Brock/Springer Series in Contemporary Bioscience

Genetics and Molecular Biology of Anaerobic Bacteria

Brock/Springer Series in Contemporary Bioscience

Series Editor: *Thomas D. Brock*
University of Wisconsin-Madison

(Continued after index)

Madeleine Sebald

Editor

Genetics and Molecular Biology of Anaerobic Bacteria

With 186 Figures

Springer-Verlag
New York Berlin Heidelberg London Paris
Tokyo Hong Kong Barcelona Budapest

Madeleine Sebald
Unité des Anaérobies
Institut Pasteur
75724 Paris, France

Cover photograph: Clostridium bifermentans DSM 631. Reproduced from *The Prokaryotes,* Second Edition, Vol. II. p. 1803. © 1992 Springer-Verlag, New York.

Library of Congress Cataloging-in-Publication Data
Genetics and molecular biology of anaerobic bacteria / Madeleine
  Sebald, editor.
      p.   cm. — (Brock/Springer series in contemporary bioscience)
  Includes bibliographical references and index.

    ISBN 978-1-4615-7089-9   ISBN 978-1-4615-7087-5 (eBook)
    DOI 10.1007/978-1-4615-7087-5

    1. Anaerobic bacteria—Genetics.   2. Anaerobic bacteria—Molecular
  Aspects.   I. Sebald, Madeleine.   II. Series.
  QR89.5.G46   1992
  589.9′01′5—dc20   91-28037
                                                            CIP

Printed on acid-free paper.

Softcover reprint of the hardcover 1st edition 1993

Production managed by Terry Kornak; manufacturing supervised by Jacqui Ashri.
Typeset by Asco Trade Typesetting Ltd., North Point, Hong Kong.

9 8 7 6 5 4 3 2 1

Preface

The field of bacterial genetics has been restricted for many years to *Escherichia coli* and a few other genera of aerobic or facultatively anaerobic bacteria such as *Pseudomonas*, *Bacillus*, and *Salmonella*. The prevailing view up to recent times has been that *anaerobic* bacteria are interesting organisms but nothing is known about their genetics. To most microbiologists, anaerobic bacteria appeared as a sort of distant domain, reserved for occasional intrusions by taxonomists and medical microbiologists. By the mid-1970s, knowledge of the genetics and molecular biology of anaerobes began to emerge, and then developed rapidly. This was the result of advances in molecular biology techniques, but also importantly because of improvements in basic techniques for culturing anaerobes and for understanding their biochemistry and other areas of interest. Investigations in this field were also stimulated by a renewal of interest in their ecology, their role in pathology and in biotransformations, and in the search for alternative renewable sources of energy.

The initial idea for this book came from Thomas D. Brock. When Dr. Brock requested my opinion about two years ago on the feasibility of publishing a book on the genetics of anaerobic bacteria, as a part of the Brock/Springer Series in Contemporary Bioscience, I answered positively but I was apprehensive about assuming the role of editor. However, I was soon reassured by the enthusiastic commitment of those I approached to contribute. Eventually, thanks to the caring cooperation of the contributors, the task became relatively easy.

I have tried to include most of the available data on the genetics and molecular biology of both developing and developed topics, and it is my hope that this book will furnish up-to-date information both to those entering and to those already involved in anaerobic microbiology. I also hope it will show that anaerobes are less distant than may be perceived.

Finally, I should like to express my gratitude to Thomas D. Brock, to the contributors, and also to the reviewers. It is a pleasure to acknowledge the assistance of Morris Goldner of the University of Toronto in the editorial process.

Madeleine Sebald

v

Contents

Contributors

Klaus Aktories Pharmakologisches Institut, Universitätsklinikum, Essen, D-4300 Essen, Germany

Jean-Paul Aubert Unité de Physiologie Cellulaire, Institut Pasteur, 75724 Paris Cedex 15, France

Michael Bagdasarian Michigan Biotechnology Institute, Lansing, MI 48909, USA

Chantal Bagnara Laboratoire de Chimie Bactérienne, CNRS, 13277 Marseille Cedex 9, France

Pierre Béguin Unité de Physiologie Cellulaire, Institut Pasteur, 75724 Paris Cedex 15, France

Anne Bélaich Laboratoire de Chimie Bactérienne, CNRS, 13277 Marseille Cedex 9, France

Jean-Pierre Bélaich Laboratoire de Chimie Bactérienne, CNRS, 13277 Marseille Cedex 9, France

George N. Bennett Department of Biochemistry and Cell Biology, Wiess School of Natural Sciences, Rice University, Houston, TX 77251, USA

Teruhiko Beppu Department of Agricultural Chemistry, Faculty of Agriculture, University of Tokyo, Bunkyo-ku, Tokyo 113, Japan

John K. Brehm Division of Biotechnology, PHLS Center for Applied Microbiology and Research, Salisbury, Wiltshire, UK

Bruno Canard Laboratoire de Génétique Moléculaire Bactérienne, Institut Pasteur, 75724 Paris Cedex 15, France

Jiann-Shin Chen Department of Anaerobic Microbiology, Virginia Polytechnic Institute and State University, Blacksburg, VA 24061, USA

Chih-Fong Chou VA Medical Center, 154J, Palo Alto, CA 94304, USA

Stewart T. Cole Laboratoire de Génétique Moléculaire Bactérienne, Institut Pasteur, 75724 Paris Cedex 15, France

Robert D. Coleman Environmental Research Division, Argonne National Laboratory, Argonne, IL 60439, USA

Peter T. Cox Center for Molecular Biology and Biotechnology, University of Queensland, Brisbane, Queensland 4072, Australia

Brian P. Dalrymple CSRIO Division of Tropical Animal Production, Longpacket Laboratories, Indooroopilly, Queensland 4068, Australia

Willem M. de Vos Bacterial Genetics Group, Department of Microbiology, Wageningen Agricultural University, Wageningen, The Netherlands

Wolfgang Dubbert Institut für Mikrobiologie, Technische Universität München, D-8000 München 2, Germany

Rik I.L. Eggen Bacterial Genetics Group, Department of Microbiology, Wageningen Agricultural University, Wageningen, The Netherlands

Christoph von Eichel-Streiber Institut für Medizinische Mikrobiologie der Johannes Gutenberg Universität, 6500 Mainz, Germany

Mel W. Eklund Utilization Research Division, Northwest Fisheries Center, NMFS, NOAA, Seattle, WA 98112, USA

Thierry Garnier Laboratoire de Génétique Moléculaire Bactérienne, Institut Pasteur, 75724 Paris Cedex 15, France

Christian Gaudin Laboratoire de Chimie Bactérienne, CNRS, 13277 Marseille Cedex 9, France

H.J. Gilbert Department of Agricultural Biochemistry and Nutrition, University of Newcastle upon Tyne, Newcastle upon Tyne, NE1 7RU, UK

Ulf B. Göbel Klinikum der Albert-Ludwigs-Universität Freiburg, Institut für Medizinische Mikrobiologie und Hygiene, Abteilung Mikrobiologie und Hygiene, D-7800 Freiburg, Germany

Howard Goldfine Department of Microbiology, University of Pennsylvania, School of Medicine, Philadelphia, PA 19104-6076, USA

Per Einar Granum Norwegian College of Veterinary Medicine, Department of Food Hygiene, Oslo 1, Norway

Donald G. Guiney UCSD Medical Center, University of California, San Diego, CA, USA

Herbert Hächler Department of Medical Microbiology, University of Zürich, CH-8028 Zürich, Switzerland

Winfried Hausner Lehrstuhl für Mikrobiologie, Universität Regensburg, D-8400 Regensburg, Germany

G.P. Hazlewood AFRC Institut of Animal Physiology and Genetic Research, Cambridge, CB2 4AT, UK

Monique Hermann Direction Biotechnologie et Environnement, Institut Français du Pétrole, 92506 Rueil Malmaison, France

Stephen M. Hinton Exxon Corporate Research Co. Clinton Township, Annandale, NJ 08801, USA

Matthew Hobbs   Center for Molecular Biology and Biotechnology, University of Queensland, Brisbane, Queensland 4072, Australia

Sueharu Horinouchi   Department of Agricultural Chemistry, Faculty of Agriculture, University of Tokyo, Bunkyo-ku, Tokyo 113, Japan

Sophie E.C. Hunter   Chemical Defence Establishment, Porton Down, Salisbury, Wiltshire SP4 OJQ, UK

Vivian Hwa   Department of Microbiology, University of Illinois, Urbana, IL 61801, USA

Phillip B. Hylemon   Department of Microbiology and Immunology, Medical College of Virginia, Virginia Commonwealth University, Richmond, VA 23298, USA

Angela Joe   University of British Columbia, Department of Microbiology, Vancouver BC, V6T 1W5, Canada

John L. Johnson   Department of Anaerobic Microbiology, Virginia Polytechnic Institute and State University, Blacksburg, VA 24061, USA

Ingo Just   Pharmakologisches Institut, Universitätsklinikum Essen, D-4300 Essen, Germany

Fritz H. Kayser   Department of Medical Microbiology, University of Zürich, CH-8028 Zürich, Switzerland

Gertrud Koch   Pharmakologisches Institut, Universitätsklinikum Essen, D-4300 Essen, Germany

Yong-Eok Lee   Department of Biochemistry, Michigan State University, East Lansing, MI 48824, USA

Chanyong Lee   Department of Biochemistry, Michigan State University, East Lansing, MI 48824, USA

Thomas Leisinger   Mikrobiologisches Institut ETH-Zentrum, CH-8092 Zürich, Switzerland

Lars G. Ljungdahl   Center for Biological Resource Recovery and Department of Biochemistry, University of Georgia, Athens GA 30602, USA

Francis L. Macrina   Department of Microbiology and Immunology, Virginia Commonwealth University, Richmond, VA 23298-0678, USA

Michel Magot   Unité de Microbiologie Industrielle, Sanofi Elf Bio Recherches, Labège Innopole, 31328 Labège, France

Darrell H. Mallonee   Department of Microbiology and Immunology, Medical College of Virginia, Virginia Commonwealth University, Richmond, VA 23298, USA

John S. Mattick   Centre for Molecular Biology and Biotechnology, University of Queensland, Brisbane, Queensland 4072, Australia

Barry C. McBride   University of British Columbia, Department of Microbiology, Vancouver, BC, V6T 1W5, Canada

Leo Meile   Mikrobiologisches Institut, ETH-Zentrum, CH-8092 Zurich, Switzerland

Hannes Melasniemi   Research Laboratories Alko Ltd., SF-00101 Helsinki, Finland

Menghsiao Meng   Michigan Biotechnology Institute, Lansing, MI 48909, USA

Jacqueline Millet   Unité de Physiologie Cellulaire, Institut Pasteur, 75724 Paris Cedex 15, France

Nigel P. Minton   Division of Biotechnology, PHLS Center for Applied Microbiology and Research, Salisbury, Wiltshire, UK

Thomas A. Morton   Center for Biological Resource Recovery and Department of Biochemistry, University of Georgia, Athens, GA 30602, USA

John D. Oultram   Division of Biotechnology, PHLS Center for Applied Microbiology and Research, Salisbury, Wiltshire, UK

John R. Palmer   Department of Microbiology, Ohio State University, Columbus, OH 43210, USA

Klaus Pelz   Klinikum der Albert Ludwigs Universität Freiburg, Institut für Medizinische Mikrobiologie und Hygiene, Abteilung Mikrobiologie und Hygiene, D-7800 Freiburg, Germany

Daniel J. Petersen   Department of Biochemistry and Cell Biology, Wiess School of Natural Sciences, Rice University, Houston, TX 77251, USA

Jesse C. Rabinowitz   Department of Biochemistry and Molecular Biology, University of California, Berkeley, CA 94720, USA

John N. Reeve   Division of Microbiology, Ohio State University, Columbus, OH 44210, USA

Gilles Reysset   Unité des Anaérobies, Institut Pasteur, 75724 Paris Cedex 15, France

Peter Roggentin   Biochemisches Institut in der Medizinischen Fakultät, Christian-Albrechts-Universität, D-2300 Kiel, Germany

Julian I. Rood   Department of Microbiology, Monash University, Clayton Victoria 3168, Australia

Abigail A. Salyers   Department of Microbiology, University of Illinois, Urbana IL 61801, USA

Roland Schauer   Biochemisches Institut in der Medizinischen Fakultät, Christian-Albrechts-Universität, D-2300 Kiel, Germany

Madeleine Sebald   Unité des Anaérobies, Institut Pasteur, 75724 Paris Cedex 15, France

R. Sellwood   AERC Institute for Animal Health, Compton, Newbury, Berkshire RG16 ONN, UK

Umadatt Singh   University of British Columbia, Department of Microbiology, Vancouver BC, V6T 1W5, Canada

C. Jeffrey Smith Department of Microbiology and Immunology, East Carolina University, Greenville, NC 27858-4354, USA

Walter L. Staudenbauer Institut für Mikrobiologie, Technische Universität München, D-8000 München 2, Germany

Gordon S.A.B. Stewart Department of Applied Biochemistry and Food Science, School of Agricultural and Food Science, University of Nottingham, Sutton Bonington, Loughborough LE12 5RD, UK

Wen-Jin Su Department of Biology, Xiamen University, 361005 Xiamen, Fujian, China

Tracy-Jane Swinfield Division of Biotechnology, PHLS Center for Applied Microbiology and Research, Salisbury, Wiltshire, UK

R.M. Teather Animal Research Centre, Research Branch, Agriculture Canada, Ottawa, Ontario, K1A OC6, Canada

Michael Thomm Institut für Biochemie Genetik und Mikrobiologie der Universität Regensburg, D-8400 Regensburg, Germany

Richard W. Titball Chemical Defence Establishment, Porton Down, Salisbury, Wiltshire SP4 OJQ, UK

Rodney K. Tweten Department of Microbiology and Immunology, University of Oklahoma, Health Sciences Center, Oklahoma City, OK 73190, USA

Shigezo Udaka Department of Food Science and Technology, Faculty of Agriculture, Nagoya University, Chikusa-ku, Nagoya 464, Japan

Kazue Ueno Institute of Anaerobic Bacteriology, Gifu University School of Medicine, 40 Tsukasa-Machi, Gifu, 500, Japan

Peter Valentine Department of Microbiology, University of Illinois, Urbana, IL 61801, USA

Jos Vanderleyden F.A. Janssens Laboratory of Genetics, University of Leuven, B-3001 Heverlee, Belgium

Peter Verhasselt F.A. Janssens Laboratory of Genetics, University of Leuven, B-3001 Heverlee, Belgium

Gerrit Voordouw Department of Biological Sciences, University of Calgary, Calgary, Alberta T2N 1N4, Canada

Judy D. Wall University of Missouri, Columbia, Department of Biochemistry, Columbia, MO USA

Shu-Zhen Wang Department of Anaerobic Microbiology, Virginia Polytechnic Institute and State University, Blacksburg, VA 24061, USA

Kunitomo Watanabe Institute of Anaerobic Bacteriology, Gifu University School of Medicine, 40 Tsukasa-Machi, Gifu 500, Japan

Sarah M. Whelan Division of Biotechnology, PHLS Center for Applied Microbiology and Research, Salisbury, Wiltshire, UK

David R. Woods   University of Cape Town, Rondebosch 7700, South Africa

Hideo Yamagata   Department of Food Science and Technology, Faculty of Agriculture, Nagoya University, Chikusa-ku, Nagoya 464, Japan

Helen Yeoman   Chemical Defence Establishment, Porton Down, Salisbury, Wiltshire SP4 OJQ, UK

Michael Young   Department of Biological Sciences, University College of Wales, Aberystwyth, Dyfed SY23 3DA, UK

J. Gregory Zeikus   Michigan Biotechnology Institute, Lansing, MI 48909, USA

Introduction

In comparison to other living organisms, obligate anaerobic microorganisms share two common characters: 1) an extreme sensitivity to molecular oxygen (or its derivatives); and a capacity to produce energy and perform essential biosyntheses without molecular oxygen. This latter property is often called *anoxybiontic* metabolism and relates to the lack of a complete cytochrome chain and oxygenases. In most anaerobes, the energy required for growth is obtained by *fermentation*, a redox process in which organic substrates are both donor and terminal electron acceptors. The major exceptions are the autotrophic methanogenic and acetogenic bacteria, the sulfur-metabolizing archaebacteria, and the sulfate-reducing bacteria which use, respectively, CO_2, elemental sulfur, and sulfate for hydrogen oxidation. Energy is transformed mostly by enzymatic processes associated with substrate level phosphorylations (Decker et al., 1970). As discussed by Gottschalk (1986), electron transport phosphorylations occur only in a few anaerobes, the terminal electron acceptor being in all cases more electronegative than oxygen. Strict anaerobes never oxidize organic substrates into CO_2 and H_2 exclusively (Gottschalk, 1986).

Anaerobes are unable to grow at the surface of a medium exposed to air (21% molecular oxygen) and the oxygen is either *bactericidal*, or *bacteriostatic*, that is, anaerobes are either *oxylabile* or *oxyduric*, respectively (Hungate, 1985). The sensivity to O_2 is extremely variable from one group to another, and evaluation of O_2 sensitivity requires that all organisms be cultivated under identical conditions, both for medium composition (including reducing substances) and gaseous environment. Most of the strict, i.e., obligate, anaerobes are procaryotes [domains Archaebacteria (synonym, Archaea) and Eubacteria (synonym, Bacteria)], but there are also some eucaryotes, such as anaerobic fungi and anaerotolerant/anaerobic protozoa (Eucarya). Only the anaerobic procaryotes are considered in this book, however.

There are no geological records of bacteria but, according to current views, the first microorganisms on earth were probably anaerobic (Brock, 1987; Levine, 1988). One may speculate that anaerobes even had a "Golden Age," living in the primordial atmosphere with no molecular oxygen, suffering "just" from overexposure to ultraviolet (UV) irradiation. But, presumably, since about 2.5×10^9 years ago, they have had to face increasing concentrations of molecular oxygen, which resulted from photosynthetic activity and which built up in the atmosphere [0.21%, 2×10^9 years ago; 1.05%, 1×10^9 years ago; 2.1%, 550×10^6 years ago; 21%, 400×10^6 years ago (Cloud, 1983)]. Anaerobes had to adapt using metabolic, ecological, phylogenetic, and genetic means. They succeeded by developing strategies such as negative aerotaxis, sporulation, adaptation to extreme conditions (temperature, pH, etc.), steno-trophic or eurytrophic metabolic capacities, nutritional interrelationships, attachment to the substrate, and living in close contact with O_2 consumers or H_2 producers. This resulted in highly adapted, or even bizarre, life styles, like living in a sulfur and saline environment at 100°C (*Staphylothermus*), in the termite hindgut, the bovine rumen, or as endosymbiotic bacteria (endosym-biotic methanogens), of anaerobic protozoa.

One may also speculate that some primordial anaerobes remained as strict anaerobes, while others, after unsuccessful trials to adapt to O_2, "returned" to an anaerobic metabolism and to an anaerobic mode of life. The anaerobes which belong to the latter category consist of those with generally low or very low levels of enzymes such as catalase and superoxide dismutase, or of those possessing a partial cytochrome chain with a terminal acceptor other than molecular O_2.

The extreme diversity of the anaerobic bacteria make them very interesting organisms for a genetic and molecular approach. These ef-forts have recently begun and look very promising from many viewpoints. The chapters of this book reflect this diversity and the interest in complementary approaches such as biotechnology and intraspecific to intergeneric gene exchanges. The latter techniques, first available for clostridia and *Bacteroides*, are now developed for archaebacteria, sulfate-reducing bacteria, rumen bacteria, and pigmented bac-teria.

The chapters of the book document recent data obtained with techniques as diverse as whole chromosome mapping, performing vectors, genetic ex-changes through phage conversion, transduction, conjugation (including con-jugative plasmids, conjugative transposons, transposonlike structures, and "elements"), protoplast transformation, electrotransformation, gene cloning and sequencing, polymerase chain reaction, and in vitro mutagenesis.

All these approaches are furnishing insight into gene organization and reg-ulation in anaerobes, permitting the characterization and isolation of enzymes from multienzymatic complexes, providing information on structure-function relationships, genetic determination of antibiotic resistance and of antibiotic transfer factors. They also permit manipulation of operons of interest, protein modeling, and the development of diagnostic probes.

The molecular techniques will make it possible to solve complex problems such as mechanisms of pathogenicity and of O_2 sensitivity, analysis of prop-erties such as thermophily or hyperthermophily, evolutionary relationship to anaerobes of puzzling structures such as organelles in anaerobic protozoa and the role of anaerobes as a natural reservoir of genes, i.e., antibiotic-resistance genes. Molecular biology is being intensively used in the taxonomy of anaerobes. GC content determinations and DNA-DNA hybridization studies have been fundamental for determining bacterial relationships at the intra-species and interspecies level. A bacterium can no longer be described as a new species unless that sort of information is available. Indeed, the establish-ment of close and remote relationships has been made possible from rRNA sequencing data, particularly from the 16S rRNA sequences, which appear as highly conserved throughout evolution. These have resulted in still-tentative but increasingly reliable phylogenetic trees. Molecular taxonomy of anaerobes as a whole has not been treated in this book because it rightly deserves a volume of its own. Indeed, excellent reviews have recently been published on this topic (Cato and Stackebrandt, 1989; Paster et al., 1985; Woese, 1987; Yang et al., 1985; Balows et al., 1991), and valuable information and comments are also given in several chapters of this book.

References

Balows, A.B., H.G. Trüper, M. Dworkin, W. Harder, and K.-H. Schleifer (editors). 1991. The Prokaryotes, 2nd edition, Springer-Verlag, New York.

Brock, T.D. 1987. Evolutionary relationships of the autotrophic bacteria. p. 499–512, in Schlegel H.G. and Bowien B. (editors), *Autotropic bacteria*, Springer-Verlag, Science Tech. Madison Wi.

Cato, E.P. and E. Stackbrandt. 1989. Taxonomy and phylogeny. p. 1-26. In Minton N.P. and Clarke D.J. (editors), *Clostridia*, Plenum Press, New York.

Cloud, P. 1983. The biosphere. *Scientific American* 249:132–144.

Decker, K., K. Jungermann and R.K. Thauer, 1970. Energy production in anaerobic organisms. *Angewandte Chemie* (International edition in English). 9:138–158.

Gottschalk, G. 1986. The anaerobic way of life of prokaryotes. p. 1415–1424. In Starr M.P., Stolp, H., Trüper, H.G. Balows, A. and Schlegel, H.G. (editors.) *The prokaryotes, a handbook on habitats, isolation and identification of bacteria*, volume 2, Springer-Verlag, Berlin.

Hungate, R.E. 1985. Anaerobic biotransformations of organic matter; p. 39–94. Poindexter J.S. (editor), *Bacteria in nature* vol. 1, Plenum Press, New York.

Levine, J.S. 1988. The origin and evolution of atmospheric oxygen, pp. 111–126, in King T.E., Mason H.S. and Morrison M. (editors). *Oxidases and related redox systems*, A.R. Liss, New York.

Paster, B.J., W. Ludwig, W.G. Weisburg, E. Stackebrandt, R.B. Hespell, C.M. Hahn, H. Reichenbach, K.O. Stetter and C.R. Woese. 1985. A phylogenic grouping of the *Bacteroides*, *Cytophagas*, and certain *Flavobacteria*. *Systematic and Applied Microbiology* 6:34–42.

Woese, C.R. 1987. Bacterial evolution. *Microbiological Reviews* 51:221–271.

Yang, D., B.P. Kaine and C.R. Woese. 1985. The phylogeny of Archaebacteria. *Systematic and Applied Microbiology* 6:251–256.

1

Plasmids, Phages, and Gene Transfer in Methanogenic Bacteria

Thomas Leisinger and Leo Meile

1.1 Introduction

Methanogenic bacteria are strictly anaerobic archaebacteria with a unique form of energy metabolism involving the generation of methane. Their substrate range for methanogenesis is limited to carbon dioxide, formate, methanol, methylamines, and acetate, but no single isolate is able to utilize all these carbon sources (Jones et al., 1987). Elucidation of the methanogenic pathway from carbon dioxide to methane has led to the discovery of several novel biochemical reactions and six new coenzymes specifically involved in methanogenesis (DiMarco et al., 1990), and various genes of the methanogenic pathway have been isolated and sequenced (Palmer and Reeve, Chapter 2, this volume). However, until recently, manipulation of these genes in vitro and their reintroduction into methanogens for functional analysis has not been possible, because methodologies for the genetic manipulation of these organisms were not available. The purpose of this chapter is to document both the genetic tools for methanogens and their potential for the development of plasmid cloning systems. This topic has recently been reviewed by Nagle (1989) and by Leisinger and Meile (1990).

The pathway of methanogenesis and its regulation are focal points of research on methanogens. Genetic methods for these organisms are also needed for the analysis of other features of the biology of methanogens, such as thermophily, halophily, and the limited oxygen tolerance of some species (Kiener and Leisinger, 1983). Recent reviews have covered the role of genetics and molecular biology for the biotechnological exploitation of methanogens (Konisky, 1989) or stressed the importance of genetic studies within these organisms for the understanding of gene expression and regulatory mechanisms (Brown et al., 1989).

1.2 Extrachromosomal Elements of Methanogens

The need for host–vector systems has led, during the past 10 years, to a search for extrachromosomal elements of methanogenic bacteria. As a result of these efforts, plasmids as well as bacteriophages or viruslike particles have now been detected in representatives of the *Methanobacteriales*, the *Methanococcales*, and the *Methanomicrobiales*, i.e., in each of the three main groups of methanogens defined by rRNA sequence comparisons (Woese, 1987). With respect to the occurrence of extrachromosomal elements, methanogens as a group are thus not different from other bacteria.

Plasmids

The plasmids of methanogens described so far are listed in Table 1.1. Each of these elements is cryptic, and the chromosome of the respective host contained no sequences homologous to plasmid DNA. Several screenings for plasmids of methanogens isolated from nature yielded a comparatively small proportion of plasmid-

Table 1.1 Plasmids of methanogenic bacteria

Host	Plasmid			Reference
	Designation	DNA (kb)	Copy number[a]	
Methanobacterium thermoauto-trophicum (Marburg)	pME2001	4.5	15–30	Meile et al. (1983)
Methanobacterium thermo-formicicum z-245	pFZ1	10.5	ND	Nölling et al. (1991)
Methanobacterium thermo-formicicum THF	pFV1	14	ND	Nölling et al. (1991)
Methanococcus sp. AG 86	pURB900	20	ND	Zhao et al. (1988)
Methanococcus jannaschii	pURB800	64	ND	Zhao et al. (1988)
	pURB801	18	ND	Zhao et al. (1988)
Methanococcus sp. C5	pURB500	8.7	3	Wood et al. (1985)
Methanolobus vulcani PL12-M	pMP1	7	ND	Thomm et al. (1983)
Methanosarcina acetivorans C2A	pC2A	5.1	6	Sowers and Gunsalus (1988)

[a]ND, not determined.

bearing strains. This may reflect the true situation or it may simply be caused by difficulties in the detection of plasmids. Only 1 among 21 newly isolated methanococci harbored a plasmid (Wood et al., 1985), and an examination of 9 acetotrophic and 3 methylotrophic methane-producing bacteria isolated from the same environment revealed only the presence of similar if not identical plasmids in 3 acetotrophic organisms (Sowers and Gunsalus, 1988). Three of seven strains of the thermophilic, formate-utilizing species *Methanobacterium thermoformicicum* contained plasmids (Table 1.1). Although these strains were isolated far from one another, their plasmids were closely related, as revealed by DNA–DNA hybridization (Nölling et al., 1991).

The circular, multicopy plasmid pME2001 of *Methanobacterium thermoautotrophicum* (Marburg) represents the best-characterized plasmid found in a methanogen. Its complete nucleotide sequence has been determined (Bokranz et al., 1990), and a short transcript, 611 base pairs in length, has been mapped (Meile et al., 1988). Figure 1.1 shows the location of this transcript on the circular map of pME2001 as well as the arrangement of four putative open-reading frames (ORFs). Sequences complementary to the 3′ end of 16S rRNA have been found in front of ORF1, ORF2, and ORF3.

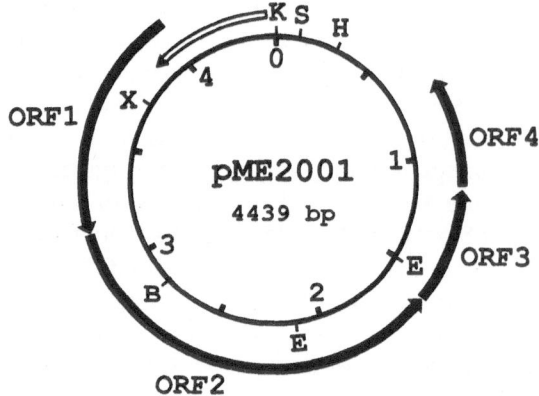

Figure 1.1 Circular map of plasmid pME2001 of *Methanobacterium thermoautotrophicum* with restriction sites for *Bal*I (B), *Eco*RI (E), *Hpa*I (H), *Kpn*I (K), *Sma*I (S), and *Xho*I (X). Mapped transcript is represented by *open arrow*; open reading frames (ORF1 to ORF4) are indicated by *solid arrows*.

They are believed to be ribosome-binding sites. The amino acid sequences derived from the ORFs did not exhibit similarity to known protein sequences. There is, however, evidence that none of the genes carried on plasmid pME2001 is essential for the host strain. A plasmid-free mutant of *M. thermoautotrophicum* (Marburg), isolated from a nitrosoguanidine-treated population, grew at a slightly reduced

Table 1.2 Bacteriophages and viruslike particles (VLPs) of methanogens

| Host | Virus of VLP | | | | Reference |
	Designation	Particle shape	DNA (kb)	Type	
Methanobacterium thermo-autotrophicum (Marburg)	ψM1	Polyhedric	27.1	Lytic phage	Meile et al. (1989a)
Methanobrevibacter smithii G	PG	Polyhedric	60 to 70	Lytic phage	Baresi and Bertani (1984, and personal communication)
Methanobrevibacter smithii PS	PMS1	Polyhedric	35	Lytic phage	Knox and Harris (1986)
Methanococcus voltae A3	VLP A3	Lemon-shaped	23	VLP	Wood et al. (1989)
Methanococcus voltae PS	VTA	ND[a]	ND	VLP	Bertani (1989)

[a] ND, not determined.

rate but to the same extent as the wild-type strain (Meile et al., 1989b).

Bacteriophages and viruslike particles

A recent review of the viruses of archaebacteria described a number of virus–host systems in the *Halobacteriales* as well as in the sulfur-dependent, extremely thermophilic *Thermoproteales* and *Sulfolobales* (Zillig et al., 1988). Some of these systems have been extensively characterized at the molecular level. In contrast, information on viruses of methanogenic bacteria is limited. Table 1.2 lists bacteriophages and related genetic elements of the methanogens. In addition, there is a preliminary report on a bacteriophage of a *Methanothrix* sp. Icosahedral particles of 100-nm diameter without tails were observed by electron microscopy in the culture fluid of this acetotrophic organism. The low growth rate of the host and its inability to grow on solid medium have prevented further characterization of the apparently infective particles (Roustan et al., 1986).

Two viruses of methanogens, bacteriophage ψM1 and the viruslike particle VLP A3 of *Methanococcus voltae*, have been characterized to some extent. Morphologically they represent the two types of virions that have been observed (Table 1.2). These two morphological types have also been observed in other

archaebacteria (Zillig et al., 1988). A third type of archaebacterial virus, rod-shaped bacteriophages with an inner envelope plus a unit membrane, from *Thermoproteus* (Zillig et al., 1988), has not been observed in methanogens.

The virulent, oxygen-resistant bacteriophage ψM1 specifically infected *Methanobacterium thermoautotrophicum* (Marburg); it infected none of the three other thermophilic representatives of the genus *Methanobacterium* that were tested. The phage was isolated from an anaerobic sludge digester operated at 55° to 60°C. Electron micrographs (Figure 1.2) show a polyhedral head and a long, noncontractile tail. In one-step growth experiments at 62°C, a latent period of 4 h and an apparent burst size of six infective particles per cell were observed. High-titer lysates [2×10^{10} plaque-forming unit (PFU) per ml] were obtained by three subsequent cycles of phage propagation. The phage particles contained linear double-stranded DNA of 30.4 ± 1.0 kb as determined by electron microscopy. About 85% of these DNA molecules represented ψM1 DNA, whereas the rest were multimers of plasmid pME2001 carried by the host (Meile et al., 1989a).

Restriction enzyme analysis of the linear phage DNA resulted in the circular map of the ψM1 genome shown in Figure 1.3. The discrepancy between the 30.4-kb physical length of ψM1 DNA and the 27.1-kb genome size on the

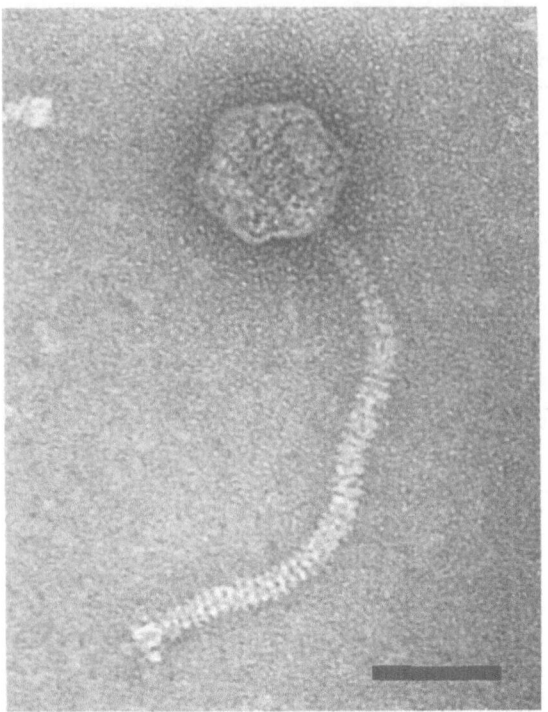

FIGURE 1.2 Electron micrograph of phage ψMl negatively stained with uranyl acetate. Bar = 50 nm.

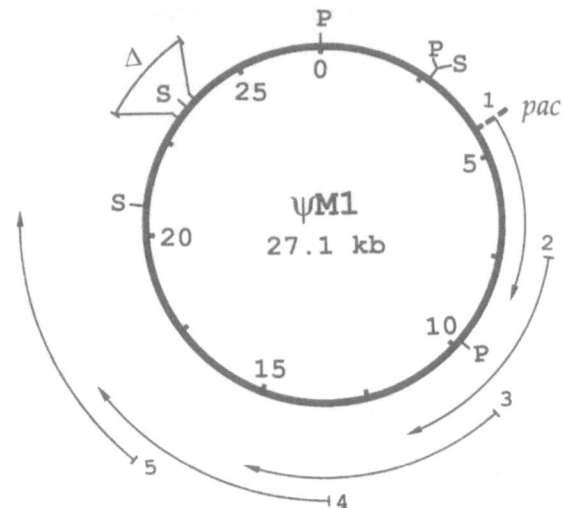

FIGURE 1.3 Circular map of phage ψM1 DNA with *Pvu*II (P) and *Sal*I (S) restriction sites. Packaging of DNA begins at *pac* site in clockwise direction. *Numbered arrows* indicate average location of left ends of phages 1 to 5 resulting from a packaging round. Marked DNA fragment at position 23.25 kb (Δ) is deleted in phage ψM2.

restriction map indicates that ψM1 is circularly permuted and exhibits terminal redundancy of approximately 3 kb. Further restriction analysis and labeling studies showed that packaging of ψM1 DNA from a concatemeric precursor initiates at a *pac* site located at coordinate 4.6 kb on the circular map and proceeds clockwise (Figure 1.3). Because evidence for at least five consecutive packaging rounds was obtained, permutation extends over a large part of the genome (Jordan et al., 1989). Efficient packaging of plasmid pME2001 and ψM1-mediated generalized transduction (Meile et al., 1990) indicate that the packaging machinery of this phage exhibits relaxed specificity.

Little is known about the structure and function of genes encoded on bacteriophage ψM1 DNA. Phage ψM2, a spontaneous mutant of ψM1, carries a deletion covering the *Sal*I site at position 23.25 kb on the circular map (see Figure 1.3). The deleted fragment spans 692 bp and encodes an ORF that, in the wild-type

phage, represents one copy of a tandem duplication. The function of this DNA region remains unknown, because ψM2 apparently does not differ from the wild type in its phenotypic properties and the amino acid sequences encoded by the duplicated DNA region do not show similarity to proteins of known sequence (R. Stettler, unpublished observations).

One gene product, presumably encoded on ψM1 DNA, has recently been purified and characterized in our laboratory (D. Stax, unpublished observations). Bacteriophage ψM1 lysates contain an oxygen-sensitive lytic enzyme capable of dissolving cell suspensions of the noninfected host strain. Thirtyfold purification of this enzyme yielded a homogeneous protein with a subunit molecular mass of approximately 33 kDa (kilodaltons). The lytic enzyme was shown to represent a pseudomurein endopeptidase hydrolyzing the ε-Ala-Lys bond of pseudomurein. The lytic enzyme in ψM1 lysates and the previously characterized pseudomurein endopeptidase from autolysates of *Methanobacterium wolfei* (Kiener et al., 1987)

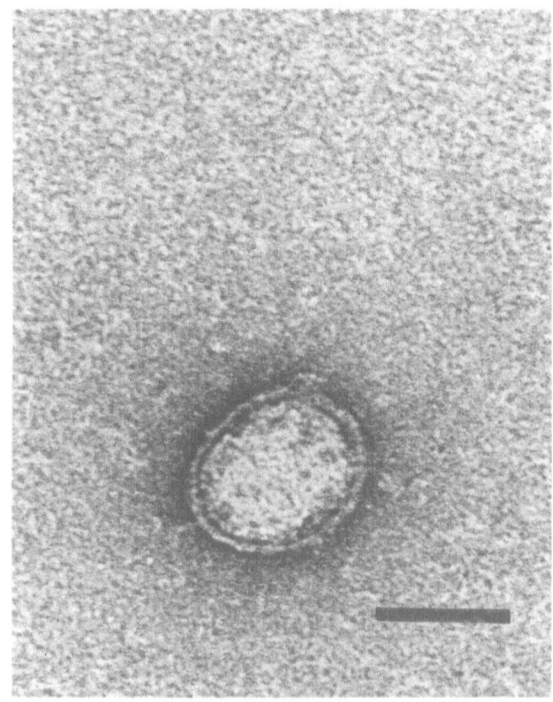

FIGURE 1.4 Electron micrograph of negatively stained viruslike particle of *Methanococcus voltae* A3. Bar = 50 nm. (From Wood et al., 1989.)

cus voltae A3 represents the other type of virus that has been characterized (Wood et al., 1989). Supernatant fluid from cultures of this organism in the late exponential phase contain as many as 10^{10} VLPs per milliliter. VLPs are lemon-shaped particles about 50 by 70 nm in size (Figure 1.4). They consist of an envelope, of which the major component is a 13-kDa protein, and of cccDNA of 23 kb (see Table 1.2). DNA hybridization indicated site-specific integration of DNA from VLP A3 into the host genome. Although *M. voltae* PS, a close relative of strain A3, did not produce VLPs, DNA homologous to that of VLP A3 was detected in its chromosome but not in the genome of other methanococci.

A spontaneous mutant of *M. voltae* A3 lost the capacity to produce VLPs and suffered a deletion of 80% of the chromosomally integrated VLP A3 DNA. Attempts to infect this mutant with purified VLPs were unsuccessful, as were efforts to demonstrate infectivity of VLPs with a range of methanococci as indicator strains. The VLP of *M. voltae* A3 thus exhibits some similarity to lysogenic bacteriophages but infectivity, inducibility, or VLP-mediated gene transfer have not yet been demonstrated.

1.3 Mutants of Methanogenic Bacteria

Availability in the same organism of suitable phage or plasmid vectors, of selectable genetic markers, and of a transformation procedure are prerequisites for establishing a gene cloning system. Considerable efforts have thus been devoted to the search for selective markers suitable for incorporation into potential vectors. This search has been successful in several methanogens, and of particular interest are mutants of *Methanococcus voltae* and *Methanobacterium thermoautotrophicum*, two organisms for which some of the other elements of a host–vector system are available. It appears that these two organisms are becoming the favored candidates for the application of recombinant DNA technology to methanogens.

Methanococcus voltae is a mesophilic metha-

are thus similar with respect to subunit molecular mass, substrate specificity, temperature optimum, and oxygen sensitivity. Each enzyme can be used for the preparation of protoplasts from *Methanobacterium* sp.. Phage ψM1 DNA has been shown to hybridize with chromosomal DNA of *M. wolfei* (Meile et al., 1989a), raising the possibility that *M. wolfei* carries a defective phage that encodes pseudomurein endopeptidase.

Protoplasts of *Methanobacterium* sp. may become useful tools for transformation provided that protocols for their regeneration are developed. In this context, it is of interest that three different protease preparations, a *Streptomyces* sp. serine protease, pronase V, and proteinase K, have been used for lysis of *M. formicicum*. Treatment with these enzymes resulted in damage to a limited region of the cell envelope and left the pseudomurein sacculi intact (Bush, 1985).

The viruslike particle VLP A3 of *Methanococ-*

Table 1.3 Auxotrophic and drug-resistant mutants useful for gener transfer experiments

Strain	Growth requirement or inhibitor	Resistance factor	Reference
Methanococcus voltae PS	L-Histidine		Bertani and Baresi (1987)
	Purines		Bertani and Baresi (1987)
	Cyanocobalamin		Bertani and Baresi (1987)
	8-Aza-2,6-diaminopurine	1,000	Bowen and Whitman (1987)
	8-Azaguanine	10,000	Bowen and Whitman (1987)
	8-Azahypoxanthine	>10,000	Bowen and Whitman (1987)
	6-Mercaptopurine	>500	Bowen and Whitman (1987)
	6-Azauracil	2,000	Bowen and Whitman (1987)
	Pseudomonic acid	40	Possot et al. (1988)
	1,2,4-Triazole-3-alanine	800	Sment and Konisky (1989)
Methanobacterium thermoautotrophicum Marburg	L-Leucine		Kiener et al. (1984)
	L-Tryptophan		Rechsteiner et al. (1986)
	Adenosine		Kiener et al. (1984)
	Thiamine		Kiener et al. (1984)
	Formate		Tanner et al. (1989)
	Pseudomonic acid	50	Kiener et al. (1986)
	5-Fluorouracil	>1,000	Nagle et al. (1987)
	6-Mercaptopurine	ND[a]	Worrell and Nagle (1990)
	8-Aza-2,6-diaminopurine	ND	Worrell and Nagle (1990)

[a] ND, not determined.

nogen that uses carbon dioxide or formate. Its genome size is $1.8 \pm 0.3 \times 10^9$ daltons (Da) (Klein and Schnorr, 1984). The biochemistry and the ultrastructure of this organism have recently been reviewed (Jarell and Koval, 1989). Natural transformation at low frequency with genomic (Bertani and Baresi, 1987) and plasmid (Gernhardt et al., 1990) DNA has been carried out in *M. voltae* PS, and a number of mutant phenotypes useful for genetic studies are available in the same strain (Table 1.3). The thermophilic rod *Methanobacterium thermoautotrophicum* grows optimally at 65°C and utilizes carbon dioxide exclusively as a substrate. It has been estimated that the total genome size of the organism is 1.2×10^9 Da (Brandis et al., 1981). Two strains of *M. thermoautotrophicum*, ΔH and Marburg, are widely used for biochemical studies. They exhibit marked differences in biochemical and physiological characteristics. DNA relatedness between these strains amounted to 46% as determined by DNA–DNA hybridization (Brandis et al., 1981). The

Marburg strain is well suited for genetic studies because mutants (Table 1.3), the potential vector plasmid pME200l (see Figure 1.1), and the generalized transducing bacteriophage ψM1 (see Figure 1.2) are available in this organism.

Mutants of *Methanococcus voltae* PS and *Methanobacterium thermoautotrophicum* Marburg with specific requirements for an amino acid, a purine, or a vitamin have been described (Table 1.3). The average frequency of spontaneous revertants per mutant bacterium amounted to 5×10^{-8} or less, which makes these auxotrophs suitable for genetic experiments. Methods for the selective enrichment of auxotrophs are available. Bacitracin has successfully been employed for this purpose in *Methanobacterium* spp. (Rechsteiner et al., 1986; Jain and Zeikus, 1987), and an enrichment procedure based on a mixture of 6-azauracil and 8-azahypoxanthine has been used for obtaining acetate auxotrophs of *Methanococcus maripaludis* (Ladapo and Whitman, 1990). A formate auxotroph of *Methanobacterium thermoautotrophicum* Marburg has recently been isolated. The metabolic function of formate is unknown. Free formate is not an intermediate in the methanogenic pathway or in CO_2 fixation, and the compound seems to fulfill an as-yet-unknown biosynthetic role in this methanogen (Tanner et al., 1989). The biochemical lesions of the auxotrophic mutants listed in Table 1.3 have not been determined. Nevertheless, because these auxotrophs allow strong selection of the dominant wild-type allele, they have proven useful in demonstrating transformation with genomic DNA and a transduction-like gene transfer in *M. voltae* PS as well as generalized transduction in *M. thermoautotrophicum* Marburg (see following).

The wild-type genes that complement auxotrophic mutants may also be used as selective markers in the construction of cloning vectors. In view of this approach and because tryptophan auxotrophs of *M. thermoautotrophicum* are available, we have cloned a 7-kb DNA fragment from this organism that encodes the enzymes for tryptophan biosynthesis. The nucleotide sequence of this fragment revealed that the seven structural genes of this pathway are arranged in a single operon, that the gene order is unique, and that no gene fusions have occurred (L. Meile, unpublished observations). The cloned tryptophan operon and a tryptophan-requiring mutant of *M. thermoautotrophicum* Marburg may now be of use for studying transformation in this organism and, at a later stage, for construction of vectors.

Methanogens, like other archaebacteria, are insensitive to many of the antibiotics effective against eubacteria (Böck and Kandler, 1985). The choice of drugs that can be used as selective agents in gene transfer experiments with methanogens is thus rather limited. Drugs of potential use in genetics should act against a specific target, and spontaneous resistance should occur at a low rate. A number of growth inhibitors effective against *Methanococcus voltae* and *Methanobacterium thermoautotrophicum* do not fulfill these criteria; the amino acid analogs azaserine, methionine sulfoximine (Possot et al., 1988), 5-methyltryptophan (Gernhardt and Klein, 1986), and ethionine (Kiener et al., 1984), as well as the coenzyme M analog bromoethane sulfonate (Kiener et al., 1984; Gern-

hardt and Klein, 1986; Dybas and Konisky, 1989), are not useful as selective agents because low-level resistance to these compounds occurs at high rates and in many cases seems to be caused by changes in cell permeability. The same holds true for the purine analogs 8-azaguanine and 8-azahypoxanthine as inhibitors of *M. thermoautotrophicum* (Worrell and Nagle, 1990).

The drug-resistant mutants included in Table 1.3 appear to be suitable for genetic experiments because they occur at low rates and exhibit high resistance factors. Some of these mutants have been biochemically characterized. Mutants of *M. voltae* and *M. thermoautotrophicum* resistant to purine and pyrimidine analogs were shown to be defective in various steps of the salvage pathways, which activate the analogs to inhibitory nucleotides (Bowen and Whitman, 1987; Nagle et al., 1987; Worrell and Nagle, 1990). Pseudomonic acid, one of the few antibiotics equally effective against eubacteria and archaebacteria, inhibits growth of *M. voltae* and *M. thermoautotrophicum* by interacting with isoleucyl-tRNA synthetase (Böck and Kandler, 1985; Kiener et al., 1986; Possot et al., 1988). The compound inhibits this enzyme competitively with respect to isoleucine, and pseudomonic acid–resistant mutants of *M. thermoautotrophicum* exhibited pseudomonic acid-resistant isoleucyl-tRNA synthetase activity (Rechsteiner et al., 1986). The structural gene encoding a resistant isoleucyl-tRNA synthetase was cloned from this organism. Comparison of its nucleotide sequence with the sequence of the wild-type gene revealed that the mutation to pseudomonic acid resistance is caused by a single base exchange in a highly conserved region (U. Jenal and T. Leisinger, unpublished observations). The cloned structural gene of the pseudomonic acid-resistant isoleucyl-tRNA synthetase represents a dominant resistance marker from *M. thermoautotrophicum*, which now can be tested for its use in transformation experiments and for the construction of potential cloning vectors.

Recruitment of biosynthesis or resistance genes from methanogens for vector construction would seem to ensure expression of the selectable phenotype in methanogen recipient strains. It bears, however, the risk of recombination between the cloned selective markers and the homologous resident genes and may lead to unstable cloning vectors. Possot et al. (1988) have therefore explored an approach that relies on the use of eubacterial drug resistances as selective markers in methanogens. They have surveyed a number of antibiotics for which cloned eubacterial resistance genes are available as growth inhibitors of *M. voltae*. Both fusidic acid and puromycin effectively inhibited this methanogen. Because the two compounds inhibited polypeptide synthesis in *M. voltae* cell extracts, their mode of action in the methanogen and in eubacteria appeared to be similar. On the basis of this information, the puromycin acetyltransferase gene of *Streptomyces alboniger* was successfully used in the construction of the first integration vector for a methanogen (Gernhardt et al., 1990).

1.4   Gene Transfer

Introduction of a recombinant vector into a methanogen host may be achieved in principle either by conjugal transfer or by transformation. The use of conjugation for plasmid delivery implies mobilization of shuttle vectors across the boundaries of primary kingdoms, that is, from a eubacterium into a methanogen recipient. Interkingdom gene transfer by conjugation has been reported from *Escherichia coli* to *Saccharomyces cerevisiae* (Heinemann and Sprague, 1989), but it is not known whether conjugation occurs also between eubacteria and archaebacteria. The only known archaebacterial conjugation system is the natural genetic transfer mechanism of the extreme halophile *Halobacterium volcanii*, which allows bidirectional conjugative exchange of genetic information within this organism (Rosenshine et al., 1989). Experimental approaches for introducing DNA into methanogens have therefore been based on transformation. Transformation procedures for methanogens are still in their infancy, whereas efficient methodologies have been developed for the transformation of *Halobacterium halobium* and *H. volcanii*. These archaebacteria

can be transformed at frequencies of up to 10^7 transformants per microgram (μg) of DNA with genomic (Cline et al., 1989), bacteriophage (Cline and Doolittle, 1987), and plasmid DNA (Charlebois et al., 1987; Hackett and DasSarma, 1989; Lam and Doolittle, 1989; Holmes and Dyall-Smith, 1990).

Transformation

The first example of a genetic transfer mechanism in a methanogen was transformation with genomic DNA of *Methanococcus voltae* PS described by Bertani and Baresi (1987). Using a simple protocol that avoided stresses which might cause autolysis of the recipient cells, they obtained up to 10^2 transformants per μg of DNA. Two genes involved in histidine and in purine biosynthesis were used as markers in these experiments. The same procedure permitted delivery of a recombinant plasmid into *M. voltae* (Gernhardt et al., 1990). The transformation efficiency per DNA mass observed with plasmid DNA was similar to the efficiency with genomic DNA. However, a comparison based on the number of transformants per gene copy demonstrated that the transformation efficiency with plasmid DNA was several orders of magnitude lower than the already low efficiencies observed with genomic DNA. The reasons for these low efficiencies remain to be explored. Restriction in *M. voltae* of plasmid DNA extracted from *E. coli* is a likely explanation for the extreme inefficiency of plasmid transformation.

A transformation system has also been reported for *Methanobacterium thermoautotrophicum* (Marburg). The wild-type strain growing on the surface of Gelrite (containing gellan gum) was transformed to 5-fluorouracil resistance with genomic DNA from a mutant resistant to this drug (Worrell et al., 1988). The yield of transformants was dependent on the amount of DNA, and transformation was only detected on the surface of medium solidified with Gelrite but not in agar plates or in liquid medium. Gellan gum is a linear heteropolysaccharide containing glucuronic acid, rhamnose, and glucose. Wider application of this method must await transformation with additional markers and substantiation of the role of gellan gum in facilitating transformation.

Transduction

Bacteriophage ψM1 of *Methanobacterium thermoautotrophicum* (Marburg) has been shown to transduce four chromosomal markers that were tested at frequencies of 6×10^{-4} to 5×10^{-6} per PFU (Meile et al., 1990). The markers used were resistance to pseudomonic acid and three genes involved in the biosynthesis of tryptophan, leucine, and adenine. ψM1 is a virulent, generalized transducing phage that may be of use for strain construction and mutant analysis in *M. thermoautotrophicum*.

Capsduction

A gene transfer agent, VTA (for *V*oltae *T*ransfer *A*gent), has been discovered by Bertani (1989) in *Methanococcus voltae* PS. Cell-free filtrates of this strain contain particles capable of transferring genetic markers. The three auxotrophic mutants of *M. voltae* PS (see Table 1.3) were transduced to wild type by VTA at efficiencies in the range of 10^3 to 10^5 transductants per milliliter of donor culture. DNA of constant size was recovered from VTA, and this DNA appeared to be degraded by restriction endonucleases as though it were highly heterogeneous in sequence (G. Bertani, personal communication). This suggests that VTA particles may carry an array of DNA fragments representing the *M. voltae* chromosome rather than a viral genome. If this is confirmed, VTA would resemble the gene transfer system described in *Rhodobacter capsulatus* (Yen et al., 1979), which was named capsduction by its discoverer, Marrs (1983). Interestingly, the VTA particles resemble in size and shape those of the *Rhodobacter* agent (G. Bertani and F. Eiserling, personal communication).

An integration vector for *Methanococcus voltae*

The construction of an integration vector for *M. voltae* PS by Gernhardt et al. (1990) represents a

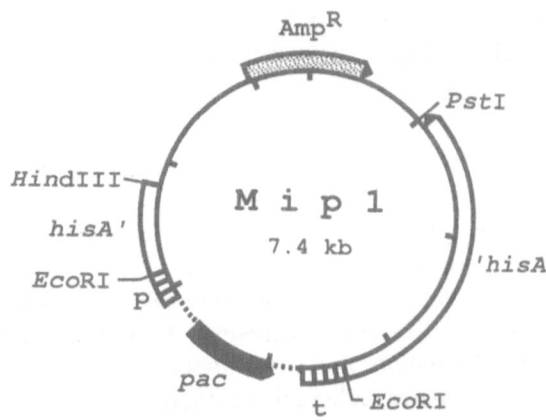

FIGURE 1.5 Map of integration vector Mip1. The *his*A gene (*open segments*) of *Methanococcus voltae*, cloned into derivative of plasmid pUC18, was interrupted by insertion of 1.8-kb *Eco*RI fragment. This fragment contained puromycin acetyltransferase (*pac*) gene (*solid segment*) and adjacent sequences (...) from *Streptomyces alboniger* flanked by promoter (p) and terminator (t) sequences of *M. voltae mcr* transcription unit. (Modified after Gernhardt et al., 1990.)

major advance in the development of a cloning system for this methanogen. The vector construct (Figure 1.5) is based on the *E. coli* vector pUC18, the puromycin acetyltransferase gene from *Streptomyces alboniger* as a selective marker in the methanogen and the cloned *his*A gene of *M. voltae* for integration of the construct into the chromosome. To ensure expression of the eubacterial resistance gene in the archaebacterial background, it was put under the control of the strong promoter and terminator sequences of the *M. voltae* methyl coenzyme M reductase transcription unit (Bokranz et al., 1988; Klein et al., 1988). Transformation of *M. voltae* wild type with this construct and selection for puromycin resistance yielded resistant clones that were analyzed for the type of integration into the chromosome the vector had undergone. In most cases integration had occurred by homologous recombination via the *his*A gene sequences, and the integrated plasmid was maintained stably under selective pressure. The integration vector makes *M. voltae* accessible to genetic manipulation. It should allow expression of foreign genes, inactivation of defined resident genes to obtain stable mutants, and analysis of gene expression signals in vivo.

1.5 Summary and Conclusions

Genetic studies in methanogenic bacteria have been initiated with *Methanococcus voltae* PS and *Methanobacterium thermoautotrophicum* (Marburg) as model organisms. Low-frequency transformation and a number of selectable genes are available in these organisms. Genetic markers are transferred in *M. voltae* by a transduction-like process and in *M. thermoautotrophicum* by generalized transduction with bacteriophage ψM1. Although these procedures may find application in mutant analysis and strain construction within methanogens, progress in molecular biology depends on the development of cloning vectors. A step in this direction has been made with the construction of an integration vector for *M. voltae*. To develop this system into a plasmid vector, a DNA fragment carrying a *M. voltae* replicon is needed. In *M. thermoautotrophicum* a potential vector plasmid and cloned selectable genes are available, but progress in vector development has been hampered by the inefficiency of the transformation procedure. Increase of the transformation frequency for this organism, as well as for *M. voltae*, is thus one of the most pressing needs for further advances in methanogen genetics. It will involve the evaluation of alternative methods for introducing DNA, such as electroporation and protoplast transformation.

Acknowledgments. We thank A.M. Cook for critically reading the manuscript, J. Konisky and D. Studer for the electron micrographs, and J. Bertani and W.M. de Vos for the communication of unpublished results. Work from this laboratory was supported by the Swiss National Foundation for Scientific Research.

References

Baresi, L. and G. Bertani. 1984. Isolation of a bacteriophage for a methanogenic bacterium. *In: Abstracts of Annual Meeting of the American Soci*ety *for Microbiology*, New Orleans, I74, p. 133. Washington, D.C.: American Society for Microbiology.

Bertani, G. 1989. Transduction-like gene transfer in a methanogen. *In: Abstracts of Annual Meeting of the American Society for Microbiology*, New Orleans, I30, p. 222. Washington, D.C.: American Society for Microbiology.

Bertani, G. and L. Baresi. 1987. Genetic transformation in the methanogen *Methanococcus voltae* PS. *J. Bacteriol.* **169**:2730–2738.

Böck, A. and O. Kandler. 1985. Antibiotic sensitivity of archaebacteria. *In: The Bacteria*, Vol. 8, C.R. Woese, and R.S. Wolfe, eds., pp. 525–544. New York: Academic Press.

Bokranz, M., A. Klein, and L. Meile. 1990. Complete nucleotide sequence of plasmid pME2001 of *Methanobacterium thermoautotrophicum* (Marburg). *Nucleic Acids Res.* **18**:363.

Bokranz, M., G. Bäumner, R. Allmansberger, D. Ankel-Fuchs, and A. Klein. 1988. Cloning and characterization of the methyl coenzyme M reductase genes from *Methanobacterium thermoautotrophicum*. *J. Bacteriol.* **170**:568–577.

Bowen, T.L. and W.B. Whitman. 1987. Incorporation of exogenous purines and pyrimidines by *Methanococcus voltae* and isolation of analog-resistant mutants. *Appl. Environ. Microbiol.* **53**:1822–1826.

Brandis, A., R.K. Thauer, and K.O. Stetter. 1981. Relatedness of strains ΔH and Marburg of *Methanobacterium thermoautotrophicum*. *Zentralbl. Bakteriol. Hyg. 1 Abt. Orig. C* **2**:311–317.

Brown, J.W., C.J. Daniels, and J.N. Reeve. 1989. Gene structure, organization, and expression in archaebacteria. *Crit. Rev. Microbiol.* **16**:287–338.

Bush, J.W. 1985. Enzymatic lysis of the pseudomurein-containing methanogen *Methanobacterium formicicum*. *J. Bacteriol.* **163**:27–36.

Charlebois, R.L., W.L. Lam, S.W. Cline, and W.F. Doolittle. 1987. Characterization of pHV2 from *Halobacterium volcanii* and its use in demonstrating transformation of an archaebacterium. *Proc. Natl. Acad. Sci. USA* **84**:8530–8534.

Cline, S.W. and W.F. Doolittle. 1987. Efficient transfection of the archaebacterium *Halobacterium halobium*. *J. Bacteriol.* **169**:1341–1344.

Cline, S.W., L.C. Schalkwyk, and W.F. Doolittle. 1989. Transformation of the archaebacterium *Halobacterium volcanii* with genomic DNA. *J. Bacteriol.* **171**:4987–4991.

DiMarco, A.A., T.A. Bobik, and R.S. Wolfe. 1990. Unusual coenzymes of methanogenesis. *Annu. Rev. Biochem.* **59**:355–394.

Dybas, M. and J. Konisky. 1989. Transport of coenzyme M (2-mercaptoethanesulfonic acid) and methylcoenzyme M [(2-Methylthio)ethanesulfonic acid] in *Methanococcus voltae*: identification of specific and general uptake systems. *J. Bacteriol.* **171**:5866–5871.

Gernhardt, P. and A. Klein. 1986. Analogue resistant mutants of *Methanococcus voltae*. In: *Archaebacteria 85*, O. Kandler, and W. Zillig, eds., p. 398. Stuttgart: Gustav Fischer Verlag.

Gernhardt, P., O. Possot, M. Foglino, L. Sibold, and A. Klein. 1990. Construction of an integration vector for use in the archaebacterium *Methanococcus voltae* and expression of a eubacterial resistance gene. *Mol. Gen. Genet.* 221:273–279.

Hackett, N.R. and S. DasSarma. 1989. Characterization of the small endogenous plasmid of *Halobacterium* strain SB3 and its use in transformation of *H. halobium. Can. J. Microbiol.* 35:86–91.

Heinemann, J.A. and G.F. Sprague, Jr. 1989. Bacterial conjugative plasmids mobilize DNA transfer between bacteria and yeast. *Nature* (London) 340:206–209.

Holmes, M.L. and M. Dyall-Smith. 1990. A plasmid vector with a selectable marker for halophilic archaebacteria. *J. Bacteriol.* 172:756–761.

Jain, M.K. and J.G. Zeikus. 1987. Method for isolation of auxotrophic mutants of *Methanobacterium ivanovii* and initial characterization of acetate auxotrophs. *Appl. Environ. Microbiol.* 53:1387–1390.

Jarell, K.F. and S.F. Koval. 1989. Ultrastructure and biochemistry of *Methanococcus voltae. Crit. Rev. Microbiol.* 17:53–87.

Jones, W.J., D.P. Nagle, Jr., and W.B. Whitman. 1987. Methanogens and the diversity of archaebacteria. *Microbiol. Rev.* 51:135–177.

Jordan, M., L. Meile, and T. Leisinger. 1989. Organization of *Methanobacterium thermoautotrophicum* bacteriophage ψM1 DNA. *Mol. Gen. Genet.* 220:161–164.

Kiener, A. and T. Leisinger. 1983. Oxygen sensitivity of methanogenic bacteria. *Syst. Appl. Microbiol.* 4:305–312.

Kiener, A., C. Holliger, and T. Leisinger. 1984. Analogue-resistant and auxotrophic mutants of *Methanobacterium thermoautotrophicum. Arch. Microbiol.* 139:87–90.

Kiener, A., T. Rechsteiner, and T. Leisinger. 1986. Mutation to pseudomonic acid resistance of *Methanobacterium thermoautotrophicum* leads to an altered isoleucyl-tRNA synthetase. *FEMS Microbiol. Lett.* 33:15–18.

Kiener, A., H. König, J. Winter, and T. Leisinger. 1987. Purification and use of *Methanobacterium wolfei* pseudomurein endopeptidase for lysis of *Methanobacterium thermoautotrophicum. J. Bacteriol.* 169:1010–1016.

Klein, A. and M. Schnorr. 1984. Genome complexity of methanogenic bacteria. *J. Bacteriol.* 158:628–631.

Klein, A., R. Allmansberger, M. Bokranz, S. Knaub, B. Müller, and E. Muth. 1988. Comparative analysis of genes encoding methyl coenzyme M reductase in methanogenic bacteria. *Mol. Gen. Genet.* 213:409–420.

Knox, M.R. and J.E. Harris. 1986. Isolation and characterization of a bacteriophage of *Methanobrevibacter smithii*. In: *Abstracts of the XIV International Congress on Microbiology*, Manchester, England. IUMS. International Union of Microbiology Societies.

Konisky, J. 1989. Methanogens for biotechnology: application of genetics and molecular biology. *Trends Biotechnol.* 7:88–92.

Ladapo, J. and W.B. Whitman. 1990. Method for isolation of auxotrophs in archaebacteria: role of the acetyl-CoA pathway of autotrophic CO_2 fixation in *Methanococcus maripaludis. Proc. Nat. Acad. Sci. USA* 87:5598–5602.

Lam, W.L. and W.F. Doolittle. 1989. Shuttle vectors for the archaebacterium *Halobacterium volcanii. Proc. Nat. Acad. Sci. USA* 86:5478–5482.

Leisinger, T. and L. Meile. 1990. Approaches to gene transfer in methanogenic bacteria. In: *Microbiology and Biochemistry of Strict Anaerobes Involved in Interspecies Hydrogen-Transfer*, J.P. Belaich, M. Bruschi, and J.L. Garcia, eds. New York: Plenum Press, pp. 11–23.

Marrs, B.L. 1983. Genetics and molecular biology. In: *Studies in Microbiology*, pp 186–214, *Vol. 4, The Phototrophic Bacteria*, J.G. Ormerod, eds., Oxford: Blackwell Scientific.

Meile, L., P. Abendschein, and T. Leisinger. 1990. Transduction in the archaebacterium *Methanobacterium thermoautotrophicum* Marburg. *J. Bacteriol.* 172:3507-3508.

Meile, L., A. Kiener, and T. Leisinger. 1983. A plasmid in the archaebacterium *Methanobacterium thermoautotrophicum. Mol. Gen. Genet.* 191:480–484.

Meile, L., J. Madon, and T. Leisinger. 1988. Identification of a transcript and its promoter region on the archaebacterial plasmid pME2001. *J. Bacteriol.* 170:478–481.

Meile, L., U. Jenal, D. Studer, M. Jordan, and T. Leisinger. 1989a. Characterization of ψM1, a virulent phage of *Methanobacterium thermoautotrophicum* Marburg. *Arch. Microbiol.* 152:105–110.

Meile, L., T. Rechsteiner, U. Jenal, M. Jordan, and T. Leisinger. 1989b. Approaches to the development of gene transfer systems in *Methanobacterium thermoautotrophicum*. In *Microbiology of Extreme Environments and Its Potential for Biotechnology*, M.S. Da Costa, J.C. Duarte, and R.A.D. Williams, eds., pp. 253–257. London: Elsevier.

Nagle, D.P., Jr. 1989. Development of genetic systems in methanogenic archaebacteria. *Dev. Ind. Microbiol.* 30:43–51.

Nagle, D.P., R. Teal and A. Eisenbraun. 1987. 5-Fluorouracil-resistant strain of *Methanobacterium thermoautotrophicum. J. Bacteriol.* 169:4119–4129.

Nölling, J., M. Frijlink, and W.M. de Vos. 1991. Isolation and characterization of plasmids DNA from different strains of *Methanobacterium thermoformicicum*. *J. Gen. Microbiol.* **137**:1981–1986.

Possot, O., P. Gernhardt, A. Klein, and L. Sibold. 1988. Analysis of drug resistance in the archaebacterium *Methanococcus voltae* with respect to potential use in genetic engineering. *Appl. Environ. Microbiol.* **54**:734–740.

Rechsteiner, T., A. Kiener, and T. Leisinger. 1986. Mutants of *Methanobacterium thermoautotrophicum*. *Syst. Appl. Microbiol.* **7**:1–4.

Rosenshine, I., R. Tchelet, and M. Mevarech. 1989. The mechanism of DNA transfer in the mating system of an archaebacterium. *Science* **245**:1387–1389.

Roustan, J.L., J.P. Touzel, G. Prensier, H.C. Dubourguier, and G. Albagnac. 1986. A bacteriophage for *Methanothrix* sp. *In Abstracts of the 5th International Symposium on Microbial Growth on C1 Compounds*, J.A. Duine, and H.W. van Verseveld, eds., p. 35, Haren: The Netherlands.

Sment, K.A. and J. Konisky. 1989. Excretion of amino acids by 1,2,4-triazole-3-alanine-resistant mutants of *Methanococcus voltae*. *Appl. Environ. Microbiol.* **55**:1295–1297.

Sowers, K.R. and R.P. Gunsalus. 1988. Plasmid DNA from the acetotrophic methanogen *Methanosarcina acetivorans*. *J. Bacteriol.* **170**:4979–4982.

Tanner, R.S., M.J. McInerney, and D.P. Nagle. 1989. Formate auxotroph of *Methanobacterium thermoautotrophicum* Marburg. *J. Bacteriol.* **171**:6534–6538.

Thomm, M., J. Altenbuchner, and K.O. Stetter.

1983. Evidence for a plasmid in a methanogenic bacterium. *J. Bacteriol.* **153**:1060–1062.

Woese, C.R. 1987. Bacterial evolution. *Microbiol. Rev.* **51**:221–271.

Wood, A.G., W.B. Whitman, and J. Konisky. 1985. A newly-isolated marine methanogen harbors a small cryptic plasmid. *Arch. Microbiol.* **142**:259–261.

Wood, A.G., W.B. Whitman, and J. Konisky. 1989. Isolation and characterization of an archaebacterial viruslike particle from *Methanococcus voltae* A3. *J. Bacteriol.* **171**:93–98.

Worrell, V.E., D.P. Nagle, Jr., D. McCarthy, and A. Eisenbraun. 1988. Genetic transformation system in the archaebacterium *Methanobacterium thermoautotrophicum* Marburg. *J. Bacteriol.* **170**:653–656.

Worrell, V.E. and D.P. Nagle. 1990. Genetic and physiological characterization of the purine salvage pathway in the archaebacterium *Methanobacterium thermoautotrophicum* Marburg. *J. Bacteriol.* **172**:3328–3334.

Yen, H.C, N.T. Hu, and B.L. Marrs. 1979. Characterization of the gene transfer agent made by an overproducer mutant of *Rhodopseudomonas capsulata*. *J. Mol. Biol.* **131**:157–168.

Zhao, H., A.G. Wood, F. Widdel, and M.P. Bryant. 1988. An extremely thermophilic *Methanococcus* from a deep sea hydrothermal vent and its plasmid. *Arch. Microbiol.* **150**:178–183.

Zillig, W., W.D. Reiter, P. Palm, F. Gropp, H. Neumann, and M. Rettenberger. 1988. Viruses of archaebacteria. *In: The Bacteriophages*, Vol. I, R. Calendar, eds., pp. 517–558. New York: Plenum.

2

Methanogen Genes and the Molecular Biology of Methane Biosynthesis

John R. Palmer and John N. Reeve

2.1   Introduction

Methanogenic bacteria (methanogens) produce methane (natural gas; sometimes called biogas) as the end product of their energy-generating metabolism. This biochemical process, termed methanogenesis, is the rate-limiting and final step in the anaerobic biodegradation of organic compounds. It occurs naturally in freshwater and marine sediments, marshes, paddy fields, geothermal springs, and in the digestive tracts of invertebrates and vertebrates. Termites and ruminants are major sources of biologically produced methane. Approximately 65% of the methane released to the atmosphere, equal to approximately 1% of the atmospheric carbon cycle, is of biological origin; the remainder is produced geologically from wells, mines, and natural vents (Ehhalt, 1974; Daniels, 1984).

Controlled anaerobic biodegradation, in digestors specifically designed for this purpose, is an established biotechnological process. It is used worldwide to reduce and detoxify agricultural, industrial, and urban wastes and to generate methane from waste biomass for use as fuel. This is already a biotechnology of major environmental and economic importance, and there are many opportunities for its further development and application. The economics of producing methane by biodegrading biomass specifically grown for this purpose has long been a topic of discussion. Only recently, however, has the concept of modifying this process by genetic engineering been incorporated into these discussions. Because methanogens have unique metabolic pathways, and many are thermophiles or hyperthermophiles (Stetter et al., 1981; Jones et al., 1983, 1987, 1989; Zhao et al., 1988; Huber et al., 1989), they are also likely to be exploited as sources of novel metabolites, pharmaceuticals, and industrial biocatalysts (Konisky, 1989). Methanogens have already proven to be a fruitful source of restriction enzymes (Schmid et al., 1984; Thomm et al., 1988a; Lunnen et al., 1989). With these future opportunities in mind, this chapter reviews the current status of the molecular biology of methanogens and our knowledge of genes whose products are directly involved in methanogenesis. Reviews of the physiology, enzymology, and biochemistry of methanogens and methanogenesis have appeared recently (Keltjens and van der Drift, 1986; Jones et al., 1987; Rouvière and Wolfe, 1988; Lancaster, 1989).

2.2   Methanogens as *Archaea*

The existence of a third biological kingdom, the *Archaebacteria*, or *Archaea*,* was recognized in the late 1970s from comparisons of the oligonucleotides generated by RNase digestions of 16S rRNAs (Woese and Fox, 1977; Woese et al.,

---

*The recent recommendation (Woese et al., 1990) that the terms *Archaea, Bacteria, Eucarya,* and Domain be used to replace *Archaebacteria, Eubacteria, Eucaryotes,* and Primary Kingdom has been followed.

1978). All known methanogens are *Archaea*, their closest nonmethanogenic relatives being the extremely halophilic *Archaea*. Although the methanogens are unified as a group by their ability to generate energy by the process of methanogenesis, they are otherwise extremely diverse with the GC content of their DNA ranging from 25 to 60 mol% (Jones et al., 1987). The phylogenetic distances (association coefficients in the range 0.2 to 0.3) among the three orders of methanogens (*Methanobacteriales, Methanococcales*, and *Methanomicrobiales*) are comparable to the distance separating the bacterial gram-negative vibrios and gram-positive bacilli (Balch et al., 1979). Methanogen genomes are circular, double-stranded DNA molecules of approximately the same size as those of *Bacteria* (Klein and Schnorr, 1984; Sitzmann and Klein, 1991), and methanogens are procaryotes as they lack a nuclear membrane but in some molecular properties they more closely resemble the *Eucarya* than the *Bacteria* (Brown et al., 1989; Iwabe et al., 1989).

2.3   Cloned Methanogen Genes

Amino acid and purine biosynthetic genes

With the discovery of the *Archaea* came the anticipation of a novel molecular biology. It was therefore somewhat of a surprise when initial studies demonstrated that cloned methanogen genes could be functionally expressed in *Escherichia coli* and *Bacillus subtilis* (Reeve et al., 1982; Wood et al., 1983; Morris and Reeve, 1984). Auxotrophic mutations in these *Bacteria* were complemented by the expression of cloned methanogen DNAs. This observation essentially eliminated the possibility that methanogens had different genetic codes, many introns, or radically different gene organizations. To date, restriction fragments of methanogen genomes have been cloned that complement mutations in the *E. coli hisA* (Wood et al., 1983; Cue et al., 1985; Weil et al., 1987), *hisI* (Beckler and Reeve, 1986), *argG* (Wood et al., 1983; Morris and Reeve, 1984, 1988), *proC* (Hamilton and Reeve,

1985a), *trpBA* (Sibold and Henriquet, 1988; Meile et al., 1991), *glnA* (Possot et al., 1989b), and *purE* genes (Hamilton and Reeve, 1985a, 1985b) and the *B. subtilis argA* genes (Morris and Reeve, 1984). Sequencing these methanogen genes has often revealed a common ancestry with the complemented bacterial gene and, in some cases, has demonstrated the presence of linked open-reading frames (ORF) with related functions. The gene adjacent to the *Methanosarcina barkeri argG* gene, for example, appears to be *carB* (encoding the enzyme carbamyl phosphate synthetase), which is also involved in arginine biosynthesis (Morris and Reeve, 1988). The *Methanosarcina barkeri* and *Methanococcus vannielii argG* genes were found to be related to each other and to the human argininosuccinate synthetase-encoding gene. This human gene, which contains at least nine introns, has apparently evolved from the same ancestral sequence as the two methanogen *argG* genes, the *E. coli argG* gene, and the *B. subtilis argA* gene.

Physical linkage and cotranscription of genes whose products are involved in the same biosynthetic pathway are common in *Bacteria*. In contrast, polycistronic transcriptional units are uncommon in *Eucarya*. Multigene transcriptional units have been found frequently in methanogens, e.g., the *mcrBDCGA* genes that encode the subunit polypeptides of methyl coenzyme M reductase (Allmansberger et al., 1986; Bokranz and Klein, 1987; Cram et al., 1987; Bokranz et al., 1988; Klein et al., 1988; Weil et al., 1988, 1989; Sherf and Reeve, 1990) and the *trpEGCFBAD* genes (Sibold and Henriquet, 1988; Meile et al., 1991) whose products catalyze different steps in the same biosynthetic pathway, but this is not always the case. The *hisA* and *hisI* genes of *Methanococcus vannielii* are not closely linked, and both the *hmfB* (Sandman et al., 1990) and *slg* genes (Bröckl et al., 1991) of *Methanothermus fervidus* appear to form monocistronic transcriptional units.

Nitrogen fixation (*nif*) genes

Dinitrogen (N_2) fixation ability is widely distributed in the *Bacteria* but has not yet been detected in the *Eucarya*. N_2 fixation has been demonstrated in *Methanosarcina barkeri* (Murray and Zinder, 1984; Bomar et al., 1985; Lobo and Zinder, 1988, 1990), *Methanolobus tindarius* (König et al., 1985), *Methanobacterium ivanovii* (Magot et al., 1986), *Methanobacterium thermoautotrophicum* (Fardeau et al., 1987), and *Methanococcus thermolithotrophicus* (Belay et al., 1984), showing that this ability is present in all three orders of methanogens. Southern hybridizations, using the cyanobacterial (*Anabaena*) *nifH* gene as the probe, demonstrated related nucleotide sequences in the genomes of 14 methanogens (Possot et al., 1986). Subsequently it was discovered that there are two distinctly different *nifH* genes in *Methanococcus thermolithotrophicus* (Souillard and Sibold, 1989). The nucleotide sequences of these two genes, *nifH1* and *nifH2*, plus the sequences of *nifD* and *nifK* (Souillard and Sibold, 1989) from this methanogen and of *nifH* from *Methanobacterium ivanovii* (Souillard et al., 1988) and *Methanococcus voltae* (Souillard and Sibold, 1986), have now been determined. The archaeal *nifH1* and *nifH2* genes of *M. thermolithotrophicus* and the *nifH* gene of *Methanobacterium ivanovii* appear to have evolved from the same common, very ancient ancestor as all bacterial *nifH* genes. The sequences of the two *nifH* genes of *Methanococcus thermolithotrophicus* are less similar to each other than they are to the sequences of many bacterial *nifH* genes. They also are differentially expressed, as only transcription of *nifH1* is inhibited by the presence of fixed nitrogen.

Transcription and translation machinery genes

tRNA genes. Genes encoding tRNAs have been cloned and sequenced from several methanogens. They occur as individual transcriptional units, linked to form clusters of tRNA genes and also linked in multigene transcriptional units to rRNA and 7S RNA encoding genes. In *Methanococcus vannielii*, 19 tRNA genes have been sequenced, 12 of which are in two transcriptional units. The larger of these contains 7 tRNA genes and a single 5S rRNA gene (Wich et al., 1984) whereas the second contains just 5 tRNA genes (Wich et al., 1986a). Substantially conserved regions from both

these transcriptional units also occur in *Methanococcus voltae* and *Methanothermus fervidus* (Wich et al., 1986b; Haas et al., 1989). Of the remaining sequenced *Methanococcus vannielii* tRNA genes, 1 is located between rRNA genes (16S-tRNA^{Ala}-23S-5S) (Jarsch and Böck, 1983) and the others are in transcriptional units containing only 1 or 2 tRNA genes (Wich et al., 1986a).

The secondary structures of the mature tRNAs in the hyperthermophile *Methanothermus fervidus* (optimum growth temperature ~83°C) are predicted from their encoding DNA sequences to have increased numbers of G-C base pairs, which presumably increase the resistance of these molecules to heat denaturation (Haas et al., 1989). Only three tRNA genes in *Methanococcus vannielii* (Wich et al., 1986a) and one tRNA gene in *M. voltae* (Wich et al., 1986b) appear to encode the 3' terminal CCA sequence present in mature tRNA molecules. Although introns have been found in tRNA genes in halophilic (Daniels et al., 1985) and sulfur-dependent *Archaea* (Kaine et al., 1983; Kaine 1987; Wich et al., 1987), they have not yet been demonstrated in tRNA genes in methanogens. In *Methanococcus vannielii*, 2-selenouridine is a natural component of several mature tRNAs (Politino et al., 1990).

The gene (*ileS*) encoding isoleucyl-tRNA synthetase in *Methanobacterium thermoautotrophicum* strain Marburg and a mutant of this gene conferring pseudomonic acid resistance have been cloned and sequenced. (Jenal et al., 1991). A gene (*purL*) located 3' to *ileS* appears to encode formylglycineamidine ribonucleotide synthase II.

rRNA genes. Methanogen rRNA genes conform to the standard bacterial arrangement, being linked in the order 16S-23S-5S. There is one 16S-23S-5S rRNA transcriptional unit in *Methanococcus voltae* (Sitzmann and Klein, 1991), two in *Methanobacterium thermoautotrophicum* (Østergaard et al., 1987), *Methanobacterium formicicum*, (Lechner et al., 1985), *Methanothermus fervidus* (Haas et al., 1989) and *Methanothrix soehngenii* (see Chapter 4, this volume) and four in *Methanococcus vannielii* (Jarsch et al., 1983). Sequencing has revealed in most, but not all cases, a tRNA^{Ala} gene in the spacer region between the 16S and 23S rRNA encoding sequences. This organization of genes is also found in the extreme halophiles and in *Archaeoglobus* and *Thermococcus celer*, two non-methanogenic *Archaea* that cluster phylogenetically with the methanogens (Achenbach-Richter and Woese, 1988). Immediately upstream of one of the two rRNA transcriptional units in *Methanothermus fervidus* and *Methanobacterium thermoautotrophicum* strain Marburg are 7S RNA and tRNA^{Ser} genes (Haas et al., 1990). The methanococcal species *Methanococcus vannielii* (Wich et al., 1984) and *Methanococcus voltae* (Wich et al., 1986b) contain one or two additional 5S rRNA-encoding genes, respectively, not linked to the rRNA genes but clustered with tRNA genes. The rRNA operon-associated 5S rRNA genes in the two methanococcal species are more closely related to each other than are the operon-linked and operon-unlinked 5S rRNA genes in the same species (Wich et al., 1986b). Complete 16S rRNA encoding sequences have been determined for *Methanobacterium formicicum* (Lechner et al., 1985), *Methanococcus vannielii* (Jarsch and Böck, 1985a, b) *Methanothrix soehngenii* (Eggen et al., 1989), *Methanobacterium thermoautotrophicum* (Østergaard et al., 1987) and *Methanothermus fervidus* (Haas et al., 1990).

7S RNA genes. All *Archaea* contain large amounts of a stable RNA molecule, designated (because of its size) as the 7S RNA, whose function is unknown. A role in a ribosome-associated activity is suggested by the discovery that the 7S RNA-encoding gene in *Methanothermus fervidus* and *Methanobacterium thermoautotrophicum* is adjacent to and possibly cotranscribed with rRNA genes (Haas et al., 1990) and by the similarities in the secondary structures predicted for archaeal 7S RNAs and for the RNA component of eucaryal signal recognition particles. Related primary sequences and secondary structures have also been detected in the small cytoplasmic scRNA of *Bacillus subtilis* and the 4.5S RNAs of many *Bacteria* (Haas et al., 1990; Kaine, 1990). Genes encoding 7S RNAs, together with their flanking regions, have also been cloned and sequenced from

Methanococcus voltae (Kaine and Merkel, 1989) and *Methanosarcina acetivorans* (Kaine, 1990) but linkage of these methanogen 7S RNA genes to rRNA genes was not reported.

Genes encoding RNA polymerases, ribosomal proteins, and elongation factors. Archaeal DNA-dependent RNA-polymerases (RNAP) are complex, multisubunit enzymes that on the basis of immunological cross-reactivities were predicted to be more similar to eucaryal RNAPs than to bacterial RNAPs (Gropp et al., 1986; Thomm et al., 1986; Zillig et al., 1988, 1989). This prediction has now been substantiated by gene cloning and sequencing. The four largest subunits of RNAP in *Methanobacterium thermoautotrophicum* Marburg are encoded by adjacent genes arranged in the sequence B", B', A, C (Berghöfer et al., 1988). The B' subunit has nucleotide-binding sequences in common with the β subunit of *E. coli* RNAP (Thomm et al., 1988b). In *E. coli*, the genes encoding the β and β' subunits of RNAP are linked to genes encoding ribosomal proteins L1, L10, L11, L12 in the order L11-L1-L10-L12-β-β' (Post et al., 1979). In *Methanococcus vannielii*, the sequence L1-L10-L12 is retained but the L11-encoding gene is replaced by an ORF that appears to encode a polypeptide unrelated to L11 (Baier et al., 1990). The 3' region of the gene which encodes the *Methanococcus vannielii* equivalent of the β' subunit of *E. coli* RNAP has also been cloned and sequenced (Auer et al., 1989a, 1989b; Lechner et al., 1989). It is located immediately upstream of the *Methanococcus vannielii* "streptomycin" operon, an operon predicted by comparison with *E. coli* sequences to encode the methanogen equivalents of ribosomal proteins S17, S10, and S12, elongation factors EF = G and EF = Tu and two unknown proteins in the sequence ORF1-ORF2-S12-S7-EF = G-EF = Tu-S10. Approximately 30 kb 3' to this transcriptional unit are the *Methanococcus vannielii* equivalents of the *E. coli* "S10" and "spectinomycin" operons. Genes encoding the methanococcal equivalents of L22-S3-L29 and L14-L24-L5-S14-S8-L6-L18-S5-L30-L15-*SecY*-L36, have been identified in these two operons, respectively (Auer et al., 1989a, 1989b). This gene organization, apart from the transfer of the S10 and S17 genes to the streptomycin and S10 operons respectively and the absence of a L16 gene in the S10 operon, is identical to that of the ribosomal protein genes in *E. coli*. There are five additional ORFs (a– e) within the *M. vannielii* spectinomycin operon that show no similarity to bacterial ribosomal protein genes. Three of these (c, d, and e) and the ORF1 of the streptomycin operon are, however, related to eucaryal ribosomal protein genes (Auer et al., 1989a, 1989b). The primary sequences of the *M. vannielii* B' RNAP subunit gene, and the EF = G and EF = Tu genes, are significantly more similar to their eucaryal equivalents than to their bacterial equivalents. The presence and conservation of the *secY* gene, promoter distal to the *M. vannielii* spectinomycin operon, suggests that protein translocation across cell membranes and protein secretion in this methanogen are likely to have features in common with the bacterial mechanism established in *E. coli*. There is, in fact, now evidence for the synthesis of flagellins and surface-layer (S-layer) proteins (Bröckl et al., 1991) in methanogens as precursors containing leader peptides very similar to those found in *E. coli* (see following).

Genes for structural proteins

Flagellins. The structures of archaeal cell walls and membranes are significantly different from those of their bacterial counterparts (Jones et al., 1987). The biosynthesis and insertion of flagella into archaeal cell envelopes may, therefore, have features not found in *Bacteria*. Archaeal flagella are, in general, thinner (10–13 nm) than bacterial flagella (~20 nm) (Jarrel and Koval, 1989) and are formed from several different flagellin subunits, which are glycosylated in *Halobacterium halobium*, *Methanospirillum hungatei* (Weiland et al., 1985), and possibly in *Methanococcus voltae*. The N-terminal amino acid sequences obtained from the two subunit flagellins of *Methanospirillum hungatei* (molecular weights of 24,000 and 25,000) and the 31,000-Da flagellin of *Methanococcus voltae* are, allowing for conservative substitutions, 84% to 90% identical to the amino acid sequence predicted for a flagellin in *H. halobium* (Kalmokoff et al., 1990). The sequence of the *Methanococcus voltae* flagellin has, however, no obvious similarities to sequences in bacterial flagellins. The homology between the N-terminal sequences of the mature methanogen flagellins and the sequence of the *H. halobium* flagellin predicted from the cloned flagellin gene begins at codon 13, suggesting that a 12-amino-acid leader peptide is removed during the maturation of these flagellin polypeptides.

Surface-layer (S-layer) proteins. Cells of *Methanothermus fervidus* and *Methanothermus sociabilis* are surrounded by a surface layer composed of subunits of a glycosylated protein. The genes (*slg*) encoding these proteins have been cloned, sequenced, and compared from these two hyperthermophiles (Bröckl et al., 1991). Both genes are 1782 bp long and apparently form monocistronic transcriptional units. There are only nine nucleotide differences in the two *slg* genes, resulting in only three conservative amino acid substitutions in the encoded polypeptides. As these S-layer glycoproteins constitute 5% to 10% of the total cellular protein, there must be frequent transcription of the *slg* genes and translation of the resulting mRNAs. Their ribosome binding sites (RBS) are, in fact, perfectly complementary to those of the 16S rRNAs of the methanogens, and the sequences suggested as the "box A" and "box B" promoter elements for transcription of these genes differ from the consensus sequences for methanogen promoters at only one position (Zillig et al., 1988; Thomm et al., 1989). Comparison of the sequences of the *slg* genes to the N-terminal amino acid sequences obtained by sequencing the mature proteins indicates that these proteins are synthesized as precursors with N-terminal leader peptides containing 22 amino acid residues that form a hydrophobic core followed, at the cleavage site, by the sequence ala-gly-ala. This sequence conforms to the ala-×-ala-ala motif recognized and cleaved by bacterial signal peptidases (Perlman and Halvorson, 1983). The secondary structures predicted for the S-layer proteins from these two *Methanothermus* species have 14% more β-sheet content than the structures predicted for S-layer glycoproteins in mesophilic species.

Histonelike proteins. Many archaeal species contain histonelike DNA-binding proteins (Brown et al., 1989). The histonelike protein in *Methanothermus fervidus*, termed HMf (*Histone M. fervidus*), is composed of two almost identical polypeptides designated HMf-1 and HMf-2, and the gene (*hmfB*) encoding HMf-2 has been cloned and sequenced (Sandman et al., 1990). HMf binding to double-stranded DNA molecules in vitro increases their resistance to thermal denaturation and results in the formation of quasi-spherical, macromolecular HMf–DNA complexes very reminiscent of nucleosomes. More than 30% of the amino acid residues in HMf-2 are conserved in the consensus sequences of eucaryal histones H2A, H2B, H3, and H4, which strongly suggests a common ancestor for HMf and eucaryal histones. The two hyperthermophilic methanogens, *Methanothermus fervidus* and *Methanopyrus kandleri*, also contain reverse gyrase, an enzyme so far found only in hyperthermophiles. (Bouthier de la Tour et al., 1990, 1991).

The amino acid sequences of chromosomal proteins from both genera of the *Methanosarcinaceae* have been determined. Methanogen chromosomal protein MC1 (previously designated HMb), the major chromosomal protein in *Methanosarcina barkeri*, contains 93 amino acid residues (molecular weight, 10,757) of which 15% are acidic and 27% basic (Laine et al., 1986; Imbert et al., 1990). The major chromosomal protein from *Methanosarcina* sp. CHT155 differs from MC1 at only nine positions, seven of which are conservative amino acid substitutions (Chartier et al., 1988, 1989b). The primary sequence of MC1 is not closely related to any eucaryal histone or any bacterial or archaeal DNA-binding protein characterized to date from organisms other than *Methanosarcinaceae*. Three structurally similar variants of MC1 (MC1a, MC1b, and MC1c) are, however, found in the related methanogen, *Methanothrix soehngenii* (Chartier et al., 1989a). Overall, the primary sequences of the different MC1 proteins are only approximately 60% identical but they all contain highly conserved residues between positions 17 and 35, a region rich in glycines and alanines, and also in the most basic region between positions 45 and 58.

Genes encoding metabolic enzymes

Glyceraldehyde-3-phosphate dehydrogenase (GAPDH). The nucleotide sequences of the GAPDH-encoding (*gap*) genes cloned from the mesophiles *Methanobacterium bryantii* and *Methanobacterium formicicum* predict amino acid sequences that are 95% identical to each other and 70% identical to GAPDH in the hyperthermophile *Methanothermus fervidus* (Fabry and Hensel, 1987, 1988; Fabry et al., 1989). Although there is only limited primary sequence identity between these archaeal GAPDHs and GAPDHs from *Bacteria* and *Eucarya*, there appears to be substantial similarity in their secondary structures. The hyperthermophilic GAPDH from *M. fervidus* has reduced numbers of glycine and serine residues and an increased number of isoleucine residues when compared with the mesophilic GAPDHs. These changes could increase the thermostability of this enzyme by increasing its hydrophobicity and primary chain rigidity and by improving the packing of internal amino acid residues.

Malate dehydrogenase (MDH). The amino acid sequence predicted for MDH from the cloned *Methanothermus fervidus* gene has a very low similarity to the amino acid sequences of this enzyme from eucaryal and bacterial sources. However, as for the GAPDHs, similar secondary structures are predicted for all MDHs (Honka et al., 1990). The same amino acid residues are also conserved at functionally important sites in all MDHs. Considerable variation also exists within archaeal MDHs. The N-terminal sequence obtained for MDH from *Sulfolobus acidocaldarius* is very different from the corresponding sequence of the *Methanothermus fervidus* MDH (Görisch and Jany, 1989).

2.4   Methane Genes

Overview of methanogenesis

Methanogenesis is a low-redox process that yields small amounts of energy during the reduction of CO, CO_2, formate, methanol, methylamines, or acetate to methane (CH_4) (Figure 2.1). In oxidative phosphorylation, the

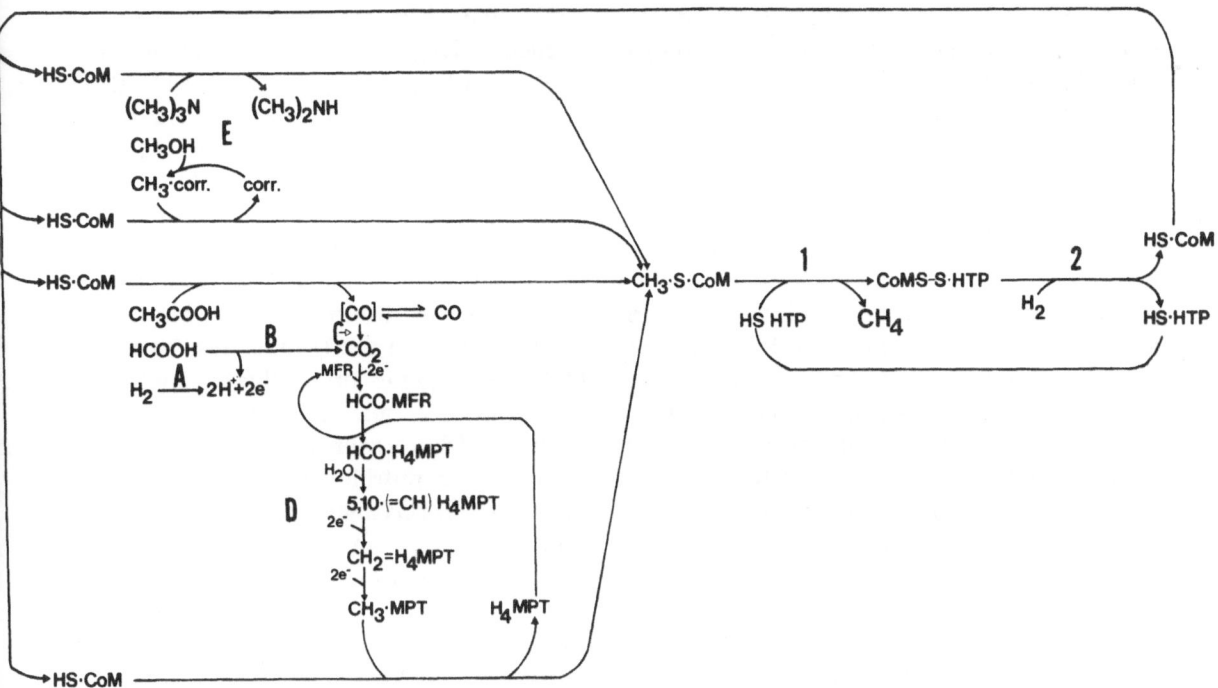

FIGURE 2.1 Outline of pathways of carbon metabolism leading to methane biosynthesis. Abbreviations: 1, methyl coenzyme M reductase (MR); 2, heterodisulfide reductase; A, hydrogenase; B, formate dehydrogenase; C, carbon monoxide dehydrogenase; D, sequence of enzymes involved in fixation of carbon dioxide and subsequent reduction of C1 moiety to the methyl level; E, methyltransferases involved in catabolism of methanol and methylamines; MFR, methanofuran; H_4MPT, tetrahydromethanopterin.

reduction of 1 mol of O_2 by H_2 to form H_2O releases 476 kJ. This is equivalent to 119 kJ per electron transferred, whereas during the reduction of 1 mol of CO_2 by H_2 to CH_4 only 16.3 kJ are generated per electron transferred.

Methane synthesis per se is catalyzed by methyl-coenzyme-M reductase (MR) (Ellermann et al., 1988; Rouvière et al., 1988; Ellermann et al., 1989). The CH_3 group of methyl-coenzyme-M (CH_3-S-CoM) is reduced to CH_4 by reductant transferred from a methanogen-specific cofactor, mercaptoheptanoylthreonine phosphate, conventionally known as HS-HTP (Noll et al., 1986, 1987). During the dissimilation of methanol (van der Meijden et al., 1983a, 1983b, 1984; Zydowski et al., 1987), methylamines (Naumann et al., 1984; Jones et al., 1987), or acetate (Eikmanns and Thauer, 1985; Krzycki et al., 1985; Terlesky et al., 1986; Krzycki and Prince, 1990) to methane, methyl groups are transferred intact from these substrates to coenzyme-M (HS-CoM).

Methanogenesis from CO, CO_2, or formate, however, requires first a series of reductive steps to generate the CH_3 group (Jones et al., 1985, 1987; Keltjens and Vogels, 1988; Rouvière and Wolfe, 1988). CO_2 is fixed initially to the cofactor methanofuran (MFR) to produce formyl-MFR (Donnelly and Wolfe, 1986). The formyl group is then transferred to a second cofactor, tetrahydromethanopterin (H_4MPT), which carries the C1 moiety during its sequential reduction through the methenyl and methylene levels to the methyl level (Keltjens and Vogels, 1988; Ma and Thauer, 1990). The methyl group is then transferred to HS-CoM to form CH_3-S-CoM, the substrate for MR (Sauer, 1986; Poirot et al., 1987). Methyl group reduction to CH_4 by MR also results in the synthesis of a heterodisulfide, designated CoM-S-S-HTP, which must be reduced to regenerate the cofactors HS-CoM and HS-HTP, needed for continued methanogenesis (Bobik et al., 1987; Ellermann et al., 1987).

An H_2-dependent CoM-S-S-HTP oxido-reductase has been described (Hedderich and Thauer, 1988; Hedderich et al., 1989). Energy conservation by methanogens could be linked to this disulfide bond reduction, rather than to the CH_4-generating reaction per se (Ellermann et al., 1987). If this is the case, methanogens may be viewed as sulfur reducers that use the CH_4-biosynthesizing system to generate an endogenous disulfide bond for subsequent use as a terminal electron acceptor. Although growth of methanogens in the absence of CH_4 biosynthesis has not been observed, methanogens can reduce elemental sulfur to H_2S, and it seems likely that some methanogens can generate energy via sulfur dissimilation (Stetter and Gaag, 1983).

Definition of methane genes

We have chosen, so that we may facilitate this review, to define all the genes that encode enzymes involved in methanogenesis as "methane genes." This is clearly a very broad definition, and the inclusion of a gene in this grouping may depend upon the methanogen being studied and the methanogenic substrate. Reactions 1 and 2 in Figure 2.1 appear, however, to be essential for methanogenesis from all substrates and to be present in all methanogens. The enzymes that catalyze these reactions, methyl coenzyme M reductase (MR) and heterodisulfide reductase are probably therefore encoded by essential methane genes in all methanogens. Two MR isoenzymes have recently been purified from *Methanobacterium thermoautotrophicum* strains ∆H and Marburg (Rospert et al., 1990), respectively, and Southern blot hybridizations confirm the presence of two related but distinct *mcr* operons in the genomes of these species and also in the genome of *Methanothermus fervidus* (J. Palmer, A. Hennigan and J. Reeve, unpublished observations). Reactions A and B in Figure 2.1, catalyzed by hydrogenases (Yamazaki, 1982; Jin et al., 1983; Fox et al., 1987; Hausinger, 1987; Muth et al., 1987; Wackett et al., 1987; Fauque, 1989; Fiebig and Friedrich, 1989; Reeve and Beckler, 1990; Shah and Clark, 1990) and formate dehydrogenase (Schauer and Ferry,

1980, 1986; May et al., 1988; Sparling and Daniels, 1990), respectively, generate the reductant and, in the latter case, possibly also the substrate needed for methanogenesis from CO_2 and formate. These enzymes are therefore encoded by methane genes in methanogens growing by oxidizing H_2 or HCOOH to supply reductant to reduce CO_2 to CH_4. The enzymes that catalyze the sequential steps leading from CO_2 to the CH_3 group of CH_3-S-CoM (group D reactions in Figure 2.1) must also be encoded by methane genes under these growth conditions (Jones et al., 1985, 1987; Keltjens and Vogels, 1988; Rouvière and Wolfe, 1988).

The most nutritionally versatile methanogens, the *Methanosarcinaceae*, are capable of methanogenesis from methanol and from mono-, di-, and trimethylamines (group E reactions in Fig. 2.1) (Hippe et al., 1979; van der Meijden et al., 1983a, 1983b, 1984; Naumann et al., 1984; Zydowski et al., 1987) and from acetate (Eikmanns and Thauer, 1985; Krzycki et al., 1985; Terlesky et al., 1986; Krzycki and Prince, 1990). The methyl transferases needed to catabolize these different substrates are substrate specific and substrate inducible and are therefore encoded by methane genes whose expression is growth-substrate dependent. A corrinoid intermediate participates in the transfer of methyl groups from methanol (van der Meijden et al., 1983b) and probably also from methylamines (Naumann et al., 1984) and acetate (Eikmanns and Thauer, 1985) to HS-CoM. The ability of *Methanosarcina* species to grow using alternative substrates for methanogenesis offers the attractive possibility of obtaining substrate-dependent, conditional mutants (Reeve, 1988). This has been hampered by the growth of these methanogens as multicellular clumps with thick heteropolysaccharide outer layers, which make protoplast formation difficult. Growth of single cells of *Methanosarcina thermophila* has, however, now been obtained by increasing the salt concentration of the growth medium, and this procedure plus the availability of a disaggregating enzyme should facilitate the isolation of substrate-dependent mutants (Sowers and Gunsalus, 1988; Xun et al., 1990). Methanogenesis using secondary alcohols as the source of reductant has also been reported (Widdel, 1986; Widdel et al., 1988; Frimmer and Widdel, 1989; Widdel and Wolfe, 1989). The syntheses of the specific secondary alcohol dehydrogenases needed for this ability are induced by their substrates, and these dehydrogenases are also encoded by substrate-regulated methane genes.

Methanogenesis from acetate requires carbon monoxide dehydrogenase (CODH) to catalyze cleavage of the $C-C$ bond (Vogels and Visser, 1983; Eikmanns and Thauer, 1984, 1985; Krzycki et al., 1985; Terlesky et al., 1986; Bhatnagar et al., 1987; Hausinger, 1987; Terlesky and Ferry, 1988a; Krzycki and Prince, 1990), and elevated amounts of CODH are found in cells of *Methanosarcina barkeri* growing on acetate as compared to cells growing on methanol (Krzycki et al., 1985; Terlesky et al., 1986; Bhatnagar et al., 1987). CODH activity is also required to generate the reductant used by acetate-catabolizing methanogens to reduce the CH_3 moiety of CH_3-S-CoM to CH_4. This complex role of CODH in methanogenesis from acetate has focused recent attention on investigating CODH-encoding genes as methane genes. As CODH is also required for the assimilation of carbon by methanogens (Jones et al., 1987; Abbanat and Ferry, 1990), it will be very interesting to determine how these alternative roles are regulated and how the expression of CODH-encoding genes is controlled. Genes encoding the subunits of CODH have been cloned and sequenced from *Methanosarcina thermophila* (K. Sowers and R. Gunsalus, personal communication) and from *Methanothrix soehngenii* (Eggen et al., 1991 and see Chapter 4, this volume).

In methanogens, energy conservation is coupled with methanogenesis. Genes encoding the catalysts involved in energy conservation and ATP synthesis therefore also fall within the broad definition of methane genes (Blaut and Gottschalk, 1985; Blaut et al., 1987; Müller et al., 1987a, 1987b, 1988a, 1988b; Lancaster, 1989; Deppenmeier et al., 1989; Kaesler and Schönheit, 1989a, 1989b). As methanogens appear to use a chemiosmotic mechanism to couple electron transfer to ATP synthesis, methane genes must also encode sodium- and proton-translocating proteins, membrane-

bound ATPases, and electron transport proteins such as cytochromes and ferredoxins. There are several unique and complex cofactors that are obligatory participants in methanogenesis (see Figure 2.1) (Wolfe, 1985; Jones et al., 1987; Dimarco et al., 1990a), and their biosyntheses are catalyzed by enzymes that must also be encoded by methane genes (White, 1989; Schwarzkopf et al., 1990).

Methane genes cloned to date

Methyl coenzyme M reductase (MR). Species that possess active MR are methanogens, and this enzyme can constitute as much as 10% of their total cellular protein (Ellefson and Wolfe, 1981). The importance and abundance of MR inevitably focused initial attention on determining its structure and the mechanisms directing its synthesis and regulation. The MR holoenzyme contains, in all methanogens so far examined, three different polypeptide subunits arranged in an α_2, β_2, γ_2 stoichiometry. The encoding genes have been designated *mcrA*, *mcrB*, and *mcrG*, respectively, and have been cloned and sequenced from two mesophilic methanococci, *M. vannielii* (Cram et al., 1987) and *M. voltae* (Allmansberger et al., 1986; Klein et al., 1988); a mesophilic methanosarcina, *M. barkeri* (Bokranz and Klein, 1987); a thermophilic methanobacterium, *M. thermoautotrophicum* (Bokranz et al., 1988), and a hyperthermophilic methanobacterium, *M. fervidus* (Weil et al., 1988). In all cases these genes form part of a tightly linked, cotranscribed cluster of five genes arranged in the order *mcrBDCGA* and conventionally termed the *mcr* operon. There appear to be two related but distinct *mcr* operons in *M. thermoautotrophicum* and *M. fervidus* (Rospert et al., 1990; J. Palmer, A. Hennigan, J. Reeve, unpublished observations). Functions have not been determined for the polypeptides encoded by the *mcrC* and *mcrD* genes, although antibodies raised against the MR holoenzyme of *M. vannielii* coprecipitate the product of the *M. vannielii mcrD* gene (Sherf and Reeve, 1990).

Because all five sequenced *mcr* operons appear to have evolved from a common ancestor (Klein et al., 1988; Weil et al., 1988, 1989),

Table 2.1 Estimation of evolutionary divergence by means of quantitative pairwise comparisons of gene products of *mcr* genes (gp*mcr*)

	Percentage identity[b]				
Strains compared[a]	gp*mcrB*	gp*mcrD*	gp*mcrC*	gp*mcrG*	gp*mcrA*
M.f. vs. M.t.	82	51	74	79	83
M.v. vs. M.vo.	82	66	83	82	91
M.f. vs. M.v.	61	34	59	59	66
M.f. vs. M.vo.	65	34	62	59	71
M.f. vs. M.b.	53	35	61	63	68
M.vo. vs. M.b.	61	35	56	53	63
M.v. vs. M.b.	53	32	50	55	59
M.t. vs. M.b.	58	33	54	57	66
M.t. vs. M.v.	58	31	51	53	61
M.t. vs. M.vo.	62	39	64	63	72

[a] Abbreviations: M.f., *Methanothermus fervidus*; M.t., *Methanobacterium thermoautotrophicum* strain Marburg; M.v., *Methanococcus vannielii*; M.vo., *Methanococcus voltae*; M.b., *Methanosarcina barkeri*.
[b] Percentage of identical amino acids in pairs of gene products based on published alignments (Weil et al., 1988, 1989) and as previously reported (Klein et al., 1988).

quantitative pairwise comparisons (Table 2.1) can be used to estimate evolutionary divergences. The results obtained from this analysis agree with the phylogenetics of methanogens established previously by analyses of 16S rRNA sequences (Woese and Fox, 1977; Balch et al., 1979; Jones et al., 1987). Table 2.1 also demonstrates that the extent of differences in overall base composition is not paralleled in the differences in the encoded amino acid sequences (Klein et al., 1988; Weil et al., 1988, 1989). The RBSs preceding the α, β, and γ subunit-encoding genes appear to be much stronger than the RBSs preceding the *mcrC* and *mcrD* genes, predicting less synthesis of the products of the latter two genes. Codon preferences in the *mcr* operons also point to a similar conclusion. The NNC codon is used preferentially in NNC/U pairs of synonymous codons in the *mcrA*, *mcrB*, and *mcrG* genes, but not in the *mcrC* or *mcrD* genes (Cram et al., 1987). The C residue in the wobble position enhances codon–anticodon interactions and has been shown to increase the rate of translation (Thomas et al., 1988). The preference for NNC codons is least evident in the *mcrA*, *mcrB*, and *mcrG* genes of the hyperthermophile *M. fervidus* (Weil et al., 1988, 1989). This subtle modulation of codon–anticodon hydrogen bonding may therefore be of only limited value against a

background of much stronger stabilizing activities, which must be present for translation to proceed accurately in cells of *M. fervidus* growing at temperatures above 80°C.

During the evolutionary adaptation of methanogens to different environments, the divergence permitted in the *mcr* genes would have been limited by the need to maintain MR function. The most highly conserved sequences in MRs in all extant methanogens are therefore likely to be active sites or coenzyme-binding sites. A block of 23 amino acid residues is conserved in the α subunit of MR, which has thus been suggested as the region that binds co-factor M (HS-CoM) (Klein et al., 1988). Conserved oligonucleotide sequences in MR genes may also be used as group-specific hybridization probes (Weil et al., 1988). Methanogens can be identified and quantified in samples taken from natural environments or anaerobic digestors by the use of these probes, thus avoiding the difficulties inherent in culturing these fastidiously anaerobic microorganisms.

Formylmethanofuran transferase (FTR). During methanogenesis from CO_2 or formate, the enzyme formylmethanofuran: tetrahydromethanopterin formyltransferase (FTR) catalyzes the transfer of a formyl group from formylmethanofuran (MFR) to tetrahydromethanopterin

(H₄MPT) yielding 5-formyltetrahydrometha-
nopterin (formyl-H₄MPT) (Donnelly and
Wolfe, 1986). The FTR-encoding gene (*ftr*) from
Methanobacterium thermoautotrophicum ΔH has
been cloned, sequenced, and functionally expressed in *E. coli* (DiMarco et al., 1990b). This
gene encodes an acidic protein (molecular
weight of 31,401), which, with the exception of
lys, asp, and glu codons, has a codon usage
very similar to that of the *mcrA*, *mcrB*, and *mcrG*
genes in *Methanobacterium thermoautotrophicum*
strain Marburg. An RBS precedes the *ftr* translation initiation codon, and a region located immediately 3' to the *ftr* gene consisting of two
inverted repeat sequences, one bounded by
oligo-T sequences, has been implicated in
transcription termination. The presence of a
truncated ORF that terminates 61 bp 5' to the
ftr gene and the apparent absence of a promoter within this intergenic region suggest that
the *ftr* gene is part of a polycistronic transcriptional unit.

Formate dehydrogenase. Approximately 50% of
the methanogens studied can use formate as a
growth substrate (Jarrel and Kalmokoff, 1988),
and under these circumstances formate dehydrogenase (FDH) accounts for 2% to 3% of the
total soluble protein (May et al., 1988). FDH
oxidizes formate and generates the reducing
equivalents needed to reduce CO_2 to CH_4. The
two genes (*fdhA* and *fdhB*) that encode the α
and β subunits of FDH, which were cloned
and sequenced from *Methanobacterium formicicum*, overlap by 1 bp and are cotranscribed as
part of a 12-kb transcript (Shuber et al., 1986;
Patel and Ferry, 1988). Transcription of the
fdhAB region is regulated by molybdenum
availability. In the absence of molybdenum, the
activity and synthesis of FDH declines but
transcription of the *fdhA* and *fdhB* genes increases (May et al., 1988). Exposure of *Methanobacterium formicicum* to tungsten results in the
synthesis of an inactive FDH that is identical in
subunit composition to the active enzyme but
which contains a metal-free molybdopterin
cofactor (May et al., 1988). In contrast, *Methanococcus vannielii* requires tungsten for optimum
growth on formate (Jones and Stadtman, 1977,
1981). Sequences as far as 140-bp upstream of

the site of transcription initiation are essential
for regulation of transcription of FDH-encoding
genes in *E. coli* (Birkmann et al., 1987), and the
nucleotide sequences upstream of the initiation
sites for *fdh* transcription in *E. coli* and *M. formicicum* are 61% identical over a 60-bp region
(Patel and Ferry, 1988). Two sequences, 8 bp
and 11 bp in length, are 100% identical (Patel
and Ferry, 1988), and it will be interesting to
learn if these sequences have the same regulatory functions in both species. In cells of *Methanococcus thermolithotrophicus* exposed to molecular hydrogen, FDH activity is reduced although
the FDH enzyme is not sensitive to H_2 (Sparling and Daniels, 1990). It appears that decreased synthesis of FDH in response to the
presence of H_2 is responsible for this decline in
FDH in *Methanococcus thermolithotrophicus* cells.

Hydrogenases and ferredoxins. Two NiFe hydrogenases, distinguished by their activity
toward the methanogen redox cofactor 8-
hydroxy-5-deazaflavin (coenzyme F_{420}), are
found in methanogens (Yamazaki, 1982; Jin et
al., 1983; Fox et al., 1987; Muth et al., 1987;
Wackett et al., 1987; Fauque, 1989; Fiebig and
Friedrich, 1989; Reeve and Beckler, 1990; Shah
and Clark, 1990): a coenzyme F_{420}-reducing
hydrogenase (FRH) and a methyl viologen-
reducing hydrogenase (MVH), which is incapable of reducing cofactor F_{420}. *Methanobacterium
thermoautotrophicum* (Hausinger, 1987) and
Methanococcus jannaschii (Shah and Clark, 1990)
possess both types of enzyme but some species, such as *Methanospirillum hungatei* (Sprott et
al., 1987), have been reported to contain only
the hydrogenase that does not reduce F_{420}.

The complete nucleotide sequences of the
genes encoding FRH (Alex et al., 1990) and
MVH (Reeve et al., 1989) have been determined from *Methanobacterium thermoautotrophicum* ΔH, and part of the MVH-encoding region
has also been cloned and sequenced from
Methanothermus fervidus (Steigerwald et al.,
1990). The FRH-encoding genes are apparently
organized in a single transcriptional unit
(*frhADGB*) in which *frhA*, *frhB*, and *frhG* encode the α, β, and γ subunits, respectively.
Although the function of the product of the
frhD gene is unknown, the *frhD* sequence indi-

cates a common evolutionary ancestry with the *hyaD* gene of *E. coli* (Menon et al., 1990). Genes encoding FRH in *Methanococcus voltae* have also been cloned (Sitzmann and Klein, 1991). The MVH-encoding genes in *Methanobacterium thermoautotrophicum* also apparently constitute a single transcriptional unit, with genes arranged in the order *mvhDGAB*. The *mvhD*, *mvhG*, and *mvhA* genes encode the δ, γ and α subunit polypeptides of MVH respectively.

A comparison of the large (*frhA* and *mvhA* gene products) and small (*frhG* and *mvhG* gene products) subunits of both methanogen hydrogenases with the corresponding subunits of NiFe hydrogenase from a range of *Bacteria* reveals conserved sequences of amino acids, suggesting a common ancestry for all these procaryotic NiFe hydrogenases (Prickril et al., 1986; Menon et al., 1987, 1990; Li et al., 1987; Leclerc et al., 1988; Reeve et al., 1989; Alex et al., 1990; Reeve and Beckler, 1990). In both *Methanobacterium thermoautotrophicum* (Reeve et al., 1989) and *Methanothermus fervidus* (Steigerwald et al., 1990), a fourth ORF, designated *mvhB*, is located promoter-distal to the MVH-encoding genes as part of the *mvh* transcriptional unit. Although the two *mvhB* gene products are only 64% identical, both are predicted to encode polypeptides containing six tandemly repeated bacterial ferredoxin-like domains, each possessing two [4Fe-4S] clusters and separated by hydrophilic α-helical regions. This polyferredoxin structure could function as an electron conduit, transferring electrons through the different [4Fe-4S] centers, possibly into the complex subcellular structures, designated methanoreductosomes, which have been implicated as the sites of methanogenesis (Wackett et al., 1987; Mayer et al., 1988).

The different domains of the polyferredoxin might supply electrons at different steps during the reduction of CO_2 to CH_4. The polyferredoxin could also act as a reservoir of reducing potential, maintaining a highly reduced intracellular environment by buffering the cell against transient exposures to oxidizing agents. Domains 2, 3, 4, and 5 of the *Methanobacterium thermoautotrophicum* polyferredoxin (Reeve et al., 1989) contain amino acid sequences that

resemble the pattern established for group 4 (archaeal) ferredoxins (Otaka and Ooi, 1987). In addition, domain 3 contains 18 amino acid residues within a 30-amino-acid sequence that are identical to a sequence of amino acids in a monoferredoxin purified from *M. barkeri* DSM800 (Hausinger et al., 1982). A 5-amino-acid sequence in domain 1 is also present at the same location within the ferredoxin-like region of the large subunit of the Fe hydrogenase of *Desulfovibrio vulgaris* (Voordouw and Brenner, 1985).

Monomeric ferredoxins have been isolated and characterized from a number of methanogens. Based on their amino acid sequences, the ferredoxins from *Methanosarcina* species resemble the clostridial ferredoxins, which contain two [4Fe-4S] centers (Hatchikian et al., 1989). The ferredoxin from *Methanosarcina barkeri* DSM800, however, contains a [3Fe-3S] cluster (Moura et al., 1982); the ferredoxin from *Methanosarcina barkeri* Fusaro contains two [4Fe-4S] clusters (Hatchikian et al., 1982); and the ferredoxin from *Methanosarcina thermophila* contains either a [3Fe-3S] or a [4Fe-4S] center (Terlesky and Ferry, 1988b). These ferredoxins appear to be involved in electron transport from pyruvate dehydrogenase (Hatchikian et al., 1982) and CODH (Terlesky and Ferry, 1988a). A ferredoxin containing two [4Fe-4S] clusters per molecule has also been characterized from *Methanococcus thermolithotrophicus*; this ferredoxin may also function as an electron acceptor for CODH (Hatchikian et al., 1989).

ATPases. The genes *atpA* and *atpB*, encoding the α and β subunits of a membrane-bound ATPase in *Methanosarcina barkeri*, have been cloned and sequenced (Inatomi et al., 1989). These genes are separated by only 3 bp with the RBS for *atpB* located within the terminal region of the *atpA* gene, indicating that both genes are transcribed as part of the same mRNA. The amino acid sequences predicted for the α and β subunits are 52% to 60% identical to the respective subunits of vacuolar H⁺-ATPases from *Neurospora crassa*, (Bowman et al., 1988a, 1988b) *Arabidopsis thaliana* (Manolson et al., 1988), *Saccharomyces cerevisiae* (Nelson et al., 1989), and to the corresponding subunits of ATPase from *Sulfolobus acidocaldarius* (Denda et al., 1988a, 1988b). The α and β subunits of *Methanosarcina barkeri* ATPase are 22% and 24% identical, respectively, to the β and α subunits of *E. coli* F_1 ATPase (Gay and Walker, 1981; Kanazawa et al., 1981, 1982; Saraste et al., 1981). Conservation of these ATPase amino acid sequences throughout the three biological domains suggests a common and ancient ancestry for this protein. Further, 126 residues are identical in an alignment of the α (578 amino acid residues) and β (459 amino acid residues) subunits of the *M. barkeri* ATPase, implying that the current *atpA* and *atpB* genes resulted from a gene duplication event (Inatomi et al., 1989). In contrast, the gene encoding a vanadate-sensitive, membrane associated ATPase in *Methanococcus voltae* shows no similarity to any previously sequenced gene (Dharmavaram et al., 1991). It encodes 565 amino acid residues including a 12 amino acid N-terminal sequence that is not present in the purified enzyme.

2.5 Gene Structure in Methanogens

Overview

The structure and organization of methanogen genes in general appears to follow the standard bacterial pattern (Reeve et al., 1986; Reeve, 1987, 1989; Brown et al., 1989; Sitzmann and Klein, 1991). As in *Bacteria*, one promoter may direct the transcription of several intron-free genes, producing a polycistronic messenger RNA (mRNA). Purine-rich sequences immediately upstream of ORFs presumably function, as in *Bacteria*, as ribosome-binding sites (RBS), positioning ribosomes for translation initiation. The structure of DNA-dependent RNA polymerases (RNAP) and the sequences of promoter regions in methanogens are, however, very different from their bacterial counterparts and more closely resemble their eucaryal equivalents (Gropp et al., 1986; Thomm et al., 1986, 1989; Berghöfer et al., 1988; Zillig

et al., 1989). In methanogens, transcription initiation is likely therefore to be an area of molecular biology that differs from the standard procaryotic model.

Codon usage

Methanogens employ the universal genetic code, with codon choice generally reflecting the overall GC content of the genome (Cue et al., 1985; Reeve et al., 1986; Brown et al., 1989). *Methanococcus voltae*, *Methanococcus vannielii*, and *Methanothermus fervidus* have low GC contents (30 to 33 mole%) and the codons in their housekeeping genes contain predominantly A and U residues, especially in the wobble position (Cue et al., 1985; Reeve et al., 1986; Brown et al., 1989). There is, however, a preference in highly expressed genes for a codon with a C in the wobble position in synonymous pairs of NNU/C codons (Cram et al., 1987; Weil et al., 1989). This has been interpreted as a mechanism to maximize translation, an argument that was advanced originally to explain the same codon preference in highly expressed genes in *Saccharomyces cerevisiae*, a eucaryote with a genome that is 70% AT (Bennetzen and Hall, 1982).

Functional reading frames in *Eucarya* and *Bacteria* are dominated by RNY codons (R, purines; Y, pyrimidines; N, any base) (Shepard, 1981), and this also holds true for most protein-encoding genes in archaeal methanogens (Brown et al., 1989). An exception is the *purE* gene of *Methanobacterium thermoautotrophicum* ΔH. Although the *purE* gene of *Methanobrevibacter smithii* conforms to the RNY rule, and these two *purE* genes have clearly evolved from a common ancestor, only the *M. smithii purE* (30.6 mol% GC) and not the *M. thermoautotrophicum purE* gene (47 mol% GC) now contains predominantly RNY codons (Hamilton and Reeve, 1985b).

Introns

Most genes in higher *Eucarya* contain intervening sequences (introns) separating coding regions (exons) (Sharp, 1985; Darnell and Doolittle, 1986; Gilbert et al., 1986). In contrast, very few introns have been found in procaryotic genes and none, so far, in methanogen genes. Introns present in the first genomes may have been lost from most procaryotic genes through an evolutionary streamlining process, or introns may have been added to originally intron-free genes during the development of the *Eucarya*. Both arguments can be supported by the observation that the *argG* genes of *Methanococcus vannielii* and *Methanosarcina barkeri* do not contain introns, whereas the ancestrally homologous human gene contains at least nine introns (Morris and Reeve, 1988). In the *nifH* region of *Methanococcus thermolithotrophicus*, an unusual nucleotide sequence is flanked by sequences that resemble the consensus sequences for intron splice sites, suggesting the vestiges of an intron (Souillard et al., 1988). Only one mobile element, ISM1, has been cloned and sequenced from a methanogen (Hamilton and Reeve, 1985a), although many such elements are known to be present, and their mobilities are known to be responsible for major genomic rearrangements that occur in the related halophilic *Archaea* (Sapienza et al., 1982; Brown et al., 1989).

Methanogen RNAP, promoters, and terminators

Methanogen RNAPs contain 10 to 12 subunit polypeptides (Thomm et al., 1986; see also Chapter 3, this volume). The largest, designated A, B', B'', and C, are most closely related to subunits of eucaryal RNAPs. Methanogen RNAPs are not inhibited by rifampicin, streptolydigin, or α-amanitin. Methanogen promoters appear to contain at least two structural elements. Transcription is initiated in vivo primarily at G residues in a weakly conserved ATGC sequence known as box B. Approximately 25 bp 5' to box B is a more conserved box A element with a consensus sequence in methanogens of 5'TTTATATA (Thomm et al., 1989). The sequence 5'TATATA-(18 or 19 bases)-TGC is essential for promoters of stable RNA operons in *Methanococcus vannielii* (Thomm and Wich, 1988). Methanogen promoters therefore resemble eucaryal promoters, which also con-

tain an upstream "TATA" box and a short conserved sequence adjacent to the site of transcription initiation. In common with bacterial but not eucaryal RNAPs, promoter binding in vitro by purified RNAPs (Thomm et al., 1989; Brown and Reeve, 1989) does not require additional transcription factors. Accurate transcription initiation does, however, involve additional factors that can be separated from the RNAP of *Methanococcus thermolithotrophicus* by phosphocellulose chromatography (Thomm and Stetter, 1985; Frey et al., 1990).

Analyses of sequences immediately downstream of methanogen genes has, in some cases, revealed inverted repeat sequences followed by an oligo-T sequence and in other cases only the presence of oligo-T sequences (Brown et al., 1989). It appears therefore that stem-loop structures in transcripts may be involved in terminating some transcripts but that such structures are not always needed. As inverted repeat sequences have not been detected downstream of genes in the hyperthermophilic species *Methanothermus fervidus*, the spontaneous formation of short regions of dsRNA may not be useful as a regulatory mechanism in species that grow above 80°C (Weil et al., 1988; Haas et al., 1989, 1990).

2.6   Conclusions

During the past decade, a substantial body of information has been accumulated that demonstrates that although methanogens are phylogenetically *Archaea* and are fastidiously anaerobic, their molecular biology is not radically different from that of organisms that have been better studied. Methanogen genes can be expressed in *Bacteria*, and there seems no doubt that both bacterial (Possot et al., 1989a) and eucaryal genes can, with the appropriate manipulations, be expressed in methanogens. The current limitations to this latter work appear to be technical rather than fundamental biological differences. Many genes have been cloned and are available, together with plasmids, phages and transformation protocols, for use with methanogens (see Chapter 1, this volume). We anticipate that during the next decade many methane genes will be cloned and the regulation of their expression determined. Using recombinant genetic techniques, methanogens will be constructed that can catalyze the biodegradation of a substantially wider range of substrates to methane. Bioremediation of polluted environments and methane generation, directly from toxic pollutants, by genetically engineered methanogens is now conceivable. Development of green plants specifically as substrates for biogas production by methanogens is also worth considering in locations where the growth of conventional crops is prevented by local soil, climatic, or economic conditions.

Acknowledgments. We thank those individuals who provided us with information before it had been published. Research in the authors laboratory is supported by the U.S. Department of Energy, the U.S. Office of Naval Research, and the National Science Foundation.

References

Abbanat, D.R. and J.G. Ferry. 1990. Synthesis of acetyl coenzyme A by carbon monoxide dehydrogenase complex from acetate-grown *Methanosarcina thermophila. J. Bacteriol.* **172**:7145–7150.

Achenbach-Richter, L. and C.R. Woese. 1988. The ribosomal gene spacer region in Archaebacteria. *Syst. Appl. Microbiol.* **10**:211–214.

Alex, L.A., J.N. Reeve, W.H. Orme-Johnson and C.T. Walsh. 1990. Cloning, sequence determination and expression of the genes encoding the subunits of the nickel-containing 8-hydroxy-5-deazaflavin reducing hydrogenase from *Methanobacterium thermoautotrophicum* ΔH. *Biochemistry* **29**:7237–7244.

Allmansberger, R., C. Bollschweiler, U. Konheiser, B. Muller, E. Muth, G. Pasti, and A. Klein. 1986. Arrangement and expression of methyl CoM reductase genes in *Methanococcus voltae. Syst. Appl. Microbiol.* **7**:13–17.

Auer, J., K. Lechner and A. Böck. 1989a. Gene organization and structure of two transcriptional units from *Methanococcus* coding for ribosomal proteins and elongation factors. *Can. J. Microbiol.* **35**:200–204.

Auer, J., G. Spicker and A. Böck. 1989b. Organization and structure of the *Methanococcus* transcriptional unit homologous to the *Escherichia coli* "spectinomycin operon." *J. Mol. Biol.* **209**:21–36.

Baier, G., W. Piendl, B. Redl, and G. Stoffler. 1990. Structure, organization and evolution of the L1 equivalent ribosomal protein gene of the archaebacterium *Methanococcus vannielii. Nucleic Acids Res.* **18**:719–724.

Balch, W.E., G.E. Fox, L.J. Magrum, C.R. Woese, and R.S. Wolfe. 1979. Methanogens: reevaluation of a unique biological group. *Microbiol. Rev.* **43**:260–296.

Beckler, G.S. and J.N. Reeve. 1986. Conservation of primary structure in the *hisI* gene of the archaebacterium, *Methanococcus vannielii,* the eubacterium *Escherichia coli,* and the eucaryote *Saccharomyces cerevisiae.* 1986. *Mol. Gen. Genet.* **204**:133–140.

Belay, N., R. Sparking and L. Daniels. 1984. Dinitrogen fixation by a thermophilic methanogenic bacterium. *Nature* (London) **312**:286–288.

Bennetzen, J. and B.D. Hall. 1982. Codon selection in yeast. *J. Biol. Chem.* **257**:3026–3031.

Berghöfer, B., L. Kröckel, C. Körtner, M. Truss, J. Schallenberg, and A. Klein. 1988. Relatedness of archaebacterial RNA polymerase core subunits to their eubacterial and eukaryotic equivalents. *Nucleic Acids Res.* **16**:8113–8128.

Bhatnagar, L., J.A. Krzycki and J.G. Zeikus. 1987. Analysis of hydrogen metabolism in *Methanosarcina barkeri*: regulation of hydrogenase and role of CO-dehydrogenase in H_2 production. *FEMS Microbiol. Lett.* **41**:337–343.

Birkman, A., F. Zinoni, G. Sawers, and A. Böck. 1987. Factors affecting transcriptional regulation of the formate-hydrogen-lyase pathway of *Escherichia coli. Arch. Microbiol.* **148**:44–51.

Blaut, M. and G. Gottschalk. 1985. Evidence for a chemiosmotic mechanism of ATP synthesis in methanogenic bacteria. *Trends Biochem. Sci.* **10**:486–489.

Blaut, M., V. Müller, and G. Gottschalk. 1987. Proton translocation coupled to methanogenesis from methanol + hydrogen in *Methanosarcina barkeri. FEBS Lett.* **215**:53–57.

Bobik, T.A., K.D. Olson, K.M. Noll, and R.S. Wolfe. 1987. Evidence that the heterodisulfide of coenzyme M and 7-mercaptoheptanoyl-threonine phosphate is a product of the methylreductase reaction in *Methanobacterium. Biochem. Biophys. Res. Commun.* **149**:455–460.

Bokranz, M., G. Baumner, R. Allmansberger, D. Ankel-Fuchs, and A. Klein. 1988. Cloning and characterization of the methyl coenzyme M reductase genes from *Methanobacterium thermoautotrophicum. J. Bacteriol.* **170**:568–577.

Bokranz, M. and A. Klein. 1987. Nucleotide sequence of the methyl coenzyme M reductase gene cluster from *Methanosarcina barkeri. Nucleic Acids Res.* **15**:4350–4351.

Bomar, M., K. Knoll, and F. Widdel. 1985. Fixation of molecular nitrogen by *Methanosarcina barkeri. FEMS Microbiol. Ecol.* **31**:47–55.

Bouthier de la Tour, C., C. Portemer, R. Huber, P. Forterre, and M. Duguet. 1991. Reverse gyrase in thermophilic eubacteria. *J. Bacteriol* 173:3921–3923.

Bouthier de la Tour, C., C. Portemer, M. Nadal. K.O. Stetter, P. Forterre, and M. Duguet. 1990. Reverse gyrase, a hallmark of the hyperthermophilic Archaebacteria. *J. Bacteriol.* 172:6803–6808.

Bowman, B.J., R. Allen, M.A. Wechser, and E.J. Bowman. 1988a. Isolation of genes encoding the *Neurospora* vacuolar ATPase. *J. Biol. Chem.* 263:14002–14007.

Bowman, E.J., K. Tenney, and B.J. Bowman. 1988b. Isolation of genes encoding the *Neurospora* vacuolar ATPase. *J. Biol. Chem.* 263:13994–14001.

Bröckl, G., M. Behr, S. Fabry, R. Hensel, H. Kaudewitz, E. Biendl, and H. König. 1991. Analysis and nucleotide sequence of the genes encoding the surface-layer glycoproteins of the hyperthermophilic methanogens *Methanothermus fervidus* and *Methanothermus sociabilis*. *Eur. J. Biochem.* 199:147–152.

Brown, J.W., C.J. Daniels, and J.N. Reeve. 1989. Gene structure, organization and expression in archaebacteria. *CRC Crit. Rev. Microbiol.* 16:287–338.

Brown, J.W. and J.N. Reeve, 1989. Transcription initiation and a RNA polymerase binding site upstream of the *purE* gene of the archaebacterium *Methanobacterium thermoautotrophicum* strain ΔH. *FEMS Microbiol. Lett.* 60:131–136.

Chartier, F., B. Laine, and P. Sautière. 1988. Characterization of the chromosomal protein MC1 from the thermophilic archaebacterium *Methanosarcina* spp. CHTI 55 and its effect on the thermal stability of DNA. *Biochim. Biophys. Acta* 951:149–156.

Chartier, F., B. Laine, D. Bélaïche, and P. Sautière. 1989a. Primary structure of the chromosomal proteins MC1a, MC1b and MC1c from the archaebacterium *Methanothrix soehngenii*. *J. Biol. Chem.* 264:17006–17015.

Chartier, F., B. Laine, D. Bélaïche, J.-P. Touzel, and P. Sautière. 1989b. Primary structure of the chromosomal protein MC1 from the archaebacterium *Methanosarcina* sp. CHTI 55. *Biochim. Biophys. Acta* 1008:309–314.

Cram, D.S., B.A. Sherf, R.T. Libby, R.J. Mattaliano, K.L. Ramachandran, and J.N. Reeve. 1987. Structure and expression of the genes, *mcrBDCGA*, which encode the subunits of component C of methyl coenzyme M reductase in *Methanococcus vannielii*. *Proc. Natl. Acad. Sci. USA* 84:3992–3996.

Cue, D., G.S. Beckler, J.N. Reeve, and J. Konisky. 1985. Structure and sequence divergence of two archaebacterial genes. *Proc. Natl. Acad. Sci. USA* 82:4207–4211.

Daniels, L. 1984. Biological methanogenesis: physiological and practical aspects. *Trends Biotechnol.* 2:91–98.

Daniels, C.J., R. Gupta, and W.F. Doolittle. 1985. Transcription and excision of a large intron in the tRNA^{Trp} gene of an archaebacterium, *Halobacterium volcanii*. *J. Biol. Chem.* 260:3132–3134.

Darnell, J.E. and W.F. Doolittle. 1986. Speculations on the early course of evolution. *Proc. Natl. Acad. Sci. USA* 83:1271–1275.

Denda, K., J. Konishi, T. Oshima, T. Date, and M. Yoshida. 1988a. The membrane-associated ATPase from *Sulfolobus acidocaldarius* is distantly related to F_1-ATPase as assessed from the primary structure of its α-subunit. *J. Biol. Chem.* 263:6012–6015.

Denda, K., J. Konishi, T. Oshima, T. Date, and M. Yoshida. 1988b. Molecular cloning of the β-subunit of a possible non-F_0F_1 type ATP synthase from the acidothermophilic archaebacterium, *Sulfolobus acidocaldarius*. *J. Biol. Chem.* 263:17251–17254.

Deppenmeier, U., M. Blaut, and G. Gottschalk. 1989. Dependence on membrane components of methanogenesis from methyl-CoM with formaldehyde or molecular hydrogen as electron donors. *Eur. J. Biochem.* 186:317–323.

Dharmavaram R., P. Gillevet, and J. Konisky. 1991. Nucleotide sequence of the gene encoding the vanadate-sensitive membrane-associated ATPase of *Methanococcus voltae*. *J. Bacteriol.* 173: 2131–2133.

DiMarco, A.A., T.A. Bobik, and R.S. Wolfe. 1990a. Unusual coenzymes of methanogenesis. *Annu. Rev. Biochem.* 59:355–394.

DiMarco, A.A., K.A. Sment, J. Konisky, and R.S. Wolfe. 1990b. The formylmethanofuran:tetrahydromethanopterin formyltransferase from *Methanobacterium thermoautotrophicum* ΔH. *J. Biol. Chem.* 265:472–476.

Donnelly, M.I. and R.S. Wolfe. 1986. The role of formylmethanofuran:tetrahydromethanopterin formyltransferase in methanogenesis from carbon dioxide. *J. Biol. Chem.* 261:16653–16659.

Eggen. R.I.L., H. Harmsen, A.C.M. Geerling, and W.M. de Vos. 1989. Nucleotide sequence of a 16S rRNA encoding gene from the archaebacterium *Methanothrix soehngenii*. *Nucleic Acids Res.* 17:9469.

Eggen R.I.L., A.C.M. Geerling, M.S.M. Jetten, and W.M. de Vos. 1991. Cloning, expression, and sequence analysis of the genes for carbon monoxide dehydrogenase of *Methanothrix soehngenii*. *J. Biol. Chem.* 266:6883–6887.

Ehhalt, D.H. 1974. The atmospheric cycle of methane. *Tellus* 26:58–70.

Eikmanns, B. and R.K. Thauer. 1984. Catalysis of an isotopic exchange between CO_2 and the carboxyl group of acetate by *Methanosarcina barkeri* grown on acetate. *Arch. Microbiol.* 138:365–370.

Eikmanns, B. and R.K. Thauer. 1985. Evidence for the involvement and role of a corrinoid enzyme in methane formation from acetate in *Methanosarcina barkeri. Arch. Microbiol.* 142:175–179.

Ellefson, W.L. and R.S. Wolfe. 1981. Component C of the methylreductase system of *Methanobacterium. J. Biol. Chem.* 256:4259–4262.

Ellermann, J., R. Hedderich, R. Böcher, and R.K. Thauer. 1988. The final step in methane formation. *Eur. J. Biochem.* 172:669–677.

Ellermann, J., A. Kobelt, A. Pfaltz, and R.K. Thauer. 1987. On the role of N-7-mercaptoheptanoyl-O-phospho-L-threonine (component B) in the enzymatic reduction of methyl-coenzyme M to methane. *FEBS Lett.* 220:358–362.

Ellermann, J., S. Rospert, R.K. Thauer, M. Bokranz, A. Klein, M. Voges, and A. Berkessel. 1989. Methyl-coenzyme-M reductase from *Methanobacterium thermoautotrophicum* (strain Marburg). *Eur. J. Biochem.* 184:63–68.

Fabry, S. and R. Hensel. 1987. Purification and characterization of D-glyceraldehyde-3-phosphate dehydrogenase from the thermophilic archaebacterium *Methanothermus fervidus. Eur. J. Biochem.* 165:147–155.

Fabry S. and R. Hensel. 1988. Primary structure of glyceraldehyde-3-phosphate dehydrogenase deduced from the nucleotide sequence of the thermophilic archaebacterium *Methanothermus fervidus. Gene* (Amst.) 64:189–197.

Fabry, S., J. Lang, T. Niermann, M. Vingron, and R. Hensel. 1989. Nucleotide sequence of the glyceraldehyde-3-phosphate dehydrogenase gene from the mesophilic methanogenic archaebacteria *Methanobacterium bryantii* and *Methanobacterium formicicum. Eur. J. Biochem.* 179:405–413.

Fardeau, M.-L., J.-P. Peillex, and J.-P. Belaïch. 1987. Energetics of the growth of *Methanobacterium thermoautotrophicum* and *Methanococcus thermolithotrophicus* on ammonium chloride and dinitrogen. *Arch. Microbiol.* 148:128–131.

Fauque, G. 1989. Properties of (NiFe) and (NiFeSe) hydrogenases from methanogenic bacteria. In: *Microbiology of Extreme Environments and Its Potential for Biotechnology*, M.S. Costa, J.C. Da Duarte, and R.A.D. Williams, eds., pp. 216–236. London: Elsevier.

Fiebig, K. and B. Friedrich. 1989. Purification of the F_{420}-reducing hydrogenase from *Methanosarcina barkeri* (strain Fusaro). *Eur. J. Biochem.* 184:79–88.

Fox, J.A., D.J. Livingston, W.H. Orme-Johnson, and C.T. Walsh. 1987. 8-hydroxy-5-deazaflavin-reducing hydrogenase from *Methanobacterium thermoautotrophicum*: 1. Purification and characterization. *Biochemistry* 26:4219–4227.

Frey, G., M. Thomm, B. Brüdigam, H.P. Gohl, and W. Hausner. 1990. An archaebacterial cell-free transcription system. The expression of

tRNA genes from *Methanococcus vannielii* is mediated by a transcription factor. *Nucleic Acids Res.* 18:1361–1387.

Frimmer, U. and F. Widdel. 1989. Oxidation of ethanol by methanogenic bacteria. *Arch. Microbiol.* 152:479–483.

Gay, N.J. and J.E. Walker. 1981. The *atp* operon: nucleotide sequence of the promoter and the genes for the membrane proteins, and the Δ subunit of *Escherichia coli* ATP-synthase. *Nucleic Acids Res.* 9:3919–3926.

Gilbert, W., M. Marchionni, and G. Knight. 1986. On the antiquity of introns. *Cell* 46:151–154.

Görisch, H. and K.-D. Jany. 1989. Archaebacterial malate dehydrogenase: the amino-terminal sequence of the enzyme from *Sulfolobus acidocaldarius* is homologous to the eubacterial and eukaryotic malate dehydrogenases. *FEBS Lett.* 247:259–262.

Gropp, F., W.D. Reiter, A. Sentenac, W. Zillig, R. Schnabel, M. Thomm, and K.O. Stetter. 1986. Homologies of components of DNA-dependent RNA polymerases of archaebacteria, eukaryotes and eubacteria. *Syst. Appl. Microbiol.* 7: 95–101.

Haas, E.S., C.J. Daniels, and J.N. Reeve. 1989. Genes encoding 5S rRNA and tRNAs in the extremely thermophilic archaebacterium *Methanothermus fervidus. Gene* (Amst.) 77:253–263.

Haas, E.S., J.W. Brown, C.J. Daniels, and J.N. Reeve. 1990. Genes encoding the 7S RNA and tRNA[Ser] are linked to one of the two rRNA operons in the genome of the extremely thermophilic archaebacterium *Methanothermus fervidus. Gene* (Amst.) 90:51–59.

Hamilton, P.T. and J.N. Reeve. 1985a. Structure of genes and an insertion element in the methane producing archaebacterium *Methanobrevibacter smithii. Mol. Gen. Genet.* 200:47–59.

Hamilton, P.T. and J.N. Reeve. 1985b. Sequence divergence of an archaebacterial gene cloned from a mesophilic and a thermophilic methanogen. *J. Mol. Evol.* 22:351–360.

Hatchikian, E.C., M. Bruschi, N. Forget, and M. Scandellari. 1982. Electron transport components from methanogenic bacteria: the ferredoxin from *Methanosarcina barkeri* (strain fusaro). *Biochem. Biophys. Res. Commun.* 109:1316–1323.

Hatchikian, E.C., M.L. Fardeau, M. Bruschi, J.P. Belaich, A. Chapman, and R. Cammack. 1989. Isolation, characterization and biological activity of the *Methanococcus thermolithotrophicus* ferredoxin. *J. Bacteriol.* 171:2384–2390.

Hausinger, R.P. 1987. Nickel utilization by microorganisms. *Microbiol. Rev.* 51:22–42.

Hausinger, R.P., I. Moura, J.J.G. Moura, A.V. Xavier, M.H. Santos, J. LeGall, and J.B. Howard. 1982. Amino acid sequence of a 3Fe:3S ferredoxin from the "Archaebacterium" *Metha-*

nosarcina barkeri (DSM 800). *J. Biol. Chem.* **257**:14192–14197.

Hedderich, R. and R.K. Thauer. 1988. *Methanobacterium thermoautotrophicum* contains a soluble enzyme system that specifically catalyzes the reduction of the heterodisulfide of coenzyme M and 7-mercaptoheptanoylthreonine phosphate with H_2. *FEBS Lett.* **234**:223–227.

Hedderich, R., A. Berkessel, and R.K. Thauer. 1989. Catalytic properties of the heterodisulfide reductase involved in the final step of methanogenesis. *FEBS Lett.* **255**:67–71.

Hippe, H., D. Caspari, Fiebig, and G. Gottschalk. 1979. Utilization of trimethylamine and other N-methyl compounds for growth and methane formation by *Methanosarcina barkeri*. *Proc. Natl. Acad. Sci. USA* **76**:494–498.

Honka, E., S. Fabry, T. Niermann, P. Palm and R. Hensel. 1990. Properties and primary structure of the L-malate dehydrogenase from the extremely thermophilic archaebacterium *Methanothermus fervidus*. *Eur. J. Biochem.* **188**:623–632.

Huber, R., M. Kurr, H.W. Jannasch, and K.O. Stetter. 1989. A novel group of abyssal methanogenic archaebacteria (*Methanopyrus*) growing at 110°C. *Nature* (London) **342**:833–834.

Imbert, M., B. Laine, N. Helbecque, J.-P. Mornon, J.-P. Hénichart, and P. Sautière. 1990. Conformational study of the chromosomal protein MC1 from the archaebacterium *Methanosarcina barkeri*. *Biochim. Biophys. Acta.* **1038**:346–354.

Inatomi, K.-I., S. Eya, M. Maeda, and M. Futai. 1989. Amino acid sequence of the α and β subunits of *Methanosarcina barkeri* ATPase deduced from cloned genes. *J. Biol. Chem.* **264**:10954–10959.

Iwabe, N., K.-I. Kuma, M. Hasegawa, S. Osawa, and T. Miyata. 1989. Evolutionary relationship of archaebacteria, eubacteria and eukaryotes inferred from phylogenetic trees of duplicated genes. *Proc. Natl. Acad. Sci. USA* **86**:9355–9359.

Jarrel, K.F. and M.L. Kalmokoff. 1988. Nutritional requirements of the methanogenic archaebacteria. *Can. J. Microbiol.* **34**:557–576.

Jarrel, K.F. and S.F. Koval. 1989. Ultrastructure and biochemistry of *Methanococcus voltae*. *CRC Crit. Rev. Microbiol.* **17**:53–87.

Jarsch, M., J. Altenbuchner, and A. Böck. 1983. Physical organization of the genes for ribosomal RNA in *Methanococcus vannielii*. *Mol. Gen. Genet.* **189**:41–47.

Jarsch, M. and A. Böck. 1983. DNA sequence of the 16S rRNA/23S rRNA intercistronic spacer of two rRNA operons of the archaebacterium *Methanococcus vannielii*. *Nucleic Acid Res.* **11**:7537–7544.

Jarsch, M. and A. Böck. 1985a. Sequence of the 16S ribosomal RNA gene from *Methanococcus vannielii*: evolutionary implications. *Syst. Appl. Microbiol.* **6**: 54–59.

Jarsch, M. and A. Böck. 1985b. Sequence of the 23S rRNA gene from the archaebacterium *Methanococcus vannielii*: evolutionary and functional implications. *Mol. Gen. Genet.* **200**:305–312.

Jenal U., T. Rechsteiner, P.Y. Tan, E. Bühlmann, L. Meile and T. Leisinger. 1991. Isoleucyl-tRNA synthetase of *Methanobacterium thermoautotrophicum* Marburg *J. Biol. Chem.* **266**:10570–10577.

Jin, S.-L.C., D.K. Blanchard, and J.-S. Chen. 1983. Two hydrogenases with distinct electron-carrier specificity and subunit composition in *Methanobacterium formicicum*. *Biochim. Biophys. Acta* **748**:8–20.

Jones, J.B. and T.C. Stadtman. 1977. *Methanococcus vannielii*: culture and effects of selenium and tungsten on growth. *J. Bacteriol.* **130**:1404–1406.

Jones, J.B. and T.C. Stadtman. 1981. Selenium-dependent and selenium-independent formate dehydrogenases of *Methanococcus vannielii*. *J. Biol. Chem.* **256**:656–663.

Jones, J.W., M.I. Donnelly, and R.S. Wolfe. 1985. Evidence of a common pathway of carbon dioxide reduction to methane in methanogens. *J. Bacteriol.* **163**:126–131.

Jones, J.W., J.A. Leigh, F. Mayer, C.R. Woese, and R.S. Wolfe. 1983. *Methanococcus jannaschii* sp. nov., an extremely thermophilic methanogen from a submarine hydrothermal vent. *Arch. Microbiol.* **136**:254–261.

Jones, W.J., D.P. Nagle, Jr., and W.B. Whitman. 1987. Methanogens and the diversity of archaebacteria. *Microbiol. Rev.* **51**:135–177.

Jones, W.J., C.E. Stugard, and H.W. Jannasch. 1989. Comparison of thermophilic methanogens from submarine hydrothermal vents. *Arch. Microbiol.* **151**:314–318.

Kaesler, B. and P. Schönheit. 1989a. The role of sodium ions in methanogenesis. *Eur. J. Biochem.* **184**:223–232.

Kaesler, B. and P. Schönheit. 1989b. The sodium cycle in methanogenesis. *Eur. J. Biochem.* **186**:309–316.

Kaine, B.P. 1987. Intron-containing tRNA genes of *Sulfolobus solfataricus*. *J. Mol. Evol.* **25**:248–256.

Kaine, B.P. 1990. Structure of the archaebacterial 7S RNA molecule. *Mol. Gen. Genet.* **221**:315–321.

Kaine, B.P., R. Gupta, and C.R. Woese. 1983. Putative introns in tRNA genes of procaryotes. *Proc. Natl. Acad. Sci. USA* **80**:3309–3313.

Kaine, B.P and V.L. Merkel. 1989. Isolation and characterization of the 7S RNA gene from *Methanococcus voltae*. *J. Bacteriol.* **171**:4261–4266.

Kalmokoff, M.L., T.M. Karnauchow, and K.F. Jarrell. 1990. Conserved N-terminal sequences in the flagellins of archaebacteria. *Biochem. Biophys. Res. Commun.* **167**:154–160.

Kanazawa, H., T. Kayano, T. Kiyasu, and M.

Futai. 1982. Nucleotide sequence of the genes for β and ϵ subunits of proton-translocation ATPase from *Escherichia coli*. *Biochem. Biophys. Res. Commun.* **105**:1257–1264.

Kanazawa, H., T. Kayano, K. Mabuchi, and M. Futai. 1981. Nucleotide sequence of the genes coding for α, β and γ subunits of the proton-translocating ATPase of *Escherichia coli*. *Biochem. Biophys. Res. Commun.* **103**:604–612.

Keltjens, J.T. and C. van der Drift. 1986. Electron transfer reactions in methanogens. *FEMS Microbiol. Rev.* **39**:259–303.

Keltjens, J.T. and G.D. Vogels. 1988. Methanopterin and methanogenic bacteria. *BioFactors* **1**:95–103.

Klein, A. and M. Schnorr. 1984. Genome complexity of methanogenic bacteria. *J. Bacteriol.* **158**:628–631.

Klein, A., R. Allmansberger, M. Bokranz, S. Knaub, B. Muller, and E. Muth. 1988. Comparative analysis of genes encoding methyl coenzyme M reductase in methanogenic bacteria. *Mol. Gen. Genet.* **213**:409–420.

König, H., E. Nusser, and K.O. Stetter. 1985. Glycogen in *Methanolobus* and *Methanococcus*. *FEMS Microbiol. Lett.* **28**:265–269.

Konisky, J. 1989. Methanogens for biotechnology: application of genetics and molecular biology. *Trends Biotechnol.* **7**:88–92.

Krzycki, J.A. and R.C. Prince. 1990. EPR observation of carbon monoxide dehydrogenase, methyreductase and corrinoid in intact *Methanosarcina barkeri* during methanogenesis from acetate. *Biochim. Biophys. Acta* **1015**:53–60.

Krzycki, J.A., L.J. Lehman, and J.G. Zeikus. 1985. Acetate catabolism by *Methanosarcina barkeri*: evidence for involvement of carbon monoxide dehydrogenase, methyl coenzyme M and methylreductase. *J. Bacteriol.* **163**:1000–1006.

Laine, B., F. Chartier, M. Imbert, R. Lewis, and P. Sautière. 1986. Primary structure of the chromosomal protein HMb from the archaebacteria *Methanosarcina barkeri*. *Eur. J. Biochem.* **161**:681–687.

Lancaster, J.R., Jr. 1989. Sodium, protons and energy coupling in the methanogenic bacteria. *J. Bioenerg. Biomembr.* **21**:717–740.

Lechner, K., G. Heller, and A. Böck. 1989. Organization and nucleotide sequence of a transcriptional unit of *Methanococcus vannielii* comprising genes for protein synthesis elongation factors and ribosomal proteins. *J. Mol. Evol.* **29**:20–27.

Lechner, K., G. Wich, and A. Böck. 1985. The nucleotide sequence of the 16S rRNA gene and flanking regions from *Methanobacterium formicicum*: the phylogenetic relationship between methanogenic and halophilic archaebacteria. *Syst. Appl. Microbiol.* **6**:157–163.

Leclerc, M., A. Colbeau, B. Cauvin, and P.M. Vignais. 1988. Cloning and sequencing of the genes encoding the large and the small subunits of the H_2 uptake hydrogenase (*hup*) of *Rhodobacter capsulatus*. *Mol. Gen. Genet.* **214**:97–107.

Li, C., H.D. Peck, Jr., J. LeGall, and A.E. Przybyla. 1987. Cloning, characterization and sequencing of the genes encoding the large and small subunits of the periplasmic [NiFe] hydrogenase of *Desulfovibrio gigas*. *DNA* **6**:539–551.

Lobo, A.L. and S.H. Zinder. 1988. Diazotrophy and nitrogenase activity in the archaebacterium *Methanosarcina barkeri* 227. *Appl. Environ. Microbiol.* **54**:1656–1661.

Lobo, A.L. and S.H. Zinder. 1990. Nitrogenase in the archaebacterium *Methanosarcina barkeri* 227. *J. Bacteriol.* **172**:6789–6796.

Lunnen, K.D., R.D. Morgan, C.J. Timan, J.A. Krzycki, J.N. Reeve, and G.G. Wilson. 1989. Characterization and cloning of *Mwo*I (GCN$_7$GC), a new type-II restriction-modification system from *Methanobacterium wolfei*. *Gene* (Amst.) **77**:11–19.

Ma, K. and R.K. Thauer. 1990. Purification and properties of N^5, N^{10}-methylene-tetrahydromethanopterin reductase from *Methanobacterium thermoautotrophicum* (strain Marburg). *Eur. J. Biochem.* **191**:187–193.

Magot, M., O. Possot, N. Souillard, M. Henriquet and L. Sibold. 1986. Structure and expression of *nif* (nitrogen fixation) genes in methanogens. *In: Biology of Anaerobic Bacteria*, H.C. Dubourguier, G. Albagnac, J. Montreuil, C. Romond, P. Sautiere, and J. Guillaume, eds., pp. 193–199. Amsterdam: Elsevier.

Manolson, M.F., B.F.F. Ouellette. M. Filion and R.J. Poole. 1988. cDNA sequence and homologies of the "57-kDa" nucleotide-binding subunit of the vacuolar ATPase from *Arabidopsis*. *J. Biol. Chem.* **263**:17987–17994.

May, H.D., P.S. Patel, and J.G. Ferry. 1988. Effect of molybdenum and tungsten on synthesis and composition of formate dehydrogenase in *Methanobacterium formicicum*. *J. Bacteriol.* **170**:3384–3389.

Mayer, F., M. Rohde, M. Salzmann, A. Jussofie, and G. Gottschalk. 1988. The methanoreductosome: a high-molecular-weight enzyme complex in the methanogenic bacterium strain Göl that contains components of the methylreductase system. *J. Bacteriol.* **170**:1438–1444.

Meile, L., R. Stettler, R. Banholzer, M. Kotik, and T. Leisinger, 1991. Tryptophan gene cluster of *Methanobacterium thermoautotrophicum* Marburg: molecular cloning and nucleotide sequence of a putative *trpEGCFBAD* operon. *J. Bacteriol.* **173**:5017–5023.

Menon, N.K., H.D. Peck, Jr., J. LeGall, and A.E. Przybyla. 1987. Cloning and sequencing of the genes encoding the large and small subunits of the periplasmic (NiFeSe) hydrogenase of

Desulfovibrio baculatus. J. Bacteriol. **169**:5401–5407.

Menon, N.K., J. Robbins, H.D. Peck, C.Y. Chatelus, E.S. Choi, and A.E. Przybyla. 1990. Cloning and sequencing of a putative *Escherichia coli* [NiFe] hydrogenase-1 operon containing six open reading frames. *J. Bacteriol.* **172**:1969–1977.

Morris, C.J. and J.N. Reeve. 1984. Functional expression of an archaebacterial gene from the methanogen *Methanosarcina barkeri* in *Escherichia coli* and *Bacillus subtilis*. In: *Microbial Growth on C1 Compounds*, R.L. Crawford, and R.S. Hanson, eds., pp. 205–209. Washington, D.C.: American Society for Microbiology.

Morris, C.J. and J.N. Reeve. 1988. Conservation of structure in the human gene encoding argininosuccinate synthetase and the *argG* genes of the archaebacteria *Methanosarcina barkeri* MS and *Methanococcus vannielii. J. Bacteriol.* **170**:3125–3130.

Moura, J., J.J.G. Moura, B.H. Huynh, and H. Santos. 1982. Ferredoxin from *Methanosarcina barkeri*: evidence for the presence of a three-iron center. *Eur. J. Biochem.* **126**:95–98.

Müller, V., M. Blaut, and G. Gottschalk. 1987a. Generation of a transmembrane gradient of Na$^+$ in *Methanosarcina barkeri. Eur. J. Biochem.* **162**:461–466.

Müller, V., M. Blaut, and G. Gottschalk. 1988a. The transmembrane electrochemical gradient of Na$^+$ as driving force for methanol oxidation in *Methanosarcina barkeri. Eur. J. Biochem.* **172**:601–606.

Müller, V., C. Winner, and G. Gottschalk. 1988b. Electron-transport-driven sodium extrusion during methanogenesis from formaldehyde and molecular hydrogen by *Methanosarcina barkeri. Eur. J. Biochem.* **178**:519–525.

Müller, V., G. Kozianowski, M. Blaut, and G. Gottschalk. 1987b. Methanogenesis from trimethylamine + H$_2$ by *Methanosarcina barkeri* is coupled to ATP formation by a chemiosmotic mechanism. *Biochim. Biophys. Acta* **892**:207–212.

Murray, P.A. and S.H. Zinder. 1984. Nitrogen fixation by a methanogenic archaebacterium. *Nature* (London) **312**:284–286.

Muth, E., E. Morschel, and A. Klein. 1987. Purification and characterization of an 8-hydroxy-5-deazaflavin-reducing hydrogenase from the archaebacterium *Methanococcus voltae. Eur. J. Biochem.* **169**:571–577.

Naumann, E., K. Fahlbusch, and G. Gottschalk. 1984. Presence of a trimethylamine: HS-coenzyme M methyltransferase in *Methanosarcina barkeri. Arch. Microbiol.* **138**:79–83.

Nelson, H., S. Mandiyan, and N. Nelson. 1989. A conserved gene encoding the 57-kDa subunit of the yeast vacuolar H$^+$-ATPase. *J. Biol. Chem.* **264**:1775–1778.

Noll, K.M., M.I. Donnelly, and R.S. Wolfe. 1987. Synthesis of 7-mercaptoheptanoylthreonine phosphate and its activity in the methylcoenzyme M methylreductase system. *J. Biol. Chem.* **262**:513–515.

Noll, K.M., K.L. Rinehart, Jr., R.S. Tanner, and R.S. Wolfe. 1986. Structure of component B (7-mercaptoheptanoylthreonine phosphate) of the methylcoenzyme M methylreductase system of *Methanobacterium thermoautotrophicum. Proc. Natl. Acad. Sci. USA* **83**:4238–4242.

Østergaard, L., N. Larsen, H. Leffers, J. Kjems, and R. Garrett. 1987. A ribosomal RNA operon and its flanking region from the archaebacterium *Methanobacterium thermoautotrophicum*, Marburg strain: transcription signals, RNA structure and evolutionary implications. *Syst. Appl. Microbiol.* **9**:199–209.

Otaka, E. and T. Ooi. 1987. Examination of protein sequence homologies: IV. Twenty-seven bacterial ferredoxins. *J. Mol. Evol.* **26**:257–267.

Patel, P.S. and J.G. Ferry. 1988. Characterization of the upstream region of the formate dehydrogenase operon of *Methanobacterium formicicum. J. Bacteriol.* **170**:3390–3395.

Perlman, D. and H.D. Halvorson. 1983. A putative signal peptidase recognition site and sequence in eukaryotic and prokaryotic signal peptides. *J. Mol. Biol.* **167**:391–409.

Poirot, C.M., S.W.M. Kengen, E. Valk, J.T. Keltjens, C. van der Drift, and G.D. Vogels. 1987. Formation of methyl coenzyme M from formaldehyde by cell free extracts of *Methanobacterium thermoautotrophicum*: evidence for the involvement of a corrinoid methyl transferase. *FEMS Microbiol. Lett.* **40**:7–14.

Politino, M., L. Tsai, Z. Veres, and T.C. Stadtman. 1990. Biosynthesis of selenium-modified tRNAs in *Methanococcus vannielii. Proc. Natl. Acad. Sci. USA* **87**:6345–6348.

Possot, O., M. Henry, and L. Sibold. 1986. Distribution of DNA sequences homologous to *nifH* among archaebacteria. *FEMS Microbiol. Lett.* **34**:173–177.

Possot, O., P. Gernhardt, M. Foglino, A. Klein, and L. Sibold. 1989a. Expression of an eubacterial puromycin resistance gene in the archaebacterium *Methanococcus voltae*. In: *Microbiology and Biochemistry of Strict Anaerobes Involved In Interspecies Hydrogen Transfer*, J.P. Belaich, M. Bruschi, and J.L. Garcia, eds., pp. 527–529. Fed. Eur. Microb. Soc.

Possot, O., L. Sibold, and J.-P. Aubert. 1989b. Nucleotide sequence and expression of the glutamine synthetase structural gene, *glnA*, of the archaebacterium *Methanococcus voltae. Res. Microbiol.* **140**:355–371.

Post, L.E., G.D. Strycharz, M. Nomura, H. Lewis, and P.P. Dennis. 1979. Nucleotide sequence of the ribosomal protein gene cluster adjacent to the gene for RNA polymerase subunit β in *Escherichia coli*. *Proc. Natl. Acad. Sci. USA* **76**:1697–1701.

Prickril, B.C., M.H. Czechowski, A.E. Przybyla, H.D. Peck, Jr., and J. LeGall. 1986. Putative signal peptide on the small subunit of the periplasmic hydrogenase from *Desulfovibrio vulgaris*. *J. Bacteriol.* **167**:722–725.

Reeve, J.N. 1987. Gene structure in methanogenic bacteria. *Poultr. Sci.* **66**:927–933.

Reeve, J.N. 1988. Structure and organization of methane genes. *In: Anaerobes Today*, J.M. Hardie, and S.P. Borriello, eds., pp. 95–104. Chichester: Wiley.

Reeve, J.N. 1989. Molecular biology of methanogens and methanogenesis. *In: Genetics and Molecular Biology of Industrial Microorganisms*, C.L. Hershberger, S.W. Queener, and G. Hegeman, eds., pp. 207–214. Washington, D.C.: American Society for Microbiology.

Reeve, J.N. and G.S. Beckler. 1990. Conservation of primary structure in procaryotic hydrogenases. *FEMS Microbiol. Rev.* **87**:419–424.

Reeve, J.N., N.J. Trun, and P.T. Hamilton. 1982. Beginning genetics with methanogens. *In: Genetic Engineering of Microorgansims for Chemicals*, A. Hollaender, R.D., DeMoss, S. Kaplan, J. Konisky, D. Savage, and R.S. Wolfe. eds., pp. 233–244. New York: Plenum.

Reeve, J.N., P.T. Hamilton, G.S. Beckler, C.J. Morris, and C.H. Clarke. 1986. Structure of methanogen genes. *Syst. Appl. Microbiol.* **7**:5–12.

Reeve, J.N., G.S. Beckler, D.S. Cram, P.T. Hamilton, J.W. Brown, J.A. Krzycki, A.F. Kolodziej, L. Alex, W.H. Orme-Johnson, and C.T. Walsh. 1989. A hydrogenase-linked gene in *Methanobacterium thermoautotrophicum* strain ΔH encodes a polyferredoxin. *Proc. Natl. Acad. Sci. USA* **86**:3031–3035.

Rospert, S., D. Lindner, J. Ellermann, and R.K. Thauer. 1990. Two genetically distinct methyl-coenzyme M reductases in *Methanobacterium thermoautotrophicum* strain Marburg and Δ. *Eur. J. Biochem.* **194**:871–877.

Rouvière, P.E. and R.S. Wolfe. 1988. Novel biochemistry of methanogenesis. *J. Biol. Chem.* **263**:7913–7916.

Rouvière, P.E., T.A. Bobik, and R.S. Wolfe. 1988. Reductive activation of the methyl coenzyme M methylreductase system of *Methanobacterium thermoautotrophicum* ΔH. *J. Bacteriol.* **170**:3946–3952.

Sandman, K., J.A. Krzycki, B. Dobrinski, R. Lurz, and J.N. Reeve. 1990. HMf, a DNA-binding protein isolated from the hyperthermophilic archaeal species *Methanothermus fervidus*, is most closely related to histones. *Proc. Natl. Acad. Sci. USA* **87**:5788–5791.

Sapienza, C., M.R. Rose, and W.F. Doolittle. 1982.

High-frequency genomic rearrangements involving archaebacterial repeat sequence elements. *Nature* (London). **295**:384–386.

Saraste, M., N.J. Gay, A. Eberle, M.J. Runswick, and J.E. Walker. 1981. The *atp* operon: nucleotide sequence of the genes for the α, β and ϵ subunits of *Escherichia coli* ATP synthase. *Nucleic Acids Res.* **9**:5287–5296.

Sauer, F.D. 1986. Tetrahydromethanopterin methyltransferase, a component of the methane synthesizing complex of *Methanobacterium thermoautotrophicum*. *Biochem. Biophys. Res. Commun.* **136**:542–547.

Schauer, N.L. and J.G. Ferry. 1980. Metabolism of formate in *Methanobacterium formicicum*. *J. Bacteriol.* **142**:800–807.

Schauer, N.L. and J.G. Ferry. 1986. Composition of the coenzyme F_{420}-dependent formate dehydrogenase from *Methanobacterium formicicum*. *J. Bacteriol.* **165**:405–411.

Schmid, K., M. Thomm, A. Laminet, F.G. Laue, C. Kessler, K.O. Stetter, and R. Schmitt. 1984. Three new restriction endonucleases MaeI, MaeII and MaeIII from *Methanococcus aeolicus*. *Nucleic Acids Res.* **12**:2619–2628.

Schwarzkopf, B., B. Reuke, A. Kiener, and A. Bacher. 1990. Biosynthesis of coenzyme F_{420} and methanopterin in *Methanobacterium thermoautotrophicum*. *Arch. Microbiol.* **153**:259–263.

Shah, N.N. and D.S. Clark. 1990. Partial purification and characterization of two hydrogenases from the extreme thermophile *Methanococcus jannaschii*. *Appl. Environ. Microbiol.* **56**:858–863.

Sharp, P.A. 1985. On the origin of RNA splicing and introns. *Cell.* **42**:397–400.

Shepard, J.C. 1981. Method to determine the reading frame of a protein from the purine/pyrimidine genome sequence and its possible evolutionary significance. *Proc. Natl. Acad. Sci. USA* **78**:1596–1600.

Sherf, B.A. and J.N. Reeve. 1990. Identification of the *mcrD* gene product and its association with component C of methyl coenzyme M reductase in *Methanococcus vannielii*. *J. Bacteriol.* **172**:1828–1833.

Shuber, A.P., E.C. Orr, M.A. Recny, P.F. Schendel, H.D. May, N.L. Schauer, and J.E. Ferry. 1986. Cloning, expression and nucleotide sequence of the formate dehydrogenase genes from *Methanobacterium formicicum*. *J. Biol. Chem.* **261**:12942–12947.

Sibold, L. and M. Henriquet. 1988. Cloning of the *trp* genes from the archaebacterium *Methanococcus voltae*: nucleotide sequence of the *trpBA* genes. *Mol. Gen. Genet.* **214**:439–450.

Sitzmann J., and A. Klein. 1991. Physical and genetic map of the *Methanococcus voltae* chromosome. *Mol. Microbiol.* **52**:505–513.

Souillard, N. and L. Sibold. 1986. Primary structure and expression of a gene homologous to *nifH* (nitrogenase Fe protein) from the archaebacterium *Methanococcus voltae*. *Mol. Gen. Genet.* **203**:21–28.

Souillard, N. and L. Sibold. 1989. Primary structure, functional organization and expression of nitrogenase structural genes of the thermophilic archaebacterium *Methanococcus thermolithotrophicus*. *Mol. Microbiol.* **3**:541–551.

Souillard, N., M. Magot, O. Possot, and L. Sibold. 1988. Nucleotide sequence of regions homologous to *nifH* (nitrogenase Fe protein) from the nitrogen-fixing archaebacteria *Methanococcus thermolithotrophicus* and *Methanobacterium ivanovii*: evolutionary implications. *J. Mol. Evol.* **27**:65–76.

Sowers, K.R. and R.P. Gunsalus. 1988. Adaptation for growth at various saline concentrations by the archaebacterium *Methanosarcina thermophila*. *J. Bacteriol.* **170**:998–1002.

Sparling, R. and L. Daniels. 1990. Regulation of formate dehydrogenase activity in *Methanococcus thermolithotrophicus*. *J. Bacteriol.* **172**:1464–1469.

Sprott, G.D., K.N. Shaw, and T.J. Beveridge. 1987. Properties of the particulate enzyme F_{420}-reducing hydrogenase isolated from *Methanospirillum hungatei*. *Can. J. Microbiol.* **33**:896–904.

Steigerwald, V.J., G.S. Beckler, and J.N. Reeve. 1990. Conservation of hydrogenase and polyferredoxin structures in the hyperthermophilic archaebacterium *Methanothermus fervidus*. *J. Bacteriol.* **172**:4715–4718.

Stetter, K.O. and G. Gaag. 1983. Reduction of molecular sulphur by methanogenic bacteria. *Nature* (London) **305**:309–311.

Stetter, K.O., M. Thomm, J. Winter, G. Wildgruber, H. Huber, W. Zillig, D. Jane-Covic, H. König, P. Palm, and S. Wunderl. 1981. *Methanothermus fervidus*, sp. nov., a novel extremely thermophilic methanogen isolated from an Icelandic hot spring. *Zentralbl. Bakteriol. Mikrobiol. Hyg. 1 Abt. Orig. C* **2**:166–178.

Terlesky, K.C. and J.G. Ferry. 1988a. Ferredoxin requirement for electron transport from the carbon monoxide dehydrogenase complex to a membrane-bound hydrogenase in acetate-grown *Methanosarcina thermophila*. *J. Biol. Chem.* **263**:4075–4079.

Terlesky, K.C. and J.G. Ferry. 1988b. Purification and characterization of a ferredoxin from acetate-grown *Methanosarcina thermophila*. *J. Biol. Chem.* **263**:4080–4082.

Terlesky, K.C., M.J.K. Nelson, and J.G. Ferry. 1986. Isolation of an enzyme complex with carbon monoxide dehydrogenase activity containing corrinoid and nickel from acetate-grown *Methanosarcina thermophila*. *J. Bacteriol.* **168**:1053–1058.

Thomas, L.K., D.B. Dix, and R.C. Thompson.

1988. Codon choice and gene expression: synonymous codons differ in their ability to direct aminoacylated-transfer RNA binding to ribosomes *in vitro*. *Proc. Natl. Acad. Sci. USA* **85**:4242–4246.

Thomm, M., G. Frey, B.J. Bolton, F. Laue, C. Kessler, and K.O. Stetter. 1988a. MvnI: a restriction enzyme in the archaebacterium *Methanococcus vannielii*. *FEMS Microbiol. Lett.* **52**:229–234.

Thomm, M., A.J. Lindner. G.R. Hartmann, and K.O. Stetter. 1988b. Affinity labelling of the active center of DNA-dependent RNA polymerases within the archaebacterial kingdom. *Syst. Appl. Microbiol.* **10**:101–105.

Thomm, M., J. Madon, and K.O. Stetter. 1986. DNA-dependent RNA polymerases of the three orders of methanogens. *Biol. Chem. Hoppe-Seyler* **367**:473–481.

Thomm, M. and K.O. Stetter. 1985. Transcription in methanogens: evidence for specific in vitro transcription of the purified DNA-dependent RNA polymerase of *Methanococcus thermolithotrophicus*. *Eur. J. Biochem.* **149**:345–351.

Thomm, M. and G. Wich. 1988. An archaebacterial consensus promoter sequence for stable RNA genes with homology to the eukaryotic TATA box. *Nucleic Acids Res.* **16**:151–163.

Thomm, M., G. Wich, J.W. Brown, G. Frey, B.A. Sherf, and G.S. Beckler. 1989. An archaebacterial promoter sequence assigned by RNA polymerase binding experiments. *Can. J. Microbiol.* **35**:30–35.

van der Meijden, P., B.W. teBrommelstroet, C.M. Poirot, C. van der Drift, and G.D. Vogels. 1984. Purification and properties of methanol: 5-hydroxybenzimidazolylcobamide methyltransferase from *Methanosarcina barkeri*. *J. Bacteriol.* **160**:629–635.

van der Meijden, P., H.J. Heythuysen, A. Pouwels, F. Houwen, C. van der Drift, and G.D. Vogels. 1983a. Methyltransferases involved in methanol conversion by *Methanosarcina barkeri*. *Arch. Microbiol.* **134**:238–242.

van der Meijden, P., L.P.J.M. Jansen, C. van der Drift, and G.D. Vogels. 1983b. Involvement of corrinoids in the methylation of coenzyme M (2-mercaptoethanesulfonic acid) by methanol and enzymes from *Methanosarcina barkeri*. *FEMS Microbiol. Lett.* **19**:247–251.

Vogels, G.D. and C.M. Visser. 1983. Interconnection of methanogenic and acetogenic pathways. *FEMS Microbiol. Lett.* **20**:291–297.

Voordouw, G. and S. Brenner. 1985. Nucleotide sequence of the gene encoding the hydrogenase from *Desulforibrio vulgaris* (Hildenborough). *Eur. J. Biochem.* **148**:515–520.

Wackett, L.P., E.A. Hartwieg, J.A. King, W.H. Orme-Johnson, and C.T. Walsh. 1987. Electron microscopy of nickel-containing methanogenic enzymes: methyl reductase and F_{420}-reducing hydrogenase. *J. Bacteriol.* **169**:718–727.

Weil, C.F., G.S. Beckler, and J.N. Reeve. 1987. Structure and organization of the *hisA* gene of the thermophilic archaebacterium *Methanococcus thermolithotrophicus*. *J. Bacteriol.* **169**:4857–4860.

Weil, C.F., B.A. Sherf, and J.N. Reeve. 1989. A comparison of the methyl reductase genes and gene products. *Can. J. Microbiol.* **35**:101–108.

Weil, C.F., D.S. Cram, B.A. Sherf, and J.N. Reeve. 1988. Structure and comparative analysis of the genes encoding component C of methyl coenzyme M reductase in the extremely thermophilic archaebacterium *Methanothermus fervidus*. *J. Bacteriol.* **170**:4718–4726.

Weiland, F., G. Paul, and M. Sumper. 1985. Halobacterial flagellins are sulfated glycoproteins. *J. Biol. Chem.* **260**:15180–15185.

White, R.H. 1989. Steps in the conversion of α-ketosuberate to 7-mercaptoheptanoic acid in methanogenic bacteria. *Biochemistry* **28**:9417–9423.

Wich, G., M. Jarsch, and A. Böck. 1984. Apparent operon for a 5S ribosomal RNA gene and for tRNA genes in the archaebacterium *Methanococcus vannielii*. *Mol. Gen. Genet.* **196**:146–151.

Wich, G., W. Leinfelder, and A. Böck. 1987. Genes for stable RNA in the extreme thermophile *Thermoproteus tenax*: introns and transcription signals. *EMBO J.* **6**:523–528.

Wich, G., L. Sibold, and A. Böck. 1986a. Genes for tRNA and their putative expression signals in *Methanococcus*. *Syst. Appl. Microbiol.* **7**:18–25.

Wich, G., L. Sibold, and A. Böck. 1986b. Divergent evolution of 5S rRNA genes in *Methanococcus*. *Z. Naturforschung.* **2**:373.

Widdel, F. 1986. Growth of methanogenic bacteria in pure culture with 2-propanol and other alcohols as hydrogen donors. *Appl. Environ. Microbiol.* **51**:1056–1062.

Widdel, F. and R.S. Wolfe. 1989. Expression of secondary alcohol dehydrogenase in methanogenic bacteria and purification of the F_{420}-specific enzyme from *Methanogenium thermophilum* strain TCI. *Arch. Microbiol.* **152**:322–328.

Widdel, F., P.E. Rouvière, and R.S. Wolfe. 1988. Classification of secondary alcohol-utilizing methanogens including a new thermophilic isolate. *Arch. Microbiol.* **150**:477–481.

Woese, C.R. and G.E. Fox. 1977. Phylogenetic structure of the procaryotic domain: the primary kingdoms. *Proc. Natl. Acad. Sci. USA* **74**:5088–5090.

Woese, C.R., O. Kandler and M.L. Wheelis. 1990. Towards a natural system of organisms: proposal for the domains archaea, bacteria and eucarya. *Proc. Natl. Acad. Sci. USA* **87**:4576–4579.

Woese, C.R., L.J. Magrum and G.E. Fox. 1978. Archaebacteria. *J. Mol. Evol.* **11**:245–252.

Wolfe, R.S. 1985. Unusual coenzymes of methanogenesis. *Trends Biochem. Sci.* **10**:396–399.

Wood, A.G., H. Redborg, D.R. Cue, W.B. Whitman, and J. Konisky. 1983. Complementation of *argG* and *hisA* mutations of *Escherichia coli* by DNA cloned from the archaebacterium *Methanococcus voltae*. *J. Bacteriol.* **156**:19–26.

Xun L., R.A. Mah, and D.R. Boone. 1990. Isolation and characterization of disaggregatase from *Methanosarcina mazei LYC*. *Appl. Environ. Microbiol.* **56**:3693–3698.

Yamazaki, S. 1982. A selenium-containing hydrogenase from *Methanococcus vannielii*. *J. Biol. Chem.* **257**:7926–7929.

Zhao, H., A.G. Wood, F. Widdel, and M.P. Bryant. 1988. An extremely thermophilic *Methanococcus* from a deep sea hydrothermal vent and its plasmid. *Arch. Microbiol.* **150**:178–183.

Zillig, W., P. Palm, W.-D. Reiter, F. Gropp, G. Pühler, and H.-P. Klenk. 1988. Comparative evaluation of gene expression in archaebacteria. *Eur. J. Biochem.* **173**:473–482.

Zillig, W., H.-P. Klenk, P. Palm, G. Pühler, F. Gropp, R.A. Garrett, and H. Leffers. 1989. The phylogenetic relations of DNA-dependent RNA polymerases of archaebacteria, eukaryotes and eubacteria. *Can. J. Microbiol.* **35**:73–80.

Zydowsky, L.D., T.M. Zydowsky, E.S. Haas, J.W. Brown, J.N. Reeve, and H.G. Floss. 1987. Stereochemical coursed of methyl transfer from methanol to methyl coenzyme M in cell-free extracts of *Methanosarcina barkeri*. *J. Am. Chem. Soc.* **109**:7922–7923.

3

Genes for Stable RNAs and Their Expression in *Archaea*

Michael Thomm and Winfried Hausner

3.1 Introduction

The investigation of stable RNA sequences has led to the discovery of the *Archaebacteria* as a second procaryotic line of descent (Woese and Fox, 1977). Archaebacteria are no more related to typical bacteria than to the eucaryotic cytoplasms. Novel designations have been proposed to express this tripartite division of the living world, thus rejecting the conventional procaryote–eucaryote dichotomy (Woese et al., 1990). According to this proposal, life on earth can be seen as comprising three domains, called the *Bacteria* (formerly eubacteria), the *Archaea* (archaebacteria), and the *Eucarya* (eucaryotes). Two major phylogenetic branches (kingdoms) can be distinguished within the domain of *Archaea*: the coherent kingdom of *Crenarchaeota* (formerly sulfur-metabolizing thermophiles) and the phenotypically diverse second kingdom of *Euryarchaeota*, comprising the three orders of methanogens (Balch et al., 1979): the order *Thermococcales* (Zillig et al., 1987), the sulfate reducer *Archaeoglobus* (Achenbach-Richter et al., 1987), the extreme halophiles, and the genus *Thermoplasma* (Woese, 1987).

The genes for stable RNAs from representatives of all phylogenetic groups of *Archaea* have been cloned and their arrangements and sequences determined. Thus, enough structural data have been accumulated to allow a significant comparison of the molecular organization of tRNA and rRNA genes in bacterial, eucaryotic, and archaeal cells. This comparison will constitute the first part of this chapter.

Although DNA sequences upstream and downstream from numerous rRNA and tRNA genes of *Archaea* have been established, little was known about the mechanisms and transcription signals regulating the expression of stable RNA genes in *Archaea*. The major reason for this paucity of information has been the lack of genetic transfer procedures and in vitro transcription systems that would allow testing and refining the predictions inferred from structural studies in functional assays. However, a convincing proposal for two promoter motives of stable RNA genes in *Methanococcus* has been derived from sequence analyses and transcription mapping experiments (Wich et al., 1986a). These sequences comprise an AT-rich sequence, called box A, located between position -40 to -20 relative to the transcription start site, and a second motif at the transcription start site (box B). Footprinting (nuclease protection) experiments showed that some of these conserved sequences are bound by the purified RNA polymerase from *Methanococcus vannielii* (Thomm and Wich, 1988). From these footprinting experiments and from the sequence analyses of RNA genes from methanogens and the thermophile *Sulfolobus*, the octanucleotide TTTA T/A ATA (TATA box) has been inferred as a general promoter element for stable RNA genes in *Archaea* (Thomm and Wich, 1988; Reiter et al., 1988a).

A modified version of this motif was found

at the same location upstream from stable RNA genes of extreme halophiles (Mankin and Kagramanova, 1988; Thomm and Wich, 1988). Further footprinting experiments and sequence analyses supported the conclusion that this sequence also comprises a constituent of the promoters of archaeal protein-encoding genes (Brown et al., 1988; Reiter et al., 1988a; Thomm et al., 1988). Although the RNA polymerase of *Methanococcus* binds to the promoter, all purified RNA polymerases of *Archaea* are unable to initiate transcription at the correct site in vitro. However, cell-free transcription systems allowing the expression of tRNA genes of *Methanococcus* (Frey et al., 1990), rRNA genes of *Sulfolobus shibatae* (Hüdepohl et al., 1990), and a protein-encoding gene of *Methanobacterium thermoautotrophicum* (Knaub and Klein, 1990) have been described. The *Methanococcus* and *Sulfolobus* systems both initiate at the same site in vitro and in vivo. The *Methanobacterium* system starts transcription at a box B-like sequence 10 nucleotides upstream of the in vivo initiation site. The low efficiency and unusual start site of this in vitro system suggest that it lacks an essential component. The availability of specific transcription systems offers the opportunity to obtain biochemical evidence for the significance of conserved DNA sequences upstream and downstream from archaeal genes.

We summarize here some of our experiments defining the DNA sequences that promote and cause the termination of transcription in *Methanococcus vannielii*. These and other experiments (Hausner et al., 1991) demonstrate that the TATA box at −25 and a second signal at the transcription initiation site are indispensable for initiation of transcription. Accurate cell-free transcription in *Methanococcus* and *Sulfolobus* is mediated by soluble transcription factors (Frey et al., 1990; Hüdepohl et al., 1990); this is reminiscent of eucaryotic transcription systems. The similarities and differences of stable RNA transcription in *Archaea*, *Bacteria*, and *Eucarya* are discussed. Stable RNA genes are highly expressed in vivo, and because only a single type of RNA polymerase appears to be present in archaeal cells, the analysis of transcription in these cells may pro-

vide a basis for understanding the mechanism of archaeal transcription in general.

3.2 Organization of Stable RNA Genes

Within the *Euryarchaeota*, ribosomal RNA genes are arranged in the sequence 5′-16S-tRNA^{Ala}-23S-5S-3′ (Figure 3.1). The tRNA gene in the intercistronic spacer between 16S and 23S RNA is missing in all *Crenarchaeota* investigated so far. According to this feature, organisms that show a thermophilic phenotype, such as *Thermococcus* and *Archaeoglobus*, are also clearly linked with their phylogenetic relatives, the methanogens and halophiles (Achenbach-Richter and Woese, 1988). In *Desulfurococcus mobilis*, *Thermoproteus tenax*, and *Thermophilum pendens*, the 5S genes are unlinked from the 16S/23S genes and are transcribed from an independent promoter (Neumann et al., 1983; Kjems and Garrett, 1987; Kjems et al., 1990). Additional unlinked 5S genes have been observed in *Sulfolobus* strain B12 (Reiter et al., 1987; strain B12 has been described recently as a new species *Sulfolobus shibatae*; Grogan et al., 1990), *Thermococcus* (Neumann et al., 1983), and *Methanococcus* (Jarsch et al., 1983; Wich et al., 1987b). The additional 5S genes of *Methanococcus vannielii* and *M. voltae* are located within tRNA operons (Wich et al., 1984, 1987b). One rRNA operon of *Methanothermus fervidus* and *Methanobacterium thermoautotrophicum* is linked to a 7S gene (Haas et al., 1990; see Figure 3.1, and also following).

In *Thermoplasma*, the genes for 16S, 23S, and 5S rRNA are physically separated by 1.5 to 7.5 kb (Tu and Zillig, 1982) and transcribed from independent promoters (Ree and Zimmermann, 1990), indicating a unique arrangement of ribosomal RNA genes in this organism into three transcription units. The number of rRNA operons varies from one in the *Crenarchaeota* to four in *Methanococcus vannielii* (see Figure 3.1). The domain *Bacteria* shares with *Archaea* the general organization of rRNA genes. In the *Eucarya* the 5S gene is separated. However, a separate 5S rRNA gene has also been disco-

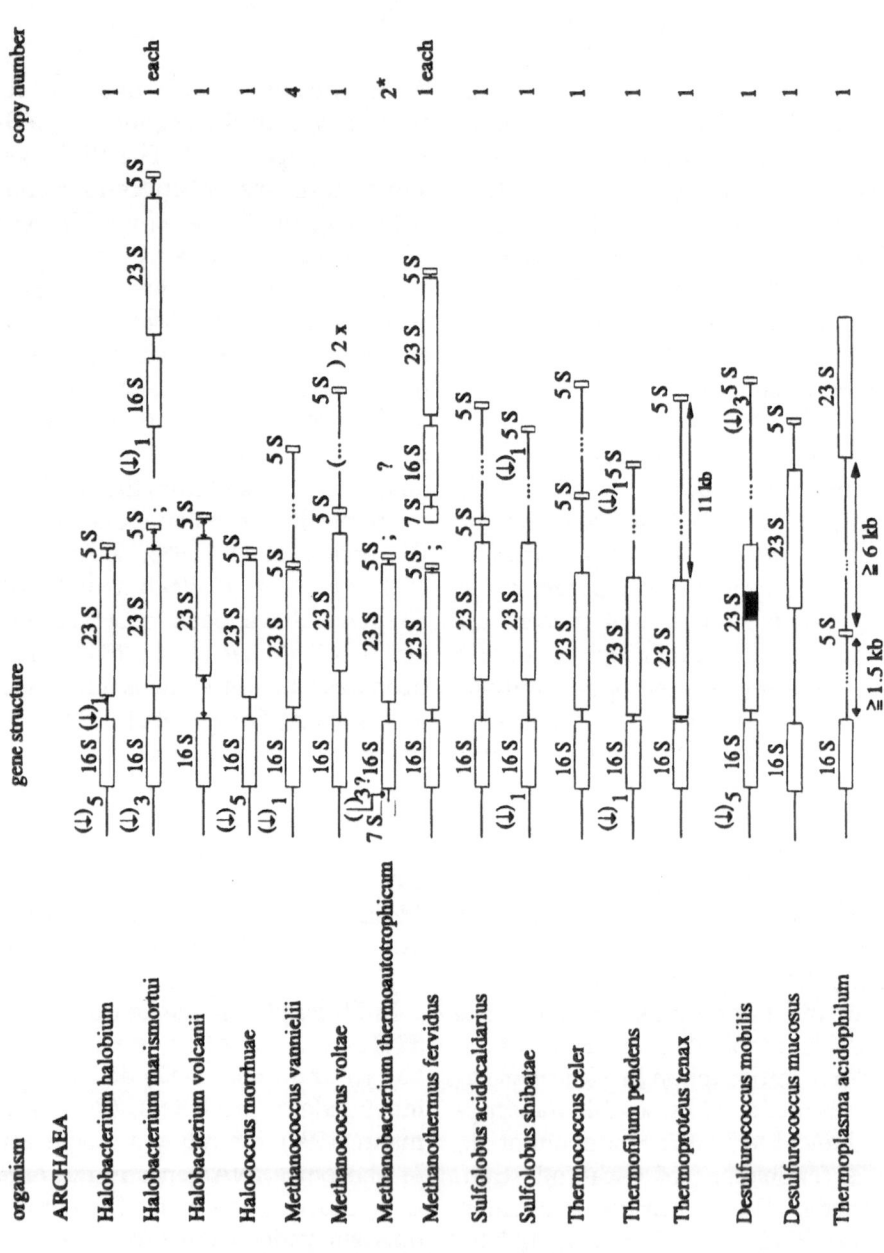

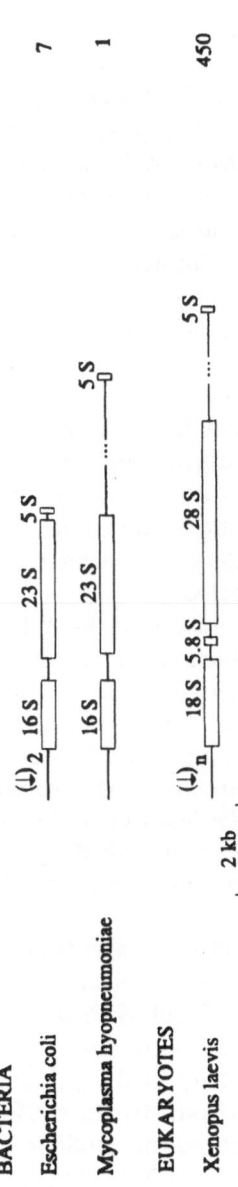

BACTERIA

Escherichia coli   7

Mycoplasma hyopneumoniae   1

EUKARYOTES

Xenopus laevis   450

2 kb

FIGURE 3.1   Arrangement and transcription of archaeal rRNA genes. Sequences encoding mature rRNA species are boxed. Dots indicate that distance of unlinked 5S gene to rRNA operon has not been precisely determined. Transcription start sites are indicated by (↓). Indices show number of mapped initiation sites; in *Xenopus*, Index n indicates multiple sites. Distance between 16S and 23S rRNA derived from mapping data has been corrected using DNA sequencing data provided by Achenbach-Richter and Woese (1988). Location of intron in 23S RNA of *Desulfurococcus mobilis* is indicated by black box.
*The organization of the second operon in *Methanobacterium thermoautotrophicum* has not yet been investigated, and interpretation of the mapping experiments for transcription start sites is difficult (see discussion in text).

vered in *Mycoplasma* (Taschke et al., 1986; Figure 3.1). Eucaryotic 18S, 5.8S, and 28S rRNA genes are cotranscribed as a large 40S or 45S precursor in lower and higher *Eucarya*, respectively. The 5.8S rRNA of *Eucarya* is homologous to the 5' end of the 23S rRNA of *Bacteria* and *Archaea*.

Desulfurococcus mobilis shares with some lower *Eucarya* the presence of an intron within the 23S rRNA gene (Kjems and Garrett, 1985). The splicing process is similar to that of the class III introns of eucaryotic tRNAs. However, in contrast to these introns, which are linear and rapidly degraded after the splicing event, the excised *Desulfurococcus* intron circularizes in a manner similar to that of self-splicing introns (Kjems and Garrett, 1988). Because the *Desulfurococcus* intron is located in a "hot spot" intron site of eucaryotic rRNAs and is spliced by a mechanism similar to that of eucaryotic tRNA introns, both rRNA and tRNA introns of *Eucarya* may have a common origin in the introns of *Archaea*.

Transfer RNA genes in *Archaea* are organized in operons or single genes and may be located within rRNA operons, in the spacer region between 16S and 23S RNA and as trailers downstream from the 5S gene. This is the typical situation encountered in *Bacteria*. In *Eucarya*, tRNA genes are clustered, but neighboring genes are rarely cotranscribed into multimeric precursors. Transfer RNA genes hitherto have not been observed within an rRNA operon in *Eucarya*. The absence of a tRNA gene in the 16S/23S RNA spacer region of extreme thermophiles has been discussed as an important argument supporting the phylogenetic tree of Lake (1989). According to this proposal, these organisms (called eocytes by Lake) are more closely related to *Eucarya* than methanogens and halophiles, which cluster together with (eu)bacteria (Lake, 1988). However, the 16S/23S RNA spacer region of *Mycobacterium* (Suzuki et al., 1988a) and *Streptomyces* (Suzuki et al., 1988b) also does not contain a tRNA gene. Because this feature is also a common property of the *Actinomycetes*, an important group in the domain *Bacteria*, the presence or absence of a tRNA gene does not argue for a specific relationship of *Crenarchaeota* to *Eucarya*. No

molecular feature of stable RNAs has yet been detected that is absolutely specific for one of the two archaeal kingdoms, and thus far, no phenotypic properties can be defined that clearly indicate a closer relationship of one of the two kingdoms of *Archaea* to *Eucarya* (see following). Also, almost all data accumulated thus far argue for the monophyletic nature of *Archaea*.

Some tRNA genes from *Sulfolobus solfataricus* (Kaine et al., 1983), *Haloferax volcanii* (Daniels et al., 1985), *Thermoproteus tenax* (Wich et al., 1987a), and *Thermofilum pendens* (Kjems et al., 1989) contain an intron. Most archaeal tRNA introns are located in the anticodon loop, as they are in eucaryotic nuclear tRNAs. The tRNA intron of *T. pendens* is located in the variable loop of the tRNA precursor. No intron has ever been detected at this position of the tRNA molecule. A tRNA^{Trp} intron endonuclease has been purified from *Haloferax volcanii* (Thompson and Daniels, 1988). Unlike eucaryotic intron endonucleases, this enzyme appears to be sequence specific and does not require a complete mature tRNA structure for substrate recognition. In stable RNA genes of bacteria, no introns have yet been detected.

Most tRNA genes of *Archaea* lack the 3' terminal CCA sequence, which must be added posttranscriptionally. These two features previously have been seen in eucaryotic tRNA genes (Melton et al., 1980), some tRNA genes of *E. coli* bacteriophages, and in some chromosomal tRNA genes of *Bacillus* (King et al., 1986).

A striking similarity to eucaryotes is the presence, in *Archaea*, of a gene encoding a stable RNA of about 300 nucleotides. This 7S RNA gene has been found in all *Archaea* examined (Moritz and Goebel, 1985; Haas et al., 1990; Kaine, 1990). The 7S RNAs of *Archaea* and *Eucarya* share a very similar secondary structure. However, the homology in the primary sequence is limited to a hairpin structure. This specific domain of about 40 nucleotides is also conserved in the 4.5S RNA (114 nucleotides) of *E. coli* and the small cytoplasmatic RNA (271 nucleotides) of *Bacillus subtilis* (Struck et al., 1988; Kaine and Merkel, 1989; Haas et al., 1990). Thus, these small bacterial RNAs and 7S RNAs might be evolutionary homologs. However, their size and secondary structure result in a greater structural resemblance between archaeal and eucaryotic 7S RNA molecules. In eucaryotic cells, 7S RNA is a major constituent of the signal recognition particle involved in translocation of secretory proteins (Zwieb, 1989). The function of 7S RNA in archaebacterial cell metabolism remains an intriguing question.

There are so many excellent reviews and original papers about the structural features of archaeal stable RNAs and the phylogenetic trees based on these sequences (Böck et al., 1986; Leffers et al., 1987; Woese, 1987; Brown et al., 1988; Kjems and Garrett, 1990) that they cannot be discussed in detail here. This chapter therefore focuses on the following aspects of the expression of stable RNA genes.

3.3 Transcription of Stable RNA Genes

Transcription initiation sites

All living cells require specific mechanisms to synthesize the tremendous amount of RNA required to constitute the RNA component of ribosomes. In bacteria, three factors seem to contribute to the high expression rate of rRNA genes. First, there are multiple copies of the genes (seven in *E. coli*; Kenerley et al., 1977). Second, the genes are expressed from tandem promoters (Young and Steitz, 1979). Finally, the RNA polymerase shows a high affinity to these promoters, most likely because they show high homology to the eubacterial consensus sequence (Hawley and McClure, 1983).

Eucaryotic cells contain 50 to 500 identical repeat units of rRNA genes per haploid genome, which are clustered at a distinct site of the chromosome and separated by a nontranscribed spacer region. In *Xenopus laevis* and probably in other eucaryotes, the spacer is composed mainly of repeated DNA sequences, some of which contain promoter-like structures. These reduplicated promoter sites are bound by transcription factors and give rise to short transcripts upstream from the true rRNA promoter (Moss, 1983). Because RNA poly-

merase I does not detach after transcription of both the 40S rRNA precursor and the short DNA sequences upstream from the gene, the multiple promoter sites appear to deliver the RNA polymerase to the primary promoter, thus ensuring a high expression rate for rRNA genes (Sollner-Webb et al., 1987).

A similar initiation, termination, and reinitiation mechanism has been demonstrated to occur upstream from the 16S/23S operon of *Desulfurococcus mobilis* (Kjems and Garrett, 1987). Four transcription initiation sites seem to direct the RNA polymerase to the primary initiation site located 134 bp upstream from the mature 16S rRNA.

Multiple transcription initiation sites have also been mapped upstream from the RNA operon of *Halobacterium cutirubrum* (Dennis, 1985) and *H. halobium* (see Figure 3.1; Mankin and Kagramanova, 1986). These two organisms are closely related and should be subsumed into the species *H. salinarium* (Larsen and Grant, 1989). However, in contrast to *Desulfurococcus*, the transcripts initiating at distant sites of halophiles do not terminate upstream from the DNA region encoding the 16S rRNA. Thus, this readthrough from several upstream promoters to the terminator of the operon resembles the mechanism encountered in *bacteria* (Boros et al., 1983). Multiple transcription initiation sites have also been observed upstream from the rRNA operon of *Halococcus morrhuae* (Larsen et al., 1986) and from one of the two operons of *Halobacterium marismortui* (Mevarech et al., 1989). The three transcription start sites mapped upstream of one of the two rRNA operons of *Methanobacterium thermoautotrophicum* have not yet been clearly established (Ostergaard et al., 1987). The situation is further complicated by the presence of a 7S RNA gene immediately upstream of this operon that was only noticed later (Haas et al., 1990). Only a single transcription start site exists upstream from rRNA genes from *Methanococcus vannielii* (Wich et al., 1986a), *Sulfolobus shibatae* (Reiter et al., 1987), *Thermofilum pendens* (Kjems et al., 1990), *Thermoplasma acidophilum* (Ree and Zimmermann, 1990), and one operon of *Halobacterium marismortui* (Mevarech et al., 1989; see also Figure 3.1). Therefore, multiple start sites upstream from rRNA operons cannot be considered a general archaeal mechanism.

An additional putative transcription start site has been located in the spacer region upstream for the 23S RNA of *Halobacterium salinarium* by S1 nuclease mapping experiments (Mankin and Kagramanova, 1988). A consensus promoter sequence is located at the correct distance upstream from this nucleotide, suggesting that this S1 signal is caused by initiation of transcription and not by processing of the rRNA precursor. This additional promoter may help to adjust cellular levels of RNAs that are located far downstream of the primary promoter. These RNAs might otherwise be expressed at lower levels because of premature termination of transcription. It is unclear whether a similar mechanism operates in other organisms.

Only a few transcription start sites of archaeal tRNA genes and operons have been mapped (Wich et al., 1986a, 1987b; Kjems and Garrett, 1988). These data indicate the presence of a single initiation site. In *Bacteria*, tRNA genes can be expressed both from a single and from tandem promoters (Caillet et al., 1985). The transcription of eucaryotic tRNA and 5S rRNA genes is controlled by an intragenic promoter (see following).

Promoter sequences

The transcription initiation sites upstream from archaeal stable RNA genes have been located by S1 mapping or primer-extension experiments. A major disadvantage of the protocols used by most investigators was that they did not allow a distinctive discrimination between initiation start sites and processing sites (especially when applied to stable RNA genes). To determine the 5' end of a primary transcript, it is important to demonstrate that: (1) the RNA initiates with a ribonucleoside triphosphate; and (2) the same initiation site is used when the RNA is synthesized in vitro in the absence of the processing machinery of the cell.

Wich et al. (1986a) were the first to locate the transcription start site of a primary transcript upstream from a rRNA operon using a guanosyl transferase capping experiment. They proposed two conserved DNA sequences as possible promoter signals for stable RNA genes of *Methanococcus*: the "box A" sequence ACCGAAA-TTTATATA-TA, extending from position 20–40 upstream from the 5' end of the primary transcript, and the "box B" motif TGCAAGT, at the transcription start site. The footprint of the *Methanococcus* RNA polymerase extends from position -30 to $+20$ relative to the transcription start site (Thomm and Wich, 1988). Hence, the 3' part of the "box A" of Wich et al. (1986a), the octanucleotide TTTA T/A ATA, is located within the RNA polymerase-binding site. This octanucleotide shows striking homology to the TATA box of eucaryotic promoters of protein-encoding genes in both location and sequence (Corden et al., 1980). A very similar sequence has been found upstream from primary transcripts of *Sulfolobus* genes (Reiter et al., 1988a). Thus, the DNA sequences upstream from transcription initiation sites appear highly conserved among phylogenetically distant *Archaea*. Sequence analyses of the DNA region upstream from stable RNA genes of a variety of further genera from both phylogenetic kingdoms of *Archaea* confirm this conclusion (Figures 3.2–3.4). However, although this strict conservation argues for the importance of these sequences in evolution, their function as promoter signals has not yet been demonstrated.

One way to investigate the significance of a conserved DNA sequence is to alter it in vitro and determine the effect of the mutation on the biological function. We used a cell-free transcription system (Frey et al., 1990) to define archaeal promoter sequences in a functional assay. As template for these experiments, we used the tRNA^{Val} gene of *M. vannielii*, which has a -25 region with perfect homology to the consensus promoter sequence; at the 3' end, it has an oligo-dT sequence, which has been proposed by Wich et al. (1986a) to be a potential terminator signal (Figure 3.5). Analysis of DNA deletion clones showed that the expression rate of this tRNA is not dramatically reduced after the DNA region from -590 to -35 is removed (Thomm et al., 1990; Hausner et al., 1991). However, removal of nucleotides extending into the TATA box or beyond leads to a complete inactivation of this template (Hausner

a)

Sequence	RNA	Organism	Reference
TACCTAAAACAATACATATTACAACACGTTTCATATTATgCAAATC	rRNA1	Mc. vannielii	(Wich et al., 1986a)
TACCTAAAACAATACATATTACAACACGTTTCATATTATgCAAATC	rRNA2	"	"
AACCGAAATATTTATATACTAGAATACCCTCCTATACTATgCTCTT	tRNA-op.	"	(Wich et al., 1986b)
TACCGAAAACTTTATATATTATAACACTAGTATTCAGTATgCGAACA	5S/tRNA-op.	"	"
CACCGAAAAGTTTATATATCATGAAATACTATGTTTAGTTTGCTCTCA	tRNA Val	"	"
CACCGAAAACTTTATATACTGTTTATTATGTATTTCATTTGGAAGTT	tRNA Thr	"	"
TACCGAAAACTTTATATAATAATTTCAATCTAATTAATGCAAGTC	tRNA Gln	"	"
TGTAAAAAGGTTTATATAGTAGAATGTAATTGTTATTGTGCGGTCA	tRNA-op.	"	"
TTACAAAAGTATATATACTAAGAAAGATATGCTTAATTGTGGATTT	tRNA Phe	"	"
GATGGAAACAATTTATATAGTTATAGTGCTTAGTCATTGTGCAAGTA	5S/tRNA-op.	Mc. voltae	"
AATAGTAAACTATATAAGCTAGAACAAGTTATGTAAATTGGCTAGG	7S RNA	"	(Kaine and Merkel, 1989)
TATCAAAAAATTTAAATAAGATTGAAAAATAAAATATAAAATGGCAG	7S RNA	Mt. fervidus	(Haas et al., 1990)
GTCCGAAAACTTTATATATGAAAAATTCAAAGGTAAATTATAGCTAA	tRNA-op.	"	(Haas et al., 1989)
AATCGAAAAAATATAAAATAGTTTTATCTAATCTATCCATTAGCTTT	tRNA-op.	"	"
-----AAGCTTAAATAAATAGAGCTGCCCTACAGTAATGG------	rRNA-op.	Mth. soehng.	(Eggen et al., 1990)
TGCCATAACCTTTATAAACTCACTGTGACAATACTTTATTTGGTGGGC	7S RNA	"	"
GTGCCAAAACTTTATATATCGACGGGGAATAGAGTAACTGGCGGCG	7S RNA	A. fulgidus	(Kaine, 1990)
CTTCGAAAGTTATATATACTGATTTGCTATTCTTTACTTgCACATA	16S RNA	Tc. celer	"
TCACGAAAATCCTATATAGATGTGTTCTATATAGTGTtCGGCAACG-	5S RNA	Tp. acidophilum	(Ree and Zimmermann, 1990)
GATCAAAATGCTTATATCCCTCTTAATGATATAGTCCATaCACGCTT	23S RNA	"	"

b)

A_{17} A_{20} A_{19} A_{12} C_9 T_{15} T_{14} T_{20} A_{20} T_{15} A_{20} T_{18} A_{17}
T_2 G_1 T_3 G_4 T_3 A_2 A_6 A_3 A_2 C_2 C_2
$-_1$ G_3 A_4 C_2 C_2 C_3 G_1
C_2 T_3 G_1 C_2

A A A A N | T T A T A T A T | A

consensus methanogens
(usually low and intermediate GC)

FIGURES 3.2 to 3.4 The promoter sequences in *Archaea* are highly conserved. Figure 3.2 shows the results for methanogens and euryarchaeota excepting halophiles, Figure 3.3 for halophiles, and Figure 3.4 for the *Crenarchaeota*. (A) DNA sequences upstream from archaeal tRNA/rRNA and 7S genes have been aligned to yield maximal homology, with TATA box at −25. This promoter element has been identified by footprinting and cell-free transcription experiments using purified components from *Methanococcus* (Thomm and Wich, 1988; Thomm et al., 1990; Hausner et al., 1991). The consensus octanucleotide (methanogens and halophiles) and hexanucleotide (*Crenarchaeota*) are boxed. Lowercase letters indicate transcription start sites. Euryarchaeota *Archaeoglobus*, *Thermococcus*, and *Thermoplasma* are listed together with methanogens in Figure 3.2. In Figure 3.3, P_i indicates putative promoter in spacer region between 16S and 23S RNA in *Halobacterium salinarium*. (B) Conserved nucleotides in −25 region of archaeal promoters. Subscripts indicate base frequency at each position. Consensus derived is shown at bottom. DNA region that has been identified as being most important for cell-free transcription in *Methanococcus* is boxed (Figure 3.2). Corresponding DNA region in halophiles (Figure 3.3) and *Crenarchaeota* (Figure 3.4) is also boxed. Note modification of consensus sequence in halophiles, which can be correlated with their high GC content. (From Thomm et al., 1989.)

a)

```
CGCCGACATATTTATCCTCCGGCCTTGTGTGTTGCATCCCaCGAAGAA        rRNA P1      H. salinarium  (Mankin and Kagramanova, 1988)
GGCGAAACTGCTTACAACGCCCCAACCCACCACGCGCTGGGT              rRNA P2            "         "
TCGACGGTGTTTATGTACCCCACCACTCGGATGAGATGCGAacGAC          rRNA P3                     (Dennis, 1985; Mankin et al., 1986)
GTCCGATGCCCTTAAGTACAACAGGGTACTTCGGTGAATGCGAaCG          rRNA P4            "         "
ATTCGATGCCCTTAAGTAATAACGGGTGTTCCGATGAGATGCGAaCG        rRNA P5            "         "
ATTCGATGCCCTTAAGTAATAACGGGGCGTTACGAGGAATTGCGaACG       rRNA P6            "         "
ATTCGATGCCCTTAAGTAATAACGGGGCGTTCGGGCGAAATGCGaACG       rRNA P7                     (Mankin and Kagramanova, 1988)
GATCGTGTCCCTTAAGTGGGAGACGGGGCAACGATGAATCgCGACGA        rRNA Pi                     (Moritz and Goebel, 1985)
CCGAAAGGCCTTAAGAACGACCCGGGTAGGATGAGATGGACTAGGC         7S  RNA      Hf. volcanii  (Daniels et al., 1985)
TCGAAACCCCTTTAAGAAAAATCGCCATACGAGAGAGTGCAGACAGA        tRNA Trp           "         (Daniels et al., 1986)
AAACGAAAGTCTTAACATAGCCAGACTCGTTTGTTGATCATG------       tRNA Met           "         "
AAAGGAAAGTCTATTTACCCACCGGCAGTACGAGAGATTGCAAGGG-        tRNA Lys           "         "
ATCGAAACGGATTAAACTATCCGAGAGAGGCAACAATGGAAGCC-          tRNA Ser           "         "
GATTCGAAAGCTTAAATTGTACCCGGACAACGGAGAGATGCGTCCGA        tRNA Val           "         "
CACCGTCAGCGTTAAGCACAAGACCCGGATATCCAGTAACTGCGCC-        tRNA Cys           "         "
-----AAGCTTAAATACAACCAGCACAACCAGGAAGTTGAGCCC           tRNA Metm     H.  morrhuae  (Datta et al., 1989)
TTCCGACGGGTTTATCCGTTACCCGGGATTCCGAATGGAAATGCGAA        rRNA P1            "         (Larsen et al., 1986)
ATCCGACGCCCTTAATTGGTACAGGGCACTCGGATGAAATGCAGAAA        rRNA P2            "         "
CTTCGAAGGGTTTATACCCTCGAACGGTGTACGAAGAGATCCGAAGG        rRNA P3                     "
AGACCGTCCCATTTATTATACTTTTTCCCATGGATGTAATgCGAAGG        rRNA P       H.  marismortui (Mevarech et al., 1989)
CTTCGACGGGCGTTAAGTGTGGCTCACCCATCGGAATGAAATgCGAAC       rRNA P1            "         "
TTCCGACGCCCTTAAGTGTAACAGGGCGTTCGGAATGAACgCAAAGG        rRNA P2            "         "
ATCCGACGCCCTTAAGTGTAACAGGGTGCTCGGAATGAACgCGAACG        rRNA P3            "         "
```

b)

```
G14 A16 C8  G11 C11 C11 C15 T23 T22 A22 A16 G12 T13 A10
A5  G4  A6  A6  G7  G7  T6  T6  A1  T1  T6  A6  A5  G6
C3  T2  T5  C4  C8  T3  A1  C1  C1  T2  C3  C5  T4  C3
-1  -1  G3  T2  A2  A2  G1  G1          A1      T2
            -1

G   A   N   G   C   C   T   T   A   A   G   T   A
                G   G

consensus halophiles
          (high GC)
```

FIGURE 3.3 See Figure 3.2 for caption.

a)

```
GCGAAAAATT TTTAAT TTAGGGTGTGTTTAGGGATGGTCgCGCCTTAATTGTTTGT    rRNA
AGCGAAAAAA TTTAAA TCGGTGAGTAAGTACGCTCGGGGCCGGTAGTCTAGCGG      tRNA Ala     Tp. tenax  (Wich et al., 1987)
ACAAAAGCTT TTTAAA TTCGCGCAAAGCTTAGACCTAGCGGGTAGGCCAGCTA       tRNA Met          "
GGCTTAAAAA TATCCTG TCATATAACGAGTTGGCGTTCCGGTAGTCTAGCG         tRNA Ala          "
GCTTGAAAAAT TTTAAAC TGAGCAGTTTATATCAGAGACGGCGGGGGTGCCCGAG     tRNA Leu          "
GAGAAAACATT TTAAT CCTGAGGAGAAAATACTGGACAGGCGGGGGTGCCCGAG      tRNA Leu          "

ATGCTAAAGG TTTATT ACCCAGGAAGTATTCCGGTCATGGGGGTTACGAAGCC       tRNA Met     Tf. pendens  (Kjems et al., 1990)
CGGGAAAAGC TTTTAAG CATGCCTTTTACTTCCTTCTAgAGGCTCAGCGGCCG       tRNA Gly          "
AAGCATAATA TTCATA TAACCCCCGTTTACTAACTAGATTgCCGCCATGGGCA       rRNA              "
TAGTTAATTT TTTATA TGTTATGAGTACTTAATTTgCCACCCGGCCACAG          5S rRNA      S. shibatae  (Reiter et al., 1987)
AGAAGTTAGA TTTATA TGGGATTTCAGAACAATATGTAATAATgCGGATGCCCC      rRNA              "
GGCATAACTC TTTTAAA GGTAACTATTTTATTTATGTTATAGTGGCCCGTAGCT      tRNA Met     S. solfataricus  (Kaine, 1987)
ACACGAAGAG TTTAAAA ACGGTAAAGATTTAAACTATTAGAGAGGGCCCGTCGT      tRNA Val          "
ATCATTAAAG TGTAAAA TAGGCTTGAAAAAGATATTAATATTGCGGCCGTCGTCT     tRNA Gly          "
CCAATAAAAC CTATAAAA GTCATATGTAAATAATAATAATGCCCGCCTAGCTCAGC    tRNA Phe          "

GCATAAAGTA TATATAAA CCCTTATCGCATAGAGTAAGATTCCAGACGCTTACAgC    rRNA P1      D. mobilis  (Kjems and Garrett, 1988)
ACCCGTCATG ATTAATA CCTTGGAGCAAATAGATTCATCaaGCCGCGGCATT        rRNA P2           "
TAGTGAACGC TTTGAAA GCAGCTGTGTTCCACGAGTGAAGCACTCTACGTGG        rRNA P3           "
AGAGAACTGG TTCAAA CACGTCAGGCTTTTCCCGACGTCATCCCGTgCTCAG        rRNA P4           "
-GAATTCATC CTTGAG GCAGTGGTGGGAACCGGGTTGAGCAGGGAGGATgCCGC      rRNA P5           "
AGAGTAAGGT TTTAAAA CCCCAGTAATAGATTATGGGGACTAcCGGTGCCCGACCC    5S rRNA P1        "
CCTAACACAC TATACAA TATATTGATGCTCGCAATAGTGGTAgCCCTAATAGTC      5s rRNA P2        "
TAAGGAGATC TTTGAA AGCGCTGAGACAACACTgAAGTATCTTGAGAAAATCAT      5s rRNA P3        "
```

b)

```
A17  A17  A13  T9   C7   T20  T20  T20  A19  A18  A18  T8   C8
T5   C3   C5   G7   T7   A1   C2   G3   T4   T4   A8   A5   T5
C1   G2   G3   A6   G5   G1   A3   T1   G3   G1   C4   A5   A5
T1   T1   T2   C1   A4   C1   A1   C1   A1   C1   G3   G3   G5

A    A    A    N    N    T    T    T    A    A    A

          consensus crenarchaeota
          (low and intermediate GC)
```

FIGURE 3.4   See Figure 3.2 for caption.

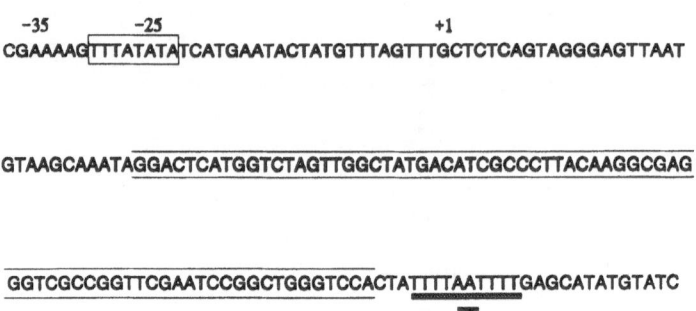

FIGURE 3.5   Genomic sequence of tRNA^Val gene of *Methanococcus vannielii*. This gene was used as template to establish the DNA sequences promoting and terminating initiation of cell-free transcription. Recombinant plasmid shown harbors the wild-type upstream region to position −35; it contains archaeal consensus promoter sequence (Thomm and Wich, 1988; Reiter et al., 1988a) upstream (boxed octanucleotide) and also terminator signal proposed by Wich et al. (1986b) downstream from the gene (indicated by grey bar and the letter T below sequence). Sequence encoding mature tRNA^Val is indicated by parallel lines above and below sequence.

et al., 1991). This finding supports the conclusion that the TATA box is necessary for initiation of transcription. To provide conclusive evidence for the significance of this sequence as a promoter signal, a series of point mutations has been introduced into the DNA region upstream from the tRNA^Val gene (Hausner et al., 1991). The effects of some of these mutations on the rate of cell-free transcription are summarized in Figure 3.6. When the T in position 2, 5, and 7 of the consensus was replaced by G, the efficiency of transcription was dramatically reduced. In contrast, single-point mutations upstream and downstream from the TATA box did not significantly affect the expression rate of the tRNA gene (Figure 3.6). From these experiments, we concluded that the TATA box is a major constituent of an archaeal promoter. To assess the importance of the second box at the transcription start site, the initiator nucleotide was mutated to a T (Figure 3.6). Analysis of the in vitro transcripts from this template showed that this nucleotide is indispensable for initiation of transcription. Two further nucleotides of this box B sequence were required for a high rate of cell-free transcription (Figure 3.6; Hausner et al., 1991). Thus, a TATA box at a distance of about 20 nucleotides to an ATGC-like motif appears to be the minimal requirement of an archaeal promoter. The sequence AAAAG up-

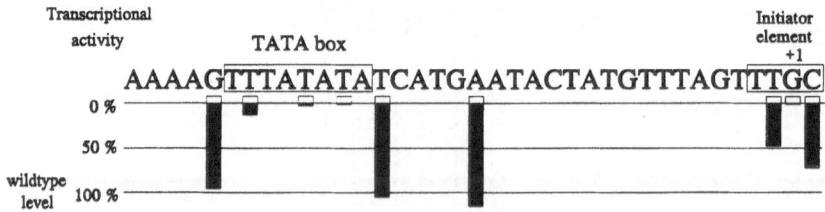

FIGURE 3.6   Analysis of effects of single-point mutations in the 5′ flanking region of the tRNA^Val gene of *Methanococcus vannielii* on the rate of cell-free transcription. DNA sequence of 5′ flanking region of tRNA^Val gene from position −35 to +2. Single-point mutations introduced into this region by in vitro mutagenesis are indicated by small open rectangles below sequence; T or A residue usually was replaced by G. G at position −31 was replaced by T; T and C residues at position −1 and +2 by A; and G at transcription start site by T. Black bars indicate template activity of plasmids containing single-point mutations. Note that promoter activity is dramatically reduced only by mutations in region of TATA box and initiator element.

stream of the TATA box (Brown et al., 1989; Figures 3.2–3.4) is not essential for initiation of transcription but appears to contribute to promoter function (Hausner et al., 1991). To avoid confusion with the well-established designations of A box and B box for the internal control regions of the RNA polymerase III promoter (see following), we suggest the designations TATA box and initiator element for the two structural elements of a typical archaeal promoter (Hausner et al., 1991). The archaeal TATA box appears homologous to the TATA box of eucaryotic polymerase II promoters at both the structural and the functional level (Hausner et al., 1991).

In contrast to organisms of the domains *Archaea* and *Bacteria*, eucaryotic cells have evolved different mechanism for the expression of rRNA and tRNA genes: These two classes of stable RNA genes are transcribed by two different types of RNA polymerase and do not share a common promoter sequence. In most eucaryotic systems, the DNA region from +10 to −40 constitutes the minimal RNA polymerase I promoter (Clos et al., 1986; Sollner-Webb et al., 1987). The sequences preceding rRNA genes from different eucaryotes do not show significant homologies. Thus, the polymerase I promoter appears to be species specific (Sommerville, 1984). In contrast to RNA polymerase I and II transcription systems, the promoter signals for eucaryotic tRNA and 5S genes reside downstream from the transcription initiation site (reviewed by Geiduschek and Tocchini-Valentini, 1988). Two sequences, A box and B box, corresponding to the region encoding the D and the TψC loops of mature tRNA, mediate initiation of transcription by RNA polymerase III. These sequences are highly conserved among all eucaryotes and are also found in the tRNA genes of bacteria. Moreover, owing to the presence of the A box and B box motives, tRNA genes of *E. coli* and the chloroplasts of *Euglena* are expressed with high efficiency by the polymerase III transcriptional machinery (Gruissem et al., 1982), although bacterial RNA polymerases initiate at upstream promoters. Inspection of the DNA sequences of archaeal tRNA genes revealed that the eucaryotic A box and B box sequences

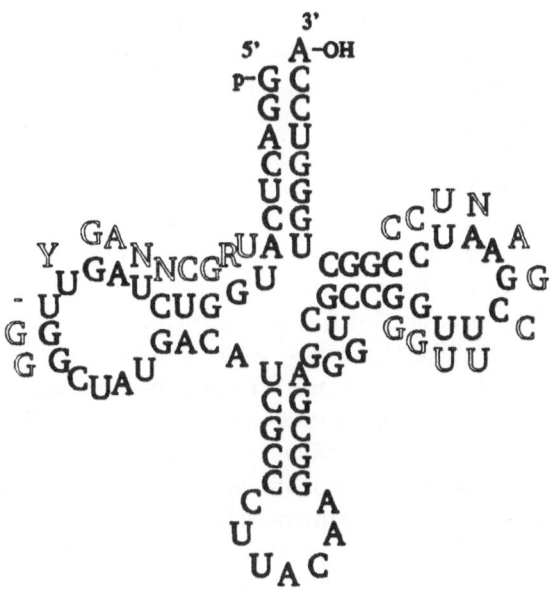

FIGURE 3.7   Archaeal tRNA sequence containing eucaryotic polIII promoter elements A box and B box: cloverleaf representation of tRNA^Val gene of *Methanococcus vannielii*. Intragenic polIII consensus promoter sequences are shown at corresponding position beside cloverleaf structure. Left: D loop, A box; right: TψC loop, B box. Abbreviations: R, purine; Y, pyrimidine; N, purine or pyrimidine.

are also highly conserved in the *Archaea* (Waldschmidt, 1989; Hausner, 1989; Figure 3.7). Deletion of internal sequences corresponding to the TψC loop did not abolish the template activity of the tRNA^Val gene (Thomm et al., 1990). However, a systematic analysis of the effects of internal deletions of an archaeal tRNA gene on the efficiency of cell-free transcription has not yet been performed. Thus, the function of internal DNA sequences in the initiation or termination of transcription remains to be elucidated.

Terminator sequences

Several DNA sequences downstream from archaeal genes have been proposed as possible terminator signals: transcripts from stable RNA genes of *Methanococcus vannielii* (Wich et al., 1986b), *Sulfolobus shibatae* (Reiter et al., 1988b), and *Thermofilum pendens* (Kjems et al., 1990) terminate within oligo-dT sequences strictly resembling the terminator sequences established for polymerase III transcription systems (Geiduschek and Tocchini-Valentini, 1988). Similar sequences have been found at the 3' end of protein-encoding genes from *Sulfolobus shibatae* (Reiter et al. 1988b). However, downstream from most protein-encoding genes of methanogens and extreme halophiles, sequences similar to rho-independent terminators of *E. coli* have been observed (see review by Brown et al., 1989). These structures can form hairpin-like structures, which are followed by an oligo-dT sequence. Transcripts from stable RNA genes from *Desulfurococcus mobilis*, *Methanobacterium thermoautotrophicum*, and *Thermofilum pendens* terminate at the end or after polypyrimidine sequences (Kjems and Garrett, 1987; Ostergaard et al., 1987; Kjems et al., 1990). However, transcription termination and 3' processing sites can barely be distinguished by mapping the 3' end of transcripts in vivo. Further, the function of these conserved oligo-dT and pyrimidine-rich sequences in the termination of transcription has not been established.

To identify the DNA sequences necessary for termination of transcription in a member of the *Archaea*, we performed a mutational analysis of the DNA region downstream from the tRNA^{Val} gene of *Methanococcus vannielii*. Various clones with deletions at the 3' end of the gene were generated and ligated to a DNA fragment harboring the intact 3' end of the tRNA^{Val} gene (Figure 3.8). These constructs contain the two putative terminator sites TTTTAATTTT (Wich et al., 1986b) in tandem. When the RNA polymerase does not stop at the first (mutated) terminator, an additional longer transcript should be synthesized. The amount of this longer RNA product is inversely correlated with the efficiency of termination at the first terminator. Thus, the effect of a mutation can be quantitated twofold by measuring the ratio of wild-type to elongated transcript. When a construct containing two wild-type terminators in tandem was used as a template in an in vitro transcription experiment, the efficiency of termination at the first and second oligo-dT sequence was 95% and 5%, repectively (Figure 3.8, top row). When six nucleotides of the first terminator have been deleted, about two-thirds of

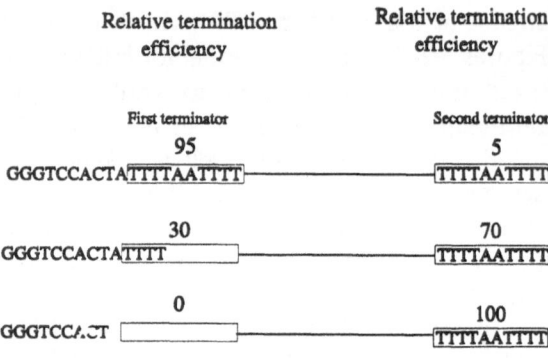

FIGURE 3.8 Oligo-dT sequences direct termination of transcription at tRNA genes of *Methanococcus vannielii*. DNA sequences of constructs containing putative terminator sequence TTTTAATTTT (Wich et al., 1986b) in tandem. DNA region between two terminator signals has been deleted by exonuclease III treatment. Efficiency of termination of transcription at first and second terminator was determined by measuring ratio of wild-type to elongated transcript. Relative amounts of corresponding transcripts are indicated above terminator signals (boxed).

the transcripts stop at the second and one-third at the first terminator (Figure 3.8, middle row). After the TTTT sequence has also been removed from the template (Figure 3.8, bottom row), transcription is terminated almost exclusively at the second terminator. These findings support the conclusion that the oligo-dT sequences downstream from stable RNA genes of M. *vannielii* are a major signal mediating termination of transcription.

3.4 Universal Features of Stable RNA Genes

No unique molecular property of stable RNA genes that is common to all the *Archaea* has been discovered so far. Some unique characteristics, such as unlinked expression of the three ribosomal RNAs (*Thermoplasma*), an intron in the variable loop of a transfer RNA (*Thermofilum pendens*), and a mixed-type splicing mechanism of a 23S rRNA intron (*Desulfurococcus mobilis*), appear to be restricted to a few genera or species.

Some features, such as the molecular organization of rRNA operons, may be *Eucarya*-like in one genus (*Thermoproteus*) and *Bacteria*-like in others (*Halobacterium*). However, both linked and unlinked expression of the 5S rRNA gene have been observed even within one genus (*Desulfurococcus mobilis* and *D. mucosus*). A universal feature of archaeal stable RNA genes that distinguishes *Archaea* from *Bacteria* is the presence of a 7S RNA.

Most similarities to *Eucarya* are found at the level of the transcriptional mechanisms. Both the polypeptides involved in the expression of stable RNA genes and the transcription signals appear to be very similar in the various genera of *Archaea*. All archaeal RNA polymerases show a multisubunit structure and genomic sequences resembling eucaryotic RNA polymerase (Huet et al., 1983; Schnabel et al., 1983; Pühler et al., 1989). The consensus promoter sequences are almost identical between *Methanococcus* and *Sulfolobus*, which represent the two phylogenetically separated kingdoms of the *Archaea* (see Figures 3.2 and 3.4). The archaeal TATA box is the major element determining the transcription start site (Haussner et al., 1991). This property, and the similarity in sequence and location to the eucaryotic TATA box, suggest homology of important parts of the archaeal and RNA polymerase II promoter. Minor differences in the consensus promoter sequences between extreme halophiles and methanogens (Figures 3.2 and 3.4) can be correlated with the high GC content of the cellular DNA of extreme halophiles (Thomm et al., 1989) and do not argue for a structural diversity of promoter structures in the *Euryarchaeota*. The existence of transcription factors provides a further similarity to eucaryotic gene transcription (Frey et al., 1990; Hüdepohl et al., 1990).

The finding that the purified RNA polymerase of *Archaea* binds to the promoter (Thomm et al., 1989) and shows semispecific initiation at initiator element-like sequences (Hüdepohl et al., 1990), unlike eucaryotic enzymes; but requires additional factors for correct initiation of transcription, as do eucaryotic RNA polymerases, suggests a novel function of the archaeal transcription factors. The further investigation of the biochemistry of archaeal

transcription might contribute to a deeper understanding of the evolution of the transcription apparatus and lead to the discovery of new mechanisms of regulation of gene expression.

3.5 Summary

The sequential and structural organization of stable RNA genes from all major groups of *Archaea* has been analyzed. The transcription start sites located upstream of many rRNA/ tRNA genes have been determined, and putative promoter and terminator sequences have been inferred from sequence comparisons. Footprinting and cell-free transcription experiments have been used to investigate the biological functions of these conserved DNA sequences, allowing the following conclusions.

Within the *Euryarchaeota*, operons encoding ribosomal RNAs show the bacteria-like organization: 5'-16S-tRNA-23S-5S-3'. *Crenarchaeota* do not contain a tRNA gene in the spacer between 16S and 23S rRNA and usually show unlinked 5S genes organized into a separate transcription unit. In *Methanococcus*, additional unlinked 5S genes exist; these are located within clusters of tRNAs. Some tRNA genes of *Archaea* and a 23S rRNA gene of *Desulfurococcus mucosus* contain an intron. However, this similarity to eucaryotes is restricted to a few tRNA species and has been found only in some genera of *Archaea*. Universal eucaryotic features are the presence of a gene encoding a 7S RNA of unknown function and the presence of eucaryotic promoter sequences. A TATA box octanucleotide at -25 and a second conserved sequence at the transcription start site are required for initiation of transcription. The archaeal TATA box element determines the transcription initiation site. Both in structure and function, this element closely resembles the TATA box of eucaryotic RNA polymerase II promoters. Oligo-dT sequences, which are conserved downstream from most archaeal genes, direct termination of transcription by the *Methanococcus* RNA polymerase. These sequences resemble the terminator signals recognized by RNA polymerase III of eucaryotes. The expression of archaeal stable

RNA genes is mediated by at least two transcription factors that might activate initiation of transcription by a hitherto unknown mechanism.

Acknowledgments. This work was supported by the Deutsche Forschungsgemeinschaft and the Fonds der Chemischen Industrie. We thank Dr. Karl Stetter for supporting parts of this work by funds from the Leibniz Preis.

References

Achenbach-Richter, L. and C.R. Woese. 1988. The ribosomal gene spacer region in archaebacteria. *Syst. Appl. Microbiol.* **10**:211–214.

Achenbach-Richter, L., K.O. Stetter, and C.R. Woese. 1987. A possible biochemical missing link among archaebacteria. *Nature* (London) **327**:348–349.

Balch, W.E., G.E. Fox., L.J. Magrum, C.R. Woese, and R.S. Wolfe. 1979. Methanogens: reevaluation of a unique biological group. *Microbiol. Rev.* **1979**:260–296.

Böck, A., H. Hummel, M. Jarsch, and G. Wich. 1986. *In: Genes for Stable RNA in Methanogens: Phylogenetic and Functional Aspects. Biology of Anaerobic Bacteria*, H.C. Doubourguier, et al., eds. pp. 206–226. Amsterdam: Elsevier.

Boros, I., E. Csordas-Toth, A. Kiss, I. Kiss, I. Török, A. Udvardy, K. Udvardy, and P. Venetianer. 1983. Identification of two new promoters probably involved in the transcription of a ribosomal RNA gene of *Escherichia coli. Biochim. Biophys. Acta* **739**:173–180.

Brown, J.W., C.J. Daniels, and J.N. Reeve. 1989. Gene structure, organization and expression in Archaebacteria. *Crit. Rev. Microbiol.* **16**:287–338.

Brown, J.W., M. Thomm, G.S. Beckler, G. Frey, K.O. Stetter, and J.N. Reeve. 1988. An archaebacterial RNA polymerase binding site and transcription initiation of the *hisA* gene in *Methanococcus vannielii. Nucleic Acids Res.* **16**:135–150.

Caillet, J., J.A. Plumbridge, and M. Springer. 1985. Evidence that *pheV*, a gene for tRNA[Phe] of *E. coli* is transcribed from tandem promoters. *Nucleic Acids Res.* **13**:3699–3710.

Clos, J., A. Normann, A. Öhrlein, and I. Grummt. 1986. The core promoter of mouse rDNA consists of two functionally distinct domains. *Nucleic Acids Res.* **14**:7581–7595.

Corden, J., B. Wasylyk, A. Buchwalder, P. Sassone-Corsi, C. Kedinger, and P. Chambon. 1980. Promotor sequences of eukaryotic protein-coding genes. *Science* **209**:1406–1414.

Daniels, C.J., S.E. Douglas, and W.F. Doolittle. 1986. Genes for transfer RNAs in *Halobacterium volcanii. Syst. Appl. Microbiol.* **7**:26–29.

Daniels, C.J., R. Gupta, and W.F. Doolitle. 1985. Transcription and excision of a large intron in the tRNA[trp] gene of an archaebacterium, *Halobacterium volcanii. J. Biol. Chem.* **260**:3132–3134.

Datta, P.K., L.K. Hawkins, and R. Gupta. 1989. Presence of an intron in elongator methionine-tRNA of *Halobacterium volcanii. Can. J. Microbiol.* **35**:189–194.

Dennis, P.P. 1985. Multiple promoters for the transcription of the ribosomal RNA gene cluster in *Halobacterium cutirubrum. J. Mol. Biol.* **186**:457–461.

Eggen, R., H. Harmsen, and M. de Vos. 1990. Organization of a ribosomal RNA gene cluster from the archaebacterium *Methanothrix soehngenii. Nucleic Acids Res.* **18**:1306.

Frey, G., M. Thomm, B. Brüdigam, H.P. Gohl, and W. Hausner. 1990. An archaebacterial cell-free transcription system. The expression of tRNA genes from *Methanococcus vannielii* is mediated by a transcription factor. *Nucleic Acids Res.* **18**:1361–1367.

Geiduschek, P.E. and G.P. Tocchini-Valentini. 1988. Transcription by RNA polymerase III. *Annu. Rev. Biochem.* **57**:873–914.

Grogan, D., P. Palm, and W. Zillig. 1990. Isolate B12, which harbours a virus-like element, represents a new species of the archaebacterial genus *Sulfolobus, Sulfolobus shibatae,* sp. nov. *Arch. Microbiol.* **154**:594–599.

Gruissem, W., D. Prescott, B.M. Greenberg, and R.B. Hallick. 1982. Transcription of *E. coli* and *Euglena* chloroplast tRNA gene clusters and processing of polycistronic transcripts in a HeLa cell-free extract. *Cell* **30**:81–92.

Haas, E.S., C.J. Daniels, and J.N. Reeve. 1989. Genes encoding 5S rRNA and tRNAs in the extremely thermophilic archaebacterium *Methanothermus fervidus. Gene* **77**:253–263.

Haas, E.S., J.W. Brown, C.J. Daniels, and J.N. Reeve. 1990. Genes encoding the 7S RNA and a tRNA[Ser] are linked to one of the two rRNA operons in the genome of the extremely thermophilic archaebacterium *Methanothermus fervidus. Gene* **90**:51–58.

Hausner, W. 1989. In vitro Mutagenese von tRNA Genen Zur Ermittlung promotor-und terminatoraktiver DNA Sequenzen. Diploma Thesis, Universität Regensburg, FRG.

Hausner, W., G. Frey, and M. Thomm. 1991. Control regions of an archaeal gene. ATATA box and an initiator element promote cell-free transcription of the tRNA[Val] gene of *Methanococcus vanniellii, J. Mol. Biol.* **L22**:495–508.

Hawley, D.K. and R. McClure. 1983. Compilation and analysis of *Escherichia coli* promoter DNA sequences. *Nucleic Acids Res.* **11**:2237–2255.

Hüdepohl, U., W.D. Reiter, and W. Zillig. 1990. In

vitro transcription of two rRNA genes of the archaebacterium *Sulfolobus* sp. B12 indicates a factor requirement for specific transcription. *Proc. Nat. Acad. Sci. Natl. USA* **87**:5851–5855.

Huet, J., R. Schnabel, A. Sentenac, and W. Zillig, 1983. Archaebacteria and eukaryotes possess DNA-dependent RNA polymerases of a common type. *EMBO J.* **2**:1291–1294.

Jarsch, M., J. Altenbuchner, and A. Böck. 1983. Physical organization of the genes for ribosomal RNA in *Methanococcus vannielii*. *Mol. Gen. Genet.* **189**:41–47.

Kaine, B.P. 1987. Intron-containing tRNA genes of *Sulfolobus solfataricus*. *J. Mol. Evol.* **25**:248–254

Kaine, B.P. 1990. Structure of the archaebacterial 7S RNA molecule. *Mol. Gen. Genet.* **221**:315–321.

Kaine, B.P. and V.L. Merkel. 1989. Isolation and characterization of the 7S RNA gene from *Methanococcus voltae*. *J. Bacteriol.* **171**:4261–4266.

Kaine, B.P., R. Gupta, and C.R. Woese. 1983. Putative introns in tRNA genes of prokaryotes. *Proc. Nat. Acad. Sci. USA* **80**:3309–3312.

Kenerley, M.E., E.A. Morgan, L. Post, L. Lindahl, and M. Nomura. 1977. Characterization of hybrid plasmids carrying individual ribosomal RNA transcription units of *Escherichia coli*. *J. Bacteriol.* **132**:931–949.

King, T.C., R. Sirdegkmukh, and D. Schlesinger. 1986. Nucleolytic processing of ribonucleic acid transcripts in procaryotes. *Microbiol. Rev.* **50**:428–451.

Kjems, J. and R.A. Garrett. 1985. An intron in the 23S ribosomal RNA gene of the archaebacterium *Desulfurococcus mobilis*. *Nature* (London) **318**:675–677.

Kjems, J. and R.A. Garrett. 1987. Novel expression of the ribosomal RNA genes in the extreme thermophile and archaebacterium, *Desulfurococcus mobilis*. *EMBO J.* **6**:3521–3530.

Kjems, J. and R.A. Garrett. 1988. Novel splicing mechanism for the ribosomal RNA intron in the archaebacterium *Desulfurococcus mobilis*. *Cell* **54**:693–703.

Kjems, J. and R.A. Garrett. 1990. Secondary structural elements exclusive to the sequences flanking ribosomal RNAs lend support to the monophyletic nature of the archaebacteria. *J. Mol. Evol.* **31**:25–32.

Kjems, J., H. Leffers, T. Olesen, and R.A. Garrett. 1989. A unique tRNA Intron in the variable loop of the extreme thermophile *Thermofilum pendens* and its possible evolutionary implications. *J. Biol. Chem.* **264**:17834–17837.

Kjems, J., H. Leffers, T. Olesen, I. Holz, and R.A. Garrett. 1990. Sequence, organization and transcription of the ribosomal RNA operon and the downstream tRNA and protein genes in the archaebacterium *Thermofilum pendens*. *Syst. Appl. Microbiol.* **13**:117–127.

Knaub, S. and A. Klein. 1990. Specific transcription of cloned *Methanobacterium autotrophicum* transcription units by homologous RNA polymerase in vitro. *Nucleic Acids Res.* **18**:1441–1446.

Lake, J. 1988. Origin of the eukaryotic nucleus determined by rate-invariant analysis of rRNA sequences. *Nature* (London) **331**:184–186.

Lake, J. 1989. Origin of the eukaryotic nucleus: eukaryotes and eocytes are genotypically related. *Can. J. Microbiol.* **35**:109–118.

Larsen, H. and W.D. Grant. 1989. Group III. Extremely halophilic archaeobacteria: Order Halobacteriales Ord. Nov. In: *Bergey's Manual of Systematic Bacteriology*, Vol. 3. J.T. Stanley, M.P. Bryant, N. Pfennig, and J.G. Holt, eds. pp. 2216–2224. Baltimore: Williams & Wilkins.

Larsen, N., H. Leffers, J. Kjems, and R.A. Garrett. 1986. Evolutionary divergence between the ribosomal RNA operons of *Halococcus morrhuae* and *Desulfurococcus mobilis*. *Syst. Appl. Microbiol.* **7**:49–57.

Leffers, H., J. Kjems, L. Ostergaard, N. Larsen, and R.A. Garrett. 1987. Evolutionary relationship amongst archaebacteria. A comparative study of 23S ribosomal RNAs of a sulphur-dependent extreme thermophile, an extreme halophile and a thermophilic methanogen. *J. Mol. Biol.* **195**:43–61.

Mankin, A.S. and V.K. Kagramanova. 1986. Complete nucleotide sequence of the single ribosomal RNA operon of *Halobacterium halobium*: secondary structure of the archaebacterial 23S rRNA. *Mol. Gen. Genet.* **202**:152–161.

Mankin, A.S. and V.K. Kagramanova. 1988. Complex promoter pattern of the single ribosomal RNA operon of an archaebacterium *Halobacterium haloblum*. *Nucleic Acids Res.* **16**:4679–4692.

Melton, D.A., E.M. De Robertis, and R. Cortese. 1980. Order and intracellular location of the events involved in the maturation of a spliced tRNA. *Nature* (London) **284**:143–148.

Mevarech, M., S. Hirsch-Twizer, S. Goldman, E. Yakobson, H. Eisenberg, and P.P. Dennis. 1989. Isolation and characterization of the rRNA gene clusters of *Halobacterium marismortui*. *J. Bacteriol.* **171**:3479–3485.

Moritz, A. and W. Goebel. 1985. Characterization of the 7S RNA and its gene from halobacteria. *Nucleic Acids Res.* **13**:6969–6979.

Moss, T. 1983. A transcriptional function for the repetitive ribosomal spacer in *Xenopus laevis*. *Nature* (London) **302**:223–228.

Neumann, H., A. Gierl, J. Tu, J. Leibrock, D. Staiger, and W. Zillig. 1983. Organization of the genes for ribosomal RNA in archaebacteria. *Mol. Gen. Genet.* **192**:66–72.

Ostergaard, L., N. Larsen, H. Leffers, J. Kjems, and R.A. Garrett. 1987. A ribosomal RNA operon and its flanking region from the archaebacterium *Methanobacterium thermoautotrophicum*,

Marburg strain: transcription signals, RNA structure and evolutionary implications. *Syst. Appl. Microbiol.* **9**:199–209.

Pühler, G., H. Leffers, F. Gropp, P. Palm, H.-P. Klenk, F. Lottspeich, R.A. Garrett, and W. Zillig. 1989. Archaebacterial DNA-dependent RNA polymerases testify to the evolution of the eukaryotic nuclear genome. *Proc. Nat. Acad. Sci. USA.* **86**:4569–4573.

Ree, H.K. and R.A. Zimmermann, 1990. Organization and expression of the 16S, 23S and 5S ribosomal RNA genes of the archaebacterium *Thermoplasma acidophilum. Nucleic Acids Res.* **18**:4471–4478.

Reiter, W., P. Palm, and W. Zillig. 1988a. Analysis of transcription in the archaebacterium *Sulfolobus* indicates that archaebacterial promoters and eukaryotic RNA pol II promoters are of the same type. *Nucleic Acids Res.* **16**:1–19.

Reiter, W., P. Palm, and W. Zillig. 1988b. Transcription termination in the archaebacterium *Sulfolobus*: signal structures and linkage to transcription initiation. *Nucleic Acids Res.* **16**:2445–2459.

Reiter, W., P. Palm, W. Voos, J. Kaniecki, B. Grampp, W. Schulz, and W. Zillig. 1987. Putative promoter elements for the ribosomal RNA genes of the thermoacidophilic archaebacterium *Sulfolobus* sp. strain B12. *Nucleic Acids Res.* **15**:5581–5595.

Schnabel, R., M. Thomm, R. Gerardy-Schahn, W. Zillig, K.O. Stetter, and J. Huet. 1983. Structural homology between different archaebacterial DNA-dependent RNA polymerases analysed by immunological comparison of their components. *EMBO J.* **2**:751–755.

Sollner-Webb, B., J. Windle, S. Henderson, J. Tower, V. Culotta, S. Kass and N. Craig. 1987. Initiation and termination of ribosomal RNA transcription and processing of the primary rRNA transcript. In: *RNA Polymerase and the Regulation of Transcription.* W.S. Reznikoff, R.R. Burgess, J.E. Dahlberg, C.A. Gross, M.T. Record and M.P. Wickens, eds., pp. 187–194. New York: Elsevier.

Sommerville, J. 1984. RNA polymerase I promotors and transcription factors. *Nature* (London) **310**:189–190.

Struck, J.C.R., H.Y. Toschka, T. Specht, and V.A. Erdmann. 1988. Common structural features between eukaryotic 7SL RNAs, eubacterial 4.5S RNA and scRNA and archaebacterial 7S RNA. *Nucleic Acids Res.* **16**:7740.

Suzuki, Y., Y. Ono, A. Nagata, and T. Yamada. 1988a. Molecular cloning and characterization of an rRNA operon in *Streptomyces lividans* TK21. *J. Bacteriol.* **170**:1631–1636.

Suzuki, Y., A. Nagata, Y. Ono, and T. Yamada. 1988b. Complete nucleotide sequence of the 16S rRNA gene of *Mycobacterium bovis* BCG. *J. Bacteriol.* **170**:2886–2889.

Taschke, C., M. Klinkert, J. Wolters, and R. Herrmann. 1986. Organization of the ribosomal RNA genes in *Mycoplasma hyopneumoniae*: The 5S rRNA gene is separated from the 16S and 23S rRNA genes. *Mol. Gen. Genet.* **205**:428–433.

Thomm, M. and G. Wich. 1988. An archaebacterial promoter element for stable RNA genes with homology to the TATA box of higher eukaryotes. *Nucleic Acids Res.* **16**:151–163.

Thomm, M., B.A. Sherf, and J.N. Reeve. 1988. RNA polymerase-binding and transcription initiations sites upstream of the methyl reductase operon of *Methanococcus vannielii. J. Bacteriol.* **170**:1958–1961.

Thomm, M., G., Frey, W. Hausner, and B. Brüdigam. 1990. An archaebacterial in vitro transcription system. In: *Microbiology and Biochemistry of Strict Anaerobes Involved in Interspecies Hydrogen Transfer,* J.-P. Belaich, M. Bruschi, and J.-L. Garcia, eds., pp 305–312. New York: Plenum Press.

Thomm, M., G. Wich, J.W. Brown, G. Frey, B.A. Sherf, and G.S. Beckler. 1989. An archaebacterial promoter sequence assigned by RNA polymerase binding experiments. *Can. J. Microbiol.* **35**:30–35.

Thompson, L.D. and C.J. Daniels. 1988. A tRNA^Trp intron endonuclease from *Halobacterium volcanicum*. Unique substrate recognition properties. *J. Biol. Chem.* **263**:17951–17959.

Tu, J. and W. Zillig. 1982. Organization of rRNA structural genes in the archaebacterium *Thermoplasma acidophilum. Nucleic Acids Res.* **10**:7231–7245.

Waldschmidt, R. 1989. Transkriptionsfuktoren der RNA Polymerase III. Ph.D. Thesis, Phillips-Universität, Marburg, FRG.

Wich, G., M. Jarsch, and A. Böck. 1984. Apparent operon for a 5S ribosomal RNA gene in the archaebacterium *Methanococcus vannielii. Mol. Gen. Genet.* **196**:146–151.

Wich, G., H. Hummel, M. Jarsch, U. Bär, and A. Böck. 1986a. Transcription signals for stable RNA genes in *Methanococcus. Nucleic Acids Res.* **14**:2459–2479.

Wich, G., L. Sibold, and A. Böck. 1986b. Genes for tRNA and their putative expression signals in *Methanococcus. Syst. Appl. Microbiol.* **7**:18–25.

Wich, G., W. Leinfelder, and A. Böck. 1987a. Genes for stable RNA in the extreme thermophile *Thermoproteus tenax*: introns and transcription signals. *EMBO J.* **6**:523–528.

Wich, G., L. Sibold, and A. Böck. 1987b. Divergent evolution of 5S rRNA genes in *Methanococcus. Z. Naturforsch. Sect. C. Biosci.* **42**:373–380.

Woese, C.R. 1987. Bacterial evolution. *Microbiol. Rev.* **51**:221–271.

Woese, C.R., and G.L. Fox. 1977. Phylogenetic structure of the procaryotic domain: the primary kingdom. *Proc. the Natl. Acad. Sci. USA* **74**:5088–5090.

Woese, C.R., O. Kandler, and M. Wheelis. 1990. Towards a natural system of organisms. Proposal of the Domain Archaea, Bacteria and Eucarya. *Proc. Natl. Acad. Sci. USA* **87**:4576–4579.

Young, R.A. and J.A. Steitz. 1979. Tandem promoters direct *E. coli* ribosomal RNA synthesis. *Cell* **17**:225–234.

Zillig, W., I. Holz, H. Klenck, J. Trent, S. Wunderl, D. Janekovic, E. Imsel, and B. Haas. 1987. *Pyrococcus woesei* sp. nov., an ultra-thermophilic marine archaebacterium representing a novel order, Thermococcales. *Syst. Appl. Microbiol.* **9**:62–70.

Zwieb, C. 1989. Structure and function of signal recognition particle RNA. *Prog. Nucleic Acid Res. Mol. Biol.* **37**:207–234.

4

Molecular Biology of the Acetoclastic Methanogen *Methanothrix soehngenii*

Rik I.L. Eggen and Willem M. de Vos

4.1 Introduction

Acetate forms about 70% of the methanogenic substrates in anaerobic digestors (Smith and Mah, 1966; Gujer and Zehnder, 1983; Zinder et al., 1984; Fukuzaki et al., 1990) and is the only dicarbon substrate that methanogenic bacteria can degrade completely (Thauer et al., 1989). The ability to catabolize acetate to methane is restricted to only two genera, *Methanosarcina* and *Methanothrix*, of the Kingdom *Euryarchaeota*, which belongs to the archaeal domain (Woese et al., 1990). *Methanosarcina* spp. are generalistic organisms, capable of degrading a variety of substrates like H_2/CO_2, methanol, methylamines, and acetate (Hutten et al., 1980; Kenealy and Zeikus, 1982; Krzycki et al., 1982; Smith and Mah, 1980). In contrast, *Methanothrix* sp. is a specialist that can use only acetate as its carbon and energy source (Huser et al., 1982; Zehnder et al., 1980). The pathway of the acetoclastic methanogenesis in *Methanosarcina* and *Methanothrix* (Figure 4.1) involves the activation of acetate to acetyl-CoA, for which different mechanisms have been found. In *Methanosarcina* spp., acetate is activated by the enzymes acetate kinase and phosphate acetyltransferase (Aceti and Ferry, 1988; Fischer and Thauer, 1988; Laufer et al., 1987; Lundie and Ferry, 1989; Terlesky et al., 1987; Thauer, 1988), whereas in *Methanothrix* spp. acetate is activated in one step by acetyl-CoA synthetase (Kohler and Zehnder, 1984; Pellerin et al., 1987; Jetten et al., 1989a). The enzyme carbon monoxide dehydrogenase catalyzes the cleavage of

acetyl-CoA into CO and CH_3 moieties and the subsequent oxidation of CO to CO_2 (Eikmanns and Thauer, 1984; Kohler and Zehnder, 1984; Krzycki and Zeikus, 1984; Terlesky et al., 1986; Grahame and Stadtman, 1987; Aceti and Ferry, 1988; Jetten et al., 1989b). The CH_3 moiety is reduced to CH_4 by methyltransferase and methyl-CoM reductase, using the electrons derived from the oxidation of CO (Lovley et al., 1984; Nelson and Ferry, 1984; Krzycki et al., 1985; Grahame and Stadtman, 1987; Laufer et al., 1987; Fischer and Thauer, 1988; van de Wijngaard et al., 1988; Jetten et al., 1990a).

In this chapter we focus on a mesophilic *Methanothrix* species, first described as *Methanothrix soehngenii* strain Opfikon (Zehnder et al., 1980; Huser et al., 1982). Various *Methanothrix* strains have since been described, and from some the G + C content has been determined (varying between 49.0% and 52.6%; Huser et al., 1982; Touzel et al., 1988; Patel and Sprott, 1990). The genus *Methanothrix* includes both mesophilic strains [*M. soehngenii* strain FE (Touzel et al., 1988) and *M. concilii* strain GP6 (Patel, 1984)] and thermophilic strains [*Methanothrix* sp., strain CALS-1 (Zinder et al., 1984) and *M. thermoacetophila*, strain Z517 (Nozhevnikova and Chudina, 1985)]. It has been proposed that the latter three axenic *Methanothrix* species should be included in the genus *Methanosaeta* (Patel and Sprott, 1990). In addition, because the purity of the type species *Methanothrix soehngenii* strain Opfikon has been disputed, it was suggested that the species *M. soehngenii* is nomenclaturally invalid (Patel

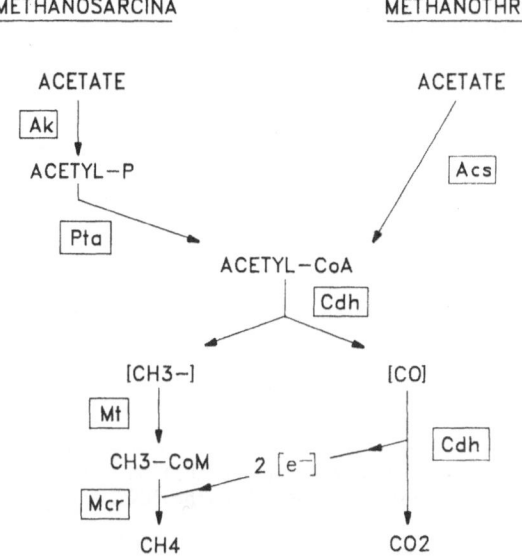

METHANOSARCINA

METHANOTHRIX

FIGURE 4.1 Pathway of acetoclastic metha-nogenesis in genera *Methanothrix* and *Methanosar-cina*. Abbreviations: Ak, acetate kinase; Pta, phos-photransacetylase; Acs, acetyl-CoA synthetase; Cdh, carbon monoxide dehydrogenase; Mt, methyl transferase; Mcr, methyl-CoM reductase; e^-, electron transfer. Protein-bound intermediates indicated in brackets.

and Sprott, 1990). However, since discussions are under way to retain the name *Methanothrix soehngenii* or at least the generic name *Metha-nothrix* (A.J.B. Zehnder, personal communica-tion) and to prevent subsequent confusion, we maintain in this chapter the originally proposed nomenclature (Huser et al., 1982).

Because of its high affinity for acetate ($K_s = 0.5$ mM) and its low acetate threshold (<10 μM), *M. soehngenii* is abundant in anaero-bic digestors where it can be stably maintained, although its growth yield (1.4 g dry wt/mol ace-tate) and growth rate (mean generation time of 9 days) are low (Zehnder et al., 1980; Huser et al., 1982; Gujer and Zehnder, 1983; Jetten et al., 1990b). Most of the key enzymes involved in the acetate metabolism of *M. soehngenii*, e.g., acetyl-CoA synthetase, carbon monoxide dehydrogenase, and methyl-CoM reductase, have been purified and studied in some detail (Jetten et al., 1989a, 1989b, 1990a, 1991). These biochemical studies have allowed for a genetic examination of the acetate pathway in this

organism. In this chapter, current knowledge about the molecular biology of *M. soehngenii* is summarized.

4.2 Ribosomal RNA Cluster and Phylogeny

16S ribosomal RNA sequence

Hybridization experiments with an oligonu-cleotide designed to detect 16S rRNA sequences indicated that the *Methanothrix soehngenii* genome contains two copies of a 16S rRNA gene. One 16S rRNA gene together with its flanking regions has been cloned, and the nucleotide sequence has been determined (Eggen et al., 1989, 1990). This sequence has been compared with available 16S rRNA se-quences of other methanogens, which resulted in homology values as indicated in Table 4.1. These results illustrate the phylogenetic differ-ences between the species and indicate that *Methanothrix* spp., both the mesophilic (*M. soehngenii*) and the thermophilic species (*Methanothrix* CALS-1), belong to the order *Methanomicrobiales*, represented by *Methano-thrix* spp., *Methanosarcina barkeri*, and *Metha-nospirillum hungatei*. This is in line with pre-viously described relationships, based on 16S rRNA oligonucleotide catalogs (Stackebrandt et al., 1982).

Ribosomal RNA operon organization

The 16S rRNA gene is part of an rRNA cluster with an operon-like organization, with the order 16S rRNA—tRNA[ala]—23S rRNA (Eggen et al., 1990). This organization is similar to that of rRNA operons from other euryarchaeal species (Figure 4.2; Achenbach-Richter and Woese, 1988; Brown et al., 1989). The spacer tRNA[ala] genes, all with an UGC anticodon, are strongly conserved. The tRNA[ala] gene from *Methanothrix soehngenii* is remarkable in that the CCA 3′ terminus is encoded by the tRNA gene, which is not a general property of archaeal tRNA genes (Achenbach-Richter and Woese, 1988). The rRNA operon is probably trans-

Table 4.1 Similarity matrix for 16S rRNA sequences from methanogenic species[a]

Organism	1	2	3	4	5	6	7	8
Methanomicrobiales								
1 *Methanothrix soehngenii* strain Opfikon	—	99.9[b]	91.2	85.5	82.0	81.0	79.8	78.0
2 *Methanothrix soehngenii* strain FE*		—	91.2	85.5	82.0	81.1	79.9	78.0
3 *Methanothrix* sp. strain CALS-1*			—	85.0	81.5	78.9	78.0	76.4
4 *Methanosarcina barkeri* strain 227*				—	82.2	79.0	78.2	77.6
5 *Methanospirillum hungatei*					—	80.0	78.8	77.2
Methanobacteriales								
6 *Methanobacterium formicicum*						—	90.8	82.3
7 *Methanobacterium thermoautotrophicum* strain Marburg							—	79.3
Methanococcales								
8 *Methanococcus vannielii*								—

[a]Sequences were obtained from literature (Eggen et al., 1989; Jarsch and Böck, 1985; Lechner et al., 1985; Østergaard et al., 1987; Yang et al., 1985) or kindly provided by Dr. C.R. Woese, University of Illinois, from the rRNA Database Project (*).
[b]Numbers represent percentages homology for various pairs of 16S rRNA sequences. Homology = matches/(length-unpaired residues).

cribed into a primary RNA molecule from which the individual mature RNAs are released by RNA processing steps. This conclusion is based on primer extension experiments that showed the location of the transcription initiation site and processing positions in the RNA (Eggen et al., 1990). The presumed transcription initiation site was located 173 residues upstream from the start of the 16S rRNA gene. This region contains boxes A and B, similar to the consensus sequence of archaeal promoters (Figure 4.3; Zillig et al., 1988; Thomm et al., 1989; Knaub and Klein, 1990; Reiter et al., 1990). Processing sites were observed 5' of the first nucleotide of the 16S rRNA gene and at a

sequence with a specific secondary structure formed by nucleotide sequences flanking the 16S rRNA gene (Eggen et al., 1990).

4.3 Genes Involved in Acetate Activation and Cleavage

Acetyl-CoA synthetase gene

As indicated in Figure 4.1, acetate is activated in *Methanothrix soehngenii* by acetyl-CoA synthetase (Acs) with CoA and ATP as cofactors, forming acetyl-CoA (Jetten et al., 1989a). Acs must therefore be able to coordinately bind

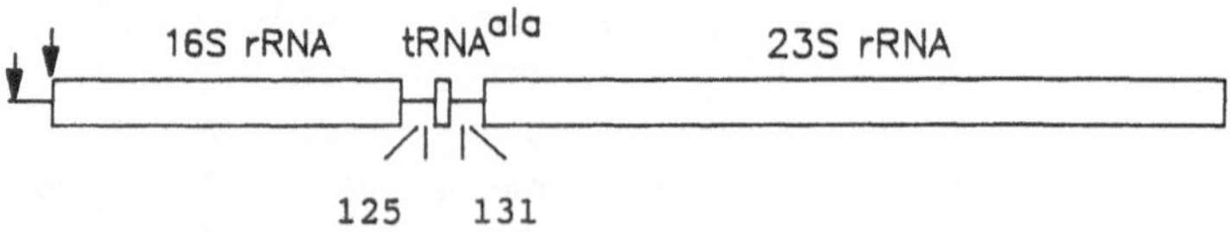

```
                    16S rRNA    tRNAᵃˡᵃ         23S rRNA

      M. soehngenii              125    131
      Msp. hungatei               68    154
      Mb. formicicum              89    220
      Mb. thermoautotrophicum ΔH 106    163
      H. cutirubrum              105    384
      H. halobium                105    351
      H. volcanii                 97    222
      Hc. morrhua                146    216
```

FIGURE 4.2 Schematic representation of genetic organization of the rRNA operon from *Methanothrix soehngenii*. Sizes of intercistronic spacer regions are compared with those of other similarly organized operons from various Archaea (Achenbach-Richter and Woese, 1988; Brown et al., 1989). Left arrow indicates putative primary transcription initiation site; right arrow, the RNA processing sites.

Consensus archaeal promoter

	Box A		Box B
	TTTATATA	18–23 bp	$\frac{A}{T}TG\frac{A}{C}$

Methanothrix soehngenii

	Box A		Box B
rRNA operon	TTTAAATA	19	ATGG
cdh operon	TTTATAAA	26	TTAA
	TCTATATT	20	TTAT
acs gene	TTATAATT	25	CTGA
	TTATTTAT	20	GTGC

FIGURE 4.3 Promoter and promoter-like structures of *Methanothrix soehngenii* genes. Shaded letters in boxes B indicate experimentally verified transcription initiation sites. Consensus sequence for archaeal promoters is obtained from literature (Zillig et al., 1988; Thomm et al., 1989; Knaub and Klein, 1990; Reiter et al., 1990).

acetate, CoA, and ATP. Acs from *M. soehngenii* constitutes up to 4% of the cytoplasmic protein and has a native molecular mass of 146 kDa, composed of two identical subunits (AcsA) of 73 kDa. On the basis of the kinetics of purified Acs, it was suggested that two binding sites for ATP occur in Acs (Jetten et al., 1989a).

Using a polyclonal antiserum raised against purified Acs, the *M. soehngenii acs* gene was isolated from a genomic library in *Escherichia coli* and cloned as a 3.5-kb *Sal*I fragment in pUC19 (Eggen et al., 1991b). After being introduced in *E. coli*, a major immunoreactive polypeptide was produced that was as much as 5% of total cellular protein with a slightly smaller molecular mass than AcsA, as purified from *M. soehn-*

genii. In *E. coli* other transcription or translation signals may be used, resulting in a shortened protein, or perhaps Acs is post-translationally modified in *M. soehngenii*, leading to a higher molecular mass. Despite this difference, Acs activity could be detected in cell-free extracts from *E. coli*, with a similar specific activity as measured in cell-free extracts from *M. soehngenii* (2.1 μmol/mg·min). This indicates that functionally active AcsA is efficiently produced in *E. coli*.

The genetic organization of the *acs* gene as deduced from its nucleotide sequence is depicted in Figure 4.4. The *acs* gene, with an A + T content of 43.8% is preceded by a stretch of nucleotides with a higher A + T percentage (57.4%), containing possible archaeal expres-

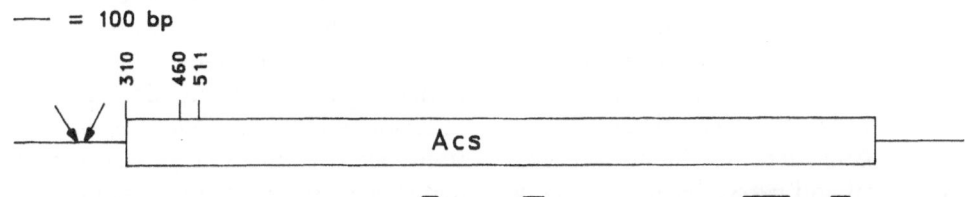

FIGURE 4.4 Genetic organization of *acs* gene. Open-reading frame (ORF) indicated by open bar, flanking regions by solid line. Arrows represent putative archaeal promoter-like structures. Positions of putative translational start signals are indicated on the ORF. Regions in Acs that showed more than 50% homology with acetyl-CoA synthetase, coumarate-CoA ligase, and luciferin monooxygenase (see Figure 4.5) are indicated below open-reading frame by black boxes.

Table 4.2 *Methanothrix soehngenii* Acs and proteins with similar functions, e.g., binding of ATP or CoA

Enzyme	Organism	Length (aminoacids)	Identity (%)[a]	Binding of: ATP	CoA
Acetyl-CoA synthetase	*Methanothrix soehngenii*	672	100	+	+
Acetyl-CoA synthetase	*Neurospora crassa*	643	42	+	+
Coumarate-CoA ligase	*Petroselinum crispum*	612	26	+	+
Luciferin monooxygenase	*Photinus pyralis*	594	24	+	−
Acetate kinase	*Escherichia coli*	472	20	+	−

[a] Percentage identity between *Methanothrix* Acs and respective proteins.

sion signals. The transcription initiation has been determined, and two possible promoter regions have been identified (see Figure 4.3) that are similar to the consensus sequence for Box A and Box B of archaeal promoters (see Figure 4.3; Thomm et al., 1989; Zillig et al., 1988; Knaub and Klein, 1990; Reiter et al., 1990). For translation initiation, three start codons, which are preceded by ribosome-binding sites, can be used in the same reading frame (Figure 4.4).

Database searches showed that the deduced amino acid sequence of *M. soehngenii* AcsA is homologous to that of proteins with similar functions found in both bacterial and eucaryal species (Figure 4.4 and Table 4.2; De Wet et al., 1987; Lozoya et al., 1988; Matsuyama et al., 1989; Connerton et al., 1990). In addition, two putative ATP-binding sequences, one A type and one B type (Albano et al., 1989) were observed.

Carbon monoxide dehydrogenase genes

Carbon monoxide dehydrogenase (Cdh) is a key enzyme in the conversion of acetate to methane in acetoclastic methanogens. The enzyme catalyzes both the cleavage of the dicarbon bond in acetyl-CoA and the oxidation of CO to CO_2 (Eikmanns and Thauer, 1984; Kohler and Zehnder, 1984; Krzycki and Zeikus, 1984; Terlesky et al., 1986; Grahame and Stadtman, 1987; Aceti and Ferry, 1988; Jetten et al., 1989b). In *M. soehngenii* the enzyme constitutes up to 4% of total cellular protein and has a native molecular mass of 190 kDa, composed of two large subunits of 79 kDa (CdhA) and two small subunits of 19 kDa (CdhB; Jetten et al.,

1989b). Metal analysis of purified Cdh indicated the presence of 2 mol Ni and 20 mol Fe per mol enzyme (Jetten et al., 1989b). Electron paramagnetic resonance analysis has provided evidence for the presence of four iron-sulfur clusters in the enzyme, being identified as two [4Fe-4S] cubanes and probably two [6Fe-6S] prismanes (Jetten et al., 1991).

The *M. soehngenii* cdhA and cdhB genes that code for the large and the small subunits of Cdh, respectively, were isolated from a genomic library in *E. coli* using polyclonal antibodies raised against purified Cdh (Eggen et al., 1991a). Subcloning of these genes on a 4.8-kb HindIII fragment into *E. coli*, resulted in the production of two immunoreactive polypeptides, which corresponded in size to the purified Cdh subunits (Jetten et al., 1989b). Carbon monoxide dehydrogenase activity could, however, not be observed in extracts prepared from both aerobically or anaerobically grown *E. coli* cells containing the cdh genes. It was shown by Jetten et al. (1989b) that Cdh activity, measured by methyl viologen coupled CO-oxidoreductase activity, was oxygen insensitive. However, recent electron paramagnetic resonance experiments showed that the iron-sulfur clusters in Cdh were extremely oxygen sensitive (Jetten et al., 1991). As an alternative anaerobic expression host, the sulfate-reducing bacterium *Desulfovibrio vulgaris* was used because its hydrogenase contains iron-sulfur clusters similar to those in *M. soehngenii* (Hagen et al., 1989; Jetten et al., 1991). Although CdhA and CdhB were produced, no Cdh activity could be detected in this host (Eggen et al., 1991a). Among several reasons for the absence of Cdh activity in the expres-

— = 100 bp

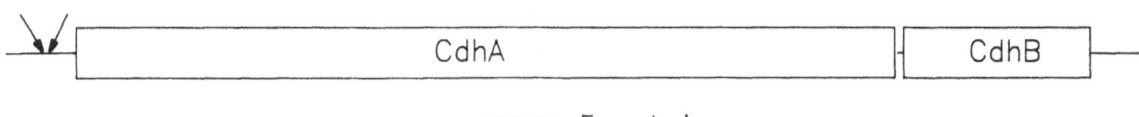

Ferredoxin

Acyl—CoA oxidase

FIGURE 4.5 Genetic organization of *cdh*A and *cdh*B genes. Open-reading frames indicated by open bars, flanking and intergenic regions by solid lines. Arrows represent putative transcription initiation sites. Location of regions in CdhA that showed homology with acyl-CoA oxidases and ferredoxins is indicated below open-reading frame by shaded boxes.

sion hosts used, one possibility could be that the iron-sulfur clusters in *E. coli* or *D. vulgaris* are not properly incorporated. To possibly circumvent this problem, other anaerobic bacteria that are known to contain iron-sulfur clusters similar to *M. soehngenii* Cdh could be used as an alternate host, such as *Clostridium* spp., for which genetic systems have been described (Blaschek, 1989).

The *M. soehngenii cdh* genes are organized in an operon-like structure with the order *cdh*A—*cdh*B, coding for proteins with molecular masses of 89.461 and 21.008 daltons, respectively (Figure 4.5). Upstream from *cdh*A, two archaeal promoter-like structures, both Box A and Box B, were observed (see Figure 4.3), which could be used for transcription initiation (Eggen et al., 1991a). The intergenic region consists of only 19 nucleotides and does not contain transcriptional initiation signals. Hence both *cdh* genes are probably organized into a single transcription unit. The presumed translation initiation sites of both *cdh* genes are preceded by sequences complementary to the 3′ end of the 16S rRNA from *M. soehngenii* (Eggen et al., 1989).

Database searches showed that CdhA contains a stretch of 110 amino acids that showed 24% identity with acyl-CoA oxidases from *Candida tropicalis* (Figure 4.5; Okazaki et al., 1986; Eggen et al., 1991a). This is compatible with the proposed function of Cdh (Jetten et al., 1989b). Another region, consisting of 64 residues, could be identified as a ferredoxin domain (Figure 4.5) similar to archaeal-type ferredoxins (Otaka and Ooi, 1987; Reeve et al., 1989;

Steigerwald et al., 1990). This confirms the proposed function of Cdh as an electron carrier (Jetten et al., 1989b). The ferredoxin domain contains 8 cysteine residues per CdhA molecule, which could bind two [4Fe-4S] clusters. However, spin-label recovery in the Cdh enzyme indicates that only one [4Fe-4S] cubane is fixed in the ferredoxin (Jetten et al., 1991). The remarkable [6Fe-6S] prismane (Jetten et al., 1991) is probably fixed by a cluster of cysteine residues in another region of the enzyme, for which the NH₂-terminal region of CdhA is a good candidate (Eggen et al., 1991a). In homology searches of the amino acid sequences of CdhB with proteins in a database, no significant homology was observed for the CdhB sequence, and its function is currently unknown. Possibly CdhB is somehow involved in the correct assembly of Cdh or its iron-sulfur clusters.

Cdhs from acetoclastic methanogens contain identical epitopes

In the absence of a suitable host that allows functional production of the *M. soehngenii* Cdh, a detailed structure–function analysis of this essential enzyme is not possible. Therefore, a comparative study of the Cdhs from acetoclastic methanogens was initiated by analyzing the immunological relationship of Cdhs from *M. soehngenii*, *Methanothrix* sp. strain CALS-1, and *Methanosarcina barkeri* strain Fusaro. Extracts were prepared from the cells, grown on acetate, fractionated by sodium dodecyl sulfate-polyacrylamide gel electrophoresis (SDS-

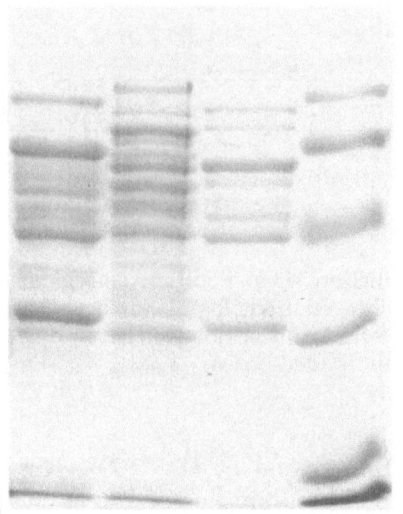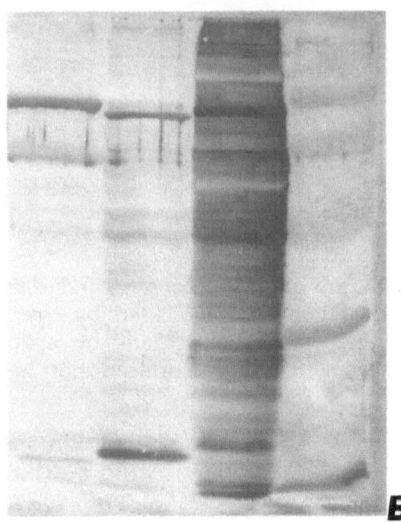

FIGURE 4.6 Coomassie brilliant blue stained **(A)** and immunoblot analysis **(B)** of proteins from acetate-grown acetoclastic methanogens that were fractionated in 12.5% SDS-polyacrylamide gel. Immunoblot analysis was performed using anti-CDH serum as primary antibody and anti-rabbit alkaline phosphatase conjugate as secondary antibody.

Lane 1, *Methanosarcina barkeri* strain Fusaro; lane 2, *Methanothrix* sp. strain CALS-1; lane 3, *Methanothrix soehngenii*. Arrows indicate position of large and small subunits of carbon monoxide dehydrogenase. M_r values of prestained marker proteins (lane 4) indicated (in kilodaltons) to right of each panel.

PAGE), blotted, and screened using anti-Cdh antiserum raised against purified Cdh from *Methanothrix soehngenii* (Jetten et al., 1989b; Eggen et al., 1991a). In all three extracts, two prominent immunoreactive proteins were observed (Figure 4.6) with molecular masses similar to that of the subunits of the purified Cdhs from *M. soehngenii* (79 and 21 kDa; Jetten et al., 1989b), *Methanosarcina barkeri* (92 and 18 kDa; Krzycki and Zeikus, 1984; Grahame and Stadtman, 1987b), and *Methanothrix* sp. strain CALS-1 (85 and 23 kDa; Allan and Zinder, 1990). These results indicate that the Cdhs from different acetoclastic methanogens contain identical epitopes, suggesting the presence of common structural elements.

4.4   Concluding Remarks

Although the physiological and biochemical knowledge of acetoclastic methanogens has already accumulated for some years, research on the molecular biology of these unique archaea has only recently started. In this review we have focused on the recent developments in the genetic research of a specialist among the acetoclastic methanogens, *Methanothrix soehngenii*. The results obtained so far are a good start for future research directed toward the structure–function analysis of acetyl-CoA synthetase and carbon monoxide dehydrogenase, the evolutionary relationship of carbon monoxide dehydrogenases from different acetoclastic methanogens, and regulation of the expression of genes involved in methanogenesis.

Acknowledgments.   This work was supported by a grant from the Programme Committee on Industrial Biotechnology (PcIB). We thank all coworkers and students who contributed to the research summarized in this review; Dr. C.R. Woese, University of Illinois, for supplying us with 16S rRNA sequences from the rRNA Database Project; and Dr. S.H. Zinder for providing *Methanothrix* sp. strain CALS-1.

References

Aceti, D.J. and J.G. Ferry. 1988. Purification and characterization of acetate kinase from acetate-grown *Methanosarcina*. Evidence for regulation of synthesis. *J. Biol. Chem.* **263**:15444–15448.

Achenbach-Richter, L. and C.R. Woese. 1988. The ribosomal gene spacer region in archaebacteria. *Syst. Appl. Microbiol.* **10**:211–214.

Albano, M., R. Breitling, and D.A. Dubnau. 1989. Nucleotide sequence and genetic organization of the *Bacillus subtilis comG* operon. *J. Bacteriol.* **171**:5386–5304.

Allan, G. and S.H. Zinder. 1990. Characterization of carbon monoxide dehydrogenase activity in the thermophilic acetotrophic methanogen: *Methanothrix* sp. strain CALS-1. *In: Abstracts*, 90th Annual Meeting of the American Society for Microbiology, Abstr. I40. Washington, D.C.: American Society for Microbiol.

Blaschek, H.P. 1989. Genetic manipulation of the clostridia. *Dev. Ind. Microbiol.* **30**:35–42.

Brown, J.W., C.J. Daniels, and J.N. Reeve. 1989. *CRC Crit. Rev. Microbiol.* **16**:287–338.

Connerton, I.F., J.R.S. Fincham, R.A. Sandeman, and M.J. Hynes. 1990. Comparison and cross-species expression of the acetyl-CoA synthetase genes of the ascomycete fungi, *Aspergillus nidulans* and *Neurospora crassa*. *Mol. Microbiol.* **4**:451–460.

De Wet, J.R., K.V. Wood, M. DeLuca, D.K. Helinski, and S. Subramani. 1987. Firefly luciferase gene: structure and expression in mammalian cells. *Mol. Cell. Biol.* **7**:725–737.

Eggen, R., H. Harmsen, and W.M. de Vos. 1990. Organization of a ribosomal RNA gene cluster from the archaebacterium *Methanothrix soehngenii*. *Nucleic Acids Res.* **18**:1306.

Eggen, R., H. Harmsen, A. Geerling, and W.M. de Vos. 1989. Nucleotide sequence of a 16S rRNA encoding gene from the archaebacterium *Methanothrix soehngenii*. *Nucleic Acids Res.* **17**:22.

Eggen, R.I.L., A.C.M. Geerling, M.S.M. Jetten, and W.M. de Vos. 1991a. Cloning, expression and sequence analysis of the genes for carbon monoxide dehydrogenase of *Methanothrix soehngenii*. *J. Biol. Chem.* **266**:6883–6887.

Eggen, R.I.L., A.C.M. Geerling, A.B.P. Boshoven and W.M. de Vos. 1991b. Cloning, sequence analysis, and functional expression of the acetyl coenzyme A synthetase gene from *Methanothrix soehngenii* in *Escherichia coli*. *J. Bacteriol.* **173**: 6383–6389.

Eikmanns, B. and R.K. Thauer. 1984. Catalysis of an isotopic exchange between CO_2 and the carboxyl group of acetate by *Methanosarcina barkeri* grown on acetate. *Arch. Microbiol.* **135**:365–370.

Fischer, R. and R.K. Thauer. 1988. Methane formation from acetyl phosphate in cell extracts of *Methanosarcina barkeri*. *FEBS Lett.* **228**:249–253.

Fukuzaki, S., N. Nishio, and S. Nagai. 1990. Kinetics of the methanogenic fermentation of acetate. *Appl. Environ. Microbiol.* **56**:3158–3163.

Grahame, D.A. and T.C. Stadtman. 1987a. In vitro methane and methyl coenzyme M formation from acetate: evidence that acetyl-CoA is the required intermediate activated form of acetate. *Biochem. Biophys. Res. Commun.* **147**:254–258.

Grahame, D.A. and T.C. Stadtman. 1987b. Carbon monoxide dehydrogenase from *Methanosarcina barkeri*. Disaggregation, purification, and physicochemical properties of the enzyme. *J. Biol. Chem.* **262**:3706–3712.

Gujer, W. and A.J.B. Zehnder. 1983. Conversion processes in anaerobic digestion. *Water Sci. Technol.* **15**:49–77.

Hagen, W.R., A.J. Pierik, and C. Veeger. 1989. Novel electron paramagnetic resonance signals from an Fe/S protein containing 6 iron atoms. *J. Chem. Soc. Faraday Trans.* **85**:4083–4090.

Huser, B.A., K. Wuhrmann, and A.J.B. Zehnder. 1982. *Methanothrix soehngenii* gen. nov. sp. nov., a new acetotrophic non-hydrogen-oxidizing methane bacterium. *Arch. Microbiol.* **132**:1–9.

Hutten, T.J., H.C.M. Bongaerts, C. van der Drift, and G.D. Vogels. 1980. Acetate, methanol and carbon dioxide as substrates for growth of *Methanosarcina barkeri*. *Antonie Leeuwenhoek J. Microbiol.* **46**:601–610.

Jarsch, M. and A. Böck. 1985. Sequence of the 16S rRNA from *M. vannielii*. *Syst. Appl. Microbiol.* **6**:54–59.

Jetten, M.S.M., A.J.M. Stams, and A.J.B. Zehnder. 1989a. Isolation and characterization of acetyl-coenzyme A synthetase from *Methanothrix soehngenii*. *J. Bacteriol.* **171**:5430–5435.

Jetten, M.S.M., A.J.M. Stams, and A.J.B. Zehnder. 1989b. Purification and characterization of an oxygen stable carbon monoxide dehydrogenase of *Methanothrix soehngenii*. *Eur. J. Biochem.* **241**:1938–1947.

Jetten, M.S.M., A.J.M. Stams, and A.J.B. Zehnder. 1990a. Purification and some properties of the methyl-CoM reductase of *Methanothrix soehngenii*. *FEMS Microbiol. Lett.* **66**:183–186.

Jetten, M.S.M., A.J.M. Stams, and A.J.B. Zehnder. 1990b. Acetate threshold values and acetate activating enzymes in methanogenic bacteria. *FEMS Microbiol. Ecol.* **73**:339–344.

Jetten, M.S.M., W.R. Hagen, A.J. Pierik, A.J.M. Stams, and A.J.B. Zehnder. 1991. Paramagnetic centers and acetyl-coenzyme A/CO exchange activity of carbon monoxide dehydrogenase from *Methanothrix soehngenii*. *Eur. J. Biochem.* **195**:385–391.

Kenealy, W.R. and J.G. Zeikus. 1982. One-carbon

metabolism in methanogens: evidence for synthesis of a two-carbon cellular intermediate and unification of catabolism and anabolism in *Methanosarcina barkeri*. *J. Bacteriol.* **151**:932–941.

Knaub, S. and A. Klein. 1990. Specific transcription of cloned *Methanobacterium thermoautotrophicum* transcription units by homologous RNA polymerase *in vitro*. *Nucleic Acids Res.* **18**:1441–1446.

Kohler, H.-P.E. and A.J.B. Zehnder. 1984. Carbon monoxide dehydrogenase and acetate thiokinase in *Methanothrix soehngenii*. *FEMS Microbiol. Lett.* **21**:287–292.

Krzycki, J. and J.G. Zeikus. 1984. Characterization and purification of carbon monoxide dehydrogenase from *Methanosarcina barkeri*. *J. Bacteriol.* **158**:231–237.

Krzycki, J.A., L.J. Lehman, and J.G. Zeikus. 1985. Acetate catabolism by *Methanosarcina barkeri*: evidence for involvement of carbon monoxide dehydrogenase, methyl-coenzyme M, and methylreductase. *J. Bacteriol.* **163**:1000–1006.

Krzycki, J.A., R.H. Wolkin, and J.G. Zeikus. 1982. Comparison of unitrophic and mixotrophic substrate metabolism by an acetate-adapted strain of *Methanosarcina barkeri*. *J. Bacteriol.* **149**:247–254.

Laufer, K., B. Eikmanns, U. Frimmer, and R.K. Thauer. 1987. Methanogenesis from acetate by *Methanosarcina barkeri*: catalysis of acetate formation from methyl iodide, CO_2, and H_2 by the enzyme system involved. *Z. Naturforsch. Teil C Biosci.* **42**:360–372.

Lechner, K., G. Wich, and A. Böck. 1985. The nucleotide sequence of the 16S rRNA gene and flanking regions from *M. formicicum*. On the phylogenetic relationship between methanogenic and halophilic archaebacteria. *Syst. Appl. Microbiol.* **6**:157–163.

Lovley, D.R., R.H. White and J.G. Ferry. 1984. Identification of methyl coenzyme M as an intermediate in methanogenesis from acetate in *Methanosarcina* spp. *J. Bacteriol.* **160**:521–525.

Lozoya, E., H. Hoffmann, C. Douglas, D. Schulz, D. Scheel, and K. Hahlbrock. 1988. Primary structures and catalytic properties of isoenzymes encoded by the two 4-coumarate:CoA ligase genes in parsley. *Eur. J. Biochem.* **176**:661–667.

Lundie, L.L., Jr., and J.G. Ferry. 1989. Activation of acetate by *Methanosarcina thermophila*. *J. Biol. Chem.* **264**:18392–18396.

Matsuyama, A., H. Yamamoto, and E. Nakano. 1989. Cloning, expression, and nucleotide sequence of the *Escherichia coli* K-12 ackA gene. *J. Bacteriol.* **171**:577–580.

Nelson, M.J.K. and J.G. Ferry. 1984. Carbon monoxide dependent methyl coenzyme M methyl reductase in acetotrophic *Methanosarcina* spp. *J. Bacteriol.* **160**:526–532.

Nozhevnikova, A.N. and V.I. Chudina. 1985. Morphology of the thermophilic acetate methane bacterium *Methanothrix thermoacetophila* sp. nov. *Microbiology* **53**:756–766.

Okazaki, K., T. Takechi, N. Kambara, S. Fukui, I. Kubota, and T. Kamiryo. 1986. Two acyl-coenzyme A oxidases in peroxisomes of the yeast *Candida tropicalis*: primary structures deduced from the genomic DNA sequence. *Proc. Natl. Acad. Sci. USA* **83**;1232–1236.

Østergaard, L., N. Larsen, H. Leffers, J. Kjems, and R. Garrett. 1987. A ribosomal RNA operon and its flanking region from the archaebacterium *Methanobacterium thermoautotrophicum*, Marburg strain: transcription signals, RNA structure and evolutionary implications. *Syst. Appl. Microbiol.* **9**:199–209.

Otaka, E. and T. Ooi. 1987. Examination of protein sequence homologies IV. Twenty-seven bacterial ferredoxins. *J. Mol. Evol.* **26**:257–267.

Patel, G.B. 1984. Characterization and nutritional properties of *Methanothrix concilii* sp. nov., a mesophilic, aceticlastic methanogen. *Can. J. Microbiol.* **30**:1383–1396.

Patel, G.B. and G.D. Sprott. 1990. *Methanosaeta concilii* gen. nov., sp. nov. ("*Methanothrix concilii*") and *Methanosaeta thermoacetophila* nom. rev., comb. nov. *Int. J. Syst. Bacteriol.* **40**:79–82.

Pellerin, P., B. Gruson, C. Prensier, G. Albagnac, and P. Debeire. 1987. Glycogen in *Methanothrix*. *Arch. Microbiol.* **146**:377–381.

Reeve, J.N., G.S. Beckler, D.S. Cram, P.T. Hamilton, J.W. Brown, J.A. Krzycki, A.F. Kolodziej, A. Alex, W.H. Orme-Johnson, and C.T. Walsh. 1989. A hydrogenase-linked gene in *Methanobacterium thermoautotrophicum* strain ΔH encodes a polyferredoxin. *Proc. Natl. Acad. Sci. USA* **86**:3031–3035.

Reiter, W.-D., U. Hüdepohl, and W. Zillig. 1990. Mutational analysis of an archaebacterial promoter: essential role of a TATA box for the transcription efficiency and start-site selection *in vitro*. *Proc. Natl. Acad. Sci. USA* **87**:9509–9513.

Smith, P.H. and R.A. Mah. 1966. Kinetics of acetate metabolism during sludge digestion. *Appl. Environ. Microbiol.* **14**:368–371.

Smith, P.H. and R.A. Mah. 1980. Acetate as sole carbon and energy source for growth of *Methanosarcina barkeri* strain 227. *Appl. Environ. Microbiol.* **39**:993–999.

Stackebrandt, E., E. Seewaldt, W. Ludwig, K.-H. Schleifer, and B.A. Huser. 1982. The phylogenetic position of *Methanothrix soehngenii*. Elucidated by a modified technique of sequencing oligonucleotides from 16S rRNA. *Zentralbl. Bakteriol. Parasitenkd. Infektionskr. Hyg.* 1. Abt. Orig. **3**:90–100.

Steigerwald, V.J., G.S. Beckler, and J.N. Reeve. 1990. Conservation of hydrogenase and polyferredoxin structures in the hyperthermophilic archaebacterium *Methanothermus fervidus*. *J. Bacteriol.* **172**:4715–4718.

Terlesky, K.C., M.J.K. Nelson, and J.G. Ferry. 1986. Isolation of an enzyme complex with carbon monoxide dehydrogenase activity containing corrinoid and nickel from acetate-grown *Methanosarcina thermophila*. *J. Bacteriol.* **168**:1053–1058.

Terlesky, K.C., M.J. Barber, D.J. Aceti, and J.G. Ferry. 1987. EPR properties of the nickel-iron-carbon center in an enzyme complex with carbon monoxide dehydrogenase activity from acetate-grown *Methanosarcina thermophila*. *J. Biol. Chem.* **262**:15392–15395.

Thauer, R.K. 1988. Citric-acid cycle, 50 years on: modifications and an alternative pathway in anaerobic bacteria. *Eur. J. Biochem.* **176**:497–508.

Thauer, R.K., D. Möller-Zinkhan, and A.M. Spormann. 1989. Biochemistry of acetate catabolism in anaerobic chemotrophic bacteria. *Ann. Rev. Microbiol.* **43**:43–67.

Thomm, M., G. Wich, J.W. Brown, G. Frey, B.A. Sherf, and G.S. Beckler. 1989. An archaebacterial promoter sequence assigned by RNA polymerase binding experiments. *Can. J. Microbiol.* **35**:30–35.

Touzel, J.P., G. Prensier, J.L. Roustan, I. Thomas, H.C. Dubourguier, and G. Albagnac. 1988. Description of a new strain of *Methanothrix soehn-genii* and rejection of *Methanothrix concilii* as a synonym of *Methanothrix soehngenii*. *Int. J. Syst. Bacteriol.* **38**:30–36.

van de Wijngaard, W.M.H., C. van der Drift, and G.D. Vogels. 1988. Involvement of a corrinoid enzyme in methanogenesis from acetate in *Methanosarcina barkeri*. *FEMS Microbiol. Lett.* **52**:165–172.

Woese, C.R., O. Kandler, and M.L. Wheelis. 1990. Towards a natural system of organisms: proposal for the domains Archaea, Bacteria and Eucarya. *Proc. Natl. Acad. Sci. USA* **87**:4576–4579.

Yang, D., B.P. Kaine, and C.R. Woese. 1985. The phylogeny of archaebacteria. *Syst. Appl. Microbiol.* **6**:251–256.

Zehnder, A.J.B., B.A. Huser, T.D. Brock, and K. Wuhrmann. 1980. Characterization of an acetate-decarboxylating non-hydrogen-oxidizing methane bacterium. *Arch. Microbiol.* **124**:1–11.

Zillig, W., P. Palm, W.-D. Reiter, F. Gropp, G. Pühler, and H.-P. Klenk, 1988. Comparative evaluation of gene expression in archaebacteria. *Eur. J. Biochem.* **173**:473–482.

Zinder, S.H., S.C. Cardwell, T. Anguish, M. Lee, and M. Koch. 1984. Methanogenesis in a thermophilic anaerobic digestor: *Methanothrix* sp. as an important aceticlastic methanogen. *Appl. Environ. Microbiol.* **47**:796–807.

5

Mutations

Madeleine Sebald

5.1 Introduction

Mutagenesis is a basic approach to bacterial genetics. Microbiologists previously have shown only limited interest in mutant strains of anaerobes, and this interest was directed toward such modified properties as increased formation of metabolites, e.g., neutral solvents of *Clostridium acetobutylicum* or hyperproduction of clostridial toxins for vaccines. Investigators often faced unwanted losses in properties of laboratory strains in which the mechanism may have been related to "strain degeneration," spontaneous mutation, or plasmid loss. Only the pioneering work was focused on obtaining mutants to inquire into such basic mechanisms as oxygen sensitivity, and those unrealistic approaches to "aerobize" anaerobes of course failed.

Most of the published data give no indication of forward mutations, or only little mention of the nonselected traits; thus, it is difficult to identify point mutations, deletions, or pleiotropic mutations. The mutations that have been described were rarely identified at the gene level, and thus we use their phenotypic designation. Mutants were quite exclusively screened and obtained in *Clostridium* and *Bacteroides*. Available data are very scarce in other gram-negative anaerobic genera and bifidobacteria.

In this chapter, we provide fundamental data on specific aspects of mutagenesis in strict anaerobes. Although limited, these may be a source of information for further investigations. Several other reviews on mutations in anaerobes have been published (Jones and Woods, 1986; Rogers, 1986; Awang et al., 1988; Reed, 1989). This review provides data on the selected markers used, the mutant phenotypes and their reversibility. The data have been ordered by genus (*Clostridia*, Tables 5.1 to 5.6; gram-negative anaerobes, Tables 5.7 to 5.9; *Bifidobacterium*, Table 5.10). We also provide data on phage and plasmid curing (Table 5.11).* Information on mutations in *Archaebacteria* and sulfate-reducing bacteria is given in Chapters 1 and 34, and transposition mutagenesis is reviewed in Chapters 6 and 38.

5.2 Spontaneous Mutations

Anaerobes are subject to spontaneous mutagenesis, defined as a "chronic disease" by Benzer (1961). The mechanism of spontaneous mutations has not been investigated in anaerobes, but virtually all DNA rearrangements (deletions, insertions, base substitutions, frameshifts) may be expected to be represented in spontaneous mutations; insertions and deletions should be particularly well represented. Among the spontaneous mutations occurring in anaerobes, changes in function in-

---

Abbreviations: The abbreviations used in the text are defined in the legend of Tables 5.1 to 5.11.

clude loss of sporulation, lack of ability to produce secondary metabolites or toxins, and loss or acquisition of antibiotic resistance. The acquisition of spontaneous rifampicin resistance (Rf^r) is particularly easy to obtain in most anaerobes at a frequency of about 10^{-8} to 10^{-9}, and the Rf^r serves as a useful selective trait. The mechanism of the Rf^r mutation is unknown, and this mutation is as a rule very stable and unable to revert to sensitivity either spontaneously or following mutagenesis (e.g., in *Bacteroides fragilis*) (M. Sebald, unpublished observations); this situation is compatible with a deletion. The existence of mutator genes, or transposition mutagenesis, should be an alternative mechanism, as suggested with *Clostridium thermocellum* ATCC2705 (Gomez et al., 1980). Spontaneous losses of antibiotic resistance have been observed or selected in *C. perfringens* (Brefort et al., 1977; Abraham et al., 1985) or *Bacteroides fragilis* (Magot et al., 1981; Welch and Macrina, 1981; Shimell et al., 1982; Tally et al., 1982; Shoemaker et al., 1985; Smith, 1987) and have been shown to correspond to plasmid deletions.

5.3 Mutagens

UV irradiation

It is more convenient to irradiate spores than vegetative cells because spores are metabolically inert and thus insensitive to oxygen. Moreover, each spore contains one copy only of the chromosome, in contrast to vegetative cells that are often filamentous and thus multinucleated, particularly in exponentially grown clostridia.

In a general way, conditions for optimal mutagenesis are those leading to a 10^{-2} to 10^{-3} spore survival. Because the effects of UV radiation are cumulative, iterative exposure of a spore suspension to UV irradiation may be performed if necessary. Vegetative cells may suffer from O$_2$ exposure, and O$_2$ may interfere with UV irradiation, as is discussed further here. The irradiation may be performed in an anaerobic chamber, but UV irradiation has a noxious effect on the plastic envelope. UV irradiation of

vegetative cells may be performed in a microaerobic environment as recommended by Daldal (1985). Wavelengths other than UV, e.g., γ rays, have been rarely used (Gomez et al., 1980).

Chemical mutagens

N-methyl-N'-nitro-N-nitrosoguanidine (MNNG) and ethyl methanesulfonate (EMS) are the mutagens most commonly used; they are most effective when concentration and time exposure lead to about 1 of 100 cells surviving in buffer in nongrowing conditions. EMS may be used in 1 M Tris buffer as described by Tessman et al. (1964). For MNNG mutagenesis, the malate buffer initially used (Adelberg et al., 1965) is inadequate for almost all clostridia and may be replaced by phosphate or citrate buffer. Other mutagens, such as nitrous acid, have been used occasionally for anaerobes (Bayen et al., 1967).

UV radiation versus chemical mutagens

There has been some controversy regarding the relative effectiveness of both types of agents as mutagens. With *C. thermocellum*, Gomez et al. (1980) reported that UV radiation was more effective than MNNG in causing forward mutations to 5-fluorouracil and rifampicin resistance. With *C. acetobutylicum* and *C. pasteurianum*, Bowring and Morris (1985) calculated the relative induced mutation frequency for rifampicin resistance. This is defined as the proportion of a mutant strain capable of surviving exposure to a mutagen based on the initial proportion of spontaneous mutants. In *C. pasteurianum*, Bowring and Morris obtained a higher frequency of induced mutation to rifampicin after MNNG exposure than after UV irradiation, particularly when MNNG exposure was performed in aerobic conditions. A high frequency of induced mutations to rifampicin resistance was obtained by EMS treatment of *C. acetobutylicum*. Rifampicin mutations are not necessarily representative of the cell response to mutagen agents, as discussed, but the data of Bowring and Morris (1985) also suggest that

UV radiations are relatively inefficient as a means of producing auxotrophic mutants in *C. acetobutylicum*. It has generally been concluded from this work that "direct mutagens" such as MNNG or EMS, which cause base mispairing and hence misreplication of DNA, are more active than "indirect mutagens" such as UV radiations, whose mutagenic activity occurs via misrepair of damaged DNA (error prone repair) by SOS processing. The tendency of anaerobes to be deficient in error-prone DNA repair is discussed later.

However, UV radiation may be a valuable method of mutagenesis, as an alternative to chemical mutagenesis, and several mutants from Tables 5.1 to 5.5 illustrate this point.

5.4   Repair of Radiation Damage

Photoreactivation

The major lesions induced by UV radiation (far-UV, 254 nm) consist of the formation of intrastrand cyclobutane-pyrimidine dimers, and the repair of these lesions occurs by three mechanisms in *Escherichia coli*: photoreactivation, excision repair, and postreplication repair.

Photoreactivation is an error-free repair process occurring in presence of visible light of wavelengths of 320 to 410 nm. The visible light activates a cellular enzyme that binds specifically at sites with tandem pyrimidines to generate a new enzyme-DNA chromophore.

The first experimental approach on photoreactivation in a strict anaerobe was made by Rambler and Margulis (1980) using *Clostridium sporogenes*. They showed unambiguously, in this bacterium, the reversion by visible light of the effects of irradiation with lethal doses of UV radiation. After an UV irradiation (1.2×10^3 J/m^2) reduced the vegetative cell survival by 99.9%, photoreactivation of more than 80% of the cells was obtained following exposure to white light at 2000 lux for 8 h in anaerobic conditions; there was no recovery when the irradiated control was maintained in the dark. Moreover, these authors observed a higher photoreactivation of

C. sporogenes than of facultative or strict aerobic bacteria tested in similar anaerobic conditions. This suggests that the *C. sporogenes* enzyme involved in photoreactivation is active in anaerobiosis, in contrast to the corresponding enzyme from the aerobes that is known to be repressed in anaerobic conditions, as in the case of *E. coli*. A photoreactivation system is apparently lacking in *C. butyricum* ATCC19398, *C. acetobutylicum* ATCC824 (Carrasco, 1989), and *B. fragilis* (Jones et al., 1980a).

Excision repair

Error-prone repair in *E. coli* gives rise to mutations after UV irradiation and requires the induction of the inducible SOS system. *B. fragilis* was shown by Jones et al. (1980a) to be more sensitive to far-UV radiation damage under aerobic conditions than under anaerobic conditions, and more sensitive to damage when in the exponential growth phase than in the stationary phase. A similar effect of oxygen was shown with *B. fragilis* cells treated by MNNG, EMS, acriflavine, or mitomycin C (Slade et al., 1984).

To determine whether the oxygen effect was caused by an increased sensitivity to UV radiation under aerobic conditions or to an increased sensitivity to oxygen after UV irradiation, Slade et al. (1981) observed the effect of H_2O_2 and quenchers of oxygen radicals on the survival of the cells after UV irradiation. Quenchers of toxic oxygen derivatives did not increase the resistance of *B. fragilis* cells to far-UV irradiation when made under aerobic conditions, and H_2O_2 did not sensitize anaerobically irradiated cells to far-UV radiation. Moreover, the differences in survival obtained in aerobic versus anaerobic conditions could not be explained in terms of numbers of cyclobutane-pyrimidine dimers present in the DNAs of the irradiated cells.

Jones and Woods (1981) investigated the liquid-holding recovery (LHR) of *B. fragilis* UV-irradiated cells by comparing the survival of the cells when plated immediately and after suspension in nutrient medium maintained aerobically or anaerobically. The results showed that LHR occurred under aerobic conditions only. Because mutations occuring at *uvr*A, *uvr*B, or *uvr*C, but not at the *rec*A block LHR of *E. coli* (Smith, 1978), the results obtained by Jones and Woods raise the hypothesis of a possible repression of *uvr* or *uvr*-like genes in *Bacteroides* in aerobic compared to anaerobic conditions.

The LHR under aerobic conditions was inhibited by chloramphenicol whereas the final level of excision repair in *B. fragilis* Bf-2 was not affected (Abratt et al., 1986). This fact again suggests that pyrimidine dimer removal is not the process responsible for increased LHR in aerobic conditions. The possibility of an excision repair process that works on a photoproduct other than cyclobutyl-pyrimidine dimers, such as pyrimidine-pyrimidone (6-4), or through modification involving DNA structural features has been reported (Brash et al., 1987, and literature cited there in). Caffeine has been reported to inhibit excision repair processes but not the inducible *rec*A-dependent repair process in *E. coli*. Sodium arsenite is thought to inhibit *rec*A-dependent steps in DNA repair in *E. coli*. In *B. fragilis* Bf-2, far-UV irradiation under anaerobic conditions resulted in the induction of a new 95-kDa protein and the increased synthesis of two proteins of 90 kDa and 70 kDa. There was no induction of a 40-kDa protein that could be a candidate for a RecA protein on the basis of the molecular mass of the *E. coli* RecA protein (Schuman et al., 1982). Caffeine at 1 mg/ml inhibited the induction of the three proteins, and sodium arsenite at 100 mM did not. By analogy with *E. coli* these results suggest that UV radiation induces three proteins involved in the excision repair process but not in the *rec*A-dependent repair process. Moreover, *B. fragilis* strain Bf-2, like *E. coli rec*A, is capable of DNA synthesis after UV irradiation, albeit at a low rate, and DNA synthesis also occurs in presence of caffeine (Schumann et al., 1984).

When *B. fragilis* Bf-2 was exposed to near-UV radiation (310–400 nm), a "V-shaped" survival curve was obtained (Slade et al., 1982); here again the *B. fragilis* response is similar to that obtained with a *E. coli rec*A mutant and differ-

ent from the corresponding wild-type strain of *E. coli* (Slade et al., 1982). All these arguments together strongly suggest the lack of recombination functions in *Bacteroides fragilis*. Nevertheless, gene cloning techniques have brought new insights in the recombination functions both in *B. fragilis* and in *C. acetobutylicum*.

We must also keep in mind that most work on repair of radiation damage was done on *B. fragilis* strain Bf-2, which may be not representative of the species because of its prolonged subcultures in air (Schumann et al., 1983). Data on mutagenesis of *C. butyricum* ATCC19398 suggest this strain has an efficient error-prone DNA repair (Carrasco and Soto, 1987); after UV irradiation, an increased rate of mutation to nalidixic acid or rifampicin resistance was observed when cells were incubated in presence of caffeine.

5.5 Recombination Functions in Anaerobes

No investigations had been reported on recombination functions in anaerobes until fairly recently, and the general feeling was that these functions were lacking because the genetic exchanges initially studied were mainly the result of plasmid transfer. Nevertheless, recombination functions are necessarily involved in such phenomena as transposition, deletion, lysogenisation, excision repair, etc.

A *rec*A-like gene was characterized in *B. fragilis* by Goodman et al. (1987). A genomic library of *B. fragilis* Bf-1 was established in *E. coli* HB101, and clones containing inserts were selected first by insertional inactivation of the EcoRI endonuclease gene of pEcoR251. The recombinant plasmids were thus reintroduced by transformation in *E. coli* HB101 *rec*A, and transformants were screened for resistance to methyl methanesulfonate (MMS). A MMS[r] transformant, HB101 [pHG100], turned out to be more resistant to killing by UV irradiation than the recipient HB101. The plasmid pHG100 contained a 5.2-kb insert; this clearly restored the repair of MMS and 4-nitro-quinoline-1-oxide- (NQO-) injured DNA and the recombination proficiency in strain HB101. However, the ability of the cloned *B. fragilis* fragment to repair UV-damaged DNA, P_1 prophage induction, and λ phage reactivation was less effective. In an *E. coli* in vitro transcription-translation system, the insert of plasmid pHG100 was shown to code for two proteins of about 37 and 39 kDa.

Antibodies raised against the *E. coli* RecA protein reacted with a UV-inducible 40-kDa polypeptide in extracts of *E. coli* C600 and with a non-UV-inducible polypeptide of 40 kDa in extracts of *E. coli* HB101 . These antibodies also reacted with polypeptides of 40, 39, and 37 kDa in UV-induced and noninduced extracts of *E. coli* HB101 [pHG100] and with polypeptides of 39 and 37 kDa in UV-induced and noninduced *B. fragilis* cells. In the experimental conditions used, however, the authors did not observe DNA homology between the insert of pHG100 and restriction endonuclease digests of *E. coli* C600 recA+ chromosomal DNA. The incomplete complementation of certain defects in the *E. coli* recA, the lack of homology between the *B. fragilis* recA-like gene and the *E. coli* recA gene, and the differences in the molecular mass of the products suggest the existence in *B. fragilis* of a repair system closely resembling, but not identical to, the rec system of *E. coli*.

A recA-like gene of *C. acetobutylicum* was also recently characterized by H. Azeddoug, J. Hubert, and G. Reysset (personal communication). After chromosomal DNA was "shotgunned" into plasmids pUC19 or pKNT11, recombinant clones restored the MMS resistance of *E. coli* HB101 recA. Other recA functions were partially or not restored, and the protein encoded by the *C. acetobutylicum* recA-like gene did not have the molecular mass expected for a RecA protein. Nevertheless, there was a significant DNA homology between the cloned gene and the recA gene of *E. coli* when analyzed by Southern hybridization.

In conclusion, these recently acquired data demonstrate clearly that the response of anaerobes to DNA-damaging agents may be somewhat different from the *E. coli* response, thus showing the need to pursue such investigations.

5.6   Mutants in Anaerobes

Selection

A positive selection is the easiest way to obtain mutants. This method is feasible for selection of phenotypes such as antibiotic resistance, radiation resistance, and ability to sporulate (heat or alcohol/chloroform treatment). A positive selection of mutants in metabolic pathways is possible through the use of suicide substrates in which the rationale for selection relies on the reversibility of enzymatic reactions (i.e., alcohol or acid analogs such as halogen derivatives), as is illustrated in Table 5.5.

In *C. thermosaccharolyticum*, low producers of acetic acid were found in fluoroacetate-resistant mutants, and their metabolism was shifted to ethanol production (Rothstein, 1986). Several enzymatic deficiencies have been characterized in *C. acetobutylicum* mutants selected as resistant to acetate and butyrate halogen analogs (Janati-Idrissi et al., 1987; Junelles et al., 1987; El Kanouni et al., 1989).

It is possible to use, for clostridia (Cueto and Mendez, 1990), the "proton suicide method" developed by Winkelman and Clark (1984) for the selection of fermentation mutants in *E. coli*. The technique is based on the lethal effect of H^+ as a-parallel to acid production when occurring in the presence of bromate and bromide sodium salts. Because the final products, elemental bromine (Br_2) and hypobromite (BrO^-) are toxic to the cell, only cells affected by diminishing acid production will survive. A rapid technique of detection of nonsolvent producers has also been developed (Clark et al., 1989) that facilitates their screening.

In some cases a direct detection of the mutants is possible, e.g., granulose-negative colonies by iodine staining, variations in toxin/enzyme production in presence of substrate added to medium (e.g., phospholipase, hemolysin, protease). Colonial variations may also be of interest because they often correspond to sporulation mutants (transparent colonies), capsule-negative mutants (nonmucoid colonies), or metabolic mutants (small or large colonies).

The replica-plating technique is useful for detecting indirectly selected characters, i.e., auxo-trophic mutants, antibiotic-sensitive mutants among a resistant population, sporulation mutants from a sporogenic population (exposure of the replicated plates to chloroform), and also for isolation of bacteriocin- or phage-cured derivatives. There is no limitation to the replica-plating technique when one is using an anaerobic chamber.

The isolation of auxotrophic mutants is generally facilitated by penicillin enrichment, associated in the case of β-lactam antibiotic resistance of the wild type with penicillinase or sulbactam. Valuable alternatives to penicillin enrichment consist of using D-cycloserine (Mendez and Gomez, 1982) (see Table 5.2) or streptozotocin according to Lengeler (1979) (also M. Sebald, unpublished observations).

Comments

The mutants that have been described in the literature are given in Tables 5.1 to 5.10. Some specific comments follow the tables.

The following list gives the abbreviations used in the tables.

Sporulation mutants and toxin formation. In bacilli, the isolation of sporulation mutants has resulted in obtention of either point mutations or deletions. The point mutations affect some step essential for the sporulation process and are capable of reversion. Among Spo mutants, pleiotropic mutants are found that are affected additionally in at least two apparently unrelated activities (Schaeffer, 1969). At least some of them revert to the Spo$^+$ condition and revert at the same time in the other altered properties, which suggests the mutation is a point mutation affecting a regulatory gene and provides insight into the biochemical events underlying, or coregulated with, the sporulation process. These observations have been fundamental to the early development of the genetics of bacilli. Similarly, several sporulation mutants have been obtained in toxigenic clostridia (see Table 5.3). In both *C. histolyticum* and *C. perfringens*, sporulation mutants blocked at an early stage were altered in α-toxin and enterotoxin production, respectively, and the Osp/Spo$^+$ revertants

clearly recovered the ability to produce toxins. Nevertheless, additional data are necessary for understanding the relationship between sporulation and enterotoxin production in *C. perfringens* (see discussion in Granum and Stewart, Chapter 16 this volume). For *C. perfringens* Θ-hemolysin or sialidase, the correlation was not so obvious, but the mutants affected in those products were frequently affected in sporulation.

Hyperproducer strains may also be expected among the mutants or the revertants and have been obtained for phospholipase C in *C. perfringens* (Sebald and Cassier, 1969) (see Table 5.3), but the interpretation is difficult because the mutants were selected as colonial variants. From a fundamental point of view, it is evident that no direct correlation between spore formation and toxin production can be made unless direct isolation of Tox$^-$ mutants led to Spo$^-$ strains.

Sporulation mutants and mutants affected in solvent production. A relationship between sporulation and solvent production was suspected early, and Weizmann and Rosenfeld (1937) maintained a high solvent productivity in *C. acetobutylicum* by heating the inoculum, thus likely counterselecting the Spo$^-$ mutants (see also Gapes et al., 1983). It was thus tempting to correlate sporulation and solvent production, two postgrowth events in batch cultures (Rogers, 1986; Woods and Jones, 1986).

Sporulation mutants have been isolated either after mutagenesis, in *C. pasteurianum* (Morris and Robson, 1973; Robson et al., 1974) and *C. acetobutylicum* (Jones et al., 1982; Rogers et al., 1986; Rogers and Palosaari, 1987), or as spontaneous mutants in *C. acetobutylicum* (continuous culture; Meinecke et al., 1984) and *C. butyricum* (chemostat; Zoutberg et al., 1989a, 1989b) (see Tables 5.4 and 5.5). Sporulation mutants have also been found among mutant strains modified in solvent production (Hermann et al., 1985; Clark et al., 1989) or selected as resistant to rifampicin (Jones et al., 1982; Clark et al., 1989). In a general way the blockage stage of these sporulation mutants and their reversion rate were not well studied and one must also consider that the granulose-

Table 5.1 Antibiotic-resistant mutants in clostridia

Original strain	Mutant designation	Phenotype	Mode of obtention	Reference
Clostridium acetobutylicum NCIB8052	Several	Rf^r ($2\mu g/ml$)	UV irradiation, mitomycin C ($10\mu g/ml$); MNNG ($50\ \mu g/ml$); EMS (2%, vol/vol)	Bowring and Morris, 1985
C. acetobutylicum				
ATCC39236 Spo$^-$		Ap^r	EMS (1%, vol/vol)	Lemmel, 1985
ATCC39236 Spo$^-$		Em^r	EMS (1%, vol/vol)	Lemmel, 1985
C. butyricum				
ATCC19398		Rf^r	UV irradiation; MMS treatment (0.01%–0.20%)	Carrasco and Soto, 1987
ATCC19398		Nal^r	UV irradiation; MMS treatment (0.01%–0.20%)	
ATCC19398		$5\text{-}FU^r$	UV irradiation; MMS treatment (0.01%–0.20%)	
C. pasteurianum				
ATCC6013	FDC1 and FDC3	Sm^r	Spontaneous	Daldal, 1985
	MAC17	Sm^r	UV	Daldal, 1985
	FDC2, FDC4	Nal^r	Spontaneous	Daldal, 1985
	MAC42	Nal^r	UV	Daldal, 1985
	FDC5	Rf^r	Spontaneous	Daldal, 1985
	MAC34	Rf^r	UV	Daldal, 1985
C. perfringens				
659-4(CP593)	659-4 Rif^r	Rf^r ($2\ \mu g/ml$)	Spontaneous	Sebald et al., 1975; Sebald and Bréfort, 1975

Strain	Markers	Mutagen	Reference
C. perfringens CP593	R^{Fr} (2 μg/ml), Nal^r (100 μg/ml); obtained by successive spontaneous mutations	Spontaneous	Bréfort et al., 1977
C. perfringens VPI11268 11268CDR	R^{Fr} (10 μg/ml)	Spontaneous	Heefner et al., 1984
C. perfringens CP601 (see Table 5.11) CP603	R^{Fr} (2 μg/ml)	Spontaneous	Bréfort et al., 1977
C. perfringens CP602 (see Table 5.11) CP602R	R^{Fr} (2 μg/ml)	Spontaneous	Bréfort et al., 1977
C. perfringens CPN52 (see Table 5.2) CPN58	Sm^r (500 μg/ml)	Spontaneous	Bréfort et al., 1977
C. perfringens CM165 (CN3870) CM198	Nal^r (50 μg/ml)	Spontaneous	Rood and Wilkinson, 1975
C. perfringens CW92 CW482	R^{fr}Spc^r	Spontaneous	Rood et al., 1978
C. perfringens CW362 CW504, CM288	Nal^{r}R^{fr}	Spontaneous	Rood et al., 1978; Sander et al., 1982
C. perfringens CM288 CM289, CM292, CM293, CM294	Nal^{r}R^{fr}Met^r	MNNG(150 μg/ml)	Sindar et al., 1982
C. tertium ATCC19405 A1	Sm^r	Not indicated	Knowlton et al., 1984
C. thermocellum ATCC2705	R^{fr} (50 μg/ml)	UV irradiation, γ irradiation or MNNG	Gomez et al., 1980

71

Table 5.2 Auxotrophic mutants in clostridia

Original strain	Mutant designation and phenotype	Mode of obtention	Reversion	Reference
Clostridium acetobutylicum				
P262	P262-1, ARG⁻Tyr⁻	EMS (2.5% vol/vol)	$<10^{-9}$	Jones et al., 1985
P262	P262-2, His⁻Met⁻	EMS (2.5% vol/vol)	$<10^{-9}$	Jones et al., 1985
N1-4 Rif^r	N1-4080, Pro⁻	EMS (2%, vol/vol) p.e.		Reysset et al., 1987
ATCC8052	SBA28, Glt⁻	UV		Oultram and Young, 1985
ATCC8052	SBA53, Leu⁻	MNNG (2%, vol/vol)		Oultram and Young, 1985
IFP903 Rif^r	903-9, Asn⁻	EMS (2%, vol/vol)		Reysset and Sebald, 1985
ATCC824 Rif^r	824-7, Asn⁻	EMS (2%, vol/vol)		Reysset and Sebald, 1985
NCIB 8052	Hist⁻Glu⁻	UV irradiation		Bowring and Morris, 1985
NCIB8052	Rib⁻; Hist⁻; Glu⁻; Pro⁻; Try⁻; Ade⁻; Ura⁻	MNNG (50 μg/ml) or EMS (2% vol/vol)	10^{-8} to 10^{-4}	Bowring and Morris, 1985
C. butyricum				
ATCC6014	Group 1, Try⁻ (TS deficient)	MNNG (500 μg/ml);p.e.	ND^a	Baskerville and Twarog, 1972
ATCC6014	Group 2, Try⁻/Ind⁻	MNNG (500 μg/ml); p.e.	ND	Baskerville and Twarog, 1972
ATCC6014	mtr-4 and mtr-8; 5 Mt^r (50 μg/ml); derepressed in AS; PRAI, InGPS, TS activities in presence of tryptophan	Spontaneous	ND	Carrasco and Soto, 1987
ATCC19398	AC 10, Ile⁻	UV	1.10^{-5}	Carrasco and Soto, 1987
ATCC19398	AC 11, Val⁻	UV	2.10^{-5}	Carrasco and Soto, 1987
ATCC19398	AC12, Ile⁻	MMS	5.10^{-6}	Carrasco and Soto, 1987
ATCC19398	AC 13, Ade⁻	MMS	1.10^{-5}	Carrasco and Soto, 1987
ATCC19398	AC 14, Val⁻	MMS	$2.5.10^{-5}$	Carrasco and Soto, 1987
ATCC19398	AC 15, Try⁻	MMS	$1.5.10^{-6}$	Carrasco and Soto, 1987
C. pasteurianum				
ATCC 6013	FDC11, Try⁻	EMS;p.e.	2.10^{-7}	Daldal, 1985
	FDC12, Gly⁻	EMS;p.e.	5.10^{-7}	Daldal, 1985
	FDC13, Ade⁻/Hypo⁻	MNNG;p.e.	6.10^{-6}	Daldal, 1985
	FDC14, Ade⁻	MNNG;p.e.	10^{-7}	Daldal, 1985
	FDC15, Ura⁻	MNNG;p.e.	2.10^{-7}	Daldal, 1985
	FDC16, Ade⁻/Hypo⁻/Gua⁻	MNNG;p.e.	7.10^{-5}	Daldal, 1985
	FDC17, Glu⁻	MNNG;p.e.	2.10^{-6}	Daldal, 1985

Strain	Mutant	Mutagenesis; selection	Frequency	Reference
	FDC18, Asp⁻	MNNG;p.e.	2.10^{-6}	Daldal, 1985
	FDC21, Asp⁻	EMS;p.e.	ND[a]	Daldal, 1985
	FDC22, Ade⁻/Hypo⁻	EMS;p.e.	ND	Daldal, 1985
	FDC28, Ade⁻	EMS;p.e.	ND	Daldal, 1985
	FDC29, Glu⁻	EMS;p.e.	ND	Daldal, 1985
	MAC 57, Ade⁻	UV;p.e.	ND	Daldal, 1985
	MAC58, Ade⁻/Hypo⁻	UV;p.e.	ND	Daldal, 1985
	MAC73, Ade⁻/Hypo⁻/Gua⁻	UV;p.e.	ND	Daldal, 1985
C. pasteurianum				
W5	1–8, Nif⁻	MNNG (100 μg/ml); p.e. + Pase; lack of growth on limiting ammonia medium	+	Simon and Brill, 1971
C. perfringens				
BP6K-N5 (CPN50)	CPN54, Pro⁻	MNNG (100 μg/ml);p.e.	$\leq 6.10^{-8}$	Sebald and Costilow, 1975
CPN50	CPN53, Lys⁻	MNNG (100 μg/ml);p.e.	10^{-7}	Brefort et al., 1977
CPN54	CPN55, Pro⁻Lys⁻	MNNG (100 μg/ml); p.e. (two-step mutagenesis)	$<10^{-9}$	Brefort et al., 1977
CPN52 (See Table 5.11)	CPN57, Pro⁻	MNNG (100 μg/ml);p.e.		Brefort et al., 1977
CP720	CP721, Ade⁻Nic⁻	MNNG (100 μg/ml);p.e.	$<10^{-9}$	Brefort et al., 1977
BP6K-N5	Six independent mutants, Ade⁻	MNNG (100 μg/ml);p.e.	$\leq 4.10^{-8}$ to 2.10^{-9}	Sebald and Costilow, 1975
BP6K-N5	352-2; 366A4, Nic⁻	MNNG (100 μg/ml);p.e.	$\leq 1.10^{-8}$ to $<3.10^{-9}$	Sebald and Costilow, 1975
BP6K-N5	352-20S, Ura⁻	MNNG (100 μg/ml);p.e.	$\leq 6.10^{-8}$	Sebald and Costilow, 1975
PX7 (see Table 5.11)	9-X; 53, Ade⁻	MNNG (100 μg/ml);p.e.	$\leq 5.10^{-9}$	Sebald and Costilow, 1975
PX7 (see Table 5.11)	38, Ura⁻	MNNG (100 μg/ml);p.e.	$\leq 10^{-9}$	Sebald and Costilow, 1975
PX7 (see Table 5.11)	7,12,33,43,57, Nic⁻	MNNG (100 μg/ml);p.e.	$\leq 10^{-9}$	Sebald and Costilow, 1975
PX7 (see Table 5.11)	3, Thia⁻	MNNG (100 μg/ml);p.e.	$\leq 10^{-9}$	Sebald and Costilow, 1975
PX7 (see Table 5.11)	10,16,51, Ade⁻Nic⁻	MNNG (100 μg/ml);p.e. (one step)	$\leq 10^{-9}$	Sebald and Costilow, 1975
PX7 (see Table 5.11)	34, Ura⁻Nic⁻	MNNG (100 μg/ml);p.e. (one step)	$\leq 10^{-7}$	Sebald and Costilow, 1975
PX7 (see Table 5.11)	9-XI, Pro⁻	MNNG (100 μg/ml);p.e.	$\leq 2.10^{-9}$	Sebald and Costilow, 1975
C. thermocellum				
ATCC27405	BM11, Ade⁻	Spontaneous;p.c.e.	+	Mendez and Gomez, 1982
	BM23, Leu⁻	Spontaneous; p.c.e.	–	Mendez and Gomez, 1982
	BM31, Ade⁻	Spontaneous; p.c.e.	–	Mendez and Gomez, 1982
	BM74, Ileu⁻	UV irradiation of bacilli; p.c.e.	–	Mendez and Gomez, 1982

[a]ND, not done.

Table 5.3 Toxinogenic clostridia: mutants altered in spore formation or toxin production

Original strain	Mutant designation	Phenotype	Mutagenesis/Mode of selection	Other properties	Reference
Clostridium botulinum type E					
ATCC9564	MSp$^+$	Spo$^+$ Tox$^-$	MNNG (150 μg/ml)/unknown mode		Emeruwa and Hawirko, 1972, 1975; Emeruwa et al., 1974; Hawirko et al., 1973
ATCC9564	RSpoIIIa	Spo$_{III}$ Tox$^-$	MNNG; pinhead translucent colonies	Sporulation blocked at stage II in presence of glucose (catabolite repression)	Emeruwa and Hawirko, 1972, 1975; Emeruwa et al., 1974; Hawirko et al., 1973
C. botulinum type A, 190L (Osp, phage induced by mitomycin C)	AT10, AT14, AT15	Increased sporulation rate	Detergent treatment (SDS, DOC, Brij58, Tween 80)/colonial variant	No phage induced by mitomycin C; lipase negative; lack of outer cell well layer	Takumi et al., 1980
C. histolyticum					
G54Spo$^+$ Tox$^+$	Sp$^-$11,12,13	Spo$_0$ Tox$^-$	UV/transparent colonies	Osp revertants were Tox$^+$	Sebald and Schaeffer, 1965; Bayen et al., 1967
G54Spo$^+$ Tox$^+$	Osp17	Osp$_0$ Tox$^\pm$	UV/transparent colonies		Sebald and Schaeffer, 1965; Bayen et al., 1967
G54Spo$^+$ Tox$^+$	Sp$^-$14	Spo$_{II}$ Tox$^+$	UV/transparent colonies		Sebald and Schaeffer, 1965; Bayen et al., 1967
G54Spo$^+$ Tox$^+$	Sp$^-$310	Spo$_{III}$ Tox$^+$	MNNG/transparent colonies	Cortex and inner coat formation	Sebald and Schaeffer, 1965; Bayen et al., 1967
G54Spo$^+$ Tox$^+$	Sp$^-$32	Spo$_{IV}$ Tox$^+$	MNNG/transparent colonies		Sebald and Schaeffer, 1965; Bayen et al., 1967
G54Spo$^+$ Tox$^+$	Sp$^-$37	Spo$_{IV}$ Tox$^+$	MNNG/transparent colonies	Hyperproduction of cortex; poor formation of inner coat; no outercoat formation	Sebald and Schaeffer, 1965; Bayen et al., 1967
G54Spo$^+$ Tox$^+$	Sp$^-$15,21,34,38	Spo$_V$ Tox$^+$	NO$_2$H; MNNG/transparent colonies	Forms cortex, inner and outer coats	Sebald and Schaeffer, 1965; Bayen et al., 1967
G54Spo$^+$ Tox$^+$	Sp$^-$31	Spo$_{ab/III}$ Tox$^+$	MNNG/transparent colonies	No cortex and outercoat; innercoat formed	Sebald and Schaeffer, 1965; Bayen et al., 1967
G54Spo$^+$ Tox$^+$	Sp$^-$18	Spo$_{ab/II}$ Tox$^+$	MNNG/transparent colonies	No cortex and outercoat; innercoat formed	Sebald and Schaeffer, 1965; Bayen et al., 1967
G54Spo$^+$ Tox$^+$	Osp325	Osp$_{ab}$ Tox$^+$	MNNG/transparent colonies	No cortex; coats formed	Sebald and Schaeffer, 1965; Bayen et al., 1967
G54Spo$^+$ Tox$^+$	Sp$^-$35	Spo$_{IV}$ Tox$^+$	MNNG/transparent colonies	Forms cortex and iner coat; no outercoat formed; no spore germination unless sodium dipicolinate added to	Sebald and Schaeffer, 1965; Bayen et al., 1967; Sebald, 1968

Organism / Type	Strain	Phenotype	Method	Reference
C. perfringens				
Type A ATCC3624		Osp^+ to Spo^-; $PLC^{++}The^+$	Spontaneous or acridine orange/colonial sectors	Sebald and Cassier, 1969
Type A ATCC3624		Osp to Spo^-; $PLC^{++}The^-$	Acridine orange/colonial sectors	Sebald and Cassier, 1969
Type A ATCC3624		Osp to Spo^-; $PLC^{+/\pm}The^{++}$	Spontaneous or acridine orange/selection; colonial sectors	Sebald and Cassier, 1969
Type A ATCC3624		Osp to Spo^-; $PLC^{+/\pm}The^-$	Spontaneous or acridine orange/selection; colonial sectors	Sebald and Cassier, 1969
C. perfringens				
Type A, BP6K PLC^+ The^+	MLH10	PLC^-, The^+ Ga^a	$MNNG/PLC^-$ on egg yolk agar	Higashi et al., 1973; 1974
Type A, BP6K PLC^+ The^+	32	PLC^+, The^- Sm^R Ga	$MNNG/The^{\pm/-}$ on sheep blood agar	Higashi et al., 1973; 1974
Type A, BP6K PLC^+ The^+	46	PLC^+, $The^\pm$ Sm^R Gb^a	$MNNG/The^{\pm/-}$ on sheep blood agar	Higashi et al., 1973; 1974
Type A, BP6K PLC^+ The^+	300-2	PLC^+, $The^\pm$ Gb	$MNNG/The^{\pm/-}$ on sheep blood agar	Higashi et al., 1973; 1974
Type A, BP6K PLC^+ The^+	300-11	PLC^-, $The^\pm$ Gb	$MNNG/The^{\pm/-}$ on sheep blood agar	Higashi et al., 1973; 1974
C. perfringens				
NCTC 8798	8-1	Spo_0^- CPE^-	Acridine orange/colonial variant	Duncan et al., 1972
8-1	R8, R42, R43	Spo^+ CPE^+	Spontaneous/heat treatment	Duncan et al., 1972
NCTC 8798	8-3	Osp_0 CPE^-	Acridine orange/colonial variant	Duncan et al., 1972
NCTC 8798	8-14, 8-15	Osp_0 CPE^-	Acridine orange/colonial variant	Duncan et al., 1972
NCTC 8798	8-16	Spo_0^- $CPE^\pm$	Acridine orange/colonial variant	Duncan et al., 1972
NCTC 8798	8-6	Sp^+ Lzd (coatless) CPE^+	Spontaneous/colonial variant	Duncan et al., 1972; Cassier and Ryter, 1971
8-6	R3	Sp^+ Lz^+ (normal coat) CPE^+	Spontaneous/heat treatment	Duncan et al., 1972
NCTC 8798	8-7	Spo_{ii}^- CPE^+	Spontaneous/colonial variant	Duncan et al., 1972
NCTC 8798	8-17	Spo_{ab}^- CPE^+	Acridine orange/colonial variant	Duncan et al., 1972
NCTC 8798	8-20	Spo_{ab}^- CPE^+	Acridine orange/colonial variant	Duncan et al., 1972
C. perfringens				
Type A, Lechien The^+	LNG5	The^-	$MNNG/The^-$ on sheep blood agar	Lapointe and Fredette, 1975
Type A, Lechien The^+	LNG11	$The^\pm$, non-pathogenic	$MNNG/The^-$ on sheep blood agar	Lapointe and Fredette, 1975

(Continued next page)

Table 5.3 (Continued)

Original strain	Mutant designation	Phenotype	Mutagenesis/Mode of selection	Other properties	Reference
C. perfringens Type A					
CN3870 (Hgn+SLD+PLC+Spo+)	CM150, 154	Hgn± SLD± PLC± Spo+	MNNG/identified as Hgn−/±		Rood and Wilkinson, 1975
CN3870 (Hgn+ SLD+ PLC+ Spo+)	CM169	Hgn± SLD±	MNNG/identified as Hgn−/±		Rood and Wilkinson, 1975
CN3870 (Hgn+ SLD+ PLC+ Spo+)	CM82	Hgn± SLD± PLC− Osp−	MNNG/identified as Hgn−/±		Rood and Wilkinson, 1975
CN3870 (Hgn+ SLD+ PLC+ Spo+)	CM171	Hgn− SLD− PLC+ Spo−	MNNG/identified as Hgn−/±		Rood and Wilkinson, 1975
CN3870 (Hgn+ SLD+ PLC+ Spo+)	CM165	Hgn− SLD± PLC++ Spo−	MNNG/identified as Hgn−/±		Rood and Wilkinson, 1975
CM198 (CM165 NalR)	Cm213, CM211	Hgn+ SLD+/± PLC+ Spo+	MNNG, Spo+ revertants selected heat treatment		Rood and Wilkinson, 1975
CM198 (CM165 NalR)	CM206, CM207	Hgn± SLD± PLC+ Spo+	MNNG except CM206 (spontaneous), Spo+ revertants selected by heat treatment		Rood and Wilkinson, 1975
CM198 (CM165 NalR)	CM209	Hgn− Spo+ PLC+ Spo+	MNNG, Spo+ revertants selected heat treatment		Rood and Wilkinson, 1975
CM198 (CM165 NalR)	CM216, CM219	Hgn− SLD± PLC++ Osp	MNNG, Spo+ revertants selected heat treatment		Rood and Wilkinson, 1975

[a]Ga, Gb: groups defined by their capacity of synergistic Θ-hemolytic activity, detected by cross-streaking on sheep blood agar plates.

Table 5.4 Sporulation and germination mutants in nontoxigenic clostridia

Original strain	Mutant designation	Phenotype	Mutagenesis/mode of selection	Other properties	Reference
Clostridium acetobutylicum DSM1731		Spo^-	Spontaneous/continuous culture under phosphate limitation; iodine staining	Stage of blockage not indicated	Meinecke et al., 1984
C. acetobutylicum P262	ds-1 and ds-2	$Spo_{O-I}{}^-$ Rf^r	EMS (2.5%, vol/vol) 10% survival/Rf^r	Granulose$^-$ capsule$^-$ blockage before stage II (no septum formation); no solvent production	Jones et al., 1982
P262	ds-3 and ds-4	$Spo_{O-I}{}^-$	EMS (2.5%, vol/vol) 10% survival/iodine staining	Granulose$^-$ capsule$^-$ blockage before stage II (no septum formation); no solvent production, but capsule$^+$	Jones et al., 1982
P262	spo-1 and spo-2	$Spo_{>II}{}^-$, Rf^r	EMS/Rf^r	Granulose$^+$; normal solvent production	Jones et al., 1982
P262	spo-3 and spo-4	$Osp_{>II}$ Rf^r	EMS/Rf^r	Granulose$^+$; deficient in solvent production	Jones et al., 1982
C. bifermentans Not indicated	1 to 5	Slow germination improved by NaOH treatment of spores	spontaneous/ungerminated spores		Wyatt and Waites, 1971
C. pasteurianum ATCC6013	FDC6	Spo^-	EMS/iodine staining		Daldal, 1985
ATCC6013	FDC7	Spo^-	MNNG/iodine staining		Daldal, 1985

Table 5.5 Metabolic mutants in clostridia

Original strain	Mutant designation	Phenotype	Mutagenesis/selection	Other properties	Reference
Clostridium acetobutylicum ATCC824	SA-1	Bol^{tol} (18.6 g/l) (WT: <15 g/l)	Spontaneous/sequential transfers in increasing concentrations of Bol	Increased growth rate; increased AMY activity; 13% increased butanol yield	Lin and Blaschek, 1983
C. acetobutylicum ATCC39236 Spo$^-$		Bol^{tol} (15 g/l) (WT: <15 g/l)	EMS/tolerant to Bol		Lemmel, 1985
C. acetobutylicum 903 Spo$^+$	904	Osp Bol^{tol} (10 g/l) (WT: 7 g/l)	MNNG/tolerant to Bol	30% increased solvent yield	Hermann et al., 1985
C. acetobutylicum ATCC824	G1	Bol^{tol} (13.5 g/l) (WT: 11.8 g/l)	Spontaneous/sequential transfers in increasing concentrations of Bol	Low Bol yield; modified carbohydrate fermentations; high production of "autobacteriocin"	Soucaille et al., 1987
C. acetobutylicum DSM1732 Aco$^+$ Bol$^+$ Eto$^+$	CA G 1 to 10	Aco$^+$ Bol$^-$ Eto$^+$	Spontaneous/sequential transfers at subinhibitory concentrations of AA (0.5 M)	BDH$^+$, BAD$^{-/\pm}$; sporulation unaffected	Dürre et al., 1986
C. acetobutylicum NRRL B643 Aco$^+$ Bol$^+$ Eto$^+$ Aca$^+$ Bua$^+$ Bud$^-$	Type 1 e.g., AA-3; AA-C4	Aco$^+$ Bol$^\pm$ Eto$^-$ Aca$^{+/++}$ Bua$^+$ Bud$^+$	Spontaneous/resistant to AA or CLA	ADH^d, BDH^d, normal AK, BK and BAD activities; PAT^{++} PBT^{++}	Rogers et al., 1986; Rogers and Palosaari, 1987
C. acetobutylicum NRRL B643 Aco$^+$ Bol$^+$ Eto$^+$ Aca$^+$ Bua$^+$ Bud$^-$	Type 2 e.g. AA-9	Aco$^\pm$ Bol$^\pm$ Eto$^-$ Aca^{++} Bua$^+$ Bud$^-$	Spontaneous/resistant to AA to CLA	ADH^d, BDH^d, PBT^{++}	Rogers et al., 1986; Rogers and Palosaari, 1987
C. acetobutylicum NRRL B643 Aco$^+$ Bol$^+$ Eto$^+$ Aca$^+$ Bua$^+$ Bud$^-$	B18	Aco$^+$ Bol$^+$ Eto^{++} Aca$^\pm$ Bua$^-$ "acid$^-$"	EMS/low acid-producing colonies on BCP plates	BDH^{++}, BAD^{++}	Rogers et al., 1986; Rogers and Palosaari, 1987
C. acetobutylicum NRRL B643 Aco$^+$ Bol$^+$ Eto$^+$ Aca$^+$ Bua$^+$ Bud$^-$	6A	Aco$^\pm$ Bol$^-$ Eto$^-$ Aca^{++} Bua^{++} Spo$^-$	EMS/granulose$^-$ by iodine staining	AK^{++}, BK^{++} BAD$^-$, BDH$^-$ PBT^{++} PAT^{++}	Rogers et al., 1986; Rogers and Palosaari, 1987

Strain	Mutant	Phenotype	Mutagenesis/selection	Properties	Reference
C. acetobutylicum ATCC824	M3, M5	Bol− Eto± Aco− Bua+/+± Aca+/± Spo−	MNNG (100 μg/ml)/resistant to 4 CLB; nonsolvent-producing strains	BDH^d, BAD−, CoA-T− BK+, PBT+	Clark et al., 1989
	Rif^r B12	Bol− Eto± Aco± Bua++ Aca+ Osp	MNNG (100 μg/ml)/resistant to 4 CLB; nonsolvent-producing strains/resistance to Rf (2 μg/ml)	BDH^d, BAD−, CoA-T− BK+, PBT+	Clark et al., 1989
	Rif^r D10	Bol− Eto− Aco± Bua++ Aca± Osp	MNNG (100 μg/ml)/resistant to 4 CLB; nonsolvent-producing strains/resistance to Rf (2 μg/ml)	BDH+, BAD±, CoA-T−, BK++, PBT++	Clark et al., 1989
	Type IV	Bol± Eto± Aco− Bua++ Aca+	Spontaneous/colonial variant	BDH^d, BAD^d, CoA-T^d, BK+, PBT+	Clark et al., 1989
	2-BBR	Bol± Eto± Aco± Bua+ Aca++	MNNG on filter paper disk/resistant to BrB	BDH++, BAD++, CoA-T^d BK+, PBT+	Clark et al., 1989
	2-BBD	Bol+ Eto+ Aco+ Bua+ Aca+	MNNG on filter paper disk/resistant to BrB	BDH+, BAD+, CoA-T+, BK+, PBT+	Clark et al., 1989
C. acetobutylicum N1-4 Aco+ Bol+ Eto± Aca± Bua± Iva− Spo+	AS2-1	Aco− Bol− Eto++ Aca− Bua− Iva− Spo+	UV irradiation on plates (3.10^−4 survival)	Altered in several carbohydrate fermentations; THL^d	Hayashida and Ahn, 1990; Ahn and Hayashida, 1990
C. acetobutylicum ATCC824 Aco+ Bol+ Eto+ Aca+ Bua+	2-Br Bu1	Aco− Bol− Eto+ Aca+ Bua+	MNNG/BrB^r	CoA-T^d, AAD^d	Junelles et al., 1987 Janati-Idrissi et al., 1987
	Acidogenic mutants	Aco± Bol± Eto± Aca++ Bua++	MNNG/BrA^r, CLA^r, FA^r or CLB^r		Junelles et al., 1987
C. acetobutylicum ATCC10132 Aco+ Bol+ Eto+ Aca+ Bua+	Group I	Aco++ Bol++ Eto+ Aca± Bua±	MNNG/resistance to NaBr + NaBrO$_3$ (proton suicide method)	Opaque colonies	Cueto and Mendez, 1990
	Group II	Aco− Bol− Eto± Aca± Bua+/++	MNNG/resistance to NaBr + NaBrO$_3$ (proton suicide method)		Cueto and Mendez, 1990
	Group III	Aco± Bol± Eto+ Aca++ Bua++	MNNG/resistance to NaBr + NaBrO$_3$ (proton suicide method)		Cueto and Mendez, 1990

(Continued next page)

Table 5.5 (Continued)

Original strain	Mutant designation	Phenotype	Mutagenesis/selection	Other properties	Reference
C. butyricum LMD77^{-11} (Buo$^-$ Ipo$^-$ Spo$^+$)	Spo$^-$ Buo^{tol}		Spontaneous/prolonged growth in chemostat; aggregated growth	Buo$^+$ Ipo$^+$	Zoutberg et al., 1989a, 1989b
C. pasteurianum W-5 (ATCC6013)	"Class 1"	DCCD^r	MNNG/decreased sensitivity to DCCD (80 μM) or butyricin 7423[a]	H$^+$-ATPase (BF$_0$F$_1$) relatively insensitive to butyricin inhibition; but with an increased sensitivity to DCCD; cultures resistant to butyricin	Clarke et al., 1982
W-5 (ATCC6013)	"Class 2"	DCCD^r	MNNG/decreased sensitivity to DCCD (80 μgM)	H$^+$-ATPase with reduced sensitivity to DCCD and butyricin; cultures with reduced sensitivity to butyricin	Clarke et al., 1982
C. pasteurianum ATCC6013	MR505	Granulose negative	MNNG (1 mg/ml) or UV irradiation at 10^{-3} survival; colorless by exposure to iodine	GSAse$^-$	Morris and Robson 1973; Robson et al., 1974; Mackey and Morris, 1974
MR505	MR505 revertant	Granulose positive	"Mutagenesis"	GSAse$^+$	Morris and Robson, 1973; Robson et al., 1974; Mackey and Morris, 1974
C. pasteurianum NCIB9486	MR30	Granulose negative	Mutagenesis (MNNG or UV?); colorless colonies by exposure to iodine	GSAse$^-$	Robson et al., 1974
MR30	MR30 revertant	Granulose positive	Mutagenesis (MNNG or UV?); colonies stained by exposure to iodine	GSAse$^+$	Robson et al., 1974
NCIB9486	MR31	Granulose negative	Mutagenesis (MNNG or UV?); colonies stained by exposure to iodine	ADP-GPPase$^-$	Robson et al., 1974
MR31	MR31 revertant	Granulose positive	Mutagenesis (MNNG or UV?); colonies stained by exposure to iodine	ADP-GPPase$^+$	Robson et al., 1974

Organism	Strain	Phenotype	Mutagenesis/method	Properties	Reference
C. saccharolyticum ATCC35040	PNA	Pyr^b	MNNG (100 μg/ml) p.e./poor growth in presence of pyruvate (vs. good growth in glucose)	Increased ethanol yield; increased ethanol tolerance (50% inhibition at 65 g/l vs. 35 g/l for the W.T.)	Murray et al., 1983
C. saccharolyticum ATCC35040	A-15	Eto^{tol} (70 g/l)	Spontaneous/sequential transfers in increasing amounts of Eto (to 75 g/l)	Improved utilization of cellobiose; improved ethanol yield in coculture with *Zymomonas anaerobia*	Asther and Khan, 1985
C. saccharolyticum ATCC12662	SmGF1, Sm10	Lfe^{-c} FR$^-$	MNNG (1 mg/ml)/small colonies		Schwartz and Stadman, 1970
	Sm4	GLR–FR++	MNNG (1 mg/ml)/small colonies		Schwartz and Stadman, 1970
	Sm7	GLR–FR–	MNNG (1 mg/ml)/small colonies		Schwartz and Stadman, 1970
	Sm13	GLR++FR+	MNNG (1 mg/ml)/small colonies		Schwartz and Stadman, 1970
	Sm39	GLR++PRR++	MNNG (1 mg/ml)/small colonies		Schwartz and Stadman, 1970
C. thermoaceticum DSM521	ATCC39289	Low pH tolerance, improved growth rate at pH 7.0	Spontaneous/growth at gradually lowered pH in chemostat		Keller et al., 1985
C. thermoaceticum YM4	γ-25a	AVI++CBA++ CMCase++, XYL++	UV, then γ irradiation (25 Rad)/ higher cellulase production	Hyperproducer of ethanol	Mori, 1990
C. thermohydrosulfuricum ATCC33223	Z67-143	CATrd	MNNG 400 μg/ml, 1 h, enrichment on 2-deoxyglucose/ colonies with large clear zones on iodine-stained glucose-starch agar plates		Hyun and Zeikus, 1985
ATCC33223	Z21-109	CAT^r; AMY++ on starch medium	MNNG 400 μg/ml, 1 h, enrichment on 2-deoxyglucose/ colonies with large clear zones on iodine-stained glucose-starch agar plates	Increased growth rate on starch; increased production of ethanol	Hyun and Zeikus, 1985

(Continued next page)

Table 5.5 (Continued)

Original strain	Mutant designation	Phenotype	Mutagenesis/selection	Other properties	Reference
C. thermohydrosulfuricum 39E (ATCC33223)	39EA	Eto^{tol} (56 g/l)	Spontaneous/sequential transfers in increased concentration of Eto (to 50 g/l)	Pyruvate not fermented; lower ethanol, higher lactate and acetate yields; lack of NAD-linked alcohol dehydrogenase	Lovitt et al., 1984
C. thermosaccharolyticum ATCC7956 Eto$^+$ Aca$^+$	8	Eto^{++} Aca$^\pm$	MNNG (500 µg/ml)/resistance to FA (30 mM)		Rothstein, 1986
Clostridium sp. *thermophilic*; 187 AK$^+$, LDH$^+$, PAT$^+$ Eto$^+$	187-102-27	AD^d, LDH^d, PAT^{++} Eto^{++} on cellulose	Two cycles of mutagenesis by MNNG at 0.6% survival, followed by growth on cellobiose 1% + ethanol 1%	Catabolite repression of ethanol by glucose or cellobiose	Kurose et al., 1988

[a] Butyricin interacts with DCCD-binding protein of membrane H$^+$-ATPase (BF$_0$F$_1$) but not with its soluble BF$_1$ component.
[b] Pyr$^-$, unable to use pyruvic acid as carbon source.
[c] Lfe$^-$, lack of lysine fermentation.
[d] Amylase synthesis of W.T. is inducible and repressed by glucose; CAT^r, resistant to catabolite repression by glucose for amylase; also, for glucose isomerase, lactase, and isomaltase in case of mutant Z21-109.

Table 5.6 Other mutants in clostridia

Original strain	Mutant designation	Phenotype	Mode of obtention/selection	Reference
Clostridium acetobutylicum P262	lyt-1	AUT^d, Bol^{tol}	EMS (0.2%, vol/vol) colonies with nolysis on solid medium added with autoclaved P262 cells	Allock et al., 1981
C. acetobutylicum NI-4080	NI-4081	AUT^d	EMS (0.2%, vol/vol)/resistance to sucrose induced autoplast formation	Reysset et al., 1987
C. acetobutylicum N1-4 His⁻	N1-441	AUT^d	EMS (0.2%, vol/vol)/colonies with nolysis on solid medium added with purified N1-4 walls	Podvin, 1989
C. pasteurianum ATCC6013	MAC53	UV^r	UV irradiation	Daldal, 1985
C. perfringens CPN50(pIP404)	CPN51(pIP404)	BCN5⁻; immune to BCN5	Acridine orange/colonies that do no produce bacteriocin	Brefórt et al., 1977
NCTC8798 R⁻M⁺[a]	Several, i.e. 8-4, 8-6 8-14, 8-15; 8-16	R±M⁺[a] R⁺M±[a]	Spontaneous/colonial variants (pleiotropic mutants, see Table 5.4) acridine orange/Spontaneous/colonial variants (pleiotropic mutants, see Table 5.4)	Drefus-Fourcade et al., 1972
NCTC8798 R⁻M⁺[a]	8-47	R⁻M⁻*	MNNG/Spontaneous/colonial variants (pleiotropic mutants, see Table 5.4)	Drefus-Fourcade et al., 1972
NCTC8798 R⁻M⁺[a]	8-65	R±M⁻*	MNNG/Spontaneous/colonial variants (pleiotropic mutants, see Table 5.4)	Drefus-Fourcade et al., 1972
C. themocellum ATCC2705		5-Fu^r (50 µg/ml)	UV irradiation; γ-ray irradiation or MNNG/resistance to 5-FU	Gomez et al., 1980

[a] R and M, DNA restriction and modification; *, indicates minute plaque formation corresponding to low burst.

Table 5.7 Antibiotic-resistant mutants in *Bacteroides*, *Mitsuokella*, *Porphyromonas*, *Prevotella*, and *Selenomonas*

Original strain	Mutant designation	Phenotype	Mode of obtention	Reference
Bacteroides				
B. fragilis 638	638R	Rf^r (MIC > 128 μg/ml) (WT:MIC:0.064 μg/ml)	Spontaneous	Privitera et al., 1979 Sebald et al., 1990
B. fragilis 638R	638Cp^r	Cip^r (25 μg/ml)	NTG mutagenesis	Guiney et al., 1989
B. fragilis 638R	IB132	Fr^r (25 μg/ml)	Unknown	Smith and Spiegel, 1987
B. fragilis TM4000 (638R)	TM429	Trs^r (10 μg/ml)	Spontaneous	Hecht and Malamy, 1989
B. fragilis ATCC29771	ATCC29762	Rf^r (MIC > 25 μg/ml)	Spontaneous	Van Tassel and Wilkins, 1978
B. fragilis AM17	AM75	Met^r (MIC 100 μg/ml)	MNNG (60 μg/ml)	Britz and Wilkinson, 1979
B. fragilis AM78	AM78.1; 78-5	Met^r (MIC 25 μg/ml)	MNNG (100 μg/ml)	Britz and Wilkinson, 1979
B. fragilis AM39	AM70	Met^r (MIC 100 μg/ml)	MNNG (80 μg/ml)	Britz and Wilkinson, 1979
B. fragilis AM21	AM72	Met^r (MIC 100 μg/ml)	MNNG (100 μg/ml)	Britz and Wilkinson, 1979
B. uniformis V528	V528Rf^r	Rf^r (40 μg/ml) (WT:MIC < 5 μg/ml)	Unknown	Macrina et al., 1981
B. uniformis 0061	BU1001	Rf^r (20 μg/ml)	Spontaneous	Shoemaker et al., 1985
B. uniformis 0061	BU1100	Requires thymidine and is thus Tp^r (200 μg/ml)	Spontaneous	Shoemaker et al., 1986
B. vulgatus BV1	BV1R	Rf^r (MIC > 128 μg/ml)	Spontaneous	Sebald et al., 1990
Mitsuokella				
M. multiacida	46/5 F100	Rf^r (20 μg/ml)	Spontaneous	Flint et al., 1989
Porphyromonas *P. gingivalis* 3079.03	3079.03R1	Rf^r (80 μg/ml)	"Multistep process"	Holt et al., 1988
Prevotella *P. buccae* 8062	8062 Rif^r	Rf^r (80 μg/ml)	Spontaneous	Guiney and Bouic, 1990
P. ruminicola ssp. *brevis*, B$_1$4	F101	Rf^r (15 μg/ml)	Spontaneous	Flint et al., 1988
P. ruminicola ssp. *brevis*, B$_1$4	F130	Fu^r (20 μg/ml)	Spontaneous	Flint et al., 1988
Selenomonas *S. ruminantium* 521C1	521CIR	Rf^r (5 μg/ml)	Spontaneous	Flint et al., 1988
FB314	FB314R	Rf^r (5 μg/ml)	Spontaneous	Flint et al., 1988

negative trait often used for detecting asporogenous clostridia is frequently unstable (Meinecke et al., 1984; M. Sebald, unpublished observations).

The data from Jones et al. (1982) (see Table 5.4) show that granulose-negative mutants blocked at an early stage of sporulation (before stage II) did not produce solvents, in contrast to mutants blocked at stage III or later, which form "typical clostridical forms." Nevertheless, the authors obtained further granulose-negative mutants that produced spores and

Table 5.8 Auxotrophic mutants in *Bacteroides fragilis*

Original strain	Mutant designation	Phenotype	Mode of obtention	Reference
ATCC29762 (2044R3)	ATCC29763, A40	Arg⁻	EMS (2%, vol/vol); p.e. + cla	Van Tassel and Wilkins, 1978
ATCC29762 (2044R3)	ATCC29764 and 5 other mutants	Hist⁻	EMS (2%, vol/vol); p.e. + cla	Van Tassel and Wilkins, 1978
ATCC29762 (2044R3)	ATCC29765, M27, M47	Met⁻	EMS (2%, vol/vol); p.e. + cla	Van Tassel and Wilkins, 1978
ATCC29762 (2044R3)	M80, Th18, Th34	Thr⁻	EMS (2%, vol/vol); p.e. + cla	Van Tassel and Wilkins, 1978
ATCC29762 (2044R3)	Th64, T23	Trp⁻	EMS (2%, vol/vol); p.e. + cla	Van Tassel and Wilkins, 1978
ATCC29762 (2044R3)	ATCC29767, ATCC29766	Ade⁻/Gua⁻	EMS (2%, vol/vol); p.e. + cla	Van Tassel and Wilkins, 1978
ATCC29768 (12256)	A8, ATCC29769, A15	Arg⁻	EMS (2.5%, vol/vol); p.e. + cla	Van Tassel and Wilkins, 1978
ATCC29768 (12256)	H4, H7	His⁻	EMS (2.5%, vol/vol); p.e. + cla	Van Tassel and Wilkins, 1978
ATCC29768 (12256)	ATCC29770, P19	Ade⁻/Gua⁻	EMS (2.5%, vol/vol); p.e. + cla	Van Tassel and Wilkins, 1978
TM4000 (638R)	TM420 (JC101)	Arg⁻Hist⁻	Unknown	Hecht and Malamy, 1989

solvents (Long et al., 1984; Jones and Woods, 1986). The early-stage sporulation mutants have been reported to revert, exhibiting a wild-type phenotype (Jones and Woods, 1986).

In conclusion, a deficiency in spore formation is frequently associated, in solventogenic clostridia, with a modified solvent formation (generally a deficient production), even if solvent formation and granulose accumulation are not sporulation-specific events.

We cannot consider that sort of mutation without pointing out the recent data relative to the old concept of "degeneration." The term degeneration or degenerative process has been given to changes observed on repeated subculturing. These changes affect the acetone-butanol-ethanol (ABE) fermentation, with a low sugar utilization and a low solvent production including a metabolic transition to acid products, and they are generally linked with morphological changes (poor sporulation, chain formation) and variations in colonial morphology. The degeneration phenomenon was interpreted in terms of S ⇄ R variations (Kutzenok and Aschner, 1952) and later of mutations, and was considered to be irreversible (Adler and Crow, 1987). This interpretation was ques-

tioned by George and Chen (1983). The reversibility of the so-called degeneration was first shown by Hartmanis et al. (1986) in *C. acetobutylicum* ATCC824 by simply increasing the initial cell density of the subcultures, and other valuable data have been reported by Jones and Woods (1986), Rogers (1986) and Awang et al. (1988).

Two confirmations of reversibility were recently reported. Barbeau et al. (1988) have studied the variations in solvent versus acid production in continuous cultures of *C. acetobutylicum* IFP918 conducted under different conditions. They observed a long-term metabolic drift toward acid production at high dilution rates, with a high sensitivity to lysis of the cells and a variation in bacterial colony form. They succeeded in reverting the overall degenerative process in a two-stage fermentation either by decreasing pH, temperature, or dilution rate, or by growth in nitrogen limitation, or by addition of butyric and acetic acids. The reversion was also obtained by batch culturing of the acid-drifted bacteria from the continuous cultures.

Hayashida and Yoshino (1990) made sequential subcultures of vegetative cells of *C. aceto-*

Table 5.9 Other mutants in *Bacteroides* and *Porphyromonas*

Original strain	Mutant designation	Mode of obtention	Selected phenotypic trait	Other properties	Reference
Bacteroides fragilis BF-2	MTC25	EMS (2%, vol/vol)	Sensitive to mitomycin C (0.06 μg/ml) (WT:0.14 μg/ml)	Unaffected in UV resistance; reduced host-cell phage reactivation	Abratt et al., 1985
Bacteroides fragilis BF-2	UVS9	EMS (2%, vol/vol)	Sensitive to UV	Moderately sensitive to mitomycin C (0.08 μg/ml); reduced host-cell phage reactivation	Abratt et al., 1985
Bacteroides fragilis BF-2	HS5, HS9	"mutagenesis"	Increased survival at 48°C	Unaffected in UV resistance; increased H_2O_2- and UV-induced host-cell phage reactivation	Goodman et al., 1985
Bacteroides fragilis TM4000(638R) SLD$^+$	TC2	Heavy MNNG Mutagenesis	Poor growth in rat granuloma pouch fluid	Cys$^-$; SLD$^-$; other unidentified mutations	Russo et al., 1990
TC2	TC1Rb	Unknown	Growth on rat granuloma pouch fluid + cysteine	SLD$^+$ revertant	Russo et al., 1990
Porphyromonas gingivalis W50	W50/BE1		Slow pigmenting	Avirulent; low trypsin-like activity; smaller external layer envelope	McKee et al., 1988; Marsh et al., 1989

Table 5.10 Auxotrophic mutants in *Bifidobacterium bifidum*

Original strain	Mutant designation	Mode of obtention	Phenotype	Reversion	Reference
Es5		MNNG (200 μg/ml); p.e.	Ade$^-$; Arg$^-$; Asp$^-$; Glu$^-$; Ileu$^-$; Leu$^-$; Met$^-$; Gua$^-$; Ura$^-$; Nic$^-$; Pyr$^-$; Thia$^-$	10^{-7} to 10^{-8}	Ueda et al., 1983
Es5	N103	MNNG (200 μg/ml); p.e.	Ant$^-$		Ueda et al., 1983
Es5	N109; N297; N411; N503	MNNG (200 μg/ml); p.e.	Ind$^-$		Ueda et al., 1983
Es5	N002; N326; N328	MNNG (200 μg/ml); p.e.	Try$^-$		Ueda et al., 1983

Table 5.11 Plasmid and phage (ϕ) curing in clostridia and *Bacteroides*

	Original strain			Derivatives			
Designation	Phenotype	Plasmid/phage harbored	Mode of obtention	Designation	Phenotype	Plasmid/phage harbored	Reference
Clostridium acetobutylicum NI-4081 (pIM13)	Em^r	pIM13	Regeneration of protoplasts under "adverse" conditions	NI-4082	Sensitive to Em	Cured of pIM13	Azeddoug et al. (in manuscript)
C. botulinum type C Stockholm, 468C	Tox C_1^+	Phage C	Acridine orange; culture in presence of phage antiserum	AO-2; AO-28	Tox$C_1^-C_3^-$	Cured of ϕC_1	Inoue and Iida, 1970; Eklund et al., 1971; Eklund, Chapter 12, this volume
C. botulinum type D 1873	ToxD$^+$	Phage D	Acridine orange culture in presence of phage antiserum	e.g., AO-20	ToxD$^-C_3^-$	Cured of ϕD	Inoue and Iida, 1970; Eklund et al., 1972; Eklund, Chapter 12, this volume
C. botulinum type G Six strains	ToxG$^+$, BOTG$^+$	Plasmid, 126 kb	Serial transfers at 44°C		ToxG$^-$, BOTG$^-$	Plasmid free	Eklund et al., 1988
C. oedematiens type A HS37	Tox α^+	ϕNA1	Heating of spores	HS37	Tox α^-	Cured of ϕ NA1	Eklund et al., 1974
C. oedematiens type B 8024	Tox α^+	ϕNB1	Acridine orange	AO26; AO52	Tox α^-	Cured of ϕ NB1	Eklund, Chapter 12, this volume
C. perfringens 659 = CP590	Tc^{r}Cm^{r}Em^{r}Cc^r	pIP401, pIP402	Acridine orange or ethidium bromide	659-1; 659-2 (= CP591)	Tc^{r}Cm^r	pIP401	Sebald and Bréfort, 1975; Sebald et al., 1975; Bréfort et al., 1977
659 = CP590	Tc^{r}Cm^{r}Em^{r}Cc^r	pIP401, pIP402	Proflavine	659-3 (= CP592)	Em^{r}Cc^r	pIP402	Sebald and Bréfort, 1975; Sebald et al., 1975; Bréfort et al., 1977

Strain	Phenotype	Plasmid	Curing agent	Derivative	Phenotype	Plasmid status	Reference
CP591	Tc^{r}Cm^r	pIP402	Proflavine	659-4 (= CP593)	Sensitive to four antibiotics	Plasmid free	Sebald and Bréfort, 1975; Sebald et al., 1975; Bréfort et al., 1977
C. perfringens CP600	Tc^{r}Cm^{r}Em^{r}Cc^r	phage φI; plasmids +	Ethidium bromide and proflavine	CP602	Sensitive to four antibiotics	Plasmid free	Bréfort et al., 1977
CP600	Tc^{r}Cm^{r}Em^{r}Cc^r	phage φI; plasmids +	proflavine in presence of serum anti-φI	CP601	Sensitive to φI	Cured of φI	Bréfort et al., 1977
C. perfringens CPN50 (BP6K-N5)	BCN5$^+$	pIP404	Acridine orange	CPN52	BCN5$^-$, sensitive to bacteriocin BC5	Cured of pIP404	Ionesco et al., 1976; Bréfort et al., 1977
C. perfringens CW55	BCN$^+$	pCW4	Acridine orange, ethidium bromide, proflavine	CW55PER	BCN$^-$	Cured of pCW14	Mihelc et al., 1978
C. perfringens CW482	Tc^{r}Cm^r	pCW2, pCW3	Spontaneous; also obtained by curing agents treatment	CW531	Tc^s	pCW2$^+$; cured of PCW3	Rood et al., 1978
C. perfringens ATCC3626B	Caseinase$^+$	pHB101, pHB102	Acriflavine; high temperature		Caseinase$^-$	Cured of pHB101	Blaschek and Solberg, 1981
C. perfringens VPI11267CDR	Tc^r	pCW3	Transfers at 46°C		Tc^s	Cured of pCW3	Heefner et al., 1984
C. perfringens NCTC8798		φX, φY	Acridine orange in presence of φY antiserum	PX7	Sensitive to φY; resistant to φX	φX; cured of φY	Sebald and Costilow, 1975; Sebald, unpublished data
C. tetani Four strains	Tox$^+$	pCL1; 75 kb	MNNG, acridine orange; plating of heated spores		Tox$^-$	Plasmid free	Hara et al., 1977; Laird et al., 1980; Finn et al., 1984
Bacteroides fragilis V479	Cc^{r}Tc^s	pBF4	Spontaneous	V479C	Tc^r	Plasmid free	Shoemaker et al., 1985

butylicum N1-4, phenotypically Spo⁺ Bol⁺ Aco±
Eto± Aca± Bua± (Table 5.5) when grown in
batch culture. From the 15th subculture, they
observed a drastic decrease in the solvent production. A degenerated Spo⁻ strain, DGN32,
was isolated from the 30th subculture. It was
phenotypically Bol± Aco± Eto± Aca⁺ Bua⁺
when the fermentation was conducted in batch
culture with no pH adjustment, and there was
no distinct transition from the initial acidogenic
phase toward the solventogenic phase. Nevertheless, when grown in batch at pH adjusted to
5.1, the phenotype of strain DGN32 was Bol⁺
Aco⁺ Eto⁺ Aca⁺ Bua⁺, thus very similar to the
wild-type strain. As a consequence, one may
suspect that reversible "degeneration" has
often been confused with mutation, particularly when no reversion rate estimation was
made.

Germination mutants. A germination mutant of
C. perfringens NCTC8798, 8-6, was shown to require lysozyme for germination after heat treatment of spores and to possess coatless spores
(Cassier and Ryter, 1971) (Table 5.3); moreover,
the mutant had a normal UV resistance, which
thus gives valuable information on the role of
coats both in UV resistance and in heat resistance. At least in this case the coats appear to
have nothing to do with spore UV resistance;
the role of coats in the heat resistance appears
here as an indirect one, such as protection of,
or repository for, a lytic enzyme. The comparative sizes of the cortex in the wild type and the
mutant, observed by Algie, have also contributed to the question of the previously
admined "expansive cortex theory" as an explanation of spore heat resistance (J.E. Algie,
personal communication; and Algie, 1980).

Mutagenesis and new metabolic pathways. Mutagenesis is also of interest in creating new
metabolic capabilities that should correspond to
derepressed pathways. An excellent example is
given by the *C. acetobutylicum* mutant AS2-1
obtained by Hayashida and Ahn (1990) and
characterized by a new ethanol-isovaleric pathway and a deficiency in thiolase (acetyl-CoA
acetyltransferase) (Ahn and Hayashida, 1990)
(Table 5.5).

5.7   Phage and Plasmid Curing

Phage and plasmid curing has been obtained in
Clostridium and *Bacteroides*, and has greatly contributed to the assessment of the role of these
elements in some of their characters and in
obtaining isogenic strains for genetic studies.
Many techniques are available, particularly in
clostridia (Table 5.11). Phage and plasmid curing nevertheless requires technical expertise
and either easily selectable markers or a high
frequency of curing. Phage or plasmid curing
cannot be achieved in many cases for reasons
that are unknown but that could be as trivial
as a high copy number of plasmids, defective
phage, or lack of indicator strain.

5.8   Concluding Remarks and Perspectives

As far as eubacteria are concerned, mutants
have been mostly isolated in *Bacteroides* and
Clostridium, but also in other gram-negative
genera, bifidobacteria, and sulfate-reducing
bacteria (see Chapter 34, this volume). Although poorly identified, they are of considerable interest for the development of genetics
of these groups, as is discussed in several
chapters of this volume. They are also useful
for tracing genetic exchanges in in vivo experimental models (Bréfort et al., 1977, Privitera et al., 1981), or for tracing a bacterium in
a complex environment (Flint et al., 1989). In
several situations, in vivo strain interactions
should be further investigated in conventional
animals to which strains carrying genetic
markers are administered, rather than in
mono- or heteroxenic animals, which can be
limited in their interpretation.

Mutagenesis plays a key role in the improvement of industrially important strains. As
documented in this chapter, it has permitted
the selection of strains with an improved
growth rate or a better adaptation to new substrates (see also Nochur et al., 1990), mutants
that are autolytic deficient, mutants resistant to
toxic products, mutants with new metabolic
pathways, and sporulation mutants bypassing

the complex cycles of sporulation and germination in clostridia (e.g., see Largier et al., 1985). For applications, see also Erickson and Fung (1988).

From an evolutionary point of view, additional fundamental data on the DNA repair mechanisms should be of the utmost interest. *Bacteroides fragilis* is relatively O_2 tolerant, has catalase and superoxide dismutase activities, and is likely less representative of "primitive" anaerobes than clostridia, for example. A great variability appears to exist among error-prone DNA repair systems in clostridia, and this shows the weakness of any prematurate generalization in terms of evolution.

Further information on the mechanisms of mutagenesis will give new insight on the recombination processes in anaerobes that are important to appreciate for genetic manipulations. Shuttle vectors allowing gene transfer both in *Clostridia/Bacteroides* and *E. coli/B. subtilis* are now available, but it will be important to appreciate whether or not those anaerobes are recombinant proficient and to be capable of modifying this trait for specific purposes. Selection of restriction-less mutants of anaerobic strains used as recipients for gene transfer is another priority for development.

This chapter has focused on radiation-induced and chemically induced mutagenesis, but transposon mutagenesis has also been performed in anaerobes, mainly in *Bacteroides* and more recently in clostridia (e.g., Bertram and Dürre, 1989). Transposon mutagenesis has many advantages over chemically induced or radiation-induced mutagenesis. Insertion mutations may be localized in a gene library, thus permitting cloning of adjacent regions; transposons, after introduction in a suitable host, should be excised with the precise restoration of the initial gene.

Both "heterologous" transposons or "indigenous" transposons, now available in clostridia and *Bacteroides*, should be used. Nevertheless, these techniques, to be efficient, require improvement of the transfer delivery frequencies and also more knowledge about the frequency of spontaneous excision and the mode of excision. In vitro site-directed mutagenesis should also be applied to anaerobes, and therefore would be of considerable interest to outline the lack of expression in *E. coli* of genes when related to an unfavorable codon bias [see the recent application to the expression of tetanus toxin in *E. coli* (Makoff et al., 1989)] and to study the structure–function relationship of proteins of interest.

WT, wild type. When WT phenotype is not specified, characters + (prototrophy, enzymatic activities, metabolic products, toxins), S (antibiotic sensitivity), or Spo+ (sporulation) may be attributed to wild type with respect to corresponding ones indicated in mutants.

Mutagenesis and postmutagenic treatment. EMS, methanesulfonate acid ethyl ester (ethyl methanesulfonate); MMS, methyl methanesulfonate; MNNG, N-methyl-N'-nitro-N-nitrosoguanidine; NO_2H, nitrous acid; p.e., p.c.e., p.e.+cla, p.e.+Pase: mutagenesis was followed by either penicillin, penicillin+D-cycloserine, penicillin+clavulinic acid, or penicillin+penicillinase enrichment. Treatment was in absence of substrate for which the requirement was searched.

Selective agents. AA, allyl alcohol; BCP, bromocresol purple; BrA, bromoacetate; BrB, 2-bromobutyrate; CA, crotyl alcohol; CLA, chloroacetate; CLB, 4-chlorobutyrate; DCCD, dicyclohexyl-carbodi-imide; FA, fluoroacetate; 5-FU, 5-fluorouracil; 5-Mt, 5-methyltryptophan.

Antibiotics. r, resistant; s, sensitive; MIC, minimal inhibitory concentration; Ap, ampicillin; Cc, clindamycin; Cip, ciprofloxacin; Cm, chloramphenicol; Em, erythromycin; Fu, fusidic acid; Met, metronidazole; Nal, nalidixic acid; Rf, rifampicin; Sm, streptomycin; Spc, spectinomycin; Tc, tetracycline; Tp, trimethoprim; Trs, trospectomycin.

Auxotrophic mutants. +, prototrophic; −, auxotrophic. Amino acids and nitrogen substrates: Ant, anthranilate; Arg, arginine; Asn, asparagine; Asp, aspartic acid; Cys, cysteine; Glu, glutamic acid; Glt, glutamine; Gly, glycine; His, histidine; Ileu, isoleucine; Ind, indole, Leu, leucine; Lys, lysine; Met, methionine; Pro, proline; Thr, threonine;

Try, tryptophan; Tyr, tyrosine; Val, valine. Bases: Ade, adenine; Gua, guanine; Hypo, hypoxanthine; Ura, uracil. Vitamins: Nic, nicotinic acid; Pyr, pyridoxamine; Rib, riboflavin; Thia, thiamine HCl. Others: Nif⁻, unable to fix nitrogen; Ade⁻/Hypo⁻, mutant requires either of these supplements.

Enzymatic activities. ++, increased activity; d, deficient; −, lack of activity. AAD, acetoacetate decarboxylase; ADH, NADP⁺ dependent alcohol dehydrogenase with acetone as substrate; ADP-GPPase, ADP-glucose pyrophosphorylase; AK, acetate kinase; AMY, amylase (glucoamylase+pullulanase); AS, anthranilate synthetase; AUT, autolysin; AVI, avicellase; BAD, NAD⁺ dependent butyraldehyde dehydrogenase; BDH, NADH-dependent alcohol dehydrogenase with butanol as substrate; BK, butyrate kinase; CBA, cellobiase; CMCase, carboxymethylcellulase; CoA-T, CoA-transferase; FR, formate-dependent triphenyltetrazolium reductase; GLR, glycine reductase; GSAse, granulose synthase; H⁺-ATPase, ATP phosphohydrolase; InGPS, indole glycerol phosphate synthetase; LDH, lactate dehydrogenase; PAT, phosphoacetyl transferase; PBT, phosphobutyryl transferase; PRAI, phosphoribosyl anthranilic acid isomerase; PRR, proline reductase; THL, thiolase (acetyl-CoA acetyltransferase); TS, tryptophan synthetase; XYL, xylanase.

Metabolic products. ++, hyperproduction; ±, low production; −, no production; tol, tolerant to product. Aca, acetic acid; Aco, acetone; Bol, butanol; Bua, butyric acid; Bud, butyraldehyde; Eto, ethanol; Ipo, isopropanol; Iva, isovaleric acid. Eto^{tol} (70g/l) indicates the strain is tolerant to ethanol at 70 g/l.

Toxins and toxin-like products (including enzymes). BCN, bacteriocin; BOT, boticin; CPE, *C. perfringens* enterotoxin; Hgn, hemagglutinin; PLC, phospholipase C; SLD, sialidase (N-acetylneuraminidase); The, thetahemolysin; Tox, toxin.

Sporulation and germination. Spo⁺, spore-forming; Spo⁻, asporogenous; Spo$_0$⁻, Spo$_{II}$⁻···Spo$_{ab}$⁻, sporulation process is blocked at stage 0, II, . . . or is abnormal.

Osp, oligosporogenous (sporulation rate <10⁻²); Lz⁺, lysozyme is not required for germination; Lzd, lysozyme-dependent for germination; DPA, dipicolinic acid.

References

Abraham, L.J., A.L. Wales, and J.I. Rood. 1985. World-wide distribution of the conjugative *Clostridium perfringens* tetracycline resistance plasmid, pCW3. *Plasmid* **14**:37–46.

Abratt, V.R., D.T. Jones, and D.R. Woods. 1985. Isolation and physiological characterization of mitomycin C-sensitive/UV-sensitive mutants in *Bacteroides fragilis. J. Gen. Microbiol.* **131**:2479–2483.

Abratt, V.R., G.L. Lindsay, and D.R. Woods. 1986. Pyrimidine dimer excision repair of DNA in *Bacteroides fragilis* wild-type and mitomycin C-sensitive/UV-sensitive mutants. *J. Gen. Microbiol.* **132**:2577–2581.

Adelberg, E.A., M. Mandel, and G.C.C. Chen. 1965. Optimal conditions for mutagenesis by N-methyl-N'-nitro-N-nitrosoguanidine in *Escherichia coli* K12. *Biochem. Biophys. Res. Commun.* **18**:788–795.

Adler, H.I. and W. Crow. 1987. A technique for predicting the solvent-producing ability of *Clostridium acetobutylicum. Appl. Environ. Microbiol.* **53**:2496–2499.

Ahn, B.K. and S. Hayashida. 1990. Metabolism mechanism of ethanol-isovaleric acid fermentation by a *Clostridium saccharoperbutylacetonicum* UV-mutant. *Agric. Biol. Chem.* **54**:353–357.

Algie, J.E. 1980. The heat resistance of bacterial spores due to their partial dehydration by reverse osmosis. *Curr. Microbiol.* **3**:287–290.

Allock, E.R., S.J. Reid, D.T. Jones, and D.R. Woods. 1981. Autolytic activity and an autolysis-deficient mutant of *Clostridium acetobutylicum. Appl. Environ. Microbiol.* **42**:929–935.

Asther, M. and A.W. Kahn. 1985. Improved fermentation of cellobiose, glucose and xylose to ethanol by *Zymomonas anaerobia* and a high ethanol tolerant strain of *Clostridium saccharolyticum. Appl. Microbiol. Biotechnol.* **21**:234–237.

Awang, G.M., G.A. Jones, and W.M. Ingledew. 1988. The acetone-butanol ethanol fermentation. *Crit. Rev. Microbiol.* **15** (Suppl. 1):S33–S67.

Barbeau, J.Y., R. Marchal, and J.-P. Vandecasteele. 1988. Conditions promoting stability of solventogenesis or culture degeneration in continuous fermentations of *Clostridium acetobutylicum. Appl. Microbiol. Biotechnol.* **29**:447–454.

Baskerville, E.N. and R. Twarog. 1972. Regulation of the tryptophan synthetic enzymes in *Clostridium butyricum. J. Bacteriol.* **112**:304–314.

Bayen, H., C. Frehel, A. Ryter, and M. Sebald. 1967. Etude cytologique de la sporulation chez Clostridium histolyticum. Ann. Inst. Pasteur (Paris) 113:163–173.

Benzer, S. 1961. Genetic fine structure. Harvey Lecture 56:1–21.

Bertram, J. and P. Dürre. 1989. Conjugal transfer and expression of streptococcal transposons in Clostridium acetobutylicum. Arch. Microbiol. 151:551–557.

Blaschek, H.P. and M. Solberg. 1981. Isolation of a plasmid responsible for caseinase activity in Clostridium perfringens ATCC3626B. J. Bacteriol. 147:262–266.

Bowring, S.N. and J.G. Morris. 1985. Mutagenesis of Clostridium acetobutylicum. J. Appl. Bacteriol. 58:577–584.

Brash, D.E., S. Seetharam, K.H. Kraemer, M.M. Seidman, and A. Bredberg. 1987. Photoproduct frequency is not the major determinant of the UV base substitution hot spots or cold spots in human cells. Proc. Natl. Acad. Sci. USA 84:3784–3786.

Brefort, G., M. Magot, H. Ionesco, and M. Sebald. 1977. Characterization and transferability of Clostridium perfringens plasmids. Plasmid 1:52–66.

Britz, M.L. and R.G. Wilkinson. 1979. Isolation and properties of metronidazole-resistant mutants of Bacteroides fragilis. Antimicrob. Agents Chemother. 16:19–27.

Carrasco, A. 1989. Photoreactivating capacity in Clostridium butyricum and Clostridium acetobutylicum. Lett. Appl. Microbiol. 8:131–134.

Carrasco, A. and C. Soto. 1987. Mutagenesis of Clostridium butyricum. J. Appl. Bacteriol. 63:539–543.

Cassier, M. and A. Ryter. 1971. Sur un mutant de Clostridium perfringens donnant des spores sans tuniques à germination lysozyme-dépendante. Ann. Inst. Pasteur (Paris) 121:717–732.

Clark, S.W., G.N. Bennett, and F.B. Rudolph. 1989. Isolation and characterization of mutants of Clostridium acetobutylicum ATCC824 deficient in acetoacetyl-coenzyme A:acetate/butyrate: coenzyme A-transferase (EC 2.8.3.9) and in other solvent pathways enzymes. Appl. Environ. Microbiol. 55:970–976.

Clarke, D.J., D.B. Kell, C.D. Morley, and J.G. Morris. 1982. Butyricin 7423 and the membrane H+-ATPase of Clostridium pasteurianum. Arch. Microbiol. 131:81–86.

Cueto, P.H. and B.S. Mendez. 1990. Direct selection of Clostridium acetobutylicum fermentation mutants by a proton suicide method. Appl. Environ. Microbiol. 56:578–580.

Daldal, F. 1985. Isolation of mutants of Clostridium pasteurianum. Arch. Microbiol. 142:93–96.

Dreyfus-Fourcade, M., M. Sebald, and M. Zavadova. 1972. Restriction et modification du

phage r par Clostridium perfringens NCTC 8798 et ses mutants. Ann. Inst. Pasteur (Paris) 122:1117–1127.

Duncan, C.L., D.H. Strong, and M. Sebald. 1972. Sporulation and enterotoxin production by mutants of Clostridium perfringens. J. Bacteriol. 110:378–391.

Dürre, P., A. Kuhn, and G. Gottschalk. 1986. Treatment with allyl alcohol selects specifically for mutants of Clostridium acetobutylicum defective in butanol synthesis. FEMS Microbiol. Lett. 36:77–81.

Eklund, M.W., F.T. Poysky, J.A. Meyers, and G.A. Pelroy. 1974. Interspecies conversion of Clostridium botulinum type C to Clostridium novyi type A by bacteriophage. Science 186:456–458.

Eklund, M.W., F.T. Poysky, L.M. Mseitif, and M.S. Strom. 1988. Evidence for plasmid-mediated toxin and bacteriocin production in Clostridium botulinum type G. Appl. Environ. Microbiol. 54:1405–1408.

Eklund, M.W., F.T. Poysky, and S.M. Reed. 1972. Bacteriophage and the toxigenicity of Clostridium botulinum type D. Nature New Biol. 235:16–17.

Eklund, M.W., F.T. Poysky, S.M. Reed, and C.A. Smith. 1971. Bacteriophage and the toxigenicity of Clostridium botulinum type C. Science 172:480–482.

El Kanouni, A., A.-M. Junelles, R. Janati-Idrissi, H. Petitdemange, and R. Gay. 1989. Clostridium acetobutylicum mutants isolated for resistance to the pyruvate halogens analogs. Curr. Microbiol. 18:139–144.

Emeruwa, A.C. and R.Z. Hawirko. 1972. Comparative studies of an asporogenic mutant and a wild type strain of Clostridium botulinum type E. Can. J. Microbiol. 18:29–34.

Emeruwa, A.C. and R.Z. Hawirko. 1975. Effect of cyclic AMP on catabolite repressed bacterial sporogenesis of an anaerobe. Arch. Microbiol. 105:67–71.

Emeruwa, A.C., R. Z. Hawirko, H. Halvorson, and I. Suzuki. 1974. Comparison of butyric type of fermentation in sporogenic and asporogenic mutants of Clostridium botulinum. J. Bacteriol. 120:74–80.

Erickson, L.E. and D.Y.-C. Fung. 1988. Handbook on Anaerobic Fermentations. New York: Dekker.

Finn, C.W., R.P. Silver, W.H. Habig, M.C. Hardegree, G. Zon, and C.F. Garon. 1984. The structural gene for tetanus neurotoxin is on a plasmid. Science 224:881–884.

Flint, H.J., J. Bisset, and J. Webb. 1989. Use of antibiotic resistance mutations to track strains of obligately anaerobic bacteria introduced into the rumen of sheep. J. Appl. Bacteriol. 67:177–183.

Flint, H.J., A.M. Thomson, and J. Bisset. 1988. Plasmid-associated transfer of tetracycline resistance in *Bacteroides ruminicola*. *Appl. Environ. Microbiol.* **54**:855–860.

Gapes, J.R., V.F. Larsen, and I.S. Maddox. 1983. A note on procedures for inoculum development for the production of solvents by a strain of *Clostridium butylicum*. *J. Appl. Bacteriol.* **55**:363–365.

George, H.A. and J.-S. Chen. 1983. Acidic conditions are not obligatory for onset of butanol formation by *Clostridium beijerinckii* (synonym, *C. butylicum*). *Appl. Environ. Microbiol.* **46**:321–327.

Gomez, R.F., B. Snedecor, and B. Mendez. 1980. Development of genetic principles in *Clostridium thermocellum*. *In: Developments in Industrial Microbiology 22*, L.A. Underkofler, and M.L. Wulf, eds., pp. 82–95. Arlington, Virginia: Society for Industrial Microbiology.

Goodman, H.J.K., E. Strydom, and D.R. Woods. 1985. Heat shock stress in *Bacteroides fragilis*. *Arch. Microbiol.* **142**:362–364.

Goodman, H.J.K., J.R. Parker, J.A. Southern, and D.R. Woods. 1987. Cloning and expression in *Escherichia coli* of a recA-like gene from *Bacteroides fragilis*. *Gene* **58**:265–271.

Guiney, D.G. and K. Bouic. 1990. Detection of conjugal transfer systems in oral, black-pigmented *Bacteroides* spp. *J. Bacteriol.* **172**:495–497.

Guiney, D.G., P. Hasegawa, K. Bouic, and B. Matthews. 1989. Genetic transfer systems in *Bacteroides*: cloning and mapping of the transferable tetracycline-resistance locus. *Mol. Microbiol.* **3**:1617–1623.

Hara, T., M. Matsuda, and M. Yoneda. 1977. Isolation and some properties of nontoxigenic derivatives of a strain of *Clostridium tetani*. *Biken J.* **20**:105–115.

Hartmanis, M.G.N., H. Ahlman, and S. Gatenbeck. 1986. Stability of solvent formation in *Clostridium acetobutylicum* during repeated subculturing. *Appl. Microbiol. Biotechnol.* **23**:369–371.

Hawirko, R.Z., K.L. Chung, A.C. Emeruwa and A.J.C. Magnusson. 1973. Ultrastructure and characterization of an asporogenic mutant of *clostridium botulinum* type E. *Canadian J. Microbiol.* **19**:281–284.

Hayashida, S. and B.K. Ahn. 1990. Isolation and characteristics of an acetone-butanol negative, ethanol-isovaleric acid-producing mutant of *Clostridium saccharoperbutylacetonicum* N1-4 ATCC 13564. *Agric. Biol. Chem.* **54**:343–351.

Hayashida, S. and S. Yoshino. 1990. Degeneration of solventogenic *Clostridium* caused by a defect in NADH generation. *Agric. Biol. Chem.* **54**:427–435.

Hecht, D.W. and M.H. Malamy. 1989. Tn4399, a conjugal mobilizing transposon of *Bacteroides fragilis*. *J. Bacteriol.* **171**:3603–3608.

Heefner, D.L., C.H. Squires, R.J. Evans, B.J. Kopp, and M.J. Yarus. 1984. Transformation of *Clostridium perfringens*. *J. Bacteriol.* **159**:460–464.

Hermann, M., F. Fayolle, R. Marchal, L. Podvin, M. Sebald, and J.-P. Vandecasteele. 1985. Isolation and characterization of butanol-resistant mutants of *Clostridium acetobutylicum*. *Appl. Environ. Microbiol.* **50**:1238–1243.

Higashi, Y., M. Chazono, K. Inoue, Y. Yanagase, T. Amano, and K. Shimada. 1973. Complementation in Θ-toxinogenicity between mutants of two groups of *Clostridium perfringens*. *Biken J.* **16**:1–9.

Higashi, Y., M. Chazono, Y. Yanagase, K. Inoue, and T. Amano. 1974. Complementation of Θ-toxin production between mutants of two groups of *Clostridium perfringens*. *Jpn. J. Med. Sci. Biol.* **27**:103–105.

Holt, S.C., J. Ebersole, J. Felton, M. Brunsvold, and K.S. Kornman. 1988. Implantation of *Bacteroides gingivalis* in nonhuman primates initiates progression of periodontitis. *Science* **239**:55–57.

Hyun, H.H. and J.G. Zeikus. 1985. Regulation and genetic enhancement of gluco-amylase and pullulanase production of *Clostridium thermohydrosulfuricum*. *J. Bacteriol.* **164**:1146–1152.

Inoue, K. and H. Iida. 1970. Conversion of toxigenicity in *Clostridium botulinum* type C. *Jpn. J. Microbiol.* **14**:87–89.

Ionesco, H., G. Bieth, C. Dauguet, and D. Bouanchaud. 1976. Isolement et identification de deux plasmides d'une souche bactériocinogène de *Clostridium perfringens*. *Ann. Microbiol. Inst. Pasteur* (Paris) **127B**:283–294.

Janati-Idrissi, R., A.-M. Junelles, A. El Kanouni, H. Petitdemange, and R. Gay. 1987. Selection de mutants de *Clostridium acetobutylicum* défectifs dans la production d'acétone. *Ann. Inst. Pasteur (Microbiol.)* **138**:313–323.

Jones, D.T. and D.R. Woods. 1981. Effect of oxygen on liquid holding recovery of *Bacteroides fragilis*. *J. Bacteriol.* **145**:1–7.

Jones, D.T. and D.R. Woods. 1986. Acetone-butanol fermentation revisited. *Microbiol. Rev.* **50**:484–524.

Jones, D.T., W.A. Jones, and D.R. Woods. 1985. Production of recombinants after protoplast fusion in *Clostridium acetobutylicum* P262. *J. Gen. Microbiol.* **131**:1213–1216.

Jones, D.T., F.T. Robb, and D.R. Woods. 1980a. Effect of oxygen on *Bacteroides fragilis* survival after far-ultraviolet irradiation. *J. Bacteriol.* **144**:1179–1181.

Jones, D.T., A. Van der Westhuizen, S. Long, E.R. Allock, S.J. Reid, and D.R. Woods. 1982. Solvent production and morphological changes in *Clostridium acetobutylicum*. *Appl. Environ. Microbiol.* **43**:1434–1439.

Junelles, A.-M., R. Janati-Idrissi, A. El Kanouni, H. Petitdemange, and R. Gay. 1987. Acetone-butanol fermentation by mutants selected for resistance to acetate and butyrate halogen analogues. *Biotechnol. Lett.* **9**:175–178.

Keller, F.A., J.S. Ganoung, and S.J. Luenser. 1985. Mutant strain of *Clostridium thermoaceticum* useful for the preparation of acetic acid. *U.S. Patent* 4, 513, 084.

Knowlton, S., J.D. Ferchak, and J.K. Alexander. 1984. Protoplast regeneration in *Clostridium tertium*: isolation of derivatives with high-frequency regeneration. *Appl. Environ. Microbiol.* **48**:1246–1247.

Kurose, N., S. Kinoshita, J. Yagyu, M. Uchida, S. Hanai, and A. Obayashi. 1988. Improvement of ethanol production of thermophilic *Clostridium* sp. by mutation. *J. Ferment. Technol.* **66**:467–472.

Kutzenok, A. and M. Aschner. 1952. Degenerative processes in a strain of *Clostridium butylicum. J. Bacteriol.* **64**:829–836.

Laird, W.J., W. Aaronson, R.P. Silver, W.H. Habig, and M.C. Hardegree. 1980. Plasmid-associated toxigenicity in *Clostridium tetani. J. Infect. Dis.* **142**:623.

Lapointe, J.-R. and V. Fredette. 1975. Mutant atténué nitrosoguanidine-induit dans la gangrène gazeuse à *Clostridium perfringens* type A. *Can. J. Microbiol.* **21**:1259–1269.

Largier, S.T., S. Long, J.D. Santangelo, D.T. Jones, and D.R. Woods. 1985. Immobilized *Clostridium acetobutylicum* P262 mutants for solvent production. *Appl. Environ. Microbiol.* **50**:477–481.

Lemmel, S.A. 1985. Mutagenesis in *Clostridium acetobutylicum. Biotechnol. Lett.* **7**:711–716.

Lengeler, J. 1979. Streptozotocin, an antibiotic superior to penicillin in the selection of rare bacterial mutations. *FEMS Microbiol. Lett.* **5**:417–419.

Lin, Y.-L. and H.P. Blaschek. 1983. Butanol production by a butanol-tolerant strain of *Clostridium acetobutylicum* in extruded corn broth. *Appl. Environ. Microbiol.* **45**:966–973.

Long, S., D.T. Jones, and D.R. Woods. 1984. The relationship between sporulation and solvent production in *Clostridium acetobutylicum* P262. *Biotechnol. Lett.* **6**:529–534.

Lovitt, R.W., R. Longin, and J.G. Zeikus. 1984. Ethanol production by thermophilic bacteria: physiological comparison of solvent effects on parent and alcohol-tolerant strains of *Clostridium thermohydrosulfuricum. Appl. Environ. Microbiol.* **48**:171–177.

Mackey, B.M. and J.G. Morris. 1974. Isolation of a mutant strain of *Clostridium pasteurianum* defective in granulose degradation. *FEBS Lett.* **48**:64–67.

Macrina, F.L., T.D. Mays, C.J. Smith, and R.A. Welch. 1981. Non-plasmid associated transfer of antibiotic resistance in *Bacteroides. J. Antimicrob. Chemother.* **8** (Suppl. D):77–86.

Magot, M., F. Fayolle, G. Privitera, and M. Sebald. 1981. Transposon-like structures in the *Bacteroides fragilis* plasmid pIP410. *Mol. Gen. Genet.* **181**:559–561.

Makoff, A.J., M.D. Oxer, M.A. Romanos, N.F. Fairweather, and S. Ballantine. 1989. Expression of tetanus toxin fragment C in *E. coli*: high level expression by removing rare codons. *Nucleic Acids Res.* **17**:10191–10202.

Marsh, P.D., A.S. McKee, A.S. McDermid, and A.B. Dowsett. 1989. Ultrastructure and enzyme activities of a virulent and an avirulent variant of *Bacteroides gingivalis* W50. *FEMS Microbiol. Lett.* **59**:181–186.

McKee, A.S., A.S. McDermid, R. Wait, A. Baskerville, and P.D. Marsh. 1988. Isolation of colonial variants of *Bacteroides gingivalis* W50 with a reduced virulence. *J. Med. Microbiol.* **27**:59–64.

Meinecke, B., H. Bahl, and G. Gottschalk. 1984. Selection of an asporogenous strain of *Clostridium acetobutylicum* in continuous culture under phosphate limitation. *Appl. Environ. Microbiol.* **48**:1064–1065.

Mendez, B.S. and R.F. Gomez. 1982. Isolation of *Clostridium thermocellum* auxotrophs. *Appl. Environ. Microbiol.* **43**:495–496.

Mihelc, V., C. Duncan, and G. Chambliss. 1978. Characterization of a bacteriocinogenic plasmid in *C. perfringens* CW55. *Antimicrob. Agents Chemother.* **14**:771–779.

Mori, Y. 1990. Isolation of mutants of *Clostridium thermocellum* with enhanced cellulase production. *Agric. Biol. Chem.* **54**:825–826.

Morris, J.G. and R.L. Robson. 1973. Granulose deposition and sporulation in *Clostridium pasteurianum. In: Regulation de la Sporulation Microbienne*, pp. 87–89. Colloques Internationaux du CNRS n° 227, CNRS, Paris.

Murray, W.D., K.B. Wemyss and A.W. Khan. 1983. Increased ethanol production and tolerance by a pyruvate-negative mutant of *Clostridium saccharolyticum. Eur. J. Appl. Microbiol. Biotechnol.* **18**:71–74.

Nochur, S.V., M.F. Roberts, and A.L. Demain. 1990. Mutation of *Clostridium thermocellum* in the presence of certain carbon sources. *FEMS Microbiol. Lett.* **71**:199–204.

Oultram, J.D. and M. Young. 1985. Conjugal transfer of plasmid pAMβ1 from *Streptococcus lactis* and *Bacillus subtilis* to *Clostridium acetobutylicum. FEMS Microbiol. Lett.* **27**:129–134.

Podvin, L. 1989. Activité Autolytique de *Clostridium acetobutylicum* Souche N1-4. Thèse doctorat, Université Paris.

Privitera, G., A. Dublanchet, and M. Sebald. 1979. Transfer of multiple antibiotic resistance between subspecies of *Bacteroides fragilis. J. Infect. Dis.* **139**:97–101.

Privitera, G., F. Fayolle and M. Sebald. 1981. Resistance to tetracycline, erythromycin and clindamycin in the *Bacteroides fragilis* group; inducible versus constitutive tetracycline resistance. *Antimicrob. Agents Chemother.* 20:314–320.

Rambler, M.B. and L. Margulis. 1980. Bacterial resistance to ultraviolet irradiation under anaerobiosis: implications for pre-Phanerozoic evolution. *Science* 210:638–640.

Reed, W.M. 1989. Mutation and genetic engineering of anaerobic bacteria. *In: Handbook on Anaerobic Fermentations,* L.E. Erickson, and D.Y.-C. Fung, eds., pp. 27–58. New York: Dekker.

Reysset, G. and M. Sebald. 1985. Conjugal transfer of plasmid mediated antibiotic resistance from streptococci to *Clostridium acetobutylicum*. *Ann. Ins. Pasteur (Microbiol.)* 136B:275–282.

Reysset, G., J. Hubert, L. Podvin, and M. Sebald. 1987. Protoplast formation and regeneration of *Clostridium acetobutylicum* strain NI-4080. *J. Gen. Microbiol.* 133:2595–2600.

Robson, R.L., R.M. Robson, and J.G. Morris. 1974. The biosynthesis of granulose by *Clostridium pasteurianum*. *Biochem. J.* 144:503–511.

Rogers, P. 1986. Genetics and biochemistry of *Clostridium* relevant to development of fermentation processes. *Adv. Appl. Microbiol.* 31:1–60.

Rogers, P. and N. Palosaari. 1987. *Clostridium acetobutylicum* mutants that produce butyraldehyde and altered quantities of solvents. *Appl. Environ. Microbiol.* 53:2761–2766.

Rogers, P., C. Quinn, and J.C. McDilda. 1986. Mutants modifying solvents produced by *Clostridium acetobutylicum. In: Abstracts,* Abstr. O–51 Washington D.C.: *Proc. Am. Soc. of Microbiology National Meet.*

Rood, J.I. and R.G. Wilkinson. 1975. Isolation and characterization of *Clostridium perfringens* mutants altered in both hemagglutinin and sialidase production. *J. Bacteriol.* 123:419–427.

Rood, J.I., V.N. Scott, and C.L. Duncan. 1978. Identification of a transferable tetracycline resistance plasmid (pCW3) from *Clostridium perfringens*. *Plasmid* 1:563–570.

Rothstein, D.M. 1986. *Clostridium thermosaccharolyticum* strain deficient in acetate production. *J. Bacteriol.* 165:319–320.

Russo, T.A., J.S. Thompson, V.G. Godoy, and M.H. Malamy. 1990. Cloning and expression of the *Bacteroides fragilis* TAL2480 neuraminidase gene, *nanH*, in *Escherichia coli. J. Bacteriol.* 172:2594–2600.

Schaeffer, P. 1969. Sporulation and the production of antibiotics, exoenzymes and exotoxins. *Bacteriol. Rev.* 33:48–71.

Schumann, J.P., D.T. Jones, and D.R. Woods. 1982. UV light induction of proteins in *Bacteroides fragilis* under anaerobic conditions. *J. Bacteriol.* 151:44–47.

Schumann, J.P., D.T. Jones, and D.R. Woods. 1983. Effect of oxygen and UV irradiation on nucleic acid and protein syntheses in *Bacteroides fragilis. J. Bacteriol.* 156:1366–1368.

Schumann, J.P., D.T. Jones, and D.R. Woods. 1984. Effect of UV irradiation on macromolecular synthesis and colony formation in *Bacteroides fragilis. J. Gen. Microbiol.* 130:771–777.

Schwartz, A.C. and T.C. Stadtman. 1970. Small colonies of *Clostridium sticklandii* resulting from nitrosoguanidine treatment and exhibiting defects in catabolic enzymes. *J. Bacteriol.* 104:1242–1245.

Sebald, M. 1968. Sur un mutant asporogène de *Clostridium histolyticum* incapable de synthétiser l'acide dipicolinique. *Ann. Inst. Pasteur* (Paris) 114:265–276.

Sebald, M. and G. Bréfort. 1975. Transfert du plasmide tetracycline-chloramphenicol chez *Clostridium perfringens. C. R. Acad. Sci.* Sér. D (Paris) 281:317–319.

Sebald, M. and M. Cassier. 1969. Sporulation and toxigenicity in mutant strains of *Clostridium perfringens. In: Spore IV,* L.L. Campbell, ed., pp. 306–316. Bethesda: American Society for Microbiology.

Sebald, M. and R.N. Costilow. 1975. Minimal growth requirements for *Clostridium perfringens* and isolation of auxotrophic mutants. *Appl. Microbiol.* 29:1–6.

Sebald, M. and P. Schaeffer. 1965. Toxinogénèse et sporulation chez *Clostridium histolyticum. C. R. Acad. Sci.* (Paris) 260:5398–5400.

Sebald, M., D. Bouanchaud, and G. Bieth. 1975. Nature plasmidique de la resistance à plusieurs antibiotiques chez *C. perfringens* type A, souche 659. *C. Acad. Sci.* Sér. D (Paris) 280:2401–2404.

Sebald, M., G. Reysset, and J. Breuil. 1990. What's new in 5-nitroimidazole resistance in the *Bacteroides fragilis* group. *In: Clinical and Molecular Aspects of Anaerobes,* S.P. Borriello, ed., pp. 217–225, Petersfield: Wrightson Biomedical.

Shimell, M.J., C.J. Smith, F.P. Tally, F.L. Macrina, and M.H. Malamy. 1982. Hybridization studies reveal homologies between pBF4 and pBFTM10, two clindamycin-erythromycin resistance transfer plasmids of *Bacteroides fragilis. J. Bacteriol.* 152:950–953.

Shoemaker, N.B., C. Getty, J.F. Gardner, and A.A. Salyers. 1986. Tn 4351 transposes in *Bacteroides* spp. and mediates the integration of plasmid R751 into the *Bacteroides* chromosome. *J. Bacteriol.* 165:929–936.

Shoemaker, N.B., E.P. Guthrie, A.A. Salyers, and J.F. Gardner. 1985. Evidence that the clindamycin-erythromycin resistance gene of *Bacteroides* plasmid pBF4 is on a transposable element. *J. Bacteriol.* 162:626–632.

Simon, M.A. and W.J. Brill. 1971. Mutant of Clostridium pasteurianum that does not fix nitrogen. J. Bacteriol. **105**:65–69.

Sindar, P., M.L. Britz, and R.G. Wilkinson. 1982. Isolation and properties of metronidazole-resistant mutants of Clostridium perfringens. J. Med. Microbiol. **15**:503–509.

Slade, H.J.K., D.T. Jones, and D.R. Woods. 1981. Effect of oxygen radicals and peroxide on survival after ultraviolet irradiation and liquid holding recovery of Bacteroides fragilis. J. Bacteriol. **147**:685–687.

Slade, H.J.K., D.T. Jones, and D.R. Woods. 1982. Effect of low fluencies of near-ultraviolet radiation on Bacteroides fragilis survival. FEMS Microbiol. Lett. **15**:257–259.

Slade, H.J.K., D.T. Jones, and D.R. Woods. 1984. Effect of oxygen and peroxide on Bacteroides fragilis cell and phage survival after treatment with DNA damaging agents. FEMS Microbiol. Lett. **24**:159–163.

Smith, C.J. 1987. Nucleotide sequence analysis of Tn4551: use of erm FS operon fusions to detect promotor activity in Bacteroides fragilis. J. Bacteriol. **169**:4589–4596.

Smith, CJ. and H. Spiegel. 1987. Transposition of Tn 4551 in Bacteroides fragilis: identification and properties of a new transposon from Bacteroides spp. J. Bacteriol. **169**:3450–3457.

Smith, K.C. 1978. Multiple pathways of DNA repair in bacteria and their roles in mutagenesis. Photochemistry and Photobiology. **28**:121–129.

Soucaille, P., G. Joliff, A. Izard, and G. Goma. 1987. Butanol tolerance and autobacteriocin production of Clostridium acetobutylicum. Curr. Microbiol. **14**:295–299.

Takumi, K., T. Kinouchi, and T. Kawata. 1980. Isolation of nontoxigenic variants associated with enhanced sporulation and alteration in the cell wall from Clostridium botulinum type A 190L by treatment with detergents. Microbiol. Immunol. **24**:469–477.

Tally, F.P., D.R. Snydman, M.J. Shimell, and M.H. Malamy. 1982. Characterization of pBFTM10, a clindamycin-erythromycin resistance transfer factor from Bacteroides fragilis. J. Bacteriol. **151**:686–691.

Tessman, I., R.K. Poddar, and S. Kumar. 1964. Identification of the altered bases in mutated single-stranded DNA. I. In vitro mutagenesis by hydroxylamine, ethylmethane sulfonate and nitrous acid. J. Mol. Biol. **9**:352–363.

Ueda, M., S. Nakamoto, R. Nakai, and A. Takagi. 1983. Establishment of a defined minimal medium and isolation of auxotrophic mutants for Bifidobacterium bifidum ES5. J. Gen. Appl. Microbiol. **29**:103–114.

Van Tassell, R.L. and T.D. Wilkins. 1978. Isolation of auxotrophs of Bacteroides fragilis. Can. J. Microbiol. **24**:1619–1621.

Weizmann, C. and B. Rosenfeld. 1937. The activation of the butanol-acetone fermentation of carbohydrates by Clostridium acetobutylicum (Weizmann). Biochem. J. **31**:619–639.

Welch, R.A. and F.L. Macrina. 1981. Physical characterization of Bacteroides R plasmid pBF4. J. Bacteriol. **145**:867–872.

Winkelman, J.W. and D.P. Clark. 1984. Proton suicide: general method for direct selection of sugar transport- and fermentation-defective mutants. J. Bacteriol. **160**:687–690.

Woods, D.R. and D.T. Jones. 1986. Physiological responses of Bacteroides fragilis and Clostridium strains to environmental stress factors. Adv. Microb. Physiol. **28**:1–64.

Wyatt, L.R. and W.M. Waites. 1971. Studies with spores of Clostridium bifermentans: comparison of germination mutants. In: Spore Research 1971, A.N. Barker, G.W. Gould, and J. Wolf, eds., pp. 123–131. London: Academic Press.

Zoutberg, G.R., R. Willemsberg, G. Smit, M.J. Teixeira de Mattos, and O.M. Neijssel. 1989a. Aggregate formation by Clostridium butyricum. Appl. Microbiol. Biotechnol. **32**:17–21.

Zoutberg, G.R., R. Willemsberg, G. Smit, M.J. Teixeira de Mattos, and O.M. Neijssel. 1 989b. Solvent production by an aggregate-forming variant of Clostridium butyricum. Appl. Microbiol. Biotechnol. **32**:22–26.

6

Conjugative Gene Transfer in Clostridia

Michael Young

6.1 Introduction

The conjugative transfer of plasmids between *Clostridium perfringens* strains was initially described by Sebald and Bréfort in 1975. This was the first demonstration of gene transfer in clostridia; it was also one of the earliest reports of conjugative gene transfer between gram-positive bacteria. During the past decade, several conjugative elements, both plasmids and transposons, have been identified in the predominantly pathogenic, amino acid-fermenting clostridia. As the following discussion demonstrates, these indigenous clostridial conjugative elements remain poorly characterized.

Today it is well known that conjugative gene transfer between gram-positive bacteria is widespread, particularly among enterococci and streptococci. A considerable body of knowledge has now accumulated concerning the biology of conjugative plasmids and transposons in this group of organisms (see review by Clewell, 1990), but much still remains to be learned, and these genetic elements are the subject of intensive current research. They have already become firmly established as powerful tools for genetic analysis in other gram-positive organisms, especially bacilli. Their essential properties are summarized in this chapter as are the ways in which they are beginning to be exploited for genetic analysis in clostridia.

6.2 Indigenous Conjugative Plasmids

Large conjugative plasmids conferring resistance to tetracycline (Tc^r) are widely distributed among different strains of *C. perfringens* (Rood, 1983; Abraham et al., 1985). Many of them are closely related. Hybridization analysis of 15 independently isolated examples indicated that they shared in common a substantial (at least 29 kb) DNA segment (Abraham and Rood, 1985a). Plasmid pIP401 (54 kb), one of those originally identified by the group at the Institut Pasteur (Sebald and Bréfort 1975; Sebald et al., 1975), also confers resistance to chloramphenicol (Cm^r). It has been transferred by conjugation to *Clostridium difficile* (Ionesco, 1980). Several plasmids belonging to this family have been characterized physically (Rood et al., 1978; Magot, 1984; Abraham and Rood, 1985b; Abraham et al., 1985), but functional analysis has not yet been undertaken. The recent development of convenient methods for electrotransformation of *C. perfringens* strains (Allen and Blaschek, 1988, 1990; Scott and Rood, 1989) will no doubt prove an invaluable aid in the functional analysis of indigenous conjugative clostridial plasmids.

During conjugative transfer, plasmid pIP401 frequently undergoes a 6-kb reduction in size, associated with loss of the Cm^r determinant (Bréfort et al., 1977, 1978). The deleted form of pIP401 is indistinguishable from the 47-kb plasmid pCW3 (Abraham et al., 1985). The Cm^r determinant of pIP401 resides on a 6.2-kb transposon denoted Tn4451 (Abraham and Rood, 1987), and a very similar transposon, Tn4452, is found on the related plasmid pJIR27 (Abraham and Rood, 1987). Tn4451 apparently has the potentially useful property of undergoing perfect excision in *Escherichia coli* (Abraham and Rood, 1988), which would be an invaluable aid to the sequencing of cloned, insertionally inactivated genes. However, a note of caution might be appropriate here. It was originally thought that streptococcal conjugative transposons behaved similarly, but careful analysis has indicated that transposon excision is often accompanied by the occurrence of base substitution or frameshift mutations in the target DNA (see Section 6.5, this chapter).

Pan-Hou et al. (1980) described another indigenous conjugative plasmid in clostridia; of unreported size and cured by acridine orange, it was implicated in the decomposition of methylmercury by *Clostridium cochlearium* strain T-2.

6.3 Conjugative Enterococcal Plasmids in Clostridia

A large number of conjugative plasmids have been described in enterococci (Clewell, 1981, 1990; Clewell et al., 1985; Horaud et al., 1985). They are placed in two groups, depending on whether transfer occurs in liquid medium or only on solid surfaces. Plasmids in the former group, of which pAD1 is the best-characterized example, are transferred at very high frequencies in response to specific peptides that act as sex pheromones produced by recipient bacteria having no associated plasmids (Clewell and Weaver, 1989; Clewell, 1990). To date, plasmids that are transferred in response to sex pheromones have only been found in enterococci and, as far as is known, their host range is rather limited. Molecular details of the conjugation mechanism are currently being elucidated (see review by Clewell and Weaver, 1989).

Plasmids in the latter group, in which trans-

fer occurs only on solid surfaces, are transferred by a process in which sex pheromones do not appear to be implicated. Many of them confer resistance to the macrolide/lincosamide/streptogramin B group of antibiotics (MLS^r). Two of these, pAMβ1 (26 kb) and pIP501 (35 kb), have been partially characterized (Behnke and Gilmore, 1981; Evans and Macrina, 1983; LeBlanc and Lee, 1984; Krah and Macrina, 1989; Brantl et al., 1990), but surprisingly little is known about the mechanism by which conjugative transfer occurs. These plasmids have an extraordinarily wide host range and are transferable to most gram-positive species that have been tested (Clewell, 1981; Schaberg et al., 1982; Horaud et al., 1985; Grigorova et al., 1988; Taketomo et al., 1989). Plasmid pAMβ1 can also be conjugatively transferred from *Enterococcus faecalis* to *Escherichia coli*, but it is unable to replicate in the latter (Trieu-Cuot et al., 1988).

Plasmids pAMβ1, pJH4, and pIP501 (as well as its derivative, pVA797) have been introduced into *Clostridium acetobutylicum* from a range of gram-positive donors (Oultram and Young, 1985; Reysset and Sebald, 1985; Yu and Pearce, 1986). The transfer frequencies observed with plasmid pAMβ1 varied with the donor organism, the highest frequencies (10^{-5} to 10^{-3} per recipient) being obtained when *Lactococcus lactis* was employed (Oultram and Young, 1985). Transfer was particularly inefficient (10^{-8} to 10^{-6} per recipient) when *Bacillus subtilis* was used as donor, which correlates with the occurrence of a major specific deletion of a DNA segment including *tra* genes from plasmid pAMβ1 in this host (van der Lelie and Venema, 1987). Structural instability of pAMβ1 had previously been reported in *Streptococcus sanguis* (Macrina et al., 1980). Plasmid pAMβ1 can also he transferred from one strain of *C. acetobutylicum* to another at a frequency of 10^{-6} to 10^{-5} per recipient (Oultram and Young, 1985; Reysset and Sebald, 1985) and as far as is known, the plasmid is structurally stable in this organism. Transfer of pAMβ1 to *Clostridium pasteurianum* and *Clostridium butyricum* has also been reported (Young et al., 1987).

6.4 Mobilization of Nonconjugative Plasmids to Clostridia

Several investigators have observed the mobilization of small nonconjugative plasmids by pAMβ1 or pIP501 derivatives to other gram-positive organisms (Schaberg et al., 1982; Lereclus et al., 1983; LeBlanc and Lee, 1984; Smith and Clewell, 1984; Smith, 1985). In one case (Smith and Clewell, 1984), mobilization occurred via the formation of plasmid cointegrates, whereas in another (Smith, 1985), cointegrate molecules were apparently not present in the donor strain, suggesting that mobilization may have occurred via a different mechanism (cf. Section 6.8). Yu and Pearce (1986) have suggested that the plasmid mobilized in this latter case (pAM610) is also conjugatively mobilized from *Lactococcus lactis* or *Streptococcus sanguis* donors containing pVA797 (a derivative of pIP501) to *Clostridium acetobutylicum*.

In spite of its observed structural instability in *B. subtilis*, pAMβ1 has been employed to mobilize small nonreplicative plasmids from this organism into *C. acetobutylicum* (Oultram et al., 1987). Mobilization was based on cointegrate formation, which was facilitated by incorporating a homologous MLS^r determinant into the small nonreplicative plasmid. Although somewhat cumbersome, this approach has been used successfully to complement a Leu$^-$ strain of *C. acetobutylicum* with a cloned *leu* gene cluster from *C. pasteurianum* (Oultram et al., 1988) and to introduce a *Pseudomonas xylE* gene into *C. acetobutylicum* (see Chapter 8, this volume). Cointegrate mobilization may also prove useful for transferring vectors containing the replication regions of many clostridial plasmids. Most of them, except those of pCB101 (Collins et al., 1985; Luczak et al., 1985) and pIP404 (Garnier and Cole, 1988b), do not apparently function in *B. subtilis* (reviewed by Young et al., 1989; see also Chapter 8, this volume), which would facilitate the construction of cointegrate molecules with pAMβ1.

Finally, a small bacteriocinogenic plasmid,

pIP404, found in certain strains of *C. perfringens* (Ionesco and Bouanchaud, 1973; Ionesco et al., 1974, 1976), can be mobilized by indigenous coresident conjugative plasmids. Nothing is currently known about the mechanism of conjugative mobilization of pIP404. The entire sequence of this plasmid has been determined (Garnier and Cole, 1988a) and functions attributed to the products of 6 of the 10 open reading frames (ORFs) it encodes (Garnier and Cole, 1986, 1988a, 1988b; Garnier et al., 1987). This mobilizable plasmid is the best-characterized of all clostridial plasmids; further information concerning pIP404 is presented in Chapters 8 and 17, this volume.

6.5 Conjugative Enterococcal and Streptococcal Transposons

Conjugative transposons are a class of genetic elements that can move from a site in one cell to a site in another cell by a process requiring close cell-to-cell contact (Clewell and Gawron-Burke, 1986). Over the years, it has become apparent that these elements, found predominantly (although not exclusively; see Hecht and Malamy, 1989) in gram-positive bacteria, show a remarkably wide host range. The best-characterized examples are Tn916 and Tn1545, both of which have been transferred to clostridia (see Section 6.6, this chapter). Their biology is briefly considered next. More detailed information may be obtained from several recent reviews (Clewell et al., 1985; Clewell and Gawron-Burke, 1986; Clewell, 1990; Trieu-Cuot et al., 1990).

Tn916 was discovered more than 10 years ago in the chromosome of *Enterococcus faecalis* strain DS16. This 16-kb genetic element confers Tc^r on its host and is transferable to other strains in the absence of detectable plasmid DNA (Franke and Clewell, 1981; Gawron-Burke and Clewell, 1982). Similar elements have been found in other enterococcal strains as well as in various species of *Streptococcus* (see review by Clewell, 1990). Generally speaking, they are located in the bacterial chromosome (but see Christie et al., 1987), and they confer

Tc^r (*tetM*) on their host. Functional analysis and sequencing of Tn916 and Tn1545 (a 25-kb element from *Streptococcus pneumoniae*; Courvalin and Carlier, 1986, 1987; Caillaud et al., 1987) is currently in progress.

When cloned on plasmids in *E. coli*, both Tn916 and Tn1545 retain the ability to transpose but are apparently unable to undergo conjugative transfer (Gawron-Burke and Clewell, 1984; Courvalin and Carlier, 1986, 1987). However, very recent evidence (Bertram et al., 1991) suggests that a plasmid-borne copy of Tn916 can in fact be transferred by conjugation from *E. coli* to a range of gram-positive bacteria. In *E. coli*, single-copy chromosomal insertions of Tn916 and Tn1545 are quite stably inherited, but when present on multicopy plasmids both transposons are lost spontaneously at high frequency. Loss by excision from an insertionally inactivated gene may restore gene function (Gawron-Burke and Clewell, 1982, 1984; Nida and Cleary, 1983; Courvalin and Carlier, 1986; Gaillard et al., 1986; Kathariou et al., 1987), suggesting that conjugative transposons might be powerful tools for cloning genes of interest inactivated by transposon insertion (Gawron-Burke and Clewell, 1984). It is now clear that transposon excision frequently generates base substitution or frameshift mutations in the target site (see following). Nevertheless, conjugative transposons are currently proving very useful for mutational cloning of clostridial genes in several laboratories (see Section 6.6).

Insertion mutagenesis of Tn916 in *E. coli* using Tn5 (Jones et al., 1987; Yamamoto et al., 1987) allowed identification of those regions necessary for excision in *E. coli*. Those required for transposition or conjugation were identified after reintroduction of mutated derivatives of Tn916 into *E. faecalis* via protoplast transformation (Senghas et al., 1988). Excision-defective mutants could be complemented in *trans*, indicating that a diffusible product was involved (Senghas et al., 1988). Excision-defective mutants were also unable to transpose after protoplast transformation, lending credence to the hypothesis (see Gawron-Burke and Clewell, 1982, 1984) that transposon excision, to generate a circular, nonreplicative intermedi-

ate, is a necessary first step in transposition. Further evidence to support this hypothesis has been provided by the detection of a supercoiled circular Tn916 intermediate in *E. coli* that can be isolated and used to transform *Bacillus subtilis* protoplasts, whereupon it becomes established in the bacterial chromosome (Scott et al., 1988).

Nucleotide sequencing has revealed the presence of genes near one end of Tn1545, the products of which are structurally homologous with those of the *int* and *xis* genes of lambdoid phages (Poyart-Salmeron et al., 1989). Moreover, functional homology between these gene products and the integrase and excisionase of lambdoid phages has recently been demonstrated using in vivo complementation assays for integration and excision of a mini-Tn1545 derivative (Poyart-Salmeron et al., 1989, 1990).

The nucleotide sequences of the extremities of Tn1545 and Tn916 are very similar for at least 250-bp (Caillaud and Courvalin, 1987; Clewell et al., 1988). These transposons have inverted, imperfectly repeated, 26-bp sequences at their termini with variable sequences at their extremities, and they do not generate a duplication of the target sequence on insertion. Novel mechanisms have been proposed to explain transposon integration and excision (Caparon and Scott, 1989; Poyart-Salmeron et al. 1989, 1990, Trieu-Cuot et al., 1990). The processes are conceived as being superficially similar to those employed by lambdoid phages, except that the putative "*att*" site has a variable centrally located "overlap" or "coupling" region. (The *att* site is generated when the transposon ends join to form the circular transposition intermediate; see earlier discussion.)

Both Tn916 and Tn1545 preferentially insert into DNA sequences that are very d(A + T)-rich, which might make these elements particularly suitable for undertaking insertional mutagenesis in the mesophilic clostridia, the DNA of which is also very rich in d(A + T). However, nucleotide sequencing of DNA adjacent to transposon insertion points has revealed that there appears to be a preferred consensus sequence (*att* site) for transposon insertion with which individual sequences show at least 50% homology (Poyart-Salmeron et al., 1990; Trieu-Cuot et al., 1990). Transposon insertion is apparently far from random. Those hosts in which conjugative transposons appear to have strongly preferred "hot spots" for integration (see Hill et al., 1985; Woolley et al., 1989; Mullany et al., 1990, 1991) may contain a sequence extremely similar or perhaps identical to the theoretical *att* site.

Transposon excision initially generates a circular intermediate containing a 6- to 7-bp heteroduplex (Caparon and Scott, 1989) that may contain as many as three mismatches in the variable overlap or coupling region of the *att* site (Trieu-Cuot et al., 1990). If the mismatched bases are removed by the mismatch-repair system of the host, then one of two additional products may be formed, depending on which strand is corrected. A similar heteroduplex left in the target DNA would presumably be repaired by replication (or mismatch repair). This would nicely account for the fact that base substitution or frameshift mutations are often left in the target sequence after transposon excision in *E. coli*. Experimental support comes from sequencing of *att* sites derived from excised circular transposition intermediates (Caparon and Scott, 1989) as well as junction sequences present before and after transposon excision from target molecules (Caillaud and Courvalin, 1987; Clewell et al., 1988; Poyart-Salmeron et al., 1990).

The 6- to 7-bp core region of the *att* site can differ at as many as three positions between recombining molecules (Poyart-Salmeron et al., 1990), and as an element moves from one location to another, the core region of the *att* site will evolve. The different derivatives thus generated might be expected to show different transposition specificities (Poyart-Salmeron et al., 1990). When using conjugative transposons as insertional mutagens, it may therefore be advantageous to employ several different donor strains. Alternatively, a single donor harbouring multiple copies of the element may suffice.

It has recently been shown that Tn916ΔE is transferred to strains of *Streptococcus pyogenes* with equal efficiency whether or not they contain a resident copy of Tn916 (Norgren and Scott, 1991). Clearly, one copy of the transposon does not repress the establishment of another introduced by conjugation. In about 10% of the transconjugants, Tn916ΔE had integrated by homologous recombination into the resident copy of Tn916, indicating that in *S. pyogenes* (and, perhaps, in other hosts) Tn916 can be used to introduce genetic information into the bacterial chromosome via homologous recombination.

Although our knowledge of several aspects of the biology of Tn916 and Tn1545 is growing rapidly, further extensive analysis will be required to unravel the details of the conjugation mechanism whereby the excised transposition intermediate "finds its way" from one cell to another.

6.6 Exploitation of Tn916 and Tn1545 in Clostridia

Tn916 has been transferred from *Enterococcus faecalis* to *Clostridium tetani, C. acetobutylicum,* and *C. difficile* (Davies et al., 1988; Volk et al., 1988; Bertram and Dürre, 1989; Woolley et al., 1989; Mattsson and Rogers, 1989; Mullany et al., 1991). Transfer of Tn916 from *E. coli* to the DSM 792 strain of *C. acetobutylicum* has also been reported (Bertram et al., 1991). As might be expected from the broad host range of Tn916 (see earlier discussion), transfer from one clostridial strain to another has also been obtained, as has transfer back to *E. faecalis* (Volk et al., 1988). In *C. tetani* and certain strains of *C. acetobutylicum* (DSM 792 and ATCC 824), Tn916 inserted into multiple sites in the bacterial chromosome (Volk et al., 1988; Bertram and Dürre, 1989; Mattsson and Rogers, 1989), whereas in *C. difficile* and in the NCIB 8052 strain of *C. acetobutylicum* there was evidence of a hot spot for Tn916 insertion (Woolley et al., 1989; Mullany et al., 1991). These latter findings should be reexamined using alternative donor strains, because the provenance of the transposon might influence insertion specificity (see earlier). Donors containing the Em[r] derivative Tn916ΔE (Rubens and Heggen, 1988) might be useful in this respect, as the NCIB 8052 strain of *C. acetobutylicum* gives rise to spontaneous Tc^r mutants at frequencies up to 10^{-6}.

The metabolism of *C. acetobutylicum* is well known for its acetone/butanol/ethanol fermentation (Jones and Woods, 1986). Strains of *C. acetobutylicum* defective in acetone and butanol production have been recovered after Tn916 mutagenesis and partially characterized (Mattsson and Rogers, 1989; Bertram et al., 1990). The structural genes of several enzymes concerned with solvent production have already been cloned (and sequenced) using complementation of suitable *E. coli* auxotrophs or oligonucleotide probes based on amino-terminal protein sequence information (Young et al., 1989; see also Chapter 23, this volume). Some Tn916-induced mutants will presumably be defective in regulatory genes. Analysis of such mutants should contribute greatly to our understanding of the control of solvent production in this organism. As far as this author is aware, no reports have yet appeared concerning the cloning of insertionally inactivated clostridial genes in *E. coli*.

Transfer of Tn1545 to *C. acetobutylicum* strain NCIB 8052 was initially reported by Davies et al. (1988). Subsequent work indicated that Tn1545 can be transferred to this strain from *B. subtilis* as well as between strains of *C. acetobutylicum* (Woolley et al., 1989). Detailed analysis of 10 independently isolated transconjugants indicated that Tn1545 became established at different chromosomal locations in every case (Woolley et al., 1989; S. R. Wilkinson and M. Young, unpublished observations). Some of these strains contain multiple copies of Tn1545. In a small-scale screening program, insertions were detected that disrupt genes concerned with solvent production (G. Hartland et al., unpublished observations), but these strains have not been further characterized. It is not known whether Tn1545 can be used for insertional mutagenesis in the DSM 792 and ATCC 824 strains of *C. acetobutylicum*.

Because Tn1545 inserts randomly into the chromosome of the NCIB 8052 strain of *C. acetobutylicum*, it is proving to be a very useful aid to the construction of a physical map of the genome of this organism (S.R. Wilkinson and M. Young, unpublished observations).

6.7 Indigenous Conjugative Transposons

Strains of *Clostridium difficile* contain a chromosomal element encoding Tc^r (*tetM*) that is transferred conjugatively from cell to cell in the apparent absence of plasmid DNA (Ionesco, 1980; Smith et al., 1981; Wüst and Hardegger, 1983). Hybridization analysis has shown that it has substantial homology with Tn916 (Hächler et al., 1987a) although not, apparently, at its termini (Mullany et al., 1990). The presence of a Tc^r determinant of the *tetM* class (Burdett et al., 1982) appears to be a common feature of many conjugative transposons (Clewell and Gawron-Burke, 1986). There are however reports suggesting the presence of conjugative transposons encoding resistances to both clindamycin (Cl^r) and erythromycin (Em^r) in strains of *Clostridium innocuum* and *C. difficile* (Magot, 1983; Hächler et al., 1987b). It is not known to what extent they are related to each other nor whether they share homologous sequences with Tn916.

Both the *C. difficile* element conferring Tc^r and the *C. difficile* and *C. innocuum* elements conferring Cl^r/Em^r share with the more extensively characterized streptococcal and enterococcal conjugative transposons (see foregoing sections) the ability to be transferred to other genera of gram-positive bacteria. The element conferring Tc^r has been transferred by conjugation between *C. difficile* and *Bacillus subtilis*, and hybridization analysis suggests that it inserts into a preferred site in the *C. difficile* chromosome, whereas insertion into the *B. subtilis* chromosome appears to be essentially random (Mullany et al., 1990). The elements conferring Cl^r/Em^r have been transferred by conjugation from *C. difficile* and *C. innocuum* to *Staphylococcus aureus* and *C. perfringens*, respectively (Magot, 1983; Hächler et al., 1987b). These elements deserve further investigation to elucidate their genetic organization and the molecular basis of their conjugative transfer. Having presumably become adapted to their clostridial hosts, they might prove to have advantages over their better-known streptococcal and enterococcal counterparts from the point of view of their exploitation as tools for genetic analysis in clostridia.

6.8 Conjugative Mobilization of Plasmids from Escherichia Coli to Clostridia

Plasmids from gram-negative bacteria belonging to incompatibility group P show an extremely broad host range (Thomas and Smith, 1987; Krishnapillai, 1988). Moreover, their conjugation machinery, like that of the IncQ plasmids, shows a remarkable lack of specificity. This enables IncP and IncQ plasmids to transfer and to mobilize nonconjugative plasmids from gram-negative bacteria to an extraordinarily wide range of organisms including both gram-negative and gram-positive bacteria and yeasts and plants (Simon et al., 1983; Buchanan-Wollaston et al., 1987; Trieu-Cuot et al., 1987; Heinemann and Sprague, 1989; Mazodier et al., 1989; Lazraq et al., 1990).

Work on several representative IncP plasmids, RP4, RK2, and R751 (Pansegrau et al., 1988, 1990; Guiney et al., 1988; Fürste et al., 1989; Ziegelin et al., 1989), has indicated that conjugative mobilization involves the interaction of a *trans*-acting protein complex with a *cis*-acting site on the plasmid denoted *oriT* (*origin of transfer*). A nick is made in the *oriT* region of the supercoiled plasmid to generate the "relaxosome" complex, after which a single DNA strand is transferred to the recipient starting with the 5' terminus at the nick site, accompanied by rolling-circle replication of the plasmid in the donor (Willets and Wilkins, 1984). Any plasmid containing the appropriate *oriT* site may be mobilized from a donor strain expressing the IncP *tra* functions.

Trieu-Cuot et al., (1987) constructed a mobilizable *oriT*-containing vector, pAT187, based on the pBR322 backbone, the replication region from pAMβ1 and the *aphA-3* gene conferring resistance to kanamycin (Km^r) from plasmid pIP1433 of *Campylobacter coli*. These last two components were known from previous work to be expressed and function normally in a wide range of gram-positive organisms. Trieu-

Cuot et al. that showed pAT187 became established in a wide range of aerobic gram-positive bacteria following conjugative mobilization from an *E. coli* donor containing an IncP helper plasmid, which provided the requisite *tra* gene products in *trans*.

Williams et al. (1990a) have shown that plasmid pAT187 can also become established in the NCIB 8052 strain of *C. acetobutylicum* after conjugative mobilization from an appropriate *E. coli* donor under anaerobic conditions. Plasmid pAT187 replicates at low copy number in *C. acetobutylicum* and the Km^r gene is difficult to select under anaerobic conditions. An alternative *oriT*-containing vector, pCTC1, was therefore constructed. It is a pUC-like plasmid into which has been incorporated the *oriT* region of RK2, the Em^r gene of pAMβ1 (Brehm et al., 1987), and a minimal fragment of the replication region of pAMβ1, that promotes replication at high copy number in gram-positive organisms (Swinfield et al., 1990). Plasmid pCTC1 was employed to optimize the conditions for obtaining IncP-mediated conjugative mobilization of plasmids from *E. coli* to *C. acetobutylicum* (Williams et al., 1990a). In subsequent work (Williams et al., 1990b), the construction and transfer of a range of *oriT*-containing vectors based on the replication regions of pAMβ1 (Swinfield et al., 1990), pCB101 (Minton and Morris, 1981; Collins et al., 1985) and pWV01 (Vosman and Venema, 1983) was documented (see also Chapter 8, this volume). Some of these plasmids replicated at high copy number and others at low copy number in *C. acetobutylicum* after conjugative mobilization from *E. coli* donors carrying either an autonomous IncP plasmid (pRK24 or R702) or a chromosomally integrated copy of RP4-2(Tc::Mu).

This convenient method for transferring plasmids directly from *E. coli* to *C. acetobutylicum* NCIB 8052 should prove to be a useful adjunct to (electro-)transformation (see also Chapter 7, this volume) for introducing DNA into this organism. Given the remarkable lack of specificity of the conjugation apparatus of the IncP plasmids, it is possible that the method will prove to be generally applicable throughout the clostridia. However, this remains to be tested experimentally.

6.9 Future Prospects

Conjugation has proved to he a versatile method for effecting gene transfer to and among clostridia. In all cases studied so far, whether transfer involves exotic or indigenous elements, and whether they are large conjugative plasmids, small mobilized nonconjugative plasmids, or chromosomally located conjugative transposons, little or nothing is known about the conjugation mechanism. There is always a necessity for the establishment of close cell-to-cell contact, which is usually effected by harvesting a mixed suspension of donor and recipient bacteria onto a membrane filter. It is not even known whether surface pili are involved in the establishment of initial contacts between mating pairs. Equally, it is not known whether a single or a double strand of DNA is transferred from donor to recipient.

Much more is known about conjugative elements originating in enterococci and streptococci than about indigenous clostridial elements. Concerted nucleic acid sequencing of indigenous elements would amply reward the effort involved and would provide a wealth of essential information concerning their structure and organization. This in turn would provide a basis for proceeding with functional analysis using the rudimentary genetic tools already available (Young et al., 1989). It is to be expected that the conjugation mechanism(s) employed by streptococcal and enterococcal elements will be fully characterized long before those employed by indigenous clostridial elements have been elucidated. Nevertheless, there is a strong case to be made for pressing forward with the characterization of indigenous elements, not only for comparative purposes, but also with a view to their development and exploitation as genetic tools. They are likely to be better adapted to their natural hosts than those currently being employed, which originated in enterococci and streptococci.

As far as the exploitation of conjugative elements in clostridia is concerned, there is a growing realization of the potential utility of some of the methodology that has already been developed. Considerable progress has been made using an oligonucleotide-based approach for cloning structural genes concerned with substrate assimilation, primary fermentative metabolism, toxigenicity, and antibiotic resistance from several clostridia (see other chapters in this volume). The potential usefulness of conjugative transposons for pinpointing and then isolating regulatory genes implicated in these cellular activities has already been recognized (Volk et al., 1988; Bertram et al., 1990).

Several other applications can also be envisaged. For example, conjugative transposons are extremely valuable aids for physical mapping of clostridial genomes (see Section 6.6). Whether they can be employed in clostridia to effect allelic replacement of chromosomal genes with mutated copies constructed in *vitro* (cf. Norgren et al., 1989) or conjugative mobilization of small nonconjugative plasmids (cf. Naglich and Andrews, 1988) remains to be assessed.

Plasmid-mediated conjugative mobilization of small nonconjugative plasmids is likely to prove a useful alternative to (electro-)transformation for transferring genes to clostridia. The method may be particularly valuable for those organisms that prove recalcitrant to (electro-)transformation (see Chapter 7, this volume). Two different mobilization strategies, both depending on large conjugative plasmids, either indigenous or exotic, may be used. The first relies on the formation of plasmid cointegrates (see Section 6.4), whereas the second relies on the provision of *trans*-acting factors able to recognize a transfer origin (*oriT*) incorporated into the plasmid to be mobilized (see Section 6.8).

References

Abraham, L.J. and J.I. Rood. 1985a. Molecular analysis of transferable tetracycline resistance plasmids from *Clostridium perfringens*. *J. Bacteriol.* **161**:636–640.

Abraham, L.J. and J.I. Rood. 1985b. Cloning and analysis of the *Clostridium perfringens* tetracycline resistance plasmid, pCW3. *Plasmid* **13**:155–162.

Abraham, L.J. and J.I. Rood. 1987. Identification of Tn*4451* and Tn*4452*, chloramphenicol resistance transposons from *Clostridium perfringens*. *J. Bacteriol.* **169**:1579–1584.

Abraham, L.J. and J.I. Rood. 1988. The *Clostridium perfringens* chloramphenicol resistance transposon Tn*4451* excises precisely in *Escherichia coli*. *Plasmid* 19:164–168.

Abraham, L.J., A.J. Wales, and J.I. Rood. 1985. Worldwide distribution of the conjugative *Clostridium perfringens* tetracycline resistance plasmid, pCW3, *Plasmid* 14:37–46.

Allen, S.P. and H.P. Blaschek. 1988. Electroporation-induced transformation of intact cells of *Clostridium perfringens*. *App. Environ. Microbiol.* 54:2322–2324.

Allen, S.P. and H.P. Blaschek. 1988. Electroporation-induced transformation of intact cells of *Clostridium perfringens*. *Appl. Environ. Microbiol.* 54:2322–2324.

Behnke, D. and M.S. Gilmore. 1981. Location of antibiotic resistance determinants, copy control, and replication functions on the double-selective cloning vector pGB301. *Mol. Gen. Genet.* 184:115–120.

Bertram, J. and P. Dürre. 1989. Conjugal transfer and expression of streptococcal transposons in *Clostridium acetobutylicum*. *Arch. Microbiol.* 151:551–557.

Bertram, J., A. Kuhn, and P. Dürre. 1990. Tn*916*-induced mutants of *Clostridium acetobutylicum* defective in regulation of solvent formation. *Arch. Microbiol.* 153:373–377.

Bertram, J., M. Strätz, and P. Dürre. 1991. Natural transfer of conjugative transposon Tn*916* between Gram-positive and Gram-negative bacteria. *J. Bacteriol.* 173:443–448.

Brantl, S., D. Behnke, and J.C. Alonso. 1990. Molecular analysis of the replication region of the conjugative *Streptococcus agalactiae* plasmid pIP501 in *Bacillus subtilis*—comparison with plasmids pAMβ1 and pSM19035. *Nucleic Acids Res.* 18:4783–4790.

Bréfort, G., M. Magot, H. Ionesco, and M. Sebald. 1977. Characterization and transferability of *Clostridium perfringens* plasmids. *Plasmid* 1:52–66.

Bréfort, G., M. Magot, H. Ionesco, and M. Sebald. 1978. Characterization and transferability of *Clostridium perfringens* plasmids. *In: Microbiology 1978*, D. Schlessinger, ed., pp. 242–245. Washington, D.C.: American Society for Microbiology.

Brehm, J., G. Salmond, and N.P. Minton. 1987. Sequence of the adenine methylase gene of the *Streptococcus faecalis* plasmid pAMβ1. *Nucleic Acids* Res. 15:3177.

Buchanan-Wollaston, V., J.E. Passiatore, and F. Cannon. 1987. The *mob* and *oriT* mobilization functions of a bacterial plasmid promote its transfer to plants. *Nature* (London) 328:172–175.

Burdett, V., J. Inamine, and S. Rajagopalan. 1982. Heterogeneity of tetracycline resistance determinants in *Streptococcus*. *J. Bacteriol.* 149:995–1004.

Caillaud, F. and P. Courvalin. 1987. Nucleotide sequence of the ends of the conjugative shuttle transposon Tn*1545*. *Mol. Gen. Genet.* 209:110–115.

Caillaud, F., C. Carlier, and P. Courvalin. 1987. Physical analysis of the conjugative shuttle transposon Tn*1545*. *Plasmid* 17:58–60.

Caparon, M.G. and J.R. Scott. 1989. Excision and insertion of the conjugative transposon Tn*916* involves a novel recombination mechanism. *Cell* 59:1027–1034.

Christie, P.J., R.Z. Korman, S.A. Zahler, J.C. Adsit, and G.M. Dunny. 1987. Two conjugation systems associated with *Streptococcus faecalis* plasmid pCF10: identification of a conjugative transposon that transfers between *S. faecalis* and *Bacillus subtilis*. *J. Bacteriol.* 169:2529–2536.

Clewell, D.B. 1981. Plasmids, drug resistance, and gene transfer in the genus *Streptococcus*. *Microbiol. Rev.* 45:409–436.

Clewell, D.B. 1990. Movable genetic elements and antibiotic resistance in enterococci. *Eur. J. Clin. Microbiol. Infect. Dis.* 9:90–102.

Clewell, D.B. and C. Gawron-Burke. 1986. Conjugative transposons and the dissemination of antibiotic resistance in streptococci. *Annu. Rev. Microbiol.* 40:635–659.

Clewell, D.B. and K.E. Weaver. 1989. Sex pheromones and plasmid transfer in *Enterococcus faecalis*. *Plasmid* 21:175–184.

Clewell, D.B., G.F. Fitzgerald, L. Dempsey, L.E. Pearce, F.Y. An, B.A. White, Y. Yagi, and C. Gawron-Burke. 1985. Staphylococcal conjugation: plasmids, sex pheromones and conjugative transposons. *In: Molecular Basis of Oral Microbial Adhesion*, S.E. Mergenhagen and B. Rosan, eds., pp. 194–203. Washington, D.C.: American Society for Microbiology.

Clewell, D.B., S.E. Flannagan, Y. Ike, J.M. Jones, and C. Gawron-Burke. 1988. Sequence analysis of the termini of conjugative transposon Tn*916*. *J. Bacteriol.* 170:3046–3052.

Collins, M.E., J.D. Oultram, and M. Young. 1985. Identification of restriction fragments from two cryptic *Clostridium butyricum* plasmids that promote the establishment of a replication-defective plasmid in *Bacillus subtilis*. *J. Gen. Microbiol.* 131:2097–2105.

Courvalin, P. and C. Carlier. 1986. Transposable multiple antibiotic resistance in *Streptococcus pneumoniae*. *Mol. Gen. Genet.* 205:291–297.

Courvalin, P. and C. Carlier. 1987. Tn*1545*: a conjugative shuttle transposon. *Mol. Gen. Genet.* 206:259–264.

Davies, A., J.D. Oultram, A. Pennock, D.R. Williams, D.F. Richards, N.P. Minton, and M. Young. 1988. Conjugal gene transfer in *Clostri-*

dium acetobutylicum. In: Genetics and Biotechnology of Bacilli, Vol. 2, A.T. Ganesan and J.A. Hoch, eds., pp. 391–395. Orlando: Academic Press.

Evans, R.P. Jr. and F.L. Macrina. 1983. Streptococcal R plasmid pIP501: endonuclease site map, resistance determinant location, and construction of novel derivatives. *J. Bacteriol.* **154**:1347–1355.

Franke, A.E. and D.B. Clewell. 1981. Evidence for a chromosome-borne resistance transposon (Tn*916*) in *Streptococcus faecalis* that is capable of "conjugal" transfer in the absence of a conjugative plasmid. *J. Bacteriol.* **145**:494–502.

Fürste, J.P., W. Pansegrau, G. Ziegelin, M. Kroger, and E. Lanka. 1989. Conjugative transfer of promiscuous IncP plasmids: interaction of plasmid-encoded products with the transfer origin. *Proc. Nat. Acad. Sci. USA* **86**:1771–1775.

Gaillard, J.L., P. Berche, and P.J. Sansonetti. 1986. Transposon mutagenesis as a tool to study the role of hemolysin in the virulence of *Listeria monocytogenes. Infect. Immun.* **52**:50–55.

Garnier, T. and S.T. Cole. 1986. Characterization of a bacteriocinogenic plasmid from *Clostridium perfringens* and molecular genetic analysis of the bacteriocin-encoding gene. *J. Bacteriol.* **168**:1189–1196.

Garnier, T. and S.T. Cole. 1988a. Complete nucleotide sequence and genetic organization of the bacteriocinogenic plasmid, pIP404, from *Clostridium perfringens. Plasmid* **19**:134–150.

Garnier, T. and S.T. Cole. 1988b. Identification and molecular genetic analysis of replication functions of the bacteriocinogenic plasmid pIP404 from *Clostridium perfringens. Plasmid* **19**:151–160.

Garnier, T., W. Saurin, and S.T. Cole. 1987. Molecular characterization of the resolvase gene, *res*, carried by a multicopy plasmid from *Clostridium perfringens*: common evolutionary origin for prokaryotic site-specific recombinases. *Mol. Microbiol.* **1**:371–376.

Gawron-Burke, C. and D.B. Clewell. 1982. A transposon in *Streptococcus faecalis* with fertility properties. *Nature* (London) **300**:281–284.

Gawron-Burke, C. and D.B. Clewell. 1984. Regeneration of insertionally inactivated streptococcal DNA fragments after excision of transposon Tn*916* in *Escherichia coli*: strategy for targeting and cloning of genes from gram-positive bacteria. *J. Bacteriol.* **159**:214–221.

Grigorova, R., L. Michailova, V. Miteva, N. Peneva, L. Ganova, and T. Takova. 1988. Conjugal transfer of streptococcal plasmids to strains of *Bacillus sphaericus. FEMS Microbiol. Lett.* **49**:289–294.

Guiney, D.G., C. Deiss, and V. Simnad. 1988. Location of the relaxation complex nick site within the minimal origin of transfer region of RK2. *Plasmid* **20**:259–265.

Hächler, H., F.H. Kayser, and B. Berger-Bächi. 1987a. Homology of a transferable tetracycline resistance determinant of *Clostridium difficile* with *Streptococcus faecalis* transposon Tn*916. Antilmicrob. Agents Chemother.* **31**:1033–1038.

Hächler, H., B. Berger-Bächi, and F.H. Kayser. 1987b. Genetic characterization of a *Clostridium difficile* erythromycin-clindamycin resistance determinant that is transferable to *Staphylococcus aureus. Antilmicrob. Agents Chemother.* **31**:1039–1045.

Hecht, D. and M.H. Malamy. 1989. Tn*4399*, a conjugal mobilising transposon of *Bacteriodes fragilis. J. Bacteriol.* **171**:3603–3608.

Heinemann, J.A. and J.F. Sprague, Jr. 1989. Bacterial conjugative plasmids mobilize DNA transfer between bacteria and yeast. *Nature* (London) **340**:205–209.

Hill, C., C. Daly, and G.F. Fitzgerald. 1985. Conjugative transfer of the transposon Tn*919* to lactic acid bacteria. *FEMS Microbiol. Lett.* **30**:115–119.

Horaud, T., C. LeBougenec, and K. Pepper. 1985. Molecular genetics of resistance to macrolides, lincosamides and streptogramin B (MLS) in streptococci. *J. Antimicrob. Chemother.* **16**(Suppl. A):111–135.

Ionesco, H. 1980. Transfert de la résistance à la tétracycline chez *Clostridium difficile. Ann. Inst. Pasteur Microbiol.* (Paris) **131**:171–179.

Ionesco, H. and D.H. Bouanchaud. 1973. Production de bactériocine liée a la presence d'un plasmide chez *Clostridium perfringens* type A. *C. R. Acad. Sci.* (Paris) Ser. D **276**:2855–2857.

Ionesco, H., A. Wolff, and M. Sebald. 1974. Production induite de bactériocine et d'un bacteriophage par la souche BP6K-N5 de *Clostridium perfringens. Ann. Inst. Pasteur Microbiol.* (Paris) **125B**:335–346.

Ionesco, H., G. Bieth, C. Dauguet, and D.H. Bouanchaud. 1976. Isolement et identification de deux plasmides d'une souche bacteriocinogène de *Clostridium perfringens. Ann. Inst. Pasteur Microbiol.* (Paris) **127B**:283–294.

Jones, D.T. and D.R. Woods. 1986. Acetone-butanol fermentation revisited. *Microbiol. Rev.* **50**:484–524.

Jones, J.M., C. Gawron-Burke, S.E. Flannagan, M. Yamamoto, E. Senghas, and D.B. Clewell. 1987. Structural and genetic studies of the conjugative transposon Tn*916. In: Staphylococcal Genetics*, J.J. Ferretti and R. Curtiss III, eds., pp. 54–60. Washington, D.C.: American Society for Microbiology.

Kathariou, S., P. Metz, H. Hof, and W. Goebel. 1987. Tn*916*-induced mutations in the hemolysin determinant affecting virulence of *Listeria monocytogenes. J. Bacteriol.* **169**:1291–1297.

Krah, E.R. and F.L. Macrina. 1989. Genetic analysis of the conjugal transfer determinants encoded by the streptococcal broad-host-range plasmid pIP501. *J. Bacteriol.* **171**:6005–6012.

Krishnapillai, V. 1988. Molecular genetic analysis of bacterial plasmid promiscuity. *FEMS Microbiol. Rev.* **54**:223–238.

Lazraq, R., S. Clavel-Sérès, H.L. David, and D. Roulland-Dussoix. 1990. Conjugative transfer of a shuttle plasmid from *Escherichia coli* to *Mycobacterium smegmatis FEMS Microbiol. Lett.* **69**:135–138.

LeBlanc, D.J. and L.N. Lee. 1984. Physical and genetic analyses of streptococcal plasmid pAMβ1 and cloning of its replication region. *J. Bacteriol.* **157**:445–453.

Lereclus, D., G. Menou, and M.-M. Lecadet. 1983. Isolation of a DNA sequence related to several plasmids from *Bacillus thuringiensis* after a mating involving the *Streptococcus faecalis* plasmid pAMβ1. *Mol. Gen. Genet.* **191**:307–313.

Luczak, H., H. Schwarzmoser, and W.L. Staudenbauer. 1985. Construction of *Clostridium butyricum* plasmids and transfer to *Bacillus subtilis*. *Appl. Microbiol. Biotechnol.* **23**:114–122.

Macrina, F.L., C.L. Keeler, K.R. Jones, and P.H. Wood. 1980. Molecular characterization of unique deletion mutants of the streptococcal plasmid pAMβ1. *Plasmid* **4**:8–16.

Magot, M. 1983. Transfer of antibiotic resistances from *Clostridium innocuum* to *Clostridium perfringens* in the absence of detectable plasmid DNA. *FEMS Microbiol. Lett.* **18**:149–151.

Magot, M. 1984. Physical characterisation of the *Clostridium perfringens* tetracycline-chloramphenicol resistance plasmid pIP401. *Ann. Inst. Pasteur Microbiol.* **135**:269–282.

Mattsson, P.M. and P. Rogers. 1989. Identification of some transposon insertion sites in the chromosome of *Clostridium acetobutylicum* after transfer of the tetracycline resistance transposon Tn916 from *Streptococcus faecalis*. In: Abstract, American Society for Microbiology Annual Meeting (New Orleans), p. 152. Washington, D.C.: American Society for Microbiology.

Mazodier, P., R. Petter, and C. Thompson. 1989. Intergenetic conjugation between *Escherichia coli* and *Streptomyces* species. *J. Bacteriol.* **171**:3583–3585.

Minton, N.P. and J.G. Morris. 1981. Isolation and partial characterization of three cryptic plasmids from strains of *Clostridium butyricum*. *J. Gen. Microbiol.* **127**:325–331.

Mullany, P., M. Wilks, and S. Tabaqchali. 1991. Transfer of Tn916 and Tn916ΔE into *Clostridium difficile*: demonstration of a hot-spot for these elements in the C. *difficile* genome. *FEMS Microbiol. Lett.* **79**:191–194.

Mullany, P., M. Wilks, T. Lamb, C. Clayton. B. Wren, and S. Tabaqchali. 1990. Genetic analysis of a tetracycline resistance element from *Clostridium difficile* and its conjugal transfer to and from *Bacillus subtilis*. *J. Gen. Microbiol.* **136**:1343–1349.

Naglich, J.G., and R.E. Andrews. 1988. Tn916-dependent conjugal transfer of pC194 and pUB110 from *Bacillus subtilis* into *Bacillus thuringiensis* subsp. *israelensis*. *Plasmid* **20**:113–126.

Nida, K. and P.P. Cleary. 1983. Insertional inactivation of streptolysin S expression in *Streptococcus pyogenes*. *J. Bacteriol.* **155**:1156–1161.

Norgren, M., M.G. Caparon, and J.R. Scott. 1989. A method for allelic replacement that uses the conjugative transposon Tn916: deletion of the *emm* 6.1 allele in *Streptococcus pyogenes* JRS4. *Infect. Immun.* **57**:3846–3850.

Norgren, M. and J.R. Scott. 1991. The presence of conjugative transposon Tn916 in the recipient strain does not impede transfer of a second copy of the element. *J. Bacteriol.* **173**:319–324.

Oultram, J.D. and M. Young. 1985. Conjugal transfer of plasmid pAMβ1 from *Streptococcus lactis* and *Bacillus subtilis* to *Clostridium acetobutylicum*. *FEMS Microbiol. Lett.* **27**:129–134.

Oultram, J.D., A. Davies, and M. Young. 1987. Conjugal transfer of a small plasmid from *Bacillus subtilis* to *Clostridium acetobutylicum* by cointegrate formation with plasmid pAMβ1. *FEMS Microbiol. Lett.* **42**:113–119.

Oultram, J.D., H. Peck, J.K. Brehm, D. Thompson, T.J. Swinfield, and N.P. Minton. 1988. Introduction of genes for leucine biosynthesis from *Clostridium pasteurianum* into *Clostridium acetobutylicum*. *Mol. Gen. Genet.* **214**:177–179.

Pan-Hou, H.S.K., M. Hosono, and N. Imura. 1980. Plasmid-controlled mercury biotransformation by *Clostridium cochlearium* T-2 *Appl. Environ. Microbiol.* **40**:1007–1011.

Pansegrau, W., G. Ziegelin, and E. Lanka. 1988. The origin of conjugative IncP plasmid transfer: interaction with plasmid-encoded products and the nucleotide sequence at the relaxation site. *Biochim. Biophys. Acta* **951**:365–374.

Pansegrau, W., D. Balzer, V. Kruft, R. Lurz, and E. Lanka. 1990. In *vitro* assembly of relaxosomes at the transfer origin of plasmid RP4. *Proc. Nat. Acad. Sci. USA* **87**:6555–6559.

Poyart-Salmeron, C., P. Trieu-Cuot, C. Carlier, and P. Courvalin. 1989. Molecular characterization of two proteins involved in the excision of the conjugative transposon Tn1545: homologies with other site-specific recombinases. *EMBO J.* **8**:2425–2433.

Poyart-Salmeron, C., P. Trieu-Cuot, C. Carlier, and P. Courvalin. 1990. The integration-excision system of the conjugative transposon Tn1545 is structurally and functionally related to those of lambdoid phages. *Mol. Microbiol.* **4**:1513–1521.

Reysset, G. and M. Sebald. 1985. Conjugal transfer of plasmid-mediated antibiotic resistance from streptococci to *Clostridium acetobutylicum*. *Ann. Inst. Pasteur Microbiol*. (Paris) **136**:275–282.

Rood, J.I. 1983. Transferable tetracycline resistance in *Clostridium perfringens* strains of porcine origin. *Can. J. Microbiol*. **29**:1241–1246.

Rood, J.I., V.N. Scott, and C.L. Duncan. 1978. Identification of a transferable tetracycline resistance plasmid (pCW3) from *Clostridium perfringens*. *Plasmid* **1**:563–570.

Rubens, C.E. and L.M. Heggen. 1988. Tn*916ΔE*: a Tn*916* transposon derivative expressing erythromycin-resistance. *Plasmid* **20**:137–142.

Schaberg, D.R., D.B. Clewell, and L. Glatzer. 1982. Conjugative transfer of R-plasmids from *Streptococcus faecalis* to *Staphylococcus aureus*. *Antimicrob. Agents Chemother*. **22**:204–207.

Scott, P.T. and J.I. Rood. 1989. Electroporation-mediated transformation of lysostaphin-treated *Clostridium perfringens*. *Gene* **82**:327–333.

Scott, J.R., P.A. Kirchman, and M.G. Caparon. 1988. An intermediate in transposition of the conjugative transposon Tn*916*. *Proc. Natl. Acad. Sci. USA* **85**:4809–4813.

Sebald, M. and G. Bréfort. 1975. Transfert du plasmide tétracycline-chloramphenicol chez *Clostridium perfringens*. *C. R. Acad. Sci. Paris Serie D*. **281**:317–319.

Sebald, M., D. Bouanchaud, and G. Bieth. 1975. Nature plasmidique de la résistance à plusieurs antibiotiques chez *C. perfringens* type A, souche 659. *C. R. Acad. Sci. Paris Ser. D* **280**:2401–2404.

Senghas, E., J.M. Jones, M. Yamamoto, C. Gawron-Burke, and D.B. Clewell. 1988. Genetic organization of the bacterial conjugative transposon Tn*916*. *J. Bacteriol*. **170**:245–259.

Simon, R., U. Priefer, A. Pühler. 1983. A broad host range mobilization system for in vivo genetic engineering: transposon mutagenesis in Gram negative bacteria. *BioTechnology* **2**:784–791.

Smith, M.D. 1985. Transformation and fusion of *Streptococcus faecalis* protoplasts. *J. Bacteriol*. **162**:92–97.

Smith, C.J., S.M. Markowitz, and F.L. Macrina. 1981. Transferable tetracycline resistance in *Clostridium difficile*. *Antimicrob. Agents Chemother*. **19**:997–1003.

Smith, M.D. and D.B. Clewell. 1984. Return of *Streptococcus faecalis* DNA cloned in *Escherichia coli* to its original host via transformation of *Streptococcus sanguis* followed by conjugative mobilization. *J. Bacteriol*. **160**:1109–1114.

Swinfield, T.J., J.D. Oultram, D.E. Thompson, J.K. Brehem, and N.P. Minton. 1990. Physical characterisation of the replication region of the *Streptococcus faecalis* plasmid pAMβ1. *Gene* **87**:79–90.

Taketomo, N., Y. Sasaki, and T. Sasaki. 1989. A new method for conjugal transfer of plasmid pAMβ1 to *Lactobacillus plantarum* using polyethylene glycol. *Agric. Biol. Chem*. **53**:3333–3334.

Thomas, C.M. and C.A. Smith. 1987. Incompatibility group P plasmids: genetics, evolution and use in genetic manipulation: *Annu. Rev. Microbiol*. **41**:77–101.

Trieu-Cuot, P., C. Carlier, and P. Courvalin. 1988. Conjugative plasmid transfer from *Enterococcus faecalis* to *Escherichia coli*. *J. Bacteriol*. **170**:4388–4391.

Trieu-Cuot, P., C. Carlier, P. Martin, and P. Courvalin. 1987. Plasmid transfer by conjugation from *Escherichia coli* to Gram-positive bacteria. *FEMS Microbiol. Lett*. **48**:289–294.

Trieu-Cuot, P., C. Poyart-Salmeron, C. Carlier, and P. Courvalin. 1990. Conjugative transposons of gram-positive cocci. In: *Proceedings of the Sixth International Symposium on Genetics of Industrial Microorganisms*, GIM '90, pp. 195–205. Strasbourg: Société Française de Microbiologie.

van der Lelie, D. and G. Venema. 1987. *Bacillus subtilis* generates a major specific deletion in pAMβ1. *Appl. Environ. Microbiol*. **53**:2458–2463.

Volk, W.A., B. Bizzini, K.R. Jones, and F.L. Macrina. 1988. Inter- and intrageneric transfer of Tn*916* between *Streptococcus faecalis* and *Clostridium tetani*. *Plasmid* **19**:255–259.

Vosman, B. and G. Venema. 1983. Introduction of a *Streptococcus cremoris* plasmid in *Bacillus subtilis*. *J. Bacteriol*. **156**:920–921.

Willets, N. and B. Wilkins. 1984. Processing of plasmid DNA during bacterial conjugation. *Microbiol. Rev*. **48**:24–41.

Williams, D.R., D.I. Young, J.D. Oultram, N.P. Minton, and M. Young. 1990a. Development and optimisation of conjugative plasmid transfer from *Escherichia coli* to *Clostridium acetobutylicum* NCIB 8052. In: *Clinical and Molecular Aspects of Anaerobes*, S.P. Borriello, ed., pp. 242–245. Petersfield, UK: Wrightson.

Williams, D.R., D.I. Young, and M. Young. 1990b. Conjugative plasmid transfer form *Escherichia coli* to *Clostridium acetobutylicum*. *J. Gen. Microbiol*. **136**:819–826.

Woolley, R.C., A. Pennock, R.J. Ashton, A. Davies, and M. Young. 1989. Transfer of Tn*1545* and Tn*916* to *Clostridium acetobutylicum*. *Plasmid* **22**:169–174.

Wüst, J. and U. Hardegger. 1983. Transferable resistance to clindamycin, erythromycin, and tetracycline in *Clostridium difficile*. *Antimicrob. Agents Chemother*. **23**:784–786.

Yamamoto, M., J.M. Jones, E. Senghas, C. Gawron-Burke, and D.B. Clewell. 1987. Generation of Tn*5* insertions in streptococcal conjugative transposon Tn*916*. *Appl. Environ. Microbiol*. **53**:1069–1072.

Young, M., W.L. Staudenbauer, and N.P. Minton.

1989. Recent advances in the genetics of the clostridia. *FEMS Microbiol. Rev.* **63**:301–325.

Young, M., J.D. Oultram, A. Pennock, and D.F. Richards. 1987. Gene transfer in clostridia. *In: Genetics of Industrial Microorganisms*, Part A, M. Alačević, D. Hranueli, and Ž. Toman, eds., pp. 403–413. GIM '86, Zagreb: Faculty of Food Technology and Biotechnology.

Yu, P.-L. and L.E. Pearce. 1986. Conjugal transfer of streptococcal antibiotic resistance plasmids into *Clostridium acetobutylicum. Biotechnol. Lett.* **8**:469–474.

Ziegelin, G., J.P. Fürste, and E. Lanka. 1989. TraJ protein of plasmid RP4 binds to a 19-base pair invert sequence repetition within the transfer origin. *J. Biol. Chem.* **264**:11989–11994.

7

Transformation and Electrotransformation in Clostridia

Gilles Reysset

7.1 Introduction

The introduction of DNA into bacteria by transformation remains an important step for the development of genetic and recombinant DNA technology. In addition to studying the structure and expression of exogenous genes either in *Escherichia coli* or *Bacillus subtilis*, it is important to reintroduce these genes in their original host. Genetic background, physiological, and environmental conditions are highly relevant to gene expression and regulation. This is particularly true for strict anaerobic bacteria such as clostridia whose general physiology is very different than that of *E. coli* or *B. subtilis*. Further, cloning foreign genes in strains of medical or economic significance with the purpose of modifying their genotype or improving their metabolic potentiality requires efficient gene transfer systems.

For these reasons, much experimental work has been done to develop transformation procedures for clostridia during the past 10 years. Because no natural or physiological transformation has been observed in the genus *Clostridium*, this review focuses on the artificial bacterial transformation systems. Two general techniques have been developed, protoplast transformation and electrotransformation; the latter appears to be the simplest method for DNA uptake developed to date. Two excellent reviews on this subject have been already published by Young et al. (1989a, 1989b).

7.2 Transformation of Wall-Less Cells

The first report on plasmid transformation of bacterial protoplasts was published in 1978 for streptomyces (Bibb et al., 1978) and 1 year later for *Bacillus subtilis* (Chang and Cohen, 1979). These authors observed that among protoplasts exposed to plasmid DNA in the presence of polyethylene glycol (PEG), a high proportion of the regenerated colonies was transformed. These results were of great interest to those developing genetic systems for gram-positive bacteria because the introduction of foreign DNA by this chemical transformation was shown to be independent of any natural stage of competence. The development of this transformation procedure, based on the ability of wall-less cells to regenerate, was also very exciting because the technology opened the way to various genetic applications. These included intra- and interspecific protoplast fusion (with the aim of genetic mapping), plasmid transfer by protoplast fusion, loss of plasmid by protoplasting, etc. (for review, see Hopwood, 1981).

Since 1980, many laboratories have endeavored to produce and regenerate protoplasts of a large variety of clostridia; these have included *Clostridium acetobutylicum* (Allcock et al., 1982), *C. pasteurianum* (Minton and Morris, 1983), *C. tertium* (Knowlton et al., 1984), *C. saccharoperbutylacetonicum* (later referred as to *C. acetobutylicum*) (Yoshino et al., 1984; Reysset et al.,

1987), *C. pefringens* (Heefner et al., 1984; Stal and Blaschek, 1985), *C. thermohydrosulfuricum* (Soutschek-Bauer et al., 1985), and more recently two thermophilic species (Kurose et al., 1989; Pettinari et al., 1989). Although clostridial protoplasts were relatively easy to obtain, the general conditions for an efficient regeneration of cell wall were in all cases difficult to define. Added to this problem, the conditions were highly dependent on the choice of strain. The major difficulty was probably that many spore-forming bacteria, such as clostridia, are highly autolytic, and thus the cell-wall components would be hydrolyzed as soon as they were synthetized by the protoplasts.

These activities generated the production of L-form regenerant cells unable to revert to the bacillary form. This was obviously the case for *C. perfringens* (Heefner et al., 1984), *C. tertium* (Knowlton et al., 1984), and also *C. acetobutylicum* strain NI-4 (Reysset et al., 1987). Attempts to overcome this problem have been partially successful, as was suggested for *B. subtilis* by Landman and Forman (1969): inhibitors of the autolytic activity, such as gelatin or *N*-acetylglucosamine (Heefner et al., 1984; Stal and Blaschek, 1985), were added to the regeneration medium; mutant derivatives with higher regeneration frequency (Knowlton et al., 1984) were isolated; or the osmotic stabilizers and the concentration of required ionic constituents, such as Mg^{2+} and/or Ca^{2+}, were varied for each strain.

For *C. acetobutylicum* strain NI-4, a reproducible and efficient regeneration was achieved when taking into account *all* the following recommendations: (i) a protocol for the production of wall-less cells using lysozyme and penicillin G was optimized to yield virtually 100% efficiency; (ii) a partially autolysin-deficient strain NI-4081 was used instead of the wild type and a strong autolysin inhibitor such as choline was added to the regeneration medium, thus limiting the residual autolytic activity; (iii) use of a preferential osmotic stabilizer (xylose 0.25 *M* rather than sucrose 0.3 *M*) and a minimal concentration of ionic constituents (Ca^{2+} and Mg^{2+}), which limits the multiplication of the L-form cells. These conditions enabled 20% to 40% of the protoplasts to re-

generate vegetative cell colonies when diluted in soft agar before plating (Reysset et al., 1987; Podvin et al., 1988).

Whatever the frequency of regeneration obtained, Polyethylene glycol-mediated transfection and transformation has in fact been reported for some clostridial strains (Table 7.1). Reid et al. (1983) succeeded in the transfection of *C. acetobutylicum* strain P261 using phage CA1 DNA, and Lin and Blascheck (1984) achieved transformation of strain SA1 with the *Staphylococcus aureus* plasmid pUB110 as DNA source (30 transformants per μg DNA). The transformation of stable L-phase variants or autoplasts (protoplasts produced by the autolytic activity of the strain) of *C. perfringens* strain 11268 has been reported by Heefner et al. (1984) using the PEG-mediated procedure (10^2 transformants per μg DNA). However, it was found that the stable L-phase variants were unable to revert to rod-shaped cells, although some autoplasts were both transformed and regenerated in liquid medium. Using the Heefner method, Mahony et al. (1988) succeeded in the transformation of L-phase variants of *C. perfringens* strain L13 at a high transformation rate (3.9×10^{-5} transformants per viable cell), but here again the L-form cells did not regenerate. Kurose et al. (1989) obtained 1 to 7×10^2 transformants per μg DNA of cellulolytic clostridia using the same protocol. None of these methods has been widely exploited, because in all cases the frequency of transformation was low, and regeneration to bacillary form was irregularly achieved, particularly in the case of *C. perfringens*. In fact, as observed for *C. acetobutylicum* strain NI-4081, the limiting step of the PEG-mediated transformation was the regeneration procedure. When this difficulty was finally overcome, transformation with appropriate DNA sources was achieved at an efficiency greater than 1×10^6 transformants per μg DNA (Azeddoug et al., 1989; Truffaut et al., 1989). A detailed technique used for the transformation of this particular strain of *C. acetobutylicum* has been published by Reysset et al. (1988).

The other genetic applications using protoplast regeneration have not been widely exploited. Protoplast fusion of *C. acetobutylicum*

Table 7.1 PEG-mediated transfection and transformation of clostridia

Organism	DNA (size in kb., selected marker)	Transformation frequency	Transformation procedure	Reference
Clostridium acetobutylicum				
Strain P262	Phage CA1	Not reported	Protoplast tfm	Reid et al., 1983
Strain SA-1	pUB110 (4.5; Km^r)	3×10^{1a}	Protoplast tfm	Lin and Blaschek, 1984
Strain NI-4081[b]	Phage HM3	5×10^4	Protoplast tfm	Reysset et al., 1988
Strain NI-4081	pVA677 (7.6, Em^r)	1×10^5	Protoplast tfm	Reysset et al., 1988
Strain NI-4081	pIM13 (2.2, Em^r)	1×10^5	Protoplast tfm	Truffaut et al., 1989
Strain NI-4081	pKNT14 (4.3, Em^r)	1×10^5–5×10^5	Protoplast tfm	Truffaut et al., 1989
Strain NI-4081	pKNT11 (6.5, Em^r)	1×10^6–5×10^6	Protoplast tfm	Truffaut et al., 1989
Strain n° 220	pTY10 (8.1, Cm^r)	≤ 1	TRIS-permeabilization	Yoshino et al., 1990
Strain n° 220	pTYD101 (4.0, Cm^r)	$> 1 \times 10^3$	TRIS-permeabilization	Yoshino et al., 1990
Clostridium thermohydrosulfuricum				
Strain DSM568	pGS13 (5.5, Km^r, Cm^r)	2.8×10^{-6c}	TRIS-permeabilization	Soutschek-Bauer et al., 1985
Strain DSM568	pUB110 (4.5, Km^r)	4×10^{-6c}	TRIS-permeabilization	Soutschek-Bauer et al., 1985
Clostridium Sp. (*thermophilic*)				
Strain 008-118	pCS1 (7.2, Cm^r)	1×10^2–7×10^2	Spheroplast tfm	Kurose et al., 1989
Clostridium perfringens				
Strain 11268 CDR	pJU124 (38.8, Tc^r)	1×10^2	L-form tfm	Heefner et al., 1984
Strain 11268 CDR	pJU12 (11.6 Tc^r)	1×10^2	L-form tfm	Squires et al., 1984
Strain L13 (L-form variant)	pJU16 (12.2, Tc^r)	4×10^3	L-form tfm	Mahony et al., 1988
Strain L13	pHR106 (7.9, Cm^r)	1×10^{6c}	L-form tfm	Roberts et al., 1988

[a] Transformation frequency is expressed per μg DNA.
[b] This strain derived from NI-4
[c] Transformation frequency per viable cell.

has been obtained in strain P262 (Jones et al., 1985) and also in strain NI-4081 (Reysset, unpublished observations). The problems encountered during the analysis of fusion products of *B. subtilis* protoplasts (Hotchkiss and Gabor, 1980; Levi-Meyrueis et al., 1980; Sanchez-Rivas, 1982) led to the conclusion that it was perhaps unwise to develop this technique for genetic mapping in clostridia whose gene organization is totally unknown. Heterologous protoplast fusion as direct plasmid transfer has not been reported for clostridia, but curing of a resident plasmid from protoplasts of *C. acetobutylicum* strain NI-4 derivatives was achieved (Azeddoug et al., unpublished observations) using the regeneration

procedure described by Novick et al. (1980) for *Staphylococcus aureus*.

As protoplast formation and regeneration was shown to be problematic as well as time consuming and strain dependent, it is thought that the use of this technique will be gradually phased out.

7.3 Transformation of Vegetative Cells

By protoplast transformation, the cell wall which constitutes the physical barrier to the uptake of DNA is removed by enzymatic treatment. However, suitable chemical or physical techniques have also been developed to *"permeabilize"* the cell envelope.

Uptake of DNA by chemical permeabilization

Although many workers tried to permeabilize clostridial strains with the intent of creating artificial competence, only two positive reports have so far been published. In both cases, the transformation procedure was derived from the technique initially described by Takahashi et al. (1983) for *Bacillus brevis*. Vegetative cells of *Clostridium thermohydrosulfuricum* treated with 50 mM Tris-HCl pH 8.3 were first transformed by Soutschek-Bauer et al. (1985) using pUB110 and PGS13 plasmid DNAs at a reasonable frequency of 4×10^{-6} per viable bacteria, in presence of PEG (final concentration 35%). More recently, Yoshino et al. (1990) transformed *C. acetobutylicum* strain no. 220 by the shuttle vector pTY10 but with a poor efficiency (1 cfu per μg DNA), while a deleted derivative, pTYD101, only able to replicate in *C. acetobutylicum*, transformed this host at a much higher level ($>10^3$ cfu per μg DNA).

Uptake of DNA by electric field permeabilization

The membranes of vegetative cells can be reversibly permeabilized for the uptake of DNA molecules by a high-strength electric field, as has been assessed by Chassy et al. (1988) and Shigekawa and Dower (1988). This technique, referred to as electrotransformation, is an excellent alternative to the previously described methods; it is less time consuming and more universally applicable. Nevertheless, a high efficiency of transformation by this process was shown to be more easily achieved for gram-negative than for gram-positive bacteria, and critical parameters of the technique appeared also to be either genus, species, or strain dependent.

The report of Scott and Rood (1989) on the electroporation-mediated transformation of *C. perfringens* strain L13 is certainly the one best documented for this species. Biological factors as well as variables in the electric field have been evaluated. Only bacteria harvested early in the logarithmic stage of growth and pretreated with lysostaphin (20 μg ml$^-$ for 1 h at 37°C) were competent for DNA uptake. A density ranging from 1 to 5×10^8 bacteria per ml was reported to be optimal for electrotransformation. There was a linear relationship between transformation frequency and DNA concentration from 0.1 to 1 μg DNA, and a saturating level was reached at 5 to 10 μg DNA per ml.

The transformation efficiency of *C. perfringens* strain L 13 appeared to be directly proportional to both the field strengh and the time constant of the electric pulse. With the general conditions of 25 μF, 6.25 kV cm^{-1}, and 7.5 ms, 3.0×10^5 transformants per μg DNA were obtained using vector pHR106 as the DNA source. Two main limitations of this technique should be mentioned. Because the phenotypic expression required an overnight liquid growth of the entire electroporation mixture, non independent transformed clones could be obtained. On the other hand, three other *C. perfringens* strains (CN504, JIR81, and ATCC3626B) failed to be electrotransformed using the standard conditions, underlining the strain dependency.

Allen and Blaschek (1988) succeeded in electrotransformation of *C. perfringens* strain 3624A with the *Streptococcus faecalis* plasmid pAMβ1 and the shuttle vector pHR106, using the osmotically protecting electroporation *"buffer E"* (270 mM sucrose, 1 mM MgCl$_2$ and 5 mM Na$_2$HPO$_2$4, adjusted to pH 7.4). The electric constants were approximately those of the preceding work (i.e., 25 μF, 6.25 kV cm^{-1}, 6 ms). After electroporation, the cell suspension was mixed with 5 vol of trypticase-glucose-yeast (TGY-) expression medium (which contains 25 mM CaCl$_2$, 25 mM MgCl$_2$, and 0.075% agar) and incubated for 3 h at 37°C before plating on selective agar medium. Transformation efficiency increased with voltage (from 0 transformant per μg DNA at 1.0 kV cm^{-1} to 1.4×10^2 transformants per μg DNA at 6.25 kV cm^{-1}) in parallel with the percentage of cell killing (from 6.4% to 81.8%). It should be noted that these results were obtained with bacteria harvested at mid-log phase of growth, and that the cells were not pretreated with any cell-wall-degrading agent.

Kim and Blaschek (1989) subsequently described a new, more efficient protocol for a closely related strain of *C. perfringens* (3626B). Cells harvested in late stationary phase of growth in TGY medium were washed once in electroporation "buffer B," which consists of 15% glycerol in distilled water. The pellet was then suspended in this same solution to give an unusually high concentration of bacteria per milliliter. About 1 μg of pAK201 plasmid DNA was mixed with 0.8 ml of the cell suspension. After application of the electric field (25 μF, 6.25 kV cm^{-1}), the mixture was diluted into 9 volumes of TGY medium, adding neither $CaCl_2$ nor $MgCl_2$, then incubated for 1 h at 37°C before plating on selective solid medium. The use of distilled water in "buffer B" instead of 5 mM Na_2HPO_4 increased the time constant (~40 ms versus ~6 ms) and as a consequence enhanced the transformation efficiency (10^4 transformants per μg DNA).

This last procedure, with slight modifications, has been used by Phillips-Jones (1990) for strain P90-2-2 of *C. perfringens*. Using the shuttle vector pSB92A2 as the DNA source, a level of 10^4 transformants per μg of DNA was also obtained. In her studies, the two electrotransformation buffers most often used have been tested: the SMP buffer initially described for *C. acetobutylicum* by Oultram et al. (1988), composed of 270 mM sucrose, 1 mM $MgCl_2$, 7 mM Na_2HPO_4, pH 7.4, and the "buffer B" described previously (Kim and Blaschek, 1989). Electroporation performed with solution B gave rise to a much higher transformation efficiency as compared to the SMP buffer (from 5- to 15-fold). It was also demonstrated that the DNA uptake was always elevated using stationary-phase cells: the later the cells were collected for electroporation, the higher the frequencies that were obtained. Taken together, these results may indicate that *C. perfringens* cells must be harvested for electroporation when the cell walls are as thin and weakly structured as possible; this occurs both in the early exponential growth stage and in the very late stage of stationary culture, when the cell-wall components have undergone partial degradation by autolysins.

A method for the introduction of plasmids into *Clostridium* acetobutylicum ATCC8052 by electroporation has also been published by Oultram et al. (1988). Cells were grown in 2 × YT broth [per liter: yeast extract, 10 g; bactotryptone, 16 g; NaCl, 4 g; supplemented with 0.5% (wt/vol) glucose] until mid-log phase and then harvested by centrifugation, washed once with the cold SMP electroporation buffer, and concentrated to 1:20 of the initial volume. About 0.5 μg DNA was added to 0.8 ml of the cell suspension in a Bio-Rad electroporation cuvette, and the cells were cooled on ice for 8 min before electroporation in a Bio-Rad Gene Pulser™. Standard conditions for efficient electroporation were 25 μF, 5 KV cm^{-1}, 4.7 to 6.9 ms. The cells were then held on ice for 10 min before being diluted into 8 volumes of 2 × YT broth and incubated at 37°C for 1 h before selective and nonselective plating. Using plasmid pMTL500E, transfer frequency varied between $8.0 × 10^1$ and $2.9 × 10^3$ transformants per μg plasmid DNA, with total cell survival between 0.7% and 23%.

In the author's laboratory, electrotransformation was also performed for NI-4 derivative strains of *C. acetobutylicum* (J. Hubert, personal communication). As the results have not yet been published, the general protocol is outlined here. Cells were cultivated in TYA medium [per liter: bacto-tryptone, 6 g; yeast extract, 2 g; ammonium acetate, 3 g; KH_2POH, 0.5 g; $MgSO_4 \cdot 7H_2O$, 0.4 g; $FeSO_4 \cdot 7H_2O$, 0.01 g; pH 6.5, supplemented with 1% (wt/vol) glucose] (Ogata and Hongo, 1973) until mid-log phase (~1 × 10^8 bacteria per ml). They were then harvested by centrifugation and washed four times with distilled water to which was added 100 mg l^{-1} cysteine HCl, and 30 mg l^{-1} ammonium acetate, at pH 5.5. The cells were resuspended in 1/25 to 1/50 of the initial volume in the electroporation solution (300 mM sucrose, 1 mM $MgCl_2$ in the initial washing solution) and kept on ice for 10 min. Plasmid DNA was added to 50 μl of the concentrated cells and transferred to a cold Bio-Rad electroporation cuvette (0.2 cm long). In the standard protocol, the electric constants were as follows: 25 μF, 400 Ω, 2.5 kV cm^{-1} giving a time constant varying from 6.5 to 8.5 ms. The electroporated mixture was anaerobically diluted with

1 ml of TYA medium and incubated for 1.5 h at 35°C to allow gene expression before plating on selective medium. Using pKNT11 vector as the DNA source, about 5×10^4 transformants per μg DNA were obtained. During the course of these experiments different conditions were tested. For strain NI-4 derivatives, only cells harvested at the beginning of the exponential growth phase ($\leq 10^8$ cells per ml) were electrotransformed with efficiency. The washing liquid and the electrotransformation solution must be readily reduced and the pH always higher than 5.0 (bacterial death occurs when the pH falls below 5). The percentage of cell survival after the electroporation stop depends on the voltage field for strain NI-4081. The results were as follows: 97%, 6.4%, 1%, and 0.7% for 2.5, 5, 7.5, and 10 kV cm^{-1}, respectively. Transformation was only achieved at 2.5 or 5 kV cm^{-1}. At 5 kV cm^{-1}, there was a linear relationship between DNA concentration and efficiency of transformation for 10 to 500 ng. Moreover, as already observed for C. perfringens strain 13 (Scott and Rood, 1989) only a limited concentration of cells was suitable for transformation in the electroporation cuvette (about 5×10^7 to 10^8 cells per experiment).

The general procedures of electrotransformation described for C. perfringens and C. acetobutylicum are very similar and, in fact, not so different from those defined for other grampositive microorganisms (Luchansky et al., 1988). Nevertheless, the efficiency of transformation varies dramatically depending on the recipient strain. Although an upper maximum threshold voltage exists for each microorganism, the electric field conditions defined are closely related from one strain to another. In contrast, the physiological stage of the cells that undergo electroporation largely influences the DNA uptake. In fact, the wall is not a true permeability barrier because it does sieve out large molecules such as DNA. This explains why growth conditions (i.e., medium, division rate, pH, etc.) that modified the composition and the complexity of the cell wall structure were so important in the electrotransformation process of vegetative cells. For example, Calvin and Hanawalt (1988) reported that for E. coli strain MC1061 the efficiency of transformation varied nearly by three orders of magnitude, depending on the growth conditions of the cells. These observations suggested that improvement in electrotransformation of Clostridium should be feasible in the near future. Results that have published so far on the electrotransformation both in C. perfringens and C. acetobutylicum are summarized in Table 7.2.

7.4   Perspectives

When the electroporation technique is used, it is not always necessary to purify the plasmid DNA to obtain transformants. Summer and Withers (1990) demonstrated that plasmid DNA in E. coli can be directly transferred from a plasmid-bearing strain to one without plasmids under the influence of an electric pulse. This situation has not yet been tested for the clostridia species, but the technique should be particularly useful for the movement of shuttle vectors between E. coli and clostridia. These techniques would lessen the need for the construction of large conjugative plasmids or the purification of genomic libraries. As transfer of DNA across the membrane is not oriented, some authors have used electroporation as a rapid method for plasmid curing (Heery et al., 1989). This technique was first developed for E. coli, in which curing frequencies of almost 80% were obtained. Such results have not yet been published for clostridia, but these new curing techniques would have great advantage if such a high efficiency can be achieved.

Electroporation has been also used by Calvin and Hanawalt (1988) to extract intact plasmid from transformed E. coli cells with efficiencies comparable to those obtained by traditional alkaline lysis or CsCl equilibrium density gradient techniques. As plasmid DNA preparations from clostridia are sometimes difficult to achieve with high yields, mainly because of the presence of nonspecific nucleases, this procedure may be a good alternative to the classical techniques. Kim et al. (1990) have already reported the purification by electroporation of the plasmid pDM6 of Clostridium acetobutylicum NCIB6444. The yield of DNA recovery was found to be much higher than with the alkaline procedure. However electroporation-induced plasmid release generated greater amounts of

Table 7.2 Electrotransformation of clostridia

Organism	DNA (size in kb; selected marker)	Transformation frequency[a]	Voltage (kV cm^{-1})/time constant (ms)[b]	Reference
Clostridium perfringens				
Strain 3624A	pAMβ_1 (26.5, Em^r)	1.4×10^2	6.25/6	Allen and Blaschek, 1988
Strain 3624A	pHR106 (7.9, Cm^r)	1.2×10^3	6.25/6	Allen and Blaschek, 1988
Strain 3624A	pAK201 (8, Cm^r)	1.0×10^4	6.25/40	Kim and Blaschek, 1989
Strain 13	pHR106 (7.9, Cm^r)	3.0×10^5	6.25/7.5	Scott and Rood, 1989
Strain P90.2.2	pSB92A2 (7.9, Cm^r)	1.2×10^4	6.25/5.7	Philips-Jones, 1990
Clostridium acetobutylicum				
Strain ATCC8052	pMTL500E (6.4, Em^r)	4×10^2–4×10^4	5/4.7–6.9	Oultram et al., 1988
Strain NI-4081	pKNT11 (6.5, Em^r)	1×10^3–5×10^4	5/6.5–8	Hubert et al., unpublished data (see Section 3.2)

[a] Transformation frequency is expressed per μg DNA.
[b] In all cases capacitance was 25 μFd.

single-stranded DNA than double-stranded DNA, possibly because only DNA molecules not tightly bound to the cell membrane of gram-positive bacteria may easily pass through the transient pore.

In conclusion, electroporation appears at present to be a convenient and promising technique for introducing foreign DNA into clostridial microorganisms. It can be assumed that electrotransformation efficiency will be enhanced for all bacterial species when the mechanism of electric field-induced DNA uptake will be completely elucidated. Basic studies, as for example that of Xie et al. (1990), should be of great interest for limiting the failure of experiments in this field and adapting the technique to almost all clostridial strains.

Acknowledgments. I thank M. Sebald and M. Goldner for their valuable assistance in preparing this manuscript and F. Georges for the secretarial assistance.

References

Allcock, E.R., S.J. Reid, D.T. Jones, and D.R. Woods. 1982. *Clostridium acetobutylicum* protoplast formation and regeneration. *Appl. Environ. Microbiol.* **43**:719–721.

Allen, S.P. and H.P. Blaschek. 1988. Electroporation-induced transformation of intact cells of *Clostridium perfringens*. *Appl. Environ. Microbiol.* **54**:2322–2324.

Azeddoug, H., J. Hubert, and G. Reysset. 1989. Characterization of a methyl-specific restriction system in *Clostridium acetobutylicum* strain NI-4081. *FEMS Microbiol. Lett.* **65**:323–326.

Bibb, M.J., J.M. Ward, and D.A. Hopwood. 1978. Transformation of plasmid DNA into Streptomyces at high frequency. *Nature* (London) **274**:398–400.

Calvin, N.M. and P.C. Hanawalt. 1988. High efficiency transformation of bacterial cells by electroporation. *J. Bacteriol.* **170**:2796–2801.

Chang, S. and S.N. Cohen. 1979. High frequency transformation of *Bacillus subtilis* protoplasts by plasmid DNA. *Mol. Gen. Genet.* **168**:111–115.

Chassy, B.M., A. Mercenier, and J. Flickinger. 1988. Transformation of bacteria by electroporation. *Trends Biotechnol.* **6**:303–309.

Heefner, D.L., C.H. Squires, R.J. Evans, B.J. Kopp, and M.J. Yarus. 1984. Transformation of *Clostridium perfringens*. *J. Bacteriol.* **159**:460–464.

Heery, D.M., R. Powell, F. Gannon, and L.K. Dunican. 1989. Curing of a plasmid from *E. coli* using high-voltage electroporation. *Nucleic Acids Res.* **17**:10131.

Hopwood, D.A. 1981. Genetic studies with bacterial protoplasts. *Annu. Rev. Microbiol.* **35**:237–272.

Hotchkiss, R.D. and M.H. Gabor. 1980. Biparental products of bacterial protoplast fusion showing unequal parental chromosome expression. *Proc. Natl. Acad. Sci. USA* **77**:3553–3557.

Jones, D.T., W.A. Jones, and D.R. Woods. 1985. Production of recombinants after protoplast fusion in *Clostridium acetobutylicum* P262. *J. Gen. Microbiol.* **131**:1213–1216.

Kim, A.Y. and H.P. Blaschek. 1989. Construction of an *Escherichia coli–Clostridium perfringens* shuttle vector and plasmid transformation of *Clostridium perfringens. Appl. Environ. Microbiol.* **55**:360–365.

Kim, A.Y., A.A. Vertes, and H.P. Blaschek. 1990. Isolation of a single-stranded plasmid from *Clostridium acetobutylicum* NCIB6444. *Appl. Environ. Microbiol.* **56**:1725–1728.

Knowlton, S., J.D. Ferchak, and J.K. Alexander. 1984. Protoplast regeneration in *Clostridium tertium*: isolation of derivatives with high frequency regeneration. *Appl. Environ. Microbiol.* **48**:1246–1247.

Kurose, N., T. Miyazaki, T. Kakimoto, J. Yagyu, M. Uchida, A. Obayashi, and Y. Murooka. 1989. Isolation of plasmids from thermophilic clostridia and construction of shuttle vectors in *Escherichia coli* and cellulolytic clostridia. *J. Ferment. Bioeng.* **68**:371–374.

Landman, O.E. and A. Forman. 1969. Gelatin-induced reversion of protoplasts of *Bacillus subtilis* to the bacillary form: biosynthesis of macromolecules and wall during successive steps. *J. Bacteriol.* **99**:576–589.

Levi-Meyruels, C., C. Sanchez-Rivas, and P. Schaeffer. 1980. Formation de bactéries diploïdes stables par fusion de protoplastes de *Bacillus subtilis* et effet de mutations *rec⁻* sur les produits de fusion formés. *C. R. Acad. Sci. Paris Sér. D* **291**:67–70.

Lin, Y. and H.P. Blaschek. 1984. Transformation of heat-treated *Clostridium acetobutylicum* with pUB110 plasmid DNA. *Appl. Environ. Microbiol.* **48**:737–742.

Luchansky, J.B., P.M. Muriana, and T.R. Klaenhammer. 1988. Application of electroporation for transfer of plasmid DNA to *Lactobacillus, Lactococcus, Leuconostoc, Listeria, Pediococcus, Bacillus, Staphylococcus, Enterococcus,* and *Propionibacterium. Mol. Microbiol.* **2**:637–646.

Mahony, D.E., J.A. Mader, and J.R. Dubel. 1988. Transformation of *Clostridium perfringens* L forms with shuttle plasmid DNA. *Appl. Environ. Microbiol.* **54**:264–267.

Minton, N.P. and J.G. Morris. 1983. Regeneration of protoplasts of *Clostridium pasteurianum* ATCC6013. *J. Bacteriol.* **155**:432–434.

Novick, R., C. Sanchez-Rivas, A. Gruss, and I. Edelman. 1980. Involvement of the cell envelope in plasmid maintenance: plasmid curing during the regeneration of protoplasts. *Plasmid* **3**:348–358.

Ogata, S. and M. Hongo. 1973. Bacterial lysis of *Clostridium species*. I. Lysis of *Clostridium* species by univalent cation. *J. Gen. and Appl. Microbiol.* **19**:251–261.

Oultram, J.D., M. Loughlin, T.J. Swinfield, J.K. Brehm, D.E. Thompson, and N.P. Minton. 1988. Introduction of plasmids into whole cells of *Clostridium acetobutylicum* by electroporation. *FEMS Microbiol. Lett.* **56**:83–88.

Pettinari, M.J., S.E. Ivanier, and B.S. Méndez. 1989. Protoplast formation and regeneration of a thermophilic *Clostridium* sp. *FEMS Microbiol. Lett.* **58**:255–258.

Phillips-Jones, M.K. 1990. Plasmid transformation of *Clostridium perfringens* by electroporation methods. *FEMS Microbiol. Lett.* **66**:221–226.

Podvin, L., G. Reysset, J. Hubert, and M. Sebald. 1988. Recent developments in the genetics of *Clostridium acetobutylicum. In: Anaerobes Today,* J.M. Hardie and S.P. Borriello, eds., pp. 135–140. Chichester: Wiley.

Reid, S.J., E.R. Allcock, D.T. Jones, and D.R. Woods. 1983. Transformation of *Clostridium acetobutylicum* protoplasts with bacteriophage DNA. *Appl. and Environ. Microbiol.* **45**:305–307.

Reysset, G., J. Hubert, L. Podvin, and M. Sebald. 1987. Protoplast formation and regeneration of *Clostridium acetobutylicum* strain NI-4080. *J. Gen. Microbiol.* **133**:2595–2600.

Reysset, G., J. Hubert, L. Podvin, and M. Sebald. 1988. Transfection and transformation of *Clostridium acetobutylicum* strain NI-4081 protoplasts. *Biotechnol. Techniq.* **2**:199–204.

Roberts, I., W.M. Holmes, and P.B. Hylemon. 1988. Development of a new shuttle plasmid system for *Escherichia coli* and *Clostridium perfringens. Appl. Environ. Microbiol.* **54**:268–270.

Sanchez-Rivas, C. 1982. Direct selection of complementing diploids from PEG-induced fusion of *Bacillus* protoplasts. *Mol. Gen. Genet.* **185**:329–333.

Scott, P.T. and J.I. Rood. 1989. Electroporation-mediated transformation of lysostaphin treated *Clostridium perfringens. Gene* **82**:327–333.

Shigekawa, K. and W.J. Dower. 1988. Electroporation of eukaryotes and procaryotes: a general approach to the introduction of macromolecules into cells. *Biotechniques* **6**:742–751.

Soutschek-Bauer, E., L. Hartl, and W.L. Staudenbauer. 1985. Transformation of *Clostridium thermohydrosulfuricum* DSM 568 with plasmid DNA. *Biotechnol. Lett.* **7**:705–710.

Squires, C.H., D.L. Heefner, R.J. Evans, B.J. Kopp, and M.J. Yarus. 1984. Shuttle plasmids for *Escherichia coli* and *Clostridium perfringens. J. Bacteriol.* **159**:465–471.

Stal, M.H. and H.P. Blaschek. 1985. Protoplast formation and cell wall regeneration in *Clostridium perfringens. Appl. Environ. Microbiol.* **50**:1097–1099.

Summer, D.K. and H.L. Withers. 1990. Electrotransfer: direct transfer of bacterial plasmid DNA by electroporation. *Nucleic Acids Res.* **18**:2192.

Takahashi, W., H. Yamagata, K. Yamaguchi, N. Tsukagoshi, and S. Udaka. 1983. Genetic transformation of *Bacillus brevis* 47, a protein-secreting bacterium, by plasmid DNA. *J. Bacteriol.* **156**:1130–1134.

Truffaut, N., J. Hubert, and G. Reysset. 1989. Construction of shuttle vectors useful for transforming *Clostridium acetobutylicum*. *FEMS Microbiol. Lett.* **58**:15–20.

Xie, T.D., L. Sun, and T.Y. Tsong. 1990. Study of mechanisms of electric field-induced DNA transfection. I. DNA entry by surface binding and diffusion through membrane pores. *Biophys. J.* **58**:13–19.

Yoshino, S., S. Ogata, and S. Hayashida. 1984. Regeneration of protoplasts of *Clostridium saccharoperbutylacetonicum*. *Agric. Biol. Chem.* **48**:249–250.

Yoshino, S., T. Yoshino, S. Hara, S. Ogata, and S. Hayashida. 1990. Construction of shuttle vector plasmid between *Clostridium acetobutylicum* and *Escherichia coli*. *Agri. Biol. Chem.* **54**:437–441.

Young, M., N.P. Minton, and W.L. Staudenbauer. 1989a. Recent advances in the genetics of the clostridia. *FEMS Microbiol. Rev.* **63**:301– 326.

Young, M., W.L. Staudenbauer, and N.P. Minton. 1989b. Genetics of *Clostridium*. *In*: *Clostridia*, Biotechnology Handbooks 3, N.P. Minton, and D.J. Clarke, eds., pp. 63–95. New York: Plenum.

8

Vectors for Use in *Clostridium acetobutylicum*

Nigel P. Minton, Tracy-Jane Swinfield, John K. Brehm, Sarah M. Whelan, and John D. Oultram

8.1 Introduction

In the last two decades, interest in using surplus/waste biomass in biocatalytic processes to produce chemicals and fuels of high added value, as an alternative to chemical synthesis from petroleum feedstocks, has exhibited considerable fluctuation. The underlying argument for such developments seems irrefutable. Fossil fuels are a finite resource. It follows that alternative technologies based on a renewable feedstock must eventually be developed. Impetus for undertaking the necessary research has until relatively recently been ruled by economic factors. Thus, although the world oil crisis of the 1970s saw a surge in the fortunes of potential biological processes, a subsequent reduction in petroleum prices dampened initial enthusiasm. Two factors have now refocused the world's attention on biocatalytic processes. First, from the economic standpoint, the Gulf Crisis of 1990–1991 serves to illustrate that the industrial world has become complacent in its reliance on access to plentiful and cheap petroleum reserves. Second, increased public awareness of the environmental pollution caused by fossil fuel transportation and exploitation suggests that economic considerations may soon relinquish their dominant role in policy decisions.

A group of bacteria that exhibits considerable potential in their biocatalytic capabilities is the genus *Clostridium* (Minton and Clarke, 1989). One species, *Clostridium acetobutylicum*, is of particular interest, having been employed in a pre-oil-rich world economy to produce the industrial solvents acetone and butanol. The possibility that the fermentation process it produces could experience a revival on an industrial scale is now gathering momentum. A prerequisite for its future use is the generation of strains with improved fermentation characteristics, e.g., growth on an extended range of substrates and improved product yields. Realization of this goal depends absolutely on the derivation of recombinant cloning vehicles and ancillary gene transfer methodologies with which to effect the desired manipulations. Much of the groundwork for establishing these enabling technologies has now been accomplished, and in this chapter we review the current status of the vector systems available for recombinant work in *C. acetobutylicum*.

8.2 Plasmid pAMβ1, a Tool for Host–Vector Development

During the period that the first procedures for transforming *Clostridium acetobutylicum* were being formulated, a major concern was that the plasmid DNA employed both was capable of replicating in this particular clostridial host and conferred on the cell a readily detectable phenotype. In contrast to proteolytic clostridia such as *Clostridium perfringens*, no function had been assigned to any plasmid isolated from *C. aceto-butylicum*. The cryptic nature of their plasmids appears to be a general property of all saccharo-lytic clostridia examined to date (Minton and Thompson, 1990). Apparent options included: (i) deriving a plasmid chimaera by endowing a cryptic saccharolytic clostridial plasmid with an antibiotic resistance gene known to function in clostridia; or (ii) using a gram-positive plasmid isolated from another bacterium that was known to, or was likely to, replicate and ex-press its antibiotic resistance gene in *C. aceto-butylicum*. Both strategies have been successful-ly adopted to develop host–vector systems for *C. acetobutylicum*. During the course of these studies one plasmid, pAMβ1, has proved to be of particular utility.

Plasmid pAMβ1 is a large (26.5 kb) conjuga-tive R factor, originally isolated from *Enterococ-cus* (formerly *Streptococcus*) *faecalis* (see Clewell, 1981), which mediates resistance to the macro-lide, lincosamide, and streptogramin B group of antibiotics (MLS) of which erythromycin (Em) is an example. The promiscuous nature of this conjugative plasmid among gram-positive bacteria (reviewed by Horaud et al., 1985) was substantiated by Oultram and Young (1985) when it was demonstrated that pAMβ1 could be transferred to *C. acetobutylicum* from a num-ber of donors (e.g., *Streptococcus lactis*, *Entero-coccus faecalis*, *Bacillus subtilis*, and other strains of *C. acetobutylicum*) at frequencies between 1.4×10^{-3} and 4.1×10^{-5} transconjugants per donor cell (see Chapter 6, this volume, for more details). Transfer of pAMβ1 had three major consequences: (i) it provided identifica-tion of a functional resistance gene that could be employed in the construction of *C. acetobuty-licum* vectors; (ii) it provided a replication origin that could replicate in *C. acetobutylicum*; and (iii) it formed the basis of a potential gene transfer system for introducing genetic information into *C. acetobutylicum*. In the latter case, this poten-tial was subsequently exploited to mobilize a coresident cloning vector from *B. subtilis* into *C. acetobutylicum* in the form of a cointegrate molecule (Oultram et al., 1987). During these mobilization experiments, a second antibiotic resistance gene that could function in clostri-dia, the staphylococcal plasmid pC194 chloram-phenicol (Cm) resistance gene, was identified.

Construction of replicon probe vectors

The finding that both the pAMβ1 Em^r gene and the pC194 Cm^r gene expressed in *C. acetobutyli-cum* was exploited by constructing some gen-eral purpose gram-positive replicon probe vec-tors (Swinfield et al., 1990). Based on the multipurpose *E. coli* cloning vectors pMTL20 and pMTL21 (Chambers et al., 1988), each gram-positive resistance gene was isolated as a specific DNA fragment (Em^r gene, 1.12-kb *Hha*I; Cm^r, 1.07-kb *Hpa*II), sequenced (Brehm et al., 1987; Minton et al., 1990a), and inserted into pMTL20 and pMTL21 such that the en-coded *lacZ'* gene was not inactivated. The re-sultant vectors pMTL20E/C and pMTL21E/C (Figure 8.1) therefore retain all the cloning advantages of their parental plasmids. Thus putative clostridial replicons may be cloned into 1 of the 18 unique cloning sites of the polylinker region encompassed by *lacZ'*, and recombinant clones are detected in an appro-priate *E. coli* host as colorless colonies on agar medium supplemented with 5-bromo-4-chloro-3-indolyl-β-D-galactoside (XGal). The plasmid chimaerae identified may then be tested for their ability to replicate (Rep$^+$) in a gram-positive bacterium, following transformation and selection for Em^r or Cm^r transformants.

Transformation of *Clostridium acetobutylicum* with pAMβ1-derived vectors

The broad host range of plasmid pAMβ1 has been studied by a number of laboratories, cul-minating in, among other details, the deriva-tion of smaller deletion variants and localiza-tion of its various determinants, including replication functions. Two particular pAMβ1-derived plasmids, pVA1 and pVA677, were used by Reysset et al. (1988) to develop plasmid transformation methodology for protoplasts of an autolysin mutant of *Clostridium saccharoper-butylacetonicum* strain N1-4080 (equivalent to *C.*

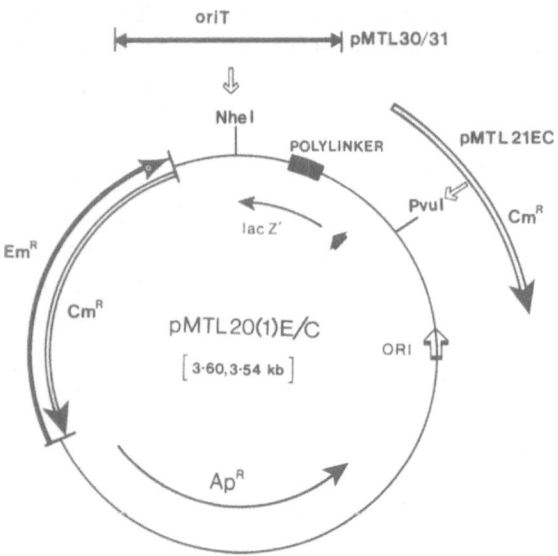

FIGURE 8.1 The pMTL gram-positive replicon cloning vectors and their "conjugative" derivatives. Antibiotic resistance genes were carried on a 1.12-kb *Hha*I fragment (pAMβ1 Em^r; Brehm et al., 1987) and a 1.07-kb *Hpa*II fragment (pC194 Cm^r; Minton et al., 1990a). Open arrow on plasmid circumference represents ColE1 origin of replication; solid, heavy black arrow within plasmid circle is *lac* promoter. Polylinker regions, containing multiple cloning sites, are those of pMTL20 and pMTL21 (Chambers et al., 1988). Indicated *Nhe*I site represents position at which pAMβ1 replicon was inserted in derivation of pMTL500E (see Figure 8.2), and the *oriT* fragment of RK2 used to generate pMTL30/pMTL31 (Williams et al., 1990a). Also illustrated is the *Pvu*II site into which pC194 *cat* gene was inserted to generate cointegrate cloning vector pMTL21EC (Oultram et al., 1988b).

acetobutylicum). The frequencies attained were between 10^4 and 10^5 transformants per µg DNA. The isolation of this particular mutant, N1-4081, was central to the success of the devised transformation procedure, forming part of a series of measures taken to limit autolysin activity (see Chapter 7 for more details).

In our own laboratory we built on previous data that had localized the pAMβ1 replicon to an approx. 5.0-kb *Eco*RI fragment (Leblanc and Lee, 1984) to clone and physically characterize the pAMβ1 replication functions (Swinfield et al., 1990). Thereafter, a *Clostridium/Escherichia coli* shuttle vector, pMTL500E, was derived

(Figure 8.2) by cloning an appropriate fragment into pMTL20E. This vector was then employed to develop an electroporation procedure for *C. acetobutylicum* NCIB 8052 (Oultram et al., 1988a), with pMTL500E transforming at frequencies of about 10^3 per µg DNA. The methodology employed has been slightly modified since its publication; melibiose is now routinely used in preference to sucrose in the electroporation buffer and the electric pulse is applied to a smaller volume of cells (0.3 ml) using 0.2-cm Bio-Rad cuvettes (field strength, 6.25 kV cm^{-1}, 25 µF, 400 Ω). In addition, the DNA of plasmids based on the pAMβ1 replicon is routinely prepared from *Bacillus subtilis*, the transformation frequencies obtained being reproducibly higher than when the same DNA is prepared from *E. coli* (J.D. Oultram and T.-J. Swinfield, unpublished observations).

Other sources of replication regions

In recent years, a significant number of clostridial cloning vectors not based on pAMβ1 have been derived using plasmid replication functions native to various different gram-positive bacteria. In our laboratory, clostridial replicons were identified by inserting restriction fragments derived from cryptic clostridial plasmids into the polylinker region of pMTL20E. The plasmids utilized (Table 8.1) were the three cryptic *Clostridium butyricum* plasmids pCB101, pCB102, and pCB103 (Minton and Morris, 1981), and two cryptic plasmids, pCP1 and pCP2, isolated from *Clostridium paraputrificum* (Swinfield and Minton, unpublished data). In the case of pCB102, pCB103, pCP1, and pCP2, the choice of restriction fragment was made on a random basis. In the case of pCB101, earlier work had demonstrated that a 3.3-kb *Sau*3A restriction fragment of pCB101 promoted the establishment of a replication deficient plasmid in *B. subtilis* (Collins et al., 1985). With the exception of pCP2, all clostridial plasmids gave at least one derivative able to transform our strain with varying degrees of efficiency (Table 8.1). The failure of the pCP2-derived plasmid to transform is assumed

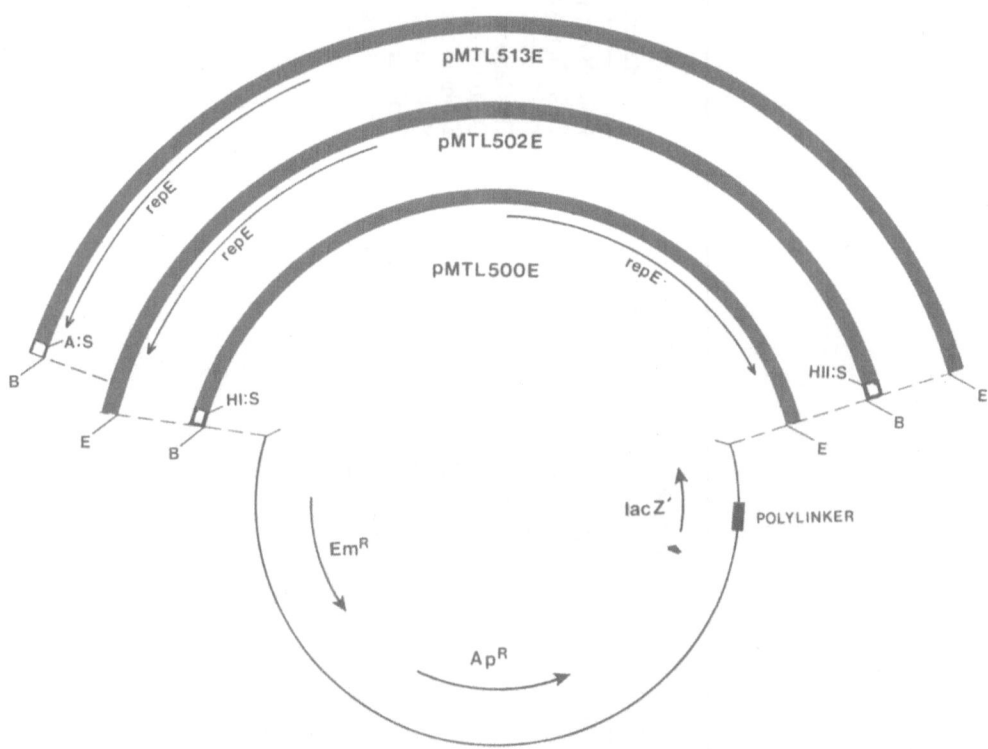

FIGURE 8.2 Cloning vectors based on the pAMβ1 replicon. The three plasmids illustrated were generated by insertion of indicated pAMβ1-derived DNA (see Swinfield et al., 1990) fragment (heavy black line) into the *Nhe*I site of pMTL20E (light line). The *lacZ'* is therefore functional (blue colonies in presence of XGal) unless inactivated by subsequent insertion of heterologous DNA into one restriction site of the polylinker region. Plasmid pMTL500E is a high-copy-number plasmid; pMTL502E and pMTL513E have a low copy number. Restriction sites are A, *Acc*I; B, *Bam*HI; E, *Eco*RI; HI, *Hpa*I; HII, *Hpa*II; S, *Sma*I: colon indicates fusion. General-purpose cloning vectors pMTL500E and pMTL502E exhibit moderate segregational stability; plasmid pMTL513E is a prototype segregational stability determinant cloning vector exhibiting extreme instability in both *Bacillus subtilis* and *Clostridium acetobutylicum*.

to result from inactivation of an essential part of the plasmid replicon during cloning.

Yoshino et al. (1990) have constructed cloning vectors for *C. acetobutylicum* by combining a 3.0-kb cryptic plasmid isolated from *C. acetobutylicum* strain No. 86 with the *E. coli* promoter probe vector pKK232-8. Transformation of *C. acetobutylicum* with the constructed vectors was elicited by a procedure similar to that previously devised for *C. thermohydrosulfuricum* (Soutschek-Bauer et al., 1985). Harvested cells were subjected to treatment with alkaline-Tris buffer before being incubated with plasmid DNA in the presence of polyethylene glycol 6000. Transformants (selected on the basis of

Cm resistance) arose at frequencies of 10 to 10^3 per μg DNA, dependent on the plasmid construction employed (see Table 8.1).

Other workers have constructed vectors based on plasmids isolated from aerobic gram-positive bacteria. Of greatest promise are those vectors based on the *B. subtilis* plasmid pIM13. This small, 2.2 kb-plasmid encodes *erm*C. Initially shown to transform *C. acetobutylicum* N1-4081 protoplasts at frequencies of 1 to 6×10^5 transformants per μg DNA (Truffaut et al., 1989), this plasmid has been used to construct a number of bifunctional plasmids (Table 8.1). These include a chimaera with pBR322 (pKNT11) and a plasmid in which the pT127

Table 8.1 Clostridial cloning vectors

Vector	Size (kb)	Source of replicon		Replication in		Transformation frequency[a]	Segregational stability[b]	Reference
		Plasmid	Organism	Bacillus	Clostridium			
pCB4	7.10	pCB102	B. butyricum	−	+	2.9×10^3	ND[c]	This laboratory
pCB5	9.50	pCB103	B. butyricum	−	+	1.1×10^2	ND	This laboratory
pCP3	10.40	pCP1	C. paraputrificum	−	+	3.5×10^2	6.5×10^{-2}	This laboratory
pCB3	7.03	pCB101	C. butyricum	+	+	4.0×10^4	1.4×10^{-2}	This laboratory
pMTL500E	6.43	pAMβ1	E. faecalis	+	+	2.9×10^3	4.1×10^{-2}	Oultram et al., 1988a
pTYD101	7.1	pCS86	C. acetobutylicum	ND	+	1.0×10^3	6.9×10^{-2}	Yoshino et al., 1990
pTYD104	3.5	pCS86	C. acetobutylicum	ND	+	1.0×10^1	4.3×10^{-1}	Yoshino et al., 1990
pKNT11	6.5	pIM13	B. subtilis	+	+	2.0×10^6	5.0×10^{-2}	Truffaut et al., 1989
pKNT14	4.3	pIM13	B. subtilis	+	+	3.0×10^5	1.2×10^{-2}	Truffaut et al., 1989

[a]Expressed as transformants of C. acetobutylicum per μg of DNA.
[b]Expressed as plasmid loss per C. acetobutylicum cell per generation.
[c]ND, not done.

Tc^r determinant has been inserted (pKNT14). More recently, shuttle vectors have been made by combining pIM13 with either pUC18 or pUC19 (G. Reysset, personal communication; see also Chapter 7).

8.3 Characterization of Clostridial Replicons

An emergent trend from studies on clostridial replicons concerns the host range of the constructed *E. coli/Clostridium* shuttle vectors. In our own studies, of the four clostridial replicons identified only pCB101 was able to function, albeit inefficiently, in *Bacillus subtilis*. This may indicate that clostridial replicons are functionally distinct from those found in aerobic gram-positive bacteria (Gruss and Ehrlich, 1989). Indeed, Phillips-Jones recently reported that a bifunctional vector constructed for use in *C. perfringens* was similarly unable to replicate in *B. subtilis* (Phillips-Jones, 1990). Although no other systematic studies of vector host range have apparently been undertaken, it is interesting to note that vectors based on two *Bacillus* plasmids (pBC16 and pT127) replicated poorly in *C. acetobutylicum* (Truffaut et al., 1989). A more definitive judgment as to whether clostridial plasmids generally rely on a different class of replicative apparatus will require a more detailed analysis of a greater number of plasmids. Some insight into this phenomenon has been gleaned by nucleotide sequence analysis of the plasmids pCB101 and pCB102 (J.K. Brehm et al., unpublished observations).

Physical characterization of the replicon of pCB101

Physical characterization of the pCB101 replication functions has provided some clues as to why this particular clostridial replicon is able to function in *B. subtilis*. Two previous independent studies established that restriction fragments derived from pCB101 enabled replication-deficient plasmids to transform *B. subtilis* (Collins et al., 1985; Luczak et al., 1985). In the former study, the plasmid obtained,

pJAB1, was reported to carry a 3.3-kb *Sau*3A, pCB101-derived fragment and appeared to replicate inefficiently in this particular grampositive host. Thus, autonomous plasmid DNA could only be detected in transformants using DNA–DNA hybridization techniques. The subsequent determination of the complete nucleotide sequence of pCB101 (Brehm and Minton, unpublished data) precisely sized the cloned fragment as 3.48 kb, and showed it to encode three major ORFs, designated A, B, and C (the respective M_r of encoded polypeptides were 11,383, 42,910, and 27,134).

Using this information, the 3.48-kb *Sau*3A fragment was further dissected (Brehm and Pennock, unpublished data) to show that all functions necessary for the autonomous replication of pCB101 in *B. subtilis* resided on a 2.5-kb subfragment, encoding open-reading frames (ORFs) B and C, composed of two contiguous *Taq*I fragments (Figure 8.3). Any attempt to reduce this fragment still further resulted in plasmids unable to replicate (Rep$^-$) in *B. subtilis*. Thus deletion of the small, 104-bp *Taq*I fragment encoding the 3' end of ORF B gave a Rep$^-$ phenotype, while removal of the 237 bp of DNA between the *Taq*I and *Aat*II site preceding ORF C also resulted in a Rep$^-$ plasmid. A major effect of the former deletion was to remove 9 codons from the 3' end of ORF B. One interpretation of the Rep$^-$ phenotype of the resultant plasmid is that ORF C is required for replication and that removal of the 9 C-terminal amino acids interferes with this role in replication. The latter deletion does not impinge on the structural integrity of either ORF B or C; however, it does delete a number of palindromic sequences that occur immediately 5' to the ORF C translational start codon and may therefore effect transcription of the downstream ORF(s). In a further experiment it was shown (Brehm and Minton, unpublished observations) that the introduction of a frameshift mutation at the unique *Xmn*I site of a plasmid carrying the 3.48-kb *Sau*3A fragment (such that the last 97 amino acids of the 130-long ORF C polypeptide were replaced with 13 "nonsense" amino acids) also resulted in a Rep$^-$ plasmid (see Figure 8.3). This result strongly suggests that ORF C is required for re-

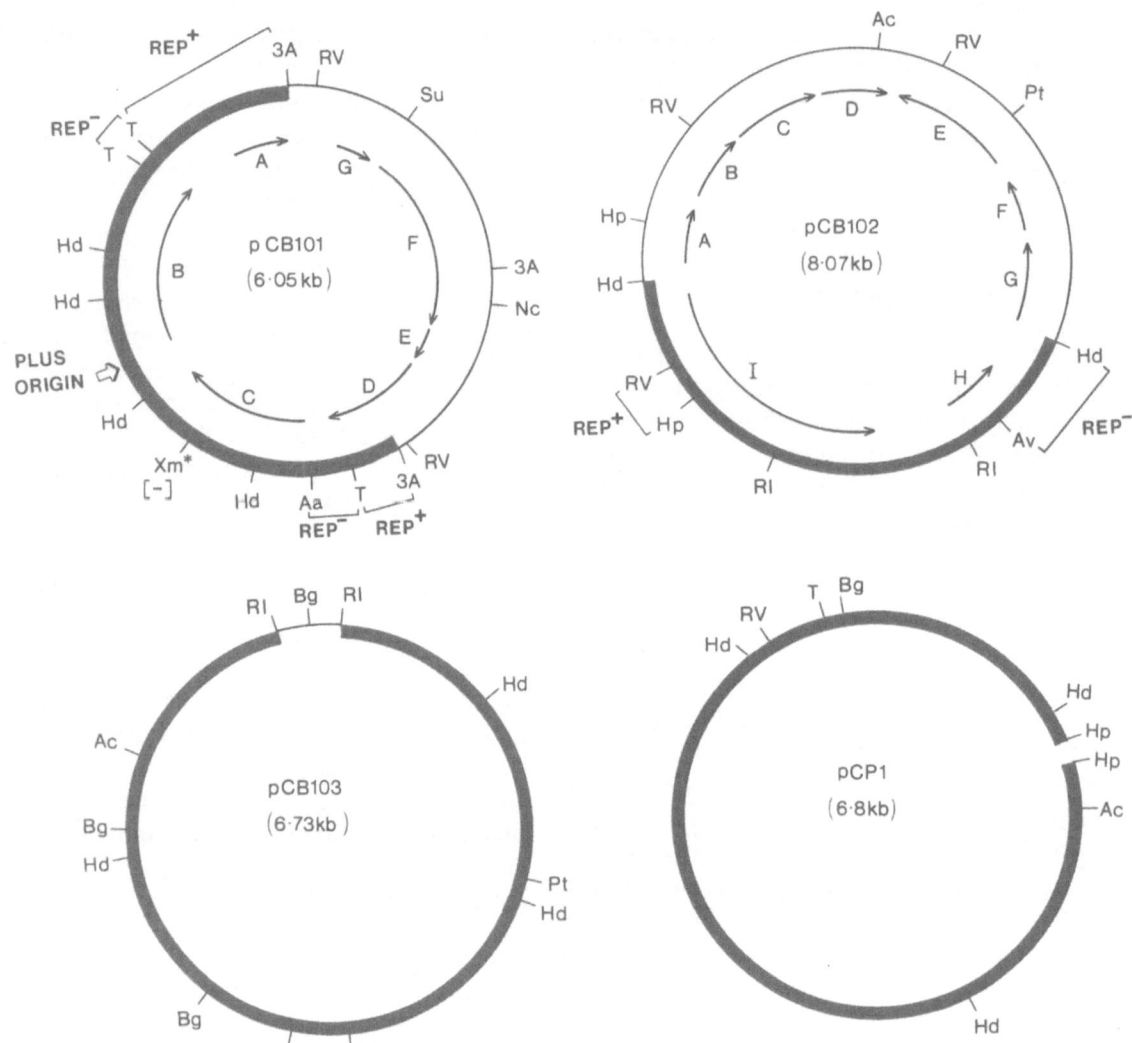

FIGURE 8.3. Sources of clostridial replicons. Extent of the illustrated plasmids used in construction of plasmids listed in Table 8.1 is indicated by heavy black line. The ORFs of plasmids pCB101 and pCB102, derived by nucleotide sequencing, are indicated by arrowed lines within the respective maps. Restriction enzyme sites are Aa, *Aat*II; Ac, *Acc*I; Av, *Ava*III; Bg, *Bgl*II; Hd, *Hind*III; Hp, *Hpa*I; Nc, *Nco*I; Pt, *Pst*I; RI, *Eco*RI; RV, *Eco*RV; Su, *Stu*I; T, *Taq*I; Xm, *Xmn*I; 3A, *Sau*3A. Regions between bracketed sites represent specifically deleted segments of DNA, which in some cases destroyed replicative ability of the plasmid (REP−). A frameshift (*) introduced at the Xm site of pCB101 also conferred a REP− phenotype on the resultant plasmid.

plication from the pCB101 origin. In parallel to these experiments, the various deletion variants were used as templates in a *E. coli*-directed *in vitro* transcription/translation system. Sodium dodecylsulfate-polyacrylamide gel electrophoresis (SDS-PAGE) analysis of the translated products confirmed the presence of polypeptides corresponding in size to the predicted ORFs B and C polypeptides. Further, the levels of both polypeptides were significantly reduced in the plasmid variant lacking the sequences 5′ to the *Aat*II site preceding ORF C. This result would be consistent with, among other factors, a reduction in transcriptional activity.

If the ORF B and C polypeptides are involved in replication, then the likelihood exists that they could share common sequences with other characterized plasmid-encoded replication proteins. A comparative analysis has indeed shown that the ORF B polypeptide exhibits significant homology to the Rep proteins of staphylococcal plasmids (pC194 and pUB110) and phage θX174 (see Minton et al., 1990b). During the past decade, a considerable amount of information has been amassed on the replication strategy of vectors of this type. It is now clear that they replicate via a rolling-circle mechanism; during the course of replication, single-stranded (ss) DNA intermediates are generated. DNA replication is initiated when the replication protein (Rep) introduces a nick in the positive DNA strand at a fixed position, termed the plus origin. As replication proceeds, the old positive strand is displaced concomitant with synthesis of a new positive strand by 3'-OH extension from the nick. Synthesis of a new negative strand, using the displaced positive strand as a template, is mediated by host factors that recognize the minus origin (for reviews, see Baas, 1985; Gruss and Ehrlich, 1989).

The significant homology between the pCB101 ORF B polypeptide and the Rep proteins of the pC194/pUB110 family of plasmids strongly suggests that pCB101 also replicates via a rolling-circle mechanism. This hypothesis is substantiated by the presence of a nucleotide sequence motif 5' to ORF B that is very similar to the plus origins of this same group of vectors (see Minton et al., 1990b). Indirect evidence that this sequence does indeed represent a plus origin of replication of pCB101 was provided by the observation that this sequence acts as a "hot spot" for deletion formation in certain plasmid constructions (Brehm and Minton, unpublished data). The compelling suggestion that pCB101 is indeed representative of the ssDNA family of plasmids is further supported by our observations that *B. subtilis* cells harboring pCB3 (pMTL20E carrying the 3.4-kb *Sau*3A fragment of pCB101) accumulate ssDNA (Brehm and Minton, unpublished observations).

Thus it appears that pCB101 possesses features that place it in the ssDNA family of gram-positive plasmids. Its replicative strategy, however, does not completely conform to dogma because a second polypeptide, that encoded by ORF C, also appears to be required for replication. The ubiquity of ssDNA plasmids in aerobic gram-positive bacteria has resulted in their almost exclusive use as the basis for cloning vehicles (Gruss and Ehrlich, 1989). It is now apparent that this type of replicative strategy is easily perturbed (i.e., by the insertion or deletion of DNA during vector construction), leading to structural and segregational instability. This general observation suggests that pCB101 may not represent the most appropriate basis for a clostridial cloning vector.

Physical characterization of the pCB102 replicon

Nucleotide sequence analysis of the entire pCB102 plasmid (Thompson et al., unpublished data) has indicated (see Figure 8.3) the presence of nine putative ORFs. All the functions necessary for the autonomous replication of pCB102 were initially localized to a 3.57-kb *Hin*dIII fragment that encodes two of the identified ORFs. The larger of the two, ORF I, encodes for a protein with a predicted M_r 68,800, while the second smaller ORF (H) encodes for a polypeptide of M_r 11,800. Not unexpectedly, neither protein exhibits any homology to any known replication protein.

Further dissection of this replication region is currently being undertaken. Data obtained to date (Thompson and Minton, unpublished observations) indicates that ORF I is not required for replication. Thus, insertion of a 2.79-kb *Hpa*I-*Hin*dIII subfragment, from which 36% of the 5' end of ORF I has been deleted, into pMTL20E results in a plasmid that is still able to replicate in *C. acetobutylicum* (see Figure 8.3). Similarly, deletion of the ORF I encoding DNA between the *Eco*RV and *Hpa*I sites has no effect on the replicative ability of a pMTL20E recombinant plasmid. In contrast, a pMTL20E plasmid recombinant derived by inserting a 2.93-kb *Hin*dIII-*Ava*III fragment was unable to transform *C. acetobutylicum*. This observation may indicate that ORF H is required for replication.

Although the identification and encoding

significance of sequences involved in pCB102 replication are not defined, a number of features of the plasmid warrant further discussion. It is notable that in contrast to the ssDNA plasmid pCB101, which is coresident with pCB102 in *C. butyricum* NCIB 7423, the ORFs of pCB102 are not encoded by the same DNA strand. The confinement of ORFs to the one DNA strand appears to be a general feature of ssDNA plasmids. The strain from which pCB101 and pCB102 were isolated is bacteriocinogenic, elaborating butyricin 7423 (Clarke and Morris, 1976). As bacteriocins are almost always encoded by plasmids, one of these two plasmids might be expected to encode butyricin 7423. Interestingly, the polypeptide encoded by ORF C has a number of features reminiscent of bacteriocins, including a high polarity index and a high content of glycine residues (7%). Further, it possesses at its extreme N terminus a 27-amino-acid motif typical of gram-positive signal peptides, implying an extracellular location for the translated protein. Although other bacteriocins do not rely on this type of signal for their secretion from the cell, this does not in itself rule against the possibility that the ORF C polypeptide corresponds to the protein previously designated butyricin 7423. Indeed, the predicted M_r of the protein following removal of the putative signal peptide corresponds almost exactly to that estimated (32,500 daltons) for purified butyricin 7423 (Clarke et al., 1975).

8.4   Features of Nonclostridial Replicons

As intimated earlier, of the clostridial vectors constructed to date the most useful have been based on replication functions of a nonclostridial origin, namely the *Enterococcus* plasmid pAMβ1 and the *Bacillus* plasmid pIM13. The replication regions of these two plasmids are now considered in turn.

The pAMβ1 replicon

As indicated, the demonstration that plasmid pAMβ1 could replicate in *Clostridium acetobuty-*

licum led us to undertake a detailed characterization of its replication region to facilitate future vector construction. Because all functions necessary for the autonomous replication of pAMβ1 had previously been shown to be localized to a 5-kb *Eco*RI fragment (Leblanc and Lee, 1984), we cloned this fragment into pMTL20E and used the resultant plasmid as a source of DNA for nucleotide sequence analysis. The fragment proved to be 5.1 kb long, to exhibit a G + C content of 34.3%, and to encode 7 putative ORFs greater than 90 amino acids (Swinfield et al., 1990). One of the most striking features immediately apparent was an area of highly repetitive DNA. Extending for 459 bp, this region was composed of two classes of directly repeated 9-bp sequences. The first 14 nonamer units conformed to consensus GTYGATCCA, while the latter 37 repeats could be represented by the sequence ACRGARCCA. This region subsequently proved to be nonessential for replication and the encompassing ORF (C) to be encoding. The polypeptide produced therefore contained reiterated tripeptide amino acid sequences of VDP (GTYGATCCA) and TEP (ACRGARCCA). The protein was also characterized by the possession of a sequence at its N terminus typical of procaryotic signal peptides (von Heijne, 1983) and an amino acid sequence motif at its C terminus that resembled the membrane-anchor sequence of a number of gram-positive proteins (see Vos et al., 1989). As all these features indicated that the ORF C polypeptide was localized to the cell surface, we suggested that the protein may play a role in the conjugation process (Swinfield et al., 1990). This finding has subsequently been reinforced by the publication of the nucleotide sequence of the region of plasmid pPD1 encoding the major pheromone-inducible protein PD78, believed to contribute to bacterial conjugation (Nakayama et al., 1990). The deduced amino acid of this conjugal protein also contains a tandem repeated sequence of XXP.

Having determined the nucleotide sequence of the 5.1-kb *Eco*RI fragment, we undertook a detailed dissection of this fragment to determine the regions of DNA important in replication (Swinfield et al., 1990). These studies showed that the largest ORF (E) encoded a polypeptide of M_r 57,380 that was essential for replication and was designated RepE. Disruption of the continuity of *repE* by either the deletion or the insertion of DNA conferred on the resultant plasmid a Rep⁻ phenotype in *B. subtilis*. Evidence was obtained indicating that RepE positively regulates plasmid copy number. Thus, deletion of a region of DNA 5' to ORF E caused a large increase in plasmid copy number concomitant with higher expression of RepE protein. A similar conclusion has been drawn for the homologous protein RepA of the related plasmid pSM19035 (Sorokin and Khazak, 1989). Although the minimum fragment capable of supporting the replication of plasmid chimaera in *B. subtilis* was shown to be a 2588-bp *HpaI-AccI* fragment, subsequent studies have indicated that the 229-bp *AccI-EcoRI* fragment adjacent to this region may be important in replication. Thus a low-copy-number plasmid carrying the entire 5.1-kb *EcoRI* fragment in which this small *Acc-EcoRI* fragment was deleted exhibited extreme segregational instability (see following).

The realization that ssDNA plasmids can suffer from structural instability when derivatized has led others to screen for plasmids that do not generate ssDNA during replication. One such plasmid was pAMβ1. Indeed, plasmids based on pAMβ1 have subsequently proven to exhibit a significantly higher degree of structural stability than vectors based on the ssDNA plasmids, e.g., pC194. The detailed dissection of the pAMβ1 replication region has allowed the construction of a number of different bifunctional *E. coli/Clostridium* shuttle vectors (see Figure 8.2), exhibiting both low and high copy number (Oultram et al., 1988a; Swinfield et al., 1990). In each case, the pAMβ1-derived replication fragment was inserted into the unique *NheI* site immediately external to the *lacZ'* gene of pMTL20E/C. The resultant vectors therefore retain all the cloning advantages of the replicon probe vectors, namely, insertion of heterologous DNA into their multiple cloning sites results in the inactivation of *lacZ'* allowing the selection of recombinant clones in appropriate *E. coli* hosts as colorless colonies on agar medium supplemented with XGal. It should be noted that, because pAMβ1 possesses such a

broad host range among gram-positive bacteria, these plasmids should prove useful in the analysis of many different bacterial species.

The pIM13 replicon

Originally isolated from *B. subtilis*, pIM13 is a 2246-bp plasmid that encodes a constitutively expressed adenine methylase highly homologous to that encoded by the *ermC* gene of the staphylococcal plasmid pE194 (Monod et al., 1986). A single 18-kDa protein (RepL) is required for replication and, although not rigorously characterized, its possession of lagging-strand conversion signals (*palA* and *palB*) and the fact that cells carrying this plasmid accumulate ssDNA (Projan et al., 1987) clearly indicate it belongs to the ssDNA class of plasmids (Gruss and Ehrlich, 1989). On the basis of similarities at the nucleotide sequence level and functional organization, staphylococcal ssDNA plasmids may be further subdivided into four distinct families, represented by the prototype plasmids pT181, pC194, pSN2, and pE194 (Novick, 1989). Plasmid pIM13 falls into the pSN2 family. Although plasmids from all four families are able to replicate in *B. subtilis*, it now seems likely that only pIM13-like vectors may be efficiently maintained in *C. acetobutylicum*. Thus pBC16Δ1 (pC194 family), pT127 (pT181 family) and pE194 (family prototype) all replicate inefficiently in this clostridial species (Truffaut et al., 1989; Reysset, personal communication). It is noteworthy that the plasmid family to which pIM13 belongs are the only ssDNA plasmids in which the direction of replication is opposite to that of the direction of transcription of the *rep* gene (Novick, 1989).

8.5 Plasmid Stability

The development of both transfer methodology and cloning vehicles for *C. acetobutylicum* NCIB 8052 has now opened up the possibility of introducing heterologous and homologous genes. In considering such experiments, it is desirable that the integrity of the introduced DNA is maintained during plasmid replication and that it is partitioned with high fidelity to

Table 8.2 Resolvase-mediated stabilization of pMTL500E

Vector	Resolvase	Segregational stability[a]	
		Bacillus subtilis	*Clostridium acetobutylicum*
pHV1431	+	1×10^{-4}	ND[b]
pHV1432	−	5×10^{-2}	ND
pMTL500E	−	8×10^{-2}	4.1×10^{-2}
pMTL531E	+	5×10^{-3}	4.2×10^{-4}
pMTL500E*res*[+]	+	4×10^{-3}	ND
pMTL500E*res*[−]	−	2×10^{-1}	ND

[a] Expressed as plasmid loss per cell per generation.
[b] ND, not determined.

each successive generation of cells. The recombinant plasmid used thus must exhibit both structural and segregational stability.

Structural stability

To date no systematic studies have been designed to test the structural stability of plasmid vectors in *C. acetobutylicum*. The data available are therefore only qualitative. In our own laboratory plasmid DNA of any of our constructed vectors, when isolated from *C. acetobutylicum*, has yet to show any gross structural rearrangements. Such observation also apply to pCB101-based vectors, even though dogma suggests that its replication strategy may result in instabilities. Plasmids based on the ssDNA plasmid pIM13 have also not been observed to exhibit any measurable structural instability (see Chapter 7). In contrast, a plasmid (pTY10) based on the cryptic *C. acetobutylicum* plasmid pCS86, although apparently stably maintained in *E. coli*, exhibited extreme structural instability when introduced into *C. acetobutylicum* (Yoshino et al., 1990). Examination of the plasmids of individual transformants revealed a range of different-sized plasmids (from 4.0 to 7.6 kb).

Segregational stability

There are currently no published examples of a *C. acetobutylicum* vector that exhibits 100% segregational stability in the absence of antibiotic selection. Thus, although the bifunctional vectors constructed in our laboratory have proven to be stably maintained in the presence of Em,

in the absence of antibiotic selection the three plasmids examined exhibited varying degrees of segregational instability (see Table 8.1). Analogous problems of instability have been observed with plasmids based on pIM13 (Reysset et al., 1989; E. Papousakis, personal communication) and pSC86 (Yoshino et al., 1990). Stabilization of introduced recombinant material could be achieved either by endowing the plasmid vehicle with a stability determinant or by promoting integration of the recombinant gene into the host genome. Although we are currently examining both possibilities, only details of former avenues of research are presented here.

pAMβ1 possesses a stability determinant

Independent studies undertaken by Janniere et al. (1990) resulted in the construction of a series of bifunctional cloning vectors for use in *B. subtilis* based on the pAMβ1 replicon. The pAMβ1-derived DNA of one of these plasmids, pHV1431, included an additional 2.12-kb of DNA to that present in pMTL500E (see Figure 8.3). Although pHV1431 exhibited a high degree of segregational stability in *B. subtilis*, a plasmid (pHV1432) derived by deletion of this extra region of DNA possessed a level of instability comparable to that exhibited by pMTL500E (Table 8.2). Subsequent subcloning of the appropriate 2.12-kb *Eco*RI-*Hind*III from pHV1431 into the polylinker region of pMTL500E resulted in a plasmid, pMTL531E, which exhibited significantly improved segre-

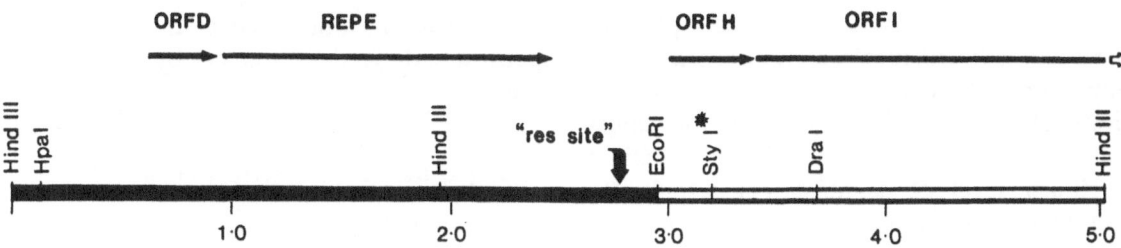

FIGURE 8.4 Schematic representation of the nucleotide sequence of the pAMβ1-derived insert of pHV1431. Heavy black line corresponds to previously identified replicon of pAMβ1 (Swinfield et al., 1990). Double line represents the identified stability determinant (ORF H, resolvase; ORF I, topoisomerase-like protein). Plasmid pMTL500E carries the region of DNA between the *Hpa*I and *Eco*RI sites. Plasmid pMTL500E*res*+ (see Table 8.2) carries a 0.7-kb subfragment between the indicated *Eco*RI and *Dra*I sites; plasmid pMTL500E*res*⁻ contains the same fragment but with frameshift (*) introduced at the *Sty*I site.

gational stability, both in *B. subtilis* and *C. acetobutylicum* (Table 8.2).

To gain insight into the nature of the identified stability determinant, the complete nucleotide sequence of the 2.12-kb *Eco*RI-*Hind*III fragment was determined (Figure 8.4). Translation of the sequence obtained indicated the presence of two putative ORFs, designated H and I (Swinfield and Minton, unpublished data). ORF H encoded a putative protein (M_r of 23,930) that demonstrates substantial identity with bacterial site-specific recombinases, including the resolvases of the gram-positive transposon Tn917 (37.4%), and the gram-positive plasmids pI524 (34.3%) and pIP404 (30.6%) (Shaw and Clewell, 1985; Garnier et al., 1987; Rowland and Dyke, 1988). The second ORF (I), which was incomplete, encoded a protein with a predicted M_r of at least 55,555 and which shares 24.3% identity with *E. coli* DNA topoisomerase I (Tse-Dinh and Wang, 1986). Circumstantial evidence, in the form of overlapping of translational stop (ORF H) and start (ORF I) codons, suggested that the expression of these two ORFs may be transcriptionally/translationally coupled.

In certain cases, segregational instability of plasmids can be caused by the predominant existence of the plasmid population as multimers, thereby reducing the overall number of plasmid units available for partition at cell division (Austin, 1988). The homologous nature of the ORF H polypeptide to bacterial resolvases may indicate that it mediates plasmid stability by maintaining the plasmid population in the monomeric state through the agency of multimer resolution. A similar role has been proposed for resolvase-like proteins encoded by the gram-positive plasmids pI524 (Rowland and Dyke, 1988) and pIP404 (Garnier et al., 1987). Indeed, examination of *B. subtilis* cleared lysates of cells harboring pMTL500E (*res*⁻) indicated that the plasmid DNA was present in a highly polymerized form. In contrast, pMTL531E (*res*+) DNA existed principally as a monomeric species (L. Janniere and S.D. Ehrlich, unpublished data). ORF I protein may possess topoisomerase activity. Additional sequence data position its translational stop codon immediately 5′ to the previously sequenced (Brehm et al., 1987) Em^r gene of pAMβ1 (Oultram, unpublished data) and reveals a "zinc finger," DNA-binding motif in the C terminus of the encoded polypeptide homologous to those present in the *E. coli* topoisomerase enzyme (Tse-Dinh and Beran-Steed, 1988). The truncated ORF I polypeptide encoded by pHV1431/pMTL531E may therefore be inactive and play no role in stability.

The pAMβ1 resolvase mediates segregational stability

To test the hypothesis that the active element of the sequenced region was indeed the identified *res* gene, a 0.73-kb *Dra*I-*Eco*RI subfragment was isolated from pMTL531E (see Figure 8.3) and inserted into pMTL500E (Swinfield et al., unpublished observations). The plasmid obtained, pMIL500E*res*+, therefore carries the entire *res* gene but encodes only 20 codons from the 5' end of ORF I. In spite of the almost total deletion of ORF I, pMTL500*res*+ exhibited a similar level of segregational stability to pMTL531E (see Table 8.2). Further, this increase in segregational stability was completely negated by the introduction of a frameshift into the coding region of *res* at a unique *Sty*I site (pMTL500E*res*−; see Table 8.2).

The site at which recombinase proteins act is generally found 5' to the structural gene and is characterized by three sequence motifs (designated I, II, and III) of dyad symmetry (see Hatfull et al., 1988). The most distal, site I, corresponds to the recombination crossover point and is characterized by the possession of a conserved central AT dinucleotide. A sequence of dyad symmetry containing the highly conserved AT dinucleotide occurs 277 bp 5' to ORF II. This putative crossover point is not encompassed by the 2.12-kb *Eco*RI-*Hind*III fragment but resides within the previously sequenced (Swinfield et al., 1990) 5.1-kb *Eco*RI fragment carrying the pAMβ1 replicon, 3' to the *repE* gene (see Figure 8.4). The position of this site correlates with the identification of a high-frequency, site-specific recombination site RFSβ in the same region by Hayes et al. (1990).

8.6 Specialist Clostridial Vectors

Developments in our laboratory and others around the world have resulted in the derivation of a number of different general-purpose cloning vectors for use in *Clostridium acetobutylicum*. Thus, vectors are available in which the detection of the insertion of foreign DNA into their unique cloning sites is performed by screening for the inactivation of either an antibiotic resistance gene or the *lacZ'*− encoded alpha peptide. In the following section, however, we describe a number of vectors of a more specialist nature that are currently under development in our laboratory. Although these vectors have principally been designed for introduction directly into the host organism by

transformation, cloning vectors have also been specifically derived for transfer by conjugative means.

"Conjugative" cloning vectors

Two different types of general-purpose cloning vectors have been constructed for introducing cloned DNA into *C. acetobutylicum* by conjugative gene transfer. The first vector, pMTL21EC, was designed to be transferred to *C. acetobutylicum* from *Bacillus subtilis* as part of a cointegrate molecule with the conjugative plasmid pAMβ1. In essence it represents a more refined version of the cointegrate pOD vectors described by Oultram et al. (1987) and was constructed by inserting a 1.07-kb *Hpa*II fragment carrying the pC194 Cm^r gene into one of the two *Pvu*II sites of pMTL21E (see Figure 8.1). As with the earlier pOD vectors, the maintenance of this Rep− plasmid on introduction into *B. subtilis* relies on its cointegration with a coresident pAMβ1 molecule by recombination between their homologous *erm* genes. Because the points of insertion of the DNA fragments encoding *erm* and *cat* resides external to *lacZ'*, pMTL21EC retains all the cloning advantages of pMTL20 with regard to ease of recombinant detection on agar medium supplemented with XGal. The utility of the vector was demonstrated by its use in the deliberate conversion of a Leu− *C. acetobutylicum* mutant strain to prototrophy by introducing a cloned heterologous *leuB* gene (Oultram et al., 1988b).

The second type of vector was described by Williams et al. (1990a). This investigation built on the observation made by Trieu-Cuot et al. (1987) that a pAMβ1-derived cloning vector (pAT187) containing the transfer origin (*oriT*) of the IncPII plasmid RK2 could be mobilized from *E. coli* to a wide variety of gram-positive recipients, provided the donor also harbored plasmid RK2. Having established that both pAT187 and pCTC1 (pMTL500E carrying *oriT*) could be transferred to *C. acetobutylicum* using this methodology, a pair of vectors (pMTL30 and pMTL31; see Figure 8.1) were constructed (Williams et al., 1990a) in which *oriT* was inserted into the *Nhe*I site of pMTL20E in both orientations. Because the *Nhe*I site is located im-

Table 8.3 Heterologous expression using the Fd cartridge

Reporter gene	Promoter	Enzyme levels (% cell-soluble protein)		
		Escherichia coli	*Bacillus subtilis*	*Clostridium acetobutylicum*
xylE	*lac*	19.86	0	ND[a]
	Fd	13.96	0.07	0.01
	None	0.22	0	ND
cat	*lac*	6.59	0.30	0
	Fd	6.81	5.1–12.6	2.9–7.2
	None	0.95	0	0

[a]ND, not determined.

mediately 3′ to *lacZ′*, both vectors retain all the cloning advantages of the progenitor plasmid pMTL20E with regard to ease of recombinant selection. These vectors subsequently proved very useful in assessing the ability of a number of gram-positive replicons to function in *C. acetobutylicum* and demonstrated, for instance, that the broad-host-range streptococcal plasmid pWV01 (Kok et al., 1984) replicates in strain NCIB 8052 (Williams et al., 1990b; see Chapter 6, this volume).

Construction of pMTL500F, a clostridial expression vector

A particularly useful refinement to a *Clostridium/E. coli* shuttle vector is the provision of a powerful promoter to elicit the expression of cloned heterologous and homologous genes. Toward this end an expression cartridge has been constructed based on the ferredoxin (Fd) gene of *Clostridium pasteurianum* (Graves et al., 1985). The Fd gene was cloned as a 650-bp *Sau*3A fragment, and site-directed mutagenesis was employed to substitute the coding region of the structural gene with the polylinker cloning region of pMTL20. These multiple cloning sites are therefore flanked at the 5′ end by the transcriptional initiation signals of the Fd gene and at the 3′ end by the Fd transcriptional termination signals (Graves and Rabinowitz, 1986). The whole cartridge is localised to a 573-bp, portable *Eco*RI fragment (Minton et al., 1990b). Into this cartridge we inserted a copy of the pseudomonad *xylE* gene that has no promoter (Zukoski et al., 1983) and the pC194 *cat* gene (Minton et al., 1990a). Both genes were

expressed with high efficiency in *E. coli*, comparing favorably with the *E. coli lac* promoter (Table 8.3). The level of expression of *xylE* in the two gram-positive organisms, on the other hand, was relatively poor and can be attributed to translational inefficiency (Leonhardt and Alonso, 1988) rather than being a consequence of poor transcription from the Fd promoter. In contrast, the *cat* gene appeared highly expressed.

Having established that the Fd transcriptional signals could be used to effect the expression of heterologous genes in *C. acetobutylicum*, we have now constructed a more refined vector that facilitates the insertion of heterologous genes 3′ to the Fd promoter and should allow the future regulatory control of transcription. A synthetic *lac* operator has been inserted immediately 3′ to the Fd + 1 (following creation of a unique *Hpa*I site by site-directed mutagenesis), and an *Nde*I site was created over the AUG start codon. Using an indirect cloning route (J.K. Brehm et al., unpublished data), the *lac* promoter of pMTL500E has been replaced with this modified Fd promoter element such that the AUG start codon of Fd becomes that of the *lacZ′* gene. The resultant plasmid has been designated pMTL500F (Figure 8.5). It is anticipated that when the plasmid is transformed into a cell that overproduces *lacI* repressor protein, transcription from the Fd promoter will be blocked. To test this supposition, the promoterless *cat* gene was inserted into both pMTL500E and pMTL500F, and the resultant plasmids were transformed into *E. coli* JM83 carrying the *lacI*q-encoding plasmid pNM52 (Gilbert et al., 1986). Exponentially growing cells carrying

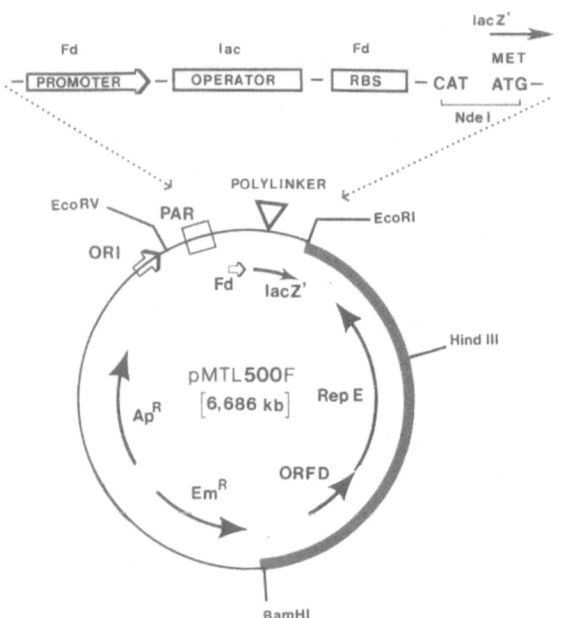

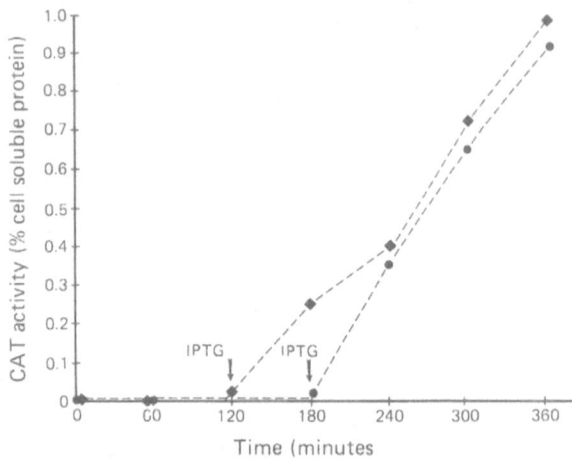

FIGURE 8.5 The *Clostridium* expression vector pMTL500F. Plasmid pMTL500F was constructed essentially by replacing the *lac po* region of pMTL500E with the indicated modified Fd promoter. During the course of this substitution, plasmid pMTL500F also acquired the pSC101 stability function, *par* (PAR). The nucleotide sequence of the Fd promoter region has been given elsewhere (Minton et al., 1990b). The ATG trinucleotide of the indicated *Nde*I restriction recognition site corresponds to the AUG translational start codon of *lacZ'*. The polylinker region is identical to pMTL500E.

FIGURE 8.6 Inducible expression of the pC194 *cat* gene cloned in pMTL500E and pMTL500F. The promoterless copy of the pC194 *cat* gene was excised from pMTL20C (Swinfield et al., 1990) as a 0.8 kb *Mnl*I fragment and inserted into the *Sma*I site of pMTL500E and pMTL500F, such that transcription was dependent on the *lac* or Fd promoter, respectively. The two recombinant plasmids were introduced into *E. coli* TG1 containing the *lacI*q-encoding plasmid pNM52 (Gilbert et al., 1986), and the two clones grown in 2XYT broth to an OD_{450} of 0.6. At this point expression was induced by addition of IPTG to a final concentration of 1 mM. CAT activity is expressed as % cell soluble protein. Symbols: (■), pMTL500E; (●), pMTL500F.

both sets of plasmids were then subjected to IPTG induction and the levels of CAT monitored. The results obtained, illustrated in Figure 8.6, clearly show that the level of repression of *cat* expression from the modified Fd promoter of pMTL500F is comparable to that obtained with the *lac* promoter of pMTL500E.

To achieve similar control in *C. acetobutylicum* it will be necessary to elicit the expression of the *lacI* in a manner similar to that obtained in *B. subtilis* (Peschke et al., 1985). To attain such a system we have constructed a plasmid, pMTL520, designed to coexist with pMTL500E, being composed of the origin of replication of the *E. coli* plasmid p15A, the Tc^r gene (Abraham et al., 1988) of the *C. perfringens* plasmid pJIR71, and the pCB101 replicon (see Minton et al., 1990b). To date, however, although this plasmid confers Tc^r on both an *E. coli* and a *B. subtilis* host, we have been unable to demonstrate that this gene functions in *C. acetobutylicum*. It may therefore prove necessary to substitute this particular *tet* gene with that of a staphylococcal plasmid such as pT127, shown to function in *C. acetobutylicum* (Truffaut et al., 1989). Similarly, we also planned to use site-directed mutagenesis and advanced genetic engineering to "fuse" the transcriptional/translational signals of the *C. pasteurianum leuB* gene to the *E. coli lacI* gene of pNM52 (Gilbert et al., 1986). Recent sequence analysis of the cloned *leuB* gene indicated that it forms part of an operon and therefore does not possess transcriptional initiation signals (Oultram, unpublished data). We therefore plan to obtain the desired promoter sequences by random cloning of such elements using the vector pMTL710 (see following).

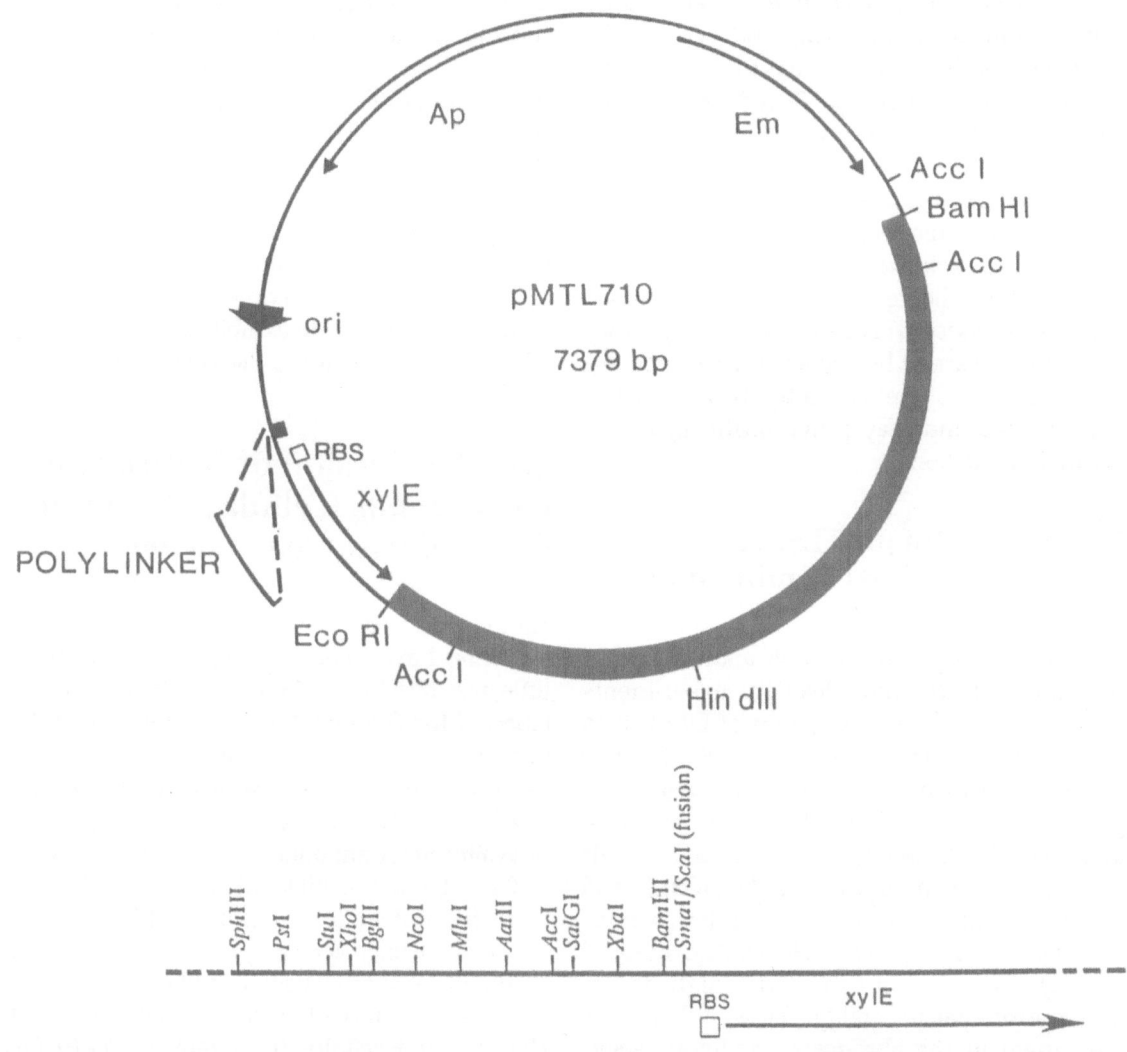

FIGURE 8.7   The promoter probe vector pMTL710. This plasmid was constructed essentially by replacing the *lac po/lacZ'* region of pMTL500E with a promoter-less copy of the pseudomonad *xylE* gene. The thick portion (heavy black line) of the plasmid is the replication origin of pAMβ1; the *xylE* region is expanded below to give details of restriction sites in the marked polylinker region. Abbreviation: RBS, ribosome-binding site.

Construction of the replicon-probe vector pMTL710

Plasmid pMTL710 (Figure 8.7) was constructed to isolate promoters active in gram-positive bacteria such as *B. subtilis* and *C. acetobutylicum* and to allow crude quantitative comparisons of their transcriptional activity in these organisms. The reporter gene used is a promoter-less copy of the pseudomonad *xylE* gene. Using an indirect route, numerous restriction enzyme cloning sites were positioned 5′ to the translational initiation codon of *xylE*, and the modified gene was inserted into a derivative of pMTL500E in which the vector *lacZ'* and the *lac po* region had been deleted. The *xylE* gene is inserted such that a low level of gene expression occurs from transcription readthrough from the upstream ColE1 RNA II promoter.

In *B. subtilis* and *C. acetobutylicum* analogous transcription does not occur, and hence cells containing pMIL710 are unable to catalyze the conversion of catechol to 2-hydroxymuconic semialdehyde and appear white when sprayed with a aqueous catechol (2% wt/vol) solution. The vector has been used to clone randomly generated genomic fragments from both *B. subtilis* and *C. acetobutylicum*, and the level of *xylE* expression of the resultant positive "yellow" clones was assessed. A number of the promoter elements carried by the appropriate recombinant plasmids are currently being further characterized and may prove useful in future vector constructions.

Construction of pMTL513E, a potential clostridial stability probe vector

During our analysis of the replication region of plasmid pAMβ1, our deletion experiments indicated that a 229-bp segment of DNA 3' to *repE* could be deleted and the resultant 4.9-kb *Eco*RI-*Acc*I fragment would still enable a pMTL20E and pMTL20C chimera to replicate in *B. subtilis*. Subsequent evaluation of cells carrying these plasmids (pMTL20Eβ13 and pMTL20Cβ13), however, revealed that its copy number was not measurable by the method employed and that it exhibited a high degree of segregational instability. Thus, when cells are grown in the absence of antibiotic selection for 10 generations and plated to give single colonies on antibiotic-free agar medium, of the 100 colonies tested in any one experiment only 1 or 2 (sometimes 0) colonies still retained antibiotic resistance. The extreme instability of this vector could be exploited as the basis of a vector for cloning elements that confer improved segregational stability. Toward this end we have inserted the 4.9-kb *Eco*RI-*Acc*I fragment into the *Nhe*I site of pMTL20E. The resultant vector, pMTL513E (Figure 8.2), therefore retains all the cloning advantages of pMTL20E with regard to ease of recombinant selection and exhibits segregational instability comparable to that of pMTL20Eβ13. It is envisaged that either random or specific restriction fragments of clostridial plasmids could be

cloned into the polylinker region of pMTL513E and the stability of the resultant vectors tested by growth of the gram-positive host in the absence of antibiotic selection. Although we have yet to obtain a plasmid-derived fragment that stabilizes this vector, stabilization by other means is known to occur. Thus an analogous vector to pMTL513E in which the copy number has been increased (from 20 to 200 copies per cell) by deletion of the copy number control region 5' to the *Hpa*I site exhibits a similar level of segregational stability to pMTL500E.

8.7   Expression of *Clostridium thermocellum* Cellulase Genes in *Clostridium acetobutylicum*

Although clostridial vectors for use in *C. acetobutylicum* have been available for some time, little use has been made of them in manipulation of the fermentation characteristics of this industrially important organism. In the last section of this chapter we describe the results of a joint study undertaken with the laboratory of Walter Staudenbauer.

To test the feasibility of using recombinant DNA technology to extend the substrate range of *C. acetobutylicum*, we have examined degradation of soluble cellulose-like polymers (β-glucans) as a model system. Unlike the well-characterized cellulolytic organism *Clostridium thermocellum*, *C. acetobutylicum* NCIB 8052 is unable to degrade cellulose. In view of the extreme complexity both of the cellulosome and of cellulose degradation in general, we have initially sought merely to engineer the organism such that soluble cellulose-like materials may be utilized as energy sources. Two genes that should enable implementation of this phenotypic change are the *celA* and *celC* genes of *C. thermocellum*. The *celA* gene product (Schwarz et al., 1986) is very efficient at the depolymerization of long-chain β-glucans, but degradation becomes far less efficient when the chain length of the polymer approaches 5 glucose residues. Although the *celC* gene product (Schwarz et al., 1988a) is markedly less efficient at depolymerization of long chains, it becomes far more efficient as chain length is shortened and is capable of reducing chains down to the level of cellobiose (two glucose units in length; Schwarz et al., 1988b). *Clostridium acetobutylicum* has been shown to derive sufficient energy for growth from the fermentation of cellobiose and the gene encoding the β-glucosidase responsible for the initial cleavage of cellobiose has been cloned in our laboratory (Oultram and Minton, unpublished data). It follows that the coordinate action of both endoglucanases may maximize the breakdown of soluble β-glucans to cellobiose and hence permit strains of *C. acetobutylicum* carrying them to grow on these β-glucans as sole sources of energy.

Cloned restriction fragments encoding both these genes were individually inserted into pMTL500E, and the resultant recombinant plasmids (pCA1 and pCC1, containing the *celA* and *celC* genes respectively) were introduced into *C. acetobutylicum* by electrotransformation. In addition, using advanced genetic engineering techniques, a fused transcriptional unit (artificial operon) was constructed consisting of, in sequential order, the *celC* transcriptional initiation signals, ribosome-binding site (RBS) and structural gene, and the *celA* RBS, structural gene, and transcriptional terminator. This operon was also cloned into pMTL500E to give the recombinant plasmid pCCA1. Recombinant clostridial clones were shown to produce the appropriate recombinant protein by a combination of *in situ* chromogenic plate tests and polyacrylamide gel activity stains of cell lysates (S. Schimming and Oultram, unpublished data). Expression of both genes was from their own promoters (in the case of the operon, *celA* expression being reliant on the *celC* promoter), and both endoglucanase activities were exported to the external medium.

To demonstrate the growth advantages afforded to the strains by these recombinant plasmids growth experiments were undertaken in a semisynthetic minimal broth medium supplemented with barley β-glucan as the sole carbon source. Preliminary results of these experiments, shown in Figure 8.8, clearly demonstrate the growth advantage conferred on *C. acetobutylicum* NCIB 8052 by the presence of the recombinant plasmids. While pCA1

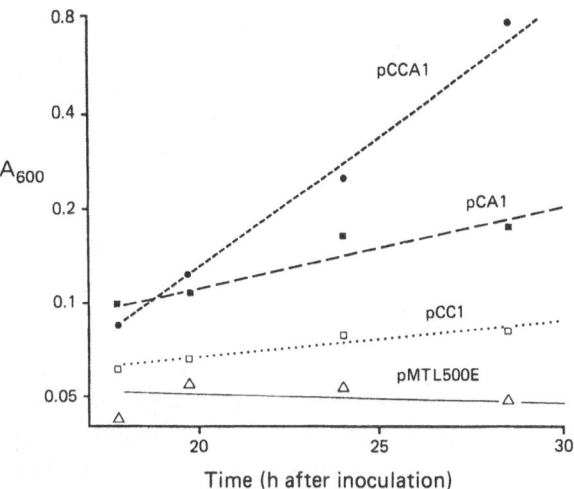

FIGURE 8.8 Growth of recombinant *C. acetobutylicum* strains on barley β-glucan. Strains were cultivated anaerobically in semisynthetic medium containing barley β-glucan as the sole source of energy. The plasmid carried by each strain is indicated next to the growth curve that it produced.

and pCC1 do allow growth on the substrate, although this was not detected in the wild-type strain, the strain carrying the *celA/celC* operon clearly shows the fastest growth rate. The result confirms the feasibility of altering the fermentation range of *C. acetobutylicum* by genetic means.

8.8 Conclusion

Proliferation in the research effort directed toward the development of genetic systems for *Clostridium acetobutylicum* is now beginning to bear fruit. Thus, rudimentary cloning vectors have been constructed and conjugative and transformation procedures formulated to effect their introduction into the cell. These enabling technologies having been established, attention is now focused on refinement of the available vectors and on their practical use in the genetic manipulation of *C. acetobutylicum*. In this chapter we have highlighted some of the refinements that are being made to improve vector utility and have described experiments which for the first time effect a beneficial change to *C. acetobutylicum* physiology by recombinant means. A number of problems still remain to be overcome, particularly with regard to recombinant segregational stability. The next few years, however, promise to be of a highly productive nature with regard elucidating the genetic and biochemical basis of the regulation of primary metabolism in this organism, and thereafter capitalising on these findings to generate strains with improved fermentation characteristics.

Acknowledgments. We wish to acknowledge the contributions of the following people to the studies described here: D.E. Thompson, A. Pennock, S. Schimming, M. Loughlin, H. Peck, S. Gunnery, R. Walmsley, A.F. Sharman, and U. Vetter. Research was carried out within the framework of the Biotechnology Action Programme of the Commission of the European Communities (BAP-0046-UK).

References

Abraham, L.J., D.I. Berryman, and J.I. Rood. 1988. Hybridisation analysis of the class P tetracycline resistance determinant from *Clostridium perfringens* R-plasmid pCW3. *Plasmid* **19**:113–120.

Austin, S.J. 1988. Plasmid partition. *Plasmid* **20**:1–9.

Baas, P.D. 1985. DNA replication of single-stranded *Escherichia coli* DNA phages. *Biochim. Biophys. Acta* **825**:111–139.

Brehm, J.K., G.P.C. Salmond, and N.P. Minton. 1987. Sequence of the adenine methylase gene of the *Streptococcus faecalis* plasmid pAMβ1. *Nucleic Acids Res.* **15**:3177.

Chambers, S.P., S.E. Prior, D.A. Barstow, and N.P. Minton. 1988. The pMTL cloning vectors: I. Improved polylinker regions to facilitate the generation of sonicated DNA for nucleotide sequencing. *Gene* **68**:139–149.

Clarke, D.J. and J.G. Morris. 1976. Butyricin 7423: a bacteriocin produced by *Clostridium butyricum* NCIB 7423. *J. Gen. Microbiol.* **95**:67–77.

Clarke, D.J., R.M. Robson, and J.G. Morris. 1975. Purification of two *Clostridium* bacteriocins by procedures appropriate to hydrophobic proteins. *Antimicrob. Agents Chemother.* **7**:256–264.

Clewell, D.B. 1981. Plasmids, drug resistance, and gene transfer in the genus *Streptococcus*. *Microbiol. Rev.* **45**:409–436.

Collins, M.E., J.D. Oultram, and M. Young. 1985. Identification of restriction fragments from two *Clostridiurn butyricum* plasmids that promote the establishment of a replication-defective

plasmid in *Bacillus subtilis*. *J. Gen. Microbiol.* **131**:2097–2105.

Garnier, T., W. Saurin, and S.T. Cole. 1987. Molecular characterisation of the resolvase gene, *res*, carried by a multicopy plasmid from *Clostridium perfringens*: common evolutionary origin for procaryotic site-specific recombinases. *Mol. Microbiol.* **1**:371–376.

Gilbert, H.J., R. Blazek, H.M.S. Bullman, and N.P. Minton. 1986. Cloning and expression of the *Erwinia chrysanthemi* asparaginase gene in *Escherichia coli* and *Erwinia carotovora*. *J. Gen. Microbiol.* **132**:151–160.

Graves, M.C. and J.C. Rabinowitz. 1986. In vivo and in vitro transcription of the *Clostridium pasteurianum* ferredoxin gene. *J. Biol. Chem.* **261**:11409–11415.

Graves, M.C., G.T. Mullenbach, and J.C. Rabinowitz. 1985. Cloning and nucleotide sequence determination of the *Clostridium pasteurianum* ferredoxin gene. *Proc. Natl. Acad. Sci. USA* **82**:1653–1657.

Gruss, A. and S.D. Ehrlich. 1989. The family of highly interrelated single-stranded deoxyribonucleic acid plasmids. *Microbiol. Rev.* **53**:231–241.

Hatfull, G.F., J.J. Salvo, E.E. Falvey, V. Rimphanitchayakit, and N.D.F. Grindley. 1988. Site-specific recombination by the γδ resolvase. *Soc. Gen. Microbiol. Symp.* **43**:149–181.

Hayes, F., C. Daly, and G.F. Fitzgerald. 1990. High frequency, site-specific recombination between lactococcal and pAMβ1 plasmid DNAs. *J. Bacteriol.* **172**:3485–3489.

Horaud, T., C. Le Bouguenec, and K. Pepper. 1985. Molecular genetics of resistance to macrolides, lincosamides and streptogramin B (MLS) in streptococci. *J. Antimicrob. Chemother.* **16**: (Suppl. A):111–135.

Janniere, L., C. Bruand, and S.D. Ehrlich. 1990. Structurally stable *Bacillus subtilis* DNA cloning vectors. *Gene* **87**:53–61.

Kok, J., J.M.B.M. Van der Vossen, and G. Venema. 1984. Construction of plasmid cloning vectors for lactic streptococci which also replicate in *Bacillus subtilis* and *Escherichia coli*. *Appl. Environ. Microbiol.* **48**:726–731.

Leblanc, D.J. and L.N. Lee. 1984. Physical and genetic analysis of streptococcal plasmid pAMβ1 and cloning of its replication region. *J. Bacteriol.* **157**:445–453.

Leonhardt, H. and J.C. Alonso. 1988. Construction of a shuttle vector for inducible gene expression in *Escherichia coli* and *Bacillus subtilis*. *J. Gen. Microbiol.* **134**:605–609.

Luczak, H., H. Schwarzmoser, and W.L. Staudenbauer. 1985. Construction of *Clostridium butyricum* plasmids and transfer to *Bacillus subtilis*. *Appl. Microbiol. Biotechnol.* **23**:114–122.

Minton, N.P and J.G. Morris. 1981. Isolation and

partial characterisation of three cryptic plasmids from strains of *Clostridium butyricum*. *J. Gen. Microbiol.* **127**:325–331.

Minton, N.P and D.J. Clarke. 1989. Clostridia. *In: Biotechnology Handbook Series, Vol. III.* New York: Plenum.

Minton, N.P. and D.E. Thompson. 1990. Genetics of anaerobes. *In: Anaerobes in Human Disease,* B.I. Duerden and B.S. Drasar, eds., pp. 38–61. London: Arnold.

Minton, N.P., T.-J. Swinfield, J.K. Brehm, and J.D. Oultram. 1990a. The gram-positive cloning vector pBD64 arose by a 1844 bp deletion of pC194-derived DNA. *Nucleic Acids Res.* **18**:1651.

Minton, N.P., J.K. Brehm, J.D. Oultram, D.E. Thompson, T.-J. Swinfield, A. Pennock, S. Schimming, S.M. Whelan, U. Vetter, M. Young, and W.L. Staudenbauer. 1990b. Vector systems for the genetic analysis of *Clostridium acetobutylicum*. *In: Clinical and Molecular Aspects of Anaerobes,* S.P. Borriello, ed., pp. 187–201. Petersfield, UK: Wrightson Biomedical.

Monod, M., C. Denoya, and D. Dubnau. 1986. Sequence and properties of pIM13, a macrolide-lincosamide-streptogramin B resistance plasmid from *Bacillus subtilis*. *J. Bacteriol.* **167**:138–147.

Nakayama, J., H. Nagasawa, A. Isogai, D.B. Clewell, and S. Akinori. 1990. Amino acid sequence of pheromone-inducible surface protein in *Enterococcus faecalis*, that is encoded on the conjugative plasmid pPD1. *FEBS Lett.* **267**:81–84.

Novick, R.P. 1989. Staphylococcal plasmids and their replication. *Ann. Rev. Microbiol.* **43**:537–565.

Oultram, J.D. and M. Young. 1985. Conjugal transfer of plasmid pAMβ1 from *Streptococcus lactis* and *Bacillus subtilis* to *Clostridium acetobutylicum*. *FEMS Microbiol. Lett.* **27**:129–134.

Oultram, J.D., A. Davies, and M. Young. 1987. Conjugal transfer of a small plasmid from *Bacillus subtilis* to *Clostridium acetobutylicum* by cointegrate formation with plasmid pAMβ1. *FEMS Microbiol. Lett.* **42**:113–119.

Oultram, J.D., M. Loughlin, T.-J. Swinfield, J.K. Brehm, D.E. Thompson, and N.P. Minton. 1988a. Introduction of plasmids into whole cells of *Clostridium acetobutylicum* by electroporation. *FEMS Microbiol. Lett.* **56**:83–88.

Oultram, J.D., H. Peck, J.K. Brehm, D.E. Thompson, T.-J. Swinfield, and N.P. Minton. 1988b. Introduction of genes for leucine biosynthesis from *Clostridium pasteurianum* into *Clostridium acetobutylicum* by cointegrate conjugal transfer. *Mol. Gen. Gen.* **214**:177–179.

Peschke, U., V. Beuck, H. Bujard, R. Gentz, and S. Le Grice. 1985. Efficient utilisation of *Escherichia coli* transcriptional signals in *Bacillus subtilis*. *J. Mol. Biol.* **186**:547–555.

Phillips-Jones, M.K. 1990. Plasmid transformation of *Clostridium perfringens* by electroporation methods. *FEMS Microbiol. Lett.* **66**:221–226.

Projan, S.J., M. Monod, C.S. Narayanan, and D. Dubnau. 1987. Replication properties of pIM13, a naturally occurring plasmid found in *Bacillus subtilis*, and its close relative pE5, a plasmid native to *Staphylococcus aureus*. *J. Bacteriol.* **169**:5131–5139.

Reysset, G., J. Hubert, L. Podvin and M. Sebald. 1988. Transfection of *Clostridium acetobutylicum* strain N1-4081 protoplasts. *Biotechnol. Techniq.* **2**:199–204.

Rowland, S.J. and K.G.H. Dyke. 1988. A DNA invertase from *Staphylococcus aureus* is a member of the Hin family of site-specific recombinases. *FEMS Microbiol. Lett.* **50**:253–258.

Schwarz, W.H., F. Grabnitz, and W.L. Staudenbauer. 1986. Properties of a *Clostridium thermocellum* endoglucanase produced in *Escherichia coli*. *Appl. Environ. Microbiol.* **51**:1293–1299.

Schwarz, W.H., S. Schimming, K.P. Rucknagel, S. Burgschwaiger, G. Kreil, and W.L. Staudenbauer. 1988a. Nucleotide sequence of the *celC* gene encoding endoglucanase C of *Clostridium thermocellum*. *Gene* **63**:23–30.

Schwarz, W.H., S. Schimming, and W.L. Staudenbauer. 1988b. Degradation of barley β-glucan by endoglucanase C of *Clostridium thermocellum*. *Appl. Microbiol. Biotechnol.* **29**:25–31.

Shaw, J.H. and D.B. Clewell. 1985. Complete nucleotide sequence of macrolide-lincosamide-streptogramin B-resistance transposon Tn*917* in *Streptococcus faecalis*. *J. Bacteriol.* **164**:782–796.

Sorokin, A.V. and V.E. Khazak. 1989. Structure of pSM19035 replication region and MLS-resistance gene. *In: Genetic Transformation and Expression,* L.O. Butler, C. Harwood, and B.E.B. Moseley, eds., pp. 269–281. Wimborne, UK: Intercept.

Soutschek-Bauer, E., L. Hartl, and W.L. Staudenbauer. 1985. Transformation of *Clostridium thermohydrosulfuricum* DSM 568 with plasmid DNA. *Biotechnol Lett.* **7**:705–710.

Swinfield, T.-J., J.D. Oultram, D.E. Thompson, J.K. Brehm, and N.P. Minton. 1990. Physical characterisation of the replication region of the *Streptococcus faecalis* plasmid pAMβ1. *Gene* **87**:79–90.

Truffaut, N., J. Hubert, and G. Reysset. 1989. Construction of shuttle vectors useful for transforming *Clostridium acetobutylicum*. *FEMS Microbiol. Lett.* **58**:15–20.

Trieu-Cuot, P., C. Carlier, P. Martin, and P. Courvalin. 1987. Plasmid transfer by conjugation from *Escherichia coli* to Gram-positive bacteria. *FEMS Microbiol. Lett.* **48**:289–294.

Tse-Dinh, Y.-C. and R.K. Beran-Steed. 1988. *Escherichia coli* DNA topoisomerare I is a zinc metalloprotein with three repetitive zinc-binding domains. *J. Biol. Chem.* **263**:15857–15859.

Tse-Dinh, Y.-C. and J.C. Wang. 1986. Complete nucleotide sequence of the *topA* gene encoding *Escherichia coli* DNA topoisomerase I. *J. Mol. Biol.* **191**:321–331.

Von Heijne, G. 1983. Patterns of amino acids near signal-peptide cleavage sites. *Eur. J. Biochem.* **133**:17–21.

Vos, P., G. Simons, R.J. Siezen, and W.M. de Vos. 1989. Primary structure and organisation of the gene for a procaryotic, cell envelope serine proteinase. *J. Biol. Chem.* **264**:13579–13585.

Williams, D.R., D.I. Young, J.D. Oultram, N.P. Minton, and M. Young. 1990a. Development and optimisation of conjugative plasmid transfer from *Escherichia coli* to *Clostridium acetobutylicum* NCIB 8052. *In: Clinical and Molecular Aspects of Anaerobes*, S.P. Borriello, ed., pp. 239–246. Petersfield, UK: Wrightson Biomedical.

Williams, D.R., D.I. Young, and M. Young. 1990b. Conjugative plasmid transfer from *Escherichia coli* to *Clostridium acetobutylicum*. *J. Gen. Microbiol.* **136**:819–826.

Yoshino, S., T. Yoshino, S. Hara, S. Ogata, and S. Hayashida. 1990. Construction of shuttle vector plasmid between *Clostridium acetobutylicum* and *Escherichia coli*. *Agric. Biol. Chem.* **54**:437–441.

Zukoski, M.M., D.F. Gaffney, D. Speck, M. Kauffman, A. Findell, A. Wisecup, and J.-P. Lecoq. 1983. Chromogenic identification of genetic regulatory signals in *Bacillus subtilis* based on expression of a cloned *Pseudomonas* gene. *Proc. Natl. Acad. Sci. USA* **80**:1101–1105.

9

Antibiotic Resistance Determinants of *Clostridium perfringens*

Julian I. Rood

9.1 Introduction

Clostridium perfringens is an anerobic microorganism that forms heat-resistant endospores in the gastrointestinal tract but does not sporulate in most routine culture media. In rich media it grows rapidly and ferments sugars to produce H_2 and CO_2, the evolution of which helps to maintain an anaerobic environment. The organism is aerotolerant; it is capable of surviving for extended periods of time in the presence of O_2. *C. perfringens* is of considerable clinical significance as the causative agent of human clostridial myonecrosis (gas gangrene), food poisoning, necrotic enteritis, and a variety of gastrointestinal diseases in domestic animals. The majority of these syndromes are mediated via the production of extracellular protein toxins (Hatheway, 1990).

Because of its cultural characteristics, *C. perfringens* is an ideal organism to use as a model for developing genetic methods of analysis in the *Clostridia*. Antibiotic-resistant isolates are relatively easy to obtain, and several resistance determinants have been shown to be plasmid determined (Bréfort et al., 1977; Rood et al., 1978b). In recent years simple, reproducible, transformation methods have been developed (Allen and Blaschek, 1988; Scott and Rood, 1989) and *Escherichia coli*–*C. perfringens* shuttle plasmids constructed (Squires et al., 1984; Roberts et al., 1988). It is now possible to clone *C. perfringens* genes into either *C. perfringens* or *E. coli* and to reintroduce *C. perfringens* genes that were originally cloned into *E. coli* back into

C. perfringens. With the possible exception of *Clostridium acetobutylicum* (Young et al., 1989), the genetic characteristics of *C. perfringens* are better known than those of any other member of this genus.

9.2 Plasmids

Many *Clostridium perfringens* isolates of all toxigenic types have been shown to carry plasmids (Rokos et al., 1978). These plasmids include cryptic elements, both conjugative and nonconjugative R-plasmids, bacteriocin plasmids, and a caseinase plasmid. There is little or no evidence that virulence factors are plasmid determined in *C. perfringens*. The properties of the various plasmids that have been identified are listed in Table 9.1. Three of these plasmids have been extensively studied, namely the conjugative R-plasmids pIP401 (Bréfort et al., 1977; Abraham and Rood, 1987) and pCW3 (Abraham and Rood, 1985b; Rood et al., 1978b) and the bacteriocin plasmid pIP404 (Ionesco et al., 1974; Garnier and Cole, 1988).

Sebald et al. (1975) described the isolation of two *C. perfringens* strains, CP590 and CP600, that were resistant to erythromycin [and the other macrolide-lincosamide-streptogramin B (MLS) antibiotics], tetracycline, and chloramphenicol. Two segregation patterns were observed in curing experiments, and the existence of two plasmids was postulated. One group of cured strains was sensitive to tetracycline and chloramphenicol and was postulated

Table 9.1 Wild-type plasmids of *Clostridium perfringens*[a]

Plasmid	Characteristics[b]	Size (kb)	Reference
pIP401	Tra$^+$Tc^{r}Cm^r	54	Sebald et al., 1975
pIP402	Em^{r}Cc^{r}Ln^r	63	Sebald et al., 1975
pIP404	Bcn5$^+$	10.2	Ionesco et al., 1974
pCW3	Tra$^+$Tc^r	47	Rood et al., 1978b
pHB101	Cas$^+$	3.1	Blaschek and Solberg, 1981
pJIR2	Tra$^+$Tc^r	67	Abraham and Rood, 1985a
pJIR4	Tra$^+$Tc^r	50	Abraham and Rood, 1985a
pJIR5,6	Tra$^+$Tc^r	47	Abraham and Rood, 1985a
pJIR19–21, pJIR 28–30, pJIR64	Tra$^+$Tc^r	47	Abraham et al., 1985
pJIR25	Tra$^+$Tc^{r}Cm^r	52	Abraham et al., 1985
pJIR27	Tra$^+$Tc^{r}Cm^r	50	Abraham et al., 1985
pJU124	Tra$^+$Tc^r	39	Heefner et al., 1984

[a] Various cryptic plasmids also have been reported by Abraham and Rood, 1985a; Blaschek and Klacik, 1984; Blaschek and Solberg, 1981; Bréfort et al., 1977; Heefner et al., 1984; Kordel and Schallehn, 1984; Mahony et al., 1986; Mahony et al., 1987; Phillips-Jones et al., 1989; Roberts et al., 1986; Rokos et al., 1978; Rood et al., 1978a, 1978b; and Smart et al., 1979.

[b] Tc^r, Cm^r, Em^r, Cc^r Ln^r: confers resistance to tetracycline, chloramphenicol, erythromycin, clindamycin, and lincomycin, respectively. Tra$^+$, conjugative; Cas$^+$, confers ability to produce caseinase; Bcn5$^+$, codes for production of bacteriocin BCN5.

to have been cured for the plasmid pIP401. The other cured derivatives were sensitive to erythromycin and no longer carried a second plasmid, pIP402 (Sebald et al., 1975). Further experiments showed that resistance to tetracycline and chloramphenicol could be transferred to a sensitive recipient strain, although some recipients only acquired tetracycline resistance. All the chloramphenicol-resistant transconjugants also acquired tetracycline resistance. Erythromycin resistance was not transferred (Sebald and Bréfort, 1975). These data appeared to indicate that pIP401 was a conjugative plasmid whereas pIP402 was nonconjugative. Further studies confirmed these findings and led to the more detailed analysis of these plasmids (Bréfort et al., 1977).

Electron microscopic analysis of plasmids from the wild-type strain, cured derivatives, and transconjugants showed that strain CP590 carried three plasmids. The conjugative plasmid pIP401 was 54 kb in size and coded for tetracycline and chloramphenicol resistance. Conjugative loss of the chloramphenicol resistance determinant was associated with the loss of 6 kb of DNA. The erythromycin resistance plasmid pIP402 was 63 kb (Bréfort et al., 1977). A

cryptic 10-kb plasmid, pIP403, that may code for the production of a bacteriocin was also observed. After in vivo mating a hybrid transconjugant, CP720, that carried both pIP401 and pIP402 was isolated. This strain appeared to be able to act as a conjugative donor of chromosomal markers in subsequent matings (Bréfort et al., 1977). It is possible that recombination between regions of homologous DNA may account for these observations, but no further experiments using this donor have been reported and the original observations have not been repeated with more modern methods of analysis.

Rood et al. (1978b) also studied a tetracycline- and chloramphenicol-resistant wild-type isolate of *C. perfringens*, strain CW92 (VPI11268). This strain was sensitive to erythromycin. Conjugative transfer of tetracycline resistance was readily observed and was associated with the transfer of a 47-kb plasmid, pCW3. All the transconjugants were sensitive to chloramphenicol. In addition, pCW3 could be cured from derivatives of CW92, with concomitant loss of the tetracycline resistance determinant. However, the cured strains were still resistant to chloramphenicol. Unlike the strains studied previously (Bréfort et al., 1977), it is clear that in CW92 the tetracycline and chloramphenicol resistance determinants are not genetically linked. A second plasmid, pCW2, was detected in CW92 but there was no evidence that this plasmid carried the latter determinant.

Abraham and Rood (1985b) cloned each of the five *ClaI* fragments which together constitute the entire pCW3 replicon. By comparative restriction analysis of both the *C. perfringens* and *Escherichia coli* plasmids, a detailed restriction map of pCW3 was constructed. At the same time, a restriction map of pIP401 also was constructed (Magot, 1984). Comparison of these maps confirmed the earlier suggestion (Rood et al., 1978b) that the plasmids were very closely related. The restriction profiles of pCW3 and the pIP401-derived deletion derivative pJIR23 were indistinguishable. Further analysis showed that pIP401 can be considered as a pCW3 replicon that contains the 6.2-kb chloramphenicol resistance transposon Tn*4451* (Abraham et al., 1985; Abraham and Rood, 1987). It could be postulated that chloramphenicol-resistant strains like CW92 resulted from the excision and subsequent transposition of Tn*4451* from pIP401-like plasmids to the chromosome. However, subsequent studies have shown that the chloramphenicol resistance determinants present on pIP401, and in strain CW92, do not hybridize with each other and therefore are not closely related (Rood et al., 1989).

Other multiply antibiotic-resistant strains of *C. perfringens* have been obtained from several studies in which organisms were isolated from animal feces (Rood et al., 1978a, 1985; Dutta and Devriese, 1980, 1981). Isolates that were resistant to tetracycline and the macrolide-lincosamide antibiotics were readily obtained, but no penicillin-resistant strains were reported. The incidence of antibiotic-resistant isolates appeared to be related to the use of antibiotics in the feed (Rood et al., 1978a, 1985). Subsequent studies (Rood, 1983; Abraham et al., 1985) showed that, in a significant number of these isolates, resistance to tetracycline could be transferred to sensitive recipient strains in mixed-plate matings. In contrast, resistance to erythromycin was not transferable.

C. perfringens isolates that were derived from human feces and which can act as donors of tetracycline resistance in conjugation experiments have also been reported (Heefner et al., 1984; Miyoshi, 1984).

Molecular studies have shown that the transfer of tetracycline resistance is plasmid determined in all the conjugative isolates examined from either human or animal sources (Abraham and Rood, 1985a; Abraham et al., 1985). Each of these conjugative R-plasmids has been extensively mapped by restriction endonuclease analysis and compared to the prototype plasmids pCW3 and pIP401. Plasmids that were indistinguishable from pCW3 were detected in isolates obtained independently from the United States, France, Belgium, Japan, and Australia. The strains carrying these plasmids were from porcine, bovine, human, and environmental sources. In addition, several plasmids that differed from pCW3 were isolated. Restriction analysis showed that each of these plasmids shared at least 17 kb of restriction map identity, in addition to DNA sequence similarity, with pCW3 (Abraham and Rood, 1985a; Abraham et al., 1985). The only additional conjugative tetracycline resistance plasmid studied in *C. perfringens* is pJU124 (Heefner et al., 1984; Squires et al., 1984). It is clear from the examination of the restriction digests of pJU124 that this plasmid is very closely related to pCW3.

From the comparative data it is concluded that all the known conjugative *C. perfringens* R-plasmids are either identical to pCW3, or closely related to this plasmid, and have presumably evolved from a pCW3-like replicon. It is surprising that no unrelated conjugative resistance plasmids have been reported in *C. perfringens* as are described for other bacterial species. It would be of interest to determine the detailed mechanism of conjugation that is encoded by pCW3 and the physical location of the transfer genes. Based on the wide geographical and environmental sources of the pCW3-like plasmids, it is concluded that human and animal strains of *C. perfringens* strains make up an overlapping, worldwide gene pool. Migration of both humans and animals presumably contributes to the distribution of genetic information within this gene pool (Abraham et al., 1985).

9.3   Tetracycline Resistance Determinants

There are many reports of the isolation of tetracycline-resistant strains of *Clostridium perfringens* (Smith, 1959; Sebald et al., 1975; Rood et al., 1978a, 1978b, 1985; Dutta and Devriese, 1980; Smart et al., 1983; Miyoshi, 1984; Miyoshi and Higa, 1984). These strains have been obtained from animal, human, and environmental sources. Although many isolates have the ability to transfer their tetracycline resistance by a conjugation-like process, the majority of the tetracycline-resistant isolates that have been studied are nonconjugative (Rood et al., 1978b, 1985; Rood, 1983; Miyoshi, 1984; Miyoshi and Higa, 1984). In all the donor strains examined, transfer is mediated via a conjugative plasmid that is either indistinguishable from, or closely related to, the prototype *C. perfringens* R-plasmid, pCW3 (Abraham et al., 1985).

The tetracycline resistance determinant from pCW3 has been cloned into the *Hind*III site of pJU1 to form the recombinant plasmid pJU7 (Squires et al., 1984) and into the *Eco*RI site of pUC18 to form pJIR39 (Figure 9.1; Abraham and Rood, 1985b). In addition, the tetracycline resistance determinant from the conjugative plasmid pJU124 was cloned into the *Eco*RI site of pJU1 to construct pJU10 (Squires et al., 1984). All three recombinant plasmids conferred tetracycline resistance in *E. coli*. The plasmids pJU7 and pJIR39 each contained the 1.9-kb and 2.1-kb *Eco*RI fragments from pCW3; pJU10 contained the equivalent 4-kb fragment from pJU124. Hybridization analysis has shown that the tetracycline resistance determinant from pJIR39 has considerable sequence similarity with the 4-kb fragment from pJU124 (Abraham et al., 1988). These *Eco*RI fragments are within the region of pCW3 that is common to all known conjugative *C. perfringens* R-plasmids. It therefore can be concluded that all these plasmids carry the same tetracycline resistance determinant (Abraham et al., 1985).

The tetracycline resistance conferred by conjugative strains of *C. perfringens* is induced by subinhibitory concentrations of tetracycline. In contrast, tetracycline resistance generally is constitutively expressed in nonconjugative isolates (Rood, 1983). The basis for this difference is not known. The cloned pCW3-derived tetracycline resistance determinant was thought to be inducible in *E. coli* (Abraham and Rood, 1985b). However, subsequent studies have shown that the observed effects probably resulted from differences in plasmid stability and reflected curing in the absence of selection for tetracycline resistance (B. Hoffmann and J. Rood, unpublished observations).

The pCW3-derived recombinant plasmid pJIR39 has been mapped (see figure 9.1) and extensively analyzed (Abraham and Rood, 1985b; Abraham et al., 1988). To determine if both the *Eco*RI fragments were essential for tetracycline resistance, each fragment was subcloned into pUC18. Recombinant plasmids that contained either the 1.9-kb fragment (pJIR40) or the 2.1-kb fragment (pJIR42) were obtained. However, neither plasmid conferred resistance to tetracycline. An *Sph*I deletion derivative of pJIR39 also was constructed. This plasmid, pJIR43, which contained the 1.9-kb *Eco*RI and the juxtaposed 0.8-kb *Eco*RI-*Sph*I fragment, also failed to confer tetracycline resistance. On the basis of these data it was concluded that the internal *Eco*RI and *Sph*I sites of pJIR39 were located within the tetracycline resistance determinant (Abraham and Rood, 1985b).

Further studies involved Tn1725-mediated mutagenesis of pJIR71, a *Pst*I-derived subclone of pJIR39. Five transposon derivatives were isolated and studied in detail. Strains carrying one of these plasmids, pJIR144, were sensitive to tetracycline (Abraham et al., 1988). These data were used to localize the resistance determinant to within a 1.4-kb region of pJIR71 (Figure 9.2). The 0.8 kb *Eco*RI- *Sph*I fragment was internal to the region and was used as a gene probe that was specific for this tetracycline resistance determinant. The gene probe hybridized with chromosomal DNA prepared from eight constitutive, tetracycline-resistant, nonconjugative *C. perfringens* isolates from diverse sources and also hybridized to DNA from a tetracycline-

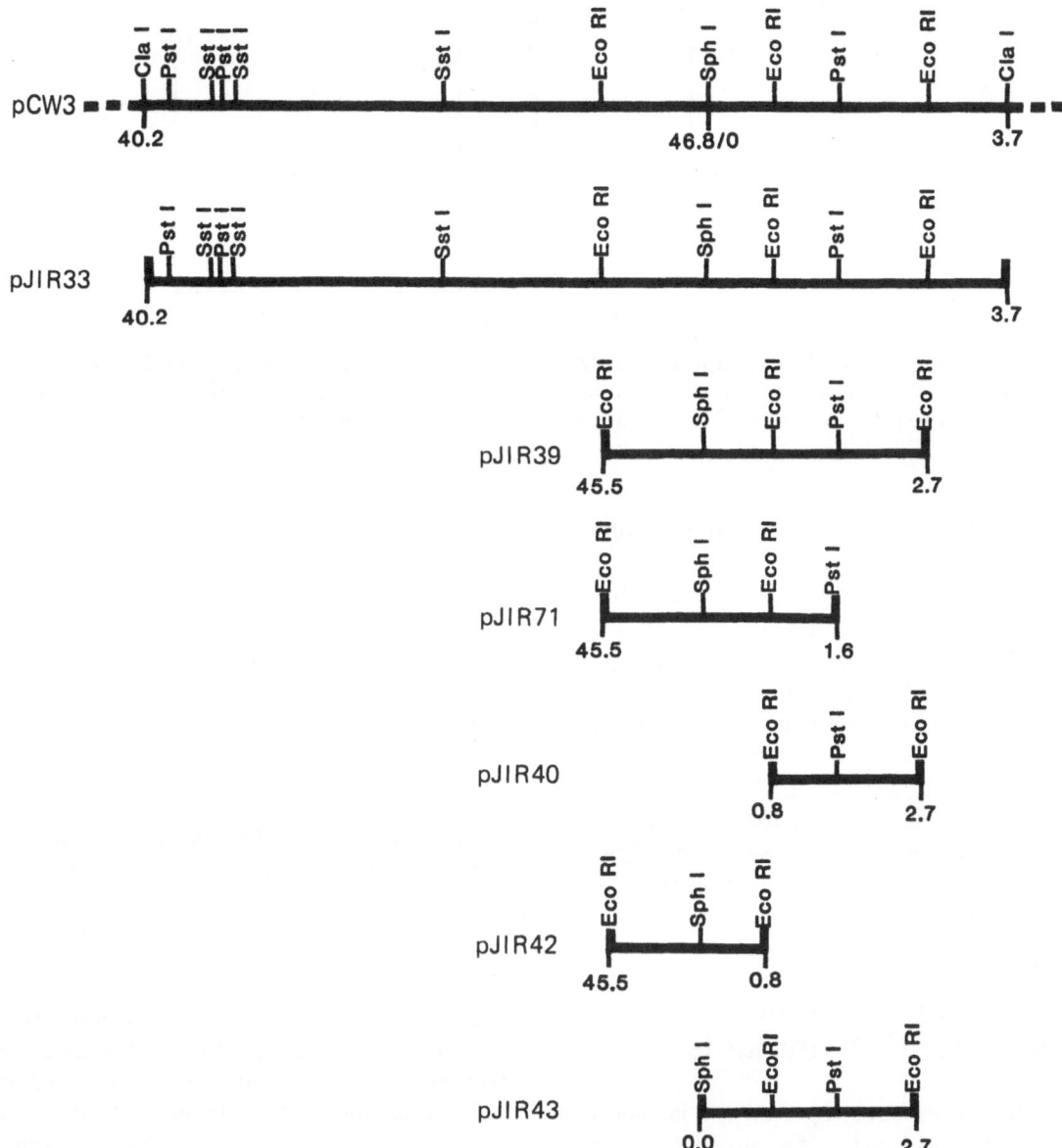

FIGURE 9.1 Derivation of Tet P recombinant plasmids. Restriction maps of plasmids derived from pCW3: plasmids pJIR33, pJIR39, and pJIR71 confer tetracycline resistance in *E coli*; pJIR40, pJIR42, and pJIR43 do not confer tetracycline resistance. pCW3 coordinates are as indicated (kb); pUC vector regions are not shown. [Reproduced from Abraham (1986) with permission of the author.]

resistant strain of *Clostridium paraputrificum*. However, it was not homologous to the tetracycline resistance determinants from the six isolates of *Clostridium difficile*, and the one *Clostridium sporogenes* strain, that were tested (Abraham et al., 1988). These results showed that the tetracycline resistance determinant from conjugative isolates was present in nonconjugative strains of *C. perfringens* and was related to a tetracycline resistance determinant from at least one other clostridial species.

Other investigators identified a number of different tetracycline resistance determinants from both gram-positive and gram-negative bacteria in earlier studies (Levy et al., 1989). Representatives of these determinants were

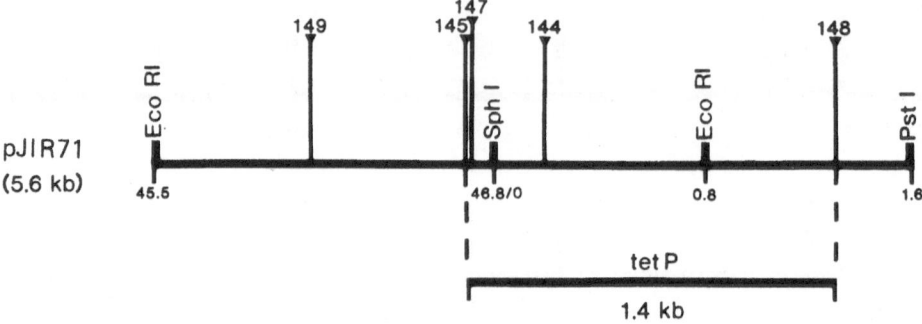

FIGURE 9.2 Sites of Tn1725 insertion into pJIR71. pCW3 coordinates indicated in kilobasepairs (kb). Numbers refer to pJIR plasmid number of particular insertion derivative. Maximum size of Tet P determinant indicated by horizontal bar. [Reproduced from Abraham et al. (1988).]

probed with the *C. perfringens* probe. None of the determinants that were tested, including the promiscuous Tet M determinant that is found in *C. difficile*, hybridized with the probe (Hächler et al., 1987b). It was therefore concluded that the *C. perfringens* tetracycline resistance determinant represented a new hybridization class, which was designated as Class P (for *perfringens*) (Abraham et al., 1988). Preliminary observations indicate that the Tet P determinant mediates an active efflux of tetracycline (L. McMurray and S.B. Levy, personal communication).

9.4 Chloramphenicol Resistance Determinants

Several studies have reported the isolation of chloramphenicol-resistant strains of *Clostridium perfringens* (Sebald et al., 1975; Rood et al., 1978b, 1985; Dutta and Devriese, 1980). However, chloramphenicol resistance is not as common as tetracycline or erythromycin resistance. Only six chloramphenicol-resistant isolates have been examined in detail, including the well-studied tetracycline-resistant isolates CP590, CP600, and CW92. The former strains, from human and bovine sources in France, carry conjugative tetracycline and chloramphenicol resistance plasmids (Bréfort et al., 1977), as does a porcine isolate from Belgium (Abraham et al., 1985). Two other porcine isolates from Belgium have been shown to carry

chloramphenicol resistance plasmids (Rood et al., 1989). In the remaining strain, CW92, chloramphenicol resistance is not plasmid determined (Rood et al., 1989). Resistance in both CP590 and CW92 is mediated via the production of chloramphenicol acetyltransferase enzymes (Rood et al., 1978b; Zaidenzaig et al., 1979).

Characterization of chloramphenicol resistance transposons

Bréfort et al. (1977) showed that conjugative transfer of pIP401 and pJIR25 [as later named by Abraham et al. (1985)] from CP590 and CP600, respectively, was often associated with a loss of the chloramphenicol resistance determinant and concomitant loss of 6 kb of DNA. These results have been confirmed and extended to a third conjugative plasmid, pJIR27 (Abraham et al., 1985). Magot (1984) constructed a restriction map of pIP401 and showed that a 6.2-kb region was lost on conjugation and that this chloramphenicol resistance determinant was located within a 10.6-kb EcoRI-PstI fragment of the parent plasmid. Further studies showed that in all three of these chloramphenicol plasmids the chloramphenicol resistance genes were located on the largest EcoRI fragments (Abraham et al., 1985).

The chloramphenicol resistance determinants from pIP401 (Abraham et al., 1985) and pJIR27 (Abraham and Rood, 1987) have been cloned to form pJIR45 and pJIR97, respectively,

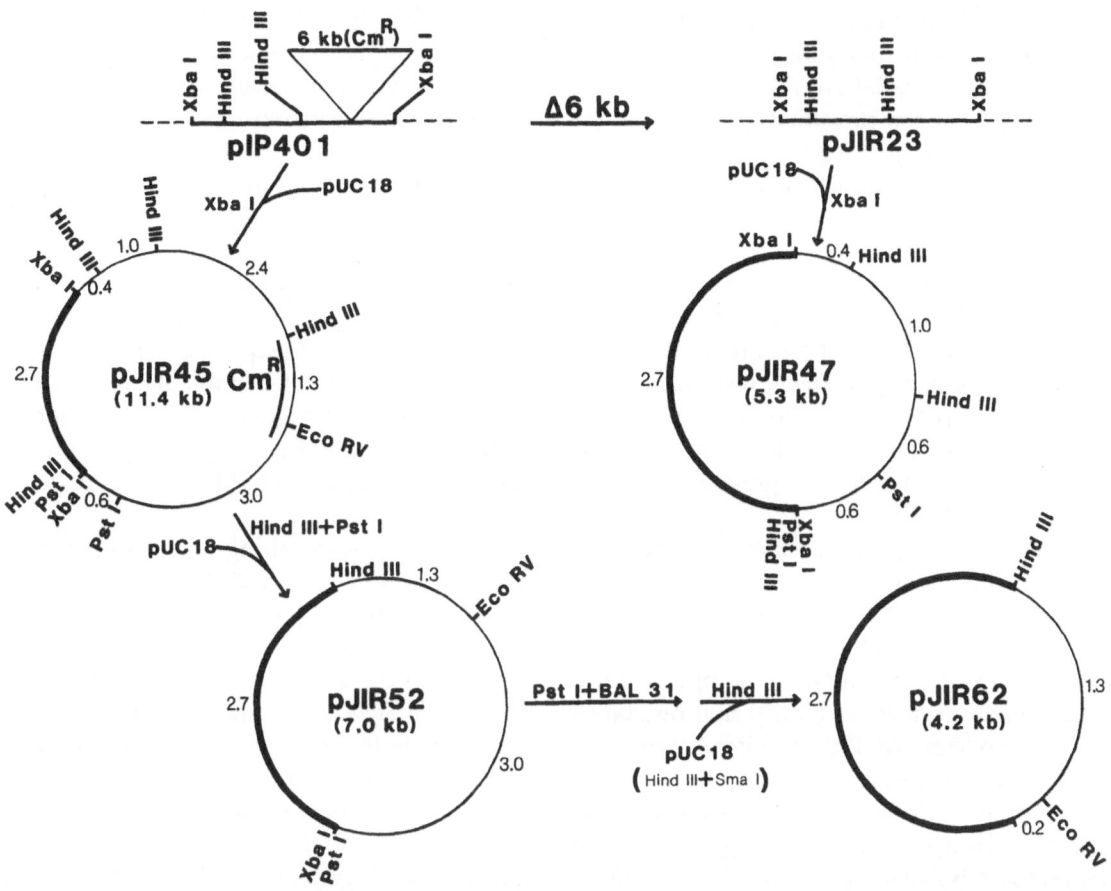

FIGURE 9.3 Derivation of recombinant plasmids derived from pIP401. Restriction maps of chloramphenicol resistance plasmids pJIR45, pJIR52, and pJIR62, and deletion derivative pJIR47. Sizes indicated in kilobasepairs (kb); pUC18 vector indicated by heavy line. [Reproduced from Abraham et al. (1985).]

and shown to be expressed in *E. coli*. The recombinant plasmids share considerable restriction identity (Abraham and Rood, 1987). Subcloning and deletion analysis of pJIR45 led to the localization of the resistance gene to within a 1.5-kb region and revealed that there was an *Eco*RV site located within the resistance gene (Figure 9.3). Hybridization analysis using this 1.5-kb region as a probe showed that all three of the conjugative *C. perfringens* plasmids carried the same chloramphenicol resistance determinant (Abraham et al., 1985). Further studies have shown that two additional chloramphenicol-resistant *C. perfringens* strains carry this gene on large plasmids (Rood et al., 1989).

The recombinant plasmids pJIR45 and pJIR97 are unstable in *recA* strains of *E. coli*, and dele-

tion plasmids are isolated spontaneously at high frequency (Abraham and Rood, 1987). These plasmids do not confer chloramphenicol resistance. The restriction profiles of six independently derived deletion derivatives of pJIR45 were analyzed and shown to be identical, indicating that essentially the same deletion event was occurring each time. The plasmids carried in these strains lack the identical 6.2-kb region that is deleted at high frequency on conjugation in *C. perfringens*. The pJIR45-derived deletion plasmid pJIR86 is therefore indistinguishable from the recombinant plasmid pJIR47, which was derived from the pIP401-derived deletion plasmid pJIR23 (see Figure 9.3). Sequence analysis has confirmed that the identical gene region is deleted in *E. coli* and *C.*

perfringens (Abraham and Rood, 1988). The excision process can therefore be regarded as a precise event. A variation of the same general deletion mechanism presumably is involved in both *E. coli* and *C. perfringens*.

Both ends of the 6.2-kb region are required for spontaneous excision in *E. coli*. Subclones that separately contain the left and right ends of this region with respect to the *EcoRV* site are therefore stable in *E. coli* (Abraham and Rood, 1987). These data, plus the similarity between the various chloramphenicol resistance plasmids and the precision of the deletion events, support the previous suggestion that the 6.2-kb regions were transposable genetic elements (Abraham et al., 1985).

Transposition of chloramphenicol resistance has been demonstrated in *E. coli* but not in *C. perfringens*. Abraham and Rood (1987) separately cloned the chloramphenicol resistance determinants from pJIR45 and pJIR97 onto a temperature-sensitive, tetracycline resistance pSC301 replicon. At the permissive temperature the spontaneous deletion phenomenon was observed in *recA* strains carrying these plasmids. However, after prolonged growth at the nonpermissive temperature, chloramphenicol-resistant, tetracycline-sensitive derivatives were isolated at a low frequency. Southern hybridization analysis of chromosomal DNA from these strains showed that they each contained a 4.2-kb *CfoI* fragment that was internal to the 6.2-kb region but did not have any homology with the chloramphenicol-sensitive deletion plasmids. It was concluded that the 6.2-kb regions had transposed from the temperature-sensitive plasmids to the *E. coli* chromosome. The regions from pIP401 and pJIR27 were therefore designated as the transposons Tn4451 and Tn4452, respectively (Abraham and Rood, 1987).

Heteroduplex analysis was used to determine the location of the Tn4451 excision site and to compare the two transposons. Delineation of both transposons revealed that Tn4451 and Tn4452 were homologous, except for a 0.4-kb region at one end of the transposons. Neither inverted nor directly repeated sequences were detected at the ends of Tn4451 by either heteroduplex analysis or hybridiza-

tion studies (Abraham and Rood, 1987). However, sequence analysis of the ends of Tn4451 showed that small, imperfect (8 of 12 bases), terminal inverted repeats were present (Abraham and Rood, 1988). Analysis of these sequences showed that they had some homology with the Tn3 family of transposable genetic elements. The first six bases of Tn4451-R (GGGGTC) were identical to the terminal sequences of several transposons, including Tn3 and Tn917. A CTAA sequence, which is often found at the internal terminus of the 38-bp inverted repeat of Tn3-like transposons (Sherrat, 1989), was found at the equivalent positions in Tn4451-L and Tn4451-R (Abraham and Rood, 1988).

Potential outward firing −35 promoter regions are found within 25 bp of the ends of many insertion sequences and transposons (Galas and Chandler, 1989). A consensus −35 sequence, TTGACA, is located 17 bp from the right end of Tn4451 (Abraham and Rood, 1988). It is possible that this sequence is involved in the formation of hybrid promoters after transposition, as occurs with other transposable elements (Galas and Chandler, 1989). The Tn3 transposons all generate a 5-bp duplication on insertion (Sherrat, 1989). Sequence analysis of the deletion plasmids pJIR47 and pJIR86, which were derived from deletion events in *C. perfringens* and *E. coli*, respectively, indicated that Tn4451 insertion may involve the duplication of a 2-bp sequence (Abraham and Rood, 1988). This suggestion must be confirmed by the sequence analysis of a transposon insertion derivative. Tn4451 and Tn4452 are the only transposons to be characterized from the clostridia. Therefore, their analysis is important for studies on the evolution of antibiotic resistance determinants. Current studies in this laboratory involve the determination of the nucleotide sequence of Tn4451.

Comparative analysis of chloramphenicol acetyltransferase genes

By use of a gene probe specific for the pIP401-derived chloramphenicol resistance determinant, hybridization analysis showed that, of six chloramphenicol-resistant *C. perfringens* strains studied, five isolates carried homologous resistance determinants that were located on large plasmids. This determinant was designated as CAT P. The remaining isolate, CW92, which also was the original parent of pCW3, did not hybridize and therefore carried a different determinant, designated CAT Q (Rood et al., 1989). This second *C. perfringens* CAT determinant has also been cloned and shown to be chromosomally located. There is no evidence that the *catQ* gene is located on a transposon (Rood et al., 1989).

CAT determinants from a variety of different bacterial species were probed with fragments specific for the *catP* and *catQ* genes (Rood et al., 1989). The *catQ*-specific probe did not hybridize with any of the other determinants. The *catQ* gene therefore represents a distinct hybridization class of chloramphenicol resistance determinant. The *catP* probe also did not hybridize to *cat* genes from other bacterial genera. However, it did exhibit homology with the *catD* gene from *Clostridium difficile* (Rood et al., 1989). Previous results suggested that the *catD* gene was present in multiple copies on the *C. difficile* chromosome (Wren et al., 1988).

Determination of the *catP* gene sequence showed that although the gene did not have significant DNA sequence homology with other *cat* genes there was notable homology at the amino acid sequence level. The region of homology included the known active site, which is at the carboxy terminal end of the molecule. The *catP* gene product shared from 42% to 45% amino acid sequence similarity with *cat* genes from other genera (Steffen and Matzura, 1989). Further analysis confirmed that the *catP* gene was more closely related to CAT monomers from *Staphylococcus aureus* and *Campylobacter coli* than to monomers from other bacteria (Bannam and Rood, 1991). However, unlike other *cat* genes from gram-positive bacteria, expression of chloramphenicol resistance was not regulated by a translation attenuation mechanism but was constitutive.

Sequence analysis has shown that the *catD* sequence is almost identical to that of *catP* (Wren et al., 1989). The two *cat* gene products differ only in two amino acids at positions 31

and 32. These conservative changes, from Ser-Leu in CATP to ThrVal in CATD, appear to have arisen by an inversion of a GCC sequence. Both CAT monomers and the *C. coli* monomer contain a 4-amino-acid deletion when compared to the other known CAT determinants. No other sequenced CAT determinants contain this deletion. It is clear from these results that the CAT P and D determinants are very closely related in terms of evolution. Transfer of pIP401 into *C. difficile* has been reported, but the resultant transconjugants were unstable (Ionesco, 1980). On the basis of these results and the sequence data it appears likely that CAT D arose from the conjugative transfer of a plasmid similar to pIP401 from *C. perfringens* into *C. difficile* and subsequent transposition of a Tn4451-like element onto the *C. difficile* chromosome.

We also have determined the nucleotide sequence of the *C. perfringens* CAT Q determinant. These results confirm the conclusions of the hybridization experiments. The *catQ* gene appears to be as closely related to the *cat* genes from *Staphylococcus aureus* as it is to the *catP* and *catD* genes; it does not contain the 4-amino-acid deletion present in the other clostridial genes. It is apparent from these results that the CAT Q determinant has evolved independently of the other clostridial genes (Bannam and Rood, 1991)

9.5   Erythromycin Resistance Determinants

There are several reports of the isolation of erythromycin-resistant strains of *Clostridium perfringens*. The first major survey was conducted in Wisconsin (United States), where it was found that erythromycin-resistant strains could be readily isolated from pig farms at which antibiotics were used in the feeds. All the erythromycin-resistant strains also were resistant to clindamycin and lincomycin (Rood et al., 1978a). Subsequent studies showed that, of the 41 isolates that were tested, none was able to transfer erythromycin resistance to sensitive recipient strains (Rood, 1983). A similar porcine survey carried out in Western Australia also led

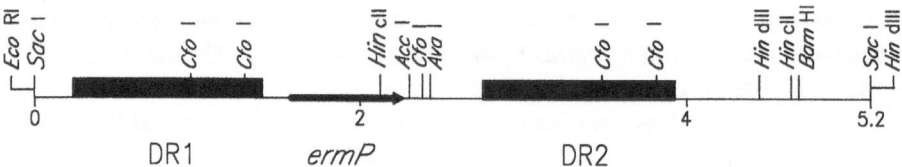

FIGURE 9.4   Genetic map of Erm P determinant. Restriction map of Erm P recombinant, pJIR 122; sizes in kb. Vector portion not shown. *erm* gene is indicated by arrow; direct repeats (DR1 and DR2) shown by heavy solid boxes.

to the ready isolation of erythromycin-resistant strains of *C. perfringens* that failed to transfer resistance in mixed-plate matings (Rood et al., 1985). Other workers have described the isolation of erythromycin-resistant strains of *C. perfringens* from a variety of sources (Dornbusch et al., 1975; Dutta and Devriese, 1980, 1981; Kordel and Schallehn, 1984).

The initial reports which depicted and characterised the conjugative tetracycline and chloramphenicol resistance plasmid pIP401 also described the nonconjugative plasmid pIP402 (Sebald and Bréfort, 1975; Sebald et al., 1975; Bréfort et al., 1977). Curing experiments showed that this plasmid conferred resistance to the macrolide-lincosamide antibiotics, i.e., erythromycin, clindamycin, and lincomycin. Throughout this chapter, this phenotype is referred to as erythromycin resistance. The plasmid pIP402 remains the only erythromycin resistance plasmid and the only nonconjugative R-plasmid ever reported from *C. perfringens*.

The erythromycin resistance determinant from a strain known to carry pIP402 has been cloned and shown to be expressed in *E. coli* (Berryman and Rood, 1989). Because we were unable to purify pIP402 consistently from the parent strain, we cloned the gene by shotgun cloning of a total DNA preparation. The resistance determinant, designated Erm P, was located on a 5.2-kb *SacI* fragment as part of the recombinant plasmid pJIR122 (Figure 9.4). This plasmid was unstable in *E. coli*. Deletion derivatives that had lost a 2.7-kb region, including the Erm P determinant, were obtained spontaneously at high frequency in *recA* strains. Subcloning and deletion analysis led to the delineation of the Erm P gene. This gene was located between two 350-bp *CfoI* fragments, one of which was retained in the

spontaneous deletion derivatives. Based on these results we suggested that these fragments form part of a directly repeated sequence that is located on either side of the resistance determinant (Berryman and Rood, 1989). Subsequent sequence analysis has confirmed that the *erm* P gene is located between two direct repeats of approximately 1300 bp and that the *CfoI* fragments are within these repeats (see Figure 9.4; D.I. Berryman and J.I. Rood, unpublished results). Although the structure of this determinant resembles that of a transposable genetic element, neither transposition of the putative element nor transposition of the direct repeats has been demonstrated in either *C. perfringens* or *E. coli*.

Hybridization experiments similar to those previously described for the *C. perfringens* Tet P, CAT P, and CAT Q determinants were carried out using an *ermP*-specific gene probe. However, the results were somewhat different. Only 5 of the 40 wild-type erythromycin-resistant *C. perfringens* strains that were probed had DNA sequences which hybridized with *ermP*. The *ermP* gene therefore is not widespread in *C. perfringens*. Nevertheless, the determinant is not restricted to this clostridial species. Five erythromycin-, clindamycin-, and lincomycin-resistant *Clostridium difficile* isolates, from diverse sources, also hybridized with the *ermP* probe as did a phenotypically similar strain of *Clostridium paraputrificum* (Berryman and Rood, 1989).

Previous workers have identified seven distinct hybridization classes of erythromycin resistance determinants, designated Erm A–G (Dubnau and Monod, 1986). Representatives of each of these classes also were probed with the *ermP-specific* gene probe. In contrast to the results obtained with the tetracycline and chloramphenicol resistance determinants, the *ermP* gene did hybridize with *erm* genes from other bacterial genera. Specifically, hybridization was observed with the plasmid pAMβ1, a representative of the Erm B/Erm AM class (Berryman and Rood, 1989). The *C. difficile* determinant, which appears to be located on a chromosomal conjugative transposon, also has been shown to hybridize with Erm B/Erm AM resistance determinants (Hächler et al., 1987a). When each of the Erm A–G plasmids was used to probe the 40 *C. perfringens* isolates previously tested, the pAMβ1 probe hybridized with the same five isolates that were positive in the *ermP* experiments. None of the other isolates hybridized with any of the probes (Berryman and Rood, 1989). The resistance determinants present in these strains therefore must represent new hybridization classes of erythromycin resistance determinants. Current studies in our laboratory involve the cloning and analysis of these genes.

We have recently determined the DNA sequence of the *ermP* gene (D.I. Berryman and J.I. Rood, unpublished observations). The sequence was compared with the previously determined sequence of the pAMβ1 gene (Brehm et al., 1987; Martin et al., 1987) and was shown to be identical. Further, the region of DNA sequence identity extends beyond the *erm* genes. Both the *ermP* and pAMβ1 *erm* genes, which are constitutively expressed, lack the leader sequence region that confers inducible resistance in other Erm B/Erm AM determinants (Brehm et al., 1987: Martin et al., 1987). The sequence identity between pAMβ1 and Erm P extends into the *C. perfringens* direct repeats in both directions (D.I. Berryman and J.I. Rood, unpublished observations). On the basis of the known broad host range of pAMβ1 (Horaud et al., 1985; Oultram and Young, 1985) it could be postulated that the Erm P determinant arose from the transfer of pAMβ1 into *C. perfringens* followed by subsequent genetic rearrangements of the surrounding regions. However, the finding that only part of the sequences present on either side of the *ermP* gene are present in pAMβ1 suggests that the clostridial determinant may represent the more primordial form. It will be of great interest to determine if a complete copy of the *C. perfringens* direct repeat can be detected in any erythromycin-resistant strains of either *Streptococcus* or *Enterococcus* spp.

9.6 *Clostridium perfringens–Escherichia coli* Shuttle Plasmids

All the well-characterized *C. perfringens* antibiotic resistance determinants have been used for the construction of *C. perfringens–E. coli* shuttle plasmids. The first shuttle plasmids that were produced carried the *tet*(P) genes from pCW3 and pJU124 (Squires et al., 1984). These plasmids utilized the *E. coli* pBR322 replicon and two small plasmids from *C. perfringens*, namely pJU121 and pJU122. The resultant shuttle plasmids pJU12, pJU13, and pJU16 encoded tetracycline resistance in *C. perfringens* and *E. coli*. Both L-phase variants and autoplasts of *C. perfringens* were transformed with these plasmids (Squires et al., 1984). Roberts et al. (1988) constructed a different shuttle plasmid, pHR106, by combining the *E. coli* replicon pSL100, the *C. perfringens* plasmid pJU122, and the *catP* gene from pJIR62 (Abraham et al., 1985). This plasmid encoded chloramphenicol resistance in *C. perfringens* and *E. coli*. L-phase variants of *C. perfringens* have been transformed with both pJU16 and pHR106 (Mahony et al., 1988; Roberts et al., 1988). Other workers have used electroporation to transform vegetative *C. perfringens* cells with pHR106 (Allen and Blaschek, 1988; Scott and Rood, 1989). Another shuttle vector, pAK201, has also been constructed (Kim and Blaschek, 1989). This plasmid carries the pBR322 replicon, the caseinase-encoding *C. perfringens* plasmid, pHB101, and the *catP* gene from pJIR62.

None of these plasmids is ideal for use in genetic manipulation between *C. perfringens* and *E. coli*. The pJU16 and pHR106 plasmids carry an *E. coli* ampicillin resistance determinant. Because *C. perfringens* isolates are always penicillin sensitive, this resistance gene should be avoided in gene cloning experiments, especially if toxin genes are involved. The pAK201 replicon contains an inactive caseinase gene and a limited number of sites suitable for cloning. None of these plasmids contain a multiple cloning site or features that enable direct screening or selection of recombinants in either *E. coli* or *C. perfringens*. In addition, these plasmids are essentially uncharacterized; very little DNA sequence information is available.

To overcome these problems we have constructed a new *C. perfringens–E. coli* shuttle plasmid, pJIR418 (Sloan et al., Plasmid, in press). This plasmid contains the replication region, *lacZ'* gene, and multiple cloning site from pUC18 but not the *bla* gene; the replication region from the *C. perfringens* plasmid pIP404; and the *ermP* and *catP* genes. Conventional XGal screening can be used to detect recombinants in *E. coli*, and insertional inactivation of the chloramphenicol or erythromycin resistance determinant can be used to detect recombinants in *C. perfringens*, if necessary. In addition, the entire sequence of this 7.4-kb plasmid has been elucidated. This plasmid should greatly facilitate the cloning of *C. perfringens* toxin genes and their subsequent reintroduction into *C. perfringens*. The plasmid is available from the author on request.

9.7   Conclusions

In recent years, our knowledge of the evolution of many bacterial antibiotic resistance determinants has greatly expanded. The role of both conjugative and nonconjugative plasmids, of conjugative chromosomal transposons, and of conventional transposons in the dissemination of antibiotic resistance determinants both within and across species and genus barriers now is more clearly understood. This knowledge has arisen primarily from cloning, hybridization, and DNA sequence analysis of antibiotic resistance genes. Most laboratories have concentrated on the detailed molecular analysis of one particular type of antibiotic resistance, for example, tetracycline resistance or aminoglycoside resistance. The approach of the researchers in this laboratory has been somewhat different as we have focused on the molecular analysis of several antibiotic resistance determinants

from one microorganism, *Clostridium perfringens*.

The results of these species-specific studies are interesting when put into context with those obtained using the complementary approach. We can certainly state that it is not possible, at least with *C. perfringens*, to make definitive species-specific generalizations. In this organism, there is a single, almost universally distributed tetracycline resistance determinant, Tet P, which is found on the ubiquitous conjugative *C. perfringens* R-plasmid, pCW3, and on closely related plasmids. Although the Tet P determinant has been found in one isolate of *Clostridium paraputrificum*, it was not detected in *Clostridium difficile* and does not hybridize with Tet determinants from other bacterial genera. Based on these data, the conclusion that *C. perfringens* did not readily exchange genetic information with either other clostridia or other genera could easily be drawn.

Different conclusions must be drawn if our studies on the *C. perfringens* chloramphenicol resistance determinants are considered. It has been shown that there are two determinants, CAT P and CAT Q. The CAT Q determinant appears to be restricted to *C. perfringens* and therefore resembles Tet P in its distribution. However, although the transposon-determined *catP* gene is located on plasmids that are very closely related to pCW3 and which contain a functional *tet*(P) gene, a gene almost identical to *catP* is found in *C. difficile* but not in other genera. Based on the latter data it would be reasonable to conclude that genetic exchange between *C. perfringens* and *C. difficile* can occur.

This conclusion must be broadened further when the erythromycin resistance determinants are considered. In *C. perfringens* there are at least two different Erm determinants and perhaps more, and one determinant, Erm P, is located on a nonconjugative plasmid between large, directly repeated sequences. The sequence identity between the *ermP* and pAMβ1 *erm* genes suggests that direct genetic exchange not only occurs between *C. perfringens* and *C. difficile* but also occurs between the clostridia and other gram-positive bacteria. It is obvious from these data that when considering the molecular epidemiology of antibiotic resistance determinants from one particular bacterium, or a group of bacteria, it is important to study more than one resistance determinant; it may be tempting to make generalizations that may not apply to all the genes that may be transferred from one microorganism to another.

The reasons for the differences in distribution of the various *C. perfringens* determinants are not known. The differences between Tet P and CAT P are not caused by variations in their ability to transfer between species because both are located on the same pCW3-like base plasmids. Perhaps Tn4451-like transposons have the ability to transpose to the *C. difficile* chromosome after conjugation, whereas pCW3 is unstable in *C. difficile* and is rapidly lost.

It is difficult to understand why there should be basically only one type of conjugative R-plasmid in such a common gastrointestinal tract inhabitant as *C. perfringens*, and thus it is important to elucidate the mechanism of conjugal transfer of pCW3-like plasmids and identify the genes involved in this transfer. Also, a conjugative chromosomal Tet M-like determinant has been found in *C. difficile* (Hächler et al., 1987b), and a Tet M transposon, Tn916, has been shown to transfer into *C. perfringens* (E. Stellwag, personal communication); it is therefore surprising that a similar determinant has not been detected in *C. perfringens*. Finally, elucidation of the pathway of evolution of the Erm P determinant must await further analysis of the Erm B/Erm AM class. Of particular importance will be the search for *C. perfringens*-like direct repeats in other genera and the detailed molecular analysis of the *C. difficile* erythromycin resistance determinant.

Molecular studies on antibiotic resistance determinants from *C. perfringens* have made an important contribution to our understanding of the molecular epidemiology and evolution of antibiotic resistance genes in gram-positive bacteria. In addition, they have led directly to the construction of *C. perfringens*–*E. coli* shuttle vectors and the development of *C. perfringens* genetics. Following the elucidation of transformation methods for *C. perfringens* (Allen and Blaschek, 1988; Scott and Rood, 1989), the use of these vectors has meant that it is now possible to clone, analyze, and manipulate *C. perfringens* genes in both *E. coli* and *C. perfringens*. These developments are the harbinger of an exciting new era in *C. perfringens* genetics.

Acknowledgment. I wish to thank the graduate students and research staff who have worked in my laboratory over the past decade for their dedication and commitment. In particular, I express my gratitude to Trudi Bannam, Dave Berryman and Joan Sloan for their assistance in the preparation of this chapter. I gratefully acknowledge the research support provided by grants from the Australian Research Council and Monash University.

References

Abraham, L.J. 1986. Molecular genetics of antibiotic resistance determinants from *Clostridium perfringens*. Ph.D. thesis, Murdoch University, Perth, Australia.

Abraham, L.J. and J.I. Rood. 1985a. Molecular analysis of transferable tetracycline resistance plasmids from *Clostridium perfringens*. J. Bacteriol. **161**:636–640.

Abraham, L.J. and J.I. Rood. 1985b. Cloning and analysis of the *Clostridium perfringens* tetracycline resistance plasmid, pCW3. *Plasmid* **13**:155–162.

Abraham, L.J. and J.I. Rood. 1987. Identification of Tn4451 and Tn4452: Two chloramphenicol resistance transposons from *Clostridium perfringens*. J. Bacteriol. **169**:1579–1584.

Abraham, L.J. and J.I. Rood. 1988. The *Clostridium perfringens* chloramphenicol resistance transposon Tn4451 excises precisely in *Escherichia coli*. *Plasmid* **19**:164–168.

Abraham, L.J., D.I. Berryman and J.I. Rood. 1988. Hybridization analysis of the class P tetracycline resistance determinant from the *Clostridium perfringens* R-plasmid, pCW3. *Plasmid* **19**:113–120.

Abraham, L.J., A.J. Wales, and J.I. Rood. 1985. Worldwide distribution of the conjugative *Clostridium perfringens* tetracycline resistance plasmid, pCW3. *Plasmid* **14**:37–46.

Allen, S.P. and H.P. Blaschek. 1988. Electroporation-induced transformation of intact cells of *Clostridium perfringens*. Appl. Environ. Microbiol. **54**:2322–2324.

Bannam, T.L. and Rood, J.I. (1991). The relationship between the *Clostridium perfringens* catQ gene product and chloramphenicol acetyltransferases from other bacteria. *Antimicrob. Agents Chemother.* **35**:471–476.

Berryman, D.I. and J.I. Rood. 1989. Cloning and hybridization analysis of *ermP*, an MLS resistance determinant from *Clostridium perfringens*. *Antimicrob. Agents Chemother.* 33:1346–1353.

Blaschek, H.P. and M. Solberg. 1981. Isolation of a plasmid responsible for caseinase activity in *Clostridium perfringens* ATCC 3626B. *J. Bacteriol.* 147:262–266.

Blaschek, HP. and M.A. Klacik. 1984. Role of DNase in recovery of plasmid DNA from *Clostridium perfringens*. *Appl. Environ. Microbiol.* 48:178–181.

Bréfort, G., M. Magot, H. Ionesco, and M. Sebald. 1977. Characterization and transferability of *Clostridium perfringens* plasmids. *Plasmid* 1:52–66.

Brehm, J., G. Salmond, and N. Minton. 1987. Sequence of the adenine methylase gene of the *Streptococcus faecalis* plasmid pAMβ1. *Nucleic Acids Res.* 15:3177.

Dornbusch, K., C.E. Nord, and A. Dahlbäck. 1975. Antibiotic susceptibility of *Clostridium* species isolated from human infections. *Scand. J. Infect. Dis.* 7:127–134.

Dubnau, D. and M. Monod. 1986. The regulation and evolution of MLS resistance. *Banbury Rep.* 24:369–387.

Dutta, G.N. and L.A. Devriese. 1980. Susceptibility of *Clostridium perfringens* of animal origin to fifteen antimicrobial agents. *J. Vet. Pharmacol. Ther.* 3:227–236.

Dutta, G.N. and L.A. Devriese. 1981. Macrolide-lincosamide-streptogramin resistance patterns in *Clostridium perfringens* from animals. *Antimicrob. Agents Chemother.* 19:274–278.

Galas, D.J. and M. Chandler. 1989. Bacterial insertion sequences. In: *Mobile DNA*, D.E. Berg and M.M. Howe, ed., pp. 109–162. Washington, D.C.: American Society for Microbiology.

Garnier, T. and S.T. Cole. 1988. Complete nucleotide sequence and genetic organization of the bacteriocinogenic plasmid, pIP404, from *Clostridium perfringens*. *Plasmid* 19:134–150.

Hächler, H., B. Berger-Bächi, and F.H. Kayser. 1987a. Genetic characterization of a *Clostridium difficile* erythromycin-clindamycin resistance determinant that is transferable to *Staphylococcus aureus*. *Antimicrob. Agents Chemother.* 31:1039–1045.

Hächler, H., F.H. Kayser, and B. Berger-Bächi. 1987b. Homology of a transferable tetracycline resistance determinant of *Clostridium difficile* with *Streptococcus (Enterococcus) faecalis* transposon Tn916. *Antimicrob. Agents and Chemother.* 31:1033–1038.

Hatheway, C.L. 1990. Toxigenic clostridia. *Clin. Microbiol. Rev.* 3:66–98.

Heefner, D.L., C.H. Squires, R.J. Evans, B.J. Kopp, and M.J. Yarus. 1984. Transformation of *Clostridium perfringens*. *J. Bacteriol.* 159:460–464.

Horaud, T., C. Le Bouguenec, and K. Pepper. 1985. Molecular genetics of resistance to macrolides, lincosamides and streptogramin B (MLS) in streptococci. *J. Antimicrob. Chemother.* 16 (Suppl. A): 111–135.

Ionesco, H. 1980. Transfert de la resistance a la tetracycline chez *Clostridium difficile*. *Ann. Microbiol. (Paris)* 131A:171–179.

Ionesco, H., G. Bieth, C. Dauget, and D. Bouanchaud. 1974. Isolement et identification de deux plasmides d'une souche bactériocinogène. *Ann. Microbiol. (Paris)* 127B:283–294.

Kim, A. Y and Blaschek, H. P. (1989). Construction of an *Escherichia coli-Clostridium perfringens* shuttle vector and plasmid transformation of *Clostridium perfringens*. *Appl. Environ. Microbiol.* 55:360–365.

Kordel, M. and G. Schallehn. 1984. Plasmid detection in a macrolide-lincosamide resistant strain of *Clostridium perfringens*. *Fed. Eur. Microbiol. Soc. Lett.* 24:153–157.

Levy, S.B., L.M. McMurray, V. Burdett, P. Courvalin, W. Hillen, M.C. Roberts, and D.E. Taylor. 1989. Nomenclature for tetracycline resistance determinants. *Antimicrob. Agents Chemother.* 33:1373–1374.

Magot. M. 1984. Physical characterization of the *Clostridium perfringens* tetracycline-chloramphenicol resistance plasmid pIP401. *Ann. Microbiol. (Inst. Pasteur)* 135B:269–282.

Mahony, D.E., J.A. Mader, and J.R. Dubel. 1988. Transformation of *Clostridium perfringens* L-forms with plasmid DNA. *Appl. Environ. Microbiol.* 54:264–267.

Mahony, D.E., M.F. Stringer, S.P. Borriello, and J.A. Mader. 1987. Plasmid analysis as a means of strain differentiation in *Clostridium perfringens*. *J. Clin. Microbiol.* 25:1333–1335.

Mahony, D.E., G.A. Clark, M.F. Stringer, M.C. MacDonald, D.R. Duchesne, and J.A. Mader. 1986. Rapid extraction of plasmids from *Clostridium perfringens*. *Appl. Environ. Microbiol.* 51:521–523.

Martin, B., G. Alloing, V. Méjean, and J.-P. Claverys. 1987. Constitutive expression of erythromycin resistance mediated by the *ermAM* determinant of plasmid pAMβ1 results from deletion of 5' leader peptide sequences. *Plasmid* 18:250–253.

Miyoshi, Y. 1984. Transferability of tetracycline resistance to *Clostridium perfringens* isolated from human feces. *Chemotherapy* 30:170–174.

Miyoshi, Y. and A. Higa. 1984. Interrelationship between drug resistance and bacteriocinogeny of *C. perfringens*. *Microbiol. Immunol.* 28:281–289.

Oultram, J.D. and M. Young. 1985. Conjugal transfer of plasmid pAMβ1 from *Streptococcus lactis* and *Bacillus subtilis* to *Clostridium acetobutylicum*. *Fed. Eur. Microbiol. Soc. Lett.* 27:129–134.

Phillips-Jones, M.K., L.A. Iwanejko, and M.S. Longden. 1989. Analysis of plasmid profiling as a method for rapid differentiation of food-associated *Clostridium perfringens*. *J. Appl. Bacteriol.* **67**:243–254.

Roberts, I., W.M. Holmes, and P.B. Hylemon. 1986. Modified plasmid isolation method for *Clostridium perfringens and Clostridium absonum*. *Appl. Environ. Microbiol.* **52**:197–199.

Roberts, I., W.M. Holmes, and P.B. Hylemon. 1988. Development of a new shuttle plasmid system for *Escherichia coli* and *Clostridium perfringens*. *Appl. Environ. Microbiol.* **54**:268–270.

Rokos, E.A., J.I. Rood, and C.L. Duncan. 1978. Multiple plasmids in different toxigenic types of *Clostridium perfringens*. *Fed. Eur. Microbiol. Soc. Lett.* **4**:323–326.

Rood, J.I. 1983. Transferable tetracycline resistance in *Clostridium perfringens* strains of porcine origin. *Can. J. Microbiol.* **29**:1241–1246.

Rood J.I., V.N. Scott, and C.L. Duncan. 1978b. Identification of a transferable tetracycline resistance plasmid (pCW3) from *Clostridium perfringens*. *Plasmid* **1**:563–570.

Rood, J.I., J.R. Buddle, A.J. Wales, and R. Sidhu. 1985. The occurrence of antibiotic resistance in *Clostridium perfringens* from pigs. *Aus. Vet. J.* **62**:276–279.

Rood, J.I., E.A. Maher, E.B. Somers, E. Campos, and C.L. Duncan. 1978a. Isolation and characterization of multiply antibiotic-resistant *Clostridium perfringens* strains from porcine feces. *Antimicrob. Agents Chemother.* **13**:871–880.

Rood, J.I., S. Jefferson, T.L. Bannam, J.M. Wilkie, P. Mullany, and B.W. Wren. 1989. Hybridization analysis of three chloramphenicol resistance determinants from *Clostridium perfringens* and *Clostridium difficile*. *Antimicrob. Agents Chemother.* **33**:1569–1574.

Scott, P.T. and J.I. Rood. 1989. Electroporation-mediated transformation of lysostaphin-treated *Clostridium perfringens*. *Gene* **82**:327–333.

Sebald, M. and G. Bréfort. 1975. Transfert du plasmide Tétracycline-Chloramphenicol chez *Clostridium perfringens*. *C. R. Acad. Sci. Paris Ser. D.* **281**:317–319

Sebald, M., D. Bouanchaud, and G. Bieth. 1975. Nature plasmidique de la resistance à plusieurs antibiotiques chez *C. perfringens* type A, souche 659. *C. R. Acad. Sci. Paris Ser. D* **280**:2401–2404.

Sherrat, D. 1989. Tn3 and related transposable elements: site-specific recombination and transposition. *In*: *Mobile DNA*, D.E. Berg and M.M. Howe, eds., pp. 163–184. Washington, D.C.: American Society for Microbiology.

Smart J.L., M.L. Truman, and M.F. Stringer. 1983. A note on the antibiotic susceptibilities of *Clostridium perfringens* serotypes isolated from meat. *J. Appl. Bacteriol.* **54**:135–139.

Smart, J.L., T.A. Roberts, M.F. Stringer, and N. Shah. 1979. The incidence and serotypes of *Clostridium perfringens* on beef, pork and lamb carcasses. *J. Appl. Bacteriol.* **46**:377–383.

Smith, H.W. 1959. The effect of continuous administration of diets containing tetracyclines and penicillins on the number of drug-resistant and drug-sensitive *Clostridium welchii* in the faeces of pigs and chickens. *J. Pathol. Bacteriol.* **77**:79–92.

Squires, C.H., D.L. Heefner, R.J. Evans, B.J. Kopp, and M.J. Yarus. 1984. Shuttle plasmids for *Escherichia coli* and *Clostridium perfringens*. *J. Bacteriol.* **159**:465–471.

Steffen, C. and H. Matzura. 1989. Nucleotide sequence analysis and expression studies of a chloramphenicol-acetyltransferase-coding gene from *Clostridium perfringens*. *Gene* **75**:349–354.

Wren, B.W., P. Mullany, C. Clayton, and S. Tabaqchali. 1988. Molecular cloning and genetic analysis of a chloramphenicol acetyltransferase determinant from *Clostridium difficile*. *Antimicrob. Agents Chemother.* **32**:1213–1217.

Wren, B.W., P. Mullany, C. Clayton, and S. Tabaqchali. 1989. Nucleotide sequence of a chloramphenicol acetyltransferase gene from *Clostridium difficile*. *Nucleic Acids Res.* **17**:4877.

Young, M., N.P. Minton, and W.L. Staudenbauer. 1989. Recent advances in the genetics of the clostridia. *Fed. of Eur. Microbiol. Soc. Rev.* **63**:301–326.

Zaidenzaig, Y., J.E. Fitton, L.C. Packman, and W.V. Shaw. 1979. Characterization and comparison of chloramphenicol acetyltransferase variants. *Eur. J. Biochem.* **100**:609–618.

10

Genetics and Molecular Biology of Antibiotic Resistance in *Clostridium difficile*: General and Specific Overview

Herbert Hächler and Fritz H. Kayser

10.1 Introduction

History, disease, epidemiology, and therapy

Clostridium difficile was first discovered in healthy infants in 1935 and referred to as *Bacillus difficilis* (Hall and O'Toole, 1935). It was not recognized as a pathogen until the late 1970s, although its capability to produce a cytotoxin was known. Pseudomembranous colitis (PMC) has also been known for a long time; Finney (1893) first described it as a complication of abdominal surgery. PMC was very rare until 1952 when, after the first report of PMC following antimicrobial therapy (Reiner et al., 1952), the incidence of antibiotic-associated PMC (AAPMC) began to increase. The association between AAPMC and *C. difficile* was not realized until investigators observed that cytotoxic activity in stools of patients with PMC could be neutralized with antisera against *C. sordellii*. In fact, unlike *C. sordellii*, *C. difficile*, which was frequently isolated from the stools, produced a cytotoxin that was cross-neutralized with the *C. sordellii* antitoxin (Larson et al., 1977; Bartlett et al., 1978; R.H. George et al., 1978; W.L. George et al., 1978). In addition to this cytotoxin, now called toxin B, an enterotoxin that caused a fluid response in the animal ileal loop assay was found soon afterward and designated toxin A (Banno et al., 1981; Taylor et al., 1981).

Clinically, *C. difficile* may cause symptoms ranging from mild antibiotic-associated diarrhoea (AAD) to (entero-) colitis (AAC) or possibly life-threatening AAPMC. Although endoscopically confirmed AAPMC is caused by *C. difficile* in more than 95% of all cases, the organism is progressively less frequently isolated with decreasing severity of the disease, i.e., in 50%–75% or 15%–25% of patients with AAC or AAD, respectively (for review, see Bartlett, 1990). Antimicrobials most often implicated in disease caused by *C. difficile* are penicillins, clindamycin, cephalosporins, and, to a lesser extent, erythromycin and trimethoprim/sulphamethoxazole (Bartlett, 1990). A small risk is posed by other drugs such as rifampin (Fekety et al., 1983), quinolones (Dan et al., 1989), or aminoglycosides (Chiou and Chiou, 1990). Virtually every drug has, however, been reported to have caused the complication at least once, even metronidazole (Saginur et al., 1980) and vancomycin (Miller and Ringler, 1987), the two major substances used for treatment of *C. difficile* disease. Besides causing the known diseases associated with antimicrobial therapy, *C. difficile* may very rarely play a direct or indirect role in other syndromes such as abscesses, osteomyelitis, peritonitis, or reactive arthritis (Levett, 1986).

After predisposition through antimicrobial therapy, *C. difficile* disease can be caused either by cells already residing in the gut or, more frequently, by strains from an exogenous source (Bartlett, 1990). The latter mode of acquisition has given rise to many nosocomial outbreaks (see Lyerly et al., 1988). This situation has led to the development of various typing schemes for the study of the epidemiology of *C. difficile*;

progress in this field has been reviewed by Tabaqchali (1990).

Treatment of AAPMC is usually started by discontinuation of the inflicting antibiotic. Then, vancomycin or metronidazole are given (Anonymous, 1989), sometimes according to sophisticated regimens to prevent relapse (Tacke and Hausman, 1988). Recently, teicoplanin (de Lalla et al., 1989) and fusidic acid (Cronberg et al., 1984) have been suggested. Prophylactic means include the use of gloves by health care personnel (Johnson et al., 1990) and isolation of patients.

Virulence and pathogenesis

Virulence factors in *C. difficile* have been reviewed by Borriello et al., (1990). The major factor is toxin A, an enterotoxin causing a fluid response in the ileal loop assay. The role of toxin B, a lethal cytotoxin, in AAPMC is not well understood. It is, however, an important diagnostic marker. Other toxins have been described, among them an actin-specific ADP-ribosyltransferase (Popoff et al., 1988; Hatheway, 1990), and three proteins, antigenically related to toxin A, that exert a cytotonic effect on hamster ovary cells (Torres and Lönnroth, 1989). However, the pathogenic significance of either of these toxins or of other virulence factors such as capsules, hydrolytic enzymes, or colonization factors is unclear. Borriello et al. (1990) suggested that they may modulate the degree of virulence in individual toxigenic strains.

The pathogenic mechanism of AAPMC is still unclear. Overgrowth of resistant *C. difficile* on antimicrobial therapy is no adequate explanation for the development of disease because the isolated *C. difficile* strain is often highly susceptible to the drug that caused the complication (Dzink and Bartlett, 1980; Bartlett, 1990). Moreover, the feces of as many as 27% of all infants are massively colonized and highly toxin positive while the carriers are asymptomatic (Viscidi et al., 1981). In contrast, only 3% of healthy adults are colonized by *C. difficile* unless they are exposed to a hospital environment with a high prevalence of *C. difficile* disease. Under these circumstances the colonization rate may reach 10%–20% (Bartlett, 1990). Additionally, older children are far less likely to develop AAPMC than adults or the elderly, but these discrepancies are unexplained.

It is known that the normal intestinal flora plays a major role in prevention of colonization and disease and may also have a profound impact on antibiotic concentrations in the colon (van der Waaij, 1989). Disturbance of this dynamic ecology by antimicrobial therapy favors *C. difficile* overgrowth and colonization. It has also been suggested that *C. difficile* might survive periods with high antibiotic concentrations through sporulation and then grow rapidly in the thus emptied ecological niches of the colon (Bartlett, 1990). Nevertheless, the mechanism by which antibiotics trigger AAD, AAC, or AAPMC as well as the role of antibiotic resistance determinants in *C. difficile* remain to be determined.

Susceptibility to antibiotics

Because *C. difficile* causes disease almost exclusively as a consequence of antimicrobial therapy, antibiotic susceptibility of the organism has been the subject of many studies (see Chow et al., 1985; Delmée and Avesani, 1986; Levett 1988; Wüst and Hardegger, 1988). Generally, collections of *C. difficile* strains react fairly homogeneously to a given drug, i.e., most strains are either susceptible, moderately resistant, or highly resistant, with only a few strains failing to fall within a relatively narrow range of minimal inhibitory concentrations (MIC). Wide variation in susceptibility of individual strains or even bimodal distribution of MIC values were observed only against chloramphenicol (Cm), clindamycin (Cc), erythromycin (Em), rifamycin (Rif), and tetracycline (Tc) (Wüst and Hardegger, 1988). Some correlation between certain serogroups and resistance to one or more of these drugs has been reported (Nakamura et al., 1987; Delmée and Avesani, 1988). In general, *C. difficile* strains are susceptible to vancomycin, metronidazole, and penicillins, moderately to highly resistant to aminoglycosides and all types of cephalosporins, and highly resistant to most quinolones.

Knowledge about the genetics and molecular biology of resistance in *C. difficile* is limited. Three determinants have been studied so far; they confer resistance to either Cc/Em, Tc, or Cm, thus covering four of the five antibiotics known to cause variable reaction of individual *C. difficile* strains. This chapter gives a broad overview of the analyzed *C. difficile* resistance determinants and the accompanying literature as well as some hypothetical perspectives.

10.2 Genetics and Molecular Biology of Resistance to Clindamycin/Erythromycin, Tetracycline, and Chloramphenicol in *Clostridium difficile*

Although considerable effort has been put into investigating the clinical, epidemiological, biochemical, pathogenic, and therapeutic aspects of *C. difficile* and its toxins, genetic and molecular biological studies are limited. Such studies are concerned primarily with three issues: (i) cloning and sequencing of toxins and secreted antigens (Muldrow et al., 1987; Wren et al., 1987; Dailey and Schloemer, 1988a; Sauerborn and von Eichel-Streiber, 1990); (ii) DNA analysis for diagnostic and epidemiological use (Clabots et al., 1988b; Tabaqchali, 1990; Wren et al., 1990), and (iii) antibiotic resistance (for references, see following sections). For a recent general overview of clostridial genetics see also Young et al. (1989).

Antibiotic resistance in *C. difficile* has so far been found to be exclusively mediated by the chromosome. Although *C. difficile* plasmids were investigated as early as 1982 (Muldrow et al., 1982) and later found to be useful for epidemiological purposes (Steinberg et al., 1987; Clabots et al., 1988a, 1988b) they have always had to be considered cryptic. Intra- and interspecific, as well as intergeneric, transfer of antibiotic resistance involving *C. difficile* was only achieved by conjugation-like events involving chromosomal sequences. Even when Tc^r, mediated by plasmid pIP401, was transferred from *C. perfringens* to *C. difficile* the phe-

notype was unstable, and plasmid DNA consistent with pIP401 could never be shown (Ionesco, 1980). The striking absence of plasmid linkage to antibiotic resistance in *C. difficile* is not understood.

Resistance to clindamycin/erythromycin

General aspects. Macrolides, lincosamides, and streptogramins (MLS), although structurally distinct groups of antibiotics, inhibit protein synthesis in many gram-positive and gram-negative bacteria by similar mechanisms. They are used for treatment of infections caused by gram-positive organisms because the gram-negative cell wall does not allow efficient accumulation of these drugs (Arthur et al., 1987). The detailed mode of action of these antibiotics is still a matter of controversy. It is known that they bind to the large (50S) subunit of the ribosome, disrupting protein synthesis. In the case of spiramycin there is evidence that the antibiotic, once bound to the 50S subunit, stimulates dissociation of peptidyl-tRNA from ribosomes during translocation from the A to the P site of the ribosome, leading to premature peptide termination. This mode of action may slightly vary if other MLS antibiotics are involved (Brisson-Noël et al., 1988b).

Three mechanisms of resistance to MLS antibiotics have been detected (see Arthur et al., 1987). The major mechanism involves chemical alteration of the ribosome target site, leading to reduced affinity to all macrolides, lincosamides, and type B steptogramins while leaving susceptibility to type A streptogramins unaltered. It is generally referred to as MLS$_B$ resistance (MLS$_B$^{r}) or simply MLS^r (Ounissi and Courvalin, 1982). The second mechanism consists of enzymatic modification or hydrolysis of one or few antibiotics of one group, and does therefore not give rise to much cross-resistance (Arthur et al., 1987). The third mechanism, an efflux system for certain macrolides, has been reported most recently (Goldman and Capobianco, 1990).

The molecular mechanism of MLS^r involves alteration of the MLS target site on the 50S ribosome subunit, which is achieved by N^6, N^6-

dimethylation of a specific adenine residue corresponding to position 2058 of the *Escherichia coli* 23S rRNA. Because this part of the 23S rRNA is highly conserved in most procaryotes, ribosomes of other species are equally well modified. Methylation is always catalyzed by an rRNA methylase that is encoded by an *erm* (*erythromycin resistance methylase*) gene located on an MLS^r determinant.

At least 13 MLS^r determinants have been described and characterized, mostly in gram-positive, and a few in gram-negative, organisms. They have been designated *erm*A–G, *erm*AM, *erm*BC, *erm*FS, *erm*FU, *erm*P, and *erm*Z (Arthur et al., 1987; Hächler et al., 1987a; Monod et al., 1987; Brisson-Noël et al., 1988a; Berryman and Rood, 1989; Halula and Macrina, 1990). Berryman and Rood (1989) have reported an additional *erm* gene in *C. perfringens* that has not been named or characterized, but it does not hybridize with either of the known determinants. The genes were classified by DNA–DNA hybridization analyses. Originally, four classes of MLS^r determinants (classes A–D) were delineated (Courvalin et al., 1985). That scheme, however, seems to have been abolished in favor of a system of seven categories, *erm*A–*erm*G, now in use. Some inconsistency of classification still remains. Although the hydridization classes are distinct, functional relationship between all seven classes of determinants is strongly implied by significant amino acid similarity as detected by sequence analysis (Arthur et al., 1987).

Expression of MLS^r is either constitutive, inducible, or zonal (Arthur et al., 1987). Induction of the phenotype by low concentrations of Em has been studied extensively. Regulation was shown to occur posttranscriptionally by translation attenuation. According to this model, *erm*C mRNA is constitutively synthesized but contains a leader sequence upstream of the structural gene, a sequence that is able to form two hairpin structures side by side. One of these hairpins sequesters, and thereby inactivates, the Shine–Dalgarno ribosome-binding site necessary for translation of the methylase gene. No protein is synthesized. In the presence of an MLS antibiotic, translation of a small leader peptide located further upstream than both

hairpins is interrupted, and the ribosome thus stalled causes a conformational change of the secondary mRNA structure. The two hairpins are disrupted in favor of a single new one. By that means, the formerly sequestered Shine–Dalgarno site is freed, leading to translation of the methylase gene (Mayford and Weisblum, 1989). Further, the ribosome stalled within the leader peptide sequence prevents degradation of the mRNA from the 5' end, thus promoting methylase expression even more (Sandler and Weisblum, 1989). Other genes, *erm*A and *erm*G, are believed to be regulated in the same way; however, those systems are even more complicated because both their leader sequences code for two leader peptides rather than one (Monod et al., 1987; Sandler and Weisblum, 1989).

Aspects in Clostridium difficile. Resistance to Em and Cc in *C. difficile* was always of particular interest because Cc often seemed to trigger the disease. Successful transfer of Em^r and Cc^r by filter mating was first achieved by Wüst and Hardegger (1983). These authors observed transconjugants at very low frequency in matings with one of two potential *C. difficile* donors. Even though Cc^r was associated with Em^r in all transconjugants obtained they could only be selected on media containing Cc, not on media containing Em. Subsequent tests revealed that all strains were also resistant to pristinamycin I, a type B streptogramin, indicating that an MLS^r determinant was involved. MLS^r was not associated with Tc^r, which was also present in both donors, because transfer of Tc^r and MLS^r always occurred independently.

The first molecular data about the *C. difficile* MLS^r determinant, obtained by further analysis of the previously characterized strains (Wüst and Hardegger, 1983), were presented by Hächler et al. (1987a). The authors demonstrated that a plasmid of 8 kilobase-pairs (kb) that was present in the donor but in none of the transconjugants was not associated with resistance; they assumed that both MLS^r and Tc^r were located on the chromosome. Hybridization experiments using Tn551 or parts of it as a probe revealed that *C. difficile* DNA of the donors and of all MLS^r transconjugants but not of the recipient, contained sequences homologous to Tn551. Subsequent experiments showed that the homology was confined to a 1.3 kb region of Tn551 covering the entire ermB determinant. Interestingly, the hybridization data demonstrated that, irrespective of the restriction enzyme used, the donor and the transconjugants always carried the determinant in one fragment of the same size. This suggested that the donor DNA might have integrated site specifically.

The C. difficle MLS^r marker was thus identified as a class ermB determinant and designated ermZ. Because Tn551 had originally been discovered in Staphylococcus aureus, attempts to transfer the C. difficile MLS^r determinant to an S. aureus recipient were undertaken, using the original donors. In one such mating experiment, indeed, one of the two donors gave rise to MLS^r staphylococcal transconjugants at a very low frequency. That was the first demonstration of C. difficile being involved in intergeneric transfer of genetic material. Interestingly, the donor was the strain that had previously failed to donate resistance intraspecifically, and, in contrast to the C. difficile matings (see foregoing), staphylococcal transconjugants could only be selected on Em- containing media.

No plasmid DNA was detected in the selected staphylococcal transconjugant. The chromosomal DNA contained an additional 12.8 kb HindIII fragment compared to the recipient (Figure 10.1). The ermZ marker, however, was actually located on a 9.5 kb fragment (Figure 10.1A). Subsequent hybridizations showed that both the 9.5 and the 12.8 kb fragments carried C. difficile DNA (Figure 10.1B). This indicated that additional C. difficile sequences, flanking ermZ, had been transferred to Staphylococcus aureus, which suggested that ermZ might be located on a C. difficile-specific transposable element. However, both S. aureus fragments showed also homology to DNA of the susceptible C. difficile recipient, indicating that the transferable C. difficile MLS^r element contained sequences to be found in both C. difficile strains, the donor and the recipient. One possible explanation for this intriguing finding would be that there are transposable elements

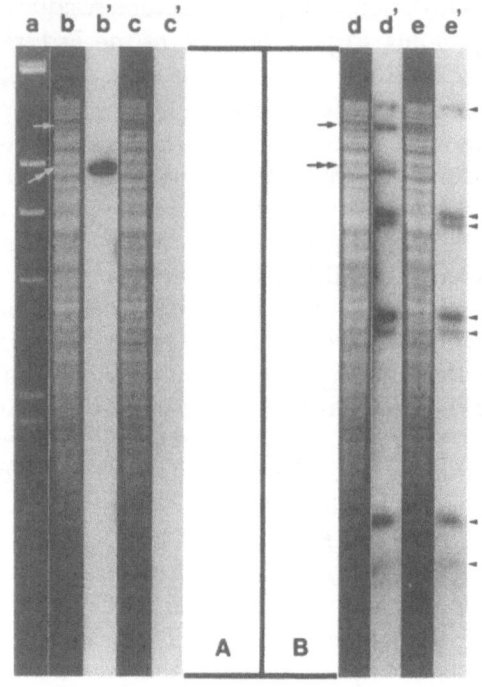

FIGURE 10.1   HindIII-digested chromosomal DNA of Staphylococcus aureus transconjugant (lanes b, d) and recipient (lanes c, e) after hybridization with ermB probe (A) or chromosomal DNA of the Clostridium difficile donor strain as probe (B). Fragment sizes according to λ-HindIII standards (a). Arrows point to additional 12.8 kb and ermZ carrying 9.5 kb bands. Arrowheads highlight hybridizing bands associated with ribosomal RNA genes (see text). Taken from Hächler et al., 1987a.

common in many C. difficile strains. In the original C. difficile donor the MLS^r element might have integrated into one such element, which might then have been transferred as a whole to S. aureus. Such hypothetical transposable elements in C. difficle, with or without an MLS^r determinant and with a tendency to site-specific integration, might be functionally similar to the Tn916 family of conjugative transposons (see section on Tc^r), because neither plasmid DNA nor transduction or transformation seemed to be involved in MLS^r transfer.

When chromosomal digests of S. aureus were hybridized with labeled whole-cell DNA of C. difficile, seven hybridizing bands were observed that clearly were not associated with

MLS^r (Figure 10.1B) (Hächler et al., 1987a). Speculations that they might be multiple IS-like elements, common in both species, could not be confirmed. Recent hybridization analysis using labeled *E. coli* ribosomal RNA as a probe showed signals identical to those shown in Figure 10.1B (lane e'). This demonstrated that the unexplained signals in fact represented ribosomal RNA genes (K. Hadorn and H. Hächler, unpublished data), which are known to be highly conserved in bacteria.

Berryman and Rood (1989) have cloned *erm*P from *C. perfringens*. An internal fragment of *erm*P was used to probe both *C. difficile* MLS^r donors used by Hächler et al. (1987a), among other strains belonging to various species of *Clostridium*. Homology between the two determinants was discovered. Not surprisingly, *erm*P also hybridized to a standard plasmid carrying *erm*B and to a strain of *C. paraputrificum*, thus adding to the already impressive list of organisms carrying *erm*B determinants including *Streptococcus pneumoniae*, *S. pyogenes*, *S. agalactiae*, *S. sanguis*, *Enterococcus faecalis*, *Staphylococcus aureus*, *E. coli*, and *Klebsiella pneumoniae* (Arthur et al., 1987; Brisson-Noël et al., 1988a). Berryman and Rood (1989) suggested that *erm*P may be located on a transposable element. If so, a detailed comparison of the regions flanking *erm*P and *erm*Z would be most desirable, because the authors have pointed out that *erm*P seemed to be more common in *C. difficile* than in *C. perfringens*. Further, in contrast to *erm*Z, *erm*P could never be transferred though it is located on a plasmid in *C. perfringens*.

Em^r in *C. difficile* has been shown to be transferable to *Bacillus subtilis* and then back to *C. difficile* (Tabaqchali, 1989). Moreover, an MLS^r determinant from a toxigenic *C. difficile* strain has been cloned into a cosmid vector (Wren et al., 1988a). From the same strain, the toxin A gene was also cloned into an independent cosmid by these authors. Interestingly, the two different inserts showed homology with each other, indicating that they might contain overlapping fragments. This led the authors to assume that the two genes were linked. On this basis they suggested that antibiotics might play a role in the regulation of virulence factors.

Resistance to tetracycline

General aspects. Tetracyclines exert their antibacterial activity by binding to the 30S ribosome subunit, thereby inhibiting the binding of aminoacyl-tRNAs to the A site and disrupting protein synthesis (Levy, 1984).

Three mechanisms of resistance to tetracyclines (Tc^r) in a wide variety of bacterial genera have been discovered and described (see Salyers et al., 1990). These mechanisms consist of reduced accumulation of Tc because of a membrane-bound Tc efflux system, protection of ribosomes against the effects of Tc, or enzymatic inactivation of the drug. An impressive number of Tc^r determinants (Tet) in many hybridization classes have been reported (Levy, 1988), and a nomenclature scheme has been proposed to avoid confusion (Levy et al., 1989).

The efflux mechanism seems to be largely confined to gram-negative rods (Tet classes A–F), and to some genera of gram-positive organisms including *Staphylococcus*, *Bacillus*, and *Streptococcus* (classes K and L). Determinants conferring Tc^r through ribosome protection (classes M and O) are found in many gram-positive and some gram-negative genera, including anaerobes (Levy, 1988), and even in *Mycoplasma* and *Ureaplasma* (Roberts, 1990). Most recently, TetM has been detected on a multiresistance plasmid in a clinical isolate of *Listeria monocytogenes* (K. Hadorn, personal communication). Tc modification has so far been discovered only in *Bacteroides* (class X), and its capacity to confer resistance in vivo has not been established (Salyers et al., 1990).

The exact mechanism of ribosome protection mediated by TetM, N, and O has not been elucidated. It is known, however, that ribosomes of resistant cells are more resistant to Tc than those of susceptible cells in an in vitro translation system (Burdett, 1986). Moreover, there is evidence that a cytoplasmic factor, presumably a protein, rather than modification of the ribosome itself is responsible for this effect. This evidence came from experiments with ribosomes of Tc-susceptible cells that could be made resistant by adding ribosome-free extracts from resistant cells. Conversely, the soluble factor could be removed from resistant ribosomes by washing them with a high salt solution (Burdett, 1988). Whether the postulated cytoplasmic factor is the product of the TetM determinant itself, or whether there is an indirect mode of action, is not known.

The gene products of TetM and TetO have been deduced from the respective nucleotide sequences (Martin et al., 1986; Sougakoff et al., 1987; Sanchez-Pescador et al., 1988a), and they have been found to be proteins of 72.5 and 72.29 kDa, respectively. Even though TetM and TetO fall into two distinct hybridization classes, their products have 77% of the amino acids in common. In addition, the N-terminal regions of both proteins share homology with the N-terminal regions of elongation factors Tu and G (Sanchez-Pescador et al., 1988b; Salyers et al., 1990), mainly within the GTP binding sites of these proteins. This led Sanchez-Pescador et al. (1988b) to the suggestion that TetM product and similar proteins might bridge Tc-impaired protein synthesis by acting as alternate elongation factors, but this hypothesis has not been confirmed.

Despite considerable effort, the mechanism of transfer and transposition of the so-called conjugative transposons has not been elucidated completely. The prototype, Tn916, has been described (Franke and Clewell, 1981; Gawron-Burke and Clewell, 1984). However, the Tn916 family has grown to contain Tn918, Tn919, Tn925, and Tn1545 (Clewell et al., 1985; Fitzgerald and Clewell, 1985; Christie et al., 1987; Courvalin and Carlier, 1987). Progress in the study of the mechanism of transposition has been achieved in several laboratories. Precise excision on transposition of Tn916 was observed (Gawron-Burke and Clewell, 1984), and the functional map was determined by analysis of numerous Tn5 insertion mutants of Tn916 (Senghas et al., 1988). Both ends of Tn916 were sequenced and found to be almost identical with the respective sequences in Tn1545 (Caillaud and Courvalin, 1987; Clewell et al., 1988); they contain neither direct nor indirect terminal repeats, possess variable base pairs (bp) at their extremities, and do not generate duplications of target DNA on insertion. A covalently closed circular intermediate of the

transposon after excision has been detected (Scott et al., 1988), and two proteins coded by sequences close to one end have been found to be essential for excision of the element (Poyart-Salmeron et al., 1989; Storrs et al., 1991). These authors also found homology of the two proteins to excision/integration-catalyzing enzymes, Xis and Int, encoded by phage λ and suggested similar functioning of λ and Tn1545 with respect to excision.

On the basis of the available knowledge and a massive amount of experimental data, an elegant model for excision and insertion of conjugative transposons has been proposed by Caparon and Scott (1989). According to this model, the transposon is flanked by variable "coupling sequences" that consist of about 5 bp and originate from the target sequences of the present and the previous insertion site of the transposon. On excision two staggered cuts, at the coupling regions on both sides of the transposon, produce two ends with 5-bp, 3' single-stranded overhangs. These ends are joined to form the postulated free circular intermediate of the transposon, and the interrupted host chromosome is reconstituted correspondingly. Because the two joined coupling sequences are not identical, 5 bp heteroduplexes are formed in the reconstituted target of the chromosome as well as in the circular intermediate. The corresponding heteroduplex region of the circular intermediate is referred to as "joint" (Caparon and Scott, 1989) or "core" (Clewell et al., 1988). After the first replication, the heteroduplex in the reconstituted chromosome segregates into two distinct subpopulations of cells representing either strand of the parental DNA. Reinsertion of the circular intermediate into a new target is thought to be the reverse of the process, leading to two new, slightly different "coupling sequences."

TetM has been found to be exceptionally widespread among many species and genera that are only slightly related (see earlier discussion in this section). A plausible explanation for this fact has been offered by Salyers et al. (1990), who pointed out that both tetM and O are preceded by excellent Shine–Dalgarno and promoter sequences, facilitating expression in diverse genetic backgrounds. Further, the product is an easily diffusible cytoplasmic protein, somehow interacting with protein synthesis, the components of which are known to be highly conserved. Additionally, the conjugative transposons carrying these determinants seem to be widespread, prompting several investigators to use these versatile elements to generate insertional mutants and facilitate genetic studies of remote organisms such as *Listeria* (Kathariou et al., 1987), *Thermus aquaticus* (Sen and Oriel, 1990), *Clostridium acetobutylicum* (Woolley et al., 1989; Bertram et al., 1990), *C. tetani* (Volk et al., 1988), or *C. difficile* (Mullany et al., 1990). Cloning and shuttle vectors are also being constructed, using parts of conjugative transposons, to allow phenotypic expression of cloned genes in those bacteria.

Aspects in Clostridium difficile. The first reports on the genetics of Tc resistance in *C. difficile* showed that Tc^r was transferable between different strains of *C. difficile* and from *C. perfringens* to *C. difficile*, but not vice versa (Ionesco, 1980; Smith et al., 1981). Later, Wüst and Hardegger (1983) confirmed these findings and showed that their *C. difficile* Tc^r was independent from other transferable markers. Because neither transduction nor transformation seemed to be involved (Smith et al., 1981; H. Hächler, unpublished data), and close cell-to-cell contact between donor and recipient cells was essential, a conjugation-like event was assumed to be responsible for the transfer.

Molecular genetic data about Tc^r in the *C. difficile* strains, originally described by Wüst and Hardegger (1983), were reported by Hächler et al. (1987b). Restriction endonuclease analysis first confirmed that all transconjugants were isogenic with the recipient, whereas both donors showed different genetic backgrounds. Subsequent hybridization experiments showed that all Tc^r *C. difficile* donors and transconjugants carried DNA homologous to *tet*M (Figure 10.2), whereas no hybridization was detected when *tet*A, B, C, or P were used as probes. Additional analyses revealed that at least six of seven internal and junction fragments of Tn*916*, representing 95% of its length, shared homology with the Tc^r determinant in *C. difficile*. The seventh fragment escaped test-

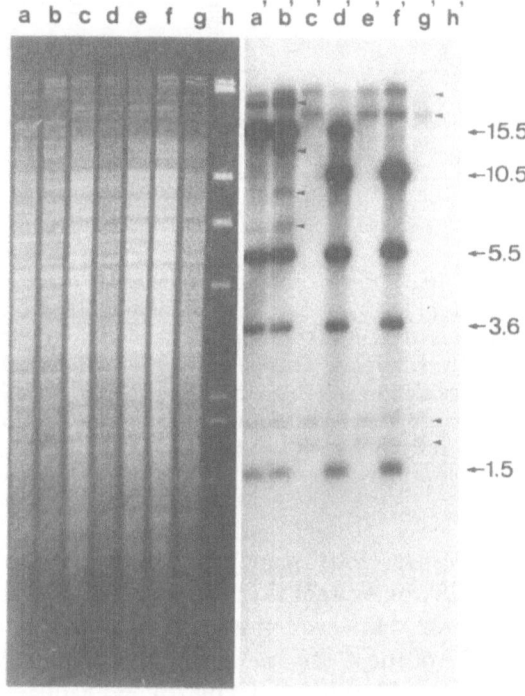

Figure 10.2 Agarose electrophoresis of *Hind*III-digested *Clostridium difficile* DNA (lanes a–g), and autoradiography after hybridization with α^{32}P-dATP-labeled Tn*916* DNA (lanes a'–g'). Lanes a, b, Tc^r MLS^r donors; c, Tc^s recipient; d, f, Tc^r transconjugants; e,g, Tc^s/ MLS^r transconjugants; h, λ-*Hind*III size standard. Strong Tc^r-associated hybridization signals are indicated by sizes in kilobases. Weak signals from weak homology with 5.4-kb Tn*916*- internal *Hinc*II fragment are indicated with arrowheads. (Taken from Hächler et al., 1987b.)

ing because of its small size (0.4 kb). Close relationship between the *C. difficile* Tc^r element and the entire Tn*916* was concluded, though different intensities of the hybridization signals had been noted along the length of the transposon, the brightest bands being the *tet*M-carrying 4.8 kb and the adjacent 5.4 kb *Hinc*II fragments of Tn*916* (Hächler et al., 1987b) [for a physical map of Tn*916*, see Clewell and Gawron-Burke (1986)]. These findings were generally confirmed by Mullany et al. (1990).

The absence of homology to *tet*P, a worldwide-distributed determinant of *C. perfringens* (Abraham et al., 1985), is somewhat surprising because transfer of *tet*P from *C. perfringens* to *C. difficile* has been demonstrated

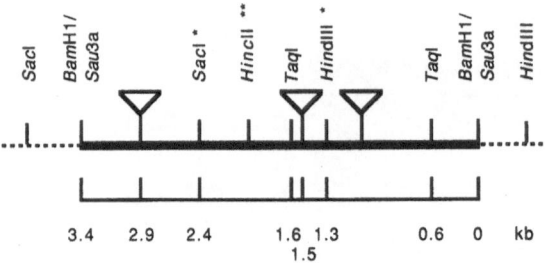

FIGURE 10.3 Restriction endonuclease map of plasmid pPPM20. Bold line, cloned DNA from *Clostridium difficile* carrying *tet*M; dashed line, vector DNA; triangles, Tn*1000* insertions, all abolish Tc^r; *, restriction sites also found in *tet*M on Tn*916*; **, restriction site not found in *tet*M on Tn*916*. (Redrawn from Figure 1 in Mullany et al., 1990.)

and because close relationship between Cm^r and MLS^r determinants of the two species was discovered. However, it should be noted that only Tc^r of the *C. difficile* strains mentioned in the work of Hächler et al. (1987b) and Mullany et al. (1990) has so far been tested by hybridization.

Significant progress was achieved when Mullany et al. (1989, 1990) cloned *tet*M from a *C. difficile* strain used earlier by Hächler et al. (1987b). The Tc^r phenotype was expressed in *E. coli* by the recombinant plasmid pPPM20 carrying a 3.4 kb *Sau*3A fragment. A restriction endonuclease map was constructed (Figure 10.3), and the Tc^r determinant was shown to cover a region of more than 1.9 kb of pPPM20 by means of Tn*1000* insertional inactivation. In vitro transcription translation revealed that the same region coded for a protein of 68 kDa, which was similar in size to the functionally related products of TetM and TetO (Sougakoff et al., 1987). This, together with the foregoing hybridization data discussed, is strong evidence that TetM in *C. difficile* and in Tn*916* are functionally related; further, TetO and M proteins showed as much as 75% amino acid homology despite the lack of hybridization of the respective genes.

The relationship of the Tc^r-carrying element of *C. difficile* with Tn*916* was further analyzed and found to be closest regarding the Tet determinant where several corresponding restriction sites were detected (Figure 10.3). The remain-

ing parts of the element showed less similarity, because two oligonucleotides directed at both ends of Tn*916* did not hybridize to the *C. difficile* element. Additionally, only two of six *Hinc*II sites known in Tn*916* seemed to be present in *C. difficile*, giving rise to a comigrating fragment of 2.8 kb. Nevertheless, Mullany et al. (1990) believed the element to be a conjugative transposon although somewhat different from Tn*916*.

Aside from intraspecific exchange, intergeneric transfer of Tc^r from *C. difficile* (donor or transconjugants) to *B. subtilis* and back to the *C. difficile* recipient could be demonstrated (Mullany et al., 1990). Further, Tc^r was successfully transferred to the *C. difficile* recipient from a Tn*916*-carrying strain of *B. subtilis*. This was the first report of *C. difficile* acting as a recipient in intergeneric matings. Subsequent hybridization analysis of donors and transconjugants using the 3.4 kb insert of pPPM20 as a probe led to some interesting discoveries. Of more than 20 *C. difficile* transconjugants tested, all carried two copies of the Tc^r element, which seemed to have integrated into the chromosome at two specific sites. This was always seen, whether *C. difficile* or *B. subtilis* had acted as donors and whether the *C. difficile* Tc^r element or Tn*916* was transferred. In contrast, *B. subtilis* transconjugants always showed one copy, integrated at various sites. Interestingly, the original *C. difficile* donor also seemed to contain only a single copy of the element.

Considering the homology of the *C. difficile* element with Tn*916* along more than 95% of its length, functional relationship was assumed. Mullany et al. (1990) suggested that the element might well carry an excision-integration system, related to that of lambdoid phages, like Tn*916* and Tn*1545*. In addition, they offered a possible explanation for a weak, yet unexplained, homology between the Tc^s *C. difficile* recipient and the 5.4 kb internal *Hinc*II fragment of Tn*916*, which had previously been reported by Hächler et al. (1987b) (see Figure 10.2). They reasoned that the Tc^s recipient could contain the degenerated and truncated remains of a once functional conjugative transposon, now allowing homologous recombination that would account for the observed site specificity. This explanation is plausible although it hardly seems to be compatible with two copies of the Tc^r element in the chromosome and with the fact that the weak homology refers to an internal fragment of Tn916 rather than to its ends. Recently, Mullany et al. (1991) reported data in support of their explanation; they found a hot-spot for the integration of Tn916 and a related derivative in the *C. difficile* chromosome.

Taking into account the model (Caparon and Scott, 1989) of excision and integration of Tn916 and the data reported by Mullany et al. (1990), this hypothesis may be considered. On conjugation of a donor with the *C. difficile* recipient CD37, Tn916 or the *C. difficile* Tc^r element undergoes excision from the donor DNA by means of two staggered cuts catalyzed by putative Xis-Int proteins, which are encoded by a region close to one end of the transposon. The partly nonhomologous, 3' single-stranded overhangs of the transposon, consisting of 4 or 5 nucleotides, join to form a free circular intermediate with a heteroduplex at its "joint." Upon replication the heteroduplex segregates to two slightly different molecules, both of which adsorb to two specific yet different attachment sites on the recipient chromosome. Integration at both sites is then achieved by a reverse mechanism, placing the "joint" region at one junction and the "coupling region" of the target at the other junction of the integrating transposon. The *C. diffcile* donor may carry only one copy of the transposon either because of excision of one copy or because only one specific attachment site is present. The fact that *B. subtilis* transconjugants always carry one single copy at variable sites may be because multiple yet less specific sites give rise to lowered transfer efficiency and one of the two circular intermediates is lost.

One shortcoming of this hypothesis is the necessity of replication of the circular intermediate for resolution of the heteroduplex into two distinct integration units, because the circular intermediates are believed not to undergo replication (Caparon and Scott, 1989). However, these authors did not offer an explanation for multiple insertions of the element, even though they had previously presented several suggestions in that context (Scott et al., 1988). In a similar model, Poyart-Salmeron et al. (1990) suggested DNA repair for resolution of the heteroduplex as an alternative to replication. Whether this could lead to multiple insertions remains to be investigated. A novel transposable element, Tn5030 carrying Tc^r and MLS^r, which is believed to be a conjugative transposon, has been detected in *Bacteroides* spp. (Halula and Macrina, 1990). The authors reported site-specific integration of the element into the chromosome of the recipient, as well as evidence for multiple insertions. These features, although found in an unrelated system, may support our hypothesis because they are strikingly similar to the properties of the transfer of Tc^r in *C. difficile*.

Resistance to chloramphenicol

General aspects. Resistance to chloramphenicol (Cm^r) is found in a wide variety of gram-positive and gram-negative bacteria (Shaw, 1984). It is usually caused by enzymatic modification and inactivation of the drug. The responsible enzyme, chloramphenicol acetyltransferase (CAT), exists in various structurally and functionally similar forms. It is a tetramer consisting of identical subunits of 210–220 amino acids and catalyzes acetyl-CoA-dependent acetylation of the antibiotic. CAT-specifying *cat* genes are categorized by their expression. Constitutive expression is generally found in gram-negative bacteria, and induction by subinhibitory concentrations of Cm is found in gram-positive organisms, although there are exceptions to the rule. Cm^r determinants have been found to be located on the chromosome, on plasmids, and on transposons. Also, many have been shown to function in various genetic backgrounds on transfer. This has drawn attention to *cat* genes to be used as selection markers in shuttle vector systems, for less studied species such as clostridia (Dubbert et al., 1988; Wren et al., 1988b), or even for the construction of fusion proteins (Harwood et al., 1983). At least 12 different *cat* genes have been cloned, 9 of which have been sequenced (see Figure 10.4 for details).

```
E. coli Tn9        MEKKITGYTTVDISQWHRKEHFEAFQSVAQCTYNQTVQLDITAFLKTVKKNKHKFYPAFIHILARLM
P. mirabilis       MDTKRVGILVVDLSQWGRKEHFEAFQSFAQCTFSQTVQLDITSLLKTVKQNGYKFYPTPIYIISLLV

B. pumilus cat-86   M FKQID ENYLRKEHFHHYMTLTRCSYSLVINLDITKLHAILKEKKLKVYPVQIYLLARAV
S. aureus pC194    MNFNKIDLDNWKRKEIFNHY LNQQTTFSITTEIDISVLYRNIKQEGYLFYPAFIFLVTRVI
S. aureus pC221     MTFNIIKLENWDRKEYFEHY FNQQTTYSITKEIDITLFKDMIKKKGYEIYPSLIYAIMEVV
S. aureus pUB112    MTFNIIKLENWDRKEYFEHY FNQQTTYSITKEIDITLFKDMIKKKGYEIYPSLIYAIMEVV
S. aureus pC223     MTFNIINLETWDRKEYFNHY FNQQTTYSVTKELDITLLKSMIKDKGYELYPALIHAIVSVI

C. perfringens catP   MVFEKIDKNSWNRKEYFDHYFASVPCTYSMSLKVDITQ        IKEKGMKLYPAMLYYIAMIV
C. difficile catD     MVFEKIDKNSWNRKEYFDHYFASVPCTYSMTVKVDITQ        IKEKGMKLYPAMLYYIAMIV
                      ccc c                      $$  cc           c        cc


E. coli Tn9        NAHPEFRMAMK DGELVIWDSVHPCYTVFHEQTETFSSLWSEYHDDFRQFLFIYSQDVACYGDNLAY
P. mirabilis       NKHAEFRMAMK DGELVIWDSVNPGYNIFHEQTETFSSLWSYYHKDINRFLKTYSEDIAQYGDDLAY

B. pumilus cat-86   QKIPEFRMDQVND ELGYWEILHPSYTILNKDTKTFSSIWTPFDENFAQFYKSCVADIETFSKSSNL
S. aureus pC194    NSNTAFRTGYNSDGDLGYWDKLEPLYTIFDGVSKTFSGIWTPVKNDFKEFYDLYLSDVDKYNGSGKL
S. aureus pC221     NKNKVFRTGINSENKLGYWDKLNPLYTVFNKQTEKFTNIWTESDNNFTSFYNNYKNDLLEYKDKEEM
S. aureus pUB112    NKNKVFRTGINSENKLGYWDKLNPLYTVFNKQTEKFTNIWTESDNNFTSFYNNYKNDLFEYKDKEEM
S. aureus pC223     NRNKVFRTGINSEGNLGYWDKLDPLYTVFNKETEKFSNIWTESNASFNSFYNSYKNDLFKYKDKNEM

C. perfringens catP NRHSEFRTAINQDGELGIYDEMIPSYTIFHNDTETFSSLWTECKSDFKSFLADYESDTQRYGNNHRM
C. difficile catD   NRHSEFRTAINQDGELGIYDEMIPSYTIFHNDTETFSSLWTECKSDFKSFLADYESDTQRYGNNHRM
                      cc        c       c c         c     c        c       c


E. coli Tn9        FPKGFI ENMFFVSANPWVSFTSFDLNVANMDNFFAP FTMGKYYTQGDKVLMPLAIQVHHAVCDGF
P. mirabilis       FPKEFI ENMFFVSANPWVSFTSFNLNMANINNFFAPVFTIGKYYTQGDKVLMPLAIQVHHAVCDGF

B. pumilus cat-86   FPKPHMPENMFNISSLPWIDFTSFNLNVSTDEAYLLPIFTIGKFKVEEGKIILPVAIQVHHAVCDGY
S. aureus pC194    FPKTPIPENAFSLSIIPWTSFTGFNLNINNNSNYLLPIITAGKFINKGNSIYLPLSLQVHHSVCDGY
S. aureus pC221     FPKKPIPENTIPISMIPWIDFSSFNLNIGNNSNFLLPIITIGKFYSENNKIYIPVALQLHHAVCDGY
S. aureus pUB112    FPKKPIPENTIPISMIPWIDFSSFNLNIGNNSSFLLPIITIGKFYSENNKIYIPVALQLHHAVCDGY
S. aureus pC223     FPKKPIPENTVPISMIPWIDFSSFNLNIGNNSRFLLPIITIGKFYSKDDKIYLPYSLQVHHAVCDGY

C. perfringens catP EGKPNAPENIFNVSMIPWSTFDGFNLNLQKGYDYLIPIFTMGKIIKKDNKIILPLAIQVHHAVCDGF
C. difficile catD   EGKPNAPENIFNVSMIPWSTFDGFNLNLQKGYDYLIPIFTMGKIIKKDNKIILPLAIQVHHAVCDGF
                      c    cc    c  cc  c c  cc          c   c cc        c   c cc cccc


E. coli Tn9        HVGRMLNELQQYCDEWQGGA           219 aa /  44.8 % identical with catD
P. mirabilis       HVGRLLNEIQQYCDEGCK            217 aa /  43.4 % identical with catD

B. pumilus cat-86   HAGQYVEYLRWLIEHCDEWLNDSLHIT    220 aa /  43.9 % identical with catD
S. aureus pC194    HAGLFMNSIQDLSDRPNDWLL          216 aa /  44.8 % identical with catD
S. aureus pC221     HASLFMNEFQDIIHKVDDWI           215 aa /  43.4 % identical with catD
S. aureus pUB112    HASLFINEFQDIINKVDDWI           215 aa /  43.4 % identical with catD
S. aureus pC223     HVSLFMNEFQNIIDNVNEWI           215 aa /  45.3 % identical with catD

C. perfringens catP HICRFVNELQELIIVTQVCL           212 aa /  99.1 % identical with catD
C. difficile catD   HICRFVNELQELIIVTQVCL           212 aa / 100  % identical with catD
                      c
```

FIGURE 10.4 Comparison of amino acid sequences of various CAT monomers. Symbols: c, conserved in all nine sequences; $, differences between *Clostridium difficile* and *C. perfringens*. Numbers indicate total of amino acids (aa) per subunit. Original references: Tn9, Marcoli et al., 1980, *Proteus mirabilis*, Charles et al., 1985; *cat-86*, Harwood et al., 1983; pC194, Horinouchi and Weisblum, 1982; pC221, Shaw et al., 1985; pUB112, Brückner and Matzura, 1985; pC223, Zyprian, 1987. [Redrawn according to Steffen and Matzura (1989), and extended to contain the sequence of *C. difficile* (Wren et al., 1989).]

Aspects in Clostridium difficile. Cm^r in *C. difficile*, although known for more than a decade, was not studied in detail until comparatively recently, and it has not been shown to be transferable either intra- or interspecifically. In 1988, a *cat* gene on a 1.9 kb chromosomal fragment of *C. difficile* was cloned in *E. coli* to form the recombinant plasmids pPPM9 and pPPM10, both of which carried the same insert in either orientation (Wren et al., 1988b). Although the phenotype was inducible in the original *C. difficile* strain by Cm, it was either constitutively expressed in the *E. coli* host (pPPM9) or not at all (pPPM10), indicating that the regulatory elements had not been cloned or were not functional in *E. coli*. This allowed the direction of transcription to be determined. The size of the gene on pPPM9 was determined by insertional inactivation using Tn*1000*, randomly introduced into the plasmid and followed by restriction endonuclease mapping. A *cat* internal fragment from pPPM9 was subsequently used as a genetic probe.

Hybridization studies revealed that several Cm^r *C. difficile* strains from different European countries carried homologous sequences, whereas all Cm^s *C. difficile* strains, as well as other known *cat* genes and Cm^r strains of other species, did not (Wren et al., 1988b). Nevertheless, subsequent studies proved that the *C. difficile cat* gene, later designated *cat*D, was not unique but showed homology with *cat*P (Rood et al., 1989), a Cm^r determinant on the 6.2 kb transposable element Tn*4451*, that had originally been found in plasmid pIP401 of *C. perfringens* (Magot, 1984; Abraham and Rood, 1987). It is also known that *cat*D and *cat*Q, a second Cm^r determinant from *C. perfringens*, do not hybridize to each other even at low stringency, although both are located on the chromosome (Rood et al., 1989). To what extent *cat*P-surrounding sequences of Tn*4451* share homology, if at all, with sequences surrounding *cat*D in *C. difficile* is not known, because *cat* internal DNA fragments were exclusively used as probes (Rood et al., 1989). More knowledge about homology of these surrounding sequences would be desirable because Cm^r in *C. difficile*, unlike the related *cat*P in *C. perfringens*, has not been shown to be transferable.

The nucleotide and deduced amino acid sequences of *cat*D recently became available (Wren et al., 1989). The *cat*D open reading frame gives rise to a polypeptide of 212 amino acids. With the exception of two amino acids (31 and 32), the sequence is identical with that reported for *cat*P from *C. perfringens* (Steffen and Matzura, 1989) (see Figure 10.4). The respective nucleotide sequences, including a stretch of 100 bp further downstream, are also identical with the exception of the two triplets coding for the amino acids 31 and 32. Further, the two nucleotide sequences are almost identical upstream of the structural gene, and promoter and Shine–Dalgarno consensus sequences are the same (Figure 10.5).

Comparison of the amino acid sequences of CATs from *C. difficile* and *C. perfringens* with those coded by seven other Cm^r determinants showed amino acid identity between 43.4% and 45.2% (see Figure 10.4), suggesting nearly the same degree of relationship of the clostridial markers with either one of these. This is supported by recent work including 4 more CAT sequences (Bannam and Rood, 1991). Considering amino acids with similar chemical structure and properties as equal, however, pointed to the staphylococcal plasmid pC194 as possibly the closest relative of clostridial CATs (71.3% amino acid similarity) (Steffen and Matzura, 1989). This assumption is supported by the fact that *cat*A from another clostridial species, *C. butyricum*, hybridizes with pC194 (Dubbert et al., 1988) and by the fact that genetic material can be exchanged between *C. difficile* and *S. aureus* by filtermating (Hächler et al., 1987a).

The presented sequence data seem to imply that all Cm^r determinants thus far investigated may be derived from a common ancestor. In all nine sequences 41 amino acids are conserved (see Figure 10.4), among them particularly the motif Q-HH-VCDG-H near the carboxyterminus that constitutes part of the active center (Steffen and Matzura, 1989). The evolutionary distance between *cat*D and *cat*P, however, is much shorter than between *cat*D and all others, suggesting that these two determinants split only fairly recently. Taking into account the fact that the only two different amino acids are bet-

```
                                      - 35                        - 10
C. perfringens catP    TCGAAGTGGGCAAGTTGAAAAATTCACAAAAATGTGGTATAATAT
C. difficile catD      TCGAAGTGGGCAAGTTGAAAAATTCACAAAAATGTGGTATAATAT


                       ---> catP mRNA                    SD
C. perfringens catP    CTTTG TTCATTAGAGCGATAAACTTGGAATTTGAGAGGGAACTT
C. difficile catD      CTTTGCTT ATTAGAGCGATAAACTT GAATTTGAGAGGGAACTT
                           $  $                      $


C. perfringens catP    AG ATG GTA TTT ---
C. difficile catD      AG ATG GTA TTT ---
                          M   V   F   ---


C. difficile antigen promoter functionning in E. coli:

                                              mRNA        mRNA
      - 35                        - 10        --->        --->
---GTTCACATCCTCCACCTAAAGCAAATCCGTTTACAGCAGACGTCTTTAATGCCGTAGTTTA---
```

FIGURE 10.5   Comparison of regulatory sequences upstream of structural genes of catP in *Clostridium perfringens* (Steffen and Matzura, 1989) and catD in *C. difficile* (Wren et al., 1989), and of an antigen promoter region from *C. difficile* (Dailey and Schloemer, 1988b). Symbols: -35, -10, promoter consensus sequences; SD, Shine–Dalgarno ribosome-binding sequence; $, differences between sequences of catP and catD. Bold letters signify starting points of mRNA synthesis.

ter conserved in *cat*D than in *cat*P, when compared to the seven other determinants ($ symbol in Figure 10.4), one might speculate that the gene was first introduced into *C. difficile* and transferred from there to *C. perfringens*.

Regulation of several *cat* genes in gram-positive organisms has been proposed to occur on the translational level, involving an inverted repeat element between promoter and coding sequence. The element is able to form a hairpin structure, thereby sequestering the Shine Dalgarno signal (Ambulos et al., 1985; Shaw et al., 1985). Because the respective clostridial sequence is much shorter and does not contain any signs of inverted repeats involving the ribosomal binding site (H. Hächler, unpublished computer analysis) (see Figure 10.5), regulation of clostridial *cat* genes is considered to be accomplished by a different, unknown mechanism. An additional problem in understanding clostridial *cat* regulation arises from the fact that *cat*P is expressed in *E. coli* from its own promoter, whereas *cat*D requires a vector promoter for expression (Wren et al., 1988b; Rood et al., 1989). It seems difficult to understand why the *cat*D promoter as well as its

Shine–Dalgarno signal should not be recognized by the *E. coli* machinery since they are identical with the respective elements of *cat*P (Figure 10.5). Furthermore, another *C. difficile* promoter, cloned and sequenced independently by Dailey and Schloemer (1988b), proved to be functional in *E. coli*, even though its −35 and −10 signals were less similar to the respective standard *E. coli* sequences than the promoters of either *cat*P or *cat*D (Figure 10.5). That the inducibility of the *cat*D gene is lost on cloning into *E. coli* is little surprising, because several cloned gram-positive *cat* genes, originally inducible, are expressed constitutively in *E. coli* (Wren et al., 1988b).

10.3   Summary and Conclusion

Investigation of the molecular genetics of antibiotic resistance in *Clostridium difficile* has shown that the organism's R determinants are related to corresponding genes in other bacteria. Moreover, bidirectional in vivo exchange of genetic material with other species and genera has been demonstrated. Interestingly, plasmids have not been shown to be involved in either resistance or resistance transfer in *C. difficile*, even though many cryptic plasmids have been detected. It is not known whether these findings are simply the result of the relatively small number of strains analyzed or whether they are relevant, perhaps even with respect to the pathogenesis of *C. difficile* disease.

DNA homology between the related species *C. difficile* and *C. perfringens* has been shown regarding MLS^r, and, to an extraordinary extent, Cm^r, but not regarding Tc^r. Tc^r in *C. difficile* shows homology to TetM, originally found in *Enterococcus faecalis*. These findings indicate that Cm^r probably was transferred in the not-too-distant past, either from *C. difficile* to *C. perfringens* or vice versa, or to both species from a common source. The acquisition of Tc^r and MLS^r by *C. difficile* cannot be traced because both TetM and *erm*B have been found in many unrelated bacteria. Resistance transfer involving *C. difficile* occurs infrequently and always requires close cell-to-cell contact. The transfer events studied to date seem to be consistent with the mechanism of conjugative transposition. However, only in the case of Tc^r, are there data supporting this model. Alternatively, the suggestions about plasmids integrated into the chromosome put forward by Odelson et al. (1987; see p. 100) should be taken into consideration.

Despite considerable effort, the role of antibiotics or antibiotic resistance of *C. difficile* in the development of AAD, AAC, or AAPMC has not been determined. Moreover, the striking differences in susceptibility to disease caused by *C. difficile* in adults and infants await clarification, although several hypothetical explanations have been offered (Lyerly et al., 1988). We believe that special attention should be exercised with these questions, because knowledge about the molecular mechanisms involved might lead to simple guidelines for prevention of the disease and will also provide insight into evolutionary and ecological aspects of the intestinal microflora. In this context, the suggestion that resistance and virulence genes might be coregulated (Wren et al., 1988a) is of particular interest.

In conclusion, PMC caused by *C. difficile* is an interesting infectious disease because it is so strongly correlated with treatment of patients with antiinfective agents. Some of these drugs can even cause and cure the disease. On account of the number and severity of cases and the massive worldwide application of antimicrobials, any effort to elucidate the exact role of antibiotic resistance in *C. difficile* and of the antibiotics themselves in the pathogenesis of the disease would be justified.

References

Abraham, L.J. and J.I. Rood. 1987. Identification of Tn*4451* and Tn*4452*, chloramphenicol resistance transposons from *Clostridum perfringens. J. Bacteriol.* **169**:1579–1584.

Abraham, L.J., A.J. Wales, and J.I. Rood. 1985. Worldwide distribution of the conjugative *Clostridium perfringens* tetracycline resistance plasmid, pCW3. *Plasmid* **14**:37–46.

Ambulos, N.P., Jr., S. Mongkolsuk, J.D. Kaufman, and P.S. Lovett. 1985. Chloramphenicol-induced translation of *cat-86* mRNA requires two cis-acting regulatory regions. *J. of Bacteriol.* **164**:696–703.

Anonymous. 1989. Treatment of *Clostridium difficile* diarrhea. *Med. Lett. Drugs Ther.* **31**:94–95.

Arthur, M., A. Brisson-Noël, and P. Courvalin. 1987. Origin and evolution of genes specifying resistance to macrolide, lincosamide and streptogramin antibiotics: data and hypotheses. *J. Antimicrob. Chemother.* **20**:783–802.

Bannam, T.L., and J.I. Rood. 1991. Relationship between the *Clostridium perfringens catQ* gene product and chloramphenicol acetyltransferases from other bacteria. *Antimicrob. Agents Chemother.* **35**:471–476.

Banno, Y., K. Kobayashi, K. Watanabe, K. Ueno, and Y. Nozawa. 1981. Two toxins (D-1 and D-2) of *Clostridium difficile* causing antibiotic-associated colitis: purification and some characterization. *Biochem. Int.* **2**:629–635.

Bartlett, J.G. 1990. *Clostridium difficile*: clinical consideration. *Rev. Infect. Dis.* **12**(Suppl.2):S243–S251.

Bartlett, J.G., T.W. Chang, M. Gurwith, S.L. Gorbach, and A.B. Onderdonk. 1978. Antibiotic-associated pseudomembranous colitis due to toxin-producing clostridia. *N. Engl. J. Med.* **298**:531–534.

Berryman, D.I. and J.I. Rood. 1989. Cloning and hybridization analysis of *ermP*, a macrolide-lincosamide-streptogramin B resistance determinant from *Clostridium perfringens. Antimicrob. Agents Chemother.* **33**:1346–1353.

Bertram, J., A. Kuhn, and P. Dürre. 1990. Tn*916*-

induced mutants of *Clostridium acetobutylicum* defective in regulation of solvent formation. *Arch. Microbiol.* **153**:373–377.

Borriello, S.P., H.A. Davies, S. Kamiya, P.J. Reed., and S. Seddon. 1990. Virulence factors of *Clostridium difficile*. *Rev. Infect. Dis.* **12** (Suppl. 2):S185–S191.

Brisson-Noël, A.M. Arthur, and P. Courvalin. 1988a. Evidence for natural gene transfer from gram-positive cocci to *Escherichia coli*. *J. Bacteriol.* **170**:1739–1745.

Brisson-Noël, A., P. Trieu-Cuot, and P. Courvalin. 1988b. Mechanism of action of spiramycin and other macrolides. *J. Antimicrob. Chemother.* **22**(Suppl. B):13–23.

Brückner, R. and H. Matzura. 1985. Regulation of the inducible chloramphenicol-acetyltransferase gene of the *Staphylococcus aureus* plasmid pUB112. *EMBO J.* **4**:2295–2300.

Burdett, V. 1986. Streptococcal tetracycline resistance mediated at the level of protein synthesis. *J. Bacteriol.* **165**:564–569.

Burdett, V. 1988. Genetic and biochemical basis of tetracycline resistance in *Streptococcus ssp.* *Genome* **30**(Suppl. 1):166 (Abstr. 31.57.6).

Caillaud, F. and P. Courvalin. 1987. Nucleotide sequence of the ends of the conjugative shuttle transposon Tn1545. *Mol. Gen. Genet.* **209**:110–115.

Caparon, M.G. and J.R. Scott. 1989. Excision and insertion of the conjugative transposon Tn916 involves a novel recombination mechanism. *Cell* **59**:1027–1034.

Charles, J.G., J.W. Keyte, and W.V. Shaw. 1985. Nucleotide sequence analysis of the *cat* gene of *Proteus mirabilis*: comparison with the type 1 (Tn9) *cat* gene. *J. Bacteriol.* **164**:123–129.

Chiou, C-Y. and H-L. Chiou. 1990. Aminoglycosides and *C. difficile* colitis. *N. Engl. J. Med.* **322**:338.

Chow, A.W., N. Cheng, and K.H. Barlett. 1985. In vitro susceptibility of *Clostridium difficile* to new β-lactam and quinolone antibiotics. *Antimicrob. Agents Chemother.* **28**:842–844.

Christie, P.J., R.Z. Korman, S.A. Zahler, J.C. Adsit, and G.M. Dunny. 1987. Two conjugation systems associated with *Streptococcus faecalis* plasmid pCF10: identification of a conjugative transposon that transfers between *S. faecalis* and *Bacillus subtilis*. *J. Bacteriol.* **169**: 2529–2536.

Clabots, C., S. Lee, D. Gerding, M. Mulligan, R. Kwok, D. Schaberg, R. Fekety, and L. Peterson. 1988a. *Clostridium difficile* plasmid isolation as an epidemiologic tool. *Eur. J. Clin. Microbiol. Infect. Dis.* **7**:312–315.

Clabots, C.R., L.R. Peterson, and D.N. Gerding. 1988b. Charcterization of a nosocomial *Clostridium diffcile* outbreak by using plasmid profile typing and clindamycin susceptibility testing. *J. Infect. Dis.* **158**:731–736.

Clewell, D.B., F.Y. An, B.A. White, and C. Gawron-Burke. 1985. *Streptococcus faecalis* sex pheromone (cAM373) also produced by *Staphyloccus aureus* and identification of a conjugative transposon (Tn918). *J. Bacteriol.* **162**:1212–1220.

Clewell, D.B. and C. Gawron-Burke. 1986. Conjugative transposons and the dissemination of antibiotic resistance in streptococci. *Annu. Rev. Microbiol.* **40**:635–659.

Clewell, D.B., S.E. Flannagan, Y. Ike, J.M. Jones, and C. Gawron-Burke. 1988. Sequence analysis of termini of conjugative transposon Tn916. *J. Bacteriol.* **170**:3046–3052.

Courvalin, P. and C. Carlier. 1987. Tn 1545. A conjugative shuttle transposon. *Mol. Gen. Genet.* **206**:259–264.

Courvalin, P., H. Ounissi, and M. Arthur. 1985. Multiplicity of macrolide-lincosamide-streptogramin antibiotic resistance determinants. *J. Antimicrob. Chemother.* **16**(Suppl. A):91–100.

Cronberg, S., B. Castor, and A. Thorén. 1984. Fusidic acid for the treatment of antibiotic-associated colitis induced by *Clostridium difficile*. *Infection* **12**:276–279.

Dailey, D.C. and R.H. Schloemer. 1988a. Cloning and expression of secreted antigens of *Clostridium difficile* in *Escherichia coli*. *Infect. Immun.* **56**:1655–1657.

Dailey, D.C. and R.H. Schloemer. 1988b. Identification and characterization of *Clostridium difficile* promoter element that is functional in *Escherichia coli*. *Gene* **70**:343–350.

Dan, M., Z. Samra, and E. Wolfson. 1989. *Clostridium difficile* colitis associated with ofloxacin therapy. *Amer. J. Med.* **87**:479.

de Lalla, F., G. Privitera, E. Rinaldi, G. Ortisi, D. Santoro, and G. Rizzardini. 1989. Treatment of *Clostridium diffcile*-associated disease with teicoplanin. *Antimicrob. Agents Chemother.* **33**:1125–1127.

Delmée, M. and V. Avesani. 1986. Comparative in vitro activity of seven quinolones against 100 clinical isolates of *Clostridium difficile*. *Antimicrob. Agents Chemother.* **29**:374–375.

Delmée, M. and V. Avesani. 1988. Correlation between serogroup and susceptibility to chloramphenicol, clindamycin, erythromycin, rifampicin and tetracycline among 308 isolates of *Clostridium difficile*. *J. Antimicrob. Chemother.* **22**:325–331.

Dubbert, W., H. Luczak, and W.L. Staudenbauer. 1988. Cloning of two chloramphenicol acetyltransferase genes from *Clostridium butyricum* and their expression in *Escherichia coli* and *Bacillus subtilis*. *Mol. Gen. Genet.* **214**:328–332.

Dzink, J. and J.G. Bartlett. 1980. In vitro susceptibility of *Clostridium difficile* isolates from patients with antibiotic-associated diarrhea or colitis. *Antimicrob. Agents Chemother.* **17**:695–698.

Fekety, R., R. O'Connor, and J. Silva. 1983. Rifampin and pseudomembranous colitis. *Rev. Infect. Dis.* **5**(Suppl. 3):S524–S527.

Finney, J.M.T. 1893. Gastro-enterostomy for cicatrizing ulcer of the pylorus. *Bull. Johns Hopkins Hosp.* **4**:53–55.

Fitzgerald, G.F. and D.B. Clewell. 1985. A conjugative transposon (Tn*919*) in *Streptococcus sanguis. Infect. Immun.* **47**:415–420.

Franke, A.E. and D.B. Clewell. 1981. Evidence for a chromosome-borne resistance transposon (Tn*916*) in *Streptococcus faecalis* that is capable of "conjugal" transfer in the absence of a conjugative plasmid. *J. Bacteriol.* **145**:494–502.

Gawron-Burke, C. and D.B. Clewell. 1984. Regeneration of insertionally inactivated streptococcal DNA fragments after excision of transposon Tn*916* in *Escherichia coli*: strategy for targeting and cloning of genes from gram-positive bacteria. *J. Bacteriol.* **159**:214–221.

George, R.H., J.M. Symonds, F. Dimock, J.D. Brown, Y. Arabi, N. Shinagawa, M.R.B. Keighley, J. Alexander-Williams, and D.W. Burdon. 1978. Identification of *Clostridium difficile* as a cause of pseudomembranous colitis. *Br. Med. J.* **1**:695.

George, W.L., E.J.C. Goldstein, V.L. Sutter, S.L. Ludwig, and S.M. Finegold. 1978. Etiology of antimicrobial-agent-associated colitis. *Lancet* i:802–803.

Goldman, R.C. and J.O. Capobianco. 1990. Role of an energy-dependent efflux pump in plasmid pNE24-mediated resistance to 14- and 15-membered macrolides in *Staphylococcus epidermidis. Antimicrob. Agents Chemother.* **34**:1973–1980.

Hächler, H., B. Berger-Bächi, and F.H. Kayser. 1987a. Genetic characterization of a *Clostridium difficile* erythromycin-clindamycin resistance determinant that is transferable to *Staphylococcus aureus. Antimicrob. Agents Chemother.* **31**:1039–1045.

Hächler H., F.H. Kayser, and B. Berger-Bächi. 1987b. Homology of a transferable tetracycline resistance determinant of *Clostridium difficile* with *Streptococcus (Enterococcus) faecalis* transposon Tn*916. Antimicrob. Agents Chemother.* **31**:1033–1038.

Hall, I.C. and E. O'Toole. 1935. Intestinal flora in new-born infants, with a description of a new pathogenic anaerobe, *Bacillus difficilis. Am. J. Dis. Child.* **49**:390–402.

Halula, M. and F.L. Macrina. 1990. Tn*5030*: A conjugative transposon conferring clindamycin resistance in *Bacteroides* species. *Rev. Infect. Dis.* **12**(Suppl. 2):S235–S242.

Harwood, C.R., D.M. Williams, and P.S. Lovett. 1983. Nucleotide sequence of a *Bacillus pumilus* gene specifying chloramphenicol acetyltransferase. *Gene* **24**:163–169.

Hatheway, C.L. 1990. Toxigenic Clostridia. *Clin. Microbiol. Rev.* **3**:66–98.

Horinouchi, S. and B. Weisblum. 1982. Nucleotide sequence and functional map of pC194, a plasmid that specifies inducible chloramphenicol resistance. *J. Bacteriol.* **150**:815–825.

Ionesco, H. 1980. Transfert de la résistance à la tétracycline chez *Clostridium difficile. Ann. Inst. Pasteur Microbiol.* **131A**:171–179.

Johnson, S., D.N. Gerding, M.M. Olson, M.D. Weiler, R.A. Hughes, C.R. Clabots, and L.R. Peterson. 1990. Prospective controlled study of vinyl glove use to interrupt *Clostridium difficile* nosocomial transmission. *Am. J. Med.* **88**:137–140.

Kathariou, S., P. Metz, H. Hof, and W. Goebel. 1987. Tn*916*-induced mutations in the hemolysin determinant affecting virulence of *Listeria monocytogenes. J. Bacteriol.* **169**:1291–1297.

Larson, H.E., J.V. Parry, A.B. Price, D.R. Davies, J. Dolby, and D.A.J. Tyrrell. 1977. Undescribed toxin in pseudomembranous colitis. *Br. Med. J.* **1**:1264–1248.

Levett, P.N. 1986. *Clostridium difficile* in habitats other than the human gastro-intestinal tract. *J. Infect.* **12**:253–263.

Levett, P.N. 1988. Antimicrobial susceptibility of *Clostridium difficile* determined by disc diffusion and breakpoint methods. *J. Antimicrob. Chemother.* **22**:167–173.

Levy, S.B. 1984. Resistance to the tetracyclines. In: *Antimicrobial Drug Resistance*, L.E. Bryan, ed., pp. 191–240. Orlando: Academic Press.

Levy, S.B. 1988. Tetracycline resistance determinants are widespread. *Amer. Soc. Microbiol. News.* **54**:418–421.

Levy, S.B., L.M. McMurry, V. Burdett, P. Courvalin, W. Hillen, M.C. Roberts, and D.E. Taylor. 1989. Nomenclature for tetracycline resistance determinants. *Antimicrob. Agents Chemother.* **33**:1373–1374.

Lyerly, D.M., H.C. Krivan, and T.D. Wilkins. 1988. *Clostridium difficile*: its disease and toxins. *Clin. Microbiol. Rev.* **1**:1–18.

Magot, M. 1984. Physical characterization of the *Clostridium perfringens* tetracycline-chloramphenicol resistance plasmid pIP401. *Ann. Inst. Pasteur Microbiol.* **135B**:269–282.

Marcoli, R., S. Iida, and T.A. Bickle. 1980. The DNA sequence of an IS1-flanked transposon coding for resistance to chloramphenicol and fusidic acid. *Fed. Eur. Biochochem. Soc. Lett.* **110**:11–14.

Martin, P., P. Trieu-Cuot, and P. Cournalin. 1986. Nucleotide sequence of the *tet*M tetracycline resistance determinant of the streptococcal conjugative shuttle transposon Tn*1545. Nucl. Acids Res.* **14**:7047–7058.

Mayford, M. and B. Weisblum. 1989. Conformational alterations in the *erm*C transcript *in vivo* during induction. *EMBO J.* **8**:4307–4314.

Miller, S.N. and R.P. Ringler. 1987. Vancomycin-induced pseudomembranous colitis. *J. Clin. Gastroenterol.* **9**:114–115.

Monod, M., S. Mohan, and D. Dubnau. 1987. Cloning and analysis of *erm*G, a new macrolide-lincosamide-streptogramin B resistance element from *Bacillus sphaericus. J. Bacteriol.* **169**: 340–350.

Muldrow, L.L., E.R. Archibold, O.L. Nunez-Montiel, and R.J. Sheehy. 1982. Survey of the extrachromosomal gene pool of *Clostridium difficile. J. Clin. Microbiol.* **16**:637–640.

Muldrow, L.L., G.C. Ibeanu, N.I. Lee, N.K. Bose, and J. Johnson. 1987. Molecular cloning of *Clostridium difficile* toxin A gene fragment in lambda gt11. *Fed. Eur. Biochem. Soc. Lett.* **213**:249–253.

Mullany, P., M. Wilks, and S. Tabaqchali. 1991. Transfer of Tn*916* and Tn*916ΔE* into *Clostridium difficile*: demonstration of a hot-spot for these elements in the *C. difficile* genome. *FEMS Microbiol. Lett.* **79**:191–194.

Mullany, P., B. Wren, M. Wilks, C. Clayton, and S. Tabaqchali. 1989. Genetic analysis of a conjugative transposon-like element from *Clostridium difficile. J. Medical Microbiol.* **28**:iii (abstr.).

Mullany, P., M. Wilks, I. Lamb, C. Clayton, B. Wren, and S. Tabaqchali. 1990. Genetic analysis of a tetracycline resistance element from *Clostridium difficile* and its conjugal transfer to and from *Bacillus subtilis. J. Gen. Microbiol.* **136**:1343–1349.

Nakamura, S., K. Yamakawa, S. Nakashio, S. Kamiya, and S. Nishida. 1987. Correlation between susceptibility to chloramphenicol, tetracycline and clindamycin, and serogroups of *Clostridium difficile. Med. Microbiol. Immunol.* **176**:79–82.

Odelson, D.A., J.L. Rasmussen, C.J. Smith, and F.L. Macrina. 1987. Extrachromosomal systems and gene transmission in anaerobic bacteria. *Plasmid* **17**:87–109.

Ounissi, H., and P. Courvalin. 1982. Heterogeneity of macrolide-lincosamide-streptogramin B-Type antibiotic resistance determinants. In: *Microbiology-1982*, Schlessinger, D., ed., pp. 167–169. Washington, D.C.: American Society for Microbiology.

Popoff, M.R., E.J. Rubin, D.M. Gill, and P. Boquet. 1988. Actin-specific ADP-ribosyltransferase produced by a *Clostridium difficile* strain. *Infect. Immun.* **56**:2299–2306.

Poyart-Salmeron, C., P. Trieu-Cuot, C. Carlier, and P. Courvalin. 1989. Molecular characterization of two proteins involved in the excision of the conjugative transposon Tn*1545*: homologies with other site-specific recombinases. *EMBO J.* **8**:2425–2433.

Poyart-Salmeron, C., P. Trieu-Cuot, C. Carlier, and P. Courvalin. 1990. The integration-excision system of the conjugative transposon Tn*1545* is structurally and functionally related to those of lambdoid phages. *Mol. Microbiol.* **4**:1513–1521.

Reiner, L., M.J. Schlesinger, and G.M. Miller. 1952. Pseudomembranous colitis following aureomycin and chloramphenicol. *Arch. Pathol.* **54**:39–67.

Roberts, M.C. 1990. Characterization of the *tet*M determinants in urogenital and respiratory bacteria. *Antimicrob. Agents Chemother.* **34**:476–478.

Rood, J.l., S. Jefferson, T.L. Bannam, J.M. Wilkie, P. Mullany, and B.W. Wren. 1989. Hybridization analysis of three chloramphenicol resistance determinants from *Clostridium perfringens* and *Clostridium difficile. Antimicrob. Agents Chemother.* **33**:1569–1574.

Saginur, R., C.R. Hawley, and J.G. Bartlett. 1980. Colitits associated with metronidazole therapy. *J. Infect. Dis.* **141**:772–774.

Salyers, A.A., B.S. Speer, and N.B. Shoemaker. 1990. New perspectives in tetracycline resistance. *Mol. Microbiol.* **4**:151–156.

Sanchez-Pescador, R., J.T. Brown, M. Roberts, and M.S. Urdea. 1988a. The nucleotide sequence of the tetracycline resistance determinant *tet*M from *Ureaplasma urealyticum. Nucl. Acids Res.* **16**:1216–1217.

Sanchez-Pescador, R., J.T. Brown, M. Roberts, and M.S. Urdea. 1988b. Homology of the TetM with translational elongation factors: implications for potential modes of *tet*M conferred tetracycline resistance. *Nucl. Acids Res.* **16**:1218.

Sandler, P. and B. Weisblum. 1989. Erythromycin-induced ribosome stall in the *erm*A leader: a barricade to 5′-to-3′ nucleolytic cleavage of the *erm*A transcript. *J. Bacteriol.* **171**:6680–6688.

Sauerborn, M. and C. von Eichel-Streiber. 1990. Nucleotide sequence of *Clostridium difficile* toxin A. *Nucl. Acids Res.* **18**:1629–1630.

Scott, J.R., P.A. Kirchman, and M.G. Caparon. 1988. An intermediate in transposition of the conjugative transposon Tn*916. Proc. Natl. Acad. Sci. USA* **85**:4809–4813.

Sen, S. and P. Oriel. 1990. Transfer of transposon Tn*916* from *Bacillus subtilis* to *Thermus aquaticus. FEMS Microbiol. Lett.* **67**:131–134.

Senghas, E., J.M. Jones, M. Yamamoto, C. Gawron-Burke, and D.B. Clewell. 1988. Genetic organization of the bacterial conjugative transposon Tn*916. J. Bacteriol* **170**:245–249.

Shaw, W.V. 1984. Bacterial resistance to chloramphenicol. *Br. Med. Bull.* **40**:36–41.

Shaw, W.V., D.G. Brenner, S.F.J. LeGrice, S.E. Skinner, and A.R. Hawkins. 1985. Chloramphenicol acetyltransferase gene of staphylococcal plasmid pC221. *Fed. Eur. Biochem. Soc. Lett.* **179**:101–106.

Smith, C.J., S.M. Markowitz, and F.L. Macrina. 1981. Transferable tetracycline resistance in *Clostridium difficile. Antimicrob. Agents Chemother.* **19**:997–1003.

Sougakoff, W., B. Papadopoulou, P. Nordmann, and P. Courvalin. 1987. Nucleotide sequence and distribution of gene *tet*O encoding tetracycline resistance in *Campylobacter coli. FEMS Microbiol. Lett.* **44**:153–159.

Steffen, C., and H. Matzura. 1989. Nucleotide sequence analysis and expression studies of a chloramphenicol-acetyltransferase-coding gene from *Clostridium perfringens*. *Gene* **75**:349–354.

Steinberg, J.P., M.E. Beckerdite, and G.O. Westenfelder. 1987. Plasmid profiles of *Clostridium difficile* isolates from patients with antibiotic associated colitis in two community hospitals. *J. Infect. Dis.* **156**:1036–1038.

Storrs, M.J., C. Poyart-Salmeron, P. Trieu-Cuot, and P. Courvalin. 1991. Conjugative transposition of Tn*916* requires the excisive and integrative activities of the transposon-encoded integrase. *J. Bacteriol.* **173**:4347–4352.

Tabaqchali, S. 1989. Molecular studies on the epidemiology and pathogenicity of *Clostridium difficile*. *Gut* (Festschrift) **30**(Suppl.):44–51.

Tabaqchali, S. 1990. Epidemiologic markers of *Clostridium difficile*. *Rev. Infect. Dis.* **12**(Suppl. 2):S192–S199.

Tacke, W. and L. Hausmann. 1988. Die pseudomembranöse Kolitis nach Antibiotikagabe–eine diagnostische und therapeutische Herausforderung. *Medizinische Klinik.* **6**:199–205.

Taylor, N.S., G.M. Thorne, and J.G. Bartlett. 1981. Comparison of two toxins produced by *Clostridium difficile*. *Infect. Immun.* **34**:1036–1043.

Torres, J.F., and I. Lönnroth. 1989. Production, purification and characterization of *Clostridium difficile* toxic proteins different from toxin A and from toxin B. *Biochim. Biophys. Acta* **998**:151–157.

van der Waaij, D. 1989. The ecology of the human intestine and its consequences for overgrowth by pathogens such as *Clostridium difficile*. *Annu. Rev. Microbiol.* **43**:69–87.

Viscidi, R., S. Willey, and J.G. Bartlett. 1981. Isolation rates and toxigenic potential of *Clostridium difficile* isolates from various patient populations. *Gastroenterology* **81**:5–9.

Volk, W.A., B. Bizzini, K.R. Jones, and F.L. Macrina. 1988. Inter- and Intrageneric transfer of Tn*916* between *Streptococcus faecalis* and *Clostridium tetani*. *Plasmid* **19**:255–259.

Woolley, R.C., A. Pennock, R.J. Ashton, A. Davies, and M. Young. 1989. Transfer of Tn*1545* and Tn*916* to *Clostridium acetobutylicum*. *Plasmid* **22**:169–174.

Wren, B.W., N. Castledine, and S. Tabaqchali. 1990. A sensitive oligonucleotide probe for the specific identification of toxigenic *Clostridium difficile* strains based of tandem repeat DNA sequence data from the toxin A gene. *J. Med. Microbiol.* **31**:XVII (abstr.).

Wren, B.W., C.L. Clayton, P.P. Mullany, and S. Tabaqchali. 1987. Molecular cloning and expression of *Clostridium difficile* toxin A in *Escherichia coli* K12. *Fed. Eur. Biochem. Soc. Lett.* **225**:82–86.

Wren, B.W., P.P. Mullany, C.L. Clayton, and S. Tabaqchali. 1988a. Molecular cloning and expression of *Clostridium difficile* toxin A and macrolide resistance determinants in *Escherichia coli*. *Gut* **29**:A723 (abstract T96).

Wren, B. W., P. Mullany, C. Clayton, and S. Tabaqchali. 1988b. Molecular cloning and genetic analysis of a chloramphenicol acetyltransferase determinant from *Clostridium difficile*. *Antimicrob. Agents Chemother.* **32**:1213–1217.

Wren, B.W., P. Mullany, C. Clayton, and S. Tabaqchali. 1989. Nucleotide sequence of a chloramphenicol acetyl transferase gene from *Clostridium difficile*. *Nucl. Acids Res.* **17**:4877.

Wüst, J. and U. Hardegger. 1983. Transferable resistance to clindamycin, erythromycin, and tetracycline in *Clostridium difficile*. *Antimicrob. Agents Chemother.* **23**:784–786.

Wüst, J. and U. Hardegger. 1988. Studies on the resistance of *Clostridium difficile* to Antimicrobial agents. *Zentralbl. Bakteriol. Mikrobiol. Hyg. Orig. A* **267**:383–394.

Young, M., N.P. Minton, and W.L. Staudenbauer. 1989. Recent advances in the genetics of the clostrida. *FEMS Microbiol. Rev.* **63**:301–326.

Zyprian, E. 1987. Analyse von Genexpressionssignalen in dem Gram-positiven Kanamycinresistenzplasmid pUB110, Entwicklung eines Gram-positiven Expressionsvektorsystems and Charakterisierung des induzierbaren Chloramphenicolacetyltransferasegens auf dem *Staphylococcus aureus*. *Plasmid* pC223. Ph.D. thesis, University of Heidelberg, Germany.

11

Genetics and Molecular Biology of Chloramphenicol Acetyltransferase of *Clostridium butyricum*

Walter L. Staudenbauer and Wolfgang Dubbert

11.1 Detoxification of Chloramphenicol

Chloramphenicol (D-threo-1-p-nitrophenyl-2-N-dichloracetamido-1,3-propanediol) (Figure 11.1) is an inhibitor of protein synthesis that is effective against a variety of *Clostridium* species. Its mode of action is to inhibit the peptidyltransferase center of procaryotic ribosomes (Hahn, 1983). Resistance of anaerobic bacteria to chloramphenicol can result from enzymatic inactivation of the antibiotic by either reduction or O-acetylation.

Reduction of the aryl nitro group of chloramphenicol to an amino group is catalyzed by nitroaryl reductases in a ferredoxin- or NADH-dependent reaction. Such enzymes are widely distributed among clostridia. Thus, Angermaier et al. (1981) reported the isolation of NADH-dependent nitroaryl reductases from *Clostridium kluyveri*. On the other hand, O'Brien and Morris (1971) described the reduction of chloramphenicol and other nitroaryl compounds via a ferredoxin-dependent reaction by *Clostridium acetobutylicum*. It was shown that the reduced chloramphenicol (100 μg/ml) was not toxic to the organism.

Acetylation of chloramphenicol results from the presence of genetic determinants (*cat*) specifying the production of chloramphenicol acetyltransferase (CAT) (acetyl-CoA:chloramphenicol O^3-acetyltransferase; EC 2.3.1.28), which catalyzes the 3-O-acetylation of chloramphenicol. The primary acetylation product, 3-acetylchloramphenicol, fails to bind to bac-

terial ribosomes and therefore lacks antibiotic activity. The inactivated antibiotic subsequently undergoes a nonenzymatic intramolecular rearrangement leading to the formation of 1-acetylchloramphenicol. The latter compound is subject to further enzymatic acetylation yielding 1,3-diacetylchloramphenicol at a slow rate (Figure 11.2). Chloramphenicol resistance from CAT production has been observed in many bacterial species, including pathogenic clostridia such as *Clostridium perfringens* (Brefort et al., 1977; Abraham and Rood, 1987) and *Clostridium difficile* (Wren et al., 1988).

The two modes of chloramphenicol resistance can be readily distinguished by employing the related antibiotic thiamphenicol in which the p-nitro group is replaced by a methylsulphonyl moiety (see Figure 11.1). Because of its altered chemical structure, thiamphenicol cannot be reduced by nitroaryl reductases but re-

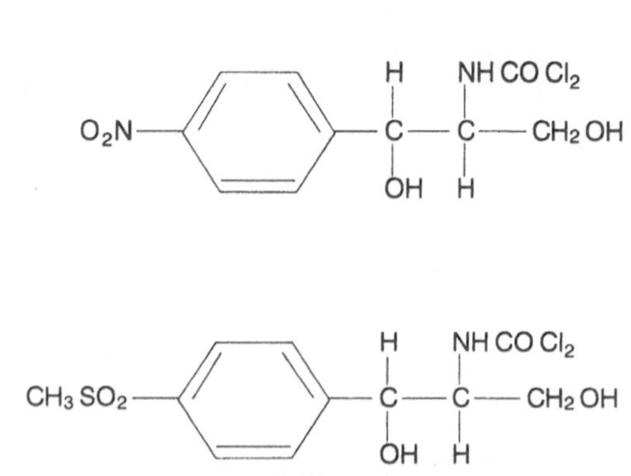

FIGURE 11.1 Chemical structure of chloramphenicol (A) and thiamphenicol (B).

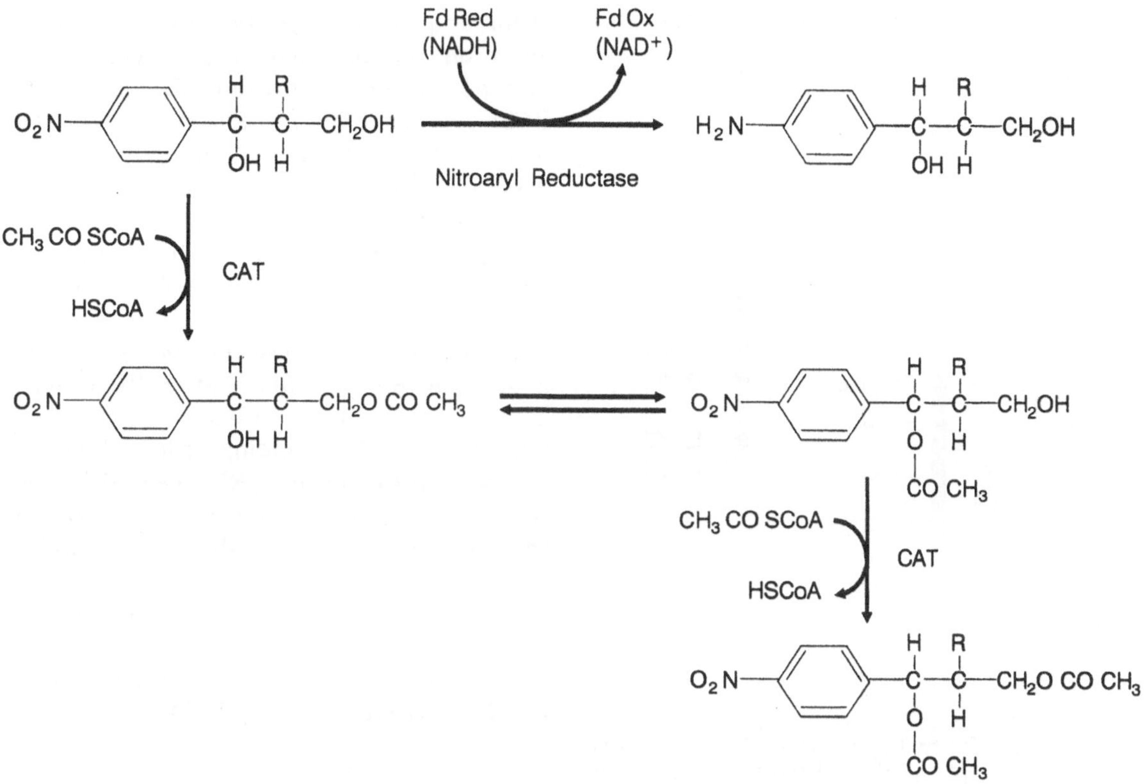

FIGURE 11.2   Detoxification of chloramphenicol by nitroaryl reductase and chloramphenicol acetyltransferase (CAT).

mains sensitive toward O-acetylation by chloramphenicol acetyltransferases.

11.2   Cloning and Expression of *Clostridium butyricum cat* Determinants

In the course of screening various *Clostridium butyricum* strains for the presence of antibiotic resistance determinants (Luczak et al., 1985), we observed the increased resistance to chloramphenicol of strains ATCC 19398 and NCIB 7423 (MIC >20 μg/ml) as compared to sensitive strains (MIC $\leq$2 μg/ml). Both strains were also insensitive to thiamphenicol, indicating a detoxification of the antibiotic by acetylation. The presence of CAT activities was confirmed by detection of the formation of monoacetylated products from chloramphenicol (Dubbert et al., 1988).

The chloramphenicol resistance determinants of *C. butyricum* ATCC 19398 and NCIB 7423 were isolated by molecular cloning in *Escherichia coli* and *Bacillus subtilis* (Dubbert et al., 1988). The location of the *cat* genes could be narrowed down to about 1.3 kb by deletion analysis (Figure 11.3). The resistance determinants showed no homology by Southern hybridization and thus represent distinct genes, designated *catA* and *catB*, respectively. Neither gene was homologous to the *cat*-86 gene of *Bacillus pumilus* (Harwood et al., 1983). The *C. butyricum catA* gene, but not the *catB* gene, showed homology to the *cat* gene of the *Staphylococcus aureus* plasmid pC194 (Horinouchi and Weisblum, 1982). Neither *catA* nor *catB* showed any homology to the endogenous plasmids of *C. butyricum* ATCC 19398 and NCIB 7423, indicating a chromosomal location of these genes.

Expression of the *C. butyricum cat* genes in *E. coli* and *B. subtilis* was independent of orienta-

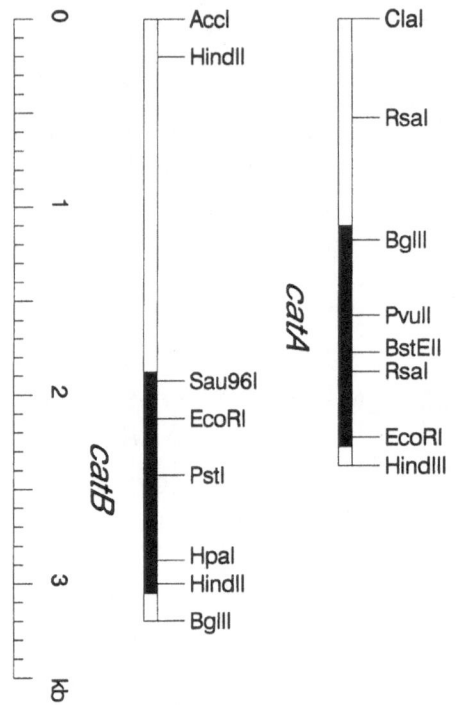

FIGURE 11.3 Restriction maps of *Clostridium butyricum* chloramphenicol resistance determinants *catA* and *catB*.

of the drug could be attributed to a stabilizing effect on the recombinant plasmids in the presence of selective pressure (Dubbert et al., 1988). A constitutive expression has also been reported for a *Clostridium perfringens cat* gene (Steffen and Matzura, 1989). The clostridial *cat* genes differ in this respect from the *cat* genes of staphylococci and bacilli, which are inducible by subinhibitory concentrations of chloramphenicol.

The expression of inducible *cat* genes from gram-positive organisms requires the destabilization of a secondary mRNA structure blocking the ribosome-binding site. Ribosome stalling on the *cat* leader mRNA caused by chloramphenicol results in the disruption of this secondary structure and thus in the induction of gene expression (Ambulos et al., 1986; Brückner et al., 1987). Such regulatory sequences are missing in the sequenced *cat* genes of *C. perfringens* (Steffen and Matzura, 1989) and *C. difficile* (Wren et al., 1989).

11.3 Properties of *Clostridium butyricum* Chloramphenicol Acetyltransferases

Chloramphenicol acetyltransferases can be readily purified to homogeneity by chromatography on an affinity matrix prepared by coupling chloramphenicol base to CNBr-activated Sepharose (Zaidenzaig and Shaw, 1976). By employing this purification method, variants of

tion, which implies that both genes are transcribed from clostridial promoters in the heterologous hosts. Gene expression was drastically increased by cloning into multicopy plasmid vectors (Table 11.1). Both genes appeared to be expressed constitutively, regardless of the addition of chloramphenicol. Enhanced CAT activities of extracts from cells grown in the presence

Table 11.1 Specific CAT activities of cell extracts from bacterial strains harboring *Clostridium butyricum cat* genes

Bacterial strain	Vector	Gene	CAT activity (U/mg)[a]	
			Cm⁻	Cm⁺
Clostridium butyricum ATCC 19398	—	*catA*	0.012	0.014
Escherichia coli JM109	pUC18	*catA*	4.009	4.105
Bacillus subtilis BD170	pUB110	*catA*	1.045	2.165
Clostridium butyricum NCIB 7423	—	*catB*	0.029	0.034
Eschorichia JM109	pUC18	*catB*	0.391	0.982
Bacillus subtilis BD170	pUB110	*catB*	0.301	0.511

[a]CAT activity was determined in crude extracts of cells grown with or without addition of chloramphenicol (Cm) (5 μg/ml).

chloramphenicol acetyltransferases have been isolated from a wide variety of bacterial species (Zaidenzaig et al., 1979; Shaw, 1983). All known variants are trimers of identical subunits with molecular weights about 25,000. The x-ray crystallographic structure of the *E. coli* enzyme has been determined (Leslie et al., 1988). The CAT monomers are highly homologous, indicating a similar tertiary structure. The highest homology among the sequenced CAT variants exists in the segment centered around the active-site histidine residue that acts as a general base catalyst in the acetyl transfer reaction.

The subunit size of *C. butyricum* CAT enzymes was determined as 22.5 kDa and 24 kDa for CAT A and CAT B, respectively. It should be noted that the apparent molecular weights of the CAT variants as determined by sodium dodecyl sulfate (SDS) gel electrophoresis may be underestimates of the true molecular weights deduced from the amino acid sequences. The *C. butyricum* enzymes have isoelectric points of 5.2 (CAT A) and 4.1 (CAT B) and are therefore acidic proteins like all other CAT variants.

Both *C. butyricum* CATs have a broad pH optimum between pH 7 and 8. The enzymes are remarkably thermoactive and show optimal activity at 70°C (CAT A) and 65°C (CAT B). They can be incubated for several hours in the absence of chloramphenicol at 60°C without loss of activity. A pronounced thermostability in the absence of substrate has also been reported for staphylococcal CATs (Shaw, 1983). Like the staphylococcal CATs, the *C. butyricum* CATs are insensitive to inactivation by thiol reagents such as 5,5'-dithiobis-2-nitrobenzoic acid (DTNB).

The Michaelis constants (K_m) of CAT A and CAT B are rather similar and correspond to those of other CAT variants (Table 11.2). Thus, the affinity of the enzymes to chloramphenicol is greater than to its analogs. Further, the enzymes show a somewhat higher affinity for chloramphenicol than for acetyl-CoA. As far as the reaction products are concerned, both *C. butyricum* CATs are capable of forming diacetyl-chloramphenicol in addition to the monoacetylated derivatives.

Table 11.2 Michaelis constants (K_m) for *Clostridium butyricum* chloramphenicol acetyltransferases

	Michaelis constant (μm)	
Substrate	CAT A	CAT B
Chloramphenicol	28.6	23.1
Thiamphenicol	55.6	40.0
Chloramphenicol base	2500	1050
Acetyl-CoA	38.0	50.1

In summary, the properties of the *C. butyricum* CATs resemble those of staphylococcal CAT variants. They differ from other gram-positive CATs by their constitutive synthesis in the absence of chloramphenicol. It remains to be seen whether this is a general feature of clostridial *cat* gene expression.

References

Abraham, L.J. and J.I. Rood. 1987. Identification of Tn4451 and Tn4452, chloramphenicol resistance transposons from *Clostridium perfringens*. *J. Bacteriol.* **169**:1579–1584.

Ambulos, N.P., E.J. Duval, and P.S. Lovett. 1986. Analysis of the regulatory sequences needed for induction of the chloramphenicol acetyltransferase gene *cat*-86 by chloramphenicol and amicetin. *J. Bacteriol.* **167**:842–849.

Angermaier, L., F. Hein, and H. Simon. 1981. Investigations on the reduction of aliphatic and aromatic nitro compounds by *Clostridium* species and enzyme systems. In: *Biology of Inorganic Nitrogen and Sulfur*, H. Boethe and A. Trebst, eds., pp. 266–275. Berlin: Springer-Verlag.

Brefort, G., H. Magot, H. Ionesco, and M. Sebald. 1977. Characterization and transferability of *Clostridium perfringens* plasmids. *Plasmid* **1**:52–66.

Brückner, R., T. Dick, and H. Matzura. 1987. Dependence of expression of an inducible *Staphylococcus aureus cat* gene on the translation of its leader sequence. *Mol. Gen. Genet.* **207**:486–491.

Dubbert, W., H. Luczak, and W.L. Staudenbauer. 1988. Cloning of two chloramphenicol acetyltransferase genes from *Clostridium butyricum* and their expression in *Escherichia coli* and *Bacillus subtilis*. *Mol. Gen. Genet.* **214**:328–332.

Hahn, F.E. 1983. Chloramphenicol. In: *Antibiotics VI*, F.E. Hahn, ed., pp. 34–45. Berlin: Springer Verlag.

Harwood, C.R., D.M. Williams, and P.S. Lovett. 1983. Nucleotide sequence of a *Bacillus pumilus* gene specifying chloramphenicol acetyltransferase. *Gene* (Amst.) **24**:163–169.

Horinouchi, S. and B. Weisblum. 1982. Nucleotide sequence and functional map of pC194, a plasmid that specifies inducible chloramphenicol resistance. *J. Bacteriol.* **150**:815–825.

Leslie, A.G.W., P.C.E. Moody, and W.V. Shaw. 1988. Structure of chloramphenicol acetyltransferase at 1.75-A resolution. *Proc. Natl. Acad. Sci. USA* **85**:4133–4137.

Luczak, H., H. Schwarzmoser, and W.L. Staudenbauer. 1985. Construction of *Clostridium butyricum* hybrid plasmids and transfer to *Bacillus subtilis*. *Appl. Microbiol. Biotechnol.* **23**:114–122.

O'Brien, R.W. and J.G. Morris. 1971. The ferredoxin-dependent reduction of chloramphenicol by *Clostridium acetobutylicum*. *J. Gen. Microbiol.* **67**:265–271.

Shaw, W.V. 1983. Chloramphenicol acetyltransferase: enzymology and molecular biology. *CRC Crit. Rev. Biochem.* **14**:1–43.

Steffen, C. and H. Matzura. 1989. Nucleotide sequence analysis and expression studies of a chloramphenicol-acetyltransferase-coding gene from *Clostridium perfringens*. *Gene* **75**:349–354.

Wren, B.W., P. Mullany, C. Clayton, and S. Tabaqchali. 1988. Molecular cloning and genetic analysis of a chloramphenicol acetyltransferase determinant from *Clostridium difficile*. *Antimicrob. Agents Chemother.* **32**:1213–1217.

Wren, B.W., P. Mullany, C. Clayton, and S. Tabaqchali. 1989. Nucleotide sequence of a chloramphenicol acetyl transferase gene from *Clostridium difficile*. *Nucleic Acids Res.* **17**:4877.

Zaidenzaig, Y. and W.V. Shaw. 1976. Affinity and hydrophobic chromatography of three variants of chloramphenicol acetyltransferases specified by R factors in *Escherichia coli*. *FEBS Lett.* **62**:266–271.

Zaidenzaig, Y., J.E. Fitton, L.C. Packman, and W.V. Shaw. 1979. Characterization and comparison of chloramphenicol acetyltransferase variants. *Eur. J. Biochem.* **100**:609–618.

12

The Role of Bacteriophages and Plasmids in the Production of Toxins and Other Biologically Active Substances by *Clostridium botulinum* and *Clostridium novyi*

Mel W. Eklund

12.1 Introduction

Clostridium botulinum and *Clostridium novyi* are pathogenic anaerobes that are characterized by their ability to produce powerful toxins lethal to man and animals. The different types of *C. botulinum* are recognized specifically for their ability to produce neurotoxins that are responsible for wound, toxicoinfectious, and food-borne forms of botulism in man and animals. Members of *C. novyi* do not produce neurotoxins, but they produce other lethal toxins and biologically active substances active in gas gangrene infections found in humans and animals.

Clostridium botulinum includes a very heterogenous group of bacterial strains that produce antigenically specific neurotoxins designated as types A, B, C1, D, E, F, and G. During the past two decades, strains of *C. botulinum* that produce more than one neurotoxin have been isolated (Sugiyama et al., 1972; Gimenez and Ciccarelli, 1978; Poumeyrol et al., 1983; Gimenez, 1984; McCroskey and Hatheway, 1984; Sakaguchi et al., 1986). Subtypes A_B, A_F, B_A, and B_F (subscript identifies the minor toxin produced) are now recognized. In addition, *Clostridium butyricum* and *Clostridium barati* have been shown to produce botulinal neurotoxins types E and F, respectively (Hall et al., 1985; Aureli et al., 1986; McCroskey et al., 1986). Both of these *Clostridium* species were isolated from infant botulism (toxicoinfectious) outbreaks.

Specific strains of *C. botulinum* types C and D also produce ADP-ribosylating enzymes designated as C2 and C3. These enzymes permit bacteria to affect eucaryotic cell functions. The C2 and C3 enzymes are ADP-ribosyltransferases that target different substrates (Aktories et al., 1987; Aktories et al., 1988; Rubin et al., 1988). C2 is a trypsin-activated binary toxin that is lethal to mice (Eklund et al., 1972; Simpson, 1982; Ohishi and Dasgupta, 1987). The C3 enzyme has several attributes of an enzymatically active portion of a toxin, but because no toxic activity has been demonstrated (with as much as 1 mg/mouse), it has been termed as exoenzyme C3. Although the different bacterial strains producing botulinal toxins differ markedly, they can be placed into six groups on the basis of their biochemical and physiological characteristics and deoxyribonucleic acid homologies. The members of these groups are shown in Table 12.1.

Clostridium novyi also includes a heterogenous group of organisms designated as types A, B, C, and D based on the production of eight different soluble antigens (Smith and Holdeman, 1968). The three pathogenic types

Table 12.1 Groups of *Clostridium* species producing botulinal neurotoxin

Group	Characteristics	Types
I	Proteolytic*	A, A_B, A_F, B, B_A, B_F, F
II	Nonproteolytic*	B, E, F
III	Nonproteolytic*	C(8 subtypes) D (6 subtypes)
IV	Proteolytic (delayed)*	G
V	*C. butyricum*	E
VI	*C. barati*	F

C. botulinum species; *C. butyricum* and *C. barati* also produce neurotoxins but not the main species producing these toxins.

include the classical type A, which has been frequently involved in gas gangrene infections; type B, the etiological agent of infectious necrotic hepatitis (black disease) that has been observed in sheep and other animals; and type D (*C. haemolyticum*), the causal organism of bacillary hemoglobinuria in animals (Betty et al., 1964; Williams, 1964; Willis, 1964). Type C is generally regarded as nonpathogenic.

The production of toxins by these different *Clostridium* species is important not only to the pathogenicity of the different strains but also to the differentiation from other closely related clostridia. This chapter describes the role of plasmids and bacteriophages in the production of toxins and other biologically active substances by *C. botulinum* and *C. novyi*. The effect of these cellular changes on the identification of these two species is also discussed.

12.2 Procedures for Determining the Relationship of Bacteriophages and Plasmids to the Production of Biologically Active Substances

Bacteriophage, usually shortened to phage, are recognized as a group of bacteria-specific viruses (Adams, 1959). Following infection, the phage nucleic acid can enter into either a lytic or nonlytic relationship with the bacterial host. During the lytic cycle, the phage nucleic acid dominates the bacterial metabolic system and new phage are replicated. Within hours after infection, the bacterial cell lyses, releasing new phage particles that are capable of infecting other bacterial cells. When the phage nucleic acid enters the nonlytic cycle, an intimate relationship is established with the host and the bacterium continues to multiply and carry the phage nucleic acid intracellularly in a noninfectious condition for an indefinite number of generations. This relationship of phage and host is referred to as the lysogenic state. Lysogenic bacteria carry the genetic units (referred to as prophage) necessary for transmission of the phage to the progeny, but they do not carry the complete phage particles. The stability of the prophage–bacterium relationship varies with the type of phage, phage–host relationship, incubation temperature and other environmental conditions, and the physiological state of the bacterium.

In any given culture, this delicate prophage–bacterium relationship is interrupted in a small percentage of bacterial cells. The phage nucleic acid enters the lytic cycle, and infective phages are released when the bacterium lyse. Most young bacterial cultures contain infective phages (Eklund et al., 1971). Bacteria-free filtrates from these young cultures therefore were used as a source of phages in the experiments described in this chapter.

A bacterial culture is generally immune to the infection by phages that it carries or to phages that are antigenically closely related and are released from other bacterial strains (Adams, 1959). To determine the relationship of phage to specific physiological and biochemical properties of a bacterial strain, phage-sensitive derivatives must be isolated, preferably from strains known to possess a specific characteristic such as neurotoxin production.

Occasionally bacterial cells within a culture will simultaneously lose their prophage genome and their immunity to the homologous phage. These cells, however, are often reinfected with free phages that occur in the culture. In experiments with *C. botulinum* types C and D and *C. novyi*, the isolation of phage-sensitive derivatives was enhanced by subculturing the neurotoxigenic strain in media containing acridine orange or acriflavine or by exposing the cells to ultraviolet light. In later experiments, phage-sensitive derivatives were isolated from untreated sporulated cultures.

Colonies isolated from these various treatments were tested for their sensitivity to the different free phages present in the young cultures of the neurotoxigenic parent strain using the agar layer procedure (Eklund and Poysky, 1974). Phage-sensitive isolates were tested to determine whether they continued to produce the different toxins and other biologically active substances. These phage-sensitive derivatives were then reinfected with specific phages, and the cultures were again tested to determine whether the phages induced the production of toxins and other substances.

Plasmids are extrachromosomal genetic elements that exist as independent replicons in a bacterial cell. They constitute from 0.1% to 5.0% of the DNA of a given cell. Plasmids are defined as nonessential elements of growth of normal cells of a host species, and under most conditions they can be gained or lost without lethal effect (Clowes, 1972). The existence of a plasmid is usually recognized by the cellular function that it determines. This may be resistance to antibiotics, production of bacteriocins, or other properties.

To determine the relationship of plasmids to a specific cellular function, such as toxin production, bacterial strains were cured of specific plasmids by repeatedly transferring them in trypticase-yeast extract-glucose (TYG) broth at 40° to 44°C. After different periods of incubation, the cultures were diluted and plated on egg-blood agar. Isolated colonies were cultured in TYG broth and assayed for toxin and bacteriocin production (Eklund et al., 1988). Both the nontoxigenic derivatives and toxigenic parent cultures were then lysed and screened for plasmids (Strom et al., 1984). A relationship between plasmid and production of toxin or bacteriocin was indicated if there was a simultaneous loss or an alteration in the molecular weight of a specific plasmid and the bacterial strain ceased to produce toxin or bacteriocin.

12.3   Phages and Their Relationship to Production of Toxins and Other Biologically Active Substances by *Clostridium botulinum* Types C and D

Electron micrographs of lysates have indicated that many different strains of *C. botulinum* types A through F harbor phages (Inoue and Iida, 1968; Vinet et al., 1968; Eklund et al., 1969; Dolman and Chang, 1972). These results coupled with observations that certain *C. botulinum* types C and D strains occasionally lost their ability to produce neurotoxin following culturing in various laboratory media suggested that phages may mediate the production of neurotoxins and other substances.

Phages and neurotoxin production

Type C and D strains are indistinguishable on the basis of physiological characteristics, biochemical reactions, and metabolic end products produced. The main characteristic that identifies the two types is the antigenic monospecificity of the neurotoxins produced: type D produces type D neurotoxin and type C produces type C1 neurotoxin.

The role that phages play in the production of neurotoxins was first demonstrated in strains of types C and D. Nonneurotoxigenic derivatives were isolated from neurotoxigenic strains following treatment with UV light or culturing in medium containing acridine orange. Each nonneurotoxigenic derivative was tested for its sensitivity to lysates of the neurotoxigenic parent and also to other nonneurotoxigenic derivatives. Different neurotoxigenic strains of types C and D were shown to carry from one to three phages. Some nonneurotoxigenic derivatives were cured of only one phage whereas others were cured of all the phages carried by the neurotoxigenic parent culture. These phages were purified by five successive single-plaque isolations using as the host the nonneurotoxigenic derivatives cured of all phages.

When the nonneurotoxigenic hosts were

Table 12.2 Relationship of specific phages to production of neurotoxins and ADP-ribosyltransferases by *Clostridium botulinum* types C and D

	Culture produced		
Bacterial strain and phage[a]	Type of neurotoxin	C_2[b]	C_3[c]
Parent 468C (1C, 2C)	C_1	+	+
Cured 468C—UV171 (2C)	−	+	−
Cured 468C—A028 (−)	−	+	−
Reinfected 468C—A028 (1C)	C_1	+	+
Parent 1873 (2D)	D	+	+
Cured 1873—AOA113 (−)	−	+	−
Reinfected 1873—AOA113 (2D)	D	+	+

[a] (TOX[+], TOX[−]) refer to phages carried by bacterium; (−) indicates no phage; 1C and 2D are TOX[+] phages; 2C is TOX[−] phage.
[b,c] C_2 and C_3 are ADP-ribosyltransferases.

reinfected with the purified phages, TOX[+] and TOX[−] phages were identified. The TOX[+] phages induced the host bacterium to produce neurotoxin whereas the TOX[−] phages did not affect toxicity (Table 12.2). When reinfected by the TOX[+] phages, the cultures simultaneously became immune to infection by the same phage and produced neurotoxins.

These same experiments were repeated numerous times to determine whether the continued participation of the phage was necessary to maintain neurotoxicity. Cultures reinfected with the TOX[+] phage were cured of their prophages by subculturing in medium containing acridine orange. In each experiment, neurotoxigenic and nonneurotoxigenic derivatives were isolated. The neurotoxigenic cultures continued to harbor the TOX[+] phage and produce neurotoxin, and they maintained their immunity to the TOX[+] phage. In contrast, the nonneurotoxigenic derivatives lost their immunity and became sensitive to the TOX[+] phage. Many of the nonneurotoxigenic strains were thus subcultured for 4 years, and in each case they maintained their nonneurotoxigenic characteristic and sensitivity to the TOX[+] phage. Numerous strains of both type C and D strains have been examined, and in all cases the continued participation of TOX[+] phage

was necessary to maintain the neurotoxigenic characteristic (Inoue and Iida, 1970, 1971; Eklund et al., 1971, 1972; Oguma et al., 1973; Hariharan and Mitchell, 1976).

Nonneurotoxigenic phage-sensitive derivatives were also isolated from strains that were subcultured in TYG broth containing antiserum against the TOX[+] phage. The number of phage-sensitive derivatives increased as the number of passages in phage antiserum medium increased. This indicated that the phage–host relationship was not true lysogeny but a pseudolysogenic relationship. Under these conditions, some progeny cells continued to lose their phages during cell division, but the antiserum protected the cured ones from being reinfected by the free phages. These phage-sensitive isolates were induced to produce neurotoxins after infection with TOX[+] phages.

These results suggested that some cells might lose their phages in TYG broth or modified egg-meat medium (MEMM) (Eklund and Poysky, 1974) without curing agents and sporulate. To test this hypothesis, cultures of types C and D isolated from many different countries were permitted to sporulate in MEMM. The spores were diluted and plated on TYG agar, and isolates were tested for phage sensitivity. Even though the degree of instability in phage–host relationship varied from strain to strain, all the neurotoxigenic strains yielded isolates that had lost their phages, immunity, and neurotoxigenicity. These phage-sensitive strains were reinfected with the TOX[+] phage from the parent, and in each case the strains were converted to neurotoxigenicity.

Structural genes for neurotoxin

To demonstrate that the TOX[+] phage carries the structural gene for the neurotoxin, the N-terminal amino acid sequence for type C1 neurotoxin was determined and a deduced oligonucleotide probe was prepared (Fujii et al., 1988). The probe hybridized DNA from five different TOX[+] phages and one TOX[−] phage. This TOX[−] phage was isolated from a toxigenic culture treated with *N*-methyl-*N*'-

nitro-*N*-nitrosoguanidine and should carry a mutation in the TOX gene; hence, cross-hybridization occurred in spite of the loss of converting ability. In the same studies, DNA from TOX$^+$ phage type D also showed positive reactions. This was caused by the similarity of N-terminal amino acid sequence of type C and D toxins. The complete type C1 DNA sequence has been reported by Hauser et al. (1990).

Phages and the production of ADP ribosyltransferase C2

In the first phases of these studies on the relationship of phages to neurotoxigenicity, type C strain 468C and type D strain 1873 ceased to produce neurotoxins when they were cured of TOX$^+$ phages. These cultures were therefore considered to be nontoxigenic. This belief, however, was changed when Eklund and Poysky (1972) demonstrated that these non-neurotoxigenic derivatives produced another toxin (later designated as C2 toxin) when the culture supernatant fluids were treated with trypsin (see Table 12.2). When the C2 toxin was inoculated intraperitoneally into mice, it produced signs that differed from those caused by the neurotoxins (Eklund and Poysky, 1972). In addition, only one of the five different type C and none of the type D antisera received from different laboratories neutralized the C2 toxin. This was further evidence that the toxin was different from the C1 and D neurotoxins. These observations were also confirmed by Nakane et al. (1978) using other type C and D strains. The subsequent studies by Ohishi et al. (1981), and Simpson (1982, 1987), described the cellular and subcellular response and the causes of these different signs in mice.

The availability of type C and D antisera without the C2 component therefore became a useful research tool for determining whether the different strains of type C and D produced C2 toxin. These antisera neutralized the C1 or D neurotoxin produced by the culture, enabling the C2 toxin that is usually masked by the dominance of neurotoxins to express itself. Of the 30 type C strains obtained from Sweden, France, Netherlands, England, South Africa,

Table 12.3 Relationship of specific phages to production of neurotoxins and ADP-ribosyltransferases by *Clostridium botulinum* types C and D

| Bacterial strain and phage[a] | Culture produced | | |
	Type of neurotoxin	C$_2$[b]	C$_3$[c]
Parent 165 (9C)	C$_1$	−	+
Cured 165—HS31 (−)	−	−	−
Reinfected 165—HS31 (9C)[d]	C$_1$	−	+
Parent 8265 (4D)	D	−	+
Cured 8265—AO12 (−)	−	−	−
Reinfected 8265—AO12 (4D)	D	−	+

[a] TOX$^+$ refers to phage carried by bacterium; (−) indicates no phage; 4D and 9C are TOX$^+$ phages.
[b,c] C$_2$ and C$_3$ are ADP-ribosyltransferases.
[d] Reinfection with TOX$^+$ phages from neurotoxigenic type C strains 165, 571, 461, Stockholm, or 468C also induced neurotoxin production.

Japan, and the United States, all except 2 produced C2 toxin (Table 12.3). The C2 toxin titer of 3 of the type C strains was very low, and the toxin was detectable only when the strains were cultured in cellophane tubes bathed in TYG broth, a procedure frequently used to detect the production of low levels of a toxin. In comparison, only one of the five type D strains produced C2 toxin.

Antisera were also produced against the C2 toxin from the phage-sensitive nonneurotoxigenic derivatives from both types C and D. These antisera neutralized the C2 toxin produced by strains of both types C and D, but they did not neutralize the neurotoxins produced by *C. botulinum* types A through G. Likewise, antiserum produced against neurotoxins types A–G (with the exception of one type C antiserum that contained the C2 component) did not neutralize the C2 toxin.

All the cultures except one continued to produce the C2 toxin following multiple transfers in laboratory media. When type C strain 164 was cured of its TOX$^+$ phage, it ceased to produce C1 neurotoxin but continued to produce C2 toxin. Of the different isolates tested, one also ceased to produce C2 toxin. When this isolate (164NT) was reinfected with the TOX$^+$

phage, it again produced only the C1 neurotoxin. Numerous unsuccessful attempts have been made to induce the production of C2 toxin by reinfection with either or both the TOX$^+$ and TOX$^-$ phages carried by the parent culture (M.W. Eklund, unpublished observation). We also studied the sporulation of strain 164 and the derivative 164NT, because Nakamura et al. (1978) reported that the C2 toxin was related to the sporulation of the culture and that titers of C2 toxin correlated with the percentage of cells forming spores. Initially, only 10% of the cells of 164NT produced spores. In comparison, approximately 35% of the cells from the parent strain 164 formed spores, suggesting that the low sporulation rate of strain 164NT might be the governing factor for C2 toxin production. The rate of sporulation of strain 164NT was therefore increased by heating the sporulated cultures at 70°C for 20 min to eliminate the vegetative cells, and the spores were subcultured again in MEMM medium. This process was continued until more than 40% of the cells formed spores. When this culture was tested for C2 toxin production, the supernatant fluid remained nontoxigenic before and after trypsin treatment. Absence of C2 toxin was also confirmed by ammonium sulfate precipitation of culture supernatants and ADP ribosylation assay (M.W. Eklund, unpublished observation). On the basis of these experiments, the production of C2 toxin by strain 164 appears to be unrelated to sporulation and to any infectious phage, although we cannot rule out the role of a defective phage or of a plasmid.

Relationship of phages to production of ribosyltransferase C3 by *C. botulinum* type C and D strains

Various research groups have reported that botulinal neurotoxins C1 and D also possess ADP-ribosyltransferase activity modifying 21- to 24-kDa proteins. Subsequent studies, however, indicated that the ADP-ribosylation activity was not caused by botulinal neurotoxins but by the action of another ADP-ribosyltransferase designated as exoenzyme C3. This exoenzyme is distinct from botulinal

neurotoxins and from C2 (Aktories et al., 1987; Rubin et al., 1988).

Type C and D strains were studied to determine the relationship of phages to production of exoenzyme C3. Production of C1 and D neurotoxins was determined by bioassay (Eklund et al., 1971). The C2 toxin was assayed by both bioassay and gel assay for ADP-ribosylation, whereas the production of C3 toxin was determined only by gel assay for ADP-ribosylation.

Bacterial strains producing botulinal C1 or D neurotoxins also produced C3 toxin (see Table 12.2). When the neurotoxigenic strains were cured of their TOX$^+$ phage, they ceased to produce both the C1 or D neurotoxin and C3 exoenzyme. Reinfection of these cured cultures with the TOX$^+$ phage induced the production of C1 and C3 components. Studies have shown that exoenzyme C3 produced by both types C and D are indistinguishable on the basis of gel assay for ADP-ribosylation (Aktories et al., 1987; Rubin et al., 1988).

Several strains of type C and D did not produce C2 toxin (see Table 12.3). This was confirmed using different culture media and also by testing the supernatant fluids (precipitated with 60% ammonium sulfate) by the gel assay for ADP-ribosylation. These strains produced only the C1 or D neurotoxin and C3 exoenzyme. After these bacteria were cured of their TOX$^+$ phages, they simultaneously ceased to produce neurotoxin C1 or D and exoenzyme C3. When these cultures were reinfected with the TOX$^+$ phage, they again produced both neurotoxins C1 or D and exoenzyme C3. The C2 toxin production was not modified during the reported experiments (see Tables 12.2 and 12.3).

The results from these studies show that specific TOX$^+$ phages govern the production of C1 and D neurotoxins and also mediate the production of C3 exoenzyme. Recent studies have shown the DNA sequence of exoenzyme C3 that is encoded by *Clostridium botulinum* type C and D phages (Popoff et al., 1990). On the basis of these data, type C and D can each produce four different combinations of toxins and ADP-ribosyltransferases. Type C can produce (1) C1,

C2, and C3; (2) C1 and C3; (3) C2; or (4) neither neurotoxin or ADP-ribosyltransferases. Similarly, type D can produce combinations of (1) D, C2, and C3; (2) D and C3; (3) C2; or (4) neither neurotoxin or ADP-ribosyltransferases.

Interconversion of *Clostridium botulinum* types C and D by phages and effect on ADP-ribosyltransferases

It was shown in the previous sections that neurotoxigenic type C and D strains are differentiated by the type of neurotoxin produced. This identity, however, was lost when the cultures were cured of their TOX^+ phages. In addition, the C2 toxin produced by type C and D strains was neutralized by C2 antiserum produced against the C2 toxin from both type C and D strains or by gel assay for ADP-ribosylation. Likewise, exoenzyme C3 produced by both type C and D strains cannot be differentiated by gel assay.

When phage-sensitive, nonneurotoxigenic derivatives of types C and D were tested for their sensitivity to the different phages of both types, some type D strains were sensitive to type C phages and vice versa. Interconversion of types C and D were observed in cultures producing C2 toxin and in cultures that did not produce C2 (Eklund and Poysky, 1974). Examples of interconversion of types C and D are shown in Table 12.4. At the beginning, type C strains produced C1 neurotoxin and ADP-ribosyltransferases C2 and C3, and type D strains produced D neurotoxin and ADP-ribosyltransferases C2 and C3. When cultures were cured of their TOX^+ phages, they both produced C2 toxin only. These nonneurotoxigenic isolates from type C and D then became sensitive to the TOX^+ phage from the parent strain and also to the TOX^+ phages of the opposite type, and they produced the corresponding neurotoxin together with the C3 component. These cultures were then cured of their phages and reinfected with the heterologous phage. In every case, the type of neurotoxin produced corresponded to the specific TOX^+ phage carried. In addition, both the type C and

Table 12.4 Interconversion of *Clostridium botulinum* types C and D and effect on ADP-ribosyltransferases and type of neurotoxin produced

Bacterial strain and phage[a]	Culture produced		
	Type of neurotoxin	C_2	C_3
Parent 153 (4C)	C_1	+	+
Cured 153—HS15 (−)	−	+	−
Reinfected 153—HS15 (2D)	D	+	+
Reinfected 153—HS15 (4C)	C_1	+	+
Parent 1873 (2D)	D	+	+
Cured 1873—AOA113 (−)	−	+	−
Reinfected 1873—AOA113 (4C)	C_1	+	+
Reinfected 1873—AOA113 (2D)	D	+	+

[a] (4C) refers to TOX^+ phage from type C strain 153; (2D) refers to TOX^+ phage from type D strain 1873; (−) indicates culture cured of phages.

D phages mediated the production of C3 exoenzyme but not of C2 toxin in all hosts.

Relationship of phages to hemagglutinin production by *C. botulinum* types C and D

Types C1 and D neurotoxins exhibit hemagglutinin (HA) and neurotoxic activity. These are different proteins and act independently under alkaline conditions. The neurotoxin has a molecular weight of approximately 150,000 and forms a complex with HA with molecular weights of 300,000 to 500,000 in acid solutions.

Oguma et al. (1976) studied the relationship of phages to the production of HA in neurotoxigenic and nonneurotoxigenic strains of types C and D. HA was produced by all three of the neurotoxigenic type C strains and one of the two neurotoxigenic type D strains. All the nonneurotoxigenic type C and D strains except C-n71 were HA negative. TOX^+ phages from type C strains and the South African type D strains induced both neurotoxin and HA production in the phage-sensitive derivatives. TOX^+ phages from the HA-negative type D strain 1873, however, induced only the production of neurotoxin type D. TOX[+] phage C-n71 isolated from a nonneurotoxigenic type C strain induced the production of HA but not neurotoxin. These results indicated that HA production can be transmitted by specific phages either separately or concomitantly with neurotoxin production.

12.4 Relationship of Plasmids to Neurotoxin and Bacteriocin Production in *Clostridium botulinum* Type G

Strains of *Clostridium botulinum* types A through G have been shown to harbor plasmids ranging in mass from 2.1 to 81 MDa (Scott and Duncan, 1978; Strom et al., 1984; Weickert et al., 1986). In each of these previous reports, no phenotypic function was ascribed to any of the plasmids.

Isolation of nonneurotoxigenic derivatives from type G

Nonneurotoxigenic derivatives were isolated from six strains of type G following daily transfers of TYG broth at 44°C, their maximum temperature of growth. Derivatives were isolated from two of the strains after only two transfers at the elevated temperatures. The toxicity of the other four strains was more stable, and 20 to 55 transfers were required before the cultures yielded nonneurotoxigenic cells. All the derivatives were transferred more than 16 times in cooked meat medium at 30°C during a 1.5-year period, and none of the isolates reverted to the parent phenotype of toxin production.

Characteristics of neurotoxigenic and nonneurotoxigenic type G cultures

All the nonneurotoxigenic isolates were tested for their sensitivity to phages of the parent strains using the agar layer procedure. Supernatant fluids from the toxigenic parent grown in TYG or CM for 6 to 18 h produced faint clearings in the confluent lawns of all nonneurotoxigenic cultures. The intensity of these clearings markedly increased when the supernatant fluid from 3-day-old cultures was used. Subsequent studies showed that these clearings were caused by bacteriocins and not bacteriophages.

Attempts were made to induce neurotoxicity by adding bacteria-free filtrates from neurotoxigenic cultures to nonneurotoxigenic cultures in the different phases of growth. Toxin production was not induced in any of the nonneurotoxigenic cultures. The results indicated that cultures concomitantly ceased to produce type G neurotoxin and to lose their immunity and ability to produce bacteriocins and that phages did not play a role in the loss of these characteristics. The neurotoxigenic strains and their derivatives were identical in all characteristics, with the exception that type G neurotoxins and bacteriocins were produced by the neurotoxigenic cultures and these characteristics were lost by the nonneurotoxigenic isolates.

Plasmids and their relationship to neurotoxin and bacteriocin production in type G

The relationship of plasmids to the production of neurotoxin and bacteriocin was studied using both nonneurotoxigenic and neurotoxigenic isolates. The cells of each isolate were lysed, extracted for plasmids, and electrophoresed in 0.7% agarose vertical slab gels in TENA buffer (Strom et al., 1984). All neurotoxic isolates continued to harbor the 81-MDa plasmid and to produce neurotoxin and bacteriocin. In contrast, the ability to produce neurotoxin and bacteriocin and the 81-MDa plasmid were concomitantly lost in all the nonneurotoxic isolates.

Bacteriocin production has previously been described in nonneurotoxigenic type E-like organisms. Kautter et al. (1966) designated this bacteriocin as boticin E. The bacteriocin from type G was therefore designated as boticin G (Eklund et al., 1988). In earlier studies, Lewis et al. (1981) and Dowell and Dezfulian (1981) reported that the only difference between *C. botulinum* type G and *C. subterminale* was the ability of type G to produce neurotoxin. In the studies of Eklund et al. (1988), the biochemical and metabolic end products produced by both neurotoxigenic and nonneurotoxigenic isolates and *C. subterminale* were the same. In recent studies on DNA homology, all strains of *C. botulinum* type G, two strains of *C. subterminale*, and one strain of *C. hastiforme* constituted one hybridization group (Suen et al., 1988). These authors proposed a new species , *Clostridium argentinense*, for this genotypically homogenous group.

12.5 Role of Phages in the Stability of Toxin Production by *Clostridium novyi* Types A and B

The main characteristic that unites types A and B is the production of the lethal alpha toxin that is neutralized by either type A or B antiserum. In contrast, types B and D are differentiated mainly in that only type B cultures produce alpha toxin (Table 12.5).

Phages and alpha toxin production by type A

The involvement of phages in the toxigenicity of *C. novyi* type A was studied by Eklund et al. (1976). In these studies, strain 5771 was permitted to sporulate in cooked meat medium and the spores were heated at 70°C for 15 min to inactivate any exogenous phage. Colonies isolated on TYG agar were tested for sensitivity to the phages of the parent culture.

Of the isolates selected, 6 simultaneously ceased to produce alpha toxin, lost their immunity, and became sensitive to the phages harbored by the toxigenic parent. The other 94 isolates continued to produce alpha toxin and were not sensitive to phages. The filtrates of supernatants from these 94 isolates, however, contained phages capable of infecting the 6 nontoxigenic derivatives.

All the nontoxigenic isolates continued to harbor a TOX⁻ phage identified as NA2, and repeated efforts to cure these bacteria of phage NA2 were unsuccessful. No evidence

Table 12.5 Biologically active substances produced by *Clostridium novyi* types A, B, C, and D[a]

Biological substance	Activity	Produced by types			
		A	B	C	D
Alpha	Necrotising, lethal	+	+	−	−
Beta	Necrotizing, lethal lecithinolytic, hemolytic	−	+	−	+
Gamma	Necrotizing, hemolytic, lecithinolytic	+	−	−	−
Delta	Oxygen labile, hemolytic	+	−	−	−
Epsilon	Lipolytic, produces pearly layer	+	−	−	−
Zeta	Hemolytic	−	+	−	−
Eta	Tropomyosinase	−	+	−	+
Theta	Lipase	−	Trace	−	+

Modified from Smith and Holdeman (1968).

was found to suggest a role of phage NA2 in alpha toxin production. The lack of involvement in alpha toxin production was also demonstrated when phage NA2 infected a nontoxigenic derivative of *C. botulinum* type C. (See later section of this chapter.) In subsequent studies, the six nontoxigenic derivatives were shown to be sensitive to only the TOX⁺ phage identified as NA1. When these cultures were infected with phages NA1 and isolated colonies were cultured, all the cultures were immune to phage NA1 and produced phage NA1 and alpha toxin (Table 12.6).

To determine whether the continued participation of phage NA1 was necessary to maintain the toxigenic characteristic, the nontoxigenic isolates infected with phage NA1 were permitted to sporulate and the spores were washed and heated to 70°C for 15 min to eliminate any free phages. When these isolates were tested, the cultures invariably either carried the TOX⁺ phage and produced alpha toxin or they concomitantly lost their TOX⁺ phage and ceased to produce alpha toxin (Eklund et al., 1976).

The relationship of phage to alpha toxin production was expanded to other strains by Schallehn et al. (1980). In these studies, five

Table 12.6 Relationship of specific phages to alpha toxin and epsilon antigen production by *Clostridium novyi* type A

Bacterial strain and phage[a]	Alpha toxin	Epsilon antigen
Parent 5771 (NA1, NA2)	+	+
Cured 5771—HS10 (NA2)	−	+
Reinfected 5771—HS10 (NA1, NA2)	+	+

[a] Phages carried by bacterium are listed inside parentheses; (−) indicates no phage. Phage NA1 is a TOX+ phage; phage NA2 is a TOX− phage.

Table 12.7 Relationship of specific phages to alpha toxin production of *Clostridium novyi* type B

Bacterial strain and phage[a]	Alpha toxin	Beta toxin
Parent 8024 (NBI)	+	+
Cured 8024—AO26 (−)	−	+
Reinfected 8024—AO26 (NBI)	+	+
Reinfected 8024—AO26 (TOX+)[b]	+	+
Reinfected 8024—AO26 (TOX−)[c]	−	+

[a] Phages carried by bacterium are listed inside parenthesis; (−) indicates no phage. NBI is a TOX+ phage.
[b] TOX+ phages from type B strains Kz391, 190, Kz394, Kz395, and Kz396.
[c] TOX− phages from type B strains Kz391, Kz395, and Kz396.

different phage-sensitive, nontoxigenic derivatives isolated from different strains of type A were studied. These nontoxigenic derivatives were tested for their sensitivity to the phage from the toxigenic parent culture and also for cross-sensitivity to phages from other toxigenic strains. Each of the five nontoxigenic strains were sensitive to the phages carried by their parent culture, and three of them were sensitive to phages from other toxigenic type A strains. In all cases, nontoxigenic derivatives were converted to alpha toxin production by the TOX+ phages.

In both studies, the TOX+ phages were very unstable in culture supernatant fluids even during overnight storage at 5° or 25°C. Numerous attempts to stabilize these phages by varying cultural conditions or by filtration in a nitrogen atmosphere were unsuccessful. All experiments therefore were performed the same day that the filtrates containing the phages were prepared.

Phages and alpha toxin production by type B

Type B strains produce the lethal alpha and beta toxins. For simplicity, cultures are referred to as nonalphatoxigenic or alphatoxigenic, depending on whether they produced alpha toxin. Nonalphatoxigenic, phage-sensitive derivatives could not be isolated from heated spores of type B strain 8024. When this strain was cultured in TYG medium containing acridine orange, 2 of the 100 isolates, designated as AO26 and AO52, simultaneously ceased to

produce alpha toxin and became sensitive to the TOX+ phage NB1 from the alphatoxigenic parent strain 8024; nevertheless, supernatant fluids from 2- and 3-day-old cultures contained levels of the lethal beta toxin. The remaining cultures were immune to phage NB1 and also continued to produce both alpha and beta toxins; strains AO52 and AO26 ceased to produce alpha toxin. Because type A strains produce alpha toxin but not beta toxin, antiserum against type A alpha toxin was used to identify alpha toxin production by type B. Type B and D (*Clostridium haemolyticum*) antisera, which contain the beta component, were used to neutralize and identify the beta toxin.

Nonalphatoxigenic derivatives AO52 and AO26 were reinfected with the phages from the parent strain 8024 and also with phages produced by five different type B strains, Both the isolates were sensitive to the TOX+ phages carried by the parent and the other five strains. All the TOX+ phages converted the nonalphatoxigenic derivatives to the alphatoxigenic state, and the alpha toxin was neutralized by both type A and B antisera (Table 12.7). TOX− phages from three of the different alphatoxigenic strains also infected AO52 and AO26, but they did not induce alpha toxin production (Table 12.7).

These curing and reinfection experiments were repeated several times and in each case the cultures followed the typical pattern of either (1) being phage sensitive and nonalphatoxigenic or (2) carrying the TOX$^+$ phage and producing alpha toxin (Eklund et al., 1976). Strains AO52 and AO26 were also tested for their sensitivity to phages from six different stype D strains. Infective phages were not demonstrated in any type D strains.

In 1975, Smith reported that deep agar colonies of *C. novyi* type B may vary from lenticular to filamentous. In the current study, filamentous colonies were observed when strains AO52 or AO26 were infected with TOX$^+$ phage NB1. Occasionally these colonies also showed evidence of "nibbling," an effect probably caused by the phage. Colonies displayed the lenticular appearance when they were cured of the TOX$^+$ phage. The TOX$^-$ phages did not have any effect on the changes in colony morphology.

This phenotypic alternation in the colonies was used to advantage in the isolation of nonalphatoxigenic derivatives. Alphatoxigenic strain 8024 was cultured in TYG broth containing antiserum against TOX$^+$ phage NB1 to determine whether phages could be isolated. After the sixth daily transfer, the cultures were plated on TYG agar. Of the 194 colonies examined, 175 were of the filamentous type and 19 were of the lenticular type. Only the lenticular colonies lost their immunity to phage NB1 and also ceased to produce alpha toxin. These data indicated that a pseudolysogenic relationship exists between phage NB1 and bacterial host 8024 in a similar manner observed with *C. botulinum* types C and D.

Conversion of *Clostridium novyi* type D to C. *C. novyi* type A by phages of type A

It was recently suggested that *C. haemolyticum* be included in the *C. novyi* species and be identified as type D because the major lethal toxin is identical with the beta toxin of *C. novyi* type B. In fact, Oakley and Warrack (1959) questioned whether *C. haemolyticum* is a member of type B or a separate type. In the previous section of this chapter, the relationship of phages to production of alpha toxin in *C. novyi* types A and B

Table 12.8 Conversion of *Clostridium novyi* type D to alpha toxin production by phage isolated from type A[a]

Bacterial strain and phage	Alpha toxin	Beta toxin
Type D		
Kz410	–	+
Kz 410 (P761)[b]	+	+
Type D		
Kz410 – AO12 (–)	–	+
Kz410 – AO12 (PSM40)[c]	+	+

[a] Modified from Schallehn and Eklund (1980).
[b] TOX$^+$ phage from alphatoxigenic *C. novyi* type A strain 761 shown in parentheses
[c] TOX$^+$ phage from alphatoxigenic *C. novyi* type A strain SM40 shown in parentheses.

was documented. When the type B strains ceased to produce alpha toxin, they closely resembled type D strains in that they produced the beta toxin.

To determine whether the type D strains could be induced to produce alpha toxin, type D strains were tested for their sensitivity to phages from 28 strains of type A. Of the 12 strains of type D tested, 6 were sensitive to four different phages of type A. These phages were able to induce alpha toxin production in at least one of the type D strains tested (examples shown in Table 12.8). Each of the alphatoxigenic type D strains was subcultured several times, and they remained stable in their ability to produce alpha toxin as well as beta toxin (Schallehn and Eklund, 1980). These converted strains of type D are indistinguishable from type B in regard to their main toxins produced. The type D strains, however, produced larger amounts of beta toxin than type B. The possibility of TOX$^+$ and TOX$^-$ phages inducing lysis and therefore releasing larger quantities of beta toxin was not established in these studies. These differences in the titer of beta toxin produced by type B and D strains agree with the results of Rutter and Collee (1969) and Nakamura et al. (1975). Phages and alpha toxin production therefore are the main characteristics that unite types A and B and differentiate types B and D.

The species *C. novyi* also includes type C strains that do not produce any of eight different soluble antigens produced by types A, B, and D. The origin and relationship of type C to other *C. novyi* types is therefore of interest. Strains of type C were not available for the studies reported in this chapter.

12.6   Interrelationship of *Clostridium botulinum* Types C and D to *Clostridium novyi* Type A

Holdeman and Brooks (1968) were the first to indicate that strains of *C. botulinum* types C and D shared some of the biochemical and physiological characteristics possessed by *C. novyi* type A. The main difference between the species was the toxins produced. In the previous section of this chapter, it was shown that alphatoxigenic strains of *C. novyi* types A and B ceased to produce alpha toxin when they were cured of their TOX$^+$ phage. The nonalphatoxigenic strains of type A then lost their identity to *C. novyi* and became biochemically and physiologically closely related to the nonneurotoxigenic strains of *C. botulinum* types C and D. Nontoxic isolates from these organisms produced the pearly layer (epsilon antigen) on egg yolk agar. This lipase activity is produced only by *C. botulinum* (with the exception of type G) and closely related organisms.

Conversion of *Clostridium botulinum* type C strain to *C. botulinum* type D and *C. novyi* type A by specific phages

The relationship of *C. novyi* type A and *C. botulinum* types C and D was also studied by testing the sensitivity of the various cured cultures to the phages of the toxic parents from both species. Nonneurotoxigenic derivatives HS37 isolated from *C. botulinum* type C strain 162 became sensitive not only to the phages of the parent type cultures but also to phages from *C. botulinum* type D and *C. novyi* type A (Eklund et al., 1974). Infection of strain H537 with phage

3C from type C strain 162 induced the production of botulinal type C1 neurotoxin and exoenzyme C3. When strain HS37 was infected with phage 1D from *C. botulinum* type D, type D neurotoxin and exoenzyme C3 were produced. *C. novyi* type A strain carried two phages. The TOX$^-$ phage NA2 had no effect on the toxigenicity of strain HS37, but TOX$^+$ phage NA1 induced the production of the alpha toxin of *C. novyi*. The alpha toxin was neutralized by both *C. novyi* type A and B antisera, but not by any of the antiserum to *C. botulinum* neurotoxins. None of these cultures produced botulinal C2 ribosyltransferase.

These studies suggest that in nature a common bacterial culture such as strain HS37 could be responsible for gas gangrene or botulism type C and D outbreaks. The specific phage that infects this common bacterial strain would therefore govern not only the toxigenicity but also the resulting disease.

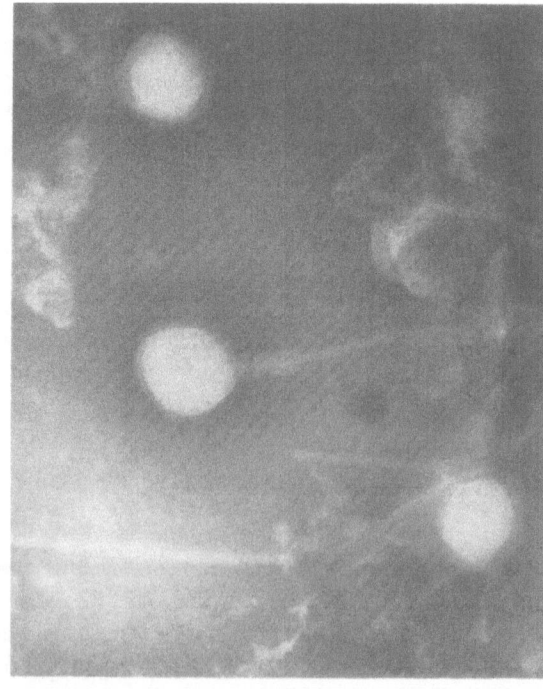

FIGURE 12.1   TOX$^+$ phage from lysates of *Clostridium botulinum* type C strain Seddon 460, which induces neurotoxin type C1 and ribosyltransferase C3 production (×177,000).

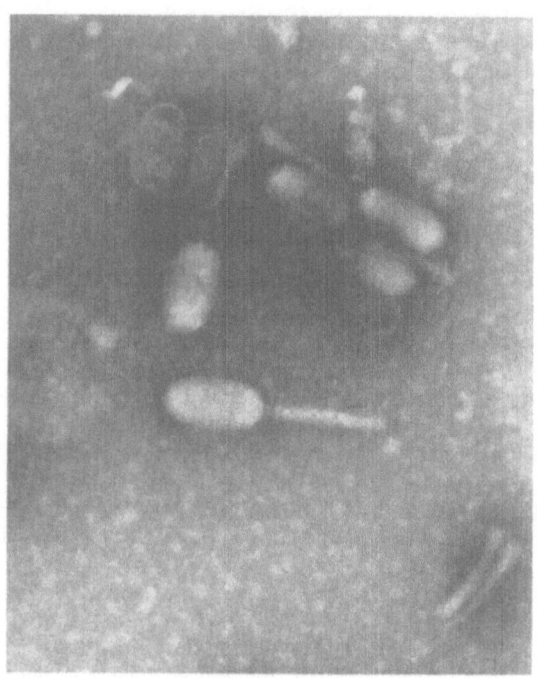

FIGURE 12.2 TOX⁺ phage from lysates of *Clostridium novyi* type B, which induces alpha toxin production (×150,000).

12.7 Morphology of TOX⁺ Phages

TOX⁺ phages inducing the production of neurotoxins in *Clostridium botulinum* types C and D and alpha toxin in *C. novyi* type A were of similar morphology. The phages exhibited polyhedral heads ranging from 65 to 100 nm in diameter and tails as long as 455 nm. In each case, the tails were surrounded by a sheath. An example of a TOX⁺ phage from *C. botulinum* type C is shown in Figure 12.1.

TOX⁺ phages inducing the production of alpha toxin in *C. novyi* type B strain 8024 differed markedly from the phages of *C. novyi* type A and *C. botulinum* types C and D. This phage exhibited an elongated head 65 nm wide and 125 nm long (Figure 12.2). The tail of the phage was 7 nm wide and 140 nm long and was surrounded by a sheath 130 nm long that covered most of the tail.

12.8 Relationship of Phages and Plasmids to Diseases and the Identification of *Clostridium botulinum*, *Clostridium novyi*, and Closely Related Organisms

Phages and plasmids have been shown to mediate the production of a number of biologically active agents. Because of this role, phages and plasmids are very important in the diseases that a bacterium can cause.

Animals and humans differ markedly in their sensitivity to the different types of botulinal neurotoxins (Smith, 1977; Smith and Sugiyama, 1988). When bacterial strains of types C and D are cured of their TOX⁺ phage, they can become sensitive to both type C and D TOX⁺ phages, and upon reinfection the bacterium will produce type C1 and D neurotoxins, respectively. The factors that determine whether a botulism outbreak will occur from types and C and D therefore depends on the sensitivity of the specific animal to the type of neurotoxin produced by the bacterium, which in turn is governed by phages. Phage-sensitive non-neurotoxigenic strains of type C have also been shown to be induced to produce the lethal alpha toxin of *C. novyi*. This bacterium and its specific TOX⁺ phage therefore determine whether the victim will develop botulism type C or D or gas gangrene.

Similar observations can be made in the *C. novyi* species. TOX⁺ phages determine whether or not alpha toxin is produced and which disease the bacterium will cause. In the case of type B, the organism can cause illness and death from either alpha toxin or beta toxin or from a combination of the two toxins.

The recent genetic studies with *C. botulinum* and *C. novyi* also emphasized the importance of phages and plasmids in the identification and classification of these pathogens. *C. botulinum* type C for many years has been recognized as two subtypes, C_{alpha} and C_{beta} (Jansen, 1971; Smith, 1977). The recent data presented in this chapter indicate that the toxins produced by different type C strains depend on the specific TOX⁺ phage that the strain carries. Type C strains can produce C1 neurotoxin, ADP-ribosyltransferases C2 and C3, *C. novyi* alpha toxin, and combinations of these toxins. *C. botulinum* type D is classified only as type D. Recent data, however, indicate type D strains can also produce D neurotoxin, C2 and C3 ribosyltransfersases, or combinations of these botulinal toxins and also the *C. novyi* alpha toxin. The current classification of these strains as C_{alpha} and C_{beta} and as type D therefore is inadequate and should be revised.

The identification of *C. botulinum* type G also is influenced by the 81 MDa plasmid that the bacterium causes. When they are cured of the 81-MDa plasmid, they cease to produce type G neurotoxin and the nonneurotoxigenic derivatives become indistinguishable from *C. subterminale* (Eklund et al., 1988).

Similar observations have been made in the classification of *C. novyi*. A type A bacterium cured of its phages ceases to produce alpha toxin and therefore resembles nonneurotoxigenic *C. botulinum* type C and D strains. Type B strains carrying the TOX$^+$ phage produce alpha toxin, but when they lose this phage they become closely related to *C. novyi* type D. In addition, type D strains can be infected with phages from type A, which induces the production of alpha toxin. This changes the identification of type D to type B.

The results of these genetic studies will undoubtedly by expanded and extended to other types of *C. botulinum* and other pathogenic clostridia. The availability of different bacterial strains that produce only one type of toxin opens new frontiers for scientists to study specific toxins and the effects of these individual or combinations of toxins in different human and animal diseases. The availability of specific cultures will also facilitate the production of specific antisera and improved toxoids by research and commercial laboratories.

Acknowledgments. These studies were supported in part by USAMRIID Contract No. 84PP4861(A-l).

References

Adams, M.H. 1959. *Bacteriophages.* New York: Wiley.

Aktories, K., S. Rösener, U. Blaschke, and G.S. Chhatwal. 1988. Botulinum ADP-ribosyl-

tranferase C3. Purification of the enzyme and characterization of the ADP-ribosylation reaction in platelet membranes. *Eur. J. Biochem.* **172**: 445–50.

Aktories, K., U. Weller, and G.S. Chhatwal. 1987. *Clostridium botulinum* type C produces a novel ADP-ribosyltransferase distinct from C2 toxin. *Fed. Eur. Biochem. Soc. Lett.* **212**:109–113.

Aureli, P., L. Fenicia, B. Pasolini, M. Gionfranceschi, L.M. McCroskey, and C.L. Hatheway. 1986. Two cases of type E infant botulism in Italy caused by neurotoxigenic *Clostridium butyricum. J. Infect. Dis.* **154**:207–211.

Betty, I., D. Buntain, and P.D. Waller. 1964. *Clostridium oedematiens*: a cause of sudden death in sheep, cattle, and pigs. *Vet. Rec.* **76**:1115–1116.

Clowes, R C. 1972. Molecular structure of bacterial plasmids. *Bacteriol. Rev.* **36**:361–405.

Dolman, C.E. and E. Chang. 1972. Bacteriophages of *Clostridium botulinum. Can. J. Microbiol.* **18**:67–76.

Dowell, V. and M. Dezfulian. 1981. Physiological characteristics of *Clostridium botulinum* and development of practical isolation and identification procedures. *In: Biomedical Aspects of Botulism,* G. Lewis, ed., pp. 205–216. New York, Academic Press.

Eklund, M.W., and F.T. Poysky. 1972. Activation of a toxic component of *Clostridium botulinum* types C and D by trypsin. *Appl. Microbiol.* **24**:108–113.

Eklund, M.W., and F.T. Poysky. 1974. Interconversion of type C and D strains of *Clostridium botulinum* by specific bacteriophages. *Appl. Microbiol.* **27**:251–258.

Eklund, M.W., F.T. Poysky, and E.S. Boatman. 1969. Bacteriophages of *Clostridium botulinum* types A, B, E, and F and nontoxigenic strains resembling type E. *J. Virol.* **3**:270–274.

Eklund, M.W., F.T. Poysky, and S.M. Reed. 1972. Bacteriophage and the toxigenicity of *Clostridium botulinum* type D. *Nature (London) New Biology* **235**:16–17.

Eklund, M.W., F.T. Poysky, S.M. Reed, and C.A. Smith. 1971. Bacteriophages and toxigenicity of *Clostridium botulinum* type C. *Science* **172**:480–482.

Eklund, M.W., F.T. Poysky, J.A. Meyers, and G.A. Pelroy. 1974. Interspecies conversion of *Clostridium botulinum* type C to *Clostridium novyi* type A by bacteriophage. *Science* **186**:456–458.

Eklund, M.W., F.T. Poysky, M.E. Peterson, and J.A. Meyers. 1976. Relationship of bacteriophages to alpha toxin production in *Clostridium novyi* types A and B. *Infect. Immun.* **14**:793–803.

Eklund, M.W., F.T. Poysky, L.M. Mseitif, and M.S. Strom. 1988. Evidence for plasmid-mediated toxin and bacteriocin production in *Clostridium botulinum* type G. *Appl. Environ. Microbiol.* **54**:1405–1408.

Fujii, M., K. Oguma, N. Yokosawa, K. Kimura, and K. Tsuzuki. 1988. Characteristics of bacteriophage nucleic acids obtained from *Clostridium botulinum* types C and D. *Appl. Environ. Microbiol.* **54**:69–73.

Gimenez, D. F. 1984. *Clostridium botulinum* subtype B$_a$. *Zentralbl. Bakteriol. Hyg.* A**257**:68–72.

Gimenez, D.F., and A.S. Ciccarelli. 1978. New strains of *Clostridium botulinum* subtype Af. *Zentralbl. Bakteriol. Parasitenk. Infektionskr. Hyg. 1 Orig.* **240**:215–220.

Hall, J.D., L.M. McCroskey, B.J. Pincomb, and C.L. Hatheway. 1985. Isolation of an organism which produces type E botulinal toxin for an infant with botulism. *J. Clin. Microbiol.* **21**:654–655.

Hariharan, H. and W.R. Mitchell. 1976. Observations on bacteriophages of *Clostridium botulinum* type C isolates from different sources and the role of certain phages in toxigenicity. *Appl. Environ. Microbiol.* **32**:145–158.

Hauser, D., M.W. Eklund, H. Kurazono, T. Binz, H. Niemann, D.M. Gill, P. Boquet, and M.R. Popoff. 1990. Nucleotide sequence of *Clostridium botulinum* C neurotoxin. *Nucleic Acids Res.* **18**:4924.

Holdeman, L.V., and J.B. Brooks. 1968. Variation among strains of *Clostridium botulinum* and related clostridia. In: *Proceedings of the First U.S.-Japan Conference on Toxic Microorganisms*, A. Herzberg, ed., pp. 278–286. Washington, DC.: U.S. Government Printing Office.

Inoue, K. and H. Iida. 1968. Bacteriophages of *Clostridium botulinum*. *J. Virol.* **2**:537–540.

Inoue, K. and H. Iida. 1970. Conversion to toxigenicity in *Clostridium botulinum* type C. *Jpn. J. Microbiol.* **14**:87–89.

Inoue, K. and J. Iida. 1971. Phage conversion of toxigenicity in *Clostridium botulinum* types C and D. *Jpn. J. Med. Sci.* **24**:53–56.

Jansen, B.C. 1971. The toxic antigenic factors produced by *Clostridium botulinum* types C and D. *Onderstepoort J. Vet. Res.* **38**:93–98.

Kautter, D.A., S.M. Harmon, R.K. Lynt, Jr., and T. Lilly, Jr. 1966. Antagonistic effect of *Clostridium botulinum* type E by organisms resembling it. *Appl. Microbiol.* **14**:616–622.

Lewis, G.E., S.S. Kalinski, D.W. Reichard, and J.F. Metzer. 1981. Detection of *Clostridium botulinum* type G toxin by enzyme-linked immunosorbent assay. *Appl. Environ. Microbiol.* **42**:1018–1022.

McCroskey, L.M., and C.L. Hatheway. 1984. Atypical strains of *Clostridium botulinum* isolated from specimens in infant botulism cases. In: *Abstracts*, Annual Meeting of the American Society for Microbiology, C159, p. 263. Washington, D.C.: American Society for Microbiology.

McCroskey, L.M., C.L. Hatheway, L. Fenicia, B. Pasolini, and P. Aureli. 1986. Characterization of an organism that produces type E botulinal toxin that resembles *Clostridium butyricum* from feces of an infant with type E botulism. *J. Clin. Microbiol.* **23**:201–202.

Nakamura, S., T. Serikawa, K. Yamakawa, S. Nishida, S. Kozaki, G. Sakaguchi. 1978. Sporulation and C2 toxin production by *Clostridium botulinum* type C strains producing no C1 toxin. *Microbiol. Immunol.* **22**:591–596.

Nakamura, S., K. Takematsu, and S. Nishida. 1975. Susceptibility to Mitomycin C and lecithinase activities of *Clostridium oedematiens* (*C. novyi*) types B and D. *J. Med. Microbiol.* **8**:289–297.

Nakane, A., K. Oguma, M. Shiozoka, and H. Iida. 1978. Production of trypsin activable toxin components by *Clostridium botulinum* types C and D. *Jpn. J. Med. Sci.* **31**:166–169.

Oakley, C.L. and G.H. Warrack. 1959. The soluble antigens of *Clostridium oedematiens* type D (*Cl. haemolyticum*). *J. Pathol. Bacteriol.* **78**:543–551.

Oguma, K., H. Iida, and K. Inoue. 1973. Bacteriophage and toxigenicity of *Clostridium botulinum*: An additional evidence for phage conversion. *Jpn. J. Microbiol.* **17**:425–426.

Oguma, K., M. Shiozaki, and H. Iida. 1976. Phage conversion to hemagglutinin production in *Clostridium botulinum* types C and D. *Infect. Immun.* **14**:597–602.

Ohishi, I. and B.R. Dasgupta. 1987. Molecular structure and biological activities of *Clostridium botulinum* C2 toxin. In: *Avian Botulism: An International Perspective*, M.W. Eklund and V.R. Dowell, Jr., eds., pp. 223–247. Springfield, IL: Thomas.

Ohishi, I., M. Iwasaka, and G. Sakaguchi. 1981. Vascular permeability activity of botulinum C2 toxin elicited by cooperation of two dissimilar protein components. *Infect. Immun.* **31**:890–895.

Popoff, M.R., P. Boquet, D.M. Gill, and M.W. Eklund. 1990. DNA sequence of exoenzyme C3, and ADP ribosyltransferase encoded by *Clostridium botulinum* C and D phages. *Nucleic Acids Res.* **18**:1291.

Poumeyrol, M., J. Billon, F. Delille, C. Haas, A. Marmonier, and M. Sebald. 1983. Intoxication botulique mortelle due a une souche de *Clostridium botulinum* de type AB. *Med. Malad. Infect.* **13**:750–754.

Rubin, E.J., D.M. Gill, P. Boquet, and M.R. Popoff. 1988. Functional modification of a 21-kilodalton G protein when ADP-ribosylated by exoenzyme C3 of *Clostridium botulinum*. *Mol. Cell. Biol.* **8**:418–426.

Rutter, J.M., and J.G. Collee. 1969. Studies on the soluble antigens of *Clostridium oedematiens* (*Cl. novyi*). *J. Med. Microbiol.* **2**:395–417.

Sakaguchi, G., S. Sakaguchi, S. Kozai, and M. Takahashi. 1986. Purification and some properties of *Clostridium botulinum* type AB toxin. *Fed. Eur. Microbiol. Soc. Microbiol. Lett.* **33**:23–29.

Schallehn, G. and M.W. Eklund. 1980. Conversion

of *Clostridium novyi* type D (*C. haemolyticum*) to alpha toxin production by phages of *C. novyi* type A. *Fed. Eur. Microbiol. Soc. Microbiol. Lett.* **7**:83–86.

Schallehn, G., M.W. Eklund, and H. Brandis. 1980. Zur phagenkonversion von *Clostridium novyi* typ A. *Zentralbl. Bakteriol. Hyg. 1 Abt. Orig. A* **247**:95–100.

Scott, V.N., and D.L. Duncan. 1978. Cryptic plasmids in *Clostridium botulinum* and *C. botulinum*-like organisms. *Fed. Eur. Microbiol. Soc. Microbiol. Lett.* **4**:55–58.

Simpson, L.L. 1982. A comparison of pharmacological properties of *Clostridium botulinum* type C1 and C2 toxins. *J. Pharmacol. Exp. Ther.* **223**:695–701.

Simpson, L.L. 1987. The pathophysiological actions of the binary toxin produced by *Clostridium botulinum*. In: *Avian Botulism: An International Perspective*, M.W., Eklund and V.R. Dowell, Jr., eds., pp. 249–264. Springfield, IL: Thomas.

Smith, L.D. 1975. *The Pathogenic Anaerobic Bacteria*. Springfield, IL: Thomas.

Smith, L. D. 1977. *Botulism: The Organism, Its Toxin, the Disease*. Springfield, IL: Thomas.

Smith, L.D., and L.V. Holdeman. 1968. *The Pathogenic Anaerobic Bacteria*. Springfield, IL: Thomas.

Smith, L.D., and H. Sugiyama. 1988. *Botulism*. Springfield, IL: Thomas.

Strom, M.S., M.W. Eklund, and F.T. Poysky. 1984. Plasmids in *Clostridium botulinum* and related species. *Appl. Environ. Microbiol.* **48**:956–963.

Suen, J.C., C.L. Hatheway, A.G. Steigerwalt, and D.J. Brenner. 1988. *Clostridium argentinense* sp. nov.: a genetically homogenous group composed of all strains of *Clostridium botulinum* toxin type G and some nontoxigenic strains previously identified as *Clostridium subterminale* or *Clostridium hastiforme*. *Int. J. Syst. Bacteriol.* **38**:375–381.

Sugiyama, H., D. Mizutani, and K.W. Yang. 1972. Basis of type A and F toxicities of *Clostridium botulinum* strain 84. *Proc. Soc. Exp. Biol. Med.* **191**:1063–1067.

Vinet, G., L. Berthiaume, and V. Fredette. 1968. Un bacteriophage dans une culture de *C. botulinum* C. *Rev. Can. Biol.* **27**:73–74.

Weickert, M.J., G.H. Chambliss, and H. Sugiyama. 1986. Production of toxin by *Clostridium botulinum* type A strains cured of plasmids. *Appl. Environ. Microbiol.* **51**:52–56.

Williams, B.M. 1964. *Clostridium oedematiens* infection (Black Disease and Bacillary Haemoglobinuria) of cattle in Mid-Wales. *Vet. Rec.* **76**:591–596.

Willis, A.T. 1964. *Anaerobic Bacteriology in Clinical Medicine*. Washington, D.C.: Butterworths.

13

Molecular Biology of Clostridial ADP-Ribosyltransferases and Their Substrates

Klaus Aktories, Gertrud Koch, and Ingo Just

13.1 Introduction

Studies in recent years have shown that several bacterial protein toxins affect the eucaryotic organism by their capacity to ADP-ribosylate target proteins of the eucaryotic cell. The underlying pathobiochemical mechanism is the toxin-catalyzed transfer of the ADP-ribose moiety of NAD onto regulatory proteins, which thereby changes the functional properties of the eucaryotic regulators. Examples of this group of toxins are diphtheria toxin, pseudomonas exotoxins A and S, cholera toxin, *Escherichia coli* heat-labile toxin, and pertussis toxin (for review, see Althaus and Richter, 1987; Moss and Vaughan, 1988, 1990).

Diphtheria toxin (Honjo et al., 1968) and *Pseudomonas* exotoxin A (Iglewski and Kabat, 1975) modify the elongation factor 2, a GTP-binding protein, which is involved in the polypeptide chain elongation on ribosomes. Cholera toxin (Cassel and Selinger, 1977; Cassel and Pfeuffer, 1978), *E. coli* heat-labile toxins (Gill and Richardson, 1980; Chang et al., 1987), and pertussis toxin (Bokoch et al., 1983; Ui, 1984) ADP-ribosylate heterotrimeric G proteins (G_s, G_i, G_o, transducin), which participate in transmembrane signal transduction, e.g., in regulation of adenylyl cyclase, cation channels, or photoperception (Pfeuffer and Helmreich, 1988). It has been reported that *Pseudomonas aeruginosa* exotoxin (exoenzyme) S ADP-ribosylates vimentin, Ras, and Ras-like proteins in vitro (Coburn et al., 1989a, 1989b). Whether these observations are of pathophysiological significance has not been clarified. The various ADP-ribosylating toxins have attracted particular attention not only for the study of pathogenetics but also as powerful tools for studying the physiological functions of their target proteins (Table 13.1).

During the past few years, several clostridial toxins have been described that possess ADP-ribosyltransferase function. These clostridial enzymes may be divided into two classes: the first class consists of toxins that catalyze the ADP-ribosylation of actin, and the second class consists of ADP-ribosyltransferases which modify the low molecular weight GTP-binding proteins of the Rho/Rac-family. In this chapter, both classes of clostridial enzymes are described in more detail.

13.2 Clostridial Toxins ADP-Ribosylating Actin

The actin ADP-ribosylating toxins are *Clostridium botulinum* C2 toxin (Aktories et al., 1986a), *C. perfringens* iota toxin (Stiles and Wilkins, 1986a; Simpson et al., 1987; Schering et al., 1988), *C. spiroforme* toxin (Popoff and Boquet, 1988; Simpson et al., 1989), and an ADP-ribosyltransferase produced by *C. difficile* (Popoff et al., 1988). The latter toxin is clearly distinct from *C. difficile* toxins A and B. All these actin ADP-ribosylating toxins are constructed according to the A-B model, i.e., the agents consist of an active enzyme component (A) possessing ADP-ribosyltransferase activity and

195

Table 13.1 ADP-ribosylating toxins and exoenzymes

Toxin or exoenzyme	Protein substrates	Effect
I	*GTP-binding proteins*	
Diphtheria toxin Pseudomonas A	Elongation factor 2	Inhibition of protein synthesis
Cholera toxin E. coli heat-labile toxins Pertussis toxin	G proteins (G_s, G_i, G_o, G_t)	Stimulation and inhibition of signal transduction
Clostridium botulinum C3 ADP-ribosyltransferase	rho (A, B, C); rac (1, 2)	Inactivation of regulatory protein
II	*ATP-binding proteins*	
C. botulinum C2 toxin	Actin (nonmuscle actin, γ- smooth muscle actin)	Inhibition of polymerization
C. perfringens iota toxin	All mammalian actin isoforms	
C. spiroforme toxin and C. difficile ADP-ribosyl- transferase	Nonmuscle actin, other actin isoforms?	
III	*Undefined*	
Pseudomonas exoenzyme S	Various in vitro substrates known, e.g., vimentin, ras-like GTP-binding proteins	Unknown

a binding component (B) that binds to the target cell and transfers the enzyme component into the cell (Ohishi et al., 1980a; Ohishi and Miyake, 1985; Stiles and Wilkins, 1986b; Popoff and Boquet, 1988). In contrast to other A and B toxins, such as diphtheria (Collier and Kandel, 1971), cholera (Gill 1977), and pertussis (Tamura et al. 1982) toxins, the two components of actin ADP-ribosylating toxins are composed of two nonlinked components. Therefore, the actin ADP-ribosylating toxins are true binary agents comparable with leucocidin (Noda et al., 1981) or anthrax toxin (Leppla, 1982).

Clostridium botulinum C2 toxin

C2 toxin is produced by *Clostridium botulinum* type C and D, which also synthesize botulinum neurotoxins C1 and D (Eklund and Poysky, 1972) and ADP-ribosyltransferase C3 (Rubin et al., 1988). Further, strains of *C. botulinum* have been described that produce C2 toxin but no neurotoxins (Eklund and Poysky, 1972; Nakamura et al., 1978). Although the neurotoxins C1 and D, and also exoenzyme C3, are phage encoded (Eklund et al., 1971, 1972; Rubin et al.,

1988; Popoff et al., 1990), C2 toxin is most probably chromosomally encoded. It has been suggested that the synthesis of C2 toxin may be associated with the sporulation of the clostridia because a smaller amount of C2 is synthesized during the vegetative period of the bacteria (Nakamura et al., 1978). The two components of the binary C2 toxin are always synthesized concomitantly. However, the ratio of the binding component to the enzyme component varies considerably with different strains (Ohishi and Okada, 1986).

The binding component of C. botulinum *C2 toxin.* The binding component (C2II) of C2 toxin has a molecular weight of about 100 kDa (Ohishi et al., 1980a, 1980b). For full activity this binding component must be proteolytically activated (Eklund and Poysky, 1972; Ohishi, 1987). Treatment of C2II with trypsin for about 30 min at 37°C releases a peptide of about 72 kDa, as determined by gel filtration, which binds to the eucaryotic cell surface and induces a binding site for the enzyme component (Figure 13.1). It has been reported that the cleaved binding components complex with each other to form pentamers with hemagglutination and hemolytic activity (Ohishi, 1987). Simpson

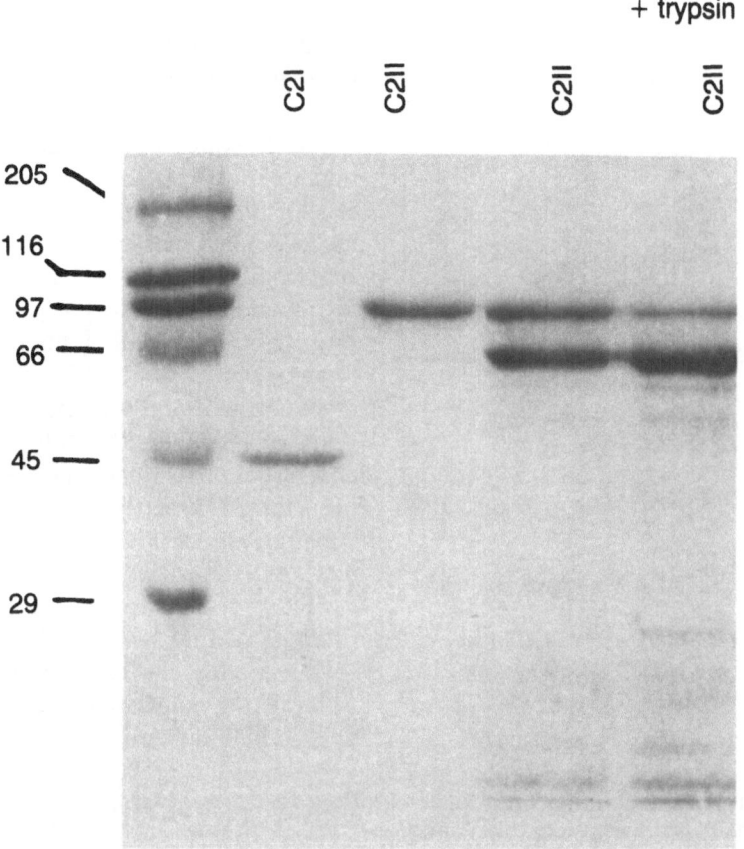

FIGURE 13.1 Components of *Clostridium botulinum* C2 toxin, which consists of component C2I and C2II. Component C2I (M_r, about 45,000) possesses ADP-ribosyltransferase activity. Component C2II is activated by partial proteolysis, which releases active peptide (M_r, about 70,000). C2II is responsible for binding of toxin to eucaryotic cell membrane.

(1989) reported that the entry of botulinum C2 toxin is blocked at low temperatures and in the presence of ammonium hydrochloride or methylamine hydrochloride. These findings were interpreted to indicate that the toxin enters the cell via an endocytotic pathway (Simpson, 1989). At present, the cell-surface receptor of C2 toxin is unknown.

The enzyme component of Clostridium botulinum *C2 toxin.* The enzyme component (C2I) of C2 toxin possesses ADP-ribosyltransferase activity (Simpson, 1984; Aktories et al., 1986b) (Figure 13.2). The substrate of the reaction is monomeric G-actin. In contrast, polymerized filamentous F-actin is apparently not modified

by the toxin (Aktories et al., 1986a, 1986b). Therefore, phalloidin, which induces polymerization of actin, inhibits toxin-induced ADP-ribosylation of actin (Aktories et al., 1986a). C2 toxin modifies actin at arginine-177, and apparently all actin-ribosylating toxins modify actin at this amino acid. This was directly demonstrated for *C. perfringens* iota toxin and C2 toxin using protein chemistry methods (Vandekerckhove et al., 1987, 1988) and for other toxins using the back ADP-ribosylating method, i.e., pretreatment of actin with C2 or iota toxin inhibited the subsequent ADP-ribosylation with other toxins.

Although iota toxin and C2 toxin modify the same amino acid of actin, these toxins differ in

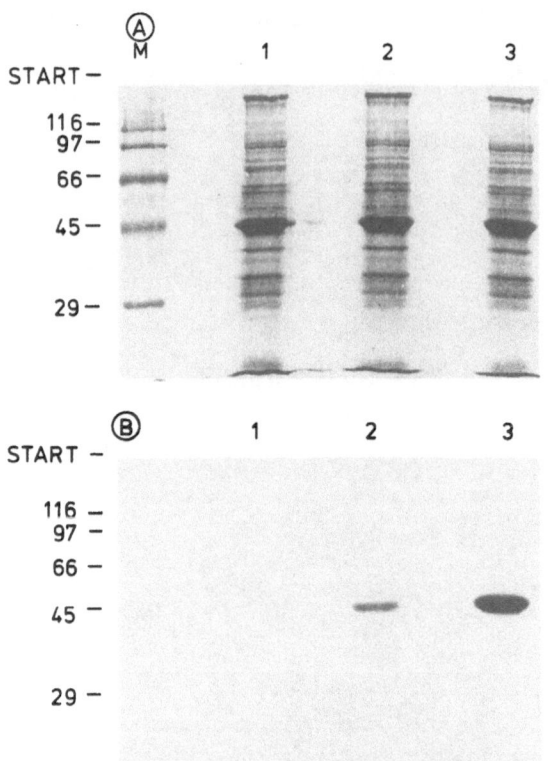

FIGURE 13.2   ADP-ribosylation of actin in platelet cytosol by *Clostridium perfringens* iota toxin and *C. botulinum* C2 toxin. Platelet cytosol was incubated without (lane 1) and with (lane 2) *C. perfringens* iota toxin, or with *C. botulinum* C2 toxin (lane 3), in presence of [^{32}P]NAD for 1 h at 37°C. Labeled proteins were analyzed by sodium dodecyl sulfate-polyacrylamide gel electrophoresis (SDS-PAGE) (A) and autoradiography (B). (Data from Schering et al., 1988.)

acids (Vandekerckhove and Weber, 1979), it has been suggested that this protein domain defines the substrate specificity towards ADP-ribosylation by C2.

The K_m-value of the ADP-ribosylation reaction for NAD is about 5 μM (Aktories et al., 1986a). Similar to other ADP-ribosylating toxins, C2 toxin possesses NAD-glycohydrolase activity and splits NAD into ADP-ribose and nicotinamide (Ohishi, 1986). As observed with other ADP-ribosylating toxins like diphtheria toxin or cholera toxin, ADP-ribosylation of actin is reversible in the presence of high concentrations of nicotinamide (>10 mM) and in the absence of NAD (Just et al., 1990). The reversal of actin ADP-ribosylation exhibits the same substrate specificity as the forward reaction, e.g., ADP-ribosylation of nonmuscle actin C2 toxin is reversed by C2 or iota toxin. These findings further support the notion that both toxins modify the same amino acid. On the other hand, skeletal muscle actin ADP-ribosylated by iota toxin is not reversed by C2.

Other actin-ADP-ribosylating toxins

The various actin-modifying toxins can be divided into two subclasses. One subclass consists of *C. perfringens* iota toxin, *C. spiroforme* toxin, and *C. difficile* ADP-ribosyltransferase; the other subclass is represented by *C. botulinum* C2 toxin. The iota-like toxins exhibit immunological cross-reactivity. The binding components of *C. perfringens* iota toxin and of *C. spiroforme* toxin are interchangeable and are able to facilitate the transfer of the *C. difficile* ADP-ribosyltransferase into the cell (note that *C. difficile* reportedly only produces the ADP-ribosyltransferase but not a binding component) (Popoff and Boquet, 1988; Popoff et al., 1988; Simpson et al., 1989). In contrast, the binding component of C2 toxin cannot transport the enzyme components of the other actin-ADP-ribosylating toxins into the cell, and the transfer of *C. botulinum* ADP-ribosyltransferase C2I into the cell is not facilitated by iota-like binding components. Accordingly, C2 toxin shows no immunological cross-reactivity with the iota-like toxins.

their substrate specificities (Schering et al., 1988). Whereas iota toxin ADP-ribosylates all actin isoforms studied so far, including α-skeletal muscle actin, α-cardiac muscle actin, α/γ-smooth muscle actin and β/γ-cytoplasmic actin, C2 toxin modifies only β/γ-cytoplasmic actin and γ-smooth muscle actin (Mauss et al., 1990). This substrate specificity of the actin ADP-ribosylating toxins is surprising if one considers that all actin isoforms are more than 95% homologous (Vandekerckhove and Weber, 1979). Because γ-smooth muscle actin that is ADP-ribosylated by C2 differs from α-smooth muscle actin only in its five N-terminal amino

Functional consequences of the ADP-ribosylation of actin

ADP-ribosylation by bacterial toxins inhibits the ability of actin to polymerize. This was first shown by the decrease in the viscosity of a solution of ADP-ribosylated actin after induction of polymerization (Aktories et al., 1986b) and confirmed by electron microscopy showing the lack of formation of actin filaments (Aktories et al., 1986a). Further, ADP-ribosylated G-actin binds to the barbed end of actin filaments, thereby inhibiting further polymerization at the fast-growing end of actin filaments (Wegner and Aktories, 1988; Weigt et al., 1989). In contrast, ADP-ribosylated actin does not bind to the pointed end of actin filaments. Actin complexed with DNAse I is a substrate of ADP-ribosylating toxins, and ADP-ribosylated actin still binds to DNAse I (Aktories et al., 1986a).

Actin binds ATP and possesses inherent ATPase activity (Pollard and Cooper, 1986; Korn et al., 1987). Polymerization of actin largely increases ATP hydrolysis by the microfilament protein. It has been shown that the ADP-ribosylation of actin inhibits its associated ATPase activity (Geipel at al., 1989). Inhibition of ATP hydrolysis by ADP-ribosylation is not caused merely by blockade of actin polymerization. At actin concentrations below its critical concentration, at low Mg^{2+} concentrations (<50 μM) and even in the monomeric actin-DNAse complex, ADP-ribosylation almost completely inhibits the actin-associated ATP hydrolysis. The cytochalasin-stimulated actin ATPase is also inhibited by ADP-ribosylation (Geipel et al., 1990). Reversal of ADP-ribosylation of modified actin restores actin ATPase activity and the ability of actin to polymerize (Just et al., 1990).

Effects of actin ADP-ribosylating toxins on intact cells. The clostridial actin ADP-ribosyltransferases are cytotoxic agents. Injection of C2 toxin into the mouse intestinal loop was shown to induce fluid accumulation (Ohishi, 1983b); the toxin largely increases vascular permeability (Ohishi, 1983a; Ohishi et al., 1980b). These toxic effects are accompanied by drastic morphological alterations of treated cells or tissues (Ohishi and Odagari, 1984). In cell culture, the actin ADP-ribosylating toxins cause cells to round up (Ohishi et al., 1984; Reuner et al., 1987). Several studies have shown that actin but no other proteins are ADP-ribosylated in intact cells, indicating that the microfilament protein is the pathophysiological substrate of the toxins. ADP-ribosylation of actin and rounding up of cells correlate in a time and dose-dependent manner. Loading of cells with [^{32}P] orthophosphate and subsequent treatment with C2 toxin caused labeling of a 43-kDa protein, which was identified as actin by protease mapping (Reuner et al., 1987). Moreover, in intact cells the actin ADP-ribosylating toxins destroy the microfilament network, as seen by staining with fluorescein-coupled phalloidin, and finally cause a complete loss of the microfilament network of cells (Reuner et al., 1987). By using the DNase inhibition test, it has been shown that the destruction of the microfilament network causes an increase in the amount of G-actin in intact cells (Aktories et al., 1989b).

A model for the cytopathic effects of the actin ADP-ribosylating toxins is shown in Figure 13.3. The binding component of the toxin induces or unmasks a binding site for the enzyme component at the surface of the target cell. The ADP-ribosyltransferase is translocated into the cell by endocytosis. In the cell, the dynamic equilibrium between G- and F-actin, which is mainly regulated by actin-binding proteins, is disturbed by the actin ADP-ribosylating toxins. The enzymes ADP-ribosylate monomeric G-actin, a modification that converts actin into a capping protein to bind to the barbed end of actin filaments. As a result, further polymerization at the fast growing end of actin filaments is blocked. Because the pointed ends of filaments are not affected by ADP-ribosylation, monomeric actin is still released, thereby becoming a substrate for the toxin. The ADP-ribosylated actin, which is incapable of polymerization, is trapped in the monomeric form and accumulates. The modified actin is therefore withdrawn from the turnover pool of actin and is no longer available for the formation of

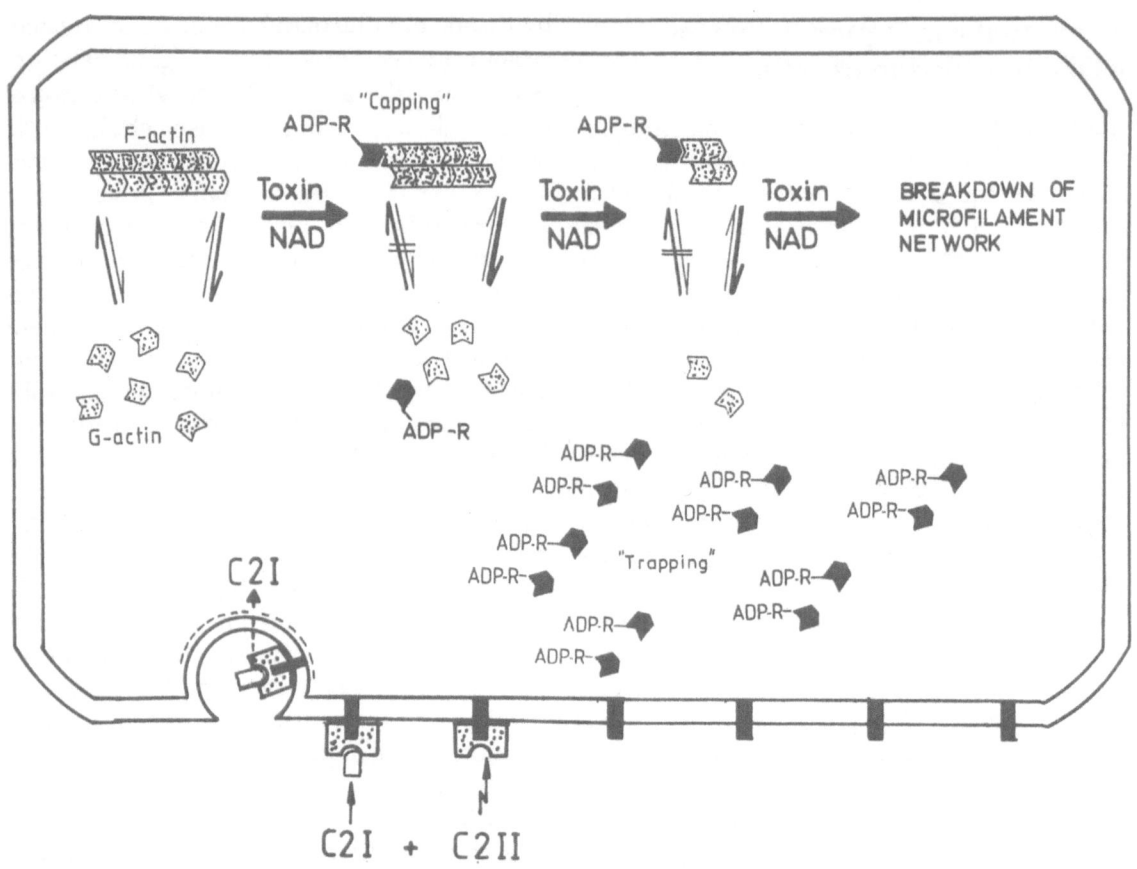

FIGURE 13.3 Model of action of actin ADP-ribosylating toxins. The binding component (e.g., C2II) of binary toxin attaches to eucaryotic cell surface, thereby inducing a binding site for enzyme component. The ADP-ribosylating component (e.g., C2II) is translocated into cell, probably via endocytosis, where toxin ADP-ribosylates G-actin. ADP-ribosylation interferes with cellular equilibrium of polymerization and depolymerization of actin. Modified actin has capability to bind to F-actin in capping protein-like manner, thereby inhibiting further growth of actin filaments at barbed end. F-actin still depolymerizes at pointed end of filaments, and released G-actin is ADP-ribosylated. ADP-ribosylated actin is not capable of polymerization and is withdrawn from pool of actin, which undergoes polymerization and depolymerization processes ("treadmilling"). Therefore, "capping" of F-actin and "trapping" of G-actin cause destruction of the microfilament network.

actin filaments. Both processes, capping of actin filaments and trapping of modified actin, will thus finally cause destruction of the microfilament network.

Actin-modifying toxins as tools for the study of cell functions. Because ADP-ribosylation blocks actin polymerization, the toxins will interfere with physiological processes that depend on the dynamic transition of G- to F-actin. For example, C2 toxin was shown to inhibit the carbachol-stimulated release of histamine from rat mast cells (Böttinger et al., 1987). On the other hand, C2 toxin affected the stimulated release of adrenaline from rat pheochromocytoma cells (PC12) in a biphasic manner (Matter et al., 1989). These findings indicate the involvement of actin in secretory events. Another model to study the role of actin is the activation of neutrophil leukocytes. Treatment of leukocytes with C2 toxin increased the superoxide anion production and enzyme release but inhibited the migration of the cells stimulated by chemotactic agents (Al-Mohanna et al., 1987; Norgauer et al., 1988; Norgauer et al., 1989).

These findings therefore support the view that actin is involved in the activation of leukocytes. Further, C2 toxin was shown to inhibit the contraction of the longitudinal smooth muscle preparation of the guinea pig ileum, most likely by a direct action on the smooth muscle cell (Mauss et al., 1989). Because the toxin modifies only monomeric G-actin but not polymerized F-actin, no interference with contractility would have occurred if only F-actin were involved in smooth muscle contraction. On the contrary, this observation suggests that G-actin or G- to F-actin transition participates in smooth muscle contraction. Thus, in addition to the mycotoxins phalloidin and cytochalasin, which act on actin (Cooper, 1987), the ADP-ribosylating toxins are novel tools to study physiological functions of their target protein actin.

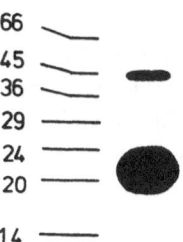

FIGURE 13.4   ADP- ribosylation of human platelet membranes by *Clostridium botulinum* type C culture filtrate. Crude human platelet membranes were incubated with filtrate of *C. botulinum* type C culture in presence of [^{32}P]NAD. Labeled proteins were analyzed by sodium dodecyl sulfate-polyacrylamide gel electrophoresis (SDS-PAGE) and autoradiography (shown). The ADP-ribosylation of 42-kDa protein (actin) was caused by C2 toxin; labeling of ~ 21-kDa protein was catalyzed by C3 ADP-ribosyltransferase.

13.3   *Clostridium botulinum* C3 ADP-Ribosyltransferase, an Exoenzyme That Modifies Low Molecular Weight GTP-Binding Proteins

Properties of *Clostridium botulinum* C3 ADP-ribosyltransferase

C3 is produced by *C. botulinum* type C and D. As is known for botulinum neurotoxins, C3 ADP-ribosyltransferase is phage encoded (Rubin et al., 1988; Popoff et al., 1990). Because commercially available botulinum neurotoxin preparations contain small quantities of C3 as a contaminant, the C3-induced ADP-ribosylation has been confused with the action of botulinum neurotoxins. However, it is now clear that C3 is neither structurally nor functionally related to the botulinum neurotoxins (Aktories and Frevert, 1987; Rösener et al., 1987; Adam-Vizi et al., 1988; Rubin et al., 1988). Final evidence for this notion comes from DNA sequencing of C3. Popoff et al. (1990) have reported that the extracellular enzyme is composed of 211 amino acids with a molecular weight of 23,546. The C

and D phage-encoded enzymes are identical in their amino acid sequences. Evidence for a precursor protein of C3 was deduced from the DNA sequence, but the complete primary structure of the precursor is not known at present. No sequence homologies have been found with any other protein, including other bacterial ADP-ribosylating toxins and enzymes (Popoff et al., 1990).

C3 has been purified from the culture filtrate of *C. botulinum* type C to apparent homogeneity (Aktories et al., 1987, 1988a). The molecular weight of C3 was determined to be about 25,000 on SDS polyacrylamide gels and about 18,000 by using gel filtration. The exoenzyme is a basic protein with a pI of about 10 (Just et al. in press). C3 is stable against short-term trypsin treatment (100 μg, 10 min at 37°C) and heating (1 min, 95°C) (Aktories et al., 1987). With [^{32}P]NAD, the ADP-ribosyltransferase labeled 22- to 24-kDa proteins in almost all tissues studied so far (Aktories et al., 1987, 1988a; Rubin et al., 1988) (Figure 13.4). The K_m of the reaction for NAD is about 0.3 μM. Isotope dilution (cold NAD), but not the addition of ADP-ribose, inhibits the labeling by C3. As is known for other ADP-ribosylating toxins, including C2 toxin, the reaction is most probably a mono-ADP-ribosylation because no further increase in the molecular weight of the labeled proteins occurs with incubation (Aktories et al., 1987).

Accordingly, ADP-ribosyltransferase activity of C3 is blocked by neither thymidine (10 mM) nor isonicotinic acid hydracid, which are inhibitors of poly-ADP-ribosylation and NAD glycohydrolases, respectively, and snake venom phosphodiesterase splits [^{32}P]AMP from ADP-ribosylated proteins. As shown for other bacterial ADP-ribosylating toxins, C3 possesses NAD glycohydrolase activity (Aktories et al., 1988a). ADP-ribosylation by C3 is reversible in the presence of nicotinamide (30 mM), in the absence of NAD, and at low pH. The optimum pH for ADP-ribosylation by C3 is about 7.5, whereas the rate of de-ADP-ribosylation is maximal at pH 5.5 (Habermann et al., 1991).

The substrates of C3 ADP-ribosyltransferase

C3 ADP-ribosylates low molecular mass GTP-binding proteins (M_r ~22,000). The Rho (Kikuchi et al., 1988; Narumiya et al., 1988; Aktories et al., 1989a; Chardin et al., 1989; Braun et al., 1989) and Rac proteins (Didsbury et al., 1989) have been identified as substrates of C3. All members of the Rho (A, B, C) protein family are modified by the exoenzyme. These GTP-binding proteins are encoded by a superfamily of homologous genes (Madaule and Axel, 1985). Other related mammalian genes are the ras (Ha, Ki, N, R) (Barbacid, 1987), ral (Chardin and Tavitian, 1986), rab (1–6) (Touchot et al., 1987), and rap (1A, 1B, 2) (Pizon et al., 1988) genes. During recent years, much information has been collected about the ras genes and their encoded proteins. They have attracted special attention because an increasing number of cell lines of rodent and human tumors contain mutants of ras genes, which are probably involved in cell transformation (Reddy et al., 1982). All the gene products, including Rho and Rac, are GTP-binding proteins possessing GTPase activity. The regulatory proteins are active in the GTP bound form and inactive after hydrolysis of the guanine nucleotide bound to GDP (Barbacid, 1987).

Proteins have been described that regulate the activity of the low molecular weight GTP-binding proteins. The GTPase-stimulating proteins (GAP) increase the GTPase activity of the GTP-binding proteins by more than 100-fold (Trahey and McCormick, 1987; Garrett et al., 1989). These GAP proteins are apparently different for the various GTP-binding proteins. Studies with mutant Ras proteins have suggested that the Ras GAP protein interacts with its target protein in the so-called effector region of Ras (amino acid sequence, 30 to 40) (Adari et al., 1988). Whether this is also true for the Rho GAP protein is not clear. It appears also that the Rho proteins are specifically controlled by other regulatory proteins that either stimulate or inhibit the GDP dissociation from and the subsequent binding of GTP to the Rho protein (Ueda et al., 1990; Mizuno et al., 1991). However, the precise functions of these interacting proteins are not known at present.

The rho genes, which apparently are ubiquitous in all eucaryotic cells, were first described in Aplysia (Madaule and Axel, 1985). In mammalian cells, at least three rho genes (A, B, C) have been observed that are about 95% homologous with each other and about 35% homologous with ras (Yeramian et al., 1987; Chardin et al, 1988). In yeast, two rho genes (RHO 1, RHO 2) have been described (Madaule et al., 1987). Although RHO 1, which is about 80% similar to the human rho A gene, is an essential gene in yeast cells, RHO 2 is not required for cell viability. Only a few data are available on the function of rho in mammalian cells. Rat-1 and NIH 3T3 mouse fibroblasts were transfected with clones containing normal and variants of the human rho A allele (Avraham and Weinberg, 1989). The mutants, encoding valine in place of glycine and leucine in place of glutamine, which are normally found in residues 14 and 64, respectively, mirror the variants of the related ras genes responsible for oncogenic activation. Although the mutant rho A did not induce focus formation or growth in soft agar, amplified colonies inoculated into nude mice were tumorigenic.

It has been reported that the Rac 1, 2 proteins are substrates of C3 (Didsbury et al., 1989). The rac genes, with 90% homology, share about 60% and 30% homology with the human rho and ras genes, respectively. Although the rac 2 transcript was mainly found in myeloid cells, the ras 1 transcript was detected in various tissues. The physiological roles of the Rho and Rac proteins are largely unknown (however, see following discussion).

Regulation of the C3-catalyzed ADP-ribosylation by guanine nucleotides and divalent cations

ADP-ribosylation of Rho or Rac proteins by C3 increases with low concentrations of Mg^{2+} (100–500 μM) whereas higher concentrations of the divalent cation decrease ADP-ribosylation (Aktories et al., 1988a; Kikuchi et al., 1988). In the absence of divalent cations, guanine nucleotides (GTP, GTPγS, GDP) stimulate ADP-ribosylation. This effect is probably caused by the stabilizing effect of nucleotides on GTP-binding proteins that lose their nucleotides in the absence of divalent cations and are denatured. In agreement with this notion is the finding that guanine nucleotides prevent the inactivation of the GTP-binding protein by heat or N-ethylmaleimide treatment (Rubin et al., 1988; Aktories et al., 1988a).

The ADP-ribosylation of the native protein depends on the type of nucleotide bound (Aktories and Frevert, 1987; Williamson et al., 1990; Narumiya et al., 1988b; Habermann et al., 1991). It appears that the GDP-bound form of the Rho protein is the primary substrate of C3. In contrast, the GTPγS-bound form is modified to a lower extent (Aktories and Frevert, 1987; Habermann et al., 1991). By analogy, whereas the inactive heterotrimeric GDP-bound form of the G_i protein or of transducin is the favored substrate of pertussis toxin, the dissociated and activated GTP-bound forms of these G proteins are very poor substrates for this toxin (van Dop et al., 1984; Tsai et al., 1984; Katada et al., 1986). This relationship explains why activation of transducin or G_i via the light receptor rhodopsin, which induces the GDP–GTP exchange and dissociation of the G protein, impairs the pertussis toxin-induced ADP-ribosylation (van Dop et al., 1984; Tsai et al., 1984). It has been shown that photoexcitation of rhodopsin reduced the C3-induced ADP-ribosylation of GTP-binding proteins of low molecular weight (Wieland et al., 1990). These findings suggest that membrane-bound receptors, which are

Table 13.2 Effector regions of low molecular weight GTP-binding proteins

c-H-Ras	27-HFVDEYDPTIEDSYRKQVV-45
c-K-Ras	27-HFVDEYDPTIEDSYRKQVV-45
c-N-Ras	27-HFVDEYDPTIEDSYRKQVV-45
Rap 1A	27-IFVEKYDPTIEDSYRKQVE-45
Rab 1	35-TYTESYISTIGVDFKIRTI-53
Ral	38-EFVEDYEPTKADSYRKKVV-56
Rho A[a]	29-QFPEVYVPTVFENYVADIE-47
Rho B	29-EFPEVYVPTVFENYVADIE-47
Rho C	29-QFPEVYVPTVFENYIADIE-47
RHO 1	34-QFPEVYVPTVFENYVADVE-52
RHO 2	31-KFPEQYHPTVFENYVTDCR-49
Rac 1	27-AFPGEYIPTVFDNYSANVM-45
Rac 2	27-AFPGEYIPTVFDNYSANVM-45

[a] The Rho A protein is ADP-ribosylated in asparagine-41. All GTP-binding proteins, which are substrates of C3 (Rho A,B,C; RHO 1, RHO 2; Rac 1 and Rac 2) share asparagine in this position. By analogy with p21-Ras, asparagine-41 is located in the so-called effector region of the GTP-binding protein.

known to cooperate with G proteins, also interact with low molecular weight GTP-binding proteins and thus assume the role of substrate of C3 in photoperception of rod outer segments.

The amino acid acceptor of C3-catalyzed ADP-ribosylation

Protein chemical analysis carried out by Narumiya and coworkers revealed that C3 modifies the Rho protein in asparagine-41 (Sekine et al., 1989) (Table 13.2). Because all substrate proteins of C3 possess asparagine at the same position, this amino acid most likely is the acceptor for ADP-ribose in these proteins. The ADP-ribose-amino-acid linkage is characterized by unique chemical stability. Treatment of ADP-ribosylated Rho protein with hydroxylamine (0.5 M) for 3 h does not split the ADP-ribose–protein bond (Aktories et al., 1988b). In comparison, the half-life of the arginine–ADP-ribose linkage formed by botulinum C2 toxin in actin has a half-life toward hydroxylamine of about 2 h (37°C). Mercury ions, which split the ADP-ribose-cysteine bond formed by pertussis toxin, do not affect the C3-induced

ADP-ribose–asparagine linkage. Even treatment with NaOH (0.5 M, 60 min), which splits the C2-toxin and pertussis toxin-catalyzed ADP-ribose–amino-acid bond, is without effect on the ADP-ribose–asparagine linkage (Aktories et al., 1988b).

Functional consequences of C3-catalyzed ADP-ribosylation in intact cells

The functional consequences of ADP-ribosylation of GTP-binding proteins are still unclear. Studies with C3 are hindered by the fact that entry of C3 into cells is poor. High concentrations of the enzyme (10–100 μg/ml) are necessary to observe effects in intact cells. In comparison, other ADP-ribosylating toxins, such as pertussis or C2 toxin, are effective at concentrations about 1000-fold lower than those of C3. It has therefore been suggested that C3 may be the enzyme component of a binary holotoxin whose binding component is not known at present (Rubin et al., 1988). However, no evidence for the existence of such a binding component has been presented.

To bypass the problem of poor cell accessibility, C3 has been microinjected into *Xenopus* oocytes (Rubin et al., 1988). *Xenopus* oocytes that are arrested in the prophase of meiosis can be stimulated to undergo maturation by progesterone (Maller and Krebs, 1977). This process can be accelerated by microinjection of C3; C3 apparently exhibits effects similar but not identical to those induced by microinjection of Ras protein into *Xenopus* oocytes (Birchmeier et al., 1985). By using the osmotic shock method, C3 has been introduced into various other cells (Rubin et al., 1988). Morphological changes in NIH 3T3 cells included rounding-up of cells and formation of binucleated cells. In PC12 cells, this treatment induced differentiation of the cells with formation of short neurites after few hours.

Similar morphological changes were observed after treatment of PC12 cells with 10 to 100 μg/ml of C3 (Nishiki et al., 1990). In the latter studies, high concentrations of C3 inhibited cell growth and induced approximately a twofold increase in choline esterase activity.

The time course of the formation of neurite-like processes, and its dependence on RNA synthesis, were similar to NGF-induced neurite formation. Treatment of intact Vero cells with C3 caused the destruction of the microfilament network without major changes of the microtubule system (Chardin et al., 1989). A similar destruction of the actin cytoskeleton was observed after treatment of hepatoma FAO cells (Aktories et al., 1990); here C3 also caused the redistribution of the intermediate filaments. Again, the microtubule system was much less affected.

Further insight into the mechanism underlying the effects of C3 on the cytoskeletal architecture was obtained by microinjection of C3 and recombinant Rho protein into 3T3 cells (Paterson et al., 1990). Injection of C3 into these cells caused rounding-up with the destruction of the microfilament network. These morphological changes were also observed with ADP-ribosylated normal Rho protein. A complete distinct and novel phenotype was observed when valine-14 Rho A protein was microinjected. This mutation of the Rho protein inhibits its GTPase activity and, by preventing hydrolysis of bound GTP, this protein thus is permanently active. Fifteen to twenty minutes after microinjection of valine-14 Rho into subconfluent cells, dramatic changes in cell morphology occurred, characterized by contraction of the cell body and fingerlike processes. Morphological changes were inhibited after ADP-ribosylation of the mutant protein, suggesting that the covalent modification renders the GTP-binding protein biologically inactive.

It has been already mentioned that C3 modifies Rho in asparagine-41 (Sekine et al., 1989). When compared with the homologous Ras protein (Pai et al., 1989), asparagine-41 is located in the so-called effector region, which covers the amino acids 32–40 in Ras (34–42 in Rho). Mutations in this region of Ras cause loss of its biological activity. Thus, if the same structure–function relationship exists in Rho, it can be assumed that modification of the protein by the incorporated ADP-ribose moiety largely hinders the interaction with a putative effector system (Figure 13.5). These studies further support the view that Rho proteins are involved in the regulation of cytoskeletal proteins, but

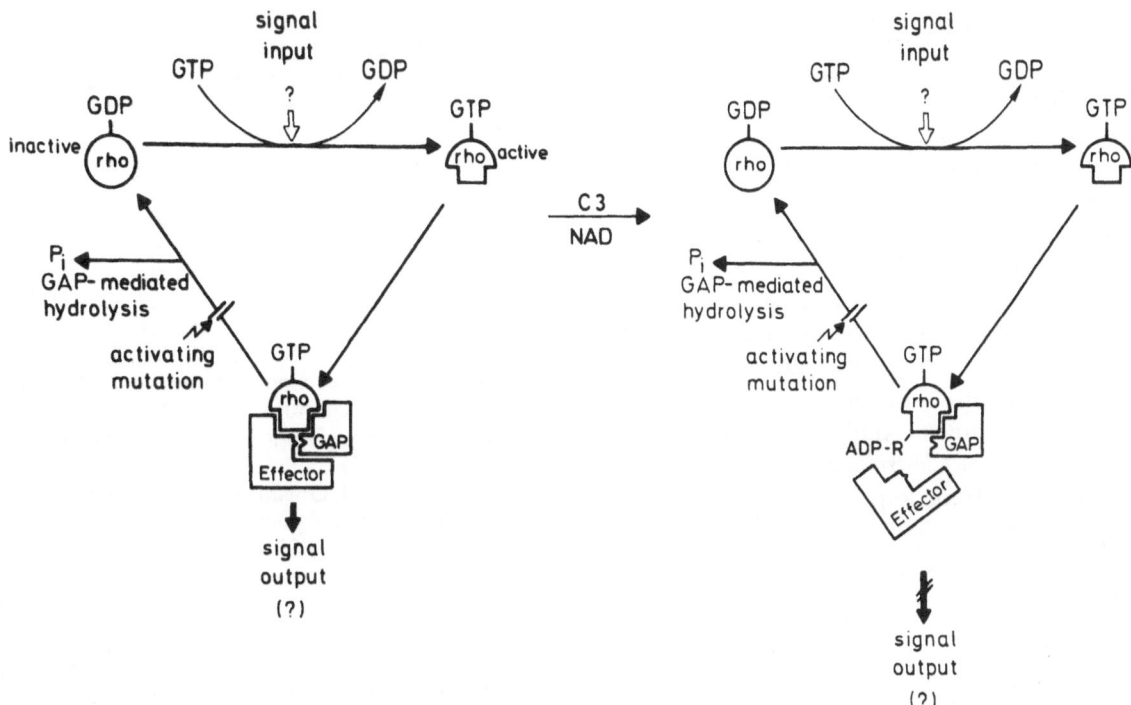

FIGURE 13.5 Model of action of *Clostridium botulinum* C3 ADP-ribosyltransferase. The GTP-binding protein, e.g., p21^{rho}, is inactive in GDP-bound form. An activating signal, carrier of which is still unknown, induces GDP–GTP exchange, and active GTP-bound p21 can interact with effector system. Hydrolysis of bound GTP to GDP by associated GTPase activity terminates active state of p21. The GTPase-activiting protein (GAP) increases deactivation of p21 severalfold. Activating mutations of p21 inhibit inherent GTPase activity and its stimulation by GAP, which results in persistently active GTP-binding protein. In the case of valine-14-p21^{rho}, the persistently active protein causes drastical morphological changes of 3T3 cells (see text). The ADP-ribosylation of valine-14-p21^{rho} by C3 inhibits cellular action of mutant protein apparently without affecting its interaction with GAP (modified according to McCormick, 1989).

whether this is a direct or indirect effect is still unclear.

No toxic or physiological effects induced by C3 in intact animals have been reported. Injection of as much as 100 μg of C3 into mice is not lethal (K. Aktories, unpublished observation). In comparison, the extremely potent botulinum neurotoxins C and D having LD$_{50}$ of about 10–20 pg/mouse. Thus, it is not clear whether C3 is really a toxin, a component of a holotoxin, or "merely" an exoenzyme.

13.4 *Clostridium limosum* ADP-ribosyltransferase

We have purified a novel ADP-ribosyltransferase which is produced by a *C. limosum* strain isolated from a human lung abscess. This ex-oenzyme is very similar to C3 having a molecular weight of about 25,000 on SDS-PAGE and an IP of 10.6. The K$_m$ for NAD is 0.5 μM similar to that found with *C. botulinum* exoenzyme C3. Deduced from the partial amino acid sequence analysis, the *C. limosum* exoenzyme exhibits an about 70% homology with C3. Also *C. limosum* enzyme modifies small GTP-binding proteins of the Rho-family. ADP-ribosylation of Rho protein by the *C. limosum* transferase is reversed by C3 indicating that both enzymes modify the same protein substrate in the identical amino acid. In contrast to *C. botulinum* ex-oenzyme C3, the *C. limosum* transferase is significantly auto-ADP-ribosylated in the presence of SDS. Thus, the ADP-ribosylation of small GTP-binding proteins of the Rho-family is not unique to C3 (Just and Schallehn, 1991; Just et al., 1992, in press).

13.5 Concluding Remarks

The discovery and subsequent characterization of various clostridial actin ADP-ribosylating toxins have resulted in significant progress in our understanding of the pathophysiological mechanisms by which these agents transmit their cytotoxic effects to eucaryotic cells. Moreover, the transferases are very instrumental tools with which to study the physiological and pathophysiological functions of their target protein actin. Similarly, with *Clostridium botulinum* ADP-ribosyltransferase C3 and the novel *Clostridium limosum* exoenzyme at hand, we hope to possess instruments to acquire further insight into the functions of the low molecular weight GTP-binding proteins of the Rho and Rac family.

References

Adam-Vizi, V., S. Rösener, K. Aktories, and D.E. Knight. 1988. Botulinum toxin-induced ADP-ribosylation and inhibition of exocytosis are unrelated events. *FEBS Lett.* **238**:277–280.

Adari, H., D.R. Lowy, B.M. Willumsen, J. der Channing, and F. McCormick 1988. Guanosine triphosphate activating protein (GAP) interacts with the p21 *ras* effector binding domain. *Science* **240**:518–521.

Aktories, K. and J. Frevert. 1987. ADP-ribosylation of a 21–24 kDa eucaryotic protein(s) by C3, a novel botulinum ADP-ribosyltransferase, is regulated by guanine nucleotide. *Biochem. J.* **247**:363–368.

Aktories, K., U. Weller, and G.S. Chhatwal. 1987. *Clostridium botulinum* type C produces a novel ADP-ribosyltransferase distinct from botulinum C2 toxin. *FEBS Lett.* **212**:109–113.

Aktories, K., T. Ankenbauer, B. Schering, and K.H. Jakobs. 1986a. ADP-ribosylation of platelet actin by botulinum C2 toxin. *Eur. J. Biochem.* **161**:155–162.

Aktories, K., M. Bärmann, I. Ohishi, S. Tsuyama, K.H. Jakobs, and E. Habermann. 1986b. Botulinum C2 toxin ADP-ribosylates actin. *Nature* (London) **322**:390-392.

Aktories, K., S. Rösener, U. Blaschke, and G.S. Chhatwal. 1988a. Botulinum ADP-ribosyltransferase C3. Purification of the enzyme and characterization of the ADP-ribosylation reaction in platelet membranes. *Eur. J. Biochem.* **172**:445–450.

Aktories, K., I. Just, and W. Rosenthal. 1988b. Different types of ADP-ribose protein bonds formed by botulinum C2 toxin, botulinum ADP-ribosyltransferase C3 and pertussis toxin. *Biochem. Biophys. Res. Commun.* **156**:361–367.

Aktories, K., U. Braun, S. Rösener, I. Just, and A. Hall. 1989a. The *rho* gene product expressed in *E. coli* is a substrate of botulinum ADP-ribosyltransferase C3. *Biochem. Biophys. Res. Commun.* **243**:70–76.

Aktories, K., K.-H. Reuner, P. Presek, and M. Bärmann. 1989b. botulinum C2 toxin treatment increases the G-actin pool in intact chicken cells: a model for the cytopathic action of actin-ADP-ribosylating toxins. *Toxicon* 27:989–993.

Aktories, K., I. Just, H. Müller, W. Wiegers, and P. Traub. 1990. Destruction of the actin cytoskeleton induced by botulinum C3 ADP-ribosyltransferase. *Naunyn-Schmiedebergs Arch. Pharmacol.* **341**(Suppl.):R331.

Al-Mohanna, F.A., I. Ohishi, and M.B. Hallett. 1987. Botulinum C_2 toxin potentiates activation of the neutrophil oxidase. *FEBS Lett.* **219**:40–44.

Althaus F.R. and C. Richter. 1987. *ADP-Ribosylation of Proteins*, pp. 1–237. Heidelberg: Springer-Verlag.

Avraham, H. and R.A. Weinberg. 1989. Characterization and expression of the human rhoH12 gene product. *Mol. Cell. Biol.* 9:2058–2066.

Barbacid, M. 1987. *ras* genes. *Annu. Rev. Biochem.* **56**:779–827.

Birchmeier, C., D. Broek, and M. Wigler. 1985. *Ras* proteins can induce meiosis in *Xenopus* oocytes. *Cell* 43:615–621.

Bokoch, G.M., T. Katada, J.K. Northup, E.L. Hewlett, and A.G. Gilmann. 1983. Identification of the predominant substrate for ADP-ribosylation by islet activating protein. *J. Biol. Chem.* **258**:2072–2075.

Böttinger, H., K.H. Reuner, and K. Aktories. 1987. Inhibition of histamine release from rat mast cells by botulinum C2 toxin. *Int. Arch. Allergy Appl. Immunol.* **84**:380–384.

Braun, U., B. Habermann, I. Just, K. Aktories, and J. Vandekerckhove. 1989. Purification of the 22-kDa protein substrate of botulinum ADP-ribosyltransferase C3 from porcine brain cytosol and its characterization as a GTP-binding protein highly homologous to the *rho* gene product. *FEBS Lett.* **243**:70–76.

Cassel, D. and Z. Selinger. 1977. Mechanism of adenylate cyclase activation by cholera toxin: inhibition of GTP hydrolysis at the regulatory site. *Proc. Natl. Acad. Sci. USA* **74**:3307–3311.

Cassel, D. and T. Pfeuffer. 1978. Mechanism of cholera toxin action: covalent modification of the guanyl nucleotide-binding protein of the adenylate cyclase system. *Proc. Natl. Acad. Sci. USA* **75**:2669–2673.

Chang, P.P., J. Moss, E. M. Twiddy, and R.K. Holmes. 1987. Type II heat-labile enterotoxins of *Escherichia coli* activates adenylate cyclase in human fibroblasts by ADP-ribosylation. *Infect. Immun.* 55:1854–1858.

Chardin, P. and A. Tavitian. 1986. The *ral* gene: a new ras related gene isolated by the use of a synthetic probe. *EMBO J.* **5**:2203–2208.

Chardin, P., P. Madaule, and A. Tavitian. 1988. Coding sequence of human *rho* cDNAs clone 6 and 9. *Nucleic Acids Res.* **16**:2717.

Chardin, P., P. Boquet, P. Madaule, M.R. Popoff, E.J. Rubin, and D.M. Gill. 1989. The mammalian G protein rho C is ADP-ribosylated by *Clostridium botulinum* exoenzyme C3 and affects actin microfilaments in Vero cells. *EMBO J.* **8**:1087–1092.

Coburn, J., S.T. Dillon, B.H. Iglewski, and D.M. Gill. 1989a. Exoenzyme S of *Pseudomonas aeruginosa* ADP-ribosylates the intermediate filament protein vimentin. *Infect Immun.* **57**:996–998.

Coburn, J., R.T. Wyatt, B.H. Iglewski, and D.M. Gill. 1989b. Several GTP-binding proteins, including p21$^{c-H-ras}$, are preferred substrates of *Pseudomonas aeruginosa* exoenzyme S. *J. Biol. Chem.* **264**:9004–9008.

Collier, R.J. and J. Kandel. 1971. Structure and activity of diphtheria toxin, I. Thiol-dependent dissociation of a fraction of toxin into enzymatically active and inactive fragments. *J. Biol. Chem.* **246**:1496–1503.

Cooper J.M. 1987. Effects of cytochalasin and phalloidin on actin. *J. Cell Biol.* **105**:1473–1478.

Didsbury, J., R.F. Weber, G.M. Bokoch, T. Evans, and R. Snyderman. 1989. rac, a novel ras-related family of proteins that are botulinum toxin substrates. *J. Biol. Chem.* **264**:16378–16382.

Eklund, M.W. and F.T. Poysky. 1972. Activation of a toxic component of *Clostridium* types C and D by trypsin. *Appl. Microbiol.* **24**:108–113.

Eklund, M.W., F.T. Poysky, and S.M. Reed. 1972. Bacteriophage and the toxigenicity of *Clostridium botulinum* type D. *Nature New Biol.* **235**:16–17.

Eklund, M.W., F.T. Poysky, S.M. Reed, and C.A. Smith. 1971. Bacteriophage and the toxigenicity of *Clostridium botulinum* type C. *Science* **172**:480–482.

Garrett, M.D., A.J. Self, C. van. Oers, and A. Hall. 1989. Identification of distinct cytoplasmic targets for ras/R-ras and rho regulatory proteins. *J. Biol. Chem.* **264**:10–13.

Geipel, U., I. Just, B. Schering, D. Haas, and K. Aktories. 1989. ADP-ribosylation of actin causes increase in the rate of ATP exchange and inhibition of ATP hydrolysis. *Eur. J. Biochem.* **179**:229–232.

Geipel, U., I. Just, and K. Aktories. 1990. Inhibition of cytochalasin D-stimulated G-actin ATPase by ADP-ribosylation with *Clostridium perfringens* iota toxin. *Biochem. J.* **266**:335–339.

Gill, D.M. and S.H. Richardson. 1980. Adenosine diphosphate-ribosylation of adenylate cyclase catalyzed by heat-labile enterotoxin of *Escherichia coli*: comparison with cholera toxin. *J. Infect. Dis.* **141**:64–70.

Gill, M.D. 1977. Mechanism of action of cholera toxin. *In: Advances in Cyclic Nucleotide Research*, Vol. 8, P. Greengard and G.A. Robison, eds., pp. 85–118. New York: Raven Press.

Habermann, B., C. Mohr, I. Just, and K. Aktories. 1991. ADP-ribosylation and de-ADP-ribosylation of the *rho* protein by *Clostridium botulinum* exoenzyme C3. Regulation by EDTA, guanine nucleotides and pH. *Biochim. Biophys. Acta* **1077**:253–258.

Honjo, T., Y. Nishizuka, O. Hayaishi and I. Kato. 1968. Diphtheria-toxin-dependent-adenosine diphosphate ribosylation of aminoacyl transferase II and inhibition of protein synthesis. *J. Biol. Chem.* **243**:3553–3555.

Iglewski, B.H. and D. Kabat. 1975. NAD-dependent inhibition of protein synthesis by *Pseudomonas aeruginosa* toxin. *Proc. Natl. Acad. Sci. USA* **72**:2284–2289.

Just, I., U. Geipel, A. Wegner, and K. Aktories. De-ADP-ribosylation of actin by *Clostridium perfringens* iota toxin and *Clostridium botulinum* C2 toxin. *Eur. J. Biochem.* **192**:723–727.

Just, I. and G. Schallehn. 1991. A novel C3-like ADP-ribosyltransferase produced by *Clostridium limosum*. *Naunyn-Schmiedeberg's Arch. Pharmacol.* Suppl. 343:152.

Just, I., C. Mohr, G. Schallehn, L. Menard, J.R. Didsbury, J. Vandekerckhove, J. van Damme, and K. Aktories. 1992. Purification and characterization of an ADP-ribosyltransferase produced by *Clostridium limosum*. *J. Biol. Chem.* in press.

Katada, T., M. Oinuma, and M. Ui. 1986. Mechanism for inhibition of the catalytic activity of adenylate cyclase by the guanine nucleotide-binding proteins serving as substrate of islet-activating protein, pertussis toxin. *J. Biol. Chem.* **261**:5215–5221.

Kikuchi, A., K. Yamamoto, T. Fujita, and Y. Takai. 1988. ADP-ribosylation of the bovine brain *rho* protein by botulinum toxin type C1. *J. Biol. Chem.* **263**:16303–16308.

Korn E.D., M.-F. Carlier, and D. Pantaloni. 1987. Actin polymerization and ATP hydrolysis. *Science* **238**:638–644.

Leppla, S.H. 1982. Anthrax toxin edema factor: a bacterial adenylate cyclase that increases cyclic AMP concentrations in eukaryotic cells. *Proc. Natl. Acad. USA Sci.* **79**:3162–3166.

Madaule, P. and R. Axel. 1985. A novel *ras*-related gene family. *Cell* **41**:31–40.

Madaule, P., R. Axel, and A.M. Myers. 1987. Characterization of two members of the rho gene family from the yeast *Saccharomyces cerevisiae*. *Proc. Natl. Acad. Sci. USA* **84**:779–783.

Maller, J.L. and E.G. Krebs. 1977. Progesterone-stimulated meiotic cell division in *Xenopus* oocytes. *J. Biol. Chem.* **252**:1712–1718.

Matter, K., F. Dreyer, and K. Aktories. 1989. Actin

involvement in exocytosis from PC12 cells: studies on the influence of botulinum C2 toxin on stimulated noradrenaline release. *J. Neurochem.* **52**:370–376.

Mauss, S., G. Koch, V.A.W. Kreye, and K. Aktories. 1989. Inhibition of the contraction of the isolated longitudinal muscle of the guinea-pig ileum by botulinum C2 toxin: evidence for a role of G/F-actin transition in smooth muscle contraction. *Naunyn-Schmiedebergs Arch. Pharmacol.* **340**:345–351.

Mauss, S., C. Chaponnier, I. Just, K. Aktories, and G. Gabbiani. 1990. ADP-ribosylation of actin isoforms by *C. botulinum* C2 toxin and *C. perfringens* iota toxin. *Eur. J. Biochem.* **194**:237–241.

McCormick, F. 1989. Gasp: not just another oncogene. *Nature* (London) **340**:678–679.

Mizuno, T., Kaibuchi, K., Yamamoto, T., Kawamura, M., Sakoda, T., Fujioka, H., Matsuura, Y. and Takai, Y. 1991. A stimulatory GDP/GTP exchange protein for smg p21 is active on the post-translationally processed form of c-Ki-ras p21 and rhoA p21. *Proc. Natl. Acad. Sci. USA* **88**:6442–6446.

Moss J. and M. Vaughan, eds. 1990. *ADP-Ribosylating Toxins and G Proteins.* Washington, D.C.: American Society for Microbiology, pp. 1–567.

Moss, J. and M. Vaughan. 1988. ADP-ribosylation of guanyl-nucleotide-binding proteins by bacterial toxins. *Adv. Enzymol.* **61**:303–379.

Nakamura, S., T. Serikawa, K. Yamakawa, S. Nishida, S. Kozaki, and G. Sakaguchi. 1978. Sporulation and C_2 toxin production by *Clostridium botulinum* type C strains producing no C_1 toxin. *Microbiol. Immunol.* **22**:591–596.

Narumiya, S., A. Sekine, and M. Fujiwara. 1988a. Substrate for botulinum ADP-ribosyltransferase, Gb, has an amino acid sequence homologous to a putative *rho* gene product. *J. Biol. Chem.* **263**:17255–17257.

Narumiya, S., Morii, N., Ohno, K., Ohashi, Y. and Fujiwara, M. 1988b. Subcellular distribution and isoelectric heterogeneity of the substrate for ADP-ribosyl transferase from *Clostridium botulinum. Biochem. Biophys. Res. Commun.* **150**:1122–1130.

Nishiki, T., S. Narumiya, N. Morii, M. Yamamoto, M. Fujiwara, Y. Kamata, G. Sakaguchi, and S. Kozaki. 1990. ADP-ribosylation of the rho/rac proteins induces growth inhibition, neurite outgrowth and acetylcholine esterase in cultured PC-12 cells. *Biochem. Biophys. Res. Commun.* **167**:265–272.

Noda, M., I. Kato, F. Matsuda, and T. Hirayama. 1981. Mode of action of staphylococcal leukocidin: relationship between binding of ^{125}I-labeled S and F components of leukocidin to rabbit polymorphonuclear leukocytes and leu-

kocidin activity. *Infect. Immun.* **34**:362–367.

Norgauer, J., I. Just, K. Aktories, and L.A. Sklar. 1989. Influence of botulinum C2 toxin on F-actin and N-formyl peptide receptor dynamics in human neutrophils. *J. Cell Biol.* **109**:1133–1140.

Norgauer, J., E. Kownatzki, R. Seifert, and K. Aktories. 1988. Botulinum C2 toxin ADP-ribosylates actin and enhances O_2- production and secretion but inhibits migration of activated human neutrophils. *J. Clin. Invest.* **82**:1376–1382.

Ohishi, I. 1983a. Lethal and vascular permeability activities of botulinum C2 toxin induced by separate injections of the two toxin components. *Infect. Immun.* **40**:336–339.

Ohishi, I. 1983b. Response of mouse intestinal loop to botulinum C2 toxin: enterotoxic activity induced by cooperation of nonlinked protein components. *Infect. Immun.* **40**:691–695.

Ohishi, I. 1986. NAD-glycohydrolase activity of botulinum C_2 toxin: a possible role of component I in the mode of action of the toxin. *J. Biochem.* (Tokyo) **100**:407–413.

Ohishi, I. 1987. Activation of botulinum C_2 toxin by trypsin. *Infec. Immun.* **55**:1461–1465.

Ohishi, I. and M. Miyake. 1985. Binding of the two components of C2 toxin to epithelial cells and brush borders of mouse intestine. *Infect. Immun.* **48**:769–775.

Ohishi, I. and Y. Odagiri. 1984. Histopathological effect of botulinum C2 toxin on mouse intestines. *Infect. Immun.* **43**:54–58.

Ohishi, I. and Y. Okada. 1986. Heterogeneities of two components of C_2 toxin produced by *Clostridium botulinum* types C and D. *J. Gen. Microbiol.* **132**:125–131.

Ohishi I., M. Iwasaki, and G. Sakaguchi. 1980a. Purification and characterization of two components of botulinum C2 toxin. *Infect. Immun.* **30**:668–673.

Ohishi, I., M. Iwasaki, and G. Sakaguchi. 1980b. Vascular permeability activity of botulinum C2 toxin elicited by cooperation of two dissimilar protein components. *Infect. Immun.* **31**:890–895.

Ohishi, I. M. Miyake, K. Ogura, and S. Nakamura. 1984. Cytopathic effect of botulinum C2 toxin on tissue — culture cell lines. *FEMS Lett. Microbiol.* **23**:281–284.

Pai, E.F., W. Kabsch, U. Krengel, K.C. Holmes, J. John, and A. Wittinghofer. 1989. Structure of the guanine-nucleotide-binding domain of the Ha-*ras* oncogene product p21 in the triphosphate conformation. *Nature* (London) **341**:209–214.

Paterson H.F., A.J. Self, M.D. Garrett, I. Just, K. Aktories, and A. Hall. Microinjection of recombinant p21-rho induces rapid changes in cell morphology. *J. Cell Biol.* (1990) **111**:1001–1007.

Pfeuffer, T. and E.J.M. Helmreich. 1988. Structural

and functional relationship of guanosine triphosphate binding proteins. *Curr. Topics Cell Regul.* **29**:129–216.

Pizon, V., P. Chardin, I. Lerosey, B. Olofson, and A. Tavitian. 1988. Human cDNAs rap 1 and rap 2 homologous to the *Drosophila* gene Dras3 encode proteins closely related to *ras* in the "effector" region. *Oncogene* **3**:201–204.

Pollard, T.D. and J.A. Cooper. 1986. Actin and actin-binding proteins. A critical evaluation of mechanism and functions. *Annu. Rev. Biochem.* **55**:987–1035.

Popoff, M.R. and P. Boquet. 1988. *Clostridium spiroforme* toxin is a binary toxin which ADP-ribosylates cellular actin. *Biochem. Biophys. Res. Commun.* **152**:1361–1368.

Popoff, M.R., P. Boquet, D.M. Gill, and M.W. Eklund. 1990. DNA sequence of exoenzyme C3, an ADP-ribosyltransferase encoded by *Clostridium botulinum* C and D phages. *Nucleic Acids Res.* **18**:1291.

Popoff, M.R., E.J., Rubin, D.M. Gill, and P. Boquet. 1988. Actin-specific ADP-ribosyltransferase produced by a *Clostridium difficile* strain. *Infect. Immun.* **56**:2299–2306.

Reddy, E., R. Reynolds, E. Santos, and M. Barbacid. 1982. A point mutation is responsible for the acquisition of transforming properties by the T24 human bladder carcinoma oncogene. *Nature* (London) **300**:149–152.

Reuner, K.H., P. Presek, C.B. Boschek, and K. Aktories. 1987. Botulinum C$_2$ toxin ADP-ribosylates actin and disorganizes the microfilament network in intact cells. *Eur. J. Cell Biol.* **43**:134–140.

Rösener, S., G.S. Chhatwal, and K. Aktories. 1987. Botulinum ADP-ribosyltransferase C3 but not botulinum neurotoxins C1 and D ADP-ribosylates low molecular mass GTP-binding proteins. *FEBS Lett.* **224**:38–42.

Rubin, E.J., D.M. Gill, P. Boquet, and M.R. Popoff. 1988. Functional modification of a 21-kilodalton G protein when ADP-ribosylated by exoenzyme C3 of *Clostridium botulinum. Mol. Cell. Biol.* **8**:418–426.

Schering, B., M. Bärmann, G.S. Chhatwal, U. Geipel, and K. Aktories. 1988. ADP-ribosylation of skeletal muscle and non-muscle actin by *Clostridium perfringens* iota toxin. *Eur. J. Biochem.* **171**:225–229.

Sekine, A., M. Fujiwara, and S. Narumiya. 1989. Asparagine residue in the rho gene product is the modification site for botulinum ADP-ribosyltransferase. *J. Biol. Chem.* **264**:8602–8605.

Simpson, L.L. 1984. Molecular basis for the pharmacological actions of *Clostridium botulinum* type C2 toxin. *J. Pharmacol. Exp. Ther.* **230**:665–669.

Simpson, L.L. 1989. The binary toxin produced by *Clostridium botulinum* enters cells by receptor-mediated endocytosis to exert its pharmacologic effects. *J. Pharmacol. Exp. Ther.* **251**:1223–1228.

Simpson, L.L., B.G. Stiles, H.H. Zepeda, and T.D. Wilkins. 1987. Molecular basis for the pathological actions of *Clostridium perfringens* Iota toxin. *Infect. Immun.* **55**:118–122.

Simpson, L.L., B.G. Stiles, H. Zepeda, and T.D. Wilkins. 1989. Production by *Clostridium spiroforme* of an iotalike toxin that possesses mono(ADP-ribosyl)transferase activity: identification of a novel class of ADP-ribosyltransferases. *Infect. Immun.* **57**:255–261

Stiles, B.G. and T.D. Wilkins. 1986a. *Clostridium perfringens* iota toxin: synergism between two proteins. *Toxicon* **24**:767–773.

Stiles, B.G. and T.D. Wilkins. 1986b. Purification and characterization of *Clostridium perfringens* iota toxin: dependence on two nonlinked proteins for biological activity. *Infect. Immun.* **54**:683–688.

Tamura, M., K. Nogimuri, S. Murai, M. Yajima, K. Ito, T. Katada, M. Ui, and S. Ishii. 1982. Subunit structure of islet-activating protein, pertussis toxin, in conformity with the A-B model. *Biochemistry* **21**:5516–5522.

Touchot, N., P. Chardin, and A. Tavitian. 1987. Four additional members of the ras gene superfamily isolated by an oligonucleotide strategy: molecular cloning of YPT-related cDNAs from a rat brain library. *Proc. Natl. Acad. Sci. USA* **84**:8210–8214.

Trahey, M. and F. McCormick. 1987. A cytoplasmic protein stimulates normal N-ras p21 GTPase, but does not affect oncogenic mutants. *Science* **238**:542–545.

Tsai, S.-C., R. Adamik, Y. Kanaho, E.L. Hewlett, and J. Moss. 1984. Effects of guanyl nucleotides and rhodopsin on ADP-ribosylation of the inhibitory GTP-binding component of adenylate cyclase by pertussis toxin. *J. Biol. Chem.* **259**:15320–15323.

Ueda, T., A. Kikuchi, N. Ohga, J. Yamamoto, and Y. Takai. 1990. Purification and characterization from bovine brain cytosol of a novel regulatory protein inhibiting the dissociation of GDP from and the subsequent binding of GTP to *rho*B p20, a ras p21-like GTP-binding protein. *J. Biol. Chem.* **265**:9373–9380.

Ui, M. 1984. Islet-activating protein, pertussis toxin: a probe for functions of the inhibitory guanine nucleotide regulatory component of adenylate cyclase. *Trends Pharmacol. Sci.* **5**:277–279.

Vandekerckhove, J. and K. Weber. 1979. The complete amino acid sequence of actins from bovine aorta, bovine heart, bovine fast skeletal muscle and rabbit slow skeletal muscle. *Differentiation* **14**:123–133.

Vandekerckhove, J., B. Schering, M. Bärmann, and K. Aktories. 1987. *Clostridium perfringens* iota toxin ADP- ribosylates skeletal muscle actin in Arg-177. *FEBS Lett.* **225**:48–52.

Vandekerckhove, J., B. Schering, M. Bärmann, and K. Aktories. 1988. Botulinum C2 toxin ADP-ribosylates cytoplasmic β/γ-actin in arginine 177. *J. Biol. Chem.* **263**:696–700.

Van Dop, C., G. Yamanaka, F., Steinberg, R.D. Sekura, C.R. Manclark, L. Stryer, and H.R. Bourne. 1984. ADP-ribosylation of transducin by pertussis toxin blocks the light-stimulated hydrolysis of GTP and cGMP in retinal photoreceptors. *J. Biol. Chem.* **259**:23–26.

Wegner, A. and K. Aktories. 1988. ADP-ribosylated actin caps the barbed ends of actin filaments. *J. Biol. Chem.* **263**:13739–13742.

Weigt, C., I. Just, A. Wegner, and K. Aktories. 1989. Nonmuscle actin ADP-ribosylated by botulinum C2 toxin caps actin filaments. *FEBS Lett.* **246**:181–184.

Wieland, T., I. Ulibarri, K. Aktories, P. Gierschik, and K.H. Jakobs. 1990. Interaction of small G proteins with photo-excited rhodopsin. *FEBS Lett.* 263:195–198.

Williamson, K.C., Smith, L.A., Moss, J. and Vaughan, M. 1990. Guanine nucleotide-dependent ADP-ribosylation of soluble rho catalyzed by *Clostridium botulinum* C3 ADP-ribosyltransferase. *J. Biol. Chem.* **265**:20807–20812.

Yeramian, P., P. Chardin, P. Madaule, and A. Tavitian. 1987. Nucleotide sequence of human rho cDNA clone 12. *Nucleic Acids Res.* **15**:1869.

14

Gene Cloning and Organization of the Alpha-Toxin of *Clostridium perfringens*

Richard W. Titball, Helen Yeoman, and Sophie E.C. Hunter

14.1 Introduction

Clostridium perfringens is widely distributed in the environment. In the soil, the bacterium is well adapted to survival by producing an array of hydrolytic enzymes capable of breaking down organic materials to yield essential nutrients (McDonel, 1986). It is not clear why many of these enzymes are also potent toxins. At least 12 toxic proteins have been shown to be produced by *C. perfringens* (McDonel, 1986), but most are considered minor toxins of questionable significance in the pathogenesis of disease.

The pattern of production of the four major toxins can be used to type strains of *C. perfringens* into one of five groups, and of the major toxins, the phospholipase C (alpha-toxin) has received the most attention. The alpha-toxin was the first bacterial toxin shown to possess enzymatic activity (MacFarlane and Knight, 1941), and the biophysical properties of this toxin have been studied extensively by numerous workers. Even so, varied results concerning such fundamental properties as the molecular weight of the toxin have been reported (Mollby, 1978; McDonel, 1986). Studies on the precise mechanisms of toxicity have required cautious interpretation because many of the alpha-toxin preparations used for these studies were probably contaminated with other *C. perfringens* toxins and enzymes (Mollby et al., 1974).

The toxin is produced by most strains of *C. perfringens* and in especially large amounts by type A strains (Mollby et al., 1976). The association of type A strains with gas gangrene in human has lead to the suggestion that this toxin plays an important role in the pathogenesis of this disease (MacFarlane, 1955; Willis, 1969). There is some experimental evidence in support of this; the toxin is known to be cytolytic for erythrocytes (MacFarlane and Knight, 1941; Titball et al., 1989) and human fibroblasts (Mollby et al., 1974; Thelestam and Mollby, 1975) and to be capable of inducing inflammatory responses, smooth muscle contraction (Fujii et al., 1986; Fujii and Sakurai, 1989), and platelet aggregation (Sugahara et al., 1976; Ohsaka et al., 1978), all of which would promote anoxia in tissues and facilitate growth of *C. perfringens*. Prior immunization with a formaldehyde toxoid has been shown to protect mice not only from alpha-toxin but also from infection with *C. perfringens* (Kameyama et al., 1975). In contrast to this evidence, attempts to correlate the level of alpha-toxin production *in vitro* with virulence of different strains of *C. perfringens* have not been successful (Mollby et al., 1976, Mollby, 1978).

Gas gangrene in humans is a serious but usually treatable disease in civilian populations, but during conflicts infection of battlefield wounds remains an important cause of incapacitation and death. For example, during World War I it is estimated that 100,000 German soldiers died of the disease, and attempts to treat the disease with antiserum met with variable results (MacPherson et al., 1922). In addition to gas gangrene in humans, it has been

211

suggested that alpha-toxin plays an important role in several other diseases including necrotic enteritis in chickens (Al-Sheikhly and Truscott, 1977).

To investigate the mode of action of the alpha-toxin and to determine factors that regulate expression of the gene *in vivo*, we have examined the molecular genetics of the encoding gene. In particular, the relationship of gene structure to the function of the protein has been investigated, and homologies found between the alpha-toxin and other procaryotic and eucaryotic lipid-metabolizing enzymes have provided further insight into the mode of action of this toxin. This approach has also permitted the precise determination of some biophysical properties of the toxin, and gene cloning has provided a source of alpha-toxin that is free of other *C. perfringens* toxins and enzymes. Alpha-toxin prepared in this way could be used for toxinological studies and may find increasing use as a reagent for the selective modification of eucaryotic membranes. In future applications, this approach should facilitate a greater understanding of the role of this toxin in the disease process and permit the development of prophylactic and therapeutic reagents against alpha-toxin and *C. perfringens* type A infections.

14.2   Gene Cloning, Location and Detection of Homologous Nucleotide Sequences in Other Bacteria

Gene cloning

Molecular cloning and the nucleotide sequence of the alpha-toxin were first reported by Titball et al. (1989), and following this, several other workers reported the results of similar experiments (Leslie et al., 1989; Okabe et al., 1989; Saint-Joanis et al., 1989; Tso and Siebel, 1989). A high alpha-toxin-producing type A strain of *Clostridum perfringens* (ATCC 13124 = NCTC 8237 = CN 1491) was used as the source of DNA in all studies except for the work of Saint-Joanis et al. (1989), who used strain 8-6, a mutant derived from the heat-resistant type A

strain NCTC 8798. The isolation of intact DNA caused problems for some workers because of the high level of deoxyribonuclease also produced by *C. perfringens* (Okabe et al., 1989). DNA fragments encoding the alpha-toxin were cloned into *Escherichia coli* using plasmids (pUC18, Titball et al., 1989; pEMBL8+, Tso and Siebel, 1989; pACYC184, Leslie et al., 1989) or bacteriophage lambda vectors (gt10, Okabe et al., 1989; NM1149, Saint-Joanis et al., 1989). Recombinant *E. coli* containing the alpha-toxin gene were isolated at a frequency of 0.2% to 1%. Expression of the gene was detected by examining the growth medium surrounding colonies or plaques for egg yolk phospholipid hydrolysis (Titball et al., 1989) or hemolysis of sheep erythrocytes (Leslie et al., 1989; Okabe et al., 1989; Saint-Joanis et al., 1989; Tso and Siebel, 1989). No instability of the cloned DNA was reported and the produced alpha-toxin did not appear to be detrimental to the host, possibly because of the low levels of phosphatidylcholine and sterols and the preponderance of odd-number branch-chain fatty acids found in procaryotic cell membranes (Tso and Siebel, 1989).

Chromosomal location of the alpha-toxin gene

It has been reported that *C. perfringens* plasmids play no role in the genetic control of alpha-toxin production (Kramer and Schallehn, 1978), and it is now known that the alpha-toxin gene is chromosomally located in strain CPN50 (Canard and Cole, 1989). The location of the gene, close to the putative origin of replication, would explain the low incidence of alpha-toxin-negative mutants because this region would be expected to be highly conserved (Canard and Cole, 1989). In addition, the observation that the gene was flanked by ribosomal RNA operons suggests that this region may be readily exchanged between strains by recombination.

By combining the reported restriction endonuclease maps of the cloned DNA fragments from strain NCTC 8237 (Leslie et al., 1989; Okabe et al., 1989; Titball et al., 1989; Tso and Siebel, 1989) a map of the alpha-toxin gene and flanking DNA has been constructed (Figure

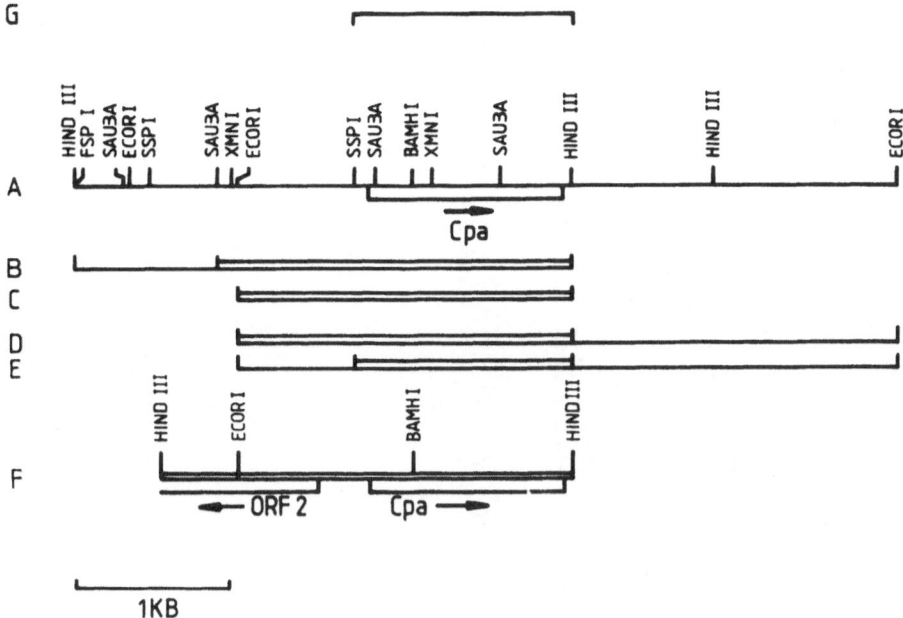

FIGURE 14.1 Genome organization adjacent to alpha-toxin gene; locations of cloned and nucleotide sequenced fragments. (A) Restriction endonuclease map of alpha-toxin gene (cpa) and flanking DNA from *Clostridium perfringens* strain NCTC 8237. Locations of fragments that were cloned (B, pT5C100, Titball et al., 1989; C, pPLC8, Tso and Siebel, 1989; D, pDAZ1, Leslie et al., 1989; E, pKM1, Okabe et al., 1989) are depicted with single line. Regions that were nucleotide sequenced are shown as double lines. Locations of restriction endonuclease sites, cpa gene, and unidentified open-reading frame (ORF 2) within DNA fragment from *C. perfringens* strain 8-6 (F, pTOX5, Saint-Joanis et al., 1989) are also shown. Location of the DNA fragment used as gene probe to detect alpha-toxin gene is also depicted (G).

14.1A). The homologous region from strain 8-6 is also shown (Figure 14.1F), with the location of the second unidentified open-reading frame identified only in this strain (Saint-Joanis et al., 1989).

Conservation of sequences resembling the alpha-toxin gene in other phospholipase C-producing clostridia

We have examined in detail the degree of conservation of the alpha-toxin gene between strains of *C. perfringens* using a DNA probe corresponding to the alpha-toxin gene (*Ssp*1 to *Hind*III fragment; Figure 14.1G). The probe was hybridized with DNA from a selection of phospholipase C-producing clostridia and other bacteria under conditions of high stringency (Table 14.1). homologous 3.1-kb *Hind*III fragment was detected in DNA isolated from *C. perfringens* types A to E, including one strain that failed to produce detectable phospholipase C on growth in liquid or on solid media (R476). In addition, sequences homologous to the alpha-toxin gene were detected in DNA isolated from one strain of *C. bifermentans* (LPHL 2825), *C. novyi* type A, and *C. barati* (formerly *C. paraperfringens*). Homologous sequences were not detected in DNA from any of the other phospholipase C-producing bacteria tested. The possibility that phospholipase C genes, which are closely related to the alpha-toxin gene, occur in other clostridial species was confirmed by the recent report that the *C. bifermentans* phospholipase C-encoding gene was 64% homologous with the alpha-toxin gene (Tso and Siebel, 1989). Previous studies have shown that the phospholipase C produced by *C. barati* was nontoxic, and there are differing reports concerning neutralization using anti-alpha-toxin serum (Naka-

Table 14.1 Detection of DNA homologous to the alpha-toxin gene in phospholipase C-producing bacteria

Origin of DNA (species and biotype)				Size of probe-reactive fragment (HindIII digest, in kb)[a]
Clostridium perfringens	type A	NCTC	8237	3.1
Clostridium perfringens	type B	NCTC	6121	3.1
Clostridium perfringens	type C	NCTC	3180	3.1
Clostridium perfringens	type D	NCTC	8346	3.1
Clostridium perfringens	type E	CN	5083	3.1
Clostridium perfringens			R476	3.1
Clostridium bifermentans		LPHL	2825	3.5
Clostridium novyi	typeA	NCTC	6737	3.1
Clostridium barati		ATCC	27639	3.3
Clostridium absonum		ATCC	27555	−
Clostridium novyi	type D	NCTC	8350	−
Bacillus cereus			B 33	−
Pseudomonas aeruginosa		NCTC	2531	−
Staphylococcus aureus		NCTC	6571	−

[a]Minus (−) indicates no probe-reactive fragment detected.

mura et al., 1973; Willis, 1977). The gamma-toxin (phospholipase C) produced by *C. novyi* is hemolytic and lethal, and it has been shown to be partially neutralized by anti-alpha-toxin serum (Willis, 1977). Unlike the alpha-toxin, the *C. novyi* enzyme was able to degrade phosphatidyl inositol and phosphatidyl glycerol. It would appear that the phospholipase C's from several clostridial species are closely related genetically and immunologically but differ in toxicity and substrate specificity. The precise nature of the proteins and of their encoding genes awaits determination.

14.3 Open-Reading Frame Nucleotide Sequence

The nucleotide sequence of the gene encoding the alpha-toxin from strain NCTC 8237 has been determined independently by four groups (Figure 14.2) (Leslie et al., 1989; Okabe et al., 1989; Titball et al., 1989; Tso and Siebel, 1989), and, with the exception of one residue (Ser^{17} to Thr^{17}; Tso and Siebel, 1989), the deduced amino acid sequences are identical. Comparison with the deduced sequence from strain 8-6 revealed several differences (Saint-

Joanis et al., 1989) that were clustered more frequently in the signal peptide.

The low GC ratio of *C. perfringens* DNA (24–27 mol%; Buchanan and Gibbons, 1974) is reflected in the codon usage of the alpha-toxin gene that is biased toward codons in which A or U predominate. This is particularly noticeable for amino acids having six codon synonyms: the codon AGA for arginine is used almost exclusively, and there is a marked preference for the codons UUA for leucine and UCA for serine. The extent of the A or U bias is highlighted by the fact that 79% of codons end in these bases, comparing well with an average of 84.5% for other *C. perfringens* genes. This figure is similar to that reported for *C. acetobutylicum* (87%; Young et al., 1989a), *C. pasteurianum* (81%; Young et al., 1989b), and *C. perfringens* plasmid-borne genes (86%; Garnier and Cole, 1988), but not to that reported for *C. thermocellum* (60%; Young et al., 1989a). Thus, the codon usage of the *C. perfringens* alpha-toxin gene is similar to the codon usage of the majority of clostridial genes. However, it differs substantially from the codon usage found in other bacteria such as *Escherichia coli* and *Bacillus subtilis* (Table 14.2).

In particular, a number of codon synonyms that are frequently found in the alpha-toxin

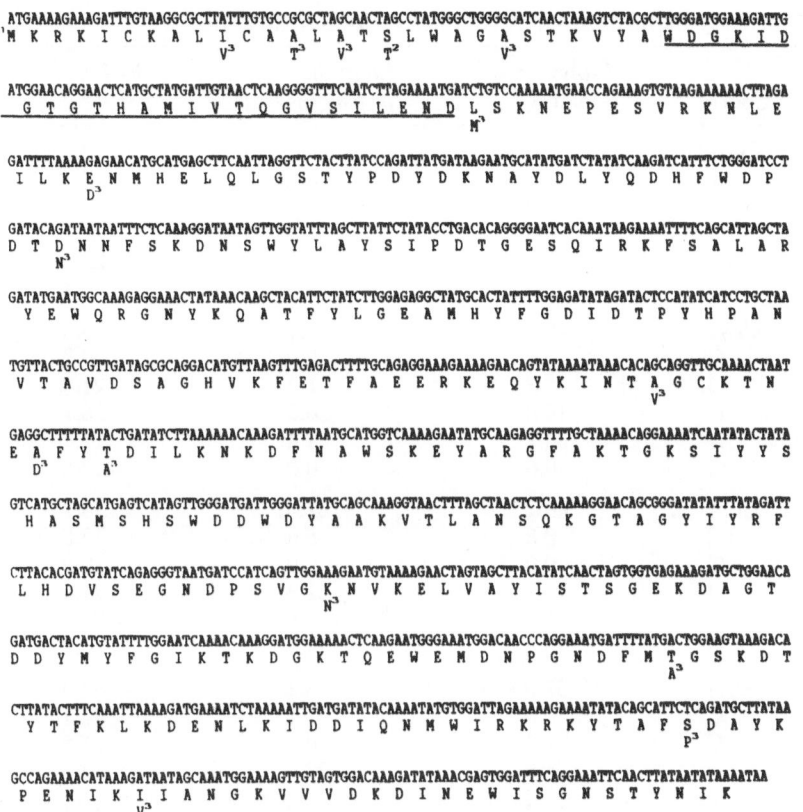

FIGURE 14.2 Open reading frame nucleotide sequence of alpha-toxin. 1, NCTC 8237 consensus sequence; 2, Tso and Siebel 1989. Strain NCTC 8237; 3, Saint-Joanis et al. 1989, strain 8-6. Experimentally determined N-terminal amino acid sequence of alpha-toxin, isolated from *Clostridium perfringens* culture supernatant fluid, is shown underlined (Titball et al., 1989, Okabe et al., 1989).

gene are found only rarely in *E. coli* genes. For example, the codons AGA (arginine), UUA (leucine), UCA (serine), ACA (threonine), GGA (glycine), and AUA (isoleucine) are found in fewer than 10% of cases in *E. coli* genes, but at least three fold more frequently in the alpha-toxin gene. It has been suggested that this difference in codon usage may present a barrier to translation of some clostridial genes, especially large genes (Garnier and Cole, 1988; Leslie et al., 1989). However, the alpha-toxin gene was efficiently expressed in *Escherichia coli*.

A wide range of values (30,000–106,000) have been reported for the experimentally determined molecular weight of the alpha-toxin (McDonel, 1986), but recently reported data from Sodium dodecyl sulfate-polyacrylamid gel electrophoresis (SDS-PAGE) analyses are similar to the values of 42,474 (Saint-Joanis et al., 1989) to 45,528 (Titball et al., 1989) for the mature toxin calculated from the deduced amino acid sequence.

14.4   Transcriptional and Translational Signals for the Alpha-Toxin Gene

Promoter sequences

Putative promoter regions have been identified for many of the clostridial genes cloned and sequenced to date. Localization of the promoter, by primer extension or S1 nuclease mapping of the transcript start site, has been reported for only a limited number of these.

Table 14.2 Codon usage of the *Clostridium perfringens* alpha-toxin gene

Amino acid	Codon	Alpha toxin gene (%)[a]	Clostridium perfringens (%)[b]	Bacillus subtilis (%)[c]	Escherichia coli (%)[c]
Arg	CGA	0.0	0.0	9.1	2.3
	C	0.0	0.0	17.5	35.0
	G	0.0	0.0	11.1	3.2
	U	0.0	21.2	25.2	58.1
	AGA	98.0	75.8	27.7	1.2
	G	2.0	3.0	9.4	0.3
Leu	CUA	26.6	9.4	6.3	1.8
	C	0.0	0.0	9.8	6.6
	G	4.3	1.2	21.8	69.1
	U	16.0	20.0	26.1	8.6
	UUA	53.2	69.4	22.4	5.8
	G	0.0	0.0	13.6	8.2
Ser	UCA	48.1	32.3	18.7	8.3
	C	3.8	5.6	12.0	25.6
	G	0.0	0.0	10.0	11.4
	U	11.3	32.3	24.4	26.5
	AGC	11.3	5.6	24.1	21.6
	U	25.5	24.2	10.7	6.5
Thr	ACA	37.0	55.3	43.3	5.9
	C	0.0	7.1	14.1	50.6
	G	1.5	0.0	27.9	19.7
	U	61.5	37.6	14.8	23.8
Pro	CCA	67.4	55.3	19.1	19.9
	C	0.0	5.3	9.8	6.0
	G	0.0	0.0	37.5	65.1
	U	32.6	39.4	33.6	9.0
Ala	GCA	39.6	47.7	27.1	22.9
	C	6.2	10.8	20.0	18.8
	G	8.0	0.0	25.4	30.5
	U	49.6	41.5	27.5	27.9
Gly	GGA	68.5	59.2	31.5	4.6
	C	0.0	11.8	29.6	40.8
	G	11.5	1.3	13.5	6.8
	U	20.0	27.7	25.4	47.8
Val	GUA	48.8	42.5	24.5	22.9
	C	6.3	2.7	24.7	12.9
	G	5.0	4.1	19.4	26.8
	U	40.0	50.7	31.4	37.5
Ile	AUA	50.0	51.0	10.6	0.5
	C	12.7	6.9	39.4	62.2
	U	37.3	42.1	50.0	37.3
Lys	AAA	72.9	70.2	75.4	76.7
	G	27.1	29.8	24.6	23.3
Asn	AAC	32.6	26.9	46.9	75.8
	U	67.4	73.1	53.1	24.2
Gln	CAA	90.0	93.7	54.2	26.6
	G	10.0	6.3	45.8	73.4
His	CAC	22.2	26.7	31.4	61.1
	U	77.8	73.3	68.6	38.9

Table 14.2 (Continued)

Amino acid	Codon	Alpha toxin gene (%)[a]	*Clostridium perfringens* (%)[b]	*Bacillus subtilis* (%)[c]	*Escherichia coli* (%)[c]
Glu	GAA	59.6	93.2	69.5	73.4
	U	40.4	6.8	30.5	26.6
Asp	GAC	14.2	20.5	36.2	49.0
	U	85.8	79.5	63.8	51.0
Tyr	UAC	14.8	9.8	38.2	59.4
	U	85.2	90.2	61.8	40.6
Cys	UGC	33.3	28.6	54.3	58.0
	U	66.7	71.4	45.7	40.6
Phe	UUC	40.0	14.3	36.0	56.5
	U	60.0	85.7	64.0	43.5
Met	AUG	100.0	100.0	100.0	100.0
Trp	UGG	100.0	100.0	100.0	100.0

[a] From Titball et al., 1989; Leslie et al., 1989; Tso and Siebel, 1989; Okabe et al., 1989; Saint-Joanis, 1989.
[b] From Roggentin et al., 1988; Tweten, 1988; van Damme-Jongsten, 1989.
[c] From Ogasawara, 1985.

The alpha-toxin gene promoter has been mapped experimentally in strain 8-6 (Saint-Joanis et al., 1989) and NCTC 8237 (Yeoman et al., in manuscript). Differences in the position of the transcript start site were observed (Figure 14.3). The promoters from each strain contained identical Pribnow boxes, which matched both the *B. subtilis* Eσ^{43} and *E. coli* Eσ^{70} consensi. However, differences between the promoters were seen in the −35 region; in strain 8-6 the final nucleotide was G, whereas in strain NCTC 8237 it was a C (see Figure 14.3). In both strains the −35 region corresponds to the *B. subtilis* Eσ^{43} consensus in the second, third, and fourth bases. Percentage conservation studies have indicated that the first four bases of this region play the most important role in polymerase recognition in gram-positive bacteria (Graves and Rabinowitz, 1986).

In addition, the "extended promoter regions" that may play a role in polymerase recognition in gram-positive bacteria (Graves and Rabinowitz, 1986) are also apparent. The degree of conservation is particularly marked for a TG motif around the −15 region. A similarly located TG motif may enhance gene expression in *E. coli* from promoters with weak −35 homology (Keilty and Rosenberg, 1987; Ponnambalam et al., 1988).

Ribosome-binding site

A large number of putative clostridial Shine–Dalgarno regions have been reported. The free energy of interaction of these ribosome-binding sites with *B. subtilis* 16S rRNA have been calculated (Tinoco et al., 1973), ranging from −10.5 to −23.5 kcal per mole. The free energy calculated for the alpha-toxin mRNA–16S rRNA complex was −10 kcal per mole, suggesting that the initiation complex is less stable than those predicted for the other clostridial genes. In view of the high level of alpha-toxin production in *C. perfringens*, it is therefore possible that the control of gene expression is not exerted at the level of translational initiation.

Like most clostridial genes the translational initiation codon of the alpha-toxin is ATG. This is located 9 nucleotides downstream of the Shine–Dalgarno purine tract and corresponds well with the clostridial consensus of 8 nucleotides.

Transcriptional stop signals

The termination of transcription of some clostridial genes has been investigated by S1 nuclease protection studies (Beguin et al., 1986; Graves and Rabinowitz, 1986; Garnier and Cole, 1988).

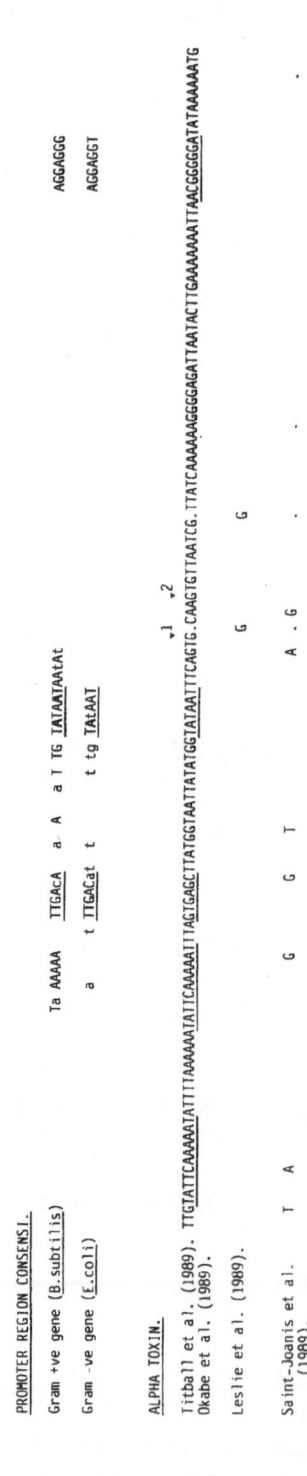

PROMOTER REGION CONSENSI.

| Gram +ve gene (B.subtilis) | Ta AAAA | TTGAcA a A a T TG TATAATAATAt | AGGAGGG |
| Gram -ve gene (E.coli) | a | t TTGACat t   t tg TAtAAT | AGGAGGT |

ALPHA TOXIN.

Titball et al. (1989). TTGTATTCAAAAATATTTAAAAAATATTCAAAAATTTAGTGAGCTTATGGTAATTATATGGTATAATTTCAGTG.CAAGTGTTAATCG.TTATCAAAAAAGGGGAGATTAATACTTGAAAAAATTAACGGGGGATATAAAAATG
Okabe et al. (1989).

Leslie et al. (1989).

Saint-Joanis et al. (1989).

FIGURE 14.3 Upstream regions of alpha-toxin gene: comparison with *Escherichia coli* Eσ70 and *Bacillus subtilis* Eσ43 consensus promoter sequences. Nucleotide differences between strains of *Clostridium perfringens* (NCTC 8237: Titball et al., 1989; Okabe et al., 1989; 8-6, Saint-Joanis et al., 1989). Dots (·) refer to absent nucleotides; transcript start points, as found by Yeoman et al. *in manuscript* and Saint-Joanis et al. (1989) are indicated by arrows. Suggested −10, −35, and upstream repeat sequences are shown underlined. Promoter region consensi of gram-positive and gram-negative genes are also shown; "extended" promoter regions postulated to play role in polymerase recognition are indicated (Graves and Rabinowitz, 1986). Bold capitals represent more than 75% conservation; capitals, 50%, lower case letters, conserved in more than 41% of cases.

These transcriptional terminators appear to be *rho* independent, with characteristic areas of dyad symmetry in the DNA, as long as 19 nucleotides and capable of hairpin loop formation. An inherent or subsequent poly-T tract, when transcribed to poly-U, leads to mRNA release.

Sequence data reported for the alpha-toxin gene included a stretch of 44 nucleotides downstream of the triple translational termination codons (Leslie et al., 1989; Okabe et al., 1989; Saint-Joanis et al., 1989; Titball et al., 1989). There are no obvious *rho*-independent transcriptional terminators in this region, and examination of the nucleotide sequence downstream of the previously published region has revealed the presence of a possible *rho*-dependent terminator located 54 nucleotides downstream of the translational stop codons (Yeoman et al., in manuscript).

Regulation of gene expression

Determination of mRNA levels in a heat-resistant strain of *C. perfringens* (Saint-Joanis et al., 1989) showed that the alpha-toxin gene was expressed constitutively and that production was about threefold higher in the stationary phase than in the logarithmic phase. This pattern of gene transcription may not be typical, because previous reports have indicated that maximal toxin production occurs in the logarithmic phase, with a reduction on entry into the stationary phase (Smyth and Arbuthnott, 1974).

Several reports have indicated the importance of factors such as iron and phosphate concentration for microbial gene expression (Pritchard and Vasil, 1986; Calderwood and Mekalanos, 1987; Guddal et al., 1989; Goldberg et al., 1990). Murata et al. (1965) demonstrated a possible relationship between environmental iron levels and alpha-toxin production by *C. perfringens*, but they were unable to show that environmental phosphate levels also affect alpha-toxin production. Environmental phosphate concentration has been shown to regulate production of the phospholipase C's of both *Pseudomonas aeruginosa* (Pritchard and Vasil, 1986) and *Bacillus cereus* (Guddal et al.,

1989), and it is possible that alpha-toxin is regulated in a similar manner.

It has been suggested that phospholipase C and alkaline phosphatase function cooperatively as part of a phosphate-scavenging mechanism within *P. aeruginosa* (Liu, 1979). This may also be the case for the *C. perfringens* alpha-toxin, as the protein shows structural motifs, such as central, catalytically essential zinc ions, that are common to both the *B. cereus* phospholipase C and alkaline phosphatase. These regulatory mechanisms may only operate in some strains of *C. perfringens*, and the reason why alpha-toxin is produced at much higher levels by type A strains has also yet to be elucidated. It is possible that global regulatory mechanisms operate in some biotypes.

14.5   Export and Posttranslational Processing

The N-terminal amino acid sequence of the alpha-toxin isolated from *Clostridium perfringens* NCTC 8237 culture supernatant fluid has been experimentally determined by two groups (see Figure 14.2) (Okabe et al., 1989; Titball et al., 1989). In both cases the results indicated that an N-terminal, 28-amino-acid peptide, with the characteristics of a signal sequence (Sarvas, 1986), was removed on export. Other observations confirmed this possibility; the cloned toxin was found predominantly in the periplasmic space of *E. coli*, and minicell analysis of the cloned gene products revealed an additional polypeptide corresponding to the precursor form of the toxin (Leslie et al., 1989).

The acquisition of three catalytically essential zinc ions probably occurs intracellularly, soon after protein synthesis, because the apoprotein is highly susceptible to proteolysis (Sato et al., 1978). The sequential acquisition (or loss) of zinc ions may be responsible for the appearance of several forms of the toxin with similar biological properties after isoelectric focusing (Krug and Kent, 1984), but we have been unable to demonstrate interconversion of the different forms after toxin incubation with 15 mM EDTA or 190 mM zinc ions. The difference in the main form of the protein expressed from the gene in *C. perfringens* (pI, 5.48) and *E. coli* (pI, 5.6) suggests that an active mechanism may be responsible for the observed heterogeneity. Phosphorylation of the alpha-toxin does not occur in *E. coli* (K. a Horics, unpublished observations), and the possibility that deamidation of the protein leads to charge differences has not been investigated.

14.6   Functional Analysis of the Alpha-Toxin

Biological activities of the alpha-toxin

Many of the early functional studies of alpha-toxin used protein preparations of dubious purity (Mollby, 1978; McDonel, 1986) and need to be interpreted with caution. The problem of obtaining alpha-toxin free of other *C. perfringens* toxins has now been circumvented by gene expression in *E. coli*. Toxin purified from this source was similar to the native protein (Titball et al., 1989) and was able to hydrolyze phosphatidylcholine and sphingomyelin (Saint-Joanis et al., 1989), cause mouse erythrocyte lysis (Titball et al., 1989), and kill mice (Tso and Siebel, 1989). These results contrast with those found in *Bacillus cereus*, a nonpathogenic aerobe in which the separate phosphatidylcholine-preferring phospholipase C (PC-PLC) and sphingomyelinase act synergistically to cause hemolysis (Gilmore et al., 1989).

In addition to the cytolytic activity of the alpha-toxin, it is now apparent that limited membrane damage by the toxin may elicit more subtle actions on cells and organs. Fujii and Sakurai (1989) have reported that alpha-toxin treatment of cells lining the rat aorta resulted in activation of the arachidonic acid cascade, leading to the production of thromboxanes which then caused an inflammatory response. The phospholipase C of *Pseudomonas aeruginosa* has also been reported to increase the levels of arachidonic acid metabolites from human granulocytes and mouse peritoneal cells (Meyers and Berk, 1990), resulting in inflammatory responses. The inflammatory responses elicited by these toxins could be reduced by using inhibitors of enzymes in the arachidonic acid pathway.

Treatment of rat aorta cells with alpha-toxin has also been reported to cause an increase in intracellular inositol-1,4,5-triphosphate, leading to increases in calcium levels and the activation of the calcium gate (Fujii et al., 1986). The resultant contraction of the aorta wall may be of significance when considering the role of alpha-toxin in the pathogenesis of disease.

Homology with other lipid-metabolizing enzymes

Significant insight into the molecular architecture and function of the alpha-toxin has been obtained by comparing the deduced amino acid sequence with those of other phospholipid-metabolizing enzymes. The most frequently reported homology is that between the N-terminal two-thirds of the alpha-toxin and the entire smaller, nontoxic PC-PLC (Leslie et al., 1989; Titball et al., 1989; Titball and Rubidge, 1990) (Figure 14.4). The possibility that the N-terminal two-thirds of the alpha-toxin represents a structural and functional homolog of the PC-PLC has recently been confirmed by expressing a truncated alpha-toxin from a recombinant plasmid (pBl5b), which contained the gene lacking the 3' coding region. The purified alpha-truncate, which terminated at proline[249], was able to digest egg yolk phospholipids as efficiently as did the alpha-toxin but did not hydrolyze sphingomyelin or cause the lysis of mouse erythrocytes. It was also at least 10 fold less toxic (Titball et al., 1991).

Calculation of the hydropathic potentials of the alpha-toxin and *B. cereus* PC-PLC did not reveal any marked similarity (Titball et al., 1989). It has been suggested that some regions of the alpha-toxin (Asp[30] to Leu[50] and Ala[145] to Pro[165]) could form surface-seeking amphipathic alpha helices (Saint-Joanis et al., 1989). These regions have no obvious functional counterparts in the *B. cereus* PC-PLC, although they would correspond to the structural helices A and D (Hough et al., 1989). It is possible that these regions promote the interaction of alpha-

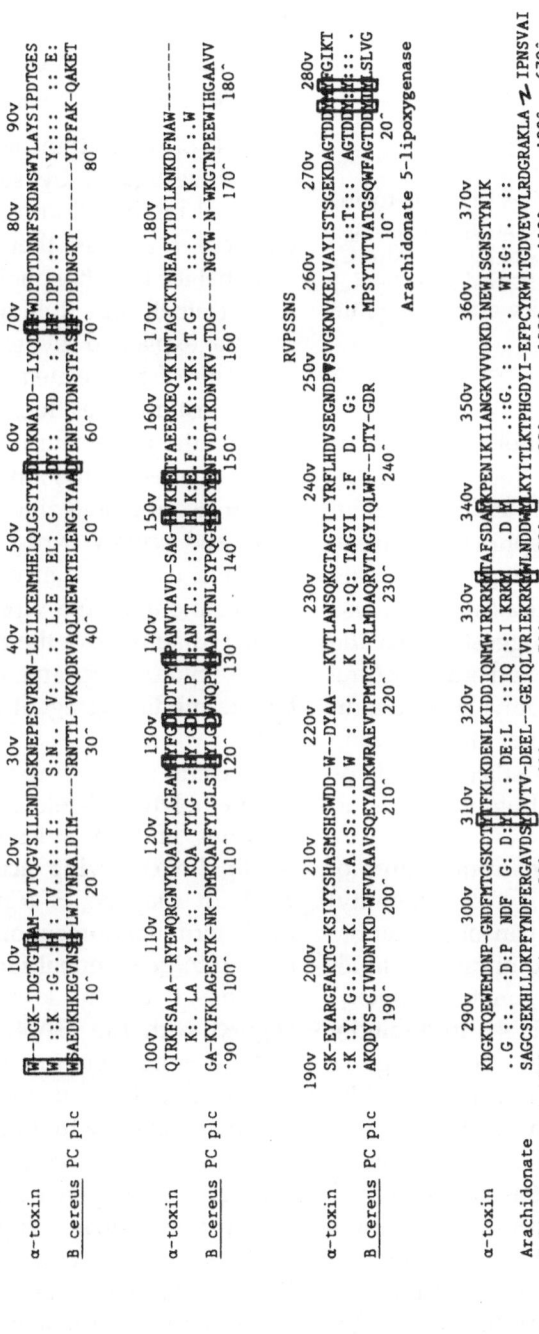

Figure 14.4 Alignment of the amino acid sequences of alpha-toxin, *Bacillus cereus* PC-PLC and arachidonate-5-lipoxygenase. Aligned histidine residues with box; PC-PLC and tyrosine residues with HA5L, point of truncation of alpha-toxin gene in plasmid pB15b, arrow.

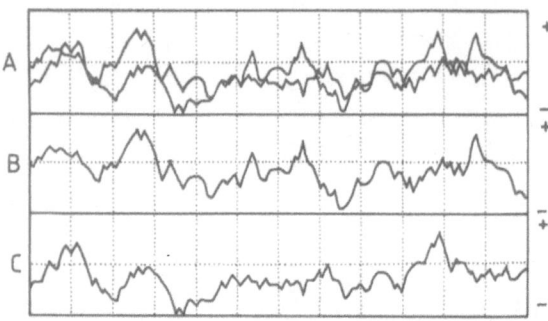

FIGURE 14.5 Hydropathy profiles of alpha-toxin, *Bacillus cereus* PC-PLC, and arachidonate-5-lipoxygenase. Hydropathy profiles of alpha-toxin (C, residues 249–370) and HA5L (B, residues 1–113) calculated using algorithm of Kyte and Doolittle (1982) are superimposed (A).

toxin with membrane phospholipids (Moreau et al., 1988).

In view of both these results and the tandem location of PC-PLC and the sphingomyelinase genes in *B. cereus*, it is tempting to speculate that the alpha-toxin represents a fusion of these two elements. It has been suggested that the C-terminal domain of alpha-toxin is a structural analog of the *B. cereus* sphingomyelinase (Saint-Joanis et al., 1989), but neither the encoding genetic elements nor the predicted proteins show significant sequence homology (Saint-Joanis et al., 1989).

In contrast, there is a significant degree of amino acid sequence homology (29% identity) between the C-terminal domain of the alpha-toxin and residues 1–113 of a eucaryotic lipid-modifying enzyme, arachidonate 5-lipoxygenase (HA5L) (see Figure 14.4). In addition, the hydropathy profiles of these regions showed a high degree of similarity (Figure 14.5), and a region of alpha helix was strongly predicted in the aligned region of both proteins (alpha-toxin 310–325, HA5L 60–70). The HA5L has not been reported to possess sphingo-myelin-hydrolyzing activities but it is required for the generation of leukotrienes from arachidonic acid (Samuelsson, 1983). It is possible that the N-terminal region of HA5L is involved in the recognition of hydrocarbon chains; the C-terminal region of alpha-toxin may perform a similar function by facilitating the recognition

of sphingomyelin. The removal of this domain in the alpha-truncate could explain the increased hydrolysis of the synthetic substrate *p*-nitrophenol phosphorylcholine because of the increased accessibility of the active site.

Characterization of functionally critical amino acids

The role of some amino acids in alpha-toxin has been investigated by several workers using chemical modification techniques (Table 14.3). Of these, it appears that histidine residues are essential for the phospholipase C activity of the toxin (Titball and Rubidge, 1990). This result can be explained by examining the tertiary structure of the related *B. cereus* PC-PLC, because histidine residues that are involved in binding the three zinc ions (Hough et al., 1989) are conserved in the alpha-toxin (Figure 14.4), which also required zinc ions for structural stability and activity (Krug and Kent, 1984). The histidine residues can only be modified after pretreatment of the alpha-toxin with EDTA, which could remove the zinc ions exposing histidine imidazole groups to the diethylpyrocarbonate modifying agent (Titball and Rubidge, 1990).

Tyrosine residues appear to be essential for the hemolytic, platelet-aggregating and lethal activities and to a lesser extent, for phospholipid-hydrolyzing activity (Sakurai et al., 1989). It may be of significance that within the C-terminal domain of alpha-toxin, which is known to be essential for hemolytic and lethal activities, five of the seven tyrosine residues are aligned with the HA5L residues (see Figure 14.4). It is possible that these C-terminal tyrosine residues play an essential role in the recognition of substrate hydrocarbon chains.

The single cysteine residue within the alpha-toxin appears to play no direct role in the activity of the protein. Superimposition of this residue onto the crystal structure of the *B. cereus* PC-PLC (Hough et al., 1989) indicates that it would be located distally to the active site. In this orientation, disulfide bridges could form with other alpha-toxin or theta-toxin molecules without masking the phospholipase C active site.

Table 14.3 Chemical modification of alpha-toxin

Residue modified	Effect on biological activity				
	EY	pNPPC	Hly	PtA	Lty
Histidine[a]	+	+	NT	NT	NT
Cysteine	−[b,c,e,]	−[e]	−[e]	−[e]	−[e]
Arginine[d]	−	NT	NT	NT	NT
Tyrosine[f]	±	−	+	+	+
Tyrosine + cysteine[g]	±	−	+	+	+
Tyrosine + cysteine[h] + histidine + NH$_2$	+	−	+	+	+

[a] Modification with EDTA + diethylpyrocarbonate (Titball and Rubidge, 1990). Inhibition (+), partial inhibition (±), or noninhibition (−) of egg yolk phospholipid (EY) hydrolysis, *p*-nitrophenol phosphorylcholine (pNNPC) hydrolysis, mouse erythrocyte lysis (Hly), platelet aggregation (PtA), or lethality (Lty). NT, not tested.
[b] Modification with *p*-hydroxymercuribenzoic acid (pHMCB) (1:1 to 1000:1 molar ratio pHMCB:alpha toxin cysteine residues, pH 6, 1 h, 22°C, ± 2m*M* EDTA).
[c] Modification with iodoacetic acid. Method as for *p*-hydroxymercuribenzoic acid modification[b].
[d] Modification with phenylglyoxal (PF) (10:1 to 100:1 molar ratio PG:alpha-toxin arginine residues, pH 8, 1h, 22°C, ±2 m*M* EDTA).
[e] Modification with N-ethylmaleimide (Sakurai et al., 1989).
[f] Modification with N-acetylimidazole (Sakurai et al., 1989).
[g] Modification with tetranitromethane (Sakurai et al., 1989).
[h] Modification with maleic anhydride (Sakurai et al., 1989).

14.7 Antigenic Analysis of the Alpha-Toxin

By using gel double diffusion techniques with polyclonal anti-alpha serum, it has been shown that the *E. coli* cloned alpha-toxin shares many or all of the epitopes found on the native protein (Titball et al., 1989). Individual epitopes have recently been investigated using monoclonal antibodies against the alpha-toxin from strains NCTC 8237 (Shuttleworth et al., 1988) and BP6K (Sato et al., 1989). In both cases of antibodies capable of neutralizing phospholipase C hemolytic and lethal activities were reported, but these activities were not coneutralized to the same extent (Sato et al., 1989), suggesting the presence of different functional sites in the alpha-toxin molecule. This is partially confirmed by our results showing that the alpha-toxin is composed of two domains (see Section 14.6).

All the monoclonal antibodies generated by Logan et al. (in manuscript) reacted equally well with the alpha-toxin or alpha-truncate (see Section 14.6), suggesting that the immunodominant region of the protein is located between amino acids 1 and 249. Only the toxin-neutralizing monoclonal antibody (3α4D10) reacted with alpha-toxin after Western blot analysis. The exact binding site of antibody 3α4D10 has been mapped more precisely after chemical cleavage of the alpha-toxin at the single cysteine residue (2-nitro-5-thiocyanobenzoic acid cleavage) or partial cleavage at methionine residues (cyanogen bromide cleavage). The results indicated that the antibody recognized a region of the protein between Cys^{197} and Met^{238} (Logan et al., in manuscript). Further characterization of the recognized epitope is currently in progress using peptides corresponding to this region.

14.8 Conclusions

A resurgence of interest in the toxins of *C. perfringens* has occurred. The cloning and nucleotide sequencing of the alpha-toxin gene has facilitated the precise determination of several fundamental biophysical properties of the protein. The role of alpha-toxin in disease is still not fully elucidated, and the regulation of gene expression *in vivo* is not understood. The failure to correlate *in vitro* alpha-toxin production with virulence should be investigated at a molecular level. We have recently cloned the gene encoding *C. perfringens* epsilon-toxin and determined its nucleotide sequence (Hunter et al., in manuscript). Parallel studies on the control of toxin gene expression may reveal the nature of regulatory systems.

Considerable progress has been made in understanding the molecular architecture and function of the toxin. It is apparent that the alpha-toxin is composed of two domains. One of these constitutes a phospholipase C capable of digesting phosphatidylcholine. The C-terminal domain is required for sphingomyelin hydrolysis, hemolysis, and lethality of the protein. The exact role of this domain has yet to be elucidated, and although this region appears to be essential for sphingomyelin hydrolysis, it seems unlikely that this domain is endowed with a catalytic site.

In the case of the two characterized phospholipase C's of *Pseudomonas aeruginosa*, it has been shown that the N-terminal two-thirds of these proteins show extensive amino acid sequence homology and that the C-terminal region may be responsible for the sphingomyelin specificity of the haemolytic enzyme (Ostroff et al., 1990). The C-terminal domain of the alpha-toxin may contain regions able to interact with membrane phospholipids. The elucidation of the function of the C-terminal domain of alpha-toxin would not only provide further insight into the molecular basis of toxicity but could also indicate the function of the homologous domain in HA5L, a eucaryotic enzyme of importance in the pathogenesis of cardiovascular and inflammatory diseases of man.

By fusing different domains from the alpha-toxin, *B. cereus* PC-PLC or *Clostridium bifermentans* phospholipase C the molecular basis of toxicity could be further investigated. If the phospholipase C-encoding genes from *C. novyi* and *C. barati* are also structurally related, then comparison of these proteins with the alpha-toxin could reveal the molecular basis of substrate specificity. The functions of some amino acids in the alpha-toxin can be tentatively assigned from comparison of their homologues in the *B. cereus* PC-PLC. A more precise determination of their functions will be possible using site-directed mutagenesis.

Expression of these native or modified phospholipase C genes in *E. coli* could provide a source of enzyme, free of other clostridial exoproteins, that could be used as phospholipid-specific probes for the investigation of eucaryotic cell membranes.

The development of a vaccine based on the alpha-toxin could proceed in a number of directions. The finding that a neutralizing antibody binds to a discrete region of the alpha-toxin suggests that a peptide vaccine could be effective. Alternatively, a vaccine based on an inactive N- or C-terminal domain of the alpha-toxin may enable a number of epitopes to be presented in the form of a nontoxic polypetide. It has been demonstrated that an antiidiotype vaccine is capable of protecting mice from *C. perfringens* epsilon toxin and infections caused by *C. perfringens* type D (Percival et al., 1990), and it is possible that this approach could also be applied to the generation of an alpha-toxin vaccine.

References

Al-Sheikhly, F. and R.B. Truscott. 1977. The interaction of *Clostridium perfringens* and its toxins in the production of necrotic enteritis of chickens. *Avian Dis.* 21:256–263.

Beguin, P., M. Rocancourt, M.C. Chebrou, and J.P. Aubert. 1986. Mapping of mRNA encoding endoglucanase A from *Clostridium thermocellum*. *Mol. Gen. Genet.* 202:251–254.

Buchanan, R.E. and N.E. Gibbons. 1974. *Bergey's Manual of Determinative Bacteriology*, 8th Ed. Baltimore: Williams & Wilkins.

Calderwood, S.B. and J.J. Mekalanos. 1987. Iron regulation of Shiga-like toxin expression in *Escherichia coli* is mediated by the *fur* locus. *J. Bacteriol.* 169:4759–4764.

Canard, B. and S. Cole. 1989. Genome organisation of the anaerobic pathogen *Clostridium perfringens*. *Proc. Natl. Acad. Sci. USA* 86:6676–6680.

Fujii, Y. and J. Sakurai. 1989. Contraction of the rat isolated aorta caused by *Clostridium perfringens* alpha-toxin (phospholipase C); evidence for the involvement of arachidonic acid metabolism. *Br. J. Pharmacol.* 97:119–124.

Fujii, Y., S. Nomura, Y. Oshita, and J. Sakurai. 1986. Excitatory effect of *Clostridium perfringens* alpha toxin on the rat isolated aorta. *Br. J. Pharmacol.* 88:531–539.

Garnier, T. and S.T. Cole. 1988. Complete nucleotide sequence and genetic organisation of the bacteriocinogenic plasmid pIP404, from *Clostridium perfringens. Plasmid* **19**:134–150.

Gilmore, M.S., A.L. Cruz-Rodz, M. Leimeister-Wachter, J. Kreft, and W. Goebbel. 1989. A *Bacillus cereus* cytolytic determinant, cereolysin AB, which comprises the phospholipase C and sphingomyelinase genes; nucleotide sequence and genetic linkage. *J. Bacteriol.* **171**:744–753.

Goldberg, M.B., V.J. DiRita, and S.B. Calderwood. 1990. Identification of an iron-regulated virulence determinant in *Vibrio cholerae*, using TnphoA mutagenesis. *Infect. Immun.* **58**:55–60.

Graves, M.C. and J.C. Rabinowitz. 1986. In vivo and in vitro transcription of the *Clostridium pasteurianum* ferredoxin gene. Evidence for "extended" promoter elements in Gram-positive organisms. *J. Biol. Chem.* **261**:11409–11415.

Guddal, P.H., T. Johansen, K. Schulstad, and C. Little. 1989. Apparent phosphate retrieval system in *Bacillus cereus. J. Bacteriol.* **171**:5702–5706.

Hough, E., L.K. Hansen, B. Birkness, K. Jynge, S. Hansen, A. Hordik, C. Little, E. Dodson, and Z. Derewenda. 1989. High resolution (1.5 A) crystal structure of phospholipase C from *Bacillus cereus. Nature* (London) **338**:357–360.

Kameyama, S., H. Sato, and R. Murata. 1975. The role of σ- toxin of *Clostrdium perfringens* in experimental gas gangrene in guinea pigs. *Jpn. J. Med. Sci. Biol.* **25**:200.

Keilty, S. and M. Rosenberg. 1987. Constitutive function of a positively regulated promoter reveals new sequences essential for activity. *J. Biol. Chem.* **262**:6389–6395.

Kramer, J. and G. Schallehn. 1978. Characterisation of plasmid DNA in a lecithinase-positive and in a lecithinase-negative strain of *Clostridium perfringens. Zentrabl. Bakteriol. Microbiol. Hyg. I Abt. Orig. A* **241**:438–447.

Krug, E.L. and C. Kent. 1984. Phospholipase C from *Clostridium perfringens*: preparation and characterisation of homogenous enzyme. *Arch. Biochem. Biophys.* **231**:400–410.

Kyte, J. and R.R. Doolittle. 1982. A simple method for displaying the hydropathic character of a protein. *J. Mol. Biol.* **157**:105–132.

Leslie, D., N. Fairweather, D. Pickard, G. Dougan, and M. Kehoe. 1989. Phospholipase C and haemolytic activities of *Clostridium perfringens* alpha-toxin cloned in *Escherichia coli*: sequence and homology with a *Bacillus cereus* phospholipase C. *Mol. Microbiol.* **3**:383–392.

Liu, P. 1979. Toxins of *Pseudomonas aeruginosa. In: Pseudomonas aeruginosa: Clinical Manifestations of Infection and Current Therapy*, R.G. Dogget, eds., pp. 63–68. New York: Academic Press.

MacFarlane, M.G. 1955. On the biochemical mechanisms of action of gas-gangrene toxins. *In: Mechanisms of Microbial Pathogenicity*, J.W. Howie and A.J. Ollea, eds., pp. 57–77. Cambridge: Cambridge University Press.

MacFarlane, M.G. and B.C.J.G. Knight. 1941. The biochemistry of bacterial toxins. I. Lecithinase activity of *C. welchii* toxins. *Biochem. J.* **35**:884–902.

MacPherson, W.G., A.A. Bowlby, C. Wallace, and C. English. 1922. *Official History of the War, Vol. 1, Medical Services Surgery of the War*, pp. 134–150. London: Her Majesty's Stationery Office.

McDonel, J.L. 1986. Toxins of *Clostridium perfringens* types A, B, C, D and E. *In: Pharmacology of Bacterial Toxins*, F. Dorner and J. Drews, eds., pp. 477–517. Oxford: Pergamon Press.

Meyers, D.J. and R.S. Berk. 1990. Characterisation of phospholipase C from *Pseudomonas aeruginosa* as a potent inflammatory agent. *Infect. Immun.* **58**:659–666.

Mollby, R. 1978. Bacterial phospholipases. *In*: J. Jeljaszewicz and T. Wadstrom, eds., pp. 367–424. *Bacterial Toxins and Cell Membranes*, London: Academic Press.

Mollby, R., M. Thelestam, and T. Wadstrom. 1974. Effect of *Clostridium perfringens* phospholipase C (alpha-toxin) on the human diploid fibroblast membrane. *J. Membr. Biol.* **16**:313–330.

Mollby, R., T. Holme, C.-E. Nord, C. Smyth, and T. Wadstrom. 1976. Production of phospholipase C (alpha-toxin), haemolysins and lethal toxins by *Clostridium perfringens* types A to D. *J. Gen. Microbiol.* **96**:137–144

Moreau, H., G. Pieroni, C. Jolivet-Raynaud, J.E. Alouf, and R. Verger, 1988. A new kinetic approach for studying phospholiopase C (*Clostridium perfringens* α-toxin) activity on phospholipid monolayers. *Biochemistry* **27**:2319–2323.

Murata, R., A. Yamamoto, S. Soda, and A. Ito. 1965. Nutritional requirements of *Clostridium perfringens* PB6K for alpha-toxin production. *Jpn. J. Med. Sci. Biol.* **18**:189–202.

Nakamura, S., T. Shimamura, M. Hayase, and S. Nishida. 1973. Numerical taxonomy of saccharolytic clostridia, particularly *Clostridium perfringens*-like strains: descriptions of *Clostridium absonum* sp.n. and *Clostridium paraperfringens. Int. J. syst. Bacteriol.* **23**:419–429.

Ohsaka, A., M. Tsuchiya, C. Oshio, M. Miyaira, K. Suzuki, and Y. Yamakawa. 1978. Aggregation of platelets in the microcirculation of the rat induced by the α-toxin (phospholipase C) of *Clostridium perfringens. Toxicon* **16**:333–341.

Okabe, A., T. Shimizu, and H. Hayashi. 1989. Cloning and sequencing of a phospholipase C gene of *Clostridium perfringens. Biochem. Biophys. Res. Commun.* **160**:33–39.

Ogasawara, N. 1985. Markedly unbiased codon usage in *Bacillus subtilis. Gene* (Amst.) **40**:145–150.

Ostroff, R.M., A.I. Vasil, and M.L. Vasil. 1990. Molecular comparison of a nonhaemolytic and a haemolytic phospholipase C from *Pseudomonas aeruginosa. J. Bacteriol.* **172**:5915–5923.

Percival, D.A., A.D. Shuttleworth, E.D. Williamson, and D.C. Kelly. (1990). Anti-idiotype protection against *Clostridium perfringens* type D. *Infect. Immun.* **58**:2487–2492.

Ponnambalam, S., B. Chan, and S. Busby. 1988. Functional analysis of different sequence elements in the *Escherichia coli* galactose operon P2 promoter. *Mol. Microbiol.* **2**:165–172.

Pritchard, A.E. and M.L. Vasil. 1986. Nucleotide sequence and expression of a phosphate-regulated gene encoding a secreted hemolysin of *Pseudomonas aeruginosa. J. Bacteriol.* **167**:291–298.

Roggentin, P., B. Rothe, F. Lottspeich, and R. Schauer. 1988. Cloning and sequencing of a *Clostridium perfringens* sialidase gene. *FEBS Lett.* **238**:31–34.

Saint-Joanis, B., T. Garnier, and S. Cole. 1989. Gene cloning shows the alpha-toxin of *Clostridium perfringens* to contain both sphingomyelinase and lecithinase activities. *Mol. Gen. Genet.* **219**:453–460.

Sakurai, J., Y. Fujii, K. Torii, and K. Kobayashi. 1989. Dissociation of various biological activities of *Clostridium perfringens* alpha toxin by chemical modification. *Toxicon* **27**:317–323.

Samuelsson, B. 1983. Leukotrienes: mediators of immediate hypersensitivity reactions and inflammation. *Science* **220**:568–575.

Sarvas, M. 1986. Protein secretion in bacilli. *Curr. Topics Microbiol. Immunol.* **125**:103–125.

Sato, H., J. Chiba, and Y. Sato. 1989. Monoclonal antibodies against alpha toxin of *Clostridium perfringens. FEMS Microbiol. Lett.* **59**:173–176.

Sato, H., Y. Yamakawa, A. Ito, and R. Murata. 1978. Effect of zinc and calcium ions on the production of alpha-toxin and proteases by *Clostridium perfringens. Infect. Immun.* **20**:325–333.

Shuttleworth, A.D., D.A. Percival, and R.W. Titball. 1988. Epitope mapping of *Clostridium perfringens* alpha toxin. *In: Bacterial Protein Toxins,* F.J. Fehrenbach, J.E. Alouf, P. Falmagne, W. Goebel, J. Jeljaszewicz, D. Jurgens and R. Rappouli, eds., pp. 65–66. *Zentralbl. Bakteriol. Microbiol. Hyg. 1. Abt.,* Suppl. 17.

Smyth, C.J. and J.P. Arbuthnott. 1974. Properties of *Clostridium perfringens* (welchii) type-A alpha-toxin (phospholipase C) purified by electrofocusing. *J. Med. Microbiol.* **7**:41–67.

Sugahara, T., T. Takahashi, S. Yamaya, and A. Ohsaka. 1976. In vitro aggregation of platelets induced by alpha-toxin (phospholipase C) of *Clostridium perfringens. Jpn. J. Med. Sci. Biol.* **29**:255–263.

Thelestam, M. and R. Mollby. 1975. Sensitive assay for the detection of toxin-induced damage to the cytoplasmic membrane of human diploid fibroblasts. *Infect. Immun.* **12**:225–232.

Tinoco, I., P.N. Borer, B. Dengler, M.D. Levine, O.C. Uhlenbeck, D.M. Crothers, and J. Gralla. 1973. Improved estimation of secondary structure in ribonucleic acids. *Nature New Biol.* **246**:40–41.

Titball, R.W. and T. Rubidge. 1990. The role of histidine residues in the alpha-toxin of *Clostridium perfringens. FEMS Microbiol. Lett.* **68**:261–266.

Titball, R.W., S.E.C. Hunter, K.L. Martin, B.C. Morris, A. D. Shuttleworth, T. Rubidge, D.W Anderson, and D.C. Kelly. 1989. Molecular cloning and nucleotide sequence of the alpha-toxin (phospholipase C) of *Clostridium perfringens. Infect. Immun.* **57**:367–376.

Titball, R.W., Lerlie, D.L., Harvey, V. and Kelly, D.C. (1991) Toxicity of *Clostridium perfringens* alpha-toxin is dependent on a domain homologous to that of an enzyme from the human arachidonic acid pathway. *Infect. Immun.* **59**:1872–1874.

Tso, J.Y. and C. Siebel. 1989. Cloning and expression of the phospholipase C gene from *Clostridium perfringens* and *Clostridium bifermentans. Infect. Immun.* **57**:468–476.

Tweten, R. 1988. Nucleotide sequence of the gene for perfringolysin O (theta toxin) from *Clostridium perfringens*: significant homology with the genes for streptolysin O and pneumolysin. *Infect. Immun.* **56**:3235–3240.

van Damme-Jongsten, M., K. Wernars, and S. Notermans. 1989. Cloning and sequencing of the *Clostridium perfringens* enterotoxin gene. *Antonie Leeuwenhoek J. Microbiol.* **56**:181–190.

Willis, T.A. 1969. *Clostridia of Wound Infection.* London: Butterworth.

Willis, T.A. 1977. *Anaerobic Bacteriology: Clinical and Laboratory Practice,* 3rd Ed., London: Butterworth.

Young, M., N.P. Minton, and W. Staudenbauer. 1989a. Recent advances in the genetics of clostridia. *FEMS Microbiol. Rev.* **63**:301–326.

Young, M., W. Staudenbauer, and N.P. Minton. 1989b. Genetics of *Clostridium. In: "Clostridia",* Biotechnology Handbooks Vol 3., N.P. Minton and D.J. Clarke, eds., pp. 63–103. New York: Plenum.

15

Gene Cloning, Organization, and Expression of θ-Toxin of *Clostridium perfringens*

Rodney K. Tweten

15.1 Introduction

The clostridia produce a plethora of extracellular products. Some of these have been termed "toxins," although not necessarily because they are lethal when administered to animals. Two toxins that are clearly toxic to animals are α-toxin and θ-toxin. Both toxins have been examined to a greater extent than most other extracellular products from *Clostridium perfringens*. The α-toxin (phospholipase C) has been termed the "lethal toxin" of *C. perfringens*; however, in the past commercial preparations of α-toxin prepared from *C. perfringens* were contaminated with θ-toxin (Smyth et al., 1975), which may have altered the results of many of these earlier studies. θ-toxin or perfringolysin O (PFO) is a thiol-activated cytolysin produced by *C. perfringens* that has been shown to be lethal to animals (Stevens et al., 1987). Clostridia other than *C. perfringens* and *C. tetani* are purported to produce a thiol-activated cytolysin, mostly on the basis of serology, although none of these cytolysins have been purified or characterized. Related toxins are produced by other pathogenic species from *Streptococcus*, *Bacillus*, and *Listeria*, which emphasizes the observation that these cytolysins have an ubiquitous place among gram-positive pathogens. The fact that similar cytolysins are found in so many species of gram-positive pathogens indicates that acquisition of this cytolysin by the various bacterial species offers a selective advantage to those organisms.

15.2 Molecular Analysis of the θ-toxin gene

The *C. perfringens* θ-toxin gene (*pfo*) has been cloned and sequenced (Tweten, 1988a, 1988b) (Figure 15.1). θ-toxin belongs to the class of cytolysins historically termed "thiol-activated cytolysins" or "oxygen-labile cytolysins." These cytolysins have several common properties; they are all secreted, with the exception of pneumolysin, each contains a highly conserved "essential" cysteine, and the cellular receptor appears to be cholesterol or closely related sterols that have an intact 3-β hydroxyl group (Prigent and Alouf, 1976). The protein encoded by the θ-toxin gene was found to have a typical structure for a secreted protein (Tweten, 1988a).

A schematic of the gene structure is shown in Figure 15.1. The leader peptide was 27 residues long (not including the amino-terminal methionine) and contained a hydrophobic core with a basic extreme amino terminus. A consensus signal peptidase cleavage site, Ser-X-Ser (Perlman and Halvorson, 1983), was present immediately preceding the amino terminus of the mature, extracellular form of the cytolysin. The amino terminus of the mature protein (secreted form) corresponded to that determined by amino-acid-sequence analysis of the purified θ-toxin produced by *C. perfringens* and by *Escherichia coli* carrying *pfo* (Figure 15.1). In *E. coli*, the recombinant-derived θ-toxin was found almost entirely in the periplasmic frac-

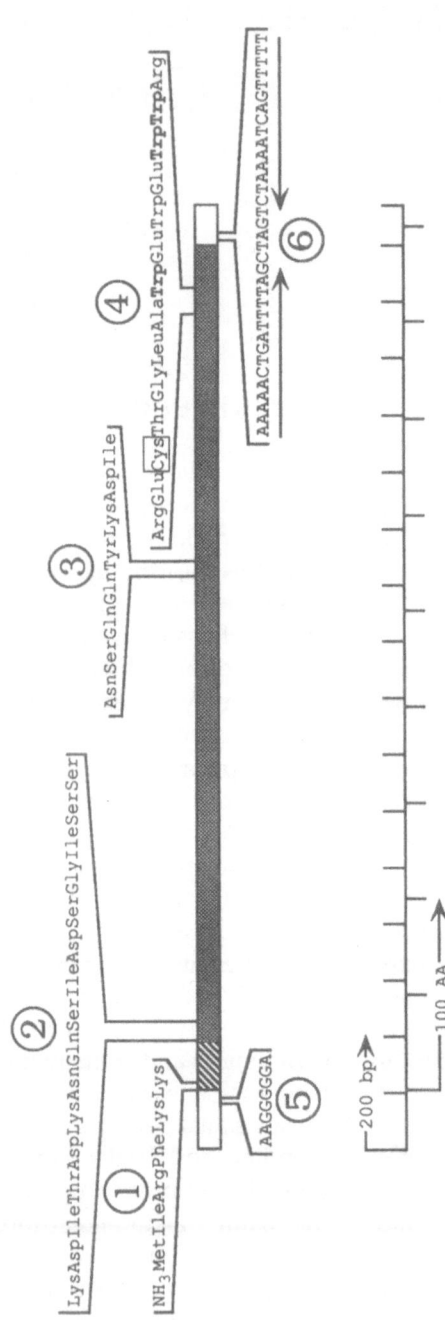

FIGURE 15.1. Schematic representation of θ-toxin gene. Coding region for secreted form of θ-toxin gene is shaded portion of the bar. Signal peptide region is designated by slanted bars. Coding regions of interest are designated by numbers: ①, DNA-derived amino-terminal sequence of θ-toxin molecule; ②, start of mature sequence of θ-toxin as derived by amino-terminal sequence analysis of perfringolysin O produced by *Clostridium perfringens* and recombinant-derived toxin from *E. coli* (both corresponded to DNA-derived sequence); ③, amino-terminal sequence of carboxy-terminal peptide generated by trypsin cleavage (Iwamoto et al., 1987); ④, 12-residue peptide that is completely conserved among all sequenced thiol-activated cytolysins and which contains "essential cysteine" (box) and three tryptophan residues (**bold**); ⑤, putative ribosome-binding site; ⑥, putative *rho*-independent transcriptional terminator. Scale for nucleotides and amino acids is shown below schematic of θ-toxin gene.

tion, indicating that it was secreted by the normal secretion machinery of *E. coli*. The recombinant-derived cytolysin from *E. coli* also exhibited characteristics identical with that isolated from *C. perfringens*. The two-dimensional peptide maps of both native and recombinant-derived θ-toxins were identical, which showed that θ-toxin synthesized in *E. coli* was not altered in its primary structure from that isolated from *C. perfringens*. The hemolytic activities of both cytolysins were also the same, within the error inherent in the hemolytic assay. Therefore, the recombinant-derived cytolysin was suitable for studies of the mechanism of action of the cytolysin and its contribution to diseases caused by *C. perfringens*.

Shimizu et al. (Shimizu et al., 1991) cloned the region upstream of *pfo* and found a regulatory sequence that stimulated a 20-fold increase in the expression the *pfo* gene in *E. coli*. The upstream region contained a structural gene that encoded a 343-residue protein, termed *pfoR*, that contained a helix-turn-helix motif typical of DNA-binding proteins. This region could apparently act in *trans* but was more efficient in *cis* for the upregulation of *pfo*. This finding was very interesting in view of the earlier findings of Imagawa et al. (1981), who found that the production of θ-toxin, collagenase, hemagglutinin, and protease appear to be coregulated. Canard and Cole (1989) have recently provided a partial genetic map of various *C. perfringens* genes and found that the genes of three putative *C. perfringens* virulence factors (θ-toxin, α-toxin, and the β-N-acetylglucosaminidase) mapped on a 200-kb fragment on the chromosome near the origin of chromosomal replication.

The primary structure of θ-toxin deduced from the DNA sequence exhibited a relatively high level of similarity (conserved residues) to the primary structure of the low molecular weight form of streptolysin O (65%) and somewhat less with pneumolysin (42%) and listeriolysin (45%). If conservative amino acid substitutions were considered, the primary structures of θ-toxin and streptolysin O (SLO) exhibited a 96% similarity. The primary difference between perfringolysin O and SLO was the presence of an additional coding region

between the coding region for the SLO signal peptide and the beginning of what is termed the low molecular weight form of SLO (Kehoe et al., 1987). Apparently this peptide is lost shortly after SLO is secreted from the cell, because the full-length form of SLO is difficult to isolate (Bhakdi et al., 1984; Kehoe and Timmis, 1984). The most striking feature of the additional sequence of SLO is the relative hydrophilicity of that region (Kehoe et al., 1987; Tweten, 1988a). When the DNA sequences for PFO and SLO were compared, the nucleotide homology began at the ATG codon for *pfo* and an out-of-phase ATG codon for *slo* (Kehoe, et al., 1987, Tweten, 1988a). We had speculated that a progenitor of *slo* might have originally more closely resembled *slo*, perhaps acquiring the additional amino-terminal coding region later. The importance of this additional peptide to either the in vitro or in vivo activity of SLO is unclear.

The most highly conserved region among the primary sequences of these cytolysins is a completely conserved region of 12 residues (see Figure 15.1) that includes the "essential" cysteine of all 4 cytolysins. This region is also conserved in its position in the carboxy-terminal region of all 4 cytolysins. The 12-residue peptide is conspicuous for its abundance of tryptophan residues, which in the case of θ-toxin constitutes 3 of the 7 tryptophan residues present in the molecule. The importance (or lack thereof) of the tryptophan residues is unknown, although they probably increase the relative hydrophobicity of this region. The term thiol-activated has been a hallmark of these cytolysins because studies have shown that activity can be restored by thiol reagents; further, when the cysteine is chemically modified the cytolytic activity is greatly reduced (Iwamoto et al., 1987). In view of studies by Pinkney et al. (1989) and Saunders et al. (1989), the term thiol-activated is probably a misnomer. These workers substituted alanine for the "essential cysteine" of pneumolysin (Pinkney et al., 1989) and streptolysin O (Saunders et al., 1989) and found that there was little change in the activity of either cytolysin.

This invariably leads one to the conclusion that the "essential cysteine" is not the same thing as the "essential sulfhydryl," because the sulfhydryl function does not appear to be required for activity. The relative hydrophobicity of the amino acid side chain is probably more important because the same workers found that the substitution of serine for the cysteine greatly reduced activity. These cytolysins probably have their activity "restored" by thiol reagents by the reduction of disulfides that have formed between the free sulfhydryl of the cytolysin and thiols in the media into which the cytolysin was secreted. We have found that highly purified perfringolysin O can be stored at 5°C for weeks without loss of activity through oxidation of the cysteine sulfhydryl (R. Tweten, unpublished results), presumably because of the absence of thiols capable of forming a disulfide with the sulfhydryl group of the cytolysin. Oxidation of the sulfhydryl group by molecular oxygen is unlikely because the oxidation with oxygen requires stringent chemical conditions and is usually irreversible (Jocelyn, 1972).

15.3   Structure/Function Analysis of θ-toxin

Modification of the cysteine with various thiol reagents generally inhibits the cytolytic action of these cytolysins, probably via a steric hindrance that interferes with an essential function of these cytolysins. Iwamoto et al. (1987) found that modification of the cysteine of θ-toxin with either N-ethylmaleimide or 5,5'-dithio-bis(2-nitrobenzoic acid) resulted in a significant reduction of hemolytic activity. The modification appeared to inhibit binding of the cytolysin to erythrocytes, which they suggested was the basis of the reduced activity. However, the modified θ-toxin appeared to undergo a structural change after modification of the cysteine. The structural change was deduced from a significant difference in the tryptic digests of the native and modified cytolysins. Native cytolysin yielded two tryptic peptides of approximately 39,000 and 15,000 daltons (see Figure 15.1 for the cleavage site), whereas the modified cytolysin exhibited only trace amounts of

these peptides after trypsin treatment. A major change in the secondary structure apparently was not indicated, because circular dichroism (CD) analysis exhibited little or no differences in the CD spectra of native and modified cytolysins. Under conditions in which the θ-toxin was reversibly modified with 5,5'-dithio-bis(2-nitrobenzoic acid), the structural change could be reversed by the removal of the modifying agent. The authors suggested that the loss of binding may have been caused by either steric hindrance of the modifying group or the conformational change induced by the modification. Another possibility, although one not mutually exclusive with the suggestions of Iwamoto et al. (1987), is that this site may also function as a trigger for structural changes in the protein on binding to a membrane surface.

Cholesterol and a number of closely related sterols inhibit these cytolysins (Prigent and Alouf, 1976), and membranes lacking cholesterol are not lysed by the cytolysins (Cowell et al., 1978). Although cholesterol is an absolute requirement for cytolysin binding, the complete story of this interaction is unclear. Rottem et al. (1982), using tetanolysin (*Clostridium tetani*), found that cholesterol concentration in the membrane of liposomes affected binding of the cytolysin to the liposome. Liposomes with more than 33 mol% cholesterol in their membrane could bind approximately four times as much tetanolysin than those with 10 mol% cholesterol, the balance being dimyristoylphosphatidylcholine in both cases. The absolute cholesterol concentration in each experiment was kept constant by adjusting the amount of total liposome so that the difference in binding resulted from the concentration of cholesterol in the membrane, not the absolute concentration of cholesterol present in the assay. At membrane cholesterol concentrations between 20 and 33 mol%, it is hypothesized that cholesterol undergoes the formation of cholesterol-rich domains that exclude most of the phospholipid (Haberkorn et al., 1977; Gershfeld, 1978). The data of Rottem et al. (1982) seem to indicate that the binding of the thiol-activated cytolysins may be sensitive to cholesterol structure in the membrane.

Membrane-dependent aggregation into supramolecular complexes is a hallmark feature of θ-toxin and related toxins. Electron micrographs (EM) of erythrocytes or erythrocyte ghost membranes that have been treated with θ-toxin or related cytolysins exhibit arc and ring structures (Mitsui et al., 1979, Smyth, et al., 1975). These structures are generally thought to be the functional lesion responsible for cell lysis; however, it is possible that a wide spectrum of aggregate sizes are present on a toxin-treated erythrocyte, and it is unclear whether these large complexes are necessary for cell lysis. Buckingham and Duncan (1983), using EM, estimated that SLO-treated erythrocyte membranes generated pores greater than 128 Å in diameter. Bhakdi et al. (1984) have estimated that these structures are composed of 25–100 monomers of SLO and that as few as 70–100 monomers of SLO per erythrocyte were sufficient to cause hemolysis. These observations did not, however, directly link aggregation to lysis. The first evidence directly linking aggregation to hemolysis was obtained by Hugo et al. (1986), using a monoclonal antibody that inhibited SLO-dependent hemolysis of erythrocytes but did not interfere with toxin binding. This antibody was found to prevent the aggregation of SLO, as determined by sucrose density centrifugation of detergent-solubilized erythrocytes that had been treated with SLO in the absence and presence of the monoclonal antibody.

More recently, we (Harris et al., 1991a) have applied the technique of fluorescence energy transfer (FET) to directly examine the real-time membrane aggregation of θ-toxin monomers on erythrocyte ghost membranes. Implicit in this technique is the fact that the fluorescent donor and acceptor molecules which participate in FET must be physically juxtaposed. The fluorescent donor used in these experiments was fluorescein, which had been coupled to θ-toxin, and the fluorescent acceptor was tetramethylrhodamine, coupled to a separate batch of θ-toxin. This donor-acceptor pair has a Förster distance (distance at which 50% FET efficiency is achieved) of 4 nm; the efficiency of FET is proportional to $1/r^6$, where r is the distance separating the donor-acceptor pair. Because FET is inversely proportional to r, doubl-

ing the Förster distance would decrease FET by more than 98%. Therefore, θ-toxin molecules that are labeled with these fluorophores must be juxtaposed for the occurrence of FET. FET was observed between labeled molecules of θ-toxin only when membranes were present, and the addition of a 1.6 molar excess of unlabeled θ-toxin completely abolished FET, indicating that only juxtaposed donor-acceptor pairs could participate in the transfer.

Using FET, it was determined that aggregation of θ-toxin is responsible for the observed lag period that occurs in the hemolytic curve before the onset of hemolysis. It had been thought that toxin binding was largely responsible for this lag period (Niedermeyer, 1985); however, our analysis showed that the time for θ-toxin membrane binding accounted for very little of the lag period. Most of the lag period was the result of the time necessary for the lateral diffusion and aggregation of toxin monomers. Only after aggregation was largely complete did the actual lysis of the erythrocytes ensue. Aggregation of θ-toxin was also responsible for the observed temperature dependence of hemolysis, and this was used to determine the activation energy for aggregation. The relationship of aggregation to the lag period was further reinforced by similar activation energies for aggregation and the lag period. The activation energy for aggregation was calculated to be 19 kcal mol^{-1}, and for the lag period, 23 kcal mol^{-1}. This approach has advantages over previous methodology in that the aggregation process can be followed kinetically. Therefore, the nature of this technique facilitates the use of a wide variety of experimental conditions in the study of the aggregation of θ-toxin. This methodology will facilitate the study of the membrane aggregation of θ-toxin-derived peptides and θ-toxin derivatives generated by in vitro mutagenesis. It will also be useful to examine the influence of membrane composition on aggregation using purified lipid systems. Finally, whether θ-toxin can form heterologous aggregates with any of the related toxins can be shown directly, because FET depends on the intimate interaction of molecules.

The general belief is that θ-toxin and related

cytolysins *do not* lyse erythrocytes by a colloid-osmotic mechanism. This hypothesis is the result of a number of studies with various thiol-activated cytolysins and has been promoted by Bhakdi et al. (1985). As early as 1947, Bernheimer (Bernheimer, 1947) reported that SLO-treated erythrocytes apparently did not undergo prelytic swelling, a characteristic of lysis by a colloid-osmotic mechanism. Duncan (1974) found that the release of rubidium-86 and hemoglobin from SLO-treated erythrocytes could not be temporally separated, suggesting that ion equilibration did not precede lysis. Similar results were found by Blumenthal and Habig (1984) when K^+ and hemoglobin release were measured from tetanolysin-treated erythrocytes. In contrast, we (Harris, et al., 1991b) have used flow cytometry to examine erythrocyte size during hemolysis by θ-toxin and found that erythrocytes increase significantly in volume before lysis, which suggests a colloid-osmotic mechanism for lysis. It was also found that 0.3 M sucrose temporarily stabilized erythrocytes from θ-toxin-induced lysis, further evidence for a colloid-osmotic mechanism of lysis. Failure to detect this prelytic stage in previous experiments may reflect the rapidity of the ion equilibration and water influx.

15.4   The Role of θ-Toxin in *Clostridium perfringens* Infection

The role of θ-toxin in *C. perfringens*-induced gangrene has not been defined, but it is unlikely that θ-toxin simply acts to cause intervascular hemolysis. Studies have suggested that this cytolysin may have a negative effect on the cells of the immune system, particularly human polymorphonuclear lymphocytes (PMNLs). Because PMNLs constitute the first line of defense against a bacterial infection, any alteration in their function may have a profound effect on the course of an infection. In a series of studies, Bremm et al. (1984, 1985, 1987) found that sublytic amounts of thiol-activated cytolysins such as θ-toxin, SLO, and alveolysin affect leukotriene release in PMNLs. Bremm et al. (1987) subsequently showed that the

cytolysin-treated PMNLs exhibited a decreased leukotriene generation on subsequent stimulation with the Ca^{2+}-ionophore A 23187, presumably because of the cytolysin-induced release of the leukotriene, which depleted available stores. The latter study also found that the conversion of LTB_4 to the less chemotactic or nonchemotactic products 20-0H-LTB_4 and 20-COOH-LTB_4 was significantly increased in the cytolysin-treated cells. LTB_4 is normally the major leukotriene released during ionophore stimulation or during phagocytosis. The significant reduction of leukotriene LTB_4 generation after cytolysin treatment may ultimately result in a reduction of PMNL infiltration into the site of infection.

Stevens et al. (1987) also investigated the effects these cytolysins have on PMNLs. Stevens and coworkers found that sublytic levels of θ-toxin (≈ 12.5 hemolytic units per 10^5 PMNLs) reduced PMNL viability and induced marked morphological changes. Increased concentrations of θ-toxin (4–32 hemolytic units per 10^5 PMNLs) inhibited PMNL chemiluminescence in a dose-dependent manner. At a concentration of 0.06 HU per 10^5 PMNLs, an increase in random migration was reported but at doses greater that 0.08 HU both random and directed migration were reduced. In contrast Stevens et al. found that α-toxin only stimulated the chemiluminescence response of PMNLs. The work of both Bremm et al. (1984, 1985, 1987) and Stevens et al. (1987) appears to show that these cytolysins can have profound adverse effects on the PMNLs, suggesting that θ-toxin is contributing to a local inhibition of the immune response of the infected host.

These findings are consistent with the observation that there is a well-demarcated zone of tissue necrosis inside which no PMNLs are present during *C. perfringens* induced gas gangrene (Robb-Smith, 1945). The PMNLs are found to be "piled-up" along the border of this zone of necrosis, suggesting that one or more soluble factors are present which inhibit PMNL migration into this area. Animal studies that make use of isogenic pairs of *C. perfringens* in which the θ-toxin gene is deleted should provide interesting insights into the contribution of this toxin in the development of gas gangrene.

Many interesting problems remain to be investigated concerning the biochemistry and genetics of θ-toxin. Such possibilities include further investigation of the highly conserved region of the cytolysin that contains the single cysteine, the domain(s) of the cytolysin which interact with the membrane surface, the minimal size of a functional membrane complex, the trigger for and nature of the conformational change that occurs after θ-toxin binds a membrane, the identification of domains involved in the formation of intermolecular contacts between membrane-aggregated molecules, and the elucidation of the crystal structure for θ-toxin. The contribution of this toxin to the establishment of gas gangrene and its effects on PMNL function should also be a subject of continued study. Finally, the recent findings concerning the regulation of θ-toxin and its chromosomal location in *C. perfringens* have provided a significant impetus to the investigation of the genetic control systems for θ-toxin expression as well as other putative extracellular virulence factors.

References

Bernheimer, A.W. 1947. Comparative kinetics of hemolysis induced by bacterial and other hemolysins. *J. Gen. Physiol.* **30**:337–353.

Bhakdi, S., M. Roth, A. Sziegoleit, and J.J. Tranum. 1984. Isolation and identification of two hemolytic forms of streptolysin-O. *Infect. Immun.* **46**:394–400.

Bhakdi, S., J.J. Tranum, and A. Sziegoleit. 1985. Mechanism of membrane damage by streptolysin-O. *Infect. Immun.* **47**:52–60.

Blumenthal, R. and W.H. Habig. 1984. Mechanism of tetanolysin-induced membrane damage: studies with black lipid membranes. *J. Bacteriol.* **157**:321–323.

Bremm, K.D., W. König, M. Thelestam, and J.E. Alouf. 1987. Modulation of granulocyte functions by bacterial exotoxin and endotoxins. *Immunology* **62**:363–371.

Bremm, K.D., H.J. Brom, J.E. Alouf, W. König, B. Spur, A. Crea, and W. Peters. 1984. Generation of leukotrienes from human granulocytes by alveolysin from *Bacillus alvei*. *Infect. Immun.* **44**:188–193.

Bremm, K.D., W. König, P. Pfeiffer, I. Rauschen, K. Theobald, M. Thelestam, and J.E. Alouf. 1985. Effect of thiol-activated toxins (streptolysin O, alveolysin, and theta toxin) on the generation of leukotrienes and leukotriene-inducing and -metabolizing enzymes from human polymorphonuclear granulocytes. *Infect. Immun.* **50**:844–851.

Buckingham, L. and J.L. Duncan. 1983. Approximate dimensions of membrane lesions produced by streptolysin S and streptolysin O. *Biochim. Biophys. Acta* **729**:115–122.

Canard, B. and S.T. Cole. 1989. Genome organization of the anaerobic pathogen *Clostridium perfringens*. *Proc. Natl. Acad. Sci. USA* **86**:6676–6680.

Cowell, J.L., K. Kim, and A.W. Bernheimer. 1978. Alteration by cereolysin of the structure of cholesterol-containing membranes. *Biochem. Biophys. Acta* **507**:230–241.

Duncan, J.L. 1974. Characteristics of streptolysin O hemolysis: kinetics of hemoglobin and [86]rubidium release. *Infect. Immun.* **9**:1022–1027.

Gershfeld, N.L. 1978. Equilibrium studies of lecithin-cholesterol interactions. I. Stoichiometry of lecithin-cholesterol complexes in bulk systems. *Biophys. J.* **22**:469–488.

Haberkorn, R.A., R.G. Griffin, M.D. Meadows, and E. Oldfield. 1977. Deuterium nuclear magnetic resonance investigation of the dipalmatoyl lecithin-cholesterol-water system. *J. Am. Chem. Soc.* **99**:7353–7355.

Harris, R.W., P.J. Sims, and R.K. Tweten. 1991a. Kinetic aspects of the aggregation of *Clostridium perfringens* theta toxin on erythrocyte membranes: a fluorescence energy transfer study. *J. Biol. Chem.* **266**:6936–6941.

Harris, R.W., P.J. Sims and R.K. Tweten. 1991b. Evidence that *Clostridium perfringens* θ-toxin induces the colloid osmotic lysis of erythrocytes. *Infect. Immun.* **59**:2499–2501.

Hugo, F., J. Reichwein, M. Arvand, S. Krämer, and S. Bhakdi. 1986. Use of a monoclonal antibody to determine the mode of transmembrane pore formation by streptolysin O. *Infect. Immun.* **54**:641–645.

Imagawa, T., T. Tatsuki, Y. Higashi, and T. Amano. 1981. Complementation characteristics of newly isolated mutants from two groups of strains of *Clostridium perfringens*. *Biken J.* **24**:13–21.

Iwamoto, M., Y. Ohno-Iwashita, and S. Ando. 1987. Role of the essential thiol group in the thiol-activated cytolysin from *Clostridium perfringens*. *Eur. J. Biochem.* **167**:425–430.

Jocelyn, P.C. 1972. *Biochemistry of the SH Group. The Occurrence, Chemical Properties, Metabolism and Biological Function of Thiols and Disulfides.* London: Academic Press.

Kehoe, M. and K.N. Timmis. 1984. Cloning and expression in *Escherichia coli* of the streptolysin O determinant from *Streptococcus pyogenes*: characterization of the cloned streptolysin O determinant and demonstration of the absence of substantial homology with determinants of other thiol-activated toxins. *Infect. Immun.* **43**:804–810.

Kehoe, M.A., L. Miller, J.A. Walker, and G.J. Boulnois. 1987. Nucleotide sequence of the streptolysin O (SLO) gene: stluctural homologies between SLO and other membrane-damaging, thiol-activated toxins. *Infect. Immun.* **55**:3228–3232.

Mitsui, K., T. Sekiya, S. Okamura, Y. Nozawa, and J. Hase. 1979. Ring formation of perfringolysin O as revealed by negative stain electron microscopy. *Biochim. Biophys. Acta* **558**:307–313.

Niedermeyer, W. 1985. Interaction of streptolysin-O with biomembranes: kinetic and morphological studies on erythrocyte membranes. *Toxicon* **23**:425–439.

Perlman, D. and H.O. Halvorson. 1983. A putative signal peptidase recognition site and sequence in eucaryotic and prokaryotic signal peptides. *J. Mol. Biol.* **167**:391–409.

Pinkney, M., E. Beachey, and M. Kehoe. 1989. The thiol-activated toxin streptolysin O does not require a thiol group for activity. *Infect. Immun.* **57**:2553–2558.

Prigent, D. and J.E. Alouf. 1976. Interaction of streptolysin O with sterols. *Biochim. Biophys, Acta* **433**:422–428.

Robb-Smith, A.H.T. 1945. Tissue changes induced by *Cl. welchii* type A filtrates. *Lancet* **ii**:362–368.

Rottem, S., R.M. Cole, W.H. Habig, M.F. Barile, and M.C. Hardegree. 1982. Strucnual charac-

teristics of tetanolysin and its binding to lipid vesicles. *J. Bacteriol.* **152**:888–892.

Saunders, K.F., T.J. Mitchell, J.A. Walker, P.W. Andrew, and G.J. Boulnois. 1989. Pneumolysin, the thiol-activated toxin of *Streptococcus pneumoniae*, does not require a thiol group for in vitro activity. *Infect. Immun.* **57**:2547–2552.

Shimizu, T., A. Okabe, J. Minami, and H. Hayashi. 1991. An upstream regulatory sequence stimulates expression of the perfringolysin O gene of *Clostridium perfringens. Infect. Immun.* **59**:137–142.

Smyth, C.J., J.H. Freer, and J.P. Arbuthnot. 1975. Interaction of *Clostridium perfringens* theta-haemolysin, a contaminant of commercial phospholipase C, with erythrocyte ghost membranes and lipid dispersions. A morphological study. *Biochim. Biophys. Acta* **382**:479–493.

Stevens, D.L., J. Mitten, and C. Henry. 1987. Effects of alpha and theta toxins from *Clostridium perfringens* on human polymorphonuclear leukocytes. *J. Infect. Dis.* **156**:324–333.

Tweten, R.K. 1988a Cloning and expression in *Escherichia coli* of the perfringolysin O (theta-toxin) gene from *Clostridium perfringens* and characterization of the gene product. *Infect. Immun.* **56**:3228–3234.

Tweten, R.K. 1988b. Nucleotide sequence of the gene for perfringolysin O (theta-toxin) from *Clostridium perfringens*: significant homology with the genes for streptolysin O and pneumolysin. *Infect. Immun.* **56**:3235–3240.

16

Molecular Biology of *Clostridium perfringens* Enterotoxin

Per Einar Granum and Gordon S.A.B. Stewart

16.1 Introduction

Clinical importance

Clostridium perfringens is a large, gram-positive, spore-forming, anaerobic and nonmotile rod belonging to the family *Bacillaceae*. It is a common inhabitant of the soil and plays an important part in the putrefication process. *C. perfringens* is also readily isolated from dust, raw meat, water, and the intestinal tract of man and animals, and is the clostridial species most frequently isolated from clinical specimens. Because this organism produces numerous toxins and enzymes, it can cause a spectrum of different diseases in man and animals (for review, see McDonel, 1986). The species is divided into five types based on the production of four major lethal toxins (Table 16.1).

Clostridium perfringens is responsible for two very different types of food poisoning in man: a mild classic form of food poisoning associated with type A strains, and a severe type of food poisoning called necrotizing enterocolitis caused by type C strains. This latter disease, while rare, is often fatal, and mainly because of β-toxin production. This toxin, however, is not addressed in this chapter (for review, see Granum, 1990). The symptoms of the *C. perfringens* type A food poisoning are acute abdominal pain and diarrhea with nausea and fever, with vomiting being rare. The symptoms usually appear 8 to 12 h (6–24 h) after ingestion of contaminated food (10^6–10^7 cells/g) and can last from 12 to 24 h (Hobbs, 1969). The food poison-

Table 16.1 The distribution of the major lethal toxins among the types of *Clostridium perfringens*

Type	Alpha	Beta	Epsilon	Iota
A	+	−	−	−
B	+	+	+	−
C	+	+	−	−
D	+	−	+	−
E	+	−	−	+

ing is seldom fatal, although a few deaths have been reported among the elderly and infirm (Parry, 1963; Sutton and Hobbs, 1965). Because the symptoms are relatively mild, the true incidence of *C. perfringens* food poisoning is not known. Nevertheless, in several countries *C. perfringens* has been reported to be the most prevalent cause of food poisoning, only exceeded in incidence by food infections caused by *Salmonella* sp. (Cliver, 1987; Reynolds, 1987). In Norway, which has a very low incidence of *Salmonella*-associated food poisoning, *C. perfringens* was by far the most important cause of bacterial food poisoning in 1988 and 1989 (Gondrosen et al., 1990).

The enterotoxin and its mode of action

Although *C. perfringens* has been associated with gastroenteritis since 1895 (Klein, 1895), the first clear demonstration of its etiological status in food poisoning was made in 1945 (McClung, 1945). Later, Stark and Duncan

(1971) showed that all clinically significant properties associated with crude extracts of sporulating enterotoxin-positive cells could be linked solely and specifically to the enterotoxin molecule. Very soon thereafter two research groups purified and partly characterized the enterotoxin protein (Hauschild and Hilsheimer, 1971; Stark and Duncan, 1972). Since publication of an improved purification method by Granum and Whitaker (1980), it has been possible to purify enough enterotoxin (200 mg in 2 days) for protein sequencing, protein modeling, and mode of action studies (Richardson and Granum, 1985).

Several studies have been carried out on the mode of action of the enterotoxin. Because we do not intend to cover this subject here, we give only some of the background on the mode of action. It has been known for some time that *C. perfringens* food poisoning results in transport alterations in the ileum. There is a net outflux of water, sodium, and chloride, and reduced uptake of glucose (McDonel, 1979), causing diarrhea; but this is the final result of the action of the enterotoxin during food poisoning. At the molecular level, a set of events are relatively apparent from the literature. The enterotoxin is produced inside the *C. perfringens* cells during sporulation in the small intestine. After cell lysis, the enterotoxin is released and processed to a more active toxin by trypsin and chymotrypsin (Granum et al., 1981, Granum and Richardson, 1991). Subsequently, it is rapidly bound to protein receptors on the brush-border membrane of epithelial cells (Wnek and McClane, 1983; Wnek and McClane, 1989), possibly through an ϵ-amino group at the lysine residue position 301 on the enterotoxin (Whitaker and Granum, 1980, Granum and Richardson 1991; Hanna et al., 1991). This is followed by rapid membrane insertion coordinated with a major configuration change from a β-sheet to an α-helix structure (Granum and Harbitz, 1985). This results in membrane permeability changes for ions and small molecules (McClane et al., 1988). As intracellular ion concentrations (particularly of Ca^{2+}) increase, morphological damage occurs, eventually leading to cell death following leakage of large molecules such as nucleotides and proteins

(Matsuda and Sugimoto, 1979; Granum, 1985; McClane et al., 1988).

16.2   Cloning of the Enterotoxin Gene

Three independent approaches have been used as cloning strategies for the enterotoxin gene. Within months of each other three reports were published in 1989 and, in consequence, were written without cross-reference. Comparison at this stage therefore is timely.

Van Damme-Jongsten et al. (1989) and Iwanejko et al. (1989) both employed a synthetic oligonucleotide probe to identify the enterotoxin gene. These probes were devised from a study of the enterotoxin protein sequence published previously (Richardson and Granum, 1985) and were chosen to represent regions of the protein sequence for which the corresponding DNA sequence could be predicted with minimum degeneracy. Van Damme-Jongsten et al. (1989) chose amino acids 125–134 while Iwanejko et al. (1989) chose amino acids 9–17 (see discussion of Figure 16.4 in later section). Degeneracy for the former is 2048 fold while that of the latter is 512 fold, and, even if equivalent lengths are employed, the former cannot have a degeneracy less than 1024 fold. Degeneracy of 1024 or 512 fold cannot be employed for the construction of effective nucleotide probes, and both groups therefore used independent strategies to minimize degeneracy. Van Damme-Jongsten et al. (1989) chose to employ the known high AT content of clostridial DNA to use only adenine or thymine in the third base position. This yielded a final degeneracy of only 4 fold. In contrast Iwanejko et al. (1989) applied the convention that G–T interactions are neutral in terms of base pairing to reduce the degeneracy of their primer also to 4 fold. Both these primers were effective probes, and it is interesting to note now that, by comparison to the known DNA sequence, the Van Damme-Jongsten primer was correct in all but one 'wobble' base while the Iwanejko primer was incorrect in three positions.

Van Damme-Jongsten et al. (1989) used total digestion of *C. perfringens* DNA with *Hind*III or

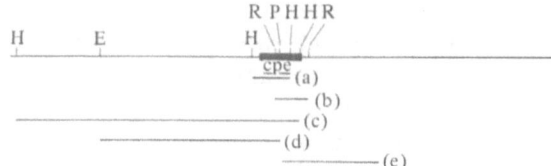

FIGURE 16.1 Diagram of extent of existing *cpe* clones with respect to their chromosomal location. (a) Van Damme-Jongsten et al. (1989), *Hind*III fragment; (b) Van Damme-Jongsten et al. (1989), *Rsa*I fragment; (c) Iwanejko et al. (1989), *Hind*III fragment; (d) Iwanejko and Stewart (unpublished data) *Eco*RI to *Pst*I fragment; (e) Hanna et al. (1989), random shear fragment. Abbreviations; *cpe*, coding region of CPE; H, *Hind*III; E, *Eco*RI; R, *Rsa*I; P, *Pst*I.

*Rsa*I to isolate DNA fragments between 0.8 and 1.2 kb in size for insertion into the *E. coli* vector pUC19 (Norrander et al., 1983). Iwanejko et al. (1989) used, by comparison, a partial *Hind*III digestion to isolate DNA fragments between 6 and 10 kb in size for insertion into the *E. coli* vector pHG165 (Stewart et al., 1986). Figure 16.1 summarizes the cloning data. Two overlapping clones obtained by Van Damme-Jongsten et al. (1989) cover the entire *Clostridium perfringens* enterotoxin gene (*cpe*) sequence. These clones are, however, only approximately 1 kb in size and do not extend far beyond the *C. perfringens* enterotoxin (CPE) sequence. A 6.8-kb DNA clone obtained by Iwanejko et al. (1989) contains substantial sequence 5' to *cpe* but fails by 88 bp to include the 3' end of the gene (L. Iwanejko, observation unpublished).

Hanna et al. (1989) used a quite different approach for cloning *cpe*. They employed a random shearing method to obtain clostridial DNA fragments and, after the addition of *Eco*RI linkers, cloned these fragments into lambda gt11. Lambda gt11 is an expression vector capable of producing a polypeptide specified by the DNA insert fragment (Young and Davis, 1983). The site used for insertion of foreign DNA is a unique *Eco*RI cleavage site located within the *lacZ* gene, 53 bp upstream from the β-galactosidase translational termination codon, and foreign DNA sequences in this vector have the potential to be expressed as β-galactosidase fusion proteins. In accordance with the foregoing, Hanna et al. (1989) obtained a single plaque that had positive affinity for an anti-CPE monoclonal antibody and, on further characterization, proved to express a β-galactosidase-CPE fusion protein. The fusion expresses the C-terminal portion of CPE from amino acid 171 to the end (320). The clostridial DNA insert is some 2.3 kb in size and consequently contains approximately 1850 bp of DNA 3' to *cpe* (Figure 16.1).

The data in Figure 16.1 indicates that none of the cloning strategies have provided an original clone containing the entire CPE gene. It is clear that the strategy employed by Hanna et al. (1989) might potentially select for partial clones but, in principle, an entire gene translationally coupled but not fused to *lacZ* could have been isolated. The strategy employed by Van Damme-Jongsten et al. (1989) was again biased against isolating the complete *cpe* because of the selected DNA size range (0.8 to 1.2kb for a 990-bp gene) and the use of total rather than partial restriction enzyme digests. Iwanejko et al. (1989), however, should have had every opportunity to isolate a complete gene sequence, yet they did not. Further, extensive efforts to reclone the entire *cpe* by complete digestion with *Eco*RI, a restriction enzyme that does not cut the *cpe* sequence and provides a 10-kb DNA fragment that can be identified as containing *cpe* by Southern hybridization (see later discussion of Figure 16.6B), have failed (L. Iwanejko and G. Stewart, unpublished observations).

An attempt has been made to reclone *cpe* as two specific restriction fragments cleaved by *Pst*I within *cpe* and flanked by *Eco*RI. The 5' *cpe* sequence has been successfully cloned on multiple independent isolates (Figure 16.1); however, the corresponding 3' terminal end of *cpe*, including some 5 kb of 3' sequence, has never been isolated (L. Iwanejko and G. Stewart, unpublished observations).

The foregoing discussion indicates there is growing, although as yet circumstantial, evidence that the cloning in *E. coli* of the entire *cpe* is highly unfavorable. The problem, however, may be associated more with cloning DNA flanking *cpe* than with *cpe* per se, as discussed next.

```
       M  L  S  N  N  L  N  P  M  V  F  E  N  I  A  K  E  V  F  L  I
                                                C
     ATGCTTAGTAACAATTTAAATCCAATGGTGTTCGAAAATGCTAAAGAAGTATTTCTTATT
            10        20        30        40        50        60

       S  E  D  L  K  T  P  I  N  I  T  N  S  N  S  N  L  S  D  G
     TCTGAGGATTTAAAAACACCAATTAATATTACAAACTCTAACTCAAATTTAAGTGATGGA
            70        80        90       100       110       120

       L  Y  V  I  D  K  G  D  G  W  I  L  G  E  P  S  V  V  S  S
     TTATATGTAATAGATAAAGGAGATGGTTGGATATTAGGGGAACCCTCAGTAGTTTCAAGT
           130       140       150       160       170       180

       Q  I  L  N  P  N  E  T  G  T  F  S  Q  S  L  T  K  S  K  E
     CAAATTCTTAATCCTAATGAACAGGTACCTTTAGCCAATCATTAACTAAATCTAAAGAA
           190       200       210       220       230       240

       V  S  I  N  V  N  F  S  V  G  F  T  S  E  F  I  Q  A  S  V
     GTATCTATAAATGTAAATTTTTCAGTTGGATTTACTTCTGAATTTATACAAGCATCTGTA
           250       260       270       280       290       300

       E  X  G  F  G  I  T  I  G  E  I  O  N  T  I  E  R  S  V  S  T
                                      G
     GAATATGGATTTGGAATAACTATAGGAGAACAAAATACAATAGAAAGATCTGTATCTACA
           310       320       330       340       350       360

       T  A  G  P  N  E  Y  V  V  Y  K  V  Y  A  T  Y  R  K  Y  Q
     ACTGCTGGTCCAAATGAATATGTATATTATAAGGTTTATGCAACTTATAGAAAGTATCAA
           370       380       390       400       410       420

       A  I  R  I  S  H  G  N  I  S  D  D  G  I  S  I  Y  K  L  T  G
                                              C
     GCTATTAGAATTTCTCATGGTAATATCTCTGATGATGGATCAATTTATAAATTAACAGGA
           430       440       450       460       470       480

       I  W  L  S  K  T  S  A  D  S  L  G  N  I  D  Q  G  S  L  I
     ATATGGCTTAGTAAAACATCTGCAGATAGCTTAGGAAATATTGATCAAGGTTCATTAATT
           490       500       510       520       530       540

       E  T  G  E  R  C  V  L  T  V  P  S  T  D  I  E  K  E  I  L
                                                        A
     GAAACTGGTGAAAGATGTGTTTTAACAGTTCCATCTACAGATATAGAAAAAGAAATCCTT
           550       560       570       580       590       600

       D  L  A  A  A  T  E  R  L  N  L  T  D  A  L  N  S  N  I  P  A
                                                     T
     GATTTAGCTGCTGCTACAGAAAGATTAAATTTAACTGATGCATTAAACTCAAACCCAGCT
           610       620       630       640       650       660

       G  I  N  L  Y  D  W  R  S  S  N  S  Y  P  W  T  Q  K  L  N  L
        T                                              K  L
     GGAAATTTATATGATTGGCGTTCTTCTAACTCATACCCTTGGACTCAAAAGCTTAATTTA
           670       680       690       700       710       720

       H  L  T  I  T  A  T  G  Q  K  Y  R  I  L  A  S  K  I  V  D
                                              C
     CACTTAACTATTACAGCTACTGGACAAAAATATAGAATTCTTAGCTAGCAAAATTGTTGAT
           730       740       750       760       770       780

       F  N  I  Y  S  N  N  F  N  N  L  V  K  L  E  Q  S  L  G  D
     TTTAATATTTATTCAAATAATTTTAATAATCTAGTGAAATTAGAACAGTCTTTAGGTGAT
           790       800       810       820       830       840

       G  V  K  D  H  Y  V  D  I  S  L  D  A  G  Q  Y  V  L  V  M
     GGAGTAAAAGATCATTATGTTGATATAAGCTTAGATGCGGACAATATGTTCTTGTAATG
           850       860       870       880       890       900

       K  A  N  S  S  Y  S  G  N  S  H  P  Y  S  I  L  F  Q  K  F
     AAAGCTAATTCATCATATAGTGGAAACTCTCACCCTTATTCAATATTATTTCAAAAATTT
           910       920       930       940       950       960

     TAA
      *
```

FIGURE 16.2 DNA sequence for *cpe* and derived amino acid sequence. Sequence from position 1 to position 873 was derived by L. Iwanejko and G. Stewart (unpublished data) and that from 874 to 963 was derived from Van Damme-Jongsten et al. (1989). Changes in base sequence between different laboratory isolates of NCTC 8239 and for *C. perfringens* stain F3686 are indicated; restriction sites for *Mbo*I (underlined); *Hind*III (overlined).

The CPE sequence

Van Damme-Jongsten et al. (1989) have reported the entire DNA sequence for *cpe*. They compared the sequence from two *C. perfringens* strains NCTC 8239 and F3686 and identified four points of variance, all in third base positions and all silent for the derived amino acid sequence (Figure 16.2). Hanna et al. (1989) sequenced 198 bp of their partial *cpe* clone derived from the NCTC 8239 strain. Surprisingly,

this sequence was identical to that derived previously for the F3686 strain, showing altered codon choices at amino acids 197, 218, and 221 [these are defined as amino acids 188, 209, and 212 in the paper by Hanna et al. (1989), because their numbering reflects the previously published amino acid sequences of Richardson and Granum (1985) that has subsequently been amended by DNA sequence data (Van Damme-Jongsten et al., 1989; see Figure 16.2 and later discussion of Figure 16.4)].

Iwanejko et al. (1989) reported the sequence for the first 30 amino acids of *cpe* cloned from *C. perfringens* 8239. These workers have extended this sequence to within 88 bp of the 3' terminus of *cpe*, the full extent of their *Hind*III clone (see Figure 16.1). This sequence is identical to that of Van Damme-Jongsten et al. (1989) except at three positions (L. Iwanejko and G. Stewart, unpublished observations): a C-to-T change at base 39 (silent), a G-to-C change at position 331 (Glu to Gln), and a C-to-T change at position 765 (silent). The only change that affects the derived amino acid sequence is at position 331, and the sequencing ladder for this region is therefore shown in Figure 16.3. Codon usage is in accord with other clostridial genes (Young et al., 1989). In a total gene length of 963 bases, there are 7 variant bases, sufficient to indicate continuing divergence not only between different strains of *C. perfringens* but between different laboratory-held isolates of the same strain, NCTC 8239. There is no evidence for multiple gene copies contributing to the sequence changes (see following).

Figure 16.4 shows the comparison of the derived amino acid sequence with that obtained from peptide analysis by Richardson and Granum (1985). The differences are as follows: peptide sequence 33–35 is not present in the derived sequence, and amino acid changes occur at peptide sequence positions 53, 61, 217, and 279. An asparagine was omitted from the peptide sequence position 261, and at position 300 a tyrosine is substituted by a serine and histidine. The major discrepancy, however, lies in the peptide sequence 88 to 103. This region was identified by Richardson and Granum (1985) as a region refractory to peptide sequencing. They suggested a 16-amino-acid sequence, but the

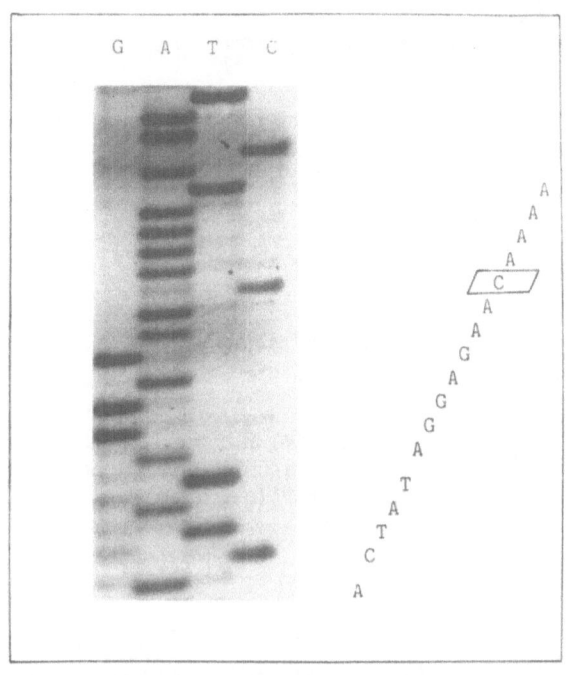

```
                10        20        30        40
MLSNNLNPMVFENAKEVFLISEDLKTPINITNNNLSNSNL
MLSNNLNPMVFENAKEVFLISEDLKTPINITN---SNSNL
                10        20        30

                50        60        70        80
SDGLYVIDKGDGLILGEPSVLSSQILNPNETGTFSQSLTK
SDGLYVIDKGDGWILGEPSVVSSQILNPNETGTFSQSLTK
   40        50        60        70

                      90       100
SKEVSIN(VVGFFTSEFIQASVEY)              TIERS
SKEVSINVNFSVGFTSEFIQASVEYGFGITIGEQNTIERS
   80        90       100       110

110       120       130       140
VSTTAGPNEYVYYKVYATYRKYQAIRISHGNISDDGSIYK
VSTTAGPNEYVYYKVYATYRKYQAIRISHGNISDDGSIYK
   120       130       140       150

150       160       170       180
LTGIWLSKTSADSLGNIDQGSLIETGERCVLTVPSTDIEK
LTGIWLSKTSADSLGNIDQGSLIETGERCVLTVPSTDIEK
   160       170       180       190

190       200       210       220
EILDLAAATERLNLTDALNSNPAGNLYDYRSSNSYPWTQK
EILDLAAATERLNLTDALNSNPAGNLYDWRSSNSYPWTQK
   200       210       220       230

230       240       250       260
LNLHLTITATGQKYRILASKIVDFNIYSNNFN-LVKLEQS
LNLHLTITATGQKYRILASKIVDFNIYSNNFNNLVKLEQS
   240       250       260       270

270       280       290       300
LGDGVKDHYVDLSLDAGQYVLVMKANSSYSGNY-PYSILF
LGDGVKDHYVDISLDAGQYVLVMKANSSYSGNSHPYSILF
   280       290       300       310

QKF
QKF
320
```

FIGURE 16.3 DNA sequencing reaction for *cpe* gene sequence between bases 319 and 342. C at position 331, previously reported as G (Van Damme-Jongsten et al., 1989), is boxed.

FIGURE 16.4 Comparison between peptide-derived amino acid sequence (Richardson and Granum, 1985) and DNA-derived amino acid sequence (differences are boxed).

derived sequence suggests it should in fact contain 28 residues.

5′ Sequence and putative promoter elements

Figure 16.5 shows the DNA sequence to 300 bp in front of *cpe*. Van Damme-Jongsten et al. (1989) and Iwanejko et al. (1989) both identified a putative Shine–Dalgarno sequence (Shine and Dalgarno, 1975) 11 bp upstream of the initiation ATG. An identical sequence was identified 14 nucleotides upstream of the *C. tetani* tetanus toxin sequence (Eisel et al., 1986). A region with significant homology to the putative −35 and −10 promoter sequences of both the *C. tetani* tetanus toxin gene (Eisel et al., 1986) and the *Bacillus cereus* penicillinase gene (Sloma and Gross, 1983) is located 5′ to the Shine–Dalgarno sequence (Figure 16.5). The homologous sequences are at −265 bp and −239 bp with respect to the enterotoxin initiation codon and represent a putative promoter sequence for the toxin gene. However, there are other regions of homology with these sequences from *C. tetani*, for example, bases −97 to −86 inclusive are identical to the −35 sequence.

A definitive identification of the transcription initiation site(s) awaits additional experimental data. Such data must unlock the genetic basis for the control of *cpe* expression and the link between sporulation and maximal synthesis (Labbe, 1980). In particular, however, there is likely to be a genetic explanation underlying those studies showing that a small amount of CPE can be detected in vegetative cells of *C. perfringens* (Granum et al., 1984; Goldner et al., 1986). In this regard, the possibility that there is a mobile genetic element 5′ to *cpe* is discussed next.

Figure 16.6A shows a detailed restriction map of the 6.8-kb *Hind*III clone containing all

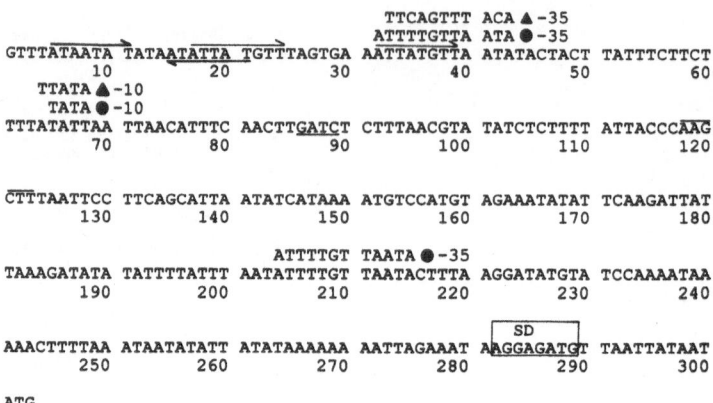

FIGURE 16.5   300-bp 5' to *cpe*: putative −35 and −10 sequences, SD sequence, inverted repeat, and direct repeat are indicated. Regions with homology to −35 and −10 promoter sequences for *Clostridium tetani* toxin gene (●) and the *Bacillus cereus* penicillinase gene (▲); restriction sequences for *Mbo*I (underlined); *Hin*dIII (overlined).

except 88 bp of the *cpe* (Iwanejko et al., 1989). There is a 330-bp *Hin*dIII fragment approximately 2.2 kb 5' to *cpe*. This *Hin*dIII fragment, and a second fragment containing 520 bp of *cpe* sequence, were used as ^{32}P-labeled 'Southern' probes against clostridial chromosomal DNA digested with *Eco*RI (Figure 17.6B). The *cpe* probe identified a single *Eco*RI band from the chromosomal digest at 10 kb, consistent with a single gene location. The 5' sequence probe, however, hybridized against a total of 10 different bands from the same digest (hybridization reactions were carried out on the same nylon membrane blot at high stringency). These multiple bands reduce to a single 330-bp band for *Hin*dIII-digested chromosomal DNA, a band equivalent in size to the hybridization probe (Figure 16.6B). These results are consistent only with multiple copies and locations for a DNA sequence homologous to a region 5' to *cpe* (experimental details to be published elsewhere).

During a *cpe* cloning program, a significant number of clones were obtained that showed positive hybridization to the 330-bp *Hin*dIII fragment but were negative to the *cpe* probe. This supports the conclusion that multiple copies and locations for the 330-bp sequence exist and can be independently cloned. A detailed structural analysis of these clones, including the extent of the multicopy sequence, has not yet been addressed. The possibility that the 330-bp *Hin*dIII sequence is a cloning artifact and not naturally contiguous with *cpe* is excluded, as the same sequence was identified in an independent clone containing DNA from *Eco*RI to *Pst*I (L. Iwanejko and G. Stewart, unpublished observations; also see Figure 16.1).

There is an interesting precedent for the presence of DNA insertion sequence (IS) elements, a type of transposable element able to insert itself into several sites in a genome, upstream of toxin gene sequences. Pritchard and Vasil (1990) have described possible IS elements upstream of the exotoxin A gene in *Pseudomonas aeruginosa*. There are intriguing comparisons between CPE and exotoxin A. Not all strains of *Pseudomonas aeruginosa* contain the *tox*A gene; there are hyper-toxin-producing strains and strains that produce only small amounts of protein and, in certain strains, sequences several kilobases 5' to the *tox*A gene are found in multiple copy. Transposable elements are known to alter regulation of toxin gene expression in *Corynebacterium diphtheriae* (Rappuoli et al., 1987), cholera toxin (Mekalanos, 1983), the α-hemolysin in *Escherichia coli* (Zabala et al., 1984) and, from the foregoing, exotoxin A in *P. aeruginosa* (Pritchard and Vasil, 1990). Given this and the evidence of multiple repeating sequences 5' to *cpe*, coupled with the knowledge that the number of IS elements in bacterial genomes can vary from 2 to more than 40 (Iida

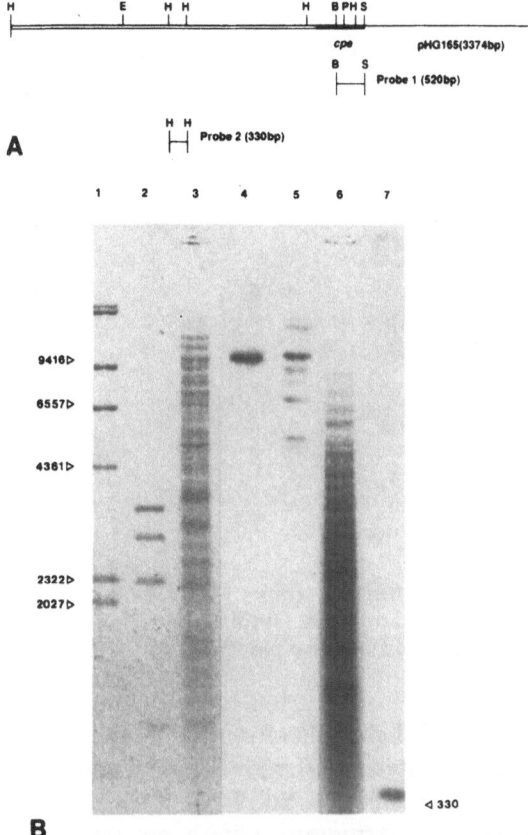

A

B

FIGURE 16.6 (**A**) Restriction enzyme map of *cpe* clone pLW1 (Iwanejko et al., 1989). Clone reaches to within 88 bp of 3' end of *cpe*. Two probes, a 330-bp *Hind*III fragment and a *Bg*III to *Sma*I fragment, used in Southern hybridization reactions, are indicated. Abbreviations: *cpe*, coding region of CPE; H, *Hind*III; E, *Eco*RI; P, *Pst*I; S, *Sma*I (located within pHG165 multiple cloning site). (**B**) Southern hybridization of restriction endonuclease-digested *Clostridium perfringens* 8239 DNA, probed sequentially with two probes identified in (**A**). Lane 1, lambda DNA digested with *Hind*III, ethidium bromide stain; lane 2, pLW1 digested with *Hind*III, ethidium bromide stain; lane 3, *C. perfringens* 8239 DNA digested with *Eco*RI, ethidium bromide stain; lane 4, lane 3 hybridized to probe 1; lane 5, lane 3 hybridized to probe 2; lane 6, *C. perfringens* 8239 DNA digested with *Hind*III, ethidium bromide stain; lane 7, lane 6 hybridized to probe 2. (Sizes indicated are in base pairs.)

et al., 1983), the possible role of such elements in the control of *cpe* expression must now be considered. At present, only two transposons have been identified and characterised in *C. perfringens* (Abraham and Rood, 1987). That one of these transposons, Tn4451, excises precisely in *E. coli* (Abraham and Rood, 1988) is of interest when considering both the possible involvement of IS elements 5' to *cpe* and the seeming difficulty in obtaining complete clones of *cpe* containing substantial 5' and 3' sequences in *E. coli*. Although all this is speculative, and could be explained equaly well by a close association with a dispersed gene family such as a small rRNA, it is perhaps timely if it encourages greater activity in the genetic analysis of *cpe*. Recently, L. Iwanejko and G. Stewart (unpublished observations) have shown that the 330 bp *Hind*III sequence does not hybridize to *C. perfringens* type C or type D chromosomal DNA, making an association with rRNA unlikely.

Gene location

Phillips-Jones et al. (1989) examined the plasmid content of enterotoxin positive *C. perfringens* food poisoning strains. These authors addressed a plasmid location for *cpe* by employing Southern hybridization with a toxin gene-specific oligonucleotide probe. Their results clearly indicated a chromosomal rather than a plasmid location. Granum (unpublished observations) has used a similar approach to confirm that *cpe* is not located on two *C. perfringens* phages. Definitive analysis of chromosome location has yet to be described; however, the recent description of a method to devise a physical map of the *C. perfringens* genome (Canard and Cole, 1989) should ensure early clarification of this question.

One aspect that is now clear is that only relatively few *C. perfringens* strains actually contain *cpe*. Van Damme-Jongsten et al. (1989) tested a total of 98 strains isolated from the feces of farm animals and found only 6% positive for hybridization against an oligonucleotide homologous with the *cpe* sequence. More recently, Van Damme-Jongsten et al. (1990) analyzed 245 strains of *C. perfringens* isolated from food and feces obtained from 186 separate confirmed outbreaks of *C. perfringens* food poisoning in the United Kingdom. Of the strains isolated from food and feces, only 59% were positive on hybridization to *cpe* probes,

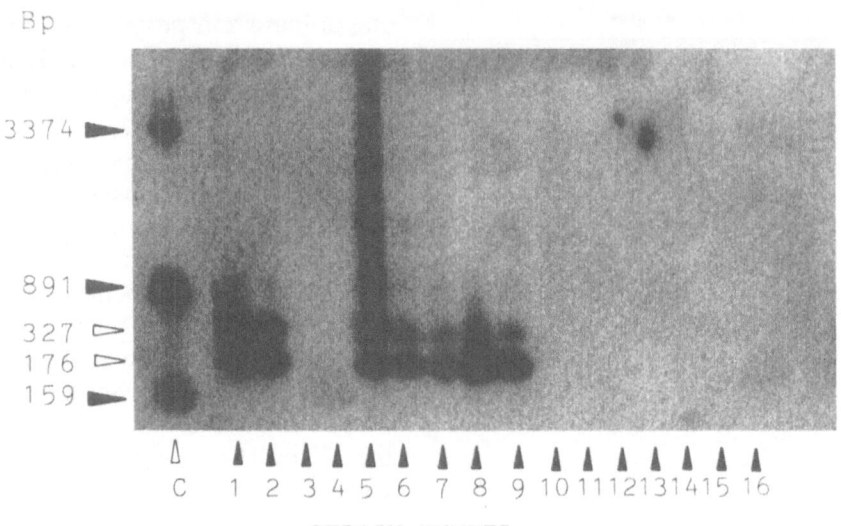

Bp

3374 ▶

891 ▶

327 ▷
176 ▷

159 ▶

△ ▲ ▲ ▲ ▲ ▲ ▲ ▲ ▲ ▲ ▲ ▲ ▲ ▲ ▲ ▲ ▲
C   1 2 3 4 5 6 7 8 9 10 11 12 13 14 15 16

STRAIN NUMBER

FIGURE 16.7 Southern hybridization of *Mbo*I-digested *Clostridium perfringens* DNA to probe I (*cpe* sequence from position 347 to 867, Figure 16.6A). *C. perfringens* strains 1–16 are identified in Table 16.2. Lane C is *Hind*III digest of plasmid pLW1; DNA sizes, base pairs.

and there was a random distribution of sero-types among these strains.

Dot blot hybridizations, such as those just described, appear to provide a reliable method for the detection of potentially enterotoxigenic *C. perfringens*. These analyses cannot, however, explore any heterogeneity in gene location or structure between different strains. Figure 16.7 shows the result of a 'Southern' analysis on 16 strains of *C. perfringens*. Nine were associated with food poisoning outbreaks, and 7 were isolated from lamb (Table 16.2). None of the lamb

Table 16.2 Source of *Clostridium perfringens* isolates analyzed for the presence of the *cpe* by Southern hybridization

Number	Isolate	Source	Presumptive food poisoning agent	*cpe*[a]	Serotype	Year	Plasmid containing
1	NCTC 8239	Salt Beef	+	+	3	1952	−
2	NCTC 10613	Minced beef	+	+	15	1968	−
3	NCTC 8449	Steamed lamb	+	−	2	1952	−
4	F 1785	Colindale (A.C. Ghosh)	+	−	38	Nk	+
5	NCTC 10239	Rissoles	+	+	12	1961	+
6	NCTC 8235	Stew	+	+	8	1951	+
7	NCTC 10612	Faeces	+	+	37	1968	+
8	NCTC 10240	Chicken	+	+	13	1959	+
9	NCTC 8679	Faeces	+	+	6	1952	+
10	L4.1	Smart et al., 1979	−	−	2, 3/4, 38, 68	1978	−
11	L53.1	Smart et al., 1979	−	−	61	1978	−
12	L60.2	Smart et al., 1979	−	−	14	1978	−
13	L5.4	Smart et al., 1979	−	−	41	1978	+
14	L99.4	Smart et al., 1979	−	−	3, 4	1978	+
15	L100.4	Smart et al., 1979	−	−	30	1978	+
16	L24.2	Smart et al., 1979	−	−	25	1978	+

[a]Data from Figure 16.7. NCTC, National Collection of Type Cultures, UK; Nk, not known.

strains contained *cpe* and two of the presumptive food poisoning isolates were negative for the gene, thus reflecting the results previously described. One of the lamb strains, L100.4, was not digested by *Mbo*I, reflecting differences in restriction modification between *C. perfringens* strains. Of the 7 toxin-positive clones, all showed identical Southern hybridization patterns for *Mbo*I digests indicating that, within the limits of a *Mbo*I restriction pattern and the probe used (probe 1, Figure 16.6a), there is no apparent heterogeneity in *cpe* location or structure between strains. This is somewhat surprising, considering that base position 459 has been shown to vary between A and C in different strains and that this variable base is within a *Mbo*I restriction sequence (see Figure 16.2). The lower of the two bands in the Southern hybridization (Figure 16.7) reflects the *Mbo*I fragment between bases 347 and 523 and is not further digested in any of the strains tested, including 8239, at the *Mbo*I site at position 459. Because this site is known to exist in strain 8239 (Figure 16.2), it would seem that it is resistant to *Mbo*I digestion, presumably as a result of adenine methylation. It is clear that continued study is necessary to elucidate, in its entirety, the molecular biology of *cpe*.

16.3   Properties of the Enterotoxin

Physicochemical properties

The physicochemical properties of CPE are summarized in Table 16.3. As discussed previously, the enterotoxin consists of 320 amino acids (Van Damme-Jongsten et al., 1989) linked together in one polypeptide chain (Granum and Skjelkvåle, 1977) with a molecular weight of 35,391. The protein is very hydrophobic (43% hydrophobic amino acids), a factor that is reflected by a low water solubility of about 4 mg/ml. The enterotoxin contains a single cysteine at residue 186 (see Figure 16.4) with no cystine and therefore one free thiol group and no stabilizing disulfide bridges. This is reflected in its relative sensitivity to heat, being rapidly

Table 16.3 Physicochemical properties of the *Clostridium perfringens* enterotoxin

One polypeptide chain (two domains?)
80% β-sheet structure and 20% random coil structure
Molecular weight 35,391 (pure protein)
Sequence known (320 amino acids, 43% hydrophobic)
Stoke's radius 2.6 nm
Heat stable to 53°C
Binds SDS anomalously (0.39 g SDS/g protein)
Solubility, 3.94 ± 0.22 mg/ml (pH 7.0, 25°C)
One free SH group
Net charge at pH 7, −10

inactivated at temperatures above 53°C (Granum and Skjelkvåle, 1977).

Protein structure–function

The primary structure of CPE is shown in Figure 16.4 and has been discussed previously in Section 16.2. Without crystallographic data (see the end of this section), it is very difficult to infer very much about secondary structure from the primary sequence. Using nine different structure prediction models (data not shown), we obtained highly conflicting putative conformations that showed consistency only in two places. N-terminal residues 10 through 25 were predicted to conform to an α-helix in eight of nine models. This prediction has support in biochemical analysis, because this section is accessible to trypsin and chymotrypsin digestion and is processed by these enzymes in the ileum to stimulate toxin activity by some threefold (Richardson and Granum, 1983, Granum and Richardson, 1991). One further putative α-helix between residue 195 and 215 was consistently predicted by all models.

The three-dimensional structure of the enterotoxin has been studied by using UV-differential spectroscopy (Granum and Whitaker, 1980) and by titration of accessible amino groups under several different conditions (Whitaker and Granum, 1980). It was concluded from these studies that the enterotoxin most probably conforms to a two-domain struc-

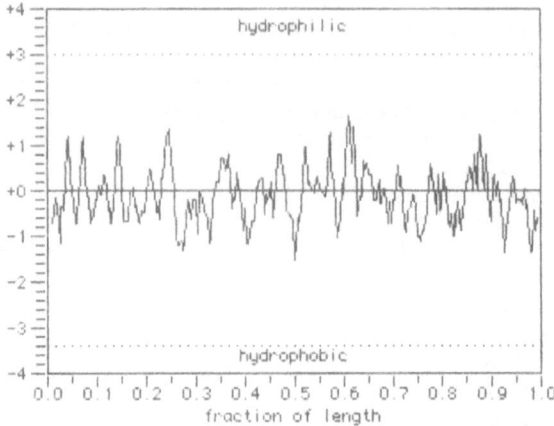

FIGURE 16.8 Hydrophilicity plot on enterotoxin from *Clostridium perfringens*, using average length of 6 amino acids. Plot is carried out using DNA Inspector program on Macintosh SE personal computer.

ture. In this respect, recent studies by Hanna et al. (1989, 1991) would appear to support the two-domain model. These workers cloned a C-terminal fragment of *cpe* (see section 16.2) and have demonstrated that the CPE amino acid sequence 290–319 completely blocks CPE specific binding to brush border membranes. This C-terminal fragment, therefore, recognizes the physiological receptor that mediates CPE cytotoxicity. Chemical modification of one or two of the ε-amino groups of the lysine residues (17 in total, 8 exposed) on the surface of the enterotoxin abolishes its biological activity (Whitaker and Granum, 1980). It has been suggested that the surface exposed and hence titratable lysine residues could be at positions 77 and 79, as these are within the most hydrophilic region of the protein molecule (Fig. 16.8). Given the work of Hanna et al. (1989, 1991) it is, however, more likely that the loss of biological activity through modification of one or two ε-amino groups are at lysine residue 301 and 319. Since the latter can be split off during chymotrypsination (together with 34 amino acids from the N-terminal end) resulting in a 3 fold increase in activity (Granum and Richardson, 1991), it seems probable that the ε-amino group of lysine 301 is the binding point for the membrane receptor.

According to Van Holde (1977) a protein containing more than 30% hydrophobic amino acids tends to associate (quaternary structure). Given the 43% hydrophobic residue content of CPE, it is not perhaps surprising to find aggregation both in vivo and in vitro. CPE has been observed as crystalline inclusions in *C. perfringens* mother cells during sporulation (Skjelkvåle and Duncan, 1975; Loffler and Labbe, 1985). Although this may be related to overproduction, as in inclusion-body formation (Williams et al., 1982), it is interesting to compare this observation with the only other naturally occurring crystalline inclusion formed from insecticidal δ-endotoxin by *Bacillus thuringiensis* (Aronson et al., 1986).

In vitro the CPE protein aggregates from solution if the concentration is maintained above 0.5 mg/ml at 4°C (P. Granum, unpublished observations). Aggregation occurs even in the presence of sodium dodecyl sulfate (SDS), where the protein binds about one-third as much compared with normal (hydrophilic) proteins (Enders and Duncan, 1976). It has also been shown that exposure to SDS gradually changes the structure of the enterotoxin from predominantly β-sheet structure (80% β-sheet structure, 20% random coil structure) to mainly α-helix structure (Granum and Harbitz, 1985). Small amounts of SDS are supposed to reflect protein behavior in a membrane environment (Robinson and Mattice, 1981). Despite this natural propensity to self-aggregate, CPE has proved to be refractory to the formation of crystals suitable for x-ray analysis. Crystals have been produced (D. Lawson and P. Artymiuk, unpublished observations) as shown in Figure 16.9. At present, however, these are too small for x-ray analysis. Nevertheless, further attempts to produce larger crystals based on the work in Artymiuk's group will hopefully succeed so that subsequent reviews can report on the detailed molecular architecture of this toxic protein.

It is clear that while substantial new information has become available on the molecular biology of CPE in recent years, there is still much to do. A detailed characterization of the genetic locus for *cpe* is awaited. With the availability of the *cpe* sequence, site-directed

FIGURE 16.9 Enterotoxin from *Clostridium perfringens* has been crystallized using method of McPherson et al. (1986). Small octahedral crystals were grown from polyethyl glycol by hanging drop method, in presence of small quantities of *n*-octyl β-glucoside. Largest bipyramidal crystals are 0.2 mm in longest dimension. (Work has been carried out by D. Lawson and P. Artymiuk, University of Sheffield.)

mutagenesis in combination with polymerase chain reaction (PCR) to provide expression-competent domain elements of the protein will continue to elucidate the binding structure and structure–function relationships for CPE. Studies on insertion of the enterotoxin into the plasma membrane could be addressed in model systems where differential sensitivity to proteolysis could establish the membrane located domain(s). Finally, with respect to much of the existing literature on CPE that reflects the sporulation-associated nature of its production, a final and definitive conclusion to the debate on the relationship between the enterotoxin and structural elements of the spore seems close at hand.

Acknowledgment. The authors would like to thank Peter Artymiuk and Askild Holck for running computer based protein structure prediction models on the enterotoxin. We would also like to thank Lesley A. Iwanejko who, as an AFRC graduate student of GSABS, has contributed much to the content of this review.

References

Abraham, L.J. and J.I. Rood. 1987. Identification of Tn445 1 and Tn4452, chloramphenicol resistance transposons from *Clostridium perfringens. J. Bacteriol.* **169**:1579–1584.

Abraham, L.J. and J.I. Rood. 1988. The *Clostridium perfringens* chloramphenicol resistance transposon Tn4451 excises precisely in *Escherichia coli. Plasmid* **19**:164–168.

Aronson, A.I., W. Bechman, and P. Dunn. 1986. *Bacillus thuringiensis* and related insect pathogens. *Microbiol. Rev.* **50**:1–24.

Canard, B. and S.T. Cole. 1989. Genome organisation of the anaerobic pathogen *Clostridium perfringens. Proc. Nat. Acad. Sci. USA* **86**:6676–6680.

Cliver, D.O. 1987. Foodborne disease in United States 1946–1986. *Int. J. Food Microbiol.* **4**:269–277.

Eisel, U., W. Jarausch, K. Goretzki, A. Henschen, J. Engels, U. Weller, M. Hudel, E. Habermann, and H. Niemann. 1986. Tetanus toxin: primary structure, expression in *E. coli* and homology with botulinum toxins. *EMBO J.* **5**:2495–2502.

Enders, G.L. Jr. and C.L. Duncan. 1976. Anomalous aggregation of *Clostridium perfringens* enterotoxin under dissociating conditions. *Can. J. Microbiol.* **22**:1410–1414.

Goldner, S.B., M. Solberg, S. Jones, and L.S. Post. 1986. Enterotoxin synthesis by nonsporulating cultures of *Clostridium perfringens*. *Appl. Environ. Microbiol.* **52**:407–412.

Gondrosen, B., G. Langeland, and N. Aas. 1990. *Norwegian Food Control Authority's Report on Food-Associated Diseases in 1988 and 1989*. SNT, Oslo.

Granum, P.E. 1985. The effect of Ca^{++} and Mg^{++} on the action of *Clostridium perfringens* enterotoxin on Vero cells. *Acta Pathol. Microbiol. Immunol. Scand. B* **93**:41–48.

Granum, P.E. 1990. *Clostridium perfringens* toxins involved in food poisoning. *Int. J. Food Microbiol.* **10**:101–112.

Granum, P.E. and O. Harbitz. 1985. A circular-dichroism study of the enterotoxin from *Clostridium perfringens* type A. *J. Food Biochem.* **9**:137–146.

Granum, P.E. and M. Richardson. 1991. Chymotrypsin treatment increases the activity of *Clostridium perfringens* enterotoxin. *Toxicon* 29, 898–900.

Granum, P.E. and R. Skjelkvåle. 1977. Chemical modification and characterization of enterotoxin from *Clostridium perfringens* type A. *Acta Pathol. Microbiol. Scand. B* **85**:89–94.

Granum, P.E. and J.R. Whitaker. 1980. Improved method for purification of enterotoxin from *Clostridium perfringens* type A. *Appl. Environ. Microbiol.* **39**:1120–1122.

Granum, P.E., J.R. Whitaker, and R. Skjelkvaåle. 1981. Trypsin activation of enterotoxin from *Clostridium perfringens* type A. Fragmentation and some physicochemical properties. *Biochim. Biophys. Acta* **668**:325–332.

Granum, P.E., W. Telle, O. Olsvik, and A. Stavn. 1984. Enterotoxin formation by *Clostridium perfringens* during sporulation and vegetative growth. *In. J. Food Microbiol.* **1**:43–49.

Hanna, P.C., A.P. Wnek, and B.A. McClane. 1989. Molecular cloning of the 3' half of the *Clostridium perfringens* enterotoxin gene and demonstration that this region encodes receptor binding activity. *J. Bacteriol.* **171**:6815–6820.

Hanna, P.C., T.A. Mietzner, G.K. Schoolnik, and B.A. McClane. 1991. Localization of the receptor-binding region of *Clostridium perfringens* enterotoxin utilizing cloned toxin fragments and synthetic peptides. The 30 C-terminal amino acids define a functional binding region. *J. Biol. Chem.* 266, 11037–11043.

Hauschild, A.H.W. and R. Hilsheimer. 1971. Purification and characteristics of the enterotoxin from *Clostridium perfringens* type A. *Can. J. Microbiol.* **17**:1425–1433.

Hobbs, B. C. 1969. *Clostridium perfringens* and *Bacillus cereus* infections. *In: Food- Borne Infections and Intoxications*, H. Reiman, ed., pp. 131–173. New York: Academic Press.

Iida, S., J. Meyer, and W. Arber. 1983. Prokaryotic IS elements. *In; Mobile Genetic Elements*, J.A. Shapiro, pp. 159–221. London: Academic Press.

Iwanejko, L.A., M.N. Routledge, and G.S.A.B. Stewart. 1989. Cloning in *Escherichia coli* of the enterotoxin gene from *Clostridium perfringens* type A. *J. Gen. Microbiol.* **135**:903–909.

Klein, E. 1895. Ueber einen pathogenen anaë roben Darmbacillus, *Bacillus enteritidis sporogenes. Centralblatt pür Bakteriologie und Parasitenkunde.* IAbt. **18**:737–743.

Labbe, R.G. 1980. Relationship between sporulation and enterotoxin production in *Clostridium perfringens* type A. *Food Technol.* **34**:88–90.

Loffler, A. and R.G. Labbe. 1985. Isolation of an inclusion body from sporulating, enterotoxin-positive *Clostridium perfringens. FEMS Microbiol. Lett.* **27**:143–147.

Matsuda, M. and N. Sugimoto. 1979. Calcium-independent and dependent steps in action of *Clostridium perfringens* enterotoxin on Hela and vero cells. *Biochem. Biophys. Res. Commun.* **91**:629–636.

McClane, B.A., A.P. Wnek, K.I. Hulkower, and P.C. Hanna. 1988. Divalent cation involvement in the action of *Clostridium perfringens* type A enterotoxin. *J. Biol. Chem.* **263**:2423–2435.

McClung, L.S. 1945. Human food poisoning due to *Clostridium perfringens* (*welchii*) in freshly cooked chicken: preliminary note. *J. Bacteriol.* **50**:229–231.

McDonel, J.L. 1979. The molecular mode of action of *Clostridium perfringens* enterotoxin. *Am. J. Clin. Nutr.* **32**:210–218.

McDonel, J.L. 1986. Toxins of *Clostridium perfringens* types A, B, C, D and E. *In: Pharmacology of Bacterial Toxins*, F. Dorner and H. Drews, eds., pp. 447–517. New York: Pergamon.

McPherson, A., S. Koszelak, H. Axelrod, J. Day, R. Williams, L. Robinson, M. McGrath, and D. Cascio. 1986. An experiment regarding crystallization of soluble proteins in the presence of β-octyl glucoside. *J. Biol. Chem.* **261**:1969–1975.

Mekalanos, J.J. 1983. Duplication and amplification of toxin genes in *Vibrio cholerae. Cell* **35**:253–263.

Norrander, J., T. Kempe, and J. Messing. 1983. Construction of improved M13 vectors using oligodeoxynucleotide-directed mutagenesis. *Gene* (Amst.) **26**:101–106.

Parry, W.H. 1963. Outbreak of *Clostridium welchii* food poisoning. *B. Med. J.* **2**:1616–1619.

Phillips-Jones, M.K., L.A. Iwanejko, and M.S. Longden. 1989. Analysis of plasmid profiling as a method for rapid differentiation of food-associated *Clostridium perfringens* strains. *J. Appl. Bacteriol.* **67**:243–254.

Pritchard, A.E. and M.L. Vasil. 1990. Possible insertion sequences in a mosaic genome organisation upstream of the exotoxin A gene in *Pseudomonas aeruginosa*. *J. Bacteriol.* **172**:2020–2028.

Rappuoli, R., M. Perugini, and G. Ratti. 1987. DNA elements of *Corynebacterium diphtheriae* with properties of insertion sequence and usefulness for epidemiological studies. *J. Bacteriol.* **169**:308–312.

Reynolds, D. 1987. Large scale purification of *Clostridium perfringens* type A enterotoxin and of *Staphylococcus aureus* enterotoxin A. Masters thesis, Public Health Laboratory Service, Porton Down, U.K.

Richardson, M. and P.E. Granum. 1983. Sequence of the aminoterminal part of the enterotoxin from *Clostridium perfringens* type A. Identification of points of trypsin activation. *Infect. Immun.* **40**:943–949.

Richardson, M. and P.E. Granum. 1985. The amino acid sequence of the enterotoxin from *Clostridium perfringens* type A. *FEBS Lett.* **182**:479–484.

Robinson, R.M. and W.L. Mattice. 1981. Conformational properties of central nervous system myelin basic protein, β-endorphin, and β-lipotropin in water and in the presence of anionic lipids. *Biopolymers* **20**:1421–1434.

Shine, J. and C.L. Dalgarno. 1975. Determination of cistron specificity in bacterial ribosomes. *Nature* (London) **254**:34–38.

Skjelkvåle, R. and C.L. Duncan. 1975. Enterotoxin formation by different toxigenic types of *Clostridium perfringens*. *Infect. Immun.* **11**:563–575.

Sloma, A. and M. Gross. 1983. Molecular cloning and nucleotide sequence of the type 1 β-lactamase gene from *Bacillus cereus*. *Nucleic Acids Res.* **11**:4997–5004.

Smart, J.L., T.A. Roberts, M.F. Stringer, and N.Shah. 1979. The incidence and serotypes of *Clostridium perfringens* on beef, pork and lamb carcasses. *J. Appl. Bacteriol.* **46**:377–383.

Stark, R.L. and C.L. Duncan. 1971. Biologic characteristics of *Clostridium perfringens* type A enterotoxin. *Infect. Immun.* **4**:89–96.

Stark, R.L. and C.L. Duncan. 1972. Purification and biochemical properties of *Clostridium perfringens* type A enterotoxin. *Infect. Immun.* **6**:662–673.

Stewart, G.S.A.B., S. Lubinsky-Mink, C.G. Jackson, A. Cassel, and J. Kuhn. 1986. pHG165: a pBR322 copy number derivative of pUC8 for cloning and expression. *Plasmid* **15**:172–181.

Sutton, F.G.A. and B.C. Hobbs. 1965. Food poisoning caused by heat-sensitive *Clostridium welchii*. A report of five recent outbreaks. *J. Hyg. Cambridge* **66**:135–146.

Van Damme-Jongsten, M., K. Wernars, and S. Notermans. 1989. Cloning and sequencing of the *Clostridium perfringens* enterotoxin gene. *Antonie Leeuwenhoek J. Microbiol.* **56**:181–190.

Van Damme-Jongsten, M., J. Rodhouse, R.J. Gilbert, and S. Notermans. 1990. Synthetic DNA probes for detection of enterotoxigenic *Clostridium perfringens* strains isolated from outbreaks of food poisoning. *J. Clin. Microbiol.* **28**:131–133.

Van Holde, K. E. 1977. Effect of amino acid composition and microenvironment on protein structure. *In: Food Proteins*, J.R. Whitaker, and S.R. Tannenbaum, (eds.,) pp. 1–13. Westport: Avi.

Whitaker, J.R. and P.E. Granum. 1980. The role of amino groups in the biological and antigenic activities of *Clostridium perfringens* type A enterotoxin. *J. Food Biochem.* **4**:201–217.

Williams, D.C., R.M. Van Frank, W.L. Muth, and J.P. Burnett. 1982. Cytoplasmic inclusion bodies in *Escherichia coli* producing biosyntetic human insulin proteins. *Science* **215**:687–688.

Wnek, A.P. and B.A. McClane. 1983. Identification of a 50,000 M_r protein from rabbit brush border membranes that binds *Clostridium perfringens* enterotoxin. *Biochem. Biophys. Res. Commun.* **112**:1094–1105.

Wnek, A.P. and B.A. McClane. 1989. Preliminary evidence that *Clostridium perfringens* type A enterotoxin is present in a 160,000-M_r complex in mammalian membranes. *Infect. Immun.* **57**:574–581.

Young, M., W.L. Staudenbauer, and N.P. Minton. 1989. Genetics of *Clostridium*. *In: Clostridia, Biotechnology Handbooks 3*, N.P. Minton and N.P. Clarke, eds., pp. 63–103. New York: Plenum.

Young, R.A. and R.W. Davis. 1983. Efficient isolation of genes using antibody probes. *Proc. Nat. Acad. Sci. USA* **80**:1194–1198.

Zabala, J.C., J.M. Garcia-Lobo, E. Diaz-Aroca, F. de la Cruz, and J.M. Ortiz. 1984. *Escherichia coli* alpha-haemolysin synthesis and export genes are flanked by a direct repetition of IS91-like elements. *Mol. Gen. Genet.* **197**:90–97.

17

Molecular Genetic Studies of UV-Inducible Bacteriocin Production in *Clostridium perfringens*

Stewart T. Cole and Thierry Garnier

17.1 Introduction

In the 1970s, Ionesco and Bouanchaud (1973) reported the first demonstration for a clostridial species that bacteriocin production was associated with the presence of a small plasmid. It was later shown that *Clostridium perfringens* CPN50 synthesized large quantities of the bacteriocin BCN5 after irradiation with ultraviolet light and that this bactericidal protein was secreted into the medium (Ionesco et al., 1974). Subsequent genetic studies revealed that the bacteriocinogenic plasmid, pIP404, was about 10 kb in size and could be mobilized by conjugation by certain cryptic plasmids commonly found in *C. perfringens* (Brefort et al., 1977). Several years later, pIP404 was adopted by our laboratory as a model system for studying DNA damage-inducible genes in the clostridia and as a source of material for vector construction.

17.2 Molecular and Functional Characterization of pIP404

After purification of pIP404 by CsCl density gradient centrifugation and appropriate subcloning experiments, a detailed restriction map was produced that served as the basis of a DNA sequencing strategy. This culminated in the determination of the complete nucleotide sequence of pIP404, that contains 10,207 bp and has a dA + dT content of 75%, which is very close to that of the *C. perfringens* chromosome (73%–76%; Smith and Hobbs, 1974). The genome of pIP404 comprises 10 open reading frames and a locus consisting of two related families of direct repeats in a dispersed tandem array, which most probably corresponds to the origin of replication, *ori* (Garnier and Cole, 1988a, 1988b). Briefly, the genes of pIP404 can be classified into two groups, one providing basic plasmid functions, the other associated with bacteriocin production.

Our current knowledge of the genetic organization is summarized in Figure 17.1. To test for biological functions, the expression of the different open-reading frames was initially studied in *Escherichia coli* and *Bacillus subtilis* and, more recently, in *C. perfringens*. *E. coli* proved to be a poor host for expressing these clostridial genes and only one of them, *orf6*, was found to be expressed at a detectable level. The reason for this is almost certainly the extraordinary difference in codon usage by these two eubacteria; this topic is developed further later in this chapter. Subcloning in *B. subtilis* enabled the *rep* gene to be identified, which, together with *ori*, is essential for replication of pIP404. Copy number control appears to be regulated by the *cop* gene, which encodes a hydrophobic protein and RNA1, a 150-nucleotide RNA molecule that may act as an antisense RNA to the *rep* gene (Garnier and Cole, 1988b).

An additional means of controlling copy number, and thus ensuring efficient distribution of plasmid molecules to daughter cells during cell division, is provided by the *res* gene, which codes for a site-specific recombinase or

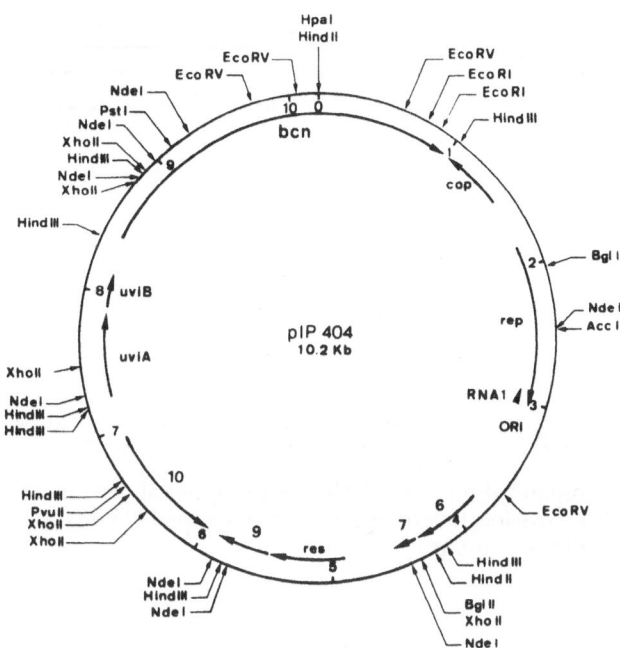

FIGURE 17.1 Organization and restriction map of bacteriocinogenic plasmid pIP404. Positions and sizes of genes indicated by arrows. Known genes and their products: *uviA*, possible bacteriocin immunity protein; *uviB*, possible bacteriocin secretion protein; *bcn*, bacteriocin BCN5; *cop*, copy number control; *rep*, DNA replicase; RNA1, antisense RNA to *rep* involved in copy number control; *res*, resolvase involved in site-specific recombination. No functions have been attributed yet to open-reading frames 6, 7, 9, and 10. ORI, putative origin of replication.

resolvase capable of resolving plasmid multimers into monomeric species (Garnier et al., 1987). Another interesting property of pIP404 is the finding that it does not employ a single-stranded replicative intermediate, unlike the majority of plasmids from gram-positive bacteria.

Identification of UV-inducible genes on pIP404

Our initial strategy for identifying the *bcn* gene encoding BCN5 was to clone overlapping restriction fragments into *E. coli* and to test for production of the bacteriocin (Garnier and Cole, 1986). This approach failed, because we had made the naive assumption that the *bcn* mRNA would not only be produced in this heterologous host but also translated. Consequently, a differential dot blot approach was employed to localize the gene; various DNA fragments were used as probes to detect mRNAs synthesized after UV irradiation of exponentially growing

C. perfringens CPN50. In this way it was established that a 4-kb segment of pIP404 was heavily transcribed after induction and that it consisted of two other genes, in addition to *bcn*, that were named *uviA* and *uviB*, for UV inducible (Figures 17.1 and 17.2).

Codon usage

The *bcn* gene was the first to be sequenced from any *C. perfringens* strain, and on inspection of its reading frame it was apparent that the codons preferentially employed were those rich in dA + dT, which is consistent with the remarkably low dG + dC content of the bacterial genome. One consequence of this "discrimination" against codons rich in dG + dC is a codon utilization diametrically opposed to that of the most popular host for cloning and expression studies, *E. coli*. Excellent examples of this polarity in codon choice are provided by the Arg and Leu families; the major codons employed by pIP404, as well as by *C. perfringens*

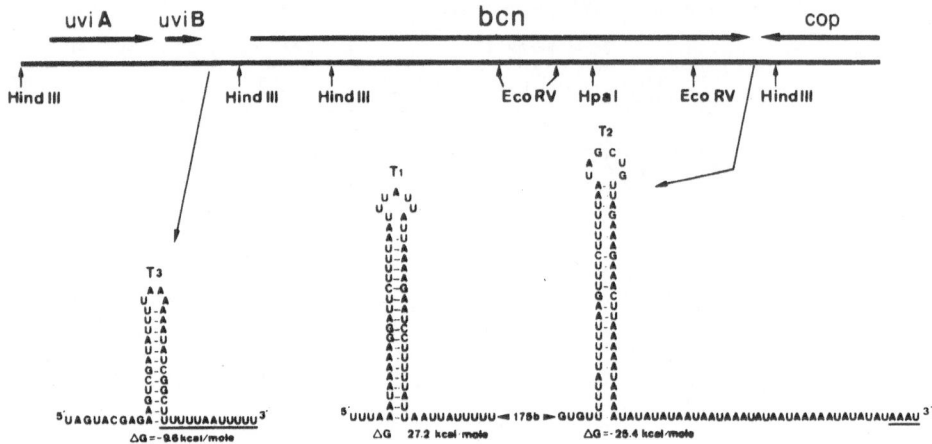

FIGURE 17.2 Detailed organization of UV-inducible region of pIP404 shows location of *uviAB* operon and *bcn* gene. Positions and structures of transcriptional terminators are shown in lower part; mapped 3' ends of mRNAs are underlined.

chromosomal genes (Saint-Joanis et al., 1989), are AGA and UUA, which account for more than 80% and 71% respectively, of the triplets employed (Garnier and Cole, 1988a).

By contrast, in *E. coli* genes these codons occur at a frequency of about 1% and 6%, respectively (Ogasawara, 1985), and in the case of AGA it is well documented that frequent use correlates with low expression levels (Chen and Inouye, 1990). Consequently, it seemed likely that expression of clostridial genes in *E. coli* could be limited by the availability of the minor tRNAs. Support for this interpretation was provided by gene fusion experiments in which the *bcn* mRNA was both transcribed and translated from *E. coli* signals. Again, no *bcn* activity was detected although in minicell experiments peptides corresponding to intermediate translational products were detected (T. Garnier, unpublished observations).

In an extended expression study of 13 genes from *C. perfringens*, only 2 were expressed at significant levels in *E. coli*, and it is clear that the expression level is a function not only of codon usage but also of gene size (Cole et al., unpublished observations). Independent confirmation of the concept that "rare" codons limit expression of clostridial genes in *E. coli* was recently provided by the elegant work of Makoff et al. (1989), who found greatly improved expression levels of a fragment of the tetanus toxin gene after harmonizing the codon usage with that of the host.

17.3 Properties and Secretion of BCN5

The *bcn* gene of pIP404 has the potential to encode a protein of 96,591 daltons, and a few hours after UV irradiation a polypeptide of this size is found in the medium (Garnier and Cole, 1986; Figure 17.3). Bacteriocins of gram-negative bacteria are not exported via the signal peptide pathway but are secreted in a semispecific manner by a controlled lysis procedure in which the "lysis" or "colicin-release protein," encoded by the bacteriocinogenic plasmid, activates a membrane-bound phospholipase to permeabilize the cell envelope (Pugsley, 1984a, 1984b).There are some parallels between this procedure and the secretion of BCN5, which also lacks an N-terminal signal sequence; after initial accumulation of bacteriocin in the bacterium (Figure 17.3), cell lysis leads to release into the medium. Although we suspect that the cell lysis associated with BCN5 production is caused by a function encoded by pIP404, it could also result from induction of the

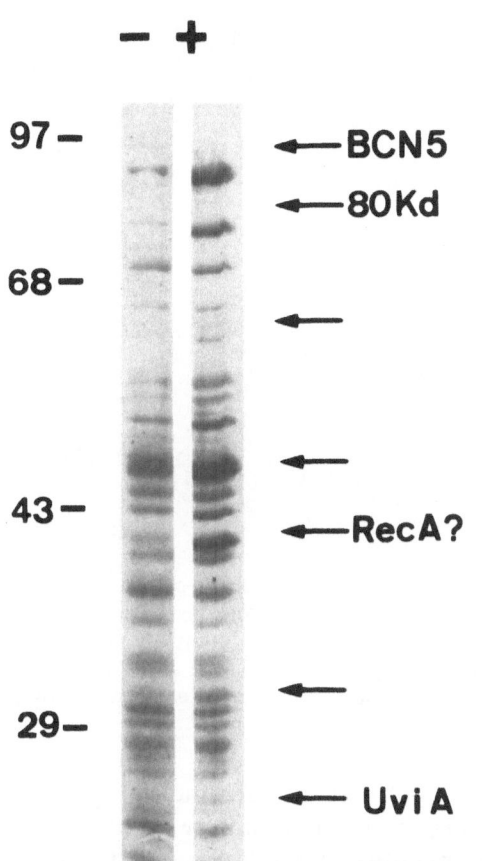

FIGURE 17.3 Sodium dodecylsulfate- polyacrylamide gel (SDS-PAGE) analysis of cytoplasmic proteins from *Clostridium perfringens* strain CPN50 produced before (−) and after (+) UV induction. Positions of molecular weight markers, (kDa); proteins induced by UV light, arrows. Known name of protein is given.

structure of BCN5 deduced from the nucleotide sequence reveals a protein of 890 amino acid residues with a high glycine content (11.5%), a property that is common to many bacteriocins and is believed to facilitate their transfer across cell membranes. A possible clue to the biological activity of BCN5 was provided by its hydropathy profile, because this predicted an extended lipophilic region (residues 815–869) near the carboxyl-terminus. Bacteriocins from gram-negative bacteria with similar hydropathy profiles possess membrane-spanning domains and function as ionophores (Pugsley, 1984a, 1984b).

17.4 Possible Roles for *uviA* and *uviB*

The *uviA* and *uviB* genes are arranged in an operon preceding the *bcn* gene on pIP404 (see Figure 17.2), and both transcriptional units are heavily transcribed in response to DNA damage. This common control mechanism suggests a possible functional relatedness and, in analogy with the colicin systems (Pugsley, 1984a, 1984b), it is likely that the *uvi* genes are involved in providing immunity to BCN5 and in facilitating its secretion.

A bacteriocin immunity gene is known to be carried by pIP404 (Brefort et al., 1977), and transformation of *C. perfringens* strain 13 (Scott and Rood, 1989) with shuttle vectors carrying the different plasmid genes has localized this to the *uvi* region although precise identification has not yet been obtained. The most likely candidate is *uviA*, because this encodes a soluble protein with a predicted molecular weight of 22,013 daltons and on induction pIP404 produces a protein of this size (see Figure 17.3). In contrast, the *uviB* gene product is a small protein of 64 residues, with a hydrophobic NH_2-terminal domain and a highly charged COOH terminus, which could play a role in BCN5 secretion by perturbing the cell membrane. Studies currently in progress should clarify the respective functions of the *uvi* cistrons.

lysogenic phage Φ29 harbored by CPN50 and most other type A strains of *C. perfringens*.

In addition to the 96-kDa form of BCN5, a second, minor protein of 80 kDa is found in the medium. This polypeptide is specified by pIP404 and probably derived by proteolysis from BCN5, because the plasmid does not carry an additional open-reading frame capable of encoding it. This finding is noteworthy; many years previously, when BCN5 was partially purified, its molecular weight was estimated as 82 kDa (Wolff and Ionesco, 1975). At present we do not know whether both species are endowed with bacteriocin activity. The primary

17.5 Transcription Signals Required for UV-Inducible Expression

The *Enterobacteriaceae* respond to DNA-damaging treatments such as UV light by inducing a family of repair functions, and accessory activities known as the SOS response (Walker, 1987) that is coordinately controlled by the LexA and RecA proteins. LexA is a repressor that binds to a site, the SOS box, near the promoters of the SOS genes. On induction of the SOS response the RecA protein is activated and catalyzes the cleavage, or autocleavage, of LexA, thus enabling RNA polymerase access to the promoters and derepressing the SOS regulon. Among the targets subject to SOS control are the plasmid-borne bacteriocin or colicin genes (Pugsley, 1984a, 1984b). To determine whether the UV-inducible genes carried by pIP404 were controlled in an analogous manner, their transcription was studied by a combination of in vivo and in vitro techniques (Garnier and Cole, 1988c). S1 nuclease protection experiments showed that the *uviA* and *uviB* genes were cotranscribed whereas the *bcn* gene was contained in a separate transcriptional unit. In both cases transcription termination occurred just downstream of stem and loop structures (see Figure 17.2).

Different transcript mapping techniques showed expression of the *bcn* gene and the *uviAB* operon to be directed by tandem promoters inducible by UV light. A representative primer extension analysis of *bcn* mRNA is shown in Figure 17.4; it can be seen that transcription initiates at three promoters (P1, P2, P3) and that promoters P1 and P2 are initially productive while P3 is activated later. Similar studies with *uviAB* revealed two promoters, P4 and P5, which were both activated simultaneously.

Inspection of the nucleotide sequences encompassing the transcriptional start points allowed the promoter elements to be identified (Garnier and Cole, 1988c). As shown in Figure 17.5, the five UV-regulated promoters from pIP404 can be classified into two groups. Three of them (P1, P3, and P5) are highly homo-

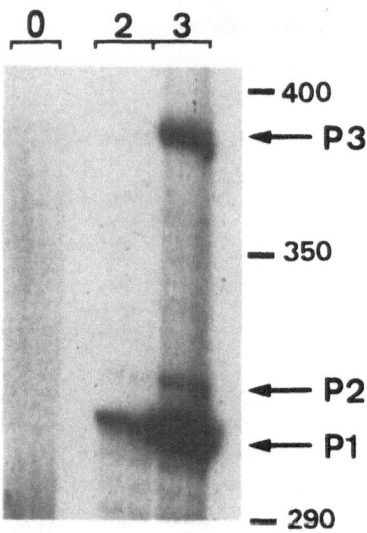

FIGURE 17.4 Kinetic analysis of production of *bcn* mRNA as judged by primer extension. Time after induction (in hours) is shown above; transcripts initiating at promoters P1, P2, and P3 indicated by arrows; positions of size markers denoted by bars.

logous but bear little resemblance to the consensus "bacterial" promoter sequence. Identical hexamers, TTTACA, and octamers, CTTTTTAT, occur in the "−35" and "−10" regions, respectively. In contrast, promoters P2 and P4 show little resemblance to one another although P4 does show reasonable homology with the consensus sequence. These regulatory regions do not contain common motifs likely to represent operator sequences nor do they possess sites homologous to the LexA recognition sequence (Walker, 1987). If expression of these UV-inducible genes were controlled at the transcriptional level by repression, removal of the repressor should lead to promoter recognition and transcription initiation by the major form of RNA polymerase.

To test this possibility, an in vitro transcription system was developed using RNA polymerases purified from *C. perfringens* or *B. subtilis*. The clostridial enzyme (Garnier and Cole, 1988c), which consisted of five subunits (β, 145 kDa; β', 135 kDa; α or σ, 52, 48 or 44 kDa), was unable to recognize the UV-inducible promoters when linear templates were employed but was highly active with constitutively expressed clostridial promoters such

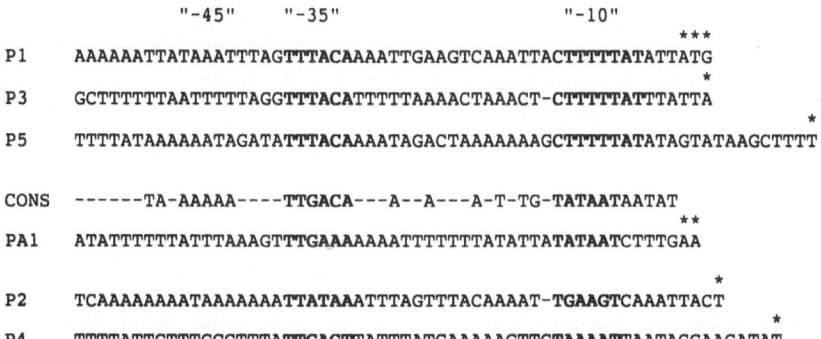

FIGURE 17.5 Nucleotide sequence of UV-inducible promoters and comparison with constitutive clostridial promoter (PA1; Garnier and Cole, 1988b) and consensus sequence for promoter from gram-positive bacterium (Graves and Rabinowitz, 1986). Transcriptional startpoints indicated by asterisk (*).

as PA1 (see Figure 17.5; Garnier and Cole, 1988b). Productive interaction with P4 could be demonstrated when supercoiled templates were used to program RNA synthesis with either the *C. perfringens* or *B. subtilis* enzymes. Taken together these findings indicate that the UV-inducible promoters of pIP404 are not negatively regulated but are more likely to be positively controlled, possibly by a novel sigma factor or a similar regulatory protein.

17.6 Concluding Remarks

Our studies of pIP404 have provided a wealth of important information about gene structure, expression, and control in *Clostridium perfringens*. Although there are many features in common between the production of BCN5 and the synthesis of colicins by gram-negative bacteria, there are also quite striking differences, notably in transcriptional control, and we would like to conclude this chapter with a speculative scenario for bacteriocin production in *C. perfringens*.

Bacteriocinogenic plasmids are common among *C. perfringens* isolates (Brefort et al., 1977; Mahony, 1979) and confer an advantage on the host that harbors them, as secretion of bacteriocin inhibits growth of related species that might inhabit the same ecological niche. This selective advantage is maximized by linking bacteriocin production to DNA damage, be-

cause this is a sensitive indicator of stressful growth conditions and one could imagine that it is precisely when the environment is least favorable that competition for nutrients and key metabolites is most fierce. Under normal, nonstressed conditions, *C. perfringens* CPN50 grows contentedly and is immune to the action of its own bacteriocin because of the low level transcription of *uviA* from the P4 promoter by vegetative RNA polymerase. Following a stimulus that provokes DNA damage, such as reduced availability of a substrate essential for replication, the clostridial equivalent of the SOS response is induced. This requires the synthesis of a new DNA-binding protein, which is probably a novel sigma factor given the abundance and diversity of sigma factors in gram-positive bacilli. In consequence RNA polymerase is modified and can now transcribe the *bcn* gene from promoters P1, P2, and P3 as well as the *uviAB* operon from P4 and P5, thereby providing the ancillary immunity and secretion functions. In addition, several chromosomal genes are switched on as suggested by the induction of the different protein species in Figure 17.3; these may include repair functions and RecA. Secretion of large quantities of BCN5 leads to elimination of possible sensitive competitors and therefore to increased availability of nutrients for the producing strain CPN50. Replication can now resume, dispensing with high-level transcription of the *bcn* and *uviAB* genes.

Acknowledgments. Work described in this article was supported by research grants 851006 and 880588 from the Institut National de la Santé et de la Recherche Médicale, and by the Fondation Roux.

References

Brefort, G., M. Magot, H. Ionesco, and M. Sebald. 1977. Characterization and transferability of *Clostridium perfringens* plasmids. *Plasmid* 1:52–66.

Chen, G.-F.T. and M. Inouye. 1990. Suppression of the negative effect of minor arginine codons on gene expression; preferential usage of minor codons within the first 25 codons of the *Escherichia coli* genes. *Nucleic Acids Res.* 18:1465–1473.

Garnier, T. and S.T. Cole. 1986. Characterization of a bacteriocinogenic plasmid from *Clostridium perfringens* and molecular genetic analysis of the bacteriocin-encoding gene. *J. Bacteriol.* 168:1189–1196.

Garnier, T. and S.T. Cole. 1988a. Complete nucleotide sequence and genetic organization of the bacteriocinogenic plasmid, pIP404, from *Clostridium perfringens*. *Plasmid* 19:134–150.

Garnier, T. and Cole, S.T. 1988b. Identification and molecular genetic analysis of replication functions of the bacteriocinogenic plasmid pIP404 from *Clostridium perfringens*. *Plasmid* 19:151–160.

Garnier, T. and S.T. Cole. 1988c. Studies of UV-inducible promoters from *Clostridium perfringens* in vivo and in vitro. *Mol. Microbiol.* 2:607–614.

Garnier, T., W. Saurin, W. and S.T. Cole. 1987. Molecular characterization of the resolvase gene, *res*, carried by a multicopy plasmid from *Clostridium perfringens*: common evolutionary origin for prokaryotic site-specific recombinases. *Mol. Microbiol.* 1:371–376.

Graves, M.C. and J.C. Rabinowitz. 1986. In vivo and in vitro transcription of the *Clostridium pasteurianum* ferredoxin gene. *J. Biol. Chem.* 261:11409–11415.

Ionesco, H. and D.H. Bouanchaud. 1973. Production de bactériocine liée à la présence d'un plasmide chez *Clostridium perfringens* type A. *C.R. Acad. Sci.* (Paris) 276:2855–2857.

Ionesco, H., A. Wolff. and M. Sebald. 1974. Production induite de bactériocine et d'un bactériophage par la souche BP6K-N5 de *Clostridium perfringens*. *Ann. Microbiol. Inst. Pasteur* (paris) 125B:335–346.

Mahony, D.E. 1979. Bacteriocin, bacteriophage and other epidemiological typing methods for the genus *Clostridium*. *Methods Microbiol.* 13:1–30.

Makoff, A.J., M.D. Oxer, M.A. Romanos, N.F. Fairweather, and S. Ballantine. 1989. Expression of tetanus toxin fragment C in *E. coli*: high level expression by removing rare codons. *Nucleic Acids Res.* 17:10191–10202.

Ogasawara, N. 1985. Markedly unbiased codon usage in *Bacillus subtilis*. *Gene* (Amst.) 40:145–150.

Pugsley, A.P. 1984a. The ins and outs of colicins. Part I: Production and translocation across membranes. *Microbiol. Sci.* 1:168–175.

Pugsley, A.P. 1984b. The ins and outs of colicins. Part II. Lethal action, immunity and ecological implications. *Microbiol. Sci.* 1:203–205.

Saint-Joanis, B., T. Garnier, and S.T. Cole. 1989. Gene cloning shows the alpha toxin of *Clostridium perfringens* to contain both sphingomyelinase and lecithinase activities. *Mol. Gen. Genet.* 219:453–460.

Scott, P.T. and J.I. Rood. 1989. Electroporation-mediated transformation of lysostaphin-treated *Clostridium perfringens*. *Gene* (Amst.) 82:327–333.

Smith, L.D.S. and G. Hobbs. 1974. Clostridium *In: Bergey's Manual of Determinative Bacteriology*, R.E. Buchanan and N.E. Gibbons, ed., p. 551–572 Baltimore: Williams & Wilkins.

Walker, G.C. 1987. The SOS response of *Escherichia coli*. *In: Escherichia coli and Salmonella typhimurium*, J.L. Ingraham, K.B. Low, B. Magasanik, M. Schaechter, and H.E. Umbarger, eds., pp. 1346–1357. Washington, D.C.: American Society for Microbiology.

Wolff, A. and H. Ionesco. 1975. Purification et caractérisation de la bactériocine N5 de *Clostridium perfringens* BP6K-N5 type A. *Ann. Microbiol.* (Paris) 126B:343–356.

18

Genome Mapping of *Clostridium perfringens* Type A Strains

Stewart T. Cole and Bruno Canard

18.1 Introduction

Among the clostridia of medical and veterinary interest, *Clostridium perfringens* has long been studied at the microbiological and biochemical levels. This important human pathogen produces a wide assortment of exotoxins that are believed to play a major role in pathogenesis and which provide the actual basis for the classification of the isolates in five serotypes, A to E (Finegold, 1977). Isolates of serotype A are predominantly associated with human disease and with gas gangrene of man and animals.

Relatively little is known at the genetic level, however, because classical approaches have been hampered by the lack of genetic tools, the most crucial being a DNA transfer system. The introduction of pulsed field gel electrophoresis (PFGE) in 1982 (Schwartz et al.) and its recent modifications (Schwartz and Cantor, 1984; Carle and Olson, 1985; Carle et al., 1986; Chu et al., 1986) opened new fields of genetic research. Bacterial chromosome organization became amenable to physical study with standard molecular biology techniques, circumventing the tedious construction of genetic maps. Moreover, linkage data could be converted into physical distances, providing a common basis for genome analysis among organisms.

Genetic mapping can be decisive in studying bacterial virulence, because pathogenesis determinants are often found associated with mobile genetic elements or clusters of related functions (Miller et al., 1989). As a first step toward a better understanding of virulence, we constructed a physical and genetic map, or genome map (Canard and Cole, 1989), of the reference strain *Clostridium perfringens* CPN50 (ATCC 27059). This chapter extends the study to two more strains, the clinical isolate A99 and the reference strain NCTC6785, both gangrene isolates from Germany and England, respectively, and establishes the extent of genome polymorphism between three type A strains.

18.2 Construction of Physical Maps

Preliminary information

The mapping strategy is imposed by the microorganism itself, i.e., by its genome size, dG + dC content and the degree of precision desired. Because the genome of *C. perfringens* is very rich in dA + dT ($\approx 75\%$; Smith and Hobbs, 1974), we tested restriction endonucleases with dG + dC-rich recognition sites. Six of these were used extensively (*Apa*I, *Fsp*I, *Mlu*I, *Nru*I, *Sac*II, *Sma*I) because they generated a limited number of fragments (6–20). In addition, assuming a random distribution of sites and a genome size typical for bacteria (4–5 Mb), this number of enzymes should generate physical intervals of 50 to 200 kb, suitable for gene mapping. Of particular interest are the restriction endonucleases giving the smallest number of cuts, as these can be used to generate end probes for indirect labeling or linking analysis (see following).

It is wise to use at least two different PFGE apparatuses because they all have their particular pros and cons, and this can help to avoid significantly underestimating the actual genome size (Grothues and Tümmler, 1987; Römling et al., 1989). Field inversion gel electrophoresis (FIGE) (Carle et al., 1986) is of particular interest for direct comparison of bacterial isolates, but great care must be taken to estimate correctly the sizes of fragments larger than 800 kb and to detect potentially very large, unresolved, scattered fragments.

Direct strategy

Many strains can be quickly and conveniently compared by PFGE analysis of their chromosomes, which often yields interesting but complex restriction patterns readily revealed by ethidium bromide staining (Figure 18.1). To generate partial genomic maps, single- and double-restriction digests can be performed, and these will give not only a reliable indication of the genome size but also information about site distribution. However, because of the large number of fragments in the 100 to 200 kb range it is sometimes difficult to attribute fragments unambiguously to a given chromosomal location. Greater precision can be obtained by Southern blotting and the use of hybridization probes, either known genes or anonymous fragments of cloned DNA, to establish physical relationships between large restriction fragments. Such an approach produces islands of overlapping fragments, often somewhat inaccurate with regard to the position of several restriction sites. It is sometimes possible to link two markers to the same fragment, but this strategy is limited by the availability of probes and its random approach. The data for a given probe may be stored conveniently as a mapping table (Table 18.1), allowing quick detection of possible overlaps.

Indirect end-labeling

This is an adaptation of the Smith and Birnstiel procedure (1976) and relies on cloning DNA fragments bearing a rare "reference" restriction

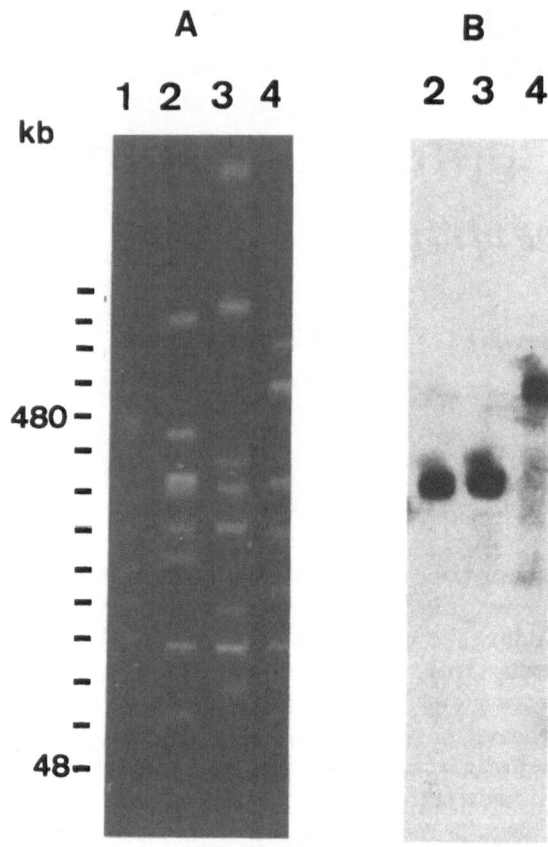

FIGURE 18.1   Comparative macrorestriction analysis of three type A *Clostridium perfringens* strains. Genomic DNA in agarose plugs was digested with *Apa*I and resolved by FIGE. Samples: CPN50, lane 2; NCTC6785, lane 3; A99, lane 4. (**A**), Ethidium bromide-stained gel; (**B**), same gel after transfer to nylon membrane and hybridization with probe for *nanH* gene. Positions of DNA size standards (multimers of lambda) are indicated (lane 1).

site at one end (Graham et al., 1987). In our case, we cloned *Mlu*I-*Eco*RI, *Mlu*I-*Hind*III and *Sma*I-*Hind*III DNA fragments for use in indirect end-labeling experiments. DNA was digested to completion with *Mlu*I (the "reference" enzyme) then partially digested with a second enzyme, such as *Nru*I, as explained in Figure 18.2. The samples were subjected to PFG electrophoresis, blotted to a membrane, and hybridized to the cloned end probe. The map for the second enzyme was then deduced from two overlapping directions from the *Mlu*I sites.

At this stage, previous "floating islands"

Table 18.1 Mapping table of *nagH*

Origin: *Clostridium perfringens* CPN50
Nature: 4.5-kb *Eco*RI fragment
Biological function: *N*-Acetylglucosaminidase

CPN50

		NluI	SmaI	ApaI	FspI	NruI
MluI		830	160	650	400	160
	SmaI		160	50	160	160
		ApaI		650	180	50
			AviII		450	160
				NruI		160

NCTC6785

		MluI	SmaI	ApaI	FspI[a]	NruI
MluI		850	170	660	—	170
	SmaI		165	55	—	165
		ApaI		660	—	50
			AviII		—	160
				NruI		165

A99

		MluI	SmaI	ApaI	FspI*	NruI
MulI		620	70	620	—	70
	SmaI		160	50	—	160
		ApaI		620	—	50
			AviII		—	160
				NruI		165

[a]Not determined.

from a given marker can be located, and internal sites for other enzymes predicted. Although it is often tedious to obtain the end clones, this procedure is extremely powerful and precise and produces internally consistent maps in a fast and nonrandom fashion.

Linking analysis and map closure

When all the reference fragments have been mapped, the end clones are used to probe a genomic library to find a clone bearing a rare reference restriction site. This clone can then be used to isolate a missing end clone or to link two fragments and join two maps. Figure 18.3 shows a typical linking analysis. Note that linking clones can be used for mapping other strains than the one from which they come, which should facilitate genome characterization of any interesting strain. If virtually all the end clones have been isolated, it is not necessary to isolate the linking clones, because direct hybridization with a genomic DNA digest will indicate their contiguity. This analysis closes maps in a linear nonrandom fashion and reveals whether the chromosome is linear or circular.

Superimposing a gene map

Having established a physical map of the genome, it is relatively simple to superimpose a gene map by hybridization. Probes can be prepared from cloned clostridial genes of known function, or a heterologous approach may be employed. It is noteworthy that any organism, even if phylogenetically very distant, is likely to contain key genes of conserved sequence useful for mapping. Obviously chances are best if the organism has a close dG + dC content, and because of the exquisite sensitivity of radioactive probes it is worth trying a simple, low-stringency hybridization.

Of particular interest are ribosomal RNA operons. These genes have an almost invariable dG + dC content regardless of the species of microorganism, are present in several copies, and also may contain rare restriction sites. In our case, each ribosomal RNA operon contains the cluster SacII-SacII-SmaI-SmaI-NruI-SmaI-NruI. By using probes specific for the 5' and 3' ends of the operons, both the genomic distribution and polarity can be established.

Good candidates for heterologous probing, as found by us and others (Filer et al., 1981), are the genes for elongation factor Tu, heat shock proteins, and subunits of RNA polymerase, DNA gyrase, and ATP synthase as well as the genes for stable RNAs.

18.3   The Genome Map of *Clostridium perfringens* CPN50

Using the combination of techniques just described, we initially established a genomic map for strain CPN50 and showed it to contain a single circular chromosome of about 3.6 Mb (Canard and Cole, 1989). More than 100 restriction sites were localized and the positions of 24 known genes and loci determined (Figure

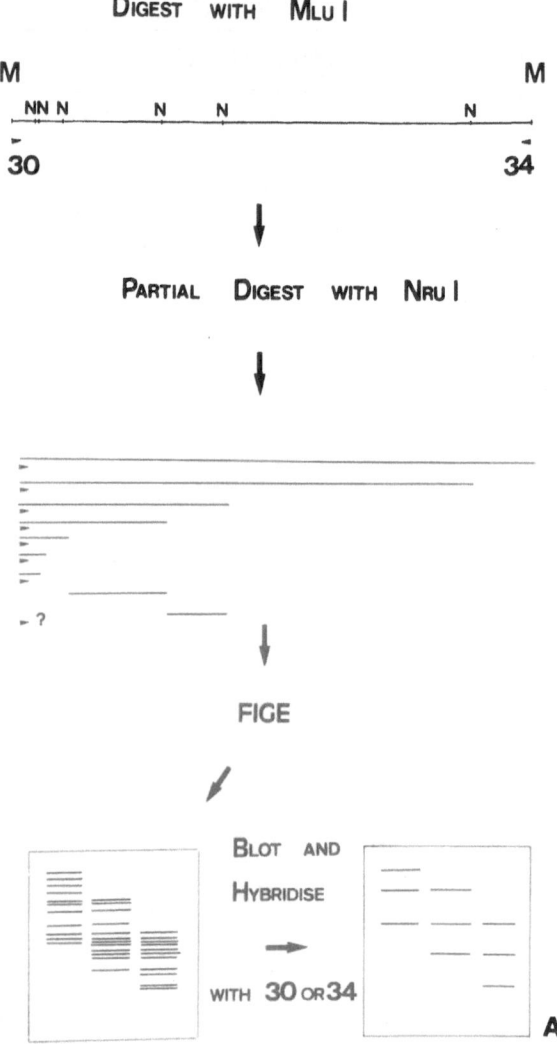

FIGURE 18.2 (A) Principle of indirect end-labeling. Genomic DNA is digested to completion with *Mlu*I then partially digested with *Nru*I and the fragments resolved by FIGE. After transfer of DNA to nylon membranes, hybridization with end probe (*Mlu*I-*Eco*RI) is performed. (B). Example of indirect end-labeling analysis of CPN50 DNA. Outermost lanes correspond to complete digests with *Mlu*I or *Nru*I; sizes of *Mlu*I fragments are indicated. Central lanes correspond to complete *Mlu*I digests, followed by digestion with increasing concentrations of *Nru*I. Hybridization was performed with probes 30 or 34, which correspond to left and right ends, respectively, of 830-kb *Mlu*I fragment of CPN50. With probe 34, two largest *Mlu*I-*Nru*I partial digestion products could only be seen after prolonged exposure of autoradiogram. Map deduced from this experiment is shown at top in **A**.

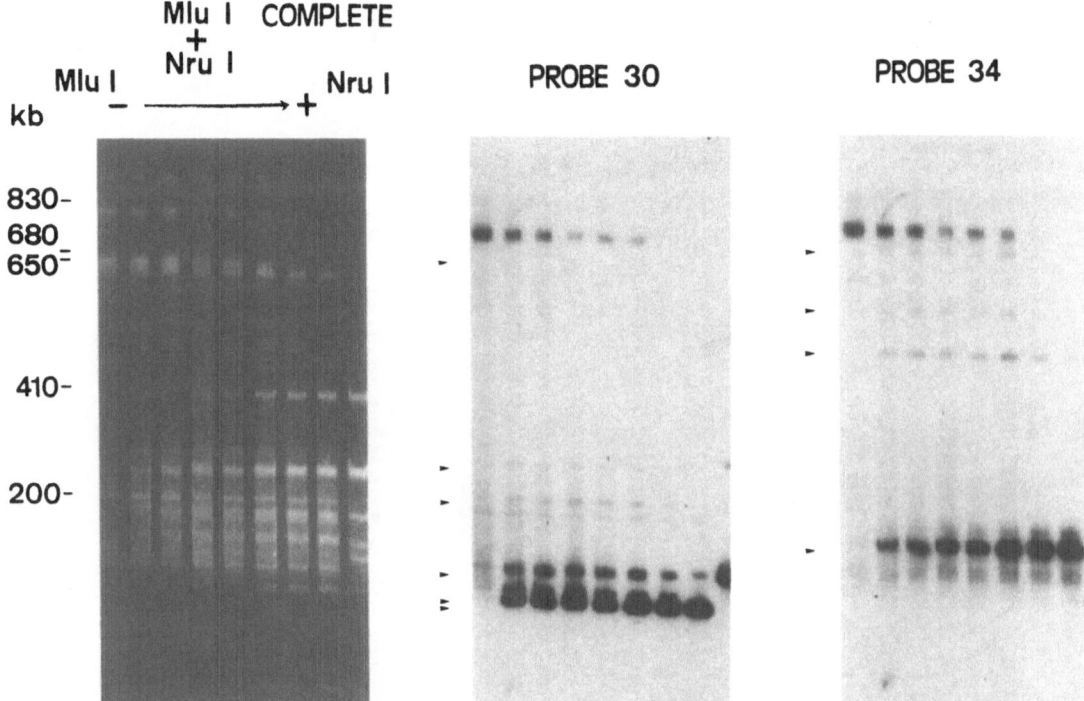

FIGURE 18.2 (B)

18.4). The distribution of the ribosomal RNA operons with respect to the putative origin of replication, defined by the *gyrA* and *gyrB* genes, is particularly informative. Seven copies are located to the right and transcribed clockwise while two more are situated to the left of *gyrAB* and transcribed counterclockwise. As is the case in *Bacillus subtilis* and *Staphylococcus aureus*, this arrangement is consistent with bidirectional replication (Henches et al., 1982; Piggot and Hoch, 1985; Hopewell et al., 1990).

The known genes located on the CPN50 genome map fall into two groups, coding for housekeeping functions or potential virulence factors. It is intriguing that three genes implicated in pathogenesis (*plc*, *pfo*, and *nagH* coding for the cytolytic toxins α and Θ and the *N*-acetyl glucosaminidase, respectively) are located in a 200-kb region near the putative origin of replication. A fourth "virulence" gene, *nanH* coding for a sialidase, is situated elsewhere but on the same *Fsp*I fragment as the lysogenic phage Φ29 (Canard and Cole, 1989, 1990).

Comparative genome mapping of *C. perfringens* strains

Having established the genomic organization for one type A strain, it was relatively simple to extend this approach to additional strains and to perform a comparative study. Consequently, we chose NCTC6785, which like CPN50 is a reference strain but not lysogenic for Φ29, and A99, which was recently isolated from a patient with gangrene (Roggentin et al., 1988).

Comparison of restriction profiles showed a reasonable degree of conservation in the size and number of fragments obtained with *Apa*I, *Mlu*I, and *Sma*I but more diversity with the other enzymes tested. A direct comparison of the *Apa*I fragments from the three strains, together with the localization of the *nagH* gene, can be seen in Figure 17.1. Accurate measurements obtained for the sizes of the fragments generated by these three enzymes are summarized in Table 18.2.

Fortunately, most of the *Mlu*I sites have

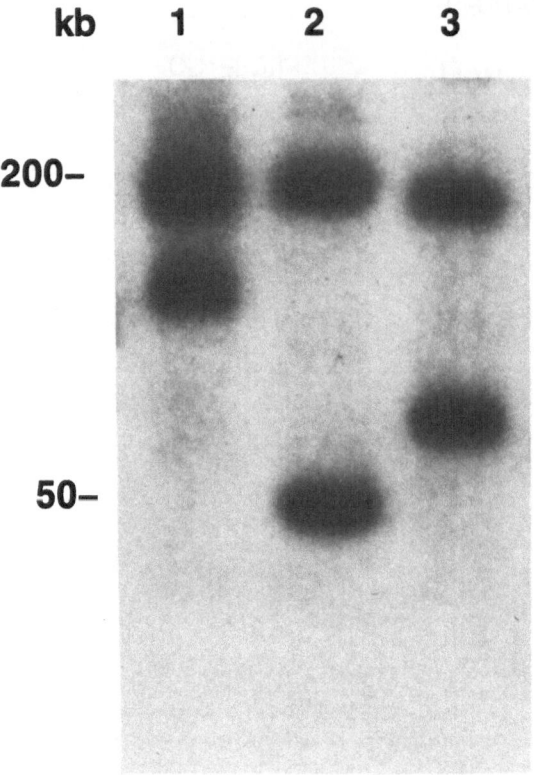

FIGURE 18.3 Linking clone analysis. *Clostridium perfringens* DNA was cleaved with *Mlu*I and *Apa*I, resolved by FIGE, then transferred to nylon membrane and hybridized with probe made from linking clone carrying *Mlu*I site (probe 8 in Figure 18.4). Samples: strain CPN50, lane 1; strain NCTC6785, lane 2; strain A99, lane 3.

Table 18.2 Restriction fragments of *Clostridium perfringens* genomic DNA

	CPN50			NCTC6785			A99		
	*Mlu*I	*Sma*I	*Apa*I	*Nlu*I	*Sma*I	*Apa*I	*Mlu*I	*Sma*I	*Apa*I
	1200	1460	650	880	1150	900	1150	1390	620
	830	440	530	850	440	660	650	440	550
	680	410	440	660	420	440	640	410	550
	650	405	420	640	410	410	630	360	450
	200	170	410	260	260	360	200	260	410
		170	360	200	260	260	190	170	360
		170	320		170	200		160	280
		120	200		160	140		110	200
		120	140		120	130		100	100
		110	100		110	50		75	40
		75	40		90			40	35
		45	35		40				
Total	3560	3695	3645	3490	3630	3550	3460	3515	3595

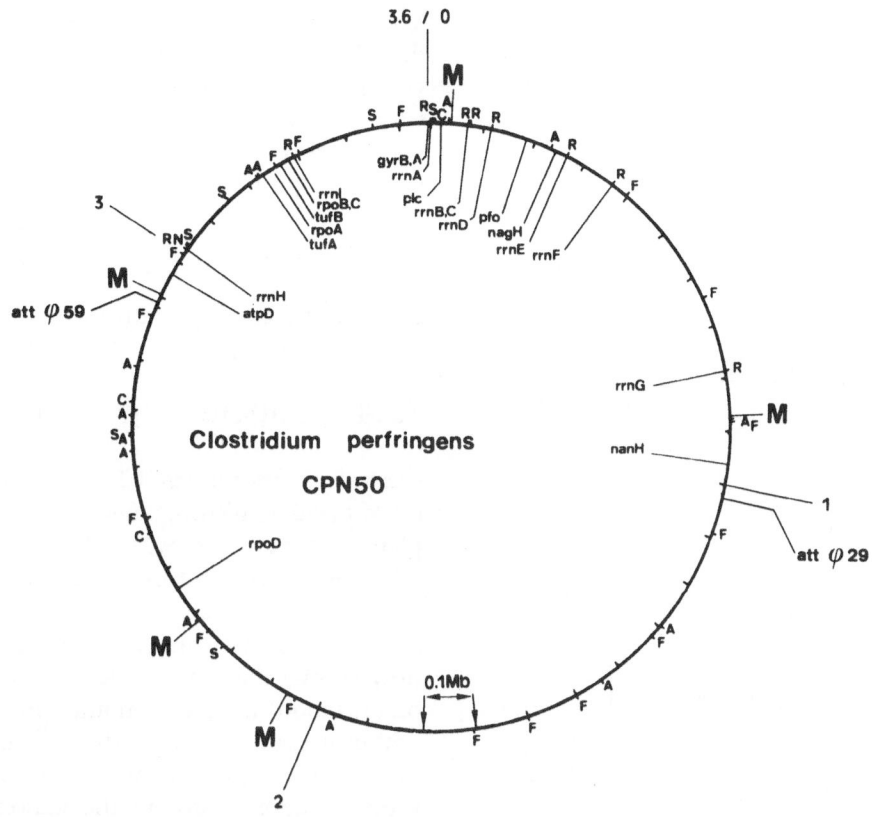

Figure 18.4 Physical and gene map of chromosome of *Clostridium perfringens* strain CPN50. Restriction sites for *Apa*I (A), *Fsp*I (F), *Mlu*I (M), *Sac*II (C), and *Sma*I (S) are positioned on outside of circle; R denotes rare site cluster, *Sac*II (2), *Sma*I (2), *Nru*I, *Sma*I, and *Nru*I, which occurs in each RNA operon. Scale in 0.1-Mb intervals is shown on inside; known genes or genetic loci are identified by cross-hybridization with heterologous probes. Known clostridial genes: *gyrA*, DNA gyrase; *nagH*, β-N-acetylglucosaminidase; *nanH*, sialidase; *pfo*, perfringolysin or Θ-toxin; *plc*, phospholipase C or alpha-toxin; *rrnA-I*, rRNA operons; *att*Φ29, integration site for Φ29; *att*Φ59, integration site for Φ59. Other genetic loci: *atpD*, β subunit of ATP synthase; *gyrB*, DNA gyrase; *rpoA*, α subunit of RNA polymerase; *rpoB*, β subunit of RNA polymerase; *rpoC*, β subunit of RNA polymerase; *rpoD*, σ⁴³ subunit of RNA polymerase, *tufA*, B, elongation factor Tu.

been conserved (although the genomes of NCTC6785 and A99 have one additional site), and this enabled maps to be generated using an indirect end-labeling and linking analysis with the probes obtained from CPN50. A typical linking experiment is illustrated in Figure 18.3. This combined approach culminated in detailed maps of the chromosomes of NCTC6785 and A99, which are shown in Figure 18.5, where they are compared with CPN50.

It is interesting to note the similarity in genome size of geographically distant isolates (Table 18.2 and unpublished observations), and this may represent a taxonomic criterium in the future. Differences are barely detectable under most single-digest conditions but a simultaneous increase, or decrease, in the size of a given restriction fragments indicates a probable insertion or deletion. In this respect, major events detected did not exceed the size of standard mobile genetic elements such as phages or transposons (see Figure 18.4, probe 2 and between probe 4 and 5). Insertion sites for bacteriophages Φ59 and Φ29 have already been mapped in CPN50 (Canard and Cole, 1990) and "RFLP-like" events can be detected around them. However, most of the variations observed lie close to ribosomal RNA-encoding

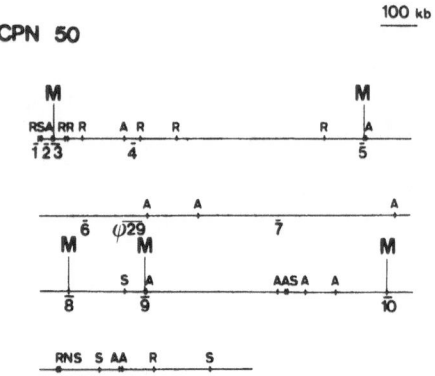

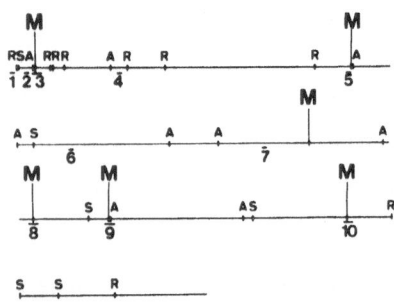

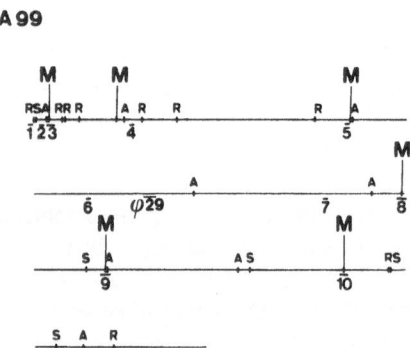

loci. We do not understand the significance of this, because the ribosomal operons do not vary in number among all *C. perfringens* strains studied (unpublished observations) and were very conserved in location and orientation relative to the putative origin of replication. Likewise, it is not clear why those rare restriction sites not located in ribosomal RNA operons, such as *Mlu*I, *Apa*I, and some *Sma*I sites, have been so well conserved during evolution.

18.4   Concluding Remarks

Clostridium perfringens CPN50 was the first gram-positive microorganism for which a complete genome map was established (Canard and Cole, 1989), and this study represents the first detailed genome map comparison of different isolates from the same species. We have now shown that linking clones and the global organization of the genome are conserved within a species (this study and unpublished results) and this, in conjunction with a set of probes scattered around the genome, should facilitate the rapid genome mapping of any *C. perfringens* strain. Although it is too early to suggest the possible future of genome comparisons, we speculate that they will provide new taxonomic tools and direct access to important genomic regions, such as variable loci associated with disease or pathogenesis. Further, such comparisons should define in molecular terms how a species is evolving and may play a role as a molecular clock. Unfortunately, so long as an insufficient number of species are mapped no information on the speed of the clock will be available.

Acknowledgments.   We would like to thank M. Popoff, R. Schauer, and M. Sebald for gifts of strains. Work described in this article was supported by research grant 880588 from the Institut National de la Santé et de la Recherche Médicale and by the Fondation Roux.

FIGURE 18.5 Comparative genomic analysis of three type A strains of *Clostridium perfringens*. Maps were deduced from data obtained by double digestions, indirect end-labeling, and linking analysis. Positions of the probes used (1–10) are indicated, and restriction sites denoted, as follows: M, *Mlu*I; A, *Apa*I; S, *Sma*I. Symbols: ribosomal RNA operons, R; approximate location of attachment site of Φ29, φ29. For comparative purposes, maps are presented in linear form.

References

Canard, B. and S.T. Cole. 1989. Genome organization of the anaerobic pathogen *Clostridium perfringens. Proc. Nat. Acad. Sci. USA* **86**:6676–6680.

Canard, B. and S.T. Cole. 1990. Lysogenic phages of *Clostridium perfringens*: mapping of the chromosomal attachment sites. *FEMS Microbiol. Lett.* **66**:323–326.

Carle, G.F. and M.V. Olson. 1985. An electrophoretic karyotype for yeast. *Proc. Nat. Acad. Sci. USA* **82**:3756–3760.

Carle, G.F., M. Frank, and M.V. Olson. 1986. Electrophoretic separations of large DNA molecules by periodic inversion of the electric field. *Science* **232**:65–68.

Chu, G., D. Vollrath, and R.W. Davis. 1986. Separation of large DNA molecules by contour-clamped homogeneous electric fields. *Science* **234**:1582–1585.

Filer, D., R. Dhar, and A.V. Furano. 1981. The conservation of DNA sequences over very big periods of evolutionary time. *Eur. J. Biochem.* **120**:69–77.

Finegold, S. 1977. *Anaerobic Bacteria in Human Disease*. New York: Academic Press.

Graham, M.Y., T. Otani, I. Boime, M.V. Olson, G.F. Carle, and D.D. Chaplin. 1987. Cosmid mapping of the human chorionic gonadotropin β subunit genes by field-inversion gel electrophoresis. *Nucleic Acids Res.* **15**:4437–4448.

Grothues, D. and B. Tümmler. 1987. Genome analysis of *Pseudomonas aeruginosa* by field inversion gel electrophoresis. *FEMS Microbiol. Lett.* **48**:419–422.

Henckes, G., F. Vannier, M. Seiki, N. Ogasawara, H. Yoshikawa, and S.J. Seror-Laurent. 1982. Ribosomal RNA genes in the replication origin region of *Bacillus subtilis* chromosome. *Nature* (London) **299**:268–271.

Hopewell, R., M. Oram, R. Briesewitz, and M. Fisher. 1990. DNA cloning and organization of the *Staphylococcus aureus gyrA* and *gyrB* genes: close homology among gyrase proteins and implications for 4-quinolone action and resistance. *J. Bacteriol.* **172**:3481–3484.

Miller, J.F., J.J. Mekalanos, and S. Falkow. 1989. Coordinate regulation and sensory transduction in the control of bacterial virulence. *Science* **243**:916–922.

Piggot, P.J. and J.A. Hoch. 1985. Revised genetic linkage map of *Bacillus subtilis*. *Microbiol. Rev.* **49**:158–179.

Roggentin, P., B. Rothe, F. Lottspeich, and R. Schauer. 1988. *FEBS Lett.* **238**:31–34.

Römling, V., D. Grothues, W. Bautsch, and B. Tümmler. 1989. A physical genome map of *Pseudomonas aeruginosa* PAO. *EMBO J.* **8**:4081–4089.

Schwartz, D.C., W. Saffran, J. Welsh, R. Hass, M. Goldenberg, and C.R. Cantor. 1982. New techniques for purifying large DNAs and studying their properties and packaging. *Cold Spring Harbor Symp. Quant. Biol.* **47**:189–195.

Schwartz, D.C. and C.R. Cantor. 1984. Separation of yeast chromosome-sized DNAs by pulsed-field gradient gel electrophoresis. *Cell* **37**:67–75.

Smith, H.O. and M.L. Birnstiel. 1976. A simple method for DNA restriction site mapping. *Nucleic Acids Res.* **3**:2387–2398.

Smith, L.D.S. and G. Hobbs. 1974. Clostridium. In: *Bergey's Manual of Determinative Bacteriology*, R.E. Buchanan and N.E. Gibbons, ed., pp. 551–572. Baltimore: Williams Wilkins.

19

Molecular Biology of the *Clostridium difficile* Toxins

Christoph von Eichel-Streiber

19.1 Introduction

In the nineteenth century a clinical entity accompanied by hemorrhagic diarrhea, designated pseudomembranous colitis (PMC), was first described in a young woman after abdominal surgery (Finney, 1893). As a result of the widespread therapeutic use of antibiotics, similar clinical cases were subsequently reported with increasing incidence (Bartlett and Gorbach, 1977). Because of its induction by antibiotics the disease in its mild form was designated antibiotic-associated diarrhea (AAD). The fatal form of the disease is still known as PMC, after the first reported case.

The search for virus particles responsible for the disease led to the discovery of a toxic factor in fecal suspensions. The activity was detected because of its cytopathic effect on different cell lines used to isolate the expected virus. The time course of development of the cytopathic effect and the fact that it could not be serially propagated led to the conclusion that the stool extracts contained a bacterial toxin (Bartlett et al., 1977; Larson et al., 1977). Soon thereafter this cytotoxic activity was associated with toxin-producing clostridia because it was neutralized by antibodies directed against *Clostridium sordellii* toxins (Larson and Price, 1977; Rifkin et al., 1977). These antibodies were part of a gas gangrene antiserum. The search for *C. sordellii* in affected patients finally led to the isolation of *Clostridium difficile* as the causative agent of PMC (Bartlett et al., 1978; George et al., 1978).

C. difficile received its name from the fact that it was initially difficult to isolate from stools. The first isolation of *C. difficile* in healthy newborn infants was reported in 1935 by Hall and O'Toole; at that time *C. difficile* was not recognized as a human pathogen. Several culture media and selective agars were introduced in the late 1970s, so that today the isolation of the bacterium is no longer a major problem (George et al., 1979).

Age plays an important role in determining the susceptibility to colonization by *C. difficile*. Infants can harbor toxigenic *C. difficile* strains without showing signs of the disease (Larson et al., 1982). The higher colonization rate of infants (50%–75% in infants, compared with 0%–11% in adults; Toma et al., 1988) is probably caused by age-associated changes in the mucous membrane and its microflora. Most *C. difficile* carriers are nevertheless healthy (George, 1986). Serotyping of *C. difficile* isolates has led to the differentiation of toxic and nontoxic serotypes, both isolated from healthy infants (Delmée et al., 1988; Toma et al., 1988). In adults, the same serotyping allows differentiation of virulent from avirulent strains.

The disease results from an antibiotic-induced modification of the intestinal ecology. The indigenous flora normally suppresses *C. difficile* growth in the colon. This has been shown in animal models, notably in rodents, where the normal colon microflora protected against *C. difficile* colonization (Wilson et al., 1985, 1986). Although first noticed after the treatment of patients with clindamycin or lincomycin, it is now apparent that virtually every antibiotic can induce AAD or PMC (Tedesco, 1979; George, 1988). The disease is either initiated by resistant *C. difficile* strains overgrowing the physiological flora in the course of treatment (e.g., clindamycin), or by sensitive *C. difficile* strains shifting from the dormant spore to the vegetative form after the end of therapy (e.g., ampicillin). Consequently, with clindamycin the onset of the disease is abrupt; with ampicillin treatment, it is delayed.

Extensive investigations led to the identification and isolation of two major toxins designated enterotoxin A or D1 and cytotoxin B or D2 (Banno et al., 1981, 1984; Taylor et al., 1981; Sullivan et al., 1982). Their role in the pathophysiology of AAD or PMC has been proven (Lyerly et al., 1985a). ToxinA (ToxA) acts locally and systemically; it induces specific symptoms in the animal model system (Chang et al., 1978a); toxinB (ToxB) only acts systemically. Intragastric application has no effect unless the ToxB preparation is contaminated with trace amounts of ToxA (Lyerly et al., 1985a) or the epithelium is mechanically altered, as occurs in the course of surgery (Bartlett and Gorbach, 1977; Keighley et al., 1978).

This chapter discusses the molecular genetics and some aspects of the molecular biology of ToxA and ToxB of *C. difficile*. Toxigenic clostridia have recently been reviewed by Hatheway (1990); the more clinical aspects of *C. difficile* and its toxins have been reviewed by Lyerly et al. (1988b).

19.2   Protein Chemistry

Characterization of *C. difficile* toxins A and B by protein chemical methods led to incomplete or even conflicting results. These studies have been impeded by the large size of these molecules. The following section briefly reviews the reported methods for toxin preparation.

Growth of bacteria

Different media have been used for selective growth of the bacterium (Iwen et al., 1989). A chemically defined medium has been described that allows growth of *C. difficile* at a rate 40% slower than that observed in complex media (Seddon and Borriello, 1989). A comparison of methods for growth and purification of *C. difficile* toxins showed that the resultant toxin preparations vary depending on the growth conditions (Torres and Lönnroth, 1988).

To isolate *C. difficile* toxins, bacteria are usually grown in freshly autoclaved brain heart infusion broth at 37°C. After inoculation with an overnight *C. difficile* culture (1:100 of the final volume) the bacteria grow even if they are not incubated under strictly anaerobic conditions. Because of difficulties associated with toxin purification, Taylor et al. (1981) introduced a dialysis flask culture system for the growth of bacteria. Growing bacteria are inoculated into physiological NaCl solution and separated from the protein-rich medium by a dialysis membrane, thus minimizing contamination with proteins of the growth medium. Similar systems had already been used for the isolation of botulinum toxin from *C. botulinum* (Sterne and Wentzel, 1950). Maximum amounts of toxin A and B were obtained after 4–5 days of culture (Ketley et al., 1984, 1986), indicating that the toxins are produced in the stationary phase of the bacterial culture. Ketley et al. (1986) showed that *C. difficile* toxin production is unrelated to sporulation.

Most of the *C. difficile* strains isolated from patients are either high or low toxin producers. Toxin-positive strains always produce both ToxA and ToxB (Wren et al., 1987a). Wren et al. monitored ToxA by an enzyme-linked immuno absorbent assay (ELISA) with specific polyclonal antisera and ToxB by its cytotoxic effect in vitro on cultured cells. It must be considered, however, that ToxA also exhibits cytotoxic activity (Tucker et al., 1990) and that low cytotoxicity titers may have resulted from ToxA alone. Differential toxin production has been reported that depended on the amino acid composition of the growth medium and on the particular strain used for toxin production; starvation of certain amino acids resulted in the production of only one of the two toxins (Haslam et al., 1986).

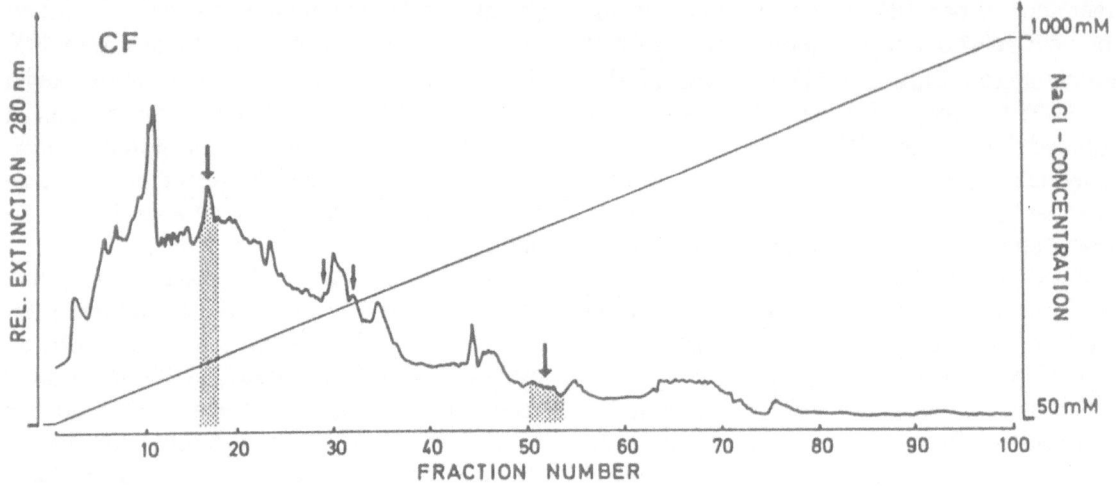

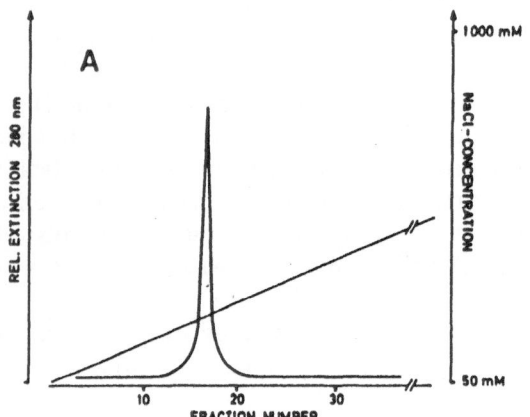

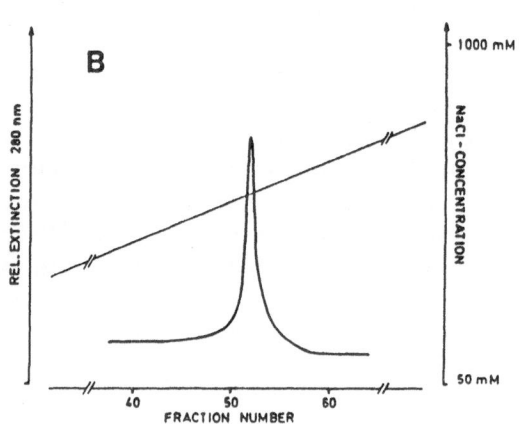

FIGURE 19.1 Mono Q anion-exchange chroma-tography to purify *Clostridium difficile* toxins. Panel **CF**: Separation of culture filtrate. A sample of 20 ml culture filtrate (20 mg/ml of protein; *C. difficile* supernatant, concentrated and depleted of low molecular weight contaminants) was loaded onto a Mono Q 10/10 column. Proteins were eluted with linear NaCl gradient (50–1000 mM) with 50 mM Tris HCl pH 7.5 running buffer. Relative extinction was recorded at 280 nm with amplification factor of 0.5. Arrows indicate four areas active in cyto-toxicity testing. Hatched areas at 149–169 mM and 475–505 mM NaCl represent ToxA and ToxB, re-spectively. Only these two fractions had cytotoxic-ity titer greater than 1:100. Panels **A** and **B**: Rechromatography of fractions with toxic activity. Active fractions of first separation (ToxA between 149 and 169 mM; ToxB between 475 and 505 mM) were separately loaded onto the Mono Q 10/10 col-umn again. As indication of their purity both pro-teins eluted as symmetrical peaks. Panel **A**: ToxA elution with 50 mM Tris HCl running buffer (pH 7.5) and linear NaCl gradient, relative extinction recorded with amplification factor of 1.0. Panel **B**: ToxB eluted with 20 mM Piperazine/HCl buffer (pH 5.0) and linear NaCl gradient; relative extinc-tion recorded with amplification factor of 2.0.

Toxin purification

To purify *C. difficile* toxins routinely, ammo-nium sulfate-concentrated, cell-free culture su-pernatants are subjected to molecular sieve chromatography, e.g., on Sepharose G100. Alternatively, concentration and ultrafiltration can be achieved by a one-step procedure by the use of an artificial kidney (D. Knautz, personal communication). In both cases the ultrafiltrate contains ToxA and ToxB, which can subsequently be separated by anion-exchange chromatography (Figure 19.1). Characteristically, ToxA elutes before ToxB in a salt gradient. Screening for both toxins is routinely done by testing the fractions of the eluate on adherent cells, such as HeLa, Y1, and CHO cells (Bowman and Riley, 1986; Maniar et al., 1987). Final purification is achieved by two chromatographic separations using a high resolution Mono Q anion exchange column in a fast protein liquid chromatography system (FPLC), which yield ToxA and ToxB in highly purified form (Eichel-Streiber et al., 1987). For ToxA, two Mono Q separations are effected with 50 mM Tris HCl pH 7.5; for ToxB the first separation is performed with 50 mM Tris HCl pH 7.5 and the second with 20 mM piperazine-HCl, pH 5.0. For highly purified ToxA and ToxB, partial N-terminal amino acid sequences have been determined (Meador and Tweten, 1988; Dove et al., 1990).

Three special procedures for ToxA purification have been described. (1) Acidic acid precipitation of ToxA at its pI of about 5.5 was first used to purify this protein (Sullivan et al., 1982). (2) Two, monoclonal antibodies (mAbs), PCG-4 and 1337C8, have been used for the purification of ToxA by affinity chromatography. Both mAbs react specifically with ToxA at its C-terminal repeat (see Section 19.6). ToxA retained by the affinity column was eluted by an alkaline or chaotropic buffer (0.02 M NaOH/ 0.15 M NaCl, Lyerly et al., 1986a; 3 M NaSCN, Eichel-Streiber et al., unpublished observations). The ToxA so prepared is highly purified but loses biological activity. The hemorrhagic toxin (HT) of *C. sordellii* is also retained by this column but may be eluted under milder conditions. The activity of HT is preserved; thus, for its isolation affinity purification is the method of choice (Martinez and Wilkins, 1988).

(3) ToxA can also be purified by interaction of the molecule with thyroglobulin (Krivan and Wilkins, 1987). The purification is based on the fact that ToxA has affinity for the carbohydrate structure Galα1-3Galβ1-4GlcNAc (see Section 19.4). The interaction of ToxA with this carbohydrate was first discovered with rabbit erythrocytes, but the trisaccharide is also pres-

ent in bovine thyroglobulin. At 4°C, ToxA binds to thyroglobulin that is covalently coupled to Affi-Gel 15. ToxA is released from the affinity column by a temperature shift to 37°C (Krivan and Wilkins, 1987). The purification is a one-step procedure, and the yield of ToxA is 56%–80%. The column is reusable, and the thermal elution is gentle; it preserves the biological activity of the ToxA, which is only slightly contaminated with thyroglobulin. Kamiya et al. (1989) reported that such ToxA preparations are contaminated with ToxB. They therefore proposed two additional separations on anion-exchange columns subsequent to the thyroglobulin-affinity column purification. The ToxA/ToxB cross-contamination is not surprising because of the 63% homology between ToxA and ToxB (see Section 19.8, isolation of the *tox* genes).

Isolation of pure ToxB on the basis of ligand structure or immune-affinity chromatography, as described for ToxA, cannot be done at present because neither a receptor molecule nor specific high-titer antibodies are known at present. Purification can be achieved by final high-resolution, anion-exchange chromatography with the use of two modified buffers that have been elaborated. The first buffer is the piperazine-HCl buffer, pH 5.0, already mentioned (Eichel-Streiber et al., 1987), and the second is a 10 mM Tris HCl pH 8.0 buffer supplemented with 50 mM CaCl$_2$ (Meador and Tweten, 1988). ToxB can be purified by either protocol without any contaminating 150-kDa polypeptide. The Meador and Tweten protocol for separation of ToxB and the 150-kDa contaminant protein depends on three conditions: the frozen storage of crude toxin preparations with the addition of glycerol; the presence of 50 mM CaCl$_2$ in the elution buffer; and the preparation of the toxin on high-resolution anion-exchange columns.

Contaminating proteins

The major contaminant in ToxA preparations is a 39- to 43-kDa polypeptide that is a better immunogen than ToxA. Immunization with a mixture of these two antigens was the basis for the development of a latex immunagglutina-

Table 19.1 Comparison of molar composition of *Clostridium difficile* proteins

Amino acid	ToxA[a]	ToxB[b]	ToxA[c]	ToxB[d]	150 kDa[e]	150 kDa[f]
asp	6.1	7.7	16.0	14.7	11.6	11.1
asn	10.7	8.2	—	—	—	—
glu	5.5	8.9	8.0	11.5	12.2	10.6
gln	2.5	2.5	—	—	—	—
gly	6.1	5.8	7.1	10.7	10.9	10.9
ala	5.0	3.8	5.7	4.5	10.3	9.7
val	4.0	5.8	5.1	6.0	7.7	9.7
leu	8.1	8.0	8.3	8.0	9.1	9.7
ile	9.6	9.8	8.6	8.6	5.4	6.2
pro	2.3	2.2	2.1	2.6	3.8	2.9
met	1.1	1.9	0.8	0.3	1.5	2.3
phe	5.8	5.6	5.6	5.0	2.7	2.7
tyr	6.3	5.7	4.3	2.8	2.9	2.6
trp	0.9	0.7	1.3	1.9	N.D.	N.D.
his	1.2	1.0	1.3	1.4	1.0	1.1
ser	7.9	8.5	7.0	7.5	4.3	5.1
thr	6.2	5.1	7.6	5.1	4.7	4.9
cys	0.3	0.3	0.3	0.3	0.9	N.D.
lys	8.2	6.6	8.6	6.4	6.3	6.7
arg	1.8	2.0	2.0	2.5	4.1	3.5

[a-f] ToxA, ToxB, and the 150-kDa contaminant of ToxB are compared with respect to their molar amino acid composition (in mol%; ND, amino acid not determined). Rows *a* and *b*: ToxA and ToxB calculated from DNA sequences; rows *c–f*, data determined from isolated proteins; *c*, Lyerly et al., 1986a; *d*, Lyerly et al., 1986b; *e*, Bisseret et al., 1989; *f*, Pothoulakis et al., 1986. Data calculated from DNA sequences of ToxA and ToxB and those derived from protein analysis are in good agreement. Data determined for 150-kDa protein differ from those of ToxB; both proteins are two separate polypeptides.

tion test (Kamiya et al., 1986). The contaminant protein, originally thought to represent ToxA, is now known to be devoid of cytotoxic activity and, by molecular biology techniques, to be a molecule distinct from ToxA (Lyerly et al., 1986c; 1988a). However, detection of this protein in stool specimens may be an aid to detection of *C. difficile* (Bowman et al., 1986). Besides *C. difficile*, some other bacteria such as *C. sporogenes*, proteoytic *C. botulinum*, and *Peptostreptococcus anaerobius* also produce a similar polypeptide and are thus detected by latex agglutination (Lyerly et al., 1988a; Miles et al., 1988). We isolated mAbs specific for this protein. In the Ouchterlony test, mAb 1346G9 precipitates the 39-kDa protein but does not cross-react with pure ToxA (see Figure 19.4, later in chapter).

ToxB preparations are regularly contaminated with a 150-kDa protein that may be reduced to 50-kDa subunits. The earlier contention that this protein is identical to ToxB (Pothoulakis et al., 1986; Bisseret et al., 1989) has been refuted (Eichel-Streiber et al., 1987; Meador and Tweten, 1988). These proteins were often copurified because of their similar physicochemical properties. Comparison of the amino acid compositions of the isolated proteins proves that the polypeptides prepared by Pothoulakis et al. and Bisseret et al., are not identical to ToxB (Table 19.1).

In addition to the presence of true contaminants, lower M_r degradation products of ToxA and ToxB are often found in toxin preparations. Such proteins are mainly seen after immune or silver staining of large amounts of toxins separated on gels and have formerly been interpreted as subunits of ToxA (Kamiya et al., 1989). Cloning of the *tox* genes (Dove et al., 1990; Eichel-Streiber et al., 1990a, 1990b; Johnson et al., 1990) showed that ToxA and ToxB are single-chain proteins of 300 and 250 kDa, respectively.

In addition to contaminant proteins, size

variations of the toxins must be considered. Molecular weight of 52 kDa (Rihn et al, 1984), 300–500 kDa (Sullivan et al., 1982) and 230 kDa (Rautenberg and Stender, 1986) has been reported for ToxA isolated from different strains. Because of the huge difference in M_r, the peptide isolated by Rihn et al. (1984) seems to be rather a contaminant than a true size variant of ToxA. Alternatively, it may be an enterotoxin different from ToxA, such as the one that described by Guiliano et al. (1988). ToxA usually is bigger than ToxB (Sullivan et al., 1982; Eichel-Streiber et al., 1987), but this is not true for the toxins reported by Rautenberg and Stender (1986). The heterogeneity of the preparations used in the analysis of the mechanism of action of *C. difficile* toxins indicates that the interpretation of some studies has to be questioned.

Characterization of ToxA and ToxB from *C. difficile* VPI10463

The *C. difficile* strain most extensively analyzed is VPI10463 from the Virginia Polytechnical Institute (Blacksburg, United States). This strain produces high titers of both toxins. Sullivan et al. (1982) reported an estimated size of 450 to 500 kDa for both ToxA and ToxB according to gel filtration. In native polyacrylamide gels, the sizes were 550 kDa and 360 kDa, respectively. In sodium dodecylsulfate-polyacrylamide gel electrophoresis (SDS-PAGE), purified toxins migrate homogenously as polypeptides of M_r 300 (ToxA) and 250 (ToxB) (Figure 19.2; Eichel-Streiber et al., 1987). Isoelectric point ranges measured for ToxA and ToxB were 4.7–5.7 and 4.1–4.7, respectively (Lyerly et al., 1986a; Eichel-Streiber et al., 1987).

According to the deduced amino acid sequences of the cloned *toxA* and *toxB* genes, their M_r are 308 and 270, and their predicted pI are 5.3 and 4.1, respectively. Thus the M_r measured by SDS-PAGE closely corresponded to the theoretical sizes, and the experimentally determined isoelectric points were in the expected range. The differences between native and denaturing conditions on the PAGE gel are not yet fully understood. These data indi-

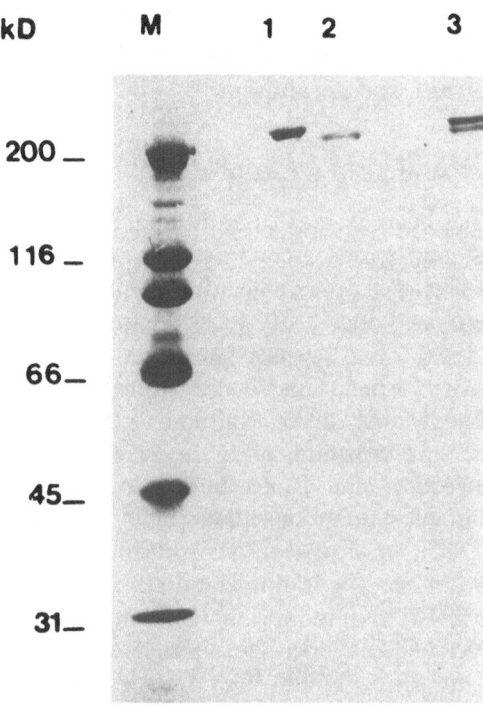

FIGURE 19.2   Gel electrophoresis of purified ToxA and ToxB. FPLC-purified ToxA and ToxB (2 μg) were separated on SDS-PAGE 5%–20% gradient gel and subsequently stained with Coomassie blue. Lane M, molecular size markers (in kilodaltons); lane 1, purified ToxA; lane 2, purified ToxB (these two lanes demonstrate purity of prepared polypeptides); lane 3, mixture of ToxA and ToxB to show size difference between the toxins.

cate that ToxA and ToxB may exist as dimers in the native state.

19.3   Immunochemical Analysis of *Clostridium difficile* Toxins

Preparation of antibodies against *C. difficile* toxins is impeded by their lethal effects (Hatheway, 1990), by the difficulties in preparing inactive but immunogenic toxoids (C.v Eichel-Streiber, unpublished observations), and by their immunosuppressive effects on monocyte functions exerted by both toxins (Däubener et al., 1988). Strongly immunogenic contaminants (e.g., the 39- to 43-kDa protein present in ToxA and the 150-kDa polypeptide in ToxB preparations) further complicate the production of high-titered, specific antisera.

Polyclonal sera against ToxA

ToxA has been used as an immunogen in mice, rabbits, and goats. Active ToxA was applied first at sublethal doses with subsequent booster injections with increasing doses (Eichel-Streiber et al., 1987). ToxA has also been used in toxoid form after formalin inactivation (Ehrich et al., 1980; Lyerly et al., 1986a; Rothman et al., 1988). A third immunization used the toxoid in the first injection and ToxA bound to agarose beads in subsequent injections (Libby and Wilkins, 1982). In all cases ToxA-specific antisera were induced. The immunization of mice gave rise to isolation of monoclonal antibodies.

Polyclonal sera block the cytotoxic activity of ToxA in vitro, inhibit the ToxA-dependent hemagglutination with rabbit erythrocytes (Krivan et al., 1986), neutralize the lethal effect of ToxA in animals (Lima et al., 1988), and also detect the ToxA-related hemorrhagic toxin (HT) of *C. sordellii* (Martinez and Wilkins, 1988). No cross-reaction with ToxB was observed (Lyerly et al., 1985b; Eichel-Streiber et al., 1989). Antibodies contained in ToxA antisera react mainly with the C-terminal repetitive structure of ToxA, which is the immunodominant part of the molecule (see Section 19.6) and which differs itself in part from ToxB. The reactivity of polyclonal ToxA sera against the N-terminal two-thirds of ToxA is very low, but titers rise against this part of ToxA and also against ToxB after long-term immunizations (Eichel-Streiber et al, unpublished observations, and Figure 19.3). The ToxA antiserum titered in Figure 19.3 reacted monospecifically with ToxA at dilution greater than 10^{-5} fold. The specificity of ToxA (and ToxB) autisera in thus dependent on the dilution of the serum used in the individual reaction.

Polyclonal sera against ToxB

ToxB is a very poor immunogen, and it is difficult to obtain satisfactory mono- or polyclonal antibodies. Immunization with formalin-

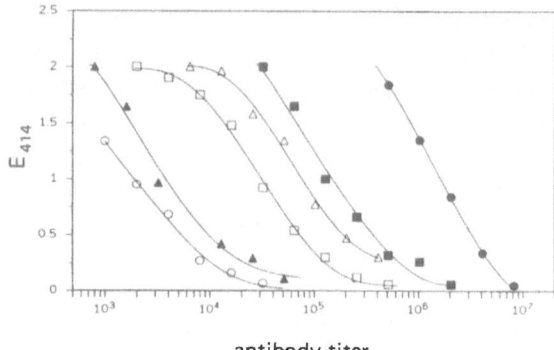

antibody titer

FIGURE 19.3   Titration of antisera by ELISA. Horse *Clostridium sordellii* LT antiserum, *C. difficile* ToxB, and ToxA rabbit antisera were titered against ToxA and ToxB (0.3 μg per well) of *C. difficile* in an ELISA. Extinction at 414 nm was measured and plotted against antibody titers. Antibody titers against *C. difficile* ToxA (open symbols) and ToxB (solid symbols) were $1:1.5 \times 10^5$ (□, ToxA) and $1:8 \times 10^5$ (■, ToxB) of ToxB antiserum; $1:1.5 \times 10^4$ (○, ToxA) and $1:5 \times 10^6$ (●, ToxB) of *C. sordellii* LT antiserum; $1:5 \times 10^5$ (△, ToxA) and $1:3 \times 10^4$ (▲, ToxB) of ToxA antiserum. *C. difficile* antitoxins thus have medium titers against both homologous toxins. Without prior adsorption these antitoxins are not specific for either toxin. A ToxA-specific antiserum is obtained in the early phase of ToxA immunization (Section 19.3). The *C. difficile* ToxB cross-reactive *C. sordellii* LT antiserum has high ToxB and low ToxA titer and thus is specific for ToxB.

inactivated ToxB did not induce high antibody titers (Eichel-Streiber et al., unpublished data). Libby and Wilkins (1982) raised a serum against a 1:9 mixture of partially purified ToxB of *C. difficile* VPI10463 (devoid of ToxA) and culture filtrate of the nontoxigenic strain VPI 2037. The resultant sera had neutralization titers of 1:256 in mouse lethality tests and 1:640 in cell tests. Meador and Tweten (1988) isolated ToxB-specific antibodies by the use of a ToxB affinity column; a serum titer was not determined.

In our laboratory the best rabbit antiserum against ToxB was raised by intravenous injection of trypsinized ToxB. In ELISA tests this serum had a ToxB titer of $1:8 \times 10^5$ and a ToxA titer of $1:1.5 \times 10^5$. For this serum no dilution can be chosen at which it reacts only with ToxB and not ToxA; the neutralization titer in in vitro cell tests was 1:320.

Characterization of *Clostridium sordellii* lethal toxin antiserum

At present the best antibody reactive with ToxB of *C. difficile* is a horse serum against *C. sordellii* lethal toxin (LT antiserum). *C. sordellii* toxins LT and HT are related to ToxB and ToxA of *C. difficile*, respectively (Popoff, 1987; Martinez and Wilkins, 1988). LT antisera cross-react with ToxB in ELISAs and Western blots, and they neutralize ToxB cytotoxic activity with a titer of 1:6400 (Eichel-Streiber et al., unpublished data). Neutralization of ToxB cytotoxicity by LT antisera is currently used for the definitive identification of ToxB within fecal specimens of patients suffering from AAD or PMC (Willey and Bartlett, 1979).

Immunological cross-reactivity between *C. difficile* ToxB and *C. sordellii* LT antisera has been reported by several groups (Larson and Price, 1977; Rifkin et al., 1977; Chang et al., 1978b; Popoff, 1987). Popoff stated that the ToxB-LT cross-reactivity is only partial. We tested a horse *C. sordellii* LT antiserum, a rabbit ToxB, and a rabbit ToxA antiserum in an ELISA with ToxB and ToxA as antigens (see Figure 19.3) (Eichel-Streiber et al., 1990b). The antibody titers of the ToxB antiserum were $1:8 \times 10^5$ and $1:1.5 \times 10^5$ against ToxB and ToxA, those of the *C. sordellii* LT antiserum $1:5 \times 10^6$ against ToxB and $1:1.5 \times 10^4$ against ToxA. The titers of the ToxA antiserum were $1:5 \times 10^5$ against ToxA and $1:3 \times 10^4$ against ToxB. To confirm that the LT antiserum detected the *C. difficile* ToxB polypeptide, we also used the serum in a Western blot analysis at a 10^{-4} dilution. The LT serum specifically detected the ToxB molecule within a culture filtrate of *C. difficile* and only weakly stained ToxA (data not shown).

The medium titer of ToxB antiserum against both ToxA and ToxB contrasted to the LT antiserum high ToxB and low ToxA titers. This great difference of reactivity between the two antisera is difficult to explain. The higher antibody titer of the LT antiserum against ToxB may in part result from a species effect of the immunization (the LT serum was raised in a horse, the ToxB serum in a rabbit); it cannot be attributed to a lower toxicity of LT compared to

ToxB (Gill, 1982; Lyerly et al., 1988b). Whether the LT toxin is also a better immunogen or lacks the immunosuppressive effects of ToxB (Däubener et al., 1988) cannot be judged at present. The analyzed reaction pattern proves that LT and ToxB have a higher degree of homology with each other than have LT and ToxA. The medium titers of ToxB antisera against ToxA and ToxB are a consequence of the 63% homology between the polypeptides (see Section 19.8). This homology is also seen in the ToxA antiserum. Nevertheless, by cross-absorption, antisera were generated that were monospecific for ToxA or ToxB and thus only reacted with the homologous toxin in counter-cross-immunelectrophoresis (Lyerly et al., 1985b). As mentioned, against ToxA this kind of monospecificity is also monitored in the early phase of the immune response. These monospecific polyclonal ToxA antibodies are exclusively directed against the ToxA repetitive C terminus (Eichel-Streiber et al., 1989), which obviously is immunologically different from ToxB repeats.

Monoclonal antibodies

Several attempts have been made to induce monoclonal antibodies (mAbs) against ToxA and ToxB of *C. difficile*. Mice were immunized with ToxA and ToxB, starting either with sublethal doses and then increasing the amount of toxin used for booster injections or by using formalin toxoid. Two groups of mAbs were isolated (Lyerly et al., 1986a; Eichel-Streiber et al., 1987). Representatives of the first group are PCG-4 and 1337C8; these two mAbs are specific for ToxA and do not cross-react with ToxB. The mAbs precipitated ToxA in the Ouchterlony test (Figure 19.4), blocked the hemagglutination of rabbit erythrocytes with ToxA (Figure 19.5), and neutralized ToxA in vivo. In a competitive ELISA, PCG-4 and 1337C8 compete for the same epitopes (Eichel-Streiber et al., unpublished data).

Both mAbs react with parts of the C-terminal repetitive structure identified by cloning of the *toxA* gene (see Section 19.6). The second group of mAbs, which all cross-react with both ToxA and ToxB, is represented by mAb G-2 (Lyerly et al., 1986a) and a series of mAbs isolated by

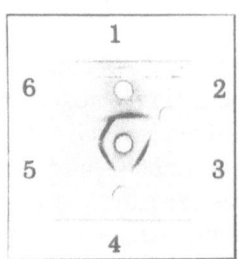 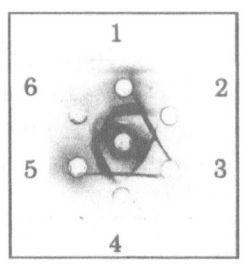 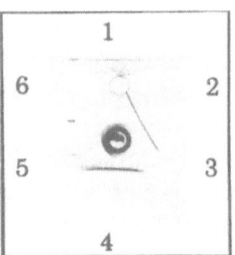

FIGURE 19.4 Ouchterlony test with antibodies specific for ToxA or its 39-kDa contaminant protein. In central well of test agar slides 600 µg each of ToxA (panel A), ToxA with 39-kDa contaminant (panel B), and 39-kDa contaminant polypeptide alone (panel C) were tested in Ouchterlony double diffusion at 37°C against following antibodies: wells 1 and 5, rabbit ToxA antiserum at 1:2 dilution; wells 3 and 6, mouse monoclonal antibody (mAb) 1337C8 specific for ToxA (ascites at 1:8 dilution); wells 2 and 4, mAb 1346G9 specific for 39-kDa contaminant (ascites at 1:8 dilution). The mAbs 1337C8 and 1346G9 react exclusively with their respective antigens. Spur formation seen with ToxA antibodies indicates that mAb 1337C8 activity is part of polyclonal ToxA rabbit serum but not vice versa. Comparable Ouchterlony test against ToxB was negative.

FIGURE 19.5 Hemagglutination of ToxA and rabbit erythrocytes. Rabbit erythrocytes (50 µl, 1×10^8/ml) and ToxA (50 µl, 20 µg/ml) were incubated at 4°C. Reaction was examined after 6–12 h. Row C: ToxA-induced hemagglutination (positive control). For inhibition of hemagglutination, ToxA was preincubated with appropriate dilutions of different antibodies at RT for 30 min before addition of erythrocytes. Row A: inhibition of hemagglutination by serial dilution of mAb 1337C8 (mouse ascites twofold stepwise diluted, from 1:5 to 1:10240); mAb 1337C8 inhibits hemagglutination to 1:320-fold dilution. Row B: same experiment as row A; inhibition by rabbit ToxA antiserum serial dilution to dilution of 1:10. Rows D–F: negative controls, sedimentation of erythrocytes incubated without ToxA; row D, mAb 1337C8 alone, serially diluted; row E, rabbit ToxA-antiserum serial dilution; row F, buffer control.

Eichel-Streiber et al. (1987) and Rothman et al. (1988). The apparent discrepancy between the monospecificity of ToxA polyclonal antisera (of the early phase) and the cross-reactivity of mAbs can now be understood on the basis of the homology of the toxin polypeptides.

Both toxins have some antigenic epitope in common; the derived antibodies were not equivalently represented within the polyclonal antisera so that the "cross-reactive epitopes" were not recognized.

Lyerly et al. (1989) recently found that different subclasses of antibodies directed against unrelated proteins unspecifically interacted with ToxA and ToxB. These experiments indicate that an unspecific interaction of *C. difficile* toxin with polyclonal sera or mAbs may exist in addition to the specific reaction of the reported mAbs.

19.4 Search for Cell Receptors

ToxA is known to interact specifically with intestinal cells. Only ToxA and not ToxB acts locally and induces the PMC-specific histological changes (Lima et al., 1988; Vernet et al., 1989). Consequently, the first cells investigated for a ToxA cell receptor were brush-border membranes (Krivan et al., 1986). The interaction between ToxA and brush-border membranes is mediated by a receptor that is threefold more highly expressed in conventional compared to axenic mice (Lucas et al., 1989). In addition, the physiological intestinal flora seems to induce a higher receptor density (Lucas et al., 1989). Hemagglutination of rabbit erythrocytes and fucosidase-treated human B-serotype erythrocytes indicated that the trisaccharide Galα1-3Galβ1-4GlcNAc was part of the receptor molecule for ToxA (Krivan et al., 1986). This trisaccharide is also present in thyroglobulin, which is commercially available and was used for some time for affinity purification of ToxA (Krivan and Wilkins, 1987; Kamiya et al., 1989).

The extent of ToxA cytotoxicity to different cells depends on the carbohydrate expressed on the cell surface and may reach levels similar to that observed for cytotoxin ToxB (Tucker et al., 1990). A receptor molecule has not yet been found for ToxB.

19.5 Molecular Genetics of the Toxins

Construction of the genetic library

Muldrow et al. (1987) were the first to report expression cloning of *toxA* fragments in lambda gt11. However, *C. difficile* DNA was not stable within these constructs and thus only a small,

0.3-kb *Taq*I fragment of *toxA* was isolated. Wren et al. (1987b) also described lambda gt11 clones containing *C. difficile* DNA; again these clones were unstable when subcultured in *Escherichia coli* host cells.

Two groups have constructed expression libraries in plasmid vectors. Price et al. (1987) cloned *Pst*I fragments of *C. difficile* chromosomal DNA into pBR322. This library allowed the isolation of the total *toxA* and *toxB* genes (Dove et al., 1990; Johnson et al., 1990). We have constructed a genetic library of partially digested chromosomal *C. difficile* DNA that is blunt end-ligated to the vector pUC12 (Eichel-Streiber et al., 1989). Size-fractionation of the partially digested chromosomal DNA was performed to avoid the expression of the entire toxin genes within *E. coli* host cells. The established genetic library also led to the isolation of the complete *toxA* and *toxB* genes (Eichel-Streiber et al., 1990b; Sauerborn and Eichel-Streiber, 1990).

All the recombinant clones carrying *C. difficile* chromosomal DNA in plasmid vectors reported so far stably transmit their DNA to the *E. coli* host. The stability of *C. difficile* chromosomal DNA in plasmid vectors as opposed to the instability in the lambda/gt11 system is striking and not yet understood.

Screening for recombinant *toxA* clones was done with antibodies against ToxA whereas *toxB* was isolated by oligodeoxyribonucleotide probing (Schulze and Eichel-Streiber, 1990), antibody screening (Eichel-Streiber et al., 1990b), or screening for cytotoxic activity (Johnson et al., 1990).

19.6 The *toxA* Gene

Isolation of *toxA*

Price et al. (1987) and Eichel-Streiber et al. (1989) screened their genetic libraries with polyclonal sera for the expression of ToxA fragments. Both groups first isolated parts of the 3′ end of the *toxA* gene. Eichel-Streiber et al. (1989) determined the direction of transcription of *toxA* and the position of the 3′ end of the gene. The expression of ToxA fragments in each of the recombinant clones was controlled by external promoters of the plasmid vectors. All clones from this screening carried parts of a sequence of the 3' end, which was therefore designated the "common region." The overlapping clones coded for a large repetitive domain of the C-terminal one-third of ToxA, and it was apparent that polyclonal antisera react predominantly with this domain of the molecule (Western blot analysis of representative *toxA* clones pCd14, pCd22, pCd13; see Figure 19.10, later in chapter).

Extension of the DNA sequence to the full length of the gene was achieved by chromosome walking (Dove et al., 1990; Eichel-Streiber et al., 1990a). Three open-reading frames (ORF) were identified and designated *toxA*, *utxA* (ORF upstream from *toxA*), and *dtxA* (ORF downstream from *toxA*; Sauerborn and Eichel-Streiber, in manuscript). Three clones (pCd122, pCd14, and pCd13) (Eichel-Streiber et al., 1990a) (Figure 19.6) or four clones (pCd19, pCd17, pCd11L, and pCd11R-6) (Dove et al., 1990) encompass the entire *toxA* gene. Southern analysis with pCd122, pCd14, and pCd13 showed that the restriction-fragment length of chromosomal DNA of *C. difficile* VPI10463 was identical to that shown in Figure 19.6 (Eichel-Streiber and Laufenberg-Feldmann, unpublished observations).

Analysis of the *toxA* gene

The nucleotide sequence of *toxA* was determined independently by Dove et al. (1990) and Sauerborn and Eichel-Streiber (1990); the respective deduced amino acid sequences are identical. The gene contains 26.9 mol% G·C, which is in good agreement with the overall G·C content of 28 mol% of *C. difficile* chromosomal DNA (Gottschalk et al., 1981).

The 5' untranslated region was analyzed for the presence of regulatory sequences (Dove et al., 1990; Eichel-Streiber and Sauerborn, 1990). SD boxes AGGAGGT were proposed directly upstream the initiation codon of *toxA*. These sequences are typical for from clostridia (Young et al., 1989). A promoter-like sequence with reasonable spacing of 20 nt is present −239 nt upstream from the *toxA* ORF. This promoter as well as the *toxB* and *utxA* promoter are highly homologous to reported clostridial promoters

(Young et al., 1989; and Figure 19.7A). Similar long spacing between the coding sequence and the −10 region is seen in the tetanus and botulinus toxin promoters (Niemann et al., 1988; Binz et al., 1990; see Figure 19.7A). One *C. difficile* promoter that is active in *E. coli* has been analyzed more extensively (Dailey and Schloemer, 1988). This promoter is active in *E. coli*, so that expression of *C. difficile* DNA in *E. coli* can be promoted by internal or external signals.

A stem loop was found 69 bp downstream from *toxA* ($\Delta G = -20.0$ kcal/mol; Tinoco et al., 1973; Sauerborn and Eichel-Streiber, 1990) that probably represents the transcription terminator for *toxA* (Figure 19.7B).

Dot matrix plot analysis of the *toxA* gene showed that the 8130-bp ORF consists of two domains, a 2499-bp repetitive structure at the 3' immunodominant end and the rest of the *toxA* gene, which has no repetitions. Base-pair comparison led to the definition of nine different short repetitive oligonucleotide sequences (SRONs), A–I. Combinations of these SRONs give rise to "in frame" oligopeptides designated CROPs (combined repetitive oligopeptides; Eichel-Streiber and Sauerborn, 1990). Fig 8A delineates a representative part of the repetitive domain of *toxA* and its deduced amino acid sequence. All 30 different CROPs defined within the repetitive area are composed of a varying number of individual SRONs. Similar short oligonucleotide sequences could not be defined for *toxB* because of its greater variability (see Section 9.7, Figure 19.6C and 19.8B).

Analysis of the ToxA polypeptide

The 8130-bp *toxA* ORF corresponds to a protein of 2710 amino acid with a calculated size of 308.057 daltons and a pI of 5.3. The N terminus of ToxA has been sequenced (Dove et al., 1990); it is not processed, and the determined amino acid sequence is identical to that deduced from DNA sequencing.

Protein chemical data had predicted the existence of repetitive structures or a subunit composition of toxin A (Lyerly et al., 1986a). Comparison of the ToxA sequence with itself showed a 833-amino-acid repetitive structure

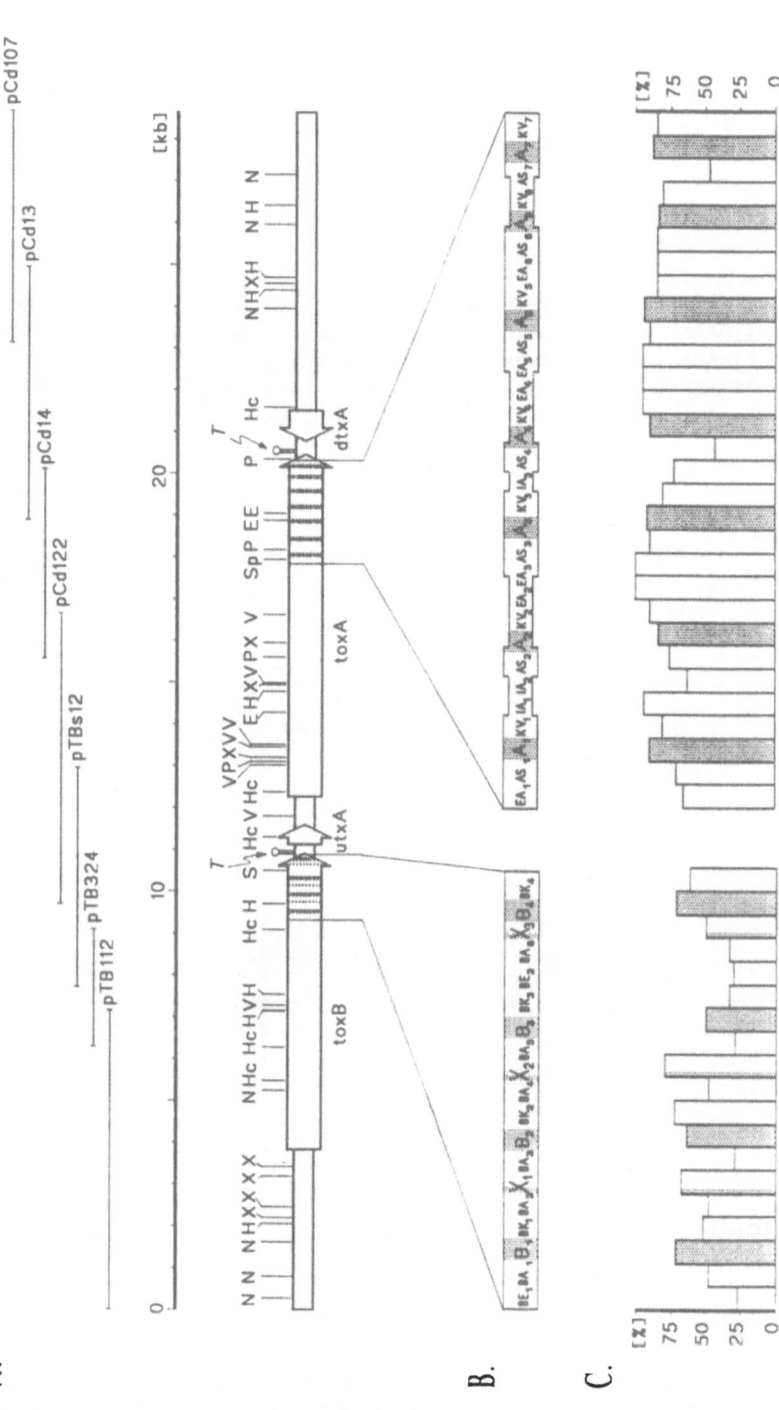

FIGURE 19.6 Organization of *tox* genes and structure of ToxA/ToxB C The five narrow boxed ToxA-CROP-sets indicate sequences that are "deletermini. (**A**) Clones pTB112, pTB324, pTBs12, pCd122, pCd14, pCd13, leted" in the ToxB repetitive domain. All IA ToxA-CROPs are deleted in and pCd107 are part of a 29-kb fragment of *C. difficile* DNA which en- ToxB; deletions of ALICE$_2$ to EA$_3$, ALICE$_4$ to EA$_5$, and ALICE$_6$ to AS$_7$ give codes for four open reading—frames, *toxB*, *utxA*, *toxA*, and *dtxA*, thus rise to ToxB-CROPs X$_{1-3}$. (**C**) Variability of CROPs: The individual CROPs far described. A Partial map of this region is shown; restriction enzymes were compared with their consensus sequences (compare Figure 19.9). The are: E, *EcoRI*; V, *EcoRV*; Hc, *HincII*; H, *HindIII*; N, *NdeI*; P, *PstI*; S, *SphI*; minimum homology of ToxB-CROPs is 27%, its maximum, 80%; ToxA-Sp, *SpeI*; X, *XbaI*. Two stem loops, probably transcription terminators of CROPs have minimal and maximal homologies of 42% and 100%, respec-*toxB* and *toxA*, are indicated by a *T* (compare also Figure 19.7B). (**B**) Suc- tively. Thus, the ToxB-CROPs are variable; those of ToxA are conserved cession of CROPs (*combined repetitive oligopeptides*) of repetitive except four areas we named "hot spots" of ToxA (see 19.6, and Eichel-C-termini: ToxB contains the following repeats: BALI$_{1-4}$; X$_{1-3}$; BA$_{1-6}$; Streiber and Sauerborn, 1990). BK$_{1-4}$; BE$_{1-2}$. ToxA contains: ALICE$_{1-7}$; KV$_{1-7}$; AS$_{1-7}$; EA$_{1-6}$; IA$_{1-3}$.

A.

		"-35"	"-10"	distance referring to ATG
GRAM[+] E. Coli		5'-aa-aaaaa----<u>TTGACA</u>	---a--a---a-t-tg- <u>TATAAT</u>aatat---•-3'	./.
	lacZ	5'-ggcaccccaggc<u>TTTACA</u>	ctttatgcttccggctcg <u>TATGTT</u>gtgtgg▲-3'	45 nt
C. botulinum				
	botA	5'-ccaattgttttaa<u>CCCTAT</u>	cttataacggtaaatata <u>TATGTT</u>tatctatg▲-3'	127 nt
C. tetani				
	tet	5'-aaatttaaattt<u>TCAGGT</u>	tacaaaaaataacctgat <u>TATGTT</u>atatgtaa<u>T</u>-3'	136 nt
C. difficile				
	P$_{DS}$	5'-g<u>TTCACA</u>tcctccacctaaagcaaatccg<u>TTTACA</u>gcagacgtctttt▲-3'		n.d.
	catD	5'-gaagtgggcaag<u>TTGAAA</u>	aattcacaaaaatgtgg <u>TATAAT</u>atctttgct-3'	48 nt
	toxA	5'-atataagatatg<u>TTAACA</u>	aattactatcagacaatctcc<u>TTATCT</u>aatagaaga-3'	239 nt
	toxB	5'-tatagaacaaag<u>TTTACA</u>	tatttatttcagacaacgtct<u>TTATTC</u>aatcgaaga-3'	169 nt
	utxA	5'-gactatgatgaa<u>TGCACA</u>	gtagttcaccttttttatatt <u>TCTAAT</u>ggtaacaaa-3'	45 nt

B.

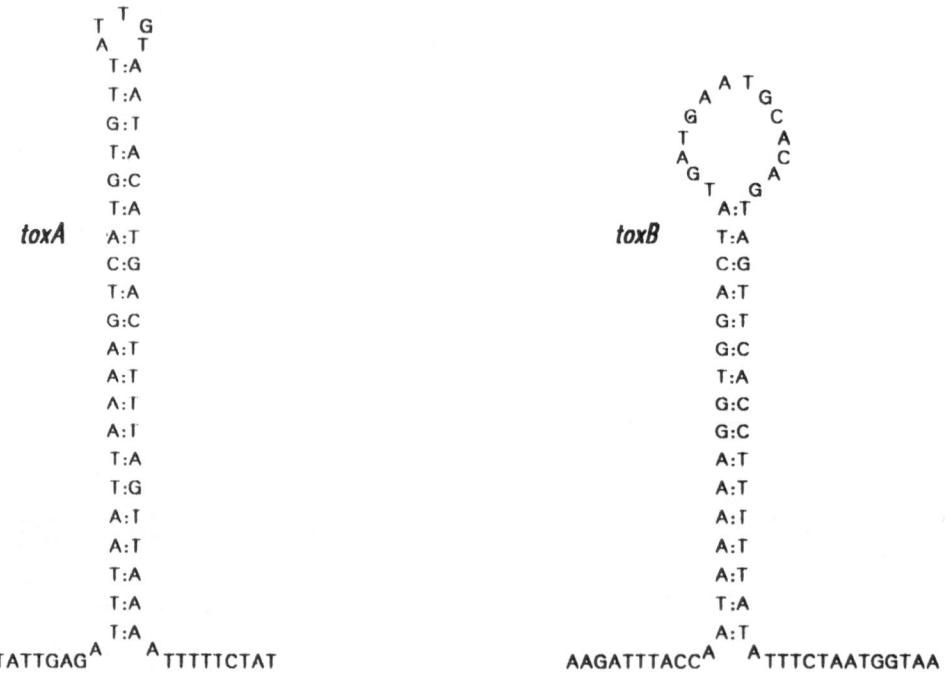

toxA   ΔG = - 20.0 Kcal/Mol

toxB   ΔG = - 18.2 Kcal/Mol

FIGURE 19.7 Analysis of regulatory sequences. (A) Comparison of putative promoter structures: Promoter-like structures of *toxB* (unpublished data), *utxA* and *toxA* (Eichel-Streiber and Sauerborn, 1990), and *catD* (Wren et al., 1989) are compared with promoters identified for: *C. difficile* (P$_{DS}$, Dailey and Schloemer, 1988); toxin promoters of *C. botulinum* neurotoxin A (*botA*, Binz et al., 1990) and *C. tetani* neurotoxin (*tet*, Niemann et al., 1988); a consensus sequence of promoter struc-

at the C terminus. This structure has been divided into five different groups of oligopeptides (Dove et al., 1990; Eichel-Streiber et al., 1990a). Both laboratories recognized five groups of repeats; the length of the individual sequences varies between 20 and 50 amino acids. Dove et al. classified 38 repeats (designated class I and class IIA-D) by the sequence homology; the 30 repeats (CROPs) of Eichel-Streiber et al. (1990a) were additionally classified according to their position within the repetitive domain. The immunodominant common region of recombinant *toxA* clones isolated by antibody screening (Eichel-Streiber et al., 1989) is part of this C-terminal repetitive domain.

The 30 different CROPs of the repetitive immunodominant C-terminal third of ToxA were designated ALICE, KV, AS, EA, and IA (Figure 19.9). In all 7 repetitions of ALICE, of the 50-amino acid repeats, 30 amino acids are identical. Within the four groups of 21 amino-acid CROPs, 9 amino acids are conserved in the 7 KV, 14 in the 6 EA, 5 in the 3 IA, and 10 (5) in the 5 (7) AS CROPs (Eichel-Streiber and Sauerborn, 1990; for CROP consensus sequences see Figure 19.9). The comparison of the individual CROPs with their repetitive consensus sequences delineated two large and two small regions of high conservation and four variable regions (see Figure 19.6C). The minimal and maximal CROP homology was 42% and 100%, respectively.

Variability of the repetitive C-terminal third

For other proteins, repetitive domains are the site of deletion and multiplication occurring by recombination events. Hollingshead et al. (1987) reported that size and sequence variations of streptococcal M6 protein occurred by intragenic recombination between repeated sequences in the type 6 gene. Deletions or multiplications are responsible for size variations of repetitive domains of streptococcal M protein (Jones et al., 1988; Miller et al., 1988). The observed four "hot spots" in ToxA may have occurred by similar events in the *toxA* repetitive sequence (Eichel-Streiber and Sauerborn, 1990). Dynamic processes at the *C. difficile* ToxA C terminus could be the basis of toxin size variants discussed here. In addition to these structural considerations, functional modifications may be mediated by similar events. A modulation of binding capacity of ToxA to its target cells (Kamiya et al., 1989) might be the biological function of the C-terminal repetitive structure. Deletions of the total repeat could also explain the loss of activity in vivo (Eichel-Streiber, 1991 in press).

Besides intragenic events, intergenic recombination may occur between *toxA* and *toxB*, which could also contribute to the toxin-variation.

Differentiation of two separate functional Parts

The fact that the clones originally isolated from the genetic library all reside at the 3' end of *toxA* indicated that ToxA is composed of two structurally different parts. The C terminus is involved in toxin–cell interaction; it contains the huge repetitive structure revealed by DNA sequence analysis. This part of ToxA induced high-titer antisera in the course of the immune reaction and thus carried the immunodominant

---

tures of other gram-positive bacteria (Graves and Rabinowitz, 1986); and P_{lacZ} of *E. coli*. The −35 and −10 boxes (underlined) with their spacing are indicated. Those transcription starts determined are marked by a bold, printed, underlined **A** (or **T** or *). The indicated tetanus and botulinus toxin promoters have extreme spacing of 127/136 nt between the ATG start codon and the transcription start. Similar spacing is evident for the proposed *C. difficile* toxin promoters (169 and 239 nt). (**B**)

Terminators of transcription of *toxA* and *toxB*: An inverted repeat of 21-bp stem size begins 69 nt downstream of the TAA translation stop signal of *toxA*. The free energy of 20.0 kcal/mol has been calculated according to Tinoco et al. (1973). Similar structure of 16 bp begins 100 nt downstream of the TAG translation stop signal of *toxB*; its free energy is 18.2 kcal/mol. No further stem loop structure is observed, especially not between *utxA* and *toxA*.

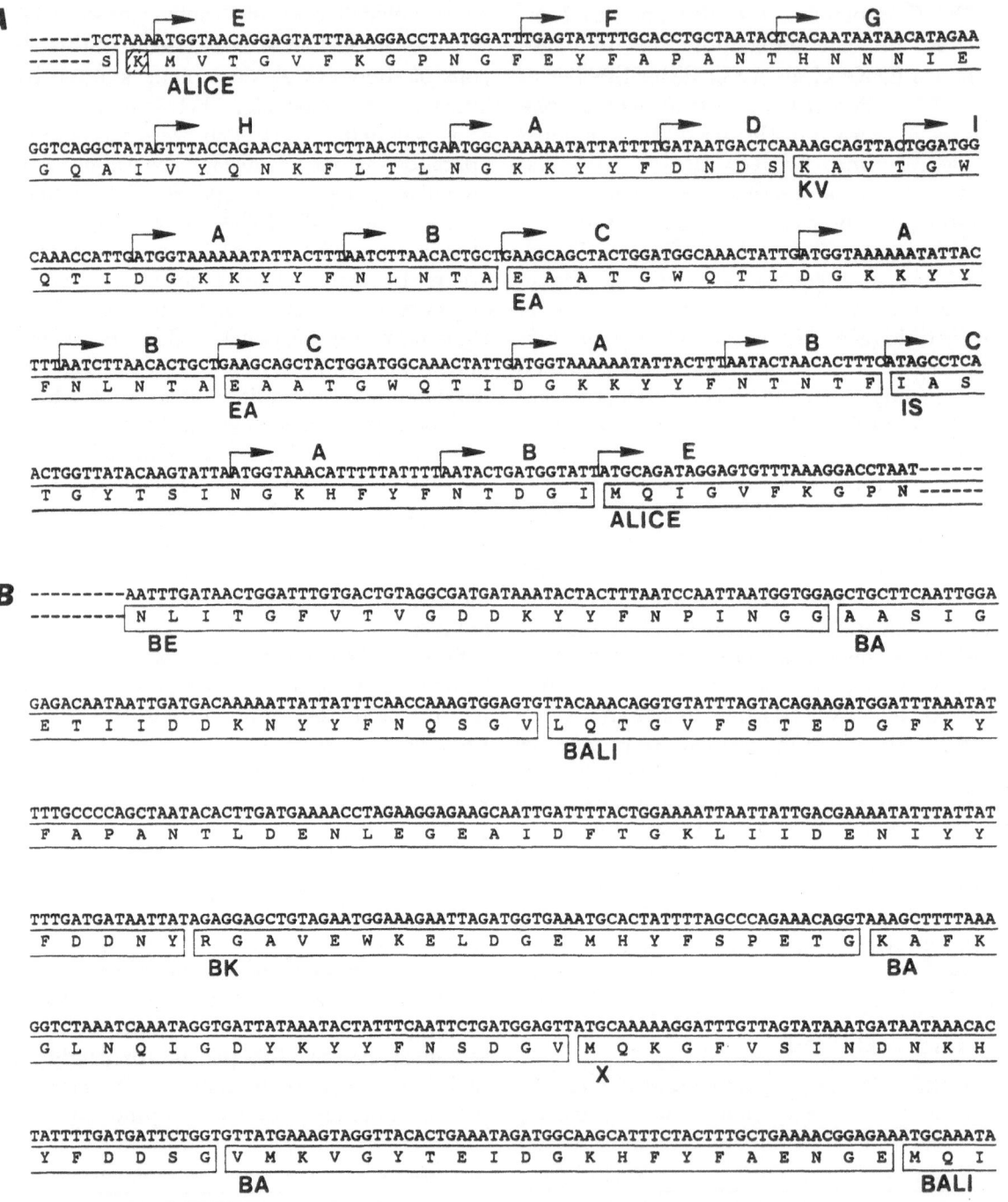

FIGURE 19.8 Delineation of characteristic parts of toxin repetitive domains. (A) Toxin A partial sequence: the sequence between ALICE 4 and ALICE 5 was chosen to demonstrate the composition of ToxA-CROPs by different SRONs (short repetitive oligonucleotides). The DNA and the derived aa sequence are shown. The arrows indicate the start of the SRONs named with capital letters (A–I). The boxed aa sequence represents individual CROPs indicated (ALICE, KV, AS, and EA; Eichel-Streiber and Sauerborn, 1990). (B) Toxin B partial sequences: The sequence between BE₁ and BALI₂ was chosen to demonstrate the succession of CROPs of ToxB. The DNA and the derived aa sequence are shown. The boxed aa sequence represents the individual ToxB-CROPs indicated. As *toxB* is less conserved than *toxA*, SRONs could not be identified as was done for *toxA* (see A).

```
ALICE   MQIGVFKGPNGFEYFAPANTDANNIEGQAIVYQNKFLTLNGKKYYFGNNS
BALI    MQIGVFNTEDGFKYFApANTLDEN*EGEAI*Y*tG*L**DEKIYYFDD*Y
X       MQ*GFV*-------------------------------INDK*FYF*DSG

KV                                        KAVTGWQTIBGKKYYFNPNTA
BK                                        TAAvGWK*LDGEKYYFDPGTA

AS                                        IASTGYKTIBGKHFYFBTDGI
BA                                        *A**GL**IDD**YVFN**G*

EA                                        EAATGWQTIBGKKYYFNTNTA
BE                                        ****G*****D*KYYF*****

IA                                        IAA*GLQTIBNBKYYFB*DTA
```

FIGURE 19.9 Comparison of ToxA and ToxB CROPs. ToxA consensus CROPs (combined repetitive oligopeptides) have been determined from ALICE$_{1-7}$, KV$_{1-7}$, AS$_{1-7}$, EA$_{1-6}$, and IA$_{1-3}$; ToxB consensus CROPs from BALI$_{1-4}$, X$_{1-3}$, BA$_{1-6}$, BK$_{1-4}$, and BE$_{1-2}$. The order of the individual CROP-repeats in the ToxA and ToxB C-terminus is seen in Figure 19.6B. Short lines (-) indicate a position where no consensus amino acid could be formulated. ToxB-CROP X has 19 aa; the long line (⸺) in its middle indicates a 31 aa deletion discussed in the text. Boxed aa are conserved in homologous ToxA- and ToxB-CROPs. For ToxA-CROPs, consensus amino acids can be given for all positions, except two in CROP-IA. ToxA-repeats are highly homologous, whereas those of ToxB are more variable (Figure 19.6C). In consequence consensus amino acids are missed at individual positions of all five ToxB-CROPs.

domains of ToxA. The N-terminal two-thirds of the molecule mediates the toxic activity. A derivative of toxin A with the repetitive domain deleted is still active in vitro but not in vivo (Eichel-Streiber 1991 in press).

Two cross-reactive mAbs (1337C8 and PCG-4) block the ToxA-induced hemagglutination of erythrocytes from rabbits and human serotype B (after fucosidase treatment). The agglutination reactions are mediated by the repetitive domain of ToxA and the trisaccharide Galα1-3Galβ1-4GlcNAc present on the cell surfaces (Section 19.4, Krivan et al., 1986). Price et al. (1987) reported that PCG-4 reacts with a 4.7-kb fragment of *toxA*. A 2.1-kb *Pst*I subclone of this fragment of the 3' repetitive part of *toxA* reacts with mAb 1337C8 (Eichel-Streiber et al., 1989). Subclones derived from the N-terminal adjacent sequences were all negative in the mAb staining. Experiments have been conducted to localize the antigenic determinants recognized by 1337C8 to a ToxA fragment, encoded by 1.8-kb of the repetitive area (Eichel-Streiber and Sauerborn, 1990). This DNA fragment was further reduced to 600 bp, the minimal sequence necessary to yield a recombinant protein reactive with mAb 1337C8 (Sauerborn et al., 1991, in press). Vaccination with a nontoxic recombinant peptide of the ToxA repetitive domain protected animals against the ToxA lethal effect (Lyerly et al., 1990).

Homology with streptococcal glucosyltransferases

The fact that the C-terminal repeat harbors a carbohydrate-binding domain is also supported by the demonstrated homology of the ToxA repeats to sequences found within the glucosyltransferases GtfB, GtfC, and GtfI of *Streptococcus mutans* and *S. sobrinus* (Ferretti et al., 1987; Shiroza et al., 1987; Ueda et al., 1988). These streptococcal enzymes contain C-terminal repetitive sequences homologous to those of ToxA (Eichel-Streiber and Sauerborn, 1990). For GtfI it could be shown that successive deletion of the C-terminal repeats eliminated the carbohydrate-binding capacity whereas the enzymatic activity, located at the N-terminal part of the enzyme, was retained (Ferretti et al., 1987). Deletion of the ToxA repetitive domain similarly eliminated the in vivo ToxA activity (Eichel-Streiber, 1991 in press).

Secretion of ToxA

Usually N-terminal signal sequences mediate the secretion of bacterial proteins (Randell and Hardy, 1989). As the ToxA N terminus is not being processed (Dove et al., 1990) and is also not composed of hydrophobic amino acids, the mediation of toxin export by an N-terminal structure is rather unlikely. Therefore, the possibility of a C-terminal signal involved in ToxA secretion merits consideration. In ToxA a hydrophobic C-terminal sequence can be found adjacent to the repetitive structure; and C-terminal secretion signals have been identified in the E. coli hemolysin family (Koronakis et al., 1989). The Hly export is dependent on the gene products of HlyB and HlyD encoded in proximity on the E. coli chromosome. In this respect the utxA gene upstream of ToxA must be considered as a candidate for a protein involved in the ToxA export. This idea is supported by the fact that there is no evident transcription termination between toxA and utxA. To evaluate this hypothesis, further studies are needed.

19.7   The toxB Gene

The toxB gene has been isolated by oligodeoxynucleotide (Schulze and Eichel-Streiber, 1990), antibody (Eichel-Streiber et al., 1990b), and cytotoxicity screening (Johnson et al., 1990).

Oligodeoxynucleotide probe screening

A 41mer and a complementary 17mer were taken to isolate the toxB 5′ end (Schulze and Eichel-Streiber, 1990; Eichel-Streiber et al., in manuscript). The oligonucleotide probes were synthesized on the basis of the reported N-terminal amino acid sequence of ToxB (Meador and Tweten, 1988). A or T were chosen for their third positions because of the A·T richness of C. difficile DNA. This proved to be a good choice, as was shown later by DNA sequencing.

Screening of the gene bank yielded eight positive clones (pTB1-8). Considering the high A·T content of C. difficile DNA hybridizations were carried out in tetramethyl ammonium chloride (TMA) solution (DiLella and Woo, 1987). To verify isolated toxB fragments, the two complementary DNA probes were used to read the DNA sequence in both directions, starting from the toxB 5′ end (Schulze and Eichel-Streiber, 1990). The derived amino acid sequence was identical to the published N-terminal amino acid sequence (Meador and Tweten, 1988) except for the extreme N-terminal amino acid (Ser not Trp; Eichel-Streiber et al., 1990b). Two restriction sites (RsaI and XbaI), asymmetric to the toxB 5′ end, helped determine the direction and orientation of the toxB sequence and thus facilitated chromosome walking for isolation of the total toxB gene.

Screening with ToxB-reactive antibodies

Because of their relatively low titers and cross-reactivity with ToxA (see Figure 19.3), C. difficile ToxB antisera could not be used to screen a genetic library. Rather, the C. sordellii horse LT antiserum was used for colony blotting of the gene bank of C. difficile VPI10463 (Eichel-Streiber et al., 1990b). This screening yielded 11 positive pTBs clones. Restriction digests indicated a similarity of some clones to pTB112 isolated by oligonucleotide screening (Schulze and Eichel-Streiber, 1990); others were similar to pCd122, which encoded for the N terminus of ToxA (Eichel-Streiber et al., 1990a; see Figure 19.6A).

Two clones, pTB112 and pTBs12, isolated in our laboratory extended furthest into the toxB gene from the 5′ and 3′ ends, respectively. A 500-bp toxB fragment situated in the center of the gene (between 6.8 and 7.4-kb on the physical map; Figure 19.6A), could not be isolated by either colony blotting, oligonucleotide screening, or chromosome walking. A HincII fragment, detected by Southern analysis of chromosomal DNA, joined and overlapped the two parts of toxB (pTB324; Eichel-Streiber et al., 1990b). Figure 19.6A delineates a partial map of the total toxB gene; the 5′ end is located 54 bp upstream of a RsaI site, and the 3′ end is 405 bp downstream of *Sph*I site at 3.8 kb and 10.9 kb on the map, respectively.

Screening for ToxB activity

Johnson et al. (1990) cloned a *Xba*I-*Sph*I fragment of chromosomal DNA (3.3–10.9 kb on the physical map; Figure 19.6A) and reported that the resulting *E. coli* clones were stable and expressed cytotoxic ToxB fragments. Thus, production of "suicide fragments" derived from parts of ToxB can be ruled out. This also could have been an explanation for the difficulty to clone the central *Hinc*II fragment of *toxB* (Figure 19.6A; pTB324). The *toxB* subclones isolated in our laboratory contained a maximum of 3-kb of *toxB* coding sequence. None of the clones we isolated exhibited cytotoxic activity (unpublished observations).

Analysis of the toxin B gene and polypeptide

Three clones (pTB112, pTB324, and pTBs12) cover the total *toxB* gene (Figure 19.6A). The position of *toxB* is 194 bp upstream of *utxA*, which itself is located upstream of *toxA* (Dove et al., 1990; Sauerborn and Eichel-Streiber, 1990). The *toxB* DNA sequence determined by us was identical to the one reported by Barrosso et al. (1990). The gene has a size of 7098 bp; the deduced protein has a M_r of 269.709 and a pI of 4.1.

The size of the cloned *C. difficile* insert DNA might have permitted isolation of overlapping clones encoding the total *toxB* gene; it is not clear why there was a gap in the center of the *toxB* gene between the two groups of clones from the primary screening. This part of the molecule contains a high number of amino acids with a molar ratio of less than 2% (cluster 2 of amino acids present in both toxins; see also Figure 19.11, later in chapter), and we predict that it encodes for the toxic domain of ToxB (Eichel-Streiber et al., in manuscript).

We analyzed the 5′ untranslated region of *toxB* for the presence of putative regulatory sequences. A Shine–Dalgarno sequence (AGGAGA) typical for clostridia (Young et al., 1989) was found upstream of the ATG start codon. A *toxB* promoter-like structure is presented in Figure 19.7A. Interestingly, all clostridial toxin promoters seem to have an extremely long spacing between their transcription and translation starts (169 nt for *toxB*). A stem loop structure that probably represents the transcription terminator for *toxB* (Sauerborn and Eichel-Streiber, 1990; see Figure 19.6A and Figure 19.7B) was found 100 bp downstream from *toxB* ($\Delta = -18.2$ kcal/mol; Tinoco et al., 1973).

Expression of ToxB fragments

In contrast to the expressed recombinant ToxA fragments, which were exclusively positioned at the immunodominant C terminus, the ToxB-positive clones were derived from both the N terminus and the C terminus of toxin B. Compared with ToxA, no part of ToxB is immunodominant. N-terminal ToxB fragments were expressed by clones that contained DNA fragments downstream of the *Xba*I site at 3.3 kb of the map presented in (Figure 19.6A). Figure 19.10 shows a Western blot of four clones, pTB1, pTB2, pTB8, and pTB112. The latter clone carried the longest *toxB* fragment and expressed the largest polypeptide of about 120 kDa (Eichel-Streiber et al., in manuscript). C-terminal fragments were expressed by clones pTBs12 and pTBs29 (see Figure 19.6A and 19.10).

The isolated *toxB* fragments derived from the 5′ end of the gene contained DNA fragments oriented in both directions (pTB27 and pTB72; Eichel-Streiber et al., 1990b); ToxB expression was obviously independent of the internal P_{lacZ} of pUC12. A strong promoter element in front of *toxB* that is active in *E. coli* must obviously exist, allowing transcription to occur opposite the P_{lacZ} of the pUC vector molecule. This assumption is supported by the findings of Dailey and Schloemer (1988), who reported a *C. difficile* promoter active in *E. coli*.

All clones derived from the 3′ end of *toxB* carried their insert DNA opposite to the external P_{lacZ}. In pTBs12 and pTBs29, 3 and 2 kb of the 3′ end of toxB were cloned, respectively (Figure 19.6A). Some bands were stained with the ToxB cross-reactive *C. sordellii* LT antiserum

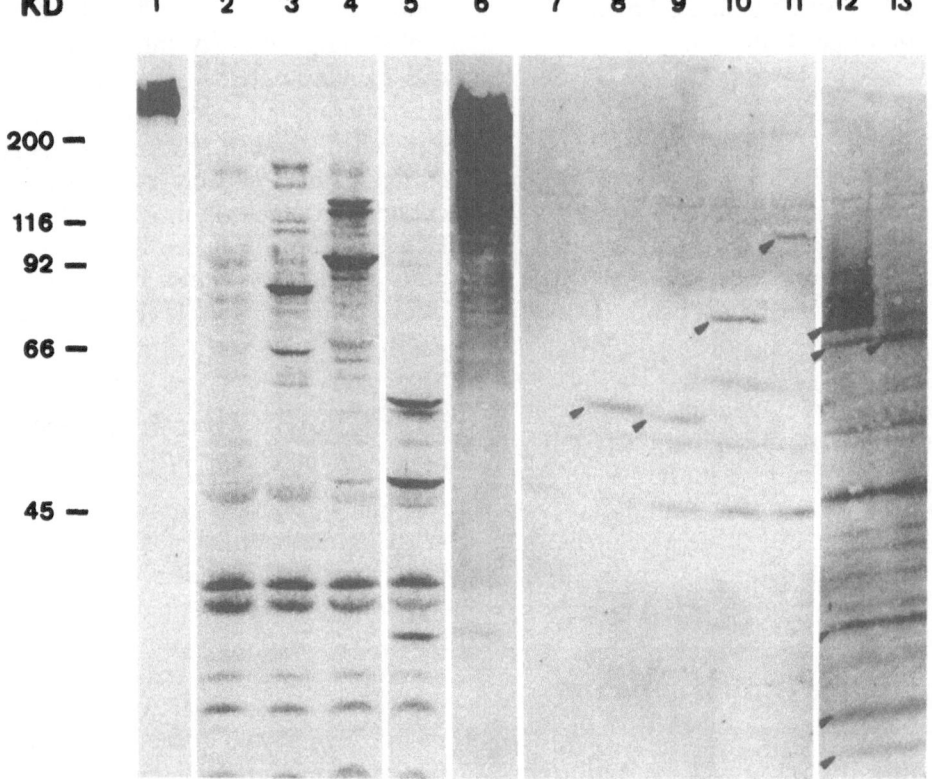

FIGURE 19.10 Western blot of recombinant proteins. Samples of ToxA and ToxB (1 μg) and total lysates of the individual recombinant clones (50 μg) were separated on a 5%–20% SDS-PAGE gel and thereafter transfered to nitrocellulose. Proteins were detected using polyclonal ToxA-specific rabbit serum (lanes 1–5) or horse serum specific for the ToxB crossreactive *C. sordellii* LT (lanes 6–13). Positive controls: ToxA, lane 1; ToxB, lane 6. Negative controls: *E. coli* JM83 transformed with pUC12, lanes 2 and 7. ToxA antiserum staining of ToxA-fragments encoded by the following recombinant plasmids: lane 3: pCd14; lane 4: pCd22; lane 5: pCd13. pCd14, pCd22, and pCd13 are three clones of the C-terminal immundominant re-

petitive domain of ToxA, originally designated "common region" (Eichel-Streiber et al., 1989). ToxB antibody staining of recombinant ToxB-fragments expressed by the clones indicated: lane 8: pTB1; lane 9: pTB2; lane 10: pTB8; lane 11: pTB112; lane 12: pTBs12; lane 13: pTBs29. Recombinant ToxB-fragments indicated by arrow. Clones pTB1 and pTB2 contain the same insert of the 5'end, but oppositely oriented. pTB1, pTB8, and pTB112 carry *toxB* fragments of increasing size. Clones pTBs12 and pTBs29 are derived from the *toxB* 3'end, they both contain their insert DNA opposite to the P_{lacZ} of the cloning vector. The indicated polypeptides are ToxB fragments (see discussion in Chapter 20.8).

even at higher dilutions (1:10⁵). Detected polypeptides had a size of 80 to 90 kDa, with some minor bands at 15 to 30 kDa (see Figure 19.10). Besides *toxB* fragments, both clones contained *utxA* and *toxA* fragments. Because diluted *C. sordellii* LT-antiserum hardly detected pure ToxA (Figure 19.3), the detected recombinant proteins were not ToxA fragments. Clone pCd138, which lacked *utxA* and the *toxB* 3' end, did not produce *C. sordellii* LT

antiserum-specific proteins (Eichel-Streiber et al., in manuscript).

The deletion of the *toxB* fragments of pTBs12 also eliminated the signals in Western blots (M. Sauerborn and C.v. Eichel-Streiber, unpublished observations). Thus, the low molecular weight polypeptides (M_r Of 16; Figure 19.10) could not correspond to the UtxA polypeptide; a minimum of *toxB* coding sequence was obviously necessary for expression to occur.

We therefore assume that expression of C-terminal ToxB fragments resulted from internal initiation of transcription within the coding sequence as a result of the AT richness of *C. difficile* DNA.

19.8 Comparison of *toxA* and *toxB*

DNA sequencing confirmed earlier findings that ToxA and ToxB are related (Barroso et al.,1990; Eichel-Streiber et al., 1990b). The basis for a first hypothesis that ToxA and ToxB are related (Eichel-Streiber et al., 1987) was the reactivity of some monoclonal antibodies, supported by their similar lethal effects (Gill, 1982). Additional support was given by the reported similar cytotoxic activity in vitro (Tucker et al., 1990).

Structural aspects

Alignment of the amino acid sequences showed a striking homology between ToxA and ToxB (Figure 19.11). ToxA and ToxB have an average of 49% identical amino acids; an additional 14% are conservative substitutions (Eichel-Streiber et al., in manuscript). Two domains exist, the N-terminal, highly conserved two-thirds and the C-terminal third with a repetitive sequence found in ToxA and ToxB Figure 19.11; Eichel-Streiber and Sauerborn, 1990).

The C-terminal repetitive oligopeptides (CROPs) of ToxA and ToxB are in part homologous (Eichel-Streiber and Sauerborn, 1990; Eichel-Streiber et al., in manuscript). ToxB CROPs analogous to ToxA repeats ALICE, KV, AS, and EA were designated BALI, BK, BA, and BE, respectively (see Figure 19.9). The conservation of the individual groups of ToxA repeats is much higher than that of the ToxB repeats (Figure 19.6C). Some sequences are missing in ToxB (IA of ToxA; Eichel-Streiber and Sauerborn, 1990; Eichel-Streiber et al., in manuscript), whereas a short sequence X is only found in ToxB. X seems to have evolved by a 93-bp deletion of BALI; it only encodes for the first 7 and final 12 amino acid of BALI (see Figure 19.9). This is a further indication that recombinational processes occur in the *C. difficile tox* genes.

The N-terminal two-thirds of ToxB and ToxA have three clusters of amino acid that are boxed in Figure 19.11 (boxes 1–3). Cluster 1 contains 150 hydrophobic amino acids conserved in both toxins (data not shown). Cluster 2 is composed of highly conserved amino acids (68% identity and an additional 7% conservative substitutions; amino acids 1149–1298 of ToxA and 1147–1297 of ToxB). This part of both toxins contains a large number of amino acids with a molar ratio of less than 2%, most of them positioned at identical sites (M; C; H; W; R; see Eichel-Streiber et al., in manuscript), e.g., three of six histidine residues are conserved. Histidine-rich regions have been discussed as motifs for the biological function of clostridial neurotoxins (Binz et al., 1990). This stretch of highly conserved amino acid of ToxB is encoded by clone pTB324, containing the 500-bp DNA fragment that could not be cloned from the genetic library. Amino acids 1450–1700 (Cluster 3: third box, see Figure 19.11) differ the most between ToxB and ToxA (32% identical amino acids, 15% conservative substitutions). This structure might be responsible for the differential ToxA/ToxB effects.

Functional aspects

As mentioned ToxA is composed of two functional parts, a N-terminal toxic domain and a C-terminal repetitive ligand domain, which can be blocked by the mAbs 1337C8 and PCG-4 (Eichel-Streiber and Sauerborn, 1990). These two mAbs react only with ToxA. Thus, the ToxA:ToxB homology leaves out those determinants within the repetitive C-terminal domain involved in ToxA targeting.

These differences in mAb binding between ToxA and ToxB indicate a functional diversity that is also reflected by their different reaction patterns in vivo and in vitro. Primarily, the differences of the in vivo activities of the toxins on individual cells and thus organs are based on different targeting mediated by the repetitive ligand domain. The fact that the traces of ToxA in ToxB

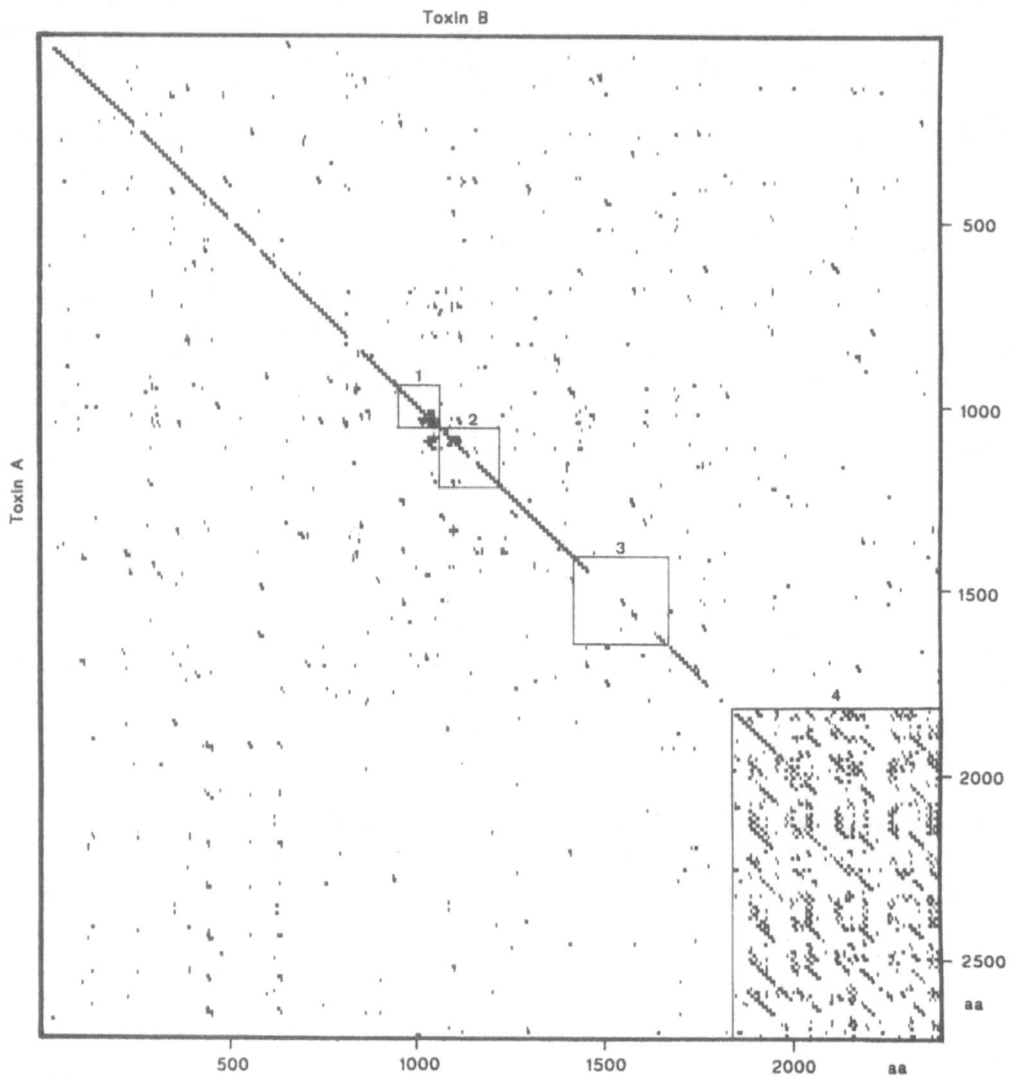

FIGURE 19.11   Dot matrix comparison of ToxA and ToxB aa sequence. ToxA and ToxB have lengths of 2710 and 2366 aa, respectively. The majority of aa are homologous in ToxA and ToxB. The boxed areas are: 1. An accumulation of hydrophobic aa; 2. A cluster of highly conserved, rare aa; 3. A stretch of aa differing between ToxA and ToxB; 4. The C-terminal repetitive domains of both toxins.

preparations are necessary to mediate an intra-gastrical activity of ToxB supports this inter-pretation (Lyerly et al., 1985a). In addition to differences mediated by toxin targeting due to their ligand domains there should be an in vivo counterpart for the different in vitro cytotoxic titers of ToxA and ToxB; even in susceptible cells ToxA is less active than ToxB by a factor of 10^3 (Tucker et al., 1990). The similar qualitative effects of both toxins in in vitro incubations with cell targets (Tucker et al., 1990) are prob-ably the result of the high homology between their toxic N-terminal domains (see Figure 19.11).

The *C. difficile* ToxA and ToxB repeats are homologous to those found in the streptococcal glucosyltransferases. However, the extended homology between ToxA repeats and the repeats of glucosyltransferases (Gtf) of *Streptococci* (Eichel-Streiber and Sauerborn, 1990) is not found in ToxB. Nevertheless, we expect a carbohydrate binding of the ToxB repetitive domain in analogy to that of ToxA and the Gtfs.

Developmental aspects

The C-terminal repetitive domains of ToxA and of some glucosyltransferases are involved in carbohydrate binding. This structure–function relationship seems to find a counterpart in the development of the two *C. difficile* toxin genes and the *gtf* genes of *Streptococci*. Ueda et al. (1988) reported that *gtfC* may have evolved by gene duplication of *gtfB* followed by recombination and mutations. The high homology suggests that toxins A and B similarly may have evolved by gene duplication and thus probably have a common ancestor.

Because *C. sordellii* LT toxin is related to ToxB (Popoff, 1987) and HT to ToxA (Martinez and Wilkins, 1988), these toxins belong to one family of clostridial toxins, classified "lethal toxins" by Hatheway (1990).

19.9 Summary and Conclusions

- ToxA and ToxB are the major virulence factors of *Clostridium difficile*.

- Purification procedures have been elaborated for both toxins; the major contaminant of ToxA is a 39-43 kDa polypeptide, and ToxB is often contaminated with a 150-kDa protein.

- Two ToxA-specific mAbs have been generated that are directed against the immunodominant C-terminal repetitive domain; the best ToxB antibody is a *C. sordellii* LT antiserum that recognizes both N-terminal and C-terminal ToxB epitopes.

- A trisaccharide Galα1-3Galβ1-4GlcNAc reacts with the ToxA C-terminal repetitive domain; this structure is involved in the

ToxA–cell-receptor binding. For ToxB, such a receptor has not yet been found.

- ToxA and ToxB of *C. difficile* are encoded by two separate genes positioned in close proximity on the chromosome; between them is located a third open-reading frame, *utxA*, whose function is unknown.

- The *toxA* gene has a length of 8130 bp, and the corresponding protein has a calculated size of 308.057 daltons and a pI of 5.3; the *toxB* gene has a size of 7098 bp, the deduced protein an M_r of 269.709 and a pI of 4.1.

- ToxA and ToxB are homologous over 63% of their amino acid sequence, with a N-terminal toxic and a C-terminal repetitive ligand domain.

- The C terminus of ToxA is composed of 30 oligopeptides (CROPs) and that of ToxB of 19, both classified into five groups; ToxA CROPs are highly conserved, ToxB CROPs less conserved, and all CROPs are mutually homologous.

- The amino acid sequence of ToxA repeats is homologous to C-terminal repeats found in streptococcal glucosyltransferases (Gtf); functionally, both structures are sites of protein–carbohydrate interaction.

- The structural and functional homologies between ToxA and ToxB indicate that they may have evolved by gene duplication with subsequent recombination and mutations; similar processes have been discussed for GtfB and GtfC.

Acknowledgments. This article is dedicated to Prof. Dr. Paul Klein, whose help and advice made possible my contribution to *Clostridium difficile* molecular biology.

The author thanks Dr. U. Hadding for his support during the beginning of the investigations on *Clostridium difficile* toxins, and Markus Sauerborn for the numerous discussions that were essential for the development of the projects. Some of the data presented herein will be contained in the doctoral theses of Donald Knautz, Sabine Sartingen, Markus Sauerborn, and Jöry Schulze. The work was supported by grants from the Deutsche Forschungsgemeinschaft, the Bundesministerium für Forschung

und Technologie, and the Naturwissenschaft-
lich-medizinisches Forschungszentrum Mainz.

References

Banno T., T. Kobayashi, K. Watanabe, K. Ueno,
and Y. Nozawa. 1981. Two toxins (D-1 and D-2)
of *Clostridium difficile* causing antibiotic-
associated colitis: purification and some characterization. *Biochem. Int.* **2**:629–635.

Banno, Y., T. Kobayashi, and K. Kono. 1984.
Biochemical characterization and biological actions of two toxins (D-1 and D-2) from *Clostridium difficile*. *Rev. Infect. Dis.* **6**:11–20.

Barroso, L.A., S.Z. Wang, C.J. Phelps, J.L. Johnson, and T.D. Wilkins. 1990. Nucleotide Sequence of the *Clostridium difficile* toxin B gene.
Nucleic Acids Res. **18**:4004.

Bartlett, J.G. and S.L. Gorbach. 1977. Pseudomembranous enterocolitis (antibiotic-related
colitis). *Adv. Intern. Med.* **22**:455–476.

Bartlett, J.G., A.B. Onderdonk, R.L. Cisneros, and
D.L. Kasper. 1977. Clindamycin-associated colitis due to a toxin-producing species of *Clostridium* in hamsters. *J. Infect. Dis.* **136**:701–705.

Bartlett, J.G., T.W. Chang, M. Gurwith, S.L. Gorbach, and A.B. Onderdonk. 1978. Antibiotic-
associated pseudomembranous colitis due to
a toxin-producing clostridia. *N. Engl. J. Med.*
298:531–534.

Binz, T., H. Kurazono, M. Wille, J. Frevert, K.
Wernars, and H. Niemann. 1990. The complete
sequence of botulinum neurotoxin A and comparison with other clostridial nerotoxins. *J. Biol.
Chem.* **265**:9153–9158.

Bisseret, F., G. Keith, B. Rihn, I. Amiri, B. Werne-
burg, R. Girardot, O. Baldacini, G. Green, V.K.
Nguyen, and H. Monteil. 1989. *Clostridium difficile* toxin B: characterization and sequence of
three peptides. *J. Chromatogr. Biomed. Appl.*
490:91–100.

Bowman, R.A. and T.V. Riley. 1986. Isolation of
Clostridium difficile from stored specimens and
comparative susceptibility of various tissue culture cell lines to cytotoxin. *FEMS Microbiol. Lett.*
34:31–35.

Bowman, R.A., S.A. Arrow, and T.V. Riley. 1986.
Latex particle agglutination for detecting and
identifying *Clostridium difficile*. *J. Clin. Pathol.*
39:212–214.

Chang, T.W., J.G. Bartlett, S.L. Gorbach, and A.B.
Onderdonk. 1978a. Clindamycin-induced enterocolitis in hamsters as a model of pseudomembranous colitis in patients. *Infect. Immun.*
20:526–529.

Chang, T.W., S.L. Gorbach, and J.G. Bartlett.
1978b. Neutralization of *Clostridium difficile* toxin by *Clostridium sordellii* antitoxins. *Infect. Immun.* **22**:418–422.

Däubener, W., E. Leiser, C.v. Eichel-Streiber, and
U. Hadding. 1988. *Clostridium difficile* toxins A
and B inhibit human immune response in vitro.
Infect. Immun. **56**:1107–1112.

Dailey, D.C. and R. Schloemer. 1988. Identification and characterization of *Clostridium difficile*
promoter element that is functional in *Escherichia coli*. *Gene* **70**:343–350.

Delmée, M., G. Verellen, V. Avesani, and G. Fran-
cois. 1988. *Clostridium difficile* in neonates:
serogrouping and epidemiology. *Eur. J. Pediatr.*
147:36–40.

DiLella, A.G. and S.L.C. Woo. 1987. Hybridization of genomic DNA to oligonucleotide probes
in the presence of tetramethylammonium
chloride. *In: Methods of Enzymology*, Vol. 156,
S.L. Berger and A.P. Kimmel, eds., pp. 447–
451. New York: Academic Press.

Dove, C.H., S.-Z. Wang, S.B. Price, C.J. Phelbs,
D.M. Lyerly, T.D. Wilkins, and J.L. Johnson.
1990. Molecular characterization of the *Clostridium difficile* toxin A gene. *Infect. Immun.*
58(2):480–488.

Ehrich, M., R.L. van Tassell, J.M. Libby, and T.D.
Wilkins. 1980. Production of *Clostridium difficile*
antitoxin. *Infect. Immun.* **28**:1041–1043.

Eichel-Streiber, C.v. and M. Sauerborn. 1990. *Clostridium difficile* toxin A carries a C-terminal
repetitive structure homologous to the carbohydrate binding region of streptococcal glyco-
syltransferases. *Gene* **96**:107–113.

Eichel-Streiber, C.v., R. Laufenberg-Feldmann,
and M. Sauerborn. 1990a. The analysis of the
Clostridium difficile toxin A reveals the 3′ end to
be composed of a 2499 bp repetitive structure.
In: Bacterial Protein toxins, pp. 115–116. Stuttgart: Fischer-Verlag.

Eichel-Streiber, C.v., U. Harperath, D. Bosse, and
U. Hadding. 1987. Purification of two high
molecular weight toxins of *Clostridium difficile*
which are antigenically related. *Microb.
Pathogen.* **2**:307–318.

Eichel-Streiber, C.v., D. Suckau, M. Wachter, and
U. Hadding. 1989. Cloning and characterisation
of overlapping DNA-fragments of the Toxin A
gene of *Clostridium difficile*. *J. Gen. Microbiol.*
135:55–64.

Eichel-Streiber, C.v., R. Laufenberg-Feldmann, S.
Sartingen, J. Schulze, and M. Sauerborn.
1990b. Cloning of *Clostridium difficile* toxin B
gene and demonstration of high N-terminal
homology between Toxin A and B. *Med. Microbiol. Immuno.* **179**:271–279.

Ferretti, J.J., M.L. Glipin, and R.R.B. Russell.
1987. Nucleotide sequence of a glycosyltrans-
ferase gene from *Streptococcus sobrinus* MFe28. *J.
Bacteriol.* **169**:4271–4278.

Finney, J.M. 1893. Gastro-enterostomy for cicatrizing ulcer of the pylorus. *Johns Hopkins Hosp.
Bull.* **11**:53–55.

George, R.H. 1986. The carrier state: *Clostridium difficile. J. Antimicrob. Chemother.* **18**:47–58.

George, R.H., J.M. Simonds, F. Dimock, J.D. Brown, Y. Arabi, N. Shinagawa, M.R.B. Keighly, L. Alexander-Williams, and D.W. Burdon. 1978. Identification of *Clostridium difficile* as a cause of pseudomembranous colitis. *Br. Med. J.* **1**:695.

George, W.L. 1988. Antimicrobial agent-associated diarrhea in adult humans. *In*: *Clostridium difficile*, R.D. Rolfe and S.M. Finegold, eds., pp. 31–44. San Diego: Academic Press.

George, W.L., V.L. Sutter, D. Citron, and S.M. Finegold. 1979. Selective and differential medium for isolation of *Clostridium difficile. J. Clin. Microbiol.* **9**:214–219.

Gill, D.M. 1982. Bacterial toxins: a table of lethal amounts. *Microbiol. Rev.* **46**:86–94.

Gottschalk, G., J.R. Andreesen, and H. Hippe. 1981. The genus *Clostridium* (nonmedical aspects). p. 1767–1803. In: M.P. Starr, H. Stolp, H.G. Trüper, A. Balows, and H.G. Schlegel (Eds), *The prokaryotes, a handbook on the habitants, isolation and identification of bacteria*, Berlin: Springer-Verlag.

Graves, C.M. and J.C. Rabinowitz. 1986. In vivo and in vitro transcription of the *Clostridium pasteurianum* ferredoxin gene. *J. Biol. Chem.* **261**:11409–11415.

Guiliano, M., F. Piemonte, and P.M. Gianfrilli. 1988. Production of an enterotoxin different from toxin A by *Clostridium difficile. FEMS Microbiol. Lett.* **50**:191–195.

Hall, I.C. and E. O'Toole. 1935. Intestinal flora in new-born infants. *Am. J. Dis. Child.* **49**:390–402.

Haslam, S.C., J.M. Ketley, T.J. Mitchell, J. Stephen, D.W. Burdon, and D.C.A. Candy. 1986. Growth of *Clostridium difficile* and production of toxins A and B in complex and defined media. *J. Med. Microbiol.* **21**:293–297.

Hatheway, C.L. 1990. Toxigenic clostridia. *Clin. Microbiol. Rev.* **3**:66–98.

Hollingshead, S.K., V.A. Fischetti, and J.R Scott. 1987. Size variation in group A streptococcal protein is generated by homologous recombination between intragenic repeats. *Mol. Gen. Genet.* **207**:196–203.

Iwen, P.C., S.J. Booth, and G.L. Woods. 1989. Comparison of media for screening of diarrheic stools for the recovery of *Clostridium difficile. J. Clin. Microbiol.* **27**:2105–2106.

Johnson, J.L., C. Phelps, L. Barroso, M.D. Roberts, D.M. Lyerly, and T.D. Wilkins. 1990. Cloning and expression of the toxin B gene of *Clostridium difficile. Curr. Microbiol.* **20**:397–401.

Jones, K.F., S.K. Hollingshead, J.R. Scott, and V.A. Fischetti. 1988. Spontaneous M6 protein size mutants of group A streptococci display variation in antigenic and opsogenic epitopes.

Proc. Nat. Acad. Sci. USA **85**:8271–8275.

Kamiya, S., P.J. Reed, and S.P. Borriello. 1989. Purification and characterisation of *Clostridium difficile* toxin A by bovine thyroglobulin affinity chromatography and dissociation in denaturing conditions or without reduction. *J. Med. Microbiol.* **30**:69–77.

Kamiya S., S. Nakamura, K. Yamakawa, and S. Nishida. 1986. Evaluation of a commercially available latex immunoagglutination test kit for detection of *Clostridium difficile* D-1 toxin. *Microbiol. Immuno.* **30**:177–181.

Keighly, M.R.B., J. Alexander-Williams, Y. Arabi, D. Youngs, D.W. Burdon, K. Shinagawa, H. Thompson, and S. Bentley. 1978. Diarrhoea and pseudomembranous colitis after gastrointestinal operations. *Lancet* **i**:1165–1167.

Ketley, J.M., S.C. Haslam, T.J. Mitchell, J. Stephen, D.C.A. Candy, and D.W. Burdon. 1984. Production and release of toxins A and B by *Clostridium difficile. J. Med. Microbiol.* **18**:385–391.

Ketley, J.M., T.J. Mitchell, S.C. Haslam, J. Stephen, D.C.A. Candy, and D.W. Burdon. 1986. Sporogenesis and toxin A production by *Clostridium difficile. J. Med. Microbiol.* **22**:33–38.

Koronakis, V., E. Koronakis, and C. Hughes. 1989. Isolation and analysis of the C-terminal signal directing export of *Escherichia coli* hemolysin protein across both bacterial membranes. *EMBO J.* **8**:595–605.

Krivan, H.C. and T.D. Wilkins. 1987. Purification of *Clostridium difficile* toxin A by affinity chromatography on immobilized thyroglobulin. *Infect. Immun.* **55**:1873–1877.

Krivan, H.C., G.F. Clark, D.F. Smith, and T.D. Wilkins. 1986. Cell surface binding site for *Clostridium difficile* enterotoxin: evidence for a glycoconjugate containing the sequence Galα1-3Galβ1-4GlcNAc. *Infect. Immun.* **53**:573–581.

Larson, H.E. and A.B. Price. 1977. Pseudomembranous colitis: presence of clostridial toxin. *Lancet* **ii**:1312–1314.

Larson, H.E., J.V. Parry, A.B. Price, D.R. Davis, J. Dolby, and D.A.J. Tyrrell. 1977. Undescribed toxin in pseudomembranous colitis. *Br. Med. J.* **1**:1246–1248.

Larson, H.E., F.E. Barcley, P. Honour, and I.D. Hill. 1982. Epidemiology of *Clostridium difficile* in infants. *J. Infect. Dis.* **146**:727–733.

Libby, J.M. and T.D. Wilkins. 1982. Production of antitoxins to two toxins of *Clostridium difficile* and immunological comparison of the toxins by cross-neutralisation studies. *Infect. Immun.* **35**:374–376.

Lima, A.A.M., D.M. Lyerly, T.D. Wilkins, D.J. Innes, and R.L. Guerrant. 1988. Effects of *Clostridium difficile* toxins A and B in rabbit small and large intestine in vivo and on cultured cells in vitro. *Infect. Immun.* **56**:582–588.

Lucas, F., G.W. Elmer, E. Brot-Laroche, and G. Corthier. 1989. Fixation of *Clostridium difficile* toxin A and cholera toxin to intestinal brush border membranes from axenic and conventional mice. *Infect. Immun.* **57**:1680–1683.

Lyerly, D.M. and T.D. Wilkins. 1986c. Commercial latex test for *Clostridium difficile* toxin A does not detect toxin A. *J. Clin. Microbiol.* **23**:622–623.

Lyerly, D.M., P.E. Carrig, and T.D. Wilkins. 1989. Nonspecific binding of mouse monoclonal antibodies to *Clostridium difficile* toxins A and B. *Curr. Microbiol.* **19**:303–306.

Lyerly, D.M., C.J. Phelps, and T.D. Wilkins. 1985b. Monoclonal and specific polyclonal antibodies for immunoassay of *Clostridium difficile* toxin A. *J. Clin. Microbiol.* **21**:12–14.

Lyerly, D.M., D.W. Ball, J. Toth, and T.D. Wilkins. 1988a. Characterization of cross-reactive proteins detected by culturette brand rapid latex test for *Clostridium difficile*. *J. Clin. Microbiol.* **26**:397–400.

Lyerly, D.M., H.C. Krivan, and T.D. Wilkins. 1988b. *Clostridium difficile*: its disease and toxins. *Clin. Microbiol. Rev.* **1**:1–18.

Lyerly, D.M., J.L. Johnson, S.M. Frey, and T.D. Wilkins. 1990. Vaccination against lethal *Clostridium difficile* enterocolitis with a nontoxic recombinant peptide of Toxin A. *Curr. Microbiol.* **21**:29–32.

Lyerly, D.M., C.J. Phelps, J. Toth, and T.D. Wilkins. 1986a. Characterization of toxins A and B of *Clostridium difficile* with monoclonal antibodies. *Infect. Immun.* **54**:70–76.

Lyerly, D.M., M.D. Roberts, C.J. Phelps, and T.D. Wilkins. 1986b. Purification and properties of toxins A and B of *Clostridium difficile*. *FEMS Microbiol. Lett.* **33**:31–35.

Lyerly, D.M., K.E. Saum, D.K. MacDonald, and T.D. Wilkins. 1985a. Effects of *Clostridium difficile* given intragastrically to animals. *Infect. Immun.* **47**:349–352.

Maniar, A.C., T.W. Williams, and G.W. Hammond. 1987. Detection of *Clostridium difficile* toxin in various tissue culture monolayers. *J. Clin. Microbiol.* **25**:1999–2000.

Martinez, R.D. and T.D. Wilkins. 1988. Purification and characterization of *Clostridium sordellii* hemorrhagic toxin and cross-reactivity with *Clostridium difficile* toxin A (enterotoxin). *Infect. Immun.* **56**:1215–1221.

Meador J., III and R.K. Tweten. 1988. Purification and characterization of toxin B from *Clostridium difficile*. *Infect. Immun.* **56**:1708–1714.

Miles, B.R., J.A. Siders, and S.D. Allen. 1988. Evaluation of a Commercial Latex test for *Clostridium difficile* for reactivity with *Clostridium difficile* and cross-reactions with other bacteria. *J. Clin. Microbiol.* **26**:2452–2455.

Miller, L., L. Gray, E. Beachey, and M. Kehoe. 1988. Antigenic variation among group A streptococcal M proteins. *J. Biol. Chem.* **263**:5668–5673.

Muldrow, L.L., G.C. Ibeanu, N.I. Lee, N.K. Bose, and J. Johnson. 1987. Molecular cloning of *Clostridium difficile* toxin A gene fragment in lambda/gt11. *FEBS Lett.* **213**:249–253.

Niemann, H., B.A. Beckh, T. Binz, U. Eisel, S. Demotz, T. Mayer, and C. Widman. 1988. Tetanus toxin: evaluation of the primary sequence and potential applications. *Zentralbl. Bakteriol. (Suppl.)* **17**:29–38.

Popoff, M.R. 1987. Purification and characterization of *Clostridium sordellii* lethal toxin and cross-reactivity with *Clostridium difficile* cytotoxin. *Infect. Immun.* **55**:35–43.

Pothoulakis, C., L.M. Barone, R. Ely, B. Faris, M.E. Clark, C. Franzblau, and J.T. LaMont. 1986. Purification and properties of *Clostridium difficile* cytotoxin B. *J. Biol. Chem.* **261**:1316–1321.

Price, S.B., C.J. Phelps, T.D. Wilkins, and J.L. Johnson. 1987. Cloning of the carbohydrate-binding portion of the toxin A gene of *Clostridium difficile*. *Curr. Microbiol.* **16**:55–60.

Randall, L.L. and S.J.S. Hardy. 1989. Unity in function in the absence of consensus in sequence: role of leader peptides in export. *Science* **243**:1156–1159.

Rautenberg, P. and F. Stender. 1986. Characterization of immunogenic p230 as the toxin A of *Clostridium difficile*. *FEMS Microbiol. Lett.* **37**:1–7.

Rifkin, G.D., F.R. Fekety, J. Silva, Jr., and R.B. Sack. 1977. Antibiotic-induced colitis implication of a toxin neutralised by *Clostridium sordellii* antitoxin. *Lancet* **ii**:1103–1106.

Rihn, B., J.M. Scheftel, R. Girardot, and H. Monteil. 1984. A new purification procedure for *Clostridium difficile* enterotoxin. *Biochem. Biophys. Res. Commun.* **124**:690–695.

Rothman, S.W., M.K. Gentry, J.E. Brown, D.A. Foret, M.J. Stone, and M.P. Strickler. 1988. Immunochemical and structural similarities in toxin-A and toxin-B of *Clostridium difficile* shown by binding to monoclonal antibodies. *Toxicon* **26**:583–599.

Sauerborn, M. and C.v. Eichel-Streiber. 1990. Nucleotide sequence of *Clostridium difficile* toxin A. *Nucleic Acids Res.* **18**:1629–1630.

Sauerborn, M., M. Hofstetter, R. Laufenberg-Feldmann, and C.v. Eichel-Streiber. 1991. The C-terminal repetitive domain of *Clostridium difficile* Toxin A and its significance for the induction of disease. *Zentralbl. Bakteriol. Hyg.* Supplement, in press.

Schulze, J. and C. Eichel-Streiber. 1990. Cloning and characterization of the *Clostridium difficile* toxin B gene. *Zentralbl. Bakteriol. Mikrobiol. Hyg.* A **273**:126.

Seddon, S.V. and S.P. Borriello, 1989. A chemically defined and minimal medium for *Clostridium difficile. Lett. Appl. Microbiol.* **9**:237–239.

Shiroza, T., S. Ueda, and H.K. Kuramitsu. 1987. Sequence analysis of the *gtfB* gene from *Streptococcus mutans. J. Bacteriol.* **169**:4263–4270.

Sterne, M. and L.M. Wentzel. 1950. A new method for the large scale production of high-titered botulium formal-toxoid types C and D. *J. Immunol.* **65**:175–183.

Sullivan, N.M., S. Pellett, and T.D. Wilkins. 1982. Purification and characterization of toxin A and B of *Clostridium difficile. Infect. Immun.* **35**:1032–1040.

Taylor, N.S., G.M. Thorne, and J.G. Bartlett. 1981. Comparison of two toxins produced by *Clostridium difficile. Infect. Immun.* **34**:1036–1043.

Tedesco, F.J. 1979. Antibiotic associated pseudomembranous colitis with negative proctosigmoidoscopy examination. *Gastroenterology* **77**:295–297.

Tinoco, I., Jr., P.N. Borer, B. Dengler, M.D. Levine, O.C. Uhlenbeck, D.M. Crothers, and J. Gralla. 1973. Improved estimation of secondary structure in ribonucleic acid. *Nature New Biol.* **246**:40–41.

Toma, S., G. Lesniak, M. Magnus, H.-L. Lo, and M. Delmée. 1988. Serotyping of *Clostridium difficile. J. Clin. Microbiol.* **26**:426–428.

Torres, J.F. and I. Lönnroth. 1988. Comparison of methods for the production and purification of toxin A from *Clostridium difficile. FEMS Microbiol. Lett.* **52**:41–47.

Tucker, K.D., P.E. Carrig, and T.D. Wilkins. 1990. Toxin A of *Clostridium difficile* is a potent cytotoxin. *J. Clin. Microbiol.* **28**:869–871.

Ueda, S., T. Shiroza, and H.K. Kuramitsu. 1988. Sequence analysis of the *gtfC* gene from *Streptococcus mutans* GS-5. *Gene* (Amst.) **69**:101–109.

Vernet, A., G. Corthier, F. Dubos-Ramaré, and A.L. Parodi. 1989. Relationship between levels of *Clostridium difficile* toxin A and toxin B and cecal lesions in gnotobiotic mice. *Infect. Immun.* **57**:2123–2127.

Willey, S.H. and J.G. Bartlett. 1979. Cultures for *Clostridium difficile* in stools containing a cytotoxin neutralized by *Clostridium sordellii* antitoxin. *J. Clin. Microbiol.* **10**:880–884.

Wilson, K.H., J.V. Sheagren, and R. Freter. 1985. Population dynamics of ingested *Clostridium difficile* in the gastrointestinal tract of the Syrian hamster. *J. Infect. Dis.* **151**:355–361.

Wilson, K.H., J.V. Sheagren, R. Freter, L. Weatherbee, and L. Lyerly. 1986. Gnotobiotic models for the study of the microecology of *Clostridium difficile* and *E. coli. J. Infect. Dis.* **153**:547–551.

Wren, B., S.R. Heard, and S. Tabaqchali. 1987a. Association between production of toxins A and B and the types of *Clostridium difficile. J. Clin. Pathol.* **40**:1397–1401.

Wren, B.W., C.L. Clayton, P.P. Mullany, and S. Tabaqchali. 1987b. Molecular cloning and expression of *Clostridium difficile* toxin A in *Escherichia coli* K 12. *FEBS Lett.* **225**:82–86.

Wren, B.W., P. Mullany, C. Clayton, and S. Tabaqchali. 1989. Nucleotide sequence of a chloramphenicol acetyl transferase gene from *Clostridium difficile. Nucleic Acids Res.* **17**:4877.

Yamakawa, K., S. Nishida, and S. Nakamura. 1983. C2 toxicity in extract of *Clostridium botulinum* type C spores. *Infect. Immun.* **41**:858–860.

Young, M., N.P. Minton, and W.L. Staudenbauer. 1989. Recent advances in the genetics of clostridia. *FEMS Microbiol. Rev.* **63**:301–326.

20

Gene Cloning and Expression in *Escherichia coli* of Clostridial Sialidases

Peter Roggentin and Roland Schauer

20.1 Introduction

Sialidases (*N*-acetylneuraminosyl-glycohydro-lases; EC 3.2.1.18) regularly occur in higher animals of the deuterostomate lineage from the echinoderms to the vertebrates but have not been found in lower animals and plants (Corfield et al., 1981; Corfield and Schauer, 1982b). They also occur in myxoviruses (Drzeniek et al., 1974), some *Protozoa* (Pereira, 1983; Reuter et al., 1987), fungi (Corfield et al., 1981), and in many different bacterial species (Müller, 1974; Rosenberg and Schengrund, 1976). Strikingly, most of these microorganisms have close contact to higher animals and are commensals or pathogens. Such pathogenic bacteria, for instance, are clostridia, a few species of which were described to secrete sialidase (Fraser, 1978; *Clostridium perfringens*, *C. septicum*, *C. chauvoei*, *C. sordellii*, *C. tertium*). Popoff and Dodin (1985) also detected sialidase in a few strains (8%) of *C. butyricum*. These facts were used for the studies described here.

In vertebrates, sialidases occur together with their substrates, α-glycosidically linked, terminal sialic acids of sialooligosaccharides and sialoglycoconjugates (Corfield et al., 1981; Corfield and Schauer, 1982b), and are responsible for the catabolism of these macromolecules. Sialic acids, comprising a family of about 30 derivatives of neuraminic acid (Schauer, 1985), are usually $\alpha(2\text{-}3)$- or $\alpha(2\text{-}6)$-linked to a subterminal hexose, mostly galactose, of the glycan chains

of glycoproteins or gangliosides; less frequently, they are $\alpha(2\text{-}8)$-linked to another sialic acid in gangliosides or glycoproteins. Sialic acids have also been detected in bacteria, in colominic acid of *Escherichia coli* K as polysialic acid (Barry, 1959), in heteropolysaccharides, e.g., of *Neisseria meningitidis* (Apicella and Robinson, 1970; Liu et al., 1971), *Citrobacter* sp. (Barry et al., 1963; Keleti et al., 1971), group B *Streptococcus* sp. (Baker and Kasper, 1976), *Salmonella* sp. (Kedzierska, 1978), *Corynebacterium* sp. (Dawes et al., 1974), *Actinomyces viscosus* (Jones et al., 1986), *Lactobacillus plantarum* (Sakellaris et al., 1988), and *Rhizobium meliloti* (Defives et al., 1989), or linked to KDO in the core region of lipopolysaccharides, e.g., *Rhodobacter* sp. (Krauss et al., 1988). Sialic acids are often $\alpha(2\text{-}8)$- or $\alpha(2\text{-}9)$-linked in bacterial sialoglycoconjugates. These sialic acids can be hydrolyzed by sialidases, although at reduced rates when compared with the $\alpha(2\text{-}3)$-linkages to galactose of animal glycoconjugates (Corfield and Schauer, 1982a).

For the elucidation of the origin and evolution of sialidases, the following questions are of considerable interest:

1. Why is sialidase and its substrate produced only in very distinct phylae in organisms of recent phylogenetic origin?

2. In which organism did sialidase originate?

3. Are sialidases related to each other or did they develop separately in different phylae?

With regard to the evolution of sialic acids and sialidases, a hypothesis has been put forward by Schauer and Vliegenthart (1982) in which the possibility of gene transfer from animals to the pathogenic microorganisms is considered. As a first step in the investigation on the evolution of sialidase, we examined the enzyme properties and gene sequences of several clostridial sialidases to gain insight into the relatedness of these enzymes and to be able to compare their structures with those of viral sialidases.

20.2 Isolation and Characterization of Clostridial Sialidases

Four sialidases were isolated from three clostridial species in our laboratory. The enzyme secreted by *Clostridium sordellii* G12 was purified to homogeneity from the culture medium by applying ion-exchange chromatography, gel filtration, isoelectric focusing, and fast performance liquid chromatography (FPLC)-techniques (Roggentin et al., 1987). Comparable methods were used for the isolation of the sialidase from *C. chauvoei* NC08596 with the exception of isoelectric focusing, which was replaced by hydrophobic interaction chromatography (Heuermann et al., 1991). *C. perfringens* A99 produces two sialidases, one being secreted ("large" enzyme, 61 kDa) and the other being retained by the cells ("small" enzyme, 43 kDa). The secreted enzyme was purified from the culture medium by the methods described. The cellular sialidase was solubilized by treating the cells with lysozyme and gentle sonication, followed by the same purification procedure. The pure enzymes were used for characterization and in some cases for N-terminal sequencing of the proteins.

The properties of these sialidases, which were characterized as described (Roggentin et al., 1987), are summarized in Table 20.1. Remarkably, relatively large differences were observed for the molecular masses of the proteins. The isoelectric points (pI) of all enzymes are at an acidic pH and in one case (*C. chauvoei*) even at a denaturing pH value. The sialidase from *C. chauvoei* is different from the other three enzymes with respect not only to pI and molecular weight, but also in hydrophobicity, which may give rise to the dimeric form of the native enzyme. Further, its activity is increased by Ca^{2+} and it shows a high affinity to gangliosides and a low specific activity with all substrates tested. An even higher hydrophobicity was observed with the sialidase from *C. septicum*, which has so far prevented its isolation because of aggregation and interaction with other proteins. An increase of activity by the addition of Ca^{2+} is also described for the sialidases of *Vibrio cholerae* (Rosenberg and Schengrund, 1976), *Streptococcus* sp. K 6646 (Berg et al., 1983), and *Bacteroides fragilis* (Von Nicolai, 1975); however, the mechanism of activation is still unknown. The low specific activity of *C. chauvoei* sialidase may be caused by steric hindrance or a higher sensitivity to damage during the isolation procedure. Further, the high temperature optimum (55°C) of the "large," secreted sialidase from *C. perfringens* differs from that of the other enzymes, the optima of which are around 37°C and correspond to the body temperature of higher animals. Similarities among the four sialidases were only found in the acidic pH optima and the preference for α(2-3)-linked sialic acids when compared to α(2-6) and α(2-8)-bonds.

Immunological investigations of clostridial sialidases were carried out with antibodies raised against the partially purified enzymes from *C. perfringens* ("large" enzyme), *C. sordellii*, *C. septicum*, and the cloned "small" sialidase from *C. perfringens* expressed in *E. coli*. The four antisera prepared showed inhibitory potency with the homologous enzymes and, to a lesser extent, with the heterologous clostridial sialidases (Roggentin et al., 1988a). On the other hand, a direct immunoassay developed for the species-specific detection of *C. perfringens* sialidase (Roggentin et al., 1990) and used to measure only those sialidase molecules that remain active after antibody binding exhibits no cross-reactivity with 12 heterologous sialidases. These results indicate that the antigenic sites at

Table 20.1 Comparison of the properies of clostridial sialidases

Sialidase from:	Molecular weight (kDa) Native[a]	Denatured[b]	Sub-units	Hydro-phobic	pI (pH)	Temperature (optimal °C)	pH (optimal)	Stable on freezing	Ca^{2+} require-ment	Substrates preferred α(2−3) > (2−6)	Ganglio-sides	Specific activity (U/mg)
Clostridium perfringens A99												
"Small" enzyme	33	43	1	−	5.15	36	6.1	−	−	+	−	470
"Large" enzyme	63	61	1	−	5.2	55	4.8	n.d.[c]	−	+	n.d.	n.d.
Clostridium sordellii G12	38.5	40	1	−	4.4	36	6.0	+	−	+	(+)	480
Clostridium chauvoei NC08596	300	150	2	+	<3.5	37	5.5	+	+	+	++	25

[a]Determined by gel filtration.
[b]Determined by SDS-PAGE.
[c]n.d., not determined.

the active center are similar in different sialidases, while the remaining parts of the molecules are quite different.

In conclusion, the four clostridial sialidases investigated are heterogenous with regard to various characteristics. From the information obtained by comparison of these properties no conclusions can be drawn as to whether these enzymes are genetically related. We therefore investigated and compared the nucleotide sequences of the genes encoding clostridial sialidases as well as the predicted amino acid sequences.

20.3 Cloning of Sialidase Genes

For the detection of cloned sialidase genes, two strategies were applied. The first is based on labeled DNA probes representing nucleotide sequences that were derived from N-terminal amino acid sequences of the pure enzymes. Sequencing was carried out in the laboratory of Dr. F. Lottspeich (Max-Planck-Institut für Biochemie, D-8033 Martinsried) as described (Eckerskorn et al., 1988). These probes were used in screening the sialidase genes from *C. sordellii* and the "large" enzyme from *C. perfringens*, the investigation on the latter being in progress. For the sialidase gene from *C. septicum*, a DNA probe was successfully applied that represents a conserved sequence present in all bacterial sialidases that have been sequenced (Roggentin et al., 1989). The second strategy is based on the expression of sialidase activity by *E. coli* clones transformed with clostridial DNA fragments ligated to pUC vectors. At the beginning of the project, it was uncertain whether the structural genes of sialidases follow their own promoters and are recognized by the host's RNA polymerase. Fortunately, the sialidase genes cloned so far have their own promoter and are expressed in *E. coli*. Active clones were detected by spraying the plates with the fluorogenic sialidase substrate 4-methylumbelliferyl-α-D-N-acetylneuraminic acid (MU-Neu5Ac) and monitoring a blue-white fluorescence under UV-light at 360 nm.

In addition to the genes coding for the siali-

dases of *C. sordellii* and *C. septicum*, a previously unknown sialidase of *C. perfringens* was detected by cloning and expression of the corresponding gene in *E. coli*. This gene encodes a "small" protein (43 KDa) that is quite different from that of the known "large?" secreted enzyme of *C. perfringens* (61 KDa). The small sialidase was also found in the cells of *C. perfringens*. The three clostridial sialidase genes cloned were sequenced by the method of Sanger et al. (1977), and the nucleotide and the predicted amino acid sequences were compared.

20.4 Comparison of Sialidase Genes and Amino Acid Sequences

The promoter regions upstream from the structural genes of the sialidases of *Clostridium septicum* (Rothe et al., 1991) and *C. sordellii* (Rothe et al., 1989) and of the small sialidase from *C. perfringens* (Roggentin et al., 1988b) were completely different but always contained a Shine–Dalgarno sequence 3 to 9 base pairs upstream of ATG and regions with dyad symmetry probably representing recognition or termination sites.

An alignment of the deduced amino acid sequences of the three sialidases is shown in Figure 20.1. Because of the size of the proteins, which were 111 kDa and 1014 amino acids for the *C. septicum* sialidase, 45 kDa and 404 amino acids for the enzyme of *C. sordellii*, and 43 kDa and 382 amino acids for the small *C. perfringens* sialidase, large, nonaligning regions at both the C- and N-terminal sites were present only in the *C. septicum* protein. However, those regions where the three sequences overlapped showed significant similarities. A leader peptide, containing mainly hydrophobic amino acids, was found in the enzymes of *C. septicum* and *C. sordellii*, sialidases that are known to be secreted (Fraser, 1978). The absence of a leader peptide in the "small" sialidase of *C. perfringens* may explain its presence in the cell where it is believed to have no physiological function. From comparison of the three sequences, values for homology of nucleotides or amino acids were

FIGURE 20.1   Alignment of amino acid sequences of sialidases from *Clostridium septicum* (Rothe et al., 1991), *C. sordellii* (Rothe et al., 1989), and *C. perfringens* (Roggentin et al., 1988b) based on gene structures. Symbols: −, no amino acid at this position; double bold print, homologous amino acids at certain positions; framed, repeated sequences.

```
Ala Asp Arg Tyr Leu Lys Glu Arg Thr Gly Glu Thr Thr Ser Lys Asp Leu Pro Phe Pro  376
 -   -   -   -   -   -   -   -   -   -   -   -   -  Ser Asn Leu Asn Thr
 -   -   -   -   -   -   -   -   -   -   -   -   -  Lys Asn Leu Asp Ile

Glu Gly Ala Val Lys Thr Glu Pro Val Asp Ile Phe Thr Pro Gly Glu Leu Gly Ser Asn  396
Thr Asn Glu Pro Gln Lys Thr Thr Val Phe Asn Lys Asn Asp Asn Thr Trp Asn Ala Gln
Ser His Lys Pro Glu Pro Leu Ile Leu Phe Asn Lys Asp Asn Asn Ile Trp Asn Ser Lys

Asn Phe Arg Ile Pro Ala Leu Tyr Thr Thr Lys Asp Gly Thr Val Leu Ala Ser Ile Asp  416
Tyr Phe Arg Ile Pro Ser Leu Gln Thr Leu Ala Asp Gly Thr Met Leu Ala Phe Ser Asp
Tyr Phe Arg Ile Pro Asn Ile Gln Leu Leu Asn Asp Gly Thr Ile Leu Thr Phe Ser Asp

Val Arg Lys Gly Gly Gly His Asp Ala Pro Asn Asn Ile Asp Thr Gly Ile Lys Arg |Ser  436
Ile Arg Tyr Asn Gly Ala Glu Asp  -  His Ala Tyr Ile Asp Ile Gly Ala Ala Lys |Ser
Ile Arg Tyr Asn Gly Pro Asp Asp  -  His Ala Tyr Ile Asp Ile Ala Ser Ala Arg |Ser

Thr Asp Gly Gly Val Thr Trp| Asp Glu Gly Lys Ile Ile Leu Asp Tyr Pro Gly Ala Ser  456
Thr Asp Asn Gly Gln Thr Trp| Asp  -   -   -   -   -   -   -   -   -   -   -   -
Thr Asp Phe Gly Lys Thr Trp| Ser  -   -   -   -   -   -   -   -   -   -   -   -

Ser Ala Ile Asp Thr Ser Leu Leu Gln Asp Asp Glu Thr Gly Arg Ile Phe Leu Ile Val  476
 -   -   -   -   -   -   -   -   -   -   -   -   -   -   -   -   -   -   -   -
 -   -   -   -   -   -   -   -   -   -   -   -   -   -   -   -   -   -   -   -

Thr His Phe Ala Glu Gly Tyr Gly Phe Gly Asn Ser Lys Thr Gly Ser Gly Tyr Val Glu  496
 -   -   -   -   -   -   -   -   -   -   -   -   -   -   -   -   -   -   -   -
 -   -   -   -   -   -   -   -   -   -   -   -   -   -   -   -   -   -   -   -

Ile Glu Gly Lys Arg Tyr Leu Lys Leu Leu Gly Ala Asn Asp Thr Ile Tyr Thr Val Arg  516
 -   -   -   -   -  Tyr  -  Lys Thr Val Met Glu Asn Asp Arg Ile Asp Ser Thr Phe
 -   -   -   -   -  Tyr  -  Asn Ile Ala Met Lys Asn Asn Arg Ile Asp Ser Thr Tyr

Glu Gly Val Val Tyr Asp Ser Asn Gly Glu Ala Thr Asn Tyr Thr Val Asp Asn Asn Asn  536
Ser Arg Val Met  -  Asp Ser Thr Thr Val Val Thr Asp  -  Thr Gly Arg Ile Ile Leu
Ser Arg Val Met  -  Asp Ser Thr Thr Val Ile Thr Asn  -  Thr Gly Arg Ile Ile Leu

Glu Leu Tyr Glu Asn Gly Asn Arg Ile Gly Asn Val Leu Leu Ser Asn Ser Pro Leu Lys  556
Ile Ala Gly Ser Trp  -  Asn Lys Asn Gly Asn Trp Ala Ser Ser Thr Thr Ser Leu Arg
Ile Ala Gly Ser Trp  -  Asn Thr Asn Gly Asn Trp Ala Met Thr Thr Ser Thr Arg Arg

Val Mat Gly Thr Ser Phe Leu Ser Leu Ile Tyr |Ser Asp Asp Asp Gly Gln Thr Trp| Ser  576
Ser Asp Trp  -  Ser  -  Val Gln Met Val Tyr |Ser Asp Asp Asn Gly Glu Thr Trp| Ser
Ser Asp Trp  -  Ser  -  Val Gln Met Ile Tyr |Ser Asp Asp Asn Gly Leu Thr Trp| Ser

Asp Pro Ile Asp Leu  -   -  Asn Lys  -  Glu Val Lys Thr Asp Trp Met Arg Phe Leu  596
Asp Lys Val Asp Leu Thr Thr Asn Lys Ala Arg Ile Lys Asn Gln Pro Ser Asn Thr Ile
Asn Phe Ile Asp Leu Thr Lys Asp Ser Ser Lys Val Lys Asn Gln Pro Ser Asn Thr Ile

Gly  -   -  Thr Gly Pro Gly Lys Gly His Gln Ile Lys Thr Gly Arg Tyr Ala Gly Arg  616
Gly Trp Leu Ala Gly Val Gly Ser Gly Ile Val Met Ser Asp Gly Thr Ile Val Met Pro
Gly Trp Leu Gly Gly Val Gly Ser Gly Ile Val Met Asp Asp Gly Thr Ile Val Met Pro

Leu Leu Phe Pro Val Tyr Leu Thr Asn Ala Ser Gly Phe Gln Ser Ser Ala Val Ile Tyr  636
Ile Gln Ile Ala Leu Arg Glu Asn Asn Ala Asn Asn Tyr Tyr Ser Ser  -  Val Ile Tyr
Ala Gln Ile Ser Leu Arg Glu Asn Asn Glu Asn Asn Tyr Tyr Ser Leu  -  Ile Ile Tyr

|Ser Asp Asp Asn Gly Ala Thr Trp| Asn Ile Gly Glu Thr Ala Thr Asp Gly Arg Leu Met  656
|Ser Lys Asp Asn Gly Glu Thr Trp| Thr Met Gly Asn Lys Val Pro Asp Pro Lys  -   -
|Ser Lys Asp Asn Gly Glu Thr Trp| Thr Met Gly Asn Lys Val Pro Asn Ser Asn  -   -

Asp Asn Gly Asp Arg Ala Ser Ala Glu Thr Ile Thr Thr Asn Thr Ser Gly Gly Val Gly  676
 -   -   -   -   -   -   -   -   -   -   -   -   -   -   -   -   -   -   -   -
 -   -   -   -   -   -   -   -   -   -   -   -   -   -   -   -   -   -   -   -

Gln Leu Thr Glu Cys Gln Val Val Glu Met Pro Asn Gly Gln Leu Lys Met Phe Met Arg  696
 -   -  Thr Ser Glu Asn Met Val Ile Glu Leu Asp Gly Ala Leu Ile Met Ser Ser Arg
 -   -  Thr Ser Glu Asn Met Val Ile Glu Leu Asp Gly Ala Leu Ile Met Ser Thr Arg
```

FIGURE 20.1 (cont.)

obtained (Table 20.2). A greater similarity on the DNA and protein level between the sialidases of C. sordellii and the "small" enzyme of C. perfringens can be seen, but the similarity is less when comparing these enzymes with C. septicum. This behavior does not conform with the evolutionary distances of the three species delineated from 16S rRNA sequences (Dorsch, 1990), because C. septicum and C. perfringens are more closely related to each other than to C. sordellii.

In each sialidase protein, a short amino acid

```
Asn Thr Gly Gly Asn Ser Gly Arg Val Arg Ile Ala Thr│Ser Phe Asp Gly Gly Ala Thr│716
Asn Asp Gly  -   -  Lys Asn Tyr Arg Ala Ser Tyr Ile│Ser Tyr Asp Met Gly Ser Thr│
Tyr Asp Tyr  -   -  Ser Gly Tyr Arg Ala Ala Tyr Ile│Ser His Asp Leu Gly Thr Thr│

Trp│Glu Asp Asp Val Val Arg Asp Glu Asn Ile Lys Glu Pro Tyr Cys Gln Leu Ser Val 736
Trp│Glu  -   -  Val Tyr Asp Pro Leu His Asn Lys Ile Ser Thr Gly Asn Gly Ser Gly
Trp│Glu  -   -  Ile Tyr Glu Pro Leu Asn Gly Lys Ile Leu Thr Gly Lys Gly Ser Gly

Ile Asn Tyr Ser  -  Gln Lys Ile Asp Gly Lys Asp Ala Ile Ile Phe Ala Ile Pro Asp 756
Cys Gln Gly Ser Phe Ile Lys Val Thr Ala Lys Asp Gly His Arg Leu Gly Phe Ile Asp
Cys Gln Gly Ser Phe Ile Lys Ala Thr Thr Ser Asn Gly His Arg Ile Gly Leu Ile Ser

Ala Asn Tyr Pro Asn Arg Val Asn Gly Thr Val Arg Val Gly Leu Ile Thr Glu Asn Gly 776
Ala  -  Pro Lys Asn Thr Lys Gly Gly Tyr Val Arg Asp Asn Ile Thr Val Tyr Met Ile
Ala  -  Pro Lys Asn Thr Lys Gly Gly Tyr Ile Arg Asp Asn Ile Ala Val Tyr Met Ile

Ser Tyr Glu Asn Gly Glu Pro Arg Tyr Asp  -  Ile Glu Trp Arg Tyr Asn Lys Val Val 796
Asp Phe Asp Asp Leu Ser Lys Gly Ile Arg Glu Leu Cys Ser Pro Tyr Pro Glu Asp Gly
Asp Phe Asp Asp Leu Ser Lys Gly Val Gln Glu Ile Cys Ile Pro Tyr Pro Glu Asp Gly

Ala Pro Gly Thr Tyr Gly Tyr Ser Cys Leu Ser Glu Met Pro Asn Gly Glu Ile Gly Leu 816
Asn Ser Ser Gly Gly Gly Tyr Ser Cys Leu Ser Phe Asn  -  Asp Gly Lys Leu Ser Ile
Asn Lys Leu Gly Gly Gly Tyr Ser Cys Leu Ser Phe Lys  -  Asn Asn His Leu Gly Ile

Phe Tyr Glu Gly Arg Gly Ser Arg Gln Met Ser Phe Thr Arg Met Asn Ile Asp Tyr Leu 836
Leu Tyr Glu Ala Asn Gly Asn Ile Glu Tyr Lys Asp Leu Thr Asp Tyr Tyr Leu Ser Ile
Val Tyr Glu Ala Asn Gly Asn Ile Glu Tyr Gln Asp Leu Thr Asp Tyr Tyr Ser Leu Ile

Lys Ala Asp Leu Leu Gln Asp Val Pro Ala Ala Asn Ile Lys Ser Tyr Thr Thr Asn Ser 856
 -   -   -   -   -   -   -   -   -   -   -   -   -   -   -   -   -   -   -   -
Asn Lys Gln END

Glu Asn Asn Ile Tyr Asp Pro Gly Asp Lys Ile Ser Leu Asn Val Thr Phe Asp Gln Thr 876
 -   -   -   -   -   -   -   -   -   -   -   -   -   -   -   -   -   -   -   -


Val Ser Leu Ile Gly Asp Arg Thr Ile Thr Ala Asp Ile Gly Gly Lys Glu Val Leu Leu 896
 -   -   -   -   -   -   -   -   -   -   -   -   -   -   -   -   -   -   -   -


Thr Leu Ala Asn Ser Lys Gly Gly Ser Glu Tyr Thr Phe Glu Gly Thr Val Pro Ala Asp 916
 -   -   -   -   -   -   -   -   -   -   -   -   -   -   -   -   -   -   -   -


Ile Ser Asn Gly Asn Tyr Thr Ile Thr Ile Lys Gly Lys Ser Gly Leu Lys Ile Val Asn 936
 -   -   -   -   -   -   -   -   -   -   -   -   -   -   -   -   -   -   -   -


Val Val Asn Lys Val Thr Asp Ile Thr Glu Asp Arg Asn Thr Gly Leu Asn Val Gln Val 956
 -   -   -   -   -   -   -   -   -   -   -   -   -   -   -   -   -   -   -   -


Gly Glu Glu Val Gln Ser Val Asp Lys Thr Leu Leu Gln Asp Leu Val Asp Ser Thr Ser 976
 -   -   -   -   -   -   -   -   -   -   -   -   -   -   -   -   -   -   -   -


Asn Leu Ile Lys Glu Asp Tyr Thr Glu Glu Ser Trp Ile Leu Tyr Glu Lys Ala Leu Glu 996
 -   -   -   -   -   -   -   -   -   -   -   -   -   -   -   -   -   -   -   -


Val Ala Asn Lys Phe Leu Val Asn Glu Ile Ala Val Gln Glu Glu Val Asp Ala Ala Lys 1016
 -   -   -   -   -   -   -   -   -   -   -   -   -   -   -   -   -   -   -   -


Pro Thr Leu Glu Asn Ala Tyr Lys END
 -   -   -  Glu Asn Asn Lys Lys Leu Lys END 1027
```

FIGURE 20.1 (cont.)

sequence is found to be repeated at four positions. Examination of these regions in the three clostridial sialidases revealed that five amino acids are conserved in all 12 repeats: Ser-X-Asp-X-Gly-X-Thr-Trp. It is remarkable that this sequence is also present four times in all other bacterial sialidase sequences so far determined (Roggentin et al., 1989; Rothe et al., 1991; Henningsen et al., 1991) and in a less similar form in influenza A virus sialidase (Air et al., 1987) (Figure 20.2). Further, some of the distances between the four blocks of repeated sequences

Table 20.2 Percent proportions of identical nucleotides (upper part) and amino acids (lower part), between sialidases from three clostridial species, after amino acid alignment

	C. septicum	C. sordellii	C. perfringens
C. septicum	100	48.3	43.6
C. sordellii	32.5	100	78.0
C. perfringens	27.5	72.0	100

```
Clostridium        +71  Ala Arg Ser Thr Asp Phe Gly Lys Thr Trp Ser Tyr
perfringens A99    +140 Ile Tyr Ser Asp Asp Asn Gly Leu Thr Trp Ser Asn
"small" sialidase  +208 Ile Tyr Ser Lys Asp Asn Gly Glu Thr Trp Thr Met
                   +255 Tyr Ile Ser His Asp Leu Gly Thr Thr Trp Glu Ile

                   +89  Ala Lys Ser Thr Asp Asn Gly Gln Thr Trp Asp Tyr
Clostridium        +158 Val Tyr Ser Asp Asp Asn Gly Glu Thr Trp Ser Asp
sordellii G12      +226 Ile Tyr Ser Lys Asp Asn Gly Glu Thr Trp Thr Met
                   +273 Tyr Ile Ser Tyr Asp Met Gly Ser Thr Trp Glu Val

                   +431 Lys Arg Ser Thr Asp Gly Gly Val Thr Trp Asp Glu
Clostridium        +563 Ile Tyr Ser Asp Asp Asp Gly Gln Thr Trp Ser Asp
septicum NC0054714 +627 Ile Tyr Ser Asp Asp Asn Gly Ala Thr Trp Asn Ile
                   +700 Ala Thr Ser Phe Asp Gly Gly Ala Thr Trp Glu Asp

                        Lys Arg Ser Thr Asp Gly Gly Lys Thr Trp Ser Ala
Actinomyces             Val Tyr Ser Asp Asp His Gly Lys Thr Trp Gln Ala
viscosus DSM 43798      Ala His Ser Thr Asp Gly Gly Gln Thr Trp Ser Glu
                        Ser Met Ser Cys Asp Asp Gly Ala Ser Trp Thr Thr

                   +263 Arg Thr Ser Arg Asp Gly Gly Ile Thr Trp Asp Thr
Vibrio             +585 Ile Tyr Ser Asp Asp Gly Gly Ser Asn Trp Gln Thr
cholerae 395       +653 Phe Leu Ser Lys Asp Gly Gly Ile Thr Trp Ser Leu
                   +718 Trp Phe Ser Phe Asp Glu Gly Val Thr Trp Lys Gly

                   +71  Ala Arg Ser Thr Asp Gly Gly Lys Thr Trp Asn Lys
Salmonella         +145 Tyr Lys Ser Thr Asp Asp Gly Val Thr Phe Ser Lys
typhimurium LT-2   +210 Ile Tyr Ser Thr Asp  -  Gly Ile Thr Trp Ser Leu
                   +254 Phe Glu Thr Lys Asp Phe Gly Lys Thr Trp Thr Glu

Influenza A virus  +353 Phe Ser Tyr Leu Asp Gly Gly Asn Thr Trp Leu Gly
(H13N9)

Consensus sequence:     Ile Tyr Ser Thr Asp Gly Gly Lys Thr Trp Ser Thr

Frequency (%):       28  36  92  32 100  32 100  20  92  96  36   8
```

FIGURE 20.2 Comparison of conserved amino acid sequences in bacterial and viral sialidases.

are also similar (Figure 20.3). Such conserved and repeated sequences in proteins are thought to have been derived from one sequence by multiplication and to be involved in the three-dimensional structure of the protein (Sambrook et al., 1989). The considerable similarity observed in the primary structures of sialidases suggests that they may have a common origin.

These unexpected similarities of the three clostridial sialidases examined are contrary to the evolution of the species, suggesting that sialidase is encoded by a mobile gene. Although nothing is known about the evolution of sialidases, first insights into possible mechanisms of gene transfer were obtained during physical mapping of the chromosome of C. perfringens by Canard and Cole (1989; Chap-

```
Clostridium
                Start-M - 87 -|AKSTDNGQTWDY|- 57 -|VYSDDNGETWSD|- 56 -|IYSKDNGETWTM|- 35 -|YISYDMGSTWEV|
sordellii G12


Clostridium
                Start-M - 69 -|ARSTDFGKTWSY|- 57 -|IYSDDNGLTWSN|- 56 -|IYSKDNGETWTM|- 35 -|YISHDLGTTWEI|
perfringens A99


Clostridium
                Start-M - 429 -|KRSTDGGVTWDE|- 120 -|IYSDDDGQTWSD|- 52 -|IYSDDNGATWNI|- 61 -|ATSFDGGATWFD|
septicum NC0054714


Salmonella
                Start-M - 69 -|ARSTDGGKTWNK|- 62 -|YKSTDDGVTFSK|- 53 -|IYSTD-GITWSL|- 32 -|FETKDFGKTWTE|
typhimurium LT-2


Vibrio
                Start-M - 261 -|RTSRDGGITWDT|- 310 -|IYSDDGGSNWQT|- 56 -|FLSKDGGITWSL|- 53 -|WFSFDEGVTWKG|
cholerae 395
```

FIGURE 20.3   Comparison of distances (number of amino acids) between conserved regions in five bacterial sialidases.

ter 18 this volume). Probing with the gene encoding the "small" sialidase of *C. perfringens* revealed its position near the attachment site of a lysogenic phage. If this phage or an "earlier" phage transported the sialidase gene, many unexplained facts would become understandable. This would contribute to an explanation as to why only one species, i.e., *C. sordellii*, of closely related species such as *C. sordellii* and *C. bifermentans* produces this enzyme (Roggentin et al., 1985), and why even strains of one species can differ in this property, as described for *C. butyricum* (Popoff and Dodin, 1985) and *Salmonella typhimurium* (Vimr et al., 1988). These observations indicate that the process of gene transfer is still in progress. The mechanism of transfer by phages may explain the presence of sialidases in different bacterial species, but the connection with viruses, protozoa, fungi, and higher animals remains unknown. It is possible that the sialidase first evolved in animals and then spread to microorganisms via transfection with DNA liberated during infectious processes. To confirm this hypothesis, we are presently attempting to clone and sequence sialidases of higher animals.

Acknowledgments. We thank the Deutsche Forschungsgemeinschaft for support of this project (grant Scha 202/13-1).

References

Air, G.M., R.G. Webster, P.M. Colman, and W.G. Laver. 1987. Distribution of sequence differences in influenza N9 neuraminidase of tern and whale viruses and crystallisation of the whale neuraminidase complexed with antibodies. *Virology* **160**:346–354.

Apicella, M.A. and J.A. Robinson. 1970. Physicochemical properties of *Neisseria meningitidis* group C and Y polysaccharide antigens. *Infect. Immun.* **2**:392–397.

Baker, C.J. and D.L. Kasper. 1976. Identification of sialic acid in polysaccharide antigens of group B *Streptococcus. Infect. Immun.* **13**:284–288.

Barry, G.T. 1959. Detection of sialic acid in various *Escherichia coli* strains and in other species of bacteria. *Nature* (London) **183**:117–118.

Barry, G.T., F. Chen, and E. Roark. 1963. Isolation of N-acetylneuraminic acid and 4-oxonorleucine from a polysaccharide obtained from *Citrobacter freundii. J. Gen. Microbiol.* **33**:97–116.

Berg, J.O., L. Lindqvist, G. Andersson, and C.E. Nord. 1983. Neuraminidase in *Bacteroides fragilis. Appl. Environ. Microbiol.* **46**:75–80.

Canard, B. and S.T. Cole. 1989. Genome organization of the anaerobic pathogen *Clostridium perfringens. Proc. Natl. Acad. Sci. USA* **86**:6676–6680.

Corfield, A.P. and R. Schauer. 1982a. Metabolism of sialic acids. *In: Sialic Acids—Chemistry, Metabolism and Function*, Cell Biology Monographs, Vol. 10, R. Schauer, ed., pp. 195–261. Wien: Springer

Corfield, A.P. and R. Schauer. 1982b. Occurrence

of sialic acids. *In: Sialic Acids—Chemistry, Metabolism and Function*, Cell Biology Monographs, Vol. 10, R. Schauer, ed., pp. 5–50. Wien: Springer.

Corfield, A.P., J.-C. Michalski, and R. Schauer. 1981. The substrate specificity of sialidases from microorganisms and mammals. *In: Sialidases and Sialidosis, Perspectives in Inherited Metabolic Diseases*, G. Tettamanti, P. Durand, and S. DiDonato, eds., pp. 3–70. Milano: Edi Ermes.

Dawes, J., S.J. Tuach, and W.H. McBride. 1974. Properties of an antigenic polysaccharide from *Corynebacterium parvum*. *J. Bacteriol.* **120**:24–30.

Defives, C., R. Bouslamti, J.C. Derieux, O. Kol, and B. Fournet. 1989. Characterization of sialic acids containing lipopolysaccharide from *Rhizobium meliloti* M 11 S. *FEMS Lett.* **57**:203–208.

Dorsch, M. 1990. Ein Beitrag zur Phylogenie der Gram-positiven Eubakterien. Dissertation, Kiel.

Drzeniek, R., R. Rott, H.E. Müller, R. Carubelli, and W. Gielen. 1974. Myxovirus neuraminidases. *Behring Inst. Mitt.* **55**:1–10.

Eckerskorn, C., W. Mewes, H. Goretzki, and F. Lottspeich. 1988. A new siliconized-glass fiber as support for protein-chemical analysis of electro-blotted proteins. *Eur. J. Biochem.* **176**:509–519.

Fraser, A.G. 1978. Neuraminidase production by Clostridia. *J. Med. Microbiol.* **11**:269–280.

Henningsen, M., P. Roggentin, and R. Schauer. 1991. Cloning and sequencing of an *Actinomyces viscosus* sialidase gene. *Glyoconj. J.* **8**:151.

Heuermann, D., P. Roggentin, R.G. Kleineidam, and R. Schauer. 1991. Purification and characterization of a sialidase from *Clostridium chauvoei* NC08596. *Glycoconjugate J.* **8**:95–101.

Jones, A.H., C.C. Lee, B.J. Moncla, M.R. Robinovitch, and D.C. Birdsel. 1986. Surface localization of sialic acid on *Actinomyces viscosus*. *J. Gen. Microbiol.* **132**:3381–3391.

Kedzierska, B. 1978. N-Acetylneuraminic acid: a constituent of the lipopolysaccharide of *Salmonella toucra*. *Eur. J. Biochem.* **91**:545–552.

Keleti, J., O. Lüderitz, D. Mlynarcik, and J. Sedlak. 1971. Immunochemical studies on *Citrobacter* O antigens (Lipopolysaccharides). *Eur. J. Biochem.* **20**:237–244.

Krauss, J.H., G. Reuter, R. Schauer, J. Weckesser, and H. Mayer. 1988. Sialic acid-containing lipopolysaccharides in purple nonsulfur bacteria. *Arch. Microbiol.* **150**:584–589.

Liu, T.Y., E.C. Gotschlich, F.T. Dunne, and E.K. Jonssen. 1971. Studies on the meningococcal polysaccharides II. Composition and chemical properties of the group B and group C polysaccharide. *J. Biol. Chem.* **246**:4703–4712.

Müller, H.E. 1974. Neuraminidases of Bacteria and Protozoa and their pathogenic role. *Behring Inst. Mitt.* **55**:34–56.

Pereira, M.E.A. 1983. A developmentally regulated neuraminidase activity in *Trypanosoma cruzi*. *Science* **219**:1444–1446.

Popoff, M.R. and A. Dodin. 1985. Survey of neuraminidase production by *Clostridium butyricum*, *Clostridium beijerinckii* and *Clostridium difficile* strains from clinical and nonclinical sources. *J. Clin. Microbiol.* **22**:873–876.

Reuter, G., R. Schauer, R. Prioli, and M.E.A. Pereira. 1987. Isolation and properties of a sialidase from *Trypanosoma rangeli*. *Glycoconjugate J.* **4**:339–348.

Roggentin, P., W. Berg, and R. Schauer. 1987. Purification and characterization of sialidase from *Clostridium sordellii* G12. *Glycoconjugate J.* **4**:349–359.

Roggentin, P., G. Gutschker-Gdaniec, R. Schauer, and R. Hobrecht. 1985. Correlative properties for a differentiation of two *Clostridium sordellii* phenotypes and their distinction from *Clostridium bifermentans*. *Zentralbl. Bakteriol. Mikrobiol Hyg. 1 Abt. Orig. A Med. Mikrobiol. Infektionskr. Parasitol.* **260**:319–328.

Roggentin, P., G. Gutschker-Gdaniec, R. Hobrecht, and R. Schauer. 1988a. Early diagnosis of clostridial gas gangrene using sialidase antibodies. *Clin. Chim. Acta* **173**:251–262.

Roggentin, P., B. Rothe, F. Lottspeich, and R. Schauer. 1988b. Cloning and sequencing of a *Clostridium perfringens* sialidase gene. *FEBS Lett.* **238**:31–34.

Roggentin, T., R.G. Kleineidam, P. Roggentin, and R. Schauer. 1990. An improved immunoassay for a quick detection of *Clostridium perfringens*. *Forum Mikrobiol.* **13**:92.

Roggentin, P., B. Rothe, J.B. Kaper, J. Galen, L. Lawrisuk, E.R. Vimr, and R. Schauer. 1989. Conserved sequences in bacterial and viral sialidases. *Glycoconjugate J.* **6**:349–353.

Rosenberg, A. and C.L. Schengrund. 1976. Sialidases. *In: Biological Roles of Sialic Acid*, A. Rosenberg, and C.L. Schengrund, eds. New York: Plenum.

Rothe, B., B. Rothe, P. Roggentin, and R. Schauer. 1991. The sialidase gene from *Clostridium septicum*: Cloning, sequencing, expression in *Escherichia coli* and identification of conserved sequences in sialidases and other proteins. *Mol. Gen. Genet.* **226**:190–197.

Rothe, B., P. Roggentin, R. Frank, H. Blöcker, and R. Schauer. 1989. Cloning, sequencing and expression of a sialidase gene from *Clostridium sordellii* G12. *J. Gen. Microbiol.* **135**:3087–3096.

Sakellaris, G., F.N. Kolisis, and A.E. Evangelopoulos. 1988. Presence of sialic acids in *Lactobacillus plantarum*. *Biochem. Biophys. Res. Commun.* **155**:1126–1132.

Sambrook, J., E.F. Fritsch, and T. Maniatis. 1989. *Molecular Cloning—A Laboratory Manual*, 2nd Ed. Vol. 2. Cold Spring Harbor, New York:

Cold Spring Harbar Laboratory.

Sanger, F., S. Nicklen, and A.R. Coulson. 1977. DNA sequencing with chain terminating inhibitors. *Proc. Natl. Acad. Sci. USA* **74**:5463–5467.

Schauer, R. 1985. Sialic acids and their role as biological masks. *Trends Biochem. Sci.* **10**:357–360.

Schauer, R. and J.F.G. Vliegenthart. 1982. Introduction. *In: Sialic Acids—Chemistry, Metabolism and Function*, Cell Biology Monographs, Vol. 10. R. Schauer ed., pp. 1–3. Wien: Springer.

Vimr, E.R., L. Lawrisuk, J. Galen, and J.B. Kaper. 1988. Cloning and molecular analysis of neuraminidase genes from *Vibrio cholerae* and *Salmonella typhimurium*. *In: Sialic Acids 1988, Proceedings of the Japanese-German Symposium on Sialic Acids*, R. Schauer, and T. Yamakawa, eds., pp. 140–141. Kiel: Verlag Wissenschaft und Bildung.

Von Nicolai, H. 1975. Vergleichende Untersuchungen zur Substratspezifität mikrobieller Neuraminidasen. Bonn: Habilitationsschrift.

21

Cloning and Expression of *Clostridium acetobutylicum* Genes Involved in Carbohydrate Utilization

Peter Verhasselt and Jos Vanderleyden

21.1 Introduction

The bioconversion of plant raw materials into refined chemicals has become a stimulating field of biotechnology. The expanding studies on *Clostridium acetobutylicum*, a gram-positive anaerobic bacterium that produces solvents such as acetone, butanol, and ethanol from carbohydrates, illustrate this interest (Zeikus, 1980; Blaschek, 1986; Jones and Woods, 1986a; McNeil and Kristiansen, 1986; Rogers, 1986; Bahl and Gottschalk, 1988; Young et al., 1989).

The first industrial production of acetone and butanol was by fermentation of starch, usually from potatoes, wheat, or maize. These biotransformations were run according to the Weizmann process (Weizmann, 1915, 1919). Because of difficulties with cereals as a fermentation substrate and the low prices of molasses, which were initially regarded as waste products, investigations on the use of molasses soon began. The acetone-butanol-ethanol (ABE) fermentation of molasses rapidly superseded the previous grain process. After World War II, the rapidly expanding petrochemical industry began to produce acetone and butanol at low prices from oil derivatives. This, together with increasing prices for carbohydrate substrates, resulted in closing down the ABE fermentation plants in Western countries.

Because of the oil crisis in 1973, the synthetic route to butanol became much more expensive, resulting in a renewed interest for the ABE fermentation. However, so far this interest has been sustained only at the laboratory level; the economic feasibility of the ABE fermentation still depends highly on the cost of raw materials. In 1980 these costs were estimated at 57% to 116% of the selling price of the solvents, using conventional carbohydrate sources. The high cost of conventional substrates combined with the ability of *C. acetobutylicum* to ferment many different carbohydrates has stimulated research into the use of alternative substrates of both cellulosic or noncellulosic nature.

During recent years much progress has been made in molecular biotechnology. The application of gene technology to *C. acetobutylicum* should allow an amplification of substrate degradation capabilities and an extension of the spectrum of usable substrates. This would make future fermentations more flexible in available substrates, thus allowing the exploitation of relatively short-term surpluses and the use of the least expensive source of carbohydrates.

This chapter focuses on *Clostridium acetobutylicum* genes related to carbohydrate metabolism.

21.2 Utilization of Monosaccharides

Early studies on the utilization of various carbohydrates in the ABE fermentation by *Clostridium acetobutylicum* indicated that nearly all sugars occurring in plant biomass can be fermented (Robinson, 1922; Nakhmanovich and

$$\text{Enzyme I} + \text{PEP} \rightleftarrows \text{P-enzyme I} + \text{pyruvate}$$

$$\text{P-enzyme I} + \text{HPr} \rightleftarrows \text{P-HPr} + \text{enzyme I}$$

$$\text{P-HPr} + \text{enzyme III} \rightleftarrows \text{HPr} + \text{P-enzyme III}$$

$$\text{P-enzyme III} + \text{glucose}_{ext} \rightleftarrows \text{enzyme III} + \text{glucose 6-P}_{int}$$

$$\text{P-enzyme III} + \text{fructose}_{ext} \rightleftarrows \text{enzyme III} + \text{fructose 1-P}_{int}$$

FIGURE 21.1   Uptake of glucose and fructose by phosphotransferase system.

Reaction	Enzyme
maltose + phosphate $\rightleftarrows$ glucose + glucose 1-P	maltose phosphorylase
glucose + ATP $\rightleftarrows$ glucose 6-P + ADP + H$^+$	glucose kinase

FIGURE 21.2   Uptake of maltose by *Clostridium acetobutylicum*.

Shcheblykina, 1959; Compere and Griffith, 1979). Hydrolysis of wood components for example results in a mixture of lignin, oligosaccharides, and monosaccharides, including pentoses and hexoses. Except for lignin, all these compounds can be converted into solvents by *C. acetobutylicum*. Oligosaccharides can be converted into monosaccharides by endogenous or exogenous enzyme activities. While pentose sugars are somewhat less favored, hexose sugars and particularly glucose serve as the best substrates for microbial growth and activity. However, the highest amount of fermented sugars in batch cultures is achieved with mixtures of glucose and xylose. This successful fermentation results from a fast growth and a rapid transition to solventogenesis on glucose and from a strong acid reconsumption during xylose metabolism after glucose exhaustion (Fond et al., 1986).

Uptake of sugars by *C. acetobutylicum* is not very well documented. Glucose and fructose appear to be taken up by a phosphotransferase system as in *Escherichia coli* and *Bacillus subtilis* (Figure 21.1). A first enzyme from this system (enzyme I) transfers the phosphoryl group from phosphoenolpyruvate (PEP) to a protein HPr. The phosphorylated form of HPr transfers its phosphoryl group to a second enzyme, called enzyme III, which interacts with enzyme II. Enzyme II is a membrane protein that forms a channel through the membrane and catalyzes the transfer of the phosphoryl group from enzyme III to the sugar that is translocated from the culture medium to the cytoplasm. In *E. coli* and *B. subtilis*, enzymes I and III are cytoplasmic; in *C. acetobutylicum*, enzyme III is found both in the cytoplasm and the membrane (Bahl and Gottschalk, 1988; W. Mitchell, personal communication).

It is suggested that other substrates are taken up by the symport mechanism driven by the transmembrane proton gradient (Bahl and Gottschalk, 1988). The substrate, together with protons, enter the cell via a substrate-specific permease. Disaccharides such as sucrose, maltose, or lactose that enter the cell in this way are cleaved by appropriate phosphorylases (Figure 21.2). Once the substrates are taken up by the cell, they are fed into the metabolic pathways that lead to cell growth, cell division, and acid and solvent production.

For the utilization of hexose sugars, *C. acetobutylicum*, like all the clostridia, uses the Embden–Meyerhof pathway almost exclusively (Rogers, 1986). The net result of the glycolysis is the conversion of 1 mole glucose into 2 moles pyruvate, generating 2 moles ATP and 2 moles NADH. The anaerobic catabolism of pentose sugars is accomplished by the pentose phosphate pathway (Volesky and Szczesny, 1983) (Figure 21.3). Another possible pathway for degrading pentose sugars usually exists in the phosphorolytic cleavage of xylose 5-P into

Reaction	Enzyme
ribulose 5-P $\rightleftarrows$ ribose 5-P	Phosphopentose isomerase
ribulose 5-P $\rightleftarrows$ xylulose 5-P	Phosphopentose epimerase
xylulose 5-P + ribose 5-P $\rightleftarrows$ glyceraldehyde 3-P + sedoheptulose 7-P	transketolase
glyceraldehyde 3-P + sedoheptulose 7-P $\rightleftarrows$ fructose 6-P + erythrose 4-P	transaldolase
xylulose 5-P + erythrose 4-P $\rightleftarrows$ fructose 6-P + glyceraldehyde 3-P	transketolase

FIGURE 21.3   Pentose phosphate pathway.

acetyl-P and glyceraldehyde 3-P. However, as in most other obligate anaerobes, the phosphoketolase that catalyzes this reaction is not present in *C. acetobutylicum* (Rogers, 1986). As a net result, 3 mole pentose 5-P are converted into 2 mole fructose 6-P and 1 mol glyceraldehyde 3-P and these products enter the glycolytic pathway. For ribose, fermentation of 3 moles pentose to pyruvate provides the cell with 5 mol ATP and 5 mol NADH. To use xylose, the microorganism must convert it into xylulose and then to xylulose 5-P. There are no data available for the enzymes catalyzing these reactions in *C acetobutylicum*. In *B. subtilis*, this pathway involves two enzymes, the D-xylose isomerase and the xylulose kinase (Klier and Rapoport, 1988).

Although metabolism of most sugars gives rise to the production of mainly acetate, butyrate, ethanol, acetone, and butanol, utilization of some substrates leads to the production of other end products. In the case of arabinose, propionic and valeric acids were also detected (Forsberg et al., 1987). Utilization of rhamnose leads to the production of 1,2-propanediol, propionic acid, and *n*-propanol in addition to the expected acids and alcohols. In *E. coli*, L-rhamnose is metabolized in a separate pathway, yielding dihydroxyacetone-P, which is further metabolized in the Embden–Meyerhof pathway, and L-lactaldehyde, which is converted to L-1,2-propanediol under anaerobiosis. Propanediol probably, is produced by the same pathway in the clostridia (Forsberg et al., 1987).

Intermediates of sugar metabolism evidently can be fermented if they can enter the cell. *Clostridium acetobutylicum* can utilize pyruvate as the sole carbon source, but only acids are produced. When pyruvate is added to a glucose fermentation medium in the acidogenic phase, glucose utilization decreases and no solvents are produced throughout the fermentation. Addition of pyruvate during solventogenesis yields simultaneous utilization of both pyruvate and glucose, but solventogenesis ceases and acidogenesis starts again (Junelles et al., 1987; Janati-Idrissi et al., 1989). The former illustrates that pyruvate is a key intermediate in the carbohydrate metabolism of *C. acetobutylicum*. In the oxidative decarboxylation of pyruvate to acetyl-CoA, electrons are transferred to ferredoxin, catalyzed by pyruvate ferredoxin oxidoreductase. The NADH formed by the glycolysis and pentose phosphate pathway is also reoxidized by transfer of the electrons to ferredoxin by NADH-ferredoxin oxidoreductase. NADPH-ferredoxin oxidoreductase catalyzes the electron transfer from ferredoxin to NADP$^+$, the only way in which *C. acetobutylicum* can generate NADPH required for biosynthesis (Jungermann et al., 1973). However, ferredoxin is mainly reoxidized by electron transfer from reduced ferredoxin to protons, generating molecular hydrogen in the culture medium: an iron-containing hydrogenase catalyzes this reaction (Jungermann et al., 1973; Petitdemange et al., 1976; Adams et al., 1980). This electron transfer system plays a key role in the *C. acetobutylicum* metabolism, and ferredoxin seems to be the limiting factor in pyruvate oxidation (Volesky and Szczesny, 1983).

During the first growth phase of *C. acetobutylicum*, acids are produced, generating additional ATP. However, acidogenesis consumes only a portion of the reducing power produced during sugar catabolism. The electron carriers are regenerated by production of molecular hydrogen. Because of release of acids, the pH of the medium decreases and it becomes progressively more difficult to reoxidize NADH to NAD⁺. Excess electrons are no longer used in the production of hydrogen but are taken up in the reassimilation of the produced acetic and butyric acid, leading to the production of solvents (Jones and Woods, 1986a) (see also Chapter 22, this volume). It remains uncertain whether decrease in hydrogen production in the solventogenic phase results from the regulation of hydrogenase production or from inactivation of enzyme activity (Andersch et al., 1983; Kim and Zeikus, 1985).

The observation that solvent production is associated with decreased hydrogen production stimulated research on hydrogenase activity in C. acetobutylicum. Hydrogenase activity can be inhibited by increasing the partial pressure of hydrogen in the medium or by gassing with carbon monoxide. Under appropriate conditions, these modulations of hydrogenase activity increase solvent production (Kim et al., 1984; Datta and Zeikus, 1985; Doremus et al., 1985; Yerushalmi and Volesky, 1985; Yerushalmi et al., 1985; Maddox, 1989). In fermentations with CO sparging, lactic acid becomes the predominant organic acid in the acidogenic phase and is produced by the reduction of pyruvate. Inhibition of hydrogenase activity results in a decreased regeneration of oxidized ferredoxin and in a decreased rate of oxidative decarboxylation of pyruvate. Under these conditions, reduction of pyruvate is catalyzed by lactate dehydrogenase. The gene encoding this enzyme has recently been cloned (Young et al., 1989). During solventogenesis lactate is reassimilated and fermented into solvents (Datta and Zeikus, 1985).

In spite of the key role of hydrogenase activity in directing the metabolism towards solvent production, clostridial hydrogenase genetics are not well developed. The hydrogenase gene from C. butyricum has been cloned and expressed in E. coli HK16, a hyd mutant of E. coli C600 (Karube et al., 1983), but no further characterization or nucleotide sequence is yet available. A gene from Clostridium pasteurianum encoding another important protein involved

in electron flow, ferredoxin, has been cloned in E. coli HB101 and the nucleotide sequence has been determined (Graves et al., 1985) (see also Chapter 25, this volume).

Another phenomenon besides the altered electron flow that is attended by solvent production is intracellular and extracellular polymer synthesis (Junelles et al., 1989). The extracellular polymer is most likely a highly acetylated polysaccharide and is produced during growth and acidogenesis, when butanol is formed from glucose and reassimilated butyrate. It is suggested that this extracellular polymer can act as a sink for storage of non-reduced compounds when excess reducing power is required (Häggström and Förberg, 1986). For instance, it is observed that this polymer is reutilized when butyrate and butanol are produced simultaneously.

The intracellular polymer is granulose, a glycogen-like reserve polymer composed of linear chains of $\alpha(1\text{-}4)$-linked D-glucopyranose units. In C. acetobutylicum, P262 production of granulose is dependent on ADP-glucose pyrophosphorylase and granulose synthase. These enzyme activities and corresponding granulose accumulation were only detected in cells at the end of the exponential growth phase, which is also accompanied by the onset of solvent production. Further, isolation of pleiotropic mutants defective in solvent production and granulose synthesis and resumption of these processes in revertants of these mutants suggest a common pathway of regulation (Reysenbach et al., 1986). These regulatory components are not known, but both processes have been shown to be associated with the accumulation of threshold concentrations of acids produced during acidogenic growth phase (Häggström and Molin, 1980; Gottschal and Morris, 1981, 1982; Bahl et al., 1982; Martin et al., 1983; Yu and Saddler, 1983; Long et al., 1984). During sporulation, granulose is reconsumed, supporting the suggestion that granulose may act as a source of carbon and energy during sporulation. However, evidence has been obtained that granulose is not essential for sporulation, because mutants defective in granulose synthesis are still able to produce mature spores, although at a lower frequency.

Granulose production therefore may result in more effective sporulation under appropriate conditions (Reysenbach et al., 1986).

21.3   Utilization of Starchy Materials

Clostridium acetobutylicum degrades starch by the action of two types of amylase, α-amylase and glucoamylase. α-Amylase [α-D-(1,4)-glucan glucanohydrolase (EC 3.2.1.1)] is an endoacting enzyme that hydrolyzes internal α-1,4 bonds and can by-pass α-1,6 linkages. Oligosaccharides and low molecular weight dextrins are produced. Glucoamylase [α-D-(1,4)-glucan glucanohydrolase (EC 3.2.1.3)] is an exoacting amylase. It splits off glucose monomers from the nonreducing ends and degrades both α-1,4 and α-1,6 linkages (Ensley et al., 1975). The combined action of both enzymes enables *C. acetobutylicum* to degrade starch completely to glucose. Surprisingly, these enzymes have long not been studied (Bahl and Gottschalk, 1988), although the first ABE fermentations were based on starchy materials (see section 21.1.). Moreover, it was suggested that amylase activity may be a limiting factor of solvent production by *C. acetobutylicum* ATCC 824 on starch (Lin and Blaschek, 1983; Hermann et al., 1985).

Ensley et al. (1975) investigated the influence of carbon sources on the synthesis of α-amylase and glucoamylase by *Clostridium acetobutylicum* NRRL B 592. Amylase production was monitored in media containing 0.8% yeast extract, 0.05% sodium thioglycolate, and 0.5% glucose, maltose, fructose or starch. When starch was the carbon source, 0.01% glucose was added with 0.49% starch to decrease the length of the lag phase. At specific intervals, samples were centrifuged and the amylolytic enzyme activities in the supernatant and in the pellet were measured. It was found that between 67% and 78% of glucoamylase activity and from 56% to 59% of α-amylase activity is secreted, while the remaining activities are loosely associated with the cells. Production of both enzymes occurs in the logarithmic growth phase of *C. acetobutylicum*, together with high glucose consumption and acidogenesis. Because spore formation

starts later, it is suggested that neither formation nor secretion of amylolytic enzymes is attributed to sporulation. Synthesis of amylolytic enzymes and growth are both influenced by carbon source. While glucose, maltose, and fructose give comparable growth rates, growth is slowest with starch as sole carbon source as demonstrated by the growth curves (Ensley et al., 1975). α-Amylase production is highest with starch as substrate and decreases with glucose, maltose, and fructose, respectively. High levels of glucoamylase occur with growth on glucose and maltose, while lower levels are attained with growth on fructose and starch (Ensley et al., 1975).

It is concluded that both enzyme activities are under separate biosynthetic control, because glucoamylase activity is induced by growth on glucose and α-amylase activity by growth on starch. Rather than starch itself, hydrolytic products of starch, probably produced by small amounts of constitutively expressed α-amylase activity, are responsible for the induction of α-amylase. Induction of glucoamylase by glucose is rather surprising because the presence of this readily usable substrate does not require secretion of starch-hydrolyzing enzymes. However, the induction of glucoamylase on glucose is still a conflicting issue, because Chojecki and Blaschek (1986) reported absence of all amylolytic enzymes on glucose (see also following). Further, hydrolytic enzyme synthesis is generally repressed by glucose, a regulation mechanism that is called catabolite repression. In *Aspergillus niger* induction of glucoamylase has been attributed to maltose or isomaltose (Barton et al., 1972).

Chojecki and Blaschek (1986) investigated the influence of carbon source on the formation of amylolytic enzymes by the butanol-tolerant strain SA-1 of *C. acetobutylicum*. For growth on starch, they obtained values for total activities of α-amylase and glucoamylase activity that were respectively 3.7- and 28.5 fold higher than the values obtained by Ensley et al. (1975). While Ensley et al. observed roughly simultaneous production of both enzymes, Chojecki and Blaschek (1986) observed a lag of about 15 h between production of α-amylase and glucoamylase. They suggest that breakdown pro-

```
ATGCACACAATTTCTTAAAAGAAGTGCAATAATATATTTAAAATTCCTAATGTTATATTTTATAATATTA
TATTAATATTAACATTACTTTATTTTTTTTCAATATATGCTATAATATGTATGAAAGGGAGGTATTATTA
         -35                                      -10     +1    S.D.
```

ATG	AAA	AAA	ATT	TTT	ACT	AAT	GTA	TTA	TTA	GTA	GGC	TTT	TTA	GTC
M	K	K	I	F	T	N	V	L	L	V	G	F	L	V

TTT	ATT	AGT	GCA	ATT	TTA	TAT	GGA	CAC	AGT	GTA	CAA	GCT	AGT	GAT
F	I	S	A	I	L	Y	G	H	S	V	Q	A	S	D

FIGURE 21.4 Nucleotide sequence and deduced amino acid sequence of 5′ end of *Clostridium aceto-butylicum* α-amylase gene. The −10 and −35 regions and Shine–Dalgarno sequence, underlined; transcription start, +1; putative signal sequence processing site, arrow.

ducts resulting from α-amylase action on starch are responsible for induction of glucoamylase synthesis.

To get a better understanding of the regulatory mechanisms of extracellular amylolytic enzyme production by *C. acetobutylicum*, genes encoding α-amylase activity were recently cloned (Blaschek, 1989; Verhasselt et al., 1989). The α-amylase gene from *C. acetobutylicum* ATCC 824 was cloned and expressed in *E. coli* HB101 using the plasmid pEcoR251 (Verhasselt et al., 1989). The presence of glucose, maltose, maltotriose, and maltotetraose on thin-layer chromatograms of products released from starch by the cloned enzyme demonstrated that this cloned *C. acetobutylicum* gene codes for an α-amylase. Distribution of the α-amylase activity in *E. coli* has been investigated. It was observed that only a minor fraction (8%) of the activity could be detected in the culture fluid, while 44% and 48% of the enzyme activity was localized intracellularly and within the periplasmic space, respectively (P. Verhasselt, unpublished results). Whether the α-amylase activity found in the culture fluid results from active secretion or is simply the result of cell lysis is not clear.

Nevertheless, because it appears that α-amylase activity is present in the periplasm, it is tempting to speculate that the signal sequence for secretion from *C. acetobutylicum* is promoting the transfer of the amylase through the inner membrane of *E. coli*. This was clearly demonstrated for the α-amylase of *B. coagulans* by Cornelis et al. (1982). The α-amylase of *C. acetobutylicum* has been partially purified by hydrophobic interaction chromatography and the

molecular weight, estimated by SDS-PAGE, is about 65 kDa. SDS-PAGE analysis of proteins produced by minicells containing the cloned *C. acetobutylicum* α-amylase gene revealed a protein with a molecular weight of about 60 (P. Verhasselt, unpublished results). Temperature and pH optima of the partially purified *C. acetobutylicum* α-amylase have been determined; optimal activity occurred at 37°C and at pH 5. Roughly the same pH activity profile has been observed with the cloned α-amylase in *E. coli* (Verhasselt et al., 1989).

All these results suggest that the purified *C. acetobutylicum* α-amylase is the same as the α-amylase encoded by the *C. acetobutylicum* DNA in *E. coli*. Nucleotide sequence of the cloned gene has been determined, with the transcription start. Figure 21.4 shows promoter region, transcription start, Shine–Dalgarno sequence, N-terminal amino acid sequence, and putative signal sequence processing site. The promoter shows good homology with the *B. subtilis* σ^{43} promoter (i.e., TTGACA . . . . TATAAT). The Shine–Dalgarno sequence shows the characteristics of a purine stretch and is spaced from the start codon by 8 nucleotides.

As in *B. licheniformis*, α-amylase is temporally expressed and subject to catabolite repression in *C. acetobutylicum* (Blaschek, 1989; Laoide and McConnell, 1989). Although *amyL*, the structural α-amylase gene of *B. licheniformis*, is temporally activated at the onset of the stationary phase under nonrepressing growth conditions when cells presumably initiate sporulation, initiation of sporulation is not a prerequisite for either temporal activation or derepression of α-amylase synthesis. It was shown that *spoO* mutants, which are defective in sporulation initiation, have enzyme activity profiles similar to that of the parent strain when the cultures are grown in nonrepressing medium. The *cis* sequences essential for mediating catabolite repression lie downstream from the promoter region and upstream from the signal sequence cleavage site, a sequence that is 108 bp long (Laoide and McConnell, 1989; Laoide et al., 1989).

No homology has yet been found between the *C. acetobutylicum* α-amylase gene and the α-amylase genes from *Bacillus* species. Moreover, no detailed studies on expression of the α-amylase gene in *C. acetobutylicum* have been carried out. Thus, it remains unclear whether expression of the α-amylase gene in *C. acetobutylicum* is subject to a regulation mechanism similar to that in *B. licheniformis*.

21.4 Utilization of Lignocellulosic Materials

Cellulose is considered to be the most abundant carbon source on earth. Unfortunately, cellulose is not available as a pure compound in biomass. Wood, for instance, is composed of three major components: cellulose, hemicellulose, and lignin. Cellulose and hemicellulose are fermented by some anaerobic bacteria. Cellulose occurs as long crystalline microfibrils surrounded by the amorphous hemicellulose. This complex is embedded in a matrix of lignin. Because lignin is recalcitiant to degradation by anaerobic bacteria, this component limits cellulose and hemicellulose degradation by steric hindrance (Zeikus, 1980; Volesky and Szczesny, 1983).

Utilization of cellulose

Cellulose is a linear polymer composed of D-glucose monomers linked by β-1,4 bonds. Cellulose can be degraded by three classes of hydrolytic enzymes: exoglucanase or 1,4-β-D-glucanase (EC 3.2.1.91) cleaves specifically cellobiose or glucose units from the nonreducing end of cellulose, depending on the exogluca-

nase concerned; endo-1,4-β-D-glucanase (EC 3.2.1.4) cleaves internal cellulosic bonds, and cellobiase or 1,4-β-D-glucanase (EC 3.2.1.21) cleaves glucose monomers from the nonreducing ends of cellooligosaccharides. Cellulose-degrading organisms generally produce multiple forms of these enzymes, which seem to act synergistically and are organized into a cellulosome in *C. thermocellum* (Bayer et al., 1985). In gene banks of *C. thermocellum*, at least 15 different endoglucanase genes and 2 different cellobiase genes have been identified (Béguin et al., 1983, 1985, 1987; Cornet et al., 1983a, 1983b; Millet et al., 1985; Schwartz et al., 1985, 1986, 1988a, 1988b; Grépinet and Béguin, 1986; Joliff et al., 1986a, 1986b; Pétré et al., 1986; Romaniec et al. 1987a, 1987b; Soutschek-Bauer and Staudenbauer , 1987; Gräbnitz and Staudenbauer, 1988; Hall et al., 1988; Hazlewood et al., 1988; Kadam et al., 1988; Gräbnitz et al., 1989).

Although *C. acetobutylicum* produces cellulolytic enzymes, levels of cellulase activities are low in comparison with other cellulolytic bacteria and fungi, and no crystalline cellulose is degraded (Allcock and Woods, 1981). This results in poor solvent production on lignocellulosic substrates. Complete degradation of cellulose is believed to require at least two synergistically acting cellulase components, namely endoglucanase and exoglucanase. Endoglucanases initiate the degradation process, creating new ends from which the exoglucanases release cellobiose or glucose (Lee et al., 1985a). The fact that hydrolysis of crystalline cellulose in *C. thermocellum* can be accomplished by two proteins of the cellulosome complex contributes to this model. The molecular weight of one of these proteins is close to that of a purified *C. thermocellum* endoglucanase. Whether this endoglucanase is one of the two enzymes mentioned remains to be determined (Wu et al., 1988).

One approach to drive solvent production by *C. acetobutylicum* on crystalline cellulose is that of hydrolysis of lignocellulosic substrates before fermentation. Lignocellulosic substrates are pretreated by steam explosion, alkali, or SO_2 and are hydrolyzed by *Trichoderma reesei* cellulase before or simultaneous with fermenta-

Table 21.1 Comparison of endoglucanase and cellobiase activities of *Clostridium acetobutylicum* strains P270, NRRL B 527, and ATCC 824

	P270	NRRL B 527	ATCC 824
Endoglucanase			
Localization	Extracellular	Extracellular and cell associated (5%)	Extracellular and cell associated (10%–20%)
pH optimum	4.6	5.2	5.2
Expression	Inducible	Catabolite repression	Catabolite repression
Cellobiase			
Localization	Extracellular	Cell associated for 75%	Cell associated for 80%
pH optimum	N.D.[a]	4.8	4.8–6.0
Expression	Constitutive	Catabolite repression	Catabolite repression
Inhibition	N.D.	1% cellobiose	1% cellobiose

[a] N.D., not determined.

tion (Marchal et al., 1984, 1985b, 1986; Pareck et al., 1988).

A second approach consists of fermentation of chemically and physically pretreated lignocellulosic residues by sequential coculture (Fond et al., 1983; Yu et al., 1985). *C. thermocellum*, grown anaerobically at 60°C for 3 days, produces active cellulolytic and xylanolytic enzymes through which it can convert crystalline cellulose and hemicellulose to ethanol. When it is grown on lignocellulosic substrates, glucose and pentoses are accumulated in the culture fluid. After lowering the temperature of the culture fluid to 37°C, *C. acetobutylicum* is added. *C. acetobutylicum* starts to utilize glucose and pentoses and to produce acetate and butyrate. However, solventogenesis is only obtained with addition of extra butyrate to the fermentation fluid. This sequential coculture approach, with added butyric acid, leads to the production of butanol, acetone, and ethanol by *C. acetobutylicum* in addition to the ethanol produced by *C. thermocellum* (Yu et al., 1985). Fond et al. (1983) chose the mesophilic *Clostridium* H10 for hydrolysis of cellulose before solvent production by *C. acetobutylicum*. In this co culture system, acids are also produced instead of solvents because the rate of cellulolysis and sugar supply by the cellulolytic strain are too low. Again, solventogenesis is restored by addition of butyric acid.

A third approach is the genetic manipulation of *C. acetobutylicum* to enable this bacterium to directly ferment cellulosic materials to solvents. Allcock and Woods (1981) reported the production of an extracellular cellobiase and an inducible endoglucanase by the industrial *C. acetobutylicum* strain P270, a close derivative of strain P262. Lee et al. (1984, 1985a) described cellulolytic activity of *C. acetobutylicum* strains NRRL B 527 and ATCC 824. Of 21 solvent-producing *Clostridium* strains tested, none were able to grow on cellulose. However, *C. acetobutylicum* NRRL B 527 and ATCC 824 exhibited endoglucanase activity as indicated by the production of clearing zones on carboxymethyl cellulose-containing medium. In addition, both strains also produced cellobiase activity. Differences in cellulolytic activities among the three strains P270, NRRL B 527, and ATCC 824 are summarized in Table 21.1.

Zappe et al. (1986) were able to clone a DNA fragment of *C. acetobutylicum* P262 in *E. coli* HB101 using pEcoR251 as cloning vector. This DNA fragment encoded endoglucanase and cellobiase activity. Both enzyme activities were expressed in *E. coli*, as indicated by degradation of carboxymethyl cellulose and by growth of the recombinant *E. coli* HB101(pHZ100) on cellobiose as sole carbon source. The lack of hydrolysis of Avicel, a crystalline cellulose, indicated the absence of exoglucanase activity. In

E. coli, as much as 75% of the total endoglucanase activity was associated with the periplasmic space, while no activity could be detected in the supernatant. The conditions for optimum activity of the endoglucanases of *C. acetobutylicum* P270 and *E. coli* HB101(pHZ100) differed; temperature optima were 37°C and 50°C, respectively. The cloned endoglucanase was optimally active at pH between 5.0 and 7.0 while the P270 endoglucanase was optimally active at pH 4.6. In *E. coli* HB101(pHZ100), the endoglucanase gene is transcribed from its own promoter and was found to be expressed in a constitutive way. In *C. acetobutylicum* P270, endoglucanase activity is induced by a small molecule present in molasses. Because of these significant differences, the option remains available that the cloned gene does not represent the gene in *C. acetobutylicum* that codes for the major endoglucanase activity.

The nucleotide sequence of the cloned *C. acetobutylicum* P262 endoglucanase gene has been determined (Zappe et al., 1988). The putative promoter sequence, with TTGTATT as a −35 region and TACAAT as a −10 region, resembles the consensus sequence of the σ^{43} promoter of gram-positive bacteria. The deduced amino acid sequence shows higher homology with endoglucanases from *Bacillus* species as compared to endoglucanases from *C. thermocellum*, *Cellulomonas fimi*, and *Trichoderma reesei*. The *Clostridium acetobutylicum* endoglucanase shows little homology with endoglucanases synthesized by cellulolytic species that form a true cellulolytic system, but significant homology with endoglucanases originating from species that produce only a few cellulolytic enzymes. Moreover, recent studies on bacterial evolution (Woese, 1987) indicate a separation of the gram-positive bacteria with a low mole percent of GC into several branches that divide the mesophilic clostridia, including *C. acetobutylicum*, and the bacilli from the thermophilic clostridia, including *C. thermocellum*.

Degradation of xylan

Xylan is a major constituent of plant hemicellulose (Biely, 1985). Xylan composition varies considerably with plant species and consists of hexose and pentose sugars. The major monomers are xylose, arabinose, mannose, glucose, galactose, and glucuronic acid, most of which can be fermented by *C. acetobutylicum* (Compere and Griffith, 1979). The backbone of xylan is composed of D-xylose units linked by β-1,4 bonds. Side chains consist of L-arabinofuranose and D-glucuronic acid or 4-0-methyl-D-glucuronic acid. The degree of branching varies with the source of xylan. Acetylation occurs most frequently at the 0-3 position of the D-xylose unit and to a lesser extend at the 0-2 position. The degree of acetylation also depends on the source of xylan.

Complete degradation of xylan involves the synergistic action of several hydrolytic enzymes. Endo-1,4-β-xylanase or xylanase (EC 3.2.1.8) is an endoacting enzyme and hydrolyzes internal β-1,4 linkages. In general, linkages adjacent to side-chained units are not hydrolyzed. β-Xylosidases (EC 3.2.1.37) hydrolyze xylooligosaccharides to produce D-xylose. α-L-Arabinofuranosidase (EC 3.2.1.55) and α-glucuronidase liberate side chains that are linked to the main chain by an α-1,2 bond. Acetyl residues are removed from the main chain by acetylesterase (EC 3.1.1.6) or acetyl xylan esterase (see figure 1 in Biely, 1985). Xylan can be degraded by acid or enzyme hydrolysis before fermentation by *C. acetobutylicum*. However, this results in additional costs of enzyme for the removal of toxic by-products produced during acid hydrolysis. Alternatively, xylan could be directly fermented by a *C. acetobutylicum* strain producing active xylanolytic enzymes (Lemmel et al., 1986). Different xylanolytic enzymes of *C. acetobutylicum* have been purified and characterized (Lee et al., 1985b, 1987; Lee and Forsberg, 1987a, b) (Table 21.2).

Interestingly, in *C. acetobutylicum* ATCC 824 both xylanases show a different activity on xylan as well as on xylooligosaccharides. Xylanase A hydrolyzes xylan randomly, yielding oligosaccharides with chain lengths between 2 and 6. Hydrolysis of xylan by xylanase B yields end products with chain lengths of 2 and 3. Because no antigenic similarity was exhibited by the two xylanases, they are presumably encoded by two different genes. Further experiments have also shown that the xylosidase is

Table 21.2 Properties of *Clostridium acetobutylicum* xylanolytic enzymes

		Xylanase			Xylosidase	Arabino-furanosidase
			ATCC 824			
	P262	NRRL B 527	A	B	ATCC 824	ATCC 824
pH optimum	6	5.8	5.0	6.0	6.0–6.5	5.0–5.5
Localization		Extracellular for 67%–95%	Extracellular for 90%		Cell-associated for 98%	Cell associated for 91%
T optimum	37°–43°C	N.D.[a]	50°C	60°C	45°C	N.D.[a]
MW	28 kDa	N.D.	65 kDa	29 kDa	224 kDa	94 kDa
pI	10	N.D.	4.45	8.5	5.85	8.15

[a]N.D., not determined.

able to hydrolyze the oligosaccharides, resulting from the degradation of xylan by the two xylanases, and yielding xylose.

A common feature of all the *C. acetobutylicum* xylanolytic enzymes is the low stability of the enzymes at a pH that is optimal for solvent production (Bahl et al., 1982). Therefore, the instability of these enzymes represents a limiting factor in the direct conversion of xylan to solvents. This problem could be solved by solvent production at neutral pH (Holt et al., 1984) or by genetic manipulation of *C. acetobutylicum* for xylan degradation at low pH. Clearly, there is a need for an appropriate combination of synergistically acting hydrolytic enzymes for the degradation of plant cell wall hemicellulosic material (Lee and Forsberg, 1987b). Zappe et al. (1987) reported the cloning and expression of a *C. acetobutylicum* P262 xylanase gene in *E. coli* HB101 using pEcoR251 as cloning vector. After expression in *E. coli* HB101(pHZ300), xylanase activity was mainly localized in the cytoplasm (98%). Xylanases were purified from *E. coli* HB101(pHZ300) and from *C. acetobutylicum* and were found to have similar characteristics; *C. acetobutylicum* P262 xylanase characteristics are indicated in Table 21.2.

No total digestion of xylan by *C. acetobutylicum* has been obtained. Resistance of xylan to enzymatic degradation may be caused by end product inhibition or the structure of xylan itself or both. Elucidation of the factors limiting the complete metabolism of xylan by *C. acetobutylicum* is a prerequisite for the efficient use of plant cell hemicellulose as fermentation substrate (Lee et al., 1985b).

21.5 Utilization of Other Substrates

In addition to starch and lignocellulosic materials, inulin is another possible polymeric carbohydrate source. Marchal et al. (1985a) demonstrated that Jerusalem artichokes can serve as substrate for ABE fermentation by *C. acetobutylicum* after acidic or enzymatic hydrolysis. Consequently, strains that exhibit higher inulolytic activity were selected, for instance, *C. acetobutylicum* strain IFP 904. However, even with this strain still better yield was obtained by adding inulinase. Efstathiou et al. (1986) and Looten et al. (1987) studied the inulinase [2,1-β-D-fructan fructano hydrolase, (EC 3.2.1.7)] activities in *C. acetobutylicum* strains ABKn8 and IFP 912. Most of the inulinases are β-fructosidases that split off fructose units from polymers with a fructose unit at the terminal β-2,1 position. They are usually distinguished from invertases, which hydrolyze sucrose and related low molecular weight sugars. Characteristics of the inulinases are described in Table 21.3. So far, no evidence for the existence of an endoacting inulolytic enzyme, such as observed with some *Aspergillus* inulinases, has been found.

Cheese whey, a waste product of the dairy industry, has attracted interest as an alternative

Table 21.3 Characteristics of *Clostridium acetobutylicum* ABKn8 and IFP 912 inulinases

Inulinase	ABKn8	IFP 912
pH optimum	4.6	5.0–6.0
Induction	Inulin	Inulin
Repression	Glucose	Glucose, fructose
Localization	Extracellular for 90%	Extracellular for 84%
T optimum	N.D.[a]	47°C

[a] N.D., not determined.

substrate for ABE fermentation because of its abundance and high lactose content. Because butanol toxicity limits the amount of sugar consumed, the relatively low sugar content (4%–5% lactose) makes whey permeate suitable for ABE fermentation. Interestingly, fermentation of whey by *C. acetobutylicum* yields butanol and acetone in a ratio of approximately 100:1. This feature is desirable for the industrial production because butanol is the highest priced solvent. The reason for this altered ratio as compared with the 6:3:1 butanol:acetone:ethanol ratio in a normal batch fermentation is not well understood (Bahl et al., 1986). In spite of this favorable solvent ratio, overall reactor productivities in batch fermentations with whey are low compared with starch and molasses as substrates (Jones and Woods, 1986a). By using immobilized *C. acetobutylicum* cells in continuous solvent production from whey, productivity could be increased more than 10 fold compared with batch fermentation (Qureshi and Maddox, 1987). Ennis and Maddox (1987) observed an inverse relationship between solvent yield and lactose utilization rate. Higher culture pH values resulted in improved lactose utilization rates and acid production at the expense of solventogenesis. On the contrary, addition of lactose to the culture fluid or lowering the pH improved solventogenesis at the expense of acid production and lactose utilization. This phenomenon could explain why in batch cultures lactose is only partially utilized (Qureshi and Maddox, 1987).

Yu et al. (1987) investigated β-galactosidase and phospho-β-galactosidase activities in the fermentation of whey permeate by different *C. acetobutylicum* strains. All strains showed both enzyme activities, except strain ATCC 824, which contained only phospho-β-galactosidase activity. Lactose is taken up by some clostridia as lactose-P via a phosphoenolpyruvate-dependent phosphotransferase system (Dills et al., 1980). Phospho-β-galactosidase hydrolyzes lactose-P to glucose and galactose 6-P. Glucose enters the glycolytic pathway while galactose 6-P is metabolized to triose phosphates. Alternatively, lactose is taken up by some clostridia as such and hydrolyzed intracellularly into glucose and galactose by β-galactosidase. Yu et al. (1987) observed that in *C. acetobutylicum* P262 phospho-β-galactosidase induction is associated with acidogenic phase, while β-galactosidase is associated with solventogenesis. Both enzymes were found to be repressed by glucose. Generally, phospho-β-galactosidase is present at much higher levels than β-galactosidase. This would suggest that lactose is mainly taken up as lactose-P and that both lactose assimilation pathways are associated with a different growth phase of *C. acetobutylicum* P262. The relative importance of these two potential pathways for lactose utilization remains to be determined.

The *C. acetobutylicum* fermentation products usually include acetate, butyrate, acetone, butanol, and ethanol. Investigations have been made to determine whether other economically valuable products could be produced by using alternative substrates. Fermentation of L-rhamnose resulted in the production of 1,2-propanediol, propionic acid, and *n*-propanol (Forsberg et al., 1987). Forsberg (1987) demonstrated that *C. acetobutylicum* grows with glycerol as sole carbon source and that 1,3-propanediol is produced as the major fermentation product. When 1,2-propanediol is included in this medium, this substrate is reduced to *n*-propanol, and less 1,3-propanediol is produced. Jewell et al. (1986) demonstrated that addition of propionic, valeric, and 4-hydroxybutyric acid results in the production of the corresponding alcohols (1-propanol, 1-pentanol, and 1,4-butanediol, respectively) by *C. acetobutylicum*. These acids are used as electron acceptors in a way similar to that of acetate and butyrate during solventogenesis.

21.6   Future Prospects

The utilization of nearly all carbohydrates occurring in biomass by *Clostridium acetobutylicum* offers great potential for the acetone-butanol-ethanol fermentation. Although *C. acetobutylicum* produces different hydrolytic enzymes, most of these activities are clearly not optimal, as demonstrated by the relatively poor substrate consumption during fermentation of polymeric substrates. Substrate catabolism of *C. acetobutylicum* could be amplified by application of genetic techniques via improvement of the existing hydrolytic activities or via introduction of foreign genes, encoding more active hydrolytic enzymes. This approach will also increase the spectrum of usable substrates. Increased flexibility with respect to substrates used will allow future fermentation units to use the most inexpensive carbon sources and to exploit relatively short-term surpluses (Blaschek, 1986; McNeil and Kristiansen, 1986).

These improvements in substrate utilization should be accompanied by improvement of the performance of *C. acetobutylicum* in other areas, including (1) elucidation of the molecular mechanisms involved in the metabolic control and diversity, to produce novel economically valuable products whose chemical synthesis is difficult and cost-prohibitive (Young et al., 1989), and (2) construction of strains with enhanced solvent yields along with the construction of strains with higher end product tolerance (Jones and Woods, 1986b).

The production of strains with improved performance in substrate utilization and solvent production should go hand in hand with developments in the field of process technology, leading to (1) improvements in the processing of feedstocks yielding fermentable substrates; (2) development of adapted continuous solvent production systems; (3) improvement of by-product utilization, and (4) improvement of product recovery.

These improvements in the overall performance of *Clostridium acetobutylicum* and in the process technology seem to be a very real possibility to make the fermentation route economically competitive with chemical solvent synthesis (Jones and Woods, 1986b).

Acknowledgments.   The authors wish to thank numerous colleagues who contributed to this chapter by sending recent reprints or preprints of their work. P.V. is the recipient of a fellowship of the Instituut voor Wetenschappelijk Onderzoek in de Nijverheid en de Landbouw (I.W.O.N.L.).

References

Adams, M.W.W., L.E. Mortensen, and J.S. Chen. 1980. Hydrogenases. *Biochim. Biophys. Acta* **594**:105–176.

Allcock, E.R. and D.R. Woods. 1981. Carboxymethylcellulase and cellobiase production by *Clostridium acetobutylicum* in an industrial fermentation medium. *Appl. Environ. Microbiol.* **41**:539–541.

Andersch, W., H. Bahl, and G. Gottschalk. 1983. Levels of enzymes involved in acetate, butyrate, acetone and butanol formation by *Clostridium acetobutylicum*. *Eur. J. Appl. Microbiol. Biotechnol.* **18**:327–332.

Bahl, H., W. Andersch, K. Braun, and G. Gottschalk. 1982. Effect of pH and butyrate concentration on the production of acetone and butanol by *Clostridium acetobutylicum* grown on continuous culture. *Eur. J. Appl. Microbiol. Biotechnol.* **14**:17–20.

Bahl, H. and G. Gottschalk. 1988. Microbial production of butanol/acetone. *In: Biotechnology*, Vol. 6B, H.J. Rehm, and G. Reed, eds., pp. 1–30. Weinheim: VCH Verlagsgesellschaft.

Bahl, H., M. Gottwald, A. Kühn, V. Rale, W. Andersch, and G. Gottschalk. 1986. Nutritional factors affecting the ratio of solvents produced by *Clostridium acetobutylicum*. *Appl. Environ. Microbiol.* **52**:169–172.

Barton, L.L., C.E. Georgi, and D.R. Lineback. 1972. Effect of maltose on glucoamylase formation by *Aspergillus niger*. *J. Bacteriol.* **111**:771–777.

Bayer, E.A., E. Setter, and R. Lamed. 1985. Organization and distribution of the cellulosome in *Clostridium thermocellum*. *J. Bacteriol.* **163**:552–559.

Béguin, P., P. Cornet, and J.P. Aubert. 1985. Sequence of a cellulase gene of the thermophilic bacterium *Clostridium thermocellum*. *J. Bacteriol.* **162**:102–105.

Béguin, P., P. Cornet, and J. Millet. 1983. Identification of the endoglucanase encoded by the *celB* gene of *Clostridium thermocellum*. *Biochimie* **65**:495–500.

Béguin, P., J. Millet, and J.P. Aubert. 1987. The cloned *cel* (cellulose degradation) genes of *Clostridium thermocellum* and their products. *Microbiol. Sci.* **4**:277–280.

Biely, P. 1985. Microbial xylanolytic systems. *Trends Biotechnol.* **3**:286–290.

Blaschek, H.P. 1986. Genetic manipulation of *Clostridium acetobutylicum* for production of butanol. *Food Technol.* **40**:84–87.

Blaschek, H.P. 1989. Genetic manipulation of the clostridia. *Dev. Ind. Microbiol.* **30**:35–42.

Chojecki, A. and H.P. Blaschek. 1986. Effect of carbohydrate source on alpha-amylase and glucoamylase formation by *Clostridium acetobutylicum* SA-1. *J. Ind. Microbiol.* **1**:63–67.

Compere, A.L. and W.L. Griffith. 1979. Evaluation of substrates for butanol production. *Dev. Ind. Microbiol.* **20**:509–517.

Cornelis, P., C. Digneffe, and K. Willemot. 1982. Cloning and expression of a *Bacillus coagulans* amylase gene in *Escherichia coli. Mol. Gen. Genet.* **186**:507–511.

Cornet, P., J. Millet, P. Béguin, and J.P. Aubert. 1983a. Characterization of two *cel* (cellulose degradation) genes of *Clostridium thermocellum* coding for endoglucanases. *Bio/Technology* **1**: 589–594.

Cornet, P., D. Tronik, J. Millet, and J.P. Aubert. 1983b. Cloning and expression in *Escherichia coli* of *Clostridium thermocellum* genes coding for amino acid synthesis and cellulose hydrolysis. *FEMS Microbiol. Lett.* **16**:137–141.

Datta, R. and J.G. Zeikus, 1985. Modulation of acetone-butanol-ethanol fermentation by carbon monoxide and organic acids. *Appl. Environ. Microbiol.* **49**:522–529.

Dills, S.S., A. Apperson, M.R. Schmidt, and M.H. Saier, Jr. 1980. Carbohydrate transport in bacteria. *Microbiol. Rev.* **44**:385–418.

Doremus, M.G., J.C. Linden, and A.R. Moreira. 1985. Agitation and pressure effects on acetone-butanol fermentation. *Biotechnol. Bioeng.* **27**: 852–860.

Efstathiou, I., G. Reysset, and N. Truffaut. 1986. A study of inulinase activity in *Clostridium acetobutylicum* strain ABKn8. *Appl. Microbiol. Biotechnol.* **25**:143–149.

Ennis, B.M. and I.S. Maddox. 1987. The effect of pH and lactose concentration on solvent production from whey permeate using *Clostridium acetobutylicum. Biotechnol. Bioeng.* **29**:329–334.

Ensley, B., J.J. McHugh, and L.L. Barton. 1975. Effect of carbon sources on formation of α-amylase and glucoamylase by *Clostridium acetobutylicum. J. Gen. Appl. Microbiol.* **21**:51–59.

Fond, O., J.M. Engasser, G. Matta-El-Amouri, and H. Petitdemange. 1986. The acetone-butanol fermentation on glucose and xylose. I. Regulation and kinetics in batch cultures. *Biotechnol. Bioeng.* **28**:160–166.

Fond, O., E. Petitdemange, H. Petitdemange, and J.M. Engasser. 1983. Cellulose fermentation by a coculture of a mesophilic cellulolytic *Clostridium*

and *Clostridium acetobutylicum. Biotechnol. Bioeng. Symp.* **13**:217–224.

Forsberg, C.W. 1987. Production of 1,3-propanediol from glycerol by *Clostridium acetobutylicum* and other *Clostridium* species. *Appl. Environ. Microbiol.* **53**:639–643.

Forsberg, C.W., L. Donaldson, and L.N. Gibbins. 1987. Metabolism of rhamnose sugars by strains of *Clostridium acetobutylicum* and other *Clostridium* species. *Can. J. Microbiol.* **33**:21–26.

Gottschal, J.C. and J.G. Morris. 1981. The induction of acetone and butanol production in cultures of *Clostridium acetobutylicum* by elevated concentrations of acetate and butyrate. *FEMS Microbiol. Lett.* **12**:385–389.

Gottschal, J.C. and J.G. Morris. 1982. Continuous production of acetone and butanol by *Clostridium acetobutylicum* growing in a turbidostat culture. *Biotechnol. Lett.* **4**:477–482.

Gräbnitz, F. and W.L. Staudenbauer. 1988. Characterization of two β-glucosidase genes from *Clostridium thermocellum. Biotechnol. Lett.* **10**:73–78.

Gräbnitz, F., K.P. Rücknagel, M. Seiss, and W.L. Staudenbauer. 1989. Nucleotide sequence of the *Clostridium thermocellum bglB* gene encoding thermostable β-glucosidase B: homology to fungal β-glucosidases. *Mol. Gen. Genet.* **217**:70–76.

Graves, M.C., G.T. Mullenbach, and J.C., Rabinowitz. 1985. Cloning and nucleotide sequence of the *Clostridium pasteurianum* ferredoxin gene. *Proc. Nat. Acad. Sci. USA* **82**:1653–1657.

Grépinet, O. and P. Béguin. 1986. Sequence of the cellulase gene of *Clostridium thermocellum* coding for endoglucanase B. *Nucleic Acids Res.* **14**:1791–1799.

Häggström, L. and C. Förberg. 1986. Significance of an extracellular polymer for the energy metabolism in *Clostridium acetobutylicum*: a hypothesis. *Appl. Microbiol. Biotechnol.* **23**:234–239.

Häggström, L. and N. Molin. 1980. Calcium alginate immobilized cells of *Clostridium acetobutylicum* for solvent production. *Biotechnol. Lett.* **2**:241–246.

Hall, J., G.P. Hazlewood, P.J. Barker, and H.J. Gilbert. 1988. Conserved reiterated domains in *Clostridium thermocellum* endoglucanases are not essential for catalytic activity. *Gene* (Amst.) **69**:29–38.

Hazlewood, G.P., M.P.M. Romaniec, K. Davidson, O. Grépinet, P. Béguin, J. Millet, O. Raynaud, and J.P. Aubert. 1988. A catalogue of *Clostridium thermocellum* endoglucanase, β-glucosidase and xylanase genes cloned in *Escherichia coli. FEMS Microbiol. Lett.* **51**:231–236.

Hermann, M., F. Fayolle, R. Marchal, L. Podvin, M. Sebald, and J.P. Vandecasteele, 1985. Isolation and characterization of butanol-resistant mutants of *Clostridium acetobutylicum. Appl. Environ. Microbiol.* **50**:1238–1243.

Holt, R.A., G.M. Stephens, and J.G. Morris, 1984. Production of solvents by *Clostridium acetobutylicum* cultures maintained at neutral pH. *Appl. Environ. Microbiol.* **48**:1166–1170.

Janati-Idrissi, R., A.M. Junelles, A. El Kanouni, H. Petitdemange, and R. Gay. 1989. Pyruvate fermentation by *Clostridium acetobutylicum*. *Biochem. Cell. Biol.* **67**:735–739.

Jewell, J.B., J.B. Coutinho, and M. Kropinski. 1986. Bioconversion of propionic, valeric, and 4-hydroxybutyric acids into the corresponding alcohols by *Clostridium acetobutylicum* NRRL 527. *Curr. Microbiol.* **13**:215–219.

Joliff, G., P. Béguin, and J.P. Aubert. 1986a. Nucleotide sequence of the cellulase gene *celD* encoding endoglucanase D of *Clostridium thermocellum*. *Nucleic Acids Res.* **14**:8605–8613.

Joliff, G., P. Béguin, M. Juy, J. Millet, ,A. Ryter, R. Poljak, and J.P. Aubert. 1986b. Isolation, crystallization and properties of a new cellulase of *Clostridium thermocellum* overproduced in *Escherichia coli*. *Bio/Technology* **4**:896–900.

Jones, D.T. and D.R. Woods. 1986a. Acetone-butanol fermentation revisited. *Microbiol. Rev.* **50**:484–524.

Jones, D.T. and D.R. Woods. 1986b. Gene transfer, recombination and gene cloning in *Clostridium acetobutylicum*. *Microbiol. Sci.* **3**:19–22.

Junelles, A.M., A. El Kanouni, H. Petitdemange, and R. Gay. 1989. Influence of acetic and butyric acid addition on polysaccharide formation by *Clostridium acetobutylicum*. *J. Ind. Microbiol.* **4**:121–125.

Junelles, A.M., R. Janati-Idrissi, H. Petitdemange, and R. Gay. 1987. Effect of pyruvate on glucose metabolism in *Clostridium acetobutylicum*. *Biochimie* (Paris) **69**:1183–1190.

Jungermann, K., R.K. Thauer, G. Leimenstoll, and K. Decker. 1973. Function of reduced pyridine nucleotide-ferredoxine oxidoreductases in saccharolytic clostridia. *Biochim. Biophys. Acta* **305**:268–280.

Kadam, S., A.L. Demain, J. Millet, P. Béguin, and J.P. Aubert. 1988. Molecular cloning of a gene for a thermostable β-glucosidase from *Clostridium thermocellum* into *Escherichia coli*. *Enzyme Microb. Technol.* **10**:9–13.

Karube, I., N. Urano, T. Yamada, H. Hirochika, and K. Sakaguchi. 1983. Cloning and expression of the hydrogenase gene from *Clostridium butyricum* in *Escherichia coli*. *FEBS Lett.* **158**:119–122.

Kim, B.H. and J.G. Zeikus. 1985. Importance of hydrogen metabolism in regulation of solventogenesis by *Clostridium acetobutylicum*. *Dev. Ind. Microbiol.* **26**:1–14.

Kim, B.H., P. Bellows, R. Datta, and J.G. Zeikus. 1984. Control of carbon and electron flow in *Clostridium acetobutylicum* fermentations: utilization of carbon monoxide to inhibit hydrogen production and to enhance butanol yields. *Appl. Environ. Microbiol.* **48**:764–770.

Klier, A.F. and G. Rapoport. 1988. Genetics and regulation of carbohydrate catabolism in *Bacillus*. *Annu. Rev. Microbiol.* **42**:65–95.

Laoide, B.M. and D.J. McConnell. 1989. *cis* Sequences involved in modulating expression of *Bacillus licheniformis amyL* in *Bacillus subtilis*: effect of sporulation mutations and catabolite resistance mutations on expression. *J. Bacteriol.* **171**:2443–2450.

Laoide, B.M., G.H. Chambliss, and D.J. McConnell. 1989. *Bacillus licheniformis* α-amylase gene, *amyL*, is subject to promoter-independent catabolite repression in *Bacillus subtilis*. *J. Bacteriol.* **171**:2435–2442.

Lee, S.F. and C.W. Forsberg. 1987a. Isolation and properties of a β-D-xylosidase from *Clostridium acetobutylicum* ATCC 824. *Appl. Environ. Microbiol.* **53**:651–654.

Lee, S.F. and C.W. Forsberg. 1987b. Purification and characterization of an α-L-arabinofuranosidase from *Clostridium acetobutylicum* ATCC 824. *Can. J. Microbiol.* **33**:1011–1016.

Lee, S.F., C.W. Forsberg, and N.L. Gibbins. 1984. Carboxymethylcellulase and cellobiase activities of *Clostridium acetobutylicum* and *Clostridium butylicum* strains. *In: Fifth Canadian Bioenergy Research and Development Seminar*, S. Hasnain, ed., pp. 569–572. Barking, England: Elsevier.

Lee, S.F., C.W. Forsberg, and L.N. Gibbins. 1985a. Cellulolytic activity of *Clostridium acetobutylicum*. *Appl. Environ. Microbiol.* **50**:220–228.

Lee, S.F., C.W. Forsberg, and L.N. Gibbins. 1985b. Xylanolytic activity of *Clostridium acetobutylicum*. *Appl. Environ. Microbiol.* **50**:1068–1076.

Lee, S.F., C.W. Forsberg, and J.B. Rattray, 1987. Purification and characterization of two endoxylanase genes from *Clostridium acetobutylicum* ATCC 824. *Appl. Environ. Microbiol.* **53**:644–650.

Lemmel, S.A., R. Datta, and J.R. Frankiewicz. 1986. Fermentation of xylan by *Clostridium acetobutylicum*. *Enzyme Microb. Technol.* **8**:217–221.

Lin, Y.L. and H.P. Blaschek. 1983. Butanol production by a butanol tolerant strain of *Clostridium acetobutylicum* in extruded corn broth. *Appl. Environ. Microbiol.* **45**:996–973.

Long, S., D.T. Jones, and D.R. Woods. 1984. The relationship between sporulation and solvent production in *Clostridium acetobutylicum* P262. *Biotechnol. Lett.* **6**:529–534.

Looten, P., D. Blanchet, and J.P. Vandecasteele. 1987. The β-fructofuranosidase activities of a strain of *Clostridium acetobutylicum* grown on inulin. *Appl. Microbiol. Biotechnol.* **25**:419–425.

Maddox, I. 1989. The acetone-butanol-ethanol fermentation: recent progress in technology. *Biotechnol. Genet. Eng. Rev.* **7**:189–220.

Marchal, R., D. Blanchet, and J.P. Vandecasteele. 1985a. Industrial optimization of acetone-butanol fermentation: a study of the utilization of Jerusalem artichokes. *Appl. Microbiol. Biotechnol.* **23**:92–98.

Marchal, R., M. Rebeller, F. Fayolle, J. Pourquie, and J.P. Vandecasteele. 1985b. Acetone butanol fermentation of hydrolysates obtained by enzymatic hydrolysis of agricultural lignocellulosic residues. *In: Energy from Biomass*, W. Palz, J. Coombs, and D.O. Hall, eds., pp. 692–696. Barking, England: Elsevier.

Marchal, R., M. Rebeller, and J.P. Vandecasteele. 1984. Direct bioconversion of alkali-pretreated straw using simultaneous enzymatic hydrolysis and acetone-butanol fermentation. *Biotechnol. Lett.* **6**:523–528.

Marchal, R., M. Ropars, and J.P. Vandecasteele. 1986. Conversion into acetone and butanol of lignocellulosic substrates pretreated by steam explosion. *Biotechnol. Lett.* **8**:365–370.

Martin, J.R., H. Petitdemange, J. Ballongue, and R. Gay. 1983. Effects of acetic and butyric acids on solvent production by *Clostridium acetobutylicum*. *J. Gen. Microbiol.* **68**:307–318.

McNeil, B. and B. Kristiansen. 1986. The acetone butanol fermentation. *Adv. Appl. Microbiol.* **31**:61–92.

Millet, J., D. Pétré, P. Béguin, O. Raynaud, and J.P. Aubert. 1985. Cloning of ten distinct DNA fragments of *Clostridium thermocellum* coding for cellulases. *FEMS Microbiol. Lett.* **29**:145–149.

Nakhmanovich, B.M. and N.A. Shcheblykina. 1959. Fermentation of pentoses of corn cob hydrolysates by *Clostridium acetobutylicum*. *Mikrobiologiya* **28**:99–104.

Pareck, S.R., R.S. Pareck, and M. Wayman. 1988. Ethanol and butanol production by fermentation of enzymatically saccharified SO_2-prehydrolysed lignocellulosics. *Enzyme Microb. Technol.* **10**:660–668.

Petitdemange, H., C. Cherrier, G. Raval, and R. Gay. 1976. Regulation of the NADH and NADPH-ferredoxin oxidoreductases in saccharolytic clostridia of the butyric group. *Biochim. Biophys. Acta* **421**:334–347.

Pétré, D., J. Millet, R. Longin, P. Béguin, H. Girard, and J.P. Aubert. 1986. Purification and properties of the endoglucanase C of *Clostridium thermocellum* produced in *Escherichia coli*. *Biochimie* (Paris) **68**:687–695.

Qureshi, N. and I.S. Maddox. 1987. Continuous solvent production from whey permeate using cells of *Clostridium acetobutylicum* immobilized by adsorption onto bonechar. *Enzyme Microb. Technol.* **9**:668–671.

Reysenbach, A.L., N. Ravenscroft, S. Long, D.T. Jones, and D.R. Woods. 1986. Characterization, biosynthesis, and regulation of granulose in *Clostridium acetobutylicum*. *Appl. Environ. Microbiol.* **52**:185–190.

Robinson, G.C. 1922. A study of the acetone and butyl alcohol fermentation of various carbohydrates. *J. Biol. Chem.* **52**:125–155.

Rogers, P. 1986. Genetics and biochemistry of *Clostridium* relevant to development of fermentation processes. *Adv. Appl. Microbiol.* **31**:1–60.

Romaniec, M.P.M., N.G. Clarke, and G.P. Hazlewood. 1987a Molecular cloning of *Clostridium thermocellum* DNA and the expression of further novel endo-β-1,4-glucanase genes in *Escherichia coli*. *J. Gen. Microbiol.* **133**:1297–1307.

Romaniec, M.P.M., K. Davidson, and G.P. Hazlewood. 1987b. Cloning and expression in *Escherichia coli* of *Clostridium thermocellum* DNA encoding β-glucosidase activity. *Enzyme Microb. Technol.* **9**:474–478.

Schwartz, W., K. Bronnenmeier, and W.L. Staudenbauer. 1985. Molecular cloning of *Clostridium thermocellum* genes involved in β-glucan degradation in bacteriophage lambda. *Biotechnol. Lett.* **7**:859–864.

Schwartz, W., F. Gräbnitz, and W.L. Staudenbauer. 1986. Properties of a *Clostridium thermocellum* endoglucanase produced in *Escherichia coli*. *Appl. Environ. Microbiol.* **51**:1293–1299.

Schwartz, W.H., S. Schimming, K.P. Rücknagel, S. Burgschwaiger, G. Kreil, and W.L. Staudenbauer. 1988a. Nucleotide sequence of the *celC* gene encoding endoglucanase C of *Clostridium thermocellum*. *Gene* **63**:23–30.

Schwartz, W.H., S. Schimming, and W.L. Staudenbauer. 1988b. Isolation of a *Clostridium thermocellum* gene encoding thermostable β-1,3-glucanase (laminarase). *Biotechnol. Lett.* **10**:225–230.

Soutschek-Bauer, E. and W.L. Staudenbauer. 1987. Synthesis and secretion of a heat-stable carboxymethylcellulase from *Clostridium thermocellum* in *Bacillus subtilis* and *Bacillus stearothermophilus*. *Mol. Gen. Genet.* **208**:537–541.

Verhasselt, P., F. Poncelet, K. Vits, A. Van Gool, and J. Vanderleyden. 1989. Cloning and expression of a *Clostridium acetobutylicum* α-amylase gene in *Escherichia coli*. *FEMS Microbiol. Lett.* **59**:135–140.

Volesky, B. and T. Szczesny. 1983. Bacterial conversion of pentose sugars to acetone and butanol. *Adv. Biochem. Engi.* **27**:101–118.

Weizmann, C. 1915. U.K. patent 4845.

Weizmann, C. 1919. U.S. patent 1,315,585.

Woese, C.R. 1987. Bacterial evolution. *Microbiol. Rev.* **51**:221–271.

Wu, J.H.D., W.H. Orme-Johnson, and A.L. Demain. 1988. Two components of an extracellular protein aggregate of *Clostridium thermocellum* together degrade crystalline cellulose. *Biochimie* (Paris) **27**:1703–1709.

Yerushalmi, L. and B. Volesky. 1985. Importance of agitation in acetone-butanol fermentation. *Biotechnol. Bioeng.* **27**:1297–1305.

Yerushalmi, L., B. Volesky, and T. Szczesny. 1985. Effect of increased hydrogen partial pressure on the acetone-butanol fermentation by *Clostridium acetobutylicum. Appl. Microbiol. Biotechnol.* **22**:103–107.

Young, M., N. Minton, and W.L. Staudenbauer. 1989. Recent advances in the genetics of the clostridia. *FEMS Microbiol. Rev.* **63**:301–326.

Yu, E.K.C., and J.N. Saddler. 1983. Enhanced acetone-butanol fermentation by *Clostridium acetobutylicum* grown on D-xylose in the presence of acetic or butyric acid. *FEMS Microbiol. Lett.* **18**:103–107.

Yu, E.K.C., M.K.H. Chan, and J.N. Saddler. 1985. Butanol production from cellulosic substrates by sequential co-culture of *Clostridium thermocellum* and *Clostridium acetobutylicum. Biotechnol. Lett.* **7**:509–514.

Yu, P.L., J.B. Smart, and B.M. Ennis. 1987. Differential induction of β-galactosidase and phospho-β-galactosidase activities in the fermentation of whey permeate by *Clostridium acetobutylicum. Appl. Microbiol. Biotechnol.* **26**:254–257.

Zappe, H., D.T. Jones, and D.R. Woods. 1986. Cloning and expression of *Clostridium acetobutylicum* endoglucanase, cellobiase and amino acid biosynthesis genes in *Escherichia coli. J. Gen. Microbiol.* **132**:1367–1372.

Zappe, H., D.T. Jones, and D.R. Woods. 1987. Cloning and expression of a xylanase gene from *Clostridium acetobutylicum* P262 in *Escherichia coli. Appl. Microbiol. Biotechnol.* **27**:57–63.

Zappe, H., W.A. Jones, D.T Jones, and D.R. Woods. 1988. Structure of an endo-β-1,4-glucanase gene from *Clostridium acetobutylicum* P262 showing homology with endoglucanase genes from *Bacillus* spp. *Appl. Environ. Microbiol.* **54**:1289–1292.

Zeikus, J.G. 1980. Chemical and fuel production by anaerobic bacteria. *Ann. Rev. Microbiol.* **30**:423–464.

22

Cloning and Expression of *Clostridium acetobutylicum* Genes Involved in Solvent Production

George N. Bennett and Daniel J. Petersen

22.1 Introduction

Practical importance of solvent production

Clostridium acetobutylicum has been used in various circumstances for the production of acetone and butanol from starches for much of this century (for historical review, see Jones and Woods, 1986). Although the availability of petrochemicals has limited the commercial use of the fermentation in the bulk production of butanol and acetone, interest continues in the process from several perspectives. One is the broad substrate specificity of the organism, which allows growth on a wide variety of feedstocks including some that would otherwise present disposal problems (e.g., cheese whey, apple pomace, and liquors from the paper inductry). Other economic factors include the concomitant production of CO_2 and H_2, which can be recovered, and the value of the bacterial biomass as animal feed. These factors plus the potential of genetic engineering to further improve the product pattern yield and fermentation characteristics of the organism have renewed scientific interest in the solvent-forming clostridia.

Metabolic pathways

The solvent-producing pathway of *C. acetobutylicum* has been discussed in detail in recent reviews (Jones and Woods, 1986, 1989) and will only be summarized here. Acetyl-CoA serves as the starting compound for the pathways to be discussed here (Figure 22.1). Side pathways can compete with the conversion of pyruvate to acetyl-CoA: these are responsible for the production of acetoin and lactate found in certain fermentations. Acetyl-CoA can be converted to ethanol via the intermediate acetaldehyde during solvent production by the enzymes acetaldehyde dehydrogenase and alcohol (ethanol) dehydrogenase. The NAD(P)H built up during the earlier metabolic steps serves as the reducing cofactor. During the acid-producing phase, acetyl-CoA is converted to acetate and butyrate. The former compound is synthesized through conversion of acetyl-CoA to acetyl phosphate by phosphotransacetylase (PTA) followed by hydrolysis to acetate by acetate kinase (AK). One molecule of ATP can be formed during this process.

The formation of butyrate requires the action of a thiolase (acetyl-CoA acetyltransferase) to condense acetyl-CoA to yield acetoacetyl-CoA. This branch point compound serves three major functions: (i) the formation of acetone during solvent production; (ii) the energetically neutral butyrate and acetate uptake by acetoacetyl-CoA:acetate/butyrate:CoA transferase (CoAT); and (iii) the formation of butyryl-CoA. Pathway (i) requires the formation of acetoacetate, usually produced as an integral part of taking up acids by the CoAT (pathway ii). Acetoacetate decarboxylase (AADC) then acts to remove CO_2 from acetoacetate to yield acetone. Pathway (iii) reduces the acetoacetyl-CoA in a process like fatty acid

317

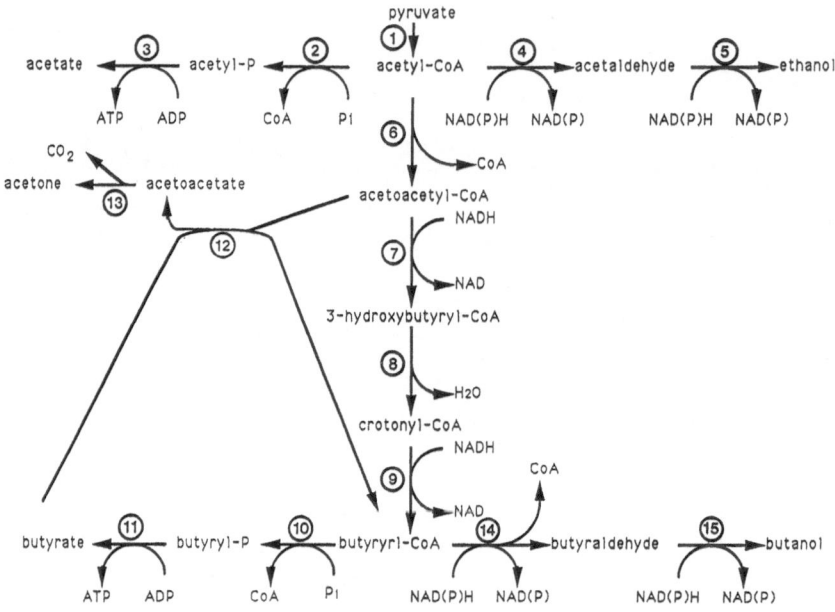

FIGURE 22.1 Solventogenic pathways in clostridia. Arrows indicate direction of carbon flow and cofactor utilization. Enzymes indicated by numbers as follows: (1) pyruvate-ferredoxin oxidoreductase; (2) phosphotransacetylase; (3) acetate kinase; (4) acetaldehyde dehydrogenase; (5) ethanol dehydrogenase; (6) acetyl-CoA acetyltransferase (thiolase); (7) 3-hydroxybutyryl-CoA dehydrogenase; (8) crotonase; (9) butyryl-CoA dehydrogenase; (10) phosphotransbutyrylase; (11) butyrate kinase; (12) acetoacetyl-CoA:acetate/butyrate:CoA transferase (CoA-transferase); (13) acetoacetate decarboxylase; (14) butyraldehyde dehydrogenase; (15) butanol dehydrogenase. Reduction of pyruvate to lactate by lactate dehydrogenase and uptake of acetate by CoA transferase are not shown.

metabolism by successive action of 3-hydroxybutyryl-CoA dehydrogenase, crotonase, and butyryl-CoA dehydrogenase, utilizing NADH as the reducing cofactor (Ljungdahl et al., 1989). At the position of butyryl-CoA the pathway again branches into either that of solvent formation via the reduction to butyraldehyde, and then butanol by butyraldehyde dehydrogenase (BAD) and butanol dehydrogenase (BDH). The production of the acidic metabolite butyrate proceeds analogously to that of acetate with conversion to a phosphorylated form (butyryl phosphate) and subsequent hydrolysis by the enzymes phosphotransbutyrylase (PTB) and butyrate kinase (BK), respectively. This set of reactions also results in the production of one molecule of ATP.

With this overview of the metabolic pathways we now consider the role of the various enzymes in metabolic regulation of solvent formation and the genetic aspects of regulation of levels of those enzymes which are induced during solvent formation.

Regulation of the metabolic pathway

This is a fruitful time to examine the basic mechanism of the shift to solvent production. Recent advances in understanding the complicated series of regulatory events leading to sporulation in the related bacterium *Bacillus subtilis* (Doi, 1989; Losick and Kroos, 1990) as well as the current focus on unraveling the basis of a variety of stress responses (e.g., heat shock, phosphate and nitrogen starvation, anaerobic and osmotic shifts, DNA damage) have developed the required technical methodology and an intellectual context in which to propose and evaluate more detailed mechanisms for the solventogenic switch. In this review we cover the biochemistry of the enzymes involved in the solvent-producing pathway and the cloning and genetic analysis that has begun. Where possible, we try to relate what is known about the formation of solvent-related enzymes to other global regulatory processes that the cells may be undergoing during the state when solvents are being formed.

The general aspects of solvent production have been the subject of recent reviews (Rogers, 1986; Bahl and Gottschalk, 1988; Jones and Woods, 1989). In this section, questions of which enzymes are induced during the solvent-producing phase and which enzymes are the key or limiting ones for solvent production are considered. From experiments analyzing the specific activities of various enzymes during vegetative and solventogenic phases, information as to which enzymes are induced has been derived. The following enzymes appear to be induced: butyraldehyde dehydrogenase (BAD, Dürre et al., 1987; Palosaari and Rogers, 1988; Yan et al., 1988), acetoacetate decarboxylase (AADC, Andersch et al., 1983; Yan et al., 1988; Hüsemann and Papoutsakis, 1989, butanol dehydrogenase (BDH, Dürre et at., 1987; Yan et al., 1988; Palosaari and Rogers, 1988; Welch et al., 1992), acetoacetyl-CoA: acetate/butyrate: CoA: transferase (CoAT, Andersch et al., 1983; Hartmanis and Gatenbeck, 1984; Hüsemann and Papoutsakis, 1989a,b; Welch et al., 1992). The enzymes involved primarily in the formation of butyrate [butyrate kinase (BK) and phosphotransbutyrylase PTB)] are not induced (Palosaari and Rogers, 1988; Hüsemann and Papoutsakis, 1989a,b; Welch et al., 1992), although levels have been reported to vary after the onset of solventogenesis (Andersch et al., 1983; Hartmanis and Gatenbeck, 1984). This pair of enzymes can serve as an alternative pathway for the uptake of acids (Meyer et al., 1986; Cary et al., 1988; Hüsemann and Papoutsakis, 1989a; Wiesenborn et al., 1989a, 1989b).

Studies employing blockage of protein synthesis by transcriptional inhibitors (rifampin) or translational inhibitors (chloramphenicol) have revealed a reduction of new synthesis of BDH, CoAT, and AADC (Ballongue et al., 1985; Palosaari and Roger, 1988; Welch et al., 1992). These experiments have further shown

that BAD activity is relatively unstable (Palosaari and Rogers, 1988; Welch et al., 1992), declining very rapidly after blockage of further synthesis in contrast to the other enzymes, which appear relatively stable.

The induction of these enzymes has been examined with regard to the culture conditions necessary to promote solvent production (George and Chen, 1983; Ballongue et al., 1985, 1989; Hüsemann and Papoutsakis, 1989b). In general the appearance of the induced enzymes requires low pH and the presence of butyrate, leading to a close correlation with the internal concentration of undissociated butyric acid (Monot et al., 1984; Gottwald and Gottschalk, 1985; Terracciano and Kashket, 1986; Hüsemann and Papoutsakis, 1988; Hüsemann and Papoutsakis, 1990). This seems to be a mechanism to avoid the toxicity of excess acid (Zerner et al., 1966; Huang et al., 1985, 1986; Ballongue et al., 1987).

Nutrient limitation and growth conditions

When the carbon source is limited, only acids are produced and the cells do not reach the solventogenic stage (Monot and Engasser, 1983; Monot et al., 1983). High glucose concentrations appear to enhance solvent production (Monot et al., 1982; Roos et al., 1985). These effects are mediated by ATP levels with the lower glucose lessening ATP levels while ATP levels may be high in the case of other nutrient limitation, allowing solvent formation (Meyer and Papoutsakis, 1989). The effect of limitation of other compounds has also been examined. Nitrogen limited cultures at pH below 5.2 can lead to solvent formation; however, the conditions were not optimal for production of solvents (Gottschal and Morris, 1981; Andersch et al., 1982; Jöbses and Roels, 1983; Monot and Engasser, 1983; Roos et al., 1985). Phosphate and iron limitation have been successfully used to produce solvents (Bahl et al., 1982a, 1986); however, iron limitation can favor increased lactate levels (Hanson and Rodgers, 1946). Sulfate limitation has allowed solvent production in continuous culture (Bahl and Gottschalk,

1985), while magnesium limitation has given varying results (Gottschal et al., 1983; Bahl and Gottschalk, 1985; Stephens et al., 1985). Phosphate or sulfate is usually chosen as the limiting factor in engineering studies.

In attempts to alter the distribution of products from a fermentation, the effect of growth conditions and various additives have been examined. A high pressure of H_2 in the fermentor diminished the production of H_2 and increased the yield of butanol and its proportion to total solvents (Gapes et al., 1982; Doremus et al., 1985; Yerushalmi et al., 1985). This can be explained by the effect of H_2 in inhibiting the hydrogenase system and thus increasing the NAD(P)H levels available for the required reductions during solvent production (Jungermann et al., 1973). The addition of CO also stimulates butanol production and reduces the production of acetone (Kim et al., 1984; Datta and Zeikus, 1985; Meyer et al., 1985, 1986; Meyer and Papoutsakis, 1989). This appears to be caused by selective inhibition of certain enzymes (e.g., acetoacetate decarboxylase and hydrogenase). Inactivation of the hydrogenase would affect intracellular nucleotide levels. When glucose limited cultures were sparged with CO, the resulting switch to solvent production was accompanied by threefold to fourfold increases in NADH and ATP levels (Meyer and Papoutsakis, 1989), correlating with the increased presence of NAD(P)H in solventogenic cells (Reardon et al., 1987; Srinivas and Mutharasan, 1987). An analogous type of solvent production pattern has been observed when certain electron acceptors, such as viologen dyes or methylene blue, are added to the medium (Ballongue et al., 1986a; Rao and Mutharasan, 1986, 1987, 1988a, 1988b).

Clostridium acetobutylicum strains and mutants

Although early work used various natural isolates, and detailed inoculation preparation regimens were developed for producing stocks (Gabriel, 1928; Rodgers et al., 1946), work has now focused on a few specific characterized stocks available from established culture collec-

tions. The strains most widely used in current studies include C. acetobutylicum P262, NCIB 8052, ATCC 824, DSM 792, DSM 1732, and C. acetobutylicum B643. Work on Clostridium beijerinckii B592 and B593 is also frequently reported. Efforts to increase the specificity of the fermentation have led to the isolation and characterization of mutants of C. acetobutylicum deficient in enzymes related to solvent production. Several strains exhibiting lower acid production and enhanced formation of butanol have been identified (Rogers and Palosaari, 1987).

Mutants with little butanol dehydrogenase activity and which produce quantities of butyraldehyde have been characterized (Rogers and Palosaari, 1987). A mutant resistant to 2-bromobutyrate that seems to have lost the ability to produce acetone was reported by Junelles et al. (1987). This may be cased by a loss of the CoAT activity, as was found in a nonsolvent-producing mutant (Clark et al., 1989).

A number of other mutants that are defective in production of solvents have been isolated. Many of these have lost more than one enzymatic activity (Clark et al., 1989; Petersen et al., 1989; Petersen and Bennett, 1991; Bertram et al., 1990). Among the most frequently lost are the CoAT, BAD, and AADC. Some mutants are further discussed in regard to individual enzymes or pathways. The use of transposon mutagenesis to produce and analyze nonsolvent-producing mutants has recently been pursued (Mattson and Rogers, 1989; Bertram et al., 1990), and this route shows promise in the future molecular analysis of the regulation of solvent production.

Relationship to other stress responses

Because the solvent-forming clostridial stage is related to physiological changes associated with the stationary phase (e.g., very slow growth, activation of the initial stages of sporulation, pH, and acidic metabolite stress), connections have been sought between the process of solvent induction and other stresses. The general topic has been reviewed by Woods and Jones (1986). Of particular interest has been the relation to the heat or alcohol shock response and the possible role of regulatory nucleotides.

Clostridium acetobutylicum ATCC 4259 responded to a temperature shift from 28° to 45°C by inducing the synthesis of 11 proteins as identified by two-dimensional gel electrophoresis (Terracciano et al., 1988). These heat shock proteins (hsp) were induced by heat shifts throughout the acid-producing phase and in the early stages of the solvent-producing phase (up to 35 mM butanol present). Some variation in individual hsp levels was noted in the duration of the heat shock response (Terracciano et al., 1988) and the return to normal levels of synthesis (Pich et al., 1990). Protein bands at 83, 74, 68, 22, 18, and 16 kDa were also found to be induced by treatment of the cells with 0.1% (13.5 mM) butanol. Both of these stresses gave a less pronounced effect and inhibited overall protein synthesis when they were given during the latter portion of the solvent-producing phase. Exposure to air also induced the 68- and 22-kDa bands. Heat shock (28°→ 45°C) gave greater induction of the similar proteins than the butanol or air treatments. The 74-kDa protein was found to be immunologically related to the DnaK protein, a common heat shock protein in many species (Tilly et al., 1983). The 22-kDa protein was found to be induced by a change in growth rate, not the onset of solvent formation (Pich et al., 1990). Using a phosphate-limited chemostat, the relation of expression of heat shock genes to solvent formation in *C. acetobutylicum* DSM 792 (ATCC 824) and DSM 1731 (ATCC 4259) was studied (Pich et al., 1990). The response was similar in these two strains. Several proteins including Hsp 72, similar to DnaK (Tilly et al. 1983), and Hsp 67, similar to Gro EL (Goloubinoff et al., 1989), were found to be synthesized at higher levels during the shift to solvent formation. This induction occurs several hours before solvents actually appear and therefore does not appear to be a case of induction by the alcohols. The heat shock response of *E. coli* can be mimicked by ethanol exposure (Neidhardt et al., 1984). Pich et al. (1990) suggested that because solvent phase-associated proteins were induced concomitant-

ly with the heat shock proteins, "conditions responsible for the metabolic switch are also able to trigger the heat shock response or vice versa." These workers have proposed the isolation and use of mutants to see if the two phenomenon are induced by the same signal or another connected regulatory mechanism. In this regard, the *dnaK*-homologous gene of *C. acetobutylicum* has been cloned (Giebeler et al., 1990).

During the acidogenic phase, the internal pH of *C. acetobutylicum* decreased with the decline in external pH but did not fall below pH 5.5 (Terracciano and Kashket, 1986). Butanol decreased the transmembrane pH gradient but had little effect on the membrane potential. In growing cells, the switch to solvent production occurred when the internal undissociated butyric acid concentration was 13 mM and total intracellular undissociated acids (acetic plus butyric) were 40 to 45 mM (Terracciano and Kashket, 1986). The rate of butanol formation was at its peak at a concentration of 50 mM butanol, but enzyme synthesis then decreased and eventually ended. *C. acetobutylicum* accumulates diadenosine-5',5'''-P'-P4-tetraphosphate (AppppA) and AppppG to high levels in response to an elevation in temperature or the addition of butanol (Balodimos et al., 1988). The adenylated nucleotides are also formed when the cultures switch from acid production to solvent production. Most of the adenylated nucleotides were found in the extracellular media, however, and the clostridia did not seem able to enzymatically degrade these compounds. Although the influence of ppGpp during the slow growth conditions occurring in the early solvent stage has been considered (Ahmed et al., 1988), no detailed measurement of this regulatory nucleotide or cAMP during the solventogenic switch has been published.

Degeneracy of cultures

Degenerative changes in clostridia strains have been long reported. If continuously subcultured, the bacteria undergo changes affecting morphological as well as physiological features (see Kutzenok and Aschner, 1952). The maintenance of industrial stocks of solvent-producing strains therefore employs a treatment to select for good spore-forming organisms. In the acidogenic phase, motile rod forms are present, while around the onset of solventogenesis motility declines and granulose synthesis begins as the cells assume the appearance of the swollen clostridial form (Preiss, 1984; Long et al., 1984a; Reysenbach et al, 1986; Woods and Jones, 1986). The production of an extracellular capsule (Jones et al., 1982; Long et al., 1984a; Häggström and Förberg, 1986) occurs before the initiation of endospore formation.

The relationship of mutant strains that have lost some functions required for solvent production to the appearance of so-called degenerate strains unable to sporulate or produce solvents is unclear. Asporogenous mutants which cannot produce solvents, granulose, or capsule have been isolated but some mutants defective in these properties appear able to produce solvents (Jones et al., 1982; Long et al., 1984b). Also, DNA synthesis inhibitors inhibit the initiation of endospore formation but do not block solvent formation (Long et al., 1984b). Colonies that are smooth and free of the characteristic hard center surrounded by outgrowths generally are poor producers of solvents, and various colony types have been defined (Adler and Crow, 1987). The enzyme profile of these colonies was not reported. No reversion of nonsolventogenic type IV colonies to the solvent-producing type I has been noted (Adler and Crow, 1987), although earlier some reversion of asporogenous cultures was reported (Jones et al., 1982; Long et al., 1984b).

Another related phenomena is the loss of motility in certain mutant strains lacking solvent production. The degenerate strain DP-4 is not motile, and the flagellin produced by this strain is smaller than that produced by the wild-type *C. acetobutylicum* ATCC 824 conferring antigenic distinction of the two types (Petersen et al., 1989; Petersen and Bennett, 1991). The major flagellin gene has been cloned (D. Petersen, unpublished observations) in order to analyze any genetic rearrangements that may take place in this gene which could be related to the regulation during the solvent-producing stage or to the loss of the large molecular size flagellin protein.

22.2 Strategies and Methodology for Cloning and Identification of Solvent-Stage Genes

The isolation of DNA from *Clostridium acetobutylicum* presented some difficulties in early work, but this was largely overcome through the isolation of DNA from protoplasts (Zappe et al., 1986). Using glycine-supplemented media to aid protoplast formation (Allcock et al., 1982) and anaerobic conditions in the early DNA preparation stage refined the procedure. The age (growth stage) of the culture was also found to be critical because of the presence of extracellular nuclease. Using a variation of the standard bacterial lysis and extraction methods, it has been possible to isolate large DNA (>40 kb) suitable for making lambda phage and cosmid libraries (J. Cary, unpublished observations).

A summary of the solvent-related genes cloned from *C. acetobutylicum*, properties of the encoded enzymes, and the cloning strategy used is presented in Table 22.1. Earlier reviews of clostridial genetics have included summaries of cloned clostridial genes (Young et al., 1989a, 1989b). In work designed to clone a gene from another bacterium into *Escherichia coli*, three methods have been used (Cary et al., 1990b). The first method requires functional protein expression in the host, allowing selection of the desired recombinant clone by complementation of a suitable *E. coli* host strain. Specifically tailored *E. coli* mutant strains are used in combination with appropriate selection or indicator media either directly or with replica plating. A suitable mutant strain can be prepared by mutagenesis, selection, and characterization of a commonly used recipient strain as was done when cloning the *adh* gene using an *adh*⁻ derivative of HB101 (Youngleson et al., 1988).

As an alternative, *E. coli* strains bearing mul-

Table 22.1 Solvent-related enzymes and genes of *Clostridium acetobutylicum*

Enzyme	Native size	Subunits	Purification	Cloning	Cloning method	Library vector/ C. acetobutylicum strain
AK	Unknown		Not done	Not done		
PTA	Unknown		Not done	Not done		
BK	85,000	2 (39 kDa)	Hartmanis, 1987	Cary et al., 1988	Complementation	pBR322/ATCC 824
PTB	264,000	8 (31 kDa)	Wiesenborn et al., 1989a	Cary et al., 1988	Complementation	pBR322/ATCC 824
Thiolase		4 (44 kDa)	Wiesenborn et al., 1988	Petersen and Bennett, 1990a	Immunoscreening	λEMBL3/ATCC 824
BHBD	Unknown		Not done	Youngleson et al., 1989b	Complementation-associated	pEcoR251/P 262
Crotonase	158,000	4 (43 kDa)	Waterson et al., 1972	Not done		
CoA Transferase	93,000	4 (2–26 kDa) (2–28 kDa)	Wiesenborn et al., 1989b	Cary et al., 1990a	Oligonucleotide probe	λEMBL3/ATCC 824
AADC	260,000	8 (28 kDa)	Zerner et al., 1966	Petersen and Bennett, 1990b; Gerischer and Dürre, 1990	Oligonucleotide probe	λEMBL3/ATCC 824 pUC9/DSM 792
BDH-NADH dep	82,000	2 (42 kDa)	Welch et al., 1989	Petersen et al., 1991	Oligonucleotide probe	λEMBL3/ATCC 824
ADH-NADPH dep	Unknown	43,274	Not done	Youngleson et al., 1988	Complementation	pEcoR251/P 262
BAD	115,000	56,000	Palosaari and Rogers, 1988	Not done		
Lactate dehydrogenase	159,000	4 (40 kDa)	Freier and Gottschalk, 1987	Contag et al., 1990	Complementation	pBR322/B 643

tiple mutations prepared for other purposes can be used. However, one must then overcome the lower transformation efficiency of the specialized host strain by an intermediate amplification step in a strain with a DNA modification system like that of the specialized host, e.g., cloning BK and PTB genes in strain DH5 followed by transformation into strain LJ32 (Cary et al., 1988). In this complementation procedure not only must the gene be expressed but it must function adequately in the different cellular environment of *E. coli*, often under conditions in which its expression is not optimal, the pH is not at an optimum, and substrate levels are less than ideal. These factors can lead to a failure to detect an initial clone by this method, even though the cloned gene can complement the reaction in *E. coli*. An example of this sort of complication was noticed in the cloning of the PTB and BK genes (which allow uptake of butyrate and growth of an *E. coli ato* mutant on this short-chain fatty acid); the CoAT gene, which should carry out the same activation and uptake of butyrate, was not cloned readily by this route because of its low expression levels. Once clones of the CoAT were available on plasmids, it was found that those, in which the enzyme was sufficiently expressed could complement the *ato* mutant and give rise to smaller colonies than those produced by complementation conferred by plasmids bearing the highly expressed BK and PTB genes.

The second technique (immunological detection) requires protein expression but does not require enzymatic activity or even production of the full intact protein. This method has been widely used in cloning genes from many species where an antibody to the desired protein is available. However, to employ this technique a relatively pure preparation of the protein that can serve as a good antigen for antibody preparation must be available. The antibody preparation must also be free of components that cross-react with *E. coli* proteins, vector-encoded proteins, or other proteins from the original organism used in the cloning experiment. When these conditions are met, screening of plaques from a lambda library can yield positive recombinants (Cary et al., 1990b; Petersen and Ben-

nett, l990a). Most commonly used are rabbit or sheep antibodies to the purified protein. The visualization can then be performed by using a second antibody conjugate (e.g., commercially available goat anti-rabbit Ab conjugated to alkaline phosphatase) or radioactively labeled protein A.

The third and most technologically demanding approach is the use of an oligonucleotide probe hybridization screening experiment to identify the recombinant λ phage plaque that bears the complementary sequence. This method requires the purification of the protein and amino acid sequence analysis of either the N-terminal portion (if the protein is not N-terminally blocked) or an internally derived peptide obtained from specific protease digestion. After the amino acid sequence is obtained, reverse translation produces a sequence showing all possible nucleotides at each position which could encode that amino acid sequence. *C. acetobutylicum* is very AT rich (72%; Cummins and Johnson, 1971), and previous sequence analysis of genes has shown a high preference for A or T in the codon "wobble" position as well as favored codons for certain amino acids (e.g., Arg; compiled in Young et al., 1989b, and Youngleson et al., 1989b). This factor reduces the number of ambiguities one needs to consider in designing the appropriate probe. This favorable feature of *C. acetobutylicum* is somewhat compromised by the necessity to make a longer probe to obtain high specificity since the sequence is composed primarily of weaker A-T base pairs and the T_m is consequently lower. To obtain specific hybridization with the probe [even using quarternary ammonium salts (DiLella and Woo, 1987)] a longer probe is often required. This is especially the case if the amino acid sequence available contains several amino acids that are encoded by many possible codons.

To avoid probe degeneracy resulting from three- or fourfold ambiguities, inosine can be placed at a few positions as a nondestabilizing base (or spacer) to allow a longer probe to be designed and synthesized (Martin and Castro, 1985). This approach has worked successfully in cloning the NADH-dependent BDH (Petersen et al., 1991). In that case the probe

Table 22.2 Enzyme-specific activities in whole cell extracts[a]

Enzyme	Specific activity in *Clostridium acetobutylicum* extracts	Reference	Specific activity of cloned gene product in *Escherichia coli* extracts	Reference	Expression system[b]
PTB	7.2	Wiesenborn et al., 1989a	22.66	Cary et al., 1988	C
BK	5.2	Hartmanis, 1987	1.44	Cary et al., 1988	C
Thiolase	3.1	Wiesenborn et al., 1988	3.47	Petersen and Bennett, 1990a	C
BHBD	8*	Hartmanis and Gatenbeck, 1984	4330*	Youngleson et al., 1989b	V
CoAT	0.36	Wiesenborn et al., 1989b	3.2	Cary et al., 1990a	V
AADC	0.033	Fridovich, 1963	3.11	Petersen and Bennett, 1990b	C
BDHI	0.031	Welch and Rudolph, in manuscript	3.1	Petersen et al., 1991	C
BDHII	0.042	Welch et al., 1989	0.38	Petersen et al., 1991	C
ADH	0.027*	Bertram et al., 1990	1.34*	Youngleson et al., 1989a	V

[a]See individual references for unit definitions; assays are comparable except as indicated by*, where activity values were assayed by significantly different methods.
[b]Gene was expressed either by sequences within clostridial insert (C) or from promoter on vector (V).

used was a 32-fold degenerate 62-mer containing 8 inosine residues. Even so, some probes fail to give specific positive hybridization signals because of hybridization to the λ vector arms or nonspecific hybridization with other fragments of *C. acetobutylicum* DNA. Once a suitable oligonucleotide is designed, labeled, and tested as a probe in Southern hybridizations to show that it gives a unique hybridization to a fragment in the chromosomal DNA, it can be used to screen plaques from a λ library (Cary et al., 1990b). Phage giving positive signals are rescreened and tested by Southern hybridization to see they contain a restriction fragment similar in size to that of the chromosome. Once found, smaller fragments can be subcloned into plasmid vectors to help limit the size of the DNA fragment bearing the gene as well as allowing for assay of the enzyme activity in growing host cells for positive confirmation of the presence of the gene on the plasmid.

Several genes from *C. acetobutylicum* have been found to be expressed well in *E. coli*. This is interesting in light of the dramatic difference in AT composition between the two organisms and the effect this causes in codon usage. Thus, even though relatively rare *E. coli* codons are used frequently in *C. acetobutylicum* genes, this does not prevent their high-level expression in *E. coli* (Youngleson et al., 1989b). In contrast to the apparently minor difficulty in translating the coding sequence, transcription of the genes has been a limitation on expression levels in certain cases. This is not particularly surprising, because solvent-stage genes may require special factors for transcription not present in *E. coli*. In some cases the entire operon or transcribed region may not have been cloned so the flanking promoter element is not present on the insert analyzed. Expression can still be attained if the insert is suitably oriented with respect to a vector promoter. The few known clostridial promoter sequences have been compared to those of *E. coli* and other organisms by Young et al. (1989b). Table 22.2 includes a list of solvent pathway enzymes expressed in *E. coli* and compares the enzymatic activities found in extracts to those reported for *C. acetobutylicum*. This assessment may be limited by the possibility that not all the protein produced in *E. coli* was enzymatically active.

22.3 Biochemical and Genetic Analysis of Noninduced Enzymes

Phosphotransbutyrylase and butyrate kinase

The enzymes phosphotransbutyrylase (E.C.2.3.1.19) and butyrate kinase (E.C.2.7.2.7) play a major role in the energy metabolism of *Clostridium acetobutylicum* and other saccharolytic and proteolytic clostridia. The PTB enzyme is capable of catalyzing the direct transfer of the butyryl moiety of butyryl CoA to inorganic phosphate, while the BK carries out the subsequent dephosphorylation with the resulting production of ATP and butyrate. The reactions are quite similar to the acetate kinase pathway.

The presence of PTB in *C. acetobutylicum* was first reported by Gavard et al. (1957). The enzyme subsequently, was found in extracts of *C. butyricum* and *C. sporogenes* (Valentine and Wolfe, 1960). The enzyme from *C. butyricum* was partially purified by Valentine and Wolfe (1960). Purification of the enzyme confirmed its distinction from the phosphotransacetylase (PTA), and the specificity of the enzyme for butyryl-CoA was established. Purification of the *C. acetobutylicum* enzyme to homogeneity proved the PTB to be an octamer of identical 31,000-kDa subunits (Wiesenborn et al., 1989a). Although PTA has not been isolated from *C. acetobutylicum*, the PTB does not resemble PTA from other clostridial species in size or number of subunits (Robinson and Sagers, 1972; Drake et al., 1981). The specificity of the PTB for butyryl-CoA marks its importance in the energy metabolism of this organism. The enzyme exhibits less than 2% relative activity with acetyl-CoA as compared to butyryl-CoA, while isovaleryl-CoA and *n*-valeryl-CoA were 95% and 73% as effective as substrates, respectively.

Thompson and Chen (1990) reported the purification of PTB from *Clostridium beijerinckii* B593. It resembled PTB from *C. acetobutylicum* ATCC 824 (Wiesenborn et al., 1989a), and they noted the high K_m for phosphate and suggested that low phosphate levels late in growth (during solvent phase) may limit the activity of PTB in the direction of formation of butyric acid, rendering more butyryl-CoA available for butanol formation. This is a possible factor as phosphate limitation favors production of solvents (Bahl et al., 1982a, 1986). Thompson and Chen also noted the ability of this PTB to react with acetoacetyl-CoA ($K_m = 1.1$ mM) in addition to its reaction with butyryl-CoA ($K_m = 0.04$ mM). This raises some interesting questions as to whether any acetoacetyl phosphate is formed in vivo. It would be expected to be quite unstable, and the possible role of this compound or its breakdown products in the cell is unknown.

As a branch point enzyme, it might be expected that PTB competes with BAD for butyryl-CoA; however, high levels of these two enzymes do not coincide. BAD is an inducible enzyme, whereas significant levels of PTB are found throughout the cell cycle. Some workers found specific activities of PTB (and BK) decreased at the end of the acidogenic phase by about 30% to 60% with respect to their peak activities (Andersch et al., 1983). Also, Hartmanis and Gatenbeck (1984) reported a nearly complete loss of PTB activity coinciding with the onset of solventogenesis. Other enzyme assays have reported little change in PTB levels after the solvent shift (Palosaari and Rogers, 1988; Hüsemann and Papoutsakis, 1989a). The enzyme was shown to be very sensitive to pH changes within the range of typical fermentations. The optimal pH was about 8.0, with sharp declines in activity with both increasing and decreasing pH (Wiesenborn et al., 1989a). The internal pH of the cell can decrease from 7 to as low as 5.5 during a typical fermentation because of the accumulation of butyric and acetic acids. The decrease in PTB activity with decreasing pH is in agreement with in vivo observations (Andersch et al., 1983; Hartmanis and Gatenbeck, 1984). Because the PTB/BK pathway serves to produce butyrate, the pH control of PTB can be viewed as feedback inhibition by butyric acid.

The butyrate kinase catalyzes the reversible formation of butyrate from butyryl phosphate with the concomitant production of ATP. The enzyme is present in many species of clostridia, including *C. acetobutylicum*, *C. pasteurianum*, *C.*

sporogenes, and *C. butyricum*. The enzyme has also been purified from *C. butyricum* (Twarog and Wolfe, 1962), *C. tetanomorphum* (Twarog and Wolfe, 1963), and *C. acetobutylicum* (Hartmanis, 1987). As the enzyme constitutes a site for ATP synthesis, it might be expected to be tightly regulated (Ballongue et al., 1986b). Indeed, during the acid formation stage of growth, both the PTB and the BK exhibit high specific activity. Phosphate-limited cultures might limit the amounts of butyryl-P formed, thus limiting the BK substrate or shifting equilibrium towards butyrate uptake, i.e., solvent production (Bahl et al., 1982a). The increased ATP levels appearing during solventogenesis (Meyer and Papoutsakis, 1989) would shift the equilibrium toward acid uptake. The reduced activity of PTB in the presence of excess ATP may account for the limited use of the pathway in most solvent-producing states.

Purified BK was shown to be a dimer of identical 39,000-Da subunits with a pI value of 5.6. It exhibits broad substrate specificity, utilizing C_2 to C_6 straight- and branched-chain carboxylic acids, with butyrate and valerate showing the highest relative activity (Hartmanis, 1987). The enzyme was only 43% as active with propionate as the substrate and could not utilize acetate. The apparent K_m values for butyrate and valerate were 14 mM, and it was inactive with dicarboxylic acids. The kinase from *C. tetanomorphum* phosphorylated butyrate, valerate, isobutyrate, and propionate, with propionate being 30% as effective as butyrate, whereas the *C. butyricum* kinase phosphorylated propionate and butyrate at equal rates. Amino acid composition analysis of BK showed only 4 Cys residues were present per subunit. Although 1 mM dithiothreitol as reducing agent was necessary for optimal activity, treatment of the enzyme with a variety of sulfhydryl-modifying reagents did not result in inhibition, even at high concentrations (Hartmanis, 1987).

The pH optimum of BK was only determined in the direction of butyryl phosphate formation wherein it showed a maximum activity around pH 7.5. The enzyme retained greater than 60% activity at pH 9.0, however a sharp decline was noted at pH values below 7.0, decreasing to approximately 10% at pH 5.5 (Hartmanis, 1987). This pH dependency is quite similar to that of the PTB. Unlike PTB, the BK has been reported to increase at the onset of solventogenesis (Hartmanis and Gatenbeck, 1984); thus, an increase in level of the enzyme may compensate for its inactivation by low pH. Other studies have not shown a large change in BK activity (Ballongue et al., 1986a; Hüsemann and Papoutsakis, 1989a).

Under certain conditions, the PTB/BK pathway was shown to be reversible (Valentine and Wolfe, 1960; Meyer et al., 1986; Hüsemann and Papoutsakis, 1989a). Taking advantage of this ability, Cary et al. (1988) succeeded in cloning the genes encoding PTB + BK by complementation of *E. coli ato fadR* mutants grown on butyrate as a sole carbon source. The PTB gene was located close to the BK gene, and the coordinate function of both of these genes allowed the uptake of butyrate in the *E. coli* mutants. The genes may form an operon as they were found to be adjacent on the *C. acetobutylicum* chromosome within a region of about 2.5 kb, separated by no more than 250 bp. Their direction of transcription was identical. Such an arrangement may account for the synchronous production of these enzymes. A clostridial promoter was highly active in the expression of these genes in *E. coli*. PTB and BK production yielded about 1% of total cellular protein as evidenced by sodium dodecyl sulfate-polyacrylamide gel electrophoresis (SDS-PAGE) of whole cell extracts of *E. coli* harboring pBR322 derivatives bearing these genes.

Phosphotransacetylase and acetate kinase

Although the genes encoding these enzymes have not been cloned from clostridia, some physiological studies and assays have been performed. Molecular weights among PTAs (E.C.2.3.1.8) of different species vary greatly, from 88,000 in *Clostridium thermoaceticum* (Drake et al., 1981), 63,000 and 75,000 in *C. acidiurici* (Robinson and Sagers, 1972), and 60,000 in *C.*

kluyveri (Klotsch, 1969), whereas the native enzyme from *E. coli* B has been estimated at 450,000 (Shimizu et al., 1969). The gene encoding the *E. coli* K-12 PTA was cloned and shown to encode an 81,000-Da subunit (Yamamoto-Otake et al., 1990).

Phosphotransacetylase has not been characterized from *C. acetobutylicum*. Stadtman (1955) purified the enzyme from *C. kluyveri*. The K_m for acetyl-CoA was 0.64 mM (Bergemeyer et al., 1963). The enzyme was 10% to 50% as effective with propionyl-CoA but could not utilize butyryl-CoA (Stadtman, 1955). A broad pH optimum activity range, from 7.4 to 8.4, was found. Enzyme activity was lost on dialysis and could not be restored by addition of known cofactors. Similar properties were found for the enzyme purified from *E. coli* B. The enzyme was less specific for acetyl-phosphate (K_m of 3 mM); however, no values were reported using other acyl-CoA compounds.

Carbon monoxide gassing of glucose-limited cultures of *C. acetobutylicum* resulted in increases in PTA and PTB. The highest specific activities for pH controlled batch fermentations were seen at pH 4.25, with essentially no activity for cultures maintained at pH 5.0 or 6.0 (Hüsemann and Papoutsakis, 1989a). However, in phosphate-limited chemostats grown under acidogenic or solventogenic conditions, the specific activity of PTA at pH 6.0 was 0.27 while only 0.04 at pH 4.3 (Andersch et al., 1983). The sevenfold drop in activity is not a reflection of the pH optimum of the enzyme as all assays were done under similar conditions. Levels of PTA in batch fermentations were found to decline gradually throughout the fermentation, with a sharp decline corresponding to the onset of solventogenesis (Hartmanis et al., 1984). Other metabolic controls for the regulation of this enzyme are indicated by these experiments. Increased reducing power is also important in regulation of *C. acetobutylicum* PTB. Although increased levels of NADH occur simultaneously with the onset of solventogenesis, a corresponding decrease in PTA activity with increasing NADH concentrations has not yet been shown. Partial inactivation of the *C.*

kluyveri PTA at relatively high ATP concentrations (0.01 M) has been noted (Stadtman, 1955).

The AK enzyme (E.C.2.7.2.1) has been purified from several species of bacteria, *Desulfovibrio vulgaris* (Mannens et al., 1988), *Methanosarcina thermophila* (Aceti and Ferry, 1988), and *E. coli* (Rose, 1955), but the enzyme has not been characterized in clostridia. This may be caused by only low levels of the enzyme: the specific activity in crude extracts of *C. kluyveri* was about 80 fold less than for *E. coli* (Rose, 1955). The gene encoding the *E. coli* K-12 AK has been cloned and shown to code for a protein of about 43,000 subunits molecular weight (Matsuyama et al., 1989). The gene is located adjacent to the *pta*, and the two genes may form an operon transcribed from the *ackA* promoter (Yamamoto-Otake et al., 1990).

In general, a Mg^{2+} or Mn^{2+} requirement has been reported for the enzyme (Rose, 1955; Aceti and Ferry, 1988; Mannens et al., 1988). The *E. coli* enzyme exhibited a K_m for acetate of 300 mM and 470 mM for propionate. V_{max} in the acetate-forming direction was five times greater than that in the reverse direction, thus favoring the formation of ATP from acetyl phosphate and ADP.

Studies on the enzyme levels of acetate kinase in *C. acetobutylicum* revealed large declines to about 15% maximal activity at the onset of solventogenesis (Andersch et al., 1983; Hartmanis et al., 1984). Acetate kinase activity decreased as the acetic acid concentration increased in the medium (Ballongue et al., 1986a). Although reversal of the AK/PTA pathway was suggested to allow uptake of acetate under certain fermentation conditions (Valentine and Wolfe, 1960), the decline in both these activities at this stage of growth suggested that the majority of the uptake does not occur by a reversal of the acid-forming pathways (Hartmanis et al., 1984). Phosphate-limited batch fermentations where pH was controlled at either 6.0 or 4.3 showed specific activities of acetate kinase of 3.61 and 1.12, respectively. Acetate kinase levels remained fairly constant in both butyrate pulsing and CO gassing experiments using *C. acetobutylicum* (Hüsemann and Papoutsakis, 1989a).

Thiolase

In bacterial systems the *ato* thiolase (acetyl-CoA:acetyl-transferase, E.C.2.3.1.9) catalyzes the CoA-dependent cleavage of acetoacetyl-CoA and the acetylation of acetyl-CoA to form acetoacetyl-CoA. A separate *fad* thiolase (thiolase I, E.C.2.3.1.16) has been distinguished from the *ato* thiolase (thiolase II) by both physical properties, subcellular localization, and substrate specificity concerning its ability to cleave both long-chain and short-chain 3-ketoacetyl-CoA substrates. Different types of thiolase have been reported in yeasts and higher eucaryotes as well.

Thiolases have been isolated in three species of *Clostridium*: *C. pasteurianum* (Berndt and Schlegel, 1975), *C. kluyveri* (Hartmanis and Stadtman, 1982), and *C. acetobutylicum* (Wiesenborn et al., 1988). These three species are significantly different, and it might be expected that the thiolase from each has different specificities relating to its cellular function. For example, *C. pasteurianum* produces mostly acetate and butyrate while *C. kluyveri* produces mostly butyrate, caproate, or succinate. During acidogenesis, the thiolase acts as a branch-point enzyme, competing with the phosphotransacetylase for the available pool of acetyl-CoA, and with the two branch pathways leading to the formation of butyric and acetic acids. The formation of acetate yields 4 mol of ATP; 3 mol of ATP is theoretically made using acetyl-CoA to produce butyrate. Butyrate formation is neutral in NADH consumption and formation whereas acetate formation leads to a buildup in NADH levels. During solventogenesis, the thiolase competes with acetaldehyde dehydrogenase for the available acetyl-CoA, and this could influence the ratio of butanol plus acetone to ethanol.

Although two thiolases have been found in *C. pasteurianum* (Berndt and Schlegel, 1975), only one has been reported in *C. acetobutylicum* (Wiesenborn et al., 1988). This enzyme appears to be coordinately expressed with at least two of the three other enzymes that, together with the thiolase, convert acetyl-CoA to butyryl-CoA (Hartmanis and Gatenbeck, 1984). In *C. acetobutylicum*, this enzyme is thought to play a

key role in the uptake of acids as it carries out the thermodynamically unfavorable condensation of two molecules of acetyl-CoA to yield the acetoacetyl-CoA used by CoAT for uptake of acetate and butyrate. However, the thiolase in *C. acetobutylicum* is not induced during solventogenesis (Yan et al., 1988).

The thiolase found in *C. acetobutylicum* is similar to other bacterial thiolases in terms of size and number of subunits. The native enzyme is composed of four subunits of equal molecular weight of 44,000 (Wiesenborn et al., 1988). The thiolase from *C. acetobutylicum* has relatively high activity throughout the physiological range of internal pH of 5.5 to 7.0 (Huang et al., 1985), unlike the enzyme of *C. pasteurianum*, which has a sharper pH optimal at pH 8.0 (Berndt and Schlegel, 1975). It is not surprising that changes in internal pH do not appear important for the regulation of the thiolase, because the thiolase is essential to the formation of the solvents acetone and butanol as well as butyrate. Other factors may play a role in modulating the net condensation. The condensation reaction is extremely sensitive to micromolar amounts of CoASH, and the relative amounts of acetyl-CoA and CoASH may be the most important factor regulating the flux through this branch-point enzyme, and hence the ratio of C2 to C3 + C4 end products (Wiesenborn et al., 1988).

Because the thiolase I found in *E. coli* is constitutively expressed whereas thiolase II is inducible, a possibility existed that both types of thiolase might be present in *C. acetobutylicum*. Using antibodies raised to the purified *C. acetobutylicum* thiolase, Petersen and Bennett (1990a) succeeded in cloning the thiolase from *C. acetobutylicum*. A lambda phage clone was identified by immunoblots of plaques transferred onto nitrocellulose as described previously. Southern hybridization confirmed the gene is encoded within an approximately 4.3-kb clostridial insert. Western blots demonstrate high expression of the thiolase protein in *E. coli*, and the molecular weight of the cloned thiolase corresponded to that observed for the purified enzyme. Hybridization studies using the cloned thiolase gene as a probe have identified only one thiolase gene in the genome and showed that it is not adjacent to the CoAT genes (D. Petersen, unpublished observations) as is the arrangement in *E. coli* (Jenkins and Nunn, 1987). Thus, if another thiolase gene is present it would be expected to have a very different nucleotide sequence.

Enzymes involved in the conversion of acetoacetyl-CoA to butyryl-CoA

The enzymes from *C. acetobutylicum* involved in conversion of acetoacetyl-CoA to butyryl-CoA have not been studied extensively. The similarity of these reactions with those of fatty acid metabolism suggest that possible structural or mechanistic conservation may be found. Activities of the enzymes in the pathway from acetyl-CoA to butyryl-CoA have been examined during the solvent switch. The thiolase, 3-hydroxybutyryl dehydrogenase, and crotonase were coordinately expressed and had maximal activity at the end of growth (Hartmanis and Gatenbeck, 1984).

It has been found that the gene encoding β-hydroxybutyryl-CoA dehydrogenase (BHBD) (E.C.1.1.1.35) is located adjacent to the *adh*1 gene cloned previously (Youngleson et al., 1988). The gene encoding this enzyme was identified by amino acid sequence homology obtained from translation of an open-reading frame sequence upstream of the *adh* gene (Youngleson et al., 1989b). The 282 amino acids indicated a M_r of 31435, and the sequence was 46% identical in amino acids to the mitochondrial fatty acid β-oxidation enzyme from the pig (Bitar et al., 1980) and 38% identical to that of the 3-hydroxyacyl-CoA dehydrogenase (HAD) part of the bifunctional HAD-enoyl-CoA hydratase enzyme from rat peroxisomes (Osumi et al., 1985; Ishii et al., 1987). A segment of six amino acids (82–87) in the *C. acetobutylicum* enzyme was conserved exactly in the three proteins. A lower level of similarity ($\approx 24\%$) was found to the λ-crystallin of rabbit lens. Some similarity was found within the β-α-β structural arrangement of the nucleotide-binding fold of other dehydrogenases to a segment of the BHBD of *C. acetobutylicum*. Secondary structure prediction programs indicated the BHBD to be α-helix rich and to exhibit some relationship to

the known structure of pig mitochondrial 3-hydroxyacyl-CoA dehydrogenase.

The BHBD gene was separated from the *adh* gene by a 354-bp intergenic region without a classical transcription terminator. Regions that could form some stem loop structures were identified earlier (Youngleson et al., 1989a). There is also an upstream Orf not completely within the pCADH100 clone. The possible function or relation of this Orf to other *C. acetobutylicum* proteins has not been determined, but it is intriguing to consider that it may encode part of a metabolically related enzyme because neither the BHBD nor the ADH appear to be expressed from a promoter of clostridial origin, and the upstream Orf may constitute another protein expressed from this operon. The location of the hydroxybutyryl-CoA dehydrogenase gene near that of the NADPH-dependent alcohol dehydrogenase (Youngleson et al., 1989a) indicates an example of clustering of metabolic genes in *C. acetobutylicum* and hints that the metabolically related crotonase and the butyryl-CoA dehydrogenase genes might also be encoded in some cluster.

The crotonase (E.C.4.2.1.17) from *C. acetobutylicum* has been purified (Waterson et al., 1972) and is stable in a lyophilized form. The enzyme has a native molecular weight of 158 kDa and a subunit size estimated at 40 to 43 kDa, as determined from SDS gels and sedimentation studies in 6 M guanidine hydrochloride, suggesting a tetrameric form for the native protein. The enzyme exhibits an extremely high turnover rate and has a K_m and apparent equilibrium constant similar to that of the enzyme from bovine liver. The *C. acetobutylicum* enzyme is specific for short-chain fatty acyl-CoA compounds, with only crotonyl-CoA and hexenoyl-CoA being hydrated. This is in contrast to the broad specificity (C4–C16) of crotonase from bovine liver (Waterson and Hill, 1972). In crude extracts, activity toward the longer chain forms was found, indicating that more than one hydratase is present in the organism. The enzyme was very sensitive to high concentrations of crotonyl-CoA, also unlike bovine liver crotonase. This is especially interesting because no inhibition was observed with hexenoyl-CoA, even at levels of substrate where the enzyme is essentially inactivated by crotonyl-CoA. The overall amino acid composition is, however, rather similar between the bacterial and bovine enzymes. It will be interesting to know the relatedness of these enzymes, especially because the hydroxybutyryl-CoA dehydrogenase has been sequenced and compared (Youngleson et al., 1989b).

The butyryl-CoA dehydrogenase (E.C.1.3.99.2) from another microorganism, *M. elsdenii*, has been isolated and the flavoprotein dissociated (Van Berkel et al., 1988). The characterization of the enzyme from *C. acetobutylicum* has not been reported although Hartmanis and Gatenbeck (1984) did observe a low activity of butyryl-CoA dehydrogenase in extracts.

22.4 Induced or Solvent-Stage Enzymes

CoA-transferase

CoA-transferases (E.C.2.8.3.9) typically activate carboxylic acids to the respective CoA-thioester at the expense of the CoA thioester of another species of carboxylic acid. This transfer of the CoA moiety serves to conserve the energy of the thiol ester and is easily reversed. During the acetone-butanol fermentation of *Clostridium acetobutylicum*, the uptake of acetate and butyrate from the medium occurs via the CoA-transferase and is directly coupled to the production of acetone by AADC, which serves to pull the CoAT reaction in the direction of acid uptake (Hartmanis et al., 1984). Different types of CoA-transferases have been identified in several species of bacteria. CoA-transferases typically show relatively broad substrate specificities, but each enzyme has characteristic preferred substrates. In most bacteria the enzyme is involved in the uptake of substrates for energy and structural use, while in butanol-forming clostridia the CoA-transferase acts mainly to detoxify the medium (Wiesenborn et al., 1989b).

The metabolic role of CoAT in *C. acetobutylicum* is fundamentally different from that of other CoA-transferases that have been characterized (the CoAT of *C. kluyveri* functions for butyrate formation), and this is reflected in the substrate specificity of the enzyme. The preferred substrates were shown to be acetate, propionate, and butyrate (4:2:1 ratio), and the K_m values were very high, ranging from 660 to 1200 mM. In contrast, the CoAT in *Clostridium* sp. SB4 exhibited K_m values ranging from 0.8 to 29 mM. The intracellular concentrations of acetate and butyrate do not exceed about 300 and 700 mM, respectively. At these subsaturating levels, the CoAT activity in vivo would be expected to be very sensitive to changes in the intracellular concentrations of acetate and butyrate. It has been noted that during solventogenesis the level of butyrate often drops much more than that of acetate. As the relative conversion rate is fourfold higher with acetate than for butyrate, it has been suggested that some regulatory mechanism may change the preference of CoAT to butyrate. Such a regulatory response at the onset of solventogenesis would account for the induction of activity observed in batch cultures of the enzyme. The CoAT has been shown to be inducible with the onset of solventogenesis (Andersch et al., 1983; Yan et al., 1988; Hüsemann and Papoutsakis, 1989b), although low levels of this enzyme are present in acid-producing cells (Andersch et al., 1983; Yan et al., 1988).

The CoAT of *C. acetobutylicum* has been purified (Wiesenborn et al., 1989b) and exhibits only superficial similarities to other bacterial CoA-transferases. The protein is a heterotetramer, consisting of two subunits of M_r 26,000 and 28,000 (Cary et al., 1990a) compared with 23,000 and 26,000 in *E. coli* (Sramek and Frerman, 1975) and 23,000 and 25,000 in *Clostridium* sp. strain SB4 (Barber et al., 1978).

Relative activity measurements in the range of pH 5.4 to 7.8 indicate greater than 90% activity from pH 6.4 to 7.8, but dropping to 56% at pH 5.4 (Wiesenborn et al., 1989b). The decrease in activity at pH 5.4 corresponding to its induction of activity in vivo was not expected, but may serve to prolong the fermentation by the cell by preventing acid uptake until the AADC is induced.

The gene encoding the CoAT from *C. acetobutylicum* has been cloned and expressed in *E. coli* (Cary et al., l990a). Cloning was achieved using hybridization of a genomic phage library with oligonucleotides designed to the N-terminal amino acid sequence obtained from the purified protein subunits. Oligonucleotides were generated to each of the two subunits and DNA from recombinant phage, which gave positive hybridization signals with both subunit probes, was subcloned and E. coli cells harboring plasmids containing these DNA inserts were analyzed for activity. The enzyme was not expressed well from its own promoter in E. coli but was expressed when oriented appropriately with a vector promoter. In E. coli the two genes encoding the two subunits of the CoAT are part of an operon (ato operon) that is involved in the degradation of short-chain fatty acids (Jenkins and Nunn, 1987). The cloned genes encoding the two clostridial CoAT subunits were adjacent to one another, with an intervening region of only 27 bp (J. Cary, unpublished observations), and thus may be expressed in a coordinate fashion in an operon type of arrangement similar to that found in E. coli. However, hybridization studies have failed to locate the C. acetobutylicum thiolase close to the cloned CoAT gene. Although the C. acetobutylicum CoAT functions in E. coli as evidenced by complementation of E. coli ato mutants, the genes encoding the CoAT and the enzyme itself appear to share little homology with their E. coli counterparts as evidenced by both Southern and Western blot analysis (Cary et al., 1990a).

Acetoacetate decarboxylase

The enzyme acetoacetate decarboxylase (E.C. 4.1.1.4) plays a key role in solvent production as it catalyzes the production of acetone by decarboxylation of acetoacetate produced by the CoAT. The kinetics of the reaction are such that it effectively pulls the less favorable CoAT reaction towards the uptake of acids. The enzyme was found to be highly inducible in fermentative cultures, with activity correlating to the onset of solvent production (Andersch et al., 1983; Ballongue et al., 1985; Hüsemann and Papoutsakis, 1989a). Although the CoAT is also highly inducible at the onset of solventogenesis, its activity is present earlier in the fermenta-

tion (Andersch et al., 1983; Janati-Idrissi et al., 1988). Because very little solvent is normally produced before the AADC is induced, it appears likely that the AADC is indeed energetically necessary for the uptake of acids from the medium.

The enzyme was first studied by Davies (1943) who was involved in determining the metabolic precursors in the production of acetone and butanol from glucose. Acetoacetate was the only substrate whose decarboxylation was effected by the AADC (Davies, 1943). Rigorous analysis produced no prosthetic groups associated with the enzyme. Crystallization of the enzyme from C. acetobutylicum B-527 (Zerner et al., 1966) allowed further enzymological studies aimed at determining the mechanism of the reaction. As in several other decarboxylases, the reaction proceeds by Schiff base formation at an active site lysine (Warren et al., 1966). No cofactors are necessary for the reaction to occur. Amino acid sequencing of the active site revealed no homology with the active site of other decarboxylases (Laursen and Westheimer, 1966).

The protein consists of an octamer of identical subunits of about 28 kDa each. Inhibition studies have shown that either two subunits combine to form a single active site or that half the number of subunits in the protein are inactive (Davies, 1943; Zerner et al., 1966). Heat activation of the enzyme at 70° for 1 h results in an approximately twofold increase in activity, whereas a biphasic inactivation is observed at temperatures at or above 75°C (Pomeroy, 1970). The enzyme shows a pH optimum at about 5.0, well suited for the production of acetone at acidic pH values (Davies, 1943). It is insensitive to oxygen and can withstand high concentrations of acetone. This latter property has been incorporated into purification procedures by extracting the enzyme from acetone powders (Fridovich, 1963; Petersen and Bennett, 1990b). Interestingly, loss of activity is observed with sonicated cell extracts.

The gene encoding the AADC was cloned by hybridization screening of a genomic phage library using oligonucleotide probes generated to the N terminus based on the amino acid sequence of the purified protein (Petersen and

Bennett, 1990b; Gerischer and Dürre, 1990). The enzyme was well expressed in *E. coli* from a promoter of clostridial origin. Deletion mapping and sequence analysis has confined the gene within a 1.07 kb *Eco*RI/*Nco*I fragment. Sequencing of the 5' upstream region has allowed the promoter to be defined, and the transcription initiation site has been identified by RNA experiments (U. Gerischer and P. Dürre, unpublished observations). The sequence encoding the previously characterized active-site peptide (Laursen and Westheimer, 1966) has also been identified, and is located about 350 bp from the N-terminal methionine. Recent sequence information has shown that the CoAT genes and the AADC gene are closely linked and are transcribed in a convergent manner on the *C. acetobutylicum* chromosome (Gerischer and Dürre 1990 Cary and Petersen, unpublished observations). The C termini are separated by only 65 bp. This small intergenic region is characterized by a large 38-bp stem loop structure that appears to function as a transcription terminator for both genes (Petersen, unpublished observations). Such an unusual arrangement has not previously been found in clostridia. The independent transcription of each gene may account for their independent expression levels during cell growth. Comparison of the transcriptional promoter regions with those of acidogenic enzymes such as BK and PTB may provide a key for induction mechanisms.

The proximity of these two genes may account for their simultaneous loss in mutants and degenerate cultures (Clark et al., 1989; Petersen et al., 1989; Petersen and Bennett 1991). Cloning of the *C. acetobutylicum* gene encoding acetoacetate decarboxylase coupled with the introduction of AADC-gene bearing plasmids into *C. acetobutylicum* mutants (Clark et al., 1989) will enable genetic analysis of the mechanism of regulation in vivo as well as allow optimization of expression of this gene for metabolic engineering purposes. The AADC gene has already been reintroduced into a *C. acetobutylicum* AADC mutant and shown to be functionally expressed in vivo (L. Mermelstein, personal communication).

Butyraldehyde dehydrogenase

The enzyme (E.C.1.2.1.10) catalyzing the formation of acetaldehyde or butyraldehyde from the CoA derivative using NADH or NADPH as a cofactor was observed in *C. kluyveri* (Burton and Stadtman, 1953; Lurz et al., 1979). Early purifications did not completely distinguish if it were a separate protein from the alcohol dehydrogenase activity as these are often combined in other organisms [e.g., *E. coli* (Rudolph et al., 1968; Shone and Fromm, 1981)]. The *C. kluyveri* enzyme has been characterized further (Smith and Kaplan, 1980). The enzyme that converts acetyl-CoA and butyryl-CoA to acetaldehyde and butyraldehyde, respectively, has been purified from *C. acetobutylicum* NRRL B643 (Palosaari and Rogers, 1988) and from *C. beijerinckii* 592 (Yan and Chen, 1990). The *C. acetobutylicum* NRRL B643 enzyme had a native M_r of 115,000 and subunit M_r of 56,000, while the enzyme from *C. beijerinckii* B592 had a native M_r of 100,000 and subunit M_r of 55,000. Acetaldehyde and butyraldehyde dehydrogenase activities copurified. The enzyme was more effective for production of butyraldehyde considering the relative K_m and V_{max} for the two substrates. Alcohol dehydrogenase and butyraldehyde dehydrogenase activities were separated by anion exchange chromatography. No alcohol dehydrogenase copurified, suggesting that in *C. beijerinckii* B592 the two enzymes are independent.

Although Palosaari and Rogers (1988) proposed the enzyme was involved in ethanol and butanol formation, the observation of ethanol formation in certain mutants (Dürre et al., 1986) and variations in the relative amounts of ethanol and butanol may suggest the need for further study. Both purified aldehyde dehydrogenases preferred NADH as a cofactor but utilized NADPH with a higher K_m. The *C. beijerinckii* enzyme exhibited even higher NADH activity compared to NADPH activity when the pH was lowered to pH 6, further suggesting that NADH is the appropriate physiological cofactor under the acidic environment existing during solvent production. The pH optimum was 6.5 to 7.0 in the aldehyde-forming reaction while it was much higher for the acyl-CoA-forming reverse reaction. This is consistent with the prevailing physiological role in aldehyde formation under acidic conditions and in agreement with Andersch et al. (1983). The K_m declined when longer chain aldehydes were used (7.2 mM for acetoaldehyde to 0.37 mM decaproaldehyde). Palosaari and Rogers (1988) observed relative turnover rates (V_{max}) for butyryl-CoA and acetyl-CoA of 5.5 to 1, perhaps indicative of its importance in specifying the ratio of solvents produced [6:1 for butanol to ethanol (Jones and Woods, 1986, 1989)]. C. acetobutylicum (strain 6A), a non-sporulating, nonsolvent-producing strain, contained no BAD activity (Palosaari and Rogers, 1988). Other mutants lacking BAD activity have also been reported (Clark et al., 1989; Petersen et al., 1989; Petersen and Bennett, 1991).

The enzyme from C. beijerinckii was affected by exposure to thiol compounds such as CoA or DTT and these were required to convert the enzyme to an active form. Yan and Chen (1990) postulated that binding CoA to the enzyme puts the enzyme in a conformation which is conducive to activation by thiols and noted the decline in activity of the enzyme when exposed to air. This lability in the presence of oxygen was not noted in the purification of the enzyme from C. acetobutylicum B643 although DTT was used in those assays (Palosaari and Rogers, 1988). The use of CHES or glycylglycine buffer after preparation of the initial extracts in phosphate buffer was stated by Palosaari and Rogers (1988) to be important for assaying the enzyme. They also noted that Tris buffer destabilized the enzyme.

Palosaari and Rogers (1988) observed that BAD was induced simultaneously with NADP-specific butanol dehydrogenase. This induction was blocked by chloramphenicol or rifampin, and the activity of BAD dropped rapidly after blockage of further synthesis, while PTB and BK remained relatively constant over the entire course of fermentation (with only a two- to fourfold variation). This rapid decline in BAD activity distinguishes it from the other solvent pathway enzymes that are stable under similar circumstances (Welch et al., 1992).

Butanol dehydrogenase

Like the AADC, the uniqueness of this enzyme has brought attention directed toward the understanding of its regulation and enzymatic properties. Only a few bacteria are known to produce butanol as a major fermentation product (e.g., C. aurantibutyricum, C. beijerinckii, C. tetanomorphum, and C. puniceum). Conflicting results as to the nature of the enzyme were first reported, depending on the method of preparation or the type of assay used. The BDH is an inducible enzyme, although even after induction the specific activity reported was rather low (Andersch et al., 1983). Not until Dürre et al. (1987) showed that two enzyme activities could be separated by ultracentrifugation was the situation clarified. Two distinct BDHs with different cofactor requirements and different pH ranges have been detected in C. acetobutylicum. The NADPH-dependent BDH has a broad pH optimum between pH 7.8 and 8.5 and is stable in oxygen (Dürre et al., 1987). The NADH-dependent BDH gave similar activities at both pH 7.8 and 6.0.

Once the existence of two BDHs was established the question remained as to their physiological roles. The lower pH optimum of the NADH-dependent BDH makes this the most likely candidate as the major butanol-forming enzyme. However, its preference for butyraldehyde was only 1.7 fold higher than for acetaldehyde, whereas the NADPH-dependent BDH was 2.4 fold higher with butyraldehyde (Dürre et al., 1987) and was shown to be induced with BAD (Palosaari and Rogers, 1988).

An NADH-dependent BDH was purified and shown to be a dimer composed of identical 42-kDa subunits (Welch et al., 1989). The K_m for butyraldehyde was 16 mM and the activity was 50 fold greater in the butanol-forming direction. As intracellular levels of butyraldehyde have been estimated to be about 1 mM (Hartmanis et al., 1984), the NADH-dependent BDH would be very sensitive to changes in the level of this substrate. The purified enzyme was reported to have about 46-fold greater activity for the utilization of butyraldehyde as a substrate than for acetaldehyde. Even greater relative activity was demonstrated with longer chain aldehydes (Welch et al., 1989). The substrate specificity and subunit composition found for this enzyme was closer to that of mammalian dehydrogenases than that found for other bacterial ADHs.

Cloning studies and further enzymatic work have revealed more complexity in the BDH enzymes. An *E. coli* HB 101 allyl alcohol-resistant (*adh⁻*) mutant was transformed with a plasmid library of *C. acetobutylicum* P262 DNA (Youngleson et al., 1988). These cells, which exhibited enhanced sensitivity, were analyzed for alcohol dehydrogenase activity. ADH activity was found to be encoded within a 3.2-kb DNA insert. However, its properties were quite different; most importantly, the enzyme was very active with ethanol as a substrate, showing just fourfold greater activity with butanol, and thus it was classified as an ADH. This ADH was also found to be NADPH specific, and its cellular role remains to be identified. The *adh1* gene of *C. acetobutylicum* P262 was analyzed by DNA sequencing. A region of 1164 bp encoded an ADH of 388 amino acid residues (M_r 43,274; Youngleson et al., 1989a), which is quite similar in size to the purified BDH enzyme (Welch et al., 1989). The amino acid sequence of the *adh* gene most closely resembled a type 3 ADH as defined by Jörnvall et al. (1987). The *C. acetobutylicum* ADH contained 39% identical residues when compared to the iron-containing *Z. mobilis* ADH and 37% with ADH4 of *S. cerevisiae*.

The sequence did not contain the conserved glycine-rich region within the β-α-β fold typically found in the Zn- containing ADHs (Jörnvall et al., 1987). Because the *C. acetobutylicum adh1* gene encodes a NADPH-dependent enzyme, this is not surprising. The fold structure (Rossmann et al., 1974) is more prominently found in NADH-dependent enzymes and the sequence comparisons with NADPH-dependent ADHs are less informative. The secondary structure prediction of the ADH indicates that it is unusually high in α-helix structure and very low in β-sheet structure. This

also correlates with its classification as a type 3 ADH and is markedly different from the type 1 and type 2 ADHs from horse liver or from *Drosophila melanogaster*, which have about equal amounts of α-helix and β-sheet structures. Downstream of the *adh* gene, a 25-bp stem loop structure ($\Delta G = -16$ kca/mol), followed by a sequence of five U's, was found. This resembles a classical rho-independent transcription terminator (Rosenberg and Court, 1979).

The gene encoding the NADH-dependent BDH has also been isolated (Petersen et al., 1991). N-terminal sequencing of the purified NADH-dependent BDH provided an amino acid sequence to which an appropriate oligonucleotide was designed. A 32-fold degenerate inosine containing 62-mer oligonucleotide was synthesized and used to screen a genomic phage library. A DNA fragment from phage exhibiting positive hybridization was cloned into pUC19. *E. coli* harboring this recombinant plasmid demonstrated NADH-dependent BDH activity. Maxicell analysis and Western blotting of whole-cell extracts showed only a single band of M_r about 43,000, which gave rise to the BDH activity.

During this time it was found that a second NADH-dependent BDH could be purified using a dye resin (Welch and Rudolph, in manuscript). This second enzyme (BDH I) is more reactive toward acetaldehyde than the previously reported BDH II, indicating that BDH I may play a much greater role in the production of ethanol. After purification to homogeneity, N-terminal sequence studies showed several differences between BDH I and BDH II within the first 20 amino acids. Only limited amino acid sequence homology was found with the ADH. To discern whether both enzyme activities were the result of a single protein or of multiple forms of a 43,000-kDa protein, the cloned enzyme was purified from *E. coli* using the same method used for preparation of the enzyme from *C. acetobutylicum*. Two separate protein fractions were obtained that closely correspond to the BDHs isolated from *C. acetobutylicum*. The two BDHs differed in pH optimum, cofactor requirements, $K_{m\text{butyraldehyde}}$, $K_{m\text{NADH}}$, and pI (Welch, 1990). These two may be considered as isozymes of BDH. The two enzymes are both encoded within the same cloned 11.9-kb clostridial DNA fragment (Petersen et al. 1991). Their close proximity may belie their coordinate regulation. Each isozyme is well expressed in *E. coli*, although one appears to be found in approximately six to sevenfold higher activity. The exact function of each enzyme and their induction during the solventogenic switch remain a mystery. The use of antibodies or nucleotide segments specific for each enzyme/gene will be essential for elucidating their roles.

Hybridization studies of both the cloned *adh* and *bdh* genes showed no homology between the three genes. This indicates that the ADH and BDH are distinct as expected from enzymological experiments. The contribution of each dehydrogenase in the production of ethanol and butanol remains to be elucidated.

22.5 Production of Other Solvent-Related Metabolites

Lactate

Although *C. acetobutylicum* normally produces relatively little lactate, species of clostridia (*C. thermolacticum* sp. nov.) exist that yield lactate as a major product. Conditions are known under which lactate is a major product of metabolism in *C. acetobutylicum* (Simon, 1947). The most favorable conditions are pH values above 5.0 with carbon monoxide or potassium cyanide or when growth is limited by low concentrations of iron or sulfate (Kubowitz, 1934; Hanson and Rodgers, 1946; Katagiri et al., 1960). Carbon monoxide diverts the metabolism of pyruvate from production of butyric acid to formation of lactate, and lactate can be formed from hexose diphosphate in the presence of glutathione. These experiments suggest a heavy metal complex is involved in the transformations of pyruvate. In continuous culture under iron or sulfate limitation, one sees butanol and acetone production at pH 4.5, but at above pH 5.0 lactate formation is favored (Bahl and Gottschalk, 1984; Bahl et al., 1982b, 1986). Under these conditions the concentra-

tion of fructose-1,6-bisphosphate, an activator of lactate dehydrogenase, is elevated (Freier and Gottschalk, 1987), and metabolism can favor production of lactate. Mutants of *C. acetobutylicum* resistant to pyruvate halogen analogs 3-fluoropyruvate of 3-bromopyruvate were selected and found to accumulate more acetoin and lactate in normal batch cultures regulated at pH 4.8 (El Kanouni et al., 1989).

Lactate dehydrogenase (E.C.1.1.1.27) was purified from *C. acetobutylicum* DSM 1731 and existed as a tetramer of 159 kDa with a pH optimum of 5.8 (Freier and Gottschalk, 1987). The enzyme catalyzed only the reduction of pyruvate and was activated by the presence of fructose-1,6-bisphosphate with calcium or magnesium ions as positive effectors. This enzyme was very unstable and had little activity below pH 5.0. The highest specific activity (0.81 U/mg) was found in cells grown under iron limitation where lactate was produced in quantity or in cells grown on complex media. Only very little was found in solvent-producing cells (0.01 U/mg) (Freier and Gottschalk, 1987). Two thermostable lactate dehydrogenases were purified from *C. thermohydrosulfuricum*; the isoenzymes were activated differently by fructose-1,6-bisphosphate (Turunen et al., 1987). The gene encoding LDH was recently cloned (Contag et al., 1989, 1990). Cloning was achieved by complementation of *E. coli* acetaldehyde dehydrogenase mutants grown anaerobically on glucose. *E. coli* cells harboring the plasmid containing the cloned LDH produced about 10-fold more lactate than *E. coli* wild type, but produced no ethanol. The cloned LDH has a subunit M_r of about 40,000.

Acetoin

Acetoin (3-hydroxy-2- butanone) is produced in a wide range of bacteria (Störmer, 1975; Asada and Fujimori, 1986; Montville et al., 1987; Petit et al., 1989). In many batch fermentations, acetoin is detected in considerable amount (as much as 10% of total solvents). In *C. acetobutylicum*, the higher level of acetoin is correlated with a concomitant reduction in acetone production (Doremus et al., 1985). Acetoin production was increased by l-bar headspace pressure and low agitation rates, and occurred mainly during the acid-producing phase and declined on production of acetone. Addition of acetoin also lowered the final concentration of acetone produced by the culture. It has been proposed that acetoin could serve as an inhibitor of acetoacetate decarboxylase, although it has not been compared to other inhibitors of the enzyme (Davies, 1943; Fridovich, 1963). The pathway for production of acetoin has not been established, but it has been proposed that it is related to the formation of lactate. Indeed, lactate concentrations decreased while acetoin production increased with increasing iron concentrations (Hanson and Rodgers, 1946). However, in this case increases in both acetone and acetoin concentrations were found. In butanediol-producing organisms, it is formed from pyruvate by the successive action of acetolactate synthetase and acetolactate decarboxylase (Störmer, 1975). Alternatively, it could be formed from a condensation with activated pyruvate formed as a part of the pyruvate-ferredoxin oxidoreductase system that produces acetyl-CoA in clostridia. Another route would postulate formation of acetoin by reaction of lactic acid with acetyl-CoA (Doremus et al., 1985).

22.6 Conclusions

Unraveling the mystery of the solventogenic switch has become a priority for increasing the commercial viability of fermentations using *Clostridium acetobutylicum*. This intrinsic feature of the organism is also of great interest as it represents a fascinating example of gene regulation in response to environmental stimuli. The cloning of genes involved in the production of acids and solvents has only recently begun to shed light on the mechanisms of gene regulation in this organism. Sequence analysis of these genes, coupled with the isolation of regulatory mutations via transposon mutagenesis and the means of introducing a gene back into *C. acetobutylicum*, will lead to an understanding of those regions within the DNA necessary for this type of regulation and the action of the proteins involved.

Acknowledgments. We would like to thank H. Bahl, P. Dürre, E. Kashket, W. Chesbro, and J.-S. Chen for copies of recent results cited here. Research on *C. acetobutylicum* at Rice University has been sponsored by the National Science Foundation and by the U.S. Department of Agriculture. D.J. Petersen has been supported by a National Science Foundation Predoctoral Fellowship.

References

Aceti, D.J. and J.G. Ferry. 1988. Purification and characterization of acetate kinase from acetate-grown *Methanosarcina thermophila*. *J. Biol. Chem.* **263**:15444–15448.

Adler, H.I. and W. Crow. 1987. A technique for predicting the solvent-producing ability of *C. acetobutylicum*. *Appl. Environ. Microbiol.* **53**:2496–2499.

Ahmed, I., R.A. Ross, V.K. Mathur, and W.R. Chesbro. 1988. Growth rate dependence of solventogenesis and solvents produced by *Clostridium beijerinckii*. *Appl. Microbiol. Biotechnol.* **28**:182–187.

Allcock, E.R., S.J. Reid, D.T. Jones, and D.R. Woods. 1982. *Clostridium acetobutylicum* protoplast formation and regeneration. *Appl. Environ. Microbiol.* **43**:719–721.

Andersch, W., H. Bahl, and G. Gottschalk. 1982. Acetone-butanol production by *Clostridium acetobutylicum* in an ammonium-linked chemostat at low pH values. *Biotechnol. Lett.* **4**:29–32.

Andersch, W., H. Bahl, and G. Gottschalk. 1983. Level of enzymes involved in acetate, butyrate, acetone and butanol formation by *Clostridium acetobutylicum*. *Eur. J. Appl. Microbiol. and Biotechnol.* **18**:327–332.

Asada, Y. and Y. Fujimori. 1986. Production of acetoin by *Escherichia coil* 32 and its regulation. *J. Agric. Chem. Soc. Jpn.* **60**:369–376.

Bahl, H. and G. Gottschalk. 1984. Parameters affecting solvent production by *Clostridium acetobutylicum* in continuous culture. *Biotechnol. Bioeng. Symp.* **14**:215–223.

Bahl, H. and G. Gottschalk. 1985. Parameters affecting solvent production by *Clostridium acetobutylicum* in continuous culture. *Biotechnol. Bioeng.* **514**:217–223.

Bahl, H. and G. Gottschalk. 1988. Microbial production of butanol/acetone. *In: Biotechnology*, Vol. 6b, H.-J. Rehm and G. Reed, eds., pp. 1–30. Weinheim: VCH Verlagsgesellschaft.

Bahl, H., W. Andersch, and G. Gottschalk. 1982a. Continuous production of acetone and butanol by *Clostridium acetobutylicum* in a two-stage phosphate limited chemostat. *Eur. J. Appl. Microbiol. Biotechnol.* **15**:201–205.

Bahl, H., W. Andersch, and G. Gottschalk. 1982b. Effect of pH and butyrate concentration on the production of acetone and butanol by *Clostridium acetobutylicum* grown in continuous culture. *Eur. J. Appl. Microbiol. Biotechnol.* **14**:17–20.

Bahl, H., M. Gottwald, A. Kuhn, V. Rale, W. Andersch, and G. Gottschalk. 1986. Nutritional factors affecting the ratio of solvents produced by *Clostridium acetobutylicum*. *Appl. Environ. Microbiol.* **52**:169–172.

Ballongue, J., R. Janati-Idrissi, H. Petitdemange, and R. Gay. 1989. Correlation between solvent production and level of solventogenic enzymes in *Clostridium acetobutylicum*. *J. Appl. Bacteriol.* **67**:611–617.

Ballongue, J., J. Amine, H. Petitdemange, and R. Gay. 1986a. Enhancement of solvent production by *Clostridium acetobutylicum* cultivated on a reducing compounds depletive medium. *Biomass* **10**:121–129.

Ballongue, J., J. Amine, H. Petitdemange, and R. Gay. 1986b. Regulation of acetate kinase and butyrate kinase by acids in *Clostridium acetobutylicum*. *Fed. Eur. Microbiol. Soc. Lett.* **35**:295–301.

Ballongue, J., J. Amine, E. Masion, H. Petitdemange, and R. Gay. 1985. Induction of acetoacetate decarboxylase in *Clostridium acetobutylicum*. *Fed. Eur. Microbiol. Soc. Microbiol. Lett.* **29**:273–277.

Ballongue, J., E. Masion, J. Amine, H. Petitdemange, and R. Gay. 1987. Inhibitor effects of products of metabolism on growth of *Clostridium acetobutylicum*. *Appl. Microbiol. Biotechnol.* **26**:568–573.

Balodimos, I.A., E.R. Kashket, and E. Rapaport. 1988. Metabolism of adenylated nucleotides in *Clostridium acetobutylicum*. *J. Bacteriol.* **170**:2301–2305.

Barker, H.A., I.-M. Jeng, N. Neff, J.M. Robertson, F.K. Tam, and S. Hosaba. 1978. Butyryl-CoA:acetoacetate CoA-transferase from a lysine-fermenting clostridium. *J. Biol. Chem.* **253**:1219–1225.

Bergemeyer, H.U., G. Holz, and G. Lang. 1963. Phosphate acetyltransferase from *Clostridium kluyveri*. Culture of the bacteria, isolation, crystallization, and properties of the enzyme. *Biochem. Z.* **338**:114–121.

Berndt, H. and H.G. Schlegel. 1975. Kinetics and properties of β-ketothiolase from *Clostridium pasteurianum*. *Arch. Microbiol.* **103**:21–30.

Bertram, J., A. Kuhn, and P. Dürre. 1990. Tn*916*-induced mutants of *C. acetobutylicum* defective in regulation of solvent formation. *Arch. Microbiol.* **153**:373–377.

Bitar, K.G., A. Perez-Aranda, and R.A. Bradshaw. 1980. Amino acid sequence of L-3-hydroxyacyl-CoA dehydrogenase from pig heart muscle. *Fed. Eur. Biochem. Soc. Lett.* **116**:196–198.

Burton, R.M. and E.R. Stadtman. 1953. The oxidation of acetaldehyde to acetyl coenzyme A. *J. Biol. Chem.* **20**:873–890.

Cary, J.W., D.J. Petersen, E.T. Papoutsakis, and G.N. Bennett. 1990a. Cloning and expression of *Clostridium acetobutylicum* ATCC 824 Acetoacetyl-coenzyme A:acetate/butyrate:coenzyme A-transferase in *Escherichia coli*. *Appl. Environ. Microbiol.* **56**:1576–1583.

Cary, J.W., D.J. Petersen, G.N. Bennett, and E.T. Papoutsakis. l990b. Methods for cloning key primary metabolic enzymes and ancillary proteins associated with the acetone-butanol fermentation of *Clostridium acetobutylicum*. *Ann. N.Y. Acad. Sci.* **589**:67–81.

Cary, J.W., D.J. Petersen, E.T. Papoutsakis, and G.N. Bennett. 1988. Cloning and expression of *Clostridium acetobutylicum* phosphotransbutyrylase and butyrate kinase genes in *Escherichia coli*. *J. Bacteriol.* **170**:4613–4618.

Clark, S.W., G.N. Bennett, and F.B. Rudolph. 1989. Isolation and characterization of mutants of *Clostridium acetobutylicum* ATCC 824 deficient in acetoacetyl coenzyme A:acetate/butyrate: coenzyme A-transferase (EC.2.8.3.9) and in other solvent pathway systems. *Appl. Environ. Microbiol.* **55**:970–976.

Contag, P.R., M.G. Williams, and P. Rogers. 1989. Cloning and expression of the *Clostridium acetobutylicum* lactate dehydrogenase genes in *Escherichia coli*. In: *Abstracts*, American Society for Microbiology National Meeting, Abstr. 0-22. Washington, D.C.: American Society for Microbiology.

Contag, P.R., M.G. Williams, and P. Rogers. 1990. Cloning of a lactate dehydrogenase gene from *Clostridium acetobutylicum* B643 and expression in *Escherichia coli*. *Appl. Environ. Microbiol.* **56**:3760–3765.

Cummins. C.S. and J.L. Johnson. 1971. Taxonomy of the clostridia: wall composition and DNA homologies in *Clostridium butyricum* and other acid-producing clostridia *J. Gen. Microbiol.* **67**:33–46.

Datta, R. and J.G. Zeikus. 1985. Modulation of acetone-butanol-ethanol fermentation by carbon monoxide and organic acids. *Appl. Environ. Microbiol.* **49**:522–529.

Davies, R. 1943. Studies on the acetone-butanol fermentation. 4. Acetoacetic acid decarboxylase of *Cl. acetobutylicum* (BY). *Biochem. J.* **37**:230–238.

DiLella, A.G. and S.L.C. Woo. 1987. Hybridization of genomic DNA to oligonucleotide probes in the presence of tetramethyl ammonium chloride. *Methods Enzymol.* **152**:447–451.

Doi, R.H. 1989. Sporulation and germination. In: *Biotechnology Handbook*, Vol. 2, *Bacillus*, C.R. Harwood, ed., pp. 169–215. New York: Plenum.

Doremus, M.G., J.C. Linden, and A.R. Moreira. 1985. Agitation and pressure effects on acetone-butanol fermentation. *Biotechnol. Bioeng.* **27**: 852–860.

Drake, H.L., S.-I. Hu, and H.G. Wood. 1981. Purification of five components from *Clostridium thermoaceticum* which catalyze synthesis of acetate from pyruvate and methyltetrahydrofolate. *J. Biol. Chem.* **256**:11137–11144.

Dürre, P., A. Kuhn, and G. Gottschalk. 1986. Treatment with allyl alcohol selects specifically for mutants of *Clostridium acetobutylicum* defective in butanol synthesis. *Fed. Eur. Microbiol. Soc. Microbiol. Lett.* **36**:77–81.

Dürre, P., A. Kuhn, M. Gottwald, and G. Gottschalk. 1987. Enzymatic investigations on butanol dehydrogenase and butyraldehyde dehydrogenase in extracts of *Clostridium acetobutylicum*. *Appl. Microbiol. Biotechnol.* **26**:268–272.

El Kanouni, A., A.-M. Junelles, R. Janati-Idrissi, H. Petitdemange, and R. Gay. 1989. *Clostridium acetobutylicum* mutants isolated for resistance to the pyruvate halogen analogs. *Curr. Microbiol.* **18**:139–144.

Freier, D. and G. Gottschalk. 1987. L(+)-lactate dehydrogenase of *Clostridium acetobutylicum* is activated by fructose-1,6,bisphosphate. *Fed. Eur. Microbiol. Soc. Microbiol. Lett.* **43**:229–233.

Fridovich, I. 1963. Inhibition of acetoacetic decarboxylate by anions. *J. Biol. Chem.* **238**:592–598.

Gabriel, C.L. 1928. Butanol fermentation process. *Ind. Eng. Chem.* **20**:1063–1067.

Gapes, J.R., V.F. Larsen, and I.S. Maddox. 1982. Microbial production of n-butanol: some problems associated with the fermentation. *In: Energy from Biomass. Proceedings of the 14th Biotechnology Conference*, pp. 90–101, Palmerston North, New Zealand.

Gavard, R., B. Hautecoeur, and H. Descourtieux. 1957. Phosphotransbutyrylase de *Clostridium acetobutylicum*. *C.R. Acad. Sci.* *(Paris)* **244**:2323–2326.

George, H.A. and J. S. Chen. 1983. Acidic conditions are not obligatory for onset of butanol formation by *Clostridium beijerinckii* (synonym. *C. butylicum*). *Appl. Environ. Microbiol.* **467**:321–327.

Gerischer, U. and P. Dürre. 1990. Cloning, sequencing and molecular analysis of the acetoacetate decarboxylase gene region from *Clostridium acetobutylicum*. *J. Bacteriol.* **172**: 6907–6918.

Giebeler, K., F. Narberhaus, and H. Bahl. 1990. Cloning of the *dna*K-homologous gene of *Clostridium acetobutylicum*. *In: Abstracts*, American Society for Microbiology National Meeting. Washington, D.C.: American Society for Microbiology.

Goloubinoff, P., A.A. Gatenby, and G.H. Lori-mer. 1989. GroE heat-shock proteins promote assembly of foreign prokaryotic ribulose bisphosphate carboxylase oligomers in *Escherichia coli*. *Nature* (London) **337**:44–47.

Gottschal, J.C. and J.G. Morris. 1981. Nonproduction of acetone and butanol by *Clostridium acetobutylicum* during glucose- and ammonium-limitation in continuous culture. *Biotechnol. Lett.* **3**:525–530.

Gottschal, J.C., R. Holt, G.M. Stephens, and J.G. Morris. 1983. Postscript: continuous production of acetone and butanol by *Clostridium acetobutylicum* growing in turbidostat cultures. *In: Acetone Butanol Fermentation and Related Topics 1980–1983*, J.D. Bu'Lock, and A.J. Bu'Lock, eds., p. 74. London: Chamelon Press.

Gottwald, M. and G. Gottschalk. 1985. The internal pH of *Clostridium acetobutylicum* and its effect on the shift from acid to solvent production. *Arch. Microbiol.* **143**:42–46.

Häggström, L. and C. Förberg. 1986. Significance of an extracellular polymer for the energy metabolism in *Clostridium acetobutylicum*: a hypothesis. *Appl. Microbiol. Biotechnol.* **23**:234–239.

Hanson, A.M. and N.E. Rodgers. 1946. Influence of iron concentration and attenuation on the metabolism of *Clostridium acetobutylicum*. *J. Bacteriol.* **51**:568–569.

Hartmanis, M.G.N. 1987. Butyrate kinase from *Clostridium acetobutylicum*. *J. Biol. Chem.* **262**:617–621.

Hartmanis, M.G.N. and S. Gatenbeck. 1984. Intermediary metabolism in *Clostridium acetobutylicum*: levels of enzymes involved in the formation of acetate and butyrate. *Appl. Environ. Microbiol.* **47**:1277–1283.

Hartmanis, M.G.N. and E.R. Stadtman. 1982. Isolation of a selenium-containing thiolase from *Clostridium kluyveri*-identification of the selenium moiety as selenomethionine. *Proc. Na. Acad. Sci. USA* **79**:4912–4916.

Hartmanis, M.G.N., T. Klason, and S. Gatenbeck. 1984. Uptake and activation of acetate and butyrate in *Clostridium acetobutylicum*. *Appl. Microbiol. Biotechnol.* **20**:66–71.

Huang, L., C.W. Forsberg, and L.N. Gibbins. 1986. Influence of external pH and fermentation products on *Clostridium acetobutylicum* intracellular pH and cellular distribution of fermentation products. *Appl. Environ. Microbiol.* **51**:1230– 1234.

Huang, L., L.N. Gibbins, and C.W. Forsberg. 1985. Transmembrane pH gradient and membrane potential in *Clostridium acetobutylicum* during growth under acetogenic and solventogenic conditions. *Appl. Environ. Microbiol.* **50**:1043–1047.

Hüsemann, M.H.W. and E.T. Papoutsakis. 1988. Solventogenesis in *Clostridium acetobutylicum* fermentations related to carboxylic acid and proton concentrations. *Biotechnol. Bioeng.* **32**:843–852.

Hüsemann, M.H.W. and E.T. Papoutsakis. 1989b. Enzymes limiting butanol and acetone formation in continuous and batch cultures of *Clostridium acetobutylicum. Appl. Environ. Microbiol.* **31**:435–444.

Hüsemann, M.H.W. and E.T. Papoutsakis. 1990. Effects of propionate and acetate additions on solvent production in batch cultures of *Clostridium acetobutylicum. Appl. Environ. Microbiol.* **56**:1497–1500.

Husemann, M.H.W. and E.T. Papoutsakis. 1989a Comparison between in vivo and in vitro enzyme activities in continuous and batch fermentations of *Clostridium acetobutylicum. Appl. Microbiol. Biotechnol.* **30**:585–595.

Ishii, N., M. Hijkata, T. Osumi, and T. Hashimoto. 1987. Structural organization of the gene for rat enoyl-CoA hydratase: 3-hydroxyacyl-CoA dehydrogenase bifunctional enzyme. *J. Biol. Chem.* **262**:8144–8150.

Janati-Idrissi, R., A.M. Junelles, H. Petitdemange, and R. Gay. 1988. Regulation of coenzyme A transferase and acetoacetate decarboxylase activities in *Clostridium acetobutylicum. Ann. Inst. Pasteur Microbiol.* **139**:683–688.

Jenkins, L.S. and W.D. Nunn. 1987. Genetic and molecular characterization of the genes involved in short chain fatty acid degradation in *Escherichia coli. J. Bacteriol.* **169**:42–52.

Jöbses, I.M.L. and J.A. Roels. 1983. Experience with solvent production by *Clostridium beijerinckii* in continuous culture. *Biotechnol. Bioeng.* **25**:1187–1194.

Jones, D.T. and D.R. Woods. 1986. Acetone-butanol fermentation revisited. *Microbiol. Rev.* **50**:484–524.

Jones, D.T. and D.R. Woods. 1989. Solvent production. *In: Biotechnology Handbooks. Vol. 3. Clostridia,* N.P. Minton and D.J. Clarke, eds., pp. 105–144. New York: Plenum.

Jones, D.T., A. van der Westhuizen, S. Long, E.R. Allcock, S.J. Reid, and D.R. Woods. 1982. Solvent production and morphological changes in *Clostridium acetobutylicum. Appl. Environ. Microbiol.* **43**:1434–1439.

Jörnvall, H., B. Persson, and J. Jeffery. 1987. Characteristics of alcohol/polyol dehydrogenases: The zinc-containing long-chain alcohol dehydrogenases. *Eur. J. Biochem.* **167**:195–201.

Jungermann, K., R.K. Thauer, G. Leimenstoll, and K. Decker. 1973. Function of reduced pyridine nucleotide-ferredoxin oxidoreductases in saccharolytic clostridia. *Biochim. Biophys. Acta* **305**:268–280.

Junelles, A.M., R. Janati-Idrissi, A. El Kanouni, H. Petitdemange, and R. Gay. 1987. Acetone-

butanol fermentation by mutants selected for resistance to acetate and butyrate halogen analogues. *Biotechnol. Lett.* **9**:175–178.

Katagiri, H., K. Imai, and T. Sugimori. 1960. On the metabolism of organic acids by *Clostridium acetobutylicum.* Part I. Formation of lactic acid and racemiase. *Bull. Agric. Chem. Soc. Jpn.* **24**:163–172.

Kim, B.H., P. Bellows, R. Datta, and J.G. Zeikus. 1984. Control of carbon and electron flow in *Clostridium acetobutylicum* fermentations: utilization of carbon monoxide to inhibit hydrogen production and to enhance butanol yields. *Appl. Environ. Microbiol.* **48**:764–770.

Klotsch, H.R. 1969. Phosphotransacetylase from *Clostridium kluyveri. Methods Enzymol.* **13**:381–386.

Kubowitz, F. 1934. Inhibition of butyric acid fermentation by carbon monoxide. *Biochem. Z.* **274**:285–298.

Kutzenok, A., and M. Aschner. 1952. Degenerative processes in a strain of *Clostridium butylicum. J. Bacteriol.* **64**:829–836.

Laursen, R.A. and F.H. Westheimer. 1966. The active site of acetoacetate decarboxylase. *J. Am. Chem. Soc.* **88**:3426–3430.

Ljungdahl, L.G., J. Hugenholtz, and J. Wiegel. 1989. Acetogenic and acid-producing Clostridia. *In: Biotechnology Handbooks, Vol. 3, Clostridia,* N.P. Minton, and D.J. Clarke, eds., pp. 145–191. New York: Plenum.

Long, S., D.T. Jones, and D.R. Woods. 1984a. Initiation of solvent production, clostridial stage and endospore formation in *Clostridium acetobutylicum* P262. *Appl. Microbiol. Biotechnol.* **20**:256–261.

Long, S., D.T. Jones, and D.R. Woods. 1984b. The relationship between sporulation and solvent production in *Clostridium acetobutylicum* P262. *Biotechnol. Lett.* **6**:529–534.

Losick, R. and L. Kroos. 1990. Dependence pathways for the expression of genes involved in endospore formation in *Bacillus subtilis. In: Regulation of Procaryotic Development,* I. Smith, R.A. Slepecky, and P. Setlow, ed., pp. 223–241. Washington, D.C.: American Society for Microbiology.

Lurz, R., F. Mayer, and G. Gottschalk. 1979. Electron microscopic study on the quaternary structure of the isolated particulate alcohol-acetaldehyde dehydrogenase complex and on its identity with the polygonal bodies of *Clostridium kluyveri. Arch. Microbiol.* **120**:255–262.

Mannens, G., G. Slegers, and A. Claeys. 1988. Purification and immobilization of acetate kinase from *Desulfovibrio vulgaris. Biotechnol. Lett.* **10**:563–568.

Martin, F.H. and M.M. Castro. 1985. Base pairing involving deoxyinosine: Implications for probe design. *Nucleic Acids Res.* **13**:8927–8938.

Matsuyama, A., H. Yamamoto, and E. Nakano. 1989. Cloning, expression, and nucleotide sequence of the *Escherichia coli* K-12 *ackA* gene. *J. Bacteriol.* **171**:577–580.

Mattson, D.M. and P. Rogers. 1989. Identification of some transposon insertion sites in the chromosome of *C. acetobutylicum* following transfer of the tetracycline resistant transposon Tn916 from *Streptococcus faecalis. In: Abstracts*, American Society for Microbiology National Meeting. Washington D.C.: American Society for Microbiology, Abstr. 0–39.

Meyer, C.L. and E.T. Papoutsakis. 1989. Increased levels of ATP and NADH are associated with increased solvent production in continuous cultures of *C. acetobutylicum. Appl. Microbiol. Biotechnol.* **30**:450–459.

Meyer, C.L., J.K. McLaughlin and E.T. Papoutsakis. 1985. The effect of CO on growth and product formation in batch cultures of *Clostridium acetobutylicum. Biotechnol. Lett.* **7**:37–42.

Meyer, C.L., J.W. Roos, and E.T. Papoutsakis. 1986. Carbon monoxide gassing leads to alcohol production and butyrate uptake without acetone formation in continuous cultures of *Clostridium acetobutylicum. Appl. Microbiol. Biotechnol.* **24**:159–167.

Monot, F. and J.M. Engasser. 1983. Production of acetone and butanol by batch and continuous culture of *Clostridium acetobutylicum* under nitrogen limitation. *Biotechnol. Lett.* **5**:213–218.

Monot, F., J.M. Engasser, and H. Petitdemange. 1983. Regulation of acetone butanol production in batch and continuous cultures of *Clostridium acetobutylicum. Biotechnol. Bioeng. Symp.* **13**:207–216.

Monot, F., J.M. Engasser, and H. Petitdemange. 1984. Influence of pH and undissociated butyric acid on the production of acetone and butanol in batch cultures of *Clostridium acetobutylicum. Appl. Microbiol. Biotechnol.* **19**:422–426.

Monot, F., J.R. Martin, H. Petitdemange, and R. Gay. 1982. Acetone and butanol production by *Clostridium acetobutylicum* in a synthetic medium. *Appl. Environ. Microbiol.* **44**:1318–1324.

Montville, T.J., A.H.-M. Hsu, and M.E. Meyer. 1987. High-efficiency conversion of pyruvate to acetoin by *Lactobacillus plantarum* during pH-controlled and fed-batch fermentations. *Appl. Environ. Microbiol.* **53**:1798–1802.

Neidhardt, F.C., R.A. Van Bogelen, and V. Vaughn. 1984. The genetics and regulation of heatshock proteins. *Annu. Rev. Genet.* **18**:295–329.

Osumi, T., N. Ishii, M. Hijikata, K. Kamijo, H. Ozasa, S. Furuta, S. Miyazawa, K. Kondo, K. Inoue, H. Kagamiyama, and T. Hashimoto. 1985. Molecular cloning and nucleotide sequence of the cDNA for rat peroxisomal enoyl-CoA hydratase:3-hydroxyacyl-CoA dehydrogenase bifunctional enzyme. *J. Biol. Chem.* **260**:8905–8910.

Palosaari, N.R. and P. Rogers. 1988. Purification and properties of the inducible coenzyme A-linked butyraldehyde dehydrogenase from *Clostridium acetobutylicum. J. Bacteriol.* **170**:2971–2976.

Petersen, D.J. and G.N. Bennett. 1990a. Cloning of *Clostridium acetobutylicum* ATCC 824 thiolase gene in *Escherichia coli. In; Abstracts*, American Society for Microbiology National Meeting. Washington, D.C.

Petersen, D.J. and G.N. Bennett. 1990b. Purification of acetoacetate decarboxylase from *Clostridium acetobutylicum* ATCC 824 and cloning of the acetoacetate decarboxylase gene. *Appl. Environ. Microbiol.* **56**:3491–3498.

Petersen, D.J., J.W. Cary, and G.N. Bennett. 1989. Isolation of a flagellar mutant of *Clostridium acetobutylicum* ATCC 824. *In: Abstracts*, American Society for Microbiology National Meeting. Washington, D.C.: American Society for Microbiology.

Petersen, D.J. and G.N. Bennett. 1991. Enzymatic characterization of a nonmotile, nonsolventogenic *Clostridium acetobutylicum* ATCC 824 mutant. *Current. Microbiol.* **23**:253–258.

Petersen D.J., R.W. Welch, F.B. Rudolph, and G.N. Bennett. 1991. Molecular cloning of an alcohol (butanol) dehydrogenase gene cluster from *Clostridium acetobutylicum* ATCC 824 *J. Bacteriology* **173**:1531–1534.

Petit, C., F. Vilchez, and R. Marczak. 1989. Formation and stabilization of diacetyl and acetoin concentration in fully grown cultures of *Streptococcus lactis* subsp. *diacetylactis. Biotechnol. Lett.* **11**:53–56.

Pich, A., F. Narberhaus, and H. Bahl. 1990. Induction of heat shock proteins during initiation of solvent formation in *C. acetobutylicum Appl. Microbiol. Biotechnol.* **33**:697–704.

Pomeroy, A. 1970. An investigation of acetoacetate decarboxylase. Ph.D. Thesis, Duke University, North Carolina.

Preiss, J. 1984. Bacterial glycogen synthesis and its regulation. *Annu. Rev. Microbiol.* **38**:419–458.

Rao, G. and R. Mutharasan. 1986. Alcohol production by *Clostridium acetobutylicum* induced by methylviologen. *Biotechnol. Lett.* **8**:893–896.

Rao, G. and R. Mutharasan. 1987. Altered electron flow in continuous cultures of *Clostridium acetobutylicum* induced by viologen dyes. *Appl. Environ. Microbiol.* **53**:1232–1235.

Rao, G. and R. Mutharasan. 1988a. Altered electron flow in a reducing environment in *Clostridium acetobutylicum, Biotechnol. Lett.* **10**:129–132.

Rao, G. and R. Mutharasan. 1988b. Directed metabolic flow with high butanol yield and selectivity in continuous cultures of *Clostridium acetobutylicum. Biotechnol. Lett.* **10**:313–318.

Reardon, K.F., T.-H. Scheper, and J.E. Bailey. 1987. Metabolic pathway rates and culture fluorescence in batch fermentations of *Clostridium acetobutylicum. Biotechnol. Prog.* **3**:153–167.

Reysenbach, A.L., N. Ravenscroft, S. Long, D.T. Jones, and D.R. Woods. 1986. Characterization, biosynthesis, and regulation of granulose in *Clostridium acetobutylicum. Appl. Environ. Microbiol.* **52**:275–281.

Robinson, J.R. and R.D. Sagers. 1972. Phosphotransacetylase from *Clostridium acidiurici. J. Bacteriol.* **112**:465–473.

Rodgers, N.E., R.H. Henika, and A.M. Hanson. 1946. Relation of strain variation and culture history to the synthesis of riboflavin by *Clostridium acetobutylicum* in whey. *J. Bacteriol.* **51**:569–570.

Rogers, P. 1986. Genetics and biochemistry of *Clostridium* relevant to development of fermentation processes. *Adv. Appl. Microbiol.* **31**:1–60.

Rogers, P. and N. Palosaari. 1987. *Clostridium acetobutylicum* mutants that produce butyraldehyde and altered quantities of solvents. *Appl. Environ. Microbiol.* **53**:2761–2766.

Roos, J.W., J.K. McLaughlin, and E.T. Papoutsakis. 1985. The effect of pH on nitrogen supply, cell lysis and solvent production in fermentations of *Clostridium acetobutylicum. Biotechnol. Bioeng.* **27**:681–694.

Rose, I.A. 1955. Acetate kinase of bacteria. *Methods Enzymol.* **1**:591–595.

Rosenberg, M. and D. Court. 1979. Regulatory sequences involved in the promotion and termination of RNA transcription. *Annu. Rev. Gen.* **13**:319–353.

Rossmann, M.G., D. Moras, and K.W. Olsen. 1974. Chemical and biological evolution of a nucleotide-binding protein. *Nature.* London **250**:194–199.

Rudolph, F.B., D.L. Purich, and H.J. Fromm. 1968. Coenzyme A-linked aldehyde dehydrogenase from *Escherichia coli. J. Biol. Chem.* **243**:5539–5545.

Shimizu, M., T. Suzuki, K.-Y. Kameda, and Y. Abiko. 1969. Phosphotransacetylase of *Escherichia coli* B, purification and properties. *Biochim. Biophy. Acta* **191**:550–558.

Shone, C.C. and H J. Fromm. 1981. Steady-state and pre-steady-state kinetics of coenzyme A linked aldehyde dehydrogenase from *Escherichia coli. Biochemistry* **20**:7494–7501.

Simon, E. 1947. The formation of lactic acid by *Clostridium acetobutylicum* (Weizman). *Arch. Biochem.* **13**:237–243.

Smith, L.M. and N.D. Kaplan. 1980. Purification, properties and kinetic mechanism of coenzyme

A linked aldehyde dehydrogenase from *Clostridium kluyveri. Arch. Biochem. Biophy.* **203**:663–675.

Sramek, S.J. and F.E. Frerman. 1975. Purification and properties of *Escherichia coli* coenzyme A-transferase. *Arch. Biochem. Biophy.* **171**:14–26.

Srinivas, S.P. and R. Mutharasan. 1987. Culture fluorescence characteristics and its metabolic significance in batch cultures of *Clostridium acetobutylicum. Biotechnol. Lett.* **9**:139–142.

Stadtman, E.R. 1955. Phosphotransacetylase from *Clostridium kluyveri. Methods Enzymol.* **1**:596–599.

Stephens, G.M., R.A. Holt, J.C. Gottschal, and J.G. Morris. 1985. Studies on the stability of solvent production by *Clostridium acetobutylicum* in continuous culture. *J. Appl. Bacteriol.* **59**:597–605.

Störmer, F. 1975. 2,3-Butanediol biosynthetic system in *Acetobacter aerogenes. Methods Enzymol.* **41**:518–533.

Terracciano, J.S. and E.R Kashket. 1986. Intracellular conditions required for initiation of solvent production by *Clostridium acetobutylicum. Appl. Environ. Microbiol.* **52**:86–91.

Terracciano, J.S., E. Rapaport, and E.R. Kashket. 1988. Stress- and growth phase-associated proteins of *Clostridium acetobutylicum. Appl. Environ. Microbiol.* **54**:1989–1995.

Thompson, D.K and J.-S. Chen. 1990. Purification and properties of an acetoacetyl coenzyme A-reacting phosphotransbutyrylase from *Clostridium beijerinckii* ("*Clostridium butylicum*") NRRL B593. *Appl. Environ. Microbiol.* **56**:607–613.

Tilly, K., N. McKittrick, M. Zylicz, and G. Georgopoulos. 1983. The *dna*K protein modulates the heat-shock response of *Escherichia coli. Cell* **34**:641–646.

Turunen, M., E. Parkkinen, J. Londesborough, and M. Korhola. 1987. Distinct forms of lactate dehydrogenase purified from ethanol-producing and lactate-producing cells of *Clostridium thermohydrosulfuricum. J. Gen. Microbiol.* **133**:2865–2873.

Twarog, R. and R.S. Wolfe. 1962. Enzymatic phosphorylation of butyrate. *J. Biol. Chem.* **237**:2472–2477.

Twarog, R. and R.S. Wolfe. 1963. Role of butyryl phosphate in the energy metabolism of *Clostridium tetamorphum. J. Bacteriol.* **86**:112–117.

Valentine, R.C. and R.S.. Wolfe. 1960. Purification and role of phosphotransbutyrylase. *J. Biol. Chem.* **235**:1948–1956.

Van Berkel, W.J.H., W.A.M. Van den Berg, and F. Mueller. 1988. Large-scale preparation and reconstitution of apo-flavoproteins with special reference to butyryl-CoA dehydrogenase from *Megasphaera elsdenii.* Hydrophobic-interaction chromatography. *Eur. J. Biochem.* **178**:197–207.

Warren, S., B. Zerner, and F.H. Westheimer. 1966. Acetoacetate decarboxylase, identification of lysine at the active site. *Biochemistry* **5**:817–823.

Waterson, R.M. and R.L. Hill. 1972. Enoyl coenzyme A hydratase (crotonase)-catalytic properties of crotonase and its possible regulatory role in fatty acid oxidation. *J. Biol. Chem.* **247**:5258–5265.

Waterson, R.M., F.J. Castellino, G.M. Hass, and R.L. Hill. 1972. Purification and characterization of crotonase from *Clostridium acetobutylicum*. *J. Biol. Chem.* **247**:5266–5271.

Welch, R.W., F.B. Rudolph, and E.T. Papoutsakis. 1989. Purification and characterization of the NADH-dependent butanol dehydrogenase from *Clostridium acetobutylicum* (ATCC 824). *Arch. Biochem. Biophy.* **273**:309–318.

Welch, R.W. 1990 Purification and studies of two butanol (ethanol) dehydrogenases and the effects of rifampicin and chloramphenicol on other enzymes important in the production of butyrate and butanol in *Clostridium acetobutylicum* ATCC 824. Ph.D. Thesis, Rice University, Texas.

Welch, R.W., S.W. Clark, G.N. Bennett, and F.B. Rudolph. 1992. Effects of rifampicin and chloramphenicol on product and enzyme levels of the acid- and solvent-producing pathways of *Clostridium acetobutylicum* (ATCC 824). *Enzyme Microb. Technol.* **14**:277–283.

Wiesenborn, D.P., F.B. Rudolph and E.T. Papoutsakis. 1988. Thiolase from *Clostridium acetobutylicum* ATCC 824 and its role in the synthesis of acids and solvents. *Appl. Environ. Microbiol.* **54**:2717–2722.

Wiesenborn, D.P., F.B. Rudolph, and E.T. Papoutsakis. 1989a. Phosphotransbutyrylase from *Clostridium acetobutylicum* ATCC 824 and its role in acidogenesis. *Appl. Environ. Microbiol.* **55**:317–322.

Wiesenborn, D.P., F.B. Rudolph, and E.T. Papoutsakis. 1989b. Coenzyme A transferase from *Clostridium acetobutylicum* ATCC 824 and its role in the uptake of acids. *Appl. Environ. Microbiol.* **55**:323–329.

Woods, D.R. and D.T. Jones. 1986. Physiological responses of *Bacteroides* and *Clostridium* strains to environmental stress factors. *Adv. Microb. Physiol.* **27**:1–64.

Yamamoto-Otake, H., A. Matsuyama, and E. Nakano. 1990. Cloning of a gene coding for phosphotransacetylase from *Escherichia coli*. *Appl. Microbiol. Biotechnol.* **33**:680–682.

Yan, R.-T. and J.-S. Chen. 1990. Coenzyme A-acylating aldehyde dehydrogenase from *Clostridium beijerinckii* NRRL B592 *Appl. Environ. Microbiol.* **56**:2591–2599.

Yan, R.-T., C. Zhu, C. Golemboski, and J. Chen. 1988. Expression of solvent-forming enzymes and onset of solvent production in batch cultures of *Clostridium beijerinckii* ("*Clostridium butylicum*"). *Appl. Environ. Microbiol.* **54**:642–648.

Yerushalmi, L., B. Volesky, and T. Szezesny. 1985. Effect of increased hydrogen partial pressure on the acetone-butanol fermentation by *Clostridium acetobutylicum*. *Appl. Microbiol. Biotechnol.* **22**:103–107.

Young, M., N.P. Minton, and W.L. Staudenbauer. 1989a. Recent advances in the genetics of the clostridia. *Fed. Eur. Microbiol. Soc. Microbiol. Rev.* **63**:301–326.

Young, M., W.L. Staudenbauer, and N.P. Minton. 1989b. Genetics of clostridia. *In: Biotechnology Handbooks, Vol. 3, Clostridia*, N.P. Minton and D.J. Clarke, eds., pp. 63–103. New York: Plenum.

Youngleson, J.S.. W.A. Jones, D.T. Jones, and D.R. Woods. 1989a. Molecular analysis and nucleotide sequence of the *adh*1 gene encoding an NADPH-dependent butanol dehydrogenase in the gram-positive anaerobic *Clostridium acetobutylicum*. *Gene* **78**:355–364.

Youngleson, J.S., D.T. Jones, and D.R. Woods. 1989b. Homology between hydroxybutyryl and hydroxyacyl coenzyme A dehydrogenase enzymes from *Clostridium acetobutylicum* and vertebrate fatty acid β-oxidation pathways. *J. Bacteriol.* **171**:6800–6807.

Youngleson, J.S., J.D. Santangelo, D.T. Jones, and D.R. Woods. 1988. Cloning and expression of a *Clostridium acetobutylicum* alcohol dehydrogenase gene in *Escherichia coli*. *Appl. Environ. Microbiol.* **54**:676–682.

Zappe, H., D.T. Jones, and D.R. Woods. 1986. Cloning and expression of *Clostridium acetobutylicum* endoglucanase, cellobiase, and amino acid biosynthesis genes in *Escherichia coli*. *J. Gen. Microbiol.* **132**:1367–1372.

Zerner, B., S.M. Coutts, F. Lederer, H.H. Waters, and F.H. Westheimer. 1966. Acetoacetate decarboxylase. Preparation of the enzyme. *Biochemistry* **5**:813–816.

23

Molecular Analysis of Glutamine Synthetase Genes and Enzymes of *Clostridium* and *Bacteroides*

David R. Woods

23.1 Introduction

Although relatively little is known about the amino acid biosynthetic pathways in anaerobes, it is often anticipated that they will be similar to those already described for aerobes. An aim of this chapter is to illustrate that this is not necessarily true, and an investigation of glutamine synthetase (GS), an enzyme that has been extensively studied in aerobes, has revealed interesting and exciting differences and novel genetic regulatory mechanisms. The glutamine synthetase gene (*glnA*) is the only example of an amino acid biosynthetic gene from two anaerobes, *Clostridium acetobutylicum* and *Bacteroides fragilis*, which has been analyzed at the molecular level.

In the biosynthesis of nitrogenous compounds the amino acids, glutamine and glutamate have a pivotal role. They provide cellular nitrogen for the synthesis of purines, pyrimidines, amino sugars, most of the amino acids, and metabolites such as NAD and p-aminobenzoate. Three enzymes GS(EC 6.3.1.2), glutamate synthetase (GOGAT, EC 1.4.1.13), and glutamate dehydrogenase (GDH, EC 1.4.1.4) catalyze the three main reactions generally used by bacteria for ammonia assimilation. GS catalyzes the reaction:

$$\text{L-glutamate} + \text{NH}_4^+ + \text{ATP} \rightarrow$$
$$\text{L-glutamine} + \text{ADP} + \text{P}_i$$

GS enzymes of eucaryotic origin are octamers (Prusiner and Stadtman, 1973) whereas the GS enzymes of eubacteria and archaebacteria are dodecamers of M_r of approximately 600,000 composed of a single type of subunit, the M_r of which falls in the range of 44,000 to 59,000 (Streicher and Tyler, 1980; Bhatnagar et al., 1986).

Procaryotes have two forms of GS, termed GSI and GSII. The majority of bacteria investigated have GSI enzymes, but members of the *Rhizobiaceae* contain both GSI and GSII enzymes. GSI is the typical procaryotic GS whereas GSII is similar to eucaryotic GS enzymes (Carlson and Chelm, 1986). Procaryotic GSI subunits investigated vary in length from 444 to 474 amino acids (Janssen et al., 1988), whereas the GSII subunits of eucaryotes investigated vary in length from 355 to 373 amino acids (Gebhardt et al., 1986; Hayward et al., 1986; Tischer et al., 1986). The eucaryotic GSII subunits lack the C-terminal portion of the procaryotic GSI subunit, including the adenylylation site. The GSII subunit from *Bradyrhizobium japonicum* is 329 amino acids long and also lacks the terminal portion of the GSI subunit and the adenylylation site (Carlson and Chelm, 1986).

In a comparison of procaryotic and eucaryotic GSI and GSII enzymes, Rawlings et al. (1987) showed that although the amino acid similarity between the enzymes was only approximately 15%, the major part of this similarity was located in five regions corresponding to β sheets in the *Salmonella typhimurium* enzyme (Almassy et al., 1986). These regions are strongly conserved in all GS enzymes analyzed to date (Janssen et al., 1988). A feature of the five conserved amino acid regions is that they are all associated with the proposed GS activity site (Almassy et al., 1986). Regions II to V are β strands closely associated with two Mn^{2+} cations of one subunit, while region I contains the Trp residue which is thought to complete the active site formed between adjacent subunits.

The structure and regulation of the GS structural gene, glnA, has been extensively studied in the enteric bacteria (Magasanik and Neidhardt, 1987; Reitzer and Magasanik, 1987; Magasanik, 1988). The glnA gene of these gram-negative aerobic bacteria is part of a complex operon (glnALG) (Pahel et al., 1982) whose transcription is regulated by the products (NR_I, NR_{II}) of the glnG (ntrC) and glnL (ntrB) genes, respectively (Ninfa and Magasanik, 1986). The NR_I and NR_{II} proteins also regulate the expression of other operons involved in nitrogen metabolism (global nitrogen regulatory system) (Kustu et al., 1979). A sigma factor (σ^{54}) encoded by the glnF (ntrA) gene is specifically required for the transcription of nitrogen-regulated (ntr) promoters (Hirschman et al., 1985; Hunt and Magasanik, 1985). These promoters do not contain canonical −35 and −10 sequences, but instead have the consensus (−26) $CTGGYAYR-N_4-TTGCA$(−10) (Beynon et al., 1983; Ausubel, 1984; Gussin et al., 1986). In Escherichia coli, σ^{54} has been implicated in the regulation of two anaerobically inducible enzymes, formate dehydrogenase and hydrogenase isoenzyme 3, which are part of the formate hydrogenlyase pathway (Birkmann et al., 1987).

The glnALG operon contains three promoters, glnAp1 and glnAp2, located upstream of the glnA gene (Reitzer and Magasanik, 1985), and glnLp situated within the glnAL intergenic region (Ueno-Nishio et al., 1983, 1984). glnAp1 and glnLp are $E\sigma^{70}$ promoters repressed by NR_I and serve to maintain basal levels of GS, whereas the σ^{54} dependent promoter glnAp2 is activated by phosphorylated NR_I when nitrogen becomes limiting.

There is no evidence for the presence of global-regulatory ntr genes in the gram-positive, endospore-forming aerobe, Bacillus subtilis. It appears that GS regulates its own expression in B. subtilis (Schreier et al., 1985;

Strauch et al., 1988). The structural glnA gene is preceded by an open-reading frame (ORF) encoding a 136-amino-acid polypeptide that contains a DNA-binding, helix-turn-helix motif and shows similarities to diverse regulatory proteins (Cro, TrpR). Although the precise mechanism is unknown, it has been suggested that the glnR protein acts in concert with GS to regulate expression of the glnA gene in B. subtilis (Strauch et al., 1988; Schreier et al., 1989; Zhang et al., 1989). It appears that the Bacillus cereus glnRA operon has a similar regulatory system (Nakano et al., 1989).

23.2   GS and glnA gene of Bacteroides fragilis

Yamamoto et al. (1984) investigated the enzymes in the ammonia assimilation pathways of the gram-negative obligate anaerobe B. fragilis and concluded that GS may not be important for ammonia assimilation into amino acids. They were unable to detect GS activity in B. fragilis cell extracts by the τ-glutamyl transferase (GGT) assay. Cell growth was not inhibited by methionine sulfoximine, a GS inhibitor of GSI enzymes. This conclusion about the lack of importance of GS in nitrogen assimilation in B. fragilis is based on the assumption that the GS of this anaerobe is similar to the GSI enzymes.

To investigate the role, structure, and regulation of the B. fragilis GS, Southern et al. (1986) cloned a glnA gene on the recombinant plasmid pJS139. This plasmid enabled E. coli glnALG deletion mutants to utilize $(NH_4)_2SO_4$ as a sole source of nitrogen. The cloned B. fragilis glnA gene was expressed from its own promoter and was subject to nitrogen repression in E. coli glnALG mutants. The GS subunit produced by pJS139 in E. coli was purified and had an apparent M_r of approximately 75,000 (Southern et al., 1986). The GS subunit from B. fragilis cells also had an apparent M_r of approximately 75,000. DNA sequencing indicated that the B. fragilis glnA gene encoded a GS subunit of 729 amino acids with a calculated M_r of 82,827 (Hill et al., 1989). The B. fragilis GS subunit is larger than that of any other known bacterial GS; E. coli, B. subtilis, Rhizobium japonicum, and Vibrio alginoly-

ticus have subunits with M_r values of 50,000, 56,000, 60,000 and 60,000, respectively (Deuel et al., 1970; Stadtman and Ginsburg, 1974; Bhandari et al., 1983; Bodasing et al., 1985).

Southern et al. (1987) determined the M_r of the *B. fragilis* GS holoenzyme and reported that the GS enzyme had an apparent M_r of approximately 490,000. In contrast to the GS of other bacteria, the GS of *B. fragilis* is a hexamer. Other procaryote GSI enzymes normally consist of 12 subunits (Streicher and Tyler, 1980; Bhatnagar et al., 1986). The GS from *Clostridium pasteurianum* has been reported to consist of 20 subunits (Krishnan et al., 1986).

Electron microscopy of negatively stained, purified, cloned *B. fragilis* GS revealed molecules with a central hole and hexagonal shape. However, tetragonal structures, which indicate side views of double-layered ring structures, were not observed.

The GS enzymes from gram-negative bacteria have been shown to be covalently modified by adenylylation, whereas there is no evidence for GS regulation by adenylylation in gram-positive bacteria (Tronick et al., 1973; Wolhueter et al., 1973). The cloned *B. fragilis* GS was not regulated by adenylylation and was unique in that it was inactivated by snake venom phosphodiesterase (Southern et al., 1987). The GGT activity of the cloned *B. fragilis* GS was inhibited by Mg^{2+} and it is interesting that the GS from another anaerobe, *C. acetobutylicum*, was also inhibited by Mg^{2+} (Usdin et al., 1986). The marked stimulation in GGT activity of the cloned GS by low concentrations of Mn^{2+} also appears to be a unique feature of the cloned *B. fragilis* GS (Southern et al., 1987).

Cloning of the *B. fragilis glnA* gene enabled the biochemical characterization of the *B. fragilis* GS, since Southern et al. (1987) showed that cell extracts of *B. fragilis* specifically and irreversibly inactivated the GS from *B. fragilis*. *E. coli* GS was not inactivated by *B. fragilis* cell extracts. The inactivation of *B. fragilis* GS by its own cell extract indicates that it is not possible to study GS activity in *B. fragilis*. This inactivation could account for the absence of GGT activity in *B. fragilis* cell extracts reported by Yamamoto et al. (1984), and would affect their conclusion that GS may not be important for ammonia assimilation into amino acids. Western blotting indicated that GS was present in *B. fragilis* cell extracts and that its production was regulated by nitrogen (Southern et al., 1987).

Because the structure and biochemical properties of the *B. fragilis* exhibited novel features, Hill et al. (1989) determined the nucleotide sequence of a 2,777-bp DNA segment containing the *B. fragilis glnA* gene. The *B. fragilis glnA* ORF of 2,187 bp encoded a GS subunit of 729 amino acids. Primer extension experiments with RNA isolated from *B. fragilis* cells showed a single major transcription start point 18 bp from the first nucleotide of the ATG start codon. Although promoter consensus sequences for *B. fragilis* have not been defined, it is interesting that an *E. coli* promoter sequence containing −35 and −10 RNA-polymerase-binding consensus sequences separated by 22 bp is situated 32 bp upstream of the *B. fragilis* transcription start point. The ATG start codon is preceded by an AAGAGA sequence 8 nucleotides upstream. A feature of the upstream region is the presence of two near-perfect, direct-repeat sequences of 46 and 45 bp from positions −127 to −182 and −74 to −30, respectively.

Hill et al. (1989) compared the nucleotide-derived amino acid sequence of the *B. fragilis* GS with the GS amino acid sequences of diverse procaryotes and eucaryotes (GSI and GSII enzymes, respectively). The *B. fragilis* GS subunit is approximately 270 and 400 amino acids longer than the GSI and GSII subunits, respectively.

The GSI and GSII enzymes can be separated into two groups on the basis of amino acid homology. Although there is only limited amino acid similarity between these two groups, a high degree of similarity exists between GS enzymes within each group. The major regions of amino acid sequence similarity between the procaryote GS enzymes involve five β sheets, all of which are associated with the proposed active site (Almassy et al., 1986; Janson et al., 1986; Rawlings et al., 1987). These regions are also conserved in GS from eucaryotes (Janssen et al., 1988). In the GS enzyme from *B. fragilis*, these five regions show far less amino acid similarity although certain conserved amino

REGION I

Hs	60	N	F	D	G	S	S	T	L	Q	S	E	G	S	N	-	S	D	M	Y	L	V	P	A	A	-	-	M	F	R	D	P	F	R
Af	54	N	Y	D	G	S	S	T	G	Q	A	P	G	E	D	-	S	E	V	I	I	Y	P	Q	A	-	-	I	F	K	D	P	F	R
Bj	42	G	F	D	G	S	S	T	Q	Q	A	E	G	H	S	-	S	D	C	V	L	K	P	V	A	-	-	V	F	P	D	A	A	R
An	50	P	F	D	G	S	S	I	R	G	W	K	A	I	N	E	S	D	M	T	M	V	L	D	P	N	T	A	W	I	D	P	F	M
St	49	M	F	D	G	S	S	I	G	G	W	K	G	I	N	E	S	D	M	V	L	M	P	D	A	S	T	A	V	I	D	P	F	F
Tf	50	S	F	D	G	S	S	I	A	G	W	K	G	I	N	E	S	D	M	I	L	L	P	D	P	D	S	A	V	L	D	P	F	M
Ca	51	M	F	D	G	S	S	I	D	G	F	V	R	I	E	E	S	D	M	N	L	R	P	N	L	D	S	F	V	I	F	P	W	R
Sc	48	A	F	D	G	S	S	I	R	G	F	Q	A	I	H	E	S	D	M	S	L	R	P	D	L	S	T	A	R	V	D	P	F	R
Bs	51	M	F	D	G	S	S	I	E	G	F	V	R	I	F	E	S	D	M	Y	L	L	Y	P	D	L	N	T	F	V	I	F	P	W	T
Bf	149	A	W	D	G	S	S	P	A	F	V	V	D	T	T	L	C	I	P	T	I	F	I	S	Y	T	G	E	A	L	D	Y	K	T

REGION II

Hs	195	A	E	V	M	P	-	A	Q	W	E	F	Q	I
Af	191	G	E	V	M	P	-	G	Q	W	E	F	Q	V
Bj	171	A	E	V	A	K	-	G	Q	W	E	F	Q	I
An	215	H	E	V	A	T	G	G	Q	C	E	L	G	F
St	215	H	E	V	A	T	A	G	Q	N	E	V	A	T
Tf	211	H	E	V	A	T	A	G	Q	H	E	I	G	V
Ca	187	H	E	V	A	E	-	G	Q	N	E	I	D	F
Sc	212	H	E	V	G	T	A	G	Q	A	E	I	N	Y
Bs	188	H	E	V	A	P	-	G	Q	H	E	I	D	F
Bf	285	N	E	V	A	P	-	N	Q	F	E	L	A	P

REGION III

Hs	237	T	F	D	P	K	P	I	P	G	N	-	W	N	G	A	G	C	H	T	N	F	S	T
Af	233	S	F	D	P	K	P	I	K	G	D	-	W	N	G	A	G	A	H	T	N	Y	S	T
Bj	213	E	F	H	C	K	P	L	G	D	T	D	W	N	G	S	G	M	H	A	N	F	S	T
An	258	T	F	M	P	K	P	I	F	G	D	-	-	N	G	S	G	M	H	C	H	Q	S	I
St	255	T	F	M	P	K	P	M	F	G	D	-	-	N	G	S	G	M	H	C	H	M	S	L
Tf	253	T	F	M	P	K	P	V	V	G	D	-	-	N	G	S	G	M	H	V	H	Q	S	L
Ca	201	S	F	M	P	K	P	I	F	G	I	-	-	N	G	S	G	M	H	V	N	M	S	L
Sc	252	T	F	M	P	K	P	I	F	G	D	-	-	N	G	S	G	M	H	V	H	O	S	L
Bs	230	T	F	M	P	K	P	L	F	G	V	-	-	N	G	S	G	M	H	C	N	L	S	L
Bf	327	L	F	H	E	K	P	Y	N	G	V	-	-	N	G	S	G	K	H	N	N	W	S	L

REGION IV

Hs	317	A	N	R	S	A	S	I	R	I	P
Af	309	A	N	R	G	A	S	I	R	V	G
Bj	290	A	S	D	G	A	S	I	R	V	P
An	341	G	N	R	S	A	S	I	R	I	P
St	338	R	N	R	S	A	S	I	R	I	P
Tf	336	K	N	R	S	A	S	I	R	I	P
Ca	285	K	N	R	T	A	L	I	R	V	P
Sc	337	R	N	R	S	A	A	M	R	I	P
Bs	314	Q	N	R	S	P	L	I	R	I	P

REGION V

Hs	337	F	E	D	R	R	P	S	A	N	C	D	P	F	S	V	T	E
Af	329	F	E	D	R	R	P	S	S	N	M	D	P	Y	V	V	T	S
Bj	311	L	E	D	R	R	P	N	S	Q	G	D	P	Y	Q	I	V	R
An	361	L	E	V	R	C	P	D	A	T	S	N	P	Y	L	A	F	S
St	357	I	E	V	R	F	P	D	P	A	A	N	P	Y	L	C	F	A
Tf	355	I	E	V	R	F	P	D	S	T	A	N	P	Y	L	A	F	S
Ca	303	V	E	L	R	C	P	D	P	S	S	N	P	Y	L	V	L	A
Sc	357	V	E	F	R	A	P	D	A	S	G	N	P	Y	L	A	F	S
Bs	332	V	E	V	R	S	V	D	P	A	A	N	P	Y	L	A	L	S
Bf	471	F	E	F	R	A	A	G	S	S	A	N	C	A	A	A	M	I

FIGURE 23.1 Comparison of amino acid sequences of five regions of homology of GS enzymes from *Homo sapiens* (Hs) (Gibbs et al., 1987), alfalfa (Af) (Tischer et al., 1986), *Bradyrhizobium japonicum* (Bj) (Carlson and Chelm, 1986), *Anabaena* sp. strain 7120 (An) (Tumer et al., 1983), *Salmonella typhimurium* (St) (Janson et al., 1986), *T. ferrooxidans* (Tf) (Rawlings et al., 1987), *Clostridium acetobutylicum* (Ca) (Janssen et al., 1988), *Streptomyces coelicolor* (Sc) (Wray and Fisher, 1988), *Bacillus subtilis* (Bs) (Strauch et al., 1988) and *Bacteroides fragilis* (Bf) (Hill et al., 1989). Amino acids are identified by single-letter code; positions of first amino acids in each region are indicated. Numbering of residues begins with start methionine at N-terminal end; identical residues are boxed. No region IV was found in *B. fragilis*.

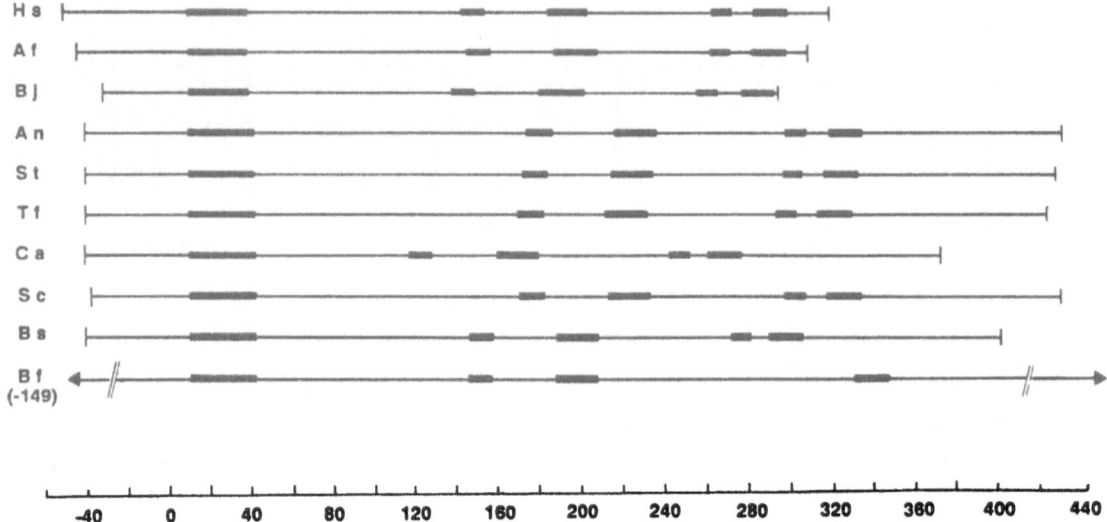

FIGURE 23.2   Relationship of GS regions I to V. GS enzymes compared are same as in Figure 23.1. Position 0 is first amino acid residue of region I, size and relative positions of regions I to V, and start and end, of each GS enzyme are indicated. No region IV was found in *Bacteroides fragilis*.

acids can be identified in four of these five regions (Figure 23.1) (Hill et al., 1989). The relative positions of the best-conserved regions in the *B. fragilis* GS (regions I, II, III and V) are similar to the positioning of these regions in the other GS enzymes (Figure 23.2).

In region I, the *S. typhimurium* GS has a Trp residue at position 58, which is thought to complete the active site formed between adjacent subunits (Almassy et al., 1986). In the GS from three gram-positive bacteria, *B. subtilis, C. acetobutylicum,* and *Streptomyces coelicolor,* the Trp residue is replaced by the functionally similar Phe residue. The *B. fragilis* GS differs from all the other procaryotic GS in that the active site Trp in region I is replaced by Val, which shows no functional similarity with Trp (Hill et al., 1989). Because the *B. fragilis* GS is not regulated by adenylylation, it is perhaps not surprising that it differs from all the other gram-negative GS enzymes in that it does not contain a conserved 18-amino-acid sequence adjacent to a Tyr residue (at position approximately 398).

The GS from *B. fragilis* differs markedly from the GSI and GSII enzymes found in other procaryotes and eucaryotes. Classification of *Bacteroides* species by ribosomal RNA sequence analysis indicated that the genus *Bacteroides* belongs to a distinct assemblage of genera including *Bacteroides, Flavobacterium,* and *Cytophaga* (Woese, 1987). The taxonomic separation of *Bacteroides* is supported by the structure of the *B. fragilis* GS. It will be interesting to determine whether the GS enzymes of other members of this group are similar to the GS of *B. fragilis*. If this is the case, then GS enzymes with large subunits arranged as hexamers should be grouped together to form a third class of GS enzymes (GSIII). Genetic studies in *Agrobacterium* and *Rhizobiaceae* have suggested the possible occurrence of a third GS by the cloning of a gene *glnT* that complements an *E. coli glnA* deletion, allowing growth in the absence of glutamine (Rossbach et al., 1988). However, no enzyme activity has yet been ascribed to the *glnT* gene product.

23.3   GS and *glnA* gene of *Clostridium acetobutylicum*

Clostridium acetobutylicum has been exploited for the industrial production of acetone and butanol (Jones and Woods, 1986). However, biological limitations (e.g., butanol toxicity) and economic factors (high substrate and solvent recovery costs) have rendered the industrial acetone and butanol fermentation unprofitable in comparison with the synthesis of solvents from petrochemicals. To make the fermentation more profitable and in the long-term interests of mankind in producing solvents from renewable resources, research is being undertaken worldwide to improve the production of solvents by *C. acetobutylicum*. The fermentation is complex in that the bacterium exhibits three distinct physiological stages: an actively growing acidogenic stage, a nongrowing solventogenic stage, and a sporulation stage. Because Long et al. (1984) had reported that nitrogen levels were important in the transitions from the acidogenic to the solventogenic stage and from the solventogenic to the sporulation stage, it was natural that studies would be carried out on the structure and regulation of GS from *C. acetobutylicum*.

Usdin et al. (1986) cloned, purified, and investigated the regulation of the *C. acetobutylicum* GS. A 6.5-kb DNA fragment from *C. acetobutylicum* cloned in the recombinant plasmid pHZ200 complemented the *glnA* lesion in an *E. coli glnALG* deletion mutant. The cloned *C. acetobutylicum glnA* gene and gene product functioned very efficiently in *E. coli* and enabled the *glnA* deletion strain to grow approximately 1.7-fold faster than a wild-type *E. coli* strain under nitrogen-limiting conditions.

The *C. acetobutylicum* GS has an apparent subunit M_r of approximately 59,000. Electron microscopy indicated that the GS had a number of features characteristic of the dodecamer assembly of the GSI subunits from other bacteria. The GS was inhibited by Mg^{2+} in the τ-glutamyltransferase assay, but there was no evidence that the GS was adenylylated. The *C. acetobutylicum* GS appears to be structurally and functionally similar to GS in other gram-positive bacteria.

The cloned *C. acetobutylicum glnA* gene was expressed from its own promoter and was subject to nitrogen regulation in *E. coli*. However, the cloned *C. acetobutylicum glnA* DNA fragment was unable to complement certain nitrogen-regulatory gene functions in *E. coli ntrB* and *ntrC* deletion strains. pHZ200 did not activate histidase production or allow growth on arginine or low concentrations of glutamine

in *E. coli glnA ntrB ntrC* deletion strains. It appears that the *C. acetobutylicum glnA* region does not contain genes that can complement *E. coli ntrB* and *ntrC* genes, which are able to regulate diverse genes involved in nitrogen metabolism. There is no evidence for the presence of a global nitrogen-regulatory system in *C. acetobutylicum*.

Janssen et al. (1988) reported the nucleotide sequence of a 2.0-kb DNA segment containing the *C. acetobutylicum glnA* gene. The DNA sequence contained an ORF that contains 1,332 nucleotides encoding 444 amino acid residues from the presumptive start codon (ATG) to the stop codon (TAA). The M_r of the predicted GS polypeptide was 49,630, which was much lower than the apparent M_r of 59,000 estimated by gel electrophoresis of the purified GS subunit (Usdin et al., 1986). *C. acetobutylicum* DNA has a GC content of 28% (Cummins and Johnson, 1971), and the *glnA* structural gene exhibited a codon usage that was strongly biased toward the use of codons in which A and U predominate.

Comparisons of the amino acid sequence of the *C. acetobutylicum* GS with other GSI and GSII enzymes indicated that the *C. acetobutylicum* GS was a type 1 enzyme and contained the five conserved amino acid regions associated with the proposed GS active site (Almassy et al., 1986; Janson et al., 1986; Janssen et al., 1988).

Studies on the crystallized GS from *S. typhimurium* (Almassy et al., 1986) indicated that the subunit contained a central loop between amino acids 156 and 173. This region is cleaved by several proteases with different specificities (Dautry-Varsat et al., 1979; Monroe et al., 1985). It is interesting that the cloned *C. acetobutylicum* GS appears to lack 26 amino acid residues involved in the formation of the central loop in *S. typhimurium* (Janson et al., 1986). The absence of a central loop may be a feature of GS from gram-positive bacteria, because the *B. subtilis* GS also appears to lack a central loop (Strauch et al., 1988). The cloned *C. acetobutylicum* GS is not subject to adenylylation (Usdin et al., 1986), and it not only lacks the tyrosine residue associated with GS adenylylation in

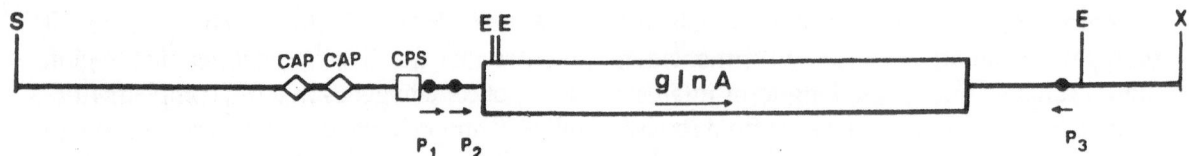

FIGURE 23.3 Diagram of *Clostridium acetobuylicum glnA* gene region. Abbreviations: putative upstream catabolite activator protein binding sites, CAP; complex palindromic sequence, CPS; promoters: P₁, P₂, P₃; direction of transcription, (→); restriction enzyme sites: S, *Sma*I; E, *Eco*RV; X, *Xba*I. Region with very high DNA curvature score and five copies of direct repeat and consensus sequence 5'—ATATTGTAA—3' is located immediately upstream of CAP sites.

E. coli, but also shows very little similarity with the 18 amino acids adjacent to the tyrosine residue in the *E. coli* GS.

23.4 Regulation of *C. acetobutylicum glnA* promoters and antisense RNA

Molecular analysis of the upstream region of the *C. acetobutylicum glnA* gene showed that it contained two putative extended promoter consensus sequences (P₁ and P₂) characteristic of gram-positive bacteria (Figure 23.3) (Janssen et al., 1988). The upstream regulatory region was also characterized by a complex palindromic sequence immediately upstream of P₁ between nucleotides 130 to 167. A 380-bp region between nucleotides 160 to 540 had a very high DNA curvature score of 8.2 (calculated according to Plaskon and Wartell, 1987) and contained five copies of a direct repeat, with the consensus sequence 5'-ATATTGTAA-3'. Similar 9-bp sequences (consensus 5'-ATATTGTTT-3') occur nine times in the *B. subtilis glnA/gltC* regulatory region (Bohannon and Sonenshein, 1989), five times in the *B. cereus glnRA* intergenic region (Nakano et al., 1989) and in the target DNAs of several *LysR* family proteins in *E. coli* (Kölling and Lother, 1985; Henikoff et al., 1988). The significance of the complex palindromic sequence and the region with the high DNA curvature score is unknown at present.

The sequences containing the putative promoter regions P₁ and P₂ were shown to have promoter activity by subcloning into promoter probe vectors (Janssen et al., 1988). Studies on the initiation of transcription of the *C. acetobutylicum glnA* gene indicated that in *C. acetobutylicum glnA* transcripts were initiated at two start points controlled by P₁ and P₂, respectively (Janssen et al., 1990). Initiation of transcription also occurred from these two sites in *E. coli glnA* deletion mutants containing the *C. acetobutylicum glnA* gene. In *E. coli*, the initiation of transcription of the *C. acetobutylicum glnA* gene was regulated by nitrogen, and a downstream region was implicated in the regulation of transcription by nitrogen. However, regulation of the *C. acetobutylicum glnA* gene is complex, and Janssen et al., (1990) suggested the involvement of regulatory mechanisms operating at the level of translation.

Janssen et al. (1988) reported that the region immediately downstream of the *glnA* structural gene consisted of a 158-bp stretch of inverted repeat sequences. mRNA transcribed from this region would have the potential to form a number of stem loop structures, including a long 59-bp stem loop with a △G value of −60.8 kcal/mol. Further downstream from the inverted-repeat region, Janssen et al. (1988) identified a promoter, P₃, oriented toward the *glnA* gene, and showed that P₃ was functional by subcloning it into a promoter probe vector. A putative antisense RNA, complementary to 43 bases overlapping the ribosome-binding site and the start of the *glnA* gene, is transcribed from a single downstream transcript start site under the control of P₃ in *C acetobutylicum* and *E. coli* (Janssen et al., 1988, 1990).

The role of antisense RNA in gene regulation is to act as an inhibitor of gene expression as a consequence of either transcriptional interference or specific RNA–RNA intermediates (Green et al., 1986). An increase in the production of antisense RNA would be expected to result in a decrease in the production of a specific gene product. To establish the involvement of the putative antisense RNA in the regulation of the C. acetobutylicum glnA gene, Janssen et al. (1990) mutated P_3 and showed that the mutation was an up-promoter mutation that resulted in a decrease in C. acetobutylicum GS activity in E. coli cells. Although it appears that the putative antisense RNA does have a role in down-regulating GS expression, it does not seem to be involved in regulation by nitrogen. GS activities in E. coli cells containing the C. acetobutylicum glnA region with the up-promoter mutation in P_3 were lower in both nitrogen-limiting and nitrogen-excess media, but GS activity was still regulated by nitrogen (Janssen et al., 1990). Previously, Janssen et al. (1988) showed that when P_3 and the putative antisense RNA region were deleted the activity of GS was still regulated by nitrogen.

The involvement of the downstream 158-bp inverted-repeat sequences (located between P_3 and the end of the glnA gene) in the regulation of the C. acetobutylicum glnA gene was shown by the construction of deletion plasmids that lacked both the inverted repeat sequences and either retained or lacked P_3 (Janssen et al., 1988,1990). Gene constructs without the inverted repeat sequences produced very low levels of GS activity. The removal of the inverted-repeat sequences will affect the secondary structures of the downstream mRNA, and these changes could play a role in termination of transcription (Morgan et al., 1985) or in the stability of the mRNA (Belasco et al., 1985; Wong and Chang, 1986). In the deletion mutant that retained P_3, this promoter is closer to the end of the glnA gene, and termination of transcription from P_3 may also be affected by the deletion of the inverted-repeat sequences. This could result in transcription from P_3 into the glnA gene, which would affect normal transcription and production of glnA mRNA. Although the levels of GS activity in E. coli containing the downstream-deleted plasmids were low, they were still regulated by nitrogen.

The regulation of the C. acetobutylicum glnA gene is complex and involves mechanisms operating at the level of transcription and translation. The regulatory systems differ from those of other gram-negative and gram-positive bacteria studied to date. Future molecular genetic studies will enable a complete understanding of the regulation of the C. acetobutylicum glnA gene and GS activity.

References

Almassy, R.J., C.A. Janson, R. Hamlin, N.H. Xuong, and D. Eisenberg. 1986. Novel subunit-subunit interactions in the structure of glutamine synthetase. Nature (London) 323:304–309.

Ausubel, F.M. 1984. Regulation of nitrogen fixation genes. Cell 37:5–6.

Belasco, J.G., J.T. Beatty, C.W. Adams, A. von Gabain, and S.N. Cohen. 1985. Differential expression of photosynthesis genes in R. capsulata results from segmental differences in stability within the polycistronic rxcA transcript. Cell 40:171–181.

Beynon, J., M. Cannon, V. Buchanan-Woolaston, and F. Cannon. 1983. The nif gene promoters of Klebsiella have a characteristic primary structure. Cell 34:665–671.

Bhandari, B., F. Vairinhos, and D.J.D. Nicholas. 1983. Some properties of glutamine synthetase from Rhizobium japonicum CC705 and CC723. Arch. Microbiol. 136:84–88.

Bhatnagar, L., J.G. Zeikus, and J-P. Aubert. 1986. Purification and characterization of glutamine synthetase from the archaebacterium Methanobacterium ivanovi. J. Bacteriol. 165:638–643.

Birkmann, A., R.G. Sawers, and A. Böck. 1987. Involvement of the ntrA gene product in the anaerobic metabolism of Escherichia coli. Mol. Gen. Genet. 210:535–542.

Bodasing, S.J., P.W. Brandt, F.T. Robb, and D.R. Woods. 1985. Purification and regulation of glutamine synthetase in a collagenolytic Vibrio alginolyticus strain. Arch. Microbiol. 140:369–374.

Bohannon, D.E. and A.L Sonenshein. 1989. Positive regulation of glutamate biosynthesis in Bacillus subtilis. J. Bacteriol. 171(4):4718–4727.

Carlson, T.A. and B.K Chelm. 1986. Apparent eukaryotic origin of glutamine synthetase II from the bacterium Bradyrhizobium japonicum. Nature 322:568–570.

Cummins, C.S. and J.L Johnson. 1971. Taxonomy of the clostridia: wall composition and DNA homologies in Clostridium butyricum and other butyric acid-producing clostridia. J. Gen. Microbiol. 67:33–46.

Dautry-Varsat, A., G.N. Cohen, and E.R. Stadtman. 1979. Some properties of *Escherichia coli* glutamine synthetase after limited proteolysis by subtilisin. *J. Biol. Chem.* **254**:3124–3128.

Deuel, T.F., A. Ginsburg, J. Yen, E. Shelton, and E.R. Stadtman. 1970. *Bacillus subtilis* glutamine synthetase: purification and physical characterization. *J. Biol. Chem.* **245**:5195–5205.

Gebhardt, C., J.E. Oliver, B.G. Forde, R. Saarelainen and B. Miflin. 1986. Primary structure and differential expression of glutamine synthetase genes in nodules, roots and leaves of *Phaseolus vulgaris*. *EMBO J.* **5**:1429–1435.

Gibbs, C.S., K.E. Campbell, and R.H. Wilson. 1987. Sequence of a human glutamine synthetase cDNA. *Nucleic Acids Res.* **15**:6293.

Green, P.J., O. Pines, and M. Ihouye. 1986. The role of antisense RNA in gene regulation. *Annu. Rev. Biochem.* **55**:569–597.

Gussin, G.N., C.W. Ronson, and F.M. Ausubel. 1986. Regulation of nitrogen fixation genes. *Annu. Rev. Gen.* **20**:567–591.

Hayward, B.E., A. Hussain, R.H. Wilson, A. Lyons, V. Woodcock, B. McIntosh, and T.J.R. Harris. 1986. The cloning and nucleotide sequence of cDNA for an amplified glutamine synthetase gene from the Chinese hamster. *Nucleic Acids Res.* **14**:999–1008.

Henikoff, S., G.W. Haughn, J.M. Calvo, and J.C. Wallace. 1988. A large family of bacterial activator proteins. *Proc. Natl. Acad. Sci. USA* **85**:6602–6606.

Hill, R.T., J.R. Parker, H.J.K Goodman, D.T. Jones, and D.R. Woods. 1989. Molecular analysis of a novel glutamine synthetase of the anaerobe *Bacteroides fragilis*. *J. Gen. Microbiol.* **135**:3271–3279.

Hirschman, J., P.K. Wong, K.S. Keener, and S. Kustu. 1985. Products of nitrogen regulatory genes *ntrA* and *ntrC* of enteric bacteria activate *glnA* transcription in vitro: evidence that the *ntrA* product is a sigma factor. *Proc. Natl. Acad. Sci. USA* **82**:7525–7529.

Hunt, T.P. and B. Magasanik. 1985. Transcription of *glnA* by purified *Escherichia coli* components: core RNA polymerase and the products of *glnF*, *glnG* and *glnL*. *Proc. Natl. Acad. Sci. USA* **82**:8453–8457.

Janson, C.A., P.S. Kayne, R.J. Almassy, M. Grunstein, and D. Eisenberg. 1986. Sequence of glutamine synthetase from *Salmonella typhimurium* and implications for the protein structure. *Gene* (Amst.) **46**:297–300.

Janssen, P.J., D.T. Jones, and D.R. Woods. 1990. Studies on *Clostridium acetobutylicum glnA* promoters and antisense RNA. *Mol. Microbiol.* **4**:1575–1583.

Janssen, P.J., W.A. Jones, D.T. Jones, and D.R. Woods. 1988. Molecular analysis and regulation of the *glnA* gene of the gram-positive anaerobe *Clostridium acetobutylicum*. *J. Bacteriol.* **170**:400–408.

Jones, D.T. and D.R. Woods. 1986. Acetone butanol fermentation revisited. *Microbiol. Rev.* **50**:484–524.

Kölling, R. and H. Lother. 1985. *AsnC*: an autogenously regulated activator of asparagine synthetase A transcription in *Escherichia coli*. *J. Bacteriol.* **164**:310–315.

Krishnan, I.S., R.K. Singhal, and R.D. Dua. 1986. Purification and characterization of glutamine synthetase from *Clostridium pasteurianum*. *Biochemistry* **25**:1589–1599.

Kustu, S.G., N.C. McFarland, S.P. Hui, B. Esmon, and G.F. Ames. 1979. Nitrogen control in *Salmonella typhimurium*: co-regulation of synthesis of glutamine synthetase and amino acid transport systems. *J. Bacteriol.* **138**(1):218–234.

Long, S., D.T. Jones, and D.R. Woods. 1984. Initiation of solvent production, clostridial stage and endospore formation in *Clostridium acetobutylicum* P262. *Appl. Microbiol. Biotechnol.* **20**:256–261.

Magasanik, B. 1988. Reversible phosphorylation of an enhancer binding protein regulates the transcription of bacterial nitrogen utilization genes. *TIBS* **13**:475–479.

Magasanik, B. and F.C. Neidhardt. 1987. Regulation of carbon and nitrogen utilization. *In: Escherichia coli and Salmonella typhimurium: Cellular and Molecular Biology*, F.C. Neidhardt ed., pp. 1318–1325. Washington, D.C.: American Society for Microbiology.

Monroe, D.M., C.M. Noyes, M.J. Griffith, R.L. Lundblad, and H.S. Kingdon. 1985. Structural and enzymatic properties of *Escherichia coli* glutamine synthetase subjected to limited proteolysis. *Curr. Top. Cell. Regul.* **27**:361–372.

Morgan, W.D., D.G. Bear, B.L Litchman, and P.H. von Hippel. 1985. RNA sequence and secondary structure requirements for rho-dependent transcription termination. *Nucleic Acids Res.* **13**:3739–3754.

Nakano, Y., C. Kato, E. Tanaka, K. Kimura, and K Horikoshi. 1989. The nucleotide sequence of the glutamine synthetase gene (*glnA*) and its upstream region from *Bacillus cereus*. *J. Biochem.* **106**:209–215.

Ninfa, A.J. and B. Magasanik. 1986. Covalent modification of the *glnG* product, NR_I, by the *glnL* product, NR_{II}, regulates the transcription of the *glnALG* operon in *Escherichia coli*. *Proc. Natl. Acad. Sci. USA* **85**:5909–5913.

Pahel, G., D.M. Rothstein, and B. Magasanik. 1982. Complex *glnA-glnL-glnG* operon of *Escherichia coli*. *J. Bacteriol.* **150**:202–213.

Plaskon, R.R. and R.M. Wartell. 1987. Sequence distributions associated with DNA curvature are found upstream of strong *E. coli* promoters. *Nucleic Acids Res.* **15**:785–796.

Prusiner, S. and E.R. Stadtman. 1973. *The Enzymes of Glutamine Metabolism.* New York: Academic Press.

Rawlings, D.E., W.A. Jones, E.G. O'Neill, and D.R. Woods. 1987. Nucleotide sequence of the glutamine synthetase gene and its controlling region of the acidophilic autotroph *Thiobacillus ferrooxidans. Gene* (Amst.) **53**:211–217.

Reitzer, L.J. and B. Magasanik. 1985. Expression of *glnA* in *Escherichia coli* is regulated at tandem promoters. *Proc. Natl. Acad. Sci. USA* **82**:1979–1983.

Reitzer, L.J. and B. Magasanik. 1987. Ammonia assimilation and the biosynthesis of glutamine, glutamate aspartate, asparagine, L-alanine, and D alanine. *In: Escherichia coli and Salmonella typhimurium: Cellular amd Molecular Biology,* F.C. Neidhardt, ed., pp. 302–320. Washington, D.C.: American Society for Microbiology.

Rossbach, S., J. Schell, and F.J. De Bruijn. 1988. Cloning and analysis of *Agrobacterium tumefaciens* V58 loci involved in glutamine biosynthesis: neither the *glnA* (GSI) nor the *glnII* (GSII) gene plays a special role in virulence. *Mol. Gen. Genet.* **212**:38–47.

Schreier, HJ., S.H. Fisher, and A L Sonenshein. 1985. Regulation of expression from the *glnA* promoter of *Bacillus subtilis* requires the *glnA* gene product. *Proc. Natl. Acad. Sci. USA* **82**: 3375–3379.

Schreier, H.J., S.W. Brown, K.D. Hirschi, J.F. Nomellini, and A.L. Sonenshein. 1989. Reguladon of *Bacillus subtilis* glutamine synthetase gene expression by the product of the *glnR* gene. *J. Mol. Biol.* **210**:51–63.

Southern, J.A., J.R. Parker, and D.R. Woods. 1986. Expression and purification of glutamine synthetase cloned from *Bacteroides fragilis. J. Gen. Microbiol.* **132**:2827–2835.

Southern, J.A., J.R. Parker, and D.R. Woods. 1987. Novel structure. properties and inactivation of glutamine synthetase cloned from *Bacteroides fragilis. J. Gen. Microbiol.* **133**:2437–2446.

Stadtman, E.R. and A. Ginsburg. 1974. The glutamine synthetase of *Escherichia coli*: structure and control. *In: The Enzymes,* Vol. X, P.D. Boyer, ed., pp. 755–807. New York: Academic Press.

Strauch, M.A., A.I. Aronson, S.W. Brown, H.J. Schreier and A.L. Sonenshein. 1988. Sequence of the *Bacillus subtilis* glutamine synthetase gene region. *Gene* **71**:257–265.

Streicher, S.L and B. Tyler. 1980. Purification of glutamine synthetase from a variety of bacteria. *J. Bacteriol.* **142**:69–78.

Tischer, E., S. DasSarma, and H.M. Goodman. 1986. Nucleotide sequence of an alfalfa glutamine synthetase gene. *Mol. Gen. Genet.* **203**: 221–229.

Tronick, S.R., J.E. Ciardi, and E.R. Stadtman. 1973. Comparative biochemical and immunological studies of bacterial glutamine synthetases. *J. Bacteriol.* **115**:858–868.

Tumer, N.E., SJ. Robinson, and R. Haselkorn. 1983. Different promoters for the *Anabaena* glutamine synthetase gene during growth using molecular or fixed nitrogen. *Nature* (London) **306**:337–342.

Ueno-Nishio, S., K.C. Backman, and B. Magasanik. 1983. Regulation at the *glnL*-operator-promoter of the complex *glnALG* operon of *Escherichia coli. J. Bacteriol.* **153**:1247–1251.

Ueno-Nishio, S., S. Mango, L.J. Reitzer, and B. Magasanik. 1984. Identification and regulation of the *glnL* operator-promoter of the complex *glnALG* operon of *Escherichia coli. J. Bacteriol.* **160**:379–384.

Usdin, K.P., H. Zappe, D.T. Jones, and D.R. Woods. 1986. Cloning, expression, and purification of glutamine synthetase from *Clostridium acetobutylicum. Appl. Environ. Microbiol.* **52**:413–419.

Woese, C.R. 1987. Bacterial evolution. *Microbiol. Rev.* **51**:221–271.

Wolheuter, R.M., H.˙ Schutt, and H. Holzer. 1973. Regulation of glutamine synthetase in vivo in *E. coli. In: The Enzymes of Glutamine Metabolism,* S. Prusiner and E.R. Stadtman, eds., pp. 45–66. New York: Academic Press.

Wong, H.C. and S. Chang. 1986. Identification of a positive retroregulator that stabilizes mRNAs in bacteria. *Proc. Natl. Acad. Sci USA* **83**:3233–3237.

Wray, L.V. and S.H. Fisher. 1988. Cloning and nucleotide sequence of the *Streptomyces coelicolor* gene encoding glutamine synthetase. *Gene* **71**:247–256.

Yamamoto, I., A. Abe, H. Saito, and M. Ishimoto. 1984. The pathway of ammonia assimilation in *Bacteroides fragilis. J. Gen. Appl. Microbiol.* **30**:499–508.

Zhang, J., M. Strauch, and A.I. Aronson. 1989. Glutamine auxotrophs of *Bacillus subtilis* that overproduce glutamine synthetase antigen have altered conserved amino acids in or near the active site. *J. Bacteriol.* **171**:3572–3574.

24

Phospholipid Biosynthetic Enzymes of Butyric Acid-Producing Clostridia

Howard Goldfine

24.1 Introduction

Among the clostridia, the most intensively studied group of organisms with respect to lipid composition and metabolism are the saccharolytic, butyric acid-producing species. Because the cellular physiology, molecular biology, and genetics of these organisms are currently of considerable interest, a brief review of the status of knowledge of the polar lipids of these organisms is provided. These compounds are intrinsically significant as structural components of the cell membrane; they are also important in terms of modulating membrane protein function. Recent work has shown that these cells have considerable flexibility in regulating their membrane lipids over a wide range of potential environmental and nutritional conditions. Because the membranes of these organisms play a key role in energy production, solute uptake, and solvent excretion, future work to modify these aspects of cellular physiology will require a deep and extensive understanding of the biochemistry and regulation of their lipids.

24.2 Lipid Composition

The phospholipid composition of the three species of butyric acid-producing clostridia *Clostri-*

dium beijerinckii (Baumann et al., 1965; Khuller and Goldfine, 1974; Goldfine et al. 1977), *C. butyricum*, (Goldfine et al. 1982; Johnston and Goldfine, 1983), and *C. acetobutylicum* (Johnston and Goldfine, 1983; Thiele et al., 1985; Oulevey et al., 1986; Lepage et al., 1987), has been thoroughly studied. All three species contain phosphatidylethanolamine (PE), phosphatidylglycerol (PG) (Figure 24.1) and diphosphatidylglycerol (cardiolipin; CL). In addition to these lipids, which are also typical of the gram-positive genus *Bacillus* (Goldfine, 1982), *C. beijerinckii* has phosphatidyl-*N*-monomethylethanolamine (PME), which largely replaces PE, and *C. acetobutylicum* has mono- and diglycosyl diacylglycerols (Oulevey et al., 1986). These glycolipids represent nearly half the polar lipid in cells grown in a low-phosphate medium (Table 24.1). In all three species, each of these phospholipids and glycolipids is present in both the diacyl and 1-alk- 1'-enyl-2-acyl forms, with the latter often predominating (see Figure 24.1). These ether lipids are historically referred to as plasmalogens because they give rise to long-chain aldehydes on treatment with mild acid. In addition to alk-1-enyl ether lipids, all three species contain unusual glycerol acetals of the plasmalogen form of PE and of the plasmalogen form of PME in *C. beijerinckii* that have not been found in other organisms (Figure 24.1). It should also be noted that

Abbreviations: ACP, acyl carrier protein; DGDG, diglycosyldiacylglycerol; CL, cardiolipin; GDG, glycosyldiacylglycerol; GAPlaE, glycerol acetal of plasmenylethanolamine; GAPlaME, glycerol acetal of plasmenyl-*N*-monomethylethanolamine; MGDG, monoglycosyldiacylglycerol; PE, phosphatidylethanolamine; PG, phosphatidylglglycerol; PME, phosphatidyl-*N*-monomethylethanolamine; PS, phosphatidylserine; PlaE, plasmenylethanolamine; PlaME, plasmenyl-*N*-monomethylethanolamine.

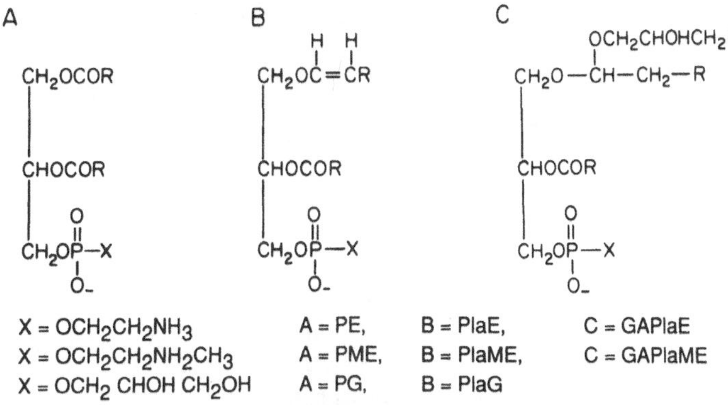

A

CH$_2$OCOR

CHOCOR

$$CH_2O\overset{O}{\underset{O_-}{\overset{\|}{P}}}-X$$

B

$$CH_2O\overset{H\ \ H}{\underset{}{\overset{|\ \ |}{C}}}=CR$$

CHOCOR

$$CH_2O\overset{O}{\underset{O_-}{\overset{\|}{P}}}-X$$

C

OCH$_2$CHOHCH$_2$OH

$$CH_2O-\overset{|}{C}H-CH_2-R$$

CHOCOR

$$CH_2O\overset{O}{\underset{O_-}{\overset{\|}{P}}}-X$$

X = OCH$_2$CH$_2$NH$_3$ A = PE,   B = PlaE,   C = GAPlaE

X = OCH$_2$CH$_2$NH$_2$CH$_3$ A = PME,   B = PlaME,   C = GAPlaME

X = OCH$_2$ CHOH CH$_2$OH   A = PG,   B = PlaG

FIGURE 24.1 Structures of major phospholipids of butyric acid-producing clostridia.

TABLE 24.1 Major polar lipids of butyric acid-producing clostridia in percentages of lipid phosphorus

Species	PE[a]	PME	GAPlaE	GAPlaME	PG	CL	GDG	References
Clostridium acetobutylicum	46(ND)[b]	0	10	0	29[c]	15[d]	+[e]	Lepage et al., 1987 Oulevy et al., 1986
C. beijerinckii	12(55)	34(78)		29.2[f]	25(38)	+[g]		Baumann et al., 1965 Khuller and Goldfine, 1974 Goldfine et al., 1977
C. butyricum	48(78)	0	16	0	21 (~1/3)	13 (~1/3)	+[h]	Goldfine et al., 1982 Johnston and Goldfine, 1983

[a] Abbreviations: CL, cardiolipin; GAPlaE, glycerol acetal of plasmenylethanolamine; GAPlaME, glycerol acetal of plasmenyl-N-monomethylethanolamine; GDG, glycosyldiacylglycerols; PE, phosphatidylethanolamine; PG, phosphatidylglycerol; PME, phosphatidyl-N-monomethylethanolamine.
[b] Numbers in parentheses represent percentage plasmalogen in lipid class. ND = not determined.
[c] Almost entirely plasmalogen (Thiele et al., 1985).
[d] Plasmalogen present (Thiele et al., 1985).
[e] Glycosyl diacylglycerols: in C. acetobutylicum grown in low-phosphate medium, mono- and diglycosyldiacylglycerol represent nearly 50% of total polar lipid, largely replacing PE and PlaE (Thiele et al., 1985).
[f] Sum of GAPlaE and GAPlaME, mostly GAPlaME (Johnston and Goldfine, 1983).
[g] Cardiolipin accumulates in cells in stationary phase of growth (Baumann et al., 1965).
[h] A small amount of glycosyldiglyceride is present (Matsumoto et al., 1971).

although phosphatidylserine and phosphatidic acid are formed in these cells as intermediates in phospholipid synthesis (see following), they do not accumulate and are usually present in nearly undetectable amounts (Koga and Goldfine, 1984; MacDonald and Goldfine, 1990). C. saccharoperbutylacetonicum ATCC 13564 appears to have a lipid composition similar to that of C. acetobutylicum (Ogata et al., 1982).

The hydrocarbon chains of these lipids are composed largely of 14-, 16-, and 18-carbon saturated and monounsaturated fatty acids and aldehydes. In addition there are variable amounts of 17- and 19-carbon cyclopropane fatty acids and aldehydes (Goldfine, 1964; Khuller

and Goldfine, 1974; Johnston and Goldfine, 1983). These are derived from the corresponding 16- and 18-carbon monounsaturated hydrocarbon chains, and the cellular content of cyclopropane chains tends to increase with culture age (Law, 1971).

24.3 Biosynthetic Pathways to Phospholipids

Phospholipid biosynthesis has been studied in both Clostridium butyricum and C. beijerinckii. As in other bacteria, the process begins with the transfer to sn-glycero-3-phosphate of the end

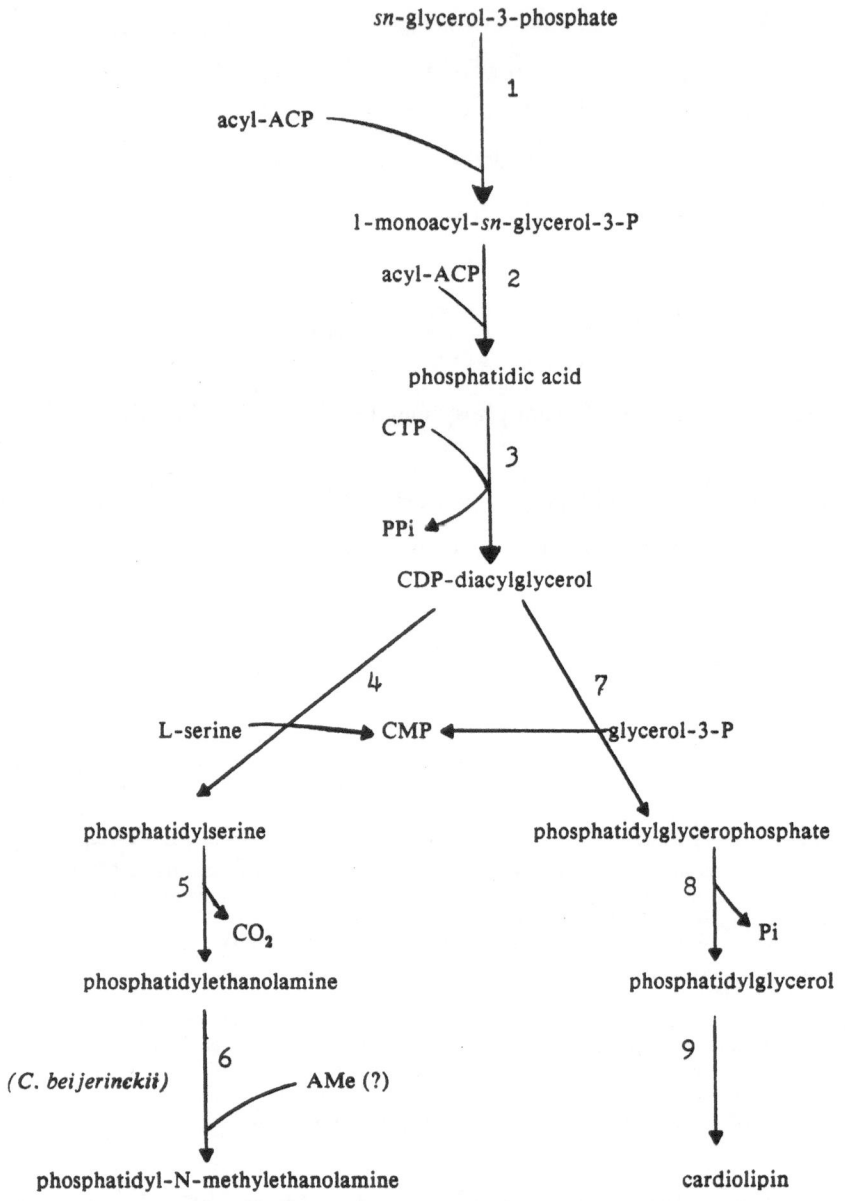

FIGURE 24.2 Pathways of phospholipid synthesis in *Clostridium beijerinckii* and *C. butyricum*. Enzymes: (1) *sn*-glycero-3-phosphate acyltransferase (Ailhaud et al., 1967; Goldfine et al., 1967; Goldfine and Ailhaud, 1971); (2) acyl-glycero-3-phosphate acyltransferase (Goldfine and Ailhaud, 1971); (3) CDP-diglyceride synthetase (Silber et al., 1980); (4) phosphatidylserine synthase (Silber et al., 1980); (5) phosphatidylserine decarboxylase (Verma and Goldfine, 1985); (6) phosphatidylethanolamine *N*-methyltransferase; (7) phosphatidylglycerophosphate synthase (Silber et al., 1980); (8) phosphatidylglycerophosphate phosphatase (Silber et al., 1980); (9) cardiolipin synthetase (Walton and Goldfine, 1987; Morii and Goldfine, 1990).

products of fatty acid synthesis, saturated or monounsaturated chains esterified as thiolesters to the 4'-phosphopantetheine prosthetic group of the acyl carrier protein. This protein has been purified from *C. beijerinckii* ATCC 6015 (formerly *C. butyricum*) and shown to have a mass of 8600 daltons (Ailhaud et al., 1967). The reaction, catalyzed by an acyl-acyl carrier protein:glycerol-3-P acyltransferase, has been characterized (Goldfine et al., 1967; Goldfine

TABLE 24.2 Enzymes of phospholipids synthesis in butyric acid-producing clostridia

Reaction	Enzyme	References
1	sn-Glycero-3-phosphate acyltransferase	Ailhaud et al., 1967
		Goldfine et al., 1967
		Goldfine and Ailhaud, 1967
2	Acyl-Glycero-3-phosphate acyltransferase	Goldfine and Ailhaud, 1971
3	CDP-diglyceride synthetase	Silber et al., 1980
4	Phosphatidylserine synthase	Silber et al., 1980
5	Phosphatidylserine decarboxylase	Verma and Goldfine, 1985
6	Phosphatidylethanolamine N-methyltransferase	
7	Phosphatidylglycerophosphate synthase	Silber et al., 1980
8	Phosphatidylglycerophosphate phosphatase	Silber et al., 1980
9	Cardiolipin synthetase	Walton and Goldfine, 1987
		Morii and Goldfine, 1990

and Ailhaud, 1971), but the membrane-bound acyltransferases have not been purified. The enzyme from *C. butyricum* has also been studied and it has been cloned into *E. coli* (see following). After the initial acylation of glycerol-3-P, which yields monoacyl glycerol-3-P (lysophosphatidic acid), a second acyltransferase that has been studied in extracts of *C. beijerinckii* utilizes either acyl-ACP or acyl-CoA in vitro (Goldfine and Ailhaud, 1971), producing phosphatidic acid (Figure 24.2). Reaction of phosphatidic acid with CTP leads to the liponucleotide CDP-diacylglycerol. Phosphatidic acid cytidylyltransferase has been studied in a number of gram-negative and gram-positive bacteria, including *C. beijerinckii* (Silber et al., 1980).

As in *E. coli*, the pathway branches (see Figure 24.2) by reaction of CDP-diacylglycerol either with L-serine to yield phosphatidylserine (PS) or with glycerol-3-P to yield phosphatidylglycerophosphate (PGP). These in turn give rise to PE by decarboxylation of PS and to PG by removal of the terminal phosphate from PGP. In *C. beijerinckii*, PE gives rise to PME by an N-methyltransferase (Goldfine, 1962; Goldfine and Ellis, 1964; Baumann et al., 1965). This reaction, however, has not been demonstrated in vitro. Finally, cardiolipin is formed by condensation of two molecules of PG. The enzymes catalyzing these reactions are listed in Table 24.2 as are references to studies of these enzymes in clostridia. The pathways to the plasmalogens and their glycerol acetals in anaerobes have not been elucidated. It is clear that they differ markedly from the oxygen-dependent mammalian pathway for the synthesis of alkyl and alkenyl ethers (Goldfine & Hagen, 1972; Paltauf, 1983).

24.4 Regulation of Lipid Synthesis in Clostridia

Regulation of membrane fluidity

Studies on *Clostridium beijerinckii* ATCC 6015 demonstrated the control of aliphatic chain composition as a function of growth temperature. As the growth temperature was lowered, the acyl chains were shown to become more unsaturated and slightly shorter. The alkenyl chains of the plasmalogens and the alkyl chains of the glycerol acetals as a group, however, were seen to become significantly more saturated at 30°C than at 37°C and slightly more saturated at 25°C than at 37°C (Khuller and Goldfine, 1974). Changes were also observed in lipid class composition, with plasmenylglycerol increasing at the expense of the ethanolamine and N-methylethanolamine plasmalogens and their glycerol acetals. The effects of these compositional changes on phospholipid thermotropic phase transitions were studied with an ESR probe, 2,2,6,6-tetramethylpiperidine-1-oxyl (TEMPO), and a fluorescent probe, diphenylhexatriene. The presumed onset of the liquid crystalline- to gel-phase transition, as observed with both probes, was 4° to 5°C lower in the phospholipids from cells grown at 25°C compared to those from cells grown at 37°C (Goldfine et al., 1977). Both probes indicated that the onset of this transition was 2° to 4°C below the growth temperature of cells grown at 37°C and 4° to 7°C above the growth temperature for the cells grown at 25°C. It was concluded that at 37°C almost all the phospholipid is potentially disordered and at 25°C most is above its liquid crystalline-to-gel transition temperature. The completion of the transitions to the gel phase were observed to be −5°C for both types of cells, which is well below the minimum growth temperature for this organism (Goldfine et al., 1977). Thus, the changes in acyl and alkenyl chains and in polar lipid composition at lower growth temperatures appear to lower the content of higher melting lipids without necessarily producing species that extend the low end of the transition to the gel phase. In *C. acetobutylicum*, lowering the growth temperature also resulted in shorter acyl chains, but there was no change in the unsaturated/saturated ratio (Lepage et al., 1987). The effects on ether-linked chains were not studied.

Regulation of lipid polymorphism

The ability of bacteria to regulate the potential for transitions from the lamellar phase to such nonlamellar phases as the reversed hexagonal and cubic phases has only recently been recognized. Initial studies with *Acholeplasma laidlawii* revealed that this natural fatty acid auxotroph was able to adjust its membrane lipids dramatically. It was shown that increased growth temperature, acyl chain unsaturation of the membrane lipids, or sterol content led to specific changes in the ratio of two major membrane glucolipids, monoglucosyl diacylglycerol (MGDG) and diglucosyl diacylglyerol (DGDG). As the lipid unsaturation or growth temperature was increased, the cellular content of DGDG increased at the expense of MGDG (Wieslander et al., 1980, 1981a). It was further shown that unsaturated MGDG tended to readily form cubic phases at physiological temperatures, but DGDG formed a stable lamellar phase (Wieslander et al., 1981b; Rilfors et al., 1984). The elevated content of DGDG observed in cells with highly unsaturated membranes served to maintain, within a narrow range, the temperature at which the mixed lipids begin to form nonlamellar phases. The

temperature of these transitions was approximately 20°C. higher than the growth temperature (Lindblom et al., 1986).

Unsaturated phosphatidylethanolamines readily undergo transitions from the lamellar to the reversed hexagonal (H_{II}) phase (Cullis et al., 1986). The presence of plasmenylethanolamine in clostridial membranes lowers the temperature of these transitions (Lohner et al., 1984). When *C. beijerinckii* (Khuller and Goldfine, 1975) or *C. butyricum* (Goldfine et al., 1981; Goldfine and Johnston, 1985) were grown with oleic acid in the absence of biotin, leading to membranes with more than 95% unsaturated plus cyclopropane chains, there was an increase in the glycerol acetals of plasmenylethanolamine and a decrease in the PE fraction. Physical studies on the phase behavior of the hydrated lipids have shown that the increased amounts of the glycerol acetal of plasmenylethanolamine served to stabilize the bilayer amounts of the glycerol acetal of plasmenylethanolamine served to stabilize the bilayer formed a lamellar phase at temperatures 50°C. above that of the similarly enriched PE fraction from these cells. As noted, *C. acetobutylicum* has both glycolipids and phospholipids in its membrane. The major glycolipids, as in *A. laidlawii*, are mono- and diglycosyldiacylglycerols, specifically galactosyldiacylglycerol and glucosylgalactosyldiacylglycerol (Oulevey et al., 1986). The fatty acid composition of this species is also readily altered by growth on fatty acids in the absence of biotin, and recent studies have shown that both the phospholipids and glycolipids are regulated in a manner similar to that seen in *C. butyricum* and in *A. laidlawii*, respectively (H. Goldfine and N.C. Johnston, unpublished observations).

Effects of solvents

Several studies have shown that *C. acetobutylicum* responds to the addition of butanol to the growth medium by increasing the ratio of saturated to unsaturated (plus cyclopropane) acyl chains in the membrane lipids (Vollherbst-Schneck et al., 1984; Baer et al., 1987; Lepage et al., 1987). Similar changes were also observed during the acetone-butanol, fermentation or on addition of ethanol, hexanol, octanol, or acetone to the growth medium (Lepage et al., 1987). Because butanol is known to increase membrane fluidity, as measured by the motions of spin-labeled or fluorescent probes, the changes observed in fatty acid ratios were viewed as a mechanism to decrease membrane fluidity in the face of solvent challenge. A butanol-tolerant strain was found to maintain relatively constant membrane fluidity at various butanol concentrations and temperature combinations, whereas the wild-type strain showed increased membrane fluidity with increasing butanol challenge (Baer et al., 1987, 1989).

Polar lipid composition was also observed to change in C. acetobutylicum during an acetone-butanol fermentation. Cardiolipin and a polar lipid presumed to be the glycerol acetal of plasmenylethanolamine (see Figure 24.1) increased. This was accompanied by decreases in phosphatidylglycerol and phosphatidylethanolamine. These changes occurred largely after growth had ceased (Lepage et al., 1987).

The effects of solvents on the polar lipids of C. butyricum have been studied in cells with controlled aliphatic chain composition. Addition of cyclohexane, hexanol or octanol to cells grown on elaidic acid in the absence of biotin, resulted in increased ratios of the glycerol acetal of plasmenylethanolamine to the sum of phosphatidylethanolamine and its plasmalogen. Since the aliphatic chains were the same with or without solvent addition, the effects on lipid class composition were seen to be independent of changes in the apolar portions of the lipid molecules. These studies suggest that cells respond to solvents to regulate lipid polymorphism in addition to regulating membrane fluidity (MacDonald and Goldfine, 1991).

24.5 Cloning of the sn-Glycerol-3-Phosphate Acyltransferase of Clostridium butyricum

As has been discussed, acyl-acyl carrier protein:sn-glycero-3-phosphate acyltransferase catalyzes the first committed step in phospholi-pid synthesis in bacteria and is considered to be important in the regulation of lipid synthesis (Cronan and Rock, 1987). In animal tissues, the corresponding enzyme uses acyl thiolesters of CoA as acyl donors (Brindley, 1985); however, in procaryotes acyl-acyl carrier protein derivatives appear to be the preferred donors in the relatively few organisms studied (Pieringer, 1989). An exception is Escherichia coli, in which both acyl-CoA and acyl-ACP esters are used by the same enzyme (Green et al., 1981). The gene plsB has been identified as the structural gene for the glycerol-3-P acyltransferase of E. coli and has been cloned and sequenced (Lightner et al., 1980, 1983). The membrane-bound enzyme has been purified to near homogeneity from cells containing the amplified gene (Green et al., 1981).

To clone the corresponding gene from C. butyricum, advantage was taken of the glycerol (or glycerol-3-P) auxotrophy of mutants in plsB in which the glycerol-3-P acyltransferase has a high K_m for glycerol-3-P (Bell, 1974). A Sau3A digest of C. butyricum DNA was ligated into the plasmid pBR327. After transformation of E. coli HB101, pools of amp^r transformants were used to transform E. coli TL400 (plsB) to glycerol-3-P prototrophy. Twenty amp^r glycerol-3-P prototrophs were obtained of which 12 gave high-frequency transformation when plasmid DNA was isolated and used to retransform E. coli TL400. Restriction digest analysis of the plasmids revealed a 6.7-kb insert. One of these hybrid plasmids, pRS25, is illustrated in Figure 24.3. The gene for sn-glycero-3-phosphate acyltransferase was localized to a 4.4-kb fragment by subcloning into pBR327. Southern hybridization with a ^{32}P-labeled probe prepared from pRS25 digested with BamHI and EcoRI showed binding to a 24.5-kb fragment from BamHI-digested C. butyricum DNA and to 5.8- and 5.0-kb fragments from an EcoRI digest. No hybridization was seen with E. coli or Bacillius subtilis DNA.

When tested in vitro, the glycerol-3-P acyltransferase expressed in one of the transformants, pRS47d, was not active with acyl-CoA as acyl donor at low concentrations of glycerol-3-P. At high concentrations (>0.4 mM), some activity was observed, consistent with the con-

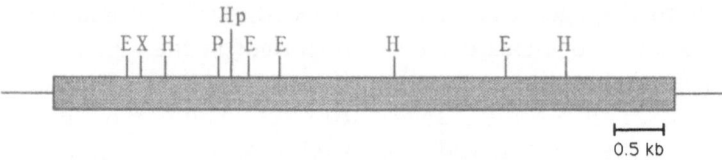

FIGURE 24.3   Restriction endonuclease map of insert in plasmid pRS25, which contains 6.7-kb Sau3A fragment of *Clostridium butyricum* DNA.

Abbreviations: *E, EcoRI; H, HindIII: Hp, HpaI; P, PstI; X, XbaI.*

tinued presence of the mutated enzyme of the host strain. With saturated and unsaturated acyl derivatives of the acyl carrier protein, however; activity was seen at lower concentrations of glycerol-3-P with a K_m of about 0.1 mM, which was similar to the K_m for glycerol-3-P of the acyltransferase in membrane particles of *C. butyricum* (Rosenthal et al., in manuscript).

Expression of the acyltransferase in *E. coli* had no effect on cellular phospholipid composition, the fatty acid composition of the principal phospholipid, phosphatidylethanolamine, or on the placement of the fatty acids on the glycerol-3-P backbone of the lipid. The largely *sn*-1-saturated, *sn*-2-unsaturated placement typical of *E. coli* was retained, despite the fact that the placement tends to be opposite in *C. butyricum*; i.e., more unsaturated acyl chains on the *sn*-1 position and more saturated chains at the *sn*-2 position. These results suggest that the enzyme may be influenced by cellular factors such as membrane lipids, other proteins in the membrane, or the cellular concentrations of reactants. Another possibility is that the existing *E. coli* acyltransferase, despite its requirement for elevated glycerol-3-P, continues to play a role in lipid synthesis in the growing transformant cell. Further studies on the function of the clostridial glycerol-3-P are currently in progress.

24.6   Prospects

The import of nutrients into the cell, the flow of the products of metabolism, including the solvents produced by clostridia, out of the cell, and the generation of proton gradients and ATP are regulated by components of the cell membrane. Because membrane lipids provide the structural basis of the membrane and modulate the functions of the active proteins, it is important to understand their structures, biosynthesis, and the regulation of their composition (Russel, 1989). Much of the work on clostridial lipids and membranes has been at the level of the whole-cell and subcellular fractions. As documented in this volume, rapid progress is being made in the introduction of genetic elements into clostridia and the transfer of genetic information between clostridial and other species. Many genes have been cloned into other species, including at least one for a key enzyme in phospholipid synthesis, as described in this chapter. Extension of these advances to other genes of lipid synthesis and metabolism is expected, and these studies should lead to a much more complete understanding of biosynthesis and regulation of membrane lipids in clostridia. This in turn holds out the prospect of controlling the structure and function of clostridial membranes.

References

Ailhaud, G.P., P.R. Vagelos, and H. Goldfine. 1967. Involvement of acyl carrier protein in acylation of glycerol 3-phosphate in *Clostridium butyricum*. I. Purification of *Clostridium butyricum* acyl carrier protein and synthesis of long-chain derivatives of acyl carrier protein. *J. Biol. Chem.* **242**:4459–4465.

Baer, S.H., H.P. Blaschek, and T.L. Smith. 1987. Effect of butanol challenge and temperature on lipid composition and membrane fluidity of butanol-tolerant *Clostridium acetobutylicum*. *Appl. Environ. Microbiol.* **53**:2854–2861.

Baer, S.H., D.L. Bryant, and H.P. Blaschek. 1989. Electron spin resonance analysis of the effect of butanol on the membrane fluidity of intact cells of *Clostridium acetobutylicum*. *Appl. Environ. Microbiol.* **55**:2729–2731.

Baumann, N.A., P-O. Hagen, and H. Goldfine. 1965. Phospholipids of *Clostridium butyricum*: Studies on plasmalogen composition and biosynthesis. *J. Biol. Chem.* **240**:1559–1567.

Bell, R.M. 1974. Mutants of *Escherichia coli* defective in membrane phospholipid synthesis: macromolecular synthesis in an *sn*-glycerol-3-phosphate acyltransferase K_m mutant. *J. Bacteriol.* **117**:1065–1076.

Brindley, D.N. 1985. Metabolism of triacylglycerols. *In: Biochemistry of Lipids and Membranes*, D.E. Vance and J.E. Vance, eds., pp. 211–241. Menlo Park: Benjamin/Cummings.

Cronan, J.E., Jr. and C.O. Rock. 1987. Biosynthesis of membrane lipids. *In: Escherichia coli* and *Salmonella typhimurium. Cellular and Molecular Biology*, Vol. 1, F. Neidhardt, J.L. Ingraham, K.B. Low, B. Magasanik, M. Schaechter, and H.E. Umbarger, eds., pp. 474–497. Washington, D.C.: American Society for Microbiology.

Cullis, P.R., M.J. Hope, and C.P.S. Tilcock. 1986. Lipid polymorphism and the roles of lipids in membranes. *Chem. Phys. Lipids* **40**:127–144.

Goldfine, H. 1962. The characterization and biosynthesis of an N-methylethanolamine phospholipid from *Clostridium butyricum*. *Biochim. Biophys. Acta* **59**:504–506.

Goldfine, H. 1964. Composition of the aldehydes of *Clostridium butyricum* plasmalogens: cyclopropane aldehydes. *J. Biol. Chem.* **239**:2130–2134.

Goldfine, H. 1982. Lipids of procaryotes—structure and distribution. *Current Topics Membr. Trans.* **17**:1–43.

Goldfine, H. and G.P. Ailhaud. 1971. Fatty acyl-acyl carrier protein and fatty acyl-CoA in the biosynthesis of phosphatidic acid in *Clostridium butyricum*. *Biochem. Biophy. Res. Commun.* **45**:1127–1133.

Goldfine, H. and M. Ellis. 1964. N-methyl groups in bacterial lipids. *J. Bacteriol.* **87**:8–15.

Goldfine, H. and P.-O. Hagen, 1972. Bacterial plasmalogens. *In: Ether Lipids: Chemistry and Biology*, F. Snyder, ed., pp. 329–350. New York: Academic Press.

Goldfine, H., G.P. Ailhaud, and P.R. Vagelos. 1967. Involvement of acyl carrier protein in acylation of glycerol 3-phosphate in *Clostridium butyricum*. II Evidence for the participation of acyl thioesters of acyl carrier protein. *J. Biol. Chem.* **242**:4466–4475.

Goldfine, H. and N.C. Johnston. 1985. Phospholipid aliphatic chain composition modulates lipid class composition, but not lipid asymmetry in *Clostridium butyricum*. *Biochim. Biophys. Acta* **813**:10–18.

Goldfine, H., N.C. Johnston, and D.G. Bishop. (1982) Ether phospholipid asymmetry in *Clostridium butyricum*. *Biochem. Biophys. Res. Commun.* **108**:1502–1507.

Goldfine, H., N.C. Johnston, and M.C. Phillips, 1981. Phase behavior of ether lipids from *Clostridium butyricum*. *Biochemistry* **20**:2908–2916.

Goldfine, H., N.C., Johnston, J. Mattai, and G.G. Shipley. 1987. The regulation of bilayer stability in *Clostridium butyricum*: Studies on the polymorphic phase behavior of the ether lipids. *Biochemistry* **26**:2814–2822.

Goldfine, H., G.K. Khuller, R.P. Borie, B. Silverman, H. Selick, N.C. Johnston, J.M. Vanderkooi, and A.F. Horwitz. 1977. Effects of growth temperature and supplementation with exogenous fatty acids on some physical properties of *Clostridium butyricum* phospholipids. *Biochim. Biophys. Acta* **488**:341–352.

Green, P.R., A.H. Merrill, Jr., and R.M. Bell. 1981. Membrane phospholipid synthesis in *Escherichia coli*. Purification, reconstitution, and characterization of *sn*-glycerol-3-phosphate acyltransferase. *J. Biol. Chem.* **256**:11151–1159.

Johnston, N.C. and H. Goldfine. 1983. Lipid composition in the classification of the butyric acid-producing clostridia. *J. Gen. Microbiol.* **129**:1075–1081.

Khuller, G.K. and H. Goldfine. 1974. Phospholipids of *Clostridium butyricum*. V. Effects of growth temperature on fatty acid, alk-1-enyl group, and phospholipid composition. *J. Lipid Res.* **15**:500–507.

Khuller, G.K. and H. Goldfine. 1975. Replacement of acyl and alk-1-enyl groups in *Clostridium butyricum* phospholipids by exogenous fatty acids. *Biochemistry* **14**:3642–3647.

Koga, Y. and H. Goldfine. 1984. Biosynthesis of phospholipids in *Clostridium butyricum*: the kinetics of synthesis of plasmalogens and the glycerol acetal of ethanolamine plasmalogen. *J. Bacteriol.* **159**:597–604.

Law, J.H. 1971. Biosynthesis of cyclopropane rings. *Accounts Chem. Res.* **4**:199–203.

Lepage, C., F. Fayolle, M. Hermann, and J.-P. Vandecasteele. 1987. Changes in membrane lipid composition of *Clostridium acetobutylicum* during acetone-butanol fermentation: effects of solvents, growth temperature and pH. *J. Gen. Microbiol.* **133**:103–110.

Lightner, V.A., R.M. Bell, and P. Modrich. 1983. The DNA sequences encoding *plsB* and *dgk* loci of *Escherichia coli*. *J. Biol. Chem.* **258**:10856–10861.

Lightner, V.A., T.J. Larson, P. Tailleur, G.D. Kantor, C.R.H. Raetz, R.M. Bell, and P. Modrich. 1980. Membrane phospholipid synthesis in *Escherichia coli*. Cloning of a structural gene (*plsB*) of the *sn*-glycerol-3-phosphate acyltransferase. *J. Biol. Chem.* **255**:9413–9420.

Lindblom, G., I. Brentel, M. Sjolund, G. Wikander, and Å. Wieslander, 1986 Phase equilibria of membrane lipids from *Acholeplasma laidlawii*: importance of a single lipid forming nonlamellar phases. *Biochemistry* **25**:7502–7510.

Lohner, K., A. Hermetter, and F. Paltauf. 1984. Phase behavior of ethanolamine plasmalogen. *Chem. Phys. Lipids* **34**:163–170.

MacDonald, D.L. and H. Goldfine. 1991. Effects of solvents and alcohols on the polar lipid composition of *Clostridium butyricum* under conditions of controlled lipid chain composition. *Appl. Environ. Microbiol.* **57**:3517–3521.

MacDonald, D.L. and H. Goldfine. 1990 Phosphatidylglycerol acetal of plasmenylethanolamine as an intermediate in ether lipid formation in *Clostridium butyricum*. *Biochem. Cell Biol.* **68**:225–230.

Matsumoto, M., K. Tamiya, and K. Koizumi. 1971. Studies on neutral lipids and a new type of aldehydogenic ethanolamine phospholipid in *Clostridium butyricum*. *J. Biochem.* (Tokyo) **69**:617–620.

Morii, H. and H. Goldfine. 1990. Phosphatidyltransferase of *Clostridium butyricum*: specificity for diacylphosphoglycerides. *Biochim. Biophys. Acta* **1044**:394–398.

Ogata, S., S. Yoshino, Y. Okuma, and S. Hayashida. 1982. Chemical composition of autoplast membrane of *Clostridium saccharoperbutylacetonicum*. *J. Gen. Appl. Microbiol.* **28**:293–301.

Oulevey, J., H. Bahl, and O.W. Thiele. 1986. Novel alk-1-enyl ether lipids isolated from *Clostridium acetobutylicum*. *Arch. Microbiol.* **144**:166–168.

Paltauf, F. 1983. Biosynthesis of 1-*O*-(1′Alkenyl)-glycerolipids (plasmalogens). *In: Ether Lipids. Biochemical and Biomedical Aspects*, (H.K. Mangold and F. Paltauf, eds.) pp. 107–128. New York: Academic Press.

Pieringer, R.A. 1989. Biosynthesis of nonterpenoid lipids. *In: Microbial Lipids, Vol. 2*, C. Ratledge and S.G. Wilkinson, eds. pp. 51–114. London: Academic Press.

Rilfors, L., G. Lindblom, Å. Wieslander, and A.

Christiansson. 1984. Lipid bilayer stability in biological membranes. *Biomembranes* **12**:205–245.

Russel, N.J. 1989. Functions of lipids: structural roles and membrane functions. *In: Microbial Lipids, Vol. 2*, C. Ratledge and S.G. Wilkinson, eds., pp. 277–365. London: Academic Press.

Silber, P., R.P. Borie, and H. Goldfine. 1980. The enzymes of phospholipid synthesis in *Clostridium butyricum*. *J. Lipid Res.* **21**:1022–1031.

Thiele, O.W., J. Oulevey, and H. Bahl. 1985. Neuartige Alkenylether-lipide aus anaeroben bakterien. *Fette Seifen Anstrichm.* **87**:551–556.

Verma, J.N. and H. Goldfine. 1985. Phosphatidylserine decarboxylase from *Clostridium butyricum*. *J. Lipid Res.* **26**:610–616.

Vollherbst-Schneck, K., J.A. Sands, and B.S. Montenecourt, 1984. Effect of butanol on lipid composition and fluidity of *Clostridium acetobutylicum* ATCC 824. *Appl. Environ. Microbiol.* **47**, 193–194.

Walton, P.A. and H. Goldfine. 1987. Transphosphatidylation activity in *Clostridium butyricum*. Evidence for a secondary pathway by which membrane phospholipids may be synthesized and modified. *J. Biol. Chem.* **262**:10355–10361.

Wieslander, Å., A. Christiansson, L. Rilfors, and G. Lindblom. 1980. Lipid bilayer stability in membranes. Regulation of lipid composition in *Acholeplasma laidlawii* as governed by molecular shape. *Biochemistry* **19**:3650–3655.

Wieslander, Å., A. Christiansson, L. Rilfors, A. Khan, L.B.-Å. Johansson, and G. Lindblom. 1981a Lipid phase structure governs the regulation of lipid composition in membranes of *Acholeplasma laidlawii*. *FEBS Lett.* **124**:273–278.

Wieslander, Å., L. Rilfors, L.B.-Å. Johansson, and G. Lindblom. 1981b. Reversed cubic phase with membrane glucolipids from *Acholeplasma laidlawii*. ^{1}H, ^{2}H, and diffusion nuclear magnetic resonance measurements. *Biochemistry* **20**:730–735.

25

The *Clostridium pasteurianum* Ferredoxin Gene

Jesse C. Rabinowitz

25.1   Iron-Sulfur Proteins

Ferredoxin (Fd) is a name first applied to a brown, iron-containing protein isolated from *Clostridium pasteurianum* (Mortenson et al., 1962). The unusually low redox potential of the protein and its participation in the pyruvate phosphoroclastic reaction and the hydrogenase reaction of *C. pasteurianum* suggested its function in electron transport in this organism (Valentine, 1964). The requirement for clostridial Fd in nitrogen fixation by extracts of *C. pasteurianum* was also demonstrated at this time, as well as its function as an electron donor in the oxidation of α-ketoglutarate, hypoxanthine, formate, and dithionite by various anaerobic bacteria (Valentine, 1964). The term Fd was also applied to a protein isolated from photosynthetic tissue because it could be replaced functionally by clostridial Fd (Tagawa and Arnon, 1962).

At the time, neither compound had been chemically characterized. They were subsequently shown to contain different Fe-S cluster types and unrelated proteins. Many other proteins related to clostridial Fd were discovered and described. These proteins were characterized as a previously unrecognized class of electron-transferring proteins containing iron and "inorganic" sulfur and are now referred to as the iron-sulfur proteins. They were also recognized by physical properties associated with the Fe-S clusters that they contain. The use of electron paramagnetic resonance spectroscopy was especially important in their characteriza-

tion. The Fe-S proteins are composed of a protein moiety associated with one or more Fe-S clusters. The clusters occur in a variety of structures that are simply represented as [xFe-yS]. These are used for classification of the compounds. Several different Fe-S proteins may be present in a particular organism. Progress in our knowledge of these proteins has been summarized at relatively regular intervals through the publication of reviews and books (Valentine, 1964; Lovenberg, 1973a; Lovenberg, 1973b; Lovenberg, 1976; Yoch and Carithers, 1979; Spiro, 1982; Beinert, 1990).

Clostridial Fds contain two Fe-S clusters of the general formula [4Fe-4S] associated with a protein of about 55 amino acids. Each iron atom is ligated to a cysteine residue of the polypeptide and three inorganic sulfide atoms that interconnect the iron atoms (Figure 25.1; [4Fe-4S]). The clusters of the clostridial type Fds have also been designated [8Fe-8S]. Additional classes of Fe-S proteins containing clusters that differ from the [4Fe-4S] cluster but also function as electron carriers have been recognized. Structures of several of these proteins and the Fe-S clusters have been determined by x-ray analysis (Stout, 1982). Schematic representations of the structures are shown in Figure 25.1. These include the [1Fe-0S] rubredoxin, which is sometimes omitted from the classification because it does not contain inorganic sulfide; the [2Fe-2S] chloroplast Fd, and other [2Fe-2S] proteins involved in methylene hydroxylation reactions (putidaredoxin) and steroid hydroxylation (adrenodoxin); the [3Fe-4S] recently dis-

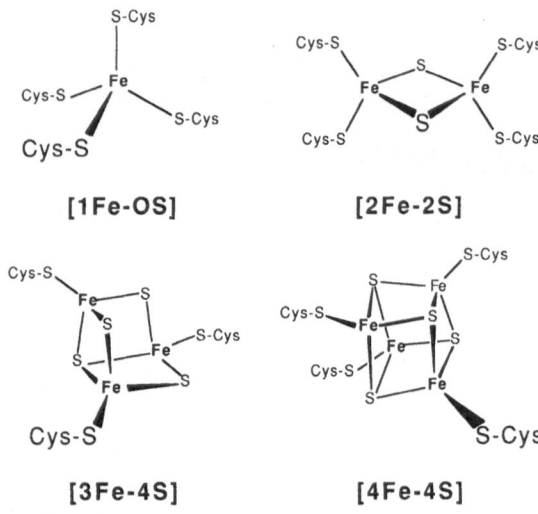

FIGURE 25.1 Schematic representation of structures of Fe-S clusters. More precise models are available from x-ray crystallographic studies (Stout, 1982).

covered in succinic and fumarate dehydrogenases (Johnson et al., 1985; Beinert, 1990); the [4Fe-4S] HIPIP (high potential iron protein) containing a single cluster, and the [8Fe-8S] clostridial Fds containing two [4Fe-4S] clusters.

In addition to the Fe-S proteins of relatively low molecular weights containing the clusters just described that act as electron carriers when associated with other proteins that have enzymic activities, the Fe-S clusters also occur as components of proteins that possess enzymic activity. They include hydrogenase and nitrogenase in reactions that involve electron transfer, but also in glutamine phosphoribosyl-pyrophosphate amidotransferase in *Bacillus subtilis*, which does not appear to be a reaction involving electron transfer (Grandoni et al., 1989). Fe-S clusters containing additional elements have also been described in hydrogenase (Ni, or Ni and Se) and nitrogenase (Mo). Fe-S clusters also occur in association with FAD or FMN in xanthine dehydrogenase and enzymes of the respiratory chain.

Fe-S proteins of *Clostridium pasteurianum*

Clostridium pasteurianum contains a number of different Fe-S proteins including representa-

tives of the [1Fe-0S], [2Fe-2S], and [8Fe-8S] classes. Rubredoxin, a [1Fe-0S] protein, was first isolated from *C. pasteurianum* (Lovenberg and Sobel, 1965). The specific function of this protein remains unknown, although it can replace clostridial Fd in some electron transfer reactions. The amino acid sequence of the protein has been determined (Lovenberg and Williams, 1969). The protein contains 54 amino acid residues with a molecular weight of 6127. *C. pasteurianum* rubredoxin is unusual relative to other rubredoxins and Fe-S proteins in retaining the N-formylmethionine group involved in initiation of synthesis of the N-terminal end (McCarthy and Lovenberg, 1970). The crystal structure of *C. pasteurianum* rubredoxin, first reported in 1970 (Herriott et al., 1970), has been further refined (Watenpaugh et al., 1980).

A protein containing a [2Fe-2S] cluster has also been found in *C. pasteurianum* (Cardenas et al., 1976). It has been named the paramagnetic protein of *C. pasteurianum*. Many Fe-S proteins containing the [2Fe-2S] cluster are known in both anaerobic and aerobic organisms, where they function in photosynthesis and a variety of enzymatic reactions. However, no function has been attributed to this *C. pasteurianum* protein, although it has been sequenced and characterized (Meyer et al., 1986). This most recent analysis shows that the protein is a dimer of two identical peptides each associated with a single [2Fe-2S] cluster; the monomeric peptide is composed of about 102 amino acid residues with a molecular weight of 11,415 (not including the [2Fe-2S]). These analytical data are not in agreement with those reported previously (Cardenas et al., 1976).

The first Fe-S protein isolated and characterized from *C. pasteurianum* was called Fd (Mortenson et al., 1962). It was found to contain 2[4Fe-4S] clusters. This protein and others containing two similar [4Fe-4S] clusters are called clostridial Fds. They may constitute as much as 2% of the cell protein. The polypeptide of the *C. pasteurianum* Fd contains 55 amino acids and functions in a variety of enzymic reactions as the electron donor or acceptor (Mortenson and Nakos, 1973). The structure of a clostridial-type Fd was first determined by x-ray analysis of the Fd from

Micrococcus aerogenes (reclassified as *Peptococcus aerogenes*) (Adman et al., 1973, 1976).

[4Fe-4S] cluster

The distinguishing feature of the Fe-S proteins is the presence of inorganic sulfide in a chromophore coordinated with iron. (As noted, rubredoxin, is stricly speaking, therefore not an Fe-S protein.) The failure to recognize this group of proteins until relatively recently probably results from the instability of the Fe-S cluster when the protein is treated with acid or denaturing agents. Under such circumstances, the sulfur is usually released as a sulfide ion or hydrogen sulfide. Earlier interpretations of these results were that the inorganic sulfide was formed by β elimination of the cysteine sulfur atoms of the polypeptide by the addition of acid to the protein. That the sulfide in the Fd was in fact inorganic sulfide was shown by reconstitution of Fd from the apoprotein and inorganic sulfide supplied as an uncharacterized water soluble complex formed from 2-mercaptoethanol, a ferric salt and sodium sulfide (Malkin and Rabinowitz, 1966; Hong and Rabinowitz, 1970). Later expenments with well-characterized, synthetic Fe-S complexes confirmed this conclusion (Holm, 1977). Later discovery of the spontaneous assembly of Fe-S clusters of known structures $[Fe(SR)_4]^Z$, $[Fe_2X_2(SR)_4]^Z$, and $[Fe_4X_4(SR)_4]^Z$ representing 1-, 2-, and 4-Fe types led to their use in "extrusion" reactions for the characterization of the Fe-S clusters of natural products of both simple and complex nature (Berg and Holm, 1982).

The apoprotein

The amino acid sequence of *C. pasteurianum* Fd was determined by Tanaka et al., (1966), and the sequences of many other clostridial Fds have also been determined (Yasunobu and Tanaka, 1973). Clostridial Fd apoproteins show the following characteristics:

1. They contain 54–56 amino acids;
2. The sequence consists of about a 28-amino-acid domain that is duplicated;

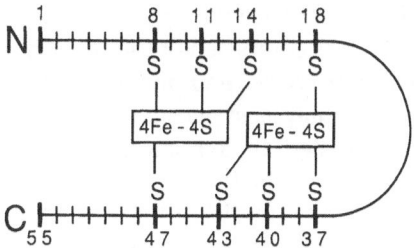

FIGURE 25.2 Positions of cysteine residues in clostridial ferredoxins (Fd). Exact numbering in polypeptide may vary by 1 or 2 depending on additions or omissions of amino acids in Fd of particular species.

3. Each domain contains 4 cysteine residues with a total of 8 cysteine residues in the apoprotein;
4. The cysteine residues occur in the following amino acid positions in the polypeptide: -8,11,14,18-proline- and -37,40,43,47-proline- (see Figure 25.2);
5. The cysteine residues ligated to each of the iron atoms occur in both domains: -8,11,14,47-proline- and -37,40,43,18-proline- (see Figure 25.2);
6. They contain 1 or 2 aromatic amino acids;
7. They generally do not contain methionine, arginine, or leucine;
8. They contain a high proportion of hydrophobic amino acids; and
9. They contain an excess of acidic amino acids.

The bacterial Fe-S proteins have been used as the basis of "trees" relating the evolution of bacterial species. When the various Fe-S proteins are considered in groups based on cluster structure, various broad classes of bacteria are recognized that are similar to the traditional classification (Fitch and Bruschi, 1987) but some differences are observed from the relationships proposed by Woese and his associates who based their tree on the ribosomal 16S RNA sequences (Fox et al., 1980). In a more recent report on attempts to determine the relationship among Fds, the [2Fe-2S], [4Fe-4S], and [8Fe-8S] were divided into five classes that included the low-potential [8Fe-8S] clostridial Fds, the high-potential [4Fe-4S] ferredoxins, and three classes of [2Fe-2S] Fds composed of the plant type, the hydroxylation type, and the *C. pasteurianum* types. This analysis suggested that each of the classes consisted of a homologous group of proteins but that the five classes of proteins are phylogenetically unrelated (Meyer, 1988). This result is of particular interest in providing evidence that the low-potential [8Fe-8S] and the plant [2Fe-2S] proteins are not related, because it has frequently been proposed that these two types of Fds are related by duplication and triplication of the very primitive clostridial type sequence to yield the plant type Fd.

The Fe-S clusters

There is general agreement that the [2Fe-2S], the [4Fe-4S], and [8Fe-8S] clusters are associated with specific polypeptides and that a particular polypeptide will only accommodate a particular type of Fe-S cluster. Thus, when clostridial apoFd is reconstituted to form the holoprotein it always forms a product with the two [4Fe-4S] clusters characteristic of the native protein, and no evidence has been obtained for the formation of a compound with a single cluster or a combination of [2Fe-2S] clusters. It has been concluded that the polypeptide chain determines the nature of the Fd-S cluster formed. On the other hand, changes in the chemical structure of the Fe-S clusters isolated in the "extrusion" analysis of complex Fe-S proteins have been reported, and it has been suggested that caution be exercised in the characterization of Fe-S clusters of complex centers by this method (Beinert, 1990). Although the peptide determines the nature of the associated Fe-S cluster, the biological mechanism by which the Fe-S cluster is formed and appears in the holoprotein is not known. It has been proposed that it is formed spontaneously without the intervention of an enzyme activity, but it has also been proposed that enzymatic activity is required for formation of the Fe-S cluster of the holoprotein.

It was shown that the clostridial apoFd could be converted to Fd in the presence of a water-soluble but uncharacterized mixture of ferric chloride and sodium sulfide in 2-mercapto-ethanol (Malkin and Rabinowitz, 1966; Hong and Rabinowitz, 1967). The same method has also been successful in the reconstitution of the [2Fe-2S] proteins putidaredoxin (Tsibris et al., 1968) and parsley Fd (Fee et al., 1971) from their respective apoproteins. This reaction has proven of value in providing Fds with isotopic Fe, S, or Se useful in studies of the cluster structure.

It is of interest that the *C. pasteurianum* apoFd can be proteolytically split into two roughly equal portions, and that the separated polypeptides can be reconstituted by the procedure described (Hong and Rabinowitz, 1967) to yield a protein that shows an absorption spectrum characteristic of Fd (Skjeldal et al., 1989). The reconstituted split protein is not stable for more than 1 h. This behavior contrasts with the marked stability of the native holoprotein, which is also strongly resistant to proteolytic digestion.

Although Fe-S clusters are formed spontaneously in solution from Fe, S^{2-}, and RS^- under anaerobic conditions, it may be questioned whether these components are present in reactive forms in all cells and under conditions that favor "spontaneous" cluster formation. It has been suggested that enzymes are involved in the formation of the components required for Fe-S cluster formation. For example, it has been suggested that the major function of the enzyme rhodanese is the formation of sulfide for this purpose, including the formation of the [4Fe-4S] cluster in *C. pasteurianum* (Bonomi et al., 1985; Cerletti, 1986). Although it seems reasonable that highly reactive molecules such as S^{2-} and Fe^{3+} might not be readily available in the cell for cluster formation and that they might be formed and used under regulated conditions such as could be provided by the action of an enzyme such as rhodanese, it remains to be demonstrated that cluster formation does indeed require an enzyme. It is noteworthy that although rhodanese is extensively distributed in bacteria, animals, and plants, its presence in *C. pasteurianum* cannot be demonstrated (Sandberg et al., 1987).

Further experimental evidence interpreted as showing that the Fe-S clusters of several proteins are formed spontaneously without the intervention of enzymatic activity has been based on several cloning experiments. The gene for human Fd, a mitochondrial [2Fe-2S] protein similar structurally and functionally to bovine adrenodoxin, was cloned and introduced into *E. coli*. A functional Fd containing the expected [2Fe-2S] cluster was formed (Coghlan and Vickery, 1989). Similar results have been reported in the expression of *B. subtilis* gene for glutamine phosphoribosylpyrophosphate amidotransferase in *E. coli*. The *B. subtilis* enzyme, but not that in *E. coli*, contains a [4Fe-4S] cluster. The cloned *B. subtilis* gene when expressed in *E. coli* was found to produce a protein that contained the expected Fe-S cluster (Makaroff et al., 1983). Although these experiments indicated that neither the human nor the *Bacillus* Fe-S protein required a homologous enzyme for the formation of the Fe-S clusters, they do not prove that enzymatic reactions were not involved. Because *E. coli* is known to contain Fe-S proteins with both [2Fe-2S] and [4Fe-4S] clusters, it is possible that endogenous *E. coli* enzymes were responsible for the formation of the Fe-S clusters in the foreign proteins that were isolated in these experiments.

Attempts to produce Fd in vitro led to the development of fractionated protein synthesizing systems from *Clostridia* and several other Gram-positive organisms (Himes et al., 1972; Stallcup et al., 1974). No synthesis of Fd or apoFd, however, was detected in the fractionated systems. Some progress was made in the purification of Fd mRNA from *C. pasteurianum*, but the amount obtained was insufficient for further studies (Liao and Rabinowitz, 1980).

25.2 Cloning the Clostridial Fd Gene

As described in the preceding sections, Fe-S proteins are a class of proteins that have only relatively recently been recognized despite the fact that they participate in a large number of biologically important functions and occur in all living cells. Our interest was both in the proteins, which are unusual with respect to size, amino acid composition, and repeat structure,

and the Fe-S clusters they contain, which are stable only when bound in the protein and are unique in containing "inorganic" sulfide. This led us to study the biosynthesis of Fd from *C. pasteurianum*. We chose to study this protein because it has served as a valuable model compound for this class of proteins in previous studies.

At the time this work was undertaken, relatively few clostridial genes had been cloned and sequenced (Young et al., 1989). The methods chosen for this particular problem were based on the use of standard DNA manipulations. A library of *C. pasteurianum* DNA was cloned in the plasmid pBR322 (Graves et al., 1985). Based on the amino acid sequence of *C. pasteurianum* Fd (Tanaka et al., 1966), a probe was synthesized (Urdea et al., 1983) corresponding to the sequence of 6 amino acids starting with Asp-39 (Figure 25.3; Met = 0). The heptadeanucleotide pool, representing a 64-fold degeneracy, had the following sequence:

5'-G-C-R-C-A-R-T-T-N-C-C-R-C-A-R-T-C-3'

where R is either purine and N is any nucleotide. Colony hybridization of the transformants revealed two positive colonies, and further analysis with products of *Sau*3A1 digestion yielded only a single fragment hybridizing with the probe. This finding, and other experimental data, are consistent with the presence of a single copy of the Fd gene in the *C. pasteurianum* genome. This fragment was sequenced with the results shown in Figure 25.3. It was found to contain the entire *C. pasteurianum* Fd sequence.

Codon usage

The fragment contains only 30% G + C, the same as the G + C content determined for the *C. pasteurianum* genomic DNA (Tonomura et al., 1965), which is one of the lowest values reported for genomic DNA (Klein and Schnorr, 1984). The G + C content of the Fd coding region is also very low (37%). Analysis of the codon usage reveals the expected strong preferred usage of A or U in the degenerate position.

5'
GATCGAGATAGTATATGATGCATATTCTTTAAATATAGATAAAGTTATAGAAGCA 55

ATAGAAGATTTAGGATTTACTGTAATATAAATTACACTTTTAAAAAGTTTAAAAA 110

CATGATACAATAAGTTATGGTAAACTTATGATTAAAATTTTAAGGAGGTGTATTT 165

	Met	Ala	Tyr	Lys	Ile	Ala	Asp	Ser	Cys	Val	Ser	Cys	Gly
TTC	ATG	GCA	TAT	AAA	ATC	GCT	GAT	TCA	TGT	GTA	AGC	TGT	GGC	207

	Ala	Cys	Ala	Ser	Glu	Cys	Pro	Val	Asn	Ala	Ile	Ser	Gln	Gly
GCT	TGT	GCT	TCA	GAA	TGT	CCA	GTT	AAT	GCT	ATA	AGT	CAA	GGA	249

	Asp	Ser	Ile	Phe	Val	Ile	Asp	Ala	Asp	Thr	Cys	Ile	Asp	Cys
GAT	TCT	ATA	TTC	GTT	ATA	GAT	GCC	GAT	ACT	TGT	ATC	GAC	TGT	291

	Gly	Asn	Cys	Ala	Asn	Val	Cys	Pro	Val	Gly	Ala	Pro	Val	Gln
GGT	AAC	TGT	GCT	AAC	GTT	TGT	CCA	GTT	GGA	GCA	CCA	GTA	CAA	333

Glu	*
GAA	TAA	TTTATAATTAAAACATAAAAATAAGAAGCCTGCAAATGCAGGCTTCT 386

TATTTTTATTGTAACTGAATTTTATTATGTTATACTATAAATTGAAGAAAATATT 441

AGGGGGAAATTGGTATGCTTAATATATTTAGGGACACTTTTCAAGTTATGTCGCC 496

GGTTAATGGAAATATAGTAAATTTAACTAATGTTCCAGACAGAATGTTTTCAGAG 551

GAAATAGTAGGAAAAGGAATAGCTGTAGACCCATTAGAGGATATAATAAGATC 604
                                                        3'

FIGURE 25.3 Nucleotide sequence of *Sau*3A1 0.6-kb fragment containing ferredoxin (Fd) gene of *Clostridium pasteurianum*. Sequence shown is nontranscribed strand corresponding to mRNA sequence. Short vertical solid arrows indicate major sites of beginning and ending of transcription. Translated amino acid sequence is shown above coding region of DNA. Asterisk (*) indicates Fd termination codon; potential RBS is boxed. Dotted lines above bases denote −35 and Pribnow regions of promoter. Dyad symmetry elements are underlined with arrows; dot indicates center of symmetry.

Translation signals

Because the total amino acid sequence of the *C. pasteurianum* Fd was known, the base sequence of the gene coding for the polypeptide was evident (see Figure 25.3). The DNA sequence also showed that the N-terminal alanine residue is preceded by a methionine residue coded for by AUG. Removal of this methionine residue appears to be the only processing that the *C. pasteurianum* Fd polypeptide undergoes. Preceding this start codon is the sequence -GGAGG-, the full complement of the potential ribosome-binding site (RBS) -CCTCC-, assuming that the *C. pasteurianum* rRNA is homologous to other eubacterial 16S rRNAs. The bases enclosed in the box are those showing complementarity to the *C. pasteurianum* 16S ribosomal RNA. The calculated free energy of binding for this sequence is −22.2 kca/mol (Tinoco et al., 1973). This value is consistent with those found for mRNAs from other Gram-positive organisms (McLaughlin et al., 1981; Hager and Rabinowitz, 1985) and is characteristically higher than the normal value found for the RBS associated with mRNA translated by *E. coli* and other Gram-negative organisms. The spacer sequence separating the RBS from the initiation codon is 12 nucleotides, which falls within the range observed in all bacteria studied (Hager and Rabinowitz, 1985; Stormo et al., 1982).

The initiation codon for *C. pasteurianum* Fd is AUG. Although this is not unexpected, it is worth noting that codons that normally code for amino acids other than methionine have been found to function as initiation codons and to specify a methionine residue under these conditions. It has been reported that 30% of the known initiation sites of Gram-positive organisms are either UUG (17%) or GUG (12%) (Hager and Rabinowitz, 1985). These codons normally code for leucine and valine, respectively. The incidence of non-AUG initiation codons is much lower in *E. coli* where only 3%

of initiation sites are non-AUG. Whether the occurrence of a high proportion of non-AUG initiation codons characteristic of Gram-positive microorganisms also occurs in clostridial proteins remains to be determined.

Transcriptional signals

The cloned *C. pasteurianum* Fd gene was used to map Fd transcripts synthesized in vivo and in vitro (Figure 25.3) (Graves and Rabinowitz, 1986). The in vivo mRNA, sized by Northern hybridization analysis and directly from the known DNA sequence after identification of the 5' and 3' termini, were identified. The 5' end was determined by primer extension-dideoxy sequencing and the 3' end by S1 nuclease mapping. The monocistronic Fd mRNA contains about 255 nucleotides and is one of the shortest bacterial mRNAs that has been described. The promoter recognition and RNA initiation sites as determined with *C. pasteurianum* in vivo are indicated in Figure 25.3. It was found that RNA polymerase from *C. pasteurianum* (in vivo), *E. coli* (in vivo and in vitro), and *B. subtilis* (in vitro) each produces a transcript of similar size in response to the Fd gene, suggesting that the same transcriptional signals are recognized. The exact nucleotide use for initiation varies slightly with the source of enzyme. *E. coli* RNA polymerase recognizes a second promoter region with efficiency about equal to that of the first, although this promoter is very weakly utilized by *C. pasteurianum* (Graves and Rabinowitz, 1986). This result confirms a previous example of a promoter being utilized by *E. coli* polymerase but not by the enzyme from a gram-positive organism (Davison et al., 1980).

The clostridial Fd gene promoter and 28 others that have been reported from Gram-positive organisms were compared and analyzed to determine any conserved features among them (Graves and Rabinowitz, 1986). Promoters of *Streptomyces*, a subgroup of Gram-positive organisms, probably do not consist of typical transcription signals and were thus not included in the analysis. The combined list demonstrates the expected conserved sequences. The classic consensus sequences of *E. coli* in the −10 and −35 regions are present in these promoters as well. Additional conserved regions are also evident in the Gram-positive promoters that are not observed in *E. coli*: an A cluster in positions −41 to −45, TG in positions −16 and −15, A residues at positions −7, −6, and −4, and T residues weakly conserved at −5 and −3. Additional data on the promoter regions of four other clostridial promoters (Young et al., 1989) does not show the poly-A region near −45 of the proposed extended consensus sequence but does show additional conserved area adjacent to the −10 region. Very few clostridial promoters have been reported since this overall analysis, and the identification of additional promoter sequences will be required to determine whether the proposed "extended" promoter sequence occurs in clostridia and other Gram-positive organisms.

Expression of the clostridial Fd gene and other clostridial genes in *E. coli*

In spite of the success in cloning the *C. pasteurianum* Fd gene and in detecting Fd transcripts *in vivo*, it has not been possible to detect the formation of the *C. pasteurianum* Fd apoprotein or holoprotein in *E. coli* transformed with the plasmid containing the clostridial Fd gene. A variety of protein staining methods and tests with antibodies to the *C. pasteurianum* Fd were employed for detection of the product without success (M. Graves and J. Rabinowitz, unpublished results). Whether the inability to detect clostridial Fd in *E. coli* bearing the clostridial Fd gene results from instability of the gene, the message, or the product or the insensitivity of the methods used for its detection is not known.

The expression of recombinant *C. pasteurianum* Fd in *E. coli* has been reported by Baur et al. (1990). The results demonstrated the formation of the Fd protein in *E. coli* in an active form with the same electronic, magnetic, redox, and enzymatic properties as the Fd expressed in *C. pasteurianum*. When expressed in *E. coli* strain JM109 from the original construct, pCP1 (Graves et al., 1985), Fd was estimated to represent about 0.14% of the cell protein. By cloning the gene into pUC9 and generating the plasmid pCP14P, the Fd represented about 0.7% of the cell protein when expressed in *E. coli* strain JM109.

Other clostridial genes have been cloned and expressed in *E. coli* to varying degrees. These include *Clostridium acetobuylicum* genes encoding enzymes involved in the acetone-butanol fermentation with subunit molecular weights ranging from 31,000 to 39,000 (Cary et al., 1988; Youngleson et al., 1988) and the *glnA* gene (Janssen et al., 1988). In the latter case it appears that cloned regions downstream from the gene were effective as promoters and regulators of the cloned gene. A sialidase gene of *Clostridium sordellii* was successfully cloned and expressed in *E. coli* (Rothe et al., 1989). Two genes from *C. pasteurianum* have also been reported to have been cloned and expressed in *E. coli*: the galactokinase gene (Daldal and Applebaum, 1985) and the gene for a molybdenum-binding protein, *mop*, with a molecular weight of about 6,000 (Hinton and Freyer, 1986). The human mitochondrial [2Fe-2S] Fd gene introduced into *E. coli* through an expression vector under the control of the $\lambda\,P_L$ promoter resulted in the production of the human Fd protein containing the [2Fe-2S] cluster in high yield with full biological activity (Coghlan and Vickery, 1989).

The lack of expression of some clostridial genes in *E. coli* has been attributed to differences in codon usage between the two organisms (Garnier and Cole, 1986; Eisel et al., 1986). It has also been suggested that expression of large genes is more seriously impaired by unfavorable codon usage than that of average-sized clostridial genes (Garnier and Cole, 1986; Leslie et al., 1989), yet the gene for Fd is exceptionally small.

The varied experiences with expression of clostridial genes in *E. coli* suggests that the basic principles governing the heterologous gene expression remain to be determined.

References

Adman, E.T., L.C. Sieker, and L.H. Jensen. 1973. The structure of a bacterial ferredoxin. *J. Biol. Chem.* 248:3987–3996.

Adman, E.T., L.C. Sieker, and L.H. Jensen. 1976. Structure of *Peptococcus aerogenes* ferredoxin:

refinement at 2Å resolution. *J. Biol. Chem.* 251:3801–3806.

Baur, J.R., M.C. Graves, B.A. Feinberg, and S.W. Ragsdale. 1990. Characterization of the recombinant *Clostridium pasteurianum* ferredoxin and comparison of its properties with those of the native protein. *BioFactors* 2:197–203.

Beinert, H. 1990. Recent developments in the field of iron-sulfur proteins. *FASEB J.* 4:2483–2491.

Berg, J.M. and R.H. Holm. 1982. Structures and reactions of iron-sulfur protein clusters and their synthetic analogs. In: *Iron-Sulfur Proteins*, T.G. Spiro, ed., p. 1–66. New York: Wiley.

Bonomi, F., S. Pagani, and D.M. Kurta. 1985. Enzymic-synthesis of the 4Fe- 4S clusters of *Clostridium pasteurianum* ferredoxin. *Eur. J. Biochem.* 148:67–73.

Cardenas, J., L.E. Mortenson, and D.C. Yoch. 1976. Purification and properties of paramagnetic protein from *Clostridium pasteurianum*. *Biochim. Biophy. Acta* 434:244–257.

Cary, J.W., D.J. Petersen, E.T. Papoutsakis, and G.N. Bennett. 1988. Cloning and expression of *Clostridium acetobuylicum* phosphotransbutyrylase and butyrate kinase genes in *Escherichia coli. J. Bactcriol.* 170:4613–4618.

Cerletti, P. 1986. Seeking a better job for an underemployed enzyme: rhodanese. *Trends Biochem. Sci.* 11:369–372.

Coghlan, V.M. and L.E. Vickery. 1989. Expression of human ferredoxin and assembly of the [2Fe-2S] center in *Escherichia coli. Proc. Natl. Acad. Sci. USA* 86:835–839.

Daldal, F. and J. Applebaum. 1985. Cloning and expression of *Clostridium pasteurianum* galactokinase gene in *Escherichia coli* K-12 and nucleotide-sequence analysis of a region affecting the amount of the enzyme. *J. Mol. Biol.* 186:533–545.

Davison, B.L., C.L. Murray, and J.C. Rabinowitz. 1980. Specificity of promoter site utilization *in vitro* by bacterial RNA polymerases on *Bacillus* phage ϕ29 DNA. *J. Biol. Chem.* 255:8819–8830.

Eisel, U., W. Jarausch, K Goretzki, A. Henschen, J. Engels, U. Weller, M. Hudel, E. Habermann, and H. Niemann. 1986. Tetanus toxin: primary structure, expression in *Escherichia coli*, and homology with botulinum toxins. *EMBO J.* 5:2495–2502.

Fee, J.A., S.G. Mayhew, and G. Palmer. 1971. The oxidation-reduction potentials of parsley ferredoxin and its selenium-containing homolog. *Biochim. Biophys. Acta* 245:196–200.

Fitch, W.M. and M. Bruschi. 1987. The evolution of prokaryotic ferredoxins — with a general method correcting for unobserved substitutions in less branched lineages. *Mol. Biol. Evol.* 4:381–394.

Fox, GE., E. Stackebrandt, R.B. Hespell, J. Gibson, J. Maniloff, T.A. Dyer, R.S. Wolfe, W.E. Balch,

R.S. Tanner, L.J. Magrum, L.B. Zablen, R. Blakemore, R. Gupta, L. Bonen, B.J. Lewis, D.A. Stahl, K.R. Luehrsen, K.N. Chen, and C.R. Woese. 1980. The phylogeny of prokaryotcs. *Science* 209:457–463.

Garnier, T. and S.T. Cole. 1986. Characterization of a bacteriocinogenic plasmid from *Clostridium perfringens* and molecular genetic analysis of the bacteriocin-encoding gene. *J. Bacteriol.* 168:1189–1196.

Grandoni, J.A., R.L. Switzer, C.A. Makaroff, and H. Zalkin. 1989. Evidence that the iron-sulfur cluster of *Bacillus subtilis* glutamine phosphoribosylpyrophosphate amido-transferase determines stability of the enzyme to degradation *in vivo*. *J. Biol. Chem.* 264:6058–6064.

Graves, M.C. and J.C. Rabinowitz. 1986. *In vivo* and *in vitro* transcription of the *Clostridium pasteurianum* ferredoxin gene. Evidence for "extended" promoter elements in gram-positive organisms. *J. Biol. Chem.* 261:11409–11415.

Graves, M.C., G.T. Mullenbach, and J.C. Rabinowitz. 1985. Cloning and nucleotide-sequence determination of the *Clostridium pasteurianum* ferredoxin gene. *Proc. Natl. Acad. Sci. USA* 82:1653–1657.

Hager, P.W., and J.C. Rabinowitz. 1985. Translational specificity in *Bacillus subtilis*. In: *The Molecular Biology of the Bacilli*, Vol. II, D.A. Dubnau, ed., pp. 1–32. New York: Academic Press.

Herriott, J.R., L.C. Sieker, L.H. Jensen, and W. Lovenberg. 1970. Structure of rubresocin: an x-ray study to 2.5 Å resolution. *J. Mol. Biol.* 50:391–406.

Himes, R.H., M.R. Stallcup, and J.C. Rabinowitz 1972. Translation of synthetic and endogenous messenger ribonucleic acid *in vitro* by ribosomes and polyribosomes from *Clostridium pasteurianum*. *J. Bacteriol.* 112:1057–1069.

Hinton, S.M. and G. Freyer. 1986. Cloning, expression and sequencing the molybdenumpterin binding-protein (*mop*) gene of *Clostridium pasteurianum* in *Escherichia coli*. *Nucleic Acids Res.* 14:9371–9380.

Holm, R.H. 1977. Synthetic approaches to the active sites of iron-sulfur proteins. *Accounts Chem. Res.* 10:427–434.

Hong, J.-S. and J.C. Rabinowitz. 1967. Preparation and properties of clostridial apoferredoxins. *Biochem. Biophys. Res. Commun.* 29:246–252.

Hong, J.-S. and J.C. Rabinowitz. 1970. The all-or-none mode of the reconstitution and the reaction of α, α'-bipyridyl and mercurials with clostridial ferredoxin. *J. Biol. Chem.* 245:6574–6581.

Janssen, P.J., W.A. Jones, D.T. Hones, and D.R. Woods. 1988. Molecular analysis and regulation of the *glnA* gene of the gram-positive anacrobe *Clostridium acetobutylicum*. *J. Bacteriol.* 170:400–408.

Johnson, M.K, J.E. Morningstar, D.E. Bennett,

B.A.C. Ackrel, and E.B. Kearney. 1985. Magnetic circular dichroism studies of succinate dehydrogenase: evidence for [2Fe-2S], [3Fe-xS], and [4Fe-4S] centers in reconstitutively active enzyme. *J. Biol. Chem.* 260:7368–7378.

Klein, A. and M. Schnorr. 1984. Genome complexity of methanogenic bacteria. *J. Bacteriol.* 158:628–631.

Leslie, D., N. Fairweather, D. Pickard, G. Dougan, and M. Kehoe. 1989. Phhospholipase C and haemolytic activities of *Clostridium perfringens* alpha-toxin cloned in *Escherichia coli*: sequence and homology with a *Bacillus cereus* phospholipase C. *Mol. Microbiol.* 3:383–392.

Liao, H.H. and J.C. Rabinowitz. 1980. Clostridial apoferredoxin messenger ribonucleic acid assay and partial purification. *Biochim. Biophys. Acta* 608:301–314.

Lovenberg, W., ed. 1973a. *Iron-Sulfur Proteins. Biological Properties*, Vol. I. New York: Academic Press.

Lovenberg, W., ed. 1973b. *Iron-Sulfur Proteins. Molecular Properties*, Vol. II. New York: Academic Press.

Lovenberg, W., ed. 1976. *Iron-Sulfur Proteins. Structure and Metabolic Mechanisms*, Vol. III. New York: Academic Press.

Lovenberg, W. and B.E. Sobel. 1965. Rubredoxin: a new electron transfer protein from *Clostridium pasteurianum*. *Proc. Natl. Acad. Sci. USA* 54:193–199.

Lovenberg, W. and W.M. Williams. 1969. Further observations on the chemical nature of rubredoxin from *Clostridium pasteurianum*. *Biochemistry* 8:141–148.

Makaroff, C.A., H. Zalkin, R.L. Switzer, and S.J. Vollmer. 1983. Cloning of the *Bacillus subtilis* glutamine phosphoribosylpyrophosphate amidotransferase gene in *Escherichia coli*. Nucleotide sequence determination and properties of the plasmid-encoded enzyme. *J. Biol. Chem.* 258:10586–10593.

Malkin, R. and J.C. Rabinowitz. 1966. The reconstitution of clostridial ferredoxin. *Biochem. Biophys. Res. Commun.* 23:822–827.

McCarthy, K. and W. Lovenberg. 1970. N-Formylmethionine: the N terminus of *Clostridium pasteurianum* rubredoxin. *Biochem. Biophys. Res. Commun.* 40:1053–1057.

McLaughlin, J.R., C.L. Murray, and J.C. Rabinowitz. 1981. Unique features in the ribosome binding site sequence of the Gram-positive *Staphylococcus aureus* β-lactamase gene. *J. Biol. Chem.* 256:11283–11291.

Meyer, J. 1988. The evolution of ferredoxins. *Trends Ecol. Evol.* 3:222–226.

Meyer, J., M.H. Bruschi, J.J. Bonicel, and G.E. Bovierlapierre. 1986. Amino-acid sequence of [2Fe-2S] ferredoxin from *Clostridium pasteurianum*. *Biochemistry* 25:6054–6061.

Mortenson, L.E., and G. Nakos. 1973. Bacterial ferredoxins and/or iron-sulfur proteins as electron carriers. *In: Iron-Sulfur Proteins, Biological Properties*, Vol. I, W. Lovenberg, ed., pp. 37–110. New York: Academic Press.

Mortenson, L.E., R.C. Valentine, and J.E. Carnahan. 1962. An electron transport factor from *Clostridium pasteurianum*. Biochem. *Biophy. Res. Commun.* **7**:448–452.

Rothe, B., P. Roggentin, R. Frank, H. Blöcker, and R. Schauer. 1989. Cloning, sequencing and expression of a sialidase gene from *Clostridium sordellii* G12. *J. Gen. Microbiol.* **135**:3087–3096.

Sandberg, W., M.C. Graves, and J.C. Rabinowitz. 1987. Role for rhodanese in Fe-S formation is doubtful. *Trends Biochem. Sci.* **12**:56.

Skjeldal, L., K Draget, and T. Ljones. 1989. Enzymatic cleavage of *Clostridium pasteurianum* apoferredoxin and reconstitution of the cleaved products. *Biochim. Biophys. Acta* **995**:59–63.

Spiro, T.G., ed. 1982. *Iron-Sulfur Proteins*. New York: Wiley.

Stallcup, M.R., W.J. Sharrock, and J.C. Rabinowitz. 1974. Ribosome and messenger specificity in protein synthesis by bacteria. *Biochem. Biophys. Res. Commun.* **58**:92–98.

Stormo, G.D., T.D. Schneider, and L.M. Gold. 1982. Characterization of translational initiation sites in *E. coli. Nucleic Acids Res.* **10**:2971–2996.

Stout, C.D. 1982. kon-sulfur crystallography. *In: Iron-Sulfur Proteins*, T.G. Spiro, ed., pp. 97–146. New York: Wiley.

Tagawa, K. and D.I. Arnon. 1962. Ferredoxins as electron carriers in photosynthesis and in the biological production and consumption of hydrogen gas. *Nature* (London) **195**:537–543.

Tanaka, M., T. Nakashima, A.M. Benson, H. Mower, and K.R. Yasunobu. 1966. The amino acid sequence of *Clostridium pasteurianum* ferredoxin. *Biochemistry* **5**:1666–1681.

Tinoco, I., P.N. Borer, B. Dengler, M.D. Levine, O.C. Uhlenbeck, D.M. Crothers, and J. Gralla. 1973. Improved estimation of secondary structure in ribonucleic acids. *Nature New Biol.* **246**:40–41.

Tonomura, B.L, R. Malkin, and J.C. Rabinowitz. 1965. Deoxyribonucleic acid base composition of clostridial species. *J. Bacteriol.* **89**:1438–1439.

Tsibris, J.C.M., R.L. Tsai, I.C. Gunsalus, W.H. Orme-Johnson, and H. Beinert. 1968. The number of iron atoms in the paramagnetic center (G = 1.94) of reduced putidaredoxin, a nonheme iron protein. *Proc. Natl. Acad. Sci. USA* **59**:959–965.

Urdea, M.S., J.P. Merryweather, G.T. Mullenbach, D. Coit, U. Heberlein, P. Valenzuela, and P.J. Barr. 1983. Chemical synthesis of a gene for human epidermal growth factor urogastrone and its expression in yeast. *Proc. Natl. Acad. Sci. USA* **80**:7461–7465.

Valentine, R.C. 1964. Bacterial ferredoxin. *Bacteriol. Rev.* **28**:497–517.

Watenpaugh, K.D., L.C. Sieker, and L.H. Jensen. 1980. Crystallographic refinement of rubredoxin at 1.2 Å resolution. *J. Mol. Biol.* **138**:615–633.

Yasunobu, K.T. and M. Tanaka 1973. The types, distribution in nature, structure-function, and evolutionary data of the iron-sulfur proteins pp. 27–130. *In: Iron-Sulfur Proteins*, Vol. II, W. Lovenberg, ed., New York: Academic Press.

Yoch, D.C. and R.P. Carithers. 1979. Bacterial iron-sulfur proteins. *Microbiol. Rev.* **43**:384–421.

Young, M., N.P. Minton, and W.L. Staudenbauer. 1989. Recent advances in the genetics of the clostridia. *FEMS Microbiol. Rev.* **63**:301–326.

Youngleson, J.S., J.D. Santangelo, D.T. Jones, and D.R. Woods. 1988. Cloning and expression of *Clostridium acetobutylicum* alcohol dehydrogenase gene in *Escherichia coli. Appl. Environ. Microbiol.* **54**:676–682.

26

Organization of the Nitrogen Fixation Genes in *Clostridium pasteurianum*

John L. Johnson, Shu-Zhen Wang, and Jiann-Shin Chen

26.1 Introduction

Clostridium pasteurianum is an obligately anaerobic and free-living, nitrogen-fixing organism. It appears to be the first nitrogen-fixing organism isolated, being described by Winogradsky in the mid-1890s (Winogradsky, 1895). Consistent nitrogen-fixing activities using cell-free preparations were first obtained using this organism (Carnahan et al., 1960), which ushered in modern studies on the biochemistry of nitrogen fixation (Burris, 1988). Although the nitrogenase from *C. pasteurianum* has been extensively studied (see references in Chen et al., 1986), the genetics of the nitrogen fixation system of this organism was not actively investigated when genetic manipulation systems became available in *Klebsiella pneumoniae* and in other free-living and symbiotic species. The nitrogen-fixation (*nif*) genes of *C. pasteurianum* have been studied only in recent years.

The organization of *nif* genes was first determined in *K. pneumoniae* by genetic and DNA sequence analyses (Ausubel and Cannon, 1980; Arnold et al., 1988). Twenty *nif* genes, which form a cluster, have been identified in *K. pneumoniae* (Figure 26.1A), and a function has been proposed for most of these genes. However, *nif* genes do not occur in a single cluster in several other diazotrophs that have been characterized. For example, *nifABQ* occur separately from the major *nif* cluster in *Azotobacter vinelandii* (Figure 26.1B; Bennett et al., 1988; Joerger and Bishop, 1988; Jacobson et al., 1989). The major *nif* cluster in these two organ-

isms has an additional difference in the occurrence of many open-reading frames (ORFs) that separate the *nif* genes in *A. vinelandii* (Figure 26.1A,B). A number of these ORFs have *nif*-related properties and have been numbered (Jacobson et al., 1989). It remains to be determined if the apparently *nif*-related ORFs from *A. vinelandii* occur elsewhere on the *K. pneumoniae* chromosome.

26.2 Genetics

Genetic diversity and unique phenotypic properties

Nearly all the studies on *Clostridium pasteurianum* have used strain W5 (ATCC 6013). This strain was obtained from Winogradsky by members of the Bacteriology Department at the University of Wisconsin in the mid-1920s (McCoy et al., 1930). Two other strains, isolated from spoilt canned pineapple by Spiegelberg (1940), are in the American Type Culture Collection (ATCC). The German Culture Collection (DSM) includes another strain (DSM 526) identified as the strain Donker. An additional strain, NRRL-B598 (Boonsermsuwong and Blaschek, 1986) from the U.S. Department of Agriculture Northern Regional Research Center, has been traced to strain A77 of the McCoy collection (L.K. Nakamura, personal communication), which was there identified as *C. pasteurianum*. Witz et al. (1967) have reported the isolation of six additional strains of *C. pasteurianum*, but these no longer appear to be ex-

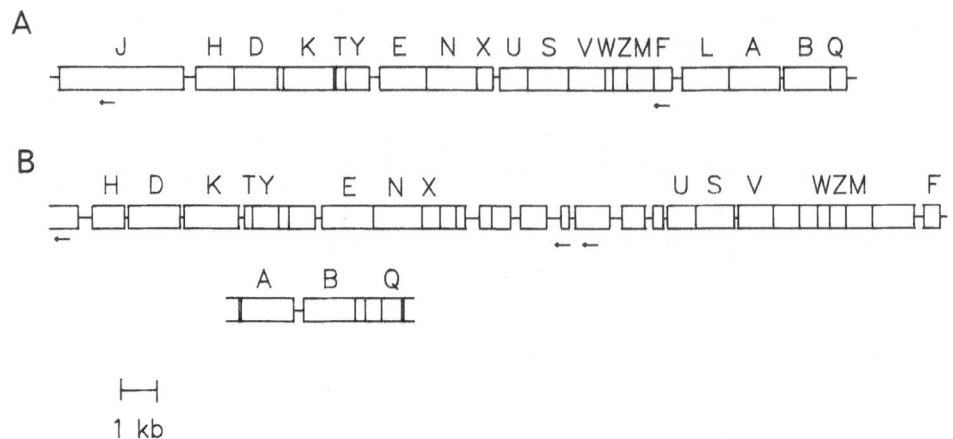

FIGURE 26.1 *nif* genes and flanking open-reading frames (ORFs; boxed) of (**A**) *Klebsiella pneumoniae* and (**B**) *Azotobacter vinelandii*. Direction of transcription is indicated by arrow for those genes and ORFs that differ from rest.

Table 26.1 DNA sequence similarity among strains of *Clostridium pasteurianum* and *C. beijerinckii*

Species	Strain	Percent DNA Sequence Similarity		
		W5	526	5481-1
C. pasteurianum	W5 (= ATCC 6013[a], VPI 4215)	(100)[b]	3	4
	ATCC 7040	84	7	6
	ATCC 7041	83	5	5
	DSM 526	5	(100)	72
C. beijerinckii	VPI 5481-1 (= ATCC 25752[a])	6	69	(100)

[a] Type strains.

[b] Amount of duplex formed by each reference strain is defined as 100%; other values are relative to the reference.

tant (J.L. Pate, personal communication). The Russian literature has referred to the presence of large numbers of *C. pasteurianum* cells in certain soils (Mishustin and Yemtsev, 1975), but we do not know how many strains are in their culture collections.

DNA sequence similarities among the strains are shown in Table 26.1. The canned pineapple spoilage organisms are indeed strains of *C. pasteurianum*; however, strain Donker (DSM 526) is a strain of *Clostridium beijerinckii*. Spiegelberg (1940) refers to "*Cl. pasteurianum* Winogradsky and *Cl. beijerinckii* Donker" as the two species most closely allied to the pineapple isolates on morphological and physiological grounds. Perhaps this *C. beijerinckii* isolate has come to be known as strain Donker and is thus mislabeled. We have not tested the NRRL-B598 strain.

On the basis of ribosomal ribonucleic acid sequence similarities (Johnson and Francis, 1975), *C. pasteurianum* belongs with the other butyric acid-producing *Clostridium* species and has been placed in a rather heterogeneous cluster that also includes *Clostridium acetobutylicum*. The general phenotypic properties of *C. pasteurianum* can be found in Bergey's Manual of Systematic Bacteriology (Cato et al., 1986). Two properties of this organism are of interest in terms of specific enrichment and isolation procedures. First, it grows well on a mineral salts basal medium in the presence of a fermentable carbohydrate and biotin. The requirement for biotin is pronounced under nitrogen-fixing growth conditions, and para-aminobenzoic acid may further stimulate growth. Second, the organism is able to grow in high concentrations of carbohydrate; sucrose, to 44%, and glucose, to 30% (Spiegelberg, 1944). As a result, a nitrogen-free, mineral salts-biotin medium containing 15% sucrose is a very selective medium for strain isolation (Witz et al., 1967).

Mutations and transformations

There has been very little work on the genetics of *C. pasteurianum* (Young et al., 1989). Plasmids have not been detected in strain WS (Lee et al., 1987). The NRRL-B598 strain has been reported to contain a 5.2-MDa plasmid by Truffaut and Sebald (1983) and a 2.3-MDa plasmid by Boonsermsuwong and Blaschek (1986). A number of mutants have been obtained with strain W5 by Simon and Brill (1971), Robson et al. (1974), and Daldal (1985).

Clostridium pasteurianum appears to be a bit recalcitrant in terms of gene transfer systems. No transformations of any type have been reported. Minton and Morris (1983) have developed procedures that result in about 10% regeneration of *C. pasteurianum* protoplasts. We have attempted to introduce plasmids into *C. pasteurianum* by electroporation and have not been successful with the *Clostridium perfringens* shuttle vector pHR106 (Roberts et al., 1988) or the *Staphylococcus* plasmids pCT20 (Keller et al., 1983) and pDL216, a derivative of pAMβ 1 (LeBlanc and Lee, 1984).

We have conjugatively transformed *C. pasteurianum* W5 (unpublished data) with the streptococcal plasmid pAM180 (Volk et al., 1988), following the procedure of Bertram and Dürre (1989). This plasmid contains the transposon Tn916 (Gawron-Burke and Clewell, 1982), which carries a tetracycline resistance gene.

26.3 Organization of the *nif* Genes in *Clostridium pasteurianum*

Genetic studies with *C. pasteurianum* have been limited by the lack of usable genetic markers and procedures for genetic manipulations. Because nitrogenase genes are conserved among nitrogen-fixing organisms (Mazur et al., 1980; Ruvkun and Ausubel, 1980), the cloned *nifHDK* genes of *Klebsiella pneumoniae* have been useful probes for the detection and cloning of *nif* genes from other organisms. For our initial cloning of the nitrogenase structural genes from *C. pasteurianum*, we utilized a *K. pneumoniae nifHD* DNA fragment as a probe (Chen et al., 1986).

The *nif* genes and flanking open-reading frames (ORFs) that have been characterized in *C. pasteurianum* W5 are shown in Figure 26.2. In a region spanning 17.5 kb of DNA (Figure 26.2A), we have identified genes that correspond to *nifH*, *nifD*, *nifK*, *nifE*, *nifN*, *nifB*, and *nifV* of *K. pneumoniae* and other organisms. The nitrogenase structural genes (*nifH1*, *nifD*, and *nifK*) of *C. pasteurianum* were positively identified from a comparison of the deduced amino acid sequences with protein sequences (Tanaka et al., 1977; Chen et al., 1986; Wang et al., 1988b). Additional *nif* genes (*nifE*, *nifN-B*, *nifVω*, and *nifVα*) of *C. pasteurianum* were identified through comparison of the deduced amino acid sequences with those of *Azotobacter vinelandii*, *K pneumoniae*, and other organisms. More recently, an ORF preceding *nifV* of *C. pasteurianum* was found to have sequence similarity to that of *chlJ* of *Escherichia coli*, and we designated this ORF *nifC* (Wang et al., 1990a).

Clostridium pasteurianum contains six *nifH*-like genes, more than any other nitrogen-fixing organism that has been studied (Chen et al., 1986 and Wang et al., 1988a). Two, *nifH1* and *nifH2*, are located on the major *nif* gene cluster (Figure 26.2A), while each of the others is located on noncontiguous *Hind*III fragments (Figure 26.2B). The organization of *nif* genes in *C. pasteurianum* differs significantly from those found in several gram-negative organisms (e.g., Arnold et al., 1988; Joerger and Bishop, 1988; Jacobson et al., 1989). The *nif* genes of *C. pasteurianum* seem to be organized according to their functions. In *C. pasteurianum*, three consecutive groups (operons) of *nif* genes show the following arrangements: (1) the first group consists of structural genes (*nifH1DK*) for nitrogenase, (2) the second group contains genes (*nifEN-B*) required for the synthesis of the iron-molybdenum cofactor (FeMoco) of

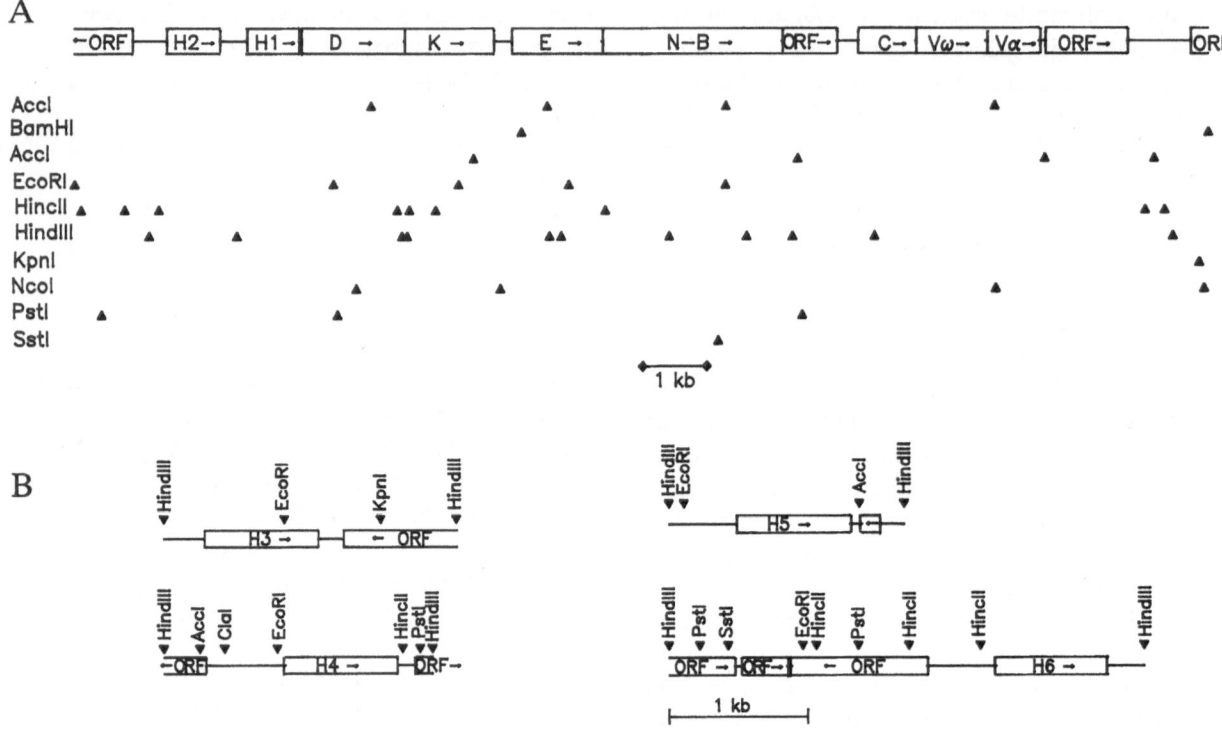

FIGURE 26.2   *nif* genes and flanking open-reading frames of *Clostridium pasteurianum*. **A**, major *nif* cluster; **B**, fragments containing additional *nifH*-like genes.

nitrogenase, where *nifN* and *nifB* are fused into one gene *nifN-B*, and (3) the third group contains a gene (*nifC*) possibly involved in molybdenum transport and additional genes (*nifVωVα*) for FeMoco synthesis, where the conserved amino acid sequence of *nifV* is encoded by two separate genes. ORFs with unknown functions are also present in groups 2 and 3. Properties of *nif* genes and ORFs in a contiguous region in *C. pasteurianum* (Figure 26.2A) are listed in Table 26.2.

Structural genes for nitrogenase

The three structural genes, *nifH1*, *nifD*, and *nifK*, constitute a transcription unit. Although *nifH2* occurs in tandem with *nifH1*, it is not part of the *nifH1DK* operon (see Figure 26.2; Wang et al., 1988a). The *nifH1* gene starts 410 bases downstream of *nifH2*, and the Fe protein has 273 amino acid residues, which is about 20 residues shorter than most *nifH*-encoded sequences studied to date. The *nifH*-encoded se-

quence of *Desulfovibrio gigas* (Kent et al., 1989) is only one residue longer than that of *C. pasteurianum*. These two polypeptides have a similarity coefficient (S_{AB}) of 0.71, which is only slightly higher than the next highest value (0.69) found between *C. pasteurianum* and *A. vinelandii nifH1* genes (with the extended C-terminal region of *A. vinelandii* NifH excluded from comparison). However, *C. pasteurianum nifH1* and *D. gigas nifH* encode long segments (to 28 residues) of identical amino acid sequences. It would be interesting to compare properties of nitrogenases purified from *C. pasteurianum* and *D. gigas*, because the *C. pasteurianum* nitrogenase has a number of unique properties that may be related to the size of its component proteins (Wang et al., 1988b).

The *nifD* gene starts 42 bases downstream of *nifH1*. It has the less common start codon GUG, and its stop codon (UAA) overlaps with the start codon (AUG) for *nifK* by 1 base. It has about 150 extra bases in the 1140- to 1290-bp region of the gene (in comparison to *nifD*

Table 26.2 Properties of *nif*[a] genes and ORFs in a cluster in *Clostridium pasteurianum*

nif Gene	Coding region	Start codon	Stop codon	Estimated pI	Encoded polypeptide	
					Number of amino acid residues	M_r
ORF	887–1	AUG	—	—	(Partial)	—
H2	1395–2213	AUG	UAA	4.93	272	29,580
H1	2623–3444	AUG	UAA	4.85	273	29,666
D	3486–5087	GUG	UAA	5.57	533	58,990
K	5086–6462	AUG	UAA	5.08	458	50,115
E	6747–8117	AUG	UAG	5.64	456	50,397
N-B	8140–10929	UUG	UAA	6.09	929	103,088
ORF	10892–11734	AUG	UAG	8.70	280	30,516
C	12084–12944	AUG	UAG	9.30	286	31,681
Vω	12962–14020	AUG	UGA	5.52	352	41,574
Vα	14030–14839	AUG	UAA	5.28	269	29,862
ORF	14953–16224	AUG	UAA	7.46	423	47,010
ORF	17207–17471	AUG	—	—	(Partial)	—

[a]References: *nifH2H1*, Chen et al. (1986); *nifD,* Chen et al. (1986), Wang et al. (1987); *nifK*, Wang et al. (1988b); *nifE*, Wang et al. (1989); *nifN-B*, our unpublished; *nifC*, Wang et al. (1990a); *nifVωVα*, our unpublished.

sequences from other organisms), and the deduced amino acid sequence in this region appears to have periodicity and is hydrophilic (Wang et al., 1988b).

The *nifK* gene of *C. pasteurianum* is about 180 bp shorter than the *nifK* genes from other organisms, and most of the missing bases are at or near the 5'-end of the gene (Wang et al., 1988b). There are 279 bp between *nifK* and the next operon [(*nifEN-B*-(ORF)]. There are two inverted repeats in this region, involving 16 and 10 bp each.

Genes involved in FeMoco synthesis or Mo transport

The *nifE* gene contains 1371 bp (Wang et al., 1989), and the deduced gene product has S_{AB} values of 0.33 with the *nifE* gene product of *A. vinelandii* and 0.3 with the *C. pasteurianum nifD* gene product. The *nifN-B* gene is located 22 nucleotides downstream of *nifE*. The predicted start codon is UUG and the gene codes for 929 amino acid residues ($M_r = 103,088$). The gene is so named because the first 480 deduced amino acid residues have sequence similarity to the predicted *nifN* product of other organisms, while the remaining 449 residues have sequence similarity to the predicted *nifB* product

of other organisms (our unpublished observations).

Using antiserum raised against a 33,000-dalton polypeptide (corresponding to the C-terminal 298 residues of Cp NifN-B), Western blot analysis detected a band with a M_r of 102,000 in nitrogen-fixing but not in ammonia-grown *C. pasteurianum* cells, indicating the presence of an intact NifN-B product. The fusion of *nifN* and *nifB* genes into *nifN-B* and the presence of the NifN-B product suggest that the *nifB* product interacts directly with the NifEN complex in other organisms.

The C-terminal 143 residues of the predicted *nifN-B* also has sequence similarity to the predicted *nifX* products of other organisms (our unpublished observations). The function of this *nifX*-like domain and the ORF downstream of *nifN-B* is not known.

The *nifC-nifVω-nifVα-* (ORF) group

The first ORF (*nifC*) of this cluster (see Figure 26.2) occurs 330 nucleotides downstream of the ORF of the *nifE-nifN-B-(ORF)* cluster. Inverted repeats, which may form a stem-and-loop structure with a stem of 23 bp, are present from 22 to 77 nucleotides downstream of *nifEN-B(ORF)*.

The predicted polypeptide (286 residues) from *nifC* has sequence similarity to that of *chlJ* of *Escherichia coli* (Wang et al., 1990a). Because the region of *chlJ* cloned and sequenced covers only the C-terminal 200 residues (Johann and Hinton, 1987), the comparison is limited to this region, and it has an S_{AB} of 0.325. The *chlJ* of *E. coli* is part of the *chlD* locus, and the *chlD* locus is involved in molybdenum transport. Mo uptake by *C. pasteurianum* and by *K. pneumoniae* is coregulated with nitrogen fixation (Elliott and Mortenson, 1976; Pienkos and Brill, 1981). Sequences similar to the proposed consensus *nif* promoter and upstream sequences of *C. pasteurianum* precede the *nifC* gene. We thus propose that *nifC* is involved in Mo uptake and is coregulated with nitrogen fixation.

Although the *nif* cluster of *K. pneumoniae* does not contain a gene equivalent to *nifC*, nitrogen fixation in *K. pneumoniae* requires the function of a *mol* locus, which is equivalent to the *chlD* locus of *E. coli* (Ugalde et al., 1985). Further, it was shown (Kennedy and Postgate, 1977) that for the *nif* genes of *K. pneumoniae* to function in *E. coli* grown at low Mo concentrations, a functional *chlD* locus is required in *E. coli*. Thus, a *nifC*- or *chlJ*-like gene may exist in the *K. pneumoniae* genome.

The two genes *nifVω* and *nifVα* are separated by 9 nucleotides, with *nifV* occurring 17 nucleotides downstream of *nifC*. *nifVω* has a coding capacity of 352 amino acid residues, whereas *nifVα* has a coding capacity of 269 amino acid residues. These two genes are so named because (1) the C-terminal 195 residues of NifVω are similar to the C-terminal region of the predicted NifV sequence of *A. vinelandii* and *K. pneumoniae*, and (2) the predicted sequence of NifVα is similar to the N-terminal region of the predicted NifV of *A. vinelandii* and *K. pneumoniae*. These relationships, for the deduced amino acids, are shown diagrammatically in Figure 26.3. From Figure 26.3, one can see that there is a region in each of the *C. pasteurianum nifV* gene products that has sequence similarity to a common region in the *nifV* product of *A. vinelandii*. What is particularly interesting about this is that the conserved amino acid residues in this region from *nifVω* are not the same as those conserved in the region from

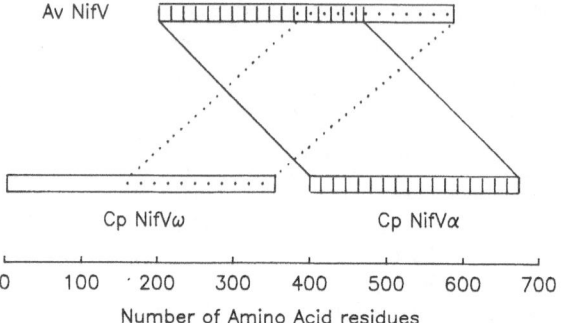

FIGURE 26.3   Schematic illustration of NifVω and NifVα of *Clostridium pasteurianum* and corresponding regions in *Azotobacter vinelandii* NifV.

nifVα. As a result, there is no significant similarity between the *nifVω* and *nifVα* genes.

The *C. pasteurianum nifV* genes are also functional in *A. vinelandii*. A plasmid containing both *nifVω* and *nifVα*, but not either alone, rescued a *nifV*-deleted strain of *A. vinelandii* (Wang et al., 1990b). The function of the ORF following *nifVα* is not known.

NifH-like genes

In addition to *nifH1*, there are five more *nifH*-like genes (*nifH2* to *nifH6*) in *C. pasteurianum* (Wang et al., 1988a); *nifH2* is upstream of *nifH1* (Figure 26.2A), but they belong to different transcriptional units. The cloned fragments, containing *nifH3* through *nifH6*, are shown in Figure 26.2B. The genomic locations of these fragments have not been determined. Among the *nifH*-like genes, similarities (S_{AB} values) of the predicted polypeptides range from 0.99 to 0.65. These relationships are illustrated in Figure 26.4. The mRNA for these *nifH*-like genes, with the exception of *nifH3*, has been detected in nitrogen-fixing *C. pasteurianum* cells grown with normal amounts of Mo (Wang et al., 1988a), but their functions remain to be determined.

A striking similarity has been observed between the predicted amino acid sequences from *nifH3* of *C. pasteurianum* and *nifH3* (*anfH*) of *A. vinelandii*, with an S_{AB} of 0.82 (Joerger et al., 1989). Because *C. pasteurianum* can grow under nitrogen-fixing conditions with low concentrations of Mo or with vanadium substitut-

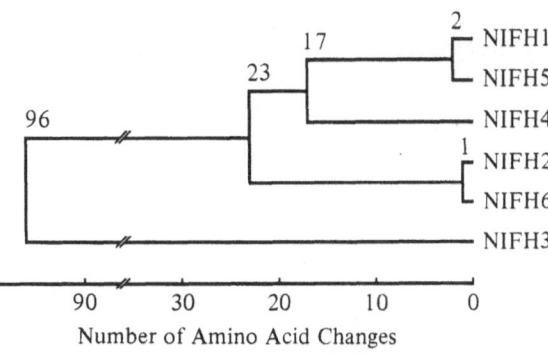

FIGURE 26.4 Deduced amino acid sequence similarities among NifH-like products.

ing for Mo (Cardenas and Mortenson, 1975; Dilworth et al., 1987; our unpublished results), it appears to contain one or more of the alternative nitrogen-fixing systems. Therefore, the *C. pasteurianum nifH3* product may be involved in an alternative nitrogenase system.

26.4 *Nif* Promoters and Upstream Sequences

The consensus *nif* promotor and activator sequences found in *Klebsiella pneumoniae* and other nitrogen-fixing organisms (Ausubel, 1984; Dixon, 1984) were not found in *Clostridium pasteurianum*. Following the determination of the transcription start site for *nifH1*, *nifE*, and *nifH*-like genes (excluding *nifH3*), similar nucleotide sequences were observed in the −100 region relative to the transcription start (Wang et al., 1988a). The proposed consensus sequence is ATCAatat-N$_{6-10}$-ATGGattc, which appears to be orientation independent. The −35 and −10 regions are more heterogeneous, with sequences similar to TTG and TATAAT preceding some *nif* operons (Wang et al., 1988a). The regulatory sequences for *nif* operons in *C. pasteurianum* will require further studies.

26.5 Concluding Remarks

Clostridium pasteurianum was one of the first nitrogen-fixing organisms isolated, and it has been used extensively in the study of the

biochemistry of nitrogen fixation. However, the genetics of the nitrogen fixation system of *C. pasteurianum* has been studied only in recent years. The organization of *nif* genes in *C. pasteurianum* W5 has several distinct features. They include (1) an overlap between *nifD* and *nifK*, (2) the fusion of *nifN* and *nifB* into *nifN-B*, (3) the division of *nifV* into *nifVα* and *nifVω*, and (4) the presence of *nifC* upstream of *nifV*. The fused and divided *nif* genes offer a unique opportunity for studies of the structure-function relationship in these gene products. Also, there appears to be a clustering of genes involved in FeMoco synthesis and possibly Mo uptake in *Clostridium pasteurianum*, which, in conjunction with a proximity of these genes to *nifH1DK*, may be an important aspect for studies concerned with the evolution and distribution of *nif* genes and with the maximization of the efficiency of nitrogen fixation. Studies with additional strains of *C. pasteurianum* may generate further leads toward these goals.

Acknowledgments. This work was supported by USDA competitive grants and by projects from the Commonwealth of Virginia.

References

Arnold, W., A. Rump, W. Klipp, U.B. Priefer, and A. Puhler. 1988. Nucleotide sequence of a 24,206-base-pair DNA fragment carrying the entire nitrogen fixation gene cluster of *Klebsiella pneumoniae*. *J. Mol. Biol.* **203**:715–738.

Ausubel, F.M. 1984. Regulation of nitrogen fixation genes. *Cell* **37**:5–6.

Ausubel, F.M. and F.C. Cannon. 1980. Molecular genetic analysis of *Klebsiella pneumoniae* nitrogen-fixation (*nif*) genes. *Cold Spring Harbor Symp. Quant. Biol.* **45**:487–499.

Bennett, L.T., F. Cannon, and D.R. Dean. 1988. Nucleotide sequence and mutagenesis of the *nifA* gene from *Azotobacter vinelandii*. *Mol. Microbiol.* **2**:315–321.

Bertram, J. and P. Dürre. 1989. Conjugal transfer and expression of streptococcal transposons in *Clostridium acetobutylicum*. *Arch. Microbiol.* **151**:551–557.

Boonsermsuwong, A. and H.P. Blaschek. 1986. Localization and inactivation of DNase activity in *Clostridium pasteurianum* NRRL-B598. *J. Ind. Microbiol.* **1**:265–270.

Burris, R.H. 1988. 100 years of discoveries in biological N$_2$ fixation. *In: Nitrogen Fixation: Hundred*

Years After, H. Bothe, F.J. de Bruijn, and W.E. Newton, eds., pp. 21–30. New York: Fisher.

Cardenas, J. and L.E. Mortenson. 1975. Role of molybdenum in dinitrogen fixation by *Clostridium pasteurianum*. *J. Bacteriol.* **123**:978–984.

Carnahan, J.E., L.E. Mortenson, H.F. Mower, and J.E. Castle. 1960. Nitrogen fixation in cell-free extracts of *Clostridium pasteurianum*. *Biochim. Biophys. Acta* **44**:520–535.

Cato, E.P., W.L. George, and S.M. Finegold. 1986. Genus *Clostridium* Prazmowski 1880, 23[AL]. *In: Bergey's Manual of Systematic Bacteriology*, Vol. 2, P.H.A. Sneath, N.S. Mair, M.E. Sharpe, and J.G. Holt, eds., pp. 1141–1200. Baltimore: Williams & Wilkins.

Chen, K.C.-K., J.-S. Chen, and J.L. Johnson. 1986. Structural features of multiple *nif*H-like sequences and very biased codon usage in nitrogenase genes of *Clostridium pasteurianum*. *J. Bacteriol.* **166**:162–167.

Daldal, F. 1985. Isolation of mutants of *Clostridium pasteurianum*. *Arch. Microbiol.* **142**:93–96.

Dilworth, M.J., R.R. Eady, R.L Robson, and R.W. Miller. 1987. Ethane formation from acetylene as a potential test for vanadium nitrogenase in vivo. *Nature* (London) **327**:167–168.

Dixon, R.A. 1984. The genetic complexity of nitrogen fixation. *J. Gen. Microbiol.* **130**:2745–2755.

Elliott, B.B. and L.E. Mortenson. 1976. Regulation of molybdate transport by *Clostridium pasteurianum*. *J. Bacteriol.* **127**:770–779.

Gawron-Burke, C. and D.B. Clewell. 1982. A transposon in *Streptococcus faecalis* with fertility properties. *Nature* (London) **300**:281–284.

Jacobson, M.R., K.E. Brigle, L.T. Bennett, R.A. Setterquist, M.S. Wilson, V.L. Cash, J. Beynon, W.E. Newton, and D.R. Dean. 1989. Physical and genetic map of the major *nif* gene cluster from *Azotobacter vinelandii*. *J. Bacteriol.* **171**:1017–1027.

Joerger, R.D. and P.E. Bishop. 1988. Nucleotide sequence and genetic analysis of the *nif*B-*nif*Q region from *Azotobacter vinelandii*. *J. Bacteriol.* **170**:1475–1487.

Joerger, R.D., M.R. Jacobson, R. Premakumar, E.D. Wolfinger, and P.E. Bishop. 1989. Nucleotide sequence and mutational analysis of the structural genes (*anf*HDGK) for the second alternative nitrogenase from *Azotobacter vinelandii*. *J. Bacteriol.* **171**:1075–1086.

Johann, S. and S. Hinton. 1987. Cloning and nucleotide sequence of the *chl*D locus. *J. Bacteriol.* **169**:1911–1916.

Johnson, J.L. and B. Francis. 1975. Taxonomy of the clostridia: ribosomal ribonucleic acid homologies among the species. *J. Gen. Microbiol.* **88**:229–244.

Keller, G., K.H. Schleifer, and F. Götz. 1983. Construction and characterization of plasmid vectors for cloning in *Staphylococcus aureus* and *Staphylococcus carnosus*. *Plasmid* **10**:270–278.

Kennedy, C. and J.R. Postgate. 1977. Expression of *Klebsiella pneumoniae* nitrogen fixation genes in nitrate reductase mutants of *Escherichia coli*. *J. Gen. Microbiol.* **98**:551–557.

Kent, H.M., M. Buck, and D.J. Evans. 1989. Cloning and sequencing of the *nif*H gene of *Desulfovibrio gigas*. *FEMS Microbiol. Lett.* **61**:73–78.

LeBlanc, D.J. and L.N. Lee. 1984. Physical and genetic analyses of streptococcal plasmid pAMβ1 and cloning of its replication region. *J. Bacteriol.* **157**:445–453.

Lee, C.-K., P. Dürre, H. Hippe, and G. Gottschalk. 1987. Screening for plasmids in the genus *Clostridium*. *Arch. Microbiol.* **148**:107–114.

Mazur, B.J., D. Rice, and R. Haselkorn. 1980. Identification of blue-green algal nitrogen fixation genes by using heterologous DNA hybridization probes. *Proc. Natl. Acad. Sci. USA* **77**:186–190.

McCoy, E., E.B. Fred, W.H. Peterson, and E.G. Hastings. 1930. A cultural study of certain anaerobic butyric acid-forming bacteria. *J. Infect. Dis.* **46**:118–137.

Minton, N.P. and J.G. Morris. 1983. Regeneration of protoplasts of *Clostridium pasteurianum* ATCC 6013. *J. Bacteriol.* **155**:432–434.

Mishustin, E.N. and V.T. Yemtsev. 1975. Anaerobic nitrogen-fixing bacteria of different soil types. *In: Nitrogen Fixation by Free-Living Microorganisms*, W.D.P. Stewart, ed., pp. 29–38. Cambridge: Cambridge University Press.

Pienkos, P.T. and W.J. Brill. 1981. Molybdenum accumulation and storage in *Klebsiella pneumoniae and Azotobacter vinelandii*. *J. Bacteriol.* **145**: 743–751.

Roberts, I., W.M. Holmes, and P.B. Hylemon. 1988. Development of a new shuttle plasmid system for *Escherichia coli* and *Clostridium perfringens*. *Appl. Environ. Microbiol.* **54**:268–270.

Robson, R.L., R.M. Robson, and J.G. Morris. 1974. The biosynthesis of granulose by *Clostridium pasteurianum*. *Biochem. J.* **144**:503–511.

Ruvkun, G.B. and F.M. Ausubel. 1980 Interspecies homology of nitrogenase genes. *Proc. Natl. Acad. Sci. USA* **77**:191–195.

Simon, M.A and W.J. Brill. 1971. Mutant of *Clostridium pasteurianum* that does not fix nitrogen. *J. Bacteriol.* **105**:65–69.

Spiegelberg, C.H. 1940. *Clostridium pasteurianum* associated with spoilage of an acid canned fruit. *Food Res.* **5**:115–130.

Spiegelberg, C.H. 1944. Sugar and salt tolerance of *Clostridium pasteurianum* and some related anaerobes. *J. Bacteriol.* **48**:13–30.

Tanaka M., M. Haniu, K.T. Yasunobu, and L.E. Mortenson. 1977. The amino acid sequence of *Clostridium pasteurianum* iron protein, a component of nitrogenase. *J. Biol. Chem.* **252**:7093–7100.

Truffaut, N. and M. Sebald. 1983. Plasmid detection and isolation in strains of *Clostridium acetobutylicum* and related species. *Mol. Gen. Genet.* **189**:178–180.

Ugalde, R.A., J. Imperial, V.K. Shah, and W.J. Brill. 1985. Biosynthesis of the iron-molybdenum cofactor and the molybdenum cofactor in *Klebsiella pneumoniae*: effect of sulfur source. *J. Bacteriol.* **164**:1081–1087.

Volk, W.A., B. Bizzini, K.R. Jones, and F.L. Macrina. 1988. Inter- and intrageneric transfer of Tn*916* between *Streptococcus faecalis* and *Clostridium tetani. Plasmid* **19**:255–259.

Wang, S.-Z., J.-S. Chen, and J.L. Johnson. 1987. Nucleotide and deduced amino acid sequences of *nifD* encoding the α-subunit of nitrogenase MoFe protein of *Clostridium pasteurianum. Nucleic Acids Res.* **15**:3935.

Wang, S.-Z., J.-S. Chen, and J.L. Johnson. 1988a. The presence of five *nifH*-like sequences in *Clostridium pasteurianum*: sequence divergence and transcription properties. *Nucleic Acids Res.* **16**:439–454.

Wang, S.-Z., J.-S. Chen, and J.L. Johnson. 1988b. Distinct structural features of the α and β subunits of nitrogenase molybdenum-iron protein of *Clostridium pasteurianum*: an analysis of amino acid sequences. *Biochemistry* **27**:2800–2810.

Wang, S.-Z., J.-S. Chen, and J.L. Johnson. 1989. Nucleotide and deduced amino acid sequences of *nifE* from *Clostridium pasteurianum. Nucleic Acids Res.* **17**:3299.

Wang, S.-Z., J.-S. Chen, and J.L. Johnson. 1990a. A nitrogen-fixation gene (*nifC*) in *Clostridium pasteurianum* with sequence similarity to *chlJ* of *Escherichia coli. Biochem. Biophys. Res. Commun.* **169**:1122–1128.

Wang, S.-Z., D.R. Dean, J.-S. Chen, and J.L. Johnson. 1990b. The N- and C-terminal portions of nifV are encoded by two different genes in *Clostridium pasteurianum. In: Nitrogen Fixation: Achievements and Objectives*, P.M. Gresshoff, L.E. Roth, G. Stacey, and W.E. Newton, eds. New York: Chapman and Hall. p.602.

Winogradsky, M.S. 1895. Recherches sur l'assimilation de l'azote libre de l'atmosphére par les microbes. *Arch. Sci. Biol.* (St. Petersburg) **3**:297–352.

Witz, D.F., R.W. Detroy, and P.W. Wilson. 1967. Nitrogen fixation by growing cells and cell-free extracts of the *Bacillaceae. Arch. Mikrobiol.* **55**:369–381.

Young, M., N.P. Minton, and W.L. Staudenbauer. 1989. Recent advances in the genetics of the clostridia. *FEMS Microbiol. Rev.* **63**:301–326.

27

Cloning and Sequencing of the *Clostridium pasteurianum* Genes Encoding Molybdenum-Pterin-Binding Proteins

Stephen M. Hinton

27.1 Introduction

The importance of molybdenum (Mo) first became apparent at the molecular level when the two proteins required for nitrogen-fixing activity were isolated (Bulen and LeComte, 1966), and subsequently, one of the nitrogenase proteins was shown to contain molybdenum and iron (MoFe protein). One of the established metabolic functions of Mo is a component of two distinct "cofactors" of various redox enzymes. The MoFe cofactor is unique to the nitrogenase system, and the Mo cofactor, a Mo–pterin complex, is common to all other known Mo enzymes. What is not known is the biosynthetic pathway for either Mo cofactor or the process by which Mo metabolism and apoenzyme biosynthesis conjoin.

Molybdenum metabolism has been studied in *Clostridium pasteurianum*, a nitrogen-fixing anaerobe, for more than a decade. *C. pasteurianum* has been used extensively in the biochemical studies of biological nitrogen fixation mainly through the research efforts of my mentor, Dr. L.E. Mortenson. It was in Dr. Mortenson's laboratory as a doctoral student that I gained an interest in nitrogen fixation and, more specifically, in the metabolic process by which microbes transform the inorganic oxyanion molybdate into an essential component of biological catalysts. In one sense, it was curiosity about the "cleverness" of microbes as (bio)inorganic chemists, but in a more practical vein it was considered as a strategy to seize the "holy grail" of molybdenum enzyme studies, which is to determine the structure and function of the catalytic Mo center.

Most structural studies on the Mo centers in various enzymes were spectroscopic studies on the holoenzyme and isolated Mo cofactors. These biophysical techniques provided detailed information about the immediate environment and possible ligands to Mo but left many more structure–function questions unanswered. We choose a converse approach to study Mo centers, which was to determine how molybdate is processed in vivo for insertion into an apoenzyme. Our research goal in this area has been to elucidate the Mo-processing pathway by a complementary biochemical and genetic study of proteins directly involved in Mo metabolism, the Mo-binding proteins. The following is a brief account of the biochemical and genetic studies that have preceded the cloning and sequencing of the clostridial molybdenum-pterin binding protein.

27.2 Biochemical Studies of Molybdenum Metabolism in *Clostridium pasteurianum*

Elliott and Mortenson (1977) demonstrated that MoO_4^{2-} is actively transported into N_2-fixing *C. pasteurianum* cells. The Mo that accumulated is evenly distributed between two cellular fractions that were separated by ion-exchange chromatography. One fraction contained two Mo enzymes, the MoFe protein of nitrogenase and formate dehydrogenase (Hinton and Mortenson, 1985b), and the other Mo fraction contained a Mo-binding storage protein (45,000 molecular weight) that was named for its postulated function (Elliott and Mortenson, 1977). This hypothesis suggested that it might be possible to study the biogenesis of the Mo cofactors if stabilized protein-bound intermediates could be identified and characterized. The hypothesis also predicted that Mo centers are not biosynthesized on the apoenzyme but rather are constructed elsewhere, possibly on small molecular weight proteins, Mo-binding proteins; and this inactive Mo precursor further requires insertion into the apoenzyme for attaining the active configuration. Further, structural information on the precursor(s) of the Mo centers could also lead to a better understanding of structure–function relationships of the active Mo-catalyst. Our goal was to identify and characterize low molecular weight Mo-binding proteins in the hope they bound a series of intermediates in the transformation of MoO_4^{2-} to Mo cofactors.

27.3   Anaerobic Electrophoresis and the Identification of Clostridial Mo Proteins

Initially, the author chose a biochemical approach to study the Mo species or intermediates involved in Mo metabolism. The first question that arose was what technology would be advantageous to identifying unknown Mo proteins. The various techniques of biochemistry available in the mid-1970s were evaluated, and the most theoretically promising were the three separation principles of electrophoresis: isotachophoresis, isoelectric focusing, and multiphasic zone electrophoresis. Isotachophoresis (today called capillary zone electrophoresis) is in principle the electrophoretic technique of greatest resolution, but the equipment was expensive and the technology for resolving pro-

teins at that time was in a developmental stage. High-resolution isoelectric focusing was also a promising electrophoretic technique, but technical problems were encountered in adapting it for anaerobic separation of acidic proteins.

The anaerobic agarose isoelectric focusing strategy we used was based on the preparative methods of Chrambach and Nguyen (1979; Chrambach, 1980) and the agarose electrofocusing technical procedures developed by Saravis and Zamcheck (1979). Anaerobic electrophoresis chambers were designed and constructed (Hinton, 1982) because the Mo proteins of *Clostridium pasteurianum* are oxygen-labile, acidic proteins that require strict anaerobic conditions to maintain their integrity. A previous study (O'Donnell and Smith, 1980) of the MoFe protein of *Klebsiella pneumoniae* demonstrated that the enzyme is oxidatively damaged even under anoxic conditions when incubated at potentials above +200 mV with respect to the standard hydrogen electrode. Therefore, the addition of the reducing agent dithionite ($Na_2S_2O_4$) was required to maintain the stability of the Mo proteins but it disrupted the acidic portion (pH < 7) of the pH gradient established with ampholytes (Pharmacia). Park (1973) reported that sodium dithionite can be used as a reducing agent in conjunction with isoelectric focusing by preelectrofocusing the dithionite through the gel into the catholyte. In our experiments, dithionite (in general, sulfur oxyanions) added to the agarose gel or to the catholyte disrupted the formation of the acidic portion of the electrofocused pH gradient.

Multiphasic zone electrophoresis (MZE) yielded the most promising results for separation and identification of Mo proteins. MZE was adapted for anaerobic conditions by adding dithionite to a polyacrylamide gel to establish reducing conditions required for integrity of oxygen-labile Mo proteins. It is important to note that highly oxidative conditions exist in a freshly polymerized gel (Morris and Morris, 1971), and reducing agents inhibit polymerization and become fully oxidized themselves. This problem was overcome using the following procedures. A gradient gel of polyacrylamide was used to resolve Mo species as large as membrane fragments and as small as MoO_4^{2-}. The polymerized pore gradient gel was preelectrophoresed in an enclosed anaerobic chamber with an electrolyte of resolving gel buffer containing dithionite. Dithionite is an anion; therefore, it migrates through the gel establishing reducing conditions.

The "stacking" gel, which contains a unique buffer, is applied on top of the resolving gel. The electrolyte and the stacking gel buffers are an isotachophoretic system designed to· concentrate the sample to a minimum bandwidth to achieve maximum resolution. With the stacking gel in place, preelectrophoresis would destroy the isotachophoretic buffer system; therefore, another strategy was used to reduce the stacking gel. The supporting medium for the stacking gel was agarose, which does not require oxidative agents or conditions for polymerization. The use of the agarose supporting medium allowed us to apply a prereduced stacking gel to the anaerobic resolving gel. It was in this manner we devised a method to prepare an anaerobic multiphasic electrophoresis gel that maintained the discontinuous buffer system (Hinton and Mortenson, 1985a).

The nondenaturing electrophoretic conditions we used were based on the strategy of Chrambach et al. (1976) to define the optimum fractionation conditions required to separate a macromolecule from its nearest neighbor. With the use of ^{99}Mo-labeled clostridial cell extracts, we were able to separate and identify 10 distinct ^{99}Mo species and free $^{99}MoO_4^{2-}$. Three of the ^{99}Mo zones represented effective electrophoretic mobility variants of the Mo-binding storage protein (Hinton and Mortenson, 1985b). With the anaerobic electrophoretic technique we were able to show that during the derepression of the nitrogenase system Mo accumulation increased 1.5 h before Mo enzyme activity was detectable (Hinton and Mortenson, 1985c). The increase in Mo accumulation is predominantly in the form of proteins, specifically the Mo-binding storage protein. The apparent order of appearance of the Mo-binding protein, before the Mo enzymes, and its relative small size compared to Mo enzymes makes it a likely candidate for a Mo-processing protein.

27.4   Purification and Identification of the Mo-Pterin-Binding Protein

A large-scale purification scheme was devised for the Mo-binding protein to isolate and characterize the putative Mo-processing protein (Hinton and Merritt, 1986). In the initial steps of purification the apparent molecular weight of the protein was in the 30,000 to 50,000 range, as determined by ultrafiltration. The addition of ammonium sulfate to fractions containing the Mo-binding storage protein resulted in a decrease to greater than 5,000 but less than 10,000 in the apparent molecular exclusion limit of the Mo protein. One possible explanation was a dissociation of an aggregate or a protein multimer into monomers. It is the small molecular weight protein, estimated to be 5,700 by gel electrophoresis, that is referred to as Mop, the Mo- (pterin-) binding protein. The isolated protein had a peak absorbancy at 293 nm and the denatured protein releases a fluorescent chromophore as well as Mo.

Spectral analysis of the chromophore suggested it is an unidentified pterin derivative. If the Mo and pterin are associated, that complex may be precursor to the molybdopterin cofactor, Moco. X-ray absorption spectroscopy (XAS) has shown that the Mo environment (a tri-oxo species) is unusual. The Mo center has no sulfur ligands, unlike most Mo enzymes; there is one exception, the clostridial formate dehydrogenase (S. Cramer and S. Hinton, unpublished results). Further studies showed that the Mo and pterin derivative in Mop were not able to reconstitute the activity of Moco-deficient nitrate reductase, the biological assay for active Moco. The pterin derivative in Mop, unlike the molybdopterin found in functional Mo enzymes, is not phosphorylated as is required for catalytic activity. These results suggest that Mop contains an inactive or precursor form of Moco, awaiting activation, possibly by phosphorylation, and insertion into the apoenzyme.

27.5   Cloning and Immunoscreening Genomic Libraries

The biochemical studies discussed here revealed a potential candidate for a Mo-processing protein. The size of Mop alone was cause for excitement for the simple reason that known molybdoenzymes are large proteins, and the relative abundance of the Mo cofactor in an enzyme is less than 1% with respect to weight. Therefore, Mop is highly enriched for Mo and pterin. Also, Mo cofactors released from the protein were very labile, making their direct study most difficult. A small molecular weight protein binding and stabilizing Mo and a chromophore seemed ideal for direct structural studies of a protein-bound Mo center. Cloning the structural gene for this Mo-binding protein was vital to determine the metabolic role of Mop and possibly locate the Mo-binding site in the protein.

The strategy used to clone the *mop* gene(s) was suggested during discussions with Drs. Maggie So and Fred Hefferon (Hinton and Freyer, 1986). *C. pasteurianum* chromosomal DNA was partially digested with the endonuclease *Sau*3A1, which resulted in restriction DNA fragments 8 to 12 kilobases (kb) in length. A 6-base pair (bp) recognition sequence is expected to be spaced approximately every 5,000 bp on a fragment of DNA; therefore, a complete *Eco*RI restriction digest was preformed to generate *Sau*3A1-*Eco*RI restriction fragments 4 to 6-kb long. The vector plasmid pBR322 was digested with three endonucleases, *Eco*RI, *Cla*I, and *Bam*HI, to ligate the clostridial restriction fragments into the vector and to remove the 26-bp *Eco*RI-*Cla*I restriction fragment by molecular sieve filtration to prevent vector self-religation. The transformed bacteria were directly plated onto nitrocellulose filters and selected for carbenicillin resistance (Cb^r). Carbenicillin, an analog of ampicillin, is recommended for library antibiotic selection because it is very effective in reducing the number of untransformed satellite colonies arising around true antibiotic resistance clones.

Polyclonal antibody raised against isolated Mop was used to screen the clones for protein expression. Antibody preparation is critical to achieve a sensitivity of detection of the few nanograms of diluted antigen protein required for successful immunoscreening. Two replica blots of the initial or master filter were immunoscreened to avoid false-positives. Following replica plating, the master filter was placed on an agar plate containing Cb and glycerol and incubated for a few hours to regrow the clones before the master filter was stored at −20°C. This method proved to be a convenient way to store and regenerate genomic libraries.

Approximately 30,000 transformed bacterial colonies constituting our clostridial genomic library were immunoscreened for Mop protein expression (Helfman et al., 1983). Two groups of clones cross-reacting with the Mop antibody were identifiable; one group had a stronger signal than the other. The positive cross-reacting clones were isolated, grown overnight, and replated on nitrocellulose filters for a second round of immunoscreening. The second round of immunoscreening is important to verify true-positives by demonstrating that all colonies of a particular clone cross-react in replating. Further, each clone was shown to produce a protein cross-reacting with Mop antisera, which had an identical effective electrophoretic mobility as purified Mop. One positive clone that carried a plasmid, pSHC50, with a 5-kb insert was chosen for further study. Plasmid-directed Mop expression in *Escherichia coli* was independent of the cloned insert orientation in the vector; therefore, expression appeared to be directed by the clostridial promoter to an estimated level of 5% of the total cellular protein.

27.6   Localization and Nucleotide Sequencing of the *mop* Gene Family

The initial evidence for the existence of the *mop* gene family was in the ambiguity of localizing the *mop* gene(s) on the expression clones and analyzing the N-terminal amino acid sequence of the purified protein preparation. To establish the *mop* gene family, a physical map of pSHC50 first was obtained by restriction endonuclease analysis to construct subclones to further localize the *mop* gene. Next, Western blot analysis of the cellular extract of several pSHC50 subclones indicated two coding regions at either end of the insert responsible for the expression of protein(s) cross-reacting with Mop antisera. The confusion about the location of the *mop* gene was partially resolved when the nucleotide sequence was obtained from the subclone, pSHC55, which produced the strongest antibody cross-reaction.

DNA sequence analysis of pSHC55 revealed an open-reading frame (ORF) that encodes for a protein of 68 amino acid residues with a predicted molecular weight of 7,038. The identity of this ORF was verified as the *mop*(I) gene by (Western blot) immunoelectrophoresis, the size of the predicted coding region, and the sequence homology of the deduced amino acid sequence of the ORF and the first 14 amino acid residues of purified protein (Hinton et al., 1987). The nucleotide sequence data confirmed we had identified a *mop* gene, but the protein sequence analysis also suggested a gene family because of the apparent existence of multiple Mop proteins.

Amino acid sequence analysis of Mop preparations from *C. pasteurianum* cells grown under nitrogen-fixing conditions identified a single amino acid in 38 of the first 44 positions. In the other 6 positions, namely 1, 15, 16, 20, 21, and 35, two amino acids were observed. The deduced amino acid sequence of the *mop*(I) gene located on pSHC55 agrees with the protein sequence except that the gene sequence predicts only 1 of the two amino acids at the 6 ambiguous positions. At that time the results suggested there were two or more closely related proteins in the purified preparation. Together the genetic and biochemical evidence suggested the possibility of duplicate *mop* genes, and the slight genetic variations among the *mop* genes are reflected in the amino acid substitutions in the copurified variant proteins.

Southern blot analysis of *C. pasteurianum* genomic DNA revealed the *mop* gene family by showing that the sequenced *mop*(I) gene used as a probe hybridizes to two *Eco*RI chromosomal restriction fragments of 6 and 2.5-kb length.

Further restriction, subcloning, and Southern hybridization analysis confirmed that the large *Eco*RI restriction fragment contained two distinct *mop* (I and II) genes: one was the original *mop*I gene isolated by immunoscreening the genomic library with a second *mop*II gene genetically linked to it. The smaller *Eco*RI chromosomal restriction fragment carried the third member of the *mop* gene family *mop*III. Establishing the existence of multiple *mop*I, *mop*II, and *mop*III genes resolved the ambiguity observed in immunoscreening subclones and protein sequence determination. For example, in immunoscreening the genomic library the clones with the strong cross-reaction signal carried two linked *mop* genes while the clones with the weaker cross-reaction carried only one copy.

Subsequent nucleotide sequence analysis of the three *mop* genes I, II, III accounted for all the ambiguities seen in the amino acid sequence analysis. It is interesting to note that most of the differences at the nucleotide level are silent substitutions. The amino acid differences that do appear in the three variants of Mop are chemically conservative substitutions, which accounts for the copurification of these proteins. It was estimated by quantitative amino acid sequencing that in the purified protein preparation the relative abundance of the *mop*I, *mop*II, and *mop*III gene products was 4:1:5, respectively. The overall amino acid composition indicates that the Mop proteins are hydrophobic (>50% of the residues are nonpolar) and lack both aromatic and cysteine residues. Therefore, the ultraviolet absorbancy at 293 nm of the purified proteins is attributed to the pterin chromophore. The lack of cysteine residues in the protein eliminates potential sulfur ligands for coordinating Mo, which is in agreement with the XAS data.

27.7 Comparison of the *mop* Gene Family

A consensus sequence of the *mop* gene family shows a homology greater than 90% at the nucleotide level, which predicts a common origin. The data are not shown, but the *mop*I and *mop*II sequences appear more closely related at the nucleotide level compared to the third gene. One explanation for this would be the duplication of the primordial gene followed by parallel independent mutations in the multiple gene family. The strong conservation of nucleotide sequence extends upstream the coding region to the predicted Shine–Dalgarno (S–D) sequence. This conserved S–D region supports the hypothesis that *C. pasteurianum*, a grampositive organism and requires a strong S–D interaction for efficient translation (Moran et al., 1982). The codon preference for A or T in the degenerate positions and intergenic regions reflects the low G + C content of clostridial genomic DNA.

The sequence similarity in the promoter regions is greatest when the putative promoter consensus sequences are used for alignment, again suggesting a common origin; however, the different amount of protein product recovered in the purified preparation may reflect varying gene expression. The three putative transcriptional binding sequences are identical. The transcriptional recognition sequences and nucleotide distance to the start codons do vary among the gene family, however, which may contribute to differences in transcriptional efficiency.

Another speculative factor that may regulate *mop* gene expression would be an autoregulatory function of the protein product. The primary structure of the Mop proteins suggests that the molecule may have two domains. The N-terminal region (amino acid residues 1–20) is hydrophilic with a predominance of positively charged amino acids (~30%). The C-terminal region is mainly composed of hydrophobic residues interrupted by negatively charged residues (i.e., glutamic acid). The secondary structure predictions (Chou and Fasman, 1978) suggest that the N-terminal domain would be a helix-turn-helix structure, and the amino acid sequence in that region is similar to DNA-binding proteins (Shaw et al., 1983). The evidence here is the basis for the following hypothesis. The N-terminal portion of *Mop* may function as a DNA-binding domain while the C-terminal region may be involved in binding the Mo–pterin complex. The expression of

mop gene products appears to be regulated in response to the level of Mo available to the cell. Therefore, Mop may be a metal-responsive regulatory protein "sensing" the availability of the anabolic source of Mo.

References

Bulen, W.A. and J.R. LeComte. 1966. The nitrogenase system from *Azotobacter*: two-enzyme requirement for N_2 reduction, and ATP hydrolysis. *Proc. Natl. Acad. Sci. USA* **56**:979–986.

Chou, P. and G.D. Fasman. 1978. Prediction of the secondary structure of proteins from their amino-acid sequence. *Adv. Enzymol.* **47**:45–147.

Chrambach, A. 1980. Electrophoresis and electrofocusing on polyacrylamide gel in the study of native macromolecules. *Mol. Cell. Biochem.* **29**:23–46.

Chrambach, A. and N.Y. Nguyen. 1979. Preparative electrophoresis, isotachophoresis and electrofocusing on polyacrylamide gel. In: *Electrokinetic Separation Methods*, P.G. Righetti, G.J. Van Oss, and J.W. Vanderhoff, eds., pp. 337–368. New York: Elsevier/North Holland.

Chrambach, A., T.M. Jovin, P.J. Svendsen, and D. Rodbard. 1976. Analytical and preparative polyacrylamide gel electrophoresis. An objectively defined fractionation route, apparatus, and procedures. In: *Methods of Protein Separations*, Vol. 2, N. Catsimpoolas, ed., pp 27–144. New York: Plenum. N.Y.

Elliott, B.B. and L.E. Mortenson. 1977. Molybdenum storage component from *Clostridium pasteurianum*. In: *Recent Developments in Nitrogen Fixation*, W. Newton, J.R. Postgate, and C. Rodriquez-Barrueco, eds., pp. 205–217. London: Academic Press.

Helfman, D.M., J.R. Feramisco, J.C. Fiddes, G.P. Thomas, and S.H. Hughes. 1983. Identification of clones that encode chicken tropo myosin by direct immunological screening of a complimentary DNA expression library. *Proc. Natl. Acad. Sci. USA* **80**:31–35.

Hinton, S.M. 1982. The processing of molybdenum during the synthesis of the molybdenum-iron protein of nitrogenase. Ph.D. thesis, West Lafayette. In: Purdue University.

Hinton, S.M., and G. Freyer. 1986. Cloning, expression and sequencing the molybdenum-pterin binding protein (*mop*) gene of *Clostridium pasteurianum* in *Escherichia coli*. *Nucleic Acids Res.* **14**:9371–9380.

Hinton, S.M. and B. Merritt. 1986. Purification and characterization of a molybdenum-pterin-binding protein (Mop) in *Clostridium pasteurianum* W5 *J. Bacteriol.* **168**:688–693.

Hinton, S.M. and L.E. Mortenson. 1985a. Anaerobic multiphasic gel electrophoresis of the molybdoproteins in extracts of *Clostridium pasteurianum*. *Anal. Biochem.* **145**:222–229.

Hinton, S.M. and L.E. Mortenson. 1985b. Identification of molybdoproteins in *Clostridium pasteurianum*. *J. Bacteriol.* **162**:477–484.

Hinton, S.M. and L.E. Mortenson. 1985c. Regulation and order of involvement of molybdoproteins during synthesis of molybdoenzymes in *Clostridium pasteurianum*. *J. Bacteriol.* **162**:485–493.

Hinton, S.M., C. Slaughter, W. Eisner, and T. Fisher. 1987. The molybdenum-pterin binding protein is encoded by a multigene family in *Clostridium pasteurianum*. *Gene* **54**:211–219.

Moran, C.P., N. Lang, S.F.J. LeGrice, G. Lee, M. Stephens, A.L. Sonenshein, J. Pero, and R. Losick. 1982. Nucleotide sequences that signal initiation of transcription and translation in *Bacillus subtilis*. *Mol. Gen. Genet.* **186**:339–346.

Morris, C.J.O.R. and P. Morris. 1971. Molecular-sieve chromatography and electrophoresis in polyacrylamide gels. *Biochem. J.* **124**:517–528.

O'Donnell, M.J. and B.E. Smith. 1980. On the nature of anaerobic oxidative damage to the MoFe protein of *Klebsiella pneumoniae* nitrogenase. *FEBS Lett.* **120**:251–254.

Park, C.M. 1973. Isoelectric focusing and the study of interacting protein systems: ligand binding, phosphate binding, and subunit exchange in hemoglobin. *Ann. N.Y Acad. Sci.* **209**:237–257.

Saravis, C.A. and N. Zamcheck. 1979. Isoelectric focusing in agarose. *J. Immunol. Methods* **29**:91–96.

Shaw, D.J., D.W. Rick, and J.R. Guest. 1983. Homology between catabolite gene activator protein and FNR a regulator of anaerobic respiration in *Escherichia coli*. *J. Mol. Biol.* **166**:241–247.

28

Cloning, Sequencing, and Expressions of Genes Encoding Enzymes of the Autotrophic Acetyl-CoA Pathway in the Acetogen *Clostridium thermoaceticum*

Thomas A. Morton, Chih-Fong Chou, and Lars G. Ljungdahl

28.1 Introduction

Clostridium thermoaceticum, first described by Fontaine et al. (1942), is a thermophilic anaerobic bacterium that ferments glucose, fructose and xylose to acetate as the only product (Reactions 1 and 2).

$$C_6H_{12}O_6 \rightarrow 3CH_3COOH \qquad [1]$$
$$2C_5H_{10}O_5 \rightarrow 5CH_3COOH \qquad [2]$$

Bacteria carrying out this type of fermentation have been called homoacetate fermenting or acetogenic. The latter term now designates bacteria capable of a total synthesis of acetate from CO_2 and CO. Fontaine et al. (1942) suggested the fermentation occurs as a two-step process; the fermentation of glucose via the Embden–Meyerhof–Parnas pathway to pyruvate, which is then converted to acetate and CO_2 (Reaction 3), and a total synthesis of acetate from CO_2 (Reaction 4).

$$C_6H_{12}O_6 + 2H_2O \rightarrow 2CH_3COOH +$$
$$2CO_2 + 8H^+ + 8e^- \qquad [3]$$
$$2CO_2 + 8H^+ + 8e^- \rightarrow CH_3COOH + 2H_2O \qquad [4]$$

Barker and Kamen (1945) and Wood (1952) conclusively showed this to be correct.

It is now well established that *C. thermoaceticum* carries out the synthesis of acetate from CO_2 via an unique autotrophic pathway called the Wood acetyl-CoA pathway. The name refers to the fact that acetyl-CoA is the first two-carbon product formed by fixation of CO_2, and it is to honor Professor H.G. Wood, who is the main contributor of the elucidation of the pathway. The Wood acetyl-CoA pathway, or variations thereof, has been shown to be present in many anaerobic bacteria, including acetogens, methanogens, and sulfate reducers. Among recent reviews covering the pathway are those by Fuchs (1986), Ljungdahl (1986), Wood et al. (1986), Wood (1989), and Wood and Ljungdahl (1991).

28.2 The Wood Acetyl-CoA Pathway and Metabolic Potentials of *Clostridium thermoaceticum*

In the Wood acetyl-CoA pathway, CO_2 is fixed in two reactions catalyzed by formate dehydrogenase (Reaction 5) and the multifunctional enzyme carbon monoxide dehydrogenase/acetyl-CoA synthase (CODH) (Reaction 6).

$$CO_2 + NADPH \rightleftharpoons HCOO^- + NADP^+ \qquad [5]$$
$$CO_2 + 2H^+ + 2e^- \rightleftharpoons CO + H_2O \qquad [6]$$

CO_2, when reduced to formate, is the precursor of the methyl group of acetyl-CoA. The reduction of formate occurs via one-carbon intermediates of the tetrahydrofolate (H_4folate) pathway and involves four enzymes: formyl-H_4folate synthetase (Reaction 7); methenyl-H_4folate cyclohydrolase (Reaction 8); methylene-H_4folate dehydrogenase (Reaction 9); and methylene-H_4folate reductase (Reaction 10).

$$HCOO^- + H_4folate + ATP \rightleftharpoons HCO\text{-}H_4folate + ADP + Pi^- \quad [7]$$

$$HCO\text{-}H_4folate + H^+ \rightleftharpoons CH\text{-}H_4folate^+ + H_2O \quad [8]$$

$$CH\text{-}H_4folate^+ + NADPH \rightleftharpoons CH_2\text{-}H_4folate + NADP^+ \quad [9]$$

$$CH_2\text{-}H_4folate + 2H^+ + 2e^- \rightleftharpoons CH_3\text{-}H_4folate \quad [10]$$

In Reaction 6, CO becomes the carbonyl group of acetyl-CoA. In addition to catalyzing the reduction of CO_2 to CO, CODH also catalyzes the final step of the synthesis of acetyl-CoA. It is a nickel/iron-sulfur/zinc protein, and evidence has been presented that CO binds to a metal center consisting of Ni and Fe (Ragsdale et al., 1988). The synthesis of acetyl-CoA involves the condensation of the methyl group of methyl-H_4folate, the CO bound to the Ni-Fe center of CODH, and CoA. Proteins involved, in addition to CODH, are a corrinoid/iron-sulfur protein (C/Fe-SP) (Ragsdale et al., 1987; Harder et al., 1989) and a methyltransferase (Drake et al., 1981; Hu et al., 1984). C/Fe-SP contains 5-methoxybenzimidazolylcobamide and a Fe_4S_4 cluster. It functions as a carrier of the methyl group from methyl-H_4folate to CODH. To accept the methyl group, the cobalt of the corrinoid must be reduced to Co^+. In vivo H_2, CO, or pyruvate may serve as reductants in reactions involving hydrogenase, CODH, or pyruvate ferredoxin oxidoreductase, respectively. The transfer of the methyl group of methyl-H_4folate to the cobalt of the corrinoid is catalyzed by the methyltransferase (Reaction 11). CODH then catalyzes the formation of acetyl-CoA (Reaction 12).

$$CH_3\text{-}H_4folate + C/Fe\text{-}SP \rightleftharpoons CH_3\text{-}C/Fe\text{-}SP + H_4folate \quad [11]$$

$$CH_3\text{-}C/Fe\text{-}SP + CO + CoA \rightleftharpoons CH_3CO\text{-}CoA + C/Fe\text{-}SP \quad [12]$$

Properties of the enzymes catalyzing Reactions 5 through 12, all of which have been purified, are covered in the review by Wood and Ljungdahl (1991).

It should be noted that acetogenic bacteria, including C. thermoaceticum, metabolize a rather wide range of organic compounds with the formation of acetate. These include methanol, formate, several sugars, alcohols, organic acids, and amino acids (see reviews by Ljungdahl, 1986 and Fuchs, 1986). Of special interest was a finding by Bache and Pfennig (1981) that Acetobacterium woodii, in the presence of CO_2, uses the O-methyl group of a number of methoxylated aromatic acids that constitute the products of lignin degradation. Several acetogens have now been found to use O-methyl groups of phenylmethyl ethers. Work by Wu et al. (1988) established with C. thermoaceticum that CO or pyruvate, rather than CO_2, was needed for the metabolism of the phenylmethyl ethers. Also, carboxyl groups of benzoic acid and similar acids as well as aromatic aldehydes are metabolized by acetogenic bacteria (Hsu et al., 1990; Lux et al., 1990). The autotrophic character of C. thermoaceticum, i.e., that it grows on CO_2 and H_2 or CO as sole carbon and energy sources, was established by Kerby and Zeikus (1983). A prerequisite for this was the demonstration by Drake (1982) that C. thermoaceticum contains hydrogenase.

The new autotrophic Wood acetyl-CoA pathway was established mostly through work with C. thermoaceticum using conventional approaches of isotopic labeling, enzymatic analyses, and protein purification. Much remains to be learned, however, about this pathway and the metabolic potentials of acetogens. For instance, chemosmotic generation of ATP must occur when acetogens grow autotrophically, synthesizing acetate with CO_2 and H_2 or CO as sole sources of carbon and energy (Hugenholtz and Ljungdahl, 1990). How is the Wood acetyl-CoA coupled to electron transport and ATP synthesis? There is almost no knowledge about the active sites of the enzymes of the acetyl-CoA pathway. Why do the tetrahydrofolate-dependent enzymes from acetogens have from 50 to 100 times higher specific activities than corresponding enzymes from other sources? How are metal centers of formate dehydrogenase, CODH, methylene-H$_4$folate reductase, hydrogenase, and the C/Fe-SP constructed and bound to the enzymes? What regulations occur of the synthesis of the enzymes and of their activities, etc.? A genetic approach, in addition to the biochemical studies, is needed to answer these questions. In this chapter we review the cloning into *Escherichia coli* and sequencing of the genes of *C. thermoaceticum* for formyl-H$_4$folate synthetase, methylene-H$_4$folate reductase, CODH, and some properties of the enzymes. The primary aims were to obtain the DNA sequences of the genes and from those deduce the amino acid sequences of the enzymes. The work has also yielded some other features related to the genetics of *C. thermoaceticum*.

28.3   Properties of Formyl-H$_4$Folate Synthetase, Methylene-H$_4$Folate Reductase, and Carbon Monoxide Dehydrogenase/Acetyl-CoA Synthase from *Clostridium thermoaceticum*

Four H$_4$folate-dependent enzymes are required to reduce formate to the methyl group of methyl-H$_4$folate (Reactions 7–10). These enzymes are present in high concentrations in homoacetogenic and purinolytic acetogenic bacteria (Buttlaire, 1980; MacKenzie, 1984; Ljungdahl, 1986). The purified enzymes of the acetogens have also very high specific activities when compared with corresponding enzymes from other bacteria, e.g., *Escherichia coli* and eucaryotic sources. Additional interest in the H$_4$folate-dependent enzymes depends on the fact that in eucaryotes (yeast and liver) the synthetase, cyclohydrolase, and dehydrogenase activities (Reactions 7, 8 and 9) are catalyzed by a single trifunctional enzyme called C$_1$-H$_4$folate synthase (Villar et al. 1985; Barlowe et al. 1989). *C. thermoaceticum* contains a bifunctional enzyme with cyclohydrolase and dehydrogenase activities (Ljungdahl et al.,

1980), whereas the synthetase is a separate enzyme. In *Clostridium formicoaceticum*, separate enzymes catalyze the three reactions. We now discuss properties of formyl-H$_4$folate synthetase, methylene-H$_4$folate reductase, and CODH, the genes of which from *C. thermoaceticum* recently were cloned into *E. coli* and sequenced.

Formyl-H$_4$folate synthetase

Formyl-H$_4$folate synthetase has been purified from the acetogens *C. thermoaceticum* (Ljungdahl et al., 1970) and *C. formicoaceticum* (O'Brien et al., 1976), as well as from the purine-fermenting *Clostridium acidiurici* and *Clostridium cylindrosporum* (Rabinowitz and Pricer, 1962). The enzymes from the four clostridia are very similar. They have molecular weights of about 240,000, consist of four subunits identical within each enzyme, and require magnesium and monovalent cations for activity, best supplied as K$^+$ or NH$_4$$^+$. Electron microscopic studies of the enzymes from *C. thermoaceticum* and *C. cylindrosporum* show that in the tetrameric enzymes monomers bind together forming dimers, which in turn combine forming the active tetrameric enzymes (Mayer et al., 1982). The formation of the dimers depends on the presence of the monovalent cations (Harmony et al., 1975). The *C. thermoaceticum* enzyme has a higher thermostability than the enzymes from the mesophilic bacteria (O'Brien et al., 1976). Elliott and Ljungdahl (1982) have obtained evidence with the *C. thermoaceticum* enzyme that a tyrosine residue is located at or near the H$_4$folate-binding site and that two cysteine residues of each subunit are in the region of interaction between the subunits. Mejillano et al. (1989) have proposed a catalytic mechanism for the formyl-H$_4$folate synthetase reactions, which involves formyl phosphate as an intermediate.

The gene encoding for formyl-H$_4$folate synthetase has been cloned into *E. coli* SK1592 using pBR322 containing a 9.5-kb *Eco*RI fragment of DNA from *C. thermoaceticum* (Lovell et al., 1988). The *E. coli* clone (CRL47) expressed catalytically active and thermostable formyl-H$_4$folate synthetase at a much higher level than

C. thermoaceticum. Cell-free extracts of CRL47 had specific activities ($\mu mol \cdot min^{-1} \cdot mg^{-1}$) of 28 to 89 when assayed at 50°C and pH 8, which is 3 to 10 fold higher than in the cell-free extract of *C. thermoaceticum*. Extracts of *E. coli* SK1592 containing native pBR322 had less than 0.008 $U \cdot mg^{-1}$, confirming reports that *E. coli* has very low or no formyl-H_4folate synthetase activity (Whiteley et al., 1959; Dev and Harvey, 1978). The enzyme from the clone CRL47 reacted with antibodies prepared against formyl-H_4folate synthetase from *C. thermoaceticum*. After purification it was found to have the same thermostability, specific activity, molecular weight, and subunit composition as the *C. thermoaceticum* enzyme. This indicates that the high thermostability (70°C) of formyl-H_4folate synthetase from *C. thermoaceticum* is inherent and that the four subunits required to form active enzymes assemble in a thermodynamically controlled reaction.

Similar results have been obtained for the leucine dehydrogenase of *C. thermoaceticum*, which has been cloned in *E. coli* (Shimoi et al., 1987). This enzyme is expressed in *E. coli* to 8.3% of the cell protein. The gene product is immunochemically identical to the *C. thermoaceticum* enzyme and shows the same thermostability. The ease of growing *E. coli*, the high level of expression of the enzyme, and its thermostability were used to produce large amounts of enzyme that could be easily purified by heat treatment and is more stable than the corresponding mesophilic enzyme (Shimoi et al., 1987). These results clearly demonstrate the advantages of cloning thermophilic genes in *E. coli*.

The complete nucleotide sequence of the gene of the *C. thermoaceticum* formyl-H_4folate synthetase has been obtained and from it the amino acid sequence (Lovell et al., 1990). The gene, 1680 nucleotides long, encodes a protein of 559 amino acids. The calculated molecular weight is 59,983, which corresponds very well with 60,000, the established molecular weight of the subunit. The amino acid sequence has been compared with the sequences of formyl-H_4folate synthetase of *C. acidiurici* sequenced by Whitehead and Rabinowitz (1988) and the formyl-H_4folate synthetase domains of cyto-

plasmic and mitochondrial C_1-H_4folate synthase from yeast (Staben and Rabinowitz, 1986; Shannon and Rabinowitz, 1988). The *C. acidiurici* enzyme has 556 amino acid residues of which 66.4% are identical with the residues of the *C. thermoaceticum* enzyme. Also, the synthetase domains of the yeast enzymes have high homology with the *C. thermoaceticum* enzyme by sharing 47% of the residues.

It is not possible at present to identify a H_4folate-binding site based on the amino acid sequences of the synthetases. There was no homology observed with other H_4folate-binding enzymes with the exception of the methylene-H_4folate reductases from *E. coli* (Saint-Girons et al., 1983) and *Salmonella typhimurium* (Stauffer and Stauffer, 1988). Residues 145–150 (F-D-I-T-V-A) of the reductases correspond with residues 197–202 (F-D-I-S-V-A) of the *C. thermoaceticum* formyl-H_4folate synthetase and residues 194–199 and 589–594 (F-D-I-T-V-A) of the *C. acidiurici* synthetase and yeast C-1 synthases, respectively. These residues are situated in a highly conserved region found in both the clostridial synthetases and the yeast C-1 synthases, indicating that they may have some function.

By comparing sequences from a number of proteins having known ATP-binding sequences, it is, however, possible to suggest that residues 58–85 and 113–124 of the *C. thermoaceticum* formyl-H_4folate synthetase constitute a two-domain binding site for ATP. The regions containing the domains are very much conserved (Figure 28.1). Of special interest is the sequence 71–78 (G-E-G-K-T-T-T-S) of the first domain containing residues 58–85. Similar sequences have been observed in many ATP-binding proteins, including the β subunit of nitrogenase (Robson, 1984), adenylate kinases, and *ras* proteins (Fry et al., 1986), ATP synthases, myosin, and several kinases (Walker et al., 1982; Duncan et al., 1986). These sequences, which are referred to as the P-loop motif, have been discussed by Higgins et al. (1986) and Saraste et al. (1990). The second domain is characterized by glycine (115), aspartic acid (124), and hydrophobic residues between these residues. It has been suggested that the aspartate is interacting with Mg^{2+} of MgATP

	58		85	113			124
Ct	K L I L V T A I T P T P A G E G K T T T S V G L T D A L			G G G Y	A Q V V		P M E D
Ca	K L V L V T A I N P T P A G E G K T T T N I G L S M G L			G G G Y	A Q V V		P M A D
Scc	K Y I L V S G I T P T P L G E G K S T T T M G L V Q A L			G G G Y	S Q V I		P M D E
Scm	K Y V L V A G I T P T P L G E G K S T T T M G L V Q A L			G G G Y	A Q V I		P M D E

FIGURE 28.1 *Clostridium thermoaceticum* formyl-H_4folate synthetase sequence (Ct) of amino acid residues 58–85 and 113–124 containing the putative binding site of ATP and comparison with similar sequences from formyl-H_4folate synthetase of *C. acidiurici* (Ca) (Whitehead and Rabinowitz, 1988) and synthetase domains of cytoplasmic (Scc) (Staben and Rabinowitz, 1986) and mitochondrial (Scm) (Shannon and Rabinowitz, 1988) C_1-H_4folate synthase from yeast. Bold letters are residues common to many ATP-binding proteins (see text). Abbreviations: Ct, *Clostridium thermoaceticum* FTHFS, residues 58–85; (Lovell et al., 1990). Ca, *C. acidiurici* FTHFS, residues 55–82 (Whitehead and Rabinowitz, 1988); Scc, *Saccharomyces cerevisiae* cytoplasmic C_1-THFS, residues 374–401 (Staben and Rabinowitz, 1986); Scm, *S. cerevisiae* mitochondrial C1-THFS, residues 398–425 (Shannon and Rabinowitz, 1988).

(Duncan et al., 1986; Fry et al., 1986). Although the evidence for the location of the ATP-binding site of formyl-H_4folate synthetase obtained by sequencing is encouraging, it must still be confirmed by biochemical analyses.

The binding between subunits of formyl-H_4folate synthase of *C. thermoaceticum* may involve a hydrophobic domain containing two cysteine residues that were found to be reactive only when the subunits were dissociated (Elliott and Ljungdahl, 1982). A Kyte and Doolittle (1982) hydropathy plot revealed a hydrophobic domain comprising residues 201–212 (V-A-S-E-V-M-A-C-L-C-L-A) containing cysteine residues 208 and 210. A similar sequence exists in the enzyme from *C. acidiurici*, which contains cysteine 207, corresponding to cysteine 210 of the *C. thermoaceticum* enzyme sequence. It was proposed by Lovell et al. (1990) that this domain may be involved in subunit interaction.

The knowledge of the protein sequences of the same enzyme from a thermophile and a mesophile can provide some insight into the features that confer thermostability, especially as there is high sequence similarity between the two enzymes. It has been proposed previously that the higher thermostability of the formyl-H_4folate synthetase of *C. thermoaceticum* over that of the enzyme from *C. acidiurici* was a result of a higher proportion of hydrophobic residues in the former enzyme (Ljungdahl et al., 1970). This was found not to be the case when the number of hydrophobic residues in each

protein was compared. However, the effect may just be limited to specific residues at critical points in the protein, which would not be seen in such an analysis. On the other hand, the *C. thermoaceticum* enzyme has 26 arginine per subunit, whereas the *C. acidiurici* enzyme has only 17. It is possible that the arginines form ionic interactions and salt bridges that may contribute to the higher thermostability of the *C. thermoaceticum* enzyme.

Methylene-H_4folate reductase

Methylene-H_4folate reductase has been purified from the acetogens *C. thermoaceticum* (Han, 1987), *C. formicoaceticum* (Clark and Ljungdahl, 1984), and *Peptostreptococcus productus* (Wohlfarth et al., 1990). The *C. thermoaceticum* enzyme was reported to be an octamer with a molecular weight of 347,000 composed of eight identical subunits. This is in contrast to the enzyme from *C. formicoaceticum* which is an octamer of two different subunits of 26,000 and 46,000 and has an $\alpha_4\beta_4$ structure. Both enzymes contain noncovalently bound flavin, nonheme iron in the form of iron-sulfur clusters, and zinc. The iron content and electron paramagnetic resonance (EPR) spectrum of the *C. thermoaceticum* enzyme indicate that each subunit contains a single Fe_4S_4 cluster. The clostridial enzymes differ from the NADPH-linked methylene-H_4folate reductase of pig liver, which is a flavoprotein but does not contain an iron-sulfur center (Daubner and Matthews,

1982; Jencks and Matthews, 1987), and the NADH-dependent enzyme from *P. productus*, also a flavoprotein and lacking iron-sulfur centers but being an octamer of identical subunits with a mass of 32 kDa. They differ also from the enzymes of *Escherichia coli* and *Salmonella typhimurium*, which apparently are $FADH_2$ dependent but couple with NADH via a FAD reductase (Katzen and Buchanan, 1965). The *E. coli* and *S. typhimurium* enzymes have been sequenced (Saint-Girons et al., 1983; Stauffer and Stauffer, 1988).

In the Wood acetyl-CoA pathway, the function of methylene-H_4folate reductase is to reduce methylene-H_4folate to methyl-H_4folate, which then is the precursor of the methyl group of acetyl-CoA. This reduction is thermodynamically favorable, and it has been suggested that it is coupled to electron transport phosphorylation, which will allow acetogens to grow autotrophically (Fuchs, 1986). In support of this was the finding that methylene-H_4folate reductase in *C. thermoautotrophicum* is associated with the cytoplasmic membrane (Hugenholtz et al., 1987). At present, the natural electron donor of the clostridial methylene-H_4folate reductase reaction is not known. It has been suggested that it may be a ferredoxin, cytochrome b_{560}, or a flavoprotein, all of which are found in the membrane fractions of *C. thermoaceticum* and *C. thermoautotrophicum* (Hugenholtz and Ljungdahl, 1990; Park et al., 1991).

The gene for methylene-H_4folate reductase has been cloned in *E. coli*, strain TG1, using the multifunctional phagemid pTZ19R containing a 2.3-kb DNA fragment obtained from a double digest, *Bam*HI-*Kpn*I, of chromosomal DNA of *C. thermoaceticum* (Chou, 1990; Chou and Ljungdahl, 1990). The *E. coli* clone formed two gene products that were observed in SDS-PAGE gels using antimethylene-H_4folate reductase antibodies; the polypeptides had the masses 36.1 kDa and 39.1 kDa. Only the 36.1-kDa protein was detected in *C. thermoaceticum*, and none was found in the *E. coli*/pTZ19R lacking the *C. thermoaceticum* DNA fragment. Sequencing of the cloned DNA fragment revealed an open-reading frame of 990 nucleotides encoding a protein of 329 amino acids, corre-

sponding to a mass of 36,143 daltons. The first 31 amino acids of the deduced sequence corresponded to the sequence determined for the purified protein.

The second polypeptide expressed in *E. coli* clone with a mass of 39.1-kDa is probably a result of translational initiation 91 nucleotides upstream of the start of the protein identified by N-terminal sequencing of the methylene-H_4folate reductase subunit. This extra peptide fragment contains 27 amino acid residues of the following sequence: M-G-A-N-I-I-H-L-I-Q-Y-T-G-G-E-F-L-A-G-F-F-R-V-R-S-K-V. This peptide fragment, having the mass of 3,026, is very hydrophobic. Thus, we suggested that the 39.1-kDa polypeptide constitutes an alternate form of the reductase subunit containing a membrane anchor. However, there is no evidence for the presence of this form of the enzyme in *C. thermoaceticum*, and it may just be the result of fortuitous translation initiation in *E. coli*.

Han (1987) found by gel filtration the mass of the active methylene-H_4folate reductase from *C. thermoaceticum* to be about 347-kDa and postulated it to be an octamer of identical subunits of about 42-kDa. The subunit expressed in *E. coli* and in *C. thermoaceticum* has a mass of 36,143 daltons calculated from the amino acid sequence. An octamer of this subunit would have a mass of 283,144 daltons. If the mass of the native enzyme determined by gel filtration is correct, it may in fact contain 10 subunits.

Methylene-H_4folate reductase from *C. thermoaceticum* contains, in moles per mole of native enzyme (347-kDa), about 34 Fe, 12 Zn, 42 sulfide, and 2.3 FAD (Park et al., 1991). It was assumed that there are 4 Fe per subunit, and EPR spectra indicate the presence of Fe_4S_4 clusters, although not of the type normally found in clostridial ferredoxins. Conventional ferredoxin type Fe_4S_4 clusters require four cysteines for ligation to the protein. The sequence of the methylene-H_4folate reductase from *C. thermoaceticum* has only two cysteines per subunit. If a single cluster were found between two subunits, it would only give half the number of iron atoms seen by iron analysis. Instead, it is more likely that two residues other than cysteines are used to bind the cluster. The two-

Protein

MTHFR	18	K-I-A-V-M-G-Y-G-S-Q-G-H-S-Q-A-Q-N-L-K-D-S-G-L-D-V-V-V-G-L-R-P-E-S-K	51
GR	22	D-Y-L-V-I-G-G-G-S-G-G-L-A-S-A-R-R-A-A-E-L-G-A- - -R-A-A-V- -V-E-S-H	52
PHBH	4	Q-V-A-I-I-G-A-G-P-S-G-L-L-L-G-Q-L-L-H-L-A-G-I- - -D-N-V-I- -L-E-R-Q	34

FIGURE 28.2 Putative binding site for flavin adenine dinucleoide (FAD) of methylene-H$_4$folate reductase (MTHFR) from *Clostridium thermoaceticum*; comparison with sequence from FAD-depending human erythrocyte glutathion reductase (GR) (Krauth-Sieghel et al., 1982) and *p*-hydroxybenzoate hydroxylase (PHBH) from *Pseudomonas fluorescens* (Weijer et al., 1982). Analysis, including secondary structure, is based on discussion by Wierenga et al. (1985). Symbols: ∞, invariant hydrophilic residue; ♦, neutral or hydrophobic residues; *, invariant glycine; ⊖, invariant negative residue, glutamic acid in FAD-binding proteins.

iron cluster in Rieske-type proteins apparently is ligated through cysteine and histidine residues (Beckman et al., 1989). There are seven histidines in the methylene-H$_4$folate reductase, and it is conceivable that two of these are involved in cluster ligation.

In the search for a possible binding site for methylene-H$_4$folate, the sequence of methylene-H$_4$folate reductase was compared with sequences of other H$_4$folate-dependent enzymes, including methylene-H$_4$folate reductases from *E. coli* (Saint-Girons et al., 1983), and *S. typhimurium* (Stauffer and Stauffer, 1988), as well as formyl-H$_4$folate synthetase from *C. thermoaceticum* (Lovell et al., 1990). The search was negative and further demonstrated that there was no homology between the clostridial methylene-H$_4$folate reductase and the enzyme from *E. coli* and *S. typhimurium*. A 32% homology was found with acetohydroxy acid isomeroreductases from *E. coli* (Wek and Hatfield, 1986) and yeast (Petersen and Holmberg, 1986). The significance, if any, of this homology is not yet understood. The hexapeptide mentioned, suggested as a possible folate-binding site on the basis of the comparison of the methylene-H$_4$folate reductases from these two organisms and found in the formyl-H$_4$folate synthetase, was not found in the *C. thermoaceticum* reductase sequence. Varalakshmi et al. (1986), studying methylene-H$_4$folate reductase from sheep liver, found that the enzyme was inactivated by phenylglyoxal and *N*-bromosuccinimide, reagents for arginine and

tryptophan residues, respectively. Methyl-H$_4$folate protected against the inactivation. The *C. thermoaceticum* enzyme has only two tryptophans, residues 300 and 326, located at the C-terminal end. This area has also 5 arginine residues. Clearly, in view of the results by Varalakshmi et al. (1986) the tryptophans could be part of a tetrahydrofolate binding site.

A putative binding site for FAD may involve residues 18 to 51, as shown in Figure 28.2. This is based on a fingerprint $\beta\alpha\beta$ nucleotide-binding unit for NAD, NADP, or FAD discussed by Wierenga and Hol (1983) and Wierenga et al. (1985). Although this sequence of methylene-H$_4$folate reductase has a strong resemblance to the sequences of other FAD-dependent proteins, biochemical proof is needed. It should also be pointed out that the native *C. thermoaceticum* reductase consists of perhaps 8 or more subunits containing 2, perhaps 3, FAD. A binding site for FAD may, therefore, involve more than 1 polypeptide.

Carbon monoxide dehydrogenase/acetyl-CoA synthase (CODH)

Of the enzymes of the Wood acetyl-CoA pathway, CODH is the best studied; the obvious reasons include its key functions as carbon monoxide dehydrogenase (Reaction 6) and acetyl-CoA synthase (Reaction 12), dependence on nickel and iron for activity, and involvement in the generation of energy. An extensive discussion of CODH is found in the review by Wood and Ljungdahl (1991).

Diekert and Thauer (1978) were first to discover carbon monoxide dehydrogenase activity in *C. thermoaceticum*. They also showed that the presence of nickel in the growth medium stimulated the formation of this activity (Diekert and Thauer, 1980). Hu et al. (1982) made the important discovery that an enzyme preparation from *C. thermoaceticum* containing CO dehydrogenase catalyzes the synthesis of acetyl-CoA from CO, methyl-H$_4$folate, and CoA. Ragsdale and Wood (1985) finally established that CODH catalyzes synthesis of acetyl-CoA. With a purified enzyme, they demonstrated that it catalyzes an exchange reaction between CO and the carbonyl group of acetyl-CoA according to the following reaction (Reaction 13).

$$\begin{array}{ccccc}
CH_3 & x & x\text{-}CH_3 & x\text{-}CH_3 & {}^{14}CO \\
| & & & & \\
{}^{14}CO \;+\; y & \rightleftharpoons & y - {}^{14}CO \rightleftharpoons & y \;+\; & \Updownarrow \quad [13] \\
| & & & & \\
SCoA & z & z - SCoA & z\text{-}ScoA & CO
\end{array}$$

This reaction involves a cleavage of the bonds between the methyl, carbonyl, and CoA moieties and demonstrates the three binding sites for these moieties, designated x, y, and z. Subsequently, exchange reactions were observed between free CoA and acetyl-CoA (Pezacka and Wood, 1986) and between the methyl groups of CH$_3$-C/Fe-SP (see Reaction 12) and methylated CODH, as well as between the latter and the methyl moiety of acetyl-CoA (Lu et al., 1990). An additional important observation is that CODH is activated by a special enzyme, CODH disulfite reductase, which can be partly replaced by dithioerythritol (Pezacka and Wood, 1986). The reduction of CODH to obtain full activity has been extensively studied by Lu et al. (1990).

CODH has now been isolated from many bacterial sources (see review by Wood and Ljungdahl, 1991). The enzyme from *C. thermoaceticum* was purified by Ragsdale et al. (1983a) and by Diekert and Ritter (1983). Ragsdale et al. (1983a) found the enzyme to be a hexamerous with the composition $\alpha_3\beta_3$ and an M_r of

440,000 whereas Diekert and Ritter (1983) proposed a tetrameric structure. It also exists as a dimer, which contains 2 Ni, 1 Zn, 11 Fe, and 14 acid-labile S per mole. Ramer et al. (1989) reported that the *C. thermoaceticum* CODH is a 3-subunit enzyme with the composition $\alpha_2\beta_2\gamma_2$, which is in opposition to all previous reports of the composition of the enzyme. As is discussed next, genetic work is consistent with a CODH containing only 2 subunits of molecular weights of 78,000 and 71,000, as described by Ragsdale et al. (1983a) and Diekert and Ritter (1983). It should be noted, however, that CODH tends to copurify with other proteins, notably CODH disulfide reductase and C/Fe-SP, that may influence the activity of the CODH. This is consistent with the observation that CODH is membrane associated and may form complexes with other membrane-associated proteins.

Considerable work has been directed toward finding the roles of the metals and the binding sites for the methyl, carbonyl, and CoA groups. Consider the exchange shown in Reaction 13. The y site binds CO; EPR studies initiated by Ragsdale et al. (1983b, 1985) indicate that the y site consists of Ni and Fe, which interact to bind the CO group, possibly via a carbonyl bridge. EXAFS studies suggest that the Ni has four sulfur ligands, is not situated in a mixed Fe-Ni cubane-type cluster, and has the ability to bind axially CO, methyl, or acetyl-CoA ligands (Cramer et al., 1987; Bastian et al., 1988). Results of additional spectroscopic investigations have suggested that four additional iron-sulfur centers besides the Ni-CO-Fe cluster are present in CODH (Lindahl et al., 1990).

The methyl moiety of acetyl-CoA (Reaction 13) occupies the x site of CODH. At present some controversy exists about this site. Pezacka and Wood (1988) labeled CODH using [^{14}C]CH$_3$I, showed that the methyl was bound to a cysteine residue of the β subunit and was converted to [^{14}C]acetyl-CoA. Lu et al. (1990) proposed that the x site is a metal, presumably a Ni of CODH, and that acetyl-Ni-CODH may form as an enzyme-bound intermediate. They based this proposal on the exchange between methyl-corrinoid enzyme and methylated CODH described earlier, that methylation of

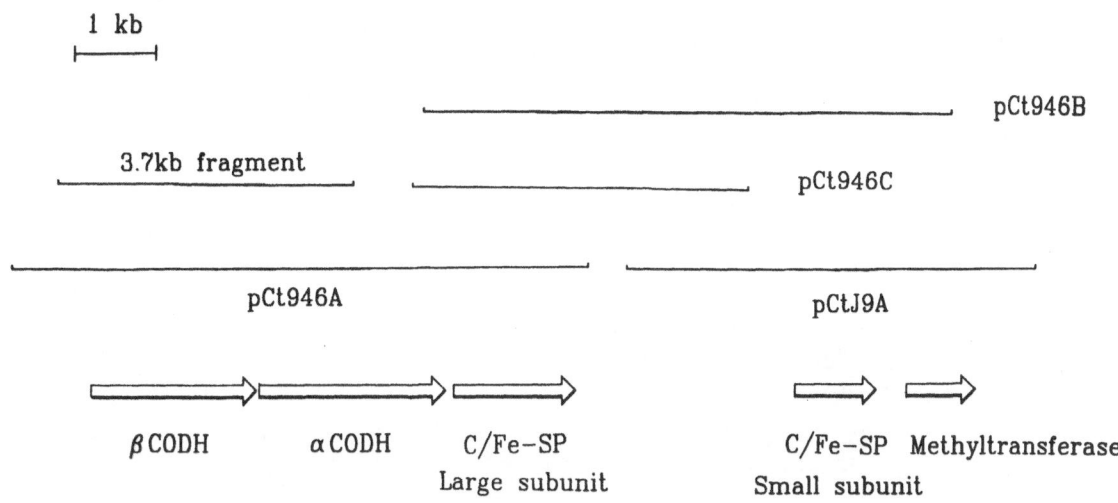

FIGURE 28.3   Arrangement of genes for CODH, C/ Fe-SP, and methyltransferase in *Clostridium thermoaceticum*. Lines indicate position of clones obtained by Roberts et al. (1989) (pCt946A, B, C and pCtJ9A) and Morton et al. (1991) (3.7-kb fragment). [Adapted from Roberts et al. (1989) and S. W. Ragsdale (personal communication).]

CODH requires a low redox potential and that carbonyl insertion occurs easily on metal centers. They postulated that the methyl cysteine may form in a side reaction.

CoA binds to the y site. This site is close to the nickel site as evidenced by an effect by CoA on the Ni-CO-Fe EPR signal (Ragsdale et al., 1985). In addition, Shanmugasundaram et al. (1988, 1989) have presented evidence that coenzyme A protects some tryptophan and arginine residues from reacting with N-bromosuccinimide and phenylglyoxal, respectively. They suggested that these residues are in the CoA-binding site.

The two genes coding for the α (mass 78,000) and β (mass 71,000) subunits of CODH are found in a cluster with the genes for the α subunits of the C/Fe-SP and the methyltransferase (Figure 28.3). The CODH genes have been cloned in *E. coli* K-12 strain JM109 using pUC9 containing a 10-kb DNA insert from *C. thermoaceticum* (pCt946A) (Roberts et al., 1989). A second fragment (3.7-kb) of *C. thermoaceticum* DNA containing the β subunit and part of the α subunit of CODH has also been inserted into pBR322, which was used to transform *E. coli* strain HB101 (Morton et al., 1991). The genes for both subunits were sequenced. The α gene codes for a protein with 729 amino acids

and a molecular weight of 81,730 and the β gene for a protein with 674 amino acids with a molecular weight of 72,928. The gene for the α subunit follows the gene for the β subunit by 23 bases and is certainly cotranscribed. There was no evidence for an open-reading frame adjacent to the genes for the α and β subunits, indicating a third subunit of CODH. Upstream, expression of the β subunit from the 3.7-kb fragment (Morton et al., 1991) indicates that there is no additional subunit ahead of the β gene that is cotranscribed, and downstream the α gene is followed by the genes for the C/Fe-SP and methyltransferase (Roberts et al., 1989).

As discussed, CODH contains perhaps as many as five different iron-sulfur clusters of which one interacts with nickel. Some of these are known to be low-potential, ferredoxin-like Fe_4S_4 clusters, but there are no C-X-X-C-X-X-C. . . .CP sequences in the CODH typical of the cysteine ligation of such clusters (Elliott et al., 1982). The cysteine-proline combination has been proposed to be essential for a low potential cluster; it is not present. There are 31 cysteine residues in the 2 subunits of CODH and, as shown in Figure 28.4, the α and β subunits each contain two cysteine-rich domains ranging from residues 506 to 528 and 583 to 608 of the α subunit and 59 to 90 and 316 to 366 of the

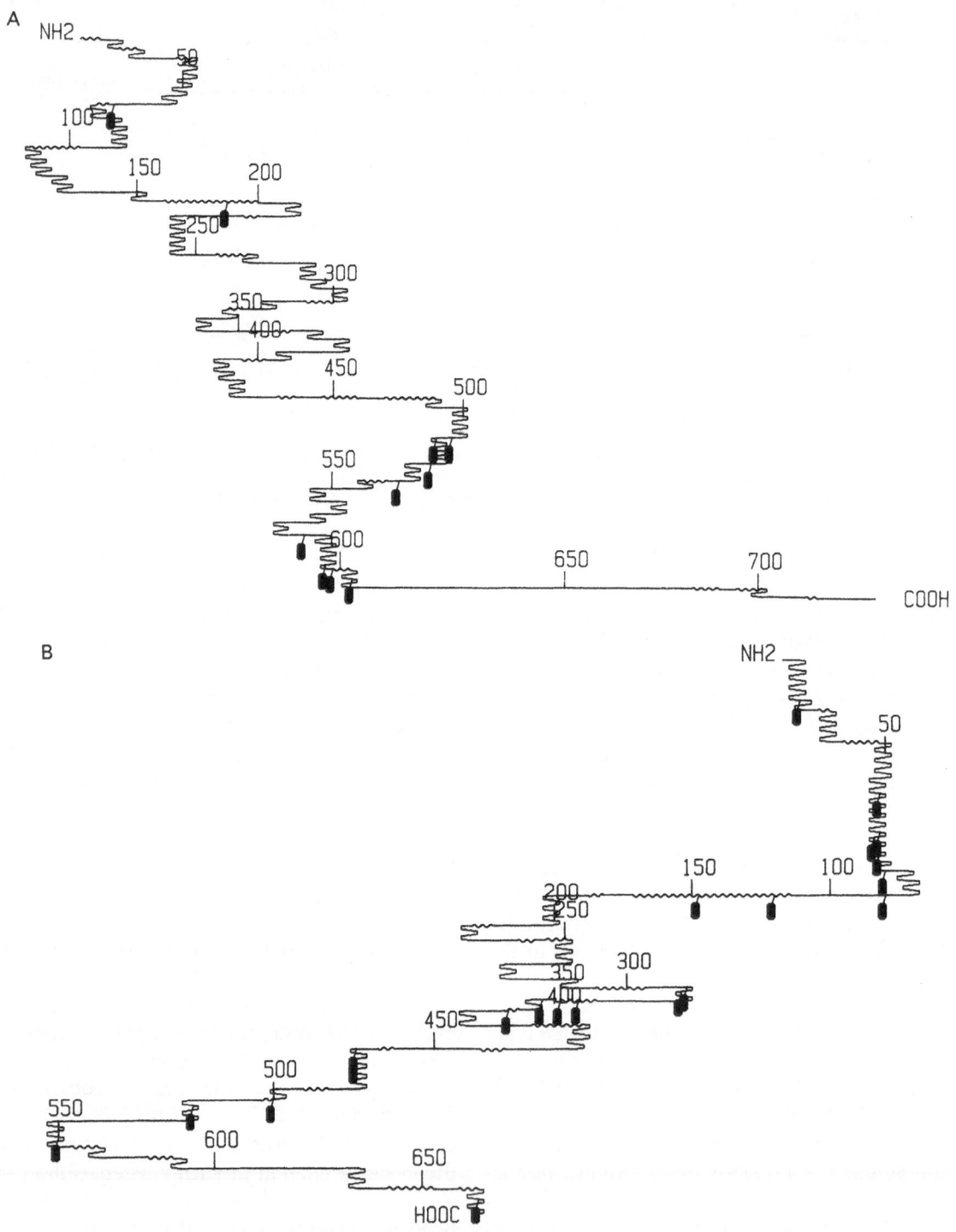

FIGURE 28.4 Predicted secondary structure of subunits of CODH according to method of Garnier et al. (1978). Sine waves represent α helices; turns are shown by 180° turns; closed circles point to position of cysteine residues. (A) α subunit; (B) β subunit.

β subunit. These domains may be sites for binding metal clusters. It is noteworthy that each of the domains of the β subunit contains a pair of vicinal cysteines (residues 67, 68, and 316, 317). Poston et al. (1966) noticed that reagents reacting with vicinal sulfhydryls inhibit synthesis of acetate from methyl-B_{12} and pyruvate as catalyzed by cell-free extracts of *Clostridium thermoaceticum*. This suggests the possible involvement of these vicinal sulfhydryls in the catalytic process, and they may be the target for CODH disulfide reductase that activates CODH (Pezacka and Wood, 1986).

Morton et al. (1991) have labeled CODH by reacting tryptophan residues with 2,4-dinitrophenylsulfenyl chloride (DNPS-Cl) and have isolated the labeled peptides after tryptic digestion. The reaction of tryptophan with DNPS-Cl in one of the peptides was prevented by coenzyme A. This peptide of the α subunit (residues 407–423) contains the tryptophan 418, and it was suggested that this may reside in the coenzyme A-binding domain of CODH. This sequence, however, has no similarity to the consensus coenzyme A-binding site proposed by Buck et al. (1985) for citrate synthase.

28.4 Transcription and Translation Control Elements in *Clostridium thermoaceticum*

All the C. *thermoaceticum* proteins discussed are expressed in *E. coli* to varying degrees. The methyltransferase and C/Fe-SP are expressed at about 1% of the soluble *E. coli* proteins, CODH at 5% (Roberts et al. 1989), and formyl-H_4folate synthetase to 30% (Lovell et al., 1988). The methylene-H_4folate reductase is also expressed, but the level was not measured (Chou and Ljungdahl, 1990). The level of expression of the genes inserted into the Lac operon was unchanged in the presence of the Lac inducer IPTG (Roberts et al., 1989). For the formyl-H_4folate synthetase (Lovell et al., 1988), which was cloned in pBR322, and the methylene-H_4folate reductase (Chou and Ljungdahl, 1990), the expression was independent of the direction of insertion in the plasmid. All this is

evidence that *E. coli* is using transcription and translation control elements present on the C. *thermoaceticum* DNA inserts. Potential candidates for these sites have been identified by homology to the *E. coli* consensus sequences.

Comparison of the sequence 50 bases from the start of transcription of 29 genes from gram-positive organisms has identified an extended promoter sequence that is similar to the -35 and -10 sequences of *E. coli* vegetative promoters (Graves and Rabinowitz, 1986). The sequences in front of each C. *thermoaceticum* gene were therefore searched for the sequences with the greatest similarity to *E. coli* promoters using a program similar to that described by Mulligen et al. (1984). Figure 28.5 shows the sequences found and their homology scores based on comparison to the consensus. The level of homology has been found to be correlated with promoter strength in vitro (Mulligen et al., 1984). Consistent with this is observation that the sequence with the highest homology, formyl-H_4folate synthetase, is expressed at up to 32% of the cell protein in *E. coli* (Lovell et al. 1988).

A homology level of 72% is close to the highest level seen for promoters found in *E. coli* (Mulligen et al., 1984). In addition, there were several other possible candidates found in the same region with less, but still significant, homology. This may result in enhancement of transcription (McClure, 1985) and account for the high level of expression in *E. coli*. In C. *thermoaceticum*, however, formyl-H_4folate represents only 2% of the cell protein (Lovell et al., 1988). There are several possible explanations for this discrepancy: the gene may be under strong repression in C. *thermoaceticum*, the plasmid containing the gene may be in high copy number in *E. coli*, or C. *thermoaceticum* may recognize a slightly different promoter sequence, with the similarity to *E. coli* promoters being coincidental.

The promoter in front of the genes for CODH with the highest homology score is very close to the translation start site. The transcription start site would be near the middle of the proposed ribosome-binding site. In this case there is another candidate further upstream with less homology, but still higher than the

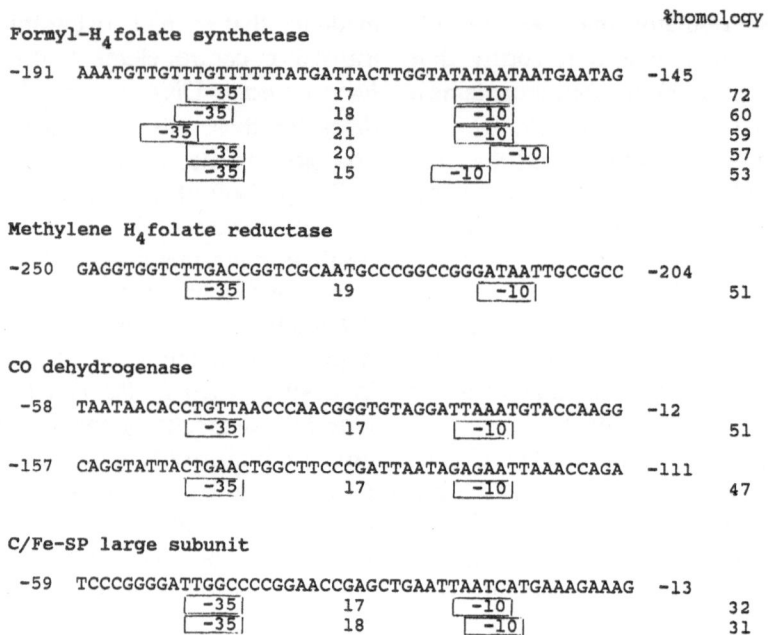

FIGURE 28.5 Potential promoter regions in front of several genes from *Clostridium thermoaceticum*. Sequences were searched for regions with highest homology to *E. coli* promoter sequence. Homology was calculated using method of Mulligen et al. (1984), shown to right. Boxes labeled −35 and −10 indicate position of sequences that correspond to principal features of *E. coli* promoters (TTGACA and TATAAT, respectively); number between them is gap size. Numbers at ends of each sequence are distances from translation start site. For CO dehydrogenase, both sequences shown precede β subunit gene; distances are relative to start of β subunit.

45% considered necessary for a functional promoter (Mulligen et al., 1984). In the case of the C/Fe-SP large subunit, however, there is no alternate because the start of this gene follows the stop codon of the α subunit of CODH by only 85 bases (Morton et al., 1991). In addition to its close proximity to the start of translation, this region also has a poor homology score. This raises the possibility that the gene for the C/Fe-SP large subunit is cotranscribed with those for CODH. This possibility seems to be excluded by the work of Roberts et al. (1989), who have a clone containing only the large subunit of the C/Fe-SP (pCt946C in Figure 28.3) and have found that it was expressed in *E. coli*.

DNA sequences following each of the genes were searched for transcription terminators. Rho factor-independent terminators are generally characterized by inverted repeats, which form a "stem-loop" structure, followed by an oligo(T) tract. Such sequences have been found

following the genes for formyl-H$_4$folate synthetase (Lovell et al., 1990) and methylene H$_4$folate reductase (Chou and Ljungdahl, 1990). The sequence between the end of the CODH genes and the start of the C/Fe-SP large subunit gene only contains a small inverted repeat, again suggesting that the genes are cotranscribed. Resolution of this question, as well as definitive identification of transcription control sites, will require RNA analysis.

Figure 28.6 shows the sequences of several *C. thermoaceticum* genes in the region around the translation start site. As in *E. coli*, most genes use ATG as the start codon. The exceptions are methylene-H$_4$folate reductase and formyl-H$_4$folate synthetase, which use GTG and TTG, respectively. This is surprising because, in terms of expression in *E. coli*, ATG > GTG > TTG (Stormo, 1986) and yet formyl-H$_4$folate synthetase is the most highly expressed of these proteins. The N-terminal amino acid of methylene-H$_4$folate, as deter-

```
C/Fe-SP large subunit      AAAGAA  AGGAGT   CAACAAAT  ATG   CCTTTGACGGGA...TAG
C/Fe-SP small subunit      GTTGAA  AGGAGT   GTTTTAAA  ATG   GCCGTCCAGATT...TAA
Methyltransferase          TTTCAA  AGGAGG   CAATCCC-  ATG   CTCATTATCGGT...TAG
Formyl-H₄folate synth.     AATGAG  AGGAGT   GGATTACA  TTG   TCCAAGGTACCC...TAG
CODH α subunit             GTACCA  AGGAGG   TAAGCAGT  ATG   CCCAGGTTCCGC...TGA
CODH β subunit             AGGATG  AGGAGG   GTCAGTTA  ATG   ACTGATTTTGAT...TAG
Methylene-H₄folate reduct. AGGGTC  AGGAGC   AAA-----  GTG   GCAGTTGTATAC...TAG

              Shine-Dalgarno          Start                Stop
                                              Translation
```

FIGURE 28.6 Sequence around translation start site of several genes from *Clostridium thermoaceti-* *cum* showing probable ribosome-binding site.

mined by protein sequencing, is alanine, indicating that the formyl-methionine has been removed following translation initiation.

There is a conserved sequence (AGGAGt/g) in front of each initiation codon, which is very similar to the *E. coli* ribosome-binding site (AGGAGG) and undoubtedly has the same function. In *E. coli* it is found 7 to 9 bases from the start codon, although most frequently at 8. This is true also for *C. thermoaceticum*, with the exception of the methylene-H₄folate reductase gene, where there are only 3 bases between the ribosome binding site and the start codon. A statistical analysis of a collection of *E. coli* translation start sites has shown there is a biased distribution of bases in the range -20 to $+13$ (relative to the first base of the initiation codon) (Stormo, 1986). The *C. thermoaceticum* genes have some features in common in this region. Between -20 and -1 the C content is half that expected on the basis of GC content. This is also seen in *E. coli* and is attributed to the need to avoid strong secondary structure formation with the ribosome-binding site. In *E. coli*, there is also a reduced guanine content, presumably to avoid confusion with the true ribosome-binding site (Hager and Rabinowitz, 1985); this is not seen in *C. thermoaceticum*.

Beyond the start codon, *E. coli* shows a preference for AAA or GCU as the second codon, A at the $+6$ position, and UUAA from $+10$ to $+13$ (Hager and Rabinowitz, 1985). In *C. thermoaceticum*, however, the second and third bases of the second codon are almost exclusively C. The only other consensus seems to be a preference for U at position $+10$.

The codon usage by *C. thermoaceticum* for the formyl-H₄folate synthetase, methylene-H₄folate reductase, and CODH is given in Table 28.1; there is a preference for G- and C-containing codons. The GC content of the *C. thermoaceticum* coding regions that have been sequenced is similar to that for the genome as a whole (55.5% and 53.6%, respectively). This value is also similar to the GC content of *E. coli* (50 to 52%), which could help explain the high level of expression of these genes in *E. coli*. Unfortunately, although they are all expressed, only the methyltransferase and formyl-H₄folate synthetase are active. The activity of the clone seems to be correlated with the absence of metal clusters in the enzyme; neither methyltransferase nor formyl-H₄folate synthetase contain metal clusters. This is unfortunate because it precludes, for the present, identifying the location and properties of the active sites of the metalloenzymes through molecular genetic techniques. Such studies would require either the isolation of the factors responsible for metal insertion or the development of a genetics system in *C. thermoaceticum*.

28.5 Plasmid

As a step toward the development of a genetics system, a plasmid was isolated from *C. thermoaceticum* and partially characterized (Chou et al., 1984; Chou, 1990). The plasmid contains about 88-kb, and a restriction map has been constructed. The plasmid was cured using novobiocin, but no effect was found in either substrate utilization, metal or drug resistance, growth, or two-dimensional protein electrophoresis pattern. As a consequence, the function of the plasmid remains unknown. Attempts to transform *C. thermoaceticum* using pBR322 containing fragments of the 88-kb plas-

Table 28.1 Codon usage in mRNAs encoding carbon monoxide dehydrogenase (**a**), formyl-THF synthetase (**b**), and methylene-THF reductase (**c**) of *Clostridium thermoaceticum*

Amino acid	Codon	Usage a	b	c	Amino acid	Codon	Usage a	b	c
Phe	UUU	20	7	3	Tyr	UAU	24	10	8
	UUC	35	9	7		UAC	18	7	6
Leu	UUA	5	0	2	Ter	UAA	0	0	0
	UUG	11	3	5		UAG	1	1	1
	CUU	9	6	2		UGA	1	0	0
	CUC	23	13	6	His	CAU	16	1	3
	CUA	2	2	1		CAC	12	8	4
	CUG	55	33	11	Gln	CAA	5	3	1
Ile	AUU	38	10	8		CAG	33	7	15
	AUC	61	25	13	Asn	AAU	15	7	1
	AUA	8	2	1		AAC	35	16	5
Met	AUG	48	14	8	Lys	AAA	39	21	8
Val	GUU	16	5	7		AAG	38	15	14
	GUC	42	22	9	Asp	GAU	43	14	9
	GIA	31	7	8		GAC	35	19	9
	GUG	15	9	2	Glu	GAA	68	15	15
Ser	UCU	4	0	1		GAG	38	17	13
	UCC	18	9	5	Cys	UGU	10	0	0
	UCA	1	1	1		UGC	21	7	2
	UCG	4	3	0	Trp	UGG	15	2	2
	AGU	8	3	1	Arg	CGU	20	5	2
	AGC	18	5	5		CGC	22	8	5
Pro	CCU	14	2	2		CGA	3	0	0
	CCC	32	13	4		CGG	19	9	4
	CCA	4	1	0		AGA	1	0	0
	CCG	24	9	5		AGG	5	4	4
Thr	ACU	9	4	2	Gly	GGU	53	17	9
	ACC	52	24	8		GGC	45	26	14
	ACA	4	2	1		GGA	7	3	3
	ACG	8	4	4		GGG	15	8	6
Ala	GCU	22	7	4					
	GCC	95	50	25					
	GCA	3	3	5					
	GCG	9	3	1					

mid were unsuccessful, mostly because ampicillin, which was used for selection of transformants, is unstable at the higher temperatures required for growth of *C. thermoaceticum*.

Acknowledgments. We thank Stephen W. Ragsdale, Jennifer A. Runquist, and Charles R. Lovell for supplying information not yet published. The work was supported by grant DK27323 from the National Institute of Diabetes and Digestive and Kidney Diseases. Support for a Georgia Power Distinguished Professorship in Biotechnology (L.G.L.) is also gratefully acknowledged.

References

Bache, R. and N. Pfennig. 1981. Selective isolation of *Acetobacterium woodii* on methoxylated aromatic acids and determination of growth yields. *Arch. Microbiol.* **130**:255–261.

Barker, H.A. and M.D. Kamen. 1945. Carbon

dioxide utilization in the synthesis of acetic acid by *Clostridium thermoaceticum*. *Proc. Natl. Acad. Sci. USA* **31**:219–225.

Barlowe, C.K., M.E. Williams, J.C. Rabinowitz, and D.R. Appling. 1989. Site-directed mutagenesis of yeast C_1-tetrahydrofolate synthase: analysis of an overlapping active site in a multifunctional enzyme. *Biochemistry* **28**:2099–2106.

Bastian, N.R., G., Diekert, E.C. Niederhoffer, B.-K. Teo, C.T. Walsh, and W.H. Orme-Johnson. 1988. Nickel and iron EXAFS of carbon monoxide dehydrogenase from *Clostridium thermoaceticum* strain DSM. *J. Am. Chem. Soc.* **110**:5581–5582.

Beckmann, J.D., P.O. Ljungdahl, and B.L. Trumpower. 1989. Mutational analysis of mitochondrial Rieske iron-sulfur protein of *Saccharomyces cerevesiae*. *J. Biol. Chem.* **264**:3713–3722.

Buck, D., M.E. Spencer, and J.R. Guest. 1985. Primary structure of the succinyl-CoA synthetase of *Escherichia coli*. *Biochemistry* **24**:6245–6252.

Buttlaire, D.H. 1980. Purification and properties of formyltetrahydrofolate synthetase. *Methods Enzymol.* **66**:585–599.

Chou, C.-F. 1990. Studies of a large plasmid and the gene encoding 5,10-methyl enetetrahydrofolate reductase from *Clostridium thermoaceticum*. Dissertation, Department of Biochemistry, University of Georgia, Athens, Georgia.

Chou, C.-F. and L.G. Ljungdahl. 1990. Cloning, expression and sequencing of the gene encoding the methylenetetrahydrofolate reductase from *Clostridium thermoaceticum*. In: *Abstracts, 90th Annual Meeting of the American Society for Microbiology*, Abstr. K-99, 236. Washington, D.C.: American Society for Microbiology.

Chou, C.-F., L.H. Carreira, and L.G. Ljungdahl. 1984. Isolation of a large plasmid from *Clostridium thermoaceticum*. In: *Abstracts*, Plasmid in Bacteria Conference, University of Illinois, Urbana.

Clark, J.E. and L.G. Ljungdahl. 1984. Purification and properties of 5,10-methylenetetrahydrofolate reductase, an iron-sulfur flavoprotein from *Clostridium formicoaceticum*. *J. Biol. Chem.* **259**:10845–10849.

Cramer, S.P., M.K. Eidsness, W.-H. Pan, T.A. Morton, S.W. Ragsdale, D.V. DerVartanian, L.G. Ljungdahl, and R.A. Scott. 1987. X-ray absorption spectroscopic evidence for a unique nickel site in *Clostridium thermoaceticum* carbon monoxide dehydrogenase. *Inorg. Chem.* **26**:2477–2479.

Daubner, S.C. and R.G. Matthews. 1982. Purification and properties of methylenetetrahydrofolate reductase from pig liver. *J. Biol. Chem.* **257**:140–145.

Dev, I.K. and R.J. Harvey. 1978. A complex of N^5,N^{10}-methylene-tetrahydrofolate dehydrogenase and N^5,N^{10}-methenyltetrahydrofolate cyclohydrolase in *Escherichia coli*. *J. Biol. Chem.* **253**:4245–4253.

Diekert, G. and M. Ritter. 1983. Purification of the nickel protein carbon monoxide dehydrogenase of *Clostridium thermoaceticum*. *FEBS Lett.* **15**:41–44.

Diekert, G.B. and R.K. Thauer. 1978. Carbon monoxide oxidation by *Clostridium thermoaceticum* and *Clostridium formicoaceticum*. *J. Bacteriol.* **136**:597–606.

Diekert, G.B. and R.K. Thauer. 1980. The effect of nickel on carbon monoxide dehydrogenase formation in *Clostridium thermoaceticum* and *Clostridium formicoaceticum*. *FEMS Microbiol. Lett.* **7**:187–189.

Drake, H.L. 1982. Demonstration of hydrogenase in extracts of the homoacetate-fermenting bacterium *Clostridium thermoaceticum*. *J. Bacteriol.* **150**:702–709.

Drake, H.L., S.-I. Hu, and H.G. Wood. 1981. Purification of five components from *Clostridium thermoaceticum* which catalyze synthesis of acetate from pyruvate and methyltetrahydrofolate. Properties of phosphotransacetylase. *J. Biol. Chem.* **256**:11137–11144.

Duncan, T.M., D. Parsonage, and A.E. Senior. 1986. Structure of the nucleotide-binding domain in the β-subunit of *Escherichia coli* F_1-ATPase. *FEBS Lett.* **208**:1–6.

Elliott, J. and L.G. Ljungdahl. 1982. Chemical modification of cysteine and tyrosine residues in formyltetrahydrofolate synthase from *Clostridium thermoaceticum*. *Arch. Biochem. Biophys.* **215**:245–252.

Elliott, J.I., S.-S. Yang, L.G. Ljungdahl, J. Travis, and C.F. Rielly. 1982. Complete amino acid sequence of the 4Fe-4S thermostable ferredoxin from *Clostridium thermoaceticum*. *Biochemistry* **21**:3294–3298.

Fontaine, F.E., W.H. Peterson, E. McCoy, M.J. Johnson, and G.J. Ritter. 1942. A new type of glucose fermentation by *Clostridium thermoaceticum* n. sp. *J. Bacteriol.* **43**:701–715.

Fry, D.C., S.A. Kuby, and A.S. Mildvan. 1986. ATP-binding site of adenylate kinase: mechanistic implications of its homology with ras-encoded p21, F_1-ATPase, and other nucleotide-binding proteins. *Proc. Natl. Acad. Sci. USA* **83**:907–911.

Fuchs, G. 1986. CO_2 fixation in acetogenic bacteria: variations on a theme. *FEMS Microbiol. Rev.* **39**:181–213.

Garnier, J., D.J. Osguthorpe, and B. Robson. 1978. Analysis of the accuracy and implications of simple methods for predicting the secondary structure of globular proteins. *J. Mol. Biol.* **120**:97–120.

Graves, M.C. and J.C. Rabinowitz. 1986. In vivo

and in vitro transcription of the *Clostridium pasteurianum* ferredoxin gene. *J. Biol. Chem.* **261**:11409–11415.

Hager, P.W. and J.C. Rabinowitz. 1985. Translational specificity in *Bacillus subtilis*. In: *The Molecular Biology of the Bacilli*, Vol. II, D.A. Dubnau, ed., pp. 1–32. New York: Academic Press.

Han, E.Y. 1987. Purification and characterization of methylenetetrahydrofolate reductase from *Clostridium thermoaceticum*. Thesis, Department of Biochemistry, University of Georgia, Athens, Georgia.

Harder, S.R., W.-P. Lu, B.A. Feinberg, and S.W. Ragsdale. 1989. Spectroelectrochemical studies of the corrinoid/iron-sulfur protein involved in acetyl coenzyme A synthesis by *Clostridium thermoaceticum*. *Biociemistry* **28**:9080–9087.

Harmony, J.A.K., R.H. Himes, and R.L. Schowen. 1975. The monovalent cation-induced association of formyltetrahydrofolate synthetase subunits: a solvent isotope effect. *Biochemistry* **14**:5379–5386.

Higgins, C.F., I.D. Hiles, G.P.C. Salmond, D.R. Gill, J.A. Downie, I.J. Evans, I.B. Holland, L. Gray, S.D. Buckel, A.W. Bell, and M.A. Hermodson. 1986. A family of related ATP-binding subunits coupled to many distinct biological processes in bacteria. *Nature* (London) **323**:448–450.

Hsu, T., S.L. Daniel, M.F. Lux, and H.L. Drake. 1990. Biotransformation of carboxylated aromatic compounds by the acetogen *Clostridium thermoaceticum*. Generation of growth-supportive CO_2 equivalents under CO_2-limited conditions. *J. Bacteriol.* **172**:212–217.

Hu, S.-I., H.L. Drake, and H.G. Wood. 1982. Synthesis of acetyl-coenzyme A from carbon monoxide, methyl-H_4folate and coenzyme A by enzymes from *Clostridium thermoaceticum*. *J. Bacteriol.* **149**:440–448.

Hu, S.-I., E. Pezacka, and H.G. Wood. 1984. Acetate synthesis from carbon monoxide by *Clostridium thermoaceticum*. Purification of the corrinoid protein. *J. Biol. Chem.* **259**:8892–8897.

Hugenholtz, J. and L.G. Ljungdahl. 1990. Metabolism and energy generation in homoacetogenic clostridia. *FEMS Microbiol. Rev.* **87**:383–390.

Hugenholtz, J., D.M. Ivey, and L.G. Ljungdahl. 1987. Carbon monoxide-driven electron transport in *Clostridium thermoautotrophicum* membranes. *J. Bacteriol.* **169**:5845–5847.

Jencks, D.A. and R.G. Matthews. 1987. Allosteric inhibition of methylenetetrahydrofolate reductase by adenosylmethionine. *J. Biol. Chem.* **262**:2485–2493.

Katzen, H.M. and J.M. Buchanan. 1965. Enzymatic synthesis of the methyl group of methionine. Repression-derepression, purification, and properties of 5,10-methylenetetrahy-drolate reductase from *Escherichia coli*. *J. Biol. Chem.* **240**:825–835.

Kerby, R. and J.G. Zeikus. 1983. Growth of *Clostridium thermoaceticum* on H_2/CO_2 or CO as energy source. *Curr. Microbiol.* **8**:27–30.

Krauth-Siegel, R.L., R. Blatterspiel, M. Saleh, E. Schiltz, R.H. Schirmer, and R. Untucht-Grau. 1982. Glutathione reductase from human erythrocytes. The sequences of the NADPH domain and of the interface domain. *Eur. J. Biochem.* **121**:259–267.

Kyte, J. and R.F. Doolittle. 1982. A simple method for displaying the hydropathic character of a protein. *J. Mol. Biol.* **157**:105–132.

Lindahl, P.A., S.W. Ragsdale, and E. Münck. 1990. Mössbauer study of CO dehydrogenase from *Clostridium thermoaceticum*. *J. Biol. Chem.* **265**:3880–3888.

Ljungdahl, L.G. 1986. The autotrophic pathway of acetate synthesis in acetogenic bacteria. *Ann. Rev. Microbiol.* **40**:415–450.

Ljungdahl, L.G., J.M. Brewer, S.H. Neece, and T. Fairwell. 1970. Purification, stability and composition of the formyltetrahydrofolate synthetase from *Clostridium thermoaceticum*. *J. Biol. Chem.* **245**:4791–4797.

Ljungdahl, L.G., W.E. O'Brien, M.R. Moore, and M.-T. Liu. 1980. Methylenetetrahydrofolate dehydrogenase from *Clostridium formicoaceticum* and methylenetetrahydrofolate dehydrogenase, methenyltetrahydrofolate cyclohydrolase (combined) from *Clostridium thermoaceticum*. *Methods Enzymol.* **66**:599–609.

Lovell, C.R., A. Przybyla, and L.G. Ljungdahl. 1988. Cloning and expression in *Escherichia coli* of the *Clostridium thermoaceticum* gene encoding thermostable formyltetrahydrofolate synthetase. *Arch. Microbiol.* **149**:280–285.

Lovell, C.R., A. Przybyla, and L.G. Ljungdahl. 1990. Primary structure of the thermostable formyltetrahydrofolate synthetase from *Clostridium thermoaceticum*. *Biochemistry* **29**:5687–5694.

Lu, W.-P., S.R. Harder, and S.W. Ragsdale. 1990. Controlled potential enzymology of methyl transfer reactions involved in acetyl-CoA synthesis by CO dehydrogenase and the corrinoid/iron-sulfur protein from *Clostridium thermoaceticum*. *J. Biol. Chem.* **265**:3124–3133.

Lux, M.F., E. Keith, T. Hsu, and H.L. Drake. 1990. Biotransformation of aromatic aldehydes by acetogenic bacteria. *FEMS Microbiol. Lett.* **67**:73–78.

MacKenzie, R.E. 1984. Biogenesis and interconversion of substituted tetrahydrofolates. In: *Folates and Pterins*, Vol. 1, *Chemistry and Biochemistry of Folates*, R.L. Blakley and S.J. Benkovic, eds., pp. 255–306. New York: Wiley.

Mayer, F., J.I. Elliott, D. Sherod, and L.G. Ljungdahl. 1982. Formyltetrahydrofolate synthetase from *Clostridium thermoaceticum*. An electron microscopic study and specific interaction of the enzyme with ATP and ADP. *Eur. J. Biochem.* **124**:397–404.

McClure, W.R. 1985. Mechanism and control of transcription initiation in procaryotes. *Ann. Rev. Biochem.* **54**:171–204.

Mejillano, M.R., H. Jahansouz, T.O. Matsunaga, G.L. Kenyon, and R.H. Himes. 1989. Formation and utilization of formyl phosphate by N^{10}-formyltetrahydrofolate synthetase: evidence for formyl phosphate as an intermediate in the reaction. *Biochemistry* **28**:5136–5145.

Morton, T.A., J.A. Runquist, S.W. Ragsdale, T. Shanmugasundaram, H.G. Wood, and L.G. Ljungdahl. 1991. The primary structure of the subunits of carbon monoxide dehydrogenase/acetyl-CoA synthase from *Clostridium thermoaceticum*. *J. Biol. Chem.* **266**:23824–23828.

Mulligen, M.E., D.K. Hawley, R. Entriken, and W.R. McClure. 1984. *Escherichia coli* promoter sequences predict in vitro RNA polymerase selectivity. *Nucleic Acids Res.* **12**:789–800.

O'Brien, W.E., J.M. Brewer, and L.G. Ljungdahl. 1976. Chemical, physical, and enzymatic comparisons of formyltetrahydrolate synthetase from thermo- and mesophilic clostridia. In: *Enzymes and Proteins from Thermophilic Microorganisms, Structure and Functions*, H. Zuber, ed., pp. 249–262. Basel: Birkhäuser.

Park, E.Y., J.E. Clark, D.V. DerVartanian, and L.G. Ljungdahl. 1991. 5,10-Methylene-tetrahydrofolate reductases: iron-sulfur-zinc flavoproteins of two acetogenic clostridia. In: *Chemistry and Biochemistry of Flavoenzymes*, Vol. I. F. Müller, ed., Boca Raton, Florida: CRC Press.

Petersen, J.G.L., and S. Holmberg. 1986. The ILV5 gene of *Saccharomyces cerevisiae* is highly expressed. *Nucl. Acid Res.* **14**:9631–9651.

Pezacka, E. and H.G. Wood, 1986. The autotrophic pathway of acetogenic bacteria: Role of CO dehydrogenase disulfide reductase. *J. Biol. Chem.* **261**:1609–1615.

Pezacka, E. and H.G. Wood. 1988. Acetyl-CoA pathway of autotrophic growth. Identification of the methyl-binding site of the CO dehydrogenase. *J. Biol. Chem.* **263**:16000–16006.

Poston, J.M., K. Kuratomi, and E.R. Stadtman. 1966. The conversion of carbon dioxide to acetate. I. The use of cobalt-methylcobalamin as a source of methyl groups for the synthesis of acetate by cell-free extracts of *Clostridium thermoaceticum*. *J. Biol. Chem.* **241**:4209–4216.

Rabinowitz, J.C. and W.E. Pricer, Jr. 1962. Formyltetrahydrofolate synthetase. I. Isolation and crystallization of the enzyme. *J. Biol. Chem.* **237**:2898–2902.

Ragsdale, S.W. and H.G. Wood. 1985. Acetate biosynthesis by acetogenic bacteria. Evidence that carbon monoxide dehydrogenase is the condensing enzyme that catalyzes the final step of the synthesis. *J. Biol. Chem.* **260**:3970–3977.

Ragsdale, S.W., J.E. Clark, L.G. Ljungdahl, L.L. Lundie, and H.L. Drake, 1983a. Properties of purified carbon monoxide dehydrogenase from *Clostridium thermoaceticum*, a nickel, iron-sulfur protein. *J. Biol. Chem.* **258**:2364–2369.

Ragsdale, S.W., L.G. Ljungdahl, and D.V. DerVartanian. 1983b. ^{13}C and ^{61}Ni isotope substitutions confirm the presence of a nickel (III)-carbon species in acetogenic CO dehydrogenase. *Biochem. Biophys. Res. Commun.* **115**:658–665.

Ragsdale, S.W., H.G. Wood, and W.E. Antholine. 1985. Evidence that an iron-nickel-carbon complex is formed by reaction of CO with the CO dehydrogenase from *Clostridium thermoaceticum*. *Proc. Natl. Acad. Sci. USA* **82**:6811–6814.

Ragsdale, S.W., P.A. Lindahl, and E. Münck. 1987. Mössbauer, EPR and optical studies of the corrinoid/iron-sulfur protein involved in the synthesis of acetyl coenzyme A by *Clostridium thermoaceticum*. *J. Biol. Chem.* **262**:14289–14297.

Ragsdale, S.W., H.G. Wood, T.A. Morton, L.G. Ljungdahl, and D.V. DerVartanian. 1988. Nickel in CO dehydrogenase. In: *The Bioinorganic Chemistry of Nickel*, J.R. Lancaster, Jr., ed., pp. 311–332. New York: VCH Publishers.

Ramer, S.E., S.A. Raybuck, W.H. Orme-Johnson, and C.T. Walsh. 1989. Kinetic characterization of the [3′-^{32}P] coenzyme A/acetyl coenzyme A exchange catalyzed by a three-subunit form of the carbon monoxide dehydrogenase/acetyl-CoA synthase from *Clostridium thermoaceticum*. *Biochemistry* **28**:4675–4680.

Roberts, D.L., J.E. James-Hagstrom, D.K. Garvin, C.M. Gorst, J.A. Runquist, J.R. Bauer, F.L. Haase, and S.W. Ragsdale. 1989. Cloning and expression of the gene cluster encoding key proteins involved in acetyl-CoA synthesis in *Clostridium thermoaceticum*: CO dehydrogenase, the corrinoid/Fe-S protein, and methyltransferase. *Proc. Natl. Acad. Sci. USA* **86**:32–36.

Robson, R.L. 1984. Identification of possible adenine nucleotide-binding sites in nitrogenase Fe- and MoFe-proteins by amino acid sequence comparison. *FEBS Lett.* **173**:394–397.

Saint-Girons, I., N. Duchange, M.M. Zakin, I. Park, D. Margarita, P. Ferrara, and G.N. Cohen. 1983. Nucleotide sequence of *metF*, the *E. coli* structural gene of 5-10-methylene tetrahydrofolate reductase and of its control region. *Nucleic Acids Res.* **11**:6723–6732.

Saraste, M., P.R. Sibbald, and A. Wittinghofer. 1990. The P-loop—a common motif in ATP- and GTP-binding proteins. *Trends Biochem. Sci.* **15**:430–434.

Shanmugasundaram, T., G.K. Kumar, and H.G. Wood. 1988. Involvement of tryptophan residues at the coenzyme A binding site of carbon monoxide dehydrogenase from *Clostridium thermoaceticum*. *Biochemistry* **27**:6499–6503.

Shanmugasundaram, T., G.K. Kumar, B.C. Shenoy, and H.G. Wood. 1989. Chemical modification of the functional arginine residues of carbon monoxide dehydrogenase from *Clostridium thermoaceticum*. *Biochemistry* **28**:7112–7116.

Shannon, K.W. and J.C. Rabinowitz. 1988. Isolation and characterization of the *Saccharomyces cerevisiae* M1S1 gene encoding mitochondrial C_1-tetrahydrofolate synthase. *J. Biol. Chem.* **263**:7717–7725.

Shimoi, H., S. Nagata, N. Esaki, H. Tanaka, and K. Soda. 1987. Leucine dehydrogenase of a thermophilic anaerobe, *Clostridium thermoaceticum*: gene cloning, purification and characterization. *Agric. Biol. Chem.* **51**:3375–3381.

Staben, C. and J.C. Rabinowitz. 1986. Nucleotide sequence of the *Saccharomyces cerevisiae* ADE3 gene encoding C_1-tetrahydrofolate synthase. *J. Biol. Chem.* **261**:4629–4637.

Stauffer, G.V. and L.T. Stauffer. 1988. Cloning and nucleotide sequence of the *Salmonella tyhimurium* LT2 *metF* gene and its homology with the corresponding sequence of *Escherichia coli*. *Mol. Gen. Genet.* **212**:246–251.

Stormo, G.D. 1986. Translation initiation. *In*: *Maximizing Gene Expression*, W. Reznikoff and L. Gold, eds., pp. 195–224.

Varalakshmi, K., H.S. Savithri, and N.A. Rao. 1986. Identification of amino acid residues essential for enzyme activity of sheep liver 5,10-methylenetetrahydrofolate reductase. *Biochem. J.* **236**:295–298.

Villar, E., B. Schuster, D. Peterson, and V. Schitch. 1985. C_1-tetrahydrofolate synthase from rabbit liver: structural and kinetic properties of the enzyme and its two domains. *J. Biol. Chem.* **260**:2245–2252.

Walker, J.E., M. Saraste, M.J. Runswick, and N.J. Gay. 1982. Distantly related sequences in the α- and β-subunits of ATP synthase, myosin, kinases and other ATP-requiring enzymes and a common nucleotide binding fold. *EMBO J.* **1**:945–951.

Weije, W.J., J. Hofsteenge, J.M. Vereijken, P.A. Jekel, and J.J. Beintema. 1982. Primary structure of p-hydroxybenzoate hydroxylase from *Pseudomonas fluorescens*. *Biochim. Biophys. Acta* **704**:385–388.

Wek, R.C., and G.W. Hatfield. 1986. Nucleotide sequence and *in vivo* expression of the *ilvY* and *ilvC* genes in *Escherichia coli* K12: Transcription from divergent overlapping promotors. *J. Biol. Chem.* **261**:2441–2450.

Whitehead, T.R. and J.C. Rabinowitz. 1988. Nucleotide sequence of the *Clostridium acidiurici* ("*Clostridium acidi-urici*") gene for 10-formyltetrahydrofolate synthetase shows extensive amino acid homology with the C_1-tetrahydrofolato synthase from *Saccharomyces cerevisiae*. *J. Bacteriol.* **170**:3255–3261.

Whiteley, H.R., M.J. Osborn, and F.M. Huennekens. 1959. Purification and properties of the formate activating enzyme from *Micrococcus aerogenes*. *J. Biol. Chem.* **234**:1538–1543.

Wierenga, R.K. and W.G.J. Hol. 1983. Predicted nucleotide-binding properties of p21 protein and its cancer-associated variant. *Nature* (London) **302**:842–844.

Wierenga, R.K., M.C.H. DeMaeyer, and W.G.J. Hol. 1985. Interaction of pyrophosphate moieties with α-helixes in dinucleotide binding proteins. *Biochemistry* **24**:1346–1357.

Wohlfarth, G., G. Geerligs, and G. Diekert. 1990. Purification and properties of a NADH-dependent 5,10-methylenetetrahydrofolate reductase from *Peptostreptococcus productus*. *Eur. J. Biochem.* **192**:411–417.

Wood, H.G. 1952. A study of carbon dioxide fixation by mass determination of the types of C^{13}-acetate. *J. Biol. Chem.* **194**:905–931.

Wood, H.G. 1989. Past and present of CO_2 utilization. *In*: *Autotrophic Bacteria*. H.G. Schlegel and B. Bowien, eds., pp. 33–52. Madison: Science Technology.

Wood, H.G. and L.G. Ljungdahl. 1991. Autotrophic character of the acetogenic bacteria. *In*: *Variations in Autotrophic Life*, J.M. Shively and L.L. Barton, eds., pp. 201–250. London: Academic Press.

Wood, H.G., S.W. Ragsdale, and E. Pezacka. 1986. The acetyl-CoA pathway: a newly discovered pathway of autotrophic growth. *Trends Biochem. Sci.* **11**:14–18.

Wu, Z., S.L. Daniel, and H.L. Drake. 1988. Characterization of a CO-dependent O-demethylating enzyme system from the acetogen *Clostridium thermoaceticum*. *J. Bacteriol.* **170**:5705–5708.

29

Genes and Proteins Involved in Cellulose Degradation by Mesophilic Clostridia

Jean-Pierre Bélaich, Anne Bélaich, Christian Gaudin, and Chantal Bagnara

Mesophilic cellulolytic clostridia are widespread in all biotopes with moderate temperatures and play an important role in the anaerobic mineralization of lignocellulosic matter. Numerous species of these bacteria have been isolated and characterized, including *Clostridium cellobioparum* (Hungate, 1944), *C. papyrosolvens* (Madden et al., 1982), *C. cellulolyticum* (Petitdemange et al., 1984), *C. cellulovorans* (Sleat et al. 1984), *C. celerecrescens* (Palop et al., 1989), *C. longisporum* (Varel, 1989) and *C. josui* (Sukhumavasi et al., 1988); the latter is thought to be moderately thermophilic. All these bacteria are true cellulolytic organisms and grow on native cellulose as carbon and energy source. The cellulolytic machinery of clostridia, like that of many cellulolytic bacteria, is embedded in parietal protuberances that were called cellulosomes by their discoverers (Lamed and Bayer, 1988). Specialized cell-surface structures of this kind have been found to exist in *C. cellobioparum*, *C. cellulovorans*, and *C. thermocellum* (Lamed et al. 1987; Shoseyov and Doy, 1990). The best-known example of a cellulosome is that of the thermophilic bacterium *C. thermocellum*, which is described in detail (see Chapter 30, this volume); the biochemical structure of cellulosome therefore is not discussed here. This chapter reviews the latest progress in the study of the biochemistry and genetics of cellulases from mesophilic clostridia.

29.1 Biochemistry

The cellulase complex from *Clostridium cellulovorans* has been recently purified and analyzed (Shoseyov and Doi, 1990). The extracellular cellulase complex was purified using its extremely high affinity for cellulose, and the main cellulase activity was found to be associated with a 900-kDa protein cluster. Three main subunits were found to exist in this cluster, with molecular masses of 170, 100, and 70 kDa. The 100-kDa component was found to be associated with a carboxymethyl cellulase (CMCase) activity whereas the 170-kDa subunit was devoid of any detectable enzymatic activity. However, the 170-kDa subunit was absolutely required for hydrolysis of crystalline cellulose, but not for CMCase or cellobiohydrolase (CBHase) activity. It was therefore concluded that the 170-kDa subunit plays a crucial role in crystalline cellulose hydrolysis and should have many interesting biochemical properties, including binding cellulose and enzymatic components, organization of the cellulase complex, and conversion of the crystalline part of the cellulose into a structure that can be more easily degraded by the accompanying hydrolytic subunits.

Similar to those from the thermophilic *C. thermocellum*, the endoglucanases* from mesophilic clostridia are firmly bound to cellulo-

---

For definition of endoglucanases and xylanes, see Chapter 30, this volume.

somes and are accordingly difficult to purify. Partial purification and characterization of a CMCase and a xylanase from *C. papyrosolvens* has been achieved, however (Garcia et al., 1989). The Michaelis constants (K_m) of CMCase and xylanase were 3.3 g CMC per liter and 2.7 g xylan per liter, respectively. The kinetic parameters obtained with CMCase from *C. papyrosolvens* were found to be similar to those reported on a crude enzyme solution from *C. thermocellum* (Shinmyo et al., 1979). Also, a 45-kDa CMCase was purified to homogeneity from *C. josui* (Fujino et al., 1989). This enzyme is able to hydrolyze small cellodextrins, and it was also observed that crystalline cellulose (Avicel) was degraded to a significant but very low extent in comparison with the hydrolysis of CMC.

Some noncellulolytic bacteria synthesize endoglucanases or xylanases. *C. acetobutylicum*, a very potent solvent producer, is one of the best-known examples of this kind of organism. Two endoxylanases were purified to homogeneity from the culture supernatant of *C. acetobutylicum* ATCC 824 (Lee et al., 1987). Because these enzymes are not clustered into complexes with high molecular masses, they have been purified from culture supernatants using the classical biochemical techniques generally applied to soluble proteins.

29.2   Genetics

The difficulties encountered in separating and purifying the cellulases from the highly structured cellulosomes suggested that gene cloning would be a useful tool for analyzing these enzymes. This research strategy was proposed and used in studies on cellulases from *Clostridium thermocellum* (Cornet et al., 1983) and then generalized to all the cellulolytic organisms. To date, however, only two mesophilic clostridia, namely *C. acetobutylicum* (Zappe et al. 1986) and *C. cellulolyticum* (Faure et al. 1988), have been studied using this approach.

An endoglucanase gene from *C. acetobutylicum* was cloned in *Escherichia coli* and sequenced (Zappe et al. 1986). Analysis of the deduced amino acid sequence showed the exis-

tence in the N-terminal region of a putative signal sequence resembling signal sequences of gram-positive bacteria. According to the classification into seven families based on hydrophobic cluster analysis proposed by Henrissat et al. (1989), this endoglucanase, which shows strong homologies with those from the *Bacillus* group, can be classified as belonging to family A, subfamily A2 (Béguin, 1990). The nucleotide sequence of the xylanase gene *xynB* from *C. acetobutylicum* was published by Zappe et al. (1990). This gene encodes for the enzyme purified by Lee et al. (1987). Analysis of the deduced amino acid sequence showed a signal sequence and strong homologies (Figure 29.1) with xylanases from *Bacillus* species, more precisely with those from *B. pumilus* (Fukusaki et al. 1984) and *B. subtilis* BB (Paice et al., 1986). These two enzymes were classified by Henrissat and Mornon (1990) in family G, and we propose to also place the XynB enzyme from *C. acetobutylicum* in this family.

Clostridium cellulolyticum is to date the mesophilic cellulolytic *Clostridium* that has been most fully studied from the molecular biology aspect. Several endoglucanases were cloned in *E. coli* (Faure et al. 1988, Pérez-Martinez et al. 1988). The nucleotide sequence of the endoglucanase-(EGCCA) encoding gene (*celCCA*) and its flanking regions was determined (Faure et al. 1989). Sequence analyses of two other genes, *celCCC* and *celCCG*, encoding for endoglucanase EGCCC and EGCCG respectively, are to be completed soon. The EGCCA-deduced amino acid sequence showed a typical signal sequence of extracellular proteins of gram-positive bacteria. The mature protein consists of 449 amino acids with a calculated M_r of 50715 daltons.

The most noteworthy feature of the sequence is the presence of a repeated stretch of 21 amino acid residues at the C-terminal of the protein (Figure 29.2) that is strongly homologous with equivalent conserved regions detected in the same position, as reported for the first time by Béguin et al. (1985) and in Chapter 30 of this volume, in practically all the endoglucanases and xylanases from *C. thermocellum*. The exception is EGE (Hall et al. 1988) and XYNZ (Grépinet et al. 1988), where the reiter-

```
1-MLRRKVIFTVLATLVMTSLTIVDNTAFAATNLNTTESTFSKEVLSTQKTYSAFNT
2-MNLRKLRLLFVMCIGLTLILTA--------------------------------V
3-MFKFKKNFLVGLSAALMSISLFSATASA--------------------------

1-QAAPKTITSNEIGVNGGYDY-ELWKDYGNTSMTLKNG--GAFSCQWSNIGNALFR
2-PAHARTITNNEMGNHSGYDY-ELWKDYGNTSMTLNNG--GAFSAGWNNIGNALFR
3---------------AST-DYWQNWTDGGGIVNAV-NGSGGNYSVNWSNTGN--FV

1-KGKKFNDTQTYKQLGNISVNYDCN-YQPYGNSYLCVYGWTSSPLVEYYIVDSWGS
2-KGKKFDSTRTHHQLGNISINYNAS-FNPSGNSYLCVYGWTQSPLAEYYIVDSWGT
3-VGKGW--TTGSPFR---TINYNAGVWAPNGNGYLTLYGWTRSPLIEYYVVDSWGT

1-WRPPGGTSKGTITVDGGIYDIYETTRINQPSIQGN-TTFKQYWSVRRTKRTSG--
2-YRPT-GAYKGSFYADGGTYDIYETTRVNQPSIIGIA-TFKQYWSVRQTKRTSG--
3-YRPT-GTYKGTVKSDGGTYDIYTTTRYNAPSIDGDRTTFTQYWSVRQSKRPTGSN

1--TISVSKHFAAWESKGMPLGKMHETAFNI-EGYQSSGKADVNSMSINIGK
2--TVSVSAHFRKWESLGMPMGKMYETAFTV-EGYQSSGSANVMTNQLFIGN
3-ATITFSNHVNAWKSHGMNLGSNWAYQVMATEGYQSSGSSNV-T--VW
```

FIGURE 29.1 Alignment of xylanase sequences from family G: 1, *Clostridium acetobutylicum* xylanase 2; 2, *Bacillus pumilus* xylanase A; 3, *Bacillus subtilis* xylanase. Identical and conserved residues are underlined.

DOMAIN I

EGCCA-**DYNNDGNVDALDFAGLKKYIM**AADHAYVKN

EGCCC-**DVNNDGAIDALDIAALKKAIL**TQTTSNISLTN

DOMAIN II

EGCCA-**DVNLDNEVNAFDLAILKKYLL**GMVSKLPSN-COOH

EGCCC-**DMNNDGNIDAIDFAQLKVNCL**TRINKIIE-COOH

FIGURE 29.2 Reiterated domains (underlined) observed in C-terminal regions of endoglucanoses EGCCA and EGCCC from *Clostridium cellulolyticum*.

ated sequence is located in the center of the protein, and EGC (Schwartz et al. 1988), which does not contain this pattern. It was recently observed that a second endoglucanase from *C. cellulolyticum* (EGCCC) also contained a reiterated amino acid sequence at the C-terminal of the protein (see Figure 29.2) (Bagnara unpublished data).

These reiterated domains, the role of which is not yet known, therefore seem to be typical of the cellulases belonging to the genus *Clostridium*. EGCCA is strongly homologous to the N-terminal moiety of EGE from *C. thermocellum*. It is noteworthy that the N-terminal sequence of an endoglucanase purified from *C. josui* also exhibits strong homologies with the putative N-terminal sequence of the mature EGCCA (Fujino et al. 1989). The three endoglucanases from *C. cellulolyticum*, EGCCA, EGCCC, and EGCCG, belong to families A, D, and E, respectively, in Henrissat's classification.

The *celCCA* gene is poorly expressed in *E. coli*, so that to purify the cloned protein it was necessary to amplify the gene expression by setting the gene behind a strong promoter. In this way the EGCCA content of the host was increased 2000 fold (Fierobe et al., 1991, in press). The protein produced in *E. coli* has been purified; the maximum CMCase activity was 95 IU/mg. In addition to CMCellulose, the enzyme hydrolyzes lichenan and barley glucan (Table 29.1) and to a lesser extent laminarin and xylan (Fierobe et al., 1991, in press). The molecular mass determined by mean of the sodium dodecylsulfate-polyacrylamide gel electrophoresis, (SDS-PAGE) technique was 49900 ± 600 daltons.

The *celCCA* gene constitutes one monocistro-

Table 29.1 Hydrolysis of various substrates by the endoglucanase EGCCA

Substrate	Bond	Monomer	Activity (%)
CMCellulose	β-1,4	Glucose	100.0
Avicel PH101	β-1,4	Glucose	0.3
Laminarin	β-1,3	Glucose	46.2
Lichenan	β-1,3 & β-1,4	Glucose	115.4
Barley glucan	β-1,3 & β-1,4	Glucose	111.5
Xylan	β-1,4	Xylose	10.0
pNPC	β-1,4	—	0.2
pNPG	β-1,4	—	<0.004

nic transcription unit; the *celCCC* and *celCCG* genes are contiguous on the chromosome, because they are separated only by a short 59-bp region. Northern mapping of transcripts by intragenic *celCCC* and *celCCG* probes revealed a similar pattern with major transcripts of approximately 5×10^3 and 6×10^3 nt in length. *celCCC* and *celCCG* therefore seem to be organized in a polycistronic transcription unit (unpublished data). This is the first time this type of organization has been reported among bacterial cellulase genes.

29.3 Conclusion

Because most cellulose-fermenting biotopes are mesophilic, it is rather paradoxical that the number of reports on mesophilic clostridial cellulases is so small in comparison with the large body of data published on cellulases from *Clostridium thermocellum*. It seems to be relevant to the purposes of both fundamental and applied research to elucidate the mechanisms of cellulose breakdown under moderate temperature conditions. This mechanism apparently involves cooperation between numerous proteins that are embedded in parietal polycellulosomes, and a knowledge of the processes underlying the morphogenesis of these structures thus would be of great use in biotechnological processes involving cellulases. The techniques of molecular biology have proved to be very efficient for analysing cellulolytic complexes, but further research is still required in this field.

One of the main problems that arises here is

the lack of gene transfer in these clostridia. These techniques are necessary to determine the role of each component of cellulosomes. The availability of appropriate genetic tools would make it possible to rapidly elucidate the breakdown mechanism of natural celluloses and would open the way to numerous biotechnical applications.

Acknowledgments. Research from our laboratory reported here was partly supported by a grant from the EEC and by a research contract from the company "Gaz de France."

References

Béguin, P. 1990. Molecular biology of cellulose degradation. *Ann. Rev. Microbiol.* **44**:219–248.

Béguin, P., P. Cornet, and J.P. Aubert. 1985. Sequence of a cellulase gene of the thermophilic bacterium *Clostridium thermocellum*. *J. Bacteriol.* **162**:102–105.

Cornet, P., D. Tronik, J. Millet, and J.P. Aubert. 1983. Cloning and expression in *Escherichia coli* of *Clostridium thermocellum* genes coding for amino acid synthesis and cellulose hydrolysis. *FEMS Microbiol. Lett.* **16**:137–141.

Faure, E., C. Bagnara, A. Bélaich, and J.P. Bélaich. 1988. Cloning and expression of two cellulase genes of *Clostridium cellulolyticum* in *Escherichia coli*. *Gene* **65**:51–58.

Faure, E., A Bélaich, C. Bagnara, C. Gaudin, and J.P. Bélaich. 1989. Sequence analysis of the *Clostridium cellulolyticum* endoglucanase-A-encoding gene, *celCCA*. *Gene* **84**:39–46.

Fierobe, H.-P., C. Gaudin, A. Bélaich, M. Loutfi, E. Faure, C. Bagnara, D. Baty and J.-P. Belaich. 1991. Characterization of endoglucanase A from *Clostridium cellulolyticum*. *J. Bacteriol.* in press.

Fujino, T., J. Sukhumavasi, T. Sasaki, K. Ohmiya,

and S. Shimizu. 1989. Purification and properties of an endo-beta 1-4glucanase from *Clostridium josui. J. Bacteriol.* **171**:4076–4079.

Fukusaki, E., W. Panbangred, A. Shinmyio, and H. Okada. 1984. The complete nucleotide sequence of the xylanase gene (*xynA*) of *Bacillus pumilus. FEBS Lett.* **171**:197–201.

Garcia, V., A. Madarro, J.L Pena, F. Pinaga, S. Valles, and A. Flors. 1989. Purification and characterization of cellulases from *Clostridium papyrosolvens. J. Chem. Technol. Biotechnol.* **46**:49–60.

Grépinet, O., M.C. Chebrou, and P. Béguin. 1988. Nucleotide sequence and deletion analysis of the xylanase gene *xynZ* of *Clostridium thermocellum. J. Bacteriol.* **170**:4582–4588.

Hall, J., G.P. Hazlewood, P.J. Barker, and H.J. Gilbert. 1988. Conserved reiterated domains in *Clostridium thermocellum* are not essential for activity. *Gene* **69**:29–38.

Henrissat, B. and J.P. Mornon. 1990. Comparison of *Trichoderma* cellullases with other β-glucanases. *In: Trichoderma reesei Cellulases: Biochemistry, Genetics, Physiology and Applications.* C.P. Kubicek, D.E. Eveleigh, H. Esterbauer, W. Steiner, and E.M. Kubicek-Pranz, eds., pp. 12–29. Vienna: Springer-Verlag.

Henrissat, B., M. Claeyssens, P. Tomme, L. Lemesle, and J.-P. Mornon. 1989. Cellulase families revealed by hydrophobic cluster analysis. *Gene* **81**:83–95.

Hungate, R.E. 1944, Studies on cellulose fermentation I. The culture and physiology of an anaerobic cellulose-digesting bacterium. *J. Bacteriol.* **48**:499–513.

Lamed, R., J. Naimark, E. Morgenstern, and E.A. Bayer. 1987. Specialized cell surface structures in cellulolytic bacteria. *J. Bacteriol.* **169**:3792–3800.

Lamed, R. and E.A. Bayer. 1988. The cellulosome concept: exocellular/extracellular enzyme reactor centers for efficient binding and cellulosis. *In: Biochemistry and Genetics of Cellulose Degradation, FEMS Symp.* 43, J.P. Aubert, P. Béguin, and J. Millet, eds. pp. 101–116. Symposium No. 43, London: Academic Press.

Lee, S.F., C.W. Forsberg, and J.B. Rattray. 1987. Purification and characterization of two endoxylanases from *Clostridium acetobutylicum* ATCC 824. *Appl. Environ. Microbiol.* **53**:644–650.

Madden, R.H., M.J. Bryder, and N.J. Poole. 1982. Isolation and characterization of an anaerobic cellulolytic bacterium, *Clostidium papyrosolvens* sp.nov. *Int. J. Syst. Bacteriol.* **32**:87–91.

Paice, G.M., R. Bourbonnais, M. Desrochers, L. Jurasek, and M. Yaguchi. 1986. A xylanase gene from *Bacillus subtilis*: nucleotide sequence and comparison with *B. pumilus* gene. *Arch. Microbiol.* **144**:201–206.

Palop, M.L., S. Valles, F. Pinaga, and A. Flors. 1989. Isolation and characterization of an anaerobic, cellulolytic bacterium, *Clostridium celerecresens* sp. nov. *Int. J. Syst. Bacteriol.* **39**:68–71.

Pérez-Martinez, G., L. Gonzalez-Candelas, J. Polaina, and A. Flors. 1988. Expression of an endoglucanase gene from *Clostridium cellulolyticum* in *Escherichia coli. J. Ind. Microbiol.* **3**:365–371.

Petitdemange, E., F. Caillet, J. Giallo, and C. Gaudin. 1984. *Clostridium cellulolyticum* sp.nov. a cellulolytic mesophilic species from decayed grass. *Int. J. Syst. Bacteriol.* **34**:155–159.

Schwartz, W.H., S. Schimming, K.P. Rûcknagel, S. Burgschwaiger, G. Kreil, and W.L Staudenbauer. 1988. Nucleotide sequence of the *celC* gene encoding endoglucanase C of *Clostridium thermocellum. Gene* **63**:23–30.

Shinmyo, A., D.V. Garcia-Martinez, and A.M. Demain. 1979. Studies on the extracellular cellulolytic enzyme complex produced by *Clostridium thermocellum. J. appl. Biochem.* **1**:202–209.

Shoseyov, O. and R.H. Doi. 1990. Essential 170-kDa subunit for degradation of crystalline cellulose by *Clostridium cellulovorans* cellulase. *Proc. Natl. Acad. Sci. USA* **87**:2192–2195.

Sleat, R., R.A. Mah, and R. Robinson. 1984. Isolation and characterization of an anaerobic cellulolytic bacterium *Clostridium cellulovorans* sp.nov. *Appl. Environ. Microbiol.* **48**:88–93.

Sukhumavasi, J., K. Ohmiya, S. Shimizu, and K Ueno. 1988. *Clostridium josui* sp.nov., a cellulolytic, moderate thermophilic species from Thai compost. *Int. J. Syst. Bacteriol.* **38**:179–182.

Varel, V.H. 1989. Reisolation and characterization of *Clostridium longisporum*, a ruminal spore forming cellulolytic anaerobe. *Arch. Microbiol.* **152**:209–214.

Zappe, H., D.T. Jones, and D.R. Woods. 1986. Cloning and expression of *Clostridium acetobutylicum* endoglucanase cellobiase and amino acid biosynthesis genes in *Escherichia coli. J. Gen. Microbiol.* **132**:1367–1372.

Zappe, H., A.J. Winsome, and D.R. Woods. 1990. Nucleotide sequence of a *Clostridium acetobutylicum* P262 xylanase gen (xynB). *Nucleic Acids Res.* **18**:2179.

30

Genes and Proteins Involved in Cellulose and Xylan Degradation by *Clostridium thermocellum*

Jean-Paul Aubert, Pierre Béguin, and Jacqueline Millet

30.1   Introduction

The thermophilic gram-positive anaerobe *Clostridium thermocellum* produces a highly active cellulolytic and hemicellulolytic enzyme complex (Ng et al., 1977; Garcia-Martinez et al., 1980; Johnson et al., 1982; Johnson and Demain, 1984). This complex, termed the cellulosome (Lamed et al., 1983a, 1983b), is located on the surface of the bacteria during growth and mediates their adherence to the substrate. It is released into the medium when growth ends.

The characteristics of the cellulosome raise many issues for the molecular biologist and the geneticist. Interesting questions include its structural organization and its mode of action; the characterization and regulation of the genes involved, in particular the *cel* (cellulose degradation) and *xyn* (xylan degradation) genes; the mechanism by which the proteins that constitute the complex are secreted and organized into cellulosomes; and how these remain anchored at discrete loci at the surface of the cell wall.

This chapter discusses both the questions that have been at least partially answered and those which are still open.

30.2   The Cellulosome

A schematic representation, based on electron microscope observations, of the cellulolytic process of *Clostridium thermocellum* (Lamed et al.,

1988) is shown in Figure 30.1. Growing cells are covered with multicellulosome-containing protuberances (**A**). Adherence is mediated via the cellulosomes located on the surface of the protuberances. On contact with the substrate, protuberances protract forming an extended corridor, and cellulose is hydrolyzed, mainly to cellobiose (**B**). The cells subsequently desorb from the substrate (**C**), but the cellulosomes remain bound to the cellulose substrate (**D**) and the hydrolytic process continues.

Cellulosomes have been purified from culture supernatants and sonicated cells (Lamed et al., 1983a). The major form appears as a polypeptide complex of M_r of 2.1 MDa. However, much larger complexes have also been isolated, as much as 100 MDa, which are probably polycellulosomes constituting the protuberances at the cell surface (Coughlan et al., 1985). Cellulosomes exhibit both carboxymethylcellulose-hydrolyzing (CMCase, endoglucanase) activity and Avicel-hydrolyzing (Avicelase) activity, which is considered to be "true" cellulase activity, because Avicel is a crystalline form of cellulose. Avicelase activity in crude culture supernatants is enhanced by Ca^{2+} ions and thiol-reducing agents (Johnson and Demain, 1984; Lamed et al. 1985).

As judged from sodium dodecyl sulfate-polyacrylamide gel electrophoresis (SDS-PAGE), the cellulosomes comprise at least 14 polypeptides with M_r ranging from 48 to 210 kDa (Lamed and Bayer, 1988). Two major bands, S1 (M_r, 210 kDa) and S8 (M_r, 75 kDa)

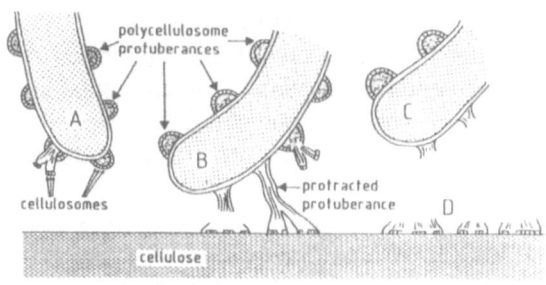

FIGURE 30.1 Schematic representation of interaction of *Clostridium thermocellum* with insoluble cellulose (see text for details). [From Lamed et al. (1988), courtesy of the Société Française de Microbiologie.]

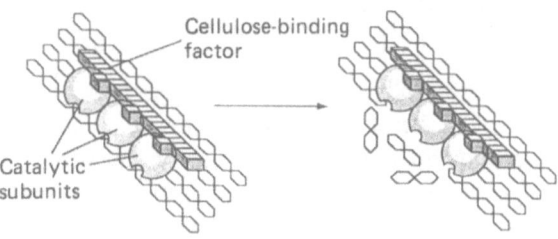

FIGURE 30.2. Schematic representation of the hydrolysis of cellulose by the cellulosome of *Clostridium thermocellum* (see text for details). [From Béguin et al. (1988), courtesy of the Société Française de Microbiologie.]

account for 24% and 28%, respectively, of the total polypeptide content of the cellulosome. Calculations based on densitometry measurements of Coomassie-stained SDS-PAGE gels suggest a complex stoichiometry between the various components. Zymograms performed with SDS-PAGE gels revealed CMCase activity in at least eight bands including S8. The S1 subunit has no CMCase activity, is highly glycosylated (Gerwig et al., 1989), and may play a role in the organization of the complex.

In addition to the cellulolytic complex, a yellow affinity substance (YAS), which attaches to the cellulose fibers, is also produced (Ng et al., 1977). According to Ljungdahl et al. (1983), YAS facilitates the binding of cellulosomes to cellulose. The structure of YAS, whose M_r is about 1 kDa, is not yet fully established but seems to be related to carotenoids (Ljungdahl et al., 1988).

Attempts to resolve the cellulosome into its active subunits by classical biochemical techniques have been hampered by the resistance of the complex to dissociation agents. Until recently only two endoglucanases had been purified from culture supernatants. One, termed endoglucanase A (EGA), is a 56-kDa protein, and was purified in the presence of 6 M urea from strain NCIB 10682 (ATCC 24405) (Pètre et al., 1981); the other, a 96-kDa protein, was purified from strain LQRI (Ng and Zeikus, 1981). Both enzymes display displayed CMCase activity but were devoid of Avicelase activity.

Wu et al. (1988) were able to isolate two components, termed S_L (M_r, 250 kDa) and S_S (M_r, 82 kDa), by gel filtration and preparative SDS-PAGE. Study of these components gave a first insight into the possible organization of the cellulosome. S_L, a highly glycosylated protein, has no CMCase nor Avicelase activity. S_S exhibited CMCase activity but no Avicelase activity. However, Avicelase activity was restored by mixing S_L and S_S. Binding experiments showed that S_L was required for the attachment of S_S to Avicel (Wu and Demain, 1988). It is likely that S_L is the S1 subunit described by Lamed and Bayer (1988) and that S_S is S8. From these observations, a model was proposed (Wu and Demain, 1988) in which S_L is an anchorage protein that mediates the binding of the S_S–S_L complex to crystalline cellulose in such a way that S_S hydrolyzes the substrate. The mechanism by which S_S displays Avicelase activity only when associated with S_L is unknown.

Electron microscopy studies of the interaction between cellulosomes or polycellulosomes and cellulose fibers led to a diagrammatic representation of the structural concept of the cellulosome (Mayer et al., 1987). According to the model, the subunits of the cellulosome are arranged in parallel rows, which allow several attachments to the cellulose fibers. As a result, a simultaneous multicutting hydrolysis of cellulose chains takes place, liberating cellodextrins, which are subsequently hydrolyzed to cellobiose. A simplified version of the same model (Figure 30.2) has also been proposed by Béguin et al. (1988).

30.3   The Cloned Genes

No methodology for transferring genetic material into *Clostridium thermocellum* is yet available. Consequently, our knowledge of the number of genes involved in cellulose and hemicellulose degradation and their characterization is based only on cloning and molecular genetics techniques.

Although *Escherichia coli* and *C. thermocellum* are unrelated, genes of this gram-positive anaerobe seem to be easily expressed in *E. coli*, and the use of appropriate substrates has allowed the screening of gene banks on plates. Endoglucanase activity is usually detected by CMC hydrolysis and the Congo red assay (Teather and Wood, 1982; Cornet et al., 1983a). Chromogenic and fluorogenic substrates such as *p*-nitrophenyl-β-D-cellobioside (pNPC) (Deshpande et al., 1984) or methylumbelliferyl-β-D-cellobioside (MUC) (Van Tilbeurgh et al., 1982) have also been used but are less specific and may lead to the detection of exoglucanases, endoglucanases, xylanases and β-glucosidases. Several gene banks of *C. thermocellum* have been constructed in *E. coli* (Cornet et al., 1983b; Millet et al., 1985; Piruzian et al., 1985; Schwarz et al., 1985; Romaniec et al., 1987). The two largest banks (Millet et al., 1985; Romaniec et al., 1987), obtained from strain NCIB 10682, were compared. This led to the characterization of 18 different DNA fragments, carrying 15 endoglucanase genes, 2 xylanase genes, and 1 β-glucosidase gene (Hazlewood et al., 1988). An additional β-glucosidase gene has been identified by Gräbnitz and Staudenbauer (1988).

30.4   The Functional Organization of the *cel* and *xyn* Genes

General organization

Attempts to detect plasmids in strain NCIB 10682 have been unsuccessful. It is therefore likely that *cel* and *xyn* genes are located on the chromosome. Hybridization experiments between *Clostridium thermocellum* total DNA, digested by various restriction enzymes, and DNA fragments carrying *cel* or *xyn* genes used as probes, showed that the cloned genes did not hybridize with each other under stringent conditions, were not reiterated, and were not clustered (Hazlewood et al., 1988). It is worth noting, however, that this study revealed the presence of transposable elements that are likely to confer some plasticity to the *C. thermocellum* genome. In addition, all the genes whose nucleotide sequences have been established (see following) seem to be organized in monocistronic transcription units.

Features of the coding regions

To date, seven *cel* genes coding for endoglucanases and one *xyn* gene coding for a xylanase have been sequenced: *celA* (Béguin et al., 1985), *celB* (Grépinet and Béguin, 1986), *celC* (Schwarz et al., 1988a), *celD* (Joliff et al., 1986a), *celE* (Hall et al., 1988), *celF* (Navarro et al., 1991), *celH* (Yagüe et al., 1990), and *xynZ* (Grépinet et al., 1988b). A partially sequenced open-reading frame termed *celX*, located upstream from *celE*, was described as probably coding for an endoglucanase (Hall et al., 1988). *CelG* is being sequenced in our laboratory. The *bglB* gene, which codes for an intracellular β-glucosidase, has also been sequenced (Gräbnitz et al., 1989).

ORF*celA* starts with a GTG codon and ends with TAG, ORF*celB*, *D*, and *H* start with ATG and end with TAA; ORF*celC* and *xynZ* start with ATG and end with TGA; ORF*celE* starts with GTG and ends with TAG; and ORF*celF* starts with TTG and ends with TGA. Because EGA had been purified from a *C. thermocellum* culture supernatant (Pètre et al., 1981), it was possible to align the nucleotide sequence of *celA* with the N-terminal amino acid sequence. This led to the identification of a 32-amino-acid signal peptide sequence (Béguin et al., 1985). Similar sequences, although not strictly identical, were found in the peptide sequences deduced from the nucleotide sequences of the other genes. In view of the putative signal sequences, the endoglucanases and xylanases of *C. thermocellum* appear to be classic secreted proteins.

EGA	417	D V N V D G N V N S T D L T M L K R Y L L K	438
	449	D V N R D G A I N S S D M T I L K R Y L I K	470-Y477
EGB	502	D V N G D G R V N S S D V A L L K R Y L L G	523
	534	D V N V S G T V N S T D L A I M K R Y V L R	555-K563
EGD	586	D V N D D G K V N S T D L T L L K R Y V L K	607
	621	D V N R D G R V N S S D V T I L S R Y L I R	642-I649
EGE	415	D V N G D G K I N S T D C T M L K R Y I L R	436
	451	D V N A D L K I N S T D L V L M K K Y L L R	472-W814
EGF	670	D V N F D G R I N S T D Y S R L K R Y V I K	691
	709	D V D G N G R I N S T D L Y V L N R Y I L K	730-G736
EGH	832	D L N F D N A V N S T D L L M L K R Y I L K	853
	871	D L N R D N K V D S T D L T I L K R Y L L Y	882-I899
XYNZ	431	D L N G D G N I N S S D L Q A L K R H L L G	452
	464	D V N R S G K V D S T D Y S V L K R Y I L K	485-Y837
EGX	?	D V N L D G Q V N S T D F S L L K R Y I L K	?+21
	?+34	D M N N D G N I N S T D I S I L K R I L L R	?+55-N?+56

FIGURE 30.3. 22-amino-acid reiterated sequences in endoglucanases and in xylanase of *Clostridium thermocellum*. Boxes with heavy outlines include amino acids with identical or closely related chemical properties (I, L, M, V; K, R; S, T); boxes with thin outlines single out nonhomologous amino acids. Numbers refer to position of first or last residue of reiterated sequences; numbers preceded by amino acid indicate nature and position of C-terminal residue of each protein.

The most striking feature in the polypeptide deduced from ORF*celA* was the existence of a 22-amino-acid reiterated sequence located at the C terminus of the protein (Béguin et al., 1985). As shown in Figure 30.3, similar reiterations were subsequently found in all the peptide sequences deduced from all *C. thermocellum* *cel* and *xyn* genes except *celC*. In EGE and XYNZ, the reiteration was found in the central region of the polypeptide; in the other enzymes, it was found, as in EGA, near the C terminus. The reiteration of a stretch of 22 amino acids is an unusual feature not commonly found in other fungal or bacterial cellulases. However, a sequence highly similar to the reiterations of *C. thermocellum* was found in a *C. cellulolyticum* endoglucanase (Faure et al., 1989). A reiterated stretch of 60 amino acids, without any similarity to the reiterated sequence of *C. thermocellum*, was found in the endoglucanase B of the alkalophilic *Bacillus* sp. N-4 (Fukumori et al., 1986).

The conservation of these reiterated regions in cellulases or xylanases of *C. thermocellum*, which otherwise display a very weak or no similarity of sequence, is presumably signifi-cant (see following discussion). These regions are often separated from the rest of the protein by a stretch of about 20 amino acids rich in Pro-Thr-Ser residues. Similar Pro-Thr-Ser-rich segments have been described as connecting separate domains in both fungal and bacterial cellulases (O'Neill et al., 1986; Teeri et al., 1987; Ong et al., 1989). In the classification of cellulases on the basis of hydrophobic cluster analysis (Henrissat et al., 1989), the proteins encoded by sequenced *C. thermocellum* genes fall into four of six families: EGB, EGC, EGE, and EGH in family A; EGA in family D; EGD and EGF in family E; XYNZ in family F (Béguin, 1990). In addition, the β-glucosidase encoded by *blgB* (Gräbnitz et al., 1989) is related to fungal β-glucosidases. EGD and EGF are related to EGA from *Pseudomonas fluorescens* subsp. *cellulosa* (Hall and Gilbert, 1988) and to an endoglucanase responsible for avocado (*Persea americana*) ripening (Tucker et al., 1987); XYNZ shows more than 40% similarity with the exoglucanase of *Cellulomonas fimi* (Grépinet et al., 1988b), and EGB shows 34% similarity with an endoglucanase from *Bacillus polymyxa* (Baird et al., 1990).

Features of the noncoding regions and expression of *cel* and *xyn* genes

Clostridium thermocellum NCIB 10682 can utilize cellulose, cellobiose, glucose, fructose, and sorbitol as carbon sources. Growth on glucose, fructose, and sorbitol follows a lag of 4 to 5 days. The doubling time is 2 h on cellobiose and is estimated at about 6 h on cellulose. In culture supernatants of cells grown in the presence of cellobiose or Avicel, Avicelase activity was found to be 8.6 and 58 U/mg cell (dry weight), respectively (Johnson et al., 1985). Because cellobiose is the main product of cellulose degradation, this suggested that the partial repression of cellulase production in culture on cellobiose resulted from a higher growth rate and might be related to the energy level in the cell.

Surprisingly, this catabolite repression-like control seems to be detectable only when Avicelase activity is measured; previous studies by the same group (Shinmyo et al., 1979) had shown that *C. thermocellum* expressed higher CMCase activity when grown on cellobiose than when grown on cellulose. The latter results were confirmed in our laboratory, and SDS-PAGE gels of both types of supernatants did not reveal major differences in the number of bands nor in their intensity (our laboratory, unpublished observations). This observation justifies the use of cells growing on cellobiose for studying the expression of individual *cel* or *xyn* genes; these cultures are much easier to manipulate than growing cells attached to cellulose fibers.

Kinetic studies of mRNA synthesis, performed with *celA*, *celD*, and *celF*, showed that *cel* mRNAs were produced only at the end of the exponential phase and at the beginning of the stationary phase in cells growing on cellobiose (Mishra et al., 1991). This is consistent with transcriptional regulation of *cel* gene expression by a catabolite-like repression mechanism. The molecular basis of regulation, however, is still unknown. Mapping of *celA* mRNA (Béguin et al., 1986) showed that in *C. thermocellum* a major transcript started 57 bp upstream from the translation initiation codon. The only consensus promoter sequences found in this region were ACGATAT at −64 and

GTAAA at −83, which are reminiscent of the −10 CCGATAT and −35 CTAAA consensus sequences of *Bacillus subtilis* σ^{28} vegetative promoter. A minor transcript started at position −134, which is preceded by a TATAAT sequence at −141 and a TTGTAT sequence at −169, reminiscent of the −10 TATAAT and −35 TTGACA consensus sequences of *E. coli* σ^{70} and *B. subtilis* σ^{43} promoters. In *E. coli* carrying *celA* cloned in pBR322, a single start site was detected at the same position as the minor transcript produced in *C. thermocellum*.

In both organisms, the size of the transcript suggested that *celA* belongs to a monocistronic transcription unit. Mapping of *celD* mRNA from *C. thermocellum* revealed the presence of two start sites, one of which was preceded by a *B. subtilis* σ^{43} consensus sequence. The other site and the start site of *celF* mRNA failed to reveal any significant similarity with tentative *celA* promoter sequences or with other identified promoter types. In addition, some *cel* genes, including *celA* and *celF*, are transcribed at a high level and others, like *celC* and *celD* at a low level (Mishra et al., 1991).

In all genes sequenced, the regions upstream from the translation start codon display a higher AT content (70%–80%) than the coding regions (50%–60%) or the genome as a whole (60%). This is probably the reason why these genes are spontaneously expressed in *E. coli*, whose RNA polymerase should recognize sequences reminiscent of its own consensus, as shown in the case of *celA*. The Shine–Dalgarno sequences found were AGGAGG in *celA*, *B*, *C*, *H* and *xynZ*, AAGGGGGA in *celD* and *celF*, and AGGAGA in *celE*. At the 3' end of all the genes, palindromes were identified that corresponded to mRNA hairpin loops with ΔG values ranging from 16 to 35 kcal/mol.

Expression in *E. coli* can be considerably increased by placing the cloned genes under the control of strong promoters. For example, a fusion of *celD* to the 5' end of the *lacZ'* gene in pUC8 resulted in hyperproduction of EGD to more than 15% of total protein. Cytoplasmic inclusion bodies were formed from which EGD has been isolated to a high degree of purity (Joliff et al., 1986b). Similar results were observed with truncated *xynZ* and *celH* genes

(Grépinet et al., 1988b; Yagüe et al., 1990). High-level expression of EGA and EGC has also been obtained by subcloning *celA* and *celC* in a temperature-regulated expression vector containing the p_L promoter of bacteriophage λ (Schwarz et al., 1987).

EGA has also been produced in other heterologous hosts. The *celA* gene, cloned in a yeast vector, was expressed in *Saccharomyces cerevisiae* at a level of about 30% of the spontaneous level of expression in *E. coli* and about 10% of that observed in *C. thermocellum* (Sacco et al., 1984). EGA is secreted from *Bacillus subtilis* and *Bacillus stearothermophilus* strains carrying *celA* cloned in the *Bacillus* vector pUB110 (Soutschek-Bauer and Staudenbauer, 1987). More recently, EGA secretion by *B. subtilis* was studied by placing the *celA* gene, with its own signal sequence or with the signal sequence of the *B. subtilis* levansucrase, under the control of the levansucrase promoter of *B. subtilis*. Expression of *celA* was found to be dependent on levansucrase regulation. The enzyme was secreted at a high rate during the log phase by a *sacU*(Hy) strain grown in the presence of sucrose (Joliff et al., 1989). Production was further increased by the introduction of the *sacB-celA* construct into the chromosome of a *B. subtilis* *sacU*(Hy) strain and amplification of the gene (Petit et al., 1990).

30.5 The Products of the *cel* and *xyn* Genes

General properties

cel and *xyn* gene products have been isolated from *E. coli* strains carrying the relevant gene: EGA (Schwarz et al., 1986), EGB (Béguin et al., 1983), EGC (Pétré et al., 1986; Schwarz et al., 1988b), EGD (Joliff et al., 1986b), and XYNZ (Grépinet et al., 1988a). Only EGA has been isolated from *C. thermocellum* culture supernatants (Pètre et al., 1981). EGD was isolated from inclusion bodies and crystallized; its three-dimensional structure is being analyzed by x-ray diffraction (Joliff et al., 1986b, 1986c). As a general rule, the enzymes purified from *E.*

coli had a slightly smaller M_r than that predicted by the nucleotide sequence or determined by Western blot of *C. thermocellum* culture supernatant. This may result from limited proteolysis, as shown in the case of EGD, which is cleaved in one of the two copies of the reiterated sequence (Chauvaux et al., 1990). In addition, it is possible that the enzymes produced by *C. thermocellum* are glycosylated, as shown for the endoglucanase purified by Ng and Zeikus (1981). However, all the enzymes are thermostable with maximum activity between 60° and 70°C. The energy of activation, when determined, was found to be high, for example, 45 kJ/mol for EGC (Pétré et al., 1986) and 26 kJ/mol for EGD (Joliff et al., 1986b); this may explain the low activities at 30°C. Determination of the M_r of native EGA, EGB, EGC, and EGD showed that these enzymes were monomeric proteins.

Specificity and catalytic properties

Comparisons of specificities of purified EGA, EGB, EGC, and EGD on various substrates are shown in Table 30.1. Specific CMCase activities of the EGs vary widely. Although it is difficult to compare determinations of CMCase activities performed in different laboratories, EGD with a specific activity of 430 IU mg^{-1} protein appears to be one of the most active cellulases described (Joliff et al., 1986b). The four EGs are also active, although to a lesser extent, on lichenan, a β-glucan containing 1,4 and 1,3 linkages. They have a very low or no Avicelase or xyla-

Table 30.1 Activity of EGA, EGB, EGC, and EGD on various substrates[a]

	CMC	Lichenan	*p*-NPC	G3	G4	G5
EGA	140	15	−	−	−	+
EGB	60	2.4	−	−	+	+
EGC	27	27	260	+	+	+
EGD	430	26	1.6	−	+	+

[a] Activity is expressed in micromole glucose equivalents or *p*-nitrophenol liberated per minute and per mg protein; CMC, carboxymethylcellulose; *p*-NPC, *p*-nitrophenyl-β-D-cellobioside; G3, cellotriose; G4, cellotetraose; G5, cellopentaose (for experimental conditions, see Pétré et al., 1986). Minus, no activity; plus, activity.

nase activity. EGs have been assayed for cellodextrinase activity. EGA hydrolyzes only cellopentaose and seems to be the most selective of the four EGs tested. EGC hydrolyzes cellotriose, cellotetraose, and cellopentaose, and appears to be the least selective EG. EGB and EGD hydrolyze cellotetraose and cellopentaose, and appear to have an intermediate selectivity.

In all cases, cellobiose, which is not hydrolyzed, is released. EGC hydrolyzes barley β-glucan, cleaving nonspecifically β-1,3 and β-1,4 bonds adjacent to cellobiose units (Schwarz et al., 1988b). Of the four EGs, EGC is the only one to have a high activity on p-NPC, which is consistent with its activity on cellotriose. The activities of EGE, EGF, and EGH have only been tested using crude extracts. EGH has a low activity on p-NPC and xylan (1:100 of the CMCase activity) (Yagüe et al., 1990); EGF has no activity on either of these two substrates (unpublished results). In contrast, EGE has a relatively high xylanase activity (1:6 of the CMCase activity) (Hall et al., 1988). Purified XYNZ has a specific activity of 166, 340, and 13 IU mg^{-1} protein for xylan, p-NP-β-D-xylobioside, and pNPC respectively, and no detectable CMCase activity (Grépinet et al., 1988a).

Analysis of the amino acid sequences deduced from the nucleotide sequences of some 50 fungal or bacterial cellulase genes led to the conclusion that most of the cellulases were organized in catalytic and noncatalytic domains linked in different orders (Béguin, 1990). In organisms such as *Trichoderma reesei* and *Cellulomonas fimi*, which are able to hydrolyze native cellulose, the noncatalytic domains, located near either the N terminus or the C terminus, have been shown to carry a substrate-binding region. In *C. thermocellum* EGs, catalytic and noncatalytic domains have also been identified (Béguin, 1990), but binding to native cellulose has not been unambiguously established. This is not surprising as individual isolated *C. thermocellum* EGs have little or no Avicelase activity.

The role of the reiterated amino acid sequence (see Figure 30.3) is still unknown. Removal of this sequence by deletion from the

genes of EGD (Chauvaux et al., 1990), EGE (Hall et al., 1988), EGH (Yagüe et al., 1990), and XYNZ (Grépinet et al., 1988b), showed that it was not required for catalytic activity. An interesting and unexpected observation was that a consensus of the first 12 residues of the reiterated sequence displayed a striking similarity with the EF-hand class of Ca^{2+}-binding sites of various Ca^{2+}-binding proteins (Chauvaux et al., 1990). Experiments with EGD showed that the enzyme bound Ca^{2+} with an association constant of 2.03×10^6 M^{-1}. However, binding was independent of the presence of the reiterated sequence (Chauvaux et al., 1990).

It is possible that the reiterated sequence plays a role in the formation of the cellulosome, by anchoring the EGs to the S_L cellulose-binding factor. This hypothesis is yet to be tested. It is in agreement with the observation that EGD (Lamed and Bayer, 1988), EGE (G.P. Hazlewood, personal communication), and XYNZ (Grépinet et al., 1988a), which carry the reiterated sequence, were shown to belong to the cellulosome; in contrast, EGC, which is devoid of the sequence, is not associated with the cellulosome (E.A. Bayer and R. Lamed, personal communication).

The cellulase whose catalytic domain is best characterized is the cellobiohydrolase CBHII of *T. reesei*. The catalytic core has been crystallized and the three-dimensional structure established (Teeri et al., 1990). The core has a diameter of 50 Å and consists of a seven-stranded, singly wound, α-β barrel. Substrate binds to a large channel formed by extended loops from the barrel, and the active site is located at the C terminus of the β sheets. From active site-modification studies of *T. reesei* enzymes, it appeared that carboxylic acid groups act as proton donor/acceptors in catalysis (Tomme and Claeyssens, 1989). Using Woodward reagent K and diethylpyrocarbonate, *C. thermocellum* EGD was also shown to contain a carboxylic and a histidyl group in the active site. Preliminary evidence, based on site-directed mutagenesis, suggests that a glutamyl residue and a histidyl residue located in the C-terminal region of the enzyme probably participate in catalysis (M. Claeyssens' group and our laboratory, unpublished observations).

30.6 Conclusion and Prospects

Although it is generally risky to make predictions, it is likely that the characterization of new *cel* or *xyn* genes of *C. thermocellum* will continue without any major difficulty. It is also likely that, by combining x-ray diffraction analysis, analysis of active sites by chemical modifications, and site-directed mutagenesis, the mechanism by which CMC or synthetic substrates are hydrolyzed by at least a few EGs from *C. thermocellum* will be established in the near future. In contrast, the problems of the fine organization, of the mechanism of action on crystalline cellulose, and of the morphogenesis of the cellulosome or polycellulosome are far from being solved.

All studies on the structural organization of the cellulosome to date have been based on biochemical approaches and electron microscopy. In vitro reconstruction experiments may be possible when all the structural genes of the constituents have been cloned and expressed in *E. coli* or another host. It is clear, however, that in vivo genetics in *C. thermocellum*, which would permit for example the inactivation of specific genes by homologous recombination, would be extremely helpful. This approach is entirely dependent on the development of tools for transferring genetic material into *C. thermocellum*. From the first promising results obtained in the past few years with several *Clostridium* species (Young et al., 1989), it is hoped that *C. thermocellum* will also be made amenable to genetic analysis.

The mechanism of action of the cellulosome on crystalline cellulose is still unclear. The properties of the individual cellulases and the experiments of partial reconstruction performed by Wu et al. (1988) indicated that the classical fungal model of synergism between endoglucanases and cellobiohydrolases freely released into the medium does not apply as such to the cellulase complex of *C. thermocellum*. A sophisticated biochemical analysis is likely to be required to understand how cellulases act, even in the simple case of the S_S component described by Wu et al. (1988), which displays only CMCase activity in the free state and acquires Avicelase activity when associated with the anchorage S_L protein. It is possible that only a few EGs are endowed with both CMCase and Avicelase activity. These EGs, when associated with the cellulosome, may hydrolyze β-1,4 bonds inside the cellulose fibers, leaving in place a cellodextrin molecule that would subsequently be degraded into cellobiose by other EGs endowed only with CMCase activity. This model, a modified version of the fungal synergistic model, could be tested experimentally by measuring the Avicelase activity of the various EGs already characterized when they are mixed with the S_L protein of Wu et al. (1988).

The most difficult question is probably the morphogenesis of the polycellulosome organelles located at the surface of the cell wall. It is likely that the rapid development of our knowledge in the field of protein secretion by gram-positive and gram-negative bacteria will furnish suggestions how the constituents of the cellulosome are translocated across the cytoplasmic membrane and the peptidoglycan layer. The main problem—how these constituents are assembled outside the cell wall into cellulosomes and polycellulosomes, while remaining attached to the cell wall during a large part of the cell cycle—will still require elucidation. Given the tools currently available to address this problem, a quick answer cannot be expected.

Acknowledgments. The research in our laboratory reported in this chapter was supported in part by a contract from Solvay & Cie Brussels, Belgium and Rhône-Poulenc Recherche, Paris, France; by grants ATP 955355 from CNRS and EN3B-0082-F from the Commission of European Communities; and by research funds from Paris 7 University.

References

Baird, S.D., D.A. Johnson, and V.L. Seligy. 1990. Molecular cloning, expression, and characterization of endo-β-1,4-glucanase genes from *Bacillus polymyxa* and *Bacillus circulans*. *J. Bacteriol.* **172**:1576–1586.

Béguin, P. 1990. Molecular biology of cellulose degradation. *Annu. Rev. Microbiol.* **44**:219–248.

Béguin, P., P. Cornet, and J. Millet. 1983. Identification of the endoglucanase encoded by the celB gene of *Clostridium thermocellum*. *Biochimie* (Paris) **65**:495–500.

Béguin, P., P. Cornet, and J.-P. Aubert. 1985. Sequence of a cellulase gene of the thermophilic bacterium *Clostridium thermocellum*. *J. Bacteriol.* **162**:102–105.

Béguin, P., O. Grépinet, J. Millet, and J.-P. Aubert. 1988. Recent aspects in the biochemistry and genetics of cellulose degradation. *In: 8th International Biotechnology Symposium*, G. Durand, L. Bobichon, and J. Florent, eds., pp. 1015–1029. Paris: Société Française de Microbiologie.

Béguin, P., M. Rocancourt, M.-C. Chebrou, and J.-P. Aubert. 1986. Mapping of mRNA encoding endoglucanase A from *Clostridium thermocellum*. *Mol. Gen. Genet.* **202**:251–254.

Chauvaux, S., P. Béguin, J.-P. Aubert, M. Bhat, L.A. Gow, T.M. Wood, and A. Bairoch. 1990. Calcium-binding affinity and calcium-enhanced activity of *Clostridium thermocellum* endoglucanase D. *Biochem. J.* **265**:261–265.

Cornet, P., J. Millet, P. Béguin, and J.-P. Aubert. 1983a. Characterization of two *cel* (cellulose degradation) genes of *Clostridium thermocellum* coding for endoglucanases. *Bio/Technology* **1**:589–594.

Cornet, P., D. Tronik, J. Millet, and J.-P. Aubert. 1983b. Cloning and expression in *Escherichia coli* of *Clostridium thermocellum* genes coding for amino acid synthesis and cellulose hydrolysis. *FEMS Microbiol. Lett.* **16**:137–141.

Coughlan, M.P., K. Hon-nami, H. Hon-nami, G. Ljungdhal, J.J. Paulin, and W.E. Rigsby. 1985. The cellulolytic enzyme complex of *Clostridium thermocellum* is very large. *Biochem. Biophys. Res. Commun.* **130**:904–909.

Deshpande, M.V., K.E. Eriksson, and L.G. Pettersson. 1984. An assay for selective determination of exo-1,4-β-glucanase in a mixture of cellulolytic enzymes. *Anal. Biochem.* **138**:481–487.

Faure, E., A. Belaich, C. Bagnara, C. Gaudin, and J.-P. Belaich. 1989. Sequence analysis of the *Clostridium cellulolyticum* endoglucanase-A-encoding gene, *celCCA*. *Gene* (Amst.) **84**:39–46.

Fukumori, F., N. Sashihara, T. Kudo, and K. Horikoshi. 1986. Nucleotide sequence of two cellulase genes from alkalophilic *Bacillus* sp. strain N-4 and their strong homology. *J. Bacteriol.* **168**:479–485.

Garcia-Martinez, D.V., A. Shinmyo, A. Madia, and A.L. Demain. 1980. Studies on cellulase production by *Clostridium thermocellum*. *Eur. J. Appl. Microbiol. Biotechnol.* **9**:189–197.

Gerwig, G.J., P. de Waard, J.P. Kamerling, J.F.G. Vliegenthart, E. Morgenstern, R. Lamed, and E.A. Bayer. 1989. Novel O-linked carbohydrate chains in the cellulase complex (cellulosome) of *Clostridium thermocellum*. *J. Biol. Chem.* **264**:1027–1035.

Gräbnitz, F. and W.L. Staudenbauer. 1988. Characterization of two β-glucosidase genes from *Clostridium thermocellum*. *Biotechnol. Lett.* **10**:73–78.

Gräbnitz, F., K.P. Rücknagel, M. Seiss, and W.L. Staudenbauer. 1989. Nucleotide sequence of the *Clostridium thermocellum bglB* gene encoding thermostable β-glucosidase B: homology to fungal β-glucosidases. *Mol. Gen. Genet.* **217**:70–76.

Grépinet, O. and P. Béguin. 1986. Sequence of the cellulase gene of *Clostridium thermocellum* coding for endoglucanase B. *Nucleic Acids Res.* **14**:1791–1799.

Grépinet, O., M.-C. Chebrou, and P. Béguin. 1988a. Purification of *Clostridium thermocellum* xylanase Z expressed in *Escherichia coli* and identification of the corresponding product in the culture medium of *C. thermocellum*. *J. Bacteriol.* **170**:4576–4581.

Grépinet, O., M.-C. Chebrou, and P. Béguin. 1988b. Nucleotide sequence and deletion analysis of the xylanase gene *xynZ* of *Clostridium thermocellum*. *J. Bacteriol.* **170**:4582–4588.

Hall, J. and H.J. Gilbert. 1988. The nucleotide sequence of a carboxymethylcellulase from *Pseudomonas fluorescens* subsp. *cellulosa*. *Mol. Gen. Genet.* **213**:112–117.

Hall, J., G.P. Hazlewood, P.J. Barker, and H.J. Gilbert. 1988. Conserved reiterated domains in *Clostridium thermocellum* endoglucanases are not essential for activity. *Gene* (Amst.) **69**:29–38.

Hazlewood, G.P., M.P.M. Romaniec, K. Davidson, O. Grépinet, P. Béguin, J. Millet, O. Raynaud, and J.-P. Aubert. 1988. A catalogue of *Clostridium thermocellum* endoglucanase, β-glucosidase and xylanase genes cloned in *Escherichia coli*. *FEMS Microbiol. Lett.* **51**:231–236.

Henrissat, B., M. Claeyssens, P. Tomme, L. Lemesle, and J.-P. Mornon. 1989. Cellulase families revealed by hydrophobic cluster analysis. *Gene* (Amst.) **81**:83–95.

Johnson, E.A. and A.L. Demain. 1984. Probable involvement of sulfhydryl groups and a metal as essential components of the cellulase of *Clostridium thermocellum*. *Arch. Microbiol.* **137**:135–138.

Johnson, E.A., F. Bouchot, and A.L. Demain. 1985. Regulation of cellulase formation by *Clostridium thermocellum*. *J. Gen. Microbiol.* **131**:2303–2308.

Johnson, E.A., M. Sakajoh, G. Halliwell, A. Madia, and A.L. Demain. 1982. Saccharification of complex cellulosic substrates by the cellulase system from *Clostridium thermocellum*. *Appl. Environ. Microbiol.* **43**:1125–1132.

Joliff, G., P. Béguin, and J.-P. Aubert. 1986a. Nucleotide sequence of the cellulase gene *celD* encoding endoglucanase D of *Clostridium thermocellum*. *Nucleic Acids Res.* **14**:8605–8613.

Joliff, G., P. Béguin, M. Juy, J. Millet, A. Ryter, R. Poljak, and J.-P. Aubert. 1986b. Isolation, crystallisation and properties of a new cellulase of *Clostridium thermocellum* overproduced in *Escherichia coli*. *Bio/Technology* **4**:896–900.

Joliff, G., P. Béguin, J. Millet, J.-P. Aubert, P. Alzari, M. Juy, and R.J. Poljak. 1986c. Crystalization and preliminary X-ray diffraction study of an endoglucanase from *Clostridium thermocellum*. *J. Mol. Biol.* **189**:249–250.

Joliff, G., A. Edelman, A. Klier, and G. Rapoport. 1989. Inducible secretion of a cellulase from *Clostridium thermocellum* in *Bacillus subtilis*. *Appl. Environ. Microbiol.* **55**:2739–2744.

Lamed, R. and E.A. Bayer. 1988. The cellulosome concept: exocellular/extracellular enzyme reactor centers for efficient binding and cellulolysis. *In: FEMS Symposium No. 43, Biochemistry and Genetics of Cellulose Degradation*, J.-P. Aubert, P. Béguin, and J. Millet, eds., pp. 101–116. London: Academic Press.

Lamed, R., E.A. Bayer, B.C. Saha, and J.G. Zeikus. 1988. Biotechnological potential of enzymes from unique thermophiles. *In: 8th International Biotechnology Symposium*, G. Durand, L. Bobichon, and J. Florent, eds., pp. 371–383. Paris: Société Française de Microbiologie.

Lamed, R., R. Kenig, E. Setter, and E.A. Bayer. 1985. Major characteristics of the cellulolytic system of *Clostridium thermocellum* coincide with those of the purified cellulosome. *Enzyme Microb. Technol.* **7**:37–41.

Lamed, R., E. Setter, and E.A. Bayer. 1983a. Characterization of a cellulose-binding, cellulase-containing complex in *Clostridium thermocellum*. *J. Bacteriol.* **156**:828–836.

Lamed, R., E. Setter, R. Kenig, and E.A. Bayer. 1983b. The cellulosome—A discrete cell surface organelle of *Clostridium thermocellum* which exhibits separate antigenic, cellulose-binding and various cellulolytic activities. *Biotechnol. Bioeng. Symp.* **13**:163–181.

Ljungdahl, L.G., B. Pettersson, K.E. Eriksson, and J. Wiegel. 1983. A yellow affinity substance involved in the cellulolytic system of *Clostridium thermocellum*. *Curr. Microbiol.* **9**:195–200.

Ljungdahl, L.G., M.P. Coughlan, F. Mayer, Y. Mori, and K. Hon-nami. 1988. Macrocellulase complexes and yellow affinity substance from *Clostridium thermocellum*. *Methods Enzymol.* **160**:483–500.

Mayer, F., M.P. Coughlan, Y. Mori. and L.G. Ljungdahl. 1987. Macromolecular organization of the cellulolytic complex of *Clostridium thermocellum* as revealed by electron microscopy. *Appl. Environ. Microbiol.* **53**:2785–2792.

Millet, J., D. Pétré, P. Béguin, O. Raynaud, and J.-P. Aubert. 1985. Cloning of ten distinct DNA fragments of *Clostridium thermocellum* coding for cellulases. *FEMS Microbiol. Lett.* **29**:145–149.

Mishra, S., P. Béguin, and J.-P. Aubert. 1991. Transcription of *Clostridium thermocellum* endoglucanase genes *celF* and *celD*. *J. Bacteriol.* **173**:80–85.

Navarro, A., M.-C. Chebrou, P. Béguin, and J.-P. Aubert. 1991. Nucleotide sequence of the cellulase gene *celF* of *Clostridium thermocellum*. *Res. Microbiol.* **142**:927–936.

Ng, T.K. and J.G. Zeikus. 1981. Purification and characterization of an endoglucanase (1,4-β-D-glucan glucanohydrolase) from *Clostridium thermocellum*. *Biochem. J.* **199**:341–350.

Ng, T.K., P.J. Weimer, and J.G. Zeikus. 1977. Cellulolytic and physiological properties of *Clostridium thermocellum*. *Arch. Microbiol.* **114**:1–7.

O'Neill, G., S.H. Goh, R.A.J. Warren, D.G. Kilburn, and R.C. Miller, Jr. 1986. Structure of the gene encoding the exoglucanase of *Cellulomonas fimi*. *Gene* (Amst.) **44**:325–330.

Ong, E., J.M. Greenwood, N.R. Gilkes, D.G. Kilburn, R.C. Miller, Jr., and R.A.J. Warren. 1989. The cellulose-binding domains of cellulases: tools for biotechnology. *Trends Biotechnol.* **7**:239–243.

Petit, M.-A., G. Joliff, J.M. Mesas, A. Klier, G. Rapoport, and S.D. Ehrlich. 1990. Hypersecretion of a cellulase from *Clostridium thermocellum* in *Bacillus subtilis* by induction of chromosomal DNA amplification. *Bio/Technology* **8**:559–563.

Pètre, J., R. Longin, and J. Millet. 1981. Purification and properties of an endo-β-1,4-glucanase from *Clostridium thermocellum*. *Biochimie* (Paris) **63**:629–639.

Pétré, D., J. Millet, R. Longin, P. Béguin, H. Girard, and J.-P. Aubert. 1986. Purification and properties of the endoglucanase C of *Clostridium thermocellum* produced in *Escherichia coli*. *Biochimie* (Paris) **68**:687–695.

Piruzian, E.S., M.A. Mogutov, G.A. Velikodov, and V.K. Akimenko. 1985. Cloning and expression of structural genes of *Clostridium thermocellum* F7 cellulolytic complex endoglucanases in cells of *Escherichia coli* K12. *Doklady Acad. Nauk. SSR* **281**:963–965.

Romaniec, M.P.M., N.G. Clarke, and G.P. Hazlewood. 1987. Molecular cloning of *Clostridium thermocellum* DNA and the expression of further novel endo-β-endoglucanase genes in *Escherichia coli*. *J. Gen. Microbiol.* **133**:1297–1307.

Sacco, M., J. Millet, and J.-P. Aubert. 1984. Cloning and expression in *Saccharomyces cerevisiae* of a cellulase gene from *Clostridium thermocellum*. *Ann. Microbiol. Inst. Pasteur* (Paris) **135A**:485–488.

Schwarz, W., K. Bronnenmeier, and W.L. Staudenbauer. 1985. Molecular cloning of *Clos-*

tridium thermocellum genes involved in β-glucan degradation in bacteriophage lambda. *Biotechnol. Lett.* **7**:859–864.

Schwarz, W.H., F. Gräbnitz, and W.L. Staudenbauer. 1986. Properties of a *Clostridium thermocellum* endoglucanase produced in *Escherichia coli. Appl. Environ. Microbiol.* **51**:1293–1299.

Schwarz, W.H., S. Schimming, and W.L. Staudenbauer. 1987. High level expression of *Clostridium thermocellum* cellulase genes in *Escherichia coli. Appl. Microbiol. Biotechnol.* **27**:50–56.

Schwarz, W.H., S. Schimming, K.P. Rücknagel, S. Burgschwaiger, G. Kreil, and W.L. Staudenbauer. 1988a. Nucleotide sequence of the *celC* gene encoding endoglucanase C of *Clostridium thermocellum. Gene* **63**:23–30.

Schwarz, W.H., S. Schimming, and W.L. Staudenbauer. 1988b. Degradation of barley β-glucan by endoglucanase C of *Clostridium thermocellum. Appl. Microbiol. Biotechnol.* **29**:25–31.

Shinmyo, A., D.V. Garcia-Martinez, and A.L. Demain. 1979. Studies on the extracellular cellulolytic enzyme complex produced by *Clostridium thermocellum. J. Appl. Biochem.* **1**:202–209.

Soutschek-Bauer, E. and W.L. Staudenbauer. 1987. Synthesis and secretion of a heat-stable carboxymethylcellulase from *Clostridium thermocellum* in *Bacillus subtilis* and *Bacillus stearothermophilus. Mol. Gen. Genet.* **208**:537–541.

Teather, R.M. and P.J. Wood. 1982. Use of Congo red-polysaccharide interactions in enumeration and characterization of cellulolytic bacteria from the bovine rumen. *Appl. Environ. Microbiol.* **43**:777–780.

Teeri, T.T., A. Jones, P. Kremlis, J. Rouvinen, M. Penttilä, A. Harkki, H. Nevalainen, S. Vanhanen, M. Saloheimo, and J.K.C. Knowles. 1990. Engineering *Trichoderma* and its cellulases. *In: Trichoderma reesei Cellulases-Biochemistry, Genetics, Physiology and Applications,* C.P.

Kubicek, D.E. Eveleigh, H. Esterbauer, W. Steiner, and E.M. Kubicek-Pranz. eds. pp. 156–167. Cambridge: The Royal Society of Chemistry, Thomas Graham House.

Teeri, T.T., P. Lehtovaara, S. Kauppinen, I. Salovuori, and J. Knowles. 1987. Homologous domains in *Trichoderma reesei* cellulolytic enzymes: gene sequence and expression of cellobiohydrolase II. *Gene* (Amst.) **51**:43–52.

Tomme, P. and M. Claeyssens. 1989. Identification of a functionally important carboxyl group in cellobiohydrolase I from *Trichoderma reesei. FEBS Lett.* **243**:239–243.

Tucker, M.L., M.L. Durbin, M.T. Clegg, and J.N. Lewis. 1987. Avocado cellulase: nucleotide sequence of a putative full-length cDNA clone and evidence for a small gene family. *Plant Mol. Biol.* **9**:197–203.

Van Tilbeurgh, H., M. Claeyssens, and C.K. De Bruyne. 1982. The use of 4-methylumbelliferyl and other chromophoric glycosides in the study of cellulolytic enzymes. *FEBS Lett.* **149**:152–156.

Wu, J.D.H. and A.L. Demain. 1988. Proteins of the *Clostridium thermocellum* cellulase complex responsible for degradation of crystalline cellulose. *In: FEMS Symposium No. 43. Biochemistry and Genetics of Cellulose Degradation,* J.-P. Aubert, P. Béguin, and J. Millet, eds., pp. 117–131. London: Academic Press.

Wu, J.D.H., W.H. Orme-Johnson, and A.L. Demain. 1988. Two components of an extracellular protein aggregate of *Clostridium thermocellum* together degrade crystalline cellulose. *Biochemistry* **27**:1703–1709.

Yagüe, E., P. Béguin, and J.-P. Aubert. 1990. Nucleotide sequence and deletion analysis of the cellulase-encoded gene *celH* of *Clostridium thermocellum. Gene* (Amst.) **89**:61–67.

Young, M., N.P. Minton, and W.L. Staudenbauer. 1989. Recent advances in the genetics of the clostridia. *FEMS Microbiol. Rev.* **63**:301–326.

31

Nucleotide Sequence of the Gene and Primary Structure of the Thermophilic β-Amylase from *Clostridium thermosulfurogenes*

Hideo Yamagata and Shigezo Udaka

31.1 Introduction

β-Amylase is an exo-type enzyme that hydrolyzes starch by removing stepwise maltose of the β-anomeric configuration from the nonreducing end of the starch molecule. Although α-amylases, endo-type enzymes that hydrolyze α-1,4 linkages of starch at random, are distributed in various kinds of organisms, β-amylases are known to be produced only by plants and certain bacteria. Extensive studies on α-amylases have revealed that α-amylases of diverse origins from mammalian to bacterial contain common well-conserved regions including active centers (Toda et al., 1982; Nakajima et al., 1986). Three-dimensional structures of α-amylases from *Aspergillus oryzae* and porcine have been determined at 3 and 2.9 Å resolution, respectively (Matsuura et al., 1984; Buisson et al., 1987). Amino acid residues stabilizing a bacillus α-amylase against irreversible thermoinactivation have been determined (Suzuki et al., 1989). In contrast, relatively little is known about the structure–function relationships of β-amylases.

The nucleotide sequences encoding the β-amylases of barley and soybean have been determined. For bacterial β-amylases, the genes encoding β-amylases of *Bacillus polymyxa* and *Clostridium thermosulfurogenes* have been cloned

and their nucleotide sequences have been determined (Kawazu et al., 1987; Kitamoto et al., 1988; Uozumi et al., 1989).

Clostridium thermosulfurogenes, a thermophilic and anaerobic bacterium, was isolated from a thermal, volcanic, algal-bacterial community via selective enrichment procedures with pectin as energy source (Schink and Zeikus, 1983). The β-amylase of *C. thermosulfurogenes* is unique among the β-amylases so far examined because it is stable and optimally active at 80° and 75°C, respectively (Hyun and Zeikus, 1985a). The enzyme should be useful for investigation of structure–function relationships of β-amylases and also for bioprocessing of starch. The *C. thermosulfurogenes* β-amylase gene cloned in *Bacillus subtilis* directed synthesis of the enzyme as thermophilic as that produced by *C. thermosulfurogenes* (Kitamoto et al., 1988).

In this chapter, we describe features of the nucleotide sequence of the cloned gene and the primary structure of the thermophilic β-amylase deduced from the nucleotide sequence, and discuss possible mechanisms conferring thermophilicity on the enzyme. In addition, we describe very efficient production of the enzyme with the aid of the *Bacillus brevis* host–vector system developed by us (Yamagata et al., 1989).

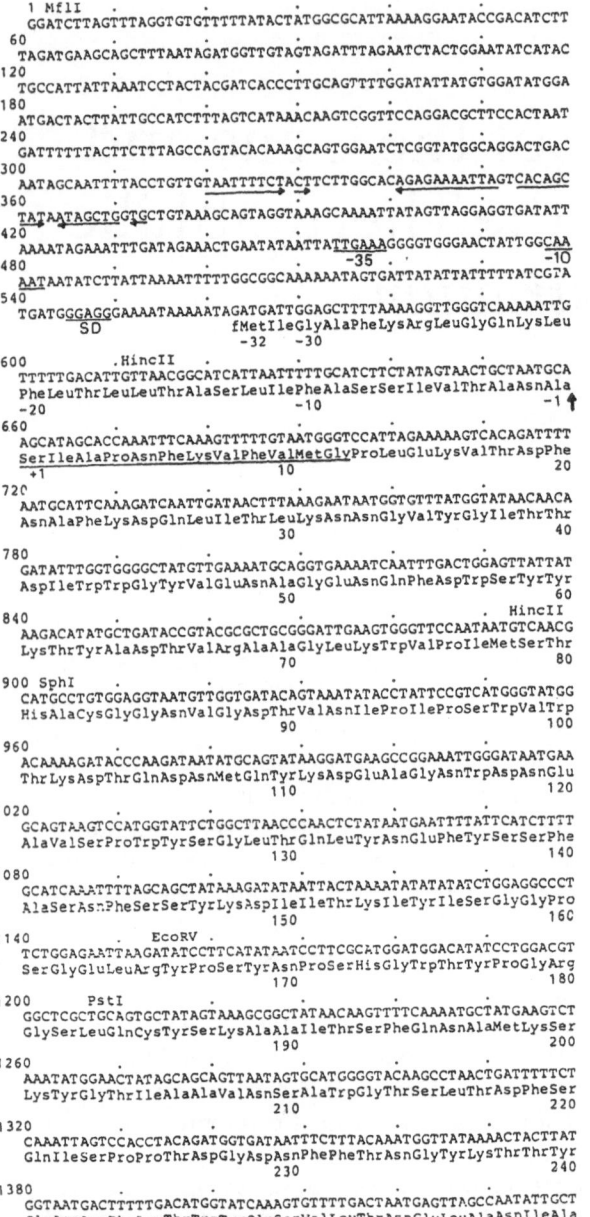

FIGURE 31.1 Nucleotide sequence of β-amylase gene and its flanking regions; only sequence of anti-sense strand is shown. Deduced amino acid sequence of the precursor β-amylase is shown below nucleotide sequence. Possible −35 and −10 sequences in promoter region and possible ribosome-binding site, SD, are underlined. Palindromic sequences shown by horizontal arrows below sequence. Amino acid sequence identical with that of NH2-terminus of purified mature *Clostridium thermosulfurogenes* β-amylase is underlined. Possible signal peptide cleavage site is indicated by vertical arrow. Restriction sites shown above nucleotide sequence. (From Kitamoto et al., 1988.)

31.2 Nucleotide Sequence of the β-Amylase Gene

The nucleotide sequence of the 2823-base pair(bp) *Mfl*I-*Eco*RI fragment containing the β-amylase gene cloned from *Clostridium thermosulfurogenes* is shown in Figure 31.1. Only one long open-reading frame (ORF) was found; it starts from ATG at nucleotide number 564 and ends in TAA at 2217. The ORF encodes a polypeptide of 551 amino acid residues. The amino acid sequence deduced from the DNA sequence contains the NH_2-terminal amino acid sequence of the mature β-amylase of *C. thermosulfurogenes* as determined chemically (amino acids +1 through +12; see underlined in Figure 31.1). The amino acid sequence from −32 to −1 shows characteristics typical of signal peptides of secretory precursors, i.e., two positively charged residues near the NH_2 terminus, followed by a hydrophobic stretch (Inouye and Halegoua, 1980).

Sequences homologous to the consensus sequences for the σ^{43}-RNA polymerase of *Bacillus subtilis* (TTGACA at −35 and TATAAT at −10 region; Moran et al., 1982) and a possible ribosomal binding site (SD), a sequence homologous to the 3′ end of *B. subtilis* 16S ribosomal RNA (Mclaughlin et al., 1981), are found upstream from the ORF (TTGAAA at nucleotide 454 through 459, CAAAAT at nucleotide 477 through 482, and GGAGGGA at nucleotide 545 through 551; underlined in Figure 31.1). A palindromic sequence that could form a stable stem and loop structure ($\Delta G = -19.5$ kcal/mol; nucleotides 2251 through 2276) followed by an AT-rich sequence are characteristics of ρ-independent transcriptional terminators of *Escherichia coli* (Rosenberg and Court, 1979). These findings suggest that the β-amylase of *C. thermosulfurogenes* is translated from a monocistronic mRNA as a secretory precursor with a signal peptide of 32 amino acid residues. The mature *C. thermosulfurogenes* β-amylase was deduced to consist of 519 amino acid residues with a molecular weight of 57,167. The molecular weight agrees well with that estimated from sodium dodecyl sulfate-polyacrylamide gel electrophoresis (SDS-PAGE) of the purified enzyme.

31.3 Codon Usage

Codon usage in the *Clostridium thermosulfurogenes* β-amylase gene is shown in Table 31.1. The G + C content of the β-amylase gene is 35.4 mol%, which agrees well with the value of 32.6 mol% for the total DNA of *C. thermosulfurogenes* (Schink and Zeikus, 1983). The G + C content of the third positions of codons used (excluding nonwobbled AUG for Met and UGG for Trp) is only 16.7%. This extreme preference for codons ending in A or T was also found in the α-amylase-pullulanase gene of *Clostridium thermohydrosulfuricum* (Melasniemi et al., 1990) and clearly discriminates the two *Clostridium* genes from the *Escherichia coli* and *B. subtilis* genes; only UUG for Leu and UGC for Cys are used with comparable frequencies as those codons ending with A or T. As for the usage of codons ending with A or T, the β-amylase gene resembles *B. subtilis* genes. Codons used frequently in *B. subtilis* but not frequently in *E. coli*, such as UCA and CUU (Leu), UCA and AGU (Ser), CCU (Pro), ACA (Thr), AGA (Arg), and GGA (Gly), are used frequently in the β-amylase gene. The exception is AUA (Ile), which is used frequently in the β-amylase gene but not in *B. subtilis* or *E. coli* genes.

31.4 Comparison of the Amino Acid Sequences of Various β-Amylases

Figure 31.2 compares the amino acid sequence of the mature β-amylase of *Clostridium thermosulfurogenes* with those of *Bacillus polymyxa* (Uozumi et al., 1989), soybean (Mikami et al., 1988), and barley (Kreis et al., 1987). Because the β-amylase of *B. polymyxa* is synthesized as a large molecule with a molecular weight higher than 100,000, the amino acid sequence of the NH_2-terminal 510 amino acid residues, which

Table 31.1   Codon usage for the *Clostridium thermosulfurogenes* β-amylase gene

Amino acid	Codon	Codon usage for C. thermosulfurogenes β-amylase gene	Bacillus subtilis proteins	Escherichia coli proteins
			Codon usage (%)[a]	
Phe	UUU	22	64	23
	UUC[b]	3	36	77
Leu	UUA	11	22	3
	UUG	9	14	4
	CUU	7	26	4
	CUC	2	10	3
	CUA	5	6	0
	CUG[b]	1	22	86
Ile	AUU	12	50	21
	AUC[b]	0	39	79
	AUA	20	11	0
Met	AUG	10	100	100
Val	GUU[b]	11	31	44
	GUC	4	25	7
	GUA[b]	15	25	31
	GUG[b]	4	19	17
Ser	UCU	16	24	44
	UCC	1	12	31
	UCA	9	19	2
	UCG	4	10	0
	AGU	13	11	6
	AGC	6	24	17
Pro	CCU	13	34	6
	CCC	0	10	3
	CCA[b]	10	19	18
	CCG[b]	3	38	74
Thr	ACU[b]	14	15	48
	ACC[b]	5	14	41
	ACA	28	43	7
	ACG	7	28	3
Ala	GCU[b]	18	28	50
	GCC	5	20	6
	GCA[b]	18	27	26
	GCG[b]	3	25	18
Tyr	UAU	36	62	17
	UAC[b]	0	38	83
His	CAU	7	69	32
	CAC	1	31	68
Gln	CAA	12	54	14
	CAG[b]	4	46	86
Asn	AAU	48	53	5
	AAC[b]	4	47	95
Lys	AAA[b]	15	75	71
	AAG	6	25	29
Asp	GAA[b]	20	64	31
	GAC	2	36	69
Glu	GAA[a]	12	70	75
	GAG	3	31	25
Cys	UGU	4	46	20
	UGC	3	54	80

Table 31.1 (cont.)

Amino acid	Codon	Codon usage for C. thermosulfurogenes β-amylase gene	Bacillus subtilis proteins	Escherichia coli proteins
			Codon usage (%)[a]	
Trp	UGG	16	100	100
Arg	CGU[b]	2	25	70
	CGC[b]	1	18	29
	CGA	0	9	0
	CGG	0	11	0
	AGA	2	28	2
	AGG	1	9	0
Gly	GGU[b]	19	25	58
	GGC[b]	5	30	40
	GGA	18	32	1
	GGG	1	14	1

[a] Percent codon usage for 21 chromosomal genes in *B. subtilis* (Ogasawara, 1985) and for typical expressed genes in *E. coli* (Guoy and Gautier, 1982).
[b] The most abundant *E. coli* tRNAs (Ikemura, 1981).

is sufficient to constitute an enzymatically active fragment (Kawazu et al., 1987), was used in this study. Of the amino acid residues of *C. thermosulfurogenes* β-amylase, 54%, 32%, and 32% are homologous to those of the *B. polymyxa*, soybean, and barley β-amylases, respectively. The homology is higher on the NH₂-terminal side than on the COOH-terminal side. Twelve regions relatively well conserved among the four β-amylases are boxed in Figure 31.2. These regions include the three highly conserved sequences of β-amylases found by Mikami et al. (1988) that might be important for expression of the enzyme activity. Nitta et al. (1989) showed that Glu-186 of soybean β-amylase is a functional group at the catalytic site. Glu at a position corresponding to Glu-186 of soybean β-amylase are conserved in all four β-amylases and located within one of the highly conserved regions. Because β-amylases are susceptible to sulfhydryl reagents, Cys-83 and Cys-322 of the *C. thermosulfurogenes* β-amylase, which located within the highly conserved regions mentioned above, may play some roles in the catalytic reaction or in maintenance of the enzyme structure. Chemical modification of Cys-95 of the soybean enzyme, corresponding to Cys-83 of the *C. thermosulfurogenes* enzyme,

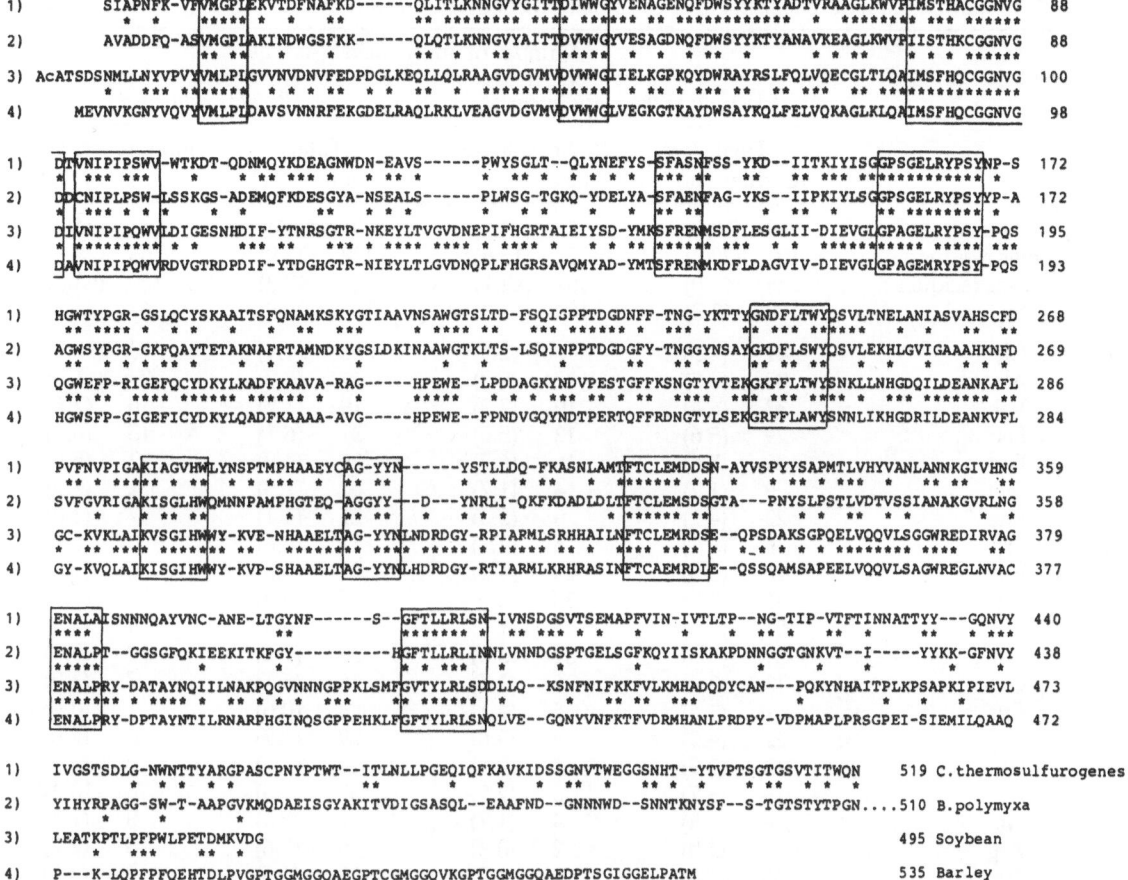

FIGURE 31.2 Comparison of amino acid sequences of mature β-amylases of (1) *Clostridium thermosulfurogenes*, (2) *Bacillus polymyxa* (Uozumi et al., 1989), (3) soybean (Mikami et al., 1988), and (4) barley (Kreis et al., 1987) are compared. For *B. polymyxa* β-amylase, amino acid sequence of 510 NH$_2$-terminal residues, which is sufficient to constitute an enzymatically active fragment, is used (see text). Sequences are aligned to obtain maximum homology; dashes in amino acid sequences denote deletion of corresponding residues; identical residues are denoted by asterisks between sequences; regions well conserved in all β-amylases are boxed. (From Kitamoto et al., 1988.)

led to loss of the activity of the enzyme (Nomura et al., 1987).

31.5 Possible Mechanisms Conferring Thermophilicity on *Clostridium thermosufurogenes* β-Amylase

On the basis of the high homology between the amino acid sequences of *C. thermosulfurogenes* and *Bacillus polymyxa* β-amylases, the structures of the two enzymes may be basically similar. The unique thermophilicity of the former enzyme should result from its nonhomologous amino acid residues. Table 31.2 compares the amino acid composition of *C. thermosulfurogenes* β-amylase to that of *B. polymyxa* β-amylase. Remarkable characteristics are as follows: (1) Higher content of Cys residue. The *C. thermosulfurogenes* β-amylase contains 7 Cys residues whereas the *B. polymyxa* β-amylase only contains 3. The increased number of Cys residues might be responsible for the heat stability of the former enzyme by generating additional disulfide bond(s), as shown by stabilization of the T4 lysozyme (Perry and Wetzel, 1984) and sub-

Table 31.2 Comparison of the amino acid compositions of mature β-amylases[a]

| | Number of amino acid residues | | | |
| | Clostridium thermosulfurogenes | | Bacillus polymyxa[b] | |
Amino acid	Total residues (%)	Nonhomologous residues (%)	Total residues (%)	Nonhomologous residues (%)
Charged residues	69 (13.3)	27 (13.3)	99 (19.4)	57 (24.8)
Hydrophilic residues	105 (20.2)	37 (15.5)	132 (25.9)	65 (28.3)
Hydrophobic residues	244 (47.0)	102 (42.7)	241 (47.3)	100 (43.5)
Neutral residues	170 (32.8)	100 (41.8)	137 (26.9)	65 (28.3)
Gly	41 (7.9)	6 (2.5)	58 (11.4)	23 (10.0)
Ala	39 (7.5)	22 (9.2)	44 (8.7)	27 (11.7)
Val	33 (6.4)	19 (7.9)	19 (3.7)	6 (2.6)
Leu	29 (5.6)	12 (5.0)	31 (6.1)	15 (6.3)
Ile	29 (5.6)	16 (6.7)	25 (4.9)	12 (5.0)
Met	9 (1.7)	4 (1.7)	7 (1.5)	2 (0.9)
Phe	22 (4.2)	8 (3.3)	21 (4.1)	7 (3.0)
Trp	16 (3.1)	5 (2.1)	13 (2.6)	2 (0.9)
Pro	26 (5.0)	10 (4.2)	23 (4.5)	6 (2.6)
Ser	46 (8.9)	30 (12.5)	47 (9.2)	29 (12.6)
Thr	51 (9.8)	28 (11.7)	32 (6.3)	9 (3.9)
Asn	51 (9.8)	31 (13.0)	39 (7.7)	19 (8.3)
Gln	15 (2.9)	6 (2.5)	16 (3.1)	7 (3.0)
Cys	7 (1.4)	5 (2.1)	3 (0.6)	1 (0.4)
Asp	22 (4.2)	9 (3.8)	29 (5.7)	16 (7.0)
Glu	15 (2.9)	6 (2.5)	18 (3.5)	9 (3.9)
Lys	19 (3.7)	6 (2.5)	37 (7.3)	24 (10.4)
His	8 (1.5)	4 (1.7)	7 (1.4)	3 (1.3)
Arg	5 (1.0)	2 (0.8)	8 (1.6)	5 (2.2)
Tyr	36 (7.0)	10 (4.2)	33 (6.5)	8 (3.5)
Total	519 (100)	239 (100)	510 (100)	230 (100)

[a] From Kitamoto et al., 1988.
[b] The 510 NH$_2$-terminal residues of the mature B. polymyxa enzyme were used (see text).

tilisin BPN' (Pantoliano et al., 1987) toward thermal inactivation. Cys-83 and Cys-91 of B. polymyxa β-amylase were shown to form a disulfide bond (Uozumi et al., 1991). (2) Lower content of hydrophilic amino acids. The number of hydrophilic amino acid residues is smaller in the C. thermosulfurogenes β-amylase (105 residues, 20% of the total residues) than in the B. polymyxa β-amylase (132 residues, 26% of the total residues). The difference is more evident when the nonhomologous residues of the two enzymes are compared; only 15% of the nonhomologous residues are hydrophilic in the former whereas 28% of those of the latter are hydrophilic. The lower content of hydrophilic amino acid residues might increase the internal hydrophobicity of the enzyme molecule so that

it folds into a heat-stable form with stronger internal packing in an aqueous solution, as is the case in the alteration of the thermostability of the Bacillus stearothermophilus neutral protease (Imanaka et al., 1986).

Figure 31.3 compares the hydropathy profiles of the C. thermosulfurogenes and B. polymyxa β-amylases. The hydrophobicity of the former is higher than that of the latter in general and especially in the several regions indicated by arrows. These regions might represent those internally packed in the β-amylase molecule, as mentioned. (3) Lower content of Gly residue; C. thermosulfurogenes β-amylase contains 41 Gly residues while B. polymyxa enzyme contains 58 Gly residues. Among the nonhomologous residues, only 2.5 % (6 residues, 5 of which locate

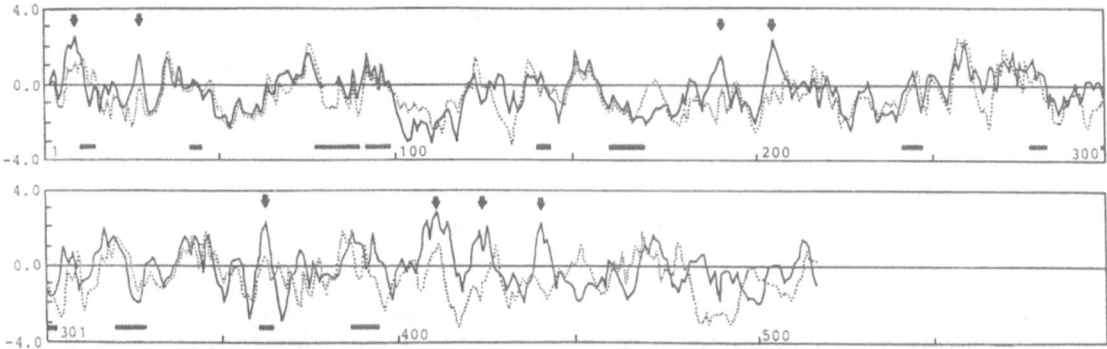

FIGURE 31.3 Comparison of hydropathy profiles of *Clostridium thermosulfurogenes* and *Bacillus polymyxa* β-amylases. Profile of mature β-amylase of *C. thermosulfurogenes* (solid line) and that of 510 NH$_2$-terminal amino acid residues of *B. polymyxa* β-amylase (dotted line) are compared. Abscissa of each panel shows amino acid number and ordinate average hydrophobicity (positive ordinate) or hydrophilicity (negative ordinate) of 5 amino acid residues calculated as described by Kyte and Doolittle (1982). Arrows indicate regions where hydrophobicity of *C. thermosulfurogenes* enzyme is markedly higher than that of *B. polymyxa* enzyme. Horizontal bars at bottom of each panel denote well-conserved regions found in amino acid sequences of 4 β-amylases of different origins (see Figure 31.2). (From Kitamoto et al., 1988.)

near the COOH terminus) are Gly in the former enzyme, whereas 10% (23 residues) are Gly in the latter enzyme. The lower content of the Gly residue which, lacks β carbon, has high backbone conformational flexibility, and destabilizes helix structure, might enhance the thermostability of the enzyme by increasing the rigidity of the molecule. (4) Higher content of Pro residue. The *C. thermosulfurogenes* β-amylase contains more Pro residues than the *B. polymyxa* β-amylase, especially in its nonhomologous region. The pyrolidine ring of proline restricts this residue to fewer conformation than are available to the other amino acids, which enhances protein stability (Matthews et al., 1987).

31.6 Efficient Production of the Thermophilic β-Amylase Using the *Bacillus brevis* Host–Vector System

The thermophilic β-amylase should be useful for starch-transforming processes at high temperature. Improved solubility of starch at high temperature should greatly facilitate its enzymatic degradation. Production of the enzyme in large amounts is important for industrial application as well as for biochemical and physicochemical investigation of the enzyme. Hyun and Zeikus (1985b) isolated a β-amylase-hyperproductive mutant of *C. thermosulfurogenes* that produced eightfold more β-amylase in starch medium than the wild type. The amount of β-amylase produced by the mutant was 46.3 units/ml of the culture medium. This value corresponds to 11 mg of the enzyme/liter, according to the specific activity of the purified enzyme (Shen et al., 1988).

We have developed an efficient system for heterologous protein production using *B. brevis* as a host. Several heterologous proteins, including human epidermal growth factor, were produced very efficiently with this system (Takagi et al., 1989; Yamagata et al., 1989; Konishi et al., 1990). Using this system, the *C. thermosulfurogenes* β-amylase gene was also expressed efficiently, and a large amount of β-amylase (1.6g/liter culture medium) was secreted into the medium. The β-amylase could be purified very easily because *B. brevis* is a mesophilic bacterium, and most of the proteins secreted by *B. brevis* could be removed by heat treatment (Mizukami et al., 1992).

References

Buisson, G., E. Duee, R. Haser, and F. Payan. 1987. Three dimensional structure of porcine pancreatic α-amylase at 2.9 Å resolution. Role of calcium in structure and activity. *EMBO Journal* **6**:3909–3916.

Gouy, M. and C. Gautier. 1982. Codon usage in bacteria: correlation with gene expressivity. *Nucleic Acids Res.* **10**:7055–7074.

Hyun, H.H. and J.G. Zeikus. 1985a. General biochemical characterization of thermostable extracellular β-amylase from *Clostridium thermosulfurogenes*. *Appl. Environ. Microbiol.* **49**:1162–1167.

Hyun, H.H. and J.G. Zeikus. 1985b. Regulation and genetic enhancement of β-amylase production in *Clostridium thermosulfurogenes*. *J. Bacteriol.* **164**:1162–1170.

Ikemura, T. 1981. Correlation between the abundance of *Escherichia coli* transfer RNAs and the occurrence of the respective codons in its protein genes: a proposal for a synonymous codon choice that is optimal for the *E. coli* translational system. *J. Mol. Biol.* **151**:389–409.

Imanaka, T., M. Shibazaki, and M. Takagi. 1986. A new way of enhancing the thermostability of proteases. *Nature* (London) **324**:695–697.

Inouye, M. and S. Halegoua. 1980. Secretion and membrane localization of proteins in *Escherichia coli*. *Crit. Rev. Biochem.* **7**:339–371.

Kawazu, T., Y. Nakanishi, N. Uozumi, T. Sasaki, H. Yamagata, N. Tsukagoshi, and S. Udaka. 1987. Cloning and nucleotide sequence of the gene coding for enzymatically active fragments of the *Bacillus polymuxa* β-amylase. *J. Bacteriol.* **169**:1564–1570.

Kitamoto, N., H. Yamagata, T. Kato, N. Tsukagoshi, and S. Udaka. 1988. Cloning and sequencing of the gene encoding thermophilic β-amylase of *Clostridium thermosulfurogenes*. *J. Bacteriol.* **170**:5848–5854.

Konishi, H., T. Sato, H. Yamagata and S. Udaka. 1990. Efficient production of human α-amylase by a *Bacillus brevis* mutant. *Appl. Microbiol. Biotechnol.* **34**:297–302.

Kreis, M., M. Williamson, B. Buxton, J. Pywell, J. Hejgaard, and I. Svendson. 1987. Primary structure and differential expression of β-amylase in normal and mutant barleys. *Eur. J. Biochem.* **169**:517–525.

Kyte, J. and R.F. Doolittle. 1982. A simple method for displaying the hydropathic character of a protein. *J. Mol. Biol.* **157**:105–132.

Matsuura, Y., M. Kusunoki, W. Harada, and M. Kakudo. 1984. Structure and possible catalytic residues of Taka-amlylase A. *J. Biochem.* **95**:697–702.

Matthews, B.W., H. Nicholson, and W.J. Becktel.

1987. Enhanced protein thermostability from site-directed mutations that decrease the entropy of unfolding. *Proc. Natl. Acad. Sci. USA* **84**:6663–6667.

McLaughlin, J.R., C.L. Murray, and J.C. Rabinowitz. 1981. Unique features in the ribosome binding site sequence of the gram-positive *Staphylococcus aureus* β-lactamase gene. *J. Biol. Chem.* **256**:11283–11291.

Melasniemi, H., M. Paloheimo, and L. Hemio. 1990. Nucleotide sequence of the α-amylase-pullulanase gene from *Clostridium thermohydrosulfuricum*. *J. Gen. Microbiol.* **136**:447–454.

Mikami, B., Y. Morita, and C. Fukazawa. 1988. Primary structure and function of β-amylase. *Seikagaku* **60**:211–216 (in Japanese).

Mikami, B., Y. Morita, and C. Fukazawa. 1988. Primary structure and function of β-amylase. *Seikagaku* **60**:211–216 (in Japanese).

Mizukami, M., H. Yamagota, K. Sakaguchi and S. Udaka 1992. Efficient production of thermostable *Clostridium thermosulfurogenes* β-amylase by *Bacillus brevis*. *J. Ferment. Bioeng.* **73**:112–115.

Moran, C.P., N. Lang, S.F. LeGrice, G. Lee, M. Stephens, A.L. Sonenshein, J. Pero, and R. Losick. 1982. Nucleotide sequences that signal the initiation of transcription and translation in *Bacillus subtilis*. *Mol. Gen. Gen.* **186**:339–346.

Nakajima, R., T. Imanaka, and S. Aiba. 1986. Comparison of amino acid sequences of eleven different α-amylases. *Appl. Microbiol. Biotechnol.* **23**:355–360.

Nitta, Y., Y. Isoda, H. Toda, and F. Sekiyama. 1989. Identification of glutamic acid 186 affinity labeled by 2,3-epoxypropyl α-D-glucopyranoside in soybean β-amylase. *J. Biochem.* **105**:573–576.

Nomura, K., B. Mikami, and Y. Morita. 1987. Partial amino acid sequences around sulfhydryl groups of soybean β-amylase. *J. Biochem.* **102**:341–349.

Ogasawara, N. 1985. Markedly unbiased codon usage in *Bacillus subtilis*. *Gene* (Amst.) **40**:145–150.

Pantoliano, M.W., R.C. Ladner, P.N. Bryan, M.L. Rollence, J.F. Wood, and T.L. Poulos. 1987. Protein engineering of subtilisin BPN': enhanced stabilization through the introduction of two cysteines to form a disulfide bond. *Biochemistry* **26**:2077–2082.

Perry, L.J. and R. Wetzel. 1984. Disulfide bond engineered into T4 lysozyme: stabilization of the protein toward thermal inactivation. *Science* **226**:555–557.

Rosenberg, M. and D. Court. 1979. Regulatory sequences involved in the promotion and termination of RNA transcription. *Annu. Rev. Genet.* **13**:319–353.

Schink, B. and J.G. Zeikus. 1983. *Clostridium ther-*

mosulfurogenes sp. nov., a new thermophile that produces elemental sulphur from thiosulphate. *J. Gen. Microbiol.* **129**:1149–1158.

Shen, G.-J., B.C. Saha, Y.-E. Lee, L. Bhatnagar, and J.G. Zeikus. 1988. Purification and characterization of a novel thermostable β-amylase from *Clostridium thermosulphurogenes. Biochem. J.* **254**:835–840.

Suzuki, Y., N. Ito, T. Yuuki, H. Yamagata, and S. Udaka. 1989. Amino acid residues stabilizing a *Bacillus* α-amylase against irreversible thermoinactivation. *J. Biol. Chem.* **264**:18933–18938.

Takagi, H., A. Miyauchi, K. Kadowaki, and S. Udaka. 1989. Potential use of *Bacillus brevis* HPD31 for the production of foreign proteins. *Agric. Biol. Chem.* **53**:2279–2280.

Toda, H., K. Kondo, and K. Narita. 1982. The complete amino acid sequence of Taka-amylase A. *Proc. Jpn. Acad.* **58**:208–212.

Uozumi, N., K. Sakurai, T. Sasaki, S. Takekawa, H. Yamagata, N. Tsukagoshi, and S. Udaka. 1989. Single gene directs synthesis of a precursor protein with β- and α-amylase activities in *Bacillus polymyxa. J. Bacteriol.* **171**:375–382.

Uozumi, N., T. Matsuda, N. Tsukagoshi, and S. Udaka. 1991. Structural and functional roles of cystein residues of *Bacillus polymyxa* β-amylase *Biochemistry* **30**:4594–4599.

Yamagata, H., K. Nakahama, Y. Suzuki, A. Kakinuma, N. Tsukagoshi, and S. Udaka. 1989. Use of *Bacillus brevis* for efficient synthesis and secretion of human epidermal growth factor. *Proc. Natl. Acad. Sci. U.S.A.* **86**:3589–3593.

32

The α-Amylase–Pullulanase (*apu*) Gene from *Clostridium thermohydrosulfuricum*: Nucleotide Sequence and Expression in *Escherichia coli*

Hannes Melasniemi

32.1 Introduction

Starch consists of two high molecular mass fractions, amylose and amylopectin, both composed of anhydro glucose units connected to each other by α-glycosidic linkages. In the straight-chain fraction, amylose, there are only α-1,4 linkages, whereas in the branched-chain fraction, amylopectin, short α-1,4 chains are linked to each other by α-1,6 linkages at the branch points. The relative amount of α-1,6 linkages depends on the source of the starch but is approximately 3%–4% in most common starches (Fogarty and Kelly, 1979). α-Amylases (EC 3.2.1.1) hydrolyze α-1,4 linkages in amylose and amylopectin in an endo fashion, producing dextrins and maltooligosaccharides and causing a rapid decrease in the viscosity and iodine-binding capacity of the starch.

α-Amylases do not hydrolyze α-1,6 linkages, although they can bypass them. Pullulanases (EC 3.2.1.41), on the other hand, hydrolyze the α-1,6 linkages in amylopectin but do not attack α-1,4 linkages. Pullulanases are distinguished from other debranching enzymes by their ability to cleave α-1,6 linkages of the linear polysaccharide pullulan, which can be conceived of as being composed of maltotriosyl units connected to each other by these linkages (Fogarty and Kelly, 1979).

From *Clostridium thermohydrosulfuricum*, a thermostable pullulanase activity was first described for the strain 39E (Hyun and Zeikus, 1985a) and a thermostable α-amylase activity together with an apparently coregulated pullulanase activity for the strain E 101-69 (Melasniemi, 1987a). As the α-amylase and pullulanase activities of the strain E 101-69 responded similarly to all parameters tested and were, in addition, physically associated, it was concluded that both activities are properties of a single protein representing a novel type of amylase (Melasniemi, 1987b). This α-amylase-pullulanase enzyme was purified (Melasniemi, 1988). Shortly after, the enzyme from the *C. thermohydrosulfuricum* strain Z 21-109, which is a derivative of 39E, was also purified (Saha et al., 1988). The enzymes produced by the strains Z 21-109 and 39E likewise have a dual α-1,4/α-1,6 activity (Saha et al., 1988; Mathupala et al., 1990), as have the enzymes from the recently isolated strains Uel 1 and Sol 1 (Klingeberg et al., 1990).

The first report on an α-amylase-pullulanase type of an enzyme seems to be that of Takasaki (1987); since then several reports on pullulanases having α-1,4 activity have appeared (Table 32.1) (see also the reviews of Saha and Zeikus (1989) and Antranikian (1990). Remarkably, enzymes of this kind are produced mainly by thermophilic bacteria. *Bacillus subtilis* and *B. circulans* are the only known nonthermophilic

Table 32.1 Bacteria producing α-amylase-pullulanase

Organism	Reference
Bacillus subtilis TU	Takasaki, 1987
Clostridium thermohydro-sulfuricum E 101-69	Melasniemi, 1987b, 1988
39E	Hyun and Zeikus, 1985a; Mathupala et al., 1990
Uel 1 and Sol 1	Klingeberg et al., 1990
Thermoanaerobium Tok6-B1	Plant et al., 1987
Thermoanaerobium brockii ATCC 33075	Coleman et al., 1987
Thermoactinomyces thal-pophilus No. 15	Odibo and Obi, 1988
Bacillus circulans F-2	Sata et al., 1989
Thermus sp. AMD33	Nakamura et al., 1989
Bacillus sp. 3183	Saha et al., 1989
Thermoanaerobacter B6A	Saha et al., 1990
Clostridium thermosufuro-genes EM1	Spreinat and Antrani-kian, 1990

producers, and even these species are not typical mesophiles but thermotolerant species capable of growth at 55°C.

32.2 Cloning of the *apu* Gene

The highest α-amylase-pullulanase yields obtained from *Clostridium thermohydrosulfuricum* E 101-69 (Melasniemi, 1987a) were about 2.1 U/ml and 0.7 U/ml (1 U = 1 μmol reducing sugar formed/min at 85°C), when assayed with pullulan and amylose, respectively. Such yields are too low for possible industrial applications. As the first step toward increasing the yield by

using recombinant DNA techniques, the α-amylase-pullulanase gene of E 101-69 was cloned in *Escherichia coli* as a 7.0-kb *Eco*RI fragment in lambda gt11 vector (Melasniemi and Paloheimo, 1989). *E. coli* JM109, harboring the fragment subcloned on the plasmids pUC18 and pUC19 to give the plasmids pALK351 and pALK353 (Figure 32.1), produced about the same amount of enzyme activity as the original host. The gene was obviously cloned together with a *C. thermohydrosulfuricum* promoter active in *E. coli*, as the amount of activity produced was not affected by the orientation of the DNA insert. The bulk (98%) of the α-amylase-pullulanase activity produced by *E. coli* carrying pALK355, derived from pALK353 by deletion of a nonessential, terminal, 0.8-kb *Eco*RI-*Bam*HI fragment (Figure 32.1), was manifested only after complete disintegration of the cells. The enzyme was judged to be intracytoplasmic because no activity was detected in inner or outer membrane vesicles separated by sucrose density gradient centrifugation and because less than 3% of the activity was released by spheroplasting the cells·by a lysozyme/EDTA treatment.

32.3 Sequence of the *apu* Gene

The nucleotide sequence of the *apu* gene (Gen-Bank accession number M28471; Melasniemi et al., 1990) contained an open-reading frame of 4425 bp. The open-reading frame started with a putative GTG initiation codon (Figure 32.2) like that of the *celA* gene of *Clostridium thermocellum* (Béguin et al., 1985), and ended with a TAG. About 10% of *E. coli* genes use a GTG initiation

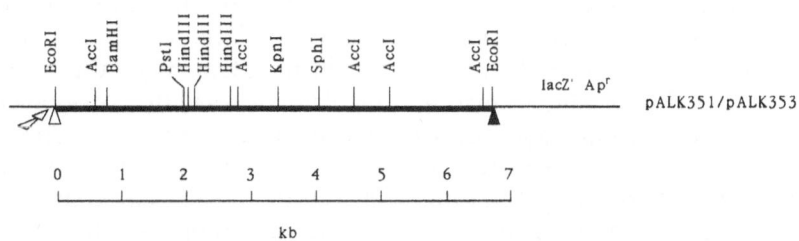

Figure 32.1 Plasmids pALK351 and pALK353. Symbols: ▲, polylinker of pALK351; △ polylinker of pALK353; arrow, *lacZ'* promoter.

```
AGATATAATTTATAGGGCTGCTTTATAGTAGCCCTATAAATTTACATATAGTGTATGTATTAACAAC   114

ATATTGACAACGAAAAAACTATAATTTATAATAGACATGAGGGAAAACGATTGCGCAAAAATTTGCA   181

CAAATAAAATGTGTTTTTATCACATTTATCTTTTTGCAATTTGTGAAAGCGATTGCCTAAAAGAAAT   248

GGGGGTGGGATTGGGAATTAATTTTTTTAAAAACATTTAAGTAAATATATGGATAAAAGGGGGATTA   315

TAGGTGTTTAAAAGGAGAGCTTTAGGATTTTTGCTGGCTTTTCTTTTAGTTTTTACAGCAGTATTTG   382
  -31 M  F  K  R  R  A  L  G  F  L  L  A  F  L  L  V  F  T  A  V  F

GCTCAATGCCTATGGAATTTGCAAAGGCTGAGACAGATACAGCGCCAGCGATAGCCAATGTTGTTGG   449
  G  S  M  P  M  E  F  A  K  A  E  T  D  T  A  P  A  I  A  N  V  V
                               +1
```

FIGURE 32.2   Nucleotide and deduced amino acid sequences of 5' end of *apu* gene and its 5' regulatory region. Putative promoter and ribosome-binding sites are denoted by double and single underlinings, respectively. Second, less likely ribosome-binding site and associated initiation codon denoted by overlinings; direct repeat of duplicated "−10" region of putative promoter indicated by dashed underlining. Two partially overlapping palindromic pairs of sequences (2 × 13, 2 × 8 nucleotides) denoted by continuous and dashed arrows, respectively. Third palindromic pair of sequences on the 3' side of putative promoter denoted by dotted arrows; N-terminal end of mature protein is underlined.

codon (Hershey, 1987). Three other TAG codons in the same reading frame occurred shortly after the TAG stop codon, securing the end of translation of the *apu* gene. The structural gene was preceded by a sequence identical to the *E. coli* and *B. subtilis* σ[43] consensus promoter sequences (Rosenberg and Court, 1979; McConnell et al., 1986). The sequence TTGACA was followed by two TATAAT sequences after 10 and 17 nucleotides, respectively (Figure 32.2), the latter distance being optimal for *E. coli* and *B. subtilis*. At least in these organisms the promoter could thus be expected to be quite effective. Seven nucleotides in front of the putative GTG initiation codon was the putative ribosomal binding sequence AAAGGGGG (see Figure 32.2) exhibiting complementarity to the 16S rRNA of *B. subtilis*, *B. stearothermophilus*, and *E. coli* (Rosenberg and Court, 1979; McConnell et al., 1986). This sequence is also found in front of the *celD* gene of *C. thermocellum* (Joliff et al., 1986). Fourteen nucleotides after the putative ribosomal binding site of *apu* there was the sequence AAAGGAG (Figure 32.2), showing likewise strong complementarity to the end of 16S rRNA. This sequence, found also in front of the *celA* and *xynZ* genes of *C. thermocellum* (Béguin et al., 1985; Grépinet et al., 1988) is, however, less likely to be the true ribosomal binding site of the *apu* gene. It is followed by the less frequent (Hershey, 1987) TTG initiation codon at a longer distance (13 nucleotides), and in addition initiation at the TTG codon would produce an abnormal signal sequence.

32.4   A Hypothetical Model for *apu* Regulation

Production of α-amylase-pullulanase by *Clostridium thermohydrosulfuricum* is not temporally regulated but occurs in parallel with growth in the presence of a stimulatory substrate such as starch. The presence of free sugars causes, however, catabolite repression: xylose, glucose, and fructose in this order, effecting an increasingly severe repression (Hyun and Zeikus, 1985b; Melasniemi, 1987a). In addition to catabolite repression, synthesis of the enzyme is also subject to induction, as shown by the observation of Hyun and Zeikus (1985b) that the catabolite repression-resistant mutant strains Z 67-143 and Z 21-109 derived from 39E still required the presence of starch for enzyme production. One might thus expect to find two regulatory sites in the 5'-region of the *apu* gene, one mediating catabolite repression and one mediating substrate induction (although models can be theoretically constructed permitting both effects to operate through a single DNA-binding site). These would most probably be palindromic sites, as regulatory proteins often

| T A A A A T G T G T T T T | * | T A T C A C A T T T | putative <u>apu</u> operator |

(Figure 32.3 diagram)

```
T A A A A T G T G T T T T * T A T C A C A T T T    putative apu operator

T A A A A T G T G T - 32nt- T T A C A T A T T T    sac Rt (B. subtilis)

T T A G T T T G T T T-3nt-C A A C A A A C T A A    xyl/xyn operator (B. subtilis)

A A A T G T G A T C T A G A T C A C A T T T        consensus CAP site (E. coli)
```

FIGURE 32.3 Similarity of *apu* sequence 3' to putative promoter, a putative operator, to some bacterial regulatory sites. One nucleotide (nt) intentional gap (*) has been left in *apu* sequence; for explanation, see text.

have their targets at such sites (Brennan and Matthews, 1989). Although it is not possible to identify these regulatory sites on the basis of their sequence alone, there are two sites that deserve attention.

Some 60 nucleotides before the putative promoter starts a region containing a palindromic pair of sequences (2×13 nucleotides, separated by 10 nucleotides; see Figure 32.2). The first 8 nucleotides of the first member of this pair, together with the 3 preceding nucleotides, have the same sequence as the duplicated "−10" region. Thus, the last 8 nucleotides of the second member of this palindromic pair have a palindromic relationship, also to 8 nucleotides, in the "−10" region. These two partially overlapping palindromic pairs of sequences are supposed to be involved in the regulation of the *apu* gene. In addition, at a classical operator site, 41 nucleotides downstream of the putative promoter and 131 nucleotides before the beginning of the *apu* structural gene, starts a 2×7 nucleotide palindrome see (Figure 32.2). This putative *apu* operator shows similarity (Figure 32.3) to a terminator sequence (*sacRt*) starting 152 nucleotides before the *B. subtilis* levansucrase structural gene (Aymerich et al., 1986; Crutz et al., 1990), as well as to three *B. subtilis xyl/xyn* operators starting some 300 or 100 nucleotides before the respective structural genes (Kreuzer et al., 1989).

All these *Bacillus* sequences are palindromic and are located 3' to the promoters of the structural genes, the induction of which they control. The 2×7 nucleotide palindrome in *apu* shows also great similarity to the consensus sequence for binding of the catabolite gene activator protein of *E. coli* (Ebright et al., 1989). Although either one of the putative *apu* regulatory sequences could be responsible for the inducibility or catabolite repression of the gene, it appears more probable that the 2×7 nucleotide palindrome 3' to the putative *apu* promoter is involved in the induction of the gene and the two partially overlapping palindromic pairs of sequences involving the "−10" region of the promoter in its catabolite repression. This is because catabolite repression in *E. coli* always involves the promoter regions of the affected genes (Crombrugghe et al., 1984), although the α-amylase gene of *Bacillus licheniformis* cloned in *B. subtilis* has been reported to be catabolically repressed independently of its promoter region (Laoide et al., 1989). In addition, as catabolite repression usually affects several genes, there should be a common target sequence in the genes, such as is provided by their promoters. In contrast, target sequences involved in induction of specific genes of an organism can be assumed to be more specialized.

32.5 Codon Usage

The 37% G+C content of the *apu* gene agrees well with the 36%–38% G+C content reported for several strains of *C. thermohydrosulfuricum* (Wiegel et al., 1979). However, the G+C content of the third ("wobble") codon position was only 27%, indicating a preference for codons ending in A or T. The codon usage of the *apu* gene (Table 32.2) resembles that observed for *C. thermocellum* genes (Grépinet et al., 1988; Schwarz et al., 1988), but is quite different from the codon usage in the structural genes of *E. coli* (Alff-Steinberger, 1984): for only 3 of 17 amino acids (excluding Met, Trp, and Cys) was the preferred codon the same as in the *E. coli*

Table 32.2 Codon usage in the *apu* structural gene[a]

Phe	*TTT	45
	TTC	13
Leu	TTA	27
	TTG	9
Leu	*CTT	28
	CTC	7
	CTA	6
	CTG	4
Ile	ATT	34
	ATC	4
	ATA	52
Met	ATG	23
Val	*GTT	44
	CTC	11
	GTA	41
	GTG	21
Ser	(*)TCT	26
	TCC	5
	TCA	26
	TCG	4
Pro	CCT	32
	CCC	4
	CCA	26
	CCG	7
Thr	ACT	29
	ACC	15
	*ACA	55
	ACG	7
Ala	GCT	29
	GCC	9
	*GCA	37
	GCG	10
Tyr	*TAT	65
	TAC	27
End	TAA	0
	TAG	0
His	*CAT	17
	CAC	5
Gln	*CAA	33
	CAG	12
Asn	*AAT	93
	AAC	20
Lys	*AAA	75
	AAG	26
Asp	*GAT	93
	GAC	39
Glu	*GAA	43
	GAG	19
Cys	TGT	0
	TGC	0
End	TGA	0
Trp	TGG	32
Arg	CGT	1
	CGC	1
	CGA	0
	CGG	0

Table 32.2 (cont.)

Ser	AGT	18
	AGC	10
Arg	*AGA	28
	AGG	11
Gly	GGT	37
	GGC	22
	*GGA	41
	GGG	17

[a] 1475 sense codons. Preference codons common with *Bacillus* genes (McConnell et al., 1986) are denoted with asterisks.

genes (data not shown). The preferred *apu* codons were more often those found in rare proteins of *E. coli* than those found in abundant proteins (Grosjean and Fiers, 1982), and in addition the codons preferred in *apu* for Arg (AGA and AGG), Ile (ATA), and Gly (GGA) have been proposed as modulators that slow down translation in *E. coli* (Grosjean and Fiers, 1982). The exceptional length of the *apu* gene together with the differences in codon usage presumably make translation of the gene in *E. coli* slow and prone to premature termination, which in turn renders the newly synthesized, untransported polypeptides susceptible to proteolytic degradation.

These factors may explain the degeneracy of the α-amylase-pullulanase polypeptides produced in *E. coli* (see following). In contrast to the differences in codon usage between *apu* and *E. coli* genes, there is a resemblance in codon usage between *Bacillus* genes (McConnell et al., 1986) and *apu*, as 14 of the 17 amino acids having more than one codon shared a common preference codon see (Table 32.2).

32.6 The *apu* Product

The preprotein deduced from the *apu* open-reading frame contained 1475 amino acids, and, like the mature α-amylase of *Bacillus amyloliquefaciens* (Takkinen et al., 1983), included no cysteine. At the N-terminal end of the preprotein was a 31-amino-acid sequence with the characteristics of a typical signal sequence (McConnell et al., 1986): three positively charged amino acids near the N terminus, a hydrophobic core, the consensus signal peptidase cleavage sequence Ala-X-Ala, and in addition a helix-breaking proline preceding the cleavage site (see Figure 32.2). The amino acid sequence after this putative signal sequence was identical to the determined N-terminal sequence of the extracellular enzyme from the native host (Melasniemi, 1988), except for position +2 at which the determined amino acid sequence had an Ile residue whereas the DNA sequence clearly indicated a Thr residue. However, the ATA codon for Ile can be changed to the ACA codon for Thr by a single point mutation.

The calculated M_r (165,600) of the deduced preprotein is very close to the M_r (165,000) observed for the longest α-amylase-pullulanase polypeptides produced in E. coli (Melasniemi and Paloheimo, 1989). This, together with the seemingly intracytoplasmic location of the enzyme, suggested that the signal sequence is not cleaved. The intracellular location, on the other hand, indicates that the signal sequence or the following sequences do not allow the transport of the enzyme in E. coli, or that a function vital for the transport of the apu product is missing. However, the possibility also exists that wrong initiation at the more 3' of the two possible ribosomal binding sites (producing a deficient signal sequence) or that a mutation in the second codon after the signal sequence cleavage site (causing an Ile to Thr change) has affected the localization of the enzyme.

The enzyme produced in E. coli was very degenerate. More than 10 α-amylase-pullulanase-specific bands were resolved by immunoblotting after sodium dodecyl sulfate-polyacrylamide gel electrophoresis. Gel filtration, on the other hand, resolved the enzyme activity into three peaks that consisted of different sets of α-amylase-pullulanase polypeptides of various lengths; the peaks eluted with approximate M_r values of 330,000, 245,000, and 200,000. The longest α-amylase-pullulanase polypeptides in each peak had approximate M_r values of 165,000, 130,000, and 100,000. The observed M_r values suggest that the enzyme produced in E. coli is dimeric, as is the extracellular enzyme

from the native host (Melasniemi, 1988). The clearly decreased length of the polypeptides in the last two peaks also indicates either that full-length polypeptides are not essential for the activity of the enzyme or that a functional enzyme can be produced by the association of shorter polypeptides.

Considering the marked degeneracy of the enzyme produced in E. coli, its thermal characteristics were suprisingly unaffected as compared with the enzyme produced by the native host (Melasniemi, 1987b): the apparent temperature optimum (80°–85°C) was only some 5°C lower and the thermal stability was the same. Because an intracellular protein in E. coli can fairly safely be assumed to be nonglycosylated, the equal stabilities of the enzymes produced in the two hosts seem to imply that the apparently covalently attached carbohydrate moiety of the native host enzyme, containing glucose, galactose, mannose, and rhamnose (Melasniemi, 1988), has no marked role in the thermostability of the enzyme. In N-glycosidically linked glycoproteins, the carbohydrate is always linked to an Asn residue in a Asn-X-Ser/Thr sequence (Sharon and Lis, 1981). The deduced amino acid sequence of the apu product revealed 13 such possible sites for N-glycosylation.

Taking into account that both forms of the extracellular enzyme purified from the native E 101-69 host contain about 10% carbohydrate (Melasniemi, 1988), the M_r of the mature polypeptide deduced from the gene sequence (162,200) is compatible with the M_r (about 180,000) observed for the lower M_r form but somewhat lower than the M_r (about 190,000) of the higher M_r form. In contrast to this, the M_r (136,500) reported for the monomeric enzyme extracted from cells of the strain Z 21-109 (Saha et al., 1988) is clearly lower than the M_r deduced for the mature E 101-69 enzyme. This implies either a difference between the strains or proteolytic degradation of the Z 21-109 enzyme during its purification.

The E 101-69 α-amylase-pullulanase occurs in cultures of the original host in three different states: at the cell surface, as an aggregated or vesicle-bound form (apparent $M_r > 10^6$) in the medium, and as two free extracellular forms (Melasniemi, 1987a, 1987b, 1988). The occurrence of the enzyme on the cell surface and as an aggregated or vesicle-bound form is reminiscent of the lipoprotein pullulanase of *Klebsiella pneumoniae* (Pugsley et al., 1986; d'Enfert et al., 1987). Neither the mechanism of anchoring of the α-amylase-pullulanase to the cell surface nor the composition of the extracellular aggregates is known. The deduced primary sequence of the enzyme shows, apart from the signal sequence, no sequence long and hydrophobic enough to serve as a membrane anchor (Kyte and Doolittle, 1982). As the cell-associated enzyme nevertheless seems to be anchored in the lipid phase, as indicated by the release of the Z 21-109 enzyme by a combined sodium choleate-lipase treatment (Saha et al., 1988), the presence of a lipid modification can be suspected at least in the enzyme on the cell surface and in the extremely high M_r aggregates. Although the total absence of cysteine from the α-amylase-pullulanase precludes the possibility of a cysteine-mediated lipid modification typical of bacterial lipoproteins (Wu, 1987), the possibility of some other kind of modification, although speculative, deserves further attention.

The extracellular E 101-69 α-amylase-pullulanase occurs in a very tight, apparently equimolar association with a small ($M_r = 24,000$) protein, the α-amylase-pullulanase satellite protein (Melasniemi, 1988), the function of which is not yet known. The N-terminal end of this protein has the sequence Ser-Thr-Tyr-Gly-Val-Phe-Glu-Tyr (H. Melasniemi, unpublished observations). The satellite protein gene was located at the other end of the 7.0-kb *Eco*RI fragment cloned (Melasniemi and Paloheimo, 1989), and the DNA sequence coding for the N-terminal end of the satellite protein started some 500 nucleotides after the *apu* open-reading frame (Melasniemi et al., 1990). Thus, the enzyme and the satellite protein are coded by two adjacent, apparently coregulated genes. Unfortunately, the sequence of the satellite protein started only just before the 3' end of the cloned fragment, which coded only for its first ten N-terminal amino acids.

32.7 Similarities Between the *apu* Product and Other Amylolytic Enzymes

Usually only little sequence similarity is found in signal sequences (McConnell et al., 1986). The putative signal sequence of the *apu* product showed, however, clear similarity to the signal sequences of the α-amylases of *Bacillus stearothermophilus* and *B. subtilis* (Melasniemi et al., 1990). The mature *apu* product, on the other hand, showed several short regions similar to sequences in several other types of amylolytic enzymes including different α-amylases, *Klebsiella pneumoniae* pullulanase (Katsuragi et al., 1987), *Pseudomonas amyloderamosa* isoamylase (Amemura et al., 1988), *K. pneumoniae* and *Bacillus macerans* maltocyclodextrin glucanotransferases (Binder et al., 1986; Takano et al., 1986), and *B. stearothermophilus* neopullulanase (Kuriki and Imanaka, 1989). In general, the *apu* product had most similarity to the neopullulanase of *B. stearothermophilus*, an enzyme that cleaves α-1,4 linkages in pullulan and also α-1,6 linkages in several oligosaccharides. Even in the case of the neopullulanase the similarities were restricted to short sequence stretches, the longest of which comprised 78 amino acids with 63% identities between the enzymes and the second longest 57 amino acids with 53% identities. Interestingly, several short sequences were also similar to sequences in *B. macerans* cyclomaltodextrin glucanotransferase; two end products (β- and γ-cyclodextrins) of this enzyme are inhibitors of the α-amylase-pullulanase.

In the sequence of α-amylases, three short regions are common to α-amylases from bacteria, fungi, plants, and mammals (Ihara et al., 1985). A further common region is evident, especially in mammalian α-amylases (Nakajima et al., 1986). According to three-dimensional α-amylase models (Matsuura et al., 1984; Buisson et al., 1987; Vihinen and Mäntsälä, 1990) the common regions come close to each other in the catalytic cleft of the enzymes to constitute the substrate binding sites and the active sites of α-amylases. These regions, or at least re-

A

```
                    2                                          4
              ------------------                        ------------------
  Cth apu    D G W R L D V A N E   602      Cth apu   T M N L L G S H D T   705
  Bsu amy    D G F R F D A A K H   180      Bsu amy   L V T W V E S H D T   270
  Bam amy    D G F R I D A A K H   235      Bli amy   A V T F V D N H D T   329
  Shy amy    D G F R I D A A K H   178      Bst amy   A V T F V D N H D T   332
  Sli amy    D G F R I D A A K H   181      BarBamy   A V T F V D N H D T   291

  Cth apu              I D A A K   676      Cth apu   A V T A V D N         970
```

B

```
  Cth apu   Ⓗ Ⓥ Ⓔ P S G Ⓓ Ⓚ N N E L T Ⓚ Ⓓ I T I T V T R E K P A R R Q N L L L L Ⓗ Ⓠ

            Q K Q Q Ⓝ Ⓗ L K K Y Ⓗ Ⓚ A K   1444   End
```

FIGURE 32.4 (**A**) Duplication of *α*-amylase regions two and four in *apu* product (boxes). Abbreviations: Cth, *Clostridium thermohydrosulfuricum* (Melasniemi et al., 1990); Bsu, *Bacillus subtilis* (Yang et al., 1983); Bam, *B. amyloliquefaciens* (Takkinen et al., 1983); Shy, *Streptomyces hygroscopicus* (Hoshiko et al., 1987); Sli, *S. limosus* (Long et al., 1987); Bli, *B. licheniformis* (Yuuki et al., 1985); Bst, *B. stearothermophilus* (Nakajima et al., 1985); Bar, barley (Rogers, 1985); apu, *α*-amylase-pullulanase; amy, *α*-amylase; Bamy, *α*-amylase B. (**B**) C terminus of *apu* product; for explanation, see text.

gions one, two, and four (numbering of Nakajima et al., 1986) are also found, although not as clearly as in *α*-amylases, in the sequences of maltocyclodextrin glucanotransferases, pullulanases, isoamylases, and in the neopullulanase.

The three regions could likewise be distinguished in the sequence of the *α*-amylase-pullulanase (Melasniemi et al., 1990). Most interestingly, considering the dual *α*-1,4/*α*-1,6 activity of the *α*-amylase-pullulanase, its second and fourth regions are in a sense duplicated. That is, characteristic motifs from each region are found at two positions separated by some 70 and 260 residues for the second and fourth regions, respectively (Figure 32.4A).

Most of the similarities between the *α*-amylase-pullulanase and other amylolytic enzymes were found at and near the common *α*-amylases regions, in the middle third part of the *α*-amylase-pullulanase sequence, the C-terminal third part of the sequence being the most specialized part of it. At the C terminus of the enzyme (Figure 32.4B), the prominent sequence of three hydroxyl-bearing Thr residues alternating with hydrophobic amino acids followed shortly after by four Leu residues (amino acids underlined in Figure 32.4), draws attention. The C terminus also contains four interesting His residues (circled amino acids), the rarest amino acid in the protein. The last two His residues occur next to Asn and Lys residues, as do the His residues at regions one,

four, and two in *α*-amylases (Nakajima et al., 1986; see Figure 32.4A), whereas a third His residue occurs next to a Gln residue. These His residues are preceded by the doublets Lys-Asp and Asp-Lys and by the triplet His-Val-Glu (circled in Figure 32.4B).

Remarkably, according to the three-dimensional *α*-amylase model of Matsuura et al. (1984) two spatially neighboring Lys and Asp residues in the catalytic cleft both bind to one glucose residue as well as individually to the glucose residues on either side of the shared one; in contrast, a Glu residue, spatially neighboring a Val residue, interacts with a glucose residue to which the His residue of region two also binds. The His residues of the Asn-His doublets at regions one and four in *α*-amylases, on the other hand, are involved in substrate binding (Matsuura et al., 1984; Buisson et al., 1987). Because of the occurrence of the *α*-amylase active region elements at the C-terminal end of the deduced *apu* product mentioned earlier, it is here hypothesized that the C terminus of the *α*-amylase-pullulanase forms a part of its active region. In accordance to this hypothesized role of the C-terminal end, the hydropathy plot (Kyte and Doolittle, 1982) of the *apu* product predicts that the whole C-terminal end of the protein is located on the surface of the molecule. In addition, according to the algorithm of Garnier et al. (1978) the extreme C terminus of the *apu* product seems to constitute one of the longest α-helical regions in the molecule.

Finally, concerning nomenclature, three different names have been used for this kind of enzyme, which cleaves α-1,4 linkages in amylose as well as α-1,6 linkages in pullulan: α-amylase-pullulanase (Melasniemi, 1988), amylopullulanase (Mathupala et al., 1990), and pullulanase type II (Antranikian, 1990). Although the term α-amylase-pullulanase requires somewhat more space than the others, I suggest its use, on the following grounds: (1) This was the first name used. (2) The *Clostridium thermohydrosulfuricum* enzyme, at least, shows as much if not more sequence similarity to α-amylases as to true pullulanases. (3) The name is compatible with the three-letter genetic symbol (*apu*) given to the *C. thermohydrosulfuricum* gene (Melasniemi et al., 1990). (4) This is the most informative of the three names proposed, because it directly indicates the dual α-1,4 (α-amylase) and α-1,6 (pullulanase) activities of the enzymes. The term amylopullulanase (Greek; *amylon*, starch), on the other hand, does not hit the mark because it fails to denote specifically the exclusively α-1,4-linked substrate, amylose, but rather denotes starch in general, which includes amylopectin, a substrate even for the true pullulanases.

Acknowledgments. I thank Dr. John Londesborough for critically reading the manuscript.

References

Alff-Steinberger, C. 1984. Evidence for a coding pattern on the non-coding strand of the *E. coli* genome. *Nucleic Acids Res.* **12**:2235–2241.

Amemura, A., R. Chakraborty, M. Fujita, T. Noumi, and M. Futai. 1988. Cloning and nucleotide sequence of the isoamylase gene from *Pseudomonas amyloderamosa* SB-15*. *J. Biol. Chem.* **263**:9271–9275.

Antranikian, G. 1990. Physiology and enzymology of thermophilic anaerobic bacteria degrading starch. *FEMS Microbiol. Rev.* **75**:201–218.

Aymerich, S., G. Gonzy-Tréboul, and M. Steinmetz. 1986. 5'-Noncoding region *sacR* is the target of all identified regulation affecting the levansucrase gene in *Bacillus subtilis. J. Bacteriol.* **166**:993–998.

Béguin, P., P. Cornet, and J.-P. Aubert. 1985. Sequence of a cellulase gene of the thermophilic bacterium *Clostridium thermocellum. J. Bacteriol.* **162**:102–105.

Binder, F., O. Huber, and A. Böck. 1986. Cyclodextrin-glycosyltransferase from *Klebsiella pneumoniae* M5al: cloning, nucleotide sequence and expression. *Gene* (Amst.) **47**:269–277.

Brennan, R.G. and B.W. Matthews. 1989. Structural basis of DNA-protein recognition. *Trends Biochem. Sci.* **14**:286–290.

Buisson, G., E. Duée, R. Haser, and F. Payan. 1987. Three dimensional structure of porcine pancreatic α-amylase at 2.9 Å resolution. Role of calcium in structure and activity. *EMBO J.* **6**:3909–3916.

Coleman, R.D., S.-S. Yang, and M.P. McAlister. 1987. Cloning of the debranching-enzyme gene from *Thermoanaerobium brockii* into *Escherichia coli* and *Bacillus subtilis. J. Bacteriol.* **169**:4302–4307.

Crombrugghe, B., S. Busby, and H. Buc. 1984. Cyclic AMP receptor protein: role in transcription activation. *Science* **224**:831–838.

Crutz, A.-M., M. Steinmetz, S. Aymerich, R. Richter, and D. Le Coq. 1990. Induction of levansucrase in *Bacillus subtilis*: an antitermination mechanism negatively controlled by the phosphotransferase system. *J. Bacteriol.* **172**:1043–1050.

Ebright, R.H., Y.W. Ebright, and A. Gunasekera. 1989. Consensus DNA site for the *Escherichia coli* catabolite gene activator protein (CAP): CAP exhibits a 450-fold higher affinity for the consensus DNA site than for the *E. coli lac* DNA site. *Nucleic Acids Res.* **17**:10295–10305.

d'Enfert, C., C. Chapon, and A.P. Pugsley. 1987. Export and secretion of the lipoprotein pullulanase by *Klebsiella pneumoniae. Mol. Microbiol.* **1**:107–116.

Fogarty, W.M. and C.T. Kelly. 1979. Starch-degrading enzymes of microbial origin. *In: Progress in Industrial Microbiology*, Vol. 15, M.J. Bull, ed., pp. 87–150. Amsterdam: Elsevier.

Garnier, J., D.J. Osguthorpe, and B. Robson. 1978. Analysis of the accuracy and implications of simple methods for predicting the secondary structure of globular proteins. *J. Mol. Biol.* **120**:97–120.

Grépinet, O., M.-C. Chebrou, and P. Béguin. 1988. Nucleotide sequence and deletion analysis of the xylanase gene (*xynZ*) of *Clostridium thermocellum. J. Bacteriol.* **170**:4582–4588.

Grosjean, H. and W. Fiers. 1982. Preferential codon usage in procaryotic genes: the optimal codon-anticodon interaction energy and the selective codon usage in efficiently expressed genes. *Gene* (Amst.) **18**:199–209.

Hershey, J.W.B. 1987. Protein synthesis. *In:*

Escherichia coli and *Salmonella typhimurium, Cellular and Molecular Biology*, Vol. 1, F.C. Neidhardt, ed., pp. 613–647. Washington, D.C: American Society for Microbiology.

Hoshiko, S., O. Makabe, C. Nojiri, K. Katsumata, E. Satoh, and K. Nagaoka. 1987. Molecular cloning and characterization of the *Streptomyces hygroscopicus* α-amylase gene. *J. Bacteriol.* **169**: 1029–1036.

Hyun, H.H. and J.G. Zeikus. 1985a. General biochemical characterization of thermostable pullulanase and glucoamylase from *Clostridium thermohydrosulfuricum. Appl. Environ. Microbiol.* **49**:1168–1173.

Hyun, H.H. and J.G. Zeikus. 1985b. Regulation and genetic enhancement of glucoamylase and pullulanase production in *Clostridium thermohydrosulfuricum. J. Bacteriol.* **164**:1146–1152.

Ihara, H., T. Sasaki, A. Tsuboi, H. Yamagata, N. Tsukagoshi, and S. Udaka. 1985. Complete nucleotide sequence of a thermophilic α-amylase gene: homology between procaryotic and eucaryotic α-amylases at the active sites. *J. Biochem.* **98**:95–103.

Joliff, G., P. Béguin, and J.-A. Aubert. 1986. Nucleotide sequence of the cellulase gene *celD* encoding endoglucanase D of *Clostridium thermocellum. Nucleic Acids Res.* **14**:8605–8613.

Katsuragi, N., N. Takizawa, and Y. Murooka. 1987. Entire nucleotide sequence of the pullulanase gene of *Klebsiella aerogenes* W70. *J. Bacteriol.* **169**:2301–2306.

Klingeberg, M., H. Hippe, and G. Antranikian. 1990. Production of novel pullulanases at high concentrations by two newly isolated thermophilic clostridia. *FEMS Microbiol. Lett.* **69**:145–152.

Kreuzer, P., D. Gärtner, R. Allmansberger, and W. Hillen. 1989. Identification and sequence analysis of the *Bacillus subtilis* W23 *xylR* gene and *xyl* operator. *J. Bacteriol.* **171**:3840–3845.

Kuriki, T. and T. Imanaka. 1989. Nucleotide sequence of the neopullulanase gene from *Bacillus stearothermophilus. J. Gen. Microbiol.* **135**: 1521–1528.

Kyte, J. and R.F. Doolittle. 1982. A simple method for displaying the hydropathic character of a protein. *J. Mol. Biol.* **157**:105–132.

Laoide, B.M., G.H. Chambliss, and D.J. McConnell. 1989. *Bacillus licheniformis* α-amylase gene, *amyL*, is subject to promoter-independent catabolite repression in *Bacillus subtilis. J. Bacteriol.* **171**:2435–2442.

Long, C.M., M.-J. Virolle, S.-Y. Chang, S. Chang, and M.J. Bibb. 1987. α-Amylase gene of *Streptomyces limosus*: nucleotide sequence, expression motifs, and amino acid sequence homology to mammalian and invertebrate α-amylases. *J. Bacteriol.* **169**:5745–5754.

Mathupala, S., B.C. Saha, and J.G. Zeikus. 1990. Substrate competition and specificity at the active site of amylopullulanase from *Clostridium thermohydrosulfuricum. Biochem. Biophys. Res. Commun.* **166**:126–132.

Matsuura, Y., M. Kusunoki, W. Harada, and M. Kakudo. 1984. Structure and possible catalytic residues of Taka-amylase A. *J. Biochem.* (Tokyo) **95**:697–702.

McConnell, D.J., B.A. Cantwell, K.M. Devine, A.J. Forage, B.M. Laoide, C. O'Kane, J.F. Ollington, and P.M. Sharp. 1986. Genetic engineering of extracellular enzyme systems of Bacilli. *In*: Biochemical Engineering IV, C.H. Lim, and K. Venkatasubramanian, eds., Ann. N.Y. Acad. Sci. **469**:1–17.

Melasniemi, H. 1987a. Effect of carbon source on production of thermostable α-amylase, pullulanase and α-glucosidase by *Clostridium thermohydrosulfuricum. J. Gen. Microbiol.* **133**:883–890.

Melasniemi, H. 1987b. Characterization of α-amylase and pullulanase activities of *Clostridium thermohydrosulfuricum*. Evidence for a novel thermostable amylase. *Biochem. J.* **246**: 193–197.

Melasniemi, H. 1988. Purification and some properties the extracellular α-amylase-pullulanase produced by *Clostridium thermohydrosulfuricum. Biochem. J.* **250**:813–818.

Melasniemi, H. and M. Paloheimo. 1989. Cloning and expression of the *Clostridium thermohydrosulfuricum* α-amylase-pullulanase gene in *Escherichia coli. J. Gen. Microbiol.* **135**:1755–1762.

Melasniemi, H., M. Paloheimo, and L. Hemiö. 1990. Nucleotide sequence of the α-amylase-pullulanase gene from *Clostridium thermohydrosulfuricum. J. Gen. Microbiol.* **136**:447–454.

Nakajima, R., T. Imanaka, and S. Aiba. 1985. Nucleotide sequence of the *Bacillus stearothermophilus* α-amylase gene. *J. Bacteriol.* **163**:401–406.

Nakajima, R., T. Imanaka, and S. Aiba. 1986. Comparison of amino acid sequences of eleven different α-amylases. *Appl. Microbiol. Biotechnol.* **23**:355–360.

Nakamura, N., N. Sashihara, H. Nagayama, and K. Horikoshi. 1989. Characterization of pullulanase and α-amylase activities of a *Thermus* sp. AMD33. *Starch Stärke* **41**:112–117.

Odibo, F.J.C. and S.K.C. Obi. 1988. Purification and characterization of a thermostable pullulanase from *Thermoactinomyces thalpophilus. J. Ind. Microbiol.* **3**:343–350.

Plant, A.R., R.M. Clemens, H.W. Morgan, and R.M. Daniel. 1987. Active-site- and substrate-specificity of *Thermoanaerobium* Tok6-Bl pullulanase. *Biochem. J.* **246**:537–541.

Pugsley, A.P., C. Chapon, and M. Schwartz. 1986. Extracellular pullulanase of *Klebsiella pneumoniae* is a lipoprotein. *J. Bacteriol.* **166**:1083–1088.

Rogers, J.C. 1985. Two barley α-amylase gene families are regulated differently in aleurone cells. *J. Biol. Chem.* **260**:3731–3738.

Rosenberg, M. and D. Court. 1979. Regulatory sequences involved in the promotion and termination of RNA transcription. *Annu. Rev. Genet.* **13**:319–353.

Saha, B.C. and J.G. Zeikus. 1989. Novel highly thermostable pullulanase from thermophiles. *Trends Biotechnol.* **7**:234–239.

Saha, B.C., S.P. Mathupala, and J.G. Zeikus. 1988. Purification and characterization of a highly thermostable novel pullulanase from *Clostridium thermohydrosulfuricum*. *Biochem. J.* **252**:343–348.

Saha, B.C., R. Lamed, C.-Y. Lee, S.P. Mathupala, and J.G. Zeikus. 1990. Characterization of an endo-acting amylopullulanase from *Thermoanaerobacter* strain B6A. *Appl. Environ. Microbiol.* **56**:881–886.

Saha, B.C., G.-J. Shen, K.C. Srivastava, L.W. LeCureux, and J.G. Zeikus. 1989. New thermostable α-amylase-like pullulanase from thermophilic *Bacillus* sp. 3183. *Enzyme Microb. Technol.* **11**:760–764.

Sata, H., M. Umeda, C.-H. Kim, H. Taniguchi, and Y. Maruyama. 1989. Amylase-pullulanase enzyme produced by *B. circulans* F-2. *Biochim. Biophys. Acta* **991**:388–394.

Schwarz, W.H., S. Schimming, K.P. Rücknagel, S. Burgschwaiger, G. Kreil, and W.L. Staudenbauer. 1988. Nucleotide sequence of the *celC* gene encoding endoglucanase C of *Clostridium thermocellum*. *Gene* (Amst.) **63**:23–30.

Sharon, N. and H. Lis. 1981. Glycoproteins: research booming on long-ignored, ubiquitous compounds. *Chem. Eng. News* **30**:21–44.

Spreinat, A. and G. Antranikian. 1990. Purification and properties of a thermostable pullulanase from *Clostridium thermosulfurogenes* EM1 which hydrolyses both α-1,6 and α-1,4-glycosidic linkages. *Appl. Microbiol. Biotechnol.* **33**:511–518.

Takano, T., M. Fukuda, M. Monma, S. Kobayashi, K. Kainuma, and K. Yamane. 1986. Molecular cloning, DNA nucleotide sequencing, and expression in *Bacillus subtilis* cells of the *Bacillus macerans* cyclodextrin glucanotransferase gene. *J. Bacteriol.* **166**:1118–1122.

Takasaki, Y. 1987. Pullulanase-amylase complex enzyme from *Bacillus subtilis*. *Agric. Biol. Chem.* **51**:9–16.

Takkinen, K., R.F. Pettersson, N. Kalkkinen, I. Palva, H. Söderlund, and L. Kääriäinen. 1983. Amino acid sequence of α-amylase from *Bacillus amyloliquefaciens* deduced from the nucleotide sequence of the cloned gene. *J. Biol. Chem.* **258**:1007–1013.

Vihinen, M. and P. Mäntsälä. 1990. Conserved residues of liquefying α-amylases are concentrated in the vicinity of active site. *Biochem. Biophys. Res. Commun.* **166**:61–65.

Wiegel, J., L.G. Ljungdahl, and J.R. Rawson. 1979. Isolation from soil and properties of the extreme thermophile *Clostridium thermohydrosulfuricum*. *J. Bacteriol.* **139**:800–810.

Wu, H.C. 1987. Posttranslational modification and processing of membrane proteins in bacteria. *In: Bacterial Outer Membrane as Model System*, M. Inouye, ed., pp. 37–71. New York: Academic Press.

Yang, M., A. Galizzi, and D. Henner. 1983. Nucleotide sequence of the amylase gene from *Bacillus subtilis*. *Nucleic Acids Res.* **11**:237–249.

Yuuki, T., T. Nomura, H. Tezuka, A. Tsuboi, H. Yamagata, N. Tsukagoshi, and S. Udaka. 1985. Complete nucleotide sequence of a gene coding for heat- and pH-stable α-amylase of *Bacillus licheniformis*: comparison of the amino acid sequences of three bacterial liquefying α-amylases deduced from the DNA sequences. *J. Biochem.* **98**:1147–1156.

33

Molecular Biology of Xylan Utilization by Thermoanaerobes

Michael Bagdasarian, Yong-Eok Lee, Chanyong Lee, Menghsiao Meng, and J. Gregory Zeikus

33.1 Introduction

Xylans are linear polymers possessing a β-1,4-linked D-xylose backbone with branches containing 4-O-methyl D-glucuronic acid and L-arabinose. In several types of wood, particularly hardwoods, some of the sugar residues are acetylated (see Figure 33.1 for the schematic representation of the structure). Xylans or hemicelluloses are the second most abundant type of polysaccharides, after cellulose, in nature. As components of plant cell walls they constitute as much as 40% of all plant biomass (Timell, 1967; Aspinall, 1980; Gong et al., 1981). It is not surprising, therefore, that they are an attractive source of carbon and energy for microbes living in natural soil and aquatic habitats and that microorganisms have developed efficient ways to degrade these polymers and use their components.

Complete breakdown of a branched acetyl xylan requires the action of several hydrolytic enzymes (Reilly, 1981; Weimer et al., 1984; Woodward, 1984; Biely, 1985; Ward and Moo-Young, 1989). The most important enzymes in xylan degradation are endo-1,4-β-D-xylanase (EC 3.2.1.8), β-D-xylosidase (EC 3.2.1.37), α-L-arabinofuranosidase (EC 3.2.1.55), α-glucuronidase (EC 3.2.1), and acetylesterase (EC 3.1.1.6). Their mode of action is schematically presented in Figure 33.1. A number of other glucosidases are essential for hydrolysis of those linkages that are present in small proportion in xylans. Thus, rhamnosidase (EC 3.2.1.40) releases L-rhamnose from the nonre-

ducing ends of the polysaccharide, and α-L-fucosidase (EC 3.2.1.51) splits the terminal fucose from xyloglucans and rhamnogalacturonans.

Optimal conditions of pH and temperature for xylan degradation vary widely depending on the source of enzymes (Dekker and Richards, 1976; Honda et al., 1985; Ward and Moo-Young, 1989). Enzymes from thermophilic anaerobes are intrinsically thermostable (i.e., their stability at elevated temperatures depends on the structure of the protein molecule) (Zeikus, 1979; Lee and Zeikus, 1991).

33.2 Isolation and Characterization of Thermophilic Xylanolytic Microorganisms

Until the mid-1980s, only one thermoanaerobic strain, *Thermoanaerobacter* B6A, had been reported to actively grow on xylan (Weimer et al., 1984). Our group has isolated several xylanolytic strains from thermal spring ecosystems in Yellowstone National Park. These include *Thermoanaerobacter* strain LX11 (unpublished), *Clostridium thermosulfurogenes* 4B (Schink and Zeikus, 1983), *Clostridium thermohydrosulfuricum* 39E (Zeikus et al., 1980), *Thermoanaerobium brockii* HTD4 (Zeikus et al., 1979), and *Clostridium thermocellum* LQRI (Ng and Zeikus, 1981), that were also studied in other laboratories (Wiegel et al., 1985). Each of these strains ex-

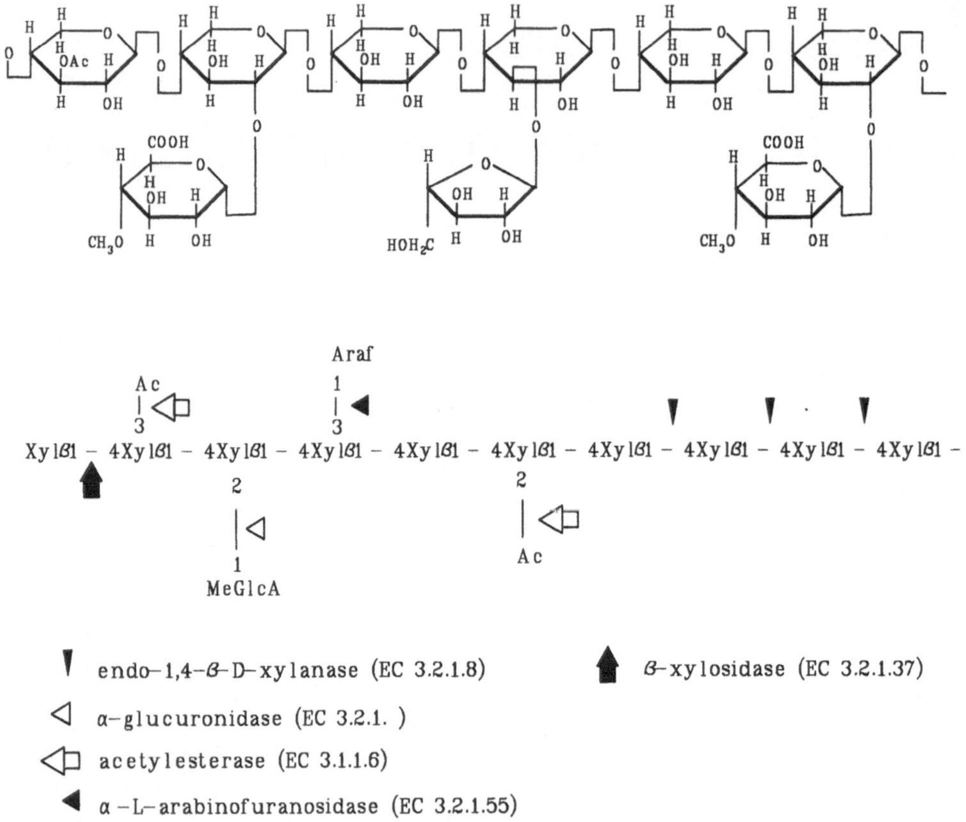

FIGURE 33.1 Schematic representation of portion of xylan chain shows types of prevalent linkages and sites of action of different xylanolytic en-zymes. Abbreviations: Ac, acetyl group; Araf, ara-binofuranose.

press xylanase and xylose isomerase activities but only *Thermoanaerobacter* sp. strain LX11 actually grows on xylan.

Although a number of saccharolytic thermo-anaerobes have been isolated and characterized to different extents, there are still no consistent taxonomic criteria for proper classification of these organisms. At present, genus and species assignments are based on spore formation abil-ity, Gram staining, substrate utilization, and end product formation (Wiegel et al., 1979; Zeikus et al., 1979; Wiegel and Ljungdahl, 1981; Schink and Zeikus, 1983; Weimer et al., 1984) Currently, species of thermoanaerobic spore formers (e.g., *Clostridium*) and non-spore formers (e.g., *Thermoanaerobacter, Thermoanaero-bium*) are designated as species of uncertain tax-onomic affiliation. However, determination of 16S rRNA sequence homologies of these organ-isms has suggested a close phylogenetic rela-

tionship among these diverse genera and only three distinct subgroups were discernible by this analytical method (Cato and Stackebrandt, 1989).

In an attempt to use a molecular approach to define the genetic relationship among xylanoly-tic thermoanaerobes and thus to facilitate the establishment of species names, we have com-pared DNA homologies of these strains by DNA–DNA hybridization. This technique has proved to be extremely useful for taxonomic studies of other species (Bradley, 1973; Schleif-er and Stackebrandt, 1983; Hurley et al., 1988), and a variation of this procedure that uses free-solution hybridization and Sl nuclease treat-ment of the hybrids was employed here. The results shown in Table 33.1 indicate the exis-tence of three groups among the xylanolytic thermoanaerobes known to date: group I in-cludes one species, *C. thermocellum*, because

Table 33.1 Comparison of DNA homology between thermoanaerobic strains that display xylanase activity

Species	Percent of relative binding with labeled DNA from							
	LQRI	B6A	LX11	4B	HTD4	JW200	39E	E100
Group I								
Clostridium thermocellum LQRI	100	13	8	5	5	5	6	6
Group II								
Thermoanaerobacter B6A	30	100	43	53	17	19	19	23
Thermoanaerobacter LX11	35	61	100	92	17	18	22	20
Clostridium thermosulfurogenes 4B	40	87	86	100	17	17	20	23
Group III								
Thermoanaerobium brockii HTD4	45	29	24	19	100	84	86	57
Thermoanaerobacter ethanolicus JW200	40	28	44	18	84	100	97	60
Clostridium thermohydrosulfuricum 39E	40	29	24	21	81	94	100	58
Clostridrium thermohydrosulfuricum E100-9	34	34	25	70	67	79	100	100

this organism exhibits less than 14% homology to other species studied; group II is composed of *Thermoanaerobacter* strains B6A and LXll and *C. thermosulfurogenes* 4B, which show more than 40% of homology among themselves; group III is composed of *C. thermohydrosulfuricum* 39E, *Thermoanaerobium brockii* HTD4, and *Thermoanaerobacter ethanolicus* JW200, because they exhibit homologies among DNA ranging from 57% to 97%. These data suggest that the organisms within each group constitute the same genus. These results are in agreement with the conclusions of Cato and Stackebrandt (1989) obtained by studying the 16S rRNA homologies, which showed that *Clostridium* species are subdivided into six separate taxonomic groups with spore-forming and nonspore-forming species included in the same group.

33.3 Thermophilic Xylanolytic Enzymes from *Thermoanaerobacter*

Thermoanaerobacter strain B6A produced high levels of xylanolytic enzymes and exhibited no cellulase activity; it was therefore used in our group as a model organism for studies of biochemistry, genetics, and molecular physiology of thermophilic xylanases.

Xylanolytic activity of the strain B6A was cell associated in the early phases of growth, but most of the activity appeared in the extracellular medium in the stationary phase. Enzyme fractions partially purified from the growth medium contained both xylose- and xylobiose-forming activities, indicating the presence of both endoxylanases and endoxylosidases. Endoxylanase of *Thermoanaerobacter* B6A was optimally active at 75°C and not active at 25°C. Native gel electrophoresis indicated multiple forms of endoxylanase. At present it is not clear whether these forms represent the products of different genes, multiple molecular forms of the same subunit polypeptides, or forms with different degree of posttranslational modification such as glycosylation. Evidence for the possible glycosylation of endoxylanases in the B6A strain was obtained by direct staining of gels and by comparisons of properties exhibited by an endoxylanase produced from the cloned gene in *Escherichia coli* (Lee and Zeikus, in manuscript).

Genetics of thermophilic xylanases

Regulation. In *Thermoanaerobacter* B6A extracellular xylanase is produced in the presence of xylan or xylose; it is not produced when glucose is used as carbon source. Moreover, the addition of glucose to the culture growing on xylose inhibits the production of xylanolytic activity (Figure 33.2). Because no xylanase activity could be detected in cultures of *Thermoanaerobacter* grown in Luria broth media (results

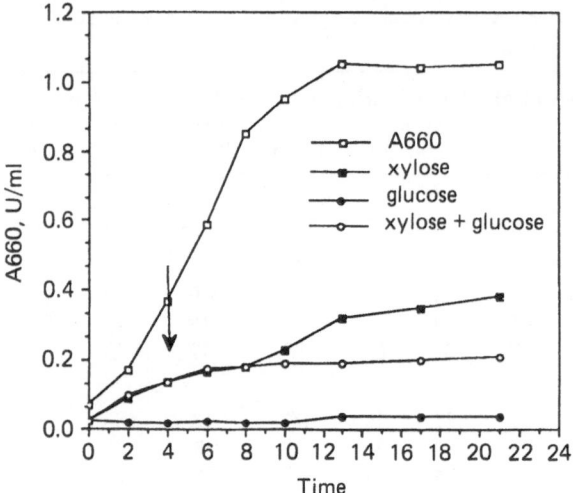

FIGURE 33.2 Production of endoxylanase by the culture of *Thermoanaerobacter* strain B6A on glucose or xylose. Optical density increase is indicated for culture grown on D-xylose as the sole carbon source. One unit is defined as the amount of enzyme that liberates 1 μmol of reducing xylose (measured as reducing sugar) per min at 60°C. Arrow indicates addition of glucose (0.3%) to culture growing on D-xylose.

not shown), we concluded that the expression of xylanase genes is regulated by induction in the presence of xylose and possibly also by catabolite repression in the presence of glucose. However, no elements of this genetic regulon have been identified.

Isolation and cloning of thermophilic xylanase genes. A gene encoding a thermophilic endoxylanase was isolated from the genomic library prepared from total DNA of the strain B6A in the cosmid vector pHC79 (C. Lee and J.G. Zeikus, in manuscript). Positive clones were identified on agar plates containing Remazol brilliant blue xylan. Restriction endonuclease analysis and Bal31-mediated deletion mapping followed by subcloning has identified a 3.9-kb fragment carrying an active endoxylanase gene. The physical map of the fragment is shown in Figure 33.3. In *E. coli*, the gene was expressed constitutively and was not subject to catabolite repression by glucose. This indicated that the regulatory gene, responsive to xylose induction, either was not present on the cloned fragment or was not functional in *E. coli*. Thermophilic en-

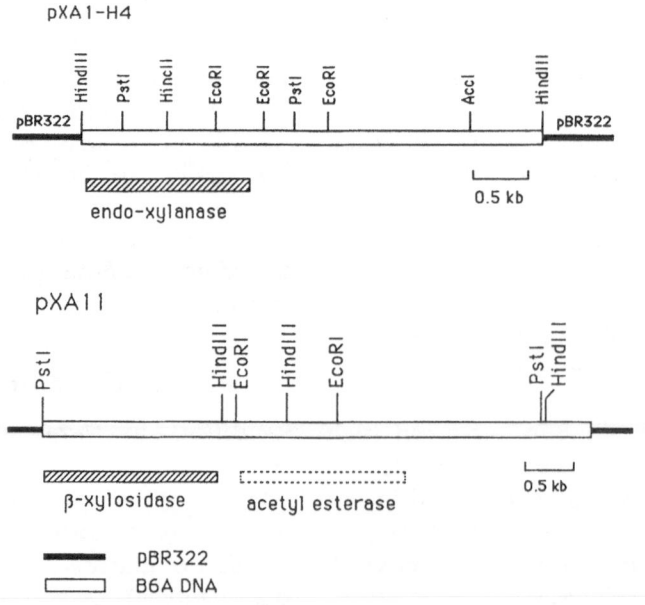

FIGURE 33.3 Physical and genetic map of DNA fragments containing endoxylanase, β-xylosidase, and acetyl esterase from *Thermoanaerobacter* B6A. Acetyl esterase gene is indicated by dotted box; precise location of its borders has not yet been determined.

doxylanase produced in *E. coli* was predominantly located in the cytoplasm and exhibited a lower thermostability than the glycosylated native enzyme by 10°–15°C.

Two other genes, one specifying β-xylosidase and another specifying acetyl esterase, were identified on the same 40-kb cosmid fragment that carried the endoxylanase gene. The β-xylosidase and acetyl esterase genes are very closely linked to one another. It is not known, however, whether the three genes are part of an operon-like structure.

Cellular location of xylanolytic enzymes in thermoanaerobes

It was observed that in the exponential phase of growth, when most of the enzyme was still cell associated, *Thermoanaerobacter* cells grown on xylan or on xylose would bind tightly to particles of insoluble xylan, whereas the cells grown on glucose remained unattached (Figure 33.4). This suggested that xylanolytic enzymes present on the surface of the cells were responsible for this binding. To visualize potential cell-surface structures in *Thermoanaerobacter*, the cells were subjected to scanning electron microscopy after treatment with cationized ferritin (CF). CF interacts with anionic groups of the cell envelope and stabilizes the structures present on the surface during dehydration (Lamed et al., 1987a, 1987b). As shown in Figure 33.5, characteristic protuberant structures (xylanosomes) were present on the surface of cells grown in xylose medium but absent from the cells grown on glucose.

It has been demonstrated that cellulolytic enzymes of *C. thermocellum* are organized into distinct multisubunit complexes called cellulosomes (Lamed et al., 1983). The cellulosome is a very large (MW, $> 2 \times 10^6$) enzyme complex containing endoglucanases and a cellulose-binding factor (Bayer and Lamed, 1986; Bayer et al., 1983; Lamed and Bayer, 1988; Kohring et al., 1990; Morag et al., 1990). It is present in both an extracellular and a cell-associated form. The latter is considered to be a discrete cell-surface organelle responsible for both the binding to the insoluble substrate and for its degradation. The occurrence of cell-associated cel-

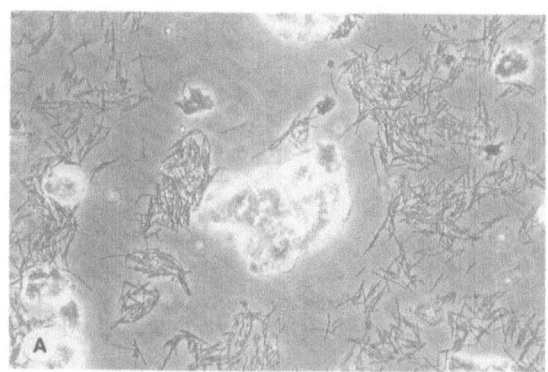

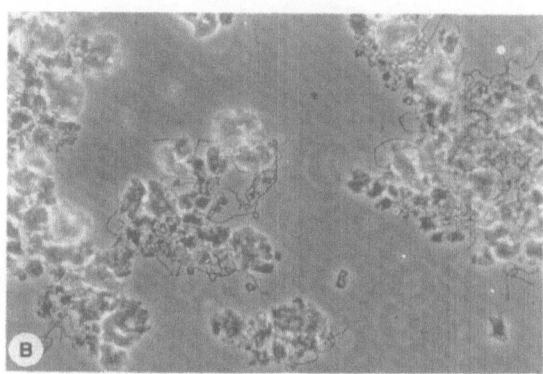

FIGURE 33.4 Phase-contrast micrographs of cells of *Thermoanaerobacter* B6A grown in D-glucose (**A**) or D-xylose (**B**) medium and mixed with insoluble xylan.

lulolytic enzymes has been suggested in several anaerobic bacteria (Groleau and Forsberg, 1981; Wood et al., 1982). It is therefore tempting to interpret the observations on xylanolytic enzymes as indicative of their location in specific xylanosomes initially present on the surface of bacterial cells and subsequently released into the medium in the stationary cultures.

It is possible that the involvement of cell-surface structures in combining multiple enzymes acting on extracellular substrates is a more general feature of anaerobic bacteria living in natural habitats.

Xylose utilization

Xylose liberated by the degradation of xylan is not phosphorylated by microbial enzyme systems; to become phosphorylated and enter the cellular metabolic pathways, it must be con-

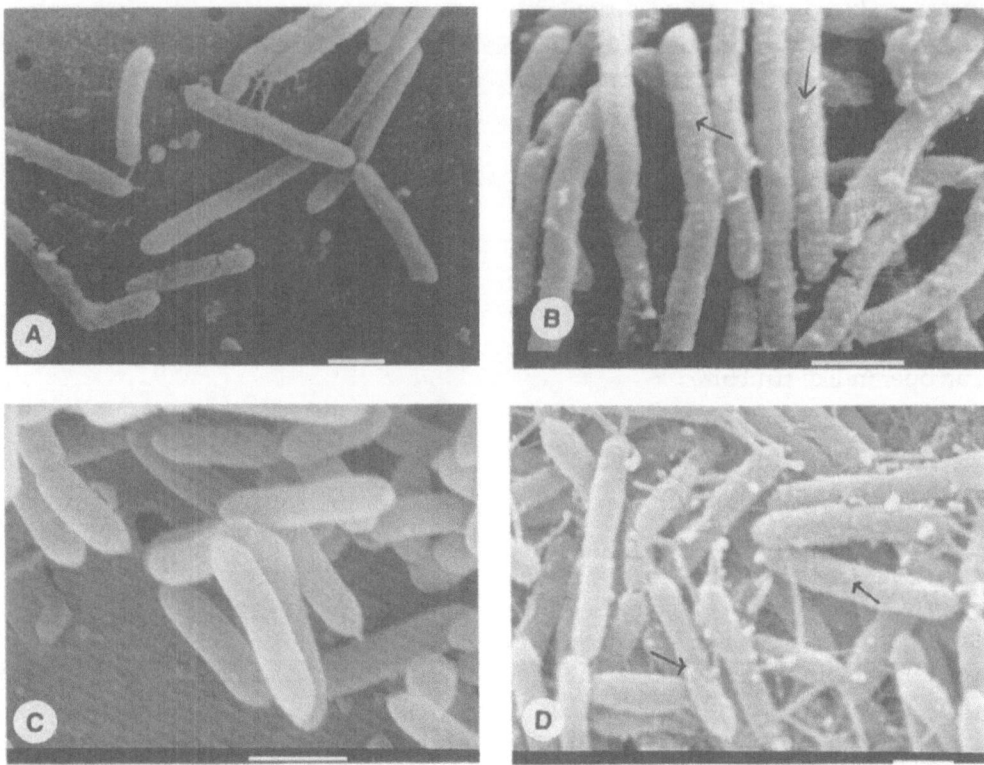

FIGURE 33.5 Scanning electron micrographs of *Thermoanaerobacter* B6A cells (**A** and **B**) and LX11 cells (**C** and **D**) labeled with cationized ferritin (CF). Cells were grown in D-glucose medium (**A, C**) or in D-xylose medium (**B, D**), stained with CF, and processed for electron microscopy. Arrows indicate cell-surface structures of putative xylanosomes. Bar, 1.0 μm.

verted to xylulose. This conversion is catalyzed by an ubiquitous intracellular enzyme, xylose isomerase [D-xylose ketol isomerase (EC 5.3.1.5)]. Increased interest in this enzyme in the past decade stems from its ability to use also D-glucose as substrate and catalyze its conversion to fructose. This enzyme is, therefore, often referred to as glucose isomerase and is widely used in industry for production of high fructose content syrups (for review, see Bucke, 1980). It is estimated that 11 billion pounds of such syrup are manufactured per year, and the increase in production has not yet reached a plateau.

Biochemical and physicochemical properties of xylose isomerase isolated from different sources have been extensively studied (Chen, 1980; Carell et al., 1984), and complete amino acid sequences of the enzyme (mostly based on the nucleotide sequences of *xyl*A, the structural gene for xylose isomerase) from *Escherichia coli* (Schellenberg et al., 1983), *Bacillus subtilis* (Wilhelm and Hollenberg, 1984, 1985), *Ampullariella* strain 3876 (Saari et al., 1987), *Arthrobacter* B3728 (Henrick et al., 1989), *Streptomyces violaceoniger* (Drocourt et al., 1988), *Streptomyces olivochromogenes* (Farber, et al., 1989), and *Streptomyces griseofuscus* (Kikuchi et al., 1990), have been reported. The crystal structure of xylose isomerase from *Actinoplanes missouriensis* (Rey et al., 1988), *Streptomyces* (Carell et al., 1989; Dauter et al., 1989; Farber et al., 1989), and *Arthrobacter* (Henrick et al., 1989; Collyer and Blow, 1990; Collyer et al., 1990) have been refined at high resolution. On the basis of this information hypotheses attempting to explain the mechanism of catalysis were proposed. However, very few experiments that could confirm the models by a functional analysis have been reported to date.

Table 33.2 Properties of thermophilic xylose isomerases

Substrate	Clostridium thermosulfurogenes[a]			Thermoanaerobacter B6A[b]		
	K_m (mM)	V_{max} (U/mg)[c]	k_{cat} (min^{-1})[d]	K_m (mM)	V_{max} (U/mg)[c]	k_{cat} (min^{-1})[d]
Xylose	20	15.7	785	16	17.6	880
Glucose	140	5.2	260	120	6.3	315
Fructose	60	2.5	125	50	2.8	140

[a] Optimum pH, 7.0–7.5; optimum temperature, 75°–80°C; stability (highest temperature at which >90% of activity remained after 1 h), 80°C.
[b] Optimum pH, 7.0–7.5; optimum temperature, 80°C; stability, 85°C.
[c] One unit (U) is amount of enzyme that produces 1 μmol of product per min.
[d] Per active site.

Thermophilic xylose isomerases

Xylanolytic thermophiles of the genus *Thermoanaerobacter* and *Clostridium* produce active xylose isomerases. As expected, the enzymes are intracellular, but they are readily purified after breakage of the cells (Lee and Zeikus, 1991). The enzymes examined to date are rich in alanine residues, which may contribute via hydrophobic interactions to their intrinsic thermostability. Pure glucose isomerases from these thermophilic organisms exhibit high thermal stability in solution and do not contain detectable glycosylation, which seems to increase the thermostability of xylanases (see: *Isolation and cloning of thermophilic xylanase genes*).

Biochemical properties of two thermophilic xylose isomerases are summarized in Table 33.2. Both enzymes require Co^{2+} for maximal stability and for activity as glucose isomerase. For the isomerization of xylose Mn^{2+} gave the highest activity, but Co^{2+} was almost as effective (Lee and Zeikus, 1991).

Genetic organization of xylose isomerases in thermoanaerobes

Regulation. In *Clostridium thermosulfurogenes* and *Thermoanaerobacter*, the synthesis of xylose isomerases was inducible by xylose. Cells grown in the presence of glucose do not produce detectable amounts of the enzyme (Lee et al., 1990a). Although both endoxylanases and xylose isomerases are induced by the same substance, we do not know whether the genes specifying these enzymes are closely linked on the chromosome of the thermophiles or even

whether the same regulatory gene might be responsible for this induction. It seems that in *C. thermosulfurogenes* the regulatory gene is not closely linked to the structural gene encoding xylose isomerase, because the cloned gene was expressed constitutively in *E. coli* and *Bacillus subtilis* and this expression was insensitive to xylose (Lee et al., 1990a, 1990b).

Structural gene xylA from C. thermosulfurogenes. The gene encoding the thermophilic xylose isomerase of *C. thermosulfurogenes* has been cloned and its nucleotide sequence determined (Lee et al., 1990b). The amino acid sequence deduced from the coding sequence of the gene exhibited considerable homology to the sequences of other xylose isomerases studied to date. In particular, amino acids that were predicted, by x-ray diffraction studies, to be part of the active center, such as His_{101}, Thr_{141}, Val_{186}, Trp_{139}, or to bind metal ions essential for the activity, i.e., Glu_{232} and Glu_{268}, were found to be highly conserved (Figure 33.6). Surprisingly, the sequence of the thermophilic enzyme from *C. thermosulfurogenes* had a much higher degree of homology to the sequences of thermolabile enzymes of *E. coli* and *B. subtilis* than to the

FIGURE 33.6 Comparison of amino acid sequence homologies of different xylose isomerases. Residues that are conserved in at least five sequences are shown in capital letters. Abbreviations: Bs, *Bacillus subtilis*; Ec, *Escherichia coli*; Ct, *Clostridium thermosulfurogenes*, Sv, *Streptomyces violaceoniger*; Amp, *Ampullariella* strain 3876; Art, *Arthrobacter* B3728; Sol, *Streptomyces olivochromogenes*; Sgr, *Streptomyces griseofuscus*; Con, consensus sequence.

```
Bs   maqshsssvn yfgsvnkvvf egkastnpla fkyynpqevi ggktmkehlr fsiaywhtfT  60
Ec   ......mqa  yfdqldrvry egskssnpla frhynpdelv lgkrmeehlr faacywhtfc
Ct   ......mnk  yfenvskiky egpksnnpys fkfynpeevi dgktmeehlr fsiaywhtfT  53
Sv   .......... .......... .......... .......... .......... ....MSfQpT
Amp  .......... .......... .......... .......... .......... ....MSlQaT
Art  .......... .......... .......... .......... .......... ....MSvQpT
Sol  .......... .......... .......... .......... .......... ....MSyQpT
Sgr  .......... .......... .......... .......... .......... ....MSfQpT
Con  ---------- ---------- ---------- ---------- ---------- ----MS-Q-T

Bs   aDgtdvFGaa TmqRp.WDhY kg.mdlArar veaAfemFek LdApffaFHD rDiaPeGstl 120
Ec   wngadmFGvg afnRp.Wqqp GEAlalAkrk advAfefFhk LhvpfycFHD vDvsPeGasl
Ct   aDgtdqFGka TmqRp.WnhY tDpmdiAkar veaAfeFFdk inApyfcFHD rDiaPeGdtl 112
Sv   PEDkFtFGLW TVGWqgRDpF GDATRpALD. PVEtVqrLAe LGAyGVTFHD dDLiPFGssd
Amp  PDDkFsFGLW TVGWqaRDaF GDATRpvLD. PIEAVhkLAe iGAyGVTFHD dDLvPFGada
Art  PaDhFtFGLW TVGWtgaDpF GvATRknLD. PVEAVhkLAe LGAyGITFHD nDLiPFdate
Sol  PEDrFtFGLW TVGWqgRDpF GDATRpALD. PVDtVqrLAg LGAhGVTFHD dDLiPFGssd
Sgr  PEDkFtFGLW TVGWqgRDpF GDATRpgLD. PVEtVr.LAe LGAyGVTFHD dDLnPFGssd
Con  P-D-F-FGLW TVGW--RD-F GDATR-ALD- PVEAV--LA- LGA-GVTFHD -DL-PFG---

Bs   kEtnqnlDii VgmikdyMrd snvkllwnTa NMFTnPrFvh GaaTscnaDV faYAaaqVkk 180
Ec   kEyinnfaqm VdvLagkqEe sGvkllwgTa NcFTnPrYga GaaTnpDpEV fsWAatqVvt
Ct   rEtnknlDti VamikdYLkt sktkVlwgTa NLFsnPrFvh GasTscnaDV faYsaaqVkk 172
Sv   tER....Esh IkrFrqALDa TGMtVPMaTT NLFTHPVFKD GgFTaNDRDV RrYAlrKtir
Amp  atR....Dgi VagFskALDe TGLiVPMvTT NLFTHPVFKD GgFTsNDRsV RrYAirKVlr
Art  aER....Eki lgdFnqALkd TGLkVPMvTT NLFsHPVFKD GgFTsNDRsI RrFAlaKVlh
Sol  tER....Esh IkrFrqALDa TGMtVPMaTT NLFTHPVFKD GgFTaNDRDV RrYAlrKtir
Sgr  tER....Esh IkrFrqALDa TGMtVPMaTT NLFTHPVFKD .rFTaNDRDV RaYAvrKtir
Con  -ER----E-- V--F--ALD- TGM-VPM-TT NLFTHPVFKD G-FT-NDRDV R-YA--KV--

Bs   glEtAkELGA enYVFWGGRE GyEtllntDl kfeLDnLarf MhMavDYake ieYtgqFlIE 240
Ec   amEathkLGg enYVLWGGRE GyEtllntDl rqerEqLgrf MqMvvEhkhk iGFqgtLlIE
Ct   alEitkELGg enYVFWGGRE GyEtllntDm efeLDnFarf LhMavDYake iGFegqFlIE 232
Sv   niDLAaELGA kTYVaWGGRE GaEsggaKDv rdALDRMkEa FdLlgEYvta qGYdlrFAIE
Amp  qmDLgaELGA kTLVLWGGRE GaEydsaKDv gaALDRyrEa LnLlaqYsed qGYglpFAIE
Art  niDLAaEMGA eTFVMWGGRE GsEydgsKDl aaALDRMrEg vdtaagYikd kGYnlriAlE
Sol  niDLAvELar kTYVaWGGRE GaEsggaKDv rvALDRMkEa FdLlgDYvts qGYdtrFAIE
Sgr  niDLAaELGA kTYVaWGGRE GaEsggaKDv rdALDRMkEa FdLlgEYvta qGYdlrFAIE
Con  --DLA-ELGA -TYV-WGGRE G-E----KD- --ALDRM-E- F-L---Y--- -GY---FAIE

Bs   PKPkEPtthq Ydtdaattia FlkqYgldnh FklNlEanHa tLAGhtFeHe lrmArvhGlL 300
Ec   PKPqEPtkhq Ydydaatvyg FlkqYgleke iklNiEanHa tLAGhsFhHe IAtAialGlF
Ct   PKPkEPtkhq YdfdVanvLA FlrkYDldky FkVNiEanHa tLAfhdFqHe lryArinGvL 292
Sv   PKPNEPRGDI LLPTVGHaLA FIerLErpEl YGVNPEvGHE QMAGLNFpHG IAQALWaGKL
mp   PKPNEPRGDI LLPTaGHaiA FVqeLErpEl FGINPEtGHE QMsnLNFtqG IAQALWhkKL
Art  PKPNEPRGDI FLPTVGHgLA FIeqLEhgDi vGlNPEtGHE QMAGLNFtHG IAQALWaeKL
Sol  PKPNqPRGDI LLPTVGHaLA FIerLErpEl YGVNPEvGHE QMAGLNFpHG IAQALWaGKL
Sgr  PKPNEPRGDI LLPTVGHaLA FIerLErpEl YGVNPEvGHE QMAGLNFpHG IAQALWaGKL
Con  PKPNEPRGDI LLPTVGH-LA FI--LE--E- -GVNPE-GHE QMAGLNF-HG IAQALW-GKL

Bs   gsVDaNqghp llgwdtdeFp tdlysttLaM yEiLqnGglg sgGlnfdakv rrssfepdDl 360
Ec   gsVDaNrgda qlgwdtdqFp nsveenALvM yEiLkaGgft tgGlnfdakv rrqstdkyDl
Ct   gsIDaNtgdm llgwdtdqFp tdirmttLaM yEvikmGgfd kgGlnfdakv rrasfepeDl 352
Sv   FHIDLNGQsG iKYDQDlrFg agdlraAFwL VDLLEsa... .Gyegprhf dfkpPrteDf
Amp  FHIDLNGQhG pKFDQDlvFg hgdllnAFsL VDLLEnGpdg gpayd.gprh fdykPsrtEd
Art  FHIDLNGQrG iKYDQDlvFg hgdltsAFft VDLLEnGfpn ggpkytgprh fdykPsrtD.
Sol  FHIDLNGQsG iKYDQDlrFg agdlraAFwL VDLLE..... saGye.gprh ldfkPprtEd
Sgr  FHaDLNGQsG iKYDQD.cgs rrrpaggviv VDLLE..... saGye.gprh fdfkPprtEd
Con  FHIDLNGQ-G -KYDQD--F- ------AF-- VDLLE-G--- --G------- ----P---D-

Bs   vyahiagmda fargLkVahk liedrvfEdv Iqhryrsfte gigleitegr anfhtleqya 420
Ec   fyGhigamdt malaLkIaar miedgelDkr Iaqrysgwns elgqqilkgq msladlakya
Ct   flGhiagmda fakgFkVayk lvkdrvfDkf Ieeryasykd gigadivsgk adfrslekya 412
Sv   dgvwasaegc mrnyLilker aaafradpev qealraarld qlaqpt..aa dglealladr
Amp  fDGvwesakd nirmYlllke rakafraDpe Vqaalaeskv delrrtptlnp getyadllad
Art  gydgvwdsak anmsmylllk eralafradp ecqeamktsg gvfelgettl nagesaadlm
Sol  idGvwasaag cmrnYlIlke raaafraDpe Vqealrrsrl delaqpt..a adgvqellad
Sgr  fDGvwasaeg cmrnYlIlkq prppsaptrr crrrasaprv wtsw.psrpl adgleallad
Con  --G------- ------I--- -------D-- I--------- ---------- ----------

Bs   lnnktiknes grqerlkpil nq........ .......... ..       465
Ec   qehhlspvhq sgrqeqlenl vnhylfdk.. .......... ..
Ct   lersqi.vnk sgrqellesi lnqylfae.. .......... ..       439
Sv   tafedfdvea Aaaaraawpfe rldqlamdhl lgarg..... ..
Amp  rsafedydad Avgakg.ygf vklnqlaidh llgar..... ..
Art  ndsasfagfd Aeaaaernfa firlnqlaie hllgsr.... ..
Sol  rtafedfdvd Aaaaraawpy erldqlamdh llgarg.... ..
Sgr  rtafedfdve Aaaargmvrt prpagdgppa grarltvapr kr
Con  ---------- A--------- ---------- ---------- --
```

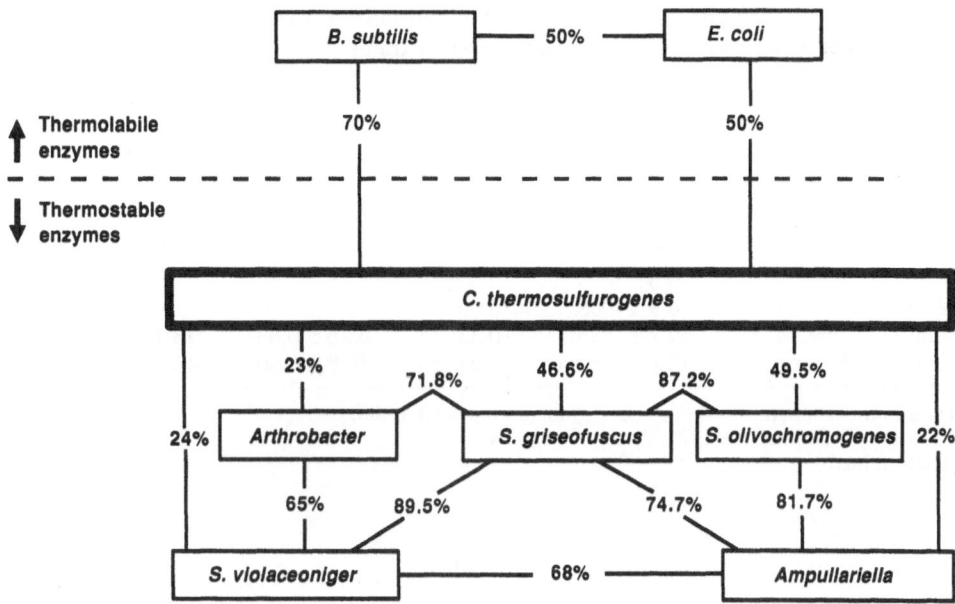

FIGURE 33.7 Summary of homology comparisons between amino acid sequences of xylose isomerases. Figures indicate percentage of identi-cal amino acids as calculated by GCG software package from University of Wisconsin.

relatively thermostable enzymes of species of *Streptomyces*, *Arthrobacter*, and *Ampulariella* (Figure 33.7). This places the *Clostridium* thermophilic enzyme in a discrete class of enzymes different from the thermolabile enzymes of the *Bacillus* type and thermostable ones of the *Streptomyces* type.

The mechanism of catalysis by xylose isomerase

Initially it was thought that enzymatic isomerization of xylose and glucose proceeds via a *cis*-endiol intermediate and is accomplished by proton transfer with a histidine residue acting as a general base and attracting the proton from the C-1 atom of the substrate (Rose et al., 1969). The analogy to triose phosphate isomerase, for which the base-catalyzed endiol mechanism has been demonstrated (Rose, 1981), and some indications from x-ray crystallographic studies (Carell et al., 1989) seemed to support this hypothesis. The early mutagenesis studies have, therefore, been interpreted as indicating that His_{101} was the residue acting as the general base in the isomerization reaction

(Batt et al., 1990). However, xylose isomerase exhibits properties very distinct from those of triose phosphate isomerase: (i) it requires divalent cations for activity and stability (Danno, 1971); (ii) it transfers the hydrogen between the C-1 and C-2 atoms of the substrate without significant exchange with the solvent (Rose et al., 1969); (iii) it exhibits anomeric specificity toward the α-configuration of the substrate indicating that the enzyme binds the closed ring form of the substrate and that a ring-opening reaction occurs as a step in the catalytic process (Feather et al., 1970; Schray and Rose, 1971); and (iv) it has a much higher K_m and approximately 1000-fold lower K_{cat} than triose phosphate isomerase (Raines et al., 1986). These differences argue that the two enzymes are unlikely to employ the same catalytic mechanisms. Moreover, crystallographic studies have indicated that the position of the substrate bound to the enzyme is different from that suggested earlier and therefore the histidine residue (His_{101} in the *Clostridium* enzyme) is positioned too far from the C-1 and C-2 atoms of the substrate to be able to attract protons from their hydroxyl groups. It was suggested instead

Table 33.3 Alterations of enzymatic properties by substitutions of amino acids in catalytic center of thermostable glucose isomerase for glucose and xylose substrates

Enzyme	Glucose			Xylose		
	K_m	V_{max}^a	Catalytic efficiency[b]	K_m	V_{max}^a	Catalytic efficiency[b]
Wild type	117 ± 6.2	13.8 ± 0.57	0.12	21.5 ± 0.58	18.7 ± 1.70	0.86
$His_{101} \rightarrow Phe$	–	N.D.[c]	–	–	N.D.	–
$His_{101} \rightarrow Gln$	138 ± 18	0.5 ± 0.08	0.004			
$His_{101} \rightarrow Asn$	168 ± 19	0.9 ± 0.06	0.005			
$Trp_{186} \rightarrow Phe$	59 ± 7.9	19.7 ± 0.38	0.33	45.5 ± 0.58	13.1 ± 0.12	0.28
$Val_{186} \rightarrow Thr$	80.3 ± 4	17.4 ± 1.10	0.22	16.8 ± 0.50	16.5 ± 0.37	0.98

[a] One unit is amount of enzyme that produces 1 μmol of product per min.
[b] Catalytic efficiency is defined as $k_{cat}(K_m^{-1})$.
[c] N.D., not determined.

that the isomerization proceeds via a metal-catalyzed hydride shift and that the essential histidine is required for either the ring opening or the binding of the substrate (Farber et al., 1987, 1989; Collyer and Blow, 1990; Collyer et al., 1990; Henrick et al., 1989).

Although the knowledge of the three-dimensional structure of a protein is required for understanding of its function, the static structure itself does not provide proof of biochemical function of individual amino acid residues. Further support is therefore needed to prove the mechanisms outlined by resolving the x-ray diffraction data.

Advances in engineering of proteins by site-specific mutagenesis of their genes have allowed verification of predictions stemming from structural analysis of xylose isomerase. Thus, it has been shown that in the *E. coli* enzyme His_{101} residue is essential for enzymatic activity by substituting it for several other amino acid residues and showing that an inactive protein was specified by the mutated gene (Batt et al., 1990; Lee et al., 1990b). Xylose isomerase from *C. thermosulfurogenes* has proved to be a particularly convenient model for protein engineering because its gene was very actively expressed in both *E. coli* and *B. subtilis* and the enzyme produced in these hosts could easily be purified in high yields (Lee et al., 1990a). It was thus found that substitutions of His_{101} by Phe leads to a complete loss of enzyme activity, whereas substitutions of several other His res-

idues has no effect on enzyme activity (Lee et al., 1990b). On the other hand, substitution of His_{101} by Gln, Glu, Asp, or Asn resulted in mutant enzymes that retained 10%–14% of the wild-type activity, whereas the K_m was not changed significantly (Table 33.3). Moreover, the resulting enzymes showed a constant activity at acidic pH down to pH 5.0, whereas the activity of the wild-type enzyme dropped sharply at pH below 6.5 (Lee et al., 1990b). This indicates that the ability of His_{101} to be protonated is essential for the activity. However, if this residue had functioned as a simple base, its substitution by the nonbasic amino acids would be expected to create an inactive enzyme. Two other possible roles for the His_{101} have been suggested: (i) catalysis of the ring opening and (ii) stabilization of the substrate bound to the enzyme.

Determination of the isotope effect of D-[2-^{2}H]glucose on the V_{max} of xylose isomerase from *C. thermosulfurogenes* xylose isomerase has demonstrated that hydrogen transfer and not the ring opening is the rate-limiting step in the isomerization reaction for both the wild type and the mutant Gln_{101} enzyme (Lee et al., 1990b). Further, the x-ray crystallographic studies of the enzyme–substrate complex revealed that only the open-chain forms of the substrate were bound to the enzyme crystals (Carell et al., 1989; Collyer et al., 1990). These findings strongly indicated that this complex is the most stable species among the reaction intermediates. We suggest, therefore, that the stabilization of this intermediate enzyme–substrate complex is the primary function of His_{101}.

The high degrees of similarity between the primary structures of xylose isomerases studied to date have indicated that certain predictions made from the structure of one species of the enzyme might be tested on a different species. With this in mind we substituted the Trp_{139} and the Val_{186} residues in the *Clostridium* thermophilic enzyme by other amino acids. In the three-dimensional structure of the *Arthrobacter* enzyme, these residues were presumed to constitute a steric hindrance for the binding of D-glucose as opposed to the binding of the smaller molecule of D-xylose. As shown in Table 33.3, substitution of each of these two residues had allowed a better binding of glucose and had created enzymes that have higher activity with glucose than with xylose as substrate. Construction of these enzymes with new catalytic properties that have not been found in nature underscores the increasing importance of engineered proteins in research and in biotechnology.

References

Aspinall, G.O. 1980. Chemistry of cell wall polysaccharides. *In: The Biochemistry of Plants*, Vol. 3, J. Preiss, ed., pp. 473–500. New York: Academic Press.

Batt, C.A., A.C. Jamieson, and M.A. Vandeyar. 1990. Identification of essential histidine residues in the active site of *Escherichia coli* xylose (glucose) isomerase. *Proc. Natl. Acad. Sci. USA* 87:618–622.

Bayer, E.A. and R. Lamed. 1986. Ultrastructure of the cell surface cellulosome of *Clostridium thermocellum* and its interaction with cellulose. *J. Bacteriol.* 167:828–836.

Bayer, E.A., R. Kenig, and R. Lamed. 1983. Adherence of *Clostridium thermocellum* to cellulose. *J. Bacteriol.* 156:818–827.

Biely, P. 1985. Microbial xylanolytic systems. *Trends Biotechnol* 3:286–290.

Bradley, S.G. 1973. Relationships among mycobacteria and nocardiae based upon DNA reassociation. *J. Bacteriol.* 113:645–651.

Bucke, C. 1980. Enzymes in fructose manufacture. *In: Enzymes and Food Processing*, G.G. Birch, N. Blackebrough, and J.K. Parker, eds., pp. 51–72. London: Applied Science.

Carrell, H.L., J.P. Glusker, V. Burger, F. Manfre,

D. Tritsch, and J.-F. Biellmann. 1989. X-ray analysis of D-xylose isomerase at 1.9 Å: native enzyme in complex with substrate and with a mechanism-designed inactivator. *Proc. Natl. Acad. Sci. USA* 86:4440–4444.

Carrell, H.L., B.H. Rubin, T.J. Hurley, and J.P. Glusker. 1984. X-ray crystal structure of D-xylose isomerase EC 5.3.1.5 at 4 Å resolution. *J. Biol. Chem.* 259:3230–3236.

Cato, E.P. and E. Stackebrandt. 1989. Taxonomy and phylogeny. *In: Clostridia*, N.P. Minton and D.J. Clarke, eds., pp. 1–26. New York: Plenum.

Chen, W. 1980. Glucose isomerase (a review). *Process Biochem.* 15:30–35.

Collyer, C.A. and D.M. Blow. 1990. Observations of reaction intermediates and the mechanism of aldose-ketose interconversion by D-xylose isomerase. *Proc. Natl. Acad. Sci. USA* 87:1362–1366.

Collyer, C.A., K. Henrick, and D.M. Blow. 1990. Mechanisms for aldose-ketose interconversion by D-xylose isomerase involving ring opening followed by a 1,2-hydride shift. *J. Mol. Biol.* 212:211–235.

Danno, G. 1971. Studies on D-glucose isomerizing enzyme from *Bacillus coagulans* strain HN-68 part 6. The role of metal ions on the isomerization of D-glucose and D-xylose by the enzyme. *Agric. Biol. Chem.* 35:997–1006.

Dauter, Z., M. Dauter, J. Hemker, H. Witzel, and K. Wilson. 1989. Crystallization and preliminary analysis of glucose isomerase from *Streptomyces albus*. *FEBS Lett.* 7:1–8.

Dekker, R.F.H. and Goeffery N. Richards. 1976. Hemicellulases: their occurrence, purification, properties and mode of action. *Adv. Carbohyd. Chem. Biochem.* 32:277–352.

Drocourt, D., S. Bejar, T. Calmels, J.P. Reynes, and G. Tiraby. 1988. Nucleotide sequence of the xylose isomerase gene from *Streptomyces violaceoniger*. *Nucleic Acids Res.* 16:9337–9337.

Farber, G.K., G.A. Petsko, and D. Ringe. 1987. The 3.0 Å crystal structure of xylose isomerase from *Streptomyces olivochromogenes*. *Protein Eng.* 1:459–466.

Farber, G.K., A. Glasfeld, G. Tiraby, D. Ringe, and G.A. Petsko. 1989. Crystallographic studies of the mechanism of xylose isomerase. *Biochemistry* 28:7289–7297.

Feather, M.S., V. Deshpande, and M.J. Lybyer. 1970. Anomeric specificity during some isomerase reactions. *Biochem. Biophys. Res. Commun.* 38:859–863.

Gong, C.-S., L.F. Chen, M.C. Flickinger, and G.T. Tsao. 1981. Conversion of hemicellulose carbohydrates. *Adv. Biochem. Eng.* 20:93–118.

Groleau, D. and C.W. Forsberg. 1981. Cellulolytic activity of the rumen bacterium *Bacteroides succinogenes*. *Can. J. Microbiol.* 27:517–530.

Henrick, K., C.A. Collyer, and D.M. Blow. 1989.

Structures of D-xylose isomerase from *Arthrobacter* strain B3728 containing the inhibitors xylitol and D-sorbitol at 2.5 Å resolution, respectively. *J. Mol. Biol.* **208**:129–157.

Honda, H., T. Kudo, and K. Horikoshi. 1985. Purification and partial characterization of alkaline xylanase from *Escherichia coli* carrying PCX-311. *Agric. Biol. Chem.* **49**:3165–3169.

Hurley, S.S., G.A. Splitter, and R.A. Welch. 1988. DNA relatedness of *Mycobacterium paratuberculosis* to other members of the family Mycobacteriaceae. *Int. J. Syst. Bacteriol.* **38**:143–146.

Kikuchi, T., Y. Itoh, T. Kasumi, and C. Fukazawa. 1990. Molecular cloning of the *xylA* gene encoding xylose isomerase from *Streptomyces griseofuscus* S-41: Primary structure of the gene and its product. *Agric. Biol. Chem.* **54**:2469–2472.

Kohring, S., J. Wiegel, and F. Mayer. 1990. Subunit composition and glycosidic activities of the cellulase complex from *Clostridium thermocellum* JW20. *Appl. Environ. Microbiol.* **56**:3798–3804.

Lamed, R. and E.A. Bayer. 1988. The cellulosome concept: exocellular/extracellular enzyme reactor centers for efficient binding and cellulolysis. *In: Biochemistry and Genetics of Cellulose Degradation*, J.P. Aubert, P. Beguin, and J. Millet, eds., pp. 101–116. New York: Academic Press.

Lamed, R., J. Naimark, E. Morgenstern, and E.A. Bayer. 1987a. Specialized cell surface structures in cellulolytic bacteria. *J. Bacteriol.* **169**:3792–3800.

Lamed, R., J. Naimark, E. Morgenstern, and E.A. Bayer. 1987b. Scanning electron microscopic delineation of bacterial surface topology using cationized ferritin. *J. Microbiol. Methods* **7**:233–240.

Lamed, R., E. Setter, R. Kenig, and E.A. Bayer. 1983. The cellulosome—a discrete cell surface organelle of *Clostridium thermocellum* which exhibits separate antigenic, cellulose-binding and various cellulolytic activities. *Biotechnol. Bioeng. Symp.* **13**:163–181.

Lee, C. and J.G. Zeikus. 1991. Purification and characterization of thermostable glucose isomerase from *Clostridium thermosulfurogenes* and *Thermoanaerobacter* strain B6A. *Biochem. J.* **273**:565–571.

Lee, C., L. Bhatnagar, B.C. Saha, Y.-E. Lee, M. Takagi, T. Imanaka, M. Bagdasarian, and J.G. Zeikus. 1990a. Cloning and expression of the *Clostridium thermosulfurogenes* glucose isomerase gene in *Escherichia coli* and *Bacillus subtilis*. *Appl. Environ. Microbiol.* **56**:2638–2643.

Lee, C., M. Bagdasarian, M. Meng, and J.G. Zeikus. 1990b. Characterization of the xylose (glucose isomerase) gene from *Clostridium thermosulfurogenes* and function of active site histidine in enzyme catalysis. *J. Biol. Chem.* **265**:19082–19090.

Morag, E., E.A. Bayer, and R. Lamed. 1990. Relationship of cellulosomal and noncellulosomal xylanases of *Clostridium thermocellum* to cellulose-degrading enzymes. *J. Bacteriol.* **172**:6098–6105.

Ng, T.K. and J.G. Zeikus. 1981. Comparison of extracellular cellulase activities of *Clostridium thermocellum* LQRI and *Trichoderma reesei* QM9414. *Appl. Environ. Microbiol.* **42**:231–240.

Raines, R.T., E.L. Sutton, D.R. Straus, W. Gilbert, and J.R. Knowles. 1986. Reaction energetics of a mutant triosephosphate isomerase in which the active-site glutamate has been changed to aspartate. *Biochemistry* **25**:7142–7154.

Reilly, P.J. 1981. Xylanases: structure and function. *In: Trends in the Biology of Fermentations for Fuels and Chemicals*, A. Hollaender, ed., pp. 111–129. New York: Plenum.

Rey, F., J. Jenkins, J. Janin, I. Lasters, P. Alard, M. Claessens, G. Matthyssens, and S. Wodak. 1988. Structural analysis of the 2.8 Å model of xylose isomerase from *Actinoplanes missouriensis*. *Proteins* **4**:165–172.

Rose, I.A. 1981. Chemistry of proton abstraction by glycolytic enzymes (aldolase, isomerases and pyruvate kinase). *Philos. Trans. R. Soc. London Ser. 1B* **29**:131–143.

Rose, I.A., E.L. O'Connell, and R.P. Mortlock. 1969. Stereochemical evidence for a cis-enediol intermediate in manganese dependent enzymes, aldose isomerases, from *Lactobacillus brevis* and *Aerobacter aerogenes*. *Biochim. Biophys. Acta* **178**:376–379.

Saari, G.C., A.A. Kumar, G.H. Kawasaki, M.Y. Insley, and P.J. O'Hara. 1987. Sequence of *Ampullariella* sp. strain 3876 gene coding for xylose isomerase. *J. Bacteriol.* **16**:612–618.

Schellenberg, G.D., A. Sarthy, A.E. Larson, M.P. Backer, J.W. Crabb, M. Lindstrom, B.C. Hall, and C.E. Furlong. 1983. Xylose isomerase from *Escherichia coli*: characterization of the protein and the structural gene. *J. Biol. Chem.* **259**:6826–6832.

Schink, B. and J.G. Zeikus. 1983. *Clostridium thermosulfurogenes* sp. nov., a new thermophile that produces elemental sulphur from thiosulphate. *J. Gen. Microbiol.* **129**:1149–1158.

Schleifer, K.H. and E. Strackebrandt. 1983. Molecular systematics of prokaryotes. *Annu. Rev. Microbiol.* **37**:143–187.

Schray, K.J. and I.A. Rose. 1971. Anomeric specificity and mechanism of 2 pentose isomerases. *Biochemistry* **10**:1058–1062.

Timell, T.E. 1967. Recent progress in the chemistry of wood hemicelluloses. *Wood Sci. Technol.* **1**:45–70.

Ward, O.P. and M. Moo-Young. 1989. Enzymatic degradation of cell wall and related plant polysaccharides. *Crit. Rev. Biotechnol.* **8**:237–274.

Weimer, P.J., L.W. Wagner, S. Knowlton, and

T.K. Ng. 1984. Thermophilic anaerobic bacteria which ferment hemicellulose. Characterization of organisms and identification of plasmids. *Arch. Microbiol.* **138**:31–36.

Wiegel, J. and L.G. Ljungdahl. 1981. *Thermoanaerobacter ethanolicus* new genus new species a new extreme thermophilic anaerobic bacterium. *Arch. Microbiol.* **128**:343–348.

Wiegel, J., L.G. Ljungdahl, and J.R. Rawson. 1979. Isolation from soil and properties of the extreme thermophile *Clostridium thermohydrosulfuricum. J. Bacteriol.* **139**:800–810.

Wiegel, J., C.P. Mothershed, and J. Puls. 1985. Differences in xylan degradation by various noncellulolytic thermophile anaerobes and *Clostridium thermocellum. Appl. Environ. Microbiol.* **49**:656–659.

Wilhelm, M. and C.P. Hollenberg. 1984. Selective cloning of *Bacillus subtilis* xylose isomerase EC 5.3.1.5 and xylulokinase EC 2.7.1.17 in *Escherichia coli* genes by insertion sequence IS-5-mediated expression. *EMBO J.* **3**:2555–2560.

Wilhelm, M. and C.P. Hollenberg. 1985. Nucleotide sequence of the *Bacillus subtilis* xylose gene: extensive homology between the *Bacillus* and *Escherichia coli* enzymes. *Nucleic Acids Res.* **13**:5717–5723.

Wood, T.M., C.A. Wilson, and C.S. Stewart. 1982. Preparation of the cellulase from the cellulolytic anaerobic rumen bacterium *Ruminococcus albus* and its release from the bacterial cell wall. *Biochem. J.* **205**:129–138.

Woodward, J. 1984. Xylanases: functions, properties and applications. *Topics Enzymol. Ferment. Biotechnol.* **8**:9–30.

Zeikus, J.G. 1979. Thermophilic bacteria: ecology, physiology and technology. *Enzyme Microb. Technol.* **1**:243–252.

Zeikus, J.G., A. Ben-Bassat, and P.W. Hegge. 1980. Microbiology of methanogenesis in thermal, volcanic environments. *J. Bacteriol.* **143**:432–440.

Zeikus, J.G., P.W. Hegge, and M.A. Anderson. 1979. *Thermoanaerobium brockii* gen. nov. and sp. nov., a new chemoorganotrophic, caldoactive anaerobic bacterium. *Arch. Microbiol.* **122**:41–48.

34

Genetics and Molecular Biology of Sulfate-Reducing Bacteria

Gerrit Voordouw and Judy D. Wall

34.1 Introduction

Because this chapter is the only one to consider the sulfate-reducing bacteria, it is appropriate to briefly review the classification, as well as some of the nutritional requirements, of the sulfate reducers before discussing their genetics and molecular biology. These topics are by no means isolated from the core of our review, because classification is now greatly advanced by the determination of 16S rRNA sequences with molecular biological methods; an understanding of nutritional requirements is also vital for rapid and efficient plating of these bacteria, which is a prerequisite for the successful application of bacterial genetics. In the context of this book it is even difficult to completely bypass a description of the ecology of these bacteria, because this provides a rationale for the existence of the various classes of sulfate reducers and clarifies their relationship with the *clostridia* and methanogens in anaerobic bacterial consortia.

34.2 Classification and Ecology of Sulfate-Reducing Bacteria

The first genera of the sulfate-reducing bacteria that were described in detail were the gram-negative *Desulfovibrio* (Postgate and Campbell, 1966) and the gram-positive *Desulfotomaculum* (Campbell and Postgate, 1965). These two genera are nutritionally very similar and are only capable of incomplete oxidation of simple organic compounds (e.g., lactate to acetate) with reduction of sulfate to sulfide. The discovery of sulfate reducers with different metabolic capacities led to the definition of the new genera *Desulfobulbus*, *Desulfobacter*, *Desulfobacterium*, *Desulfococcus*, *Desulfosarcina*, *Desulfonema*, *Thermodesulfobacterium*, and *Desulfomicrobium* (Pfennig et al., 1981; Widdel and Pfennig, 1984; Widdel, 1988). Some of these are nutritionally very limited but capable of complete oxidation; e.g., *Desulfobacter* is specific for acetate, which it oxidizes to carbon dioxide.

Although these new genera were discovered and could be distinguished by their novel nutritional requirements, sequencing of 16S rRNA, by Devereux et al. (1989) has contributed considerably to their phylogenetic classification. These authors divided the gram-negative sulfate reducers in the following seven distinct groups, based on the sequence homologies of their 16 S rRNAs: group 1, *Desulfovibrio desulfuricans* and other species (e.g., *Desulfovibrio vulgaris*); group 2,"*Desulfovibrio*" *sapovorans*; group 3, *Desulfobulbus* species; group 4, *Desulfobacter* species; group 5, *Desulfobacterium* species; group 6, *Desulfococcus* and *Desulfosarcina*; and group 7, "*Desulfovibrio*" *baarsii*. These phylogenetic studies indicated that "*Desulfovibrio*" *sapovorans* and "*Desulfovibrio*" *baarsii* require taxonomic revision because they are only distantly related to group 1 *Desulfovibrio*. In addition to these seven groups there are the gram-positive *Desulfotomaculum* already mentioned as well as the thermophilic, sulfate-reducing eubacteria

Table 34.1 Properties of selected species of distinct groups of sulfate-reducing bacteria (from Devereux et al., 1989)

Species	16S rRNA group[a]	Form	Mol% G + C	Desulfoviridin[b]	Motility	Oxidation[c]	Substrate								
							Hydrogen	Formate	Lactate	Ethanol	Actate	Fatty acids (carbon number)	Fumarate	Malate	Benzoate
Desulfovibrio desulfuricans	1	Vibrio	59	+	+	i	+	+	+	+	−	−	+	+	−
Desulfovibrio vulgaris	1	Vibrio	65	+	+	i	+	+	+	±	−	−	−	−	−
"Desulfovibrio" sapovorans	2	Vibrio	53	−	+	i	−	−	+	−	−	4–16	−	−	−
Desulfobulbus propionicus	3	Oval or onion	60	−	+	i	+	−	+	±	−	3	−	−	?
Desulfobacter postgatei	4	Oval rod	46	−	+	c	−	−	−	−	+	−	−	−	−
Desulfobacterium autotrophicum	5	Oval rod	48	−	+	c	+	+	+	+	±	3–16	+	+	−
Desulfococcus multivorans	6	Sphere	57	+	−	c	−	+	+	+	±	3–16	−	−	+
"Desulfovibrio" baarsii	7	Vibrio	66	−	+	c	−	+	−	−	+	3–18	−	−	−
Desulfotomaculum orientis	Gr+	Curved rod	45	−	+	i	+	+	+	+	−	−	−	−	−

[a] See text.
[b] Presence (+) or absence (−) of the bisulfite reductase desulfoviridin.
[c] Incomplete (i) or complete (c) oxidation of carbon substrates.

(Zeikus et al., 1983) and archaebacteria (Achenbach-Richter et al., 1987; Stetter et al., 1987). A wide variety of organisms is thus grouped under the name sulfate-reducing bacteria, and a phylogenetic tree indicating the degree of relatedness between the sulfate-reducing eubacteria and some well-known reference organisms (*Bacillis subtilis*, *Escherichia coli*, and *Myxococcus xanthus*) has been constructed from the 16S rRNA sequence data (Devereux et al., 1989; data not shown here).

Some distinctive properties for one or two representatives of the various groups are given in Table 34.1. Carbohydrates in either monomeric or polymeric form cannot in general be used by the sulfate-reducing bacteria. These bacteria thus depend on and collaborate with other genera (e.g., *Clostridium*; Chapters 29

by Bélaich and 30 by Aubert, this volume) to achieve the complete anaerobic degradation of polymeric carbohydrates, such as cellulose. The *clostridia* depolymerize and ferment cellulose to a variety of small molecular weight compounds (hydrogen, lactate, butyrate, acetate, etc.), which in the presence of sulfate can be readily utilized by the sulfate reducers (Table 34.1). This utilization can be envisaged as a two-step process in which lactate is first converted to acetate by the incompletely oxidizing species (e.g., *Desulfovibrio*) and then completely oxidized to CO_2 by the completely oxidizing species (e.g., *Desulfobacter*) in a cellulose-degrading, sulfate-reducing anaerobic consortium. In the absence of sulfate, the fermentation products will be used by methanogenic bacteria to form methane and CO_2.

34.3 Development of Genetic Systems for *Desulfovibrio*

Introduction

With the exception of the sequence determination of 16 S rRNAs discussed earlier, genetic and molecular biological studies of sulfate-reducing bacteria have been limited to group 1 species of *Desulfovibrio*. Among the reasons for this focus are the ease of cultivation of members of this genus and the wealth of biochemical information available. The criteria for the selection of species and strains for genetic development have included metabolic capabilities, plating efficiencies, and endogenous resistance or sensitivity to a variety of antibiotics.

To benefit from the power of genetic and molecular biological approaches, mutant isolation procedures and genetic exchange techniques are required. One report has appeared in which classical enrichments for mutants with nonselectable phenotypes have been demonstrated to be feasible with *D. desulfuricans* ATCC 27774 (Odom and Wall, 1987). This procedure should be equally successful with other strains that are sensitive to penicillin. More recent studies have concentrated on gene transfer techniques, reporting transduction (Rapp and Wall, 1987) and conjugation (Powell et al., 1989; van den Berg et al., 1989; Wall and Rapp-Giles, in manuscript), as discussed next. Perhaps the most facile and utilitarian form of gene transfer, transformation, has not been documented for the sulfate-reducing bacteria. Finally, although electroporation has not yet met with success for *D. desulfuricans* (J.D. Wall and B.J. Rapp-Giles, unpublished observations) a reevaluation is required in view of improvements in the equipment and techniques as applied to gram-negative bacteria, as well as the recent success with this technique in *D. fructosovorans* (Rousset et al., 1991).

Metabolic properties

Hydrogen metabolism, energy transduction, and metal corrosion top the list of subjects to be explored through development of genetic tools for the sulfate-reducing bacteria. Although these processes are generic, it should be kept in mind that the strain selected must exhibit the metabolic process to be studied and, ideally, have additional growth capabilities allowing that process to be nonessential. The capacity to use alternative substrates or terminal oxidizing agents for energy metabolism often allows the genetic dissection of the bioenergetic pathway unique to a given substrate. For example if a sulfate-reducing, bacterium can use either H_2 or organic acids as electron donors, it may be possible to obtain mutants in H_2 utilization among cells growing with organic acids.

For this reason, *D. desulfuricans* ATCC 27774 was one of the strains selected as a candidate for genetic development. This strain is capable of respiration with nitrate or sulfate as electron acceptor and can use either H_2 or organic acids as electron donors, while it can grow fermentatively on either pyruvate or choline (Peck, 1984). Consequently, a mutant of this bacterium was isolated that could no longer respire sulfate in the presence of H_2 (Odom and Wall, 1987). The successful isolation of the mutant indicated that classical procedures of enrichment could be used and that there do not appear to be multiple genomes in *D. desulfuricans* as reported for *D. gigas* (Postgate et al., 1984). Unfortunately, *D. desulfuricans* ATCC 27774 grew poorly, if at all, on minimal medium. Thus the possibility of obtaining nutritional markers is low. Prototrophic strains, capable of rapid growth in defined medium, will allow the identification of a wider array of mutations and are therefore to be preferred.

Plating efficiencies

The strictly anaerobic nature of the metabolism of these bacteria has been considered to be a potential problem for genetic manipulation. Anecdotal information suggested that O_2 sensitivity was a major obstacle in obtaining workable efficiencies for plating *Desulfovibrio* as single colonies on the surface of solidified medium. However, through the use of the anaerobic chamber and prereduced medium, most steps can now be performed without ex-

Table 34.2 Sensitivity of *Desulfovibrio* strains to antibiotics

Antibiotic[a]	*D. vulgaris* Hildenborough[b]	*D. desulfuricans* ATCC 27774[c]	*D. desulfuricans* G100A[c]
Ap	S (>1)[d]	S (5)	R (10)
Sm	R (>100)	R (100)	R (100)
Km	R (>100)	S (150)	S (150)
Tc	R (12)	S (5)	S (<1)
Rf	R (20)	S (50)	R (100)
Cm	S (>3)	S (100)	S (50)
Sp	N.R.[e]	R (>200)	S (100)
Gm	N.R.	S (50)	S (50)
Nal	N.R.	S (100)	S (100)

[a] Antibiotics are Ap, ampicillin; Sm, streptomycin; Km, kanamycin; Tc, tetracycline; Rf, rifampicin; Cm, chloramphenicol; Sp, spectinomycin; Gm, gentamycin; Nal, nalidixic acid.
[b] Data from van den Berg et al. (1989).
[c] Data from J.D. Wall and B.J. Rapp-Giles, unpublished.
[d] Numbers in parentheses are concentrations of antibiotic in μg per ml at which the cells are either resistant (R) and can grow or sensitive (S) and cannot grow.
[e] N.R., not reported.

posure of the cells to significant levels of O_2. Using these procedures, Singleton et al. (1988) calculated efficiencies of plating for *D. vulgaris* Hildenborough between 30% and 80%. Surprisingly, van den Berg et al. (1989) were able to obtain similar efficiencies after plating this same strain aerobically onto prereduced medium that was immediately incubated under anaerobic conditions. The exposure to O_2 clearly, was insufficient to cause detectable damage. However, experience with streaking of *D. desulfuricans* onto agar surfaces for growth in the anaerobic chamber confirms that colony formation is more rapid, more robust, and more efficient under a layer of prereduced soft agar (B.J. Rapp-Giles and J.D. Wall, unpublished observations).

Antibiotic sensitivities

The application of genetic and recombinant DNA techniques requires the availability of a set of selectable markers. Because of the strength of the selection pressure that can be applied to identify rare events, antibiotic resistance (and sensitivity) has played a vital role in genetic manipulation of bacteria. In general, the sulfate-reducing bacteria display resistance to high levels of a number of antibiotics (Postgate, 1984). However, the strains chosen for

genetic analysis to date have some useful sensitivities (Table 34.2). For example, chloramphenicol has proved effective in a selection against *D. vulgaris* and for acquisition of a plasmid in a conjugational cross (van den Berg et al., 1989). Classical penicillin enrichments should also be possible with this strain, as shown for *D. desulfuricans* ATCC 27774 (Odom and Wall, 1987). In contrast, *D. desulfuricans* G100A is resistant to ampicillin, cycloserine, and rifampicin, making it necessary to identify nonselectable mutations (such as those resulting in auxotrophy) by screening procedures alone. Powell et al. (1989) took advantage of the sensitivity of *D. desulfuricans* strains 8391 and 8312 to streptomycin to select for exconjugants expressing resistance from a plasmid-borne gene. When appropriate antibiotic resistances were limited or unavailable, spontaneously resistant derivatives have been readily isolated from *D. desulfuricans* (Rapp and Wall, 1987; Voordouw et al., 1990).

34.4 Desulfovibrio Bacteriophages

Genetic exchange mediated by bacteriophages, i.e., transduction, has made enormous contributions to fine-structure mapping of mutations in many bacteria. However, few instances of bacteriophages or bacteriophage-like particles have been reported for the sulfate-reducing bacteria. As early as 1973, Handley et al. reported the mitomycin C induction of phage-like particles from *D. vulgaris* Hildenborough. No plaque-forming ability was observed, and potential gene transfer activity was not explored. Recently, a lambda-like phage was isolated from an enrichment of marine sediment that could form plaques on *Desulfovibrio salexigens* (Kamimura and Araki, 1989). Again, its potential for transduction was not explored. The sensitivity of marine isolates to infection by this phage was suggested as an ecological marker for relatedness. Finally, the remaining report of a *Desulfovibrio* phage-like particle followed an observation of genetic transfer (Rapp and Wall, 1987). A small, apparently defective bacteriophage, Dd1, was found to be released during growth of *D. desulfuricans* ATCC 27774. The morphological characteristics of the released phage particles resembled those of T7 or T3 coliphages, except that the head was smaller. The head contains double-stranded DNA of uniform length, about 13.5-kb, that has been randomly packaged from the chromosome.

The inference that all areas of the chromosome have a similar probability of packaging was supported by the occurrence of similar transfer frequencies for each of four different antibiotic markers and by lack of distinctly reiterated fragments in the restriction endonuclease digestion pattern of the packaged DNA. The DNA in the particles resisted degradation by added nucleases and required divalent cations to be transferred to an appropriate recipient. Routinely, frequencies of 10^{-6} to 10^{-5} were reported in gene transfer experiments. As for *D. vulgaris* phage-like particles (Handley et al., 1973), no plaque formation by Dd1 could be detected on any host (Rapp and Wall, 1987). However, mitomycin C treatment did not increase the numbers of the bacteriophage released from *D. desulfuricans* as assayed by genetic transfer. The ability of this defective bacteriophage to transfer an antibiotic resistance marker from *D. desulfuricans* ATCC 27774 to other *D. desulfuricans* strains or to other *Desulfovibrio* species was tested without success (B.J. Rapp and J.D. Wall, unpublished observa-

tions). Thus, the practical use of Dd1 was shown to be limited to the ATCC strain 27774, although it is a valuable fine-structure mapping tool for that strain. Without an in vitro packaging system, it is not possible to transfer exogenous DNA with this vector. Thus, additional transfer techniques are required for the full analysis of *Desulfovibrio* genetics. The conjugation systems described here fill this need.

34.5 Conjugation of *Desulfovibrio*

Broad host range plasmids are capable of transfer between and stable maintenance in almost all gram-negative bacterial species. Thus, it was reasonable to expect that the *Desulfovibrio* spp. should be no exception once plating conditions and the appropriate selective markers were identified. Two reports of successful gene transfer by plasmids belonging to the Q incompatibility group appeared almost simultaneously (Powell et al., 1989; van den Berg, et al., 1989), and similar results have been obtained elsewhere (J.D. Wall and B.J. Rapp-Giles, in manuscript) (Table 34.3). Powell et al. (1989) used two *D. desulfuricans* strains as recipients in overnight matings with *E. coli* donors and reported frequencies of 5×10^{-2} and 1. Similar frequencies for IncQ plasmid transfer, about 1×10^{-2}, were obtained after overnight matings with *D. vulgaris* (van den Berg et al., 1989). Although many rounds of conjugation are possible under such conditions, the frequencies obtained indicated that a very useful level of transfer had occurred. In conjugations of 4 to 6 h with *D. desulfuricans* G200, frequencies as high as 1×10^{-3} were obtained (J.D. Wall and B.J. Rapp-Giles, in manuscript). Transfer of IncP1 plasmids from an *Alcaligenes eutrophus* donor was also reported, but this plasmid was found to be unstable in strains 8301 and 8312 (Powell et al., 1989). No instability of IncP1 plasmids was noted with *D. vulgaris* Hildenborough (van den Berg et al., 1989) or *D. desulfuricans* G200 (J.D. Wall and B.J. Rapp-Giles, in manuscript), although transfer frequencies were in the range of 10^{-5} per donor.

Table 34.3 Plasmid transfer by conjugation from *Escherichia coli* to *Desulfovibrio*

Strain	Plasmid stably transferred	Incompatibility group	Antibiotic selection	Reference
Desulfovibrio sp. NCIMB 8301	R300B	IncQ	Sm^r	Powell et al., 1989
D. desulfuricans NCIMB 8312	R300B	IncQ	Sm^r	Powell et al., 1989
D. vulgaris Hildenborough NCIMB 8303	pSUP104	IncQ	Cm^r	van den Berg et al., 1989
	pRK404 Cm	IncP	Cm^r	van den Berg et al., 1989
D. desulfuricans G200[a]	pKT230[b]	IncQ	Km^r	Wall and Rapp-Giles, unpublished
	pJRD215[c]	IncQ	Km^r	Wall and Rapp-Giles, unpublished
	pRHP20.2d[d]	IncP	Km^r	Wall and Rapp-Giles, unpublished

[a] A spontaneously nalidixic acid resistant derivative of wild-type G100A (Weimer et al., 1988).
[b] Bagdasarian et al., 1981.
[c] Davison et al., 1987.
[d] Xu et al., 1989.

The IncQ cloning vector pSUP104 has been used to introduce extra copies of the [Fe] hydrogenase genes (see following for a more detailed discussion of these genes) into *D. vulgaris* Hildenborough (van den Berg et al., 1989). Although protein levels increased, activity did not increase proportionally reflecting a complex regulation for the expression of the activity of this enzyme. When these same genes were transferred into the heterologous background of *D. desulfuricans* G200, very low levels of completely processed [Fe] hydrogenase were obtained (W.M.A.M. van Dongen, personal communication). In contrast, the introduction of the *cyc* gene encoding the *D. vulgaris* Hildenborough cytochrome c_3 into the G200 host allowed high levels of functional cytochrome to be produced (Voordouw et al., 1990).

Plasmids that can be mobilized but are not stable may serve as suicide plasmids in marker exchange experiments. Van den Berg et al. (1989) reported attempts to use pSUP5011, a pMB11 replicon, to construct [Fe] hydrogenase-negative mutants. Although not yet successful, this technique should prove useful as the limitations are identified. Preliminary experiments that apply suicide vectors for the delivery of transposons to *D. desulfuricans* have been encouraging (J.D. Wall and B.J. Rapp-Giles, in manuscript). Among factors that are likely to reduce the numbers of transconjugants obtained with recombinant plasmids are the existence of restriction endonucleases as well as nonspecific nucleases. *D. desulfuricans* Norway is the source of two restriction endonucleases, *Dde*I and *Dde*II, and the genes for *Dde*I endonuclease and methylase have been sequenced (Sznyter et al., 1987). Other type II restriction enzymes have not been documented in the sulfate-reducing bacteria but are likely to be present. Indirect evidence suggests that nonspecific nucleases may also be present in a number of strains. Chromosomal DNA preparations of *D. desulfuricans* often display evidence of significant digestion.

Attempts to isolate usable quantities of either endogenous plasmids (Postgate et al., 1984; H.M. Kent and J.R. Postgate, personal communication) or those introduced by conjugation (Powell et al., 1989; van den Berg et al., 1989; J.D. Wall and B.J. Rapp-Giles, in manuscript) have been universally discouraging, often requiring Southern analyses to visualize the isolated plasmids. One exception has been the isolation of a small endogenous plasmid from *D. desulfuricans* G200 (Wall et al., 1990). Sequencing and stability tests of this apparently cryptic plasmid should indicate its usefulness as a cloning vector in a number of *Desulfovibrio* strains.

Table 34.4   Genes cloned and sequenced or characterized for group 1 *Desulfovibrio*

Gene product	Gene name[a]	References
[Fe] hydrogenase	*hydA, B*	Voordouw et al., 1985
		Voordouw and Brenner, 1985
		Prickril et al., 1986
		Voordouw et al., 1987a
		Voordouw et al., 1987b
		van Dongen et al., 1988
		Voordouw et al., 1989b
Not identified	*hydγ, hydC*	Stokkermans et al., 1989
		Voordouw et al., 1989b
[NiFe] hydrogenase	(*hynA, hynB*)	Li et al., 1987
		Voordouw et al., 1989a
[NiFeSe] hydrogenase	(*hysA, hysB*)	Menon et al., 1987
		Voordouw et al., 1989a
Cytochrome c_3	*cyc*	Voordouw and Brenner, 1986
		Voordouw et al., 1987b
		Pollock et al., 1989
		Voordouw et al., 1990
Cytochrome c_{553}	(*cyf*)	van Rooijen et al., 1989
Flavodoxin	(*fla*)	Curley and Voordouw, 1988
		Krey et al. 1988
		Carr et al., 1990
Rubredoxin	*rub*	Voordouw, 1988
(Rubredoxin oxidoreductase)	*rbo*	Brumlik and Voordouw, 1989
Desulforedoxin	*dsr*	Brumlik et al., 1990
OMPase	*pyrF*	Li et al., 1986
Nitrogenase, Fe protein	*nifH*	Postgate et al., 1987
		Kent et al., 1989
*Dde*I, restriction endo-nuclease, methylase	*hsdM, R*	Sznyter et al., 1987

[a] The gene names in brackets are proposed names and were not used in the references indicated.

34.5.   Cloning, Sequencing, and Expression of *Desulfovibrio* Genes

Introduction

Some of the genes that have been isolated and characterized from group 1 *Desulfovibrio* are summarized in Table 34.4. The focus of this work has been on genes encoding redox proteins: hydrogenases, cytochromes, and small redox carriers like rubredoxin, flavodoxin, and desulforedoxin. Several of these redox proteins have been characterized in depth with biochemical methods. The amino acid sequences as well as the three-dimensional structures of flavodoxin (Watenpaugh et al., 1972), rubredoxin (Herriot et al., 1970; Bruschi, 1976; Adman et al., 1977; Hormel et al., 1986; Sieker et al., 1986; Frey et al., 1987), and cytochrome c_3 (Ambler, 1968; Trousil and Campbell, 1974; Pierrot et al., 1982; Higuchi et al., 1984) have been determined, and the amino acid sequences of cytochrome c_{553} and desulforedoxin are also known (Bruschi and LeGall, 1972; Bruschi et al., 1979).

These biochemical studies were often undertaken with the aim of characterizing and comparing the structure and function of a particular redox protein with the structure and function of similar proteins isolated from other bacteria. The role of a particular protein in the *Desulfovibrio* chains for transport of electrons from reduced substrates (e.g., hydrogen, lactate) to sulfate was often not clarified by, and was indeed often not the research goal of, these studies. These smaller redox carriers were selected for study not because of a perceived critical function in *Desulfovibrio* electron transport but because they were colored and thus easy to isolate, as well as because they were small, which facilitated the determination of their amino acid sequence and three-dimensional structure. The larger redox carriers, such as the hydrogenases, lactate dehydrogenase and the enzymes involved in terminal sulfate respiration, such as the sulfite reductase desulfoviridin have been isolated and characterized biochemically but were left unexplored structurally. Their structures are bound to be more informative, and molecular biological methods are currently contributing by defining the amino acid sequences of these redox carriers from the sequences of the genes determined by dideoxynucleotide chain termination technology.

These amino acid sequences, such as the one for [Fe] hydrogenase discussed here, can sometimes be interpreted in terms of a three-dimensional structure. The gene sequence also gives information on localization by providing an answer to the question whether a secretory NH$_2$-terminal signal sequence is present or absent. Precise knowledge of localization is vital for defining the electron transport chain in *Desulfovibrio*, which is partly periplasmic (hydrogenases and c-type cytochromes) and partly cytoplasmic (e.g., flavodoxin, rubredoxin, desulfoviridin). Molecular biological methods can indicate whether a given carrier is perhaps part of an operon in which other genes, possibly encoding its redox partners, are also present. Both of these two latter aspects are important to understand physiology. Finally, a most important contribution to physiology, as yet unexplored in sulfate reducers, may be expected to be a study of the possibilities of directed deletion of genes for redox proteins once they have been cloned and sequenced. In the following sections we briefly review what has been learned from the genes that have been characterized (see Table 34.4). We also indicate the possibilities of the newly developed conjugational gene transfer technology in both expression and deletion of specific genes in *Desulfovibrio*. The work on the *nifH* and *pyrF* genes (see references in Table 34.4) are not further discussed here.

Genes for [Fe] hydrogenases

Three hydrogenases have been isolated from *Desulfovibrio* and are now referred to as the [Fe], [NiFe], and [NiFeSe] hydrogenases. All three enzymes have been shown to be present in *D. vulgaris* Hildenborough (Lissolo et al., 1986; Prickril et al., 1987). The [NiFe] hydrogenase seems to be present in most representatives of the group 1 *Desulfovibrio*, while the [Fe] and [NiFeSe] hydrogenases have a more limited distribution.

The genes for [Fe] hydrogenase from *D. vulgaris* Hildenborough were the first genes that were cloned and sequenced from the genus *Desulfovibrio* (Voordouw et al., 1985; Voordouw and Brenner, 1985). Cloning was achieved by screening recombinant *E. coli* clones for expression of hydrogenase protein with immunological methods. Sequencing of the 4.7-kb *Eco*RI-*Sal*I insert of the positive recombinant clone obtained indicated that the structural genes encoding [Fe] hydrogenase form an operon: the gene (*hydA*) for the large subunit (α, 46 kDa) is immediately followed by a gene (*hydB*) for a smaller subunit (β, 13.5 kDa), as indicated in Figure 34.1. The most interesting feature of the amino acid sequence of the large subunit is that its NH$_2$ terminus is homologous to [8Fe-8S] ferredoxin. This is a well-known small (6 kDa) redox carrier that has been isolated and sequenced from a variety of bacterial sources (Bruschi and Guerlesquin, 1988). The canonical [8Fe-8S] ferredoxin sequence has two groups of 4 cysteine residues in the characteristic pattern C-I-X-C-X-X-C-X-X-X-C-P-X-X-A-I, where X is a variable amino acid residue. Together the 8 cysteine residues coordinate two [4Fe-4S] clusters.

The three-dimensional structure of [8Fe-8S] ferredoxin from *Peptococcus aerogenes* has been elucidated: it is an elegant structure that displays two-fold symmetry (Adman et al., 1973). [Fe] hydrogenase is known to contain three iron-sulfur clusters (Hagen et al., 1986). The homology with ferredoxin indicates that two of these are coordinated at the NH$_2$ terminus of the large subunit, while the third cluster must be coordinated by cysteine residues located in the COOH-terminal portion of the large sub-

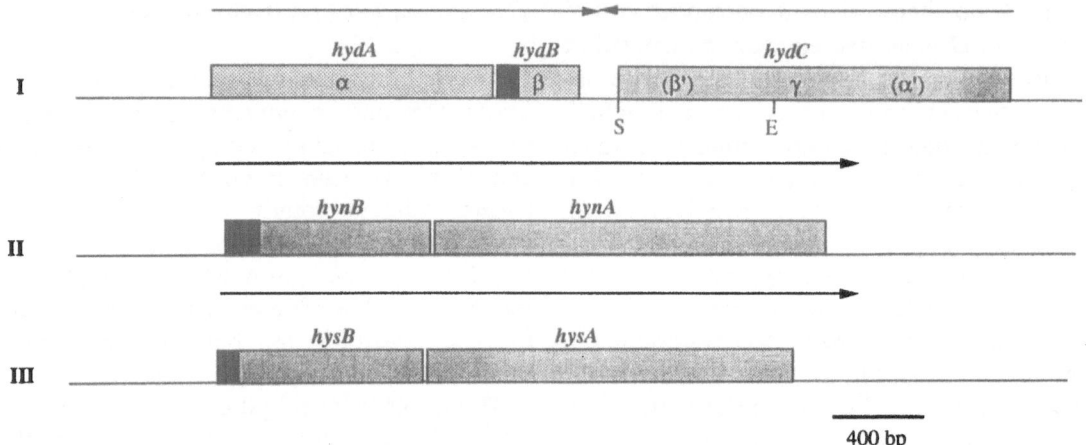

FIGURE 34.1 Survey of hydrogenase genes that have been cloned and sequenced from *Desulfovibrio* species. (I) [Fe] hydrogenase operon from *D. vulgaris* Hildenborough. Positions of *hydA* and *hydB* genes encoding large α and small β subunits of periplasmic [Fe] hydrogenase are indicated by shaded boxes. Small subunit signal sequence is represented by solid black box. Direction of transcription of the *hydA,B* operon is indicated by arrow. *hydC* gene is homologous with *hydA,B* genes as indicated (α', β') and potentially encodes a 66-kDa polypeptide when transcribed as indicated. (II) [NiFe] hydrogenase operon from *D. gigas*; small and large subunits are encoded by *hynB* and *hynA* genes, respectively. (III) [NiFeSe] hydrogenase operon of *D. baculatus*; small and large subunits are encoded by the *hysB* and *hysA* genes, respectively.

unit. The small subunit cannot participate in covalent cluster coordination because it does not contain cysteines. The large subunit can thus be regarded as being composed of two domains. The NH$_2$-terminal domain (residues 1–110) is homologous to ferredoxin and is likely to serve an electron transport function, while the COOH-terminal domain (residues 111–420) coordinates the third cluster, which has special properties (Hagen et al., 1986) and serves as the active site for hydrogen uptake or evolution.

The function of the small β subunit, which was discovered by the cloning and sequencing work, was not immediately apparent. The finding by Prickril et al. (1986), that the β subunit has a complex signal sequence of 34 amino acid residues indicated a role in the translocation of hydrogenase to the periplasm. However, this cannot be its only function, because [Fe] hydrogenase as isolated from the periplasm of *D. vulgaris* Hildenborough is a heterologous dimer of a single α (46 kDa) and a single mature β (10 kDa) subunit. Surprisingly, the large subunit lacks an NH$_2$-terminal signal sequence, and it is therefore not clear how this subunit is trans-

located. Evidence has been obtained from expression studies in *E. coli* for an unusual mechanism for the translocation of [Fe] hydrogenase. *E. coli* can express both the α and β hydrogenase subunits, but no enzyme activity is obtained. Part of the *E. coli*-expressed protein can be isolated as an $\alpha\beta$ dimer, which is lacking a properly assembled active site cluster but has the ferredoxin clusters inserted (Voordouw et al., 1987a). This fraction of the *E. coli*-expressed protein presumably resides in the periplasm, although this was not shown in the initial paper (Voordouw et al., 1987a). A more detailed study on the localization of *D. vulgaris* Hildenborough [Fe] hydrogenase expressed in *E. coli* indicated lack of translocation when either the *hydA* or the *hydB* gene was expressed separately. The pro-β polypeptide was found not to be processed under these conditions. However, when expressing the *hydA,B* operon partial processing of pro-β and partial export of both α and β polypeptides were observed.

These results suggest that assembly of an α-pro-β dimer on the cytoplasmic side of the membrane precedes and is a prerequisite for translocation. Such a mechanism, in which the pro-β signal peptide would suffice to translocate the entire complex, is unusual and cannot be considered firmly proven at present. Further evidence for an unusual translocation mechanism is provided by the fact that the presence of a long complex NH_2-terminal signal sequence on the small subunit and the absence of such a sequence from the large subunit is a universal feature in the assembly of periplasmic bacterial hydrogenases (see Figure 34.1). This applies to both the [NiFe] and [NiFeSe] hydrogenases from *Desulfovibrio* and to the periplasmic [NiFe] hydrogenases found in other bacterial genera, as discussed next.

In addition to their work on the translocation mechanism van Dongen and coworkers have found an interesting gene (*hydγ*; Stokkermans et al., 1989). This gene, referred to as *hydC* by Voordouw et al. (1989b) is immediately downstream from the *hydA,B* genes, and could encode a 66-kDa polypeptide (γ), as indicated in Figure 34.1. The NH_2-terminus of γ is homologous to the α subunit, while its COOH terminus is homologous to the β subunit of the periplasmic hydrogenase. It has not been shown that γ is expressed in *Desulfovibrio*, and its function is therefore also unknown (e.g., it could serve as a cytoplasmic [Fe] hydrogenase). Expression of a 63-kDa polypeptide in *E. coli* minicells, transformed with *hydC*, has been demonstrated. The *hydC* gene is not universally present in *Desulfovibrio* species with a periplasmic [Fe] hydrogenase and appears to be absent from *D. vulgaris* ssp. *oxamicus* Monticello, an organism for which the *hydA,B* genes were cloned and sequenced (Voordouw et al., 1989b). A detailed but speculative discussion of the possible role of *hydC* in the evolution of the genes for periplasmic hydrogenase has been given by Voordouw (1990). The γ sequence is homologous to that of hydrogenase I from *Clostridium pasteurianum* that was recently determined (Meyer and Gagnon, 1991).

Genes for [NiFe] and [NiFeSe] hydrogenases

The [NiFe] and [NiFeSe] hydrogenases that have been isolated from group 1 *Desulfovibrio* are two subunit enzymes with molecular weights of about 30 and 60 kDa for the small and large subunits, respectively. The genes encoding the [NiFe] hydrogenase of *D. gigas*, which had been well characterized by biochemical and biophysical studies (Hatchikian et al., 1978; Texeira et al., 1985; Fernandez et al., 1986; Cammack et al., 1987), have been cloned and sequenced (Li et al., 1987; Voordouw et al., 1989a). The gene for the small subunit (34 kDa for the unprocessed protein) precedes that for the large subunit (61 kDa), as indicated in Figure 34.1. No names were given to these genes in the paper of Li et al. (1987); we will refer to them here as the *hynB* and *hynA* genes, respectively, while those encoding the [NiFeSe] hydrogenase are referred to as the *hysA* and *hysB* genes (Table 34.4, Figure 34.1, and Voordouw, 1990). The small subunit has a NH_2-terminal signal sequence of 50 residues, as indicated in Figure 34.1, which after processing leaves a mature protein of 28.4 kDa. The amino acid sequences of the small and large subunits of *D. gigas* [NiFe] hydrogenase are not homologous with those of *D. vulgaris* Hildenborough [Fe] hydrogenase, with the exception of the small subunit signal sequence, which shares a consensus box R-R-X-F-X-K. This box must be critical for export of the two subunit hydrogenase via a single signal sequence, as discussed here and elsewhere (Voordouw et al., 1989b; Voordouw, 1990).

The genes for the [NiFeSe] hydrogenase of *D. baculatus* were cloned and sequenced by Menon et al. (1987; Voordouw et al., 1989a) and found to encode a small subunit of 34 kDa and a large subunit of 57 kDa. The NH_2-terminal signal sequence of the small subunit was found to be 34 amino acid residues long and contained the consensus box mentioned. The mature *hysB* gene product is therefore 31 kDa. Comparison of the two mature small subunit sequences (the *hysB* and *hynB* gene products) indicated a sequence identity of 38%, while the two large sununits can be aligned to show 34% sequence identity (Voordouw et al., 1989a). Sequence alignment indicates that the small subunit contains 10 conserved cysteine residues: C-67, C-70, C-162, C-198, C-238, C-263, C-269, C-278, C-296, and C-299 (numbering of the unprocessed *hynB* gene product), while the pairs of large subunits have only four conserved cysteines: C-65, C-68, C-530, and C-533. *D. gigas* [NiFe] hydrogenase is thought to coordinate two [4Fe-4S] clusters, requiring 4 cysteine residues each for coordination, a [3Fe-4S] cluster, which is bound by 3 cysteines (Kissinger et al., 1989), and an active site nickel. A potential nickel ligand could be identified by comparison of the nucleotide sequences of the *D. gigas hynA* and *D. baculatus hysA* genes. The TGC codon for C-530 in the *D. gigas* large subunit gene sequence is matched by a TGA in the *D. baculatus* sequence. This codon, which usually causes translation termination, has been shown to direct the incorporation of selenocysteine, when present internally in the reading frame for a selenium-containing enzyme, such as *E. coli* formate dehydrogenase (Zinoni et al., 1986).

The TGA codon identified in the *hysA* gene is also likely to encode a selenocysteine. Spectroscopic studies in the [NiFeSe] hydrogenase from *D. baculatus* have provided evidence that the selenocysteine coordinates directly to the nickel (Eidsness et al., 1989: He et al., 1989). This TGA codon, as well as the corresponding TGC codon for C-530 in the *D. gigas hynA* gene, is thus likely to specify a nickel ligand. This leaves 13 conserved cysteine residues (10 in the small and 3 in the large subunit) for the coordination of the iron-sulfur clusters. The conclusion of this analysis is, therefore, that contrary to the situation with [Fe] hydrogenase, the small subunit of the nickel-containing hydrogenases must play a prominent role in iron-sulfur cluster coordination. A pattern of cysteine residues with homology to [8Fe-8S] ferredoxin, as found in the [Fe] hydrogenase α subunit, is not present in the nickel-containing hydrogenase sequences. This precludes an assignment of any of the 13 remaining conserved cysteine residues to a particular cluster type.

The nucleotide sequence of the *D. gigas hyn-B,A* genes share considerable homology with the gene sequences for [NiFe] hydrogenase operons from other bacteria, e.g., *Bradyrhizobium japonicum* (Sayavedra-Soto et al., 1988), *Rhodobacter capsulatus* (Leclerc et al., 1988), *Rhodocyclus gelatinosus* (Uffen et al., 1990), *Azoto-*

bacter chroococcum (Ford et al., 1990), and *Escherichia coli* (Menon et al., 1990). These hydrogenases all have the conserved cysteine residues indicated as well as a signal sequence with the R-R-X-F-X-K consensus box as the NH$_2$ terminus of the small subunit. The [NiFe] hydrogenases of these latter microorganisms are generally found to be periplasmic and membrane bound and can be isolated only after trypsin or detergent treatment of the isolated membrane fraction. The COOH terminus of the small subunit of these hydrogenases appears, when compared with the *D. gigas* sequence, to have an extension of 40 to 50 amino acid residues, which is partially hydrophobic (Menon et al., 1990) and may function in the binding of these enzymes to the membrane.

An unusual hydrogenase has been isolated and characterized from *D. vulgaris* Miyazaki F by Yagi and coworkers (Yagi et al., 1976, 1985). It has subunit molecular weights of 29 and 62 kDa, similar to *D. gigas* [NiFe] hydrogenase, but appears to lack detectable nickel. The enzyme, which is thought to be periplasmic, appears to be more strongly membrane bound than *D. gigas* [NiFe] hydrogenase. Solubilization of the Miyazaki hydrogenase requires trypsin treatment of the isolated membranes just as for the [NiFe] hydrogenases from *E. coli*, *Azotobacter*, etc. Cloning and sequencing of the genes encoding the hydrogenase from *D. vulgaris* Miyazaki F, suggests it to be a bona fide [NiFe] hydrogenase, with the conserved features as outlined (Deckers et al., 1990). The small and large subunit sequences share 80% identity with those of *D. gigas*. There is no COOH-terminal extension of the small subunit that can explain its high membrane affinity. The conclusion of this work and of some recent physical studies of the enzyme are that this hydrogenase belongs also to the widespread [NiFe] hydrogenase family.

The identification of different hydrogenases in *Desulfovibrio*

The characterization of the genes for the three classes of hydrogenases found in group 1 *Desulfovibrio* allows a rapid evaluation of the potential hydrogen metabolism of deposited and newly isolated organisms. Like the genes for nitrogenase (e.g., see references for *nifH* in Table 34.4), which have been well conserved in different bacterial genera, the *hyd*, *hyn*, and *hys* genes appear sufficiently conserved to allow definitive characterization of the presence of these genes in the organism under study by Southern blotting. It would be much more difficult to gather this information by enzyme isolation and characterization. The picture that emerges from Southern blotting work is that the *hyn* genes encoding the [NiFe] hydrogenase are widespread and may even be present in every representative of the group 1 *Desulfovibrio*. The *hyd* (Voordouw et al., 1987b) and *hys* genes have a more limited distribution. This information provides important boundary conditions for theories on the role of hydrogenases in the hydrogen metabolism of group 1 *Desulfovibrio*, such as the hydrogen cycling model of Odom and Peck (1981, 1984). These theories should be tested by generating *Desulfovibrio* mutants in which one or more hydrogenase genes have been silenced by directed insertional inactivation. Application of this method has come within reach (Rousset et al., 1991).

Genes for c-type cytochromes

Sulfate-reducing bacteria harbor a variety of c-type cytochromes, e.g., cytochrome c_{553} ($M_r = 9$ kDa, 1 heme), cytochrome c_3 ($M_r = 13$ kDa, 4 hemes), and a high molecular weight cytochrome ($M_r = 75$ kDa, 16 hemes) have all been isolated from *D. vulgaris* Hildenborough (Ambler, 1968; Bruschi and LeGall, 1972; Higuchi et al., 1987). The genes encoding cytochrome c_3 (Voordouw and Brenner, 1986) and cytochrome c_{553} of *D. vulgaris* Hildenborough (van Rooijen et al., 1989) have been isolated. The nucleotide sequences of the cloned genes confirmed the known amino acid sequence of cytochrome c_3 (Ambler, 1968; Trousil and Campbell, 1974), but indicated errors in the amino acid sequence published for cytochrome c_{553} (Bruschi and LeGall, 1972). The amino acid sequence deduced from the nucleotide sequence of the gene was found to be homologous to the sequence, determined for *D. vulgaris* Miyazaki cytochrome c_{553} by protein sequencing (Nakano et al., 1983): the two proteins are both 79 amino acid residues long and have identical amino acids in 63 positions. Hildenborough cytochrome c_{553} is thus structurally similar to the protein from *D. vulgaris* Miyazaki for which the structure is being determined (Nakagawa et al., 1986).

The structure of cytochrome c_3 from *D. vulgaris* Miyazaki (amino acid sequence 88% identical to that of *D. vulgaris* Hildenborough) is also known (Higuchi et al., 1984), as well as that of cytochrome c_3 from *D. desulfuricans* Norway (Pierrot et al., 1982). The hemes in c-type cytochromes are covalently linked to two cysteine residues of the protein, which are generally present in the sequence C-X-X-C-H, via two thioether linkages. X is a variable amino acid residue in this sequence, while the histidine coordinates to the heme iron. However, in cytochrome c_3 two of the hemes connect to this generic sequence, while the other two are attached to C-X-X-X-X-C-H. The other axial heme iron ligand is also a histidine nitrogen in all four hemes.

Cytochrome c_3 thus has 8 cysteine residues for the covalent coordination and 8 histidine residues for the noncovalent coordination of the four hemes. The four hemes have low redox potentials (-400 to -200 mV). The axial heme ligands coordinating the single heme iron in cytochrome c_{553} are histidine and methionine, just as in eucaryotic cytochrome c and other monoheme bacterial cytochromes. This cytochrome has a much higher redox potential (-50 mV).

The nucleotide sequences of the genes for cytochrome c_3 and c_{553} indicate that these proteins are exported to the periplasm. The NH_2-terminal amino acid of the mature protein (alanine in both cytochromes) is preceded by a "typical" signal sequence of 22 residues for cytochrome c_3 and 24 residues for cytochrome c_{553}. These sequences have a positively charged NH_2 terminus (MRK- and MKR-, respectively) followed by a hydrophobic sequence up to the signal peptidase cleavage site (Á ↓ A), features that are typical for peptides that signal export to the periplasm in other gram-negative bacteria (Benson et al., 1985). One expects these cytochromes to be synthesized by a membrane-bound polypeptide–mRNA–ribosome complex in which the growing polypeptide chain is excreted during synthesis, resulting in a 100% periplasmic localization.

The physiological function of cytochrome c_3 is thought to be the transport of electrons to and from the enzyme hydrogenase, which is also periplasmic as discussed, while cytochrome c_{553} has been described as the electron acceptor of formate and D-lactate dehydrogenase (Odom and Peck, 1984). The gene for cytochrome c_3 is preceded by a strong putative E. coli consensus promoter and followed by a possible transcription terminator. The nucleotide sequence of the gene and its flanking regions thus suggest it to be monocistronic, although this has not been proven by direct measurement of the size of the cyc-mRNA and experimental determination of the transcription start site. No promoter has been suggested on the basis of the nucleotide sequence in the region upstream from the cyf-gene (Table 34.4), a hairpin loop structure that could serve as transcription terminator is present 3′ from the structural gene. Library searches of translated nucleotide sequences surrounding the cyf gene have not led to identification of other genes (e.g., if cytochrome c_{553} would be the exclusive redox partner of formate dehydrogenase one might hope that the two genes for these two proteins form an operon, as discussed here for rubredoxin).

Expression of the cyc gene in E. coli has been studied but does not lead to formation of functional cytochrome. Transformation of E. coli with plasmid pJ8, which has an 840-bp insert containing the cyc gene, does lead to expression of the protein (Pollock et al., 1989). Western blotting indicates that a considerable fraction of the expressed polypeptide is present in the 14-kDa precursor form from which the processed 12-kDa form is formed. The latter was shown to reside in the periplasm. Site-directed mutagenesis of the signal peptidase cleavage site abolished processing and export of the 14-kDa form. It thus appeared that E. coli can synthesize, export, and process the cyc gene product. However, E. coli does not appear to be able to covalently insert the hemes as judged by two criteria (heme fluorescence and sensitivity to proteolytic digestion) under either aerobic or anaerobic growth conditions.

Expression of the cyf gene in E. coli has been demonstrated in minicell experiments (G. Voordouw, unpublished observations). It is not known whether a functional holo-cytochrome c_{553} is formed. These results point to a certain specificity of the c-type heme insertion system in procaryotes, because E. coli can synthesize its own periplasmic c-type cytochromes (Kajie and Anraku, 1986). The Desulfovibrio system cannot be very specific because it must presumably be able to handle heme insertion in a variety of c-type cytochromes. As indicated, the Hildenborough cyc gene has recently been transferred to another Desulfovibrio host, D. desulfuricans G200, using bacterial conjugation technology, and this led to the expression of a high level of fully functional, periplasmic Hildenborough cytochrome c_3, that could readily be purified and characterized (Voordouw et al., 1990). This result does not solve the mystery of procaryotic insertion of c-type heme but demonstrates clearly that in the absence of knowledge of cofactor insertion systems which lead to functional hydrogenases and cytochromes it is prudent to select a host that is not too distant from the organism from which the gene was isolated.

Genes for cytoplasmic redox carriers

The genes for several small cytoplasmic redox proteins have been characterized by cloning and sequencing. The amino acid sequence and in some cases three-dimensional structure (flavodoxin, rubredoxin) of these carriers was known, allowing the synthesis and use of oligonucleotide probes for the cloning of these genes. The gene for flavodoxin from D. vulgaris Hildenborough was cloned and sequenced by two groups (Curley and Voordouw, 1988; Krey et al., 1988). The two sequences determined independently for the coding region of the gene are in good agreement and confirm the known amino acid sequence of flavodoxin. The absence of a signal sequence confirms flavodoxin as a cytoplasmic redox protein. Expression of functional flavodoxin in E. coli, which requires synthesis of the 148-amino-acid residue polypeptide chain and noncovalent binding of a single FMN residue, is easily achieved (Krey et al., 1988; Carr et al., 1990). Holo-flavodoxin has been purified from *E. coli*, transformed with *fla* gene containing expression vectors, and found to be indistinguishable from the protein isolated from *D. vulgaris* Hildenborough despite the presence of a nonnative NH_2 terminus in some cases. Current research on the *fla* gene focuses on the characterization of electron transport in mutant flavodoxins generated by oligonucleotide-directed mutagenesis techniques (Carr et al., 1990), but these data fall outside the scope of this chapter.

The gene for rubredoxin (*rub*) was isolated from a *D. vulgaris* Hildenborough gene library and characterized by dideoxynucleotide sequencing (Voordouw, 1988). Interestingly, sequencing of DNA upstream from the *rub* gene indicated the presence of a second larger gene of 378 bp (Brumlik and Voordouw, 1989). This gene is very likely to encode another redox protein because its NH_2 terminus is homologous to desulforedoxin, a very small redox protein isolated from *D. gigas* (19 identical residues in a stretch of 36). The two genes were shown to form an operon, and this led to the suggestion that the product of the upstream gene may be the redox partner of rubredoxin in vivo. It has therefore been tentatively named Rbo (rubredoxin oxidoreductase), product of the *rbo* gene. The *rbo-rub* operon is transcribed as a single transcript of 680 nucleotides (Brumlik and Voordouw, 1989). Expression of Rbo in *D. vulgaris* Hildenborough was recently demonstrated with the aid of antibodies directed at peptides, synthesized on the basis of the gene translated amino acid sequence of Rbo. Sodium dodecyl sulfate gel electrophoresis and Western blotting indicated Rbo to be present as a 14-kDa polypeptide in *D. vulgaris* cells, in agreement with expectations from the gene sequence. Rbo has been purified from *D. vulgaris* Hildenborough and *D. desulfuricans* (J. LeGall, personal communication), and a detailed study of its properties is now underway. These results are relevant to the elucidation of the cytoplasmic branch of the *Desulfovibrio* electron transport chain. New questions with re-

spect to the presently unknown physiological role of rubredoxin in *Desulfovibrio* can now be asked. There is no question that this approach (identification of possible redox partners by the characterization of redox protein operons) will prove important in the future, because biochemical approaches alone have not led to a complete description and understanding of the peri- and cytoplasmic electron transport chains in *Desulfovibrio*.

34.6 Summary

The genetics and molecular biology of sulfate-reducing bacteria have been fruitfully developed in the past several years. The genes for the three main types of hydrogenases, two smaller c-type cytochromes, and several well-known cytoplasmic carriers have been cloned and sequenced. Problems in the expression of some of these redox proteins caused by the failure of hosts to covalently insert redox prosthetic groups (hemes, iron-sulfur clusters) have been encountered and highlight a general deficiency in understanding how these proteins are synthesized in the functional form in vivo. More research in this area may prove to be of great general significance.

The recently developed possibilities for conjugating group 1 *Desulfovibrio* with other gram-negative bacteria have opened new research avenues. Genes for *Desulfovibrio* c-type cytochromes can now be readily expressed in other *Desulfovibrio* hosts and this will, for instance, allow a detailed study of the intramolecular electron transport pathways among the four hemes of cytochrome c_3 for which the three-dimensional structure is known in detail. Insertional inactivation of cloned genes has recently been reported (Rousset et al., 1991). This technique will allow directed mutagenesis and assessment of the physiological importance of some of the gene products indicated here, as soon as it has been made generally applicable.

In conclusion it appears that good progress has been made. However, the knowledge that there are at least eight different groups of sulfate reducers, which are not likely to be genetically related (e.g., these include the gram-positive genus *Desulfotomaculum*), offers some sobering reflections: we have only brushed the surface of a very large research topic.

Acknowledgments. G.V. acknowledges the financial support of The Natural Science and Engineering Research Council of Canada (NSERC). W. van Dongen and J. LeGall are thanked for communicating research results before publication. M.J. Brumlik, H.M. Deckers, and W.B.R. Pollock contributed considerably to this research as graduate students in G.V.'s laboratory. J.D.W. gratefully acknowledges the support of the Basic Energy Program of the U.S. Department of Energy and the research skills of B.J. Rapp-Giles.

References

Adman, E.T., L.C. Sieker, and L.H. Jensen. 1973. The structure of a bacterial ferredoxin. *J. Biol. Chem.* **259**:7045–7055.

Adman, E.T., L.C. Sieker, L.H. Jensen, M. Bruschi, and J. LeGall. 1977. A structural model of rubredoxin from *Desulfovibrio vulgaris* at 2 Å resolution. *J. Mol. Biol.* **112**:113–120.

Achenbach-Richter, L., K.O. Stetter, and C.R. Woese. 1987. A possible biochemical missing link among archaebacteria. *Nature* (London) **327**:348–349.

Ambler, R.P. 1968. The amino acid sequence of cytochrome c_3 from *Desulfovibrio vulgaris* (NCIB 8303). *Biochem. J.* **109**:47P.

Bagdasarian, M., M.R. Lurz, B. Ruckert, F.C.H. Franklin, M.M. Bagdasarian, J. Frey, and K.N. Timmis. 1981. Specific purpose cloning vectors II. Broad host range, high copy number, RSF1010-derived vectors, and a host-vector system for gene cloning in *Pseudomonas*. *Gene* (Amst.) **16**:237–247.

Benson, S.A., M.N. Hall, and T.J. Silhavy. 1985. Genetic analysis of protein export in *Escherichia coli* K12. *Annu. Rev. Biochem.* **54**:101–134.

Brumlik, M.J. and G. Voordouw. 1989. Analysis of the transcriptional unit encoding the genes for rubredoxin (*rub*) and a putative rubredoxin oxidoreductase (*rbo*) in *Desulfovibrio vulgaris* Hildenborough. *J. Bacteriol.* **171**:4996–5004.

Brumlik, M.J., G. LeRoy, M. Bruschi, and G. Voordouw. 1990. The nucleotide sequence of the gene for desulforedoxin from *Desulfovibrio gigas* indicates that the *rbo*-gene from *Desulfovibrio vulgaris* originated from a gene fusion event. *J. Bacteriol.* **172**:7289–7292.

Bruschi, M. 1976. Non-heme iron proteins: the amino acid sequence of rubredoxin from *Desulfovibrio vulgaris*. *Biochim. Biophys. Acta* **434**:4–17.

Bruschi, M. and F. Guerlesquin. 1988. Structure, function and evolution of bacterial ferredoxins. *FEMS Microbiol. Rev.* **54**:155–176.

Bruschi, M. and J. LeGall. 1972. c-Type cytochromes of *Desulfovibrio vulgaris*. The primary structure of cytochrome c_{553}. *Biochim. Biophys. Acta* **271**:48–60.

Bruschi, M., I. Moura, J. Le Gall, A.V. Xavier, and L.C. Sieker. 1979. The amino acid sequence of desulforedoxin, a new type of non heme iron protein from *Desulfovibrio gigas*. *Biochem. Biophys. Res. Commun.* **90**:596–605.

Cammack, R., D.S. Patil, E.C. Hatchikian, and V.M. Fernandez. 1987. Nickel and iron-sulphur centres in *Desulfovibrio gigas* hydrogenase: ESR spectra, redox properties and interactions. *Biochim. Biophys. Acta* **912**:98–109.

Campbell, L.L. and J.R. Postgate. 1965. Classification of the spore-forming sulfate-reducing bacteria. *Bacteriol. Rev.* **29**:359–363.

Carr, M.C., G.P. Curley, S.G. Mayhew, and G. Voordouw. 1990. Effects of substituting asparagine for glycine-61 in flavodoxin from *Desulfovibrio vulgaris* (Hildenborough). *Biochem. Int.* **20**:1025–1032.

Curley, G.P. and G. Voordouw. 1988. Cloning and sequencing of the gene encoding flavodoxin from *Desulfovibrio vulgaris* Hildenborough. *FEMS Microbiol. Lett.* **49**:295–299.

Davison, J., M. Heusterpreute, N. Chevalier, H.T. Vinh, and F. Brunel. 1987. Vectors with restriction site banks V. pJRD215, a wide-host range cosmid vector with multiple cloning sites. *Gene* (Amst.) **51**:275–280.

Deckers, H.M., F.R. Wilson, and G. Voordouw. 1990. Cloning and sequencing of a [NiFe] hydrogenase operon from *Desulfovibrio vulgaris* Miyazaki F. *J. Gen. Microbiol.* **136**:2021–2028.

Devereux, R., M. Delaney, F. Widdel, and D.A. Stahl. 1989. Natural relationships among sulfate-reducing eubacteria. *J. Bacteriol.* **171**:6689–6695.

Eidsness, M.K., R.A. Scott, B. Prickril. D.V. DerVartanian, J. LeGall, I. Moura, J.J.G. Moura, and H.D. Peck. Jr. 1989. Evidence for selenocysteine coordination to the active site nickel in the [NiFeSe] hydrogenase from *Desulfovibrio baculatus*. *Proc. Natl. Acad. Sci. USA* **86**:147–151.

Fernandez, V.M., E.C. Hatchikian, D.S. Patil, and R. Cammack. 1986. ESR detectable nickel and iron-sulphur centers in relation to the reversible activation of *Desulfovibrio gigas* hydrogenase. *Biochim. Biophys. Acta* **883**:145–154.

Ford, C.M., N. Garg, R.P. Garg, K.H. Tibelius, M.G. Yates, D.J. Arp, and L.C. Seefeldt. 1990. The identification, characterization, sequencing and mutagenesis of the genes (*hupS,L*) encoding the small and large subunits of the H_2-uptake hydrogenase of *Azotobacter chroococcum*. *Mol. Microbiol.* **4**:999–1009.

Frey, M., L. Sieker, F. Payan, R. Haser, M. Bruschi, G. Pepe, and J. LeGall. 1987. Rubredoxin from *Desulfovibrio vulgaris*. A molecular model of the oxidized form at 1.4 Å resolution. *J. Mol. Biol.* **197**:525–541.

Hagen, W.R., A. van Berkel-Arts, K.M. Kruse-Wolters, G. Voordouw, and C. Veeger. 1986. The iron-sulfur composition of the active site of hydrogenase from *Desulfovibrio vulgaris* (Hildenborough) deduced from its subunit structure and total iron-sulfur content.. *FEBS Lett.* **203**:59–63.

Handley, J., V. Adams, and J.M. Akagi. l973. Morphology of bacteriophage-like particles from *Desulfovibrio vulgaris*. *J. Bacteriol.* **115**:1205–1207.

Hatchikian, E.C., M. Bruschi, and J. LeGall. 1978. Characterization of the periplasmic hydrogenase from *Desulfovibrio gigas*. *Biochem. Biophys. Res. Commun.* **82**:451–461.

He, S.-H., M. Texeira, J. LeGall, D.S. Patil, D.V. DerVartanian, B.H. Huyn, and H.D. Peck Jr. 1989. EPR studies with ^{77}Se enriched [NiFeSe] hydrogenase of *Desulfovibrio baculatus*. Evidence for a selenium ligand to the active-site nickel. *J. Biol. Chem.* **264**:2678–2682.

Herriott, J.R., L.C. Sieker, L.H. Jensen, and W. Lovenberg. 1970. Structure of rubredoxin: an X-ray study to 2.5 Å resolution. *J. Mol. Biol.* **50**:391–406.

Higuchi, Y., M. Inaka, N. Yasuoka, and T. Yagi. 1987. Isolation and crystallization of a high molecular weight cytochrome from *Desulfovibrio vulgaris* Hildenborough. *Biochim. Biophys. Acta* **911**:341–348.

Higuchi, Y., M. Kusunoki, Y. Matsuura, W. Yasuoka, and M. Kakudo. 1984. Refined structure of cytochrome c_3 at 1.8 Å resolution. *J. Mol. Biol.* **172**:109–139.

Hormel, S., K.A. Walsh, B.C. Prickril, K. Titani, J. LeGall, and L.C. Sieker. 1986. Amino acid sequence of rubredoxin from *Desulfovibrio desulfuricans* strain 27774. *FEBS Microbiol. Lett.* **201**:147–150.

Kajie, S. and Y. Anraku. 1986. Purification of a hexaheme cytochrome c_{552} from *Escherichia coli* K12 and its properties as a nitrite reductase. *Eur. J. Biochem.* **154**: 457–463.

Kamimura, K. and M. Araki. 1989. Isolation and characterization of a bacteriophage lytic for *Desulfovibrio salexigens*, a salt requiring, sulfate-reducing bacterium. *Appl. Environ. Microbiol.* **55**:645–648.

Kent, H.M., M. Buck, and D.J. Evans. 1989. Cloning and sequencing of the *nifH* gene of *Desulfovibrio gigas*. *FEMS Microbiol. Lett.* **61**:73–79.

Kissinger, C.R., E.T. Adman, L.C. Sieker, L.H. Jensen, and J. LeGall. 1989. The crystal structure of the three-iron ferredoxin II from *D. gigas*. *FEBS Lett.* **242**:477–450.

Krey, G.D., E.F. Vanin, and R.P. Swenson. 1988.

Cloning, nucleotide sequence and expression of the flavodoxin gene from *Desulfovibrio vulgaris* (Hildenborough). *J. Biol. Chem.* **263**:15436–15443.

Leclerc, M., A. Colbeau, B. Cauvin, and P.M. Vignais. 1988. Cloning and sequencing of the genes encoding the large and the small subunits of the H_2 uptake hydrogenase (*hup*) of *Rhodobacter capsulatus*. *Mol. Gen. Genet.* **214**:97–108.

Li, C., H.D. Peck, Jr., and A.E. Przybyla. 1986. Complementation of an *Escherichia coli pyrF* mutant with DNA from *Desulfovibrio vulgaris*. *J. Bacteriol.* **165**:644–646.

Li, C., H.D. Peck. Jr., J. LeGall, and A.E. Przybyla. 1987. Cloning, characterization and sequencing of the genes encoding the large and small subunits of the periplasmic [NiFe] hydrogenase of *Desulfovibrio gigas*. *DNA (NY)* **6**:539–551.

Lissolo, T., E.S. Choi, J. LeGall, and H.D. Peck, Jr. 1986. The presence of multiple intrinsic membrane nickel containing hydrogenase in *Desulfovibrio vulgaris* (Hildenborough). *Biochem. Biophys Res. Commun.* **139**:701–708.

Menon, N.K., H.D. Peck, Jr., J. LeGall, and A.E. Przybyla. 1987. Cloning and sequencing of the genes encoding the large and small subunits of the periplasmic (NiFeSe) hydrogenase of *Desulfovibrio baculatus*. *J. Bacteriol.* **169**:5401–5407.

Menon, N.K., J. Robbins. H.D. Peck, Jr., C.Y. Chatelus, E.-S. Choi, and A.E. Przybyla. 1990. Cloning and sequencing of a putative *Escherichia coli* [NiFe] hydrogenase-1 operon containing six open reading frames. *J. Bacteriol.* **172**:1969–1977.

Meyer, J. and J. Gagnon. 1991. Primary structure of hydrogenase I from *Clostridium pasteurianum*. Biochemistry **30**:9697–9704.

Nakagawa, A., E. Nagashima, Y. Higuchi, M. Kusunoki, Y. Matsuura, N. Yasuoka, Y. Katsube, H. Chichara, and T. Yagi. 1986. Crystallographic study of cytochrome c_{553} from *Desulfovibrio vulgaris*. *J. Biochem.* **99**:605–606.

Nakano, K., Y. Kikumoto, and T. Yagi. 1983. Amino acid sequence of cytochrome c-553 from *Desulfovibrio vulgaris* Miyazaki. *J. Biol. Chem.* **258**:12409–12412.

Odom, J.M. and H.D. Peck Jr. 1981. Hydrogen cycling as a general mechanism for energy coupling in the sulfate-reducing bacteria, *Desulfovibrio* sp. *FEMS Microbiol. Lett.* **12**:47–50.

Odom, J.M. and H.D. Peck, Jr. 1984. Hydrogenase, electron transferring proteins and energy coupling in the sulfate-reducing bacteria *Desulfovibrio*. *Annu. Rev. Microbiol.* **38**:551–592.

Odom, J.M. and J.D. Wall. 1987. Properties of a hydrogen-inhibited mutant of *Desulfovibrio desulfuricans* ATCC 27774. *J. Bacteriol.* **169**:1335–1337.

Peck, H.D., Jr. 1984. Physiological diversity of the sulfate-reducing bacteria. *In: Microbial Chemoautotrophy*, W.R. Strohl and O.H. Tuovinen, eds., pp. 309–335. Columbus: Ohio State University Press.

Pfennig, N., F. Widdel, and H.G. Truper. 1981. The dissimilatory sulfate-reducing bacteria. *In: The Prokaryotes*, Vol. 1, M.P. Starr, H. Stolp, H.G. Truper, A. Balows, and H.G. Schlegel, eds., pp. 926–940. Heidelberg: Springer-Verlag.

Pierrot, M., R. Haser, M. Frey, F. Payan, and J.P. Astier. 1982. Crystal structure and electron transfer properties of cytochrome c_3. *J. Biol. Chem.* **257**:14341–14348.

Pollock, W.B.R., P.J. Chemerika, M.E. Forrest, J.T. Beatty, and G. Voordouw. 1989. Expression of the gene encoding cytochrome c_3 from *Desulfovibrio vulgaris* (Hildenborough) in *Escherichia coli*: export and processing of the apoprotein. *J. Gen. Microbiol.* **135**:2319–2328.

Postgate, J.R. 1984. *The sulphate-reducing bacteria*, 2d Ed. Cambridge: Cambridge University Press.

Postgate, J.R. and L.L. Campbell. 1966. Classification of *Desulfovibrio* species, the nonsporulating sulfate-reducing bacteria. *Bacteriol. Rev.* **30**:732–738.

Postgate, J.R., H.M. Kent, and R.L. Robson. 1987. Nitrogen fixation by *Desulfovibrio*. *In: The Nitrogen and Sulphur Cycles*, J.A. Cole and S. Ferguson, eds., pp. 457–471. Cambridge: Cambridge University Press.

Postgate, J.R., H.M. Kent, R.L. Robson, and J.A. Chesshyre. 1984. The genomes of *Desulfovibrio gigas* and *D. vulgaris*. *J. Gen. Microbiol.* **130**: 1597–1601.

Powell, B., M. Mergeay, and N. Christofi. 1989. Transfer of broad host-range plasmids to sulphate-reducing bacteria. *FEMS Microbiol. Lett.* **59**:269–274.

Prickril, B.C., M.H. Czechowski, A.E. Przybyla, H.D. Peck, Jr., and J. LeGall. 1986. Putative signal peptide on the small subunit of the periplasmic hydrogenase from *Desulfovibrio vulgaris*. *J. Bacteriol.* **167**:722–725.

Prickril, B.C., S.-H. He, C. Li, N. Menon, E.S. Choi, A.E. Przybyla, D.V. DerVartanian, H.D. Peck. Jr., G. Fauque, J. LeGall, M. Texeira, I. Moura, J.J.G. Moura, D. Patil, and B.J. Huyn. 1987. Identification of three distinct classes of hydrogenase in the genus *Desulfovibrio*. *Biochem. Biophys. Res. Commun.* **149**:369–377.

Rapp, B.J. and J.D. Wall. 1987. Genetic transfer in *Desulfovibrio desulfuricans*. *Proc. Natl. Acad. Sci. USA* **84**:9128–9131.

Rousset, M., Z. Dermoun, M. Chippaux, and J.P. Belaich. 1991. Marker exchange mutagenesis of the *hydN* genes in *Desulfovibrio fructosovorans*. *Mol. Microbiol.* **5**:1735–1740.

Sayavedra-Soto, L.A., G.K. Powell, H.J. Evans, and R.O. Morris. 1988. Nucleotide sequence of the genetic loci encoding subunits of *Bradyrhizobium japonicum* uptake hydrogenase. *Proc. Natl. Acad. Sci. USA* **85**:8395–8399.

Sieker, L.C., R.E. Stenkamp, L.H. Jensen, B. Prickril, and J. LeGall. 1986. Structure of rubredoxin from the bacterium *Desulfovibrio desulfuricans*. *FEBS Lett.* **208**:73–76.

Singleton, R., Jr., R.B. Ketchum, and L.L. Campbell. 1988. Effect of calcium cation on plating efficiency of the sulfate-reducing bacterium *Desulfovibrio vulgaris*. *Appl. Environ. Microbiol.* **54**:2318–2319.

Stetter, K.O., G. Lauerer, M. Thomm, and A. Neuner. 1987. Isolation of extremely thermophilic sulfate reducers: evidence for a novel branch of archaebacteria. *Science* **236**:822–824.

Stokkermans, J., W. van Dongen, A. Kaan, W. van den Berg, and C. Veeger. 1989. hyd γ, a gene from *Desulfovibrio vulgaris* (Hildenborough) encodes a polypeptide homologous to the periplasmic hydrogenase. *FEMS Microbiol. Lett.* **58**:217–222.

Sznyter, L.A., B. Slatko. L. Moran, K.H. O'Donnell, and J.E. Brooks. 1987. Nucleotide sequence of the *DdeI* restriction-modification system and characterization of the methylase protein. *Nucleic Acids Res.* **15**:8249–8266.

Texeira, M., I. Moura, A.V. Xavier, B.H. Huyn, D.V. DerVartanian, H.D. Peck, Jr., J. LeGall, and J.J.G. Moura. 1985. Electron paramagnetic resonance studies on the mechanism of activation and the catalytic cycle of the nickel-containing hydrogenase from *Desulfovibrio gigas*. *J. Biol. Chem.* **260**:8942–8950.

Trousil, E.B. and L.L. Campbell. 1974. The amino acid sequence of cytochrome c_3 from *Desulfovibrio vulgaris*. *J. Biol. Chem.* **249**:386–393.

Uffen, R.L., A. Colbeau, P. Richaud, and P.M. Vignais. 1990. Cloning and sequencing of the genes encoding uptake-hydrogenase subunits of *Rhodocyclus gelatinosus*. *Mol. Gen. Genet.* **221**:49–58.

van den Berg, W.A.M., J.P.W.G. Stokkermans, and W.M.A.M. van Dongen. 1989. Development of a plasmid transfer system for the anaerobic sulphate reducer, *Desulfovibrio vulgaris*. *J. Biotechnol.* **12**:173–184.

van Dongen, W., W. Hagen, W. van den Berg, and C. Veeger. 1988. Evidence for an unusual mechanism of membrane translocation of the periplasmic hydrogenase of *Desulfovibrio vulgaris* (Hildenborough), as derived from expression in *Escherichia coli*. *FEMS Microbiol. Lett.* **50**:5–9.

van Rooijen, G.J.H., M. Bruschi, and G. Voordouw. 1989. Cloning and sequencing of the gene encoding cytochrome c_{553} from *Desulfovibrio vulgaris* Hildenborough. *J. Bacteriol.* **171**:3573–3578.

Voordouw, G. 1988. Cloning of genes encoding redox proteins of known amino acid sequence from a library of the *Desulfovibrio vulgaris* (Hildenborough) genome. *Gene* (Amst.) **69**:75–83.

Voordouw, G. 1990. Hydrogenase genes in *Desulfovibrio*. *In: Microbiology and Biochemistry of Strict Anaerobes Involved in Interspecies Hydrogen Transfer*, J.P. Belaich, M. Bruschi, and J.L. Garcia, eds., pp. 37–51. New York: Plenum.

Voordouw, G. and S. Brenner. 1985. Nucleotide sequence of the gene encoding the hydrogenase from *Desulfovibrio vulgaris* (Hildenborough). *Eur. J. Biochem.* **148**:515–520.

Voordouw, G. and S. Brenner. 1986. Cloning and sequencing of the gene encoding cytochrome c_3 from *Desulfovibrio vulgaris* (Hildenborough). *Eur. J. Biochem.* **159**:347–351.

Voordouw, G., J.E. Walker, and S. Brenner. 1985. Cloning of the gene encoding the hydrogenase from *Desulfovibrio vulgaris* (Hildenborough) and determination of the NH_2-terminal sequence. *Eur. J. Biochem.* **148**:509–514.

Voordouw, G., W.R. Hagen, M. Kruse-Wolters, A. van Berkel-Arts, and C. Veeger. 1987a. Purification and characterization of *Desulfovibrio vulgaris* (Hildenborough) hydrogenase expressed in *Escherichia coli*. *Eur. J. Biochem.* **162**:31–36.

Voordouw, G., H.M. Kent, and J.R. Postgate. 1987b. Identification of the genes for hydrogenase and cytochrome c_3 in *Desulfovibrio*. *Can. J. Microbiol.* **33**:1006–1010.

Voordouw, G., N.K. Menon, J. LeGall, E.-S. Choi, H.D. Peck, Jr., and A.E. Przybyla. 1989a. Analysis and comparison of nucleotide sequences encoding the genes for [NiFe] and [NiFeSe] hydrogenase from *Desulfovibrio gigas* and *Desulfovibrio baculatus*. *J. Bacteriol.* **171**:2894–2899.

Voordouw, G., J.D. Strang, and F.R. Wilson. 1989b. Organization of the genes encoding [Fe] hydrogenase in *Desulfovibrio vulgaris subsp. oxamicus* Monticello. *J. Bacteriol.* **171**:3881–3889.

Voordouw, G., W.B.R. Pollock, M. Bruschi, F. Guerlesquin, B.J. Rapp-Giles, and J.D. Wall. 1990. Functional expression of *Desulfovibrio vulgaris* Hildenborough cytochrome c_3 in *Desulfovibrio desulfuricans* following conjugational gene transfer from *Escherichia coli*. *J. Bacteriol.* **172**:6122–6126.

Wall, J.D., B.J. Rapp-Giles, and S.P. Concannon. 1990. Identification of a small cryptic plasmid in

Desulfovibrio desulfuricans. In *Abstracts, 90th Annual Meeting of the American Society for Microbiology*, H-158. Washington, D.C.: American Society for Microbiology.

Watenpaugh, K.D., L.C. Sieker, L.H. Jensen, J. LeGall, and M. Dubourdieu. 1972. Structure of the oxidized form of a flavodoxin at 2.5 Å resolution: resolution of the phase ambiguity by animalous scattering. *Proc. Natl. Acad. Sci. USA* **69**:3185–3188.

Weimer, P.J., M.J. van Kavelaar, C.B. Michel, and T.K. Ng. 1988. Effect of phosphate on the corrosion of carbon steel and the composition of corrosion products in two-stage continuous cultures of *Desulfovibrio desulfuricans*. *Appl. Environ. Microbiol.* **54**:386–396.

Widdel, F. 1988. Microbiology and ecology of sulfate-and sulfur-reducing bacteria. *In: Biology of Anaerobic Microorganisms*, A.J.B. Zehnder, ed., pp. 469–585. New York: Wiley.

Widdel, F. and N. Pfennig. 1984. Dissimilatory sulfate- or sulfur-reducing bacteria. *In: Bergey's Manual of Systematic Bacteriology*, Vol. 1, N.R. Krieg and J.G. Holt, eds., pp. 663–679. Baltimore: Williams & Wilkins.

Xu, H.W., J. Love, R. Borghese, and J.D. Wall. 1989. Identification and isolation of genes essential for H_2 oxidation in *Rhodobacter capsulatus*. *J. Bacteriol.* **171**:714–721.

Yagi, T., K. Kimoura, and H. Inokuchi. 1985. Analysis of the active center of hydrogenase from *Desulfovibrio vulgaris* Miyazaki by magnetic measurements, *J. Biochem.* **97**:181–187.

Yagi, T., K. Kimura. H. Daidoji, F. Sakai, S. Tamura, and H. Inikuchi. 1976. Properties of purified hydrogenase from the particulate fraction of *Desulfovibrio vulgaris* Miyazaki. *J. Biochem.* **79**:661–671.

Zeikus. J.G., M.A. Dawson, T.E. Thompson, K. Ingvorsen, and E.C. Hatchikian. 1983. Microbial ecology of microbial sulphidogenesis: isolation and characterization of *Thermodesulfobacterium commune* gen. nov. sp. nov. *J. Gen. Microbiol.* **129**:1159–1169.

Zinoni, F., A. Birkman, T.C. Stadtman, and A. Bock. 1986. Nucleotide sequence and expression of the selenocysteine containing polypeptide of formate dehydrogenase (formate-hydrogen-lyase linked) from *Escherichia coli*. *Proc. Natl. Acad. Sci. USA* **83**:4650–4654.

35

Gene Transmission, MLS, and Tetracycline Resistance in *Bacteroides*

Francis L. Macrina and C. Jeffrey Smith

35.1   Overview of Conjugal Genetic Exchange in *Bacteroides*

Conjugal genetic exchange in *Bacteroides* was first described in 1979 when reports from three laboratories (Privitera et al., 1979; Tally et al., 1979; Welch et al., 1979) demonstrated conjugally transmissible macrolide-lincosamide-streptogramin (MLS) resistance in this genus. This resistance phenotype was important because it permitted bacterial growth in the presence of the clinically effective and widely used antibiotic clindamycin. In all cases this resistance was constitutively expressed. This resistance phenotype was similar to that seen in gram-positive bacteria (e.g., streptococci, staphylococci, and bacilli). In these bacteria, the MLS gene (designated *erm*) encoded an RNA methylase that modified specific adenine residues in the 23S rRNA, thus preventing MLS drugs from binding to the ribosome and inhibiting protein synthesis. The genetic basis of such conjugal transfer has been traced to both plasmids as well a nonplasmid elements. The latter appear to reside chromosomally and to resemble the conjugative transposons of the gram-positive bacteria (Clewell and Gawron-Burke, 1986; Clewell, 1990) in their behavior.

In general, conjugal transfer in *Bacteroides* has been defined on the basis of DNase resistance and the necessity for donor and recipient cells to be in close contact. Transfer requires cell-to-cell contact on a nitrocellulose or polycarbonate filter or on an agar matrix. Conjugal transfer among *Bacteroides* has not been demonstrated to occur in liquid medium. Transfer frequencies are low (10^{-4} to 10^{-6} transconjugants/donor cell), but they can be elevated one to two orders of magnitude by using isogenic mating pairs (Welch et al., 1979). Under optimal conditions the transfer process reaches maximal levels after 2 to 3 h of co-cultivation of cells on filter disks (Welch et al., 1979).

Little is known about the cellular or molecular events of conjugal transfer in *Bacteroides*. Pili or other surface appendages have not been associated with the conjugative process. The number of genes needed for conjugal transfer are not known for any of the well-characterized systems. The existence of specific surface receptors on recipient cells has not been described, and no information as to whether surface exclusion is operative in matings is available. At the cellular level, important unanswered questions about *Bacteroides* conjugal transfer include whether or not an *oriT* is needed for transfer or mobilization, whether single- or double-stranded DNA is transferred, and what macromolecular processes are essential to transfer. Evidence supporting the genetic exchange between strains of *Bacteroides* in vivo has been reported (Butler et al., 1984) and it is likely that the conjugal transfer of drug resistance plasmids occurs in the colon, resulting in the spread of antibiotic resistance genes.

35.2 Antibiotic Resistance Plasmids

Species of the intestinal *Bacteroides* have yielded three different resistance plasmids, all of which have been intensely studied: pBF4, a 41-kb plasmid from *B. fragilis*; pBFTM10, a 15-kb plasmid from *B. fragilis* (and its indistinguishable counterpart pCP1 from *Bacteroides thetaiotaomicron*); and pBI136, an 80-kb plasmid from *B. ovatus* all share some common features. All three plasmids encode high-level MLS resistance conferred by a conserved *erm* gene. The location of the *erm* gene on each of these plasmids was deduced initially by the analysis of spontaneously occurring plasmid deletion mutants (Welch et al., 1979; Magot et al., 1981; Tally et al., 1982; Smith, 1985a). The nucleotide sequence of two of the *erm* genes (*ermF* from pBF4 and *ermFS* from pBI136) has been determined, and the genes have been found to differ at only one nucleotide position. Based on restriction maps and the DNA hybridization under high stringency, it is likely that the *erm* gene of pBFTM10 is strongly homologous to *ermF* and *ermFS*. It is believed that all three of these genes share a common ancestor and that they have not diverged significantly from one another during their evolution (Odelson et al., 1987). Physical analyses of both wild-type and mutant plasmids revealed the *erm* gene of each plasmid was bordered by a similar, if not identical, sequence, which flanked the gene in a directly repeated fashion. Other than their *erm* genes and the associated directly repeated DNA sequences, pBF4, pBFTM10, and pBI136 shared no detectable homology.

The repeated sequences bordering the *erm* genes of these three plasmids have been shown to be a related, if not identical, insertion sequence (IS) elements (IS*4400* on pBFTM10, IS*4351* on pBF4, and iso-IS*4351* on pBI136). These flanking IS elements form compound transposons with their associated *erm* genes, and genetic evidence for their transposition in both *B. fragilis* and *Escherichia coli* has been obtained (Robillard et al., 1985; Shoemaker et al., 1985; Smith and Spiegel, 1987). In all three

of these transposons the *erm* genes were juxtaposed to one copy of the IS element, suggesting some interaction between these two sequences. This suggestion has been experimentally explored by the analysis of nucleotide sequence data, S1 nuclease transcription mapping studies, and the construction of gene fusions to evaluate promoter activity (Rasmussen et al., 1986, 1987; Smith, 1987). Collectively, these data indicate that the *erm* genes are under the transcriptional control of the promoters contained in the IS element. Interestingly, outward-firing promoters exist at both ends of this class of element, so it was possible to drive *erm* transcription regardless of its orientation with respect to the drug resistance gene.

Examples of the IS element being in opposite orientation with respect to the *erm* gene were found in the comparative examination of pBF4 and pBFTM10 with pBI136. Odelson et al. (1987) have proposed a model to explain the formation of these transposons that predicts that all three transposons evolved independently of one another. Following the entry of a similar *erm* gene into *Bacteroides*, transposition of the IS element next to the newly acquired gene resulted in the transcriptional control of *erm* by the IS element. This may have been required in the face of selective pressure because the invading *erm* gene could not be expressed in *Bacteroides* for several reasons [e.g., unrecognizable promoter, nonfunctional upstream regulatory sequences on the *erm* mRNA (Odelson et al., 1987)].

A second proposed transposition event then resulted in the *erm* gene being flanked by two copies of the IS element, generating a compound transposon. The differing orientation of the IS element with respect to the *erm* gene on pBF4/pBFTM10 compared to pBI136 with the qualitatively and quantitatively different intervening DNA sequences in the three transposons argues that these elements were formed as the result of three separate events. Figure 35.1 illustrates the genetic organization of these three well-characterized *Bacteroides* transposons.

Plasmid	(Size)	Transposon
pBFTM10	*(15kb)*	*Tn4400*
pBF4	*(41kb)*	*Tn4351*
pBI136	*(80kb)*	*Tn4551*

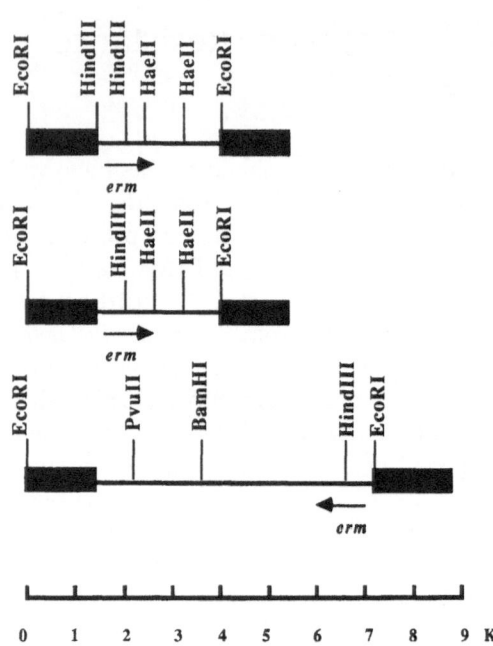

FIGURE 35.1 *erm* transposons from plasmids of *Bacteroides* origin. Restriction maps of three transposons that are discussed in text are illustrated; Arrows indicate locations of their *erm* genes (confer clindamycin resistance). Black rectangles correspond to related flanking insertion sequences that border each of these transposons in directly repeated fashion. *tetX* locus discussed in text is located adjacent to *erm* genes seen on Tn4400 and Tn4351.

35.3 Mobilization of Plasmids in *Bacteroides*

Numerous, phenotypically cryptic, plasmid molecules of varying size have been reported in *Bacteroides* (Odelson et al., 1987; Salyers et al., 1987). For plasmids of smaller size (< 6 kb) Callihan et al. (1983) have used DNA hybridization to define three distinct classes of such extrachromosomal elements. Although *Bacteroides* has proven to be a rich source of extrachromosomal DNA, the study of plasmid-linked phenotypes has been limited to antibiotic resistance traits. It is not clear at present whether other clinically, metabolically, or ecologically important phenotypes are encoded by plasmids in this genus. Certain of the phenotypically cryptic *Bacteroides* plasmids have been used as components of *Escherichia–Bacteroides* shuttle cloning vectors, providing replication origins that function in the latter genus (see following).

Available evidence supports the notion that presumed nonconjugative *Bacteroides* plasmids are capable of being mobilized by conjugative elements. Mays et al. (1982) first reported that a 5.7-kb plasmid frequently (~ 80%) was co-transferred to recipients along with a nonplasmid conjugative element conferring MLS and tetracycline resistance. The lack of a selectable marker on this plasmid prevented the unequivocal establishment of its own self-transmissibility, but its small size argued against it being conjugative. Valentine et al. (1988) confirmed and extended these observations using a phenotypically cryptic plasmid as well as the *Bacteroides* resistance plasmids pBFTM10 and pCP1. Using an intergeneric conjugal transfer system, these workers demonstrated that an IncP conjugative plasmid could mobilize these elements in both inter- (*Escherichia–Bacteroides*) and intra- (*Escherichia–Escherichia*) generic matings. These results led the authors to suggest that these *Bacteroides* plasmids contained a site that functioned operationally as an *oriT* in DNA transfer. Direct physical evidence supporting the existence of such a site is not available. Interestingly, shuttle plasmids made from either pB8-51 or pBFTM10 could be mobilized from *Bacteroides* donors containing a nonplasmid Tc^r element designated ERL. Whether the same region is being recognized by both the IncP plasmid and the ERL elements is still open to question.

The mobilization of nonreplicating circular DNA molecules presumably created by the recombination of directly repeated drug resistance transposon sequences (Tn4351) from *Bacteroides* to *E. coli* also was observed by this group (Shoemaker and Salyers, 1987). The presence of nonplasmid (i.e., chromosomal) conjugative elements in the *Bacteroides* resembling conjugative transposons appears to be a rather common occurrence (see following). Such elements potentially complicate studies aimed at defining conjugal transfer involving plasmid because integrated elements could mobilize in *trans*. In sum, the mechanism of mobilization of nonconjugative elements remains poorly defined, although available evidence suggests the process involves mechanisms similar to those operative in other gram-negative systems. The existence of a *cis*-acting region appears necessary for mobilization to occur.

35.4   Nonplasmid Conjugative Elements

It has been known since early in the 1980s that certain antibiotic-resistant strains of *Bacteroides fragilis* were capable of transferring their resistance without the involvement of detectable plasmid DNA (Macrina et al., 1981; Tally et al., 1981; Mays et al., 1982). Such transfer elements have been placed in two categories with respect to antibiotic resistance phenotypes: those that transfer tetracycline resistance alone and those which transfer tetracycline and clindamycin resistances in a linked fashion. All these systems have met the operational definition of conjugal transfer based on standard criteria (see earlier). Mating only occurs on a solid matrix (filter or agar surface), similar to the conditions seen for plasmid transfer. Donor strains carrying many of the Tc^r elements require that cells first be

grown in low concentrations of tetracycline to display maximal conjugal transfer efficiency (Smith et al., 1982; Shoemaker and Salyers, 1988).

Exceptions to this do exist, and one well-characterized donor has been shown to transfer Tc^r at comparable frequencies irrespective of treatment with tetracycline (Macrina et al., 1981). The mechanism of tetracycline enhancement of conjugal transfer proficiency remains unexplained. Usually, the Tc^r determinant itself is inducibly expressed and, as stated earlier, these elements are capable of mobilizing nonconjugative plasmids, presumably operating in a *trans* fashion. Where examined, these elements share no cross-hybridizing sequences with conjugative plasmids, although many of the MLS resistance-conferring elements carry a gene that is related to the *Bacteroides erm* gene family [e.g., *ermF*, *ermFS* (Macrina et al., 1981; Mays et al., 1982)]. The presence of insertion sequences has not been associated with these elements on the basis of either hybridization with known IS sequences or genetic structure analysis (i.e., detectable repeated sequences associated with the elements) (Halula and Macrina, 1990). Evidence suggests that related conjugal elements exist that do not carry antibiotic resistance markers (Smith and Spiegel, 1987; Shoemaker and Salyers, 1990). Moreover, the presence of such elements has been demonstrated in clinical isolates of *Bacteroides* already shown to be carrying a conjugative resistance plasmid (Smith et al., 1982).

Shoemaker and Salyers (1988) reported an interesting observation associated with the transfer of several of these types of elements, all of which conferred Tc^r and displayed greatly increased transfer efficiency following exposure to tetracycline. Specifically, when these elements were transferred to *Bacteroides uniformis* and the resulting strain was exposed to inhibitory tetracycline, two circular DNA forms (~ 10 and 11-kb) could be detected as extrachromosomal elements. Southern analyses indicated these circular molecules were normally integrated into the host cell chromosome. Certain of the Tc^r nonplasmid elements examined by these authors did not mediate the excision of the circular DNA forms.

Stevens et al. (1990) have presented evidence suggesting that the control of production of these circular DNA forms resides in a region contained within the Tc^r transfer element called ERL (confers Tc^r and MLS resistance). Stevens et al. were able to clone regions of a similar element (termed DOT) and create insertion mutations in the subcloned fragments. When these insertion mutations were introduced back into the native ERL element and its transfer evaluated, these authors found certain mutants with reduced conjugal transfer efficiencies. Two such reduced transfer mutants did not display the formation of the circular DNA molecules [termed plasmid-like forms (plf)]. One of these mutants also showed reduced efficiency of conjugative mobilization of coresident plasmids. Interestingly, these authors identified an additional mutant, also prepared by insertion mutagenesis, that displayed the plf phenotype in a constitutive fashion. Although these studies clearly established the need for this Tc^r element to produce the plf phenotype, the significance of this phenomenon remains unexplained.

Indirect evidence first suggested that nonplasmid transfer elements were large, probably in the 40- to 50-kb range or larger. Shoemaker et al. (1989) used a novel shuttle to clone segments of the DOT element (Tc^r, MLS^r). They obtained sequences that contained one or both drug resistance markers. Some of the clones that carried only Tc^r were capable of conjugal transfer in intergeneric *Bacteroides* matings. Shoemaker et al. used their cloned sequences, which overlapped, to construct a linear restriction endonuclease map. This map indicated the size of this element to be at least 50 kb. The authors could not determine whether they had cloned a chromosome–DOT element junction sequence, which would enable them to ascribe a chromosomal location to this element. Nonetheless, their work was the first report on the physical isolation of functional segments of one of these prototype elements.

Halula and Macrina (1990) have also reported the partial cloning of a Tc^r, MLS^r element, called Tn5030, from *B. fragilis*. They estimated the size of this element to be in the 40- to 50-kb range, based on overlapping clones. Interpretation of their results was complicated by the unexpected finding that sequences related to Tn5030 were present in the transconjugant strain from which their genomic libraries were prepared. Further studies are needed to fully define the molecular structure of this element.

Shoemaker and Salyers (1990) have isolated a 65-kb conjugal transfer element devoid of drug resistance determinants. This element could apparently transpose to a *Bacteroides–Escherichia* shuttle plasmid; it was designated XBU4422 and was shown by DNA hybridization experiments to be integrated into the chromosome of the host *B. uniformis* strain from which it was initially isolated. XBU4422 appeared to integrate into the host chromosome in a site-specific fashion. The availability of an apparent intact element of this type opens the door for extending the study of this class of element using both physical and genetic means.

Malamy and coworkers (Hecht and Malamy, 1989; Hecht et al., 1989) have isolated and described a 9.6-kb element designated Tn4399 that is capable of conjugally mobilizing plasmids only when it is present in *cis*. This element was discovered in experiments in which a Tc^r element was being studied in terms of its ability to restore conjugal proficiency to a transfer-defective plasmid. Transposition of Tn4399 to the transfer-defective plasmid as well as to one other plasmid in the cell occurred into several sites. One copy of this element was found on the chromosome of the donor strain used in their studies, however, a transconjugant strain isolated by them contained three copies of Tn4399 at novel chromosomal sites. These workers have further characterized the molecular events associated with the insertion of Tn4399 into a target site. Tn4399 contains 13-bp, inverted-repeat sequences at its termini, with an extra base present within the left inverted-repeat sequence. Their results suggested some specificity with a majority of independent insertions occurring in one of two locations in a single AT-rich, 14-bp target site.

In contrast to other conjugative transposons (Clewell and Gawron-Burke, 1986; Clewell, 1990), Tn4399 generates a novel 5-bp sequence at its right terminus–target junction in addition to creating a 3-bp, repeated sequence of the target site. The basis of mobilization conferred by Tn4399 is under study by Clewell's group. This element could be carrying an *oriT*-like sequence that is working in concert with additional conjugal elements within the donor cell. Tn4399 may also encode for all or part of the conjugal transfer apparatus, as is the case with conjugative transposons such as Tn916 (Clewell and Gawron-Burke, 1986; Clewell, 1990). Tn4399 was capable of mobilizing plasmids in intergeneric *Bacteroides* matings as well as in *Bacteroides*–*E. coli* matings. This transfer element provides yet another novel example of the genetic plasticity of these procaryotes, and it is likely to play a role in the dissemination of antibiotic resistance determinants.

35.5 Biochemistry of Antibiotic Resistance

Direct biochemical studies on the basis of MLS resistance in *Bacteroides* have not been reported in the literature. Early studies in the field assumed resistance to be caused by N^6 dimethylation of specific adenine residues in the 23S ribosomal RNA. The biochemistry of such methlytransferases had been well established in other bacterial systems as the sole basis of MLS resistance; only this simple enzymatic alteration of the rRNA was known to simultaneously confer resistance to three different families of antibiotics (Dubnau, 1984; Weisblum, 1985). Thus, the MLS phenotype was assumed to result from the presence of comparable enzymes in *Bacteroides*. This thinking was confirmed when DNA sequence analysis of *erm* genes from *Bacteroides* was performed. The strong resemblance of the *Bacteroides erm* gene products to the MLS methlytransferases of the staphylococci, streptococci, and bacilli clearly defined the presence of such enzymes as the basis of MLS resistance in *Bacteroides* (Rasmussen et al., 1986; Odelson et al., 1987). It is possible that there are other modes of clindamycin resistance present in the *Bacteroides*. For example, enzymes that esterify macrolides or phosphorylate lincosamides could conceivably confer resistance but direct evidence for their exis-

tence in the *Bacteroides* has not yet been reported.

Unlike the MLS phenotype, resistance to the tetracyclines in *Bacteroides* has been genetically and biochemically more complex. The first Tc^r determinant to be studied in detail was discovered by Guiney and coworkers (Guiney et al., 1984a). They cloned a fragment from plasmid pBF4 that expressed Tc^r in aerobically, but not in anaerobically, grown *E. coli*. A similar determinant was found on pBFTM10 and pBF4 but not on pBI136. In all these cases, the Tc^r was located close to the *erm* gene on these plasmids. This novel tetracycline resistance determinant was initially designated "Tc" (Speer and Salyers, 1988, 1989) but *tetX* has now been suggested by this group (Salyers et al., 1990) as a name for the gene.

Early biochemical studies (Speer and Salyers, 1988) suggested that the gene product was made in both aerobically and anaerobically grown *E. coli* cells and that expression of the phenotype was not dependent on the aerobic electron transport system. Evidence that *E. coli* strains carrying this determinant were capable of detoxifying tetracycline was also obtained in these studies (Speer and Salyers, 1989). Park and Levy (1988) have confirmed that this determinant mediated the degradation of tetracycline to a biologically less active compound, but additionally suggested the determinant present on Tn4400 was also capable of mediating the active efflux of tetracycline from the cell. The genetic resolution of these two seemingly different resistance mechanisms is underway.

Speer and Salyers (1990) did not observe an obvious efflux of tetracycline conferred by the sequences cloned from Tn4351 into *E. coli*. However, using Tn4400, they were able to clone and separate the determinants for tetracycline degradation and tetracycline efflux. The genetic determinant for efflux did not confer tetracycline resistance by itself even when present on a multicopy plasmid. The clinical significance of this class of determinants in *Bacteroides* remains open to question because the phenotype is apparently unable to be expressed in this genus. However, the expression of Tc^r in enteric species along with our knowledge of the potential for mobilization of genetic information from *Bacteroides* to *E coli* (see earlier) suggests that *Bacteroides* could be a reservoir of Tc^r determinants for other bacteria.

The cloning of a Tc^r determinant able to confer resistance in *Bacteroides* has been accomplished by Guiney and coworkers (Guiney et al., 1989). Previous efforts to clone the *Bacteroides* Tc^r determinant led to the discovery of the *tetX* type element (expressed in *E. coli* but not in *Bacteroides*), but such approaches failed to yield a Tc^r determinant which was expressed in *Bacteroides*. Accordingly, Guiney et al. (1989) created a genomic library of *B. fragilis* 1126 in *E. coli* using a *Bacteroides–E. coli* shuttle that could be mobilized back into a tetracycline-sensitive *Bacteroides* host strain.

This approach enabled these authors to isolate clones that carried recombinant plasmids which conferred tetracycline resistance in *Bacteroides*. This Tc^r determinant was mapped to a 3-kb region and when used as a DNA probe revealed the presence of related genes in other Tc^r *Bacteroides* strains capable of transferring this resistance. The 1126 strain was shown to transfer Tc^r linked to MLS^r in the absence of plasmid involvement. These authors used pulsed-field electrophoresis to demonstrate that their cloned Tc^r gene was associated with high molecular weight DNA, suggesting its chromosomal location. The availability of this cloned gene will facilitate the further analysis of nonplasmid conjugative elements in *Bacteroides* and, in addition, will help further our knowledge of the biochemical genetic basis of tetracycline resistance in the *Bacteroides*.

35.6   Gene Transmission Systems and Genetic Tools

Research on antibiotic resistance in *Bacteroides* species during the past 10 years has led to a rapid development of genetic systems suitable for manipulation of these organisms. Early work in this area was hampered by the realization that indigenous *Bacteroides* plasmids were not maintained or expressed in *E. coli*. Likewise, no *E. coli* plasmid or antibiotic resistance gene has been found to function in the *Bacter-*

oides. However, studies on transmissible MLS^r resulted in the identification of the R plasmids pBF4, pBFTM10, and pBI136 and of their affiliated *erm* transposons. Through the use of these plasmids and their antibiotic resistance genes it has been possible to construct the plasmid vectors required for routine genetic analysis. More recently, attention has centered on the novel, conjugal antibiotic resistance elements (transposons) that are not associated with plasmid DNA. The unique properties of these elements, together with the variety of antibiotic resistance genes associated with them, will undoubtedly provide the foundation for future genetic tools with innovative applications. In the following sections we describe the genetic systems and vectors that have proven useful for analysis of *Bacteroides* and, when possible, suggest areas where future development is likely to occur.

Transformation systems

The prerequisite for genetic manipulation and analysis of a bacterial species is a suitable method for the introduction of DNA into cells. For most applications, an ideal approach is through the use of genetic transformation. However, the *Bacteroides* have been found to be refractory to most commonly used transformation protocols, and it has not been until recently that electroporation has made transformation a more reliable technique for use with these organisms. The original report concerning DNA uptake by a *Bacteroides* species described a transfection system with *B. thetaiotaomicron* and a virulent phage, B1 (Burt and Woods, 1976). Transfection followed the CaCl$_2$ protocol of Mandel and Higa (Mandel and Higa, 1970) with a heat pulse at 42°C and an immediate dilution into broth. Transfection frequencies of as much as 5.5×10^8 pfu per μg of DNA were obtained, and infectivity was lost when the phage DNA was treated with DNase but not RNase or pronase. Attempts by several laboratories to apply this technique or other similar methods employing divalent cations for transformations with plasmid DNA have not been successful.

The first plasmid transformation of *Bacteroides* was demonstrated using *B. fragilis* 638

(Privitera et al., 1979) as the recipient and pBFTM10 as the transforming DNA (Smith, 1985b). Polyethylene glycol (PEG) treatment (Klebe et al., 1983) of cell suspensions in hypertonic buffer was used to induce competence for DNA uptake. Protoplast formation was not required, nor were any special precautions taken to maintain anaerobic conditions. Using pBFTM10 DNA isolated from the 638 host strain, transformation frequencies to 4×10^3 MLS^r transformants per μg of DNA were observed, and a linear relationship between DNA concentration and the number of transformations was obtained. A posttransformation expression period was required for recovery of transformants. No MLS^r colonies were observed for the first 30 min following transformation, and optimum frequencies were observed after 2.5 h of growth in nonselective broth (Smith, 1985b). The PEG transformation system was used initially to test shuttle cloning vectors for expression and maintenance in *Bacteroides*, and several useful constructs were identified (Smith, 1985c).

During the course of this work it became apparent that the source of the transforming DNA was of utmost importance for transformation. Plasmid DNA isolated from *B. fragilis* 638 transformed at high frequencies, reaching 10^{-5} transformants per viable cell per μg of DNA for some of the smaller vectors (e.g., pFD176, 7.3 kb). However, DNA isolated from *E. coli* transformed at frequencies 1,000 to 10,000 fold lower, suggesting the presence of a strong host-controlled restriction system in *B. fragilis*. *B. fragilis* 638 has proven to be the most suitable recipient strain, but other *B. fragilis* strains also can be easily transformed. In contrast, most other *Bacteroides* species have not been successfully transformed using the PEG method.

Electroporation of *Bacteroides*

Electroporation, which has proven to be a rapid and reliable method for the transformation of a wide variety of bacterial species, has been applied successfully to the *Bacteroides* (Thompson and Flint, 1989; Smith et al., 1990). Using 10% glycerol or 10% glycerol with 1 mM MgCl$_2$, strains of *B. fragilis*, *B. uniformis*, *B. ovatus*, and *B. ruminicola* have been transformed with a variety of *Bacteroides* R plasmids and shuttle vectors. Electrical field strengths of 12.5 kV/cm were required for optimum transformation of 10^5 to 10^6 transformants per μg of DNA and, as seen with other organisms, there was a direct relationship between electric field strength and transformation frequency. There was no significant effect of cell growth phase on transformation, and cell viability remained high ($> 75\%$) for all field strengths tested (Thompson and Flint, 1989; Smith et al., 1990). The requirement for a postelectroporation expression period was addressed by using the shuttle vector pFD308, containing both an MLS^r gene (*ermF*) and a chloramphenicol resistance (Cm^r) gene (Smith et al., 1990). When electroporated cells were selected on clindamycin, a 2- to 3-h recovery period was required for optimum transformation. However, in cases in which the Cm^r marker was selected, the recovery of transformants was observed almost immediately. This difference in expression period may reflect the fact that the two resistance genes differ in their mode of physiological action. The *ermF* gene presumably acts to protect the antibiotic target site (Rasmussen et al., 1986), whereas the Cm^r gene (chloramphenicol acetyl transferase from Tn9) is an antibiotic-inactivating enzyme. In the case of clindamycin, therefore, it may be that there is a minimum time required for these anaerobic bacteria to recover from stress induced during electroporation and then modify a majority of their ribosomes. Similarly, transformation experiments with *B. ruminicola* using the Tc^r plasmid pRR14 indicated that a postelectroporation expression period was not required for optimal recovery of Tc^r transformants (Thompson and Flint, 1989). Because it is unlikely that the Tc^r gene is a target-modifying enzyme, the expression period may not be required.

As suggested, the presence of strong host-controlled restriction and modification systems in the *Bacteroides* has a significant effect on transformation frequency, which is supported by studies using electroporation-induced transformation. For these experiments, *B. fragilis* 638 was transformed with the shuttle plasmid pFD288 (Smith, 1989), isolated from *E. coli*,

B. fragilis, B. thetaiotaomicron, B. ovatus, or *B. uniformis* (Smith et al., 1990). All the *Bacteroides* DNAs tested resulted in a 8- to 12-fold decrease in transformation frequency when compared to values obtained with the homologous DNA isolated from *B. fragilis* (Smith et al.1990). DNA isolated from *E. coli* resulted in even greater restriction with a decrease in frequency greater than 1000 fold. A similar 1000-fold reduction in transformation of *B. uniformis* also was noted in experiments with pDP1 isolated from *E. coli* (Thompson and Flint, 1989). This further 100-fold reduction in transformation when using *E. coli*-isolated DNA poses an interesting question concerning the putative *Bacteroides* modification systems. It is possible that multiple modification systems are present with one being common at the genus level while others operate at the species level.

In comparison to the conjugative systems, transformation of *Bacteroides* has not been used extensively as a tool for genetic analysis. However, it has proven useful in the analysis of plasmid-born antibiotic resistance genes. For example, the MLS^r gene from pBI136 (*ermFS*) was first cloned and expressed in *B. fragilis* with the aid of the PEG transformation system (Smith, 1985c), and analysis of *ermFS* expression also relied on transformation for the introduction of plasmid constructs to the *B. fragilis* host strain (Smith, 1987). More recently, the plasmid linkage of a 5-nitroimidazole resistance gene was documented using transformation into the sensitive *B. fragilis* 638 strain (Breuil et al., 1989). The feasibility of directly cloning plasmid-linked or cloned genes also exists; this was first demonstrated by construction of the *Bacteroides* vector pBI191, using *B. fragilis* 638 as the cloning host (Smith, 1985c). With the applicability of electroporation to a wider variety of *Bacteroides* species, it is likely that transformation now will find more use in routine genetic experiments, especially when plasmids must be moved between strains.

Conjugation systems

Conjugation and plasmid mobilization have been widely used as methods for the introduction of DNA into the *Bacteroides*. The essential elements required for transfer between *E. coli* and *Bacteroides* were established by Guiney et al. (Guiney et al., 1984c) in the construction of pDP1. This hybrid plasmid contains a pBR322 deletion derivative for replication and selection in *E. coli* as well as the transfer origin (*oriT*) from RK2 for high-frequency mobilization (Guiney and Yakobson, 1983). This was fused to the 15-kb MLS^r plasmid, pCP1 (pBFTM10), which contains a *Bacteroides* replication region and selective marker. Mobilization of pDP1 from *E. coli* to *Bacteroides* requires the transfer functions of a helper plasmid such as RK2 or RK231, and these can be provided in *trans* in the pDP1 host cell or by a second helper strain in triparental matings.

Filter mating transfer frequencies for pDP1 were 0.7 transconjugants per recipient cell for matings between *E. coli* strains, whereas the frequency dropped to 3×10^{-6} when *B. fragilis* was used as the recipient (Guiney et al. 1984c). No transfer was observed in the absence of an appropriate helper plasmid. Genetic and physical analyses of the *Bacteroides* transconjugants revealed that pDP1 was transferred intact but that the helper plasmid was not present. Subsequent evidence has indicated that the helper plasmid is in fact transferred into the *Bacteroides* recipients, but as with other *E. coli* plasmids it does not function in this species and is lost at high frequency (Shoemaker et al., 1986b).

Two significant improvements in the construction and use of mobilization vectors has occurred since the initial description of pDP1. The first of these was an improvement in the transfer frequencies of *E. coli* to *Bacteroides* matings. To demonstrate transposition of Tn4351 in *Bacteroides* using a suicide plasmid strategy, it was necessary to increase the transfer frequencies to levels ($> 10^{-5}$ per recipient) at which the transposition event could be observed. Salyers and coworkers (Shoemaker et al., 1986b) found that incubation of mating mixtures under aerobic conditions resulted in a 20- to 100-fold increase in the number of transconjugants as compared to identical matings performed under an anaerobic atmosphere. Under aerobic conditions transfer frequencies with typical shuttle vectors reached 1.1×10^{-3}. The *Bacter-*

oides recipients did not suffer a loss in viability, and in fact their numbers increased 3-fold in overnight matings. Presumably the *E. coli* donors in such matings act as efficient oxygen scavengers and effectively protect the *Bacteroides* recipients. Using these mating conditions it was possible to demonstrate the transposition of Tn4351 from an *E. coli* donor (see following).

The second improvement resulted from an observation that many indigenous *Bacteroides* plasmids contain regions that promote their mobilization by *E. coli* conjugative plasmids (Shoemaker et al., 1986a). As described, it was shown that shuttle vectors containing *Bacteroides* replicons such as pBFTM10 (pCP1) and the cryptic plasmid pB8-51 could be mobilized to *B. uniformis* recipients by IncP (e.g., RK231, R751) plasmids at frequencies of 10^{-8} to 10^{-5}. The IncIa plasmid R64*drd*-11 also mobilized the shuttle vectors but 1000 times less efficiently, and no mobilization was observed with the IncN, W, or F1 plasmids. Although these indigenous mobilization regions are not as efficient for transfer as the *E. coli* transfer origins, they can be used effectively for routine genetic transfer. Further, their presence on most *Bacteroides* plasmids simplifies vector construction, especially where high-frequency transfer is not required, and they facilitate transfer between *Bacteroides* when coresident with a conjugative transposon.

General-purpose cloning vectors

A wide variety of vectors have been developed in recent years, and most of these have been similarly designed with minor modifications for improved versatility and convenience (Table 35.1). The principal elements required are two distinct replicons for replication in the appropriate hosts and two antibiotic resistance markers suitable for selection of one or the other organism. An *E. coli* transfer origin is optional but desirable because it allows higher transfer frequencies between *E. coli* donors and *Bacteroides*.

Although a number of different *E. coli* plasmids have been used for these hybrids, three *Bacteroides* replicons have been used almost exclusively. Two are small cryptic plasmids that

are found widely disseminated among all the intestinal tract *Bacteroides* species and thus have the advantage of a very broad host range. One of these, pBI143, is a 2.7-kb replicon belonging the class I homology group of cryptic plasmids (Callihan et al., 1983). The second is pB8-51, a 4.23-kb plasmid in the class IIa homology group. Overall these two plasmids have similar properties; both are readily mobilized by *Bacteroides* conjugal elements and the *E. coli* IncP plasmids, they are compatible and stably maintained, and their copy number appears to be relatively high. Although neither of these plasmids has been subjected to intensive genetic analysis, some of the regions involved in replication and mobilization have been identified (Smith, 1985c; Shoemaker et al., 1986a; Valentine et al., 1988).

The 14-kb R plasmid pCP1 (pBFTM10) also has been studied intensely, and its derivatives have been developed into useful components for several shuttle vectors. Tn5 mutagenesis of pDP1 derivatives has resulted in the identification of the pCP1 regions required for replication as well as for MLS^r and the cryptic Tc^r determinant (Matthews and Guiney, 1986). These studies have led directly to the construction of plasmid vectors that are considerably smaller than pDP1 but maintain all the advantages of this vector (see Table 35.1; and Guiney et al., 1988). Generally, pCP1 has been ideal for this application because its size makes it easy to manipulate and the MLS^r marker is already in place. Like the cryptic plasmids, pCP1 appears to have a broad host range encompassing most intestinal tract *Bacteroides*.

The MLS^r genes *ermF* and *ermFS* have thus far proven to be the most useful markers for selection and maintenance of shuttle vectors in *Bacteroides* host strains. These were the first antibiotic resistance genes identified and their original plasmid location made them a logical choice for use as selectable markers. The most common source for the *erm* genes has been the 3.8-kb *Eco*RI fragment from Tn4351 (pBF4; Welch and Macrina, 1981). An unusual but extremely useful feature of this gene fragment (and also of Tn4400) is the presence of a cryptic Tc^r gene adjacent to the *erm* gene (Guiney et al., 1984a; Matthews and Guiney, 1986). As mentioned, this resistance gene is not expressed

Table 35.1 Representative *Bacteroides* plasmids and cloning vectors

Plasmid/size	Replicon[a]		Phenotype[b]		Mobilization region[c]	Reference
	EC	BA	EC	BA		
Indigenous plasmids						
pBF4/41 kb	None	pBF4	None	MLS	BA	Welch et al., 1979
pBI136/80 kb	None	pBI136	None	MLS	BA	Smith and Macrina, 1984
pCP1 (pBFTM10)/ 14 kb	None	pCP1	None	MLS	BA	Guiney et al., 1984b
pBI143/2.7 kb	None	pBI143	None	None	BA	Smith, 1985c
pB8-51/4.2 kb	None	pB8-51	None	None	BA	Shoemaker et al., 1986a
pBFTM2006/2.7 kb	None	pBFTM2006[d]	None	None	BA	Thompson and Malamy, 1990
General-purpose vectors						
pDP1/19 kb	pBR322	pCP1	Ap,Tc	MLS	EC	Guiney et al., 1984
pFD176/7.3 kb	pUC19	pBI143	Ap,lac	MLS	BA	Smith, 1985c
pFD160/5.4 kb	pUC19	pBI143	Ap,lac	None	BA	Smith, 1985c
pVAL-1/11 kb	pBR328	pB8-51	Ap,Tc	MLS	BA	Valentine et al., 1988
pDK3/8.5 kb	pBR322	pCP1	Ap	MLS	EC	Guiney et al., 1988
pFD288/8.8 kb	pUC19	pBI143	Sp,lac	MLS	EC	Smith, 1989
pBI191/5.4 kb	None	pBI143	None	MLS	NA	Smith, 1985c
Genomic cloning vectors						
pJST61/12 kb	pBR322	pBFTM2006	Ap,*endR1*	MLS	EC	Thompson and Malamy, 1990
pNJR1/14.4 kb	RSF1010	pB8-51	Km,Sm,*cos*	None	EC	Shoemaker et al., 1989
pPH6/8.9 kb	pBR322	pCP1	Ap,lac	MLS	EC	Guiney et al., 1988, 1989
Transposon and suicide vectors						
R751:Tn4351	R751	None	Tp,Tc	MLS	EC	Shoemaker et al., 1986b
pFD197/17kb	pUC19	pBI143	Ap	MLS	BA	Smith and Spiegel, 1987
pE3-1/13 kb	pBR328	None[e]	Ap,Tc	MLS	BA	Guthrie and Salyers, 1986
Expression and fusion vectors						
pFD214/6.3 kb	pUC19	pBI143	Ap	MLS[f]	BA	Smith, 1987
pFD290/8.0 kb	pUC19	pBI143	Ap	MLS	BA	Smith, in manuscript[g]

[a] Entries indicate replicon used for plasmid replication in *E. coli* (EC) or *Bacteroides* (BA)

[b] Phenotypes of plasmids in *E. coli* (EC) or *Bacteroides* (BA) are abbreviated as follows: MLS, macrolide-lincosamide-streptogramin B resistance; Ap, ampicillin resistance; lac, β-galactosidase activity; Tc, tetracycline resistance; *endR1*, *Eco*RI endonuclease; Km, kanamycin resistance; Sm, streptomycin resistance; *cos*, phage lambda *cos* site; Tp, trimethoprim resistance; Sp, Spectinomycin resistance.

[c] Entries indicate origin of mobilization region used in *E. coli* to *Bacteroides* matings. Abbreviation EC, *E. coli* transfer origins such as *oriT*; BA, *Bacteroides* mobilization regions found on indigenous plasmids; NA, not applicable.

[d] pBFTM2006 is probably identical to pBI143 (C.J. Smith unpublished observations).

[e] pE3-1 contains a functional pB8-51 mobilization region, but a region required for replication has been inactivated.

[f] MLS resistance in operon fusion vector pFD214 can only be activated by cloning in a *Bacteroides* promoter.

[g] Pertinent details concerning expression vector pFD290 are given in text.

in *Bacteroides* hosts but confers Tc^r in aerobically grown *E. coli* strains. This property has provided a positive selection for the gene fragment during vector construction in *E. coli*. Although useful, the Tc^r is not required for stable maintenance of the *erm* genes in *E. coli*, and a variety of plasmids utilizing the *erm* genes without the adjacent Tc^r gene have been described (see Table 35.1).

The Achilles heel of *Bacteroides* cloning vectors has been the sole reliance on the *erm* genes for selection in *Bacteroides* host strains. Until recently there have been no other choices available for selection of plasmids in these organisms, and this has severely restricted the approaches available for genetic analysis and strain construction. However, the *erm* vectors lately have been used to clone two new chromosomally located *Bacteroides* antibiotic resistance genes, Tc^r (Guiney et al., 1989; Shoemaker et al., 1989) and cefoxitin-imipenem (Fox-Imp; Thompson and Malamy, 1990). One of the Tc^r determinants was cloned from a conjugal transposon originally found in *B. thetaiotaomicron*, and subcloning experiments have further localized this gene to a 2.5-kb fragment that has been cloned into a shuttle plasmid (Stevens et al., 1990).

The Fox-Imp^r determinant should be extremely useful as a plasmid marker because nearly all *Bacteroides* strains are sensitive to one or both of these drugs. The Fox-Imp^r gene *cfiA*, located on a 3.6-kb *Sau*3A fragment, has been sequenced and found to encode a β-lactamase with a predicted mass of 27,260 daltons. This enzyme is highly potent with activity on carbapenems, cephamycins, and all other β-lactams tested. Fragments containing the intact gene have been subcloned, and the suitability of this gene as a selective marker for plasmids is presently being tested (M. Malamy, personal communication). These new selectable markers, together with the chloramphenicol acetyl transferase gene described next, will assist greatly in the analysis of *Bacteroides* gene expression and regulation.

Genomic cloning vectors

During the past few years several *Bacteroides* genes have been cloned directly in *E. coli* by complementation or by screening for an activity (Guthrie et al., 1985; Southern et al., 1986; Scholle et al., 1990). However, many genes are not expressed in *E. coli*, and those that have been cloned are often poorly expressed (Guthrie et al., 1985; Russo et al., 1990). Antibiotic resistance genes have been especially refractory to direct cloning in *E. coli* although some do seem to express at low levels once they are cloned. To circumvent these expres-

sion problems, several systems have been developed that are suitable for genomic cloning with expression in a *Bacteroides* host.

The basic approach is to prepare a genomic library in *E. coli* using a highly transmissible shuttle vector. The library is then mobilized into a *Bacteroides* host for screening. This strategy was used to clone the transmissible Tc^r gene from *B. fragilis* DNA in the shuttle pPH6 (Guiney et al., 1988, 1989; also see Table 35.1). The strong selective pressure afforded by the Tc^r determinant made it possible to readily obtain Tc^r isolates from the library, and a 13.2-kb insert containing the resistance gene was identified. A variation of this strategy was employed for the cloning of a Fox-Imp^r gene and a neuraminidase gene from *B. fragilis* using the positive selection vector pJST61 (Russo et al., 1990; Thompson and Malamy, 1990). The significant property of this vector is the inclusion of a target gene that must be inactivated by insertion of cloned DNA into a unique *Bgl*II restriction site. This target gene, *E. coli endR1*, is lethal to *E. coli* host strains lacking the *Eco*RI methylase; thus, only strains containing recombinant plasmids will survive. The considerable enrichment for recombinant plasmids provided by this positive selection vector significantly decreases the background of self-ligated vector, which can seriously complicate any cloning strategy especially those depending on conjugation to a secondary host for screening. The use of pJST61 should allow the rapid cloning of any *Bacteroides* gene with a measurable phenotype.

Another strategy devised by Salyers and coworkers (Shoemaker et al., 1989) is based on a shuttle cosmid vector, pNJR1 (Table 35.1), and it has been used successfully to clone large segments of a chromosomal, conjugative Tc^r-MLS^r transposon. This shuttle plasmid was derived from an RSF1010-based cosmid vector, pJRD215, which contains the *cos* site and an IncP-compatible transfer origin. The *Bacteroides* replicon, pB8-51, was inserted into the plasmid but no *Bacteroides*-selectable markers were included because this vector was intended for cloning the chromosomal Tc^r and *erm* genes. In actual cloning experiments many of the cosmids containing *Bacteroides* DNA suffered deletions, but it was possible to obtain preparations in which 40% to 60% of the cosmids had full-sized (40- to 44-kb) inserts. Using this approach, independent transconjugants were obtained containing the *erm* and Tc^r genes either separately or on apparently contiguous fragments of DNA (Shoemaker et al., 1989).

Specialized vectors

One of the first applications for the *Bacteroides* vectors and gene transmission systems was the demonstration of Tn*4351* and Tn*4551* transposition in a *Bacteroides* host. Two approaches were used to generate transposon insertions. The first required the use of a suicide delivery vehicle that was capable of replication in *E. coli* but not in *Bacteroides* and which was capable of high-frequency transfer to a *Bacteroides* host. Theoretically any *E. coli* plasmid containing a transfer origin could have been used to deliver the transposon, but because of its high rate of transfer a particularly appealing choice was the IncP plasmid R751 containing a Tn*4351* insertion (Shoemaker et al., 1986b). When this plasmid was conjugally transferred into *B. uniformis* recipients with selection for the *erm* gene, transconjugants were readily observed at frequencies of 3×10^{-6} to 5×10^{-6}. Because there was no functional *Bacteroides* replicon, all the transconjugants were the result of an insertion of Tn*4351* into the chromosome.

One unusual feature of this system was the observation that more than 50% of the transconjugants contained insertions of R751 together with the transposon. While this phenomenon has not been explained, the integrated R751 does not seem to interfere with the stability of the insertions. A second approach used for transposon delivery was to clone Tn*4551* onto a shuttle vector that was partially defective for replication in *Bacteroides* (Smith and Spiegel, 1987). The vector chosen, pFD165 (Smith, 1985c), was a pBI143::pUC19 hybrid previously shown to be extremely unstable in *B. fragilis* in the absence of selective pressure. *B. fragilis* transformed with the transposon-containing plasmid were grown in the absence of selective pressure and then plated on media containing clindamycin. MLS^r colonies were screened for plasmid content and an average of 52% of those tested contained chromosomal in-

sertions of Tn*4551*. The frequency of transposition was 7.8×10^{-5} per generation, which is in close agreement with the results for Tn*4351*.

Suicide vectors also have been used for targeted insertional inactivation of chromosomal genes in *Bacteroides*. The initial description of this technique involved the use of pE3-1, a pBR328::pB8-51 chimera containing the *ermF* gene, the pB8-51 mobilization region, but lacking a *Bacteroides* replicon (Guthrie and Salyers, 1986). A small restriction fragment bearing a portion of the cloned target gene, chondroitin lyase II, was ligated to pE3-1 and the resulting plasmid was mobilized into *B. thetaiotaomicron* with selection for MLS^r transconjugants. In the absence of a functional replicon, transconjugants can arise only by chromosomal integration of the plasmid mediated by homologous recombination. Biochemical and physical analyses of transconjugants recovered from these matings confirmed that the entire plasmid had inserted into the chromosomal chondroitin lyase II gene and abolished the enzymatic activity. This approach offers an efficient and reproducible means of mutating specific genes that have been previously cloned.

The final vector systems to be discussed here are those designed to analyze gene expression and regulation in *Bacteroides*. An extremely valuable approach for these kinds of studies is the use of operon fusions in which the regulatory regions of the gene of interest are fused to a reporter gene whose activity can be easily monitored. One operon fusion system developed for *B. fragilis* has been described in which promoters can be cloned upstream from the MLS^r structural gene, *ermFS*, and their activity is determined by selection on clindamycin-containing media (Smith, 1987). This promoter-probe vector, pFD214, was constructed by insertion of a 0.85-kb *ermFS* gene fragment bearing the entire structural gene and ribosomal-binding site into the shuttle plasmid pFD160 (MLS^s; Table 35.1). Putative promoter fragments can be cloned into the vector and then transformed or mobilized into *B. fragilis* to determine if they have activity. A major drawback of pFD214 is that the only genetic marker available for selection in a *Bacteroides* host is the *erm* reporter gene; thus, putative promoters with low activity may not be recovered. Nevertheless, it was possible to use this system to demonstrate the presence of two *Bacteroides* promoters in the insertion element IS*4351*, which bounds Tn*4551* (Smith, 1987). Improved operon fusion vectors will be rapidly developed as additional selectable *Bacteroides* genetic markers become available.

Based on the results showing promoter activity of IS*4351*, a simple expression vector suitable for the activation of transcription in *Bacteroides* has been constructed (C.J. Smith, unpublished data). The vector is based on pFD167 (Smith, 1985c), which contains the *ermFS* structural gene inserted into the *Bacteroides* (pBI143) portion of the plasmid. The *ermFS* gene is active in *Bacteroides*, presumably being transcribed from a resident promoter located on the plasmid. To provide a source of promoter activity for the activation of foreign genes, an IS*4351* gene fragment (minus the terminal 27 bases of the element) was ligated into pFD167. The resulting construct, pFD290, contained translation stop signals at the end of the IS element in all reading frames. The expression vector was tested by insertion of the Tn*9* chloramphenicol acetyl transferase (CAT) structural gene downstream of the IS promoters. This plasmid then was transformed into *B. fragilis* 638 with selection on a medium containing clindamycin, and transformants were screened for resistance to chloramphenicol. The results revealed that *B. fragilis* strains bearing the CAT plasmid were resistant to 66 μg/ml of chloramphenicol, a 16-fold increase in resistance. Evidence that CAT was under control of the IS*4351* promoter(s) was obtained by deletion of the IS sequences. Such deletion derivatives did not confer chloramphenicol resistance on *Bacteroides* strains. The phenotypic tests were confirmed by measurement of CAT activity in cell-free extracts.

These results with CAT were the first demonstration of the expression of an *E. coli* gene in *Bacteroides*, and they open several new avenues of investigation. For example, it is clear that the CAT gene may be used as a reporter gene in *Bacteroides*. This gene has proven extremely useful in other systems and may now be a valuable addition to the genetic tools available for use in the *Bacteroides*. Genetic systems

are advancing rapidly, and it is now possible to study *Bacteroides* gene function on many different levels. Although many of the genetic studies have focused on plasmid systems or antibiotic resistance, the tools are now available for detailed analysis of chromosomal genes involved in virulence and metabolism.

Acknowledgment. Work reported from the author's laboratories was supported by U.S. Public Health Service Grant no. DE09035 (F.L.M.) and a grant from the Biotechnology Research and Development Corporation (C.J.S.).

References

Breuil, J., A. Dublanchet, N. Truffaut, and M. Sebald. 1989. Transferable 5-nitroimidazole resistance in the *Bacteroides fragilis* group. *Plasmid* **21**:151–154.

Burt, S.J. and D.R. Woods. 1976. Transfection of the anaerobe *Bacteroides thetaiotaomicron* with phage DNA. *J. Gen. Microbiol.* **93**:405–409.

Butler, E., K.A. Joiner, M. Malamy, J.C. Bartlett, and F.P. Tally. 1984. Transfer of tetracycline or clindamycin resistance among strains *Bacteroides fragilis* in experimental abscesses. *J. Infect. Dis.* **150**:20–24.

Callihan, D.R., F.E. Young, and V.L. Clark. 1983. Identification of three homology classes of small, cryptic plasmids in intestinal *Bacteroides* species. *Plasmid* **9**:17–30.

Clewell, D.B. 1990. Movable genetic elements and antibiotic resistance in enterococci. *Eur. J. Clin. Microbiol. Infect. Dis.* **9**:90–102.

Clewell, D.B. and C. Gawron-Burke. 1986. Conjugative transposons and the dissemination of antibiotic resistance in streptococci. *Ann. Rev. Microbiol.* **40**:635–659.

Dubnau, D. 1984. Translational attenuation: the regulation of bacterial resistance to the macrolide-lincosamide-streptogramin B antibiotics. *CRC Crit. Rev. Biochem.* **16**:103–132.

Guiney, D.G. and E. Yakobson. 1983. Location and nucleotide sequence of the transfer origin of the broad host range plasmid RK2. *Proc. Nat. Acad. Sci. USA* **80**:3595–3598.

Guiney, D.G, Jr., P. Hasegawa, and C.E. Davis. 1984a. Expression in *Escherichia coli* of cryptic tetracycline genes from *Bacteroides* R plasmids. *Plasmid* **11**:248–252.

Guiney, D.G, Jr., P. Hasegawa, and C.E. Davis. 1984b. Homology between clindamycin resistance plasmids in *Bacteroides*. *Plasmid* **11**:268–271.

Guiney, D.G., P. Hasegawa, and C.E. Davis. 1984c. Plasmid transfer from *Escherichia coli* to *Bacteroides fragilis*: differential expression of antibiotic resistance phenotypes. *Proc. Natl. Acad. Sci. USA* **81**:7203–7206.

Guiney, D.G., P. Hasegawa, K. Bouic, and B. Mathews. 1989. Genetic transfer systems in *Bacteroides*: cloning and mapping the transferable tetracycline resistance locus. *Mol. Microbiol.* **3**:1617–1623.

Guiney, D.G., K. Bouic, P. Hasegawa, and B. Matthews. 1988. Construction of shuttle cloning vectors for *Bacteroides fragilis* use in assaying foreign tetracycline resistance gene expression. *Plasmid* **20**:17–22.

Guthrie, E.P., N.B. Shoemaker, and A.A. Salyers. 1985. Cloning and expression in *Escherichia coli* of a gene coding for chondroitin lyase from *Bacteroides thetaiotaomicron*. *J. Bacteriol.* **164**:510–515.

Guthrie, E.P. and A.A. Salyers. 1986. Use of targeted insertional mutagenesis to determine whether chondroitin lyase II is essential for chondroitin sulfate utilization by *Bacteroides thetaiotaomicron*. *J. Bacteriol.* **166**:966–971.

Halula, M. and F.L. Macrina. 1990. Tn*5030*: a conjugative transposon conferring clindamycin resistance in *Bacteroides* species. *Rev. Infect. Dis.* **12** (Suppl.2):S235–S242.

Hecht, D.W. and M.H. Malamy. 1989. Tn*4399*, a conjugal mobilizing transposon of *Bacteroides fragilis*. *J. Bacteriol.* **171**:3603–3608.

Hecht, D.W., J.S. Thompson, and M.H. Malamy. 1989. Characterization of the termini and transposition products of Tn*4399*, a conjugal mobilizing transposon of *Bacteroides fragilis*. *Proc. Natl. Acad. Sci. USA* **86**:5340–5344.

Klebe, R.J., J.V. Harriss, Z.D. Sharp, and M.D. Douglas. 1983. A general method for polyethylene-glycol-induced genetic transformation of bacteria and yeast. *Gene* (Amst.) **25**:333–341.

Macrina, F.L., T.D. Mays, C.J. Smith, and R.A. Welch. 1981. Non-plasmid associated transfer of antibiotic resistance in *Bacteroides*. *J. Antimicrob. Chemother.* **8**:77–86.

Magot, M., F. Fayolle, G. Privitera, and M. Sebald. 1981. Transposon-like structures in the B. *fragilis* MLS plasmid pIP410. *Mol. Gen. Genet.* **181**:559–561.

Mandel, M. and A. Higa. 1970. Calcium-dependent bacteriophage DNA infection. *J. Mol. Biol.* **53**:159–162.

Matthews, B.G. and D.G. Guiney. 1986. Characterization and mapping of regions encoding clindamycin resistance, tetracycline resistance, and a replication function on the *Bacteroides* R plasmid pCP1. *J. Bacteriol.* **167**:517–521.

Mays, T.D., C.J. Smith, R.A. Welch, C. Delfini, and F.L. Macrina. 1982. Novel antibiotic resistance transfer in *Bacteroides. Antimicrob. Agents Chemother.* **21**:110–118.

Odelson, D.A., J.L. Rasmussen, C.J. Smith, and F.L Macrina. 1987. Extrachromosomal systems and gene transmission in anaerobic bacteria. *Plasmid* **17**:87–109.

Park, B.H. and S.B. Levy. 1988. The cryptic tetracycline resistance determinant on Tn*4400* mediates tetracycline degradation as well as tetracycline efflux. *Antimicrob. Agents Chemother.* **32**:1797–1800.

Privitera, G., A. Dublanchet, and M. Sebald. 1979. Transfer of multiple antibiotic resistance between subspecies of *Bacteroides fragilis. J. Infect. Dis.* **139**:97–101.

Rasmussen, J.L., D.A. Odelson, and F.L. Macrina. 1986. Complete nucleotide sequence and transcription of ermF, a macrolide-lincosamide-streptogramin B resistance determinant *Bacteroides fragilis. J. Bacteriol.* **168**:523–533.

Rasmussen, J.L., D.A. Odelson, and F.L. Macrina. 1987. Complete nucleotide sequence of insertion element IS*4351* from *Bacteroides fragilis. J. Bacteriol.* **169**:3573–3580.

Robillard, N.J., F.P. Tally, and M.H. Malamy. 1985. Tn*4400*, a compound transposon isolated from *Bacteroides fragilis*, functions in *Escherichia coli. J. Bacteriol.* **164**:1248–1255.

Russo, T.A., J.S. Thompson, V.G. Godoy, and M.H. Malamy. 1990. Cloning and expression of the *Bacteroides fragilis* TAL2480 neuraminidase gene, nanH, in *Escherichia coli. J. Bacteriol.* **172**:2594–2600.

Salyers, A.A., N.B. Shoemaker, and E.P. Guthrie. 1987. Recent advances in *Bacteroides* genetics. *CRC Crit. Rev. Microbiol.* **14**:49–71.

Salyers, A.A., B.S. Speer, and N.B. Shoemaker. 1990. New perspectives in tetracycline resistance. *Mol. Microbiol.* **4**:151–156.

Scholle, R.R., H.E. Steffen, H.J.K. Goodman, and D.R. Woods. 1990. Expression and regulation of a *Bacteroides fragilis* sucrose utilization system cloned in *Escherichia coli. Appl. Environ. Microbiol.* **56**:1944–1948.

Shoemaker, N.B. and A.A. Salyers. 1987. Facilitated transfer of IncP beta R751 derivatives from the chromosome of *Bacteroides uniformis* to *Escherichia coli* by a conjugative *Bacteroides* tetracycline resistance element. *J. Bacteriol.* **169**:3160–3167.

Shoemaker, N.B. and A.A. Salyers. 1988. Tetacycline-dependent appearance of plasmid-like forms in *Bacteroides uniformis* 0061 mediated by conjugal *Bacteriodes* tetracycline resistance elements. *J. Bacteriol.* **170**:1651–1657.

Shoemaker, N.B. and A A. Salyers. 1990. A cryptic 65-kilobase-pair transposonlike element isolated from *Bacteroides uniformis* has homology with *Bacteroides* conjugal tetracycline resistance elements. *J. Bacteriol.* **172**:1694–1702.

Shoemaker, N.B., R.D. Barber, and A.A. Salyers. 1989. Cloning and characterization of a *Bacteroides* conjugal tetracycline-erythromycin resistance element by using a shuttle cosmid vector. *J. Bacteriol.* **171**: 1294–1302.

Shoemaker, N.B., C. Getty, E.P. Guthrie, and A.A. Salyers. 1986a. Regions in *Bacteroides* plasmids pBFTM10 and pB8-51 that allow *Escherichia coli-Bacteroides* shuttle vectors to be mobilized by plasmids and by a conjugative *Bacteroides* tetracycline element. *J. Bacteriol.* **166**:959–965.

Shoemaker, N.B., C. Getty, J.F. Gardner, and A.A. Salyers. 1986b. Tn4351 transposes in *Bacteroides* spp. and mediates the integration of plasmid R751 into the *Bacteroides* chromosome. *J. Bacteriol.* **165**:929–936.

Shoemaker, N.B., E.P. Guthrie, A.A. Salyers, and J.F. Gardner. 1985. Evidence that the clindamycin-erthromycin resistance gene of *Bacteroides* plasmid pBF4 is on a transposable element. *J. Bacteriol.* **162**:626–632.

Smith, C.J. 1985a. Characterization of *Bacteroides ovatus* plasmid pBI136 and of its clindamycin resistance region. *J. Bacteriol.* **161**:1069–1073.

Smith, C.J. 1985b. Polyethylene glycol-facilitated transformation of *Bacteroides fragilis* with plasmid DNA. *J. Bacteriol.* **164**:466–469.

Smith, C.J. 1985c. Development and use of cloning systems for *Bacteroides fragilis*: cloning of a plasmid-encoded clindamycin resistance determinant. *J. Bacteriol.* **164**:294–301.

Smith, C.J. 1987. Nucleotide sequence analysis of Tn4551: use of *ermFS* operon. *J. Bacteriol.* **169**:4589–4596.

Smith, C.J. 1989. Clindamycin resistance and the development of genetic systems in the *Bacteroides*. *Dev. Ind. Microbiol.* **30**:23–33.

Smith, C.J. and F.L. Macrina. 1984. Large transmissible clindamycin resistance plasmid in *Bacteroides ovatus*. *J. Bacteriol.* **158**:739–741.

Smith, C.J. and H. Spiegel. 1987. Transposition of Tn4551 in *Bacteroides fragilis*: identification and properties of a new transposon from *Bacteroides* spp. *J. Bacteriol.* **169**:3450–3457.

Smith, C.J., C.J. Parker, and M. Rogers. 1990. Plasmid transformation of *Bacteroides* spp. by electroporation. *Plasmid* **24**:100–109.

Smith, C.J., R.A. Welch, and F.L. Macrina. 1982. Two independent conjugal transfer systems operating in *Bacteroides fragilis* V479-1. *J. Bacteriol.* **151**:281–287.

Southern, J.A., J.R. Parker, and D.R. Woods. 1986. Expression and purification of glutamine synthetase cloned from *Bacteroides fragilis*. *J. Gen. Microbiol.* **132**:2827–2835.

Speer, B.S. and A.A. Salyers. 1988. Characterization of a novel tetracycline resistance that functions only in aerobically grown *Escherichia coli*. *J. Bacteriol.* **170**:1423–1429.

Speer, B.S. and A.A. Salyers. 1989. Novel aerobic tetracycline resistance gene that chemically modifies tetracycline. *J. Bacteriol.* **171**:148–153.

Speer, B.S. and A.A. Salyers. 1990. A tetracycline efflux gene on *Bacteroides* transposon Tn4400 does not contribute to tetracycline resistance. *J. Bacteriol.* **172**:292–298.

Stevens, A.M., N.B. Shoemaker, and A.A. Salyers. 1990. The region of a *Bacteroides* conjugal chromosomal tetracycline resistance element which is responsible for production of plasmidlike forms from unlinked chromosomal DNA might also be involved in transfer of the element. *J. Bacteriol.* **172**:4271–4279.

Tally, F., D. Snydman, S. Gorbach, and M. Malamy. 1979. Plasmid mediated, transferable resistance to clindamycin and erythromycin in *B. fragilis*. *J. Infect. Dis.* **139**:83–88.

Tally, F., M. Shimell, G. Carson, and M. Malamy. 1981. Chromosomal and plasmid-mediated transfer of clindamycin resistance in *Bacteroides fragilis*. In: *Molecular Biology, Pathogenicity and Ecology of Bacterial Plasmids*, S.B. Levy, and R.C. Clowes, eds., p. 51. New York: Plenum.

Tally, F., D. Snydman, M. Shimell, and M. Malamy. 1982. Characterization of pBFTM10, a clindamycin-erythromycin resistance transfer factor from *Bacteroides fragilis*. *J. Bacteriol.* **151**:686–689.

Thompson, A.M. and H.J. Flint. 1989. Electroporation-induced transformation of *Bacteroides ruminicola* and *Bacteroides uniformis* by plasmid DNA. *FEMS Microbiol. Lett.* **61**:101–104.

Thompson, J.S. and M.H. Malamy. 1990. Sequencing the gene for an imipenem-cefoxitin-hydrolyzing enzyme (CfiA) from *Bacteroides fragilis* TAL2480 reveals strong similarity between CfiA and *Bacillus cereus* β-lactamase II. *J. Bacteriol.* **172**:2584–2593.

Valentine, P.J., N.B. Shoemaker, and A.A. Salyers. 1988. Mobilization of *Bacteroides* plasmids by *Bacteroides* conjugal elements. *J. Bacteriol.* **170**:1319–1324.

Weisblum, B. 1985. Inducible resistance to macrolides, lincosamides and streptogramin type B antibiotics: the resistance phenotype, its biological diversity, and structural elements that regulate expression—a review. *J. Antimicrob. Agents Chemother.* **16**:63–90.

Welch, R.A. and F.L. Macrina. 1981. Physical characterization of *Bacteroides fragilis* R plasmid pBF4. *J. Bacteriol.* **145**:867–872.

Welch, R.A., K.R., Jones, and F.L. Macrina. 1979. Transferable lincosamide-macrolide resistance in *Bacteroides*. *Plasmid* **2**:261–268.

36

Transfer of Beta-Lactam Antibiotic Resistance in *Bacteroides*

Kunitomo Watanabe and Kazue Ueno

36.1   Introduction

It is well known that *Bacteroides fragilis* and related species produce several kinds of beta-lactamases such as oxyiminocephalosporinase, penicillinase, and metallo-beta-lactamase, with some strains showing a particularly high resistance to beta-lactam antibiotics. Until now, there have only been scattered reports about the transfer of beta-lactam resistance among *Bacteroides*. Beta-lactam resistance in *Bacteroides* seemed to be mobilized by a group of tetracycline resistance transfer elements as discussed in Chapter 35. However, we have found that *B. fragilis* could transfer beta-lactam resistances in the absence of such transfer elements.

36.2   Transfer of High-Level Ampicillin Resistance

Transfer of beta-lactam antibiotic resistance among *Bacteroides fragilis* and related species was first reported by Butler et al. (1980). They demonstrated that a highly ampicillin-resistant strain of *B. fragilis* TMP-14 could transfer its resistance determinant through a tetracycline resistance transfer element from *B. fragilis* TM 2300, or a transfer factor, plasmid pBFTM10 from *B. fragilis* TMP10. The ampicillin-resistant transconjugants acquired a new beta-lactamase as revealed by the substrate profile. The genetic location of this determinant, however, was not plasmid borne, although two plasmid species were shown in the donor *B. fragilis* TMP-14.

The transfer occurred by a filter-mating technique but not with broth-mating technique. In 1982, Sato et al. reported that *B. fragilis* GN11499 could simultaneously transfer both ampicillin and tetracycline resistances to *B. fragilis* GN11497 and *Bacteroides vulgatus* GN11496 with a frequency of 1×10^6/input donor cells. The donor strain was resistant to 800 μg/ml of benzylpenicillin and 3200 μg/ml of ampicillin and was able to express a new type of penicillinase. The ampicillin-resistant transconjugants newly acquired a penicillinase, but there is no available information on the plasmid content of the donor strain and the genetic location of this resistance determinant.

In 1990, Yamaoka et al. reported that an ampicillin-resistant *B. fragilis* GAI-10150 was able to transfer its resistance to a sensitive strain of *B. fragilis* JC-101 (Table. 36.1). *B. fragilis* GAI-10150 was resistant to 1600 μg/ml of benzylpenicillin and ampicillin, and to 400 μg/ml of cefazolin and latamoxef. Although this strain was also resistant to tetracycline, the ampicillin resistance transfer was not associated with transfer of tetracycline resistance. A tetracycline-sensitive and ampicillin-resistant transconjugant could also transfer the ampicillin resistance in a secondary mating.

It was further recognized that a plasmid of about 40-kb called pBFKW1 was always associated with the transfer of ampicillin resistance (Figure 36.1). Thus, the transconjugants acquired the ability of hydrolyzing ampicillin. The beta-lactamases from the donor and the transconjugant were compared for their sub-

Table 36.1 Transfer of ampicillin-resistance from *Bacteroides fragilis* donor cells to *B. fragilis* recipient cells by filter mating

Mating	Donor	Recipient[a]	Selective medium[b]	Transconjugants Frequency	Transconjugants Phenotype
1	GAI-10150	JC-101	Rif, Apc	1.2×10^{-8}	Rif[r], Apc[r], His[−], Arg[−]
2	GAI-10150	JC-101	Rif, Tc	$<10^{-10}$	
3	A-2[c]	TM-4000	Apc	4.0×10^{-7}	Rif[r], Apc[r], His[+], Arg[+]

[a]Strain JC-101 is His[−]Arg[−] mutant of *B. fragilis* strain TM-4000, resistant to rifampicin.
[b]Basal selective medium: GAM medium in matings 1 and 2; synthetic medium (Varel and Bryant, 1974) in mating 3. Selective media contained rifampicin (Rif, 50 μg/ml) plus approprite antibiotics at following concentrations: ampicillin (Apc, 200 μg/ml); tetracycline (Tc, 10 μg/ml).
[c]Strain A-2 is transconjugant from mating 1.

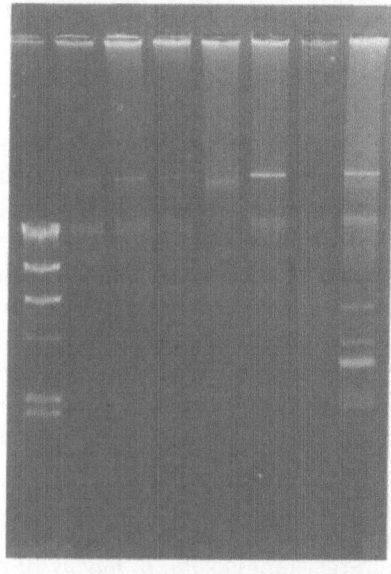

FIGURE 36.1 Plasmid content of *Bacteroides fragilis* GAI-10150, JC-101, TM-4000, and transconjugants A-2, A-2-1-3. Lane 8: GAI-10150; lane 7: JC-101 (recipient); lane 6: A-2 (transconjugant from GAI-10150 × JC-101 matings); lane 5: TM-4000 (recipient); lanes 2 to 4: A-2-1 to 3 (transconjugants from A-2 × TM-4000 matings); lane 1: *Hind*III fragments of λ DNA. (Modified from Yamaoka et al., 1990.)

Table 36.2 Beta-lactam antibiotic resistance levels of *Bacteroides fragilis* strains used in matings shown in Table 36.1

Antibiotics	Strains GAI-10150	JC-101	TM-4000	A-2
Ampicillin	1600[a]	12.5	6.25	800
Cloxacillin	3200	25	12.5	3200
Piperacillin	400	3.13	1.56	200
Cefazolin	400	12.5	6.25	200
Cefoxitin	25	3.13	1.56	12.5
Latamoxef	400	<0.39	<0.39	200

[a]MIC values expressed in μg/ml.

strate profiles and both were shown to be an oxyimino-cephalosporinase type. The minimal inhibitory concentrations (MICs) of the strains are given in Table 36.2.

In 1986, Morita et al. studied a plasmid of about 2.62-kb, called pBFSK1, that originated from *B. fragilis* strain KHMO27. The strain was isolated from human feces and was resistant to 100 μg/ml of benzylpenicillin, ampicillin, piperacillin, and cephaloridine and to 50 μg/ml of cefoperazone and cefotaxime; it was found to contain several plasmid species. The authors transformed *E. coli* HB101 with purified plasmid DNA from *B. fragilis* KHMO27 and obtained transformants showing a resistance to 50 μg/ml of ampicillin. The plasmid pBFSK1 was isolated from one such transformant. The authors characterized an oxyiminocephalosporinase (29 kDa) in *E. coli* minicells transformed by plasmid pBFSK1. This plasmid is probably not self-transferable to *B. fragilis* by reason of its small size. Whether this plasmid is cotransferable to *B. fragilis* and capable of expressing its resistance in this species was not investigated further.

36.3 Transfer of Benzylpenicillin Resistance between *Bacteroides* and *Prevotella*

In 1990, Guiney and Bouic reported that a strain of *Prevotella denticola* and a strain of *Prevotella intermedia* that were resistant to both benzylpenicillin and tetracycline could transfer both resistances to a sensitive strain of *Prevotella buccae* or a sensitive strain of *B. fragilis*. The *P. denticola* strain was plasmid free, and the *P. intermedia* strain contained two plasmid species that were presumably not associated with the resistance transfer.

36.4 Transfer of Cefoxitin Resistance

In 1986, Cuchural et al. (1986b) demonstrated the capacity to transfer a cefoxitin resistance genetic determinant from *B. fragilis* TAL-4170; the product inactivates cefoxitin. The resistance determinant was reported to be transferred by self-mobilization; this strain in itself contained no plasmid. Unlike ampicillin resistance transfer, this resistance occurred without the introduction of a tetracycline resistance transfer element from *B. fragilis* TM2300 if the donor strain was induced by growth in subinhibitory concentrations of tetracycline. The transfer frequency, however, was increased by more than 10 fold after the introduction of the tetracycline resistance transfer element from *B. fragilis* TM2300. The cefoxitin resistant transconjugants obtained from this mating were shown to acquire a new beta-lactamase with an isoelectric focusing point of 8.1.

In our laboratory, Bunai et al. (1987) also confirmed the capacity to transfer cefoxitin resistance among *B. fragilis* strains. *B. fragilis* GAI-7955 and *B. uniformis* Tm-46, which were able to inactivate cefoxitin when tested by bioassay, transferred their resistance to *B. fragilis* JC-101. These strains contained some plasmids, but the transconjugants had not acquired any CCC plasmid band from these donors. The transfer

of cefoxitin resistance in these two strains occurred independently of the transfer of tetracycline resistance.

36.5 Transfer of Carbapenem Resistance

Several authors have reported that some strains of *B. fragilis* and related species are resistant to imipenem, a carbapenem derivative. In Japan, 1 to 2% of *B. fragilis* strains isolated from clinical specimens were shown to be resistant to 25 μg/ml of imipenem. Cuchural et al. (1986a) have already reported that two isolates of imipenem-resistant *B. fragilis*, TAL-2480 and 3636, could produce a novel metallo-beta-lactamase (MW-44,000). In our laboratory, Bandoh et al. (1991) also confirmed that two Japanese isolates, *B. fragilis* GAI-30144 and GAI-30079, were resistant to 50 μg/ml of imipenem, oxyiminocephalosporins, cephamycins, and penicillins. These workers produced a metallo-beta-lactamase (MW-33,000). Attempts at transfer of imipenem resistance in both laboratories were unsuccessful.

In 1990, Thompson and Malamy succeeded in cloning and sequencing a *B. fragilis* gene (*cfi* A) encoding for an imipenem-hydrolyzing enzyme (MW-25,249). The gene seemed to be chromosomal in the resistant strain, *B. fragilis* TAL 2480. We found more recently that a new imipenem-resistant isolate, *B. fragilis* 10-73, can transfer its resistance to a *B. fragilis* sensitive strain by a conjugation-like process (unpublished observations).

36.6 Conclusion

Genetic determinants of *Bacteroides* encoding beta-lactam antibiotic resistances can be mobilized by tetracycline resistance transfer elements. But it seems that plasmid-mediated transfer of beta-lactam resistance can also occur independently of these elements. It was recognized in our Institute that a Japanese strain of *B. fragilis* that produces a carbapenem-hydrolyzing metallo-beta-lactamase could transfer its resistance genetic determinant by a filter-mating technique in the absence of the tetracycline resistance transfer element.

References

Bandoh, K., Y. Muto, K. Watanabe, K. Naoki, and K. Ueno. 1991. Biochemical properties and purification of metallo-beta-lactamase from *Bacteroides fragilis*. *Antimicrob. Agents Chemother.* **35**:371–372.

Bunai, M., K. Watanabe, Y. Hioki, M. Miyauchi, and K. Ueno. 1987. Transferable resistance to cefoxitin in the *Bacteroides fragilis* group. *Acta Sch. Med. Univ. Gifu* **35**:414–425.

Butler, T., F.P. Tally, S.L. Gorbach, and M.H. Malamy. 1980. Transferable ampicillin resistance in *Bacteroides fragilis*. *Clin. Res.* **28**:365 (abstr.).

Cuchural, G.J., Jr., M.H. Malamy, and F.P. Tally. 1986a. Beta-lactamase mediated imipenem resistance in *Bacteroides fragilis*. *Antimicrob. Agents Chemother.* **30**:645–648.

Cuchural, G.J., Jr., F.P. Tally, J.R. Storey, and M.H. Malamy. 1986b. Transfer of beta-lactamase-associated cefoxitin resistance in *Bacteroides fragilis*. *Antimicrob. Agents Chemother.* **29**:918–920.

Guiney, D.G. and K. Bouic. 1990. Detection of conjugal transfer systems in oral, black-pigmented *Bacteroides* spp. *J. Bacteriol.* **172**:495–497.

Morita, K., A. Miyano, M. Hayashi, and K. Shiga. 1986. Study on molecular genetics of beta-lactam antibiotics resistance plasmid pBFSK1 from *Bacteroides fragilis*. *J. Kyorin Med. Soc.* **17**:531–537.

Sato, K., Y. Matsuura, M. Inoue, and S. Mitsuhashi. 1982. Properties of a new penicillinase type produced by *Bacteroides fragilis*. *Antimicrob. Agents Chemother.* **22**:579–584.

Thompson, J.S. and M.H. Malamy. 1990. Sequencing the gene for an imipenem-cefoxitin-hydrolyzing enzyme (*cfi* A) from *Bacteroides fragilis* TAL-2480 reveals strong similarity between *cfi* A and *Bacillus cereus* beta-lactamase II. *J. Bacteriol.* **172**:2584–2593.

Varel, V.H. and M.P. Bryant. 1974. Nutritional features of *Bacteroides fragilis*. *Appl. Microbiol.* **28**:251–257.

Yamaoka, K., K. Watanabe, Y. Muto, N. Katoh, K. Ueno, and F.P. Tally. 1990. R-plasmid mediated transfer of beta-lactam resistance in *Bacteroides fragilis*. *J. Antibiot.* (Tokyo) **43**:1302–1306.

37

Genetics of 5-Nitroimidazole Resistance in *Bacteroides*

Gilles Reysset, Wen-Jin Su, and Madeleine Sebald

37.1 Introduction

Nitroimidazoles are imidazoles that contain a nitro group at position 2, 4, or 5 of the imidazole ring. These compounds may have various substitutions on the nitrogen at position 1 of the imidazole ring. The 5-nitroimidazoles are synthetic compounds that appear to be very potent anaerobicidal agents. Three derivatives are commonly used in clinical therapy: metronidazole [1-(2-hydroxyethyl)-2-methyl-5-nitroimidazole], ornidazole [1-(3-chloro-2-hydroxypropyl) 2-methyl-5-nitroimidazole], and tinidazole (ethyl [2-(2-methyl-5-nitro-1-imidazolyl)-ethyl] sulfone). They are used extensively for treatment of infections from anaerobic eucaryotes (*Entamoeba histolytica*, *Trichomonas vaginalis*, *Giardia*) and anaerobic procaryotes, having a bactericidal effect against all anaerobic bacteria except the genera *Bifidobacterium*, *Propionibacterium*, and *Actinomyces*. They are particularly useful in prophylaxis and therapy of infections caused by *Bacteroides* because these bacteria commonly are resistant to many antibiotics, the resistance often being transferable (as reviewed in Chapters 35, 36, and 40, this volume).

This review includes some general features on the activity of nitroimidazoles and outlines current data on the epidemiology of 5-nitroimidazole (Ni) resistance in *Bacteroides* and its relevant genetics (Breuil et al., 1989; Sebald et al., 1990; G. Reysset et al., in manuscript). The data concern mostly transferable, plasmid-borne, moderate Ni resistance and some pre-liminary results from apparently plasmid-free strains of a similar phenotype.

37.2 General Data on the Activity of the 5-Nitroimidazoles

According to current views, the 5-nitroimidazole molecule ($R-NO_2$) is inactive per se but becomes bactericidal after reductive activation leading to intermediates and active derivative(s); these have not yet been clearly identified, probably because they are highly reactive and unstable (McLafferty et al., 1982; Yeung et al., 1984; Müller, 1986). The possibility that the activity may be attributed to different reduced derivatives of the drug for different microorganisms cannot be excluded (Müller, 1986).

The first of these intermediates is a derivative whose nitro group has been univalently reduced to a nitro-free radical. As a consequence, the presence of molecular oxygen during in vitro susceptibility testing results, first, in the reoxidation of the nitro-free radical into the original form, with the concomitant reduction of molecular oxygen into superoxide anion ("futile metabolic cycle"; Perez-Reyes et al., 1980):

$$R - NO_2^- + O_2 \rightarrow R - NO_2 + O_2^-$$

Second, molecular oxygen can compete for the electrons needed for further reduction steps of the nitro-free radical. Therefore, to be fully active these compounds must be used in a complete absence of oxygen; this is particularly true for anaerobes with a superoxide dismutase activity, such as *Bacteroides*. Incomplete anaerobiosis could lead to "false resistance" (Milne et al., 1978).

The pyruvate oxidation was clearly shown to be involved in metronidazole reduction in the butyric clostridia (O'Brien and Morris, 1972; Edwards et al., 1973) and in *Trichomonas* (Müller, 1986). A pyruvate:ferredoxin oxydoreductase activity was found in most anaerobic procaryotes that are sensitive to 5-nitroimidazoles, including the *Bacteroides* (Narikawa, 1986).

The ultimate target of the activated derivatives is DNA. It has been established in vivo that the activated drug inhibits DNA synthesis as well as mediating the degradation of the existing DNA (Plant and Edwards, 1976; Knight et al., 1978). In vitro, it destabilizes the DNA helix and causes strand breakage (Knight et al., 1978, and references cited therein). The reductively activated drug binds specifically to guanine and cytosine (LaRusso et al., 1977).

37.3 The 5-Nitroimidazole Resistance in *Bacteroides*

Despite the widespread use of 5-nitroimidazoles, resistant anaerobes have been reported infrequently. A *B. fragilis* strain with a high level of resistance to metronidazole [minimal inhibitory concentration (MIC), 64 µg/ml] was first described in England (Ingham et al., 1978) and additional strains have been found in clinical specimens (Rotimi et al., 1979; Eme et al., 1983; Sprott et al., 1983; Tabaqchali et al. 1983; Dublanchet et al., 1986; Lamothe et al., 1986; Scher, 1988; Sprott and Kearns, 1988). Nevertheless, they appeared to be very rare, and may have a low potential for epidemiological dissemination because the 5-nitroimidazole resistance of the strain described by Ingham et al. (1978) was not transferable (Tally and Malamy, 1982; Breuil et al., 1989), thus excluding horizontal transmission.

Tally and Malamy (1982) showed that in the Ingham strain there were both reduced uptake of the drug and decreased nitroreductase activity. A similar mechanism was found in a N-methyl-N'-nitro-N-nitrosoguanidine (MNNG)

mutant selected for metronidazole resistance (Tally and Malamy, 1982). This is in contrast to the data of Britz and Wilkinson (1979), who found instead a reduced pyruvate dehydrogenase activity in metronidazole-resistant mutants of *B. fragilis*, but the identification of the original strain remains questionable (Tabaqchali et al., 1983). No reduced pyruvate dehydrogenase activity was found in the Ingham strain (Tally and Malamy, 1982; Tabaqchali et al., 1983) or in two other *B. fragilis* strains with a reduced sensitivity to metronidazole (Tabaqchali et al., 1983).

Strains of *Bacteroides* with a reduced sensitivity to 5-nitroimidazoles have occurred in France; these were characterized by higher levels of resistance to tinidazole (Ti) and ornidazole (Or) (MICs, 8–16 µg/ml) than to metronidazole (MICs, 2–4 µg/ml) (Dublanchet et al., 1986). The resistance was found to be transferable (Breuil et al., 1989). The MICs, although close to the admitted breakpoint of metronidazole (4 µg/ml) (Finegold, and the National Committee, 1988), were significantly different from those of the sensitive strains (Table 37.1). The strains were isolated and cultivated under strict anaerobic conditions. No interpretation of these findings appeared feasible unless assessed by a genetic approach.

37.4 Genetics of Plasmid-Mediated 5-Nitroimidazole Resistance

Transferability of the 5-nitroimidazole resistance

Our experimental approach consisted first in conjugative transfer between Ni-resistant donor strains and a Ni-sensitive *Bacteroides fragilis* recipient, strain 638R, that was plasmid free, resistant to rifampicin (Rf), and has been repeatedly shown to be a good recipient for several antibiotic resistances, e.g., tetracycline (Tc), clindamycin (Cc), erythromycin, cefoxitin, and ampicillin (Privitera et al., 1979a, 1979b; Rashtchian et al., 1982; Cuchural et al., 1986; Yamaoka et al., 1990).

Two Ni-resistant (Ni^r) strains were studied, *Bacteroides vulgatus* BV17 and *Bacteroides the-*

Table 37.1 Susceptibility levels to 5-nitroimidazoles of relevant clinical isolates of species of *Bacteroides*

Strains	MIC ranges (μg/ml)[a]			Reference
	Metronidazole	Ornidazole	Tinidazole	
B. fragilis NCTC 11295	32–64	16–32	16–32	Ingham et al., 1978
B. fragilis BF3, Tc^{r}[b]	8–16	16–32	16–32	Sebald et al., 1990
B. fragilis BF8, Tc^{r} Cc^{r}[b]	8–16	8–16	8–16	Sebald et al., 1990
B. thetaiotaomicron BT13, Tc^{r}	4–8	8–16	8–16	Sebald et al., 1990
B. vulgatus BV17, Cc^{r} Tc^{r}	2–4	4–8	4–8	Breuil et al., 1989
B. fragilis BF2, Im^{r} Tc^{r} Cc^{r}	0.25–0.5	0.5–1	0.5–1	Reysset et al., unpublished
B. fragilis 638R, Rf^{r}[b]	0.5–1	0.5–1	0.25–1	Breuil et al., 1989
B. vulgatus BV1R, Rf^{r}	0.5–1	0.5–1	0.5–1	Sebald et al., 1990

[a] Minimal inhibitory concentrations (MIC) were performed on Wilkins–Chalgren medium (Oxoid) by agar dilution method and evaluated after 48-h incubation in anaerobic chamber in 5% hydrogen, 5% CO_2, and 90% nitrogen. Ranges indicated from five independent experiments.

[b] Other relevant antibiotic resistances of strains are indicated: Cc, clindamycin (MIC $\geq$ 128 μg/ml); Tc, tetracycline (MIC $\geq$ 8 μg/ml); Rf, rifampicin (MIC $\geq$ 128 μg/ml); Im, imipenem (MIC > 32 μg/ml).

taiotaomicron BT13. When strain BV17 or BT13 was mated on plates with strain 638R and selected on medium containing tinidazole (4 μg/ml) and rifampicin (25 μg/ml), Ni^{r} Rf^{r} transconjugants were obtained with a frequency of 10^{-3} to 10^{-5} per donor (transconjugants per donor at the end of the mating), respectively. The MICs for 5-nitroimidazoles of the transconjugants were identical to or higher than those of the donor, and the resistance was stably maintained in the absence of selective pressure. The Ni^{r} transconjugants had phenotypic characters identical to those of the recipient. Because strain BV17 was also Tc^{r} Cc^{r}, and BT13 was Tc^{r}, the matings were also plated on media containing rifampicin (25 μg/ml) and tetracycline (4 μg/ml) (matings BV17 × 638R and BT13 × 638R), or rifampicin and clindamycin (10 μg/ml) (mating BV17 × 638R). The Tc and Cc resistances were not transferable, even when the donor strains were precultivated in a subinhibitory concentration of tetracycline (Privitera et al., 1979b). We therefore concluded that Ni^{r} transfer occurred independently of other antibiotic resistance transfer.

Plasmid contents of strains BV17 and BT13

Donors and transconjugants then were screened for the presence of plasmids by the alkaline lysis procedure (Birnboim and Doly, 1979), and DNAs were analyzed by elec-

trophoresis on 0.7% agarose gels. The strain BV17 was shown to harbor several major covalently closed circular (CCC) plasmid bands of 56, 7.7, 5, and 4.5-kb, with additional bands of about 15 and 9-kb, corresponding to open circular (OC) forms or dimers of the 7.7- and 4.5-kb plasmids, respectively. The Ti^{r} transconjugants obtained after mating BV17 × 638R had various plasmid patterns, but regularly harbored plasmids of 7.7 and 56-kb to which we had referred as pIP417 and pIP418, respectively (Breuil et al., 1989). Thus both plasmids were candidates to carry the gene(s) involved in Ni resistance. The strain BT13 was shown to harbor CCC plasmid bands of 10 and 4.1-kb, a plasmid band of 20-kb corresponding to the OC form or dimer of the 10-kb plasmid, and a plasmid band of about 56-kb, which was irregularly found. The Ti^{r} transconjugants from matings BT13 × 638R had the same plasmid pattern but the 4.1-kb plasmid was sometimes lacking and thus did not play a role in the resistance. Plasmids of 10-kb (pIP419) and, less likely, of 56-kb (pIP420) were candidates to carry the gene(s) involved in the Ni resistance of BT13 (Sebald et al., 1990).

Transformation of plasmids pIP417 and pIP419

The strain 638R was shown to be both chemically transformable (Smith, 1985b) and electrotransformable (Sebald et al., 1990). Experiments were performed initially by the polyethyleneglycol- (PEG-) mediated transformation procedure (Breuil et al., 1989), and then by electrotransformation (Sebald et al., 1990; Reysset et al., unpublished data), using for electroporation a Gene Pulser™ apparatus. In this instance, under the conditions described the number of transformants was a linear function of the DNA concentration over at least four orders of magnitude (0.1–500 ng DNA) with an efficiency of 1×10^7 transformants per μg DNA when pBI191 (Smith, 1985a) or pIP417 plasmid DNA originated from an isogenic strain.

In the case of plasmid DNA that originated from a 638R transconjugant containing pIP417 and pIP418 plasmids, transformants selected for tinidazole resistance were obtained at 10^6 to 10^7 transformants per μg DNA. The transformants harbored only plasmid pIP417, which then was capable of transforming strain 638R. This result suggested clearly the role of pIP417 plasmid in Ni resistance. In a similar experiment performed with a plasmid DNA preparation from strain BV17, no transformant was obtained, which might be caused by host-controlled restriction of heterologous DNA.

Similarly, Ni[r] transformants have been obtained with plasmid pIP419 when isolated from a 638R background, but not from BT13. This again might result from a 638R restriction of heterogenic DNA. Moreover, even from a 638R background the Ni[r] transformation frequencies were 1,000 to 10,000 times lower with pIP419 than with plasmid pIP417. This low transformation efficiency could be related to a less efficient expression of Ni resistance from plasmid pIP419 or to some plasmid instability immediately after transformation.

After transformation into *B. fragilis* 638R, both pIP417 and pIP419 were stably maintained, and the level of expression of Ni resistance was similar to that of the original strain. The restriction endonuclease site maps of both plasmids are given in Figures 37.1 and 37.2.

Ni[r] gene cloning

The next step consisted of cloning the Ni[r] gene. Because *Escherichia coli* has a high level of resistance to 5-nitroimidazole, except when recom-

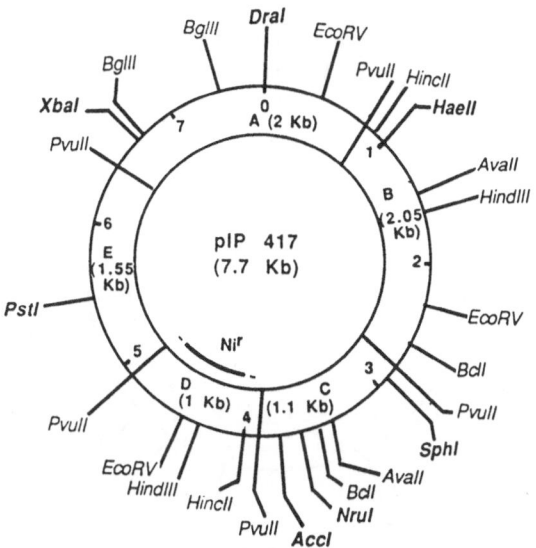

FIGURE 37.1 Physical map of plasmid pIP417. Unique site *DraI* was arbitrarily designated as 0. Plasmid pIP417 has also seven sites for *DdeI*, *MaeII*, and *MspI*, and multiple sites for *AluI*, *HaeIII*, *HhaI*, *HinfI*, *HpaII*, *MboI*, *SspI*, and *TaqI* that are not localized. No site was detected for *AvaI*, *BamHI*, *ClaI*, *EcoRI*, *HpaI*, *KpnI*, *MluI*, *NcoI*, *NdeI*, *PvuI*, *SacI*, *SalI*, *ScaI*, *SmaI*, *SpeI*, *SstII*, *StuI*, and *XhoI*. Capital letters in inner circle refer to *PvuII*-generated fragments. Fragment coding for 5-nitroimidazole resistance (Ni[r]) is shown as thick black lines. Unique restriction sites are shown in bold type.

bination deficient (Yeung et al., 1984; Hof et al., 1987), and because antibiotic resistance genes from *Bacteroides* are generally not expressed in *E. coli* (Odelson et al., 1987), the strategy decided on was to use a *Bacteroides* cloning vector, pBI191, constructed by Smith (1985a). This plasmid contains the *Bacteroides* replicon pBI143, a *B. ovatus* cryptic plasmid, to which was ligated the Cc[r] determinant from plasmid pBF4, plus a fragment of pUC19 that includes mainly a MCS (multiple cloning sequence).

We first constructed an hybrid plasmid pIP417-pBI191 to verify the feasibility of Ni[r] gene cloning. Plasmid pBI191 was linearized by *SphI* (within the MCS) and ligated with pIP417 previously linearized by the same enzyme. After electrotransformation into strain 638R, Cc[r] Ti[r] transformants were obtained and screened by 0.7% agarose gel electrophoresis. The restriction map of one of the recombinant plas-

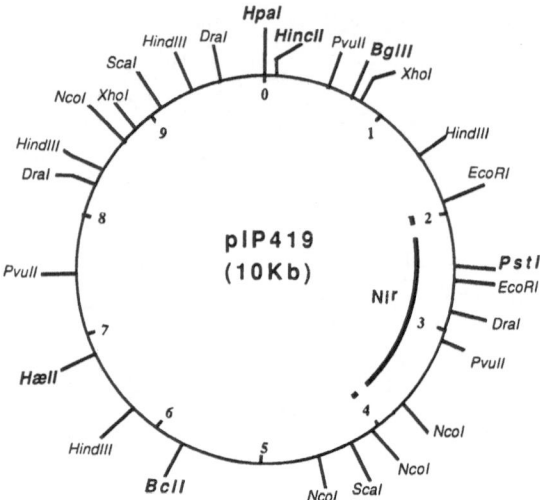

FIGURE 37.2 Physical map of plasmid pIP419. Unique site HpaI was arbitrarily designated as 0. Multiple sites, mostly not localized, were found for AccI, AvaII, DdeI, EcoRV, HinfI, MobI (Sau3A), MspI, SspI, and TaqI. Not site was detected for BamHI, BgII, BstEII, ClaI, DpnI, KpnI, MluI, NdeI, NruI, PvuI, SacI, SalI, SmaI, SpeI, SphI, or SstI. (See also the legend for Figure 37.1.)

mids, pFK700 (13-kb), is given in Figure 37.3. The pFK700 plasmid was stably maintained after 30 generations in the absence of selective pressure, and the levels of resistance expressed were identical to those expressed in strain 638R by the original plasmids. A 3.9-kb XbaI-XbaI deletion of pFK700 was constructed from this plasmid (Figure 37.3). The resultant plasmid was 9.1-kb in size, stably expressed both resistances, and may be reintroduced by transformation into strain 638R. We shall refer to this plasmid later as pFK702.

For the Ni^r gene cloning of pIP417, the vector was linearized by restriction enzyme SmaI (within the MCS), dephosphorylated, and ligated to plasmid pIP417 previously cleaved by restriction enzyme PvuII, then electrotransformed into strain 638R with selection on Wilkins–Chalgren medium containing tinidazole (4 μg/ml). The transformants were reisolated twice on selective medium; their plasmid content was screened by the alkaline lysis procedure and then electrophoresed on 0.7%

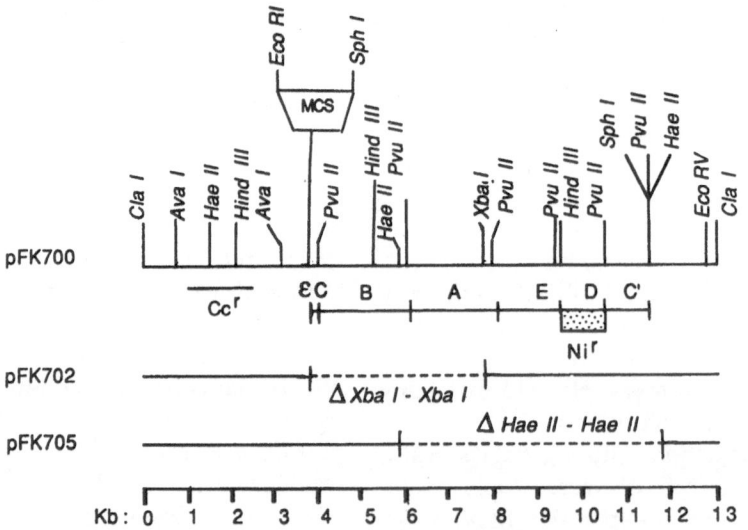

FIGURE 37.3 Maps of recombinant plasmids pFK700, pFK702, and pFK705. Plasmid pFK700 was obtained by ligation of plasmids pIP417 and pBI191, linearized by restriction enzymes SphI then transformed into 638R and selected for Ni^r Cc^r. Plasmid pFK702 was constructed from pFK700 by XbaI hydrolysis, autoligation, and transformation into B. fragilis 638R with selection for Ni^r. Note that in pFK700 fragment C' is fragment C from Figure 37.1 minus fragment PvuII-

SphI ~50 bp. Fragment ε C extending from SphI (coordinated at 3.85-kb) to PvuII (coordinated at 5.95-kb) is present in pFK700 recombinant plasmid only. Plasmid pFK705 was constructed from pFK700 by HaeII hydrolysis, autoligation, transformation into 638R, and selection for Cc^r Ti^s. Note that the insert present in construction is fragment B', corresponding to fragment B deleted from fragment HaeII-PvuII of ~150 bp, this plasmid also contains fragment ε C, as in pFK700.

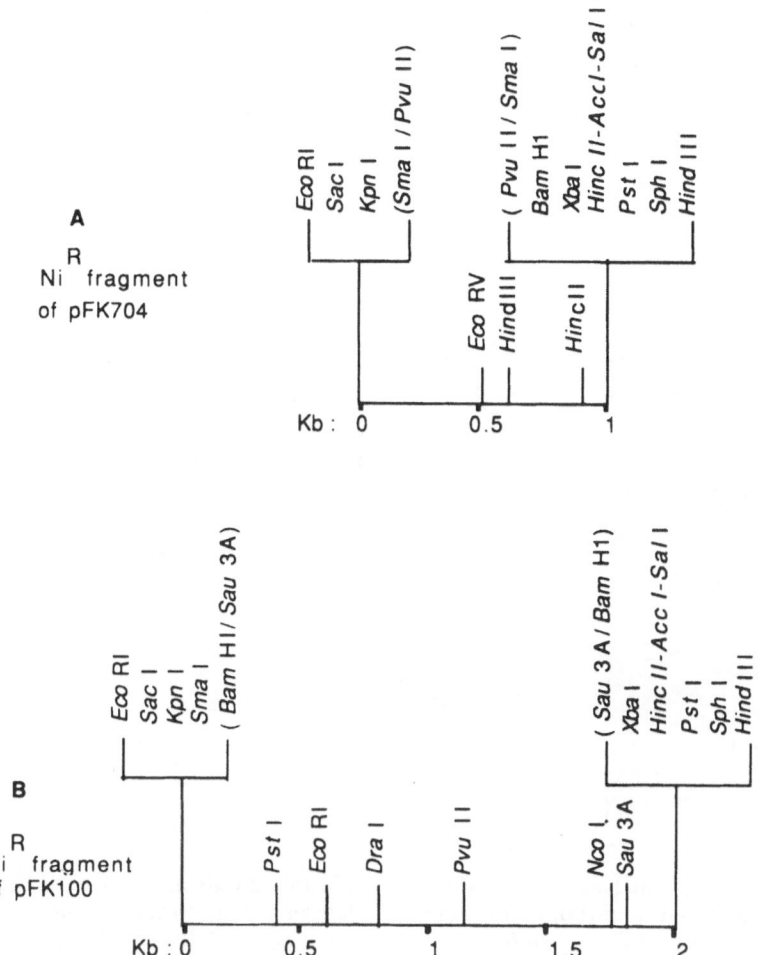

FIGURE 37.4 Restriction endonuclease maps of Ni^r inserts from plasmids pFK704 and pFK700. Ni^r from pFK704 was derived from plasmid pIP417 (**A**); Ni^r insert from pFK100 was from plasmid pIP419 (**B**).

agarose gels. A clone containing a plasmid (pFK704) of 6.3-kb was further studied; its insert of 1-kb corresponded to the *Pvu*II-*Pvu*II fragment D of pIP417 (Figures 37.1 and 37.4A)

By a similar approach, we cloned a *Sau*3A fragment of pIP419 at the *Bam*HI site (within the MCS) of pBI191. The recombinant plasmid, pFK100, was found to be 7.3-kb in size and to contain an insert of 2-kb. The *Sau*3A sites of pIP419 were not yet localized because pIP419 contains many *Sau*3A sites and only fragments whose total length was 8.2 kb were easy to detect by electrophoresis on 1% or 0.7% agarose gel. These fragments had the following sizes: 1.90, 1.48, 1.37, 1.30, 0.83, 0.81, and 0.56-kb.

The map of the insert is given in Figure 37.4B Its restriction sites, including a *Pst*I site that is unique in the pIP419 plasmid, allowed us to localize the insert on the pIP419 map as shown in Figure 37.2.

Those inserts must be subcloned and sequenced, but two categories of experimental data showed that the two Ni^r determinants are different. First, their restriction maps are different. Second, they are different enough not to cross-hybridize under stringent conditions by Southern analysis. We hybridized *Pvu*II-generated fragments of plasmid pIP417 and *Sau*3A-generated fragments of plasmid pIP419 to the ^{35}S-dATP-labeled probes pFK704 and

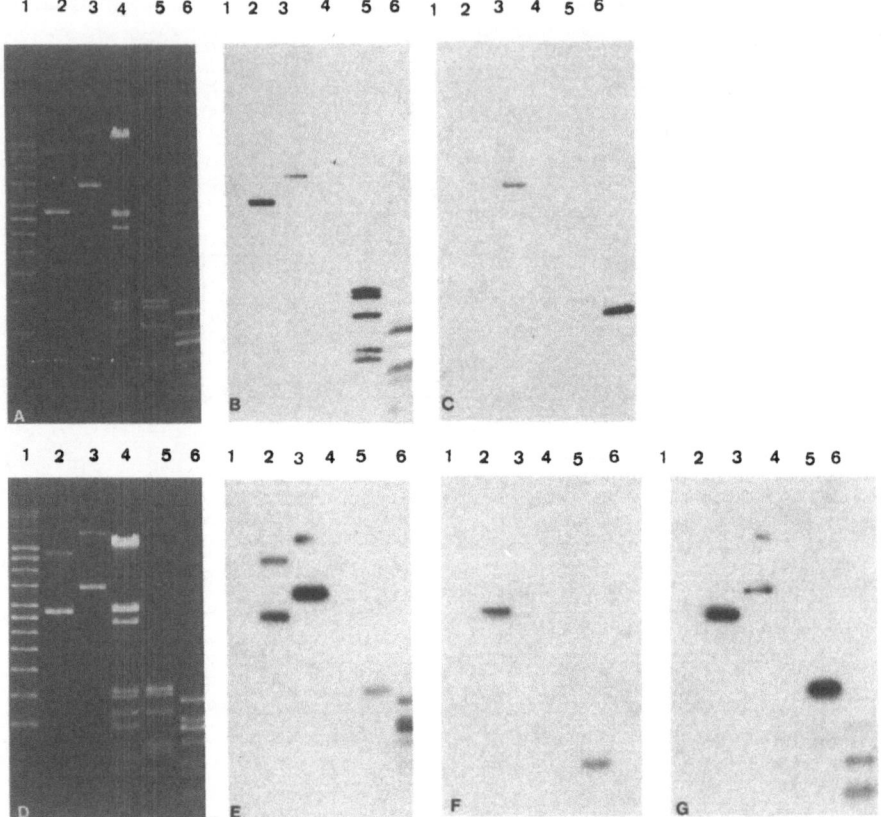

FIGURE 37.5 Southern analysis of pIP417 and pIP419 plasmid DNA. (A) and (D), identical DNA preparations run horizontally in different 0.7% agarose gels in TBE for 5 h at 50 V; (B) and (C), Southern analysis of gel (A); (E) to (G), Southern analysis of gel (D); CCC ladder (lane 1) and marker III (Boehringer Mannheim, Germany) (lane 4) are included as molecular weight standards. Preparations are plasmid pIP417 (lane 2) and same after PvuII hydrolysis (lane 5); plasmid pIP419 (lane 3) and same after Sau3A hydrolysis (lane 6). Plasmid DNAs are from transformant strains 638R (pIP417) and 638R (pIP419). Note that plasmids pIP417 and pIP419 contain two forms, CCC and OC. Sizes of plasmid bands are evaluated by reference to CCC ladder values (from bottom to top: 2.0; 2.9; 3.9; 5.0; 6.0; 7.0; 8.0; 10.1; 12.1; 14.1; and 16.2 kb) and for linear DNA to marker III (from bottom to top: 0.56; 0.83; 0.94; 1.37; 1.59; 1.90; 1.98; 3.48; 4.27; 5.00–5.15; and 21.7 kb). PvuII fragments of pIP417 (from top to bottom): B (2.05 kb), A (2 kb), E (1.55 kb), C (1.1 kb), and D (1 kb), as defined in Figure 37.1. Sau3A fragments of pIP419 (from top to bottom): 1.90; 1.48; 1.37; 1.30; 0.83; 0.81; and 0.56 kb in size. Following probes were used: (B) pIP417; (C) pFK100; (E) pIP419; (F) pFK704; and (G) pFK705.

pFK100. This was a feasible approach by virtue of the lack of homology between plasmids pBI191 and either pIP417 or pIP419 with the experimental conditions used. The results obtained showed that the pFK704 probe cross-reacted with the fragment D of pIP417 only (Figure 37.5F); similarly, the pFK100 probe cross-reacted only with a Sau3A fragment of 1.9-kb from pIP419 (Figure 37.5C). The cross-reaction found by Southern analysis between plasmids

pIP417 and pIP419 (Figures 37.5B and 37.5E) cannot be attributed to the Ni^r determinants.

Localization of the region of homology between plasmids pIP417 and pIP419

In spite of the lack of homology between the Ni^r of both plasmids, those total plasmid DNAs cross-reacted when tested by Southern analysis. Therefore, we studied the homology between both plasmids by these means, using as DNAs the complete plasmids, the *Pvu*II-generated fragments of pIP417, and the fragments of pIP419 generated by several restriction enzymes. The probes were plasmids pIP417, pIP419, pFK705 (pBI191 plus the fragments A, B', D from pIP417; Figure 37.3) containing most of the region B (see Figure 37.3), and pFK706 (pBI191 plus the fragments A C D from pIP417) (data not shown). Because of the lack of homology, under the stringent conditions used, between either Ni^r plasmids and pBI191 or pIP417 A and D fragments and pIP419, plasmids pFK705 and pFK706 may be used as probes specific for pIP417 fragments B' and C, respectively.

The results showed a strong cross-reaction between the probe pFK705 and the *Pvu*II-generated fragment B of pIP417 and a very weak reaction with its fragment C (Figure 37.5G), which may be caused by the fact that the pFK705 probe contains, in addition to the fragment *Pvu*II-*Hae*II (B'), about 50 bp of fragment C (eC). The pFK705 probe also crossreacted with several *Sau*3A fragments of pIP419 that were not localized (Figure 37.5G). By Southern analysis of plasmid pIP419 hydrolyzed by the restriction enzymes whose sites were located on the map and the pIP417 B fragment (probe pFK705), we have identified its homologous fragment on pIP419 to extend from the *Sca*I site (at 4.25-kb) to the *Hin*dIII site (at 6.25-kb) (see Figure 37.2; and data not shown). Moreover, there was no significant homology between the pIP417 fragment C (probe pFK706) and plasmid pIP419. This result confirms clearly that both plasmids have a limited region of homology localized outside the Ni^r genes.

Further studies on the plasmid transfer, particularly pIP417

The plasmid pIP417 was associated in the original *B. vulgatus* strain BV17 with a large plasmid of about 56-kb, pIP418, and this plasmid was regularly found in transconjugants but not in transformants (Breuil et al., 1989). We thus performed retransfers of plasmid pIP417 from a 638R transconjugant or transformant background into a Ni-sensitive, plasmid-free recipient strain *B. fragilis* BF2, resistant to imipenem (Im), tetracycline, and clindamycin. Preliminary experiments done by the selection of Im^r Ni^r transconjugants led us to conclude that the pIP417 plasmid was not transferable unless plasmid pIP418 was in the donor strain. Therefore, plasmid pIP418 was suggested to be a conjugative plasmid involved in the mobilization of pIP417 (Breuil et al., 1989). Further mating experiments with selection by both imipenem plus tinidazole, and tetracycline plus tinidazole, revealed that imipenem, for some unknown reason, was an inefficient selection agent, consequently leading to an underestimation of the plasmid transferability. We observed a transfer of Ni^r when a transformant 638R (pIP417) and BF2 were mated and selection made on tinidazole (4 μg/ml) plus tetracycline (4 μg/ml). Ni^r transconjugants were Rf^s Tc^r Cc^r Im^r and were additionally resistant to 638R phages (Privitera et al., 1979a) as with the recipient strain BF2. Moreover, the transconjugants contained a plasmid of the predicted size, 7.7-kb.

We verified that no transfer of resistance of tetracycline, clindamycin, or imipenem occurred from BF2 into 638R. Therefore, we may regard the genetic exchange between strains 638R and BF2 to be unidirectional. Nevertheless, realizing the small size of plasmid pIP417, we may consider that plasmid-encoded functions are involved in pIP417 mobilization rather than in self-transfer, which in turn is compatible with the possible existence in strain 638R of host-encoded transfer determinant(s) (Smith and Spiegel, 1987).

Our information on plasmid pIP419 transfer is limited, but like pIP417, plasmid pIP419 was transferable from a 638R transformant into strain BF2, and the transconjugants also harbored a plasmid of the expected size and carried the expected antibiotic resistance markers. This transfer occurs although there is no other large plasmid in the donor strain. It should also involve chromosomal transfer elements capable of mobilizing pIP419.

Table 37.2 Conjugal transfer of 5-nitroimidazoles and tetracycline in plasmid-free strains[a]

Bacteroides fragilis donor strain	*Bacteroides* recipient strain	Antibiotic resistance transfer	
		5-Nitroimidazole (Ni)	Tetracycline (Tc)
BF3 Ni[r] Tc[r]	638R	−	+
BF8 Ni[r] Tc[r]	638R	+	−
BF8 Ni[r] Tc[r]	BV1R	+	−

[a]MICs (μg/ml) of the strains are given in Table 37.1. After mating (according to Breuil et al., 1989), the selection was made on tinidazole (4 μg/ml) and tetracycline (4 μg/ml).

37.5 Genetics of Nonplasmid-Mediated 5-Nitroimidazole Resistance

During screening for the presence of plasmids in Ni[r] resistant clinical isolates of *Bacteroides*, we were not able to detect CCC DNA in two strains of *B. fragilis*, BF3 and BF8. Although the presence of a large plasmid could not be ruled out, we have suggested that both resistances are chromosomally determined. Both strains were also tetracycline resistant. The results of matings with Rf[r] Ni[s] recipient strains, either *B. fragilis* 638R or *B. vulgatus* BV1R, are given in Table 37.2. One strain, BF3, although able to transfer its Tc resistance, did not transfer the Ni resistance; however, strain BF8 was able to transfer its Ni resistance independently of Tc resistance, either homo- or heterospecifically (Sebald et al., 1990). We should note that strain BF8 has a moderate resistance to Ni akin to the strains harboring transferable plasmids. Preliminary hybridization experiments performed under stringent conditions did not show cross-hybridization between the chromosomal DNA of strains BF3, BF8, and *B. fragilis* NCTC 11295 (Ingham et al., 1978) and the plasmid DNA probe pIP419; with probe pIP417, cross-reaction was obtained with BF3 and BF8 DNA only.

37.6 Perspectives

The clear demonstration of the genetic transfer of low-level, 5-nitroimidazole resistance,

whether plasmid encoded or chromosomally determined, exemplifies the interest in these resistance patterns although moderate in terms of MIC. We must emphasize, however, that the total lack of information on the mechanism of resistance to 5-nitroimidazoles in these strains. It could differ from the two major types that have already been described in trichomonads and which are expressed at high levels (>100 μg/ml). One of these is the clinically relevant decrease of reductive activation in the presence of low pO_2 (Müller, 1986; Cerkasovova et al., 1988; Müller et al., 1988). The other type was observed in hydrogenosome-deficient mutants lacking the pyruvate:ferredoxin oxydoreductase (Müller, 1986; Cerkasovova et al., 1988).

The biochemical bases of resistance to 5-nitroimidazole in procaryotic anaerobes are obscure. First, they might involve a modification of the molecule or a loss of its reducing activation, resulting as a consequence in a reduced uptake; strong evidence in anaerobic eucaryotes and procaryotes suggests that metronidazole uptake is driven by the reductive process, but without any hint of a transport system. Second, noting that the ultimate target of 5-nitroimidazoles is the DNA, resistance to these agents might result from a modification in the interaction between the active derivative and the bacterial DNA, or from an enhanced DNA repair. The latter mechanism is a reasonable hypothesis considering that mutants of facultative anaerobes such as *Escherichia coli* and *Salmonella*, when altered in their DNA repair process, have an increased sensitivity to 5-nitroimidazoles (Yeung et al., 1984; Hof et al., 1987). This hypothesis should help to explain discrepancies between rates of reduction of various nitroimidazole derivatives, one-electron reduction potentials (E¹), and bactericidal activities of these drugs (Church et al., 1990). Investigations on the mechanism of action of the 5-nitroimidazoles is now accessible through isogenic derivatives of *Bacteroides fragilis*.

Plasmids pIP417 and pIP419 are transferable to Ni-sensitive strains such as *B. fragilis* 638R and BF2 and can be transformed into cells, particularly through electrotransformation. Therefore, they are accessible to genetic analysis. The direct gene cloning in *B. fragilis* appeared to be a useful alternative strategy to gene cloning in *E. coli*, which is highly resistant to 5-nitroimidazoles except when altered in the SOS pathway (Yeung et al., 1984; Hof et al., 1987). The recombinant plasmids pFK704 and pFK100 carry two antibiotic resistance determinants coding for Ni^r and Cc^r and are fully expressed in *B. fragilis*.

E. coli-B. fragilis shuttle vectors carrying at least two antibiotic resistance genes expressed in both hosts could be derived from our constructions, and lead to valuable tools for investigating the gene organization of *B. fragilis*, particularly because electrotransformation is now a reliable technique. Nevertheless, isolation of restriction-less *Bacteroides* strains or mutants is a prerequisite to an efficient use of such vectors.

In the near future, the development of this work includes subcloning and sequencing the plasmid-borne genes involved in the 5-nitro-imidazole resistance of these strains and cloning the chromosomally determined resistance.

The Ni^r plasmids pIP417 and pIP419, although relatively small, are transferable or likely mobilized to a sensitive strain by a conjugation-like process whose precise mechanism has not yet been elucidated. We are currently identifying the mobilization region(s) of the plasmids, and preliminary data obtained with plasmids pFK100, pFK702 (see Figure 37.3), and pIP417-pBI191 recombinant plasmids reveal the involvement of *Pvu*II fragment B in plasmid mobilization (G. Reysset et al., in manuscript). This work should lead to clarifying the mechanisms concerned in transfer of relatively small *Bacteroides* plasmids and to improving *Bacteroides* cloning vectors through the construction of efficient conjugative shuttle vectors.

Acknowledgments. We thank A. Dublanchet and C.J. Smith for the gift of strains, J. Morin for her technical assistance, F. Georges for the editing work, and M. Goldner for critical reading of the manuscript.

References

Birnboim, H.C. and J. Doly. 1979. A rapid alkaline extraction procedure for screening recombinant plasmid DNA. *Nucleic Acids Res.* 7:1513–1523.

Breuil, J., A. Dublanchet, N. Truffaut, and M. Sebald. 1989. Transferable 5-nitroimidazole resistance in the *Bacteroides fragilis* group. *Plasmid* 21:151–154.

Britz, M.L. and R.G. Wilkinson. 1979. Isolation and properties of metronidazole-resistant mutants of *Bacteroides fragilis*. *Antimicrob. Agents Chemother.* 16:19–27.

Cerkasovova, A., J. Cerkasov, and J. Kulda. 1988. Resistance of trichomonads to metronidazole. *Acta Univ. Carol. Biol.* 30:485–503.

Church, D.L., H.R. Rabin, and E.J. Laishley. 1990. Reduction of 2-, 4- and 5-nitroimidazole drugs by hydrogenase 1 in *Clostridium pasteurianum*. *J. Antimicrob. Chemother.* 25:15–23.

Cuchural, G.J., F.P. Tally, J.R. Storey, and M.H. Malamy. 1986. Transfer of β-lactamase-associated cefoxitin resistance in *Bacteroides fragilis*. *Antimicrob. Agents Chemother.* 29:918–920.

Dublanchet, A., J. Caillon, J.P. Emond, H. Chardon, and H.B. Drugeon. 1986. Isolation of *Bacteroides* strains with reduced sensitivity to 5-nitroimidazoles. *Eur. J Clin. Microbiol.* 5:346–347.

Edwards, D.I., M. Dye, and H. Carne. 1973. The selective toxicity of antimicrobial nitroheterocyclic drugs. *J. Gen. Microbiol.* 76:135–145.

Eme, M.A., J.F. Acar, and F.W. Goldstein. 1983. *Bacteroides fragilis* resistant to metronidazole. *J. Antimicrob. Chemother.* 12:523–525.

Finegold, S.M. and the National Committee for Clinical Laboratory Standards Working Group on Anaerobic Susceptibility Testing. 1988. Susceptibility testing of anaerobic bacteria. *J. Clin. Microbiol.* 26:1253–1256.

Hof, H., T. Chakraborty, R. Royer, and J.-P. Buisson. 1987. Mode of action of nitro-heterocyclic compounds on *Escherichia coli*. *Drugs Exp. Clin. Res.* 13:635–639.

Ingham, H.R., S. Eaton, C.W. Venables, and P.C. Adams. 1978. *Bacteroides fragilis* resistant to metronidazole after long-term therapy. *Lancet* i:214.

Knight, R.C., I.M. Skolimowski, and D.I. Edwards. 1978. The interaction of reduced metronidazole with DNA. *Biochem. Pharmacol.* **27**:2089–2093.

Lamothe, F., C. Fijalkowski, F. Malouin, A.M. Bourgault, and L. Delorme. 1986. *Bacteroides fragilis* resistant to metronidazole and imipenem. *J. Antimicrob. Chemother.* **18**:642–643.

LaRusso, N.F., M. Tomasz, M. Müller, and R. Lipman. 1977. Interaction of metronidazole with nucleic acids in vitro. *Mol. Pharmacol.* **13**:872–882.

McLafferty, M.A., R.L. Koch, and P. Goldman. 1982. Interaction of metronidazole with resistant and susceptible *Bacteroides fragilis*. *Antimicrob. Agents Chemother.* **21**:131–134.

Milne, S.E., E.J. Stokes, and P.M. Waterworth. 1978. Incomplete anaerobiosis as a cause of metronidazole "resistance." *J. Clin. Pathol.* **31**:933–935.

Müller, M. 1986. Reductive activation of nitroimidazoles in anaerobic microorganisms. *Biochem. Pharmacol.* **35**:37–41.

Müller, M., J.G. Lossick, and T.E. Gorrel. 1988. In vitro susceptibility of *Trichomonas vaginalis* to metronidazole and treatment outcome in vaginal trichomoniasis. *Sex. Trans. Dis.* **15**:17–24.

Narikawa, S. 1986. Distribution of metronidazole susceptibility factors in obligate anaerobes. *J. Antimicrob. Chemother.* **18**:565–574.

O'Brien, R.W. and J.G. Morris. 1972. Effect of metronidazole on hydrogen production by *Clostridium acetobutylicum*. *Arch. Mikrobiol.* **84**:225–233.

Odelson, D.A., J.L. Rasmussen, C.J. Smith, and F.L. Macrina. 1987. Extrachromosomal systems and gene transmission in anaerobic bacteria. *Plasmid* **17**:87–109.

Perez-Reyes, E., B. Kalyanarman, and R.P. Mason. 1980. The reductive metabolism of metronidazole and ronidazole by aerobic liver microsomes. *Mol. Pharmacol.* **17**:239–244.

Plant C.W. and D.I. Edwards. 1976. The effects of tinidazole, metronidazole and nitrofurazone on nucleic acid synthesis in *Clostridium bifermentans*. *J. Antimicrob. Chemother.* **2**:203–209.

Privitera, G., A. Dublanchet, and M. Sebald. 1979a. Transfer of multiple antibiotic resistance between subspecies of *Bacteroides fragilis*. *J. Infect. Dis.* **139**:97–101.

Privitera, G., M. Sebald, and F. Fayolle. 1979b. Common regulatory mechanism of expression and conjugative ability of a tetracycline resistance plasmid in *Bacteroides fragilis*. *Nature* (London) **278**:657–659.

Rashtchian, A., G.R. Dubes, and S.J. Booth. 1982. Transferable resistance to cefoxitin in *Bacteroides thetaiotaomicron*. *Antimicrob. Agents Chemother.* **22**:701–703.

Rotimi, V.O., B.I. Duerden, V. Ede, and A.E. Mackinnon. 1979. Metronidazole-resistant *Bacteroides* from untreated patient. *Lancet* i:833.

Scher, K.S. 1988. Emergence of antibiotic resistant strains of *Bacteroides fragilis*. *Surg. Gynecol. Obst.* **167**:175–179.

Sebald, M., G. Reysset, and J. Breuil. 1990. What's new in 5-nitroimidazole resistance in the *Bacteroides fragilis* group. *In: Clinical and Molecular Aspects of Anaerobes*, S.P. Borriello, ed., pp. 217–225. Petersfield: Wrightson Biomedical.

Smith, C.J. 1985a. Development and use of cloning systems for *Bacteroides fragilis*: cloning of a plasmid-encoded clindamycin resistance determinant. *J. Bacteriol.* **164**:294–301.

Smith, C.J. 1985b. Polyethylene glycol-facilitated transformation of *Bacteroides fragilis* with plasmid DNA. *J. Bacteriol.* **164**:466–469.

Smith, C.J. and H. Spiegel. 1987. Transposition of Tn*4551* in *Bacteroides fragilis*: identification and properties of a new transposon from *Bacteroides* spp. *J. Bacteriol.* **169**:3450–3457.

Sprott, M.S. and A.M. Kearns. 1988. Metronidazole-resistant *Bacteroides melaninogenicus*. *J. Antimicrob. Chemother.* **22**:951–954.

Sprott, M.S., H.R. Ingham, J.E. Hickman, and P.R. Sisson. 1983. Metronidazole-resistant anaerobes. *Lancet* i:1220.

Tabaqchali, S., A. Pantosi, and S. Oldfield. 1983. Pyruvate dehydrogenase activity and metronidazole susceptibility in *Bacteroides fragilis*. *J. Antimicrob. Chemother.* **11**:393–400.

Tally, F.P. and M.H. Malamy. 1982. Mechanism of antimicrobial resistance and resistance transfer in anaerobic bacteria. *Scand. J. Infect. Dis.* **35**:37–44(Suppl.).

Yamaoka, K., K. Watanabe, Y. Muto, N. Katoh, K. Ueno, and F.P. Tally. 1990. R-plasmid mediated transfer of β-lactam resistance in *Bacteroides fragilis*. *J. Antibiot.* (Jpn) **43**:1302–1306.

Yeung, T.-C., B.B. Beaulieu, M.A. McLafferty, and P. Goldman. 1984. Interaction of metronidazole with DNA repair mutants of *Escherichia coli*. *Antimicrob. Agents Chemother.* **25**:65–70.

38

Genetics of Polysaccharide Utilization Pathways of Colonic *Bacteroides* Species

Abigail A. Salyers, Peter Valentine, and Vivian Hwa

38.1 Introduction

The human colon supports a complex microbial ecosystem. *Bacteroides* species are prominent members of this ecosystem, accounting for approximately 25% of all isolates (Moore et al., 1978). A number of *Bacteroides* species have been found in the colon, and these species seem to have diverged considerably during evolution. Interspecies DNA–DNA homologies range from a high of 49% to a low of less than 5% (Johnson, 1978).

The *Bacteroides* species found in the human colon should be differentiated from the species originally designated to be of the genus "*Bacteroides*" that are found in the human mouth and vaginal tract and in the rumen of cattle. Shah and Collins (1988) have proposed that the genus designation *Bacteroides* should be reserved for the colonic *Bacteroides* species (Shah and Collins, 1989), and have proposed two new genera, *Prevotella* and *Porphyromonas*, to contain the oral and vaginal species (Shah and Collins, 1988, 1990). Unfortunately, the evidence used as a basis for the proposed new genera consists mainly of phenotypic traits, although Johnson and Harich have done DNA–DNA and DNA–rRNA homology studies that lend support to the new classification (Johnson and Harich, 1986). The proposed new genera will become credible when the 16S rRNA sequence analysis of oral and vaginal strains is completed. Preliminary results of 16S rRNA analysis supports the existence of at least

two genera (B. Paster, private communication). It is also likely that the *Bacteroides* found in the rumen of cattle will soon be reclassified in a different genus (D. Stahl, personal communication). Whatever the outcome of the 16S rRNA analyses, it is clear that there are deep phylogenetic divergences between the colonic *Bacteroides* species and the other species that have been classified as *Bacteroides*. Nonetheless, the species in this group are more closely related to each other than to any other known group of bacteria. This review will be limited to the colonic species of the genus *Bacteroides*.

Some human colonic *Bacteroides* species are notable for their ability to utilize a variety of polysaccharides differing in sugar composition and glycosidic linkages (Saiyers et al., 1977a, 1977b). For example, *Bacteroides thetaiotaomicron* can ferment mucopolysaccharides (charged linear polymers containing hexosamines and uronic acids), ovomucoid (branched polysaccharide containing a variety of sugars and linkages), arabinogalactan (β-galactan backbone with arabinose and galactose branches), amylose (linear α-glucan), amylopectin (branched α-glucan), pullulan (linear α-glucan with mixed α-1,4 and α-1,6 linkages), dextran (linear α-glucan), laminarin (linear β-glucan), and pectin (linear β-galacturonan). Other colonic species such as *Bacteroides ovatus*, *Bacteroides caccae*, *Bacteroides uniformis*, *Bacteroides vulgatus*, and *Bacteroides stercoris* utilize a similarly wide range of polysaccharides. In some cases, the range of polysaccharides extends to include the branched plant gums (tragacanth, guar) and the seaweed polysaccharide alginate. Growth of these *Bacteroides* species on polysaccharides is very efficient. Growth rates on most of the polysaccharides are as high as growth rates on the monosaccharide constituents.

Because polysaccharides rather than simple sugars, are probably the main form of carbohydrate available to bacteria in the colon, the ability to utilize polysaccharides may be responsible, at least in part, for the success of *Bacteroides* in colonizing the human colon. The ability to degrade polysaccharides may also contribute to the pathogenesis of *Bacteroides* opportunistic infections. *Bacteroides* can form abscesses in any part of the body, and abscess formation presumably involves breakdown of host tissue polysaccharides and glycoproteins.

Another reason for interest in polysaccharide utilization genes of the colonic *Bacteroides* species is that polysaccharide fermentation is an important activity in many bacterial ecosystems, and gram-negative anaerobes are often involved. Although these gram-negative anaerobes are usually not the same species as the ones found in the colon, they are more closely related to colonic *Bacteroides* species than to the *Escherichia coli* group. Until recently, human colonic *Bacteroides* species have been the only gram-negative anaerobes that are genetically manipulable, and the genetic tools available for use with the colonic species are consequently the most advanced. Equally important, colonic *Bacteroides* are the easiest of all the gram-negative anaerobes to work with in the laboratory, because of their limited nutritional requirements, rapid growth rates, and tolerance for oxygen.

38.2 General Features of *Bacteroides* Polysaccharide Utilization Pathways

A considerable amount of information is now available about the biochemistry and genetics of *Bacteroides* polysaccharide utilization pathways. This work is discussed in detail below

but it is useful at the outset to summarize some general features that appear to be shared by all the *Bacteroides* polysaccharide utilization pathways studied to date. The first common feature is that the polysaccharide-degrading enzymes are not extracellular but are located in the periplasm or cytoplasm (see Section 38.4). Thus, polysaccharide uptake across the outer membrane appears to be the first step in utilization. A second common feature is that the number of degradative enzymes produced is generally greater than the number of reactions to be catalyzed, and it is not uncommon to find more than one enzyme with the same activity (see Section 38.5). A third common feature of *Bacteroides* polysaccharide utilization systems is that they are regulated. That is, levels of degradative enzymes and membrane proteins are increased significantly above background only when bacteria are grown on the inducing polysaccharide (see Section 38.7).

These observations raise several immediate questions:

1. Is uptake of the polysaccharide through the outer membrane really the first step in polysaccharide utilization? If so, how many proteins are involved and what are their characteristics?

2. Why are there usually more degradative enzymes than would seem to be necessary? Does the existence of apparently redundant enzymes mean that there are some genes that are not essential for polysaccharide utilization and, if so, why are they retained by the cell?

3. Given the large number of genes involved, are all the genes in a particular polysaccharide utilization system located in a single region of the chromosome or are they scattered in separate clusters?

4. How are polysaccharide utilization genes regulated, and is there more than one level of regulation? Is regulation tied to a specific polysaccharide or can more than one type of polysaccharide induce the same genes if the polysaccharides share a particular linkage?

38.3 Current Status of *Bacteroides* Genetics

Clearly, questions such as those raised in the foregoing list cannot be answered by biochemical analyses alone. Fortunately, considerable advances in *Bacteroides* genetics have been made in recent years, and these advances have made it possible to do experiments with *Bacteroides* that are not possible with most other anaerobes. Before describing recent work on genetic analysis of *Bacteroides* polysaccharide utilization pathways, it is worthwhile to survey briefly the genetic tools that are currently available.

The first *E. coli–Bacteroides* shuttle vectors were described in 1985 (Guiney et al., 1984; Shoemaker et al., 1985). Since then, a number of shuttle vectors have been constructed (see, for example, Valentine et al., 1988; Shoemaker et al., 1989; Russo et al., 1990). The shuttle vectors make it possible to introduce cloned DNA into *Bacteroides* for purposes of testing gene expression and the effects of increased gene dosage.

Some suicide vectors have also been constructed (Guthrie and Salyers, 1986; Smith and Salyers, 1989). Suicide vectors, which replicate in *E. coli* but not in *Bacteroides*, are important tools because they make it possible to disrupt genes in the *Bacteroides* chromosome. To construct a gene disruption, a segment of DNA internal to a *Bacteroides* gene is cloned into the suicide vector (in *E. coli*) and transferred into *Bacteroides* with selection for a marker on the vector. Because the vector cannot replicate in *Bacteroides*, it can only survive by integrating in the chromosome via a single crossover recombination event between the cloned region and the homologous region on the *Bacteroides* chromosome. Such an integration event disrupts the chromosomal gene. Gene disruptions can be used to determine whether the gene is essential for growth on a particular polysaccharide. Because such insertions are polar on downstream genes, they can also be used to test for genes linked to the cloned gene in an operon.

Chemical mutagenesis of *Bacteroides* has been reported but has not proven useful. The main reason for this is the lack of a transducing phage or Hfr-type chromosomal mapping. Without such mapping tools, point mutations generated by chemical mutagenesis cannot be mapped nor can they be moved conveniently to a new genetic background (to test for multiple mutations). By contrast, transposon mutagenesis of *Bacteroides* has proven to be quite useful. The *Bacteroides* transposon Tn4351 (Shoemaker et al., 1986), has been used to generate a number of different mutants. Recently, Tn4351 has also been used as a marker to clone chromosomal loci that contain polysaccharide utilization genes (Valentine et. al., 1992).

For studies of gene regulation, it is important to be able to construct gene fusions. The first vector for constructing transcriptional fusions was based on a promoter-less erythromycin resistance gene, *ermF* (Smith, 1987). A second fusion vector was based on a promoter-less β-glucuronidase (GUS) gene (Feldhaus et al., 1991). The promoter-less GUS gene is available both on a plasmid that replicates in *Bacteroides* and on a suicide vector, which should be useful for making fusions to chromosomal genes. The GUS fusion vector has the advantage that the enzyme is easy to assay in cell extracts, but there is no positive selection for fusions. A transcriptional fusion vector that employs a promoter-less cefoxitin- imipenem resistance (*cfiA*) gene is currently under construction (M. Malamy, personal communication). This fusion vector should have the dual advantage of an easy enzyme assay and a positive selection for fusions. A potential problem with using this cefoxitin resistance gene is that it makes *Bacteroides* resistant to one of the drugs of choice for treating *Bacteroides* infections. Thus, it may be necessary to use special safety precautions.

38.4 Polysaccharide Binding and Uptake

The most familiar microbial strategy for utilization of polysaccharides involves secretion of extracellular polysaccharidases that degrade the polysaccharide into smaller units. These smaller units (usually mono- or disaccharides) are then transported into the cell. However, this strategy does not make ecological sense for organisms such as *Bacteroides*, which are found in highly competitive ecosystems. In the colon, there are many bacteria that cannot utilize polysaccharides but can use the monosaccharide or disaccharide products of polysaccharidase action. Thus, unless an organism had some very efficient method for sequestering the products of extracellular enzymes, these products would be lost to competitors.

Early studies of colonic *Bacteroides* species indicated that *Bacteroides* employed a different strategy for polysaccharide utilization. *Bacteroides* polysaccharidases are generally located in the periplasm or cytoplasm (Salyers and O'Brien, 1980; McCarthy et al., 1985; Anderson and Salyers, 1989a; Scholle et al., 1990). The one exception so far has been a galactomannanase produced by *Bacteroides ovatus* that is located in the outer membrane (Gherardini and Salyers, 1987). However, the products of this enzyme were large oligomers that would still have to be transported into the cell for further breakdown.

If the degradative enzymes are inside the cell, there should be polysaccharide-binding proteins in the outer membrane that mediate binding and uptake of the polysaccharide. Although there were some earlier reports of outer membrane proteins that were induced during growth on polysaccharides (Kotarski et al., 1985; Gherardini and Salyers, 1987), the first direct evidence that outer membrane proteins play a role in polysaccharide binding was obtained by Anderson and Salyers (1989a), who showed that radioactive starch was bound by intact cells of *Bacteroides thetaiotaomicron*. Binding was saturable and susceptible to protease activity, as expected if binding was mediated by proteins. Moreover, binding appeared to be regulated similarly to the starch-degrading enzymes. That is, binding activity was 12 fold higher if cells were grown on maltose, an inducer of synthesis of the starch-degrading enzymes, than if cells were grown on glucose.

If starch binding is essential for utilization, it should be possible to find mutants that are un-able to utilize starch because they are deficient in starch binding. Such mutants were isolated by screening colonies which had been mutagenized with Tn4351 for the ability to grow on starch (Anderson and Salyers, 1989b). Five classes of starch-minus mutants were isolated (Ms-1–Ms-5), and all were deficient in starch binding. Two of these, Ms-2 and Ms-3, appeared to be deficient solely in starch binding; i.e., the starch-degrading enzymes were properly regulated and produced at wild-type levels when cells were grown on maltose (Anderson and Salyers, 1989b). The fact that mutants that lacked the ability to bind starch were unable to grow on starch provides strong support for the hypothesis that starch binding is essential for starch utilization.

Results of biochemical and immunological analyses of membrane proteins from the starch-minus mutants indicated that at least five membrane proteins are missing in one or more of the starch-binding mutants (Tancula et al., in manuscript). Thus, the complex that mediates binding and uptake may involve a number of proteins. Some members of this complex could be periplasmic proteins, such as the periplasmic polysaccharidases. A finding that indirectly supports this hypothesis is that although intact cells bind labeled starch, membranes from disrupted cells no longer bind starch. This could be explained if loosely bound members of a complex were lost during disruption of the cells and pelleting of the membranes by centrifugation.

Some results of the starch-binding studies indicated that starch binding is separable from starch uptake. Cells of *Bacteroides thetaiotaomicron*, which had been washed and resuspended in phosphate buffer under aerobic conditions, were capable of binding starch but did not accumulate the starch. Cells resuspended in anaerobic medium, however, were able to take up and accumulate labeled starch (Anderson and Salyers, 1989a). Thus, uptake appears to involve more than simply binding the polysaccharide to the cell surface.

The outer membrane proteins that are inducible by chondroitin sulfate and galactomannan (Kotarski et al., 1985; Gherardini and Salyers, 1987; Valentine and Salyers, 1992) may be polysaccharide-binding proteins similar to those seen in the case of starch utilization. It is not yet possible to conclude whether these outer membrane proteins are essential for growth on these two polysaccharides because no mutants deficient only in these proteins have been described. However, analysis of two transposon-generated mutants of *Bacteroides ovatus* that were unable to grow on galactomannan indicates that both mutations affected membrane proteins (Valentine and Salyers, 1992; Valentine et al., 1992).

38.5 Polysaccharide-Degrading Enzymes

Neuraminidase

Neuraminic acids are 9-carbon sugars that have an *N*-acetyl group attached to the fifth carbon. Neuramic acids are ubiquitous in tissue and mucin polysaccharides, where they are covalently linked to other sugars. Neuramic acids are often the end sugar in a branch and appear to protect the rest of the polymer from degradation. Neuraminidases have proven to be a virulence factor for some pathogenic bacteria.

Russo et al. (1990) have cloned a neuraminidase gene from *Bacteroides fragilis*. This gene is of particular interest both because it appears to be important for pathogenesis and because of the clever way in which it was cloned. The gene was cloned by selecting for complementation of a mutant that lacked neuraminidase activity. This mutant had been found to be deficient in the ability to grow in a rat granuloma pouch (a model system for the study of *B. fragilis* virulence). Russo et al. constructed a clone bank in *E. coli* and mobilized it into the neuraminidase-minus mutant. The mixture of transconjugants was inoculated into granuloma pouch fluid to select for clones that complemented the growth defect of the mutant. A transconjugant that grew normally in the granuloma pouch fluid carried a cloned DNA segment which expressed neuraminidase activity.

Neuraminidase activity in wild-type *Bacter-oides fragilis* is regulated. When the clone containing the neuraminidase gene was introduced into *Bacteroides fragilis*, it was regulated normally. Thus, the *Bacteroides* promoter-operator region was cloned along with the gene. The gene was also expressed in *E. coli*, but Russo et al. (1990) showed quite clearly that expression in *E. coli* was from the lambda promoter upstream of the gene on the cloning vector and not from the *B. fragilis* promoter. The neuraminidase gene was partially sequenced and was found to share amino acid homology with regions that are highly conserved in neuraminidase genes from other pathogenic bacteria.

Chondroitin sulfate-degrading enzymes

Chondroitin sulfate is a complex, charged, host mucopolysaccharide that may be a natural substrate in the colon. Because chondroitin sulfate is also abundant in many tissues, the ability to degrade chondroitin sulfate could contribute to the pathogenesis of *Bacteroides* infections. Chondroitin sulfate consists of repeating disaccharide units containing a glucuronic acid residue and an *N*-acetyl-galactosamine residue that is sulfated either on the fourth or sixth carbon. Because of the number of linkages, several different types of enzymes are needed to degrade it to monosaccharides. The polysaccharide is first cleaved into sulfated disaccharides by chondroitinases. The sulfates are removed from the 4- and 6-sulfated disaccharides by chondro-4-sulfatase and chondro-6-sulfatase, respectively. Finally, the unsulfated disaccharides are cleaved by a glucuronidase.

Two chondroitinases have been characterized, chondroitinase I and chondroitinase II (Linn et al., 1983). Although they have similar pI values, pH optima, and substrate specificity, the two chondroitinases are antigenically and genetically distinct. The chondroitinase II gene from *Bacteroides thetaiotaomicron* has been cloned (Guthrie et al., 1985). A segment of DNA that carried the gene plus about 1 kb of adjacent sequence was expressed in *E. coli* but was not expressed in *B. thetaiotaomicron*. The reason was that the gene encoding chondro-4-sulfatase was upstream from the chondroitinase II gene. The regulated *Bacteroides* promoter that drives expression of both of these genes was found to be upstream of the chondro-4-sulfatase gene (Guthrie and Salyers, 1987; Hwa and Salyers, 1992).

The finding that there were two chondroitinases which catalyzed the same reaction raised the question of whether both were necessary for growth on chondroitin sulfate. A gene disruption that eliminated expression of the chondroitinase II gene in *B. thetaiotaomicron* was constructed using an internal segment of the gene cloned in a suicide vector. This gene disruption decreased the total chondroitinase activity of the cell by 70% but did not significantly affect the ability of the cell to grow on chondroitin sulfate (Guthrie and Salyers, 1986). Presumably, chondroitinase I alone was sufficient to support growth on chondroitin sulfate. This indicates that the initial breakdown step catalyzed by chondroitin sulfate is not a rate-limiting step in the pathway because lowering the level of enzyme does not affect the rate at which chondroitin sulfate utilization proceeds.

By contrast, a disruption that eliminated expression of the chondro-4-sulfatase gene virtually abolished growth on chondroitin sulfate (Guthrie and Salyers, 1987). This result indicates that there is only one chondro-4-sulfatase and that it is essential for growth on chondroitin sulfate. Because the disruption in the chondro-4-sulfatase gene did not affect chondro-6-sulfatase activity in cell extracts, the chondro-6-sulfatase activity must be associated with a separate enzyme. We have shown that a transposon-generated mutant of *Bacteroides thetaiotaomicron* (46-1), which was unable to grow on chondroitin sulfate, lacks chondro-6-sulfatase activity (Hwa and Salyers, 1992b). Thus, there appears to be only one chondro-6-sulfatase gene, and that gene may be essential for growth on chondroitin sulfate.

To determine if the mutations in the chondroitin sulfate utilization pathway had any effect on colonization of the colon, the three mutants just described were tested for ability to compete with the wild type for colonization of the intestinal tracts of germ-free mice, i.e., mice that have been bred and raised in a sterile environment and thus have no microflora (Salyers and Guthrie, 1988; Salyers et al., 1988; Hwa and Salyers, 1992b). An interesting result of this work is that although the chondroitinase II disruption and the chondro-4-sulfatase disruption had no effect on colonization, the chondro-6-sulfatase mutant (46-1) was rapidly outcompeted by the wild type. We have not yet established whether the in vivo effect is caused by the loss of chondro-6-sulfatase expression or by effects of the transposon insertion on some other genes, but it is clear that this locus is more important for survival in the intestinal tract than the other loci which have been tested.

The mutant 46-1 was unable to grow on heparin (Hwa and Salyers, 1992b). This phenotype is interesting because it indicates how complex the *Bacteroides* polysaccharide utilization pathways and their interconnections may be. Heparin, a polysaccharide that is structurally similar to chondroitin sulfate, does not induce expression of chondroitin sulfate utilization genes such as the chondroitinases, chondro-4(or 6)-sulfatase, or the chondroitin sulfate-associated outer membrane proteins. However, the 46-1 mutation abolishes growth on heparin as well as on chondroitin sulfate; this is not true of the disruptions in the chondroitinase II or chondro-4-sulfatase genes. Thus, even though the chondro-6-sulfatase gene is not regulated directly by heparin, it may be linked to genes that are part of the heparin utilization system.

Starch-degrading enzymes

Bacteroides thetaiotaomicron grows on three forms of starch: amylose (linear α-1,4 glucan), amylopectin (linear α-1,4 chains connected by α-1,6 branches), and pullulan (maltotriose residues connected by α-1,6 linkages). Growth on any one of these polysaccharides or on maltose induces amylase, pullulanase, and amylopectinase activity. A gene encoding a pullulanase has been cloned (Smith and Salyers, 1989). The cloned pullulanase cleaves the α-1,6 linkages of pullulan to produce maltotriose. A gene disruption that abolished expression of the cloned pullulanase gene in *Bacteroides thetaiotaomicron* did not affect the ability of *B. thetaiotaomicron* to grow on pullulan. Moreover, because the level of pullulanase in the cell was reduced by only 30%, there must be another pullulanase.

Using the pullulanase-minus mutant as the starting point, Smith and Salyers (1991) have isolated and characterized two more pullulan-degrading enzymes. Both were neopullulanases that cleaved the α-1,4 linkages of pullulan to produce trisaccharides with one α-1,4 and one α-1,6 linkage. The neopullulanases also degraded amylose. One of the neopullulanases was soluble, and one was membrane bound. An α-glucosidase that cleaved small oligomers of amylose and the mixed-linkage trisaccharides produced by the neopullulanases was also purified and characterized. Still uncharacterized is the enzyme that breaks the α-1,6 linkages in amylopectin. Also, because a transposon-generated mutant of *Bacteroides thetaiotaomicron* (Ms-1) that lacks all the enzymes listed here (including the α-glucosidase) could still grow on maltose, there must be a maltase that has not yet been detected and characterized (Anderson and Salyers, 1989b).

It is interesting to note that a screen of thousands of transposon-generated mutants of *Bacteroides thetaiotaomicron* for mutants that could not grow on starch yielded only mutants which were deficient in starch binding or appeared to be regulatory mutants (Anderson and Salyers, 1989b). No polysaccharidase-minus mutants were found in this screen. Given the number of genes encoding starch-degrading enzymes and given that the transposon Tn4351 appears to insert randomly in the chromosome, it is possible that the failure to find polysaccharidase-minus mutants by screening for mutants unable to grow on starch results from the fact that loss of a single polysaccharidase gene does not eliminate growth on starch. This was certainly the case with the cloned pullulanase.

α-Galactosidase

Many plant polysaccharides contain α-galactoside linkages. There are also some α-galactoside linkages in mucins and host glycoproteins. *Bacteroides ovatus*, a species notable for its ability to degrade a variety of branched polysaccharides, exhibited elevated levels of α-galactosidase activity when grown on polysaccharides that contain α-galactosides. This provided a good system for investigating whether the same gene can be involved in more than one polysaccharide utilization system, because in theory one α-galactosidase could act on more than one type of α-galactoside.

The results to date indicate that *Bacteroides ovatus* produces a separate α-galactosidase for each type of polysaccharide rather than one all-purpose enzyme. For example, growth on galactomannan induces one α-galactosidase (α-galactosidase I) whereas growth on melibiose, raffinose, and stachyose induces a second distinct enzyme (α-galactosidase II; Gherardini et al., 1985). Interestingly, both enzymes are capable of removing galactose residues from either galactomannan or melibiose, raffinose, and stachyose. Thus, only one enzyme would have been necessary. Recent results indicate that *Bacteroides ovatus* produces at least two other α-galactosidases. One is induced only by mucin and gum tragacanth (Valentine et al., 1991), but the inducer and natural substrate are still not known for the other α-galactosidase. This α-galactosidase was discovered in an attempt to clone α-galactosidase I or II by screening a cosmid library of *Bacteroides ovatus* DNA for expression in *E. coli* (Valentine et al. 1991). Although the activity of the cloned α-galactosidase was not detectable in cell extracts of *Bacteroides ovatus*, a mutant in which this gene was disrupted was outcompeted by the wild type in the germ-free mouse model. Thus, the gene may be important for colonization.

Sucrase

Scholle et al. (1990) cloned a 6-kb fragment from *Bacteroides fragilis* that carried a sucrase gene. The gene was expressed in *E. coli* and made it possible for *E. coli* to grow on sucrose as a sole carbohydrate source. The sucrase activity in *B. fragilis* appeared to be localized primarily in the periplasmic space. The small amount of extracellular activity was attributed to leakage of the periplasmic contents. The sucrase activity expressed in *E. coli* from the clone was also localized primarily in the periplasm. This suggested that the *B. fragilis* enzyme had a signal sequence that was recognized in *E. coli*. Scholle et al. concluded that the clone also carried gene(s) necessary for sucrose transport because the cloned fragment allowed *E. coli* to grow on sucrose as a sole carbohydrate source. Also, they were able to detect uptake of labeled sucrose in *E. coli* containing the clone. However, they did not rule out the possibility that sucrose can enter the periplasm through one of the *E. coli* porins. Because the enzyme is located in the periplasm, diffusion of sucrose through a pore would allow utilization of sucrose as the sole carbon source as sucrose would be broken down to glucose and fructose, two sugars that are readily transported by *E. coli*.

Xylanase and xylosidase

Whitehead and Hespell (1990b) cloned a 3.8-kb fragment from *Bacteroides ovatus* that expressed xylanase, xylosidase, and arabinosidase in *E. coli*. The gene encoding xylanase activity was localized to a 2-kb region of the clone. The arabinosidase and xylosidase activities were both associated with the adjacent 1.5-kb region. Genetic and biochemical evidence support the hypothesis that the arabinosidase and xylosidase activities are the result of a single bifunctional enzyme. This is unusual; none of the previously characterized xylosidases from other organisms have had arabinosidase activity. Similarly, no previously reported arabinosidases have had xylosidase activity. Whitehead and Hespell (1990b) fractionated extracts from *Bacteroides ovatus* and found, in addition to the arabinosidase associated with the cloned enzyme, there was at least one other arabinosidase that did not have xylosidase activity. Thus, the xylan utilization system of *Bacteroides ovatus* may involve enzymes other than those cloned on the 3.8-kb fragment. Recently, we have shown that a gene disruption that eliminates the xylosidase-arabinosidase grows poorly on xylan (Weaver et al., in manuscript). Disruption of the cloned xylanase gene did not have such a drastic effect. Analysis of the mutant with the disruption in the xylanase gene showed that there is a second xylanase in this strain.

38.6   Organization of *Bacteroides* Polysaccharide Utilization Genes

Only limited information is available about the organization of *Bacteroides* polysaccharidase genes, but the results to date have provided some examples of close linkage between genes in the same polysaccharide utilization system. The xylosidase and xylanase genes cloned from *Bacteroides ovatus* (Whitehead and Hespell, 1990b) were closely linked, and both genes appear to be part of a single transcriptional unit (Weaver et al., in manuscript). Similarly, the chondroitinase II and chondro-4-sulfatase genes of *Bacteroides thetaiotaomicron* are adjacent on the chromosome. These genes have been shown to form an operon that appears not to contain any other chondroitin sulfate utilization genes (Guthrie and Salyers, 1986, 1987; Hwa et al., 1992a). DNA adjacent to the transposon insertion in the mutant 46-1 (deficient in chondro-6-sulfatase activity) has been cloned (Hwa et al., unpublished data). Southern analysis indicates that this locus is not close to the chondro-4-sulfatase/chondroitinase II operon or to other transposon insertions that eliminate growth on chondroitin sulfate. Finally, recent analysis of transposon-generated mutants of *Bacteroides ovatus* that could not grow on guar gum revealed two loci which are at least 6-kb apart (and perhaps more) on the *Bacteroides* chromosome (Valentine et al., 1992). The identity of the genes in these two loci has not been established conclusively, but at least some of them encode or regulate expression of membrane proteins.

Several of the genes involved in starch utilization appear to be located in the same region of the chromosome. The starch-minus mutants Ms-1, Ms-2, Ms-3, Ms-4, and Ms-5 were generated by mutagenesis of *Bacteroides thetaiotaomicron* with Tn*4351* (Anderson and Salyers, 1989b). Using Tn*4351* as a marker, we have succeeded in cloning DNA adjacent to the transposon insertions in Ms-2 and Ms-3 (Tancula et al., in manuscript). The adjacent DNA segments were then used as probes to clone segments of wild-type *B. thetaiotaomicron* DNA that overlap the mutations in Ms-2 and Ms-3.

Southern analysis indicates that the transposon insertions in Ms-4 and Ms-5 (and possibly also in Ms-1) are within a 20-kb region that contains Ms-2 and Ms-3 (Tancula et al., in manuscript). However, the pullulanase gene cloned from *B. thetaiotaomicron* is not located in this region.

A good case could be made for the assertion that determining where genes are in the chromosome is not nearly as important as determining what the genes do and how they are regulated. Nonetheless, there are two contexts in which linkage of genes can be important. First, once a particular gene is located by cloning or transposon mutagenesis, analyzing regions of DNA adjacent to this gene may lead to discovery of genes that have not been previously detected. Second, there are already indications that some genes in a polysaccharide utilization pathway may be regulated differently from the others (see following). In such cases, determining the mechanism for the differential regulation may require knowledge of whether the genes are physically separated on the chromosome.

38.7 Regulation of *Bacteroides* Polysaccharide Utilization Genes

Many studies of *Bacteroides* polysaccharidases have noted that the polysaccharidase genes appear to be regulated. That is, the specific activities of the enzymes increase when the bacteria are grown on the inducing polysaccharide (Salyers and Kotarski, 1980; Gherardini et al., 1985; Anderson and Salyers, 1989a; Russo et al., 1990; Whitehead and Hespell, 1990b). The polysaccharide-specific membrane proteins that may be involved in polysaccharide uptake appear to be similarly regulated (Kotarski et al. 1985; Gherardini and Salyers, 1987; Valentine and Salyers, 1992). Only recently, however, has it been possible to study regulation of *Bacteroides* polysaccharide utilization genes at the molecular level.

To investigate regulation at the molecular level, it was necessary to clone a regulated *Bacteroides* promoter and show that it was properly regulated in *Bacteroides*. Because there are many

fortuitous promoters in *Bacteroides* DNA that allow expression of genes in *E. coli*, expression of a *Bacteroides* gene in *E. coli* does not prove that the real *Bacteroides* promoter is in the cloned region. To locate the real *Bacteroides* promoter, it is necessary to check expression in *Bacteroides*. To date only two regulated *Bacteroides* promoters have been cloned and shown to be regulated properly in *Bacteroides*: the promoter of the *Bacteroides fragilis* neuraminidase gene (Russo et al., 1990) and the promoter of the *Bacteroides thetaiotaomicron* operon that contains the chondro-4-sulfatase and the chondroitinase II genes (Hwa and Salyers, 1992a). A second impediment to molecular analysis of regulation of *Bacteroides* genes was the lack of an easily assayable reporter gene for generating transcriptional fusions that could be monitored in *Bacteroides*. The recent demonstration that a promoter-less *E. coli* β-glucuronidase gene (GUS) could serve as a reporter group in *Bacteroides* has solved this problem.

Hwa and Salyers (1992a) used fusions to the *E. coli* β-glucuronidase (GUS) gene to identify the region required for regulated expression of the chondro-4-sulfatase/chondroitinase II operon. The region was at least 500 bp in size. When the promoter region was cloned into a multicopy plasmid and introduced into wild-type *Bacteroides thetaiotaomicron*, there was a partial reduction of chondroitinase II activity from the chromosomal gene. This indicated that a regulatory element (possibly an activator) was being titrated by the regulatory region.

An unexpected effect of this regulatory region, when provided in multiple copies on a plasmid, was the apparent derepression of an unlinked gene in the chondroitin sulfate utilization pathway, the gene that encodes the *Bacteroides* β-glucuronidase which hydrolyzes the unsulfated disaccharides from chondroitin sulfate to monosaccharides. (Because this enzyme has a different substrate specificity from the GUS reporter group mentioned in the preceding paragraph, there was no difficulty in distinguishing between the two enzymes.) The chondroitin sulfate pathway β-glucuronidase, like the other chondroitin sulfate-regulated enzymes, usually is not detectable when the bacteria are grown on glucose. However, in cells containing multiple copies of the regulatory region of the chondro-4-sulfatase/chondroitinase II operon, β-glucuronidase activity was detectable in cells grown on glucose. None of the other genes in the chondroitin sulfate utilization system were similarly derepressed. Thus the promoter-operator region of the chondro-4-sulfatase/chondroitinase II operon may be titrating a repressor that regulates the β-glucuronidase gene but not other genes in the pathway. The region of DNA required to produce the titration effect was even larger than the region necessary for regulated expression of the operon.

Another example of differential regulation is the regulation of the galactomannanases that are induced by growth of Bacteroides ovatus on galactomannan. All the enzymes and membrane proteins so far identified in the galactomannan utilization system are expressed at high levels when cells are grown on galactomannan but are barely detectable when cells are grown on galactose or mannose, the two monosaccharide constituents of galactomannan. Surprisingly, the galactomannanases (I and II) are also induced by glucose, whereas the levels of α-galactosidase I and the outer membrane proteins remain at basal level (Valentine and Salyers, 1992).

Given the fact that the colonic Bacteroides species have diverged considerably during evolution, it is interesting to ask whether cross-expression of regulated genes can still occur among these Bacteroides species. A good example is provided by Bacteroides thetaiotaomicron and Bacteroides ovatus. Both grow on chondroitin sulfate and produce enzymes with similar characteristics, but they share a DNA–DNA homology of only 40% (Johnson, 1978). Moreover, a clone of the Bacteroides thetaiotaomicron chondroitinase II gene did not cross-hybridize with Bacteroides ovatus DNA on Southern blots (Lipeski et al., 1986). M. Feldhaus et al. (1991) have tested a fusion between the promoter-operator region of the chondroitinase II operon of Bacteroides thetaiotaomicron and the reporter gene GUS for expression in Bacteroides ovatus. The GUS fusion from Bacteroides thetaiotaomicron was expressed at the same level in Bacteroides ovatus and was regulated in the same way. Thus, despite the genetic divergence of the two species, the reg-ulatory apparatus of one has been conserved enough to be able to recognize the regulatory region from the other.

Whitehead and Hespell (1990a) have cloned a xylanase gene from the rumen species, Bacteroides ruminicola. When this clone was introduced into Bacteroides uniformis or Bacteroides fragilis, the enzyme was produced in much higher levels than in Bacteroides ruminicola. Although the xylanase activity in Bacteroides ruminicola is regulated, expression of the cloned gene in Bacteroides uniformis and Bacteroides fragilis was produced constitutively. This may indicate that the Bacteroides ruminicola gene is normally regulated by a repressor that is not produced in the two colonic Bacteroides species. Alternatively, the operator of the Bacteroides ruminicola gene may have been deleted during cloning. Neither of the colonic Bacteroides species tested in this study grows on xylan, so we cannot rule out the possibility that cross-regulation could occur. Because Bacteroides ruminicola is much more distantly related to the human colonic Bacteroides species than they are to each other, it would be interesting to learn if the same promoter-operator regions are being used in both cases.

38.8 Directions for Future Research

A number of genes involved in utilization of polysaccharides and disaccharides have now been identified. In some cases the function of the gene in Bacteroides has been established by determining the phenotype of a chromosomal insertion in the gene. Bacteroides genetics has now advanced sufficiently that when a gene has been cloned its function can be assessed by disrupting the gene in the Bacteroides chromosome or by introducing it in multiple copies. Also, the region of the chromosome surrounding the gene can be cloned. The cloned DNA can then used to generate insertional disruptions in the surrounding DNA for the purpose of locating linked genes. The cloned DNA can also be used to construct GUS fusions for the purpose of locating promoters and determining the direction of transcription. It is therefore now possible to obtain detailed information about the molecular structure and regulation of *Bacteroides* polysaccharide utilization genes.

Despite the fact that the types of experiments outlined here are now feasible, techniques for genetic manipulation of *Bacteroides* are still cumbersome and time consuming. Given the difficulty of the work and the limited amount of manpower available, it is appropriate to ask what types of questions should have the highest priority. In closing this chapter, we would like to make some suggestions. One important question, which has implications for microbial ecology as well as for application of genetic engineering to improving the efficiency of polysaccharide fermentation in economically important microbial ecosystems, is the question of what is the rate-limiting step in polysaccharide fermentation. Is it uptake of the polysaccharide, enzymatic breakdown of the polysaccharide, or fermentation of the monosaccharide products? Using genetics to identify essential genes can help to answer this question.

A second important question is, why are the polysaccharide utilization systems so complex? Do genes that are not essential for growth on the polysaccharide in laboratory medium make any contribution to survival of the organism in its natural habitat? The question of the ecological role of regulation of polysaccharide utilization genes is closely related. The assumption is generally made that regulation helps organisms to survive in shifting environments, but this hypothesis must be tested. Having well-characterized mutants to test in in vivo model systems will help to answer these questions.

Finally, there is the question of how polysaccharide utilization genes have evolved. Information about the organization and characteristics and sequences of polysaccharide utilization genes of *Bacteroides* and other polysaccharide-utilizing bacteria may shed light on this question.

Acknowledgments. This work was supported by grant AI 17876 from the U.S. National Institutes of Health.

References

Anderson, K. and A.A. Salyers. 1989a. Biochemical evidence that starch breakdown by *Bacter-oides thetaiotaomicron* involves outer membrane starch binding sites and periplasmic starch degrading enzymes. *J. Bacteriol.* **171**:3192–3198.

Anderson, K. and A.A. Salyers. 1989b. Genetic evidence that outer membrane binding of starch is required for starch utilization by *Bacteroides thetaiotaomicron*. *J. Bacteriol.* **171**:3199–3204.

Feldhaus, M., V. Hwa, Q. Cheng, and A.A. Salyers. 1991. Use of the *E. coli* β-glucuronidase (GUS) gene as a reporter gene for investigating *Bacterioides* promoters. *J. Bacteriol.* **173**:4540–4543.

Gherardini, F., M. Babcock, ard A.A. Salyers. 1985. Purification and characteristics of two alpha-galactosidases associated with catabolism of guar gum and other galactomannans. *J. Bacteriol.* **161**:500–506.

Gherardini, F.C. and A.A. Salyers. 1987. Partial purification and characterization of an outer membrane galactomannanase from *Bacteroides ovatus*. *J. Bacteriol.* **169**:2038–2043.

Guiney, D. G., P. Hasegawa, and C.E. Davis. 1984. Plasmid transfer from *Escherichia coli* to *Bacteroides fragilis*: differential expression of antibiotic resistance genes. *Proc. Natl. Acad. Sci. USA* **81**:7203–7207.

Guthrie, E.P. and A.A. Salyers. 1986. Use of targeted insertional mutagenesis to determine whether chondroitin lyase II is essential for chondroitin sulfate utilization by *Bacteroides thetaiotaomicron*. *J. Bacteriol.* **166**:966–971.

Guthrie, E.P. and A.A. Salyers. 1987. Evidence that the chondroitin lyase gene of *Bacteroides thetaiotaomicron* is adjacent to the gene for a chondro-4-sulfatase. *J. Bacteriol.* **169**:1192–1199.

Guthrie, E.P., N.B. Shoemaker, and A.A. Salyers. 1985. Cloning and expression in *Escherichia coli* of a gene coding for a chondroitin lyase from *Bacteroides thetaiotaomicron*. *J. Bacteriol.* **164**:510–515.

Hwa, V. and A.A. Salyers. 1992a. Evidence for differential regulation of genes in the chondroitin sulfate utilization pathway of *Bacteroides thetaiotaomicron*. *J. Bacteriol.* **174**:342–344.

Hwa, V. and A.A. Salyers. 1992b. Analysis of two chondroitin sulfate utilization mutants of *Bacteroides thetaotaomicron* that differ in their abilities to compete with wild type in the gestrointestinal tracts of germ free mice. *Appl. Environ. Microbiol.* **58**:869–876.

Johnson, J.L. 1978. Taxonomy of the *Bacteroides*. *Int. J. Syst. Bacteriol.* **28**:245–256.

Johnson, J.L. and B. Harich. 1986. Ribosomal ribonucleic acid homology among species of the genus *Bacteroides*. *Int. J. Syst. Bacteriol.* **36**:71–79.

Kotarski, S.F., J. Linz, D.M. Braun, and A.A. Salyers. 1985. Analysis of outer membrane polypeptides that are associated with the growth of *Bacteroides thetaiotaomicron* on chondroitin sulfate. *J. Bacteriol.* **163**:1080–1086.

Linn, S.P., T. Chan, and A.A. Salyers. 1983. Isolation and characterization of two chondroitin lyases from *Bacteroides thetaiotaomicron*. *J. Bacteriol.* **156**:859–866.

Lipeski, L., E.P. Guthrie, M. O'Brien, S.F. Kotarski, and A.A. Salyers. 1986. Comparison of protein involved in chondroitin sulfate utilization by three colonic *Bacteroides* species. *Appl. Environ. Microbiol.* **51**:978–984.

McCarthy, R., S.F. Kotarski, and A.A. Salyers. 1985. Location and characterization of enzymes involved in the breakdown of polygalacturonic acid by *Bacteroides thetaiotaomicron*. *J. Bacteriol.* **161**:493–499.

Moore, W.E.C., E.P. Cato, and L.V. Holdeman. 1978. Some current concepts of intestinal bacteriology. *Am. J. Clin. Nutr.* **31**:S33–S42.

Russo, T.A., J.S. Thompson, V.G. Codoy, and M.H. Malamy. 1990. Cloning and expression of the *Bacteroides fragilis* TAL2480 neuraminidase gene, *nanH*, in *Escherichia coli*. *J. Bacteriol.* **172**:2594–2600.

Salyers, A.A., and E.P. Guthrie. 1988. A deletion in the chromosome of *Bacteroides thetaiotaomicron* that abolishes production of chrondroitinase II does not affect the survival of this organism in the gastrointestinal tracts of exgermfree mice. *Appl. Environ. Microbiol.* **54**:1964–1969.

Salyers, A.A. and S.F. Kotarski. 1980. Induction of chrondroitin sulfate lyase activity in *Bacteroides thetaiotaomicron*. *J. Bacteriol.* **143**:780–788.

Salyers, A.A. and M. O'Brien. 1980. Cellular location of enzymes involved in the breakdown of chondroitin sulfate by *Bacteroides thetaiotaomicron*. *J. Bacteriol.* **143**:772–780.

Salyers, A.A., M. Pajeau, and R. P. McCarthy. 1988. Assessing the importance of mucopolysaccharides as substrates for *Bacteroides thetaiotaomicron* growing in the intestinal tracts of exgermfree mice. *Appl. Environ. Microbiol.* **54**:1970–1976.

Salyers, A.A., J. Vercellotti, S. West, and T.D. Wilkins. 1977a. Fermentation of mucin and plant polysaccharides by strains of *Bacteroides* from the human colon. *Appl. Environ. Microbiol.* **33**:319–322.

Salyers, A.A., S.E.H. West, J.R. Vercellotti, and T.D. Wilkins. 1977b. Fermentation of mucins and plant polysaccharides by anaerobic bacteria from the human colon. *Appl. Environ. Microbiol.* **34**:529–533.

Scholle, R.R., H.E. Steffen, H.J.K. Goodman, and D.R. Woods. 1990. Expression and regulation of a *Bacteroides fragilis* sucrose utilization system cloned in *Escherichia coli*. *J. Bacteriol.* **56**:1944–1948.

Shah, H.N. and M.D. Collins. 1988. Proposal for reclassification of *Bacteroides asaccharolyticus*, *Bacteroides gingivalis* and *Bacteroides endodontalis* in a new genus *Porphyromonas*. *Int. J. Syst. Bacteriol.* **38**:128–131.

Shah, H.N. and M.D. Collins. 1989. Proposal to restrict the genus *Bacteroides* (Castellani and Chalmers) to *Bacteroides fragilis* and closely related species. *Int. J. Syst. Bacteriol.* **39**:85–87.

Shah, H.N. and M.D. Collins. 1990. *Prevotella*, a new genus to include *Bacteroides melaninogenicus* and related species formerly classified in the genus *Bacteroides*. *Int. J. Syst. Bacteriol.* **40**:205–208.

Shoemaker, N.B., R.D. Barber, and A. A. Salyers. 1989. Cloning and characterization of a *Bacteroides* conjugal tetracycline resistance element using a shuttle cosmid vector. *J. Bacteriol.* **171**:1294–1302.

Shoemaker, N.B., C. Getty, J.F. Gardner, and A.A. Salyers. 1986. Tn*4351* transposes in *Bacteroides* and mediates the integration of R751 into the *Bacteroides* chromosome. *J. Bacteriol.* **165**:929–936.

Shoemaker, N.B., E.P. Guthrie, A.A. Salyers, and J.F. Gardner. 1985. Evidence that the clindamycin-erythromycin resistance gene of *Bacteroides* plasmid pBF4 is on a transposable element. *J. Bacteriol.* **162**:626–632.

Smith, J. 1987. Nucleotide sequence analysis of Tn*4551*: use of *ermFS* fusions to detect promoter activity in *Bacteroides fragilis*. *J. Bacteriol.* **169**:4589–4596.

Smith, K. and A. A. Salyers. 1991. Characterization of a neopullulanase and an α-glucosidase from *Bacteroides thetaiotaomicron* 95-1. *J. Bacteriol.* **173**:2962–2968.

Valentine, P.J., P. Arnold, and A.A. Salyers. 1992. Cloning and partial characterization of two loci in the Bacteroides chromosome that contain essential guar gum utilization genes. *Appl. Environ. Microbiol.* **58**: in press.

Valentine, P.J., F.C. Gherardini and A.A. Salyers. 1991. Characterization of an α-galactosidase from *Bacteroides ovatus* 0038 which may be important for colonization. *Appl. Environ. Microbiol.* **57**:1615–1623.

Valentine, P.J., and A.A. Salyers. 1992. Analysis of proteins associated with the growth of Bacteroides ovatus on quar gum. *Appl. Environ. Microbiol.* **58**: in press.

Valentine, P.J., N.B. Shoemaker, and A.A. Salyers. 1988. Mobilization of *Bacteroides* plasmids by *Bacteroides* conjugal elements. *J. Bacteriol.* **170**:1319–1324.

Whitehead, T.R. and R.B. Hespell. 1990a. Heterologous expression of the *Bacteroides ruminicola* xylanase gene in *Bacteroides fragilis* and *Bacteroides ruminicola*. *FEMS Microbiol. Lett.* **66**:61–66.

Whitehead, T.R. and R.B. Hespell. 1990b. The genes for three xylan-degrading activities from *Bacteroides ovatus* are clustered in a 3.8-kilobase region. *J. Bacteriol.* **172**:2408–2412.

39

Molecular Biology of the Fimbriae of *Dichelobacter* (Previously *Bacteroides*) *nodosus*

John S. Mattick, Matthew Hobbs, Peter T. Cox, and Brian P. Dalrymple

39.1 Background

Dichelobacter nodosus is a gram-negative anaerobe and the essential causative agent of ovine footrot (Egerton, 1977). Virulent isolates of this organism contain large numbers of fine surface filaments termed fimbriae (or common pili) (Figure 39.1), which have a diameter of about 6 nm and may extend up to several micrometers (µm) in length (Stewart, 1973). In other bacteria such fimbriae have been shown to have adhesive properties (Paranchych and Frost, 1988; Moore and Rutter, 1989; see also following) and to represent a primary mechanism for colonization of animal cell surfaces. Although the exact function of *D. nodosus* fimbriae has not yet been clearly defined, they appear to play a central role in the invasion by the bacterium of the epidermal matrix of the hoof (see Mattick et al., 1985a). Colony variants with few or no fimbriae are relatively benign or avirulent, respectively (Short et al., 1976; Skerman et al., 1981; Every and Skerman, 1983; Depiazzi and Richards, 1985; Stewart et al., 1986).

Similar fimbriae are found in a variety of other pathogenic bacteria, including *Neisseria gonorrhoeae*, *N. meningitidis*, *Moraxella bovis*, *M. lacunata*, *M. nonliquefaciens*, and *Pseudomonas aeruginosa*, among others (see Dalrymple and Mattick, 1987; also Section 39.7). These fimbriae have been collectively classified as type 4 and share a number of characteristics in common, namely a polar location on the cell, association with a phenomenon termed twitching motility (which appears to mediate surface transloca-

tion), and certain distinctive features of the structural subunit that constitutes the fimbrial strand (Ottow, 1975; Henrichsen, 1983; Dalrymple and Mattick, 1987). These include a short positively charged leader sequence in the primary translation product, an unusual modified amino acid (*N*-methylphenylalanine) as the first residue in the mature protein, a highly hydrophobic and highly conserved amino-terminal domain, and other similarities else-

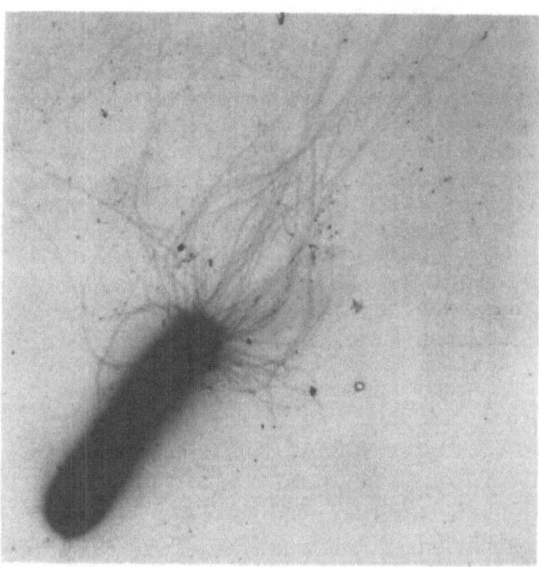

FIGURE 39.1 Electron micrograph of a *Dichelobacter nodosus* cell, showing numerous thin, flexible fimbriae projecting from the cell pole. (Photomicrograph reproduced by permission of J.R. Egerton, Department of Animal Health, University of Sydney.)

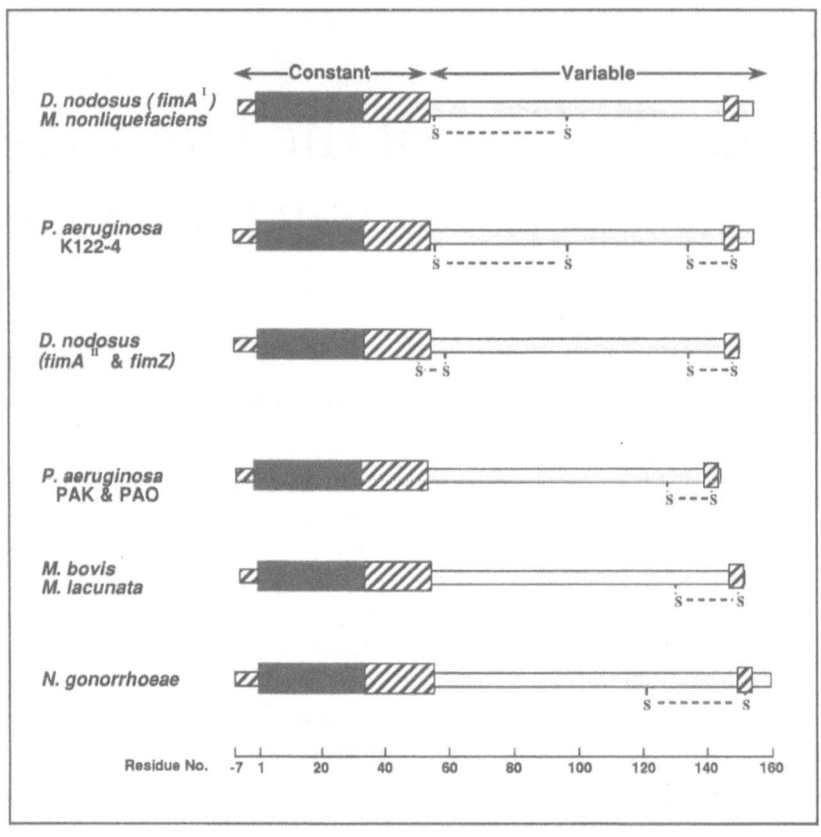

FIGURE 39.2 Schematic comparison of primary sequences of fimbrial subunits of some type 4 fimbriate pathogens. Subunit sequences are from *Dichelobacter nodosus* serogroup H (FimZ and FimA^II; Mattick et al., in press), *D. nodosus* serogroup A1, (FimA^I, Mattick et al., in press), *Neisseria gonorrhoeae* MS11 (Meyer et al., 1984), *Pseudomonas aeruginosa* PAK (Pasloske et al., 1985), *P. aeruginosa* 122-4 (Pasloske et al., 1988), *Moraxella bovis* EPP63 TfpQ (Marrs et al., 1985), *M. lacunata* ATCC 17956 TfpQ (C.F. Marrs and F. Rozsa, personal communication), and *M. nonliquefaciens* (T. Tønjum, C.F. Marrs, F. Rozsa and K. Bøvre, personal communication). In constant region, thin-striped block represents the short 6- to 7-amino-acid-leader sequence; thicker dark block, highly conserved hydrophobic 32-amino-acid sequence (see Figure 39.3), beginning with modified amino acid *N*-methylphenylalanine at position 1 in mature protein. Striped area from residue 33 to glycine at residue 54/55 represents secondary region that is conserved within species but which shows some divergence between species. Limited homology is observed between species in variable region. Pairs of cysteine residues joined by dotted lines.

where in the protein (Dalrymple and Mattick, 1987; Figures 39.2 and 39.3). Type 4 fimbrial subunits range from about 145 to 160 amino acids in length. Inter- and intraspecies variation occurs primarily in the carboxy-terminal 70% of the protein (Dalrymple and Mattick, 1987; and Figure 39.2).

It is now well established that the fimbriae of *D. nodosus* represent primary serological and immunoprotective antigens. A number of studies have shown serological differences between strains (Egerton, 1973; Schmitz and Gradin, 1980; Claxton et al., 1983), and in this respect *D. nodosus* is easily the best characterized of all the type 4 fimbriate species. Using the now-standard K-agglutination test (Egerton, 1973), extensive surveys of some thousands of field isolates have identified 9 major serogroups, termed A to I, covering at least 18 subsidiary serotypes (Claxton, 1989). The same

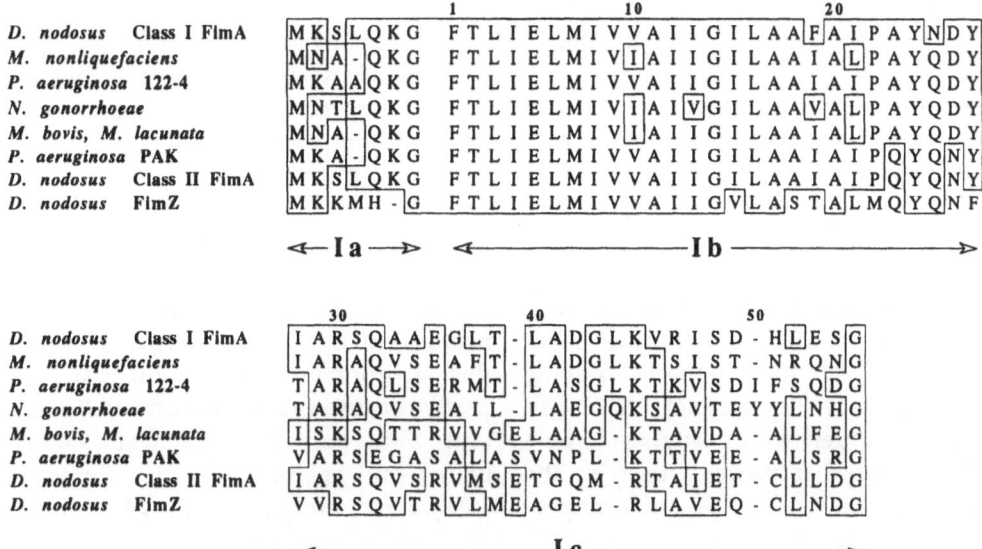

FIGURE 39.3 Sequence comparison of region I of type 4 fimbrial subunits depicted in Figure 39.2, characteristic of type 4 subunits and subdivided according to amino acid composition and degree of sequence conservation into regions Ia, Ib, and Ic (Dalrymple and Mattick, 1987). Gaps (–) introduced to improve alignment. Positions at which three or more sequences share identical amino acid residues are boxed.

range of variants is encountered in different geographical locations, although the relative prevalence of the serogroups may vary (Hindmarsh and Fraser, 1985; Kingsley et al., 1986). As far as can be presently ascertained, these groups appear to encompass essentially the entire range of serological (K-agglutination) variation in the population (Kingsley et al., 1986; Claxton, 1989), although there is some disagreement on this point, and alternative classification systems are not fully cross-referenced (Day et al., 1986; Chetwin et al., 1991). The serological profile of *D. nodosus* correlates with the range of immunity observed *in vivo*, in that effective cross-protection is a function of both the intensity of response and the degree of antigenic similarity between the challenge and vaccinating strain (see Egerton et al., 1987), and is usually limited to related variants classified within the same serogroup (Egerton, 1974; Claxton, 1981; Stewart et al., 1985; Elleman et al., 1990).

The fimbriae of *D. nodosus* are not only responsible for the K-agglutination reaction (Walker et al., 1973; Short et al., 1976; Every, 1979),

but can effectively vaccinate sheep against footrot, at a level commensurate with that obtained with whole cells, although again the extent of immunoprotection is restricted to the serogroup involved (Every and Skerman, 1982; Stewart et al., 1982, 1985; Lee et al., 1983). While isolated fimbrial preparations contain other antigens (Mattick et al., 1984), subsequent electrophoretic and matrix Western transfer analyses using serogroup-specific antisera have confirmed that the relevant serological and immunoprotective epitopes reside on the structural subunit of the fimbrial strand (Anderson et al., 1986, 1987).

Structural and antigenic variation analogous to that described in *D. nodosus* has been observed in the fimbriae of other type 4 bacteria including *M. bovis* (Lehr et al., 1985; Lepper and Hermans, 1986; Moore and Rutter, 1987; Ruehl et al., 1988), *N. gonorrhoeae* (Virji and Heckels, 1983; Meyer et al., 1984) and *P. aeruginosa* (Sastry et al., 1985; Pasloske et al., 1988), although the extent and limits of such variation remain to be defined. At least in some cases, fimbriae from these organisms have also been shown to be immunoprotective (Brinton et al., 1982; Lehr et al., 1985; Lepper, 1988; Boslego et al., 1991).

On the basis of these studies, a number of traditional whole-cell vaccines composed of mixtures of representatives of the major serogroups of *D. nodosus*, have been marketed by animal health companies since 1981. These multivalent vaccines appear to be reasonably effective against footrot infection in the field (Liardet et al., 1986; Plant and Claxton, 1986; Reed, 1986) although there have been some conflicting reports (Mulvaney et al., 1984). However, they are relatively costly, up to five-fold higher than other vaccines used in the sheep industry, and thus are a significant proportion of the value of the animal they are designed to protect. Consequently, their use and acceptance by the pastoral community has been limited, and the control of footrot is still largely dependent on the traditional and laborious methods of hoof paring and antiseptic bathing. Apart from their multivalency, the high cost of conventional footrot vaccines appears to be largely related to the fastidious requirements and sparse growth of *D. nodosus* in culture and the difficulty of obtaining stable fimbrial expression, especially under the liquid fermentation conditions required for large-scale production.

39.2 Cloning and Morphogenetic Expression of the Fimbrial Subunit Gene

Our initial molecular biological studies on *Dichelobacter nodosus* fimbriae were directed at the development of a recombinant DNA-based vaccine to circumvent the problems of fimbrial production by the natural host and to allow greater flexibility in the selection of serological representatives for inclusion in the vaccine. The prototype recombinant studies were carried out using the reference strain of *D. nodosus* serogroup A (Anderson et al., 1984; Elleman et al., 1984). The gene encoding the fimbrial subunit (*fimA*) was cloned in *Escherichia coli* by anti-

body screening of genomic libraries, where it is expressed, albeit poorly, from an associated promoter. However, in these recombinants the subunits are not assembled into mature fimbriae but rather embedded in the bacterial cell envelope, primarily in the inner membrane (Anderson et al., 1984; Elleman et al., 1986a). For both immunological and practical reasons this arrangement appeared unsuitable for vaccine production, partly because of the likely difficulties and costs of extracting the antigen from this environment on an industrial scale and partly because isolated denatured fimbrial subunits elicit poor levels of agglutinating antibodies and are not protective (Emery et al., 1984). Intact fimbriae, on the other hand, elicit good responses and are amenable to simple and industrially applicable procedures for their isolation and purification from the host cell (see Mattick et al., 1985a). Overproduction of the fimbrial subunit in *E. coli* by linking the gene to a strong promoter had little impact on the problem (Elleman et al., 1986a). Vaccination trials using sonicated cells or even purified membranes from recombinant *E. coli* containing high levels of the *D. nodosus* fimbrial subunit demonstrated some retardation of the development of foot lesions but no real reduction in the ultimate severity of the disease.

Although there are several possible explanations for the lack of formation of mature *D. nodosus* fimbriae in recombinant *E. coli* cells, the most likely is that the cloned segments lacked other ancillary genes required for the export and assembly of the fimbrial subunit in this host (see Sections 39.4 and 39.6 below), despite the fact that in some cases as much as 14-kb of *D. nodosus* genomic DNA was present (Hobbs et al., 1991). Studies on the well-characterized type 1 and related fimbrial systems found in enterotoxigenic and uropathogenic strains of *E. coli* itself have shown that fimbrial biosynthesis involves several genes, which are normally clustered (Mooi and de Graaf, 1985; Baga et al., 1987). Apart from the structural subunit, these encode a basal protein located in the outer membrane and an assembly factor located in the periplasmic space, as well as other proteins related to the subunit, which function as adhesins or influence the length and distribution of the fimbriae on the cell. However, these fimbriae are clearly distinct from those occurring in *D. nodosus*, and, not surprisingly, provision of *E. coli* fimbrial assembly genes did not result in the assembly of *D. nodosus* subunits into fimbrial structures on recombinant cells (Anderson et al., 1984). It now appears that there are a relatively large number of genes involved in the biogenesis of type 4 fimbriae in *D. nodosus* and other species, and these are located at more than one position in the genome (see Section 39.6).

The problem of morphogenetic expression of *D. nodosus* fimbriae was in the meantime solved by a different approach. It was evident by this time that the type 4 fimbriae expressed by different bacteria were structurally, functionally and evolutionarily related (Dalrymple and Mattick, 1987; Figure 39.2). In particular, the strong conservation of the hydrophobic amino-terminal region of the subunit and the unusual leader peptide (Figure 39.3) suggested to us that these proteins might share common signals for fimbrial morphogenesis in terms of the interaction of the subunit with other factors in the assembly pathway. If so, the fimbrial subunits containing these signals might be interchangeable throughout the type 4 group and thereby provide a means for the production of *D. nodosus*-type fimbriae from suitably engineered cloned genes (Mattick et al., 1985a, 1985b). *Pseudomonas aeruginosa* was selected from among the known type 4 fimbriate bacteria as the recipient host on the basis that this species is a genetically well-characterized and easily grown aerobe, highly suitable for industrial fermentation, and for which a range of plasmid shuttle vectors are available for gene cloning and transfer.

Dichelobacter nodosus fimbrial subunit genes were found to be poorly expressed in *P. aeruginosa*, a problem that was overcome by placing them under the transcriptional control of a strong promoter, either P_L or P_R from bacteriophage λ. An example of the type of construction employed is shown in Figure 39.4. In this case a 576-bp *DraI* restriction fragment of cloned *D. nodosus* genomic DNA containing the complete *fimA* coding sequence (from 30 bp upstream from the initiation codon to 69 bp downstream from the termination codon, including the 5' ribosome-binding sequence and 3' transcription termination signal), was subcloned into the *HpaI* site downstream from the P_L promoter in the plasmid pP$_L$-λ. The entire gene-promoter construction was then transferred as a *BamHI* cartridge into the corresponding site of the broad host range vector pKT240. Recombinant *P. aeruginosa* cells containing this type of construction were found to produce high yields of fimbriae that were physically, structurally, and antigenically indistinguishable from those produced by the *D. nodosus* strain from which the fimbrial subunit gene was originally derived (Mattick et al., 1985b; Elleman et al., 1986c; Mattick et al., 1987). This was shown by Western transfer analyses (Mattick et al., 1985b, 1987) and electron microscopy with immunogold labeling (Figure 39.5). These fimbriae were also equally as effective as either whole cells or isolated fimbriae from *D. nodosus* in eliciting prophylactic and therapeutic responses against footrot (Egerton et al., 1987); they formed the prototype for the development of a multivalent vaccine for use in the field, which is now in the latter stages of product registration. Quite unexpectedly, there was no trace in these fimbriae of the subunit characteristic of the *P. aeruginosa* host strain, apparently as the result of the very high level of expression of the cloned gene rather than any downregulation of the indigenous gene (Mattick et al., 1987). This also implies that the *P. aeruginosa* cells have little if any preference for assembly of the indigenous over the introduced subunit into fimbrial strands, although this is not always the case (Beard et al., 1990).

Subsequent studies have shown that similar morphogenetic expression of heterologous fimbriae may be obtained in *P. aeruginosa* from cloned subunit genes representative of other *D. nodosus* serogroups, including the quite distinctive class II strains (Elleman, 1988; Mattick et al., 1991; see also Sections 39.3 and 39.4), and more recently from *M. bovis* (Beard et al., 1990), although apparently not from *N. gonorrhoeae*, possible because of differences in the arrangement of the carboxy-terminal region of the pro-

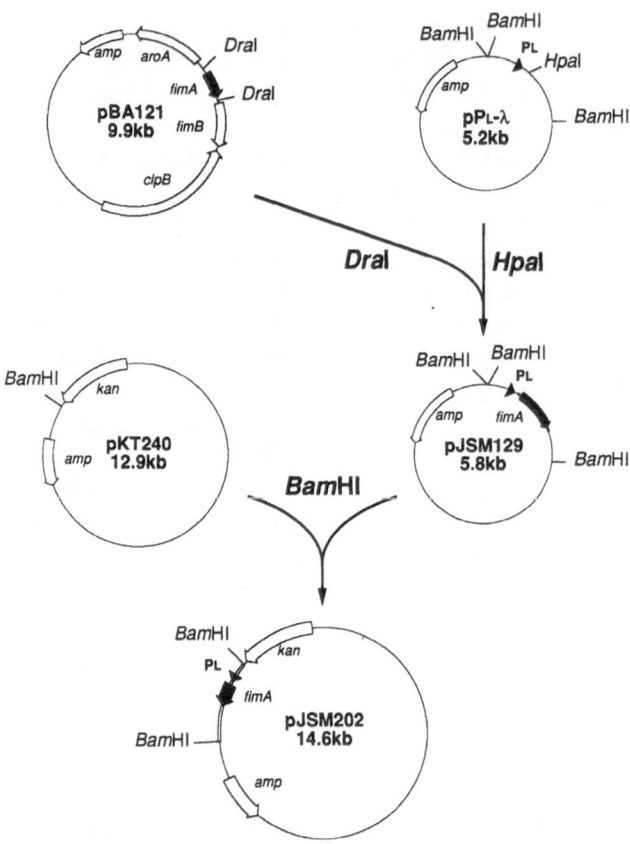

FIGURE 39.4 Construction of plasmid PJSM202 for high-level expression of *Dichelobacter nodosus* fimbrial subunit gene (*fimA*) in *Pseudomonas aeruginosa* (Mattick et al., 1985a, 1987).

tein in this species relative to others (Beard et al., 1990). This system has also been used as a vehicle for the production and export of other peptide epitopes by protein engineering (Jennings et al., 1989).

The recombinant fimbriae produced in *P. aeruginosa* have also been used to address structural and functional questions. Large-scale preparations of highly purified recombinant *D. nodosus* fimbriae have provided a sufficient amount and purity of material to generate crystals for x-ray analysis, which has allowed characterization of the space group and crystal packing (J.H. Parge, D. Christensen, J.S. Mattick, and J.A. Tainer, unpublished data) and should soon permit determination of the atomic structure of this protein. This will provide the structural context to integrate the now extensive primary sequence information (see fol-

lowing). It has also been shown that *P. aeruginosa* recombinants expressing *M. bovis* fimbriae (and containing just 624 bp of *M. bovis* DNA), but not wild-type *P. aeruginosa* controls, will adhere to bovine corneal epithelium, thereby providing direct and unequivocal evidence that the fimbrial subunit rather than some other minor component of the structure is an adhesin (W.W. Ruehl, et al., in preparation).

39.3 Molecular Basis of Antigenic Variation in *Dichelobacter nodosus* Fimbriae

Class I and class II strains

Cloning and sequencing of the fimbrial subunit genes representative of each of the major sero-

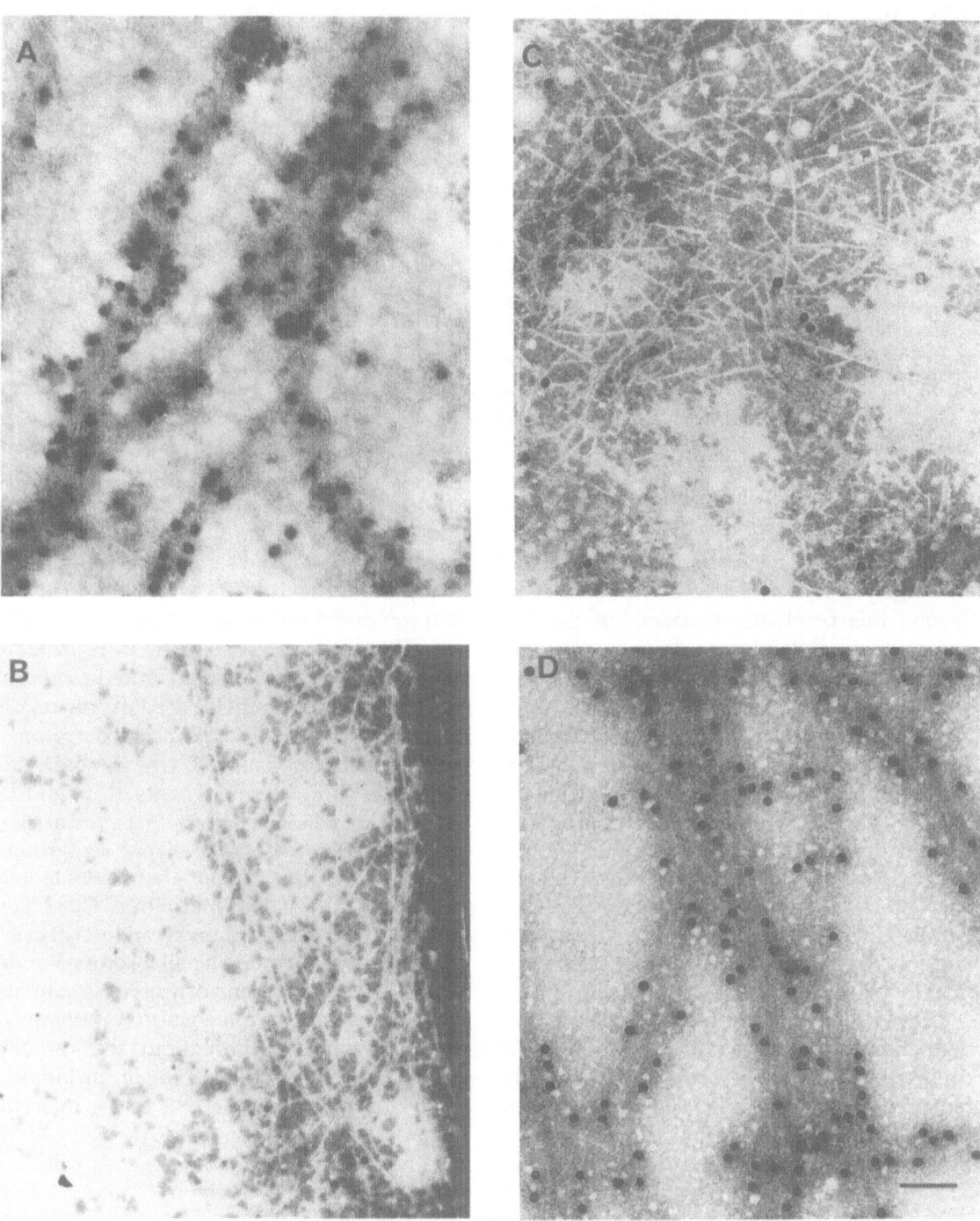

FIGURE 39.5 Immunogold labeling of host and recombinant fimbriae. *Pseudomonas aeruginosa* cells (**A, B**) and same cells harboring pJSM202 (**C, D**), were placed on parlodion-carbon-coated grids and treated with anti-*P. aeruginosa* fimbrial antiserum (**A, C**) or anti-*Dichelobacter nodosus* fimbrial antiserum (**B, D**), and then with protein A labeled with 15-nm colloidal gold particles. Grids were then negatively stained and examined by electron microscopy (Mattick et al., 1987). Figure reproduced by kind permission of the *Journal of Bacteriology*.

groups of *D. nodosus* has revealed that there are, surprisingly, two distinct classes of fimbrial subunits in the population (Elleman, 1988; Mattick, 1989; Mattick et al., 1991; Figure 39.6): class I, which contains most of the serogroups (A, B, C, E, F, G, and I), and class II, including serogroups D and H. These classes are also distinguished by a different physical and genetic organization of the fimbrial gene region (Hobbs et al., 1991; see Section 39.4). Both class I and class II *D. nodosus* fimbrial subunits show absolute conservation in protein sequence in the 7-amino-acid leader sequence and up to residue 18 of the mature protein, which includes the bulk of the highly hydrophobic amino-terminal region (Figure 39.6), both of which are characteristic of type 4 fimbrial subunits in general (Dalrymple and Mattick, 1987). Close sequence similarity extends up to residue 34, with conservative or semiconservative substitutions that are largely but not absolutely class specific. Beyond this point the sequences of the two classes of subunits diverge dramatically, with little obvious similarity except for a loose homology over the following 20 residues, up to the glycine at position 54 (which is conserved in all type 4 fimbrial subunits sequenced to date) and elsewhere in the protein, especially the carboxy-terminal region, wherein there is also some degree of conservation between species (Dalrymple and Mattick, 1987; see Figure 39.3). Surprisingly, in both of these regions the class II fimbrial subunits of *D. nodosus* appear to have more in common with those of other species than they do with class I (see following).

The class I and class II subunits also differ dramatically in the distribution and nature of conserved and variable sequence motifs in the remainder of the protein, as well as in the number and position of cysteine residues. Class I subunits contain cysteine residues at positions 56 and 100 that form a presumptive disulfide loop in the central third (region II) of the protein. Class II subunits contain two closely spaced pairs of cysteine residues, at positions 50 and 62 near the end of the conserved amino-terminal region and at positions 138 and 152 near the carboxy terminus of the protein (see Figure 39.6). There is evidence that disulfide bridges do not occur between these pairs (Elle-

man et al., 1986b), but it is not known whether such bridges are formed within them, although this is quite possibly the case. The second pair of cysteine residues occurs in a similar position and environment in the fimbrial subunits of other type 4 fimbriate species (Dalrymple and Mattick, 1987), whereas the first pair are thus far unique (see Figure 39.3).

Class I subunits: conserved and variable features

The class I fimbrial subunits of *D. nodosus* exhibit relatively close relationships in sequence and organization, essentially consisting of variable regions interspersed with more conserved domains. On the basis of sequence comparisons (Figure 39.6), these proteins may be divided into subclasses: (a) [(A,E,F)(B,I)] and (b) [G,C]. The subunits of serogroups G and C are clearly related and distinct from the others in both conserved and variable regions of the protein. The distinction among A, E, F, B, and I subunits is much less marked, and the subdivision of this larger group is based mainly on differences in the relatively conserved region Ic between residues 37 and 54, wherein the B and I subunits show close similarity to C and G. Beyond this region there are three major clusters of sequence variation between serogroups, one at each end of the putative disulfide loop and one near the carboxy terminus. These clusters not only exhibit sequence variation at every position but also require small insertions or deletions to align adjacent conserved segments. There are also other more isolated positions of variation scattered throughout the carboxy-terminal two-thirds of the protein, including a small cluster right at the end of the molecule (Figure 39.6).

Almost 30 fimbrial subunit genes from virtually all known serotypes of *D. nodosus* have now been sequenced. Integration of these data shows that, excluding interclass or intersubclass comparisons, there are more than 35 and as many as 50 amino acid changes between serogroups, but only 8 to 15 changes between different serotypes within a serogroup and 0 to 5 between independent isolates classified. This suggests that the different serogroups repre-

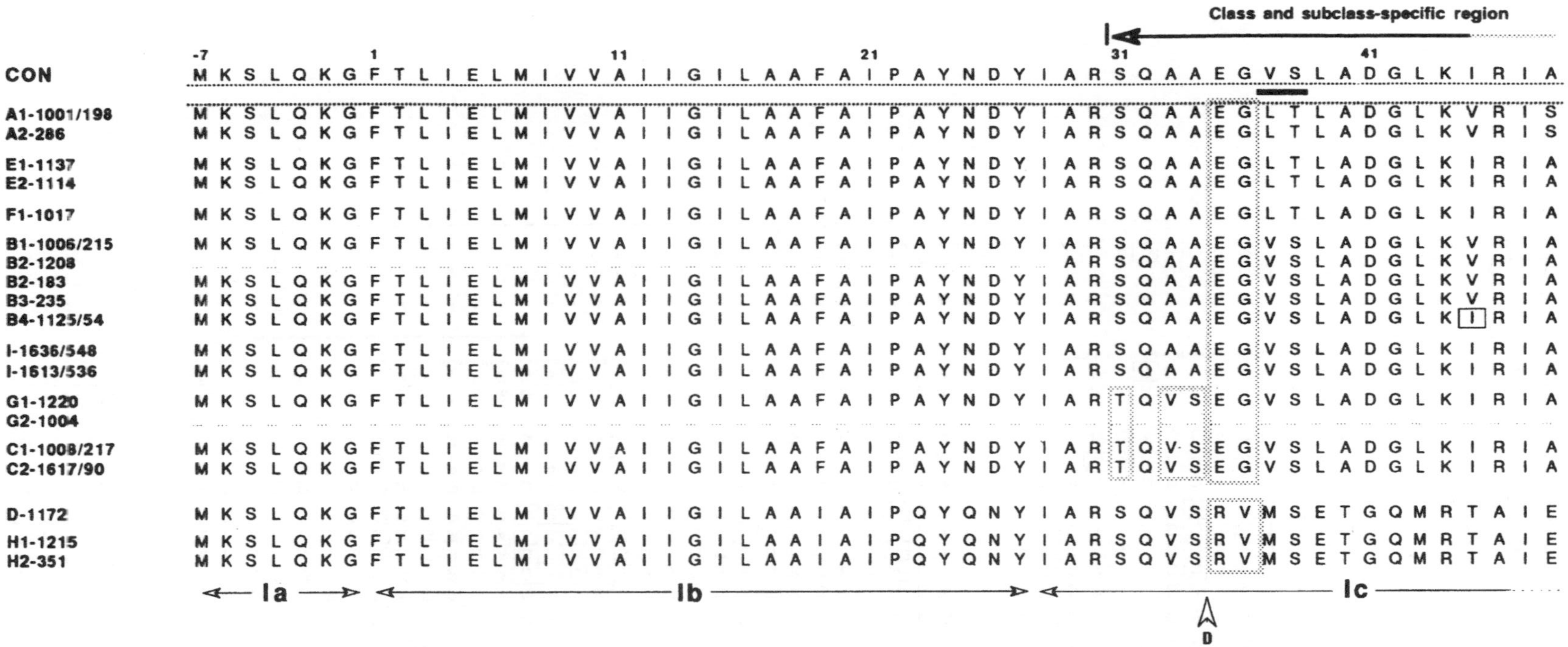

FIGURE 39.6 Predicted protein sequences of *fimA* genes representing every serogroup and serotype at present identified under system of Claxton et al. (1983), except F2 serotype, which is of doubtful validity (see Anderson et al., 1987). Sequences are numbered with respect to phenylalanine residue, which becomes amino terminus of mature subunit. Gaps have been introduced to improve alignment. Regions marked I (a,b,c), II and III are those defined by Dalrymple and Mattick (1987) in comparative analysis of type 4 fimbrial subunits from different species and are based on amino acid composition and degree of sequence conservation exhibited. Consensus sequence is presented above aligned sequences, representing the residue most commonly found at each position in class I (when two residues were found with equal frequency, the serotype A1 sequence was used). Regions where sequence is undetermined in isolates B2-1208 and G2-1004 are indicated by dots (. . .). Because we did not obtain full sequence for B2-1208, predicted amino acid sequence for another B2 isolate, 183, which was determined by Elleman et al. (1990), has been included for completeness. For isolate B4-1125/54, our data were accurate only from position 73 on. In this area, our results matched exactly those of Elleman et al. (1990) for this isolate, and thus for amino terminus of protein, data are complemented by their work. Predicted sequence of A2-286 is taken from Elleman (1988) and that of H2-351 from Hoyne et al. (1989). Possible disulfide bridges are indicated by heavy lines connecting relevant cysteine residues. Class- or subclass-specific sequence motifs are enclosed by stippled boxes. Dark bars under consensus sequence show positions at which variations are characteristic of either individual serogroups or a cluster of serogroups not coincident with class/subclass divisions. Residues that vary at level of serotype or isolate are marked by plain box in body of diagram. Three hypervariant clusters (HVI–III) (Mattick et al., 1991) are enclosed by striped boxes. Point of marked divergence of class I and class II sequences, between positions 34 and 35, is indicated by large open arrowhead.

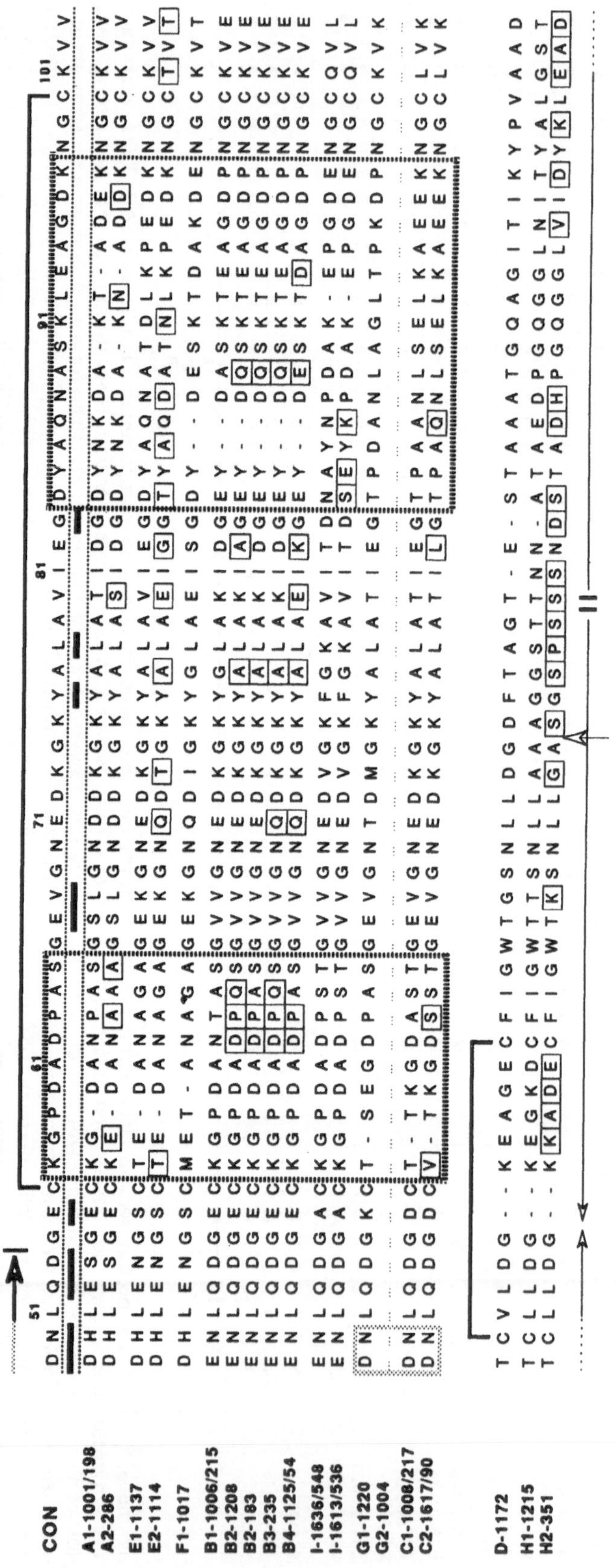

HVI HVII

CON

A1-1001/198
A2-286
E1-1137
E2-1114
F1-1017

B1-1006/215
B2-1208
B2-183
B3-235
B4-1125/54

I-1636/548
I-1613/536

G1-1220
G2-1004

C1-1008/217
C2-1617/90

D-1172
H1-1215
H2-351

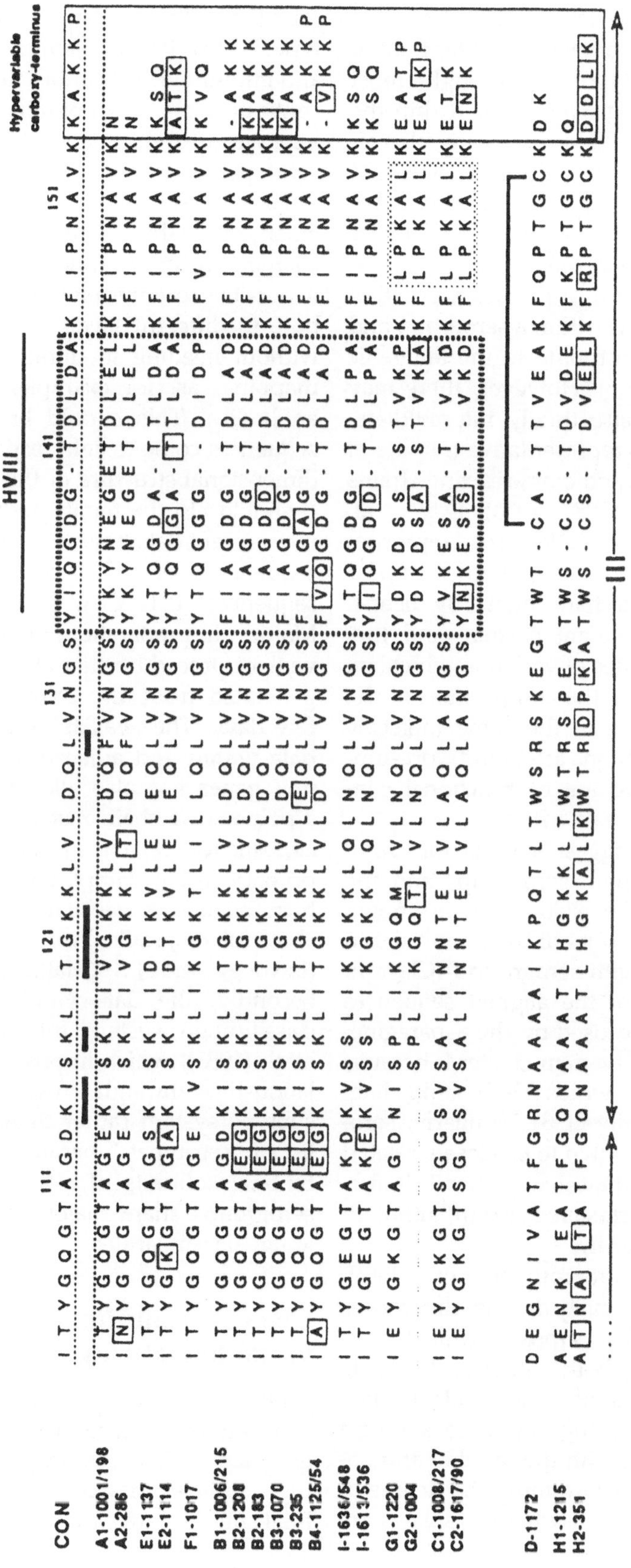

FIGURE 39.6 Continued

sent relatively discrete sets, with amino acid changes within this framework determining the subsidiary serotyping. Although most changes between serogroups occur within the hypervariant clusters, these motifs are relatively conserved within a serogroup, and would appear to be the primary antigenic signatures of serogroup classification (i.e., the major serogroup-specific epitopes). There are also other particular amino acid positions that appear to be conserved within a serogroup and which presumably contribute to the antigenic profile (see Figure 39.6). However, there may be strains that are structurally intermediate, and the serotype H2 representative (Hoyne et al., 1989; Figure 39.6) and cross-reacting strains reported by Claxton (1981, 1986), Day et al. (1986), and Chetwin et al. (1991) may represent cases in point. Nevertheless, it does appear that each serogroup is characterized by distinct motifs in the hypervariant clusters and that these represent cassettes of variation which are restricted in size or by the composition of the other variant cassettes in the same molecule (only certain combinations are found), presumably because of structural or functional constraints in the protein as a whole.

These data also suggest a basis for cross-reaction between certain serogroups and the epitopes involved. For example, it is known that serotypes B2–B4, but not B1, exhibit significant cross-reaction with serogroup I (Claxton, 1986). Examination of the aligned sequences suggests that this is caused by the asparagine-proline motif in hypervariant cluster I, because this is the only area in which B2–B4 and I share residues independently of B1. Similarly, serogroups B and F are known to be closely related (Claxton, 1981), and they share general similarities in hypervariant clusters I and III, although not so much in cluster II.

Changes within a serogroup are not as localized. Single, or occasionally double, differences occur throughout the carboxy-terminal two-thirds of the protein, both within and outside the hypervariant regions. The available data suggest that within a serogroup there is more a apparent continuum, with the number and nature of the changes determining the degree of antigenic relatedness and whether isolates are

classified with the same or a different "serotype." This is consistent with both protein and immunological studies that indicate a more continuous spectrum of minor variation within a serogroup and less discrete divisions between serotypes.

Two other general conclusions begin to emerge from these data. First, the availability of a large number of primary sequences has made it possible to determine which regions contribute to different levels of antigenic character without needing to undertake direct epitope mapping, as, for example, with monoclonal antibodies. This should be refined as more sequences come to light, especially if the three-dimensional structure of the protein is solved. It should also be mentioned that many of the more recent data have been generated by direct sequencing of PCR-amplified fimbrial gene sequences (P.T. Cox, J.R. Egerton, and J.S. Mattick, in preparation), a very rapid and convenient procedure that allows such data to be generated from any primary isolate within a few days. The availability of ample sequence data within and adjacent to the fimbrial subunit genes has also allowed development of a polymerase chain reaction- (PCR-) based diagnostic test capable of detecting D. nodosus in primary lesion samples within hours, rather than the days to weeks required by conventional microbiological culturing (P.T. Cox, S. Rachdawong, and J.S. Mattick, in preparation). Secondly, the data vindicates the serotypic classification system introduced by Claxton et al. (1983), which imposes a hierarchy of relationships through a two-tier serogroup and subtype system rather than a more extended list of "serotypes," because it is clear that the fimbrial subunits of isolates classified within serogroups share major structural as well as antigenic similarities.

Class II subunits: conserved and variable features

With only three representatives for comparison, the significance of the observed homologies and differences and their distribution is more difficult to ascertain. The D and H fimbrial subunits are almost identical until the end of region Ic, the only difference being a conservative valine-isoleucine substitution at position 51 within the first putative disulfide loop. However, the structure of the protein is clearly different from that of class I and is much more closely related to those of *Moraxella bovis* and *P. aeruginosa*, especially in the semiconserved region Ic (see Figure 39.3) and at the end of region II, where there is a disulfide bridge that has been found in all type 4 subunits analyzed to date except *D. nodosus* class I (Dalrymple and Mattick, 1987; Elleman, 1988; Mattick et al., 1991) and *M. nonliquefaciens* (T. Tonjum, C.F. Marrs, F. Rozsa, and K. Bøvre, personal communication; see Figure 39.2). This suggests that the class II subunits may have evolved in a different environment and been acquired by genetic exchange (see Section 39.5). The differences between the prototype strains of serogroup D (which has demonstrated no subsidiary serotypes) and of serotypes H1 and H2 are distributed throughout the carboxy-terminal two-thirds of the protein, with only five short segments (4–5 amino acids) of sequence conservation between them (see Figure 39.6). Although the H2 subunit has more in common with H1 than with D, it is sufficiently different from both in primary sequence to suggest it should constitute a separate serogroup (Hoyne et al., 1989). Which regions or residues may be conserved within class II serogroups, or vary between subsidiary isolates, remain to be determined.

Codon usage and distribution

Analysis of the codon usage in *D. nodosus* fimbrial subunit genes (Mattick et al., 1991), now encompassing some thousands of codons, shows a pattern consistent with that of highly expressed genes in *E. coli* (Grantham et al., 1981; Grosjean and Fiers, 1982; Sharp and Li, 1986), exhibiting third-base preferences related to optimization of codon–anticodon pairings for efficient translation with reasonable fidelity (Grosjean and Fiers, 1982). This is not surprising because the fimbrial subunit is probably the most abundant protein produced by the *D. nodosus* cell (Mattick et al., 1984), and *D. nodosus* and *E. coli* have a similar G + C content (Holdeman et al., 1984). However, there are some exceptions, such as the relative scarcity of the codons GTC and TCC, which are common in other species (Maruyama et al., 1986). Perhaps the most unusual feature is the virtual absence of the codon CTG, which is usually the predominant leucine codon in both procaryotes and eucaryotes (constituting, for example, to 60% of all leucine codons in *E. coli*), although this codon is relatively common in adjacent genes on the *D. nodosus* chromosome (see Hobbs et al., 1991; Mattick et al., 1991). The significance of these observations is unknown.

A surprising aspect of the *D. nodosus* fimbrial subunit genes is the striking imbalance in the distribution of silent base changes in codons specifying conserved amino acids in different regions of the protein (Mattick et al., 1991). Areas upstream from the gene are highly conserved, with no changes until −30 and very few thereafter. This nucleotide conservation also extends into the 5′ coding region of the gene, and all class I and class II sequences are virtually identical upto +71, encompassing the first 23 codons (Mattick et al., 1991). There are also very few base substitutions in codons specifying invariant amino acids in the remainder of the semiconserved regions Ib and Ic. In contrast, a large number of silent base changes occur in codons specifying invariant amino acids in the remainder of the protein. This is particularly evident in the class I gene, where approximately 25% of such codons have silent base substitutions (relative to the preferred codon), a figure close to random after allowance for bias in codon usage (Mattick et al., 1991). This appears to be significantly higher than that found in adjacent areas of the genome, suggesting that the 3′ sequences of the fimbrial subunit genes may represent a region of hypermutability. On the other hand, the almost absolute conservation at the nucleotide level of the 5′ coding sequences, encompassing region Ia and half of region Ib of the protein, suggests that these sequences have some key secondary function, possibly in the regulation of gene expression (at transcriptional

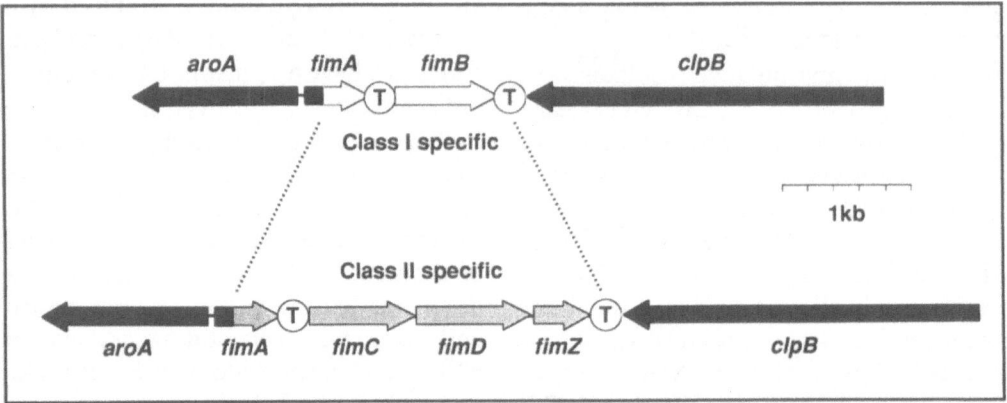

FIGURE 39.7 The *Dichelobacter nodosus* fimbrial gene region. Arrows represent genes; black shading indicates regions which in all strains examined have been shown by restriction mapping and partial DNA sequencing to be highly homologous; dashed lines represent boundaries of putative recombination event that gave rise to class II organization; symbol T represents transcription termination signals.

or translational level) or in mediating recombinational exchange (see Section 39.5).

39.4 Organization and Expression of the Fimbrial Gene Region

The two classes of *Dichelobacter nodosus* strains also exhibit different genomic organization immediately downstream from the fimbrial subunit gene (*fimA*) (Figure 39.7). This appears to be a local alteration rather than an indication of two different subspecies in the *D. nodosus* population, because the remainder of the genome, both at close range (Figure 39.7; see following) and as judged by overall restriction cleavage patterns, appears to be essentially identical.

Upstream from *fimA* in both classes in opposite transcriptional orientation is the gene *aroA*, which encodes the aromatic amino acid biosynthetic enzyme 5-enoypyruvyl shikimate-3-phosphate synthase (Hobbs et al., 1991). Downstream from *fimA* the two classes are quite different, but homology is abruptly restored about 0.9-kb (in class I) and 2.5-kb (in class II) downstream at a bidirectional transcrip-

tion termination signal, beyond which lies a large gene *clpB*, also in opposite orientation. This gene encodes the regulatory subunit of an ATP-dependent protease and is representative of a new class of protein (Gottesman et al., 1990) that has been highly conserved throughout evolutionary history in both procaryotes and eucaryotes (including yeast, trypanosomes, *Drosophila*, and plants), analogous to other well-conserved families such as heat shock proteins, membrane ATPases, and ribosomal proteins.

In both classes the sequences 5' to *fimA* are highly conserved and contain the promoter region (Hobbs et al., 1991; see following). This sequence conservation extends into the *fimA* coding sequences but then diverges within region Ib. Despite this divergence the *fimA* gene in both classes is followed closely by a transcription stop signal, yielding a major transcript 570 to 590 nucleotides in length (Hobbs et al., 1991). Sequence and transcriptional analysis have shown that the promoters of the fimbrial subunit genes in *D. nodosus*, as well as other type 4 fimbriate bacteria, are recognized by an alternative form of RNA polymerase whose sigma (initiation) factor is the *rpoN* gene product (Johnson et al., 1986; Hobbs et al., 1991). RpoN-dependent promoters are activated by *trans*-acting factors (regulators) that bind nearby DNA sequences and which only do so when activated (usually by phosphorylation) in response to an environmental signal often transduced by a sensor protein on the cell surface. Such regulators can bind to sequences in different regions of the genome to induce transcription and can also "cross-talk" to give a fuller dynamic range of responses to a range of environmental stimuli (Taylor et al., 1987; Albright et al., 1989; Deretic et al., 1989; Stock et al., 1990). Thus the fimbrial subunit genes are almost certainly controlled by a sensor-regulator system as part of the coordinate regulation of a suite of colonization factors, including possibly toxins and proteases. Further, it seems these virulence factors also share common infrastructural elements in their export pathway from the cell (see Section 39.6).

Downstream from *fimA*, class I strains contain only one other open reading frame (*fimB*) of 257 codons, whose initiation codon is in some cases only 18 bp from the end of the transcription stop signal at the *fimA*. The translation product of *fimB* is possibly an inner membrane protein involved in fimbrial assembly (Hobbs et al., 1991). It has a similar leader sequence to that of the fimbrial subunit, followed by three predicted transmembrane domains, the third of which exhibits homology to the conserved hydrophobic region of the fimbrial subunit. Sequence and PCR analyses indicate that this gene does not have a separate promoter but rather is cotranscribed with *fimA* at a level attenuated by the strength of the transcription termination signal in the intergenic region. In class II strains *fimA* is followed by a more extended region containing three genes (Hobbs et al., 1991; Figure 39.7), which appear to have a transcriptional arrangement analogous to the class I operon. The first of these genes (*fimC*) encodes a highly hydrophobic protein with no obvious similarity to other genes in these operons. The second gene, *fimD*, encodes a protein of 395 amino acids and may represent an analog of *fimB* although there is no close sequence homology. Beyond *fimD*, at the 3′ end of the class II-specific region, is a variant fimbrial subunit gene (*fimZ*) that is virtually identical in serogroups D and H and which appears to represent a duplicate, possibly redundant, gene closely related to the progenitor of the more divergent structural subunit *fimA* gene found in these strains (Hobbs et al., 1991; Figure 39.8).

```
        -7         -1   1                  10                  20                  30                  40
H fimZ  M K K M H - G   F T L I E L M I V V A I I G V L A S T A L M Q Y Q N F V V R S Q V T R V L M E A G E L R L A V
H fimA  M K S L Q K G   F T L I E L M I V V A I I G I L A A I A I P Q Y Q N Y I A R S Q V S R V M S E T G Q M R T A I
D fimA  M K S L Q K G   F T L I E L M I V V A I I G I L A A I A I P Q Y Q N Y I A R S Q V S R V M S E T G Q M R T A I
D fimZ  M K K M H - G   F T L I E L M I V V A I I G V L A S I A L M Q Y Q N F V V R S Q V T R V L M E A G E L R L A V

        50                  60                  70                  80                  90                  100
H FimZ  E Q C L N D G T T K I G N G Q N E C D P R A S G S N I I S G A S Q N P - E I V I A A N T G - - - V - V Q F P N P
H FimA  E T C L L D G - - K E - - G - K D C F I G W T T S N L L A A A G G S T T N N A T A E D P G Q G G L N I T Y A L G
D FimA  E T C V L D G - - K E A - G - - E C F I G W T G S N L L D G D F T A G T E - S T A A A T G Q A G I T I K Y P V A
D FimZ  E Q C L N D G T T T V G N G A N E C D P R A S G S N I I S G A S Q N P - E I V I A A N T G - - - V - V Q F P N P

          110                 120                 130                 140                 150
H FimZ  L T E E T A L T A T F N N S A A S I I H G K K L I W Q R Q K S G S W Y C H S N A A E K F L P S G C K Y D A S L *
H FimA  S T A E N K I E A T F G Q N A A A T L H G K K L T W T R S P E A T W S C S T D V D E K F K P T G C K - Q *
D FimA  A D D E G N I V A T F G R N A A A A I K P Q T L T W S R S K E G T W T C A T T V E A K F Q P T G C K - D K *
D FimZ  L T E E T A L T A T F N N S A A S I I H G K K L I W Q R Q K S G S W Y C H S N A A E K F L P S G C K Y D A S L *
```

FIGURE 39.8 Comparison of *fimZ*-encoded proteins with *fimA*[II]-encoded subunits. Amino acid sequences predicted from genes *fimZ* and *fimA*[II] in class II strains VCS1215 (serotype H1) and VCS1172 (serogroup D) have been aligned. Gaps (−) have been introduced to improve alignment. Pairs of cysteine residues are joined by lines below sequence blocks. Positions at which either FimA sequence matches at least one other sequence have been boxed. Figure reproduced by kind permission of *Molecular Microbiology*.

39.5 Recombinational Exchange in the Fimbrial Gene Region

The presence of a second subunit gene unique to the class II-specific region highlights the differences between the two types of fimbrial operon in *D. nodosus*, but at the same time provides insight into the evolutionary history of this part of the genome. The close proximity and similarity of the genes *fimA*^{II} and *fimZ* suggest that a gene duplication occurred in class II strains. This must have been an early event because certain motifs (e.g., FIGWT versus DPRAS at positions 66–70 in Figure 39.8) in the protein sequences encoded by these two genes in the D and H1 strains are locus specific rather than serogroup specific. This indicates that substantial divergence of *fimA*^{II} and *fimZ* occurred after this duplication but before the divergence of the D and H serogroups. Since that divergence however, there has been very little change at the *fimZ* locus (4 amino acid differences between D and H), in contrast to the *fimA* locus (46 amino acid differences between D and H).

Gene duplication is not in itself a sufficient explanation for the differences observed between the class I and class II patterns, and two particular observations point to the occurrence of a secondary event. The first is the chimeric nature of *fimA*^{II}, which is more closely related to *fimA*^{I} in region I and to *fimZ* in its remainder (Figure 39.9), and the second is the fact that class II strains lack *fimB*, but rather contain two other apparently unrelated genes, *fimC* and *fimD*. We have considered various possibilities for the origin of *fimZ* and the class II-specific sequences. The first was based on tandem duplication of the fimbrial operon in situ. However, in addition to having to invoke numerous steps to account for the loss of *fimB* and the appearance of *fimC* and *fimD*, this model does not adequately explain the relationship of *fimA*^{II} to *fimA*^{I} and *fimZ*. A second possibility, that duplication occurred as a consequence of the integration of episomal DNA, can also be eliminated when the subunit sequence relationships are considered. Although a reciprocal

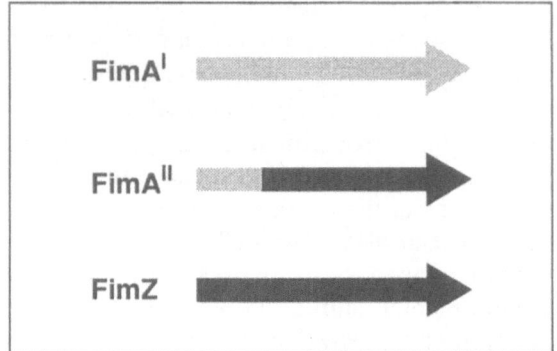

FIGURE 39.9   Relationship of FimA^{I}, FimA^{II}, and FimZ subunits. Arrows represent protein sequences. Two proteins are similarly shaded where sequences are more closely related to each other than to third sequence.

crossover between *fimA* and a second subunit gene carried on an episome would indeed give rise to an integrate comprising two recombinant subunit genes flanking residual episomal DNA, the 3' end of *fimZ* should then be related to the 3' end of *fimA*^{I} rather than *fimA*^{II}, as should also the regions downstream from *fimZ* and *fimA*^{I}. This is clearly not the case.

Because neither of these models is consistent with *fimA*^{II} and *fimZ* resulting from duplication in situ, one or both genes may have originated elsewhere. Further, because low-stringency Southern blotting analyses (data not shown) indicate that there are no other *fimZ*- or *fimA*-like genes in the *D. nodosus* genome, an exogenous source (i.e., not *D. nodosus*) of these sequences must be considered. The only explanation that accounts for the data, including the divergence of *fimA*^{II} and *fimZ* before the divergence of the D and H serogroups as well as the different organization of the class I and class II strains, is that recombination occurred between a class I ancestral genome and a piece of DNA *already* carrying a duplicated *fimZ*-like gene. In effect the 1.3-kb, class I-specific region was replaced by the 2.9-kb, class II-specific region in a direct exchange. Thus, the *fimA*^{II} gene represents a chimera, its 5' end equivalent to that of class I genes, and its 3' end a sibling to that of *fimZ* (see Figure 39.9). If the replacement had occurred in the reverse direction (class I replacing class II), *fimA*^I rather than *fimA*^{II} would have been chimeric.

Two other observations which support the conclusion that the class II-specific gene cluster originated in a species other than *D. nodosus*. First, the variable regions of *fimA*^{II} and *fimZ*-encoded subunits bear more resemblance to those found in other species than to those encoded at *fimA*^I (Elleman, 1988; Mattick et al., 1991). Second, the sequence of the leader oligopeptide (region Ia) has thus far been found to be always conserved within species but distinct between species (Dalrymple and Mattick, 1987), and so the novel leader encoded by *fimZ* is in itself suggestive of the occurrence of interspecies exchange. This signature may allow the likely source of the class II-specific region to be identified in the future as the sequences of type 4 fimbrial subunit genes from other species are determined. In this context it is also interesting to note that a fimbrial subunit gene characterized from a clinical isolate of *Pseudomonas aeruginosa* (K122-4) is quite different from other *P. aeruginosa fimA* sequences (Pasloske et al., 1988) and in fact contains cysteine residues flanking region II, an arrangement hitherto only observed in *D. nodosus* class I subunits and more recently in *Moraxella nonliquefaciens* (Figure 39.2). This subunit is also similar to *D. nodosus* class I in the semiconserved region Ic (Figure 39.3). Thus, exchange of subunit coding sequences may be relatively common among type 4 fimbriate bacteria, a possibility predicted from earlier morphogenetic complementation studies (Mattick et al., 1987), although this may not be universal (Beard et al., 1990).

Figure 39.10 shows the likely recombination sites that mark the boundaries of the introduction of class II DNA into the class I genome. Alignment of the junction sequences of class-specific regions with the right-hand constant region indicates that the right-hand recombination event occurred very near the left-hand boundary of the bidirectional transcription termination signal that separates the class-specific genes from *clpB* (see also Figure 39.7). The left-hand recombination site appears to have occurred within *fimA* between codons 17 and 32 (Figure 39.10).

Analysis of the sequences flanking *fimA* and the class-specific region provides clear evidence that recombinational exchange has also occurred between cells within the *D. nodosus* population, involving both *fimA* and the entire operon. In class I strains there are considerable differences in the intergenic sequences on either side of the transcription termination signal separating *fimA* from *fimB*. These differences fall into two or three basic motifs into which the various strains may be grouped. However, these groupings are not consistent with the relationships among the *fimA* sequences from these various strains, nor are they preserved across the intergenic terminator signal (Hobbs et al., 1991; Mattick et al., 1991). Perhaps the best example of this is provided by serotypes E1, E2, and B1 (Figure 39.11). The E1 and E2 *fimA* coding sequences are as expected closely related, with only 26 (5%) base changes between them (P.T. Cox, J.R. Egerton, and J.S. Mattick, unpublished observations), but much more distantly related to B1, which has approximately 110 (23%) base changes. However, downstream from *fimA* the E1 and E2 sequences diverge totally, and the latter becomes almost totally identical to the corresponding region from B1 (1 base difference in the following 130). These patterns cannot be explained by random drift. The only possible explanation for this observation and the more general lack of correlation of *fimA* relationships with flanking sequence motifs is that recombinational exchange has taken place between strains.

Analysis of the sequences upstream of *fimA*, although more highly conserved, suggests that recombinational exchange has also occurred between entire class I and class II operons, after the acquisition of the latter (Hobbs et al., 1991). For example, the prototype strains of class II serogroups (D and H1) have minor differences in their 5'-noncoding sequences but are identical in this region with two different strains from within the *same* serogroup from class I. The sequences immediately preceding the *clpB* terminator also appear to fall into two groups (Hobbs et al., 1991). Therefore it appears there has been exchange of both *fimA* and of the entire operon among perhaps two to three background lineages of *D. nodosus*, involving homologous sequences in the 5' coding region of

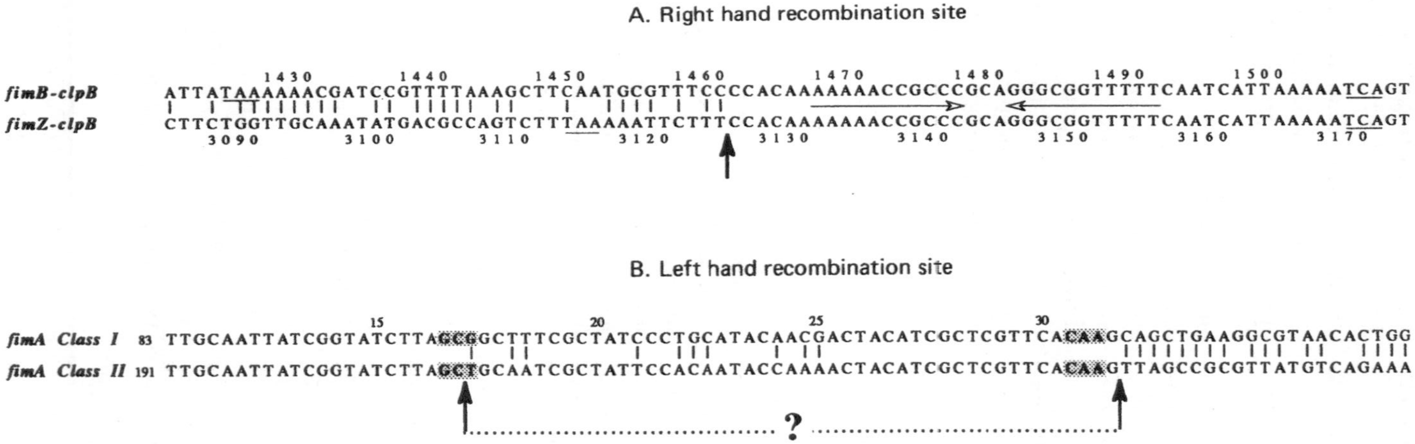

A. Right hand recombination site

```
                          1430      1440      1450      1460      1470      1480      1490      1500
fimB-clpB  ATTATAAAAACGATCCGTATTTTAAAGCTTCAATGCGTTCCCCACAAAAAACCCGCAGGGCGGTTTTCAATCATTAAAAATCAGT
           | ||||||| || ||| |||   |||| |||||||   ||||||||||  ||||||||||||||||||| |||||||||||||||
fimZ-clpB  CTTCTGGTTGCAAATATGACGCCAGTCTTTAAAAATTCTTTCCACAAAAACCGCCGCAGGGCGGTTTTTCAATCATTAAAAATCAGT
                          3090      3100      3110      3120      3130      3140      3150      3160      3170
```

B. Left hand recombination site

```
                              15            20            25            30
fimA Class I   83  TTGCAATTATCGGTATCTTAGCGGCTTTCGCTATCCCTGCATACAACGACTACATCGCTCGTTCACAAGCAGCTGAAGGCGTAACACTGG
                   ||||||||||||||||||||||| || ||| || |||  || || ||||| |||||||||||||||||||| | ||  |  ||   ||||
fimA Class II  191 TTGCAATTATCGGTATCTTAGCGCAATCGCTGCTATTCCACAATACCAAACTACATCGCTCGTTCACAAGTTAGCCGCGTTATGTCAGAAA
```

FIGURE 39.10 Alignment of sequences at boundaries of class I and class II specific regions. Mismatches indicated by vertical bars. (A) Right-hand recombination site. DNA sequences around 3' ends of *fimB* and *fimZ* (Hobbs et al., 1991) have been aligned. *fimB*, *fimZ*, and *clpB* translation termination codons are underlined. Proposed site of right-hand recombination site lies just to left of probable bidirectional transcriptional terminator (convergent arrows) and is marked with upward arrow. (B) Left-hand recombination site. DNA sequences within 5' ends of *fimA*^I and *fimA*^II (VCS1001, serotype A1, and VCS1215, serotype H1; Mattick et al., 1991) have been aligned. Numbers at left of sequences refer to nucleotide position of first residue to appear on line. Region that probably contains site of left-hand recombination site is indicated by two upward arrows joined by horizontal line. This region is approximately bounded by triplets encoding alanine and glutamine at APs 17 and 32 (highlighted). [Figure reproduced by kind permission of *Molecular Microbiology*.]

```
             450                    465                    480              S   495            510
A1-1001/198  GAA CTT AAA TTT ATT CCG AAT GCT GTT AAA AAC ... ... ... TAA TAG...CTAGCTCTTAAA
E1-1137      GAT GCT AAA TTT ATT CCG AAT GCT GTT AAA AAA ... TCA CAA TAA ......CTAGCTCTTAAA
E2-1114      GAT GCT AAA TT[C] ATT CC[A] AAT GC[A] GT[G] AAA [...] [GCA] [ACC] [AAA] TA[G] ...TATCTAGTGTAAACA
B1-1006/215  GCA GAT AAA TTT ATC CCG AAT GCA GTA AAA ... GCT AAA AAA TAG ...TATCTAGTGTAAACA


                        525          540              555              570          585
A1-1001/198  TGC........GAAAGCCTCTCTCTTGAGAGGCTTTTTT-ATGGTTTATTGTTTCTAT--CATTTAAACAAAGGAAA
E1-1137      TGT........GAAAGCCTCTCTTTTGAGAGGCTTTTTTTATGGTTTATTGTTTCTATATCATTTAAACAAAGAAA
E2-1114      T-TAGCTTACTTAAAAGCCTCTCTCTTGAGAGGCTTTTTTT....................ATTTCACCA......
B1-1006/215  T-TAGCTTACTTAAAAGCCTCTCTCTTGAGAGGCTTTTTTT....................ATTTCACCA......
                              ──────>   <──────


                360          615          630              fimB->
A1-1001/198  ATTAACTCATAATCATCTACTCTATATCTTGTC--TAA--GTAGGAGTATATATTC ATG
E1-1137      ATTAACTCATAATCATCTACTCTATATCTTGTC--TAA--GTAGGAGTATATCTTC ATG
E2-1114      ...........................................TCAATC....AGTAAAAAGCT ATG
B1-1006/215  ...........................................TTAATC....AGTAAAAAGCT ATG
                                                        ↑
```

FIGURE 39.11 DNA sequences of four isolates prototypic of serogroups A1, B1, and serotypes E1 and E2 are aligned over terminal portion of *fimA* and intergenic region between *fimA* and *fimB*. Gaps have been inserted to improve alignment of sequences. Numbering above alignment shows nucleotide positions with 1 starting at A of start codon of *fimA*. S indicates stop codon of *fimA*; start codon of *fimB* is also indicated. Convergent horizontal arrows depict position of rho-independent terminator sequence between *fimA* and *fimB*. Within coding sequence of *fimA*, individual nucleotide differences between E1 and E2 prototypes are boxed in E2 sequence. From terminal portion of *fimA* coding sequence on E2 sequence no longer resembles E1 sequence that it did earlier. However, it entirely matches (with single base exception at position 624 in this alignment, as indicated by vertical arrowhead) B1 prototype sequence in this area and into *fimB*. A1 sequence is included because it represents a serogroup closely related to E group by *fimA* sequence comparison.

fimA, in the transcription terminator 3' to *fimA* (at least within class I), and in the bidirectional terminator that forms the border of the fimbrial operon with *clpB*, the first and last of which were also involved in an interspecies exchange.

Active exchange of subunit gene sequences by self-transformation and recombination occurs in *Neisseria gonorrheae* (see Haas and Meyer, 1986; Segal et al., 1986; Seifert et al., 1988). High natural rates of self-transformation occur in this species, as well as in *M. bovis*, and appear to be related to the presence of fimbriae (Bøvre and Frøholm, 1972; Biswas et al., 1977). However, we have not yet been able to demonstrate a similar phenomenon in *D. nodosus* using a streptomycin-resistant mutant, although there may be sequence specificity in this process, as in *Neisseria gonorrhoeae* (Graves et al., 1982; Goodman and Scocca, 1988). Moreover, while each *N. gonorrheae* cell contains many partial subunit genes, we cannot detect additional fimbrial gene sequences (other than the class II variant subunit gene *fimZ*) in the *D. nodosus* genome. In low-stringency Southern blots (data not shown), using probes specific for both the variable and conserved parts of subunit genes as well as an oligonucleotide capable of detecting the conserved 5' region of the subunit gene of several different species, the only hybridizing material corresponded to the regions already cloned.

39.6 Biogenesis of Type 4 Fimbriae

An intriguing aspect of the *Dichelobacter nodosus* fimbrial operons is that the ancillary genes in the class I and class II operons are quite different. Indeed, although genes associated with the fimbrial subunit locus have been defined in other well-studied type 4 fimbriate species, no similarities have yet emerged among these in either sequence or genetic arrangement. In

Pseudomonas aeruginosa, *fimA* is cotranscribed with a downstream tRNA gene to a short mRNA (Dalrymple and Mattick, 1986; Hobbs et al., 1988). Three genes necessary for production of fimbriae (*pil B,C,D*), apparently in the same transcription unit, are located upstream from *fimA* but in the opposite orientation (Nunn et al., 1990). In *Neisseria gonorrhoeae* there are a number of silent cassettes of variant (C-terminal) coding sequences, which are used as a reservoir for antigenic variation by recombination into the *pilE* "expression locus" (Meyer et al., 1982; Siefert et al., 1988). Downstream from *pilE* are the genes *opaE1*, which encode the cell surface opacity protein and which also undergo phase and antigenic variation (although probably by the different mechanism of mutagenesis by replication slippage) (Stern et al., 1986), as well as *pilA* and *pilB*, which appear to activate and repress, respectively, transcription of the subunit gene (Taha et al., 1988). PilA is a homolog of the eucaryotic docking protein (Taha et al., 1991), suggesting it may also play a role in the export of the subunit. A new gene *pilC*, which is located elsewhere on the genome and which appears to encode a necessary basal protein for the fimbriae, has been identified (Jonsson et al., 1991), apparently as a consequence of an earlier report that such a protein might exist in *D. nodosus* (Mattick et al., 1984). In *Moraxella bovis*, the fimbrial subunit gene is associated with an invertible segment that can alternate the expression of two structurally distinct forms of the protein (Fulks et al., 1990). One of the inversion breakpoints occurs within the region of the subunit gene encoding the conserved amino-terminal domain of the protein and has homology to the recombination sites of a number of other bacterial inversion systems (Fulks et al., 1990). Within and adjacent to the invertible segment are two open-reading frames, the functions of which are presently unknown.

Thus none of the ancillary genes defined in one type 4 fimbriate bacteria appears to have a homolog defined in any of the others. This contrasts with the high degree of conservation among the structural subunits, as well as the fact that these proteins are interchangeable at least to some extent and can be effectively processed by the indigenous assembly system of another species. Thus it is difficult to imagine that the ancillary genes have diverged to the point of having no recognizable sequence relationship and yet remain functionally equivalent. The alternative is that there are a relatively large number of genes involved, and that those defined to date (as being adjacent to the subunit gene) represent only a small subset, with the different genetic arrangement of different species being largely idiosyncratic.

Our laboratory has begun to examine the question of the twitching motility phenotype (Henrichsen, 1983), which is associated with the presence of type 4 fimbriae. Twitching motility appears to mediate bacterial translocation across cell surfaces and may be an important virulence factor in vivo (Pedersen et al., 1972; Bradley, 1980; Depiazzi and Richards, 1985). Little is really known about twitching motility, although electron microscope studies have suggested that it is mediated by fimbrial extension and retraction (Bradley, 1972). To explore this phenomenon, and the biogenesis of type 4 fimbriae generally, we have increasingly turned to *P. aeruginosa* because of its physiological and genetic advantages. *P. aeruginosa* mutants that have lost twitching motility display altered colony morphology and resistance to "fimbrial-specific" bacteriophage (Bradley, 1974), and are either lacking fimbriae (fim⁻) or appear to be hyperfimbriated (fim^{NR}, nonretractile). We used phenotypic complementation of the latter to clone the corresponding sequences required for twitching motility.

These sequences were then physically mapped to a *Spe*I fragment at around 20 min on the *P. aeruginosa* chromosome, remote from the major fimbrial locus (around 75 min) where the structural subunit gene (*fimA/pilA*) and ancillary genes required for fimbrial assembly (*pilB, C, and D*) are found. A gene, *pilT*, within the twitching motility region is predicted to encode a 344-amino-acid protein that has strong homology to a variety of other bacterial proteins (Figure 39.12). These include the *P. aeruginosa* PilB gene product, the ComG ORF-1 protein from the *Bacillus subtilis comG* operon (necessary for competence), the PulE protein

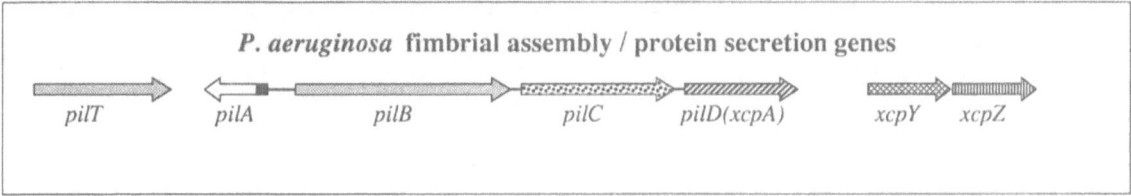

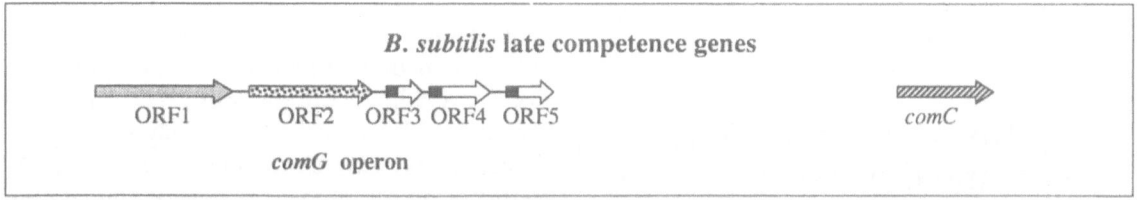

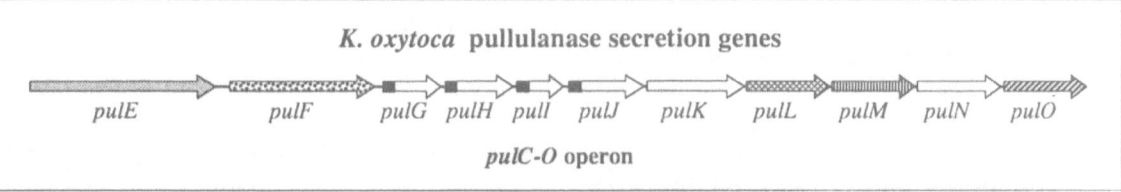

FIGURE 39.12 Homologous genes from three bacterial species, representing specialized protein export system that has been adapted to various different functions. The genetic organisation is shown of *Pseudomonas aeruginosa* fimbrial genes found at *fimA-D* (Nunn et al., 1990) and *pilT* loci, *P. aeruginosa* protein secretion genes found at *xcpl* (*xcpA*) and *xcp5* (*xcpYZ*) loci (Bally et al., 1991; Fil-loux et al., 1991), *Bacillus subtilis* late competence genes found at *comG* (Albano et al., 1989) and *comC* (Mohan et al., 1989) loci, and *Klebsiella oxytoca pulC-O* operon (d'Enfert et al. 1989; Pugsley and Reyss, 1990; Reyss and Pugsley, 1990). Arrows represent size and transcriptional orientation of genes. Similarly shaded genes encode homologous proteins.

from the *Klebsiella oxytoca* (previously *K. pneumoniae*) *pulC-O* operon (involved in pullulanase export), and the VirB-11 protein from the *virB* operon (involved in virulence), which is located on the *Agrobacterium tumefaciens* Ti plasmid. We have also identified additional sets of homologies between other *P. aeruginosa* fimbrial proteins and other *B. subtilis* Com and *K. oxytoca* Pul proteins, which suggest that these are all related members of a new and specialized protein export pathway that is widespread in the *Eubacteria* (Whitchurch et al., 1991). The substrates of this export system are predicted to be proteins which possess type 4 subunit-like leader sequences (Whitchurch et al., 1991), and which themselves are components of complexes located on the cell surface, either as organelles like fimbriae or as mediators of the uptake or export of other macromolecules.

This conclusion has been reinforced by the demonstration (Bally et al., 1991; Strom et al., 1991) that the *pilD* gene (which encodes a leader peptidase) is identical to the *xcpA* gene, which is required for export of a variety of proteins, including exotoxin A, lipase, phospholipase, elastase, and alkaline phosphatase, at least some of which are known virulence factors, and that other genes required for protein export, *xcpY/xcpZ*, share homology with *pulL/pulM* (Filloux et al., 1991). The *pul* and *xcp* operons are known to contain at least 10 to 15 genes (Lazdunski et al., 1990; Pugsley and Reyss, 1990), which suggests the same may be true for fimbrial assembly. The implication of the *xcpA/pilD* identity is that these pathways share some components in common, although there will also be system-specific factors and substrates.

We are currently using saturation transposon mutagenesis of the *P. aeruginosa* genome in an attempt to define all the genes involved in fimbrial biogenesis, as well as those encoding the sensor and regulator of this system and other coregulated targets. The ultimate objective is to obtain an integrated understanding of the mechanism of colonization of eucaryotic hosts by such bacteria and thereby to design more rational and effective strategies to interfere with infection.

39.7 Taxonomy of *Dichelobacter nodosus* and the Distribution of Type 4 Fimbriae

The correct taxonomic position of *D. nodosus* has until recently been unclear. Beveridge (1941) initially placed it in the genus *Fusiformis* (later *Fusobacterium*). It was subsequently transferred to *Ristella* (Prévot, 1948) and then *Bacteroides* (Mraz, 1963), where it had remained until very recently.

The impetus to reexamine the phylogenetic position of *D. nodosus* came essentially from the molecular genetic studies of its fimbriae (Dalrymple and Mattick, 1987), although the possibility that this species might not be a true *Bacteroides* had been foreshadowed on other criteria some years earlier (Shah and Collins, 1983). It has become evident that type 4 fimbriae constitute a distinct genetic system (Dalrymple and Mattick, 1987; Mattick et al., 1987). The type 4 fimbriate species whose subunits have been characterized (*Neisseria gonorrhoeae*, *N. meningitidis*, *Moraxella bovis*, *M. lacunata*, *M. nonliquefaciens*, and *Pseudomonas aeruginosa*) are all classified within the β and γ subdivisions of the *Proteobacteria*. A number of other species appear to possess type 4 fimbriae, based on the polar location of their fimbriae or evidence of the twitching motility phenotype (Table 39.1). Many of these are also located within the β and γ subdivisions whereas thus far no representatives of the β and δ subdivisions show clear evidence of type 4 fimbriae, indicating that this system may be largely restricted to the former taxa and be widely distributed therein. This suggested that the type 4 fimbriate organism *D. nodosus* (then *Bacteroides nodosus*) might actually

belong to one of these branches of the *Proteobacteria* (Dalrymple and Mattick, 1987).

Further, comparison of the *fimA* and *clpB* sequences among different bacteria suggested that *D. nodosus* might be more closely related to the *Neisseriaceae* and pseudomonads than the genus *Bacteroides* (Dalrymple and Mattick, 1987). This was confirmed by 16S rRNA gene sequencing, which places *D. nodosus* in the γ subdivision of *Proteobacteria* (La Fontaine and Rood, 1990). Consequently, *D. nodosus*, together with *Suttonella* (previously *Kingella*) *indologenes* and *Cardiobacterium hominis*, has recently been assigned to a new family within this subdivision, the *Cardiobacteriaceae* (Dewhirst et al., 1990).

Classification of *D. nodosus* illustrates an important problem in bacterial taxonomy, as addressed by Dewhirst et al. (1990). The members of the *Cardiobacteriaceae* are clearly genetically related to one another and also share some distinctive phenotypic characteristics. However, within this family *S. indologenes* and *C. hominis* are aerobic and saccharolytic, whereas *D. nodosus* is anaerobic and saccharolytic (Dewhirst et al., 1990). Because it appears that some phenotypes important in classical descriptive taxonomy (e.g., aerobiosis) can be lost without a correspondingly significant change in genomic content or structure, the question arises as to whether taxonomy should primarily reflect phenotypic relatedness (which is of importance from the viewpoint of clinical or applied microbiology) or genetic relatedness (which is of importance from the viewpoint of evolutionary biology). Certainly, it would appear that anaerobiosis is not necessarily an informative factor in phylogenetic classification.

Number of other gram-negative bacterial species of uncertain phylogenetic position appear to possess type 4 fimbriae, including *Eikenella corrodens*, *Bacteroides ureolyticus*, and *Bacteroides gracilis* (Tanner et al., 1981; Henrichsen, 1983). These species also have the characteristic of corroding the surface of agar plates, which is a feature of many type 4 fimbriate bacteria (see Frøholm and Bøvre, 1972; Henrichsen et al., 1972; Henriksen and Frøholm, 1975). A number of species classified as *Wolinella*

TABLE 39.1 Distribution within *Proteobacteria* of phenotypes associated with type 4 fimbriae.[a]

Taxon	Originally designated or includes	P	T	S
α subdivision				
Bradyrhizobium japonicum	*R. lupini*	⊛		
Rhizobium leguminosarum	*R. phaseoli, R. trifolii*	⊛		
Agrobacterium spp.		⊛		
P. diminuta			○	
P. vesicularis			○	
β subdivision				
Comamonas testosteroni	*P. testosteroni*	●	●	●
C. acidovorans	*P. acidovorans*	●	●	●
Pseudomonas flava			○	
P. cepacia	*P. multivorans*	○	○	
P. mallei			○	
P. solanacearum		●	●	
P. pickettii			○	
Neisseria gonorrhoeae		●	●	●
N. meningitidis		●	●	●
N. lactamica		●	●	
N. cinerea		●	●	
N. flava		●		
N. subflava	*N. pharyngis*	●		
N. perflava		●		
N. elongata		●		
Kingella kingae	*M. kingae*		●	
K. denitrificans			●	
γ subdivision				
P. stutzeeri		●	●	
P. alcaligenes		●	●	
P. pseudoalcaligenes		●	●	
P. fragi		○	○	
P. aeruginosa		●	●	●
P. chlororaphis		○		
P. aureofaciens		○		
P. putida		○	○	
P. fluorescens		○	○	
X. maltophila	*P. maltophila*	●	●	
Acinetobacter calcoaceticus		●	●	
Moraxella bovis		●	●	●
M. nonliquefaciens		●	●	●
M. atlantae		●	●	
M. lacunata		●	●	●
M. osloensis			●	
Branhamella catarrhalis	*N. catarrhalis*	●	●	
Dichelobacter nodosus	*Bacteroides nodosus*	●	●	●
Suttonella indologenes	*Kingella indologenes*	●	●	
Cardiobacterium hominis			●	
Alteromonas putrefaciens	*P. putrefaciens*	●	●	
Pasteurella canis	*Pa. multocida*		●	
Vibrio cholerae				●
Enterobacteriaceae		○	○	○
δ subdivision				
Myxococcus spp.		⊛		

[a]Abbreviations: P, polar fimbriae; T, twitching motility; S, characteristic fimbrial subunit DNA or amino acid sequence; ●, phenotype present; ○, phenotype absent; ⊛, polar fimbriae have been observed, but are probably type 5 (Ottow, 1975). In all other cases, phenotype is not known.

that are anaerobic gram-negative mammalian pathogens (Tanner et al., 1981, 1984) and presently unclassified nongonococcal urethritis strains have also been reported to have twitching motility and agar-corroding activity (Fontaine et al., 1984), and on this basis probably also have type 4 fimbriae.

A similar phenotype is also exhibited by a number of strains isolated from soft-tissue infections of cats and dogs (Love et al., 1984).

By analogy with *D. nodosus*, it is possible that the less well characterized species presently classified is the genera *Bacteroides*, *Eikenella*, and *Wolinella* may also lie within the β and γ subdivisions of the *Proteobacteria*. Certainly the available evidence suggests that type 4 fimbriae were present in the ancestor of these groups, although the observation that such fimbriae may also be found in some gram-positive species [*Streptococcus sanguis* (Henriksen and Henrichsen, 1975) and *Eubacterium fossor* (Bailey and Love, 1986)] suggests that this system may have a deeper ancestry or have been acquired on an ad hoc basis by some other species whose ecology may require colonization of eucaryotic cells. However, given the apparently complex infrastructure required for biogenesis and assembly of type 4 fimbriae, any lateral acquisition of this system must have occurred before to the dispersal of the relevant genes around the genome (as is clearly the case in those species partially characterized to date) or from a species where this has not yet occurred.

Acknowledgements. The support of the Australian Meat and Livestock Research and Development Corporation, the Australian Wool Corporation, and the National Health and Medical Research Council, during the different phases of this project, is gratefully acknowledged.

References

Albano, M., R. Breitling, and D.A. Dubnau. 1989. Nucleotide sequence and genetic organisation of the *Bacillus subtilis comG* operon. *J, Bacteriol.* 171:5386–5404.

Albright, L.M., E. Huala, and F.M. Ausubel. 1989. Prokaryotic signal transduction mediated by sensor and regulator protein pairs. *Annu. Rev. Genet.* 23:311–336.

Anderson, B.J., M.M. Bills, J.R. Egerton, and J.S. Mattick. 1984. Cloning and expression in *Escherichia coli* of the gene encoding the structural subunit of *Bacteroides nodosus* fimbriae. *J. Bacteriol.* 160:748–754.

Anderson, B.J., C.L. Kristo, J.R. Egerton, and J. Mattick. 1986. Variation in the structural subunit and basal protein antigens of *Bacteroides nodosus* fimbriae. *J. Bacteriol.* 166:453–460.

Anderson, B.J., J.S. Mattick, P.T. Cox, C.L. Kristo, and J.R. Egerton. 1987. Western blot (immunoblot) analysis of the fimbrial antigens of *Bacteroides nodosus*. *J. Bacteriol.* 169:4018–4023.

Baga, M., M. Norgren, and S. Normark. 1987. Biogenesis of *E. coli* pap pili: papH a minor pilin subunit involved in cell anchoring and length modulation. *Cell* 49:241–251.

Bailey, G.D. and D.N. Love. 1986. *Eubacterium fossor* sp. nov., an agar-corroding organism from normal pharynx and oral and respiratory tract lesions of horses. *Int. J. Syst. Bacteriol.* 36:383–387.

Bally, M., G. Ball, A. Badere, and A. Lazdunski. 1991. Protein secretion in *Pseudomonas aeruginosa*: the *xcpA* gene encodes an integral inner membrane protein homologous to *Klebsiella pneumoniae* secretion function protein PulO. *J. Bacteriol.* 173:479–486.

Beard, M.K.M., J.S. Mattick, L.J. Moore, M.R. Mott, C.F. Marrs, and J.R. Egerton. 1990. Morphogenetic expression of *Moraxella bovis* fimbriae (pili) in *Pseudomonas aeruginosa*. *J. Bacteriol.* 172:2601–2607.

Beveridge, W.I.B. 1941. Foot-rot in sheep: a transmissable disease due to infection with *Fusiformis nodosus* (n. sp.). Bull. No. 140, Council for Scientific and Industrial Research, Melbourne, Australia.

Biswas, G.D., T. Sox, E. Blackman, and P.F. Sparling. 1977. Factors affecting genetic transformation of *Neisseria gonorrhoeae*. *J. Bacteriol.* 129:983–992.

Boslego, J.W., E.C. Tramont, R.C. Chung, D.G. McChesney, J. Ciak, J.C. Sadoff, M.P. Piziak, J.D. Brown, C.C. Brinton, Jr., S.W. Wood, and J.R. Bryan, 1991. Efficacy of a parenteral gonococcal pilus vaccine in men. *Vaccine* 9:153–162.

Bradley, D.E. 1972. Evidence for the retraction of *Pseudomonas aeruginosa* RNA phage pili. *Biochem. Biophys. Res. Commun.* 47:142–149.

Bradley, D.E. 1974. The adsorption of *Pseudomonas aeruginosa* pilus-dependent bacteriophages to a host mutant with nonretractile pili. *Virology* 58:149–163.

Bradley, D.E. 1980. A function of *Pseudomonas aeruginosa* PAO pili: twitching motility. *Can. J. Microbiol.* 26:146–154.

Brinton, C.C., S.W. Wood, A. Brown, A.M. Labik, J.R. Bryan, and S.W. Lee. 1982. The development of a Neisserial pilus vaccine for gonorrhea and meningococcal meningitis. *In: Seminars in*

Infectious Disease, Vol IV: Bacterial Vaccines, J.B. Robbins, J.C. Hill, and J.C. Sadoff, eds., pp. 140–159. New York: Thieme-Stratton.

Bøvre, K. and L.O. Frøholm. 1972. Competence in genetic transformation related to colony type and fimbriation in three species of *Moraxella*. *Acta Path. Microbiol. Scand. Sect. B* **80**:649–659.

Chetwin, D.H., L.C. Whitehead, and S.E.J. Thorley. 1991. The recognition and prevalence of *Bacteroides nodosus* serotype M in Australia and New Zealand. *Aust. Vet. J.* **68**:154–155.

Claxton, P.D. 1981. Studies on *Bacteroides nodosus* vaccines with particular reference to factors associated with their efficacy in protecting against ovine virulent footrot. Ph.D. thesis, University of Sydney, Australia.

Claxton, P.D. 1986. Serogrouping of *Bacteroides nodosus* isolates. In: *Footrot in Ruminants: Proceedings of a Workshop, Melbourne, 1985.* D.J. Stewart, N.M. McKern, and D.L. Emery, eds., pp. 131–134. Syndey, Australia: CSIRO Press.

Claxton, P.D. 1989. Antigenic classification of *Bacteroides nodosus*. In: *Footrot and Foot Abscess of Ruminants.* J.R. Egerton, W.K. Yong, and G.G. Riffkin, eds., pp. 155–166. Boca Raton, Florida: CRC Press.

Claxton, P.D., L.A. Ribeiro, and J.R. Egerton. 1983. Classification of *Bacteroides nodosus* by agglutination tests. *Aust. Vet. J.* **60**:331–334.

d'Enfert, C., I. Reyss, C. Wandersman, and A.P. Pugsley. 1989. Protein secretion by Gram-negative bacteria. Characterization of two membrane proteins required for pullulanase secretion by *Escherichia coli* K-12. *J. Biol. Chem.* **264**:17462–17468.

Dalrymple, B. and J.S. Mattick. 1986. Genes encoding threonine tRNAs with the anticodon CGU from *Escherichia coli* and *Pseudomonas aeruginosa*. *Biochem. Int.* **13**:547–553.

Dalrymple, B. and J.S. Mattick. 1987. An analysis of the organization and evolution of type 4 fimbrial (MePhe) subunit proteins. *J. Mol. Evol.* **25**:261–269.

Day, S.E.J., C.M. Thorley, and J.E. Beesley. 1986. Serotyping of *Bacteroides nodosus*: proposal for 9 further serotypes (J-R) and a study of the antigenic complexity of *B. nodosus* pili. In: *Footrot in Ruminants: Proceedings of a Workshop, Melbourne, 1985.* D.J. Stewart, N.M. McKern, and D.L. Emery, eds., pp. 147–159, Sydney, Australia: CSIRO Press.

Depiazzi, L.J. and R.B. Richards. 1985. Motility in relation to virulence of *Bacteroides nodosus*. *Vet. Microbiol.* **10**:107–116.

Deretic, V., W.M. Konyecsni, C.D. Mohr, D.W. Martin, and N.S. Hibler. 1989. Common denominators of promoter control in *Pseudomonas* and other bacteria. *Bio/Technology* **7**:1249–254.

Dewhirst, F.E., B.J. Paster, S. La Fontaine, and J.I. Rood, 1990. Transfer of *Kingella indologenes* (Snell and Lapage 1976) to the genus *Suttonella* gen. nov. as *Suttonella indologenes* comb. nov.; transfer of *Bacteroides nodosus* (Beveridge 1941) to the genus *Dichelobacter* gen. nov. as *Dichelobacter nodosus* comb. nov.; and assignment of the genera *Cardiobacterium*, *Dichelobacter* and *Suttonella* to *Cardiobacteriaceae* fam. nov. in the gamma subdivision of *Proteobacteria* based upon 16S ribosomal ribonucleic acid sequence comparisons. *Int. J. Syst. Bacteriol.* **40**:426–433.

Egerton, J.R. 1973. Surface and somatic antigens of *Fusiform nodosus*. *J. Comp. Pathol.* **83**:151–159.

Egerton, J.R. 1974. Significance of *Fusiformis nodosus* serotypes in resistance of vaccinated sheep to experimental foot-rot. *Aust. Vet. J.* **50**:59–62.

Egerton, J.R. 1977. Foot-rot of sheep—pathogenesis and immunity. *Prog. Immunol.* **3**:645–650.

Egerton, J.R., P.T. Cox, B.J. Anderson, K. Kristo, M. Norman, and J.S. Mattick. 1987. The protection of sheep against footrot with a recombinant DNA-based fimbrial vaccine. *Vet. Microbiol.* **14**:393–409.

Elleman, T. 1988. Pilins of *Bacteroides nodosus*: molecular basis of serotypic variation and relationships to other bacterial pilins. *Microbiol. Rev.* **52**:233–247.

Elleman, T.C., P.A. Hoyne, D.L. Emery, D.J. Stewart, and B.L. Clark. 1984. Isolation of the gene encoding pilin of *Bacteroides nodosus* (strain 198), the causal organism of ovine footrot. *FEBS Lett.* **173**:103–107.

Elleman, T.C., P.A. Hoyne, D.L. Emery, D.J. Stewart, and B.L. Clark. 1986a. Expression of the pilin gene from *Bacteroides nodosus* in *Escherichia coli*. *Infect. Immun.* **51**:187–192.

Elleman, T.C., P.A. Hoyne, N.M. McKern, and D.J. Stewart. 1986b. Nucleotide sequence of the gene encoding the two-subunit pilin of *Bacteroides nodosus* 265. *J. Bacteriol.* **167**:243–250.

Elleman, T.C., P.A. Hoyne, D.J. Stewart, N.M. McKern, and J.E. Peterson. 1986c. Expression of pili from *Bacteroides nodosus* in *Pseudomonas aeruginosa*. *J. Bacteriol.* **168**:574–580.

Elleman, T.C., D.J. Stewart, K.G. Finney, P.A. Hoyne, and C.W. Ward. 1990. Pilins from the B-serogroup of *Bacteroides nodosus*—characterization, expression, and cross-protection. *Infect. Immun.* **58**:1545–1551.

Emery, D.L., D.J. Stewart, and B.L. Clark. 1984. The structural integrity of pili from *Bacteroides nodosus* is required to elicit protective immunity against footrot. *Aust. Vet. J.* **61**:237–238.

Every, D. 1979. Purification of pili from *Bacteroides nodosus* and an examination of their chemical physical and serological properties. *J. Gen. Microbiol.* **115**:309–316.

Every, D. and T.M. Skerman. 1982. Protection of sheep against experimental footrot by vaccination with pili purified from *Bacteroides nodosus*.

N.Z. Vet, J. **30**:156–158.

Every, D. and T.M. Skerman. 1983. Surface structure of *Bacteroides nodosus* in relation to virulence and immunoprotection in sheep. *J. Gen. Microbiol.* **129**:225–234.

Filloux, A., M. Bally, G. Ball, M. Akrim, J. Tommasson, and A. Lazdunski. 1991. Protein secretion in Gram-negative bacteria: transport across the outer membrane involves common mechanisms in different bacteria. *EMBO J.* **9**:4323–4329.

Fontaine, E.A.R., S.P. Borriello, D. Taylor-Robinson, and H.A. Davies. 1984. Differential characteristics of a small gram-negative anaerobe associated with non-gonococcal urethritis which morphologically resembles *Bacteriodes ureolyticus*. *Scand. J. Urol. Nephrol.* **86**:157–165 (Suppl.).

Frøholm, L.O.F. and K. Bøvre. 1972. Fimbriation associated with the spreading-corroding colony type in *Moraxella kingii*. *Acta Pathol. Microbiol. Scand. Sect. B* **80**:641–648.

Fulks, K.A., C.F. Marrs, S.P. Stevens, and M.R. Green. 1990. Sequence analysis of the inversion region containing the pilin genes of *Moraxella bovis*. *J. Bacteriol.* **172**:310–316.

Goodman, S.D. and J.J. Scocca. 1988. Identification and arrangement of the DNA sequence recognised in specific transformation of *Neisseria gonorrhoeae*. *Proc. Natl. Acad. Sci. USA* **85**:6982–6986.

Gottesman, S., C. Squires, E. Pichersky, M. Carrington, M. Hobbs, J.S. Mattick, B. Dalrymple, H. Kuramitsu, T. Shiroza, T. Foster, W.P. Clark, B. Ross, C.L. Squires, and M.R. Maurizi. 1990. Conservation of the regulatory subunit for the Clp ATP-dependent protease in prokaryotes and eukaryotes. *Proc. Natl. Acad. Sci. USA* **87**:3513–3517.

Grantham, R., C. Gautier, M. Gouy, M. Jacobzone, and R. Mercier. 1981. Codon catalog usage is a genome strategy modulated for gene expressivity. *Nucleic Acids Res.* **9**:43–75.

Graves, J.F., G.D. Biswas, and P.F. Sparling. 1982. Sequence-specific DNA uptake in transformation of *Neisseria gonorrhoeae*. *J. Bacteriol.* **152**:1071–1077.

Grosjean, H. and W. Fiers. 1982. Preferential codon usage in prokaryotic genes: the optimal codon-anticodon interaction energy and the selective codon usage in efficiently expressed genes. *Gene* (Amst.) **18**:199–209.

Haas, R. and T.F. Meyer. 1986. The repertoire of silent pilus genes in *Neisseria gonorrhoeae*: evidence for gene conversion. *Cell* **44**:107–115.

Henrichsen, J. 1983. Twitching motility *Annu. Rev. Microbiol.* **37**:81–93.

Henrichsen, J., L.O. Frøholm, and K. Bøvre. 1972. Studies on bacterial surface translocation. 2. Correlation of twitching motility and fimbriation in colony variants of *Moraxella nonliquefaciens, M. bovis* and *M. kingii*. *Acta. Pathol. Microbiol. Scand. Sect. B* **80**:445–452.

Henriksen, S.D. and L.O. Frøholm. 1975. A fimbriated strain of *Pasteurella multocida* with spreading and corroding colonies. *Acta Pathol. Microbiol. Scand. Sect. B* **83**:129–132.

Henriksen, S.D. and J. Henrichsen. 1975. Twitching motility and possession of polar fimbriae in spreading *Streptococcus sanguis* isolates from the human throat. *Acta. Pathol. Microbiol. Scand. Sect. B* **83**:133–140.

Hindmarsh, F. and J. Fraser. 1985. Serogroups of *Bacteroides nodosus* isolated from ovine footrot in Britain. *Vet. Res.* **116**:187–188.

Hobbs, M., B. Dalrymple, S.F. Delaney, and I.S. Mattick. 1988. Transcription of the fimbrial subunit gene and an associated transfer RNA gene of *Pseudomonas aeruginosa*. *Gene* (Amst.) **62**:219–227.

Hobbs, M., B. Dalrymple, P.T. Cox, S.P. Livingston, S.F. Delaney, and J.S. Mattick. 1991. Organisation of the fimbrial gene region of *Bacteroides nodosus*: class I and class II strains. *Mol. Microbiol.* **5**:543–560.

Holdeman, L.V., R.W. Kelley, and W.E.C. Moore. 1984. Family I. The *Bacteroidaceae*. Genus I. *Bacteroides*. In: *Bergey's Manual of Systematic Bacteriology, Vol. 1*, N.R. Kreig, and J.G. Holt, eds. pp. 604–631. Baltimore: Williams & Wilkins.

Hoyne, P.A., T.C. Elleman, N.M. McKern, and D.J. Stewart. 1989. Sequence of pilin from *Bacteroides nodosus* 351 (serogroup H) and implications for serogroup classification. *J. Gen. Microbiol.* **135**:1113–1122.

Jennings, P.A., M.M. Bills, D.O. Irving, and J.S. Mattick. 1989. Fimbriae of *Bacteroides nodosus*: protein engineering of the structural subunit for the production of an exogenous peptide. *Protein Eng.* **2**:365–369.

Johnson, K., M.L. Parker, and S. Lory. 1986. Nucleotide sequence and transcriptional initiation site of two *Pseudomonas aeruginosa* pilin genes. *J. Biol. Chem.* **261**:15703–15708.

Jonsson, A.B., G. Nyberg, and S. Normark. 1991. Phase variation of gonococcal pili by frameshift mutation in *pilC*, a novel gene for pilus assembly. *EMBO J.* **10**:477–488.

Kingsley, D.F., F.H. Hindmarsh, D.M. Liardet, and D.H. Chetwin. 1986. Distribution of serogroups of *Bacteroides nodosus* with particular reference to New Zealand and the United Kingdom. In: *Footrot in Ruminants: Proceedings of a Workshop, Melbourne, 1985*, D.J. Stewart, N.M. McKern, and D.L. Emery, eds., pp. 143–146. Sydney, Australia: CSIRO Press.

La Fontaine, S. and J.R. Rood. 1990. Evidence that *Bacteroides nodosus* belongs in subgroup gamma of the class *Proteobacteria*, not in the genus *Bacteroides*: partial sequence analysis of a *B. nodosus*

16S rRNA gene. *Int. J. Syst. Bacteriol.* **40**:154–159.

Lazdunski, A., J. Guzzo, A. Filloux, M. Bally, and M. Murgier. 1990. Secretion of extracellular proteins by *Pseudomonas aeruginosa. Biochemie* (Paris) **72**:147–156.

Lee, S.W., B. Alexander, and B. McGowan. 1983. Purification characterization and serologic characteristics of *Bacteroides nodosus* pili and the use of a purified vaccine in sheep. *Am. J. Vet. Res.* **44**:1676–1681.

Lehr, C., H.G. Jayappa, and R.A. Goodnow. 1985. Serological and protective characterization of *Moraxella bovis* pili. *Cornell Vet.* **75**:484–492.

Lepper, A.W. 1988. Vaccination against infectious bovine keratoconjunctivitis: protective efficacy and antibody response induced by pili of homologous and heterologous strains of *Moraxella bovis. Aust. Vet. J.* **65**:310–316.

Lepper, A.W.D. and L.R. Hermans. 1986. Characterisation and quantitation of pilus antigens of *Moraxella bovis* by ELISA. *Aust. Vet. J.* **63**:401–405.

Liardet, D.M., D.H. Chetwin, D.F. Kingsley, and F.H. Hindmarsh. 1986. Results of field trials in New Zealand to confirm the protective and curative effects of a 10-strain ovine footrot vaccine. *In: Footrot in Ruminants: Proceedings of a Workshop, Melbourne, 1985*, D.J. Stewart, N.M. McKern, and D.L. Emery, eds., pp. 181–184. Sydney, Australia: CSIRO Press.

Love, D.N., R.F. Jones, M. Bailey, and A. Calverly. 1984. Comparison of strains of gram-negative anaerobic agar-corroding rods isolated from soft tissue infection in cats and dogs with type strains of *Bacteroides gracilis, Wolinella recta, Wolinella succinogenes* and *Campylobacter concisus. J. Clin. Microbiol.* **20**:747–750.

Marrs, C.F., G. Schoolnik, J.M. Koomey, J. Hardy, J. Rothbard, and S. Falkow. 1985. Cloning and sequencing of a *Moraxella bovis* pilin gene. *J. Bacteriol.* **163**:132–139.

Maruyama, T., T. Gojobori, S. Aota, and T. Ikemura. 1986. Codon usage tabulated from the Genbank genetic sequence data. *Nucleic Acids Res.* **14**:r151–r197.

Mattick, J.S. 1989. The molecular biology of the fimbriae (pili) of *Bacteroides nodosus* and the development of a recombinant DNA-based vaccine. *In: Footrot and Foot Abscess of Ruminants*, J.R. Egerton, W.K. Yong, and G.G. Riffkin, eds., pp. 195–218. Boca Raton, Florida: CRC Press.

Mattick, J.S., B.J. Anderson, M.R. Mott, and J.R. Egerton. 1984. Isolation and characterization of *Bacteroides nodosus* fimbriae: structural subunit and basal protein antigens. *J. Bacteriol.* **160**:740–747.

Mattick, J.S., B.J. Anderson, J.R. Egerton. 1985a. Molecular biology and footrot of sheep. *In: Reviews in Rural Science, Vol. 6, Biotechnology and Recombinant DNA Technology in the Animal Production Industries*, R.A. Leng, J.S.F. Barker, D.B. Adams, and K.R. Hutchinson, eds., pp. 79–91. Armidale, N.S.W.: University of New England.

Mattick, J.S., B.J. Anderson, and T.C. Elleman. 1985b. Australian Patent Application 50154/85; New Zealand Patent Application 214017; European Patent Application 85905494.2; Patent Corporation Treaty Application W086/02557.

Mattick, J.S., M.M. Bills, B.J. Anderson, B. Dalrymple, M.R. Mott, and J.R. Egerton. 1987. Morphogenetic expression of *Bacteroides nodosus* fimbriae in *Pseudomonas aeruginosa. J. Bacteriol.* **169**:33–41.

Mattick, J.S., B.J. Anderson, P.T. Cox, B.P. Dalrymple, M.M. Bills, M. Hobbs, and J.R. Egerton, 1991. Gene sequences and comparison of the fimbrial subunits representative of *Bacteroides nodosus* serogroups A to I: Class I and class II strains. *Mol. Microbiol.* **5**:561–573.

Meyer, T.F., N. Mlawer, and M. So. 1982. Pilus expression in *Neisseria gonorrhoeae* involves chromosomal rearrangement. *Cell* **30**:45–52.

Meyer, T.F., E. Billyard, R. Haas, S. Storzbach, and M. So. 1984. Pilus genes of *Neisseria gonorrheae*: chromosomal organization and DNA sequence. *Proc. Natl. Acad. Sci. USA* **81**:6110–6114.

Mohan, S., J. Aghion, N. Guillen, and D.A. Dubnau. 1989. Molecular cloning and characterization of *comC*, a late competence gene of *Bacillus subtilis. J. Bacteriol.* **171**:6043–6051.

Mooi, F.R. and F.K. de Graaf. 1985. Molecular biology of fimbriae of enterotoxigenic *Escherichia coli. Curr. Topics Microbiol. Immunol.* **118**:119–138.

Moore, L.J. and J.M. Rutter. 1987. Antigenic analysis of fimbrial proteins from *Moraxella bovis. J. Clin. Microbiol.* **25**:2063–2070.

Moore, L.J. and J.M. Rutter. 1989. Attachment of *Moraxella bovis* to calf corneal cells and inhibition by antiserum. *Aust. Vet. J.* **66**:39–42.

Mraz, O. 1963. Schizomycetes. *In: Nomina und Svnonyma der Pathogenen und Saprophytaren Microben*, O. Mraz, J. Tesarcik, and F. Varejka, eds., p. 85. Jena, GDR: VEB Gustav Fischer Verlag.

Mulvaney, C.J., R. Jackson, and A.J. Jopp. 1984. Field trials with a killed nine-strain, oil adjuvanted *Bacteroides nodosus* footrot vaccine in sheep. *N.Z. Vet. J.* **32**:137–139.

Nunn, D., S. Bergman, and S. Lory. 1990. Products of three accessory genes, *pilB, pilC* and *pilD*, are required for biogenesis of *Pseudomonas aeruginosa* pili. *J. Bacteriol.* **172**:2911–2919.

Ottow, J.C.G. 1975. Ecology physiology and genetics of fimbriae and pili. *Annu. Rev. Microbiol.* **29**:79–108.

Paranchych, W. and L.S. Frost. 1988. The physiology and biochemistry of pili. *Adv. Microbiol. Physiol.* **29**:53–114.

Pasloske, B.L., B.B. Finlay and W. Paranchych. 1985. Cloning and sequencing of the *Pseudomonas aeruginosa* PAK pilin gene. *FEBS Lett.* **183**:408–412.

Pasloske, B.L., P.A. Sastry, B.B. Finlay, and W. Paranchych. 1988. Two unusual pilin sequences from different isolates of *Pseudomonas aeruginosa*. *J. Bacteriol.* **170**:3738–3741.

Pedersen, K.B., L.O. Frøholm, K. Bøvre. 1972. Fimbriation and colony type of *Moraxella bovis* in relation to conjunctival colonisation and development of keratoconjunctivitis in cattle. *Acta Pathol. Microbiol. Scand. Sect. B* **80**:911–918.

Plant, J.W. and P.D. Claxton. 1986. Efficacy of pairing, footbathing and vaccination in the treatment of footrot. *In: Footrot in Ruminants: Proceedings of a Workshop, Melbourne, 1985,* D.J. Stewart, N.M. McKern, and D.L. Emery, eds., pp. 57–61. Sydney, Australia: CSIRO Press.

Prévot, A.R. 1948. *Manual de Classification et de Détermination des Bactéries Anaérobies.* Paris: Masson.

Pugsley, A.P. and I. Reyss. 1990. Five genes at the 3′ end of the *Klebsiella pneumoniae pulC* operon are required for pullulanase secretion. *Mol. Microbiol.* **4**:365–379.

Reed, G.A. 1986. The role of footrot vaccines in Australia. *In: Footrot in Ruminants: Proceedings of a Workshop, Melbourne, 1985,* D.J. Stewart, N.M. McKern, and D.L. Emery, eds., pp. 173–176. Sydney, Australia: CSIRO Press.

Reyss, I. and A.P. Pugsley. 1990. Five additional genes in the *pulC-O* operon of the Gram-negative bacterium *Klebsiella oxytoca* UNF5023 which are required for pullulanase secretion. *Mol. Gen. Genet.* **222**:176–184.

Ruehl, W.W., C.F. Marrs, R. Fernandez, S. Falkow, and G.K. Schoolnik. 1988. Purification, characterization, and pathogenicity of *Moraxella bovis* pili. *J. Exp. Med.* **168**:983–1002.

Sastry, P.A., B.B. Finlay, B.L. Pasloske, W. Paranchych, J.R. Pearlstone, and L.B. Smillie. 1985. Comparative studies of the amino acid and nucleotide sequences of pilin derived from *Pseudomonas aeruginosa* PAK and PAO. *J. Bacteriol.* **164**:571–577.

Schmitz, J.A. and J.L. Gradin. 1980. Serotypic and biochemical characterization of *Bacteroides nodosus* isolates from Oregon. *Can. J. Comp. Med.* **44**:440–446.

Segal, E., P. Hagblom, H.S. Seifert, and M. So. 1986. Antigenic variation of gonococcal pilus involves assembly of separated silent gene segments. *Proc. Natl. Acad. Sci. USA* **83**:2177–2181.

Seifert, H.S., R.S. Ajioka, C. Marchal, P.F. Sparling, and M. So. 1988. DNA transformation leads to pilin antigenic variation in *Neisseria gonorrhoeae*. *Nature* (London) **336**:392–395.

Shah, H.N. and M.D. Collins. 1983. Genus *Bacteroides*, a chemotaxonomic perspective. *J. Appl. Bacteriol.* **55**:403–416.

Sharp, P.M. and W. Li. 1986. Codon usage in regulatory genes in *Escherichia coli* does not reflect selection for "rare" codons. *Nucleic Acids Res.* **14**:7737–7749.

Short, J.A., C.M. Thorley, and P.D. Walker. 1976. An electron microscope study of *Bacteroides nodosus*: ultrastructure of organisms from primary isolates and different colony types. *J. Appl. Bacteriol.* **40**:311–315.

Skerman, T.M., S.K. Erasmuson, and D. Every. 1981. Differentiation of *Bacteroides nodosus* biotypes and colony variants in relation to their virulence and immunoprotective properties in sheep. *Infect. Immun.* **32**:788–795.

Stern, A., M. Brown, P. Nickel, and T.F. Meyer. 1986. Opacity genes in *Neisseria gonorrhoeae*: control of phase and antigenic variation. *Cell* **47**:61–71.

Stewart, D.J. 1973. An electron microscopic study of *Fusiformis nodosus*. *Res. Vet. Sci.* **14**:132–134.

Stewart, D.J., B.L. Clark, J.E. Peterson, D.A. Griffiths, and E.F. Smith. 1982. Importance of pilus-associated antigen in *Bacteroides nodosus* vaccines. *Res. Vet. Sci.* **32**:140–147.

Stewart, D.J., B.L. Clark, J.E. Peterson, D.L. Emery, E.F. Smith, D.A. Griffiths, and I.J. O'Donnell. 1985. The protection given by pilus and whole cell vaccines of *Bacteroides nodosus* strain 198 against ovine foot-rot induced by strains of different serogroups. *Aust. Vet. J.* **62**:153–159.

Stewart, D.J., J.E. Peterson, J.A. Vaughan, B.L. Clark, D.L. Emery, J.B. Caldwell, and A.A. Kortt. 1986. The pathogenicity and cultural characteristics of virulent, intermediate and benign strains of *Bacteroides nodosus* causing ovine footrot. *Aust. Vet. J.* **63**:317–326.

Stock, J.B., A.M. Stock, and J.M. Mottonen. 1990. Signal transduction in bacteria. *Nature* **344**:395–400.

Strom, M.S., D. Nunn, and S. Lory. 1991. Multiple roles of the pilus biogenesis protein PilD: involvement of PilD in excretion of enzymes from *Pseudomonas aeruginosa*. *J. Bacteriol.* **173**:1175–1180.

Taha, M.K., B. Dupuy, W. Saurin, M. So, and C. Marchal. 1991. Control of pilus expression in *Neisseria gonorrhoeae* as an original system in the family of two-component regulators. *Mol. Microbiol.* **5**:137–148.

Taha, M.K., M. So, H.S. Seifert, E. Billyard, and C. Marchal. 1988. Pilin expression in *Neisseria gonorrhoeae* is under both positive and negative transcriptional control. *EMBO J.* **7**:4367–4378.

Tanner, A.C.R., M.A. Listgarten, and J.L. Ebersole. 1984. *Wolinella curva* sp. nov.: "*Vibrio succinogenes*" of human origin. *Int. J. Syst. Bacteriol.* **34**:275–282

Tanner, A.C.R., S. Badger, C. Lai, M.A. Listgarten, R.A. Visconti, and S.S. Socransky. 1981. *Wolinella* gen nov. *Wolinella succinogenes* (*Vibrio succinogenes* Wolin *et al.*) comb. nov. and description of *Bacteroides gracilis* sp. nov., *Wolinella recta* sp. nov., *Campylobacter concisus* sp. nov. and *Eikenella corrodens* from humans with periodontal disease. *Int. J. Syst. Bacteriol.* **31**:432–445.

Taylor, R.L., V.L. Miller, D.B. Furlong, and J.J. Mekalanos. 1987. Use of *phoA* gene fusions to identify a pilus colonisation factor coordinately regulated with cholera toxin. *Proc. Natl. Acad. Sci. USA* **84**:2833–2837.

Virji, M. and J.E. Heckels, 1983. Antigenic cross-reactivity of *Neisseria* pili: investigations with type- and species-specific monoclonal antibodies. *J. Gen. Microbiol.* **129**:2761–2768.

Walker, P.D., J. Short, R.O. Thompson, and D.S. Roberts. 1973. The fine structure of *Fusiformis nodosus* with special reference to the antigens associated with immunogenicity. *J. Gen. Microbiol.* **77**:351–361.

Whitchurch, C.B., M. Hobbs, S.P. Livingston, V. Krishnapillai, and J.S. Mattick. 1991. Characterisation of a *Pseudomonas aeruginosa* twitching motility gene and evidence for a specialised protein export system widespread in eubacteria. *Gene* **101**:33–44.

40

Genetic Exchange in Pigmented *Bacteroides*

Donald G. Guiney

40.1 Introduction

The black-pigmented *Bacteroides* constitute a significant portion of the normal microbial flora of the oral cavity of man. These organisms are particularly successful in colonizing the spaces of the gingival crevice, where they attain large numbers of organisms. In their normal ecological niche, the black-pigmented *Bacteroides* coexist with many other members of the oral flora, including streptococci, actinomycetes, fusobacteria, other oral *Bacteroides*, actinobacilli, and *Capnocytophaga*. When host defense mechanisms are compromised, the black-pigmented *Bacteroides* play a major role, often in conjunction with these other species, in both acute and chronic infections arising from the oral cavity.

Very few studies have investigated the genetics of the black-pigmented *Bacteroides*. In fact, the taxonomy of this group of organisms has been rapidly changing. A new genus for the asaccharolytic group, composed of *B. asaccharolyticus*, *B. gingivalis*, and *B. endodontalis*, has been proposed and designated *Porphyromonas* (Shah and Collins, 1988). The black-pigmented *Bacteroides* are clearly a heterogenous group of organisms that are quite distinct from the intestinal *Bacteroides*. Molecular analysis of ribosomal RNA indicates that the intestinal *Bacteroides* line diverged at a very early time in the evolution of *Eubacteria*, making these organisms unrelated to most other bacteria (Weisburg et al., 1985). These evolutionary considerations have important implications for the study of genetic

processes in *Bacteroides* species. Genetic exchange in the intestinal *Bacteroides* appears to employ both plasmids and conjugative transposons (see Chapter 35, this volume) that are quite different from the genetic systems of the facultative gram-negative bacteria. The extent to which gene transfer mechanisms discovered in the intestinal *Bacteroides* can be applied to the oral black-pigmented species is only beginning to be explored.

40.2 Antibiotic Resistance

The appearance of antibiotic resistance in previously sensitive bacteria has often been an indirect indication of genetic exchange. Early studies indicated that the black-pigmented *Bacteroides* were sensitive to both penicillin and tetracycline. These antibiotics were used for many years to treat anaerobic infections involving the oral *Bacteroides*. However, as early as 1976, 15% to 20% of the black-pigmented *Bacteroides* were reported as resistant to penicillin (Sutter and Finegold, 1976). β-lactamase production was subsequently detected in resistant strains (Murray and Rosenblatt, 1977). Although studies in the early 1980s reported a low incidence of penicillin resistance (Baker et al., 1983; Sutter et al., 1983), more recent articles have noted rates of penicillin resistance up to 50% in the oral *Bacteroides* (Brook, 1987; Crook et al., 1988). A study from Egypt found β-lactamase production in approximately 40% of strains of *B. melaninogenicus*, *B. intermedius*, and *B. gingivalis* (Wasfy et al., 1986). As expected, treatment with penicillin increases the number of resistant isolates from the subgingival space (Kinder et al., 1986). The situation appears to be similar for tetracycline resistance, with approximately 25% of isolates reported as resistant. These clinical findings have prompted investigations into the genetic basis of penicillin and tetracycline resistance in the black-pigmented *Bacteroides*.

40.3 Plasmid Analysis

Transmissible plasmids carrying antibiotic resistance genes have been isolated from multiple species of both facultative and anaerobic bacteria. Among the intestinal *Bacteroides* species, several plasmids encoding transferable clindamycin resistance have been characterized (Welch and Macrina, 1981; Tally et al., 1982; Guiney et al., 1984a; Smith and Macrina, 1984; see Chapter 35, this volume for details). In addition, multiple cryptic plasmids appear to be common in these intestinal strains. Very little information is known about the plasmid content of the black-pigmented *Bacteroides*. Bouic and Guiney (unpublished observations) examined 15 independent strains of black-

pigmented *Bacteroides* for plasmid content and sensitivity to penicillin and tetracycline. Many of these isolates were selected because of their resistance to penicillin or tetracycline. Plasmids were extracted by a modification of the Currier–Nestor procedure, which was adapted for analysis of plasmids in the intestinal *Bacteroides* (Guiney et al., 1983). The results of this study are shown in Table 40.1.

Thirteen strains were penicillin resistant (MIC > 1 μg/ml), and six were resistant to tetracycline (MIC > 1 μg/ml). Only four plasmids, all less than 10 kb in size, were detected in these strains. Although all strains with plasmids were resistant to penicillin, many resistant strains did not carry plasmids. On the basis of this study, it appears that small plasmids can be found in some strains of the black-pigmented *Bacteroides*. Whether any of these plasmids encode antibiotic resistance is unclear at this time because there are no systems for genetic manipulation of these organisms. However, most of the resistant strains in this study did not contain detectable plasmids, a situation similar to that seen in the intestinal *Bacteroides* species in which tetracycline and penicillin resistance genes are located on the chromosome.

Similar results were obtained in plasmid

Table 40.1 Antibiotic resistance and plasmid content of black-pigmented *Bacteroides*

Bacteroides strain	Antibiotic resistance (MIC, μg/ml)		Plasmid
	Penicillin	Tetracycline	
B. melaninogenicus 7135	>5	<1	None
B. melaninogenicus 108	>5	<1	None
B. melaninogenicus 542	>5	<1	None
B. melaninogenicus 2901	>5	<1	None
B. melaninogenicus 461	>5	<1	None
B. asaccharolyticus (ATCC)	5	5	None
B. gingivalis (ATCC)	<1	<1	None
B. melaninogenicus 9185	>5	2	7 kb, pSK1
			3.4 kb, pSK2
B. melaninogenicus 9639	2	>5	8 kb
B. melaninogenicus 5565	>5	>5	None
B. asaccharolyticus 4649	2	5	None
B. denticola 4636	5	<1	None
B. denticola 10553	>5	>5	None
B. melaninogenicus 8515	>5	<1	None
B. denticola 8062B	>5	<1	3.3 kb, pSK3

screening of black-pigmented strains not selected for their antibiotic resistance phenotypes. Vandenbergh et al. (1982) found three plasmids 2.7 to 2.9 MDa in size after screening 59 black-pigmented *Bacteroides* isolates; no plasmids were seen in the 11 *B. asaccharolyticus* (*gingivalis*) strains tested. Sako et al. (1988) screened 123 pigmented strains of *Bacteroides* and found a common 5.9-MDa plasmid in 5 of the isolates. Fukushima et al. (1988) detected plasmids in 14 of 98 *B. intermedius* strains and suggested that production of an extracellular viscous material was correlated with plasmid content. However, no phenotypic property has been definitively shown to be plasmid encoded in the black-pigmented *Bacteroides*.

Significantly, large plasmids appear to be quite rare in the black-pigmented strains. Because conjugation systems are generally found on plasmids greater than 10-kb in size, these results suggest that self-transmissible plasmids may be unusual in the black-pigmented *Bacteroides*. Two large plasmids, found in a strain of *B. intermedius*, did not appear to be associated with conjugal transfer of antibiotic resistance (Guiney and Bouic, 1990). Intestinal *Bacteroides* strains differ in that they can contain large plasmids, either cryptic or encoding self-transmissible clindamycin resistance. In summary, the study of plasmids in the black-pigmented *Bacteroides* has shown that these strains can carry small cryptic plasmids, but no association of these plasmids with genetic transfer has been made.

40.4 Endogenous Genetic Exchange Systems

In the intestinal *Bacteroides*, tetracycline resistance transfer systems are encoded by large genetic elements located on the chromosome (Guiney et al., 1989; Shoemaker et al., 1989). As noted, the development of widespread tetracycline and penicillin resistance in the black-pigmented *Bacteroides* prompted the examination of these organisms for antibiotic resistance transfer systems similar to those found in the intestinal *Bacteroides*. Two strains of black-pigmented *Bacteroides*, *B. denticola* 10553 and *B.*

intermedius M87-1738, were found to be capable of transferring tetracycline resistance to an antibiotic-sensitive recipient, *B. buccae* 8062 (Guiney and Bouic, 1990). The characteristics of these transfers are shown in Table 40.2. The *B. denticola* system was the most extensively studied. The transfer of resistance markers has the characteristics of a conjugation system rather than those of transformation or transduction: transfer required cell-to-cell contact and was DNase resistant. No plasmid DNA was found in the donor or recipient strains. Transfer frequencies were in the range of 10^{-4} to 10^{-5} per donor in the presence of excess recipients, and were not increased by prior treatment with tetracycline. Transfer of tetracycline resistance to the intestinal *Bacteroides* recipient, *B. fragilis*, was also observed but at a lower frequency. Retransfer was achieved between *B. fragilis* strains, indicating that the entire self-transmissible element was being transferred during the matings.

Because each of the donors, *B. denticola* and *B. intermedius*, were also resistant to penicillin and produced penicillinase, transfer of this resistance marker was also examined. *B. buccae* transconjugants selected for tetracycline resistance from matings with both *B. denticola* and *B. intermedius* were also shown to be penicillin resistant and to produce penicillinase. These results suggest that the penicillin and tetracycline resistance loci may be linked in these black-pigmented strains.

The genetic transfer systems in these black-pigmented *Bacteroides* species show a close phenotypic similarity to the tetracycline resistance transfer elements in the intestinal *Bacteroides*: (i) the tetracycline resistance loci are not encoded by detectable plasmids, and (ii) gene transfer occurs by conjugative mechanisms at frequencies of 10^{-4} to 10^{-7}. The similarity between the tetracycline resistance loci of the black-pigmented strains and *B. fragilis* was examined directly by DNA hybridization analysis (Guiney and Bouic, 1990). Using a probe containing the cloned tetracycline resistance gene from *B. fragilis* strain 1126, Southern hybridization studies demonstrated strong homology with the transferable tetracycline resistance loci of the *B. denticola* and *B. intermedius* strains.

Table 40.2 Tetracycline resistance transfer by *Bacteroides* strains[a]

Donor	Recipient	Transfer frequency, tetracycline resistance	Cotransfer of penicillin resistance
B. denticola 10553 Tet[r], Pen[r]	*B. buccae* 8062 Rif[r]	2×10^{-5}	+
B. denticola 10553 Tet[r], Pen[r]	*B. fragilis* 638 Rif[r]	3×10^{-7}	N.T.
B. intermedius M87-1738 Tet[r], Pen[r]	*B. buccae* 8062 Rif[r]	3×10^{-4}	+

[a]Abbreviations: Tet[r], tetracycline resistance; Pen[r], penicillin resistance, Rif[r], chromosomal mutation to rifampin resistance; N.T., not tested.

These results show that the tetracycline resistance genes in the oral, black-pigmented *Bacteroides* and the intestinal, *B. fragilis* group are closely related.

In summary, endogenous genetic transfer systems have been discovered in the black-pigmented *Bacteroides*. While transduction and transformation have not been reported in these species, conjugative transfer of antibiotic resistance has been detected. These conjugation systems appear closely related to the tetracycline resistance transfer elements that are common in the intestinal *B. fragilis* group of organisms. These results imply that genetic exchange occurs naturally between the black-pigmented and intestinal *Bacteroides*. The transfer of the *B. denticola* tetracycline resistance element to *B. fragilis* directly confirms that such exchange is possible. It is likely that gene transfer also occurs from *B. fragilis* to the oral strains, given the broad host range transfer properties of the intestinal tetracycline resistance elements (Shoemaker and Salyers, 1987). Two differences have been detected between the black-pigmented and *B. fragilis* systems. While most of the *B. fragilis* transfer elements are induced by prior growth in tetracycline (Privitera et al., 1979, 1981), the two black-pigmented systems are not affected by tetracycline. Of more significance, however, is the finding that tetracycline resistance is linked to penicillin resistance in the black-pigmented strains, while in intestinal strains clindamycin resistance is frequently cotransferred with tetracycline resistance. These linkages may reflect the different antibiotic-selective pressures applied to the two bacterial populations. The intestinal *Bacteroides* possess intrinsic resistance to many β-lactam antibiotics because of chromosomal cephalosporinases. Clindamycin has been used extensively to treat *B. fragilis* infections, and linkage of the clindamycin resistance locus with transferable tetracycline resistance is a selective advantage. The black-pigmented species have intrinsic susceptibility to penicillin and other β-lactams. In these strains, the linkage of a penicillinase gene to the transferable tetracycline element is of selective value because of the widespread use of penicillin and also its particular employment to treat oral infections.

40.5 Transfer of Recombinant Plasmids from *E. coli*

Genetic analysis of the intestinal *Bacteroides* has been greatly facilitated by the development of shuttle vector systems that allow transfer of recombinant plasmids from *E. coli* into *Bacteroides* species (Guiney et al., 1984b; Shoemaker et al., 1985). Because no broad host range plasmid capable of maintenance in both *Escherichia coli* and *Bacteroides* has been found, the shuttle vectors are chimeric plasmids composed of an *E. coli* replicon and a *Bacteroides* replicon. In addition, antibiotic resistance genes from *E. coli* and *Bacteroides* are not expressed well in the heterologous host, necessitating inclusion of both an *E. coli* resistance marker and the *Bacteroides* clindamycin resistance gene. Finally, either the mobilization region of RSF1010 or the transfer origin (*ori*T) of RK2 is required so that the shuttle vector can be transferred into *Bacteroides* using the conjugation system of the IncP plasmids R751 or RK2. These vectors have been used extensively to clone and manipulate *Bacteroides* DNA in *E. coli* and to return the cloned regions to *Bacteroides* for phenotypic testing and mutant construction (Guthrie and Salyers, 1986; Guiney et al., 1989).

Progulske-Fox et al. (1989) have demonstrated this same shuttle vector system could be transferred into the black-pigmented *Bacteroides* species *B. gingivalis* and *B. intermedius*. Using the plasmid pE5-2 constructed by Shoemaker et al. (1985) and R751 as the helper conjugative plasmid, they detected transfer of both the clindamycin resistance marker and the complete plasmid in matings between *E. coli* donors and *B. intermedius* and *B. gingivalis* recipients. Transfer frequencies were of the order of 10^{-7}, somewhat lower than the optimal frequencies that are obtained with shuttle vector transfer into the intestinal *Bacteroides*. A significant amount of genetic instability was found in the recipients, as the vector DNA in 4 of 20 transconjugants appeared to have suffered rearrangements. However, these results clearly indicate that the IncP-mediated shuttle vector system is applicable to the black-pigmented *Bacteroides*. The *Bacteroides* replicon on pE5-2, derived from a cryptic plasmid in intestinal *Bacteroides*, appears to replicate in *B. gingivalis*, although the stability of the replicon remains an issue.

These results are highly significant because endogenous plasmids in *B. gingivalis* appear to be rare. It is now possible to construct and test a second generation of shuttle vectors using cryptic plasmids isolated from other black-pigmented *Bacteroides*. In addition, a second selective marker, the tetracycline resistance gene, has been cloned from *B. fragilis* and is available for use in vector construction (Guiney et al., 1989). Given the extensive homology with the tetracycline resistance loci of *B. denticola* and *B. intermedius*, it is likely that the *B. fragilis* gene will be expressed in the black-pigmented *Bacteroides*. Because several genes from *B. gingivalis* have now been cloned (see Chapter 41, this volume), the availability of selective markers and shuttle vectors will facilitate the construction of specific mutants of *B.*

gingivalis as well as analysis of the expression of the cloned determinants in their native host.

40.6   Conclusions

Genetic studies of the oral black-pigmented *Bacteroides* are in their infancy. However, the genetic analysis of these black-pigmented strains has been greatly facilitated by the study of intestinal *Bacteroides*, particularly the *B. fragilis* group. Naturally occurring transformation or transduction systems have not been reported for any *Bacteroides* species. Plasmid-mediated conjugation in the black-pigmented strains has not been demonstrated and is likely, on the basis of the lack of larger plasmids detected in these species, to be rare. Instead, transfer elements in the chromosome have been detected in two strains, *B. denticola* and *B. intermedius*, and closely resemble the tetracycline resistance transfer systems that are widespread in the intestinal *Bacteroides*. In fact, genetic exchange mediated by these elements occurs between black-pigmented and intestinal species.

In an additional analogy with the intestinal strains, the IncP plasmids are able to mediate transfer of a recombinant plasmid from *E. coli* to *B. intermedius* and *B. gingivalis*. The intestinal *Bacteroides* replicon and clindamycin resistance gene are maintained and expressed in the black-pigmented strains. These results show that the shuttle vector systems developed for genetic manipulation of the *B. fragilis* group can be applied to the black-pigmented *Bacteroides*.

Acknowledgments.   Work reported from the author's laboratory was supported by U.S. Public Health Service grants AI 16463 and DE 07344 from the National Institutes of Health.

References

Baker, P.J., J.S. Slots, R.J. Genco, and R.T. Evans. 1983. Minimal inhibitory concentrations of various antimicrobial agents for human oral anaerobic bacteria. *Antimicrob. Agents Chemother.* 24:420–424.

Brook, I. 1987. Role of anaerobic beta-lactamase-producing bacteria in upper respiratory tract infections. *Pediatr. Infect. Dis. J.* 6:310–316.

Crook, D.W., G.J. Cuchural, Jr., N.V. Jacobus, and F.P. Tally. 1988. Antimicrobial resistance in oral and colonic *Bacteroides*. *Scand. J. Infect. Dis.* (Suppl.) **57**:55–64.

Fukushima, H., A. Progulske-Fox, and C.B. Walker. 1988. The presence of plasmids in oral isolates of *Bacteroides intermedius*. *J. Dent. Res.* **67**:331.

Guiney, D.G. and K. Bouic. 1990. Detection of conjugal transfer systems in oral, black-pigmented *Bacteroides* spp. *J. Bacteriol.* **172**:495–497.

Guiney, D.G., P. Hasegawa, and C.E. Davis. 1984a. Homology between clindamycin resistance plasmids in *Bacteroides*. *Plasmid* **11**:268–271.

Guiney, D.G., P. Hasegawa, and C.E. Davis. 1984b. Plasmid transfer from *Escherichia coli* to *Bacteroides fragilis*: differential expression of antibiotic resistance phenotypes. *Proc. Natl. Acad. Sci. USA* **81**:7203–7206.

Guiney, D.G., P. Hasegawa, K. Bouic, and B. Matthews. 1989. Genetic transfer systems in *Bacteroides*: cloning and mapping of the transferable tetracycline resistance locus. *Mol. Microbiol.* **3**:1617–1623.

Guiney, D.G., P. Hasegawa, D. Stalker, and C.E. Davis. 1983. Genetic analysis of clindamycin resistance in *Bacteroides* species. *J. Infect. Dis.* **147**:551–558.

Guthrie, E.P. and A.A. Salyers. 1986. Use of targeted insertional mutagenesis to determine whether chondroitin lyase II is essential for chondroitin sulfate utilization by *Bacteroides thetaiotaomicron*. *J. Bacteriol.* **166**:966–971.

Kinder, S.A., S.C. Holt, and K.S. Korman. 1986. Penicillin resistance in the subgingival microbiota associated with adult periodontitis. *J. Clin. Microbiol.* **23**:1127–1133.

Murray, P.R. and J.E. Rosenblatt. 1977. Penicillin resistance and penicillinase production in clinical isolates of *Bacteroides melaninogenicus*. *Antimicrob. Agents Chemother.* **11**:605–608.

Privitera, G., F. Fayolle, and M. Sebald. 1981. Resistance to tetracycline, erythromycin, and clindamycin in the *Bacteroides fragilis* group: inducible versus constitutive tetracycline resistance. *Antimicrob. Agents Chemother.* **20**:314–320.

Privitera, G., M. Sebald, and F. Fayolle. 1979. Common regulatory mechanism of expression and conjugative ability of a tetracycline resistance plasmid in *Bacteroides fragilis*. *Nature* (London) **278**:657–659.

Progulske-Fox, A., A. Oberste, C. Drummond, and W.P. McArthur. 1989. Transfer of plasmid pE5-2 from *Escherichia coli* to *Bacteroides gingivalis* and *B. intermedius*. *Oral Microbiol. Immunol.* **4**:132–134.

Sako, K., I. Takazoe, and K. Okuda. 1988. Isolation and characterization of plasmid DNA from *Bacteroides* strains isolated from the human oral cavity. *Oral Microbiol. Immunol.* **3**:72–76.

Shah, H.N. and M.D. Collins. 1988. Proposal for reclassification of *Bacteroides asaccharolyticus*, *Bacteroides gingivalis*, and *Bacteroides endodontalis* in a new genus, *Porphyromonas*. *Int. J. Syst. Bacteriol.* **38**:128–131.

Shoemaker, N.B. and A.A. Salyers. 1987. Facilitated transfer of IncPβ R751 derivatives from the chromosome of *Bacteroides uniformis* to *Escherichia coli* recipients by a conjugative *Bacteroides* tetracycline resistance element. *J. of Bacteriol.* **169**:3160–3167.

Shoemaker, N.B., R.D. Barber, and A.A. Salyers. 1989. Cloning and characterization of a *Bacteroides* conjugal tetracycline-erythromycin resistance element by using a shuttle cosmid vector. *J. Bacteriol.* **171**:1294–1302.

Shoemaker, N.B., E.P. Guthrie, A.A. Salyers, and J.F. Gardner. 1985. Evidence that the clindamycin-erythromycin gene of *Bacteroides* plasmid pBF4 is on a transposable element. *J. Bacteriol.* **162**:626–632.

Smith, C.J. and F. Macrina. 1984. Large transmissible clindamycin resistance plasmid in *Bacteroides ovatus*. *J. Bacteriol.* **158**:739–741.

Sutter, V.L. and S.M. Finegold. 1976. Susceptibility of anaerobic bacteria to 23 antimicrobial agents. *Antimicrob. Agents Chemother.* **10**:736–752.

Sutter, V.L., M.J. Jones, and A.T.M. Ghoneim. 1983. Antimicrobial susceptibilities of bacteria associated with periodontal disease. *Antimicrob. Agents Chemother.* **23**:483–486.

Tally, F.P., D.R. Snydman, M.J Shimell, and M.H. Malamy. 1982. Characterization of pBFTM10, a clindamycin-erythromycin resistance transfer factor from *Bacteroides fragilis*. *J. Bacteriol.* **151**:686–691.

Vandenbergh, A., S.A. Syed, C.F. Gonzalez, W.J. Loesche, and R.H. Olsen. 1982. Plasmid content of some oral microorganisms isolated from subgingival plaque. *J. Dent. Res.* **61**:497–501.

Wasfy, M.O., R.E. Bajuscak, A.C. Santos, and G.E. Minah. 1986. β-lactamase resistance of black-pigmented *Bacteroides* in gingival plaques of Egyptian children. *J. Periodontal Res.* **21**:450–454.

Weisburg, W.D., Y. Oyaizu, H. Oyaizu, and C.R. Woese. 1985. Natural relationship between *Bacteroides* and *Flavobacteria*. *J. Bacteriol.* **164**:230–236.

Welch, R.A. and F.L. Macrina. 1981. Physical characterization of *Bacteroides fragilis* R plasmid pBF4. *J. Bacteriol.* **145**:867–872.

41

Porphyromonas gingivalis: Gene Cloning of Determinants of Pathogenicity

Barry C. McBride, Umadatt Singh, and Angela Joe

41.1   Introduction

Porphyromonas gingivalis is a frequent isolate from the gingival sulcus of individuals with periodontal disease, and is believed to play a major role in the pathogenesis of chronic adult periodontitis (Slots, 1977; Tanner et al., 1979; White and Mayrand, 1981). This organism is a gram-negative, nonmotile, nonspore-forming, obligate anaerobe. Cells are generally coccobacillary (0.5 by 1 to 2 μm) in shape. Until recently, *P. gingivalis* was known as *Bacteroides gingivalis*, a member of the asaccharolytic black-pigmented *Bacteroides* group, but the species has now been reclassified as *P. gingivalis* (Shah and Collins, 1988).

This group of organisms has been extensively reviewed by Mayrand and Holt (1988). Black-pigmented *Bacteroides* species produce brown to black colonies when grown on blood agar medium. Several members of this group have been implicated in different forms of periodontal disease, usually as part of a mixed bacterial infection. A high correlation between the presence of *P. gingivalis* and advanced cases of adult periodontitis has led to an interest in its pathogenic role. Many potential virulence determinants associated with this organism may contribute to the initiation and progression of the disease process. The aim of this chapter is to present a brief overview of *Porphyromonas gingivalis*, evidence for its involvement in periodontal disease, and current work on the molecular genetic characterization of its pathogenic determinants.

41.2   A History of the Taxonomy of *Porphyromonas gingivalis*

The taxonomic grouping of this organism has undergone substantial change leading to its current classification as *Porphyromonas gingivalis* (Shah and Collins, 1988). These changes in taxonomy may cause some confusion in interpreting past literature pertaining to *P. gingivalis*. The following, therefore, serves as a brief history of the classification of this organism.

A small gram-negative anaerobic rod that produced black-pigmented colonies when grown on blood agar plates was first reported by Oliver and Wherry (1921). The dark pigment was believed to be melanin, and the organism was named *Bacterium melaninogenicum*. This designation was subsequently changed to *Haemophilus melaninogenicus* (Bergey et al., 1930) as growth of this bacterium appeared to be better on solid medium containing the X and V growth factors characteristic of the members of the genus *Haemophilus*. Roy and Kelly (1939) later reclassified this organism to the genus *Bacteroides* and named it *Bacteroides melaninogenicus*.

Only one species of black-pigmented *Bacteroides* was recognized before 1970, although Courant and Gibbons (1967) and Sawyer et al. (1962), demonstrated biochemical and immunological heterogeneity among strains of *B. melaninogenicus*. In 1970, Holdeman and Moore divided *Bacteroides melaninogenicus* into three subspecies on the basis of their fermentative activities: *B. melaninogenicus* subspecies *melaninogenicus* (strongly fermentative), *B. melaninogenicus* subspecies *intermedius* (weakly fermentative), and *B. melaninogenicus* subspecies *asaccharolyticus* (nonfermentative). Later studies (reviewed by Finegold and Barnes, 1977) indicated biochemical and genetic differences between the saccharolytic and asaccharolytic strains sufficient to elevate the asaccharolytic subspecies to the species level (*B. asaccharolyticus*).

Work conducted by Shah et al. (1976) and van Steenbergen et al. (1979) showed that *Bacteroides* species, particularly the asaccharolytic isolates, could be differentiated into oral and nonoral strains based on genetic heterogeneity. Further investigation by Coykendall et al. (1980) indicated a significant genetic difference between oral and nonoral asaccharolytic strains. The DNA base content of asaccharolytic strains isolated from the oral cavity varied from 46.5 to 48.4 mol% $G + C$, while that of nonoral asaccharolytic isolates was between 49.2 and 53.6 mol% $G + C$. It was proposed that the name *B. gingivalis* be used for the asaccharolytic strains isolated from oral sites (Coykendall et al., 1980); *B. asaccharolyticus* was retained for asaccharolytic strains isolated from nonoral sites.

In 1984, a third asaccharolytic black-pigmented *Bacteroides* species, *B. endodontalis*, was proposed (van Steenbergen et al., 1984) based on two strains isolated by Sundqvist (Sundqvist, 1976). Although originally referred to as *B. asaccharolyticus*, these strains had little or no DNA homology with either authentic *B. asaccharolyticus* or *B. gingivalis*. All three species of asaccharolytic *Bacteroides* can be rapidly differentiated by their unique protein profiles following sodium dodecyl sulfate-polyacrylamide gel electrophoresis (SDS-PAGE). In addition, several synthetic oligonucleotide DNA probes have been developed for identification of *B. ginaivalis* (Moncla et al., 1990). These probes are complementary to hypervariable regions of 16S rRNA of this organism.

Using filter hybridization, the reactivity of 6 different *B. gingivalis* DNA probes was tested against genomic DNA from a total of 77 strains of *B. gingivalis* and 105 strains that constitute 12

other *Bacteroides* species. All 6 DNA probes identified the *B. gingivalis* strains correctly; 4 of the probes were highly specific and did not hybridize with any of the other *Bacteroides* species, and 2 probes cross-reacted with 4 of the other *Bacteroides* species tested. The cross-reactivity of these *B. gingivalis* DNA probes was assessed against a panel of nucleic acids isolated from 20 microorganisms common to the oral cavity (Dix et al., 1990). Within the range of nucleic acids tested, the probes hybridized specifically to only *B. gingivalis* DNA. This indicated the potential usefulness of these probes in detection of *B. gingivalis* from within the mixed bacterial population usually present in clinical specimens.

Shah and Collins (1988) have proposed that sufficient biochemical and chemical differences exist between the asaccharolytic *Bacteroides* species and the type species of the genus *Bacteroides* (*Bacteroides fragilis*), that *B. asaccharolyticus*, *B. gingivalis*, and *B. endodontalis* be reclassified in a new genus called *Porphyromonas*. For the remainder of this review, *Bacteroides gingivalis* will be referred to as *Porphyromonas gingivalis*.

41.3 Nutrition and Physiology of *Porphyromonas gingivalis*

Porphyromonas gingivalis is an obligate anaerobe, requiring hemin and vitamin K as growth factors (Gibbons and MacDonald, 1960). A medium containing bacitracin, colistin, and nalidixic acid as selective agents has been developed for specific isolation of this organism (Hunt et al., 1986).

It has been suggested that the dark pigment produced by *P. gingivalis* is a mechanism of storage of hemin (Gibbons and MacDonald, 1960). Rizza et al. (1968) showed that *P. gingivalis* cells transferred from a growth medium containing high concentrations of hemin to a medium lacking hemin were still able to undergo 8 to 10 cell divisions. This suggested the cell accumulates and stores exogenous hemin which can be utilized during periods of hemin limitation. In 1979, Shah et al. demonstrated unequivocally that the pigment produced by *P.*

gingivalis was protohemin with traces of protoporphyrin.

Gibbons and MacDonald (1960) suggested a role for vitamin K as an electron carrier in the electron transport system of P. gingivalis. This proposal was later supported by the studies of Shah and Collins (1980). Vitamin K has also been found to stimulate synthesis of phosphophingolipids in the cell envelope, which indicates a possible role for this compound in membrane permeability (Lev and Milford, 1975).

The exact roles of hemin and vitamin K in the growth of P. gingivalis are not fully understood. The concentration of hemin available for growth appears to have an effect both on the physiology and virulence of this organism (McKee et al., 1986). Cells of strain W50 transferred to medium lacking hemin were avirulent when injected into mice, whereas cells grown in conditions of hemin excess caused 100% mortality.

Isolates of P. gingivalis from in vivo infections are generally part of a mixed bacterial population. It appears that synergistic nutritional relationships between bacteria within these infections may be important for proliferation of P. gingivalis. The presence of vitamin K-related compounds such as naphthoquinone produced by associated bacteria may enhance growth of this organism (MacDonald et al., 1963). Increased growth of P. gingivalis was observed due to protoheme produced by Wolinella recta, an organism frequently found in association with P. gingivalis in oral infections (Grenier and Mayrand, 1986).

In addition, succinate produced by fermentation of glucose by other bacteria can replace hemin as a growth factor for P. gingivalis (Mayrand and McBride, 1980; Grenier and Mayrand, 1983). Mayrand and McBride (1980) demonstrated that P. gingivalis could be cultivated in hemin-free medium provided there was an exogenous supply of succinate. In one study, succinate was produced by cocultivation with succinate-producing organisms. The P. gingivalis strain studied was not infectious when inoculated subcutaneously as a pure culture in the guinea pig model system. However, coinoculation with any one of a number of gram-positive or gram-negative succinate-producing organisms resulted in a rapidly progressing and often fatal disease. Further studies showed that incorporation of hemin or succinate into the inoculum was sufficient to replace the need for a second organism.

41.4 Porphyromas gingivalis in the Etiology of Periodontal Disease

Periodontal disease encompasses a family of inflammatory infections affecting periodontal tissue, the supporting tissue of teeth. The etiologic agents of periodontal disease are certain bacteria that play significant roles in inducing and maintaining the inflammatory process. In the healthy state, innocuous organisms colonize the tooth surface and the epithelial cells lining the gingival sulcus (Figure 41.1). The microbial population colonizing healthy tissue contains mainly facultative gram-positive rods and cocci. Periodontitis, a severe form of periodontal disease, correlates with a transition to gram-negative rods and motile organisms as the major components of the microflora inhabiting the densely populated gingival sulcus. The presence and activity of specific microbial pathogens is believed to be responsible for the extensive destruction of periodontal connective tissue supporting the attachment of teeth in the oral cavity.

A large body of evidence implicates P. gingivalis as an oral pathogen. A number of studies have shown a direct relationship between the presence of P. gingivalis and the occurrence of periodontitis (Slots, 1977; White and Mayrand, 1981; Loesche et al., 1985). White and Mayrand (1981) demonstrated that P. gingivalis was present in the gingival sulcus of patients with severe inflammation and absent from healthy sites. In a study evaluating the subgingival microflora from patients diagnosed as having rapidly progressive periodontitis, Di Murro et al. (1987) found P. gingivalis to be consistently involved. An investigation by Loesche et al. (1981) showed that isolates of P. gingivalis constituted 35%–40% of the sulcular flora of

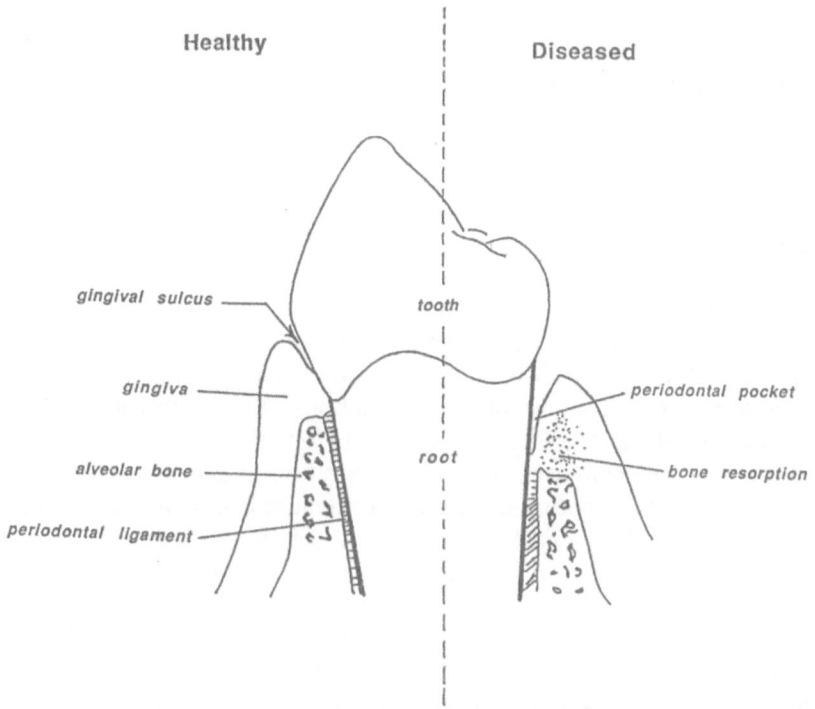

Healthy   Diseased

gingival sulcus ———

gingiva ———

alveolar bone ———

periodontal ligament ———

tooth

root

periodontal pocket

bone resorption

FIGURE 41.1 Schematic diagram of changes in periodontal tissue during periodontitis. In healthy state (left), tooth is supported in bone socket by periodontal ligament fibers that mediate attachment of root surface to alveolar bone. Innocuous bacteria, mainly facultative gram-positive rods and cocci colonize gingival sulcus. Onset of periodontitis correlates with transition to gram-negative rods and motile organisms as major components of microbial population inhabiting gingival sulcus. In initial stages of periodontitis, degradation of gingival tissue and periodontal ligament results in apical detachment of gingiva from tooth and root surface, forming periodontal pocket. As severity of disease increases, extensive destruction of periodontal ligament and alveolar bone leads to eventual loss of tooth from oral cavity.

patients with severe periodontitis. Treatment with metronidazole eliminated this organism and resolved the disease. In a subsequent study, van Dyke et al. (1988) reported a case of refractory periodontitis in which the sulcular flora contained more than 60% *P. gingivalis*. Eradication of *P. gingivalis* correlated with success in controlling the disease.

Additional evidence for the role of *P. gingivalis* in periodontal disease comes from immunological studies. Bacterial infections are frequently accompanied by an immune response that is specific for the pathogenic organism. Virtually all investigators agree that serum antibody levels to *P. gingivalis* are higher in patients diagnosed with adult periodontitis or generalized juvenile periodontitis than in other groups of individuals (Mouton et al., 1981; Altman et al.,

1982; Vincent et al., 1985; Ebersole et al., 1982, 1986; Farida et al., 1986). Several studies also show elevated levels of antibody specific to *P. gingivalis* in gingival crevicular fluid (Ebersole et al., 1985; Tew et al., 1985). Kagan (1980) investigated the local immune response and found that 32% of plasma cells from advanced periodontal lesions bound *P. gingivalis*, while only 3% of plasma cells from early lesions demonstrated the same activity. This suggests that *P. gingivalis* antigens penetrate gingival tissues where they evoke a strong local immune response.

The pathogenic potential of *P. gingivalis* has been demonstrated in animal model infections. Important early work was done by MacDonald et al. (1956) who showed that black-pigmented oral isolates were a necessary component of experimental infections in guinea pigs. In these studies it was necessary to coinfect with a minimum of two other organisms. Several investigators have subsequently shown, using the guinea pig model, that some strains of *P. gingivalis* inoculated in pure culture result in the development of a rapidly spreading infection (Takazoe et al., 1971; Grenier and Mayrand, 1987a). Neiders et al. (1988) demonstrated that eight strains of *P. gingivalis* produced lesions when injected intraperitoneally into mice. Histological examination of the tissue showed evidence of invasion of the connective tissue.

In a study by Holt et al. (1988), *P. gingivalis* was implanted into the periodontal microbiota of monkeys, resulting in ligature-induced periodontitis. Infection by *P. gingivalis* correlated with an increase in the serum antibody levels to this organism, as well as to significant bone loss at the implanted site. These results imply a direct connection between implantation of *P. gingivalis* and the clinical manifestations generally observed in periodontitis. McArthur et al. (1989) showed that immunization of squirrel monkeys with *P. gingivalis* resulted in a significant reduction in colonization of the gingival sulcus by black-pigmented *Bacteroides*.

41.5   Pathogenic Determinants of *Porphyromonas gingivalis*

The involvement of *P. gingivalis* in the pathogenesis of periodontal disease is widely accepted. However, relatively little is known about specific components of this organism that play a role in virulence. A variety of putative pathogenic determinants have been identified as factors that may contribute to the ability of *P. gingivalis* to colonize and infect host tissue.

Adhesins

Bacterial cell-surface components that mediate attachment to immobilized receptors are considered to be important pathological determinants. Accumulation of bacteria in the oral cavity is caused by binding to host tissue or to other bacteria (Slots and Gibbons, 1978; Weerkamp and McBride, 1980). A variety of different adhesive properties exhibited by a number of oral organisms facilitate binding to soft tissue, salivary pellicle, or other bacteria. Several studies show that *P. gingivalis* possesses the ability to agglutinate erythrocytes (Okuda and Takazoe, 1974) and to attach to human epithelial cells (Slots and Gibbons, 1978), saliva-coated and serum-coated hydroxyapatite (Cimasoni et al., 1987), and a number of gram-positive and gram-negative bacteria (Slots and Gibbons, 1978; Schwartz et al., 1987; Kinder and Holt, 1989).

Binding to saliva-coated hydroxyapatite and epithelium is inhibited by serum and crevicular fluid (Slots and Gibbons, 1978). Very little is known about the surface-binding proteins of *P. gingivalis* that mediate attachment to these surfaces. Boyd and McBride (1984) fractionated deoxycholate-solubilized outer membranes of *P. gingivalis* by gel filtration. A high molecular weight fraction contained the bacterial aggregating activity, while a lower molecular weight fraction contained the hemagglutinating activity. This latter fraction did not contain fimbriae. The experiment clearly showed that the two adherence activities reside on different molecular constituents.

Fimbriae from *P. gingivalis* have been isolated and the morphological, biological, immunological, and chemical properties characterized (Yoshimura et al., 1984a, 1985). These structures are composed of a monomeric subunit protein (fimbrilin) with an apparent molecular weight of 43 kDa. Native fimbriae on the cell surface of *P. gingivalis* appear as thin, curly filaments approximately 5 nm wide. It was initially believed that fimbriae were responsible for the hemagglutinating activity exhibited by *P. gingivalis* (Okuda and Takazoe, 1974; Slots and Gibbons, 1978; Okuda et al., 1981). However, subsequent studies indicate hemagglutinating activity is not an integral property of the fimbrial component of this organism (Boyd and McBride, 1984; Yoshimura et al., 1984a; Suzuki et al., 1988).

Much research has been devoted to elucidating the nature of the hemagglutinin expressed by *P. gingivalis*, but the identity of the molecule remains unclear. Inoshita et al. (1986) isolated an exohemagglutinin from the culture medium of *P. gingivalis*. Analysis of the hemagglutinin preparation by SDS-PAGE revealed 3 major protein bands with apparent molecular weights of 24, 37, and 44 kDa as well as some minor bands. A hemagglutinin has also been isolated from culture supernatant by Okuda et al. (1986). SDS-PAGE analysis showed the presence of 2 protein bands at 40 and 60 kDa. Mouton et al. (1989) identified a cell-bound hemagglutinin of *P. gingivalis* by use of crossed-immunoaffinity electrophoresis. A polyclonal antiserum with specificity restricted to the hemagglutinin was produced.

Western blot immunoassays with *P. gingivalis* cell extracts prepared in the presence of EDTA revealed two antigens with apparent molecular weights of 33 and 38 kDa. However, the antiserum detected two polypeptides of molecular weights 43 and 49 kDa from outer membrane samples of *P. gingivalis* prepared in the absence of EDTA (Deslauriers and Mouton, 1991). Deslauriers and Mouton concluded that the pair of bands at lower apparent molecular weight detected by this antiserum may be caused by either an EDTA-sensitive tertiary conformation of component polypeptides of the hemagglutinin or an EDTA-sensitive linkage of each of the polypeptides to an unknown component of 10 kDa. Further work is clearly warranted to clarify the nature of the *P. gingivalis* hemagglutinin.

A *P. gingivalis* adhesin that mediates attachment to erythrocytes was isolated in our laboratory. The isolation procedure involved gentle stirring of the cells followed by ammonium sulfate precipitation and ion-exchange and gel filtration chromotography. The native molecule had a molecular weight in excess of 10^3 kDa, and was composed of subunits with an apparent molecular weight of 43 kDa. Antisera raised to the hemagglutinin inhibited the attachment of *P. gingivalis* to erythrocytes (Table 41.1). Proteolytic enzymes destroyed binding capability of whole cells and of the purified hemagglutinin, but the molecular weight of the hemagglutinin as determined by SDS-PAGE was not altered.

Table 41.1 Effect of immune sera[a] on hemagglutinating activity of *Porphyromonas gingivalis*

IgG	Inhibitory titer[b]
Anti-*P. gingivalis*	128
Anti-HA[c]	128
Anti-43 subunit[d]	128
Preimmune serum	4

[a] IgG fractions were prepared by passage of the serum over a protein A Sepharose column. IgGs were resuspended to final concentration of 100 μg per ml for use in inhibition assay.
[b] Minimum antibody titer required to inhibit hemagglutination.
[c] Antisera raised against native hemagglutinin molecule.
[d] Antisera raised against 43-kDa subunit of hemagglutinin.

Capsule

Several strains of *P. gingivalis* appear to be encapsulated. This capsule consists of a layer of electron-dense material approximately 15 nm thick covering the outer membrane (Listgarten and Lai, 1979; Woo et al., 1979; Handley and Tipler, 1986). Woo and Holt (unpublished data) have determined that the material consists of a polysaccharide heteropolymer.

In studies relating virulence to the presence of capsule, Okuda and Takazoe (1973) found that an encapsulated virulent strain was more resistant to phagocytosis and killing by polymorphonuclear leukocytes (PMNLs) than was a similar but noncapsulated strain. Capsular material therefore appeared to serve as a protective barrier for the organism. Studies by van Steenbergen et al. (1985) supported the protective role of the bacterial capsule. They noted that virulent strains of *P. gingivalis* were more resistant to killing by human serum and by PMNLs than those strains which were less virulent. The virulent strains had a thicker capsule and were much more hydrophilic. This indicated that the observed differences in virulence may be attributed, at least in part, to differences in capsular material.

Lipopolysaccharide

The lipopolysaccharide (LPS) component of *P. gingivalis* has been reported to be atypical in that it lacks heptose and 2-keto-3-deoxy-octonate (KDO) (Hofstad, 1974; Mansheim et al., 1978; Nair et al., 1983). Studies by Johne et al. (1988), however, show that a phosphory-lated KDO derivative is present. The presence of the phosphate group may have masked the detection of KDO by the methods used in previous investigations. Bramanti et al. (1989) have reported on the chemical and biological characterization of LPS from virulent strains of *P. gingivalis*. *P. gingivalis* LPS shows very weak endotoxic activity (Mansheim and Kasper, 1977; Sveen, 1977a, 1977b).

However, several studies indicate that LPS from *P. gingivalis* may play a role in the pathogenic process. LPS inhibits growth of gingival fibroblasts (Larjava et al., 1987), a factor that could be significant in the destruction of connective tissue. LPS has also been shown to cause bone resorption in vitro (Nair et al., 1983; Millar et al., 1986) and is able to inhibit bone collagen formation (Millar et al., 1986). Its ability to induce interleukin 1 production suggests that LPS may activate the inflammatory reactions that appear to be important in the pathogenesis of adult periodontitis (Hanazawa et al., 1985).

Enzymatic activities

A number of enzymatic activities associated with *P. gingivalis* are believed to be significant in the virulence of this organism. Of particular interest is the collagenase activity elaborated by *P. gingivalis*, which may play an important role in the loss of periodontal connective tissue. This organism appears to possess a specific bacterial collagenase that degrades native type I collagen (Mayrand and Grenier, 1985; van Steenbergen and de Graaff, 1986; Sundqvist et al., 1987), the major constituent of periodontal connective tissue. Grenier and Mayrand (1987a) found that, with one exception, invasive strains had higher levels of collagenase activity. In addition to a direct involvement in collagen degradation, the proteolytic activity of *P. gingivalis* may indirectly enhance tissue destruction by inducing the production of host

collagenases by gingival fibroblasts (Uitto et al., 1989).

Other proteolytic enzymes produced by this organism are capable of degrading serum proteins involved in host defence against microbial infections. Several studies have demonstrated the ability of *P. gingivalis* to degrade immuno-globulin molecules (Kilian, 1981; Sundqvist et al., 1985; Sato et al., 1987). Grenier et al. (1989a) confirmed these results and demonstrated that the degraded molecules could serve as growth substrates. Of interest was the finding that im-munoglobulin bound to the bacterial cell could be degraded, presumably by proteases located in the outer membrane. Sundqvist et al. (1985) also showed that *P. gingivalis* can degrade complement proteins C3 and C5 from guinea pig serum both in vitro and in vivo. These enzymatic activities may contribute to the virulence of this organism by destroying or impeding the defensive response of the host immune system.

Porphyromonas gingivalis possesses proteolytic activity specific for the degradation of human plasma proteinase inhibitors (Carlsson et al., 1984a; Nilsson et al., 1985). In vivo, these proteinase inhibitors probably regulate the activity of host proteases released from PMNLs. The destruction of such inhibitors by *P. gingivalis* could therefore favor greater tissue degradation and more rapid disease progression.

Iron transport plasma proteins such as albumin, haptoglobin, hemopoxin, and transferrin are also effectively degraded by *P. gingivalis* (Carlsson et al., 1984b). The degradation of these proteins may provide a source of nutrients, including peptides, hemin, and iron, to satisfy the metabolic requirements of this asaccharolytic organism.

The proteolytic activity of *P. gingivalis* may play a role in disturbing normal tissue integrity within periodontal connective tissue. Uitto et al. (1989) have demonstrated the ability of a *P. gingivalis* protease to degrade fibroblast cell-surface glycoproteins. This includes fibro-nectin, a cell-surface and extracellular matrix protein responsible for attachment, prolifera-tion, and synthetic activity of fibroblasts in connective tissue. Significant fibrinolytic activity has also been associated with *P. gingivalis* (Nitzan et al., 1978; Wikstrom et al., 1983; Lantz et al., 1986), thereby enhancing tissue invasion by infecting microorganisms in wound sites. These combined effects would interfere with the intrinsic ability of host tissues to undergo modeling and repair.

Although it appears several enzymes produced by *P. gingivalis* are important in bacterial pathogenicity, many of the enzyme activities studied have only been demonstrated in vitro. It remains to be determined whether some of these activities play a significant role in the disease process. In particular, it is necessary to study the regulation of virulence factors and determine if they, or other factors, are produced in vivo.

Grenier et al. (1989b) showed that *P. gingivalis* possesses at least 8 proteolytic activities of different apparent molecular weight. Profiles of the proteolytic activities were found in *P. gingivalis* culture supernatants, outer membranes, vesicles, and cell extracts analyzed in SDS-polyacrylamide gels containing covalently bound bovine serum albumin (Figure 41.2). Of the 8 proteolytic bands detected, 4 were found

in the culture supernatant (P1, P2, P3, and P4). The outer membranes, vesicles, and the cell extract contained 7 major proteolytic bands each (P1, P3, P4, P5, P6, P7, and P8). No activity was found in the membrane-free extract, suggesting that the proteases were associated with the cell envelope. With the exception of P7 and P8, all the proteolytic bands were dependent on reducing agents for activity. The 8 proteolytic bands were distributed in an identical manner in all 4 strains of *P. gingivalis* studied.

Several extracellular and cell-associated proteases have been isolated and characterized (Yoshimura et al., 1984b; Abiko et al., 1985; Fujimura and Nakamura, 1987; Grenier and McBride, 1987; Ono et al., 1987; Otsuka et al., 1987; Sorsa et al., 1987; Suido et al., 1987; Tsutsui et al., 1987; Barua et al., 1989; Miyauchi et al., 1989). It appears that most of these enzymes require thiol or serine groups for activity.

Vesicles

Some oral pathogens, including *P. gingivalis*, release outer membrane fragments as vesicles (or blebs) into the growth environment (Listgarten and Lai, 1979; London et al., 1982; Haapasalo et al., 1984; Williams and Holt, 1985; McKee et al., 1986; Parent et al., 1986; Grenier and Mayrand, 1987b). The exact function of these structures is still unclear; however, specific biological activities have been associated with the vesicles produced by oral bacteria. Grenier and Mayrand (1987b) showed that vesicles produced by *P. gingivalis* are strongly proteolytic, agglutinate erythrocytes and can act as a bridge to mediate the attachment of two noncoaggregating bacterial species. Singh et al. (1989) showed that the presence of *P. gingivalis* vesicles resulted in a 10-fold increase in the binding of streptococci to serum-coated and saliva-coated hydroxyapatite. It has been postulated that the small size of vesicles may allow them to enter sites not accessible to the whole cell, and vesicles have been found in the interstitial spaces between epithelial cells (Pekovic and Fillery, 1984). These structures therefore have the potential to invade tissue and exert enzymatic damage to host cells.

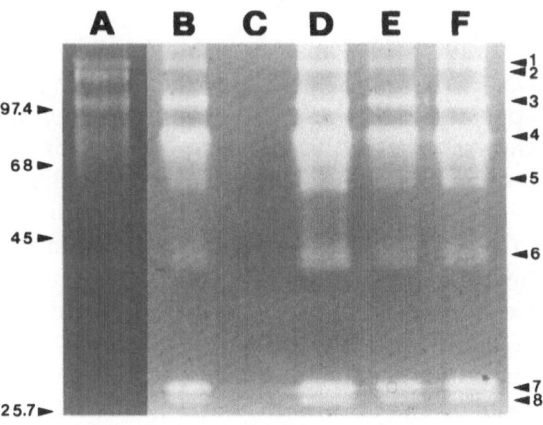

FIGURE 41.2 *Porphyromonas gingivalis* protease profiles on BSA-conjugated polyacrylamide gels developed for 2 h at 37°C in presence of 50 mM cysteine. Lanes: **A**, 15× concentrated supernatant; **B**, crude cell extract; **C**, high-speed supernatant of crude cell extract; **D**, outer membrane preparation; **E**, vesicle preparation; **F**, high-speed pellet of crude cell extract. [Reproduced from Grenier et al. (1989b) with permission.]

Metabolic end products

Other compounds that may contribute to the virulence of *P. gingivalis* are several of the physiological end products characteristic of asaccharolytic metabolism. *P. gingivalis* produces butyrate and propionate, which have been shown to have a cytotoxic effect on various human or animal cells in culture (Singer and Buchner, 1981; van Steenbergen et al., 1982; Touw et al., 1982). Volatile sulfur compounds produced by *P. gingivalis* include hydrogen sulfide, dimethyl sulfide, and methylmercaptan (Tonzetich and McBride, 1981); these have a potentially damaging effect upon host tissues because they can influence the permeability of oral mucosal tissues (Ng and Tonzetich, 1983) and reduce collagen synthesis. Indole and ammonia are also potentially toxic metabolic end products of *P. gingivalis* (MacDonald and Gibbons, 1962).

41.6   Gene Cloning of Pathogenic Determinants

With few important exceptions, current knowledge on the biology of anaerobes has resulted from classical biochemical and physiological analyses. Classical genetic approaches have only recently been applied to gram-negative anaerobes, particularly in the case of *P. gingivalis*. Standard biochemical approaches have been employed to characterize surface antigens, enzyme activities, and adhesins of *P. gingivalis*, while growth characteristics have been defined through nutritional studies. The history surrounding the taxonomic status of the organism is a vivid example of how little is known about the most fundamental properties of organisms that were recently grouped together because they formed black colonies. The advent of molecular genetics as a standard research tool has created an opportunity to define the genetic basis underlying the virulence and colonization capabilities of these organisms.

The speciation of the black-pigmented *Bacteroides* group was the first application of molecular biology in the study of these anaerobes. This was an important initial step, which defined these organisms and allowed investigators to make valid comparisons between different strains. While expertise has developed in applying standard biochemical methods to study *P. gingivalis*, the use of molecular genetics and the powerful new research possibilities it offers has been slow to develop. A number of groups however, have adopted recombinant DNA methodology to study adhesins, proteases, and immunologically important surface antigens; several researchers have recently cloned putative pathogenic determinants of *P. gingivalis*.

The gene encoding the fimbrial subunit protein (fimbrilin) of *P. gingivalis* was cloned by Dickinson et al. (1988). Chromosomal DNA was isolated from *P. gingivalis* 381 and digested with *Sac*I restriction enzyme. Fragments of 2 to 3-kb were ligated into the plasmid vector pUC13. The ligation mixture was used to transform the host cell *Escherichia coli* JM83 to generate recombinant clones. A 20-mer synthetic oligonucleotide probe corresponding to the N-terminal peptide sequence of fimbrilin was used to screen for recombinants carrying a fimbrilin gene by direct Southern blot analysis. One clone carrying a 2.5 kilobase insert fragment hybridized strongly with the probe, and was selected as a fimbrilin clone for further study.

Southern blot analysis using labelled-insert DNA as a probe showed that the fimbrilin gene appeared to be present as a single copy on the chromosome of this organism. Sequencing of the fimbrilin gene (Dickinson et al., 1988) revealed the predicted size of the mature protein to be 36 kDa, as compared to the estimated fimbrilin size of 43 kDa determined by gel electrophoresis in previous studies (Yoshimura et al., 1984a). The protein sequence had no marked similarity to other published fimbrial sequences, nor was a homologous DNA sequence found in several closely related black-pigmented *Bacteroides* species. This suggests that *P. gingivalis* fimbriae are composed of a unique protein subunit. The biological role of fimbriae, however, is not known.

McBride et al. (1990) have isolated more than 300 clones that appear to express *P. gingivalis* antigens. *P. gingivalis* ATCC 33277 chromosomal DNA was partially digested with *Pst*I endonuclease. Size-selected fragments of 3 to 9-kb were ligated to the plasmid pUC18 used as the cloning vector. Ligated DNA was used to transform *E. coli* JM83 or *E. coli* JM101 as host cells. Recombinant clones were screened for expression of *P. gingivalis* antigens by colony immunoassay, probing with antisera raised against *P. gingivalis* whole cells. The antiserum had first been extensively adsorbed with *E. coli* JM83/pUC18. More than 300 clones appeared to react with anti-*P. gingivalis* antiserum.

Three antigen clones (designated BA2, BA3, and BA4) express significant quantities of *P. gingivalis* surface antigens in *E. coli*. Analysis of these clones by (i) SDS-PAGE followed by staining with Coomassie brilliant blue R-250 and (ii) Western blot immunoassay with anti-*P. gingivalis* antiserum indicated that the apparent molecular weights of the cloned proteins expressed by BA2, BA3, and BA4 are 21, 48, and 52 kDa, respectively. The highly expressed 48-kDa *P. gingivalis* surface protein is of particular interest. The distribution and function of this protein on the *P. gingivalis* cell surface is under study.

A gene encoding a hemagglutinin from *P. gingivalis* has been cloned by Progulske-Fox et al. (1989a). A genomic library was constructed by ligating 2- to 10-kb *P. gingivalis* 381 DNA fragments from partial digestion with either *Sau*3A or *Hind*III to the plasmid pUC9, followed by transformation of host cell *E. coli* JM109. Recombinant transformants were screened by (i) colony immunoassay using antisera raised against *P. gingivalis* whole cells (the antiserum was adsorbed with *E. coli* JM109/pUC9) and (ii) a hemagglutination assay. Of the three clones demonstrating the ability to agglutinate erythrocytes, one clone carrying a 3.2-kb *P. gingivalis* DNA fragment was selected for further study.

The cloned protein is expressed on the surface of the *E. coli* host. Western blot analysis using anti-*P. gingivalis* antiserum revealed the product of the cloned gene to have an apparent molecular weight of 125 kDa. However, polyclonal monospecific antiserum to this hemagglutinin clone was found to react with two major protein bands of molecular weights 43

and 38 and minor bands of 115, 105, 32, and 30 kDa present in *P. gingivalis* cell lysate preparations. This antiserum inhibited *P. gingivalis* hemagglutinating activity, and immunoelectromicroscopic studies indicated that the cloned hemagglutinin is located on the cell surface of *P. gingivalis*.

McBride et al. (1990) have also isolated a clone that appears to carry a hemagglutinin gene from *P. gingivalis*. Recombinant clones previously analyzed for the expression of *P. gingivalis* antigens were also screened for hemagglutinating activity. A clone containing a 2.9-kb *P. gingivalis* DNA fragment was found to agglutinate human erythrocytes. The recombinant gene product is transported to the cell surface such that intact cells exhibit hemagglutinating activity. Antiserum raised against whole cells of the clone appeared to react with a 43-kDa protein present in whole-cell *P. gingivalis* lysates. This corresponds to the 43-kDa hemagglutinin purified from *P. gingivalis* ATCC 33277 (Singh and McBride, in preparation). The antiserum also inhibited *P. gingivalis* hemagglutinating activity. Restriction mapping of the *P. gingivalis* insert DNA showed that the hemagglutinin gene encoded is distinctly different from that cloned by Progulske-Fox et al. (1989a) (Figure 41.3). This introduces the possibility that more than one hemagglutinin is produced by *P. gingivalis*.

Several investigators have reported cloning of proteases from *P. gingivalis*. Roberts et al. (1989) ligated partially digested *Sau*3A fragments of *P. gingivalis* W83 chromosomal DNA into the phage cloning vector lambda EMBL4. Recombinant clones were screened for protease production by degradation of casein on skim milk overlay plates. One highly proteolytic protease clone was isolated. Further analysis showed this clone to degrade a range of synthetic substrates. Proteolytic activity appears to depend on the presence of a thiol group.

Genes encoding for type I collagenase and gelatinase activities from *P. gingivalis* have been cloned (H. Kuramitsu, personal communication). A gene bank was constructed by ligating 5-kb fragments of *Sau*3A1-partially-digested *P. gingivalis* A7A1-28 DNA to the cloning vector pPLlambda. Host cell *E. coli* HB101 was used

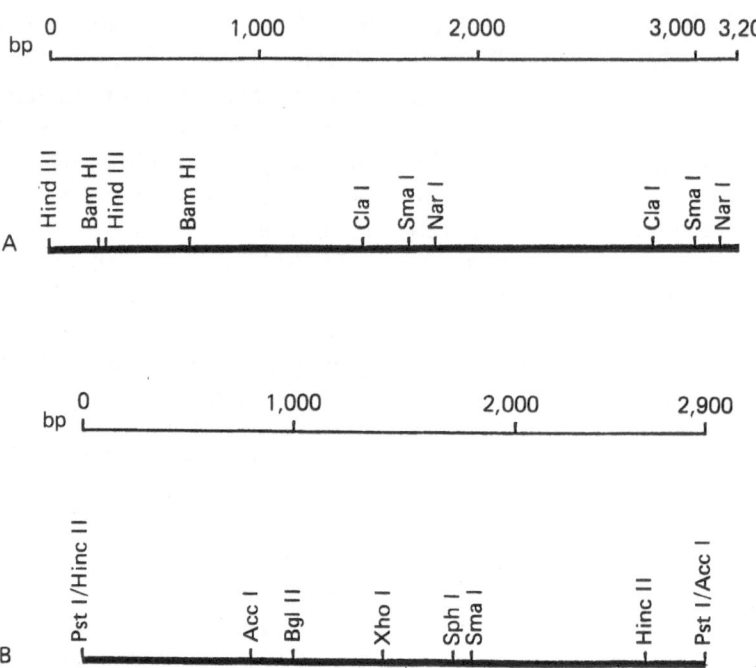

FIGURE 41.3 Restriction map of *Porphyromonas gingivalis* DNA insert containing hemagglutinin gene cloned by (**A**) Progulske-Fox et al. (1989a) and (**B**) McBride et al. (1990). Restriction maps differ significantly in restriction enzyme cleavage sites, indicating possibility of two different hemagglutinin genes. [Restriction map **A** taken from Progulske-Fox et al. (1989a) with permission.]

for isolation of recombinant clones. Two protease clones were detected by hydrolysis of casein on skim milk agar plates. Plasmids from these clones appeared to contain the same 5.9-kb insert DNA.

One clone was chosen for further characterization. Crude extracts of the clone were found to cleave type I collagen as shown by analysis of reaction mixtures by SDS-PAGE. Weak activity against type IV collagen was detected. The extracts also hydrolyzed the synthetic collagenase substrate PZ-PLGPA (phenylazobenzyl-1oxycarbonyl-L-prolyl-L-leucylglycyl-L-prolyl-D-arginine), and exhibited activity against gelatin. A deletion series of recombinant plasmids were prepared to further localize the proteolytic activity on the cloned DNA fragment. Assays for both type I collagenase and gelatinase activites indicated that the two activities appear to be expressed by two distinct genes present on the insert DNA.

Very little is known about the genetics and transfer of DNA into *P. gingivalis*. Progulske-Fox et al. (1989b) demonstrated that a shuttle plasmid, pE5-2, can be transferred via conjugation from *E. coli* to *P. gingivalis*. This plasmid appears to replicate and remain stable. It may be possible to use this shuttle vector to introduce foreign or modified genes into *P. gingivalis*.

41.7 Summary

Studies on a variety of putative pathogenic determinants elaborated by *P. gingivalis*, in addition to clinical and animal studies, implicate this organism as a periodontopathogen. However, much work is still required to specify which of the components are important in virulence. Considerable data for other pathogenic organisms show that expression of virulence determinants is a function of the growth environment; to understand the pathogenicity, it is necessary to reproduce the environment at the site of infection and assess the virulence factors.

One approach is to create genetic lesions that

modify infectivity. Decreased pathogenicity of an organism in a particular environment can then be correlated with the reduction or absence of a specific bacterial factor. While this approach may seem straightforward, it is technically difficult in the study of fastidious anaerobes. Effective systems for manipulation of DNA in these organisms have not been developed and minimal, if any, genetic information is available.

In terms of reproducing the in vivo growth environment, it is necessary to take into account specific nutritional requirements or interactions with other bacteria that may be important for the survival of the pathogen but which may prove difficult to duplicate in an experimental situation. The growth environment may be particularly critical in the study of pathogens which are isolated from bacterial infections where there is a mixed microbial population. In these situations, synergistic interactions appear to play a role in bacterial infectivity.

The presence of *P. gingivalis* in the gingival sulcus is dependent on prior colonization by a large and complex microbial population. The ability of this organism to colonize and survive in the gingival sulcus appears to rely on the support of other bacterial species for a variety of nutritional and physiological factors. Despite this problem, it is critical that attempts be made to study *P. gingivalis* under conditions which most closely mimic the natural situation. Future studies should be directed to understanding how this environment controls the expression of virulence factors. Developing appropriate genetic systems is an essential aspect of this work. To this end, it is necessary to generate useful vectors, develop protocols for efficient transfer of DNA, and therefore create the tools required to analyze the *P. gingivalis* genome and monitor gene expression in the appropriate environment.

When this has been accomplished, we will be in a position to define how this interesting oral organism manifests its pathogenicity.

Acknowledgments. We thank Howard Kuramitsu and Christian Mouton for sharing unpublished information.

References

Abiko, Y., M. Hayakawa, S. Murai, and H. Takiguchi. 1985. Glycylprolyl dipeptidylaminopeptidase from *Bacteroides gingivalis*. *J. Dent. Res.* **64**:106–111.

Altman, L.C., R.C. Page, J.L. Ebersole, and G.E. Vandesteen. 1982. Assessment of host defenses and serum antibodies to suspected periodontal pathogens in patients with various types of periodontitis. *J. Periodontal Res.* **17**:495–497.

Barua, P.K., M.E. Neiders, A. Topolnycky, J.J. Zambon, and H. Birkedal-Hansen. 1989. Purification of an 80,000 glycylprolyl peptidase from *Bacteroides gingivalis*. *Infect. Immun.* **57**:2522–2528.

Bergey, D.H., F.C. Harrison, R.S. Breed, B.W. Hammer, and F.M. Huntoon. 1930. *Bergey's Manual of Determinative Bacteriology, 3rd Ed.*, p. 314. Baltimore: Williams & Wilkins.

Boyd, J. and B.C. McBride. 1984. Fractionation of hemagglutinating and bacterial binding adhesins of *Bacteroides gingivalis*. *Infect. Immun.* **45**:403–409.

Bramanti, T.E., G.G. Wong, S.T. Weintraub, and S.C. Holt. 1989. Chemical characterization and biological properties of lipopolysaccharide from *Bacteroides gingivalis* strain W50, W83 and ATCC 33277. *Oral Microbiol. Immunol.* **4**:183–192.

Carlsson, J., J.F. Hofling, and G.K. Sundqvist. 1984a. Degradation of the human proteinase inhibitors alpha-1-antitrypsin and alpha-2-macroglobulin by *Bacteroides gingivalis*. *Infect. Immun.* **34**:644–648.

Carlsson, J., J.F. Hofling, and G.K. Sundqvist. 1984b. Degradation of albumin, hemopexin, haptoglobin and transferrin, by black-pigmented *Bacteroides* species. *J. Med. Microbiol.* **18**:39–46.

Cimasoni, G., M. Song, and B.C. McBride. 1987. Effect of crevicular fluid and lysosomal enzymes on the adherence of *Streptococci* and *Bacteroides* to hydroxyapatite. *Infect. Immun.* **55**:1484–1489.

Courant, P.R. and R.J. Gibbons. 1967. Biochemical and immunological heterogeneity of *Bacteroides melaninogenicus*. *Arch. Oral Biol.* **12**:1605–1613.

Coykendall, A.L., F.S. Kaczmarek, and J. Slots. 1980. Genetic heterogeneity in *Bacteroides asaccharolyticus* (Holdeman and Moore 1970) Finegold and Barnes 1977 (Approved Lists, 1980) and proposal of *Bacteroides gingivalis* sp. nov. and *Bacteroides macacae* (Slots and Genco) comb. nov. *Int. J. Syst. Bacteriol.* **30**:559–564.

Deslauriers, M. and C. Mouton. 1991. The hemagglutinating adhesin HA-Ag2 of *Bacteroides gingivalis* is distinct from fimbrilin. *Oral Microbiol. Immunol.* **6**:6–11.

Dickinson, D.P., M.A. Kubiniec, F. Yoshimura, and R.J. Genco. 1988. Molecular cloning and sequencing of the gene encoding the fimbrial subunit protein of *Bacteroides gingivalis*. *J. Bacteriol.* **170**:1658–1665.

Di Murro, C., R. Nisini, M. Cattabriga, A. Simonetti-D'Arca, S. Le Mole, M. Paolantonio, L. Sebastiani, and R. D'Amelio. 1987. Rapidly progressive periodontitis — Neutrophil chemotaxis inhibitory factors associated with the presence of *Bacteroides gingivalis* in crevicular fluid. *J. Periodontol.* **58**:868–872.

Dix, K., S.M. Watanabe, S. McArdle, D.l. Lee, C. Randolph, B. Moncla, and D.E. Schwartz. 1990. Species-specific oligonucleotide probes for the identification of periodontal bacteria. *J. Clin. Microbiol.* **28**:319–323.

Ebersole, J.L., M.A. Taubman, and D.J. Smith. 1985. Gingival crevicular fluid antibody to oral microorganisms. II. Distribution and specificity of local antibody responses. *J. Periodontal Res.* **20**:349–356.

Ebersole, J.L., M.A. Taubman, D.J. Smith, and D.E. Frey. 1986. Human immune responses to oral microorganisms: patterns of systemic antibody levels to *Bacteroides* spp. *Infect. Immun.* **51**:507–513.

Ebersole, J.L., M.A. Taubman, D.J. Smith, and S.S. Socransky. 1982. Humoral immune responses and diagnosis of human periodontal disease. *J. Periodontal Res.* **17**:478–480.

Farida, R., P.D. Marsh, H.N. Newman, D.C. Rule, and L. Ivanyi. 1986. Serological investigation of various forms of inflammatory periodontitis. *J. Periodontal Res.* **21**:365–374.

Finegold, S.M. and E.M. Barnes. 1977. Report of the ICSB Taxonomic Subcommittee on Gram-negative Anaerobic Rods. Proposal that the saccharolytic and asaccharolytic strains at present classified in the species *Bacteroides melaninogenicus* (Oliver and Wherry) be classified in the two species as *Bacteroides melaninogenicus* and *Bacteroides asaccharolyticus*. *Int. J. Syst. Bacteriol.* **27**:388–391.

Fujimura, S. and T. Nakamura. 1987. Isolation and characterization of a protease from *Bacteroides gingivalis*. *Infect. Immun.* **55**:716–720.

Gibbons, F.J. and J.B. MacDonald. 1960. Hemin and vitamin K compounds as required factors for the cultivation of certain strains of *Bacteroides melaninogenicus*. *J. Bacteriol.* **80**:164–170.

Grenier, D. and D. Mayrand. 1983. Etudes d'infections mixtes anaerobies comportant *Bacteroides gingivalis*. *Can. J. Microbiol.* **29**:612–618.

Grenier, D. and D. Mayrand. 1986. Nutritional relationships between oral bacteria. *Infect. Immun.* **53**:616–620.

Grenier, D. and D. Mayrand. 1987a. Selected characteristics of pathogenic and nonpathogenic strains of *Bacteroides gingivalis*. *J. Clin. Microbiol.* **25**:738–740.

Grenier, D. and D. Mayrand. 1987b. Functional characterization of extracellular vesicles produced by *Bacteroides gingivalis*. *Infect. Immun.* **55**:111–117.

Grenier, D. and B.C. McBride. 1987. Isolation of a membrane-associated *Bacteroides gingivalis* glycylprolyl protease. *Infect. Immun.* **55**:3131–3136.

Grenier, D., D. Mayrand, and B.C. McBride. 1989a. Further studies on the degradation of immunoglobulins by black-pigmented *Bacteroides*. *Oral Microbiol. Immunol.* **4**:12–18.

Grenier, D., G. Chao, and B.C. McBride. 1989b. Characterization of sodium dodecyl sulfate-stable *Bacteroides gingivalis* proteases by polyacrylamide gel electrophoresis. *Infect. Immun.* **57**:95–99.

Haapasalo, M., H. Ranta, H. Shah, K. Ranta, and K. Lounatmaa. 1984. Isolation and characterization of a new variant of black-pigmented asaccharolytic *Bacteroides*. *FEMS Microbiol. Lett.* **23**:269–274.

Hanazawa, S., K. Nakada, Y. Ohmori, T. Miyoshi, S. Amano, and S. Kitano. 1985. Functional role of interleukin 1 in periodontal disease: induction of interleukin 1 production by *Bacteroides gingivalis* lipopolysaccharide in peritoneal macrophages from C3H/HeN and C3H/HeJ mice. *Infect. Immun.* **50**:262–270.

Handley, P.S. and L.S. Tipler. 1986. An electron microscope survey of the surface structures and hydrophobicity of oral and nonoral species of the bacterial genus *Bacteroides*. *Arch. Oral Biol.* **31**:325–335.

Hofstad, T. 1974. The distribution of heptose and 2-keto-3-deoxy-octonate in *Bacteroidaceae*. *J. Gen. Microbiol.* **85**:314–320.

Holdeman, L.V. and W.E.C. Moore. 1970. *Outline of Clinical Methods in Anaerobic Bacteriology* (2nd Ed.) Blacksburg: Virginia Polytechnic Institute and State University Anaerobic Laboratory.

Holt, S.C., J. Ebersole, J. Felton, M. Brunsvold, and K. Kornman. 1988. Implantation of *Bacteroides gingivalis* in nonhuman primates initiates progression of periodontitis. *Science* **239**:55–57.

Hunt, D.E., J.V. Jones, and V.R. Dowell, Jr. 1986. Selective medium for the isolation of *Bacteroides gingivalis*. *J. Clin. Microbiol.* **23**:441–445.

Inoshita, E., A. Awano, T. Haniska, H. Tamagawa, S. Shizukishi, and A. Tsunemitsu. 1986. Isolation and some properties of exohemagglutinin from the culture medium of *Bacteroides gingivalis* 381. *Infect. Immun.* **52**:421–427.

Johne, B., I. Olsen, and K. Bryn. 1988. Fatty acids and sugars in lipopolysaccharides from *Bacteroides intermedius*, *Bacteroides gingivalis*, and *Bacteroides loescheii*. *Oral Microbiol. Immunol.* **3**:22–27.

Kagan, J.M. 1980. Local immunity to *Bacteroides gingivalis* in periodontal disease. *J. Dent. Res.* **59**(D, pt. 1):1750–1756 (special issue).

Kilian, M. 1981. Degradation of immunoglobulins A1, A2, and G by suspected principal periodontal pathogens. *Infect. Immun.* **34**:757–765.

Kinder, S.A. and S.C. Holt. 1989. Characterization of coaggregation between *Bacteroides gingivalis* T22 and *Fusobacterium nucleatum* T18. *Infect. Immun.* **57**:3425–3433.

Lantz, M., R.W. Rowland, L.M. Switalski, and M. Hook. 1986. Interactions of *Bacteroides gingivalis* with fibrinogen. *Infect. Immun.* **54**:654–658.

Larjava, H., V.-J. Uitto, E. Eerola, and M. Haapasalo. 1987. Inhibition of gingival fibroblast growth by *Bacteroides gingivalis*. *Infect. Immun.* **55**:201–205.

Lev, M. and A.F. Milford. 1975. Sensitivity of a *Bacteroides melaninogenicus* to monosaccharides: effect on enzyme induction. *J. Bacteriol.* **121**: 152–159.

Listgarten, M.A. and C.H. Lai. 1979. Comparative ultrastructure of *Bacteroides melaninogenicus* subspecies. *J. Periodontal Res.* **14**:332–340.

Loesche, W.J., S.A. Syed, E. Schmidt, and E.C. Morrison. 1985. Bacterial profiles of subgingival plaques in periodontitis. *J. Periodontol.* **56**:447–456.

Loesche, W.J., S.A. Syed, E.C. Morrison, B. Langhon, and N.S. Grossman. 1981. Treatment of periodontal infections due to anaerobic bacteria with short term treatment with metronidazole. *J. Clin. Periodontol.* **8**:29–44.

London, J., R. Celesk, and P. Kolenbrander. 1982. Physiological and ecological properties of the oral gram-negative gliding bacteria capable of attaching to hydroxyapatite. *In: Host-Parasite Interactions in Periodontal Diseases*, R.J. Genco and S.E. Mergenhagen, eds., pp. 283–298. Washington, D.C.: American Society of Microbiology.

MacDonald, J.B. and R.J. Gibbons. 1962. The relationship of indigenous bacteria to periodontal disease. *J. Dent. Res.* **41**:320–326.

MacDonald, J.B., S.S. Socransky, and R.J. Gibbons. 1963. Aspects of the pathogenesis of mixed anaerobic infections of mucous membranes. *J. Dent. Res.* **42**:529–544.

MacDonald, J.B., R.M. Sutton, M.L. Knoll, E.M. Madlener, and R.M. Grainger. 1956. The pathogenic components of an experimental fusospirochetal infection. *J. Infect. Dis.* **98**:15–20.

Mansheim, B.J. and D.L. Kasper. 1977. Purification and immunochemical characterization of the outer membrane complex of *Bacteroides melaninogenicus* subspecies *asaccharolyticus*. *J. Infect. Dis.* **135**:787–799.

Mansheim, B.J., A.B. Onderdonk, and D.L. Kasper. 1978. Immunochemical and biologic studies of the lipopolysaccharide of *Bacteroides melaninogenicus* subspecies *asaccharolyticus*. *J. Immunol.* **120**:72–78.

Mayrand, D. and B.C. McBride. 1980. Ecological relationships of bacteria involved in a simple mixed anaerobic infection. *Infect. Immun.* **27**:44–50.

Mayrand, D. and D. Grenier. 1985. Detection of collagenase activity in oral bacteria. *Can. J. Microbiol.* **31**:134–138.

Mayrand, D. and S.C. Holt. 1988. Biology of asaccharolytic black-pigmented *Bacteroides* species. *Microbiol. Rev.* **52**:134–152.

McArthur, W.P., I. Magnusson, R.G. Marks, and W.B. Clark. 1989. Modulation of colonization by black-pigmented *Bacteroides* species in squirrel monkeys by immunization with *Bacteroides gingivalis*. *Infect. Immun.* **57**:2313–2317.

McBride, B.C., A. Joe, and U. Singh. 1990. Cloning of *B. gingivalis* surface antigens involved in adherence. *Arch. Oral Biol.* **35**, Suppl:59–68.

McKee, A.S., A.S. McDermid, A. Baskerville, A.B. Dowsett, D.C. Ellwood, and P.D. Marsh. 1986. Effect of hemin on the physiology and virulence of *Bacteroides gingivalis* W50. *Infect. Immun.* **52**:349–355.

Millar, S.J., E.G. Goldstein, M.J. Levine, and E. Hausmann. 1986. Modulation of bone metabolism by two chemically distinct lipopolysaccharide fractions from *Bacteroides gingivalis*. *Infect. Immun.* **51**:302–306.

Miyauchi, T., M. Hayakawa, and Y. Abiko. 1989. Purification and characterization of glycylprolyl aminopeptidase from *Bacteroides gingivalis*. *Oral Microbiol. Immunol.* **4**:222–226.

Moncla, B.J., P. Braham, K. Dix, S. Watanabe, and D. Schwartz. 1990. Use of synthetic oligonucleotide DNA probes for the identification of *Bacteroides gingivalis*. *J. Clin. Microbiol.* **28**:324–327.

Mouton, C., D. Bouchard, M. Deslauriers, and L. Lamonde. 1989. Immunological identification and preliminary characterization of a nonfimbrial hemagglutinin adhesin of *Bacteroides gingivalis*. *Infect. Immun.* **57**:566–573.

Mouton, C., P.G. Hammond, J. Slots, and R.J. Genco. 1981. Serum antibodies to oral *Bacteroides asaccharolyticus* (*Bacteroides gingivalis*): relationship to age and periodontal disease. *Infect. Immun.* **31**:182–192.

Nair, B.C., W.R. Mayberry, R. Dziak, P.B. Chen, M.J. Levine, and E. Hausmann. 1983. Biologic effects of a purified lipopolysaccharide from *Bacteroides gingivalis*. *J. Periodontal Res.* **18**:40–49.

Neiders, M.E., P.B. Chen, H. Suido, H.S. Reynolds, J.J. Zambon, M. Schlossman, and R.J. Genco. 1988. Heterogeneity of virulence among strains of *Bacteroides gingivalis*. *J. Periodontal Res.* **24**:192–198.

Ng, W. and J. Tonzetich. 1983. Effect of H_2S on permeability of oral mucosa. *J. Dent. Res.* **62**:275. (abstr. 953)

Nilsson, T., J. Carlsson, and G. Sundqvist. 1985. Inactivation of key factors of the plasma proteinase cascade systems by *Bacteroides gingivalis*. *Infect. Immun.* **50**:467–471.

Nitzan, D., J.F. Sperry, and T.D. Wilkins. 1978. Fibrinolytic activity of oral anaerobic bacteria. *Arch. Oral Biol.* **23**:465–470.

Okuda, K. and I. Takazoe. 1973. Antiphagocytic effects of the capsular structure of a pathogenic strain of *Bacteroides melaninogenicus*. *Bull. Tokyo Dent. Coll.* **14**:99–104.

Okuda, K. and I. Takazoe. 1974. Hemagglutinating activity of *Bacteroides melaninogenicus*. *Arch. Oral Biol.* **19**:415–416.

Okuda, K., J. Slots, and R.J. Genco. 1981. *Bacteroides gingivalis*, *Bacteroides asaccharolyticus*, and *Bacteroides melaninogenicus* subspecies: cell surface morphology and adherence to erythrocytes and human buccal epithelial cells. *Curr. Microbiol.* **6**:7–12.

Okuda, K., A Yamamoto, Y. Naito, I. Takazoe, J. Slots, and R.J. Genco. 1986. Purification and properties of hemagglutinin from culture supernatant of *Bacteroides gingivalis*. *Infect. Immun.* **54**:659–665.

Oliver, W.W. and W.B. Wherry. 1921. Notes on some bacterial parasites of the human mucous membranes. *J. Infect. Dis.* **28**:341–345.

Ono, M., K. Okuda, and I. Takazoe. 1987. Purification and characterization of a thiol-protease from *Bacteroides gingivalis* strain 381. *Oral Microbiol. Immunol.* **2**:77–81.

Otsuka, M., J. Endo, D. Hinode, A. Nagata, R. Maehara, M. Sato, and R. Nakamura. 1987. Isolation and characterization of protease from culture supernatant of *Bacteroides gingivalis*. *J. Periodontal Res.* **22**:491–498.

Parent, R., C. Mouton, L. Lamonde, and D. Bouchard. 1986. Human and animal serotypes of *Bacteroides gingivalis* defined by cross immunoelectrophoresis. *Infect. Immun.* **51**:909–918.

Pekovic, D.D. and E.D. Fillery. 1984. Identification of bacteria in immunopathological mechanisms of human periodontal disease. *J. Periodontal Res.* **19**:329–351.

Progulske-Fox, A., S. Tumwasorn, and S.C. Holt. 1989a. The expression and function of a *Bacteroides gingivalis* hemagglutinin gene in *E. coli*. *Oral Microbiol. Immunol.* **4**:121–131.

Progulske-Fox, A., A. Oberste, C. Drummond, and W.P. McArthur. 1989b. Transfer of plasmid pE5-2 from *Escherichia coli* to *Bacteroides gingivalis* and *B. intermedius*. *Oral Microbiol. Immunol.* **4**:132–134.

Rizza, V., P.R. Sinclair, D.C. White, and P.R. Courant. 1968. Electron transport system of the protoheme-requiring anaerobe *Bacteroides melaninogenicus*. *J. Bacteriol.* **96**:665–671.

Roberts, I.S., M. Arnott, A. Wallace, H.N. Shah,

and D. Williams. 1989. Cloning of proteases of *Porphyromonas gingivalis*. *J. Dent. Res.* **68**:1006, abstr. 1117 (special issue).

Roy, T.E. and C.D. Kelly. 1939. Genus VIII. *Bacteroides* Castellani and Chalmers. In: *Bergey's Manual of Determinative Bacteriology*, 5th Ed., D.H., Bergey, R.S. Breed, E.G.D. Murray, and A.P. Hitchens, eds., (pp. 556–559). Baltimore: Williams & Wilkins.

Sato, M., M. Otsuka, R. Maehara, J. Endo, and R. Nakamura. 1987. Degradation of human secretory immunoglobulin A by protease isolated from the anaerobic periodontopathogenic bacterium, *Bacteroides gingivalis*. *Arch. Oral Biol.* **32**:235–238.

Sawyer, S.J., J.B. MacDonald, and R.J. Gibbons. 1962. Biochemical characteristics of *Bacteroides melaninogenicus*. A study of 31 strains. *Arch. Oral Biol.* **7**:685–691.

Schwartz, S., R.P. Ellen, and D.A. Grove. 1987. *Bacteroides gingivalis- Actinomyces viscosus*. Cohesive interactions as measured by a quantitative assay. *Infect. Immun.* **55**:2391–2397.

Shah, H.N., R.A.D. Williams, G.D. Bowden, and M.J. Hardie. 1976. Comparison of the biochemical properties of *Bacteroides melaninogenicus* from human dental plaque and other sites. *J. Appl. Bacteriol.* **41**:473–492.

Shah, H.N., R. Bonnet, B. Matteen, and R.A. Williams. 1979. The porphyrin pigmentation of subspecies of *Bacteroides melaninogenicus*. *Biochem. J.* **180**:45–50.

Shah, H.N. and M.D. Collins. 1980. Fatty acid and isoprenoid quinone composition in the classification of *Bacteroides melaninogenicus* and related taxa. *J. Appl. Bacteriol.* **48**:75–87.

Shah, H.N. and M.D. Collins. 1988. Proposal for reclassification of *Bacteroides asaccharolyticus*, *Bacteroides gingivalis*, and *Bacteroides endodontalis* in a new genus, *Porphyromonas*. *Int. J. Syst. Bacteriol.* **38**:128–131.

Singer, R.E. and B.A. Buchner. 1981. Butyrate and propionate: important components of toxic dental plaque extracts. *Infect. Immun.* **32**:458–463.

Singh, U., D. Grenier, and B.C. McBride. 1989. *Bacteroides gingivalis* vesicles mediate attachment of streptococci to serum-coated hydroxyapatite. *Oral Microbiol. Immunol.* **4**:199–203.

Slots, J. 1977. The predominant cultivable microflora of advanced periodontitis. *Scand. J. Dent. Res.* **85**:114–121.

Slots, J. and R.J. Gibbons. 1978. Attachment of *Bacteroides melaninogenicus* subsp. *asaccharolyticus* to oral surfaces and its possible role in colonization of the mouth and of periodontal pockets. *Infect. Immun.* **9**:254–264.

Sorsa, T., V.-J. Uitto, K. Suomalainen, A. Turto, and S. Lindy. 1987. A trypsin-like protease from *Bacteroides gingivalis*: partial purification and characterization. *J. Periodontal Res.* **22**:1–6.

Suido, H., M.E. Neiders, P.K. Barua, M. Nakamura, P.A. Mashimo, and R.J. Genco. 1987. Characterization of N-CBz-glycyl-glycyl-arginyl peptidase and glycyl-prolyl peptidase of *Bacteroides gingivalis*. *J. Periodontal Res.* **22**:412–418.

Sundqvist, G.K. 1976. Bacteriological studies of necrotic dental pulps. Ph.D. thesis, University of Umea, Umea, Sweden.

Sundqvist, G.K., J. Carlsson, and L. Hanstrom. 1987. Collagenolytic activity of black pigmented *Bacteroides* species. *J. Periodontal Res.* **22**:300–306.

Sundqvist, G.K., J. Carlsson, B. Herrmann, and A. Tarnvik. 1985. Degradation of human immunoglobulins G and M and complement factors C3 and C5 by black-pigmented *Bacteroides*. *J. Med. Microbiol.* **19**:85–94.

Suzuki, Y., E. Yoshimura, K. Takahasti, H. Toran, and T. Suzuki. 1988. Detection of fimbriae and fimbrial antigens on the oral anaerobe *Bacteroides gingivalis* by negative staining and serological methods. *J. Gen. Microbiol.* **134**:2713–2720.

Sveen, K. 1977a. The capacity of lipopolysaccharides from *Bacteroides*, *Fusobacterium*, and *Veillonella* to produce skin inflammation and the local and generalized Schwartzman reaction in rabbits. *J. Periodontal Res.* **12**:340–350.

Sveen, K., T. Hofstad, and K.C. Milner. 1977b. Lethality for mice and chick embryos, pyrogenicity in rabbits and ability to gelate lysate from ameobocytes of *Limulus polyphemus* by lipopolysaccharides from *Bacteroides*, *Fusobacterium* and *Veillonella*. *Acta Pathol. Microbiol. Scand. Sect. B* **85**:388–396.

Takazoe, I., M. Tanaka, and T. Homma. 1971. A pathogenic strain of *Bacteroides melaninogenicus*. *Arch. Oral Biol.* **16**:817–822.

Tanner, A.C.R., C. Haffer, G.T. Bratthall, R.A. Visconti, and S.S. Socransky. 1979. A study of the bacteria associated with advancing periodontitis in man. *J. Clin. Periodontol.* **6**:278–307.

Tew, J.G., D.R. Marshall, J.A. Burmeister, and R.R. Ranney. 1985. Relationship between gingival crevicular fluid and serum antibody titres in young adults with generalized and localized periodontitis. *Infect. Immun.* **49**:487–493.

Tonzetich, J. and B.C. McBride. 1981. Characterization of volatile sulphur production by pathogenic and nonpathogenic strains of oral *Bacteroides*. *Arch. Oral Biol.* **26**:963–969.

Touw, J.J.A., T.J.M. van Steenbergen, and J. de Graaff. 1982. Butyrate: a cytotoxin for vero cells produced by *Bacteroides gingivalis* and *Bacteroides asaccharolyticus*. *Antonie Leeuwenhoek J. Microbiol.* **48**:315–325.

Tsutsui, H., T. Kinouchi, Y. Wakano, and Y. Ohnishi. 1987. Purification and characterization of a protease from *Bacteroides gingivalis* 381. *Infect. Immun.* **55**:420–427.

Uitto, V.-J., H. Larjava, J. Heino, and T. Sorsa. 1989. A protease from *Bacteroides gingivalis* degrades cell surface and matrix glycoproteins of cultured gingival fibroblasts and induces secretion of collagenase and plasminogen activator. *Infect. Immun.* **57**:213–218.

van Dyke, T.E., S. Offenbacher, D. Place, V.R. Dowell, and J. Jones. 1988. Refractory periodontitis: mixed infection with *Bacteroides gingivalis* and other *Bacteroides* species. *J. Periodontol.* **59**:184–189.

van Steenbergen, T.J.M. and J. de Graaff. 1986. Proteolytic activity of black-pigmented *Bacteroides*. *FEMS Microbiol. Lett.* **33**:219–222.

van Steenbergen, T.J.M., F. Namavar, and J. de Graaff. 1985. Chemiluminescence of human leukocytes by black-pigmented *Bacteroides* strains from dental plaque and other sites. *J. Periodontal Res.* **20**:58–71.

van Steenbergen, T.J.M., J.J. de Soet, and J. de Graaff. 1979. DNA base composition of various strains of *Bacteroides melaninogenicus*. *FEMS Microbiol. Lett.* **5**:127–130.

van Steenbergen,T.J.M., M.D. den Ouden, J.J.A. Touw, and J. de Graaff. 1982. Cytotoxic activity of *Bacteroides gingivalis* and *Bacteroides asaccharolyticus*. *J. Med. Microbiol.* **5**:253–258.

van Steenbergen, T.J.M., A.J. van Winkelhoff, D. Mayrand, D. Grenier, and J. de Graaff. 1984. *Bacteroides endodontalis* sp. nov., an asaccharolytic black-pigmented *Bacteroides* species from infected dental root canals. *Int. J. Syst. Bacteriol.* **34**:118–120.

Vincent, J.W., J.B. Suzuki, W.A. Falkler, Jr., and W.C. Cornett. 1985. Reaction of human sera from juvenile periodontitis, rapidly progressive periodontitis, and adult periodontitis patients with selected periodontopathogens. *J. Periodontol.* **56**:464–469.

Weerkamp, A.H. and B.C. McBride. 1980. Characterization of the adherence properties of *Streptococcus salivarius*. *Infect. Immun.* **29**:459–468.

White, D. and D. Mayrand. 1981. Association of oral *Bacteroides* with gingivitis and adult periodontitis. *J. Periodontal Res.* **16**:259–265.

Wikstrom, M.B., G. Dahlen, and A. Linde. 1983. Fibrinogenolytic and fibrinolytic activity in oral microorganisms. *J. Clin. Microbiol.* **17**:759–767.

Williams, G.D. and S.C. Holt. 1985. Characteristics of the outer membrane of selected oral *Bacteroides* species. *Can. J. Microbiol.* **31**:238–250.

Woo, D.D.L., S.C. Holt, and E.R. Leadbetter. 1979. Ultrastructure of *Bacteroides* species: *Bacteroides asaccharolyticus*, *Bacteroides fragilis*, *Bacteroides melaninogenicus*, and *Bacteroides melaninogenicus* subspecies *intermedius*. *J. Infect. Dis.* **319**:534–546.

Yoshimura, F., K. Takahasti, Y. Nodasaka, and T.

Suzuki. 1984a. Purification and characterization of a novel type of fimbriae from the oral anaerobe *Bacteroides gingivalis*. *J. Bacteriol.* **160**:949–952.

Yoshimura, F., M. Nishikata, T. Suzuki, E.l. Hoover, and E. Newbrun. 1984b. Characterization of a trypsin-like protease from the bacter-

ium *Bacteroides gingivalis* isolated from human dental plaque. *Arch. Oral Biol.* **7**:559–564.

Yoshimura, F., T. Takasawa, M. Yoneyama, T. Tamaguchi, H. Shikawa, and T. Suzuki. 1985. Fimbriae from the oral anaerobe *Bacteroides gingivalis*: physical, chemical, and immunological properties. *J. Bacteriol.* **163**:730–734.

42

Cloning, Structure, and Expression of Genes of the Anaerobic Rumen Bacteria

R.M. Teather, H.J. Gilbert, and G.P. Hazlewood

42.1 Introduction

The anaerobic rumen bacteria comprise 30 or so known species of obligately anaerobic bacteria that may be reproducibly isolated from the contents of the ruminant forestomach. Together with the anaerobic protozoa and fungi also found ubiquitously in this ecosystem, these bacteria obtain carbon and energy by degrading plant biomass. In so doing, they produce acetic, butyric, and propionic acids as major fermentation products, and generate microbial biomass that, with undegraded dietary protein, serves as the protein source for the ruminant animal. The intrinsic interest in these organisms derives from their considerable metabolic, taxonomic, and phylogenetic diversity, and their economic importance may be gauged by the prominence of ruminants in world agriculture.

Through the application of conventional microbiological and biochemical techniques, a wealth of information concerning individual microorganisms, their enzyme systems, and probable contributions to the rumen fermentation has been revealed (Stewart and Bryant, 1988). The advent of molecular techniques has provided powerful new tools for investigating the genetic basis of biological function and has made available the means to study ruminal processes which in some instances have proved to be too complex for biochemical analysis.

42.2 Cellulase, Xylanase, and Related Genes

The capacity for degrading the plant structural polysaccharides cellulose and hemicellulose is an essential feature of the rumen microbial ecosystem (Chesson and Forsberg, 1988), and is a property that has attracted much interest not only because detailed knowledge of cellulolysis in vivo may lead to improved efficiency in the use of forages, but also because cellulolytic enzymes are of fundamental importance in commercial processes seeking to use plant biomass as a source of energy or chemical feedstocks.

Enzymatic hydrolysis of cellulose is a complex process requiring the cooperative action of β-(1,4)-endoglucanases (EC 3.2.1.4), β-(1,4)-exoglucanases (EC 3.2.1.91), and β-glucosidase (EC 3.1.1.21). Studies with *Clostridium thermocellum* suggest that a likely model for cellulase in anaerobic bacteria is a high molecular weight multiprotein aggregate (the cellulosome) that is composed of multiple β-glucanases encoded by multiple genes (Lamed and Bayer, 1988; Béguin et al., 1988; Hazlewood et al., 1988). Hydrolysis of xylan (the main polymeric component of hemicellulose) is principally effected by β-(1,4)-endoxylanases (EC 3.2.1.8) and β-xylosidase (EC 3.2.1.37), although other enzymes [α-arabinofuranosidase (EC 3.2.1.55), α-glucuronidase (EC 3.2.1.4), and acetyl xylan

569

esterase (EC 3.1.1.6)] are required to remove the side chains from highly branched xylans (Biely, 1985; Wong et al., 1988). Biochemical analysis has been partially successful in resolving the complexities of bacterial cellulases but has been impeded by the difficulties involved in dissociating multienzyme complexes while retaining the biological function of the individual components. The cloning of genes by in vitro genetic recombination and their expression in *Escherichia coli* has provided the means of dissecting cellulase enzyme complexes and studying the function of the constituent proteins against a cellulase-free background. Such an approach has been applied with varying degrees of success to each of the bacterial species shown to be involved in ruminal cellulose and hemicellulose degradation, and to at least two species whose role in cellulolysis in vivo is less clearly defined. To date genes coding for enzymes involved in the degradation of plant structural polysaccharides have been cloned from *Fibrobacter succinogenes, Ruminococcus albus, Ruminococcus flavefaciens, Bacteroides ovatus, Bacteroides ruminicola*, and *Butyrivibrio fibrisolvens* (see Table 42.1). In some instances, studies have progressed beyond the preliminary characterization of a gene and its encoded protein, and primary sequence data are available revealing interesting details of molecular architecture and phylogeny.

Fibrobacter succinogenes

Fibrobacter succinogenes plays a major role in the degradation of plant fiber in the rumen (Dehority and Scott, 1967; Morris and van Gylswyk, 1980; Graham et al., 1985; Stewart and Flint, 1989). Although it was only recently reclassified (Montgomery et al., 1988), there is evidence to suggest that it may yet constitute a genetically diverse group of organisms (Stahl et al., 1988; Stewart and Flint, 1989). Most of the interest in this organism has centered around its ability to efficiently degrade plant fiber, so it is not entirely surprising that all of the genes that have been cloned from it play a role in this process.

Like other cellulolytic bacteria that have been characterized (Coughlan, 1985; Robson and Chambliss, 1989) *F. succinogenes* produces a

number of β-(1,4)-glucanases (Schellhorn and Forsberg, 1984). Several of these have been characterized biochemically, including a cellodextrinase (Huang and Forsberg, 1988), two endoglucanases (McGavin and Forsberg, 1988), and a chloride-stimulated cellobiosidase (Huang et al., 1988). Gene cloning experiments have demonstrated a still greater level of complexity. The results of these studies are summarized in Table 42.1. At least nine distinct genes involved in the degradation of plant polysaccharides have been identified in this way, and it seems probable from the limited number of isolates that have been examined that the total number of genes involved may be considerably greater (Crosby et al., 1984). Each of the clones characterized to date has expressed only a single gene coding for an enzyme involved in polysaccharide degradation, and has shown no DNA sequence homology with other clones, even though most of the cloned DNA fragments have been large enough to carry a number of genes. Although it is of course possible that other genes present are not preceded by fortuitous *E. coli*-like promoter sequences, or have phenotypes that are not easily assayable, this suggests that the genes involved in cellulose degradation in this organism are widely scattered on the chromosome rather than being organized in an operon structure. The expression of four of these genes in *E. coli* has been studied in some detail.

Xylanase gene expression was independent of orientation and the level of enzyme activity in *E. coli* was comparable to that found in *F. succinogenes*. Synthesis was constitutive, with no regulation by glucose, xylan, or xylose. [Synthesis of the enzyme is apparently constitutive in *F. succinogenes* as well (Forsberg et al., 1981)]. About half of the activity was found in the periplasmic space, indicating that the enzyme possesses a recognizable signal peptide sequence (Sipat et al., 1987).

The β-(1,3)(1,4)-glucanase from *F. succinogenes* was also expressed constitutively, apparently from its own promoter sequence, in *E. coli*. Because this enzyme has not been identified in vivo or purified from *F. succinogenes* cultures, it was not possible to compare the level of gene expression between the two

Table 42.1 Genes cloned from anaerobic rumen bacteria

Cellulase, xylanase, and related genes

Fibrobacter succinogenes	Reference
β-(1,4)-Endoglucanases (six)	Crosby et al., 1984
xylanase	Sipat et al., 1987
β-(1,3)(1,4)-Glucanase	Irvin and Teather, 1988
cellodextrinase	Gong et al., 1989
Ruminococcus albus	
β-(1,4)-Endoglucanase	Kawai et al., 1987
β-(1,4)-Endoglucanase (two)	Ropmaniec et al., 1989
β-(1,4)-Endoglucanase	Ohmiya et al., 1988
β-Glucosidase	Honda et al., 1988
β-(1,4)-Endoglucanase/β-(1,4)-exoglucanase/	
β-glucosidase (total of six)	Howard and White, 1988
β-(1,4)-Endoglucanase (two)	Ware et al., 1989
β-Glucosidase	Misawa and Nakamura, 1989
Ruminococcus flavefaciens	
β-(1,4)-Endoglucanase/cellodextrinase	Barros and Thomson, 1987
Xylanase (four)	
β-(1,3)(1,4)Glucosidase	Flint et al., 1989
β-(1,4)-Endoglucanase (eight)	
β-(1,4)-Endoglucanase (four)	Huang et al., 1989
β-Glucosidase (two)	
β-(1,4)-Endoglucanase/xylanase	Howard and White, 1990
β-(1,4)-Endoglucanase	
Xylanase	White et al., 1990
Butyrivibrio fibrisolvens	
β-(1,4)-Endoglucanase	Hazlewood et al., 1990
β-(1,4)-Endoglucanase	Berger et al., 1989
cellodextrinase	Berger et al., 1990
Arabinosidase	
Xylosidase	Utt and Ingram, 1989
Xylosidase	Sewell et al., 1989
Bacteroides ruminicola	
β-(1,4)-Endoglucanase	Matsushita et al., 1989; 1990
Xylanase	Whitehead and Hespell, 1989a
β-(1,4)-Endoglucanase	Woods et al., 1989
Bacteroides ovatus	
Arabinosidase	
Xylanase	Whitehead and Hespel, 1989b
Xylosidase	

Lactose catabolic genes

Streptococcus bovis	
β-Galactosidase	
Lactose permease	Gilbert and Hall, 1987
Thiogalactoside transacetylase	

organisms, but the level of enzyme activity in *E. coli* was comparable to the total β-(1,4)-glucanase activity in *F. succinogenes* (Erfle et al., 1988). This enzyme was also partially exported to the periplasmic space, although less effi-ciently (about 10% of the total activity) than the xylanase, again indicating that the gene posses-ses a recognizable signal peptide sequence (Irvin and Teather, 1988).

Expression of *F. succinogenes* cellodextrinase

in *E. coli* was constitutive with a level of activity comparable to that found in the wild-type strain. In contrast to the xylanase and mixed linkage glucanase genes, expression of the cellodextrinase gene in *E. coli* was subject to catabolite repression by glucose. Expression was not affected by cellobiose, possibly because of the inability of cellobiose to enter the host cell. The properties of the recombinant and native enzymes were similar. Unlike the native enzyme, the recombinant enzyme was not exported to the periplasm, indicating that it did not have a signal peptide sequence that could be recognized by the host cell (Gong et al., 1989).

The product of the *cel*-3 gene, endoglucanase 3 (EG3), was also expressed constitutively in *E. coli*. Like the cellodextrinase, synthesis of the enzyme was strongly repressed by the addition of glucose to the growth medium. [Some repression of endoglucanase synthesis by glucose and cellobiose has been noted in *F. succinogenes* (Groleau and Forsberg, 1981)]. The recombinant enzyme differed in its properties from the native endoglucanase complex. It was exported to the periplasmic space, indicating that it possessed a recognizable signal peptide sequence, but gel electrophoresis revealed that at least three enzymatically active protein species were present (Taylor et al., 1987). Subsequent analysis showed that while the enzyme was efficiently exported to the periplasmic space, it remained membrane bound as a 94-kDa peptide. In aging cultures this peptide was hydrolyzed, presumably by nonspecific periplasmic proteases, to produce first a 51-kDa periplasmic peptide and finally a family of three 43-kDa peptides that differed in size by one amino acid (McGavin et al., 1989). The native enzyme has been identified in the culture supernatant, with an apparent molecular weight >94 kDa (McGavin et al., 1989).

These four examples show that genes from *F. succinogenes* can be efficiently expressed in *E. coli*. The fact that each gene has a recognizable promoter and ribosome binding site is not surprising since they were all isolated by selecting *E. coli* clones which expressed the proteins at significant levels. It has not been shown that these elements are functional in *F. succinogenes*.

All four genes were expressed constitutively in *E. coli*, suggesting that if transcription is regulated in *F. succinogenes* it probably involves a repressor rather than an activator; two of the cloned genes were apparently subject to catabolite repression, but it is possible that this effect was mediated through the vector (Taylor et al., 1987; Irvin and Teather, 1988) since the overall cellulose degradation rate of *F. succinogenes* is not affected by inclusion of glucose or cellobiose in the growth medium (Hiltner and Dehority, 1983).

Although all four of these enzymes are presumably exported from the cytoplasm in *F. succinogenes*, the proportion of enzyme activity found outside the cytoplasm in the *E. coli* clones ranged from 50% to 0%, and in the case of EG3 the mature native and recombinant enzymes had very different structures and properties. Thus while three of the four genes apparently encoded recognizable signal peptide sequences, there is evidence that further posttranslational modification occurs in *F. succinogenes* that cannot be duplicated in *E. coli*, either because *E. coli* lacks the necessary functions or because it lacks other components of the cellulase complex that interact with the peptide before or during processing.

Complete nucleotide sequences are available for two of these cloned genes, *cel*-3 (McGavin et al., 1989) and the mixed linkage glucanase gene (Teather and Erfle, 1990). The *cel*-3 gene consists of an ORF of 1,974 bp coding for a 658 amino acid protein with M_r of 73.4 kDa. The mixed linkage glucanase is encoded by an ORF of 1,047 bp, specifying a mature protein of 35.2 kDa. Both the *cel*-3 gene and the mixed linkage glucanase gene have a single consensus promoter sequence (ending at −40 and −55 bases relative to the initiation codon, respectively) and an *E. coli* type ribosome binding site (at −6 bp and −8 bp, respectively). There is a single potential rho-independent termination sequence downstream of the stop codon in the *cel*-3 gene. In the mixed linkage glucanase gene there are five potential rho-independent transcription termination sites downstream of the open reading frame, and three upstream. The noncoding regions of both these genes have a very high AT content (69% and 57% for cel-3, and 70% and 71% for the mixed linkage glucanase, respectively), as compared with 46% for the open reading frames. The existence of AT-rich regions upstream of promoter sequences has been noted as a feature of strong promoters in *Bacillus subtilis* (Reznikoff et al., 1985). Codon usage in the mixed linkage glucanase gene was unusual in that the major bacterial lysine codon, AAA, was used infrequently, and the dominant leucine codon, CTG, was not used at all.

As noted above, both EG3 and the mixed linkage glucanase are exported from the cytoplasm and processed to some extent. The amino terminus peptide sequences predicted from the DNA sequences fit the consensus for bacterial signal peptide sequences, particularly in the case of the mixed linkage glucanase (Teather and Erfle, 1990). The N-terminal amino acid sequence of the enzyme synthesized in *E. coli* has been determined in both cases. In the case of EG3, N-terminal amino acid analysis showed that the signal sequence was able to direct the translocation of the protein to the periplasmic space, but the protein was not cleaved as predicted, and remained membrane associated until it was partially degraded by periplasmic proteases (McGavin et al., 1989), resulting in a greatly truncated but active form of EG3 (M_r 43 kDa) whose N-terminus matched residue 266 of the protein encoded by the translated gene sequence. EG3 produced by *F. succinogenes* had an M_r of 118 kDa, suggesting that posttranslational glycosylation or N-terminal fatty acyl substitution occurs in the wild-type strain. Although only about 10% of the recombinant *F. succinogenes* mixed linkage glucanase could be isolated with a periplasmic cell fraction (Irvin and Teather, 1988), all of the protein was cleaved as predicted by the DNA sequence (Teather and Erfle, 1990).

Based on the analysis of early cellulase sequences, it was suggested (Robson and Chambliss, 1989) that the level of intergeneric sequence homology for bacterial cellulases is low, with the exception of relatively short conserved regions, but as more primary sequences become available, the picture that emerges is of a group of enzymes, composed of domains that are more or less conserved,

and whose order of assembly varies according to the source of the enzyme (Béguin, 1990). A seemingly valid classification based on hydrophobic cluster analysis (HCA) has recently been employed to assign the cellulases of prokaryotes and lower eukaryotes to six distinct families, some of which can be divided into subfamilies (Henrissat et al., 1989; Béguin, 1990). *F. succinogenes* cel-3 and mixed linkage glucanase genes showed no significant homology with one another or with cellulase related genes from other rumen bacteria, but each of the encoded proteins exhibited some similarity with β-glucanases from taxonomically distant genera. The cel-3 gene displayed significant homology with the *celC* gene of *C. thermocellum* (McGavin et al., 1989) and has, on that basis, been assigned to the cellulase subfamily A3 (Béguin, 1990), while the mixed linkage glucanase was identical over 36% of its residues with an enzyme of similar function produced by *Bacillus subtilis* (Murphy et al., 1984) (Figure 42.1).

An interesting feature, located toward the C-terminus of the *F. succinogenes* mixed linkage glucanase, is a series of five direct repeats of the seven amino acid sequence XxxProXxx-SerSerSerSer, where Xxx is an uncharged residue (Teather and Erfle, 1990). The last pair of repeats is separated by two additional uncharged residues. A similar feature, consisting of five direct repeats of the seven amino acid sequence ProGluProThrProValGlu, has been observed toward the C-terminus of the endoglucanase, End 1, produced by *Butyrivibrio fibrisolvens* H17c (Berger et al., 1989) (Figure 42.2). Neither the nucleic acid sequences of the two genes, nor the amino acid sequences of the respective proteins, exhibit significant homology, but in each case the repeated sequences define a similar structure, consisting of seven closely spaced turns flanked by short hydrophilic sequences. Deletion analysis has shown in the case of the *F. succinogenes* mixed linkage glucanase that the repeated sequence can be removed without inactivating the enzyme or changing its affinity for oat glucan (Teather and Erfle, 1990), suggesting that it is not directly involved in catalysis or substrate binding. Sequences such as this, lacking the repeat motif,

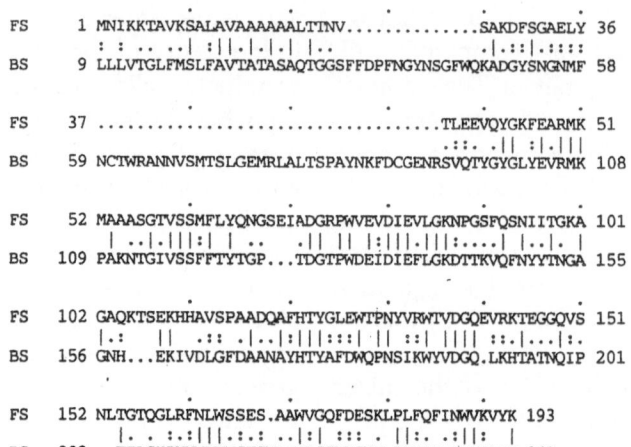

```
FS    1 MNIKKTAVKSALAVAAAAAALTINV.............SAKDFSGAELY  36
        : : ..| :||.|.| .|           |..::|..:::
BS    9 LLLVTGLFMSLFAVTATASAQTGGSFFDPFNGYNSGFWQKADGYSNGNMF  58


FS   37 ..................................TLEEVQYGKFEARMK  51
                                          .::. .|| :|.|||
BS   59 NCTWRANNVSMTSLGEMRLALTSPAYNKFDCGENRSVQTYGYGLYEVRMK 108


FS   52 MAAASGTVSSMFLYQNGSEIADGRPWVEVDIEVLGKNPGSFQSNIITGKA 101
        | ..|.|||:| | ..   .|| || :|||.|||:....| ...|. |
BS  109 PAKNTGIVSSFFTYTGP...TDGTPWDEIDIEFLGKDTTKVQFNYYTNGA 155


FS  102 GAQKTSEKHHAVSPAADQAFHTYGLEWTPNYVRWTVDGQEVRKTEGGQVS 151
        |.:   ||  .:: .|..:|:|||:::| || ::| |||| ::.|...|:.
BS  156 GNH...EKIVDLGFDAANAYHTYAFDWQPNSIKWYVDGQ.LKHTATNQIP 201


FS  152 NLTGTQGLRFNLWSSES.AAWVGQFDESKLPLFQFINWVKVYK 193
        |. . :.:|||.:.: ..|:| ::: .||:. .:||: |
BS  202 ..TTLGKIMMNLWNGTGVDEWLGSYNGVN.PLYAHYDWVRYTK 241
```

FIGURE 42.1. Comparison of the amino acid sequence of the *B. subtilis* (BS) (Murphy et al., 1984) and *F. succinogenes* (FS) mixed linkage glucanases. A vertical bar indicates identical amino acid residues, dots indicate amino acids that are closely related in evolutionary terms. [The comparison shown was prepared using the program Bestfit in the University of Wisconsin GCG package, version 6.0 (Devereux et al., 1984)].

```
B. fibrisolvens    ..psteq PDPTPVD PDPTPVD PDPTPVD PDPTPVD     PDPQPVD ptpvs..

F. succinogenes    ..rddep APQSSSS APASSSS VPASSSS VPASSSS AF VPPSSSS atnai..
```

FIGURE 42.2. Comparison of the amino acid repeats in the *B. fibrisolvens* endoglucanase End1 and the *F. succinogenes* mixed linkage glucanase.

but similarly rich in hydroxy amino acids and/or proline, and sometimes heavily glycosylated, are a recurring feature in cellulases of both bacterial and fungal origin (O'Neill et al., 1986; Penttila et al., 1986; Wong et al., 1986; Knowles et al., 1987; Teeri et al., 1987; Béguin et al., 1988; Hall and Gilbert, 1988; Hall et al., 1989; Gilbert et al., 1990). These sequences have been widely assumed to function as hinges or linkers that maintain spatial separation between the catalytic and cellulose-binding domains of certain cellulases, but a recent study has weakened this assumption and suggested a different possible origin and role. Serine-rich sequences located between the catalytic and cellulose-binding domains of a xylanase (xylA) from *Pseudomonas fluorescens* subsp. *cellulosa* were deleted without affecting substrate specificity, cellulose-binding capacity, specific activity, or K_m for xylan hydrolysis. If, as seems likely, cellulases and xylanases evolved through the acquisition and shuffling of common ancestral sequences encoding discrete domains, it is suggested that DNA sequences coding for the extended serine-rich regions may have fulfilled a role analogous to that of introns in eukaryotic genes (Ferreira et al., 1990).

Ruminococcus albus

Molecular cloning, in combination with biochemical analysis, has shown that cellulase from the Gram-positive anaerobe *Ruminococcus albus* consists of numerous enzyme activities (β-(1,4)-endoglucanase, β-glucosidase, cellobiosidase), and is encoded by multiple genes. Genes encoding cellulase-related enzymes [notably β-(1,4)-endoglucanases] have been isolated from strains SY3 (Romaniec et al., 1989), F-40 (Kawai et al., 1987; Ohmiya et al., 1988; Honda et al., 1988), AR67/68 (Ware et al., 1989), and 8 (Howard and White, 1988) (Table 42.1).

The complete nucleotide sequence of the *Eg1* gene of *R. albus* F-40 comprised 1,221 bp and coded for a mature endoglucanase (Eg1) of 363 amino acids (M_r 40.9 kDa), preceded by a putative signal peptide of 43 residues (Ohmiya et al., 1989). Palindromic sequences with potential to form hairpin loops were located upstream and downstream of the structural gene, but the deduced free energies for the proposed structures were too low for effective termination of transcription. Consensus promoter sequences and a sequence with strong Shine–Dalgarno (SD) complementarity characteristic of an *E. coli* ribosome binding site were located upstream of the *Eg1* gene. G + C content of the ORF was 46.3% and there was no bias in codon usage. At the amino acid level, Eg1 was highly homologous with endoglucanase E (EGE) from *C. thermocellum* (Hall et al., 1988). A β-glucosidase from the same strain of *R. albus*, encoded by an ORF of 2841 bp and comprising 947 amino acid residues, showed no detectable homology with other β-glucanases of rumen origin (Ohmiya et al., 1990).

Two further genes, *celA* and *celB*, encoding endoglucanase/xylanase and endoglucanase activities respectively, were isolated from a plasmid gene bank constructed from *R. albus* SY3 genomic DNA partially digested with *Sau*3A (Romaniec et al., 1989). The *celA* structural gene comprised an ORF of 1,095 bp coding for a protein with M_r of 41.2 kDa. The *celB* structural gene consisted of an ORF of 1,227 bp coding for a protein with M_r of 45.5 kDa (Poole et al., 1990). A typical Gram-positive SD sequence was located 7 bp upstream of the translational start codon of *celB*, and the deduced N-terminal region of endoglucanase B (EGB) conformed to the general pattern for the signal peptides of secreted prokaryotic proteins (positively charged N-terminus followed by a central hydrophobic core, and a flexible C-terminal region containing hydrophilic residues). The complete *celB* gene cloned in a high copy number vector was lethal to *E. coli*, but *celA* cloned in pUC18, under the control of *lac*Zp, directed high-level synthesis of endoglucanase A (EGA) in *E. coli* JM83; only 20% of total EGA was translocated to the periplasm, suggesting that the secretory signal preceding mature EGA is not efficiently recognized by *E. coli*. Gene deletion and subcloning studies with *celA* revealed

that EGA does not contain separate cellulose-binding and catalytic domains, such as have been described in an increasing number of bacterial endoglucanases and xylanases (see Béguin, 1990 for refs). Comparison of sequences revealed a high degree of homology between the *R. albus* cellulases EGA, EGB, and Eg1 (Figure 42.3). Furthermore all three endoglucanases exhibited homology with the *Butyrivibrio fibrisolvens* endoglucanase, End1 (Berger et al., 1989), and the catalytic domain of *C. thermocellum* endoglucanase E (Hall et al., 1988). A simple explanation for the evolution of multiple cellulases with a highly conserved structure is not immediately obvious but it is possible that the ancestral cellulase gene of *R. albus* may have undergone duplication so recently that the resultant multiple genes have had insufficient time to undergo substantial mutagenesis. Further factors, such as the overall structure of the multiprotein cellulase complex or the highly proteolytic rumen environment, may impose structural limitations on the constituent proteins, with the result that the multiple cellulases that have successfully evolved all conform to a common structural pattern. Extensive homology between EGA, EGB, and Eg1 of *R. albus* and End1 of *B. fibrisolvens* suggests that these four enzymes belong to the same enzyme family (subfamily A4; Béguin, 1990), and provides evidence that gene transfer may have occurred at some time between these two genera of rumen bacteria.

Further published reports and unpublished observations suggest that the true number of genes coding for cellulase-related enzymes in *R. albus* is very high. Genomic libraries of *R. albus* strain 8 (Howard and White, 1988) and strain SY3 (G.P. Hazlewood and B.A. White, unpublished) constructed in the lambda phage vector λDASH yielded large numbers of apparently different clones that were able to hydrolyze Ostazin brilliant red hydroxyethyl cellulose. Indeed, for strain 8, three different endoglucanases, three putative exoglucanases, and four clones with mixed specificity including β-glucosidase activity were identified in studies that used phage lysates as the source of enzyme activity and recombinant DNA (Howard and White, 1988). Attempts to transfer

```
EGB    1 MKLKRIAALLTAAVMSVGV MASCGGSKSDDKSKADTKSAAETSGAEGDSSESEEIPVSQTH
Eg1    1 MNS KKIGAMIAAAVLSLIVMTPAATRKIVQRQTRNSSTAVENSAAD ES ETENVPVSQTH
EGA    1 M                    RKDADRLTTLDLARSGEV              R
EGE    1 MK  K IVGLVCVLMLVS IL  GSFSVVAAS P VKGF QVSGTKLLDASGNEL VMRG
End 1  1 MHK SKCIKRVFTFLLALF VFVMA I         PATKVSAA GGT   DRSATQ V VSDMR


EGB   62 TNDPMTVTSAKDLVAKMSNGWNLGNTMDATGE GLESEISWLPTKVYTNKFMIDMLPEAGFN
Eg1   60 TNDTMTVTSAKDLVAKMSNGWNLGNTMDATGQ GLGSEVSWLPLKVTTNKYMIDMLPEAGFN
EGA   24 D    ISAM ELVGEMKTGWNLGNSLDATGAPGNASEVNWGNPKT TKE MIDAVYNKGFN
EGE   52 MRD    ISAI DLVKEIKIGWNLGNTLDA PT    ETA WGNPRT TKA MIEKVREMGFN
End1  48 V  GWNI GN SL DSFGQSYNFP Y TS   LNETY W GNPA TTKALIDEVAKAGFN
                         *         **              *     *    *** *    ***


EGB  123 VLRIPVSWGNHLI DN NYTIDPAWMDRVQEIVNYGIDDGMYVILNTHHE    EW    YMPKP
Eg1  121 VLRIPVSWGNHII DD KYTSDPAWMDRVQEIVNYGIDNGLYVILNTHHE    EW    YMPKP
EGA   78 VIRIPVSWGGHGV DAPDYKIDDEWIARVQEIVNYAYDDGAYVIINSHHE    EW    RIPDN
EGE  102 AVRVPVTWDTHIGPAP DYKIDEAWLNRVEEVVNYVLDCGMYAIINLHHD    NTW    IIPTY
End1  95 TIRIPVSWGQYTT GS DYQI PDFMNRVKEVVDYCIVNDMYVILNSHHDINSDYCFYVPNN
         *******           *      ** ** *** *       * *** ***     *   **


EGB  178 SEKDGDIEE LKAIWSQIADRFKGYDEHLIFEGLNEPRLRGEGAEWTGT       SEARE I
Eg1  176 SEKDGDIEE IKAVWAQIADRFKGYDEHLIFEGLNEPRVRGEGAEWTGT       SEARE I
EGA  135 EHIDAVDEK TAAIWKQVAERFKDYGDHLIFEGLNEPRVKGSPQEWNGGT      EEGRR  C
EGE  160 AEQRSK EK LVKVWEQIATRFKDYDDHLLFETMNEPREVGSPMEWMGGTY     EARD V
End1 155 ANKDRS EKYFKSIWTQIAKEFKNYDYHLVFETMNEPRLVGHGEEWWFPRNNPSDIREAVAC
                   ** *    ***** ***** *     **            * *


EGB  232 INEYEKAFVETVRASGGNNGDRCLMITGYAASSGYN NLSA IELPEDSDKLIISVHAYLPY
Eg1  230 INEYEKAFVETVRASGGNNGDRCLMITGYAASSGYN NLSA IELPEDSDKLIISVHAYLPY
EGA  190 VDRLNKTFLDTVRATGGNNEKRLLLMTTYASSSMSN VIKD TAIPED DHIGFSIHAYTPY
EGE  214 INRFNLAVVNTIRASGGNNDKRFIMVPT NAATGLDVALND LVIPNNDSRVIVSIHAYSPY
End1 214 INDYNQVALDAIRATGGNNATRCVMVPGYDASIE G CMTDGFKMPNDTARLILSVHAYIPY
         *        *  ********  * ***            **       ** ***** **


EGB  292 SFALDTKG T D  KYDPE DT A  IPTLFESLNELFISRDIPVIVGEFGSMNKDNID DRV
Eg1  290 SFALDTKG T D  KYDPE DT A  IPELFEHLNELFISKGIPVIVGEFGTMNKENTE DRV
EGA  249 AFTYNANA DWELFHWDDSHDG E  LVSLMTNLKENYLDRDIPVIITEYGAVNKDNNDEDRA
EGE  274 FFAMDVTG T S  YWGSDYDT ASLTSELDAIYNRFVKTGRAVIIGEFGTIDKTNLS SRV
End1 279 YFALASDTYV T  RFDDN LKYD IDSFFNDLNSKFLSRNIPVVVGETSATNRRNTA ERV
         *           *         *    *   **    *****    * *       *


EGB  346 KCLDDYL GNAAKYD IPCVWWDN   YARIGNGENFG  L LNRQEYDWYFPKLMD VFK K
Eg1  345 KCLEDYL AAAAKYD IPCVWWDN   YARIGNGENFG  L MNRADLEWYFPDLIE TFK T
EGA  308 KWVSSYI EYAELLGGIPCVWWDNG  YYSSGN ELFG  I FDRNTCTWFTDTVTD AII
EGE  330 AHAEHYAREAVSR G IAVFWWDN   YYNPGDAETYA  L LNRKTLSWYYPEIVQALMRGA
End1 335 K WADYYWGRAARYSNVAMVLWDNNIYQNNSAGSDGECHMYIDRNSLQWKDPEIIS TIM K
         *            *     ***  *                * * *       *    *


EGB  339 YAE SDPSAAAA
Eg1  397 YAE KDPASAE
EGA  361   E NAK
EGE  386 GVEPLVSPTPTPTLM
End1 394 HVD GTPA
```

Figure 42.3. Alignment of homologous regions of EGA and EGB from *R. albus* SY3 with other cellulases. The sequences are as follows: Eg1, endoglucanase I from *R. albus* (Ohmiya et al., 1989); EndI; endoglucanase I from *B. fibrisolvens* H17c (Berger et al., 1989); EGE, endoglucanase E from *C. thermocellum* (Hall et al., 1988). Residues which show identity or similarity of structure in all primary structures compared are indicated by *. (From Poole et al., 1990.)

the cloned genes to plasmid vectors have been unsuccessful, suggesting that a substantial number of the cellulase genes or their encoded proteins share a common attribute that adversely affects either plasmid replication or cell viability in *E. coli*.

Ruminococcus flavefaciens

Ruminococcus flavefaciens contributes significantly to fiber digestion in the rumen through its production of endoglucanase, exoglucanase, cellodextrinase, xylanase, and pectinase activities (Pettipher and Latham, 1979). A larger number of genes encoding cellulase-related enzymes has been cloned from this Gram-positive anaerobe than from any of the other cellulolytic rumen species, but paradoxically we know the least of all about their structure and their expression in heterologous hosts. Barros and Thomson (1987) were first to report the cloning of a β-glucanase gene from R. flavefaciens and its expression in E. coli. The encoded enzyme hydrolyzed carboxymethyl cellulose (CMC), but was most active against p-nitrophenyl-β-D-cellobioside and cellooligosaccharides and was subsequently designated a cellodextrinase. Enzyme synthesised in E. coli was largely exported to the periplasm and was not subject to catabolite repression by glucose. Subsequent studies have resulted in the isolation of genes encoding multiple endoglucanases, exoglucanases, xylanases, and β-glucosidases and a single mixed linkage, β-(1-3),(1-4)-glucanase (Flint et al., 1989; Huang et al., 1989; Howard and White, 1990; White et al., 1990) (Table 42.1). In most instances, genes have been cloned and expressed using λ phage vectors, and enzyme activities and recombinant DNA have been characterized by reference to phage lysates. The identification of clones expressing both endoglucanase and xylanase activities (Howard and White, 1990) or a combination of xylanase, xylosidase, and β-(1-3),(1-4)-glucanase activities (Flint et al., 1989) raises the interesting possibilities of multifunctional proteins or linked genes. Detailed information on the primary structure of the cloned genes and their encoded proteins is not yet available. There is evidence that some of the genes encoding cellulase-related functions of R. flavefaciens may be unstable in E. coli (B.A. White, personal communication) and this may constitute a barrier to further rapid progress.

Butyrivibrio fibrisolvens

Molecular cloning techniques have also been applied with some success by groups interested in the fiber digesting activities of Butyrivibrio fibrisolvens. Members of this genus are ubiquitous in the rumen and are active in numerous processes including the hydrolysis of plant lipids (Harfoot and Hazlewood, 1988), protein (Hazlewood et al., 1983; Cotta and Hespell, 1986), starch (Cotta, 1988), pectin (Dehority, 1969), cellulose (Shane et al., 1969), and xylan (Dehority, 1967). In some species of ruminant, butyrivibrios are the major culturable cellulolytic bacteria (Orpin et al., 1985).

Six genes coding for arabinosidase, cellodextrinase, endoglucanase, and xylosidase activities have been cloned from B. fibrisolvens (Table 42.1) and the complete nucleotide sequences have been reported for two endoglucanases and one cellodextrinase. Each of the cloned genes was expressed in E. coli, and the endoglucanase gene from B. fibrisolvens strain A46 was also expressed in Enterococcus faecalis (Mann, 1988).

Endoglucanase gene end1 from B. fibrisolvens H17c consisted of 1,641 bp and coded for a 61-kDa polypeptide of 547 residues, whose N-terminal sequence was typical of the signal peptides of secreted proteins from Gram-positive bacteria (Berger et al., 1989). The region upstream of the ORF contained three consensus promoter sequences and an E. coli type ribosome binding site that ended 8 bp upstream of the initiation codon. An 8-bp inverted repeat that could form a rho-independent terminator was found 8 bp downstream of the stop codon. An interesting feature of the End1 protein was the five perfect repeats of the sequence PDPTPVD that occurred between residues 412 and 447. Similar sequences rich in hydroxy amino acids and proline have been reported in a number of bacterial and fungal cellulases, including the mixed linkage glucanase from F. succinogenes referred to earlier in this chapter (Figure 42.2).

Comparison of amino acid sequences with those of other cellulases revealed that End 1 displays 53% similarity with EGE from C. thermocellum over 360 residues, suggesting that it belongs to cellulase subfamily A4 (Béguin, 1990). In addition, the sequence contains a region showing clear similarity with the cellulose-binding domain common to a variety of other endoglucanases and xylanases (see Gilbert et al., 1990), and although Berger et al., (1989) gave no indication as to whether End 1 binds to crystalline cellulose in the manner of other enzymes containing this sequence, it seems entirely possible that End 1 contains distinct and separate catalytic and cellulose-binding domains.

A second gene (ced1) isolated from B. fibrisolvens H17c, also comprised 1641 bp, and coded for a cellodextrinase with a calculated size of 61 kDa (Berger et al., 1990). The ced1 gene product, which was exported to the periplasm in E. coli, had some endoglucanase activity but differed from End1 with respect to the high level of activity that it displayed against cellodextrins and p-nitrophenyl-β-D-cellobioside. Amino acid sequence alignment studies revealed that Ced1 was <12% homologous with End1 from the same organism, but showed 59% and 52% similarity respectively with endoglucanase D (EGD) from C. thermocellum (Joliff et al., 1986) and endoglucanase A (EGA) from P. fluorescens subsp. cellulosa (Hall and Gilbert, 1988), suggesting that it belongs to cellulase subfamily E1 (Béguin, 1990).

A gene coding for an endoglucanase (EGA) active against a range of substrates including CMC, barley β-glucan, Avicel, filter paper, and p-nitrophenyl-β-D-cellobioside has been cloned from B. fibrisolvens strain A46, which was originally isolated as a predominant cellulolytic organism from the rumen of the high arctic Svalbard reindeer (Orpin et al., 1985). The primary structure of the gene, designated celA, comprised an ORF of 1,296 bp coding for a protein of 48.9 kDa, preceded by a potential ribosome binding site (GGGAGGA) (Hazlewood et al., 1990); the celA promoter appeared to be functional in E. coli JM83. The encoded protein, EGA, was hyperexpressed in E. coli under the control of lacZp, and had an N-terminal region of 34 amino acids that conformed to the pattern for prokaryotic signal peptides, but only 25% at most was secreted into the periplasm, suggesting that the signal peptide was not fully effective in E. coli. The G + C content of celA was 38.9% and codons having G or C in the third position occurred at frequencies of 18.6% and 15.2% respectively indicating some bias toward A and T in the "wobble" position. A comparison between this gene and end1 of B. fibrisolvens H17c revealed that the two genes were quite similar in their codon preferences. With two ex-

ceptions (AGA and GGA) both celA and end1 make little use of those codons corresponding to the minor tRNAs that in E. coli are presumed to act as modulators (Minton et al., 1986). Comparison of sequences revealed that residues 151–283 of EGA displayed >50% homology with endoglucanases from Bacillus spp. (Fukumori et al., 1989), Clostridium acetobutylicum (Zappe et al., 1988), and Erwinia chrysanthemi (Guiseppi et al., 1988) (Figure 42.4). No homology was found with End1 from B. fibrisolvens H17c. Endoglucanases showing homology with EGA display significant similarity over their 300–400 residue N-terminal catalytic domains, and have been assigned to cellulase subfamily A2 (Béguin, 1990). It would seem likely that the celA gene of B. fibrisolvens A46 has also evolved from an ancestral gene common to members of the A2 subfamily.

In contrast to the cellulases from R. albus which are highly conserved, suggesting a common ancestral gene and strong selection pressure for the retention of conserved sequences, endoglucanases EGA and End1, produced by two strains of B. fibrisolvens, clearly belong to different cellulase subfamilies. The lack of homology between B. fibrisolvens cellulases could reflect the apparent ease with which genetic exchange occurs between members of this genus and other bacteria (Teather, 1985), thus providing a mechanism for the acquisition of cellulase genes from diverse sources. Alternatively, it may be further evidence of the inadequacy of the classification applied to rumen bacteria. If rumen butyrivibrios do, in fact, comprise a genetically heterogeneous group consisting of two or more genera and numerous species (Hazlewood and Teather, 1988) it would not be surprising to find that their cellulase enzymes derive from more than one evolutionary source.

Xylan degrading enzymes, produced constitutively by many ruminal strains of B. fibrisolvens (Hespell et al., 1987), have also been investigated using cloning techniques, and genes coding for xylosidase and arabinosidase activities have been isolated (Sewell et al., 1989; Utt and Ingram, 1989). The xylB gene, coding for an enzyme which hydrolysed xylooligosaccharides from two to five residues in length,

```
EGA    151   ETIKS L RDT W GIN V I RLAM YTSDYN GY CVA
CEL-A   73   DSLKW L RDD W GIT V F RAAM YTSS  G GY IE
CEL-B   71   ESMKW L RDD W GIT V F RAAM YTSS  G GY IE
Bs1     77   DSLKW L RDD W GIT V F RAAM YTAD  G GY I
Bs2     76   DSLKW L RDD W GIT V F RAAM YTAD  G GY ID
EGZ     72   DTVAS L LKD W KSS I V RAAM GVQESG GY LQ
Cal     82   DSMKF L RDK W GMN V I RAAM YTNE  G GY ISN
CEL-C  114   NAFAA L AND W GSN V I RLAL YIGE  N GY PR

EGA    181   GKENQEKL K DI I DDA VE A A T D ND MYVII D W
CEL-A  101      D PS V K NK V KEA VE A A I D LG IYVII D W
CEL-B   99      D PS V K EK V KEA VE A A I D LG IYVII D W
Bs1    104     DNPS V K NK V KEA VE A A K E LG IYVII D W
Bs2    104     NPSVKNK K KV   KEA VE A A K E LG IYVII D W
EGZ    101      D PAGNKA K V  ERV VD A       ND MYAII G W
Cal    111     PSSQKEKI K KI V QDA ID      L N MYVII D W
CEL-C  142       N PLIE K    V YAG IE I A K E ND MYVII D W

EGA    211   H TLNDAD PN  EY              KAD A IQ FF
CEL-A  127   H ILSDND PN  IY              KEE A DE FF
CEL-B  125   H ILSDND PN  IY              KEE A KD FF
Bs1    137   H ILNDGN PN  QN              KEK A KE FF
Bs2    132   H ILNDGN PN  QN              KEK A KE FF
EGZ    126   H S HSAE N   N               RSE A IR FF
Cal    137   H ILSDNN PN  TY              KEQ A KS FF
CEL-C  168   H VHRPGD PN ADIY QGGVNEDGEEYLG A KD FF

EGA    230   GEMVRK Y KDNEN VIYEI C NEP NG DTT  W N
CEL-A  145   DEMSAL Y GDYPN VIYEI A NEP TG HNVRW D
CEL-B  143   DEMSEL Y GDYPN VIYEI A NEP NGSDVT WDN
Bs1    151   KEMSSL Y GNTPN VIYEI A NEP NG DVN  WKR
Bs2    150   KEMSSL Y GNTPN VIY I A NEP NG DVN  WKR
EGZ    141   QEMARK Y GNKPN VIYEI Y NEP L  QVS  W S
Cal    155   QEMAEE Y GKYSN VIYEI C NEP NG GTN  WAN
CEL-C  198   LHIAEK Y PNDPH LIYEL A NEP SSNSSG G P

EGA    258   D VRRYANEVIPV IR N          VD
CEL-A  173   SHIKPYAEEVIPV IR A          ND
CEL-B  172   Q IKPYAEEVIPV IR N          ND
Bs1    179   D IKPYAEEVISV IR K          ND
Bs2    177   D IKPYAEEVISV IR K          ND
EGZ    167   NTIKPYAEAVIS  IR A          ID
Cal    183   D VRPYANYIIPA IR A          ID
CEL-C  226   G I TNDEDGWEA VR EYAQPIVDALRDSGNA

EGA    275     A IILVGT PK
CEL-A  189   PNN IVIVGT A
CEL-B  188   PNN IIIVGT GT
Bs1    195   PDN IIIVGT GIW
Bs2    193   PDN IIIVGT GT
EGZ    183   PDN LIIVGT PSW
Cal    198   PNN IIIVGT S
CEL-C  255   EDN IIIVGS PNW
```

FIGURE 42.4. Amino acid sequence alignments for endoglucanases from *B. fibrisolvens* strain A46 (EGA), *Bacillus* sp. strain N-4 (CEL-A, CEL-B and CEL-C; Fukumori et al., 1989), *Bacillus subtilis* PAP115 (Bs1; Zappe et al., 1988), *B. subtilis* DLG (Bs2; Robson and Chambliss, 1987), *Erwinia chrysanthemi* (EGZ; Guiseppi et al., 1988) and *Clostridium acetobutylicum* (Cal; Zappe et al., 1988). Residues that are identical or similar in structure in all sequences compared are enclosed within a box. (From Hazlewood et al., 1990.)

was expressed in *E. coli* apparently under the control of *lacZp* (Sewell et al., 1989).

Others

It is clear from recent cloning experiments that other rumen bacteria, in addition to the acknowledged cellulolytic species, have the genetic potential to produce some of the enzymes capable of degrading plant structural polysaccharides. For example, genes coding for xylanase (Whitehead and Hespell, 1989a) and endoglucanase/xylanase (Matsushita et al., 1989 and 1990; Woods et al., 1989) activities have been cloned from *Bacteroides ruminicola*. In a further study, genes coding for three of the enzyme activities involved in xylan hydrolysis have been isolated from *Bacteroides ovatus* (Whitehead and Hespell, 1989b). In each of the above cases, active enzymes were expressed by *E. coli* and there was, in addition, some evidence for indigenous promoter activity for an endoglucanase/xylanase cloned from *B. ruminicola* (Woods et al., 1989).

In a recent study, part of the gene coding for an endoglucanase from *B. ruminocola* B₁4 has been cloned and sequenced (Matsushita et al., 1990). In *E. coli* synthesis of the functional endoglucanase with M_r of 40.5 kDa was directed by an ORF of 1,089 bp, preceded by a ribosome binding site and four closely linked promoter sequences that were shown by S1 mapping to function in *E. coli*. The same endoglucanase synthesized by *B. ruminocola* was shown by Western blotting to have an M_r of 88 kDa. Together with the observation that the ORF present on the cloned fragment of *B. ruminicola* DNA continued for at least 1,180 bp upstream of the putative *E. coli* initiation site, this suggests that the 40.5-kDa polypeptide constitutes the C-terminal catalytic domain of a much larger endoglucanase. By analogy with other bacterial endoglucanases, this raises the interesting possibility that the N-terminal region, though not essential for catalytic activity, may contain a separate functional domain. Amino acid sequence alignments indicated significant homology with endoglucanase E (EGE) from *C. thermocellum* (Hall et al., 1988) and an endoglucanase (Eg1) from *R. albus* F-40 (Ohmiya et al., 1989), suggesting that the *B. ruminicola* endoglucanase is a member of cellulase subfamily A4 (Béguin, 1990).

42.3 Lactose Catabolic Genes

Gilbert and Hall (1987) have employed a cloning approach to isolate and characterize the lactose catabolic genes of the prominent rumen bacterium *Streptococcus bovis*. In the manner of most enterobacteria, the *lac* enzymes were encoded by a cluster of three genes, with the β-D-galactosidase, lactose permease, and thiogalactoside transacetylase genes comprising an operon, controlled by the product of the closely linked *lacI* gene. The *S. bovis* promoter cloned with the lactose catabolic genes was not utilized efficiently in *E. coli* but the results of induction experiments suggested that the *lacI* repressor gene was expressed at a comparable level in *S. bovis* and *E. coli*. The primary sequence of the *S.*

bovis operator was very similar to the *E. coli lac*-Zo sequence, but the CAP binding sequence present in *E. coli lacZp* was absent from the *S. bovis* promoter, explaining the observed lack of glucose repression of the *S. bovis lac* operon in *E. coli*. Based on similarity in the organization and regulation of the *lac* operons of the two bacteria, it was suggested that the lactose catabolic genes of *S. bovis* had been acquired by genetic transfer from the enterobacteria.

42.4 Cryptic Plasmids

It is now well established that plasmids occur naturally in many of the anaerobic rumen bacteria (Hazlewood and Teather, 1988; Martin and Dean, 1989) but, with few exceptions (see for example, Flint et al., 1988), the functions encoded by plasmid borne genes have not been determined. With the object of either investigating plasmid function or of developing hybrid plasmids capable of replication in the rumen strain and in *E. coli*, cryptic plasmids from several rumen anaerobes have been cloned into *E. coli* plasmid vectors (see for example, Mann et al., 1986; Flint and Stewart, 1987). The hybrids constructed are, in general, stably maintained in *E. coli* but to date there has been no information on the nature of the proteins expressed or the primary structure of the heterologous plasmid DNA.

42.5 Conclusions

The degradation of plant structural polysaccharides is undoubtedly one of the major events occurring in the rumen, so it is not entirely surprising to find that, where molecular cloning techniques have been applied to anaerobic rumen bacteria, it has nearly always been with the object of investigating cellulase and hemicellulase enzyme systems. Each of the nine genes of rumen bacteria that have to date been sequenced encodes a protein whose function is to hydrolyze cellulose or xylan, so general observations regarding gene structure and expression within this group of bacteria are heavily biased. Notwithstanding this limitation, it is possible from the available data to see that genes from anaerobic rumen bacteria encoding cellulase-related proteins display a high degree of conformity with prokaryotic consensus sequences controlling initiation and termination of transcription and translation, and with consensus sequences regulating protein processing for membrane translocation. It could of course be argued that this is to be expected because the experimental approaches employed have selected only those genes that are readily expressed in *E. coli*, and that such regulatory sequences are either atypical, or inactive and only fortuitously present, in the rumen bacteria. However, the large number of rumen bacterial genes that have been expressed in *E. coli* makes it reasonable to conclude that there is no major deviation from consensus sequences among these bacteria at least with respect to this class of gene.

With regard to molecular architecture and the conservation of primary structure, cellulase-related proteins from the anaerobic rumen bacteria exhibit significant similarities with functionally related enzymes from other organisms. It is clear that, in common with anaerobic bacteria such as *C. thermocellum*, the cellulases of rumen bacteria, for the most part, comprise multiprotein complexes encoded by multiple genes. Furthermore, it seems probable that these proteins have evolved as a result of the acquisition of a number of ancestral genes encoding functional domains that are conserved among cellulolytic species from a variety of habitats. Although there is no conclusive evidence from protein sequence data that any of the rumen cellulases expressed in *E. coli* contain distinct and separate catalytic and cellulose-binding domains, linker sequences composed of hydroxy amino acids and proline, which are known to separate functional domains in other bacterial and fungal cellulases and xylanases, have been demonstrated in cellulases from two rumen genera. Together with biochemical data (McGavin and Forsberg, 1989) which show that an endoglucanase from *F. succinogenes* contains a region that binds specifically to cellulose, this suggests that it will not be long before a cellulase gene from a rumen bacterium is shown to encode a protein which

incorporates the consensus cellulose-binding domain.

Acknowledgements. We (H.J.G. and G.P.H.) thank the Agricultural and Food Research Council (Grant No. LRG 138) for supporting part of this work, and we gratefully acknowledge the skilled technical assistance of Judith Laurie and Keith Davidson.

References

Barros, M.E.C. and J.A. Thomson. 1987. Cloning and expression in *Escherichia coli* of a cellulase gene from *Ruminococcus flavefaciens*. *J. Bacteriol.* **169**:1760–1762.

Béguin, P., J. Millet, O. Grépinet, A. Navarro, M. Juy, A. Amit, R. Poljak, and J.-P. Aubert. 1988. The *cel* (cellulose degradation) genes of *Clostridium thermocellum. In: Biochemistry and Genetics of Cellulose Degradation*, Aubert, J.-P., P. Béguin, and J. Millet, eds., London and San Diego: Academic Press, pp. 267–282.

Béguin, P. 1990. Molecular biology of cellulose degradation. *Annu. Rev. Microbiol.* **44**:219–248.

Berger, E., W.A. Jones, D.T. Jones, and D.R. Woods. 1989. Cloning and sequencing of an endoglucanase (*end*1) gene from *Butyrivibrio fibrisolvens* H17c. *Mol. Gen. Genet.* **219**:193–198.

Berger, E., W.A. Jones, D.T. Jones, and D.R. Woods. 1990. Sequencing and expression of a cellodextrinase (*ced*1) gene from *Butyrivibrio fibrisolvens* H17c cloned in *Escherichia coli*. *Mol. Gen. Genet.* **223**:310–318.

Biely, P. 1985. Microbial xylanolytic systems. *Trends Biotechnol.* **3**:286–290.

Chesson, A. and C.W. Forsberg. 1988. Polysaccharide Degradation by rumen microorganisms. *In: The Rumen Microbial Ecosystem*, Hobson, P.N., ed., pp. 251–284. London and New York: Elsevier.

Cotta, M.A. 1988. Amylolytic activity of selected species of ruminal bacteria. *Appl. Environm. Microbiol.* **54**:878–883.

Cotta, M.A. and R.B. Hespell. 1986. Proteolytic activity of the ruminal bacterium *Butyrivibrio fibrisolvens. Appl. Environm. Microbiol.* **52**:51–58.

Coughlan, M.P. 1985. The properties of fungal and bacterial cellulases with comment on their production and application. *Biotechnol. Genet. Engin. Rev.* **3**:39–109.

Crosby, B., B. Collier, D.Y. Thomas, R.M. Teather, and J.D. Erfle. 1984. Cloning and expression in *Escherichia coli* of cellulase genes from *Bacteroides succinogenes. In: Fifth Canadian Bioenergy R & D Seminar*, Hasnain, S., ed., pp. 573–576. Amsterdam: Elsevier.

Dehority, B.A. and H.W. Scott. 1967. Extent of cellulose and hemicellulose digestion in various forages by pure cultures of rumen bacteria. *J. Dairy Sci.* **50**:1136–1141.

Dehority, B.A. 1969. Pectin-fermenting bacteria isolated from the bovine rumen. *J. Bacteriol.* **99**:189–196.

Dehority, B.A. 1967. Rate of hemicellulose degradation and utilization by pure cultures of rumen bacteria. *Appl. Microbiol.* **15**:987–993.

Devereux, J., P. Haeberli, and O. Smithies. 1984. A comprehensive set of sequence analysis programs for the VAX. *Nucleic Acids Res.* **12**:387–395.

Erfle, J.D., R.M. Teather, P.J. Wood, and J.E. Irvin. 1988. Purification and properties of a 1,3-1,4-β-D-glucanase (lichenase, 1,3-1,4-β-D-glucan 4-glucanohydrolase, EC 3.2.1.73) from *Bacteroides succinogenes* cloned in *Escherichia coli. Biochem. J.* **255**:833–841.

Ferreira, L.M.A., A.J. Durrant, J. Hall, G.P. Hazlewood, and H.J. Gilbert. 1990. Spatial separation of protein domains is not necessary for catalytic activity or substrate binding in a xylanase. *Biochemi. J.* **269**:261–264.

Flint, H.J. and C.S. Stewart. 1987. Antibiotic resistance patterns and plasmids of ruminal strains of *Bacteroides ruminicola* and *Bacteroides multiacidus. Appl. Microbiol. Biotechnol.* **23**:450–455.

Flint, H.J., C.A. McPherson, and J. Bisset. 1989. Molecular cloning of genes from *Ruminococcus flavefaciens* encoding xylanase and β-(1-3,1-4)-glucanase activities. *Appl. Environm. Microbiol.* **55**:1230–1233.

Flint, H.J., A.M. Thomson, and J. Bisset. 1988. Plasmid-associated transfer of tetracycline resistance in *Bacteroides ruminicola. Appl. Environm. Microbiol.* **54**:855–860.

Forsberg, C.W., T.J. Beveridge, and A. Hellstrom. 1981. Cellulase and xylanase release from *Bacteroides succinogenes* and its importance in the rumen environment. *Appl. Environm. Microbiol.* **42**:886–896.

Fukumori, F., T. Kudo, N. Sashihara, Y. Nagata, K. Ito, and K. Horikoshi. 1989. The third cellulase of alkalophilic *Bacillus* sp. strain N-4: evolutionary relationships within the *cel* gene family. *Gene* **76**:289–298.

Gilbert, H.J. and J. Hall. 1987. Molecular cloning of *Streptococcus bovis* lactose catabolic genes. *J. Gen. Microbiol.* **133**:2285–2293.

Gilbert, H.J., J. Hall, G.P. Hazlewood, and L.M.A. Ferreira. 1990. The N-terminal region of an endoglucanase from *Pseudomonas fluorescens* subspecies *cellulosa* constitutes a cellulose-binding domain that is distinct from the catalytic centre. *Mol. Microbiol.* **4**:759–767.

Gong, J., R.Y.C. Lo, and C.W. Forsberg. 1989. Molecular cloning and expression in *Escherichia coli* of a cellodextrinase gene from *Bacteroides succinogenes* S85. *Appl. Environm. Microbiol.* **55**:132–136.

Graham, H., P. Aman, O. Theander, N. Kolankaya, and C.S. Stewart. 1985. Influence of heat sterilization and ammonia on composition and degradation of straw by pure cultures of rumen bacteria. *Anim. Food Sci. Technol.* **12**: 195–203.

Groleau, D. and C.W. Forsberg. 1981. The cellulolytic activity of the rumen bacterium *Bacteroides succinogenes. Can. J. Microbiol.* **27**:517–523.

Guiseppi, A., B. Camii, J.-L. Aymeric, G. Ball, and N. Creuzet. 1988. Homology between endoglucanase Z of *Erwinia chrysanthemi* and endoglucanases of *Bacillus subtilis* and alkalophilic *Bacillus. Mol. Microbiol.* **2**:159–164.

Hall, J. and H.J. Gilbert. 1988. The nucleotide sequence of a carboxymethyl cellulase gene from *Pseudomonas fluorescens* subsp. *cellulosa. Mol. Gen. Genet.* **213**:112–117.

Hall, J., G.P. Hazlewood, P.J. Barker, and H.J. Gilbert. 1988. Conserved reiterated domains in *Clostridium thermocellum* endoglucanases are not essential for catalytic activity. *Gene* **69**:29–38.

Hall, J., G.P. Hazlewood, N.S. Huskisson, A.J. Durrant, and H.J. Gilbert. 1989. Conserved serine-rich sequences in xylanase and cellulase of *Pseudomonas fluorescens* subsp. *cellulosa*: internal signal sequence and unusual protein processing. *Mol. Microbiol.* **3**:1211–1219.

Harfoot, C.G. and G.P. Hazlewood. 1988. Lipid metabolism in the rumen. *In: The Rumen Microbial Ecosystem*, Hobson, P.N., ed. pp. 285–322. London and New York: Elsevier.

Hazlewood, G.P., K. Davidson, J.I. Laurie, M.P.M. Romaniec, and H.J. Gilbert. 1990. Cloning and sequencing of the *celA* gene encoding endoglucanase A of *Butyrivibrio fibrisolvens* strain A46. *J. Gen. Microbiol.* **136**:2089–2097.

Hazlewood, G.P., C.G. Orpin, Y. Greenwood, and M.E. Black. 1983. Isolation of proteolytic rumen bacteria by use of selective medium containing leaf fraction 1 protein (ribulose bisphosphate carboxylase). *Appl. Environm. Microbiol.* **45**:1780–1784.

Hazlewood, G.P. and R.M. Teather. 1988. The Genetics of Rumen Bacteria. *In: The Rumen Microbial Ecosystem*, Hobson, P.N., ed., pp. 323–341. London and New York: Elsevier.

Hazlewood, G.P., M.P.M. Romaniec, K. Davidson, O. Grépinet, P. Béguin, J. Millet, O. Raynaud, and J.-P. Aubert. 1988. A catalogue of *Clostridium thermocellum* endoglucanase, β-glucosidase and xylanase genes cloned in *Escherichia coli. FEMS Microbiol. Lett.* **51**:231–236.

Henrissat, B., M. Claeyssens, P. Tomme, L. Lemesle, and J.-P. Mornon. 1989. Cellulase families revealed by hydrophobic cluster analysis. *Gene* **81**:83–95.

Hespell, R.B., R. Wolf, and R.J. Bothast. 1987. Fermentation of xylans by *Butyrivibrio fibrisolvens* and other ruminal bacterial species. *Appl. Environm. Microbiol.* **53**:2849–2853.

Hiltner, P. and B.A. Dehority. 1983. Effect of soluble carbohydrates on digestion of cellulose by pure cultures of rumen bacteria. *Appl. Environm. Microbiol.* **46**:642–648.

Honda, H., T. Saito, S. Iijima, and T. Kobayashi. 1988. Molecular cloning and expression of a beta-glucosidase gene from *Ruminococcus albus* in *Escherichia coli*. *Enzyme Microb. Technol.* **10**:559–562.

Howard, G.T. and B.A. White. 1988. Molecular cloning and expression of cellulase genes from *Ruminococcus albus* 8 in *Escherichia coli* bacteriophage lambda. *Appl. Environm. Microbiol.* **54**:1752–1755.

Howard, G.T. and B.A. White. 1990. Cloning in *Escherichia coli* of a bifunctional cellulase/xylanase enzyme from *Ruminococcus flavefaciens* FD-1. *Anim. Biotechnol.* **1**:95–106.

Huang, L. and C.W. Forsberg. 1988. Purification and comparison of the periplasmic and extracellular forms of the cellodextrinase from *Bacteroides succinogenes*. *Appl. Environm. Microbiol.* **54**:1488–1493.

Huang, L., C.W. Forsberg, and D.Y. Thomas. 1988. Purification and characterization of a chloride stimulated cellobiosidase from *Bacteroides succinogenes* S85. *J. Bacteriol.* **170**:2923–2932.

Huang, C.-M., W.J. Kelly, R.V. Asmundson, and P.-L. Yu. 1989. Molecular cloning and expression of multiple cellulase genes of *Ruminococcus flavefaciens* strain 186 in *Escherichia coli*. *Appl. Microbiol. Biotechnol.* **31**:265–271.

Irvin, J.E. and R.M. Teather. 1988. Cloning and expression of a *Bacteroides succinogenes* mixed linkage β-glucanase (1,3-1,4-β-d-glucan 4-glucanohydrolase) gene in *Escherichia coli*. *Appl. Environm. Microbiol.* **54**:2672–2676.

Joliff, G., P. Béguin, and J.-P. Aubert. 1986. Nucleotide sequence of the cellulase gene *celD* encoding endoglucanase D of *Clostridium thermocellum*. *Nucleic Acids Res.* **14**:8605–8613.

Kawai, S., H. Honda, T. Tanase, M. Taya, S. Iijima, and T. Kobayashi. 1987. Molecular cloning of *Ruminococcus albus* cellulase gene. *Agric. Biol. Chem.* **51**:59–63.

Knowles, J., P. Lehtovaara, and T. Teeri. 1987. Cellulase families and their genes. *Trends Biotechnol.* **5**:255–261.

Lamed, R. and E.A. Bayer. 1988. The Cellulosome of *Clostridium thermocellum*. *Adv. Appl. Microbiol.* **33**:1–46.

Mann, S.P., G.P. Hazlewood, and C.G. Orpin. 1986. Characterization of a cryptic plasmid (pOM1) in *Butyrivibrio fibrisolvens* by restriction endonuclease analysis and its cloning in *Escherichia coli*. *Curr. Microbiol.* **13**:17–22.

Mann, S.P. 1988. Subcloning of beta glucanase genes from *Ruminococcus albus*, *Clostridium thermocellum* and *Butyrivibrio fibrisolvens* using the shuttle vector pSA3. *Lett. Appl. Microbiol.* **7**:119–122.

Martin, S.A. and R.G. Dean. 1989. Characterization of a plasmid from the ruminal bacterium *Selenomonas ruminantium*. *Appl. Environm. Microbiol.* **55**:3035–3038.

Matsushita, O., J.B. Russell, and D.B. Wilson. 1989. Cloning of *Bacteroides ruminicola* B14 endoglucanase gene into *Escherichia coli*. p.365. *Abstr. Annu. Meeting Am. Soc. Microbiol.*

Matsushita, O., J.B. Russell, and D.B. Wilson. 1990. Cloning and sequencing of a *Bacteroides ruminicola* B$_1$4 endoglucanase gene. *J. Bacteriol.* **172**:3620–3630.

McGavin, M.J. and C.W. Forsberg. 1988. Isolation and characterization of Endoglucanase 1 and Endoglucanase 2 from *Bacteroides succinogenes* S85. *J. Bacteriol.* **170**:2914–2922.

McGavin, M.J. and C.W. Forsberg. 1989. Catalytic and substrate-binding domains of endoglucanase 2 from *Bacteroides succinogenes*. *J. Bacteriol.* **171**:3310–3315.

McGavin, M.J., C.W. Forsberg, B. Crosby, A.W. Bell, D. Dignard, and D.Y. Thomas. 1989. Structure of the *cel*-3 gene from *Fibrobacter succinogenes* S85 and characteristics of the encoded gene product, endoglucanase 3. *J. Bacteriol.* **171**:5587–5595.

Minton, N.P., H.M.S. Bullman, M.D. Scawen, T. Atkinson, and H.J. Gilbert. 1986. Nucleotide sequence of the *Erwinia chrysanthemi* NCPPB1066 L-asparaginase gene. *Gene* **46**:25–33.

Misawa, N. and K. Nakamura. 1989. Expression and stability of a beta-glucosidase gene of *Ruminococcus albus* in *Zymomonas mobilis*. *Agric. Biol. Chem.* **53**:723–727.

Montgomery, L., B.A. Flesher, and D.A. Stahl. 1988. Transfer of *Bacteroides succinogenes* (Hungate) to *Fibrobacter* gen. nov. as *Fibrobacter succinogenes* comb. nov. and description of *Fibrobacter intestinalis* sp. nov. *Int. J. Systemat. Bacteriol.* **38**:430–435.

Morris, E.J. and N.O. van Gylswyk. 1980. Comparison of the action of rumen bacteria on cell walls from *Eragrostis tef*. *J. Agric. Sci.* **95**:313–323.

Murphy, N., D.J. McConnel, and B.A. Cantwell. 1984. The DNA sequence of the gene and genetic control sites for the excreted *B. subtilis* enzyme β-glucanase. *Nucleic Acids Res.* **12**:5355–5367.

O'Neill, G., S.H. Goh, R.A.J. Warren, D.G. Kilburn, and R.C. Miller. 1986. Structure of the gene encoding the exoglucanase of *Cellulomonas fimi*. *Gene* **44**:325–330.

Ohmiya, K., T. Kajino, A. Kato, and S. Shimizu.

1989. Structure of a *Ruminococcus albus* endo-1,4-β-glucanase gene. *J. Bacteriol.* **171**:6771–6775.

Ohmiya, K., K. Nagashima, T. Kajino, E. Goto, A. Tsukada, and S. Shimizu. 1988. Cloning of the cellulase gene from *Ruminococcus albus* and its expression in *Escherichia coli*. *Appl. Environm Microbiol.* **54**:1511–1515.

Ohmiya, K., M. Takano, and S. Shimizu. 1990. DNA sequence of β-glucosidase from *Ruminococcus albus*. *Nucleic Acids Res.* **18**:671.

Orpin, C.G., S.D. Mathiesen, Y. Greenwood, and A.S. Blix. 1985. Seasonal changes in the ruminal microflora of the high-arctic Svalbard reindeer (*Rangifer tarandus platyrhyncus*). *Appl. Environm. Microbiol.* **50**:144–151.

Penttilä, M., P. Lehtovaara, H. Nevalainern, R. Bhikhabhai, and J. Knowles. 1986. Homology between cellulase genes of *Trichoderma reesei*: complete nucleotide sequence of the endoglucanase I gene. *Gene* **45**:253–263.

Pettipher, G.L. and M.J. Latham. 1979. Characteristics of enzymes produced by *Ruminococcus flavefaciens* which degrade plant cell walls. *J. Gen. Microbiol.* **110**:21–27.

Poole, D.H., G.P. Hazlewood, J.I. Laurie, P.J. Barker, and H.J. Gilbert. 1990. Nucleotide sequence of the *Ruminococcus albus* SY3 endoglucanase genes *cel*A and *cel*B. *Mol. Gen. Genet.* **223**:217–223.

Reznikoff, W.S., D.A. Siegele, D.W. Cowing, and C.A. Gross. 1985. The regulation of transcription initiation in bacteria. *Annu. Rev. Genet.* **19**:355–387.

Robson, L.M. and G.H. Chambliss. 1987. Endo-β-1,4-glucanase gene of *Bacillus subtilis* DLG. *J. Bacteriol.* **169**:2017–2025.

Robson, L.M. and G.H. Chambliss. 1989. Cellulases of bacterial origin. *Enzyme Microb. Technol.* **11**:626–645.

Romaniec, M.P.M., K. Davidson, B.A. White, and G.P. Hazlewood. 1989. Cloning of *Ruminococcus albus* endo-β-1,4-glucanase and xylanase genes. *Lett. Appl. Microbiol.* **9**:101–104.

Schellhorn, H.E. and C.W. Forsberg. 1984. Multiplicity of extracellular β-(1,4)-endoglucanases of *Bacteroides succinogenes*. *Can. J. Microbiol.* **30**:930–937.

Sewell, G.W., E.A. Utt, R.B. Hespell, K.F. Mackenzie, and L.O. Ingram. 1989. Identification of the *Butyrivibrio fibrisolvens* xylosidase gene (*xyl*B) coding region and its expression in *Escherichia coli*. *Appl. Environm. Microbiol.* **55**:306–311.

Shane, B.S., L. Gouws, and A. Kistner. 1969. Cellulolytic bacteria occurring in the rumen of sheep conditioned to low-protein teff hay. *J. Gen. Microbiol.* **55**:445–457.

Sipat, A., K.A. Taylor, R.Y.C. Lo, C.W. Forsberg, and P.J. Krell. 1987. Molecular cloning of a xylanase gene from *Bacteroides succinogenes* and its

expression in *Escherichia coli*. *Appl. Environm. Microbiol.* **53**:477–481.

Stahl, D.A., B.A. Flesher, H.R. Mansfield, and L. Montgomery. 1988. Use of phylogenetically based hybridization probes for studies of ruminal microbial ecology. *Appl. Environm. Microbiol.* **54**:1079–1084.

Stewart, C.S. and H.J. Flint. 1989. *Bacteroides (Fibrobacter) succinogenes*, a cellulolytic anaerobic bacterium from the gastrointestinal tract. *Appl. Microbiol. Biotechnol.* **30**:433–439.

Stewart, C.S. and M.P. Bryant. 1988. The Rumen Bacteria. In: *The Rumen Microbial Ecosystem*, Hobson, P.N., ed., pp. 21–75. London and New York: Elsevier.

Taylor, K.A.B., B. Crosby, M. McGavin, C.W. Forsberg, and D.Y. Thomas. 1987. Characteristics of the endoglucanase encoded by a *cel* gene from *Bacteroides succinogenes* expressed in *Escherichia coli*. *Appl. Environm. Microbiol.* **53**:41–46.

Teather, R.M. 1985. Application of gene manipulation to rumen microflora. *Can. J. Anim. Sci.* **65**:563–574.

Teather, R.M. and J.D. Erfle. 1990. DNA sequence of a *Fibrobacter succinogenes* mixed linkage β-glucanase (1,3-1,4-β-D-glucanohydrolase) gene. *J. Bacteriol.* **172**:3837–3841.

Teeri, T.T., P. Lehtovaara, S. Kauppinen, I. Salovuori, and J. Knowles. 1987. Homologous domains in *Trichoderma reesei* cellulolytic enzymes: gene sequence and expression of cellobiohydrolase II. *Gene* **51**:43–52.

Utt, E.A. and L.O. Ingram. 1989. Cloning and sequencing of *Butyrivibrio fibrisolvens* genes for xylan degradation. p.311. *Abstr. Annu. Meeting Am. Soc. Microbiol.*

Ware, C.E., T. Bauchop, and K. Gregg. 1989. The isolation and comparison of cellulase genes from 2 strains of *Ruminococcus albus*. *J. Gen. Microbiol.* **135**:921–930.

White, B.A., J.H. Clarke, K.C. Doerner, V.K. Gupta, C.T. Helaszek, G.T. Howard, M. Morrison, A.A. Odenyo, S. Rosenzweig, and R.I. Mackie. 1990. Improving cellulase activity in *Ruminococcus* through genetic manipulation. In: *Microbial and Plant Opportunities to Improve Lignocellulose Utilization by Ruminants*, Akin, D.E., L.G. Ljungdahl, J.R. Wilson, and P.J. Harris, eds., pp. 389–400. London and New York: Elsevier.

Whitehead, T.R. and R.B. Hespell. 1989a. Cloning and expression in *Escherichia coli* of a xylanase gene from *Bacteroides ruminicola* 23. *Appl. Environm. Microbiol.* **55**:893–896.

Whitehead, T.R. and R.B. Hespell. 1989b. Cloning of genes for xylanolytic enzymes from *Bacteroides ovatus* 975. p. 258. *Abstr. Annu. Meeting Am. Soc. Microbiol.*

Wong, W.K.R., B. Gerhard, Z.M. Guro, D.G. Kilburn, R.A.J. Warren, and R.C. Miller. 1986. Characterization and structure of an endoglucanase gene celA of Cellulomonas fimi. Gene **44**:315–324.

Wong, K.K.Y., L.U.L. Tan, and J.N. Saddler. 1988. Multiplicity of β-1,4-xylanases in microorganisms: functions and applications. *Microbiological Reviews* **52**:305–317.

Woods, J.R., J.F. Hudman, and K. Gregg. 1989. Isolation of an endoglucanase gene from *Bacteroides ruminicola* subsp. *brevis. J. Gen. Microbiol.* **133**:2543–2549.

Zappe, H., W.A. Jones, D.T. Jones, and D.R. Woods. 1988. Structure of an endo-β-1,4-glucanase gene from *Clostridium acetobutylicum* P262 showing homology with endoglucanase genes from *Bacillus* spp. *Appl. Environm. Microbiol.* **54**:1289–1292.

43

Antigenic Characterization, Taxonomy, and Genetics of *Treponema hyodysenteriae*

R. Sellwood

43.1 Introduction

Spirochaetes are emerging as an important bacterial species in intestinal infections both in man and animals. These spirally shaped organisms include several genera—*Treponema, Leptospira, Borrelia, Brachyspira,* and *Spirillum.* This chapter will be concerned only with the genus *Treponema* and in particular *Treponema hyodysenteriae,* a porcine intestinal spirochaete, but work on other spirochaetes will be referred to where appropriate.

T. hyodysenteriae, an oxygen tolerant anaerobe, was first described in 1971 when Taylor and Alexander (1971) isolated a spirochaete from the feces of swine with mucohaemorrhagic diarrhoea or dysentery. Swine dysentery (SD) is a disease of the colon and reproduction of the disease by feeding conventional pigs with this spirochaete was also demonstrated. Similar studies were performed by Glock and Harris (1972) and by Harris et al. (1972), who called this spirochaete *Treponema hyodysenteriae.* It was classified as *Treponema* because it was "anaerobic, saprophytic and was pathogenic (ie. for pigs) but was not transmitted by vectors."

It soon became clear that, in experimental infection studies using both conventional and gnotobiotic pigs, the severe disease was dependant on the presence of other organisms. The most significant of these were *Bacteroides* spp.

and *Fusobacterium* spp. and it was concluded that they may play a major role in the pathogenesis of the disease (Lysons et al., 1978; Meyer et al., 1975) by facilitating colonization of the colon by *T. hyodysenteriae* and also increasing the severity of the disease (Whipp et al., 1982).

T. hyodysenteriae is characterized by a strong β-haemolysis on blood agar plates but other porcine intestinal spirochaetes have been identified that were poorly haemolytic and when inoculated into pigs were nonpathogenic (Taylor and Alexander, 1971). Nonpathogenic spirochaetes have been isolated from the intestinal tract of normal pigs and are considered to be part of the normal flora of the porcine large bowel (Hudson et al., 1976; Joens et al., 1979a,b). The weakly haemolytic, *T. hyodysenteriae*-like organism was called *Treponema innocens* (Kinyon and Harris, 1979) and 13 isolates of this type were nonpathogenic when orally inoculated into pigs. This group of nonpathogenic organisms may, however, represent more than one species because the only common characteristics are that they are isolated from the intestinal tract of pigs, are poorly haemolytic on blood agar, and are not pathogenic for pigs (see Section 43.4). The inclusion of all nonpathogens in this species should be approached with care. The type strain for *T. hyodysenteriae* is B78 and for *T. innocens* is B256 (Kinyon and Harris, 1979).

43.2 Immunity

There have been several reports of both humoral (Joens et al., 1979a; Egan et al., 1983; Burrows et al., 1984; Rees et al., 1989) and mucosal (Joens et al., 1984; Rees et al., 1989) immune responses to infection by *T. hyodysenteriae*, and also information on the degree of protective immunity of pigs that have recovered from swine dysentery. Joens et al. (1979a) demonstrated that 27 of 29 pigs, which had been infected with *T. hyodysenteriae* and recovered naturally, were refractory to the full manifestation of swine dysentery when rechallenged, although 13 of the 29 pigs had diarrhoea. Also, Olson (1974) showed that when those pigs, which recovered either naturally or had been given antibiotics, were challenged a second time they developed diarrhoea, but none of the pigs had blood in their faeces and three challenges were required before full protection was achieved. Also when Rees et al. (1989) challenged pigs with *T. hyodysenteriae* 94% developed swine dysentery after the first challenge. Of those that recovered 43% developed clinical disease after the second challenge and 10% after the third challenge. Each challenge was given after the pigs had recovered from the previous bout of clinical swine dysentery.

The immune response as a result of vaccination has been investigated. Although serotype specific (based on LPS serotypes; see Section 43.3) protection had been demonstrated using colonic loops (Joens et al., 1983, 1985), parenteral vaccination using crude preparations of strains of *T. hyodysenteriae* has not, in general, been a great success (Hudson et al., 1976; Fernie et al., 1983; Parizek et al., 1985) and other methods of vaccination have been investigated. Lysons et al. (1986, 1987) used a combination of a live oral avirulent strain of *T. hyodysenteriae* and a parenteral injection of killed virulent organisms. This regimen afforded significantly better protection than a parenteral injection alone. The conclusions drawn were that the live oral component contained antigens common to both live and killed strains and that the live component was able to deliver these antigens to the mucosal surface and antigen sampling sites of the large bowel of the parenterally primed pig.

43.3 Antigenic Characterisation

Lipopolysaccharide (LPS)

In common with other Gram-negative bacteria, strains of *T. hyodysenteriae* possess a surface coat that is predominantly composed of carbohydrate. Baum and Joens (1979) were able to extract a water-soluble antigen from a hot phenol–water extract (Westphal et al., 1952) of 13 isolates of *T. hyodysenteriae*. These antigens were investigated in immunodiffusion studies using antisera raised in rabbits against the isolates and four different serotypes were proposed. Later, other serotypes were proposed (Table 43.1) by Lemcke and Bew (1984), Mapother and Joens (1985), and Kent et al. (1989). The LPS preparations from all isolates of *T. hyodysenteriae* did not react with the LPS from *T. innocens*, the nonpathogenic intestinal spirochaete and consequently, LPS was considered important for the development of a serotype specific serological test for swine dysentery. An enzyme-linked immunoabsorbent assay (ELISA) based on LPS was used to assess the antibody status in herds with swine dysentery (Joens et al., 1982). More recently, the LPS serotyping has been modified to include five serogroups, each containing several distinct serotypes (Hampson et al., 1989a). This investigation did not analyse the LPS from all known isolates (Table 43.2) but it is clear that the serogroups/serotypes are more complex than first envisaged. However, these techniques do attempt to address the problem of a typing system based on heterogeneous LPS antigens.

Sodium dodecyl sulfate-polyacrylamide gel electrophoresis (SDS-PAGE) analysis of *T. hyodysenteriae* LPS demonstrated a pronounced difference with the LPS of other Gram-negative bacteria (Figure 43.1). The classical ladder usually observed with LPS, which is dependent on repeating oligosaccharide units, e.g., *Salmonella* LPS, was absent. By silver staining,

Table 43.1 Lipopolysaccharide (LPS) serotypes of strains of T. hyodysenteriae

Serotype	Strain	Origin	Reference
1	B78[a]	USA	Baum and Joens (1979)
	B234[b]	USA	Baum and Joens (1979)
	Dys7	USA	Baum and Joens (1979)
2	B140	USA	Baum and Joens (1979)
	B204[c]	USA	Baum and Joens (1979)
	T3	USA	Baum and Joens (1979)
	T4	USA	Baum and Joens (1979)
	9605	USA	Baum and Joens (1979)
	S75/1	GB	Lemcke and Bew (1984)
	JWPM 300/8	The Netherlands	Lemcke and Bew (1984)
3	B169	Canada	Baum and Joens (1979)
4	A1	USA	Baum and Joens (1979)
	P18A	GB	Lemcke and Bew (1984)
5[d]	B8044	USA	Mapother and Joens (1985)
6[d]	B6933	USA	Mapother and Joens (1985)
7[d]	Ack 300/8[e]	The Netherlands	Mapother and Joens (1985)
Other	KF9	GB	Lemcke and Bew (1984)
possible	VS1	GB	Lemcke and Bew (1984)
serotypes	MC52/80	GB	Lemcke and Bew (1984)
	P35/2	GB	Kent et al. (1989)

[a] ATCC 27164
[b] ATCC 31287
[c] ATCC 31212
[d] Proposed by Mapother and Joens (1985)
[e] Placed in serotype 2 by Lemcke and Bew (1984) (same strain as JWPM 300/8).

Table 43.2 Proposed serological groups of lipopolysaccharides of T. hyodysenteriae

Sero-group	Type/Organism	Serotype		
		1	2	3
A	B78	B78	—	—
B	WA1	WA1	B204	—
C	B169	B169	—	—
D	A1	A1	—	—
E	WA6	WA6	MC52/80	KF9

From Hampson et al. (1989a).

Halter and Joens (1988) detected two and Greer and Wannemuehler (1989a) four characteristic bands in the range 18 to 24kDa in pathogenic strains. *T. innocens* strains, in general, did not possess the high molecular weight band but one strain (421) possessed six bands of 17 to 26.9 kDa in the form of a partial step ladder (Halter and Joens, 1988). In contrast, *T. innocens* strain B1555a LPS did not resolve into distinct bands whereas an endotoxin preparation did (Greer and Wannemuehler, 1989a). Attenuation of two pathogenic strains resulted in the disappearance of the high molecular weight band and, in one of the strains, the appearance of three new bands of 20.4, 25.1, and 30.9 kDa (Halter and Joens, 1988). Hampson et al. (1989b) investigated the LPS of different serogroups by SDS-PAGE and immunoblotting. Most of the bands, both in silver stained gels and in immunoblotting, were in the range of 10 to 40 kDa. This contrasts with the unpublished results cited by Chatfield et al. (1988a) which gave a range of 14 to 24 kDa for the LPS components that reacted with antibody in immunoblotting experiments.

Outer membrane extracts were used to evaluate the antibody responses in vaccinated and convalescent pigs (Wannemuehler et al., 1988). Antigens in the range 14 to 19 kDa were identified. Absorption of the sera with endotoxin extracts from the homologous strain of *T. hyodysenteriae* removed the reacting antibodies.

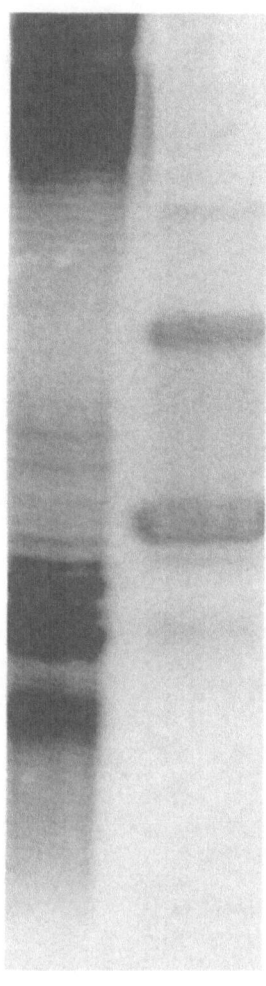

A   B

FIGURE 43.1 SDS-PAGE of lipopolysaccharide preparations from *Salmonella typhimurium* (lane A) and *Treponema hyodysenteriae* (lane B). Silver stained preparation.

These antigens were resistant to proteinase K treatment but were sensitive to periodate oxidation. Although serotyping LPS by immuno-diffusion assays has clearly been successful in the epidemiology of the disease a better understanding of the nature of the LPS is desirable. Immunoblotting of LPS preparations with hyperimmune sera, raised against LPS from different strains and convalescent pig sera, has demonstrated some cross-reactivity between different LPS serotypes (Halter and Joens, 1988; Hampson et al., 1989b). Also, immunoblotting increased the number of visible bands

when compared to silver staining and the appearance was not dissimilar to that of silver-stained LPS from Gram-negative bacteria. The authors concluded that either the silver stain did not react with some of the bands or that immunoblotting was able to reveal specific epitopes on the LPS (Halter and Joens, 1988).

Chemical analysis. The chemical composition of the LPS has been investigated. Baum and Joens (1979) reported that the carbohydrate content of the water-soluble extracts of several strains of *T. hyodysenteriae* was between 85 and 90% and the protein content was 5 to 10%. Greer and Wannemuehler (1989a) analysed the LPS and also endotoxin, prepared by a butanol–water extraction method of Morrison and Leive (1975). The carbohydrate content of the LPS of *T. hyodysenteriae* strain B204 was 80.9% and was similar to that reported by Baum and Joens (1979), who included strain B204 in their investigations. The LPS of the weakly haemolytic *T. innocens* strain B1555a had a much lower carbohydrate content of 56.3%. As would be expected the endotoxin extracts contained more protein than the purified LPS preparations (Table 43.3). An interesting result was the 2-keto-3-deoxyoctonate (KDO) content. This appeared to be low (<0.5%) when compared to the content normally encountered in *Escherichia coli* (1.0%). Gas–liquid chromatography of the phenol–water extracts revealed the presence of the fatty acids myristic

Table 43.3 Chemical analysis of treponemal LPS and endotoxin

	Percent weight of the indicated component in			
	T. hyodysenteriae		*T. innocens*	
Component	LPS	Endotoxin	LPS	Endotoxin
Protein	<1.01	11.3	<1.0	26.0
Hexose	80.9	35.2	56.3	37.8
KDO[a]	0.12	0.45	0.45	0.4
Heptose	6.4	7.5	13.4	17.8
Phosphorus	1.5	ND[b]	ND	ND

From Greer and Wannamuehler (1989a)
[a]KDO determined as a thiobarbituric acid reactive compound
[b]ND, not determined

acid, 13-methyl-myristic acid and 3-hydroxy-hexadecanoic acid. Also detected were glucosamine, KDO, heptose, rhamnose, mannose, galactose, and glucose but because of difficulties encountered with thin-layer chromatography of the KDO-like material in the LPS the authors only tentatively suggested its presence in the LPS of *T. hyodysenteriae*.

Early investigations suggested that an LPS was also present on other treponemes (Christianson, 1962; D'Alessandro and del Carpio, 1958; Jackson and Zey, 1973) and more recently LPS-like material has been reported in other spirochaetes; *Borrelia burgdorferi* (Beck et al., 1985), *Leptospira* spp. (Cinco et al., 1986; Vinh et al., 1986). However, in other investigations phenol–water extracts of *B. burgdorferi* did not contain KDO, fatty acids or glucosamine (Takayama et al., 1987). Immunoblotting of whole and protease-treated treponemes with homologous sera suggested that *Treponema pallidum* lacked a smooth LPS, but evidence for its presence in *Treponema phagedenis* (biotype Reiterii) was obtained when the characteristic Gram-negative LPS ladder was observed by immunoblotting with anti-*T. phagedenis* serum (Bailey et al., 1985) (Figure 43.2). Thus an LPS-like molecule may be present on *T. hyodysenteriae* and other spirochaetes but may also be chemically different to the LPS of gram negatives bacteria such as *Escherichia coli* and *Salmonella*.

Biological activity. Nuessen et al. (1982) investigated the effects of LPS from a strain of *T. hyodysenteriae* on a variety of cell types in vitro. They found that the LPS, like that of *E. coli*, was toxic to murine peritoneal macrophages and generated chemotactic factor(s) for leukocytes from serum. No chemotactic factor was generated when heated serum was used. The LPS was also either weakly or not mitogenic for murine spleen cells (Nuessen et al., 1982), unlike endotoxin preparations (Greer and Wannemuehler, 1989a). Preparations from *T. innocens* also behaved similarly in this assay and also in tests for adjuvanticity, pyrogenicity, and the dermal Schwartzman reaction, and the authors thus concluded that the differences in the virulence of *T. hyodysenteriae* and *T. innocens* could

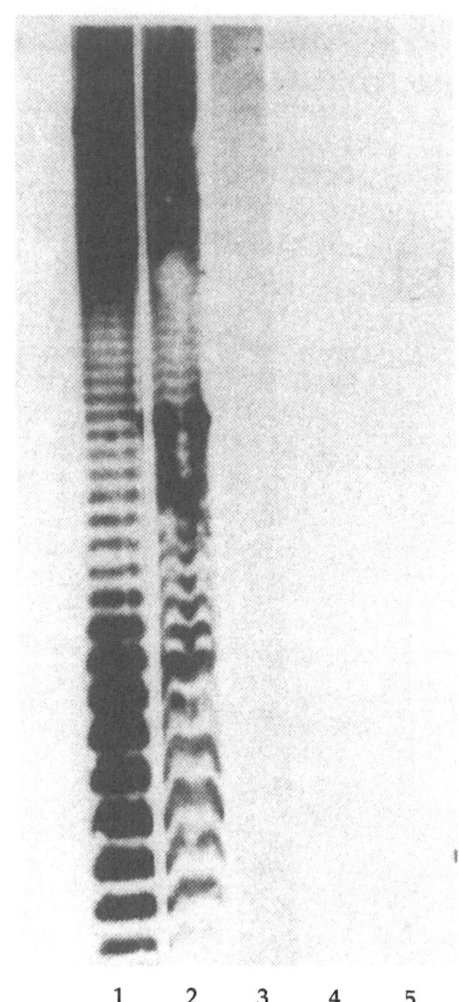

1   2   3   4   5

FIGURE 43.2 Western Blot of (1) proteinase K-treated, washed *Treponema phagedenis*; (2) washed *T. phagedenis*; (3) washed *T. pallidum*; (4) proteinase K-treated washed *T. pallidum*; (5) *Escherichia coli* LPS reacted with 1:100 dilution of rabbit anti-*T. phagedenis* serum. Note the characteristic LPS ladder pattern only in the *T. phagedenis* tracks. (From Bailey et al., 1985.)

not be attributed to the biological properties investigated (Greer and Wannemuehler, 1989a). When mice were inoculated intraperitoneally with LPS, macrophages capable of enhanced Ig-Fc and C3 receptor-mediated phagocytosis were produced. The specific opsonization of spirochaetes by convalescent sera, as measured by their attachment to mouse peritoneal cells, was inhibited also by the homologous LPS preparation (Nuessen and Joens, 1982).

Table 43.4 Pathogenicity of *T. hyodysenteriae* in C3H/HeJ and C3HeB/FeJ mice

Mouse strain	Faecal culture[a]	Gross lessions	Microscopic lesions
C3HeB/FeJ	19/21[b]	19/21	19/21
C3H/HeJ	10/15	0/15	2/15

From Nuesson et al. (1983)

[a] Data expressed represent results from culturing fecal samples at the time of necropsy

[b] Data are expressed as the number of mice positive/total number sampled

The effect of treponemal LPS in vivo has been investigated in mice (Nuessen et al., 1983). Both LPS susceptible (C3HeB/FeJ) and LPS resistant (C3H/HeJ) mice (Sultzer, 1968, 1969) were exposed orally to a strain of *T. hyodysenteriae*. Gross lesions were caused in 19 of 21 of the LPS susceptible mice but only microscopic lesions were observed in 2 of 15 of the resistant mice and colonization of the intestinal tract was greater in the susceptible mice (Table 43.4). The lethal effects of LPS with actinomycin D were also investigated by the intraperitoneal route. Nine of nine resistant mice were unaffected by injection of 1 mg of LPS (plus 14 μg actinomycin D) whereas all nine susceptible mice were killed. Using peritoneal macrophages from both C3HeB/FeJ and C3H/HeJ mice chemotaxis of the macrophages in response to LPS was observed only in the resistant (C3H/HeJ) mice. The authors concluded that treponemal LPS behaved as would be expected for Gram-negative LPS and was probably important in the pathogenesis of infection in mice. The lesion formation was probably a result of either the direct action of the LPS or indirectly involving various inflammatory responses that may result in damage. In contrast, using galactosamine-sensitized BALB/cByJ mice the lethal doses of *T. hyodysenteriae* LPS and endotoxin were 100- to 500-fold greater than similar preparations from *E. coli* (Greer and Wannemuehler, 1989a).

The ability of LPS and endotoxin preparations to stimulate production of interleukin-1 (IL-1) and tumour necrosis factor (TNF) from murine peritoneal exudate cells (PEC) has also been investigated (Greer and Wannemuehler, 1989b). LPS did not have any effect, but endotoxin induced production of both IL-1 and TNF, although only at much higher doses than the equivalent preparations from *E. coli*. The conclusions drawn were that the endotoxin may contribute to the inflammatory response by the induction of these molecules.

Haemolysin

An increasing number of haemolysins are being discovered that are "carrier" dependent. The haemolysin produced by strains of *T. hyodysenteriae* is one of this group (Picard et al., 1979; Saheb et al., 1980); others are streptolysin S (*Streptococcus pyogenes*) (Alouf, 1980) and the haemolysins of *Actinobacillus (Haemophilus) pleuropneumoniae* (Martin et al., 1985), *Streptococcus mutans* (Woltjes et al., 1981), and *Streptococcus agalactiae* (Dal and Monteil, 1983). These haemolysins require a carrier molecule, often RNA, to be present in the extraction medium before a cell-free, active preparation can be obtained. Other carriers used for streptolysin S have been α-lipoprotein, albumin, trypan blue, and some nonionic detergents (Tween 40, 60, 80, and Triton X-205) (Ginsberg, 1970).

The treponemal haemolysin is like streptolysin S in that the most efficient carrier is RNA or RNA core (a RNase resistant fraction of yeast RNA). Both haemolysins (1) are poorly antigenic, (2) cannot easily be separated from their carriers and when separation is achieved are extremely labile, (3) are difficult to visualise on SDS-PAGE, and (4) are cytotoxic to a range of tissue culture cells (Kent, 1984).

Lemcke and Burrows (1982) and Kent et al. (1988) developed improved methods for the large-scale extraction of *T. hyodysenteriae* haemolysin from liquid culture using a defined medium containing RNA-core Type IIC, which is a RNase digest of *Torula* yeast RNA. The preparation was concentrated by ultrafiltration using a low (5 kDa) mol. wt. cut-off membrane and was partially purified by DEAE-cellulose ion-exchange chromatography. The molecular weight of the component with haemolytic activity was determined by gel filtration. The size was estimated to be 19 kDa (Kent et al.,1988). This contrasts with the earlier estimates of 74 kDa (Saheb et al., 1980) and 68 kDa (Knoop, 1981). However, it seems likely that the haemolysin could be even smaller than this estimate because there is greater retention of the haemolysin by a 5-kDa cut-off membrane than a 10-kDa membrane and is at this stage still bound to RNA molecules. The most recent size estimate for streptolysin S has been 1,800 Da (Loridan and Alouf, 1986); the separation from RNA was achieved by isoelectricfocusing. This denatured haemolysin was basic (pI 9.2) and extremely labile.

Kent (1984) showed that the treponemal haemolysin was cytotoxic for a range of tissue culture cells and primary pig lymphocytes. It was also demonstrated that the yield of haemolysin from a virulent strain of *T. hyodysenteriae* (P18A) was 20× greater than the yield from a naturally occurring avirulent strain of *T. hyodysenteriae* (Kent, 1984; Kent et al., 1988) and was more cytotoxic in vitro. More recently (Lysons et al., 1991) demonstrated that the haemolysin from the virulent strain produced lesions in pig intestinal ileal and colonic loops in vivo. Therefore, it seems likely that the haemolysin from the avirulent strain is not only produced in smaller amounts but is also less toxic than the haemolysin from the virulent strain and this cytotoxin is important in virulence.

Protein antigens

In an effort to identify and characterize antigens to which an immune response has been mounted several investigators have used the techniques of SDS-PAGE and Western blotting. Joens and Marquez (1986) first described protein profiles from a strain of *T. hyodysenteriae* (B204) and *T. innocens* (B256). Forty-two distinct proteins were identified in the range 14 to >100 kDa with 12 to 16 major protein bands in each species. Five proteins were common to both but bands of 44, 46, 53, and 72 kDa were present only in strain B204. Western blotting, using serum from pigs that had been orally infected with strain B204 and had clinical signs of swine dysentery, enabled identification of 14 antigens and an additional 12 antigens, when serum from recovered pigs was used. Colonic

secretions from the same pigs enabled identification of 16-, 26-, 27-, 32-, and 59-kDa antigens. Convalescent antiserum absorbed with *T. innocens* strain B256 recognised a single 16-kDa antigen. It was concluded that this antigen could be responsible for the stimulation of a protective immune response in the convalescent pigs.

Chatfield et al. (1988a) also compared antigens from *T. hyodysenteriae* strains and *T. innocens* strains collected from healthy pigs and found the SDS-PAGE profiles similar. Hyperimmune serum raised against one of the *T. hyodysenteriae* strains (CN8368) gave similar profiles in immunoblotting experiments. Also, serum from a pig that had been vaccinated with strain CN8368 recognised bands in all strains, predominantly in the region 29 to 45 kDa. However, when the hyperimmune rabbit serum was absorbed with a strain of *T. innocens* (BL or SM), and then immunoblotted against *T. hyodysenteriae* strain CN8368 three polypeptides of 31, 36, and 68-kDa were specifically identified. The conclusions drawn were that these polypeptides may play a role in virulence.

In a similar study of intestinal treponemes from humans and swine up to 30 common antigens were identified using antiserum raised against several human and swine strains (Dettori et al., 1987). In an earlier study (Baker-Zander and Lukehart, 1984) nine cross-reacting antigens were also identified in *T. pallidum* and *T. hyodysenteriae*.

Outer envelope and endoflagella

The outer envelope of the spirochaete probably possesses polypeptide antigens. However, it has been shown that there are few surface exposed proteins on other spirochaetes. *T. pallidum* is thought to be almost devoid of surface exposed outer membrane proteins (Penn et al., 1985; Radolf et al., 1988, 1989) and in vitro incubation was necessary before three protein antigens were recognised by syphilitic rabbit sera (Stamm et al., 1987). An explanation for the limited antigenicity of *T. pallidum* has been proposed by Radolf et al. (1989).

Physical disruption of the cells has been used to obtain the outer membranes of Gram-negative bacteria (Owen et al.,1982). Chatfield et al. (1988b) used a combination of physical disruption followed by treatment with the nonionic detergent Triton X-100 containing either magnesium ions or EDTA to solubilize outer envelope proteins from a strain of *T. hyodysenteriae*. SDS-PAGE Coomassie Blue staining profiles of the crude outer envelope of *T. hyodysenteriae* strain CN8368 demonstrated polypeptides in the range 24 to 45 kDa, with a prominant band at approximately 17 kDa (Figure 43.3). The Triton X-100 treatment also selectively solubilized several polypeptides, predominantly a band at 36 kDa. The presence of EDTA (5 mM) but not Mg^{2+} considerably improved the solubilisation of this polypeptide. Immunoblotting of these preparations using serum from a vaccinated and protected pig and a hyperimmune rabbit serum identified most of the polypeptides. However, when the rabbit serum was absorbed with a strain of *T. innocens* (BL) the serum reacted strongly with the 36-kDa band but only weakly with the Triton X-100 insoluble polypeptides. In immunogold labelling experiments the rabbit hyperimmune serum reacted with the surface of the *T. hyodysenteriae* strain but reacted only poorly with *T. innocens*.

Outer membrane extracts of both *T. hyodysenteriae* and *T. innocens* prepared by Triton X-100 solubilization were also investigated by Wannemuehler et al. (1988). In contrast to the results of Chatfield et al. (1988b), many polypeptides in the 25 to 49 kDa range were found in the soluble fraction.

The endoflagella or axial filaments have also been investigated and these organelles have been, in general, easier to prepare employing usually sodium lauroyl sarkosinate (Sarkosyl) and caesium chloride isopycnic, density gradient centrifugation (Hardy et al., 1975; Miller et al., 1988; Kent et al., 1989). Analysis of the endoflagella by SDS-PAGE revealed the presence of six polypeptides of 39, 29, 27, 22, 21, and 18.5 kDa (Miller et al., 1988). Electron micrographs of the purified endoflagella revealed the presence of both thick and thin structures suggesting that the filaments do have a sheath. These observations were similar to those on *T. pallidum* (Cockayne et al., 1987).

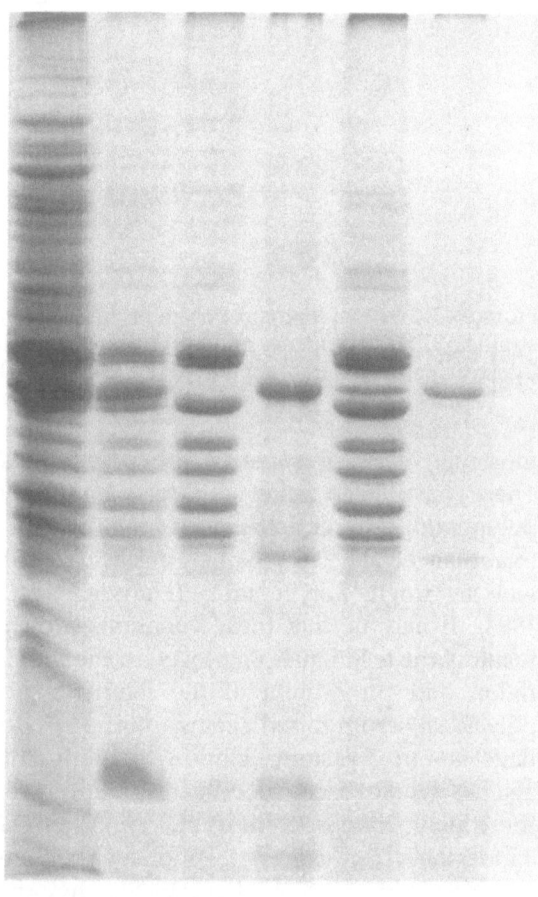

A   B   C   D   E   F

FIGURE 43.3 SDS-PAGE separation of cell envelope preparations of *T. hyodysenteriae* strain CN8368 after treatment with Triton X-100 (Coomassie blue stained). (A) Unfractionated cells; (B) Cell envelope preparation; (C) Insoluble fraction after treatment with Triton X-100 EDTA; (D) Triton X-100 EDTA soluble fraction; (E) Insoluble fraction after treatment with Triton X-100 Mg^{2+}; (F) Triton X-100 Mg^{2+} soluble fraction. (From Chatfield et al., 1988b.)

In the more extensive investigation of Kent et al. (1989) a comparison was made of the endoflagellar polypeptides of 8 serotypes of *T. hyodysenteriae* and two nonpathogenic intestinal spirochaetes (Figure 43.4). Five major polypeptides of 43.8, 38, 34.8, 32.8, and 29.4 kDa were identified for *T. hyodysenteriae* strain P18A. It was interesting that several of these bands appeared as doublets. Also, although the profiles were similar for all the spirochaetes, there were minor variations in the

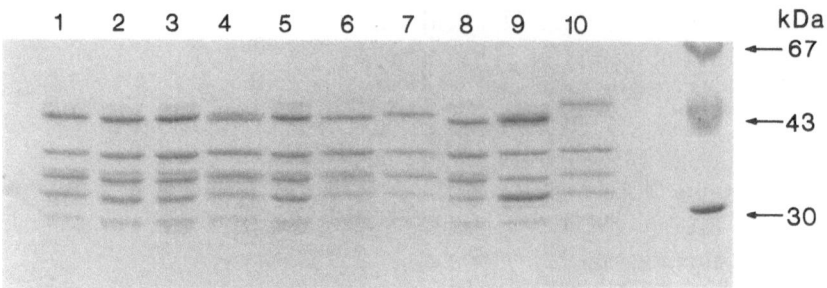

FIGURE 43.4 Axial filament polypeptides of eight strains of *T. hyodysenteriae* and two nonpathogenic intestinal spirochaetes separated by SDS-PAGE and stained with Page Blue 83. 1. B78; 2. S75/1; 3. B169; 4. P18A; 5. KF9; 6. VS1; 7. MC52/80; 8. P35/2; 9. PWS/A; 10. M1. (From Kent et al., 1989.)

molecular sizes of some of the polypetides. There have been other reports of multiple polypeptides in endoflagella of spirochaetes (Nauman et al., 1969; Blanco et al., 1986) as well as single polypeptides (Hansen et al., 1988). It may be that these variations are the result of the techniques employed in the preparation and purification of the filaments but polyclonal serum raised against purified endoflagellae in Western blotting experiments against whole bacterial cells also recognised these polypeptides (Kent et al., 1989). Similar bands in *T. hyodysenteriae* endoflagellae were also observed by Miller et al. (1988). They discussed several reasons for multiple bands including (1) contamination by extraneous proteins, (2) inclusion of hook structures of the endoflagella, (3) proteolytic degradation, and (4) the existence of several species of endoflagella, but came to no firm conclusions for their results. No information is yet available on the location of these polypeptides in the endoflagellum.

An important point in the investigation of the endoflagella was that the polypeptides contained domains that are conserved between all the *T. hyodysenteriae* strains and also the nonpathogenic spirochaetes (Kent et al., 1989). This has obvious relevance to the development of specific diagnostic assays for *T. hyodysenteriae* and swine dysentery.

These investigations of an organelle that appears to have been obtained in pure form assists the interpretation of the above outer envelope studies. Many of the polypeptides detected in these investigations (Wannemuehler

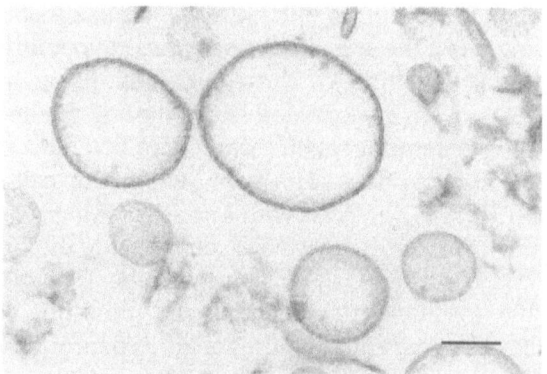

FIGURE 43.5 Electron micrograph of an outer envelope preparation of *T. hyodysenteriae* strain P18A (ultrathin sections stained with uranyl acetate; bar = 0.1 µm). (From Sellwood et al., 1989.)

et al., 1988; Chatfield et al., 1988a, 1988b) in the range 45 to 29 kDa may have been endoflagella polypeptides and not just from the outer membrane. The relatively harsh methods employed in obtaining the outer envelope may have resulted in preparations that included endoflagella.

An alternative technique using low concentrations of SDS (0.005%) produced an outer envelope preparation relatively free of endoflagella (Figure 43.5), as judged by electron microscopy (Sellwood et al., 1989) and immunoblotting with hyperimmune serum against purified endoflagella (data not shown). Hyperimmune serum raised against outer envelopes of one *T. hyodysenteriae* strain (P18A) purified in this way, reacted in immunoblotting to only two bands of 45 and 16 kDa. This

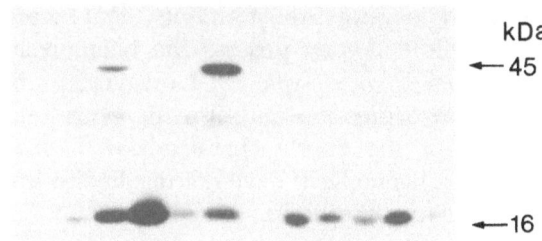

FIGURE 43.6 Western blot of whole cells from 11 strains of *T. hyodysenteriae* (lanes 3 to 13) and two non-pathogenic intestinal spirochaetes (lanes 1 and 2). Polypeptides were separated on a 12.5% SDS-PAGE gel, transfered to nitrocellulose and probed with hyperimmune gnotobiotic pig serum (B50). 1. M1; 2. PWS/A; 3. P35/2; 4. MC52/80; 5. VS1; 6. KF9; 7. P18A; 8. B169; 9. JWPM; 10. S75/1; 11. B204; 12. B78; 13. B234. (From Sellwood et al., 1989.)

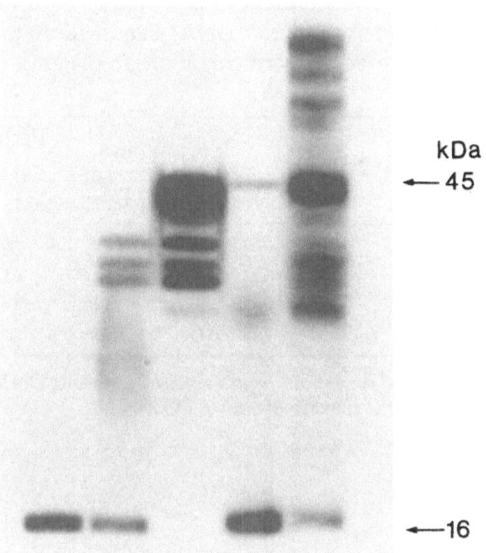

FIGURE 43.7 Western blot of polypeptides of *T. hyodysenteriae* strain P18A probed with serum from gnotobiotic pig 67A (lane 1) and 67B (lane 2) both infected with strain P18A; hyperimmune gnotobiotic pig serum (C11) raised against axial filaments of strain P18A (lane 3) or (B50) raised against outer envelopes (lane 4) or a convalescent pig serum (lane 5). (From Sellwood et al., 1989.)

was interesting because the outer envelope preparation used to raise the antiserum possessed several other polypeptides. Either the animal used for the antibody production (a gnotobiotic pig) was not able to elicit an immune response to these other polypeptides or the treatment with SDS etc. in the Western blotting experiments denatured the epitopes. The main point of interest in these studies was that the 16-kDa antigen was present in all strains of *T. hyodysenteriae* investigated but was absent from the nonpathogenic spirochaetes (Figure 43.6). The 45-kDa antigen was specific for two *T. hyodysenteriae* strains only (P18A and MC52/80). Since this serum inhibited the growth of all the *T. hyodysenteriae* strains but not the nonpathogens, it was concluded that antibodies to the 16-kDa antigen probably were responsible for this activity. This view was supported by similar growth inhibitory activity of serum from a gnotobiotic pig (67 A) that had been orally infected with P18A but the infection did not establish. Serum from this pig contained only antibodies to the 16-kDa antigen (Figure 43.7). Another pig (67B), which had also been infected, had antibodies to both the 16-kDa antigen and several axial filament polypeptides.

Immunogold labelling with the outer envelope serum suggested that the 16-kDa anti-

gen was located on the surface of the spirochaetes (Sellwood et al., 1989). Hyperimmune serum against the endoflagella had no growth inhibitory activity and was therefore thought not to be solely of importance. However, in other investigations treponemicidal activity of human serum against *T. pallidum* (Blanco et al., 1986) and hyperimmune mouse serum against a cloned polypeptide of *T. hyodysenteriae* against a heterologous strain (Boyden et al., 1989) has been correlated with antibodies against endoflagella.

The 16-kDa antigen described above may be the same as that first described by Joens and Marquez (1986) although differences encountered in periodate and protease sensitivity suggest otherwise (Sellwood, unpublished observations). The antigens in both investigations were detected by immunoblotting with convalescent pig sera which demonstrates that these antigens are not only immunogenic but are also produced in vivo.

Table 43.5 Transformations obtained from cloning *T. hyodysenteriae* genomic DNA, size fractionated by sucrose gradient centrifugation, in plasmid pAT153

Fraction no.	Size DNA (kb)	No. colonies		Percent insert >1 kb
		Amp^R	Tet^S	
1	2–6	543	437	76
2	4–8	128	122	92
3	6–11	39	39	0
4	9–15	100	100	0

*Sau*3A partial digest of *T. hyodysenteriae* genomic DNA cloned into the *Bam*H1 site of pAT153.

Cloning and expression of T. hyodysenteriae genes

There have been many attempts at cloning *T. hyodysenteriae* genes. Many of the problems encountered may have been due to the base composition and/or secondary structure of the genomic DNA. Cloning difficulties in *E. coli* may have been due to the *E. coli*-like promoter regions of the A + T rich regions of *T. hyodysenteriae*. In this laboratory we have encountered problems with cloning treponemal DNA larger than 4 kb in size. A partial *Sau*3A restriction digest of genomic DNA was size fractionated on a sucrose gradient into 4 fractions, 2–6, 4–8, 6–11 and 9–15 kb. The DNA fragments were ligated into the *Bam*H1 site of plasmid pAT153, which is in the tetracycline resistance gene, and *E. coli* were transformed. The results of the transformation are shown in Table 43.5. Many tetracycline sensitive colonies were obtained when all size fractions were used, which indicated that an insert had been successfully cloned. However, on analysis it was discovered that if the DNA size to be inserted was >6 kb the size of the insert in the recombinant plasmid was always <1 kb. When DNA (<6 kb) was ligated and *E. coli* transformed, clones were successfully obtained with stable inserts (<4.5 kb). One clone obtained had an insert of 5.3 kb but was unstable. Thus, it appeared that *T. hyodysenteriae* DNA was inherently unstable in *E. coli* plasmid vectors and one solution was to clone only small fragments.

Using this technique Sellwood et al. (1987) identified several transformants that were haemolytic and may possess the haemolysin gene from *T. hyodysenteriae*. However, problems with expression and also the poor antigenicity of the poorly characterized *T. hyodysenteriae* haemolysin have not resulted in unequivocal identification of the gene.

Boyden et al. (1989) reported the successful cloning of *T. hyodysenteriae* genes using bacteriophage λgt11, which generates fusion proteins with β-galactosidase and the pSY908, pSY909, and pSY910 plasmids, which contain the cI857 repressor and λP$_R$ promoter upstream of the T7 gene 10 promoter. Each plasmid contains a *Bam*H1 site in a different reading frame. In addition the plasmid cloning vector pUC13 was used. *T. hyodysenteriae* strain B204 was used to prepare the genomic DNA. The authors did not describe which method was most successful but it was noted that the plasmid vectors were preferred. It was also interesting that for expression the clones neither required the transcription nor translation signals of the vector which suggested that *T. hyodysenteriae* signals could function in *E. coli*. Expression of cloned genes generally were low (<2% of the *E. coli* protein) but was usually detected by seroconversion in mice. No further analysis of the genes was reported but a crude extract of one clone that expressed an endoflagellar antigen of 25 kDa (cloned antigen) provided some protection ($P < 0.2$) to CF1 mice when given intraperitoneally before being challenged with the homologous strain B204. Moreover, serum from these mice was bactericidal for the spirochaete. Although the cloned antigen was only 25 kDa the mice sera reacted in Western blots to three *T. hyodysenteriae* proteins of 37, 34, and 32 kDa. This discrepancy in size may have been due to proteolytic processing of the cloned antigen or the cloning of only part of the gene. This apparent cross-reactivity with endoflagellar proteins has been observed before (Kent et al., 1989) but this is the first indication that each of the endoflagellar polypeptides contain common epitopes. The partially purified 25-kDa antigen also protected mice ($P < 0.05$) from heterologous challenge (strain B234) and in vitro bactericidal tests using serum from mice

Table 43.6 Homology of *T. hyodysenteriae* genomic DNA with DNA from other spirochaetes

Species	Strain	Mol% G + C	Percent homology (ref. strain)	Ref.
T. hyodysenteriae	B204	25.7–25.9	100 (B204)	1
	A1	25.7–25.9	80 (B204)	1
	B78	25.6	ND	2
	B78	ND	100 (B78)	3
	B78	ND	93 (B204)	3
	10/82	26.2	100 (10/82)	4
	Dys754	25.4	100 (10/82)	4
	19/84[a]	24.7	90 (10/82)	4
	P18A	27.9	100 (P18A)	5
	P18A	28.5–30.0	100 (P18A)	6
T. innocens	B256	25.7–25.9	28 (B204)	1
	4/71	25.7–25.9	28 (B204)	1
	B256	ND	37–41 (B78)	3
	4/71	ND	40 (B78)	3
	M1	25.5	93.3 (P18A)	5
T. succinifaciens	6091	ND	2 (B78)	3
T. pallidum (Nichols)	—	52.4–53.7	<5 (B204)	1/7
		ND	17% (P18A)	5
T. phagedenis (Reiter)	—	39.0	<5 (B204)	1
T. refringens (Noguchi)	—	41.5	<5 (B204)	1
Borrelia hermsii	—	30.6	9 (B78)	2
Lyme disease spirochaete	TL0-005	28.1	16 (B78)	2
Borrelia burgdorferi	—	ND	6.4 (P18A)	5
21 Human intestinal spirochaetes	—	25.6–30.1	>70/13[b] (P18A)	5
			<50/8 (P18A)	5
Human intestinal spirochaetes	HRM2	27.3–29.7	68 (P18A)	6
	HRM2	27.7–30.0	95 (P18A)	6

[a] Identical to strain B78 (ATCC 27164).
[b] Homology to *T. hyodysenteriae* /no. of isolates.
ND Not determined.
References: (1) Miao et al. (1978); (2) Schmid et al. (1984); (3) Stanton et al. (1991); (4) Sachse and Blaha (1988); (5) Coene et al. (1989); (6) Wergifosse and Coene (1989); (7) Miao and Fieldsteel (1978)

given the partially purified antigen killed the heterologous strain B234.

43.4   Genetics

DNA composition and homology studies

The first investigation with *T. hyodysenteriae* DNA was performed by Miao et al. (1978) using DNA saturation–reassociation assays and analysis of reaction products after digestion with S1 nuclease (Table 43.6). These inves-

tigators found no reassociation (<5%) of the DNA from *T. pallidum* (Nichols), *T. phagedenis* biotype Reiter, and *T. refringens* biotype Noguchi with two *T. hyodysenteriae* strains (B204 and A1); or with two *T. innocens* strains (B256 and 4/71), whereas reassociation in the presence of homologous DNA was >95%. Control salmon sperm DNA was <2%. These workers concluded that *T. hyodysenteriae* was genetically unrelated to any of the other treponemes ie. *T. pallidum*, *T. phagedenis* and *T. refringens*. It is also interesting to note only 28% homology existed between the pathogenic *T. hyodysenteriae* strains and the closely related intestinal nonpathogens which would indicate that both pathogens and nonpathogens are within the same genus but belong to different species (Miao et al., 1978).

Investigation of the base composition also revealed major differences between the intestinal spirochaetes and the other treponemes. The base composition for *T. pallidum*, *T. phagedenis* biotype Reiter and *T. refringens* biotype Noguchi was earlier determined by Miao and Fieldsteel (1978) as 52.4 to 53.7 (G + C mol%), 39.0 mol% and 41.5 mol%, respectively. In contrast, the porcine spirochaetes had a very low G + C content of 25.7 to 25.9 mol% (Miao et al., 1978) and 24.7 to 29.7 mol% (Sachse and Blaha, 1988). The poor homology and major difference in G + C content led Miao et al. (1978) also to suggest that the porcine spirochaetes belonged to a separate group, encompassing intestinal spirochaetes and may therefore constitute a new genus. On the other hand, because the G + C content was <35 mol% for both the porcine intestinal spirochaetes and *Borrelia* spp and values did not differ by more than 5%, Sachse and Blaha (1988) proposed that *T. hyodysenteriae* and *T. innocens* should be "regarded as belonging to one species, which is to be renamed *Borrelia hyodysenteriae* (comb. nov. Blaha)."

However, a comparison by Schmid et al. (1984) of G + C content and homology between *T. hyodysenteriae* (25.6% G + C), a Lyme disease spirochaete (strain TL0-005, 28.1% G + C), and *Borrelia hermsii* (30.6% G + C) showed very low homology (16% for the Lyme disease spirochaete and 9% for *B. hermsii*). These results also suggested that whereas the Lyme disease spirochaete was considered to belong to the genus *Borrelia*, *T. hyodysenteriae* probably did not. These homology studies clearly place the intestinal, porcine spirochaetes into a separate genus or genera.

Information on the nature of spirochaetes isolated from the intestinal tract of both animals (Davis et al., 1972, 1973; Cwyk and Canale Parola, 1979; Kinyon and Harris, 1979) and man (Kaplan and Takeuchi, 1979; Douglas and Crucioli, 1981; Hovind-Hougen et al., 1982; Sanna et al., 1982, 1984) has been accumulating. However, no further genetic analysis of intestinal spirochaetes emerged until Coene et al. (1989) investigated the genomic DNA from 21 human spirochaetes isolated from faeces of patients with a variety of intestinal disorders and compared the DNA to that of one strain of *T. hyodysenteriae* (P18A) and a nonpathogenic intestinal, porcine spirochaete (Ml) (Lemcke and Burrows, 1979). The base composition was determined by a double labelling technique (Coene and Cocito, 1985) which unlike the techniques of thermal denaturation midpoint (T_m) or buoyant density was not affected by the presence of modified or unusual bases. All human isolates were very similar and had G + C contents of 25.6 to 30.1 mol%. These were similar to the G + C content determined for P18A (27.9 mol%) and Ml (25.5 mol%). DNA hybridization in liquid medium of nick-translated genomic DNA fragments of 1kb (mean size), under stringent conditions showed that hybridization varied between 31.5% (±3.4) and 104.8% (±9.7) with 13 out of a total of 21 human strains having >70% homology. These strains were considered to be in the same group as *T. hyodysenteriae*. The poor homology (<17%) of *T. hyodysenteriae* with other treponemes, *T. pallidum* (17%), *T. phagedenis* (7.9%), and *Borrelia burgdorferi* (6.4%), again suggested different genera. It is interesting that the nonpathogen M1, in this study, showed high homology with *T. hyodysenteriae* strain P18A in contrast to the results with *T. innocens* strains B256 and 4/71 (Miao and Fieldsteel, 1978; Stanton et al., 1991). This underlines the problem of including all porcine intestinal spirochaetes in the one species. It is interesting that in a further study Wergifosse and Coene (1989) observed that the highest homology (91 to 95%) was between a human strain (HRM 3) and *T. hyodysenteriae* strain P18A. They concluded that not only did these two spirochaetes fall in the same species but were also in the same subgroup. The relevance of this result is that both humans and domestic animals may be sources of intestinal spirochaetal infections; a point that did not go unnoticed by the authors.

DNA methylation and modification

Digestion of treponemal DNA with restriction endonucleases has provided some information on the composition of the genomic DNA.

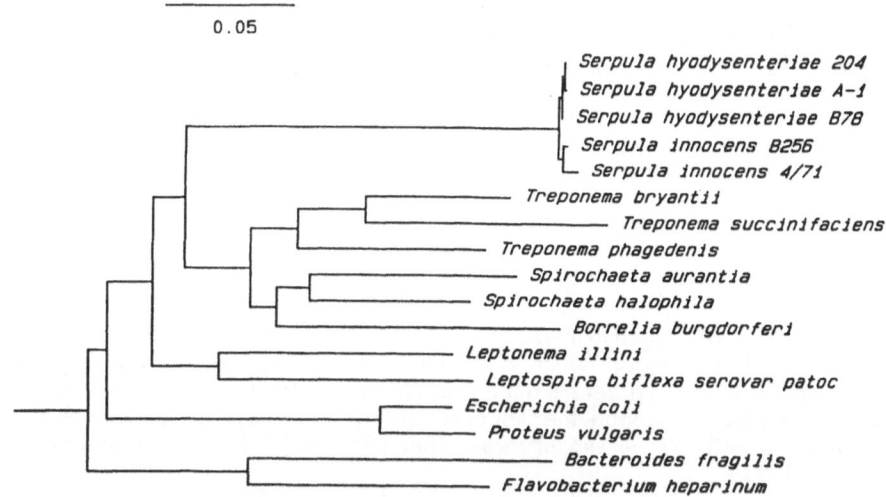

FIGURE 43.8 Dendrogram indicating phylogenetic relationship of *T. hyodysenteriae* and *T. innocens* to other spirochaetes and to distantly related bacteria. Scale bar represents a 5% difference in nucleotide sequence. Total distance between two organisms is the sum of the horizontal branch lengths. Vertical distance has no meaning. (From Stanton et al., 1991.)

Revitt and Sellwood (unpublished observations), using restriction enzymes, observed that *T. hyodysenteriae* DNA from strain P18A was cut extensively with *Sau*3A (recognition site ↓ GATC) but was only partially cut with *Mbo*1 (recognition site ↓ GATC, but only recognises the sequence when the adenine is not methylated) which suggested methylation of a proportion of the adenine bases in the GATC sequence. *Dpn*1, which cuts at a different site (GA ↓ TC) but only if the adenine on both strands is methylated (Vovis and Lacks 1977), did not digest the DNA. This is in agreement with the work of Coene et al. (1989), who found that *Mbo*1 did not cut strain P18A and also several human strains, including HRM-3. They also found that the porcine nonpathogen M1 was not cut by *Mbo*1. Although M1 was not investigated by Revitt and Sellwood, another porcine nonpathogen (4/71) also was not cut by *Mbo*1 whereas another nonpathogen (PWS/A) was cut, giving a distinct band pattern. Using a bank of nine other restriction enzymes that are influenced by methylation of cytosine Coene et al. (1989) also concluded that a proportion of these bases were also methylated.

A recent and important development in the study of genetic relationships between bacteria has resulted from the study of the nucleotide sequences of conserved and variable regions of 16S ribosomal RNA (Woese et al., 1983). Analysis of 16S RNA sequence data from strains of *T. hyodysenteriae* and *T. innocens* and data from other spirochaetes and bacteria enabled a dendrogram of genetic relationships to be constructed (Figure 43.8). These studies clearly suggested that *T. hyodysenteriae* and *T. innocens* are closely related but are both only distantly related to treponemes and other spirochaetes (Stanton et al., 1991). DNA probes have been constructed that hybridise to 16S RNA and can distinguish between *Mycoplasma* sp. (Göbel et al., 1987). Recently, Jensen et al. (1990) described the detection of *T. hyodysenteriae* and *T. innocens* with the use of 2 DNA probes complementary to 16S RNA sequences (Figure 43.9). A 17-base and a 28-base oligonucleotide probe successfully differentiated between strains of *T. hyodysenteriae* and *T. innocens*. However, the 17-base probe was less specific because under low stringency conditions it hybridised to both species. Taking into consideration SDS-PAGE profiles of whole cell proteins, DNA–DNA reassociation and 16S RNA homology, Stanton et al. (1991) proposed the new genus "*Serpula*" (Latin: little serpent) for both *T. hyodysenteriae* and *T. innocens* strains. However, because of the illegitimacy of "*Serpu-*

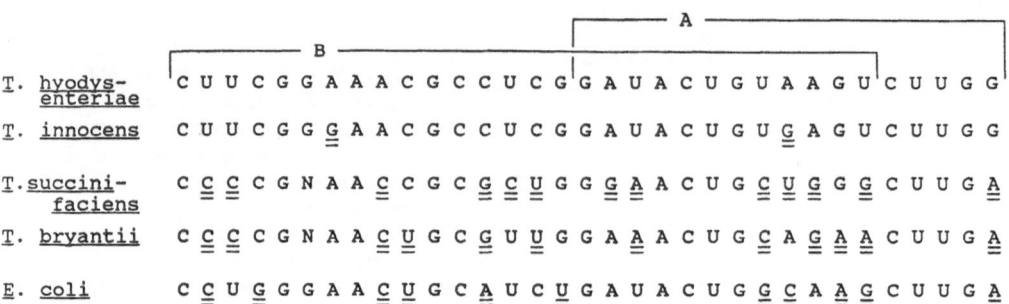

FIGURE 43.9 Comparison of *T. hyodysenteriae* 16S rRNA sequence (5'–>3') with 16S rRNA from other treponemes and *E. coli* in the region which is complementary to 17-base (A) and 28-base (8) probes. Underlined bases are those which differ from the *T. hyodysenteriae* 16S rRNA sequence. N corresponds to a variable or unidentified base. Treponeme sequences are from *T. hyodysenteriae* (strains B78, B204, A1), *T. innocens* (strains B256 and 4/71), *T. succinifaciens* (strain 6091) and *T. bryantii* (strain RUS-1); and *E. coli* (From Jensen et al., 1990.)

la" for the spirochaete genus Stanton (1992) proposed a change in the genus designation to "*Serpulina.*"

Plasmid content of *T. hyodysenteriae*

There is very little information on the plasmid content of intestinal spirochaetes. Margolin et al. (1985) reported the presence of a plasmid of approximately 8.3 kb in *T. hyodysenteriae* strain B204. However, analysis of this plasmid has not been reported.

Extrachromosomal DNA, in the form of covalently closed linear plasmids and super-coiled circular plasmids (Barbour and Garon, 1987; Barbour, 1988; Hyde and Johnson, 1988) has been found in *B. burgdorferi*. *T. pallidum* also possesses plasmid DNA (Norgard and Miller, 1981) so the observation of Margolin et al. (1985) with *T. hyodysenteriae* was not unexpected. Whether the plasmid of *T. hyodysenteriae* encodes important antigens and virulence factors or harbours antibiotic resistance genes remains to be elucidated.

43.5   Concluding Remarks

The cloning of *T. hyodysenteriae* genes is now of major interest to several groups in Europe, USA, and Australasia. The biological characterization of *T. hyodysenteriae* is being actively pursued and reagents such as monoclonal anti-bodies are becoming available. These reagents will facilitate the cloning of important genes. The investigations of other intestinal spirochaetes may also lead to greater knowledge of their importance in disease both in man and animals (Harland and Lee, 1967; Takeuchi et al., 1974). Human colonic disease and opportunistic infections in immunocompromised hosts are often associated with spirochaetes. For a review see Ruane et al. (1989). In chickens intestinal spirochaetosis has already been attributed to *T. hyodysenteriae* (Adachi et al., 1985; Davelaar et al., 1986; Dwars et al., 1989) and it is only through improvement in our knowledge of virulence and characterization of potentially protective antigens can a realistic approach to control of disease be developed. Also it remains to be seen whether all these intestinal spirochaetes fall within a single genus, but with taxanomic classification becoming increasingly questioned by DNA/RNA analysis and homology studies I believe that other genera will emerge. However there is already probably enough information to place *T. hyodysenteriae* in a new genus which needs to be defined.

References

Adachi, Y., M. Sueyoshi, E. Miyagawa, H. Minato, and S. Shoya. 1985. Experimental infection of young broiler chicks with *T. hyodysenteriae*. *Microbiol. Immunol.* **29**:683–688.

Alouf, J.E. 1980. Streptococcal toxins (Streptolysin

O, Streptolysin S, Erythrogenic toxin). *Pharmacol. Ther.* **11**:661–717.

Bailey, M.J., C.W. Penn, and A. Cockayne. 1985. Evidence for the presence of lipopolysaccharide in *Treponema phagedenis* (biotype Reiterii) but not in *Treponema pallidum* (Nichols). *FEMs Microbiol. Lett* **27**:117–121.

Baker-Zander, S.A. and S.A. Lukehart. 1984. Antigenic cross-reactivity between *Treponema pallidum* and other pathogenic members of the family *Spirochaete. Infect. Immun.* **46**:116–121.

Barbour, A.G. 1988. Plasmid analysis of *Borrelia burgdorferi*, the Lyme disease agent. *J. Clin. Microbiol.* **26**:475–478.

Barbour, A.G. and C.F. Garon. 1987. Linear plasmids of the bacterium *Borrelia burgdorferi* have covalently closed ends. *Science* **237**:409–411.

Baum, D.H. and L.A. Joens. 1979. Serotypes of beta-haemolytic *Treponema hyodysenteriae. Infect. Immun.* **25**:792–796.

Beck, G., G.S. Habricht, J.L. Benach, and J.L. Coleman. 1985. Chemical and biological characterization of a lipopolysaccharide extracted from the Lyme disease spirochete (*Borrelia burgdorferi*). *J. Infect. Dis.* **152**:108–117.

Blanco, D.R., C.I. Champion, J.N. Miller, and M.A. Lovett. 1986. Antigenic and structural characterization of *Treponema pallidum* (Nichols strain) endoflagella. *Infect. Immun.* **56**:167–175.

Boyden, D.A., F.G. Albert, and C.S. Robinson. 1989. Cloning and characterisation of *Treponema hyodysenteriae* antigens and protection in a CF-1 mouse model by immunisation with cloned endoflagellar antigen. *Infect. Immun.* **57**:3808–3815.

Burrows, M.R., R.J. Lysons, G.J. Rowlands, and R.M. Lemcke. 1984. An enzyme-linked immunosorbent assay for detecting serum antibody to *Treponema hyodysenteriae.* pp. 186. *Proceedings of the 8th Congress of the International Pig Veterinary Society*, Belgium.

Chatfield, S.N., D.S. Fernie, C. Penn, and G. Dougan. 1988a. Identification of the major antigens of *Treponema hyodysenteriae* and comparison with those of *Treponema innocens. Infect. Immun.* **56**:1070–1075.

Chatfield, S.N., D.S. Fernie, J. Beesley, C. Penn, and G. Dougan. 1988b. Characterisation of the cell envelope of *Treponema hyodysenteriae. FEMS Microbiol. Lett.* **55**:303–308.

Christianson, A.H. 1962. Studies on the antigenic structure of *Treponema pallidum. Acta Pathol. Microbiol. Scand.* **56**:166–176.

Cinco, M., E. Banfi, and E. Panfili. 1986. Heterogeneity of lipopolysaccharide banding patterns in *Leptospira* spp. *J. Gen. Microbiol.* **132**:1135–1138.

Cockayne, A., M.J. Bailey, and C.W. Penn. 1987. Analysis of sheath and core structures of the axial filament of *Treponema pallidum. J. Gen.*

Microbiol. **133**:1397–1407.

Coene, M. and C. Cocito. 1985. A microanalytical procedure for determination of the base composition of DNA. *Eur. J. Biochem.* **150**:475–479.

Coene, M., A.M. Agliano, A.T. Paques, P. Cattani, G. Dettori, A. Sanna and C. Cocito. 1989. Comparative analysis of the genomes of intestinal spirochetes of human and animal origin. *Infect. Immun.* **57**:138–145.

Cwyk, W.M. and E. Canale-Parola. 1979. *Treponema succinifaciens* sp. nov., an anaerobic spirochaete from the swine intestine. *Arch. Microbiol.* **122**:231–239.

D'Alessandro, G. and C. del Carpio. 1958. A lipopolysaccharide antigen of Treponema. *Nature (London)* **181**:991–992.

Dal, M. and H. Monteil. 1983. Haemolysin produced by group B *Streptococcus agalactiae. FEMS Microbiol. Lett.* **16**:89–94.

Davis, C. P., D. Mulcahy, A. Takeuchi, and C.D. Savage. 1972. Location and description of spiral-shaped microorganisms in the normal rat cecum. *Infect. Immun.* **6**:184–192.

Davis, C.P., J.S. McAllister and C.D. Savage. 1973. Anaerobic bacteria on the mucosal epithelium in suckling mice. *Infect. Immun.* **7**:666–672.

Dettori, G., R. Grillo, A.M. Agliano, and G. Branca. 1987. Common antigens of human intestinal treponemes and of swine *Treponema hyodysenteriae. Microbiologica* **10**:1–8.

Davelaar, F.G., H.F. Smit, K. Hovind-Hougen, R.M. Dwars, and P.C. van der Valk. 1986. Infectious typhlitis in chickens caused by spirochaetes. *Avian Pathol.* **15**:247–258.

Douglas, J.G. and V. Crucioli. 1981. Spirochaetosis a remediable cause of diarrhoea and rectal bleeding? *Br. Med. J.* **283**:1362.

Dwars, R.M., H.F. Smit, F.G. Davelaar, and W. Van'T Veer. 1989. Incidence of spirochaetal infections in cases of intestinal disorder in chickens. *Avian Pathol.* **18**:591–596.

Egan, I.T., D.L. Harris, and L.A. Joens. 1983. Comparison of the microtitration agglutination test and enzyme-linked immunoabsorbent assay for the detection of herds affected with swine dysentery. *Am. J. Vet. Res.* **44**:1323–1328.

Fernie, D.S., P.H. Ripley, and P.D. Walker. 1983. Swine dysentery: protection against experimental challenge following single dose parenteral immunisation with inactivated *Treponema hyodysenteriae. Res. Vet. Sci.* **35**:217–221.

Ginsberg, I. 1970. Streptolysin S. *In: Microbiol. Toxins*, Vol III, Montie, T. C., Kadis, S., and Ajl, S. (ed.), pp. 99–171. New York: Academic Press.

Glock, R.D. and D.L. Harris. 1972. Swine dysentery. II. Characterisation of lesions in pigs inoculated with *Treponema hyodysenteriae* in pure and mixed culture. *Vet. Med. Small Anim. Clin.* **67**:65–68.

Göbel, U.B., A. Geiser, and E.J. Stanbridge. 1987. Oligonucleotide probes complementary to variable regions of ribosomal RNA descriminate between *Mycoplasma* species. *J. Gen. Microbiol.* **133**:1969–1974.

Greer, J.M. and M.J. Wannemuehler. 1989a. Comparison of the biological responses induced by lipopolysaccharide and endotoxin of *Treponema hyodysenteriae* and *Treponema innocens*. *Infect. Immun.* **57**:717–723.

Greer, J.M. and M.J. Wannemuehler. 1989b. Pathogenesis of *Treponema hyodysenteriae*: induction of interleukin-1 and tumor necrosis factor by a treponema butanol/water extract (endotoxin). *Microb. Pathogen.* **7**:279–288.

Halter, M.R. and L.A. Joens. 1988. Lipooligosaccharides from *Treponema hyodysenteriae* and *Treponema innocens*. *Infect. Immun.* **56**:3152–3156.

Hampson, D.J., J.R.L. Mhoma, B. Combs, and J.R. Buddle. 1989a. Proposed revisions to the serological typing system for *Treponema hyodysenteriae*. *Epidemiol. Infect.* **102**:75–84.

Hampson, D.J., J.R.L. Mhoma, and B. Combs. 1989b. Analysis of lipopolysaccharide antigens of *Treponema hyodysenteriae*. *Epidemiol. Infect.* **103**:275–284.

Hansen, K., P. Hindersson, and N. Strandberg-Pedersen. 1988. Measurement of antibodies to the *Borrelia burgdorferi* flagellum improves serodiagnosis in Lyme disease. *J. Clin. Microbiol.* **26**:338–346.

Hardy, P.H., W.R. Fredericks, and E.E. Nell. 1975. Isolation and antigenic characterisation of axial filaments from the Reiter Treponeme. *Infect. Immun.* **11**:380–386.

Harland, W.A. and F.D. Lee. 1967. Intestinal spirochaetosis. *Br. Med. J.* **16**:718–722.

Harris, D.L., R.D. Glock, C.R. Christiansen, and J.M. Kinyon. 1972. Swine dysentery. I. Inoculation of pigs with *Treponema hyodysenteriae* (new species) and reproduction of the disease. *Vet. Med. Small Anim. Clin.* **67**:61–64.

Hovind-Hougen, K., A. Birch-Andersen, R. Henrik-Nielsen, M. Orholm, J.O. Pedersen, P.S. Teglbjaerg, and E.H. Thaysen. 1982. Intestinal spirochetosis: morphological characterisation and cultivation of the spirochete *Brachyspira aalborgi* gen. nov., sp. nov. *J. Clin. Microbiol.* **16**:1127–1136.

Hudson, M.J., T.J.L. Alexander, R.J. Lysons, and J.F. Prescott. 1977. Swine dysentery: protection of pigs by oral and parenteral immunisation with attenuated *Treponema hyodysenteriae*. *Res. Vet. Sci.* **21**:366–367.

Hyde, F.W. and R.C. Johnson. 1988. Characterization of a circular plasmid from *Borrelia burgdorferi*, etiologic agent of Lyme disease. *J. Clin. Microbiol.* **26**:2203–2205.

Jackson, S.W. and P.N. Zey. 1973. Ultrastructure of lipopolysaccharide isolated from *Treponema pallidum*. *J. Clin. Microbiol.* **114**:838–844.

Jensen, N.S., T.A. Casey, and T.B. Stanton. 1990. Detection and identification of *Treponema hyodysenteriae* using oligonucleotide probes complementary to 16S ribosomal RNA. *J. Clin. Microbiol.* **28**:2717–2721.

Joens, L.A., D.L. Harris, and D.H. Baum. 1979a. Immunity to swine dysentery in recovered pigs. *Am. J. Vet. Res.* **40**:1352–1354.

Joens, L.A., J.G. Songer, and D.L. Harris. 1979b. Comparison of selective culture and serologic agglutination of *Treponema hyodysenteriae* for diagnosis of swine dysentery. *Vet. Rec.* **105**:463–465.

Joens, L.A., N.A. Nord, J.M. Kinyon, and I.T. Egan. 1982. Enzyme-linked immunosorbent assay for detection of antibody to *Treponema hyodysenteriae* antigens. *J. Clin. Microbiol.* **15**:249–252.

Joens, L.A., S.C. Whipp, R.D. Glock, and M.E. Neussen. 1983. Serotype-specific protection against *Treponema hyodysenteriae* infection in ligated colonic loops of pigs recovered from swine dysentery. *Infect. Immun.* **39**:460–462.

Joens, L.A., D.W. DeYoung, J.C. Cramer and R.D. Glock. 1984. The immune response of the porcine colon to swine dysentery. *In: Proceedings of the 8th Congress of the International Pig Veterinary Society*, Belgium, p. 187.

Joens, L.A., D.W. DeYoung, R.D. Glock, M.E. Mapother, J.D. Cramer, and H.E. Wilcox III. 1985. Passive protection of segmented swine colonic loops against swine dysentery. *Am. J. Vet. Res.* **46**:2369–2371.

Joens, L.A. and R.B. Marquez. 1986. Molecular characterization of proteins from porcine spirochetes. *Infect. Immun.* **54**:893–896.

Kaplan, L.R. and A. Takeuchi. 1979. Purulent rectal discharge associated with a non-treponemal spirochaete. *J. Am. Vet. Med. Assoc.* **241**:52–53.

Kent, K.A. 1984. Haemolytic and adhesive properties of *Treponema hyodysenteriae*. PhD Thesis, University of Reading, England.

Kent, K.A., R.M. Lemcke, and R.J. Lysons. 1988. Production, purification and molecular weight determination of the haemolysin of *Treponema hyodysenteriae*. *J. Med. Microbiol.* **27**:215–224.

Kent, K.A., R. Sellwood, R.M. Lemcke, M.R. Burrows, and R.J. Lysons. 1989. Analysis of the axial filaments of *Treponema hyodysenteriae* by SDS-PAGE and immunoblotting. *J. Med. Microbiol.* **135**:1625–1632.

Kinyon, J.M. and D.L. Harris. 1979. *Treponema innocens*, a new species of intestinal bacteria and emended description of the type strain of *Treponema hyodysenteriae*. *Int. J. Syst. Bacteriol.* **29**:102–109.

Knoop, F.C. 1981. Investigation of a haemolysin produced by enteropathogenic *Treponema hyodysenteriae*. *Infect. Immun.* **31**:193–198.

Lemcke, R.M. and J. Bew. 1984. Antigenic differences among isolates of *Treponema hyodysenteriae*. In: *Proceedings of the 8th Congress of the International Pig Veterinary Society*, Belgium, p. 183.

Lemcke, R.M. and M.R. Burrows. 1979. A disc growth-inhibition test for differentiating *Treponema hyodysenteriae* from other intestinal spirochaetes. *Vet. Rec.* 104:548–551.

Lemcke, R.M. and M.R. Burrows. 1982. Studies on a haemolysin produced by *Treponema hyodysenteriae*. *J. Med. Microbiol.* 15:205–214.

Loridan, C. and J.E. Alouf. 1986. Purification of RNA-core induced streptolysin S, and isolation and haemolytic characteristics of the carrier-free toxin. *J. Gen. Microbiol.* 132:307–315.

Lysons, R.J., G.A. Hall, T.J.L. Alexander, J. Bew, and A.P. Bland. 1978. Aetiological agents and pathogenesis of swine dysentry. In: *Proceedings of the 5th Congress of the International Pig Veterinary Society*, Yugoslavia, p. 171.

Lysons, R.J., M.R. Burrows, T.G. Debney, and J. Bew. 1986. Vaccination against swine dysentery. An effective novel method. In: *Proceedings of the 9th Congress of the International Pig Veterinary Society*, Spain, p. 180.

Lysons, R.J., M.R. Burrows, P.W. Jones, and P. Collins. 1987. Parenteral priming for live oral vaccines. In: *Recent Advances in Mucosal Immunology, Part B: Effector Functions*. McGhee, J.R., Mestecky, J., Ogra, P.L., and Bienenstock, J., eds. *Adv. Exptl. Med. Biol.* 216B:1825–1829.

Lysons, R.J., K.A. Kent, A.P. Bland, R. Sellwood, W.F. Robinson, and A. J. Frost. A cytotoxic haemolysin from *Treponema hyodysenteriae*: a probable virulence determinant in swine dysentery. 1991. *J. Med. Microbiol.* 34:97–102.

Mapother, M.E. and L.A. Joens. 1985. New serotypes of *Treponema hyodysenteriae*. *J. Clin. Microbiol.* 22:161–164.

Margolin, A.B., L.A. Joens, and M.J. Hewlett. 1985. The characterization of a plasmid isolated from *Treponema hyodysenteriae* and *Treponema innocens*. Abstract No. 133 from the 66th Conference of Research Workers in Animal Science, Chicago, USA.

Martin, P.G., P. Lachance, and D.F. Niven. 1985. Production of RNA-dependent haemolysin by *Haemophilus pleuropneumoniae*. *Can. J. Microbiol.* 31:456–462.

Meyer, R.C., J. Siman, and C.S. Byerly. 1975. The etiology of swine dysentery. III. The role of selected gram-negative obligate anaerobes. *Vet. Pathol.* 12:46–54.

Miao, R. and A.H. Fieldsteel. 1978. Genetics of *Treponema*: relationship between *Treponema pallidum* and five cultivable treponemes. *J. Bacteriol.* 133:101–107.

Miao, R.M., A.H. Fieldsteel, and D.L. Harris. 1978. Genetics of *Treponema*: characterization of *Treponema hyodysenteriae* and its relationship to

Treponema pallidum. *Infect. Immun.* 22:736–739.

Miller, D.P., M. Toivio-Kinnucan, G. Wu, and G.R. Wilt. 1988. Ultrastructural and electrophoretic analysis of *Treponema hyodysenteriae* axial filaments. *Am. J. Vet. Res.* 49:786–789.

Morrison, D.C. and L. Leive. 1975. Fractions of lipopolysaccharide from *Escherichia coli* 0111:B4 prepared by two extraction procedures. *J. Biol. Chem.* 250:2912–2919.

Nauman, R.K., S.C. Holt, and C.D. Cox. 1969. Purification, ultrastructure, and composition of axial filaments from Leptospira. *J. Bacteriol.* 98:264–280.

Norgard, M.V. and J.N. Miller. 1981. Plasmid DNA in *Treponema pallidum* (Nichols): potential for antibiotic resistance by Syphilis bacteria. *Science* 213:553–555.

Nuessen, M.E. and L.A. Joens. 1982. Serotype-specific opsonization of *Treponema hyodysenteriae*. *Infect. Immun.* 38:1029–1032.

Nuessen, M.E., J.R. Birmingham, and L.A. Joens. 1982. Biological activity of a lipopolysaccharide extracted from *Treponema hyodysenteriae*. *Infect. Immun.* 37:138–142.

Nuessen, M.E., L.A. Joens, and R.D. Glock. 1983. Involvement of lipopolysaccharide in the pathogenicity of *Treponema hyodysenteriae*. *J. Immun.* 131:997–999.

Olson, L.D. 1974. Clinical and pathological observations on the experimental passage of swine dysentery. *Can. J. Comp. Med.* 38:7–13.

Owen, P.A., K.A. Graeme-Cook, B.A. Crowe, and C. Condon. 1982. Bacterial membranes: preparative techniques and criteria of purity. In: *Techniques in Lipid and Membrane Biochemistry*. Hesketh, T.R., Kornberg, H.L., Metcalf, J.C., Northcote, D.H., Pogson, C.I., and Tipton, K.F., pp. 1–69. Part 1, Amsterdam: Elsevier.

Parizek, R., R. Stewart, K. Brown, and D. Blevins. 1985. Protection against swine dysentery with an inactivated *Treponema hyodysenteriae* bacterin. *Vet. Med.* 80:82–86.

Penn, C.W., A. Cockayne, and M.J. Bailey. 1985. The outer membrane of *Treponema pallidum*: biological significance and biochemical properties. *J. Gen. Microbiol.* 131:2349–2357.

Picard, B., L. Massicotte, and S.A. Saheb. 1979. Effet du ribonucleate de sodium sur le croissance et l'activité hemolytique de *Treponema hyodysenteriae*. *Experimentia* 35:484–486.

Radolf, J.D., N.R. Chamberlain, A. Clausell, and M.V. Norgard. 1988. Identification and localization of integral membrane proteins of virulent *Treponema pallidum* subsp. *pallidum* by phase partitioning with the nonionic detergent Triton X-114. *Infect. Immun.* 56:490–498.

Radolf, J.D., M.V. Norgard, and W.W. Schulz. 1989. Outer membrane ultrastructure explains the limited antigenicity of virulent *Treponema pallidum. Proc. Natl. Acad. Sci. USA* **86**:2051–2055.

Rees, A.S., R.J. Lysons, C.R. Stokes, and F.J. Bourne. 1989. Antibody production by the pig colon during infection with *Treponema hyodysenteriae. Res. Vet. Sci.* **47**:263–269.

Ruane, P.J., M.M. Nakata, J.F. Reinhardt, and W.L. George. 1989. Spirochaete-like organisms in the gastrointestinal tract. *Rev. Infect. Dis.* **11**:184–196.

Sachse, K. and T. Blaha. 1988. Characterization of the agent of swine dysentery based on deoxyribonucleic acid homology. *Z. Bakteriol. Mikrobiol. Hygiene* **268**:8–14.

Saheb, S.A., L. Massicotte, and B. Picard. 1980. Purification and characterisation of *Treponema hyodysenteriae* haemolysin. *Biochemie* **62**:779–785.

Sanna, A., G. Dettori, R. Grillo, A. Rossi, and D. Chiarenza. 1982. Isolation and propagation of a strain of Treponema from the human digestive tract—preliminary report. *L'Igiene Moderna* **77**:287–297.

Sanna, A., G. Dettori, A.M. Agliano, G. Branca, R. Grillo, F. Leone, A. Rossi, and G. Parisi. 1984. Studies of treponemes isolated from human gastrointestinal tract. *L'Igiene Moderna* **81**:959–973.

Schmid, G.P., A.G. Steigerwalt, S.E. Johnson, A.G. Barbour, A.C. Steere, I.M. Robinson, and D.J. Brenner. 1984. DNA characterization of the spirochete that causes Lyme Disease. *J. Clin. Microbiol.* **20**:155–158.

Sellwood, R., K.A. Kent, M.R. Burrows, R.J. Lysons, and A.P. Bland. 1989. Antibodies to a common outer envelope antigen of *Treponema hyodysenteriae* with antibacterial activity. *J. Gen. Microbiol.* **135**:2249–2257.

Sellwood, R., D. Revitt, K.A. Kent, M.R. Burrows, and R.J. Lysons. 1987. The cloning and expression of the haemolysin gene from *Treponema hyodysenteriae. In: Anaerobes Today*, Hardie, J.M. and Borriello, S.P., (eds.), pp. 222–223. Chichester: John Wiley & Sons.

Stamm, L.V., R.L. Hodinka, P.B. Wyrick, and P.J. Bassford. 1987. Changes in the cell surface properties of *Treponema pallidum* that occur during in vitro incubation of freshly extracted organisms. *Infect. Immun.* **55**:2255–2261.

Stanton, T.B., N.S. Jensen, T.A. Casey, L.A. Tordoff, F.E. Dewhirst, and B.J. Paster. 1991. Reclassification of *Treponema hyodysenteriae* and *Treponema innocens* in a new genus, *Serpula*, gen. nov., as *Serpula hyodysenteriae* comb. nov. and *Serpula innocens*, comb. nov. *Int. J. Syst. Bacteriol.* **41**:50–58.

Stanton, T.B. 1992. Proposal to change the genus designation *Serpula* and *Serpulina* gen. nov. containing the species *Serpulina hyodysenteriae*, comb. nov. and *Serpulina innocens*, comb. nov. *Int. J. Syst. Bacteriol.* **42**:189–190.

Sultzer, B.M. 1968. Genetic control of leukocyte responses to endotoxin. *Nature* **219**:1253–1254.

Sultzer, B.M. 1969. Genetic factors in leukocyte responses to endotoxin: further studies in mice. *J. Immunol.* **103**:32–38.

Takayama, K., R.J. Rothenberg, and A.S. Barbour. 1987. Absence of lipopolysaccharide in the Lyme disease spirochaete, *Borrelia burgdorferi. Infect. Immun.* **55**:2311–2313.

Takeuchi, A., H.R. Jarvis, H. Nakazawa, and D.M. Robinson. 1974. Spiral-shaped organisms on the surface colonic epithelium of the monkey and man. *Am. J. Clin. Nutr.* **27**:1287–1296.

Taylor, D.J. and T.J.L. Alexander. 1971. The production of dysentery in swine by feeding cultures containing a spirochaete. *Br. Vet. J.* **127**:lviii–lxi.

Vinh, T., B. Adler, and S. Faine. 1986. Ultrastructure and chemical composition of lipopolysaccharide extracted from *Leptospira interrogans* serovar *copenhagen. J. Gen. Microbiol.* **132**:103–109.

Vovis, G.F. and S. Lacks. 1977. Complementary action of restriction enzymes EndoR.*DpnI* and EndoR.*DpnII* on bacteriophage fl DNA. *J. Mol. Biol.* **115**:525–528.

Wannemuehler, M.J., R.D. Hubbard, and J.M. Greer. 1988. Characterization of the major outer membrane antigens of *Treponema hyodysenteriae. Infect. Immun.* **56**:3032–3039.

Wergifosse P. de and M.M. Coene. 1989. Comparison of the genomes of pathogenic treponemes of human and animal origin. *Infect. Immun.* **57**:1629–1631.

Westphal, O., O. Lüderitz, and R. Bister. 1952. Uber die extraction von bakterien mit phenolwasser. *Z. Naturforschung Teil B* **7**:148–155.

Whipp, S.C., J. Pohlenz, D.L. Harris, I.M. Robinson, R.D. Glock, and R. Kunkel. 1982. Pathogenicity of *Treponema hyodysenteriae* in uncontaminated gnotobiotic pigs. In: *Proceedings of the 7th Congress of the International Pig Veterinary Society*, Mexico, p. 31.

Woese, C.R., R. Gutell, R. Gupta, and H.F. Noller. 1983. Detailed analysis of the higher-order structure of 16S-like ribosomal ribonucleic acids. *Microbiol. Rev.* **47**:621–669.

Woltjes, J., H. Legdeur-Velthuis, and J. De Graaf. 1981. Detection and characterisation of haemolysin production in *Streptococcus mutans. Infect. Immun.* **31**:850–855.

44

Nucleic Acid Hybridization for Identification and Detection of Gram-Negative Anaerobes

Ulf B. Göbel and Klaus Pelz

44.1 Introduction

Although most pathogenic clostridia and a number of nonspore-forming anaerobes such as *Bacteroides* and *Fusobacterium* species had been discovered by the end of the century (Willis, 1989), widespread knowledge was impeded by the lack of isolation and identification techniques suitable for clinical laboratories. Improved culture methods eventually lead to increasing awareness of the medical importance of anaerobic infections, but they are still considered "the most commonly overlooked of bacterial infections" (Finegold and George, 1989). The apparent lack of phenotypic markers and the fastidious nature of many anaerobes, however, makes analyses difficult and time consuming, restricted to rather few specialized laboratories. Further confusion is added by rapid changes in the taxonomy of these organisms. It is certain that molecular biology will contribute significantly not only to elucidating the role of virulence factors and host defense mechanisms in anaerobic infections but also to providing a ground for reliable classification and identification of anaerobic bacteria based on phylogenetic (evolutionary) relationships. Because identification of bacteria using nucleic acid hybridization is of increasing importance for both clinical and research laboratories, it is our objective to give the background for the understanding of basic principles and to review recent developments within the field.

44.2 Current Laboratory Diagnosis of Anaerobic Infections

Current laboratory diagnosis of anaerobic infections is based mainly on microscopic examination and isolation on appropriate culture media of the organisms involved. Isolates are then characterized by a number of phenotypic traits, which often but not always allow unambiguous identification. This examination involves detection of enzymes and their products, analysis of volatile metabolites such as fatty acids by gas-liquid-chromatography (GLC), or immuno-detection of antigens by monoclonal antibodies (Romond, 1989; Phillips, 1990). Because it is beyond the scope of this review to discuss these methods the interested reader is referred to some recent reviews and monographs (Sutter et al., 1985; Bascomb, 1987; Poxton, 1988; Finegold and George, 1989).

44.3 Nucleic Acid Hybridization

History and Principles

Single-stranded nucleic acids are able to form duplexes by the formation of hydrogen bonds between amino and carboxyl moeties of their complementary bases. Pairing typically occurs between purine and pyrimidine rings, e.g., guanine-cytosine (G-C) and adenine-thymine (A-T) or adenine-uracil (A-U) in the case of RNA. These hybrid molecules can be denatured to single strands by physical or chemical means. However, under appropriate conditions these strands reassociate to form the original duplex. This reassociation also takes place between complementary strands from different sources. The apparent usefulness of hybridization between single-stranded nucleic acids as a test for the identification of complementary sequences was recognized almost 30 years ago (Hall and Spiegelman, 1961; Gillespie and Spiegelman, 1965). It was applied some years ago to determine the genetic relatedness among organisms (McCarthy and Bolton, 1963). DNA reassociation and RNA hybridization of bacterial nucleic acids as indirect means to measure nucleotide sequence similarity provided a basis for microbial taxonomy long before sequencing techniques were available (Grimont, 1984; Johnson, 1985). However, it was 15 years before nucleic acid hybridization was used to detect pathogenic bacteria, *Listeria monocytogenes*, in experimentally infected mice (Steinman, 1975). This report did not gain widespread attention and it was a report by Moseley et. al. (1980) that sparked great interest among clinical microbiologists. A number of recent review articles have covered the broad range of applications in diagnostic microbiology (Quivey, 1987; Tenover, 1988; Yolken 1988; Albandar and Olsen, 1990; Druel and Freney, 1990; Groves, 1990).

Probe Design

Several factors determine the success of diagnostic hybridization: (i) appropriately denatured target nucleic acids, either present in solution or bound to a solid phase; (ii) a labeled probe, denatured double- or single-stranded nucleic acid of known origin, complementary to a single gene or a unique transcript (messenger or ribosomal RNA) representative for a particular organism; and (iii) proper hybridization conditions. Several strategies have been employed to develop probes. The simplest way is to use whole genomic DNA, adopting the

features of classical DNA reassociation experiments used in bacterial taxonomy. Whole genomic DNA can be labeled to high specific activity (amount of label incorporated per microgram or molecule of DNA), resulting in a rather high sensitivity. However, reassociation of probe molecules competing for the hybridization with target molecules may lead to reduced sensitivity. This is particularly true for high probe concentrations required to increase hybridization speed.

Despite the addition of hybridization rate accelerators such as dextran sulfate, hybridization usually requires overnight incubation. In filter hybridization assays, this not only results in higher background values but also leads to a loss of target molecules. Repeated isolation of chromosomal DNA is tedious, often resulting in DNA preparations of varying quality; therefore, it is difficult or impossible to standardize the labeling conditions and the use of whole chromosomal probes is not recommended. Another approach still commonly used is the application of randomly chosen fragments of genomic or extrachromosomal DNA. It is quite easy to generate specific probes this way, but because sequence and function of these fragments are usually not known they must be tested for specificity by lengthy series of hybridization experiments. With the advent of automated DNA synthesis, construction of short synthetic oligodeoxyribonucleotides complementary to known target sequences represents the most straightforward approach to develop specific probes. Oligonucleotide probes combine a number of advantages: (i) They are cheap and easy to prepare. Standard synthesis (0.2 μmol) of a probe 20 to 40 bases in length yields enough product for as many as 100,000 hybridization assays. (ii) Unlabeled or nonradioactively labeled oligonucleotides are very stable and can be kept for a long time. (iii) They can be labeled reproducibly to yield a defined specific activity. This contributes significantly to the development of standardized tests. (iv) Because the oligonucleotides are small, reaction kinetics allow for short hybridization times. (v) Under stringent conditions, the probes discriminate target sequences differing in only one nucleotide. The rapid accumulation of sequence data suggests that synthetic oligonucleotide probes will play a major role in developing rapid and accurate hybridization assays.

Target Molecules

Almost any nucleic acid sequence derived from chromosomal or extrachromosomal DNA of the organism of interest can be used for species-specific detection or identification, provided that the sequence is constantly found, that it shows no or negligible variation among members of the same taxon, and that it exhibits sufficient differences with respect to all nontarget sequences. However, it is inappropriate to select these sequences by random cloning and lengthy series of hybridization experiments. Selection should be based on a more rational approach, considering the importance of the selected sequence as a reliable phylogenetic (taxonomic) marker or its significance as a coding region for virulence factors or antibiotic resistance.

DNA–rRNA hybridization or, better, direct comparison of rRNA sequences are currently the best tools available "for determining phylogenetic relationships above the species level" (Murray et al., 1990). The same features that make rRNAs suitable markers for phylogenetic studies explain their relevance as targets for diagnostic hybridizations and microbial ecology (Woese, 1987). Ribosomal RNAs are integral constituents of ribosomes in any self-replicating cell. Common traits of these molecules, such as ubiquity, genetic stability, relative abundance, and structural conservation, result from evolutionary constraints that tend to enhance functional constancy as a basis for maximal translational efficiency. Comparing a number of rRNA sequences suggested that regions of greater sequence divergence are not scattered randomly but are confined to regions of conserved secondary structure, resulting in a unique pattern wherein stretches of conserved primary sequences are interspersed by regions of relatively high sequence variation (Figure 44.1). This observation led to the development of rRNA-based probes, which according to their respective target sequences recognize either single species or a greater number of taxa

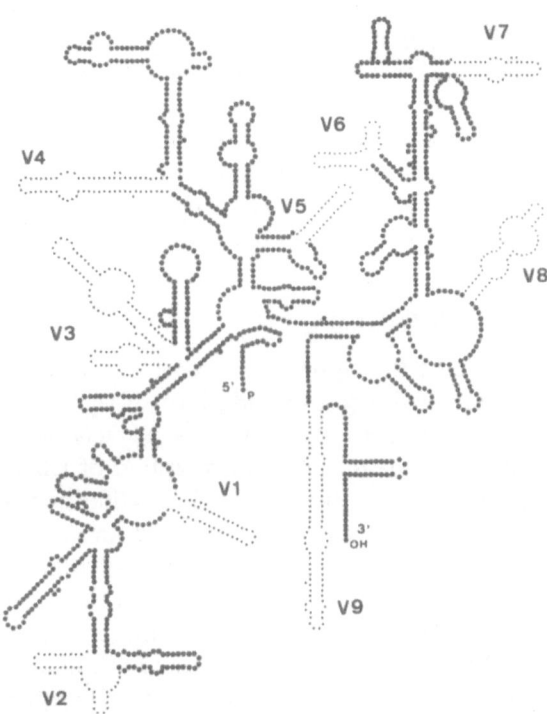

FIGURE 44.1 Secondary structure of prokaryotic consensus 16S rRNA. Variable regions (V) are numbered starting from 5' end of molecule.

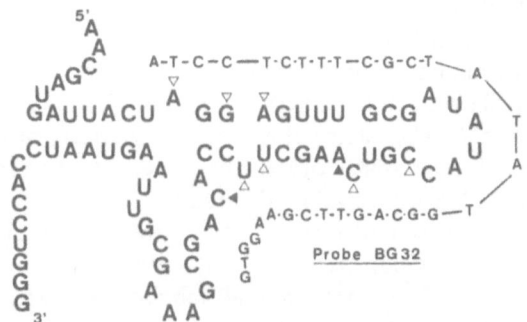

FIGURE 44.2 *Porphyromonas gingivalis* 16S rRNA molecule, variable region V5. Open triangles, base substitutions; bold triangles, insertions with respect to homologous region in *B. fragilis* 16S rRNA. The sequence of the *P. gingivalis*-specific oligonucleotide probe (BG32) is given in small uppercase letters.

(Göbel et al., 1987; Giovannoni et al., 1988); an example of a species-specific probe is given in Figure 44.2. A synthetic oligonucleotide (BG32), complementary to the "hypervariable" region (V5) of the *Porphyromonas gingivalis* 16S rRNA molecule, specifically detects bacteria of this species (Chuba et al., 1988). The ease of rRNA sequencing, together with the obvious advantages described here, has made rRNA-based probes most attractive candidates for a broad range of applications (Lane et al., 1985; Böttger, 1989; Barry et al., 1990).

Labeling Strategies

Choosing the appropriate label is critical for the sensitivity of a hybridization assay. Probes can be labeled by a variety of techniques either by enzymatical incorporation of labeled nucleotides or by chemical modification of nucleotides already incorporated in a poly-nucleotide chain. The strategy used differentiates directly labeled probes, in which a label is covalently attached, or indirectly labeled probes, where haptens are linked to the probe molecule, which are then detected by appropriate binding proteins or antibodies. Radioisotopes, e.g., ^{32}P, have been used for labeling with great success. Radiolabeled probes routinely detect about 1 to 5 attomol of target, which means they are still attractive analytical tools in a research laboratory. For most non-research laboratories, e.g., microbiology units within hospitals, nonisotopically labeled probes will be the only alternative. Although many different approaches have been reported, none achieves satisfactory levels of sensitivity. For detailed information the reader is referred to some excellent review articles (Matthews and Kricka, 1988; Dahlen et. al., 1989; Jablonski, 1989). There are several labeling strategies, such as incorporation of biotin, which is detected by avidin or streptavidin conjugates carrying enzymes or fluorochromes. Alternatively, other haptens such as digoxigenin, acetylaminofluorene, or sulfone groups are used for nonradioactive detection. Direct labeling of nucleic acids with various enzymes, such as alkaline phosphatase (AP) or horseradish peroxidase (HRP), has also been achieved. Enzymatically triggered chemiluminescence, providing sensitivities comparable to ^{32}P labeling, holds great promise to reach the detection limits needed for discovering small numbers of

bacteria present in clinical specimens (Matthews et al., 1985; Urdea et al., 1988; Schaap et al., 1989; Pollard-Knight et al., 1990a, 1990b). It should, however, be noted that substances produced by *Porphyromonas* and *Prevotella* species (formerly black-pigmented *Bacteroides* spp.) interfere with compounds of the enhancer solutions present in some chemiluminescence assays, thus leading to strong nonspecific signals (H. Gersdorf and U.B. Göbel, unpublished results).

Hybridization Format

There are basically two hybridization formats: one has either targets or probes immobilized on solid supports (mixed-phase hybridization), and the other has both free probes and target sequences (solution hybridization). Theoretical considerations concerning the hybridization kinetics are beyond the scope of this article, and interested readers may find valuable information elsewhere (Anderson and Young, 1985; Britten and Davidson, 1985; Young and Anderson, 1985). Because there are advantages and disadvantages associated with both assay formats, using one or the other largely depends on the intended application. The filter hybridization assay most commonly used in research laboratories utilizes target nucleic acids or crude samples that are spotted onto nitrocellulose or nylon membranes by using dot- or slot-shaped filtration manifolds and immobilized by baking or UV irradiation. Labeled probes are added and allowed to hybridize under the appropriate conditions, and unbound probe is removed by a series of simple washing steps. By using this format several hundred samples can be processed in parallel, and therefore this procedure lends itself to epidemiological studies or identification of culture isolates (Figure 44.3).

When a sample must be screened with several probes another format, the so-called reversed hybridization, may be advantageous. This technique employes multiple probes immobilized on filter strips and target nucleic acids labeled with photoactivatable DNA-labeling reagents or biotin before the hybridization step (Dattagupta et al., 1989a, 1989b; R. Saiki et al., 1989).

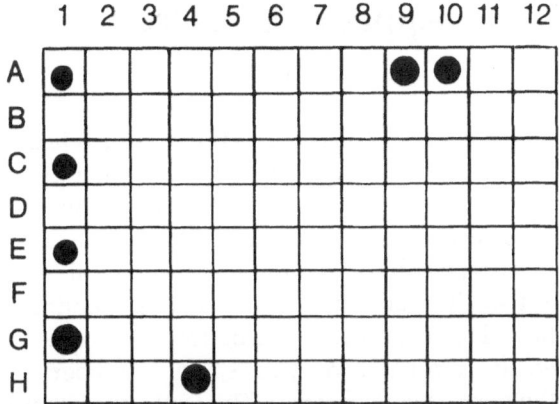

FIGURE 44.3 Autoradiography of a representative dot hybridization assay using a ^{32}P-labeled *Selenomonas sputigena* probe. Aliquots of following bacteria (clinical isolates, if not stated otherwise) were spotted: *Selenomonas sputigena* (IA-G, 2A-C); *Selenomonas flueggei* (3A-G), *Selenomonas noxia* (4A-H); *Selenomonas dianae* (5A-C); *Selenomonas infelix* (6A); *Selenomonas artemidis* (7A-H, 8A); *Selenomonas sputigena*, ATCC 35185 (9A); *Selenomonas flueggei*, ATCC 43531 (9B); *Selenomonas noxia*, ATCC 43541 (9C); *Selenomonas infelix* ATCC 43532 (9D); *Selenomonas artemidis* ATCC 43528 (9E); *Selenomonas dianae*, ATCC 43527 (9F); *Selenomonas spp.* (10A–D); *Porphyromonas gingivalis* (11A); *Prevotella intermedia* (11B); *Porphyromonas endodontalis* (11C); *Bacteroides forsythus* (11D); *Fusobacterium nucleatum* (11E); *Wolinella recta* (11F); *Peptococcus anaerobius* (11G); *Peptococcus micros* (11H); *Veillonella parvula* (12A); *Eubacterium nodosum* (12C); *Eubacterium lentum* (12D).

However, care should be taken if heterogeneous samples are to be identified by nonisotopically labeled probes. Interpretation of the results is often hampered by probes bound nonspecifically, and extensive purification of target nucleic acids is therefore indicated. This can be achieved by "sandwich hybridization," where a target nucleic acid present in a crude sample is first captured by an unlabeled probe immobilized on a solid support, washed, and then detected by a labeled second probe (Ranki et al., 1983; Morrissey and Collins, 1989).

Conventional solution hybridization required absorption to hydroxyapatite or capture using antibodies to separate bound and unbound probe molecules (Britten et al., 1974; Boguslawski et al., 1986). To circumvent this step a variety of homogeneous solution hybridization assays have been developed, where all

reactions are performed in a one-step procedure without subsequent separation (for review see Matthews and Kricka, 1988; Keller and Manak, 1989). A most promising approach involves acridinium ester-labeled DNA probes in a hybridization protection assay detecting about 10^{-16} to 10^{-17} mol of target sequences, which translates to 3×10^2 to 3×10^3 bacteria, when rRNA molecules were targeted (Arnold et al., 1989). However, by combining this assay with an additional solid-phase separation step (magnetic polycationic microspheres), the sensitivity was enhanced about 10 fold by significant reduction of the background noise. Single-cell resolution can be achieved by in situ hybridization using appropriately labeled probes. This method allows for the detection by light microscopy of specific target sequences within fixed and paraffin- or plastic-embedded tissue sections and cells (Moench, 1987; Lichter and Ward, 1990).

The apparent advantage of applying in situ hybridizations for diagnosis of bacterial infections is based on the maintenance of the cellular morphology of the infected tissue and the identification of the inflammatory cells involved. Recently in situ hybridization using fluorochrome-labeled group- and species-specific oligonucleotide probes has been described for microecological studies (Giovannoni et al., 1988; Delong et al., 1989; Amann et al., 1990). This technique holds great promise for analyzing complex anaerobic microbiota, such as resident oral or intestinal flora, inflamed periodontal pockets, and sediments from sewage treatment systems.

Amplification Strategies

A major advantage of using nucleic acid probes rather than antibodies for detection of microbial pathogens lies in the capability to amplify both target and probe molecules. Several amplification strategies have been developed allowing for amplification to a millionfold. The ultimate goal of diagnostic microbiology, detection of a single pathogenic microorganism present in a given sample, seems to become feasible. A straightforward procedure for amplifying target sequences, the so-called polymerase chain reaction (PCR), has gained widespread acceptance (Mullis and Faloona, 1987; Saiki et al., 1988; Gibbs, 1990; Innis et al., 1990). This method is based on repeated cycles of in vitro replication of a template molecule using two specific oligonucleotide primers and a thermostable DNA polymerase. Amplification of RNA molecules requires an additional reverse transcriptase step to generate cDNA templates (T. Adam and U.B. Göbel, unpublished results; E. Böttger, personal communication; Wilson et al., 1989).

In a different approach, specific DNA fragments are exponentially amplified by using two pairs of oligonucleotides in sequential rounds of template-dependent ligation (Wu and Wallace, 1989). The ligase amplification reaction (LAR) reliably discriminates single base mismatches, but the lack of a heat-stable ligase restricts its practical application to a few laboratories. A very promising alternative to PCR, simulating retroviral replication, was published by Guatelli et al. (1990). Target nucleic acid molecules are amplified in vitro under isothermal conditions by the coordinated action of three enzymes, reverse transcriptase, RNase H, and a DNA-dependent RNA polymerase. Ten millionfold amplification has been achieved in only 1 to 2 h.

In contrast to these methods designed to amplify target nucleic acids, a different concept involves amplifiable probe molecules, recombinant RNA, serving a "dual role of specific probe and amplifiable reporter" (Chu et al., 1986; Lizardi et al., 1988). Recombinant molecules of bacteriophage $Q\beta$ midivariant (MDV) RNA carrying sequences of desired target specificity are amplified by $Q\beta$ replicase, an RNA-directed RNA polymerase. Replication does not depend on oligonucleotide primers but is initiated by a unique secondary structure within the bacteriophage RNA, resulting in amplification as high as billionfold within 30 min. Other amplification schemes are based on signal amplification using probe networks. Label amplification is achieved by chemically cross-linked oligonucleotides (amplification multimers) and enzyme (HRP) labeled probes for chemiluminescent or colorimetric detection (Urdea et al., 1987).

44.4   Application of DNA Probes

There is a lack of rapid, simple, and reliable assays for identifying and enumerating a great number of anaerobic bacteria. Fermentation tests very often do not readily differentiate species and a battery of different biochemical tests is usually required for final species identification. This is particularly true for some non- or weakly fermentative species, e.g., members of the genus *Porphyromonas*. The fastidious nature of the organisms, requiring special handling and equipment, makes the speciation laborious, time consuming, and expensive. Precise description of complex microbiota found within the alimentary or genital tract is almost impossible. These technical difficulties set the stage for the introduction of nucleic acid hybridization.

Oral Cavity

The importance of periodontal infections as a major cause of tooth loss worldwide has sparked early interest to identifying the organisms associated with several forms of the disease. French et al. (1986) reported first on the application of nucleic acid hybridization for detection of periodontal pathogens using isotopically labeled, whole genomic DNA. Among the bacteria tested were two strict anaerobes, *Prevotella intermedia* (*B. intermedius*) and *Porphyromonas gingivalis*, two species that have been allocated to the genus *Prevotella* and *Porphyromonas*, respectively (Shah and Collins, 1988; Shah and Collins, 1990). The *P. intermedia* probe did not diffentiate between *P. intermedia I* and *P. intermedia II*, two strains that may represent different species (see also Roberts et al., 1987). Both *P. intermedia* types, however, have been found in periodontal pockets. The *P. gingivalis* probe seemed to be specific. Unfortunately a number of oral *Bacteroides* spp., as well as the third member of the genus *Porphyromonas*, *P. endodontalis*, were not tested. Both probes detected as few as 10^2 of the homologous bacteria in pure and mixed cultures. Subsequently, both probes have been tested against a larger panel of related and unrelated oral bacteria (Strzempko et al., 1987). The *P. intermedia* probe cross-hybridized to a number of *Bacteroides* spp., when present in high numbers (10^5 cells).

Under the assumption that most *Bacteroides* species share only 30% or less DNA homology Roberts et al. (1987) described the application of whole chromosomal DNA probes for the identification of five oral species in a dot blot assay using either pure DNA or whole bacteria as targets. Isotopically labeled probes detected about 10^4 bacteria per dot and were tested against a panel of 23 clinical isolates. The *P. asaccharolytica* and the *P. gingivalis* probe correctly identified 100% of the strains tested. All other probes, *P. loeschii*, *P. melaninogenica*, and *P. intermedia I*, missed part (3%–12%) of the biochemically identified strains. In the case of *P. intermedia*, this failure (12%) may be explained by the fact that the probe did not recognize type II strains. However, the problem has been addressed by Moncla et al. (1988).

Another species, *Selenomonas noxia*, that is possibly associated with periodontal disease has been identified by using nonisotopically labeled, whole genomic DNA (Tanner et al., 1989). In this study periodontal samples were plated, and colonies transferred from primary plates onto filters and hybridized after DNA extraction. Several randomly cloned probes were used to assess the prevalence of *Actinobacillus actinomycetemcomitans*, *P. intermedia*, *P. gingivalis*, *Eikenella corrodens*, and *Wolinella recta* in subjects with HIV- (human immunodeficiency virus-) associated periodontitis (Murray and French, 1989). Detection threshold of using ^{32}P-labeled plasmids in a slot blot procedure was about 10^2–10^3 cells for *P. gingivalis*, *P. intermedia*, *W. recta*, and *E. corrodens* after 18-h autoradiographic exposure.

A different approach for developing probes specific for *A. actinomycetemcomitans*, *Haemophilus aphrophilus*, *P. gingivalis*, *P. intermedia II*, and *P. asaccharolytica*, has been reported (Chuba et al., 1988; Göbel et al., 1989). Partial sequencing of 16S rRNA and subsequent computer-assisted sequence comparison allowed the construction of oligonucleotide probes complementary to variable regions of 16S rRNA molecules. Isotopically labeled probes were

shown to be 100% specific when tested in a dot hybridization assay; the detection limit was less than 5×10^3 organisms. A great number of oligonucleotide probes specific for various oral bacteria has been developed since and used to confirm biochemical identification (H. Gersdorf, S. Nagel, A. Meissner, I. Spiliopoulou-Sdoungou, and B. Choi, unpublished results). A representative dot blot using a *Selenomonas sputigena*-specific probe is shown in Figure 44.3. The probe did not cross-hybridize with any of the other *Selenomonas* species tested. From 61 samples applied to the nylon membrane, all except 4 strains (row 1, positions B, D, F; row 4, position H) were unambiguously identified. However, repeated biochemical testing of these isolates confirmed the probe result.

The same approach of using rRNA-based synthetic oligonucleotide probes has been adopted by another group to describe the development of probes for *A. actinomycetemcomitans*, *P. gingivalis*, *P. intermedia I* and *P. intermedia II*, *B. forsythus*, *E. corrodens*, *H. aphrophilus*, *Streptococcus intermedius*, and *Fusobacterium nucleatum* (Dix et al., 1990). All probes except the *F. nucleatum* probes were specific, having a detection limit of 10^2 bacteria. Comparing the target sequences of the *F. nucleatum* probes with that of other *Fusobacterium* species, complete sequence identity was found for *F. periodonticum*, and only one mismatch was noted for *F. mortiferum*. In an accompanying paper six probes complementary to different hypervariable regions of *P. gingivalis* 16S rRNA were tested against 77 *P. gingivalis* field strains and 105 strains of 12 other *Bacteroides* species. Four of six probes were 100% specific, and two others cross-hybridized to other species (Moncla et al., 1990). So far, DNA probes (whole genomic and randomly cloned probes) have been compared with either culture or indirect immunofluorescence.

In one study (Savitt et al., 1988), 60 periodontal samples were tested in parallel by culture and hybridization. *P. gingivalis* was found by probes in 61% and by culture in 19%; *P. intermedia* was detected by probes in 72% and by culture in 23% of the samples. Almost all culture-positive samples were also probe positive (*P. intermedia* 100%, *P. gingivalis* 91%). Probes performed equally well in the second study (Zappa et al., 1990). In 52 samples from 13 patients *P. gingivalis* and *P. intermedia* were found in 66.7% and 96.1%, respectively, by indirect immunofluorescence, and 76.5% and 78.4% by probes. The detection limit for fluorescence was greater than 10^4 organisms, whereas probes detected bacteria at levels of 10^3. However, for routine applications, especially in a clinical laboratory, radioactive markers should be replaced by nonisotopic reporter systems for probe labeling. This was the reason for Smith et al. (1989) to evaluate the application of biotinylated probes or probes directly labeled with horse radish peroxidase (HRP) for identification of seven subgingival species. Unspecific binding of labeled probes was eliminated by pretreating the membranes with proteinase K and organic solvents before hybridization. The inherent problem of cross-reactivity associated with whole genomic DNA probes could not be eliminated.

Gastrointestinal Tract

To our knowledge, the first report on the application of DNA hybridization for identification of anaerobic bacteria appeared in 1983 (Salyers et al., 1983). Randomly cloned DNA fragments were used to identify *B. thetaiotaomicron* isolates. Two of five fragments tested against 65 strains of colonic *Bacteroides* species specifically hybridized to DNA from *B. thetaiotaomicron* but not to DNA from other members of the "*B. fragilis* group." The probes detected about 50 ng of *B. thetaiotaomicron* DNA, which roughly equals about 10^7 about 10^8 bacteria. Soon after, a similar approach was chosen to develop probes for identifying and enumerating *Bacteroides vulgatus* in human feces for dietary studies (Kuritza and Salyers, 1985). A fragment of about 700 bp was shown to be specific and instrumental to determine the concentration of these bacteria in stool specimens.

This study was later complemented by a report on the enumeration of polysaccharide-degrading *Bacteroides* species in feces (Kuritza et al., 1986a). Attwood et al. (1988) used run-off transcripts in a dot-blot hybridization format to enumerate a strain of *Bacteroides ruminicola* introduced into the rumen. Randomly chosen

genomic DNA fragments from *B. uniformis, B. distasonis, Bacteroides* group 3452A (renamed *B. caccae*), and *B. ovatus* were tested for specificity and used together with the *B. thetaiotaomicron* probe to detect these species in fecal samples of two human volunteers. In an attempt to develop probes for clinically significant groups of anaerobic bacteria, Kuritza et al. (1986b) reported on a *B. fragilis*-specific probe, which was used to detect this species in spiked blood cultures, a probe specific for all members of the *B. fragilis* group, and one probe specific not only for all *Bacteroides* sp. strains tested, but also for *Fusobacterium nucleatum* and *F. necrophorum*. The latter probe contained a fragment of the 16S rRNA gene from *B. ovatus*. The overall conservation of rRNA genes explained the wide range specificity of this probe. The detection limit of radiolabeled probes was 10^6 bacteria.

Attempts to use biotinylated probes were unsuccessful because of the unspecific binding of the detection reagent (see also Smith et al., 1989). Additional *B. fragilis*-specific probes were generated by random cloning into run-off transcription vectors claiming that RNA transcripts would be more sensitive in detecting the target bacteria (Groves and Clark, 1987). Information on detection limits is, however, lacking. A battery of whole chromosomal DNA probes has been used by Morotomi et al. (1988) to develop a procedure allowing investigation of the interrelationships between colon cancer and the composition of the fecal flora. Sixty-five *Bacteroides* strains from 19 individuals were biochemically identified and subsequently tested with ^{32}P-labeled probes. The probes correctly identified 62 (95%) of these strains.

A most interesting approach was chosen by Stahl et al. (1988) using species- and group-specific 16S rRNA-targeted oligonucleotide probes to study the ecology of the ruminal flora by quantitative filter hybridization. One "signature" probe detected all but one strain of *B. succinogenes* (renamed *Fibrobacter succinogenes*). Two other oligonucleotides hybridized to the ruminal or cecal subgroup, respectively of this rather diverse species. The 16S rRNA target sequence chosen to detect *Lachnospira multiparus*, was found to be identical in a related organism *Roseburia cecicola*, an organism that has not yet been found in the rumen. Amann et al.

(1990) have shown the use of fluorescent oligonucleotide probes for in situ hybridization of cell smears from washed rumen samples. 16S rRNA-targeted oligonucleotide probes were designed to specifically discriminate between the two *Fibrobacter succinogenes* and *F. intestinalis* (exhibiting 92% rRNA sequence similarity) as well as to distinguish between *F. succinogenes* ssp. *succinogenes* and ssp. *elongata*. However, because the autofluorescence of compounds present in the samples was intense, direct microscopy of the rumen was not possible. A review of the use of gene or antibody probes to identify and enumerate rumen bacterial species has been published by Brooker et al. (1990).

Genital Tract

Although closely related to the genus *Actinomyces*, *Mobiluncus* spp. should be included in this review. These curved, motile bacteria appear mostly gram-negative and are frequently isolated from females with bacterial vaginosis. The difficulties in cultivating these organisms make them good candidates for nucleic acid hybridization. The first probes reported were whole-cell DNAs from the species *Mobiluncus curtisii* and *M. mulieris* (Roberts et al., 1984). The ^{32}P-labeled probes detected about 10^4 organisms in a filter hybridization assay; 70 strains, 39 isolates of *M. mulieris*, and 31 strains of *M. curtisii* were correctly identified. Biotin labeling did not affect specificity, but the detection limit was lowered to 10^5 to 10^6 bacteria. In a subsequent report current methods for the primary detection of these bacteria, Gram stain and culture, were compared with hybridization (Roberts et al., 1985). Gram stain detected 90% of positive samples, cultures 77% to 83%, and probes 52% to 83%.

44.5 Conclusions and Outlook

A major advantage of using nucleic acid hybridization is the ease with which probes can be developed. This is particularly true for short synthetic oligonucleotide probes complementary to defined sequences. Construction of these probes by definition requires sequenc-

ing of target nucleic acids, which has been considered tedious and time consuming by those investigators advocating the use of whole genomic or randomly cloned DNA probes. However, with the advent of rapid sequencing protocols and automated sequence analysis, primary nucleotide sequences are readily available. Target sequences of desired specificity are selected on the basis of computer-assisted sequence comparison. This allows not only for the design of probes of predefined specificity but also the accumulation of phylogenetically relevant data, provided an appropriate evolutionary marker, e.g., rRNA, has been chosen.

Synthetic oligonucleotide probes targeting defined regions within rRNA molecules have been used for a broad range of applications, including the analysis of bacteria that could not yet be cultured but which were classified by comparative rRNA analysis. The latter approach lends itself to studying complex microbial communities found in the environment or resident flora colonizing epithelial layers in animals and humans. Thus the development and systematic application of rRNA-based probes may revolutionize our understanding of microbial ecology and of the role of the resident flora in preventing or provoking exogenous or endogenous infections, respectively.

There are, however, some aspects of practical importance. A major problem of nearly all hybridization assays described here is the relative lack of sensitivity. Detection limits of 10^5 to 10^6 bacteria are acceptable for confirmatory tests of cultured isolates but are definitively not sufficient for primary detection of organisms in clinical specimens. However, optimization of sampling, sample transport and processing, hybridization conditions, and selection of appropriate reporter molecules will enhance the sensitivity significantly. Detection problems very often also arise from interfering background noise, which is either caused by intrinsic factors, such as interaction of the label with other assay components, or by extrinsic factors, e.g., autofluorescence of sample constituents (Jablonski, 1989). This interference can be reduced or completely eliminated by application of different capture assays (Thompson et al., 1989). Finally, ultimate sensititivity may be reached by combining nucleic acid hybridization with one of the amplification strategies described, leading to the desired detection limit of one target organism in a given specimen. However, interpretation of these data will be rather difficult, because pathogenic bacteria are often at least temporarily found in low numbers as part of the resident flora.

What then does detection of a single pathogenic bacterium mean? The fact that most anaerobic infections are mixed infections involving more than one microorganism represents another difficulty. A single highly specific probe may fail to detect other bacteria involved. In consequence a whole battery of different probes would be required for complete analysis. The "reverse hybridization" procedure will be most appropriate to solve this problem (Dattagupta et al., 1989a, 1989b; Saiki et al., 1989). One should be aware, however, that unusual new agents or familiar pathogens in unusual sites might still be missed (Anonymus, 1989). Nevertheless, having a number of synthetic hybridization probes or amplification primers for specific detection of most prevalent pathogens would allow for a timely initiation of therapy. There is great hope that rapid development of improved sampling and labeling techniques, as well as the evolution of new hybridization or amplification stratagies, will lead to a widespread use in both research and clinical laboratories.

Acknowledgments. We are indebted to Dr. Iris Spiliopoulou-Sdoungou and Ms. Susanne Nagel for providing the *S. sputigena*-specific oligonucleotide probe and the original autoradiography (Figure 44.3), respectively. This work has been supported by grants from the Deutsche Forschungsgemeinschaft and the Bundesministerium für Forschung und Technologie.

References

Albandar, J.M. and I. Olsen. 1990. Nucleic acid probes as potential tools in oral microbial epidemiology. *Community Dent. Oral Epidemiol.* **18**:88–94.

Amann, R.I., L. Krumholz, and D. Stahl. 1990. Fluorescent-oligonucleotide probing of whole cells for determinative, phylogenetic, and environmental studies in microbiology. *J. Bacteriol.* **172**:762–770.

Anderson, M.L.M. and B.D. Young. 1985. Quantitative filter hybridization. *In: Nucleic Acid Hybridisation*, B.D. Hames and S.J. Higgins, eds., pp. 73–111. Oxford: IRL Press.

Anonymus. 1989. DNA technology and rapid diagnosis of infection. *Lancet* ii:897–898.

Arnold, Jr., L.J., P.W. Hammond, W.A. Wiese, and N.C. Nelson. 1989. Assay formats involving acridinium-ester-labeled DNA probes. *Clin. Chem.* **35**:1588–1594.

Attwood, G.T., R.A. Lockington, G.P. Xue, and J.D. Brooker. 1988. Use of a unique gene sequence as a probe to enumerate a strain *Bacteroides ruminicola* introduced into the rumen. *Appl. Environ. Microbiol.* **54**:534–539.

Barry, T., R. Powell, and F. Gannon. 1990. A general method to generate DNA probes for microorganisms. *BioTechnology* **8**:233–236.

Bascomb, S. 1987. Enzyme tests in bacterial identification. *Methods Microbiol.* **19**:105–160.

Boguslawski, S.J., D.E. Smith, M.A. Michalak, K.E. Mickelson, C.O. Yehle, W.L. Patterson, and R.J. Carrico. 1986. Characterization of monoclonal antibody to DNA:RNA and its application to immunodetection of hybrids. *J. Immunol. Methods* **89**:123–130.

Böttger, E.C. 1989. Rapid determination of bacterial ribosomal RNA sequences by direct sequencing of enzymatically amplified DNA. *FEMS Microbiol. Lett.* **65**:171–176.

Britten, R.J. and E.H. Davidson. 1985. Hybridisation strategy. *In: Nucleic Acid Hybridisation*, B.D. Hames and S.J. Higgins, eds., pp. 3–15, Oxford: IRL Press.

Britten, R.J., D.E. Graham, and B.R. Neufeld. 1974. Analysis of repeating DNA sequences by reassociation. *In: Methods in Enzymology*, Vol. 29, L. Grossman and K. Moldave, eds., pp. 363–418. London: Academic Press.

Brooker, J.D., R.A. Lockington, G.T. Attwood, and S. Miller. 1990. The use of gene and antibody probes in identification and enumeration of rumen bacterial species. *In: Gene Probes for Bacteria*, A.J.L. Macario, and E. Conway de Macario, eds., pp. 390–415. San Diego: Academic Press.

Chu, B.C.F., F.R. Kramer, and L.E. Orgel. 1986. Synthesis of an amplifiable reporter RNA for bioassays. *Nucleic Acids Res.* **14**:5591–5603.

Chuba, P.J., K. Pelz, G. Krekeler, T.S. De Isele, and U. Göbel. 1988. Synthetic oligodeoxynucleotide probes for the rapid detection of bacteria associated with human periodontitis. *J. Gen. Microbiol.* **134**:1931–1938.

Dahlen, P.O., P.J. Hurskainen, and T.N.E. Lövgren. 1989. Alternative labels in DNA hybridization. In: *Rapid Methods and Automation in Microbiology and Immunology*, A. Balows, R.C. Tilton, and A. Turano, eds., pp. 213–221. Brescia: Brixia Academic Press.

Dattagupta, N., E. Huguenel, P. Rae, and D. Crothers. 1989a. Nucleic acid hybridization: a rapid method for the diagnosis of infectious diseases. In: *Perspectives in Antiinfective Therapy*, G.G. Jackson, H.D. Schlumberger, and H.J. Zeiler, eds., pp. 241–247. Braunschweig: Vieweg.

Dattagupta, N., P.M.M. Rae, E.D. Huguenel, E. Carlson, A. Lyga, J.A. Shapiro, and J.P. Albarella. 1989b. Rapid identification of microorganisms by nucleic acid hybridization after labeling the test sample. *Anal. Biochem.* **177**:85–89.

DeLong, E.F., G.S. Wickham, and N.R. Pace. 1989. Phylogenetic stains: ribosomal RNA-based probes for the identification of single cells. *Science* **243**:1360–1363.

Dix, K., S.M. Watanabe, S. McArdle, D.I. Lee, C. Randolph, B. Moncla, and D.E. Schwartz. 1990. Species-specific oligodeoxynucleotide probes for the identification of periodontal bacteria. *J. Clin. Microbiol.* **28**:319–323.

Druel, B. and J. Freney. 1990. Utilisation des sondes nucléiques en bactériologie. *L'Eurobiologiste* **24**:127–154.

Finegold, S.M. and W.L. George, eds. 1989. *Anaerobic Infections in Humans*. San Diego: Academic Press.

French, C.K., E.D. Savitt, S.L. Simon, S.M. Eklund, M.C. Chen, L.C. Klotz, and K.K. Vaccaro. 1986. DNA probe detection of periodontal pathogens. *Oral Microbiol. Immunol.* **1**:58–62.

Gibbs, R.A. 1990. DNA amplification by the polymerase chain reaction. *Anal. Chem.* **62**:1202–1214.

Gillespie, D. and S. Spiegelman. 1965. A quantitative assay for DNA-RNA hybrids with DNA immobilized on membranes. *J. Mol. Biol.* **12**:829–842.

Giovannoni, S.J., E.F. DeLong, G.J. Olsen, and N.R. Pace. 1988. Phylogenetic group-specific oligodeoxynucleotide probes for identification of single microbial cells. *J. Bacteriol.* **170**:720–726.

Göbel, U.B., A. Geiser, and E.J. Stanbridge. 1987. Oligonucleotide probes complementary to variable regions of ribosomal RNA discriminate between *Mycoplasma* species. *J. General Microbiology* **133**:1969–1974.

Göbel, U.B., G. Krekeler, K. Pelz, and P.J. Chuba. 1989. rRNA probes for rapid identification of bacteria associated with periodontal disease. In: *Rapid Methods and Automation in Microbiology and Immunology*, A. Balows, R.C. Tilton, and A. Turano, eds., pp. 833–838. Brescia: Brixia Academic Press.

Grimont, P. 1984. DNA/DNA hybridization in bacterial taxonomy. In: *New Horizons in Microbiology* A. Sanna and G. Morace, eds., pp. 11–19. Amsterdam: Elsevier.

Groves, D.J. 1990. Nucleic acid probes for *Bacteroides* species. In: *Gene Probes for Bacteria*, A.J.L. Macario and E. Conway de Macario, eds., pp. 233–254. San Diego: Academic Press.

Groves, D.J. and V. Clark. 1987. Preparation of ribonucleic acid probes specific for *Bacteroides fragilis*. *Diagn. Microbiol. Infect. Dis.* **7**:273–278.

Guatelli, J.C., K.M. Whitfield, D.Y. Kwoh, K.J. Barringer, D.D. Richman, and T.R. Gingeras. 1990. Isothermal, in vitro amplification of nucleic acids by a multienzyme reaction modeled after retroviral replication. *Proc. Natl. Acad. Sci. USA* **87**:1874–1878.

Hall, B.D. and S. Spiegelman. 1961. Sequence complementarity of T2-DNA and T2-specific RNA. *Proc. Natl. Acad. Sci. USA* **47**:137–146.

Innis, M.A., D.H. Gelfand, J.J. Sninsky, and T.J. White, eds. 1990. *PCR Protocols. A Guide to Methods and Applications*. San Diego: Academic Press.

Jablonski, E.G. 1989. Detection systems for hybridization reactions. In: *DNA Probes for Infectious Diseases*, F.C. Tenover, ed., pp. 15–30. Boca Raton, Florida: CRC Press.

Johnson, J.L. 1985. DNA Reassociation and RNA hybridisation of bacterial nucleic acids. *Methods Microbiol.* **18**:33–74.

Keller, G.H. and M.M. Manak. 1989. *DNA probes*. New York: Stockton Press.

Kuritza, A.P. A.A. and Salyers. 1985. Use of a species-specific DNA hybridization probe for enumerating *Bacteroides vulgatus* in human feces. *Appl. Environ. Microbiol.* **50**:958–964.

Kuritza, A.P., P. Shaughnessy, and A.A. Salyers. 1986a. Enumeration of polysaccharide-degrading *Bacteroides* species in human feces by using species-specific DNA-probes. *Appl. Environ. Microbiol.* **51**:385–390.

Kuritza, A.P., C.E. Getty, P. Shaughnessy, R. Hesse, and A.A. Salyers. 1986b. DNA probes for identification of clinically important *Bacteroides* species. *J. Clin. Microbiol.* **23**:343–349.

Lane, D.J., B. Pace, G.J. Olsen, D.A. Stahl, M.L. Sogin, and N.R. Pace. 1985. Rapid determination of 16S ribosomal RNA sequences for phylogenetic analyses. *Proc. Natl. Acad. Sci. USA* **82**:6955–6959.

Lichter, P. and D.C. Ward. 1990. Is non-isotopic in situ hybridization finally coming of age? *Nature* (London))**345**:93–94.

Lizardi, P.M., C.E. Guerra, H. Lomeli, I. Tussie-Luna, and F.R. Kramer. 1988. Exponential amplification of recombinant-RNA hybridization probes. *Biotechnology* **6**:1197–1202.

Matthews, J.A. and L.J. Kricka. 1988. Analytical strategies for the use of DNA probes. *Anal. Biochem.* **169**:1–25.

Matthews, J.A., A. Batki, C. Hynds, and L.J. Kricka. 1985. Enhanced chemiluminescent method for the detection of DNA dot-hybridization assays. *Anal. Biochem.* **151**:205–209.

McCarthy, B.J. and E.T. Bolton. 1963. An approach to the measurement of genetic relatedness among organisms. *Proc. Natl. Acad. Sci. USA* **50**:156–164.

Moench T.R. 1987. *In situ* hybridization. *Mol. Cell. Probes* **1**:195–205.

Moncla, B.J., P. Braham, K. Dix, S. Watanabe, and D. Schwartz. 1990. Use of synthetic oligonucleotide DNA probes for the identification of *Bacteroides gingivalis*. *J. Clin. Microbiol.* **28**:324–327.

Moncla, B.J., L. Strockbine, P. Braham, J. Karlinseg, and M.C. Roberts. 1988. The use of whole-cell DNA probes for the identification of *Bacteroides intermedius* isolates in a dot blot assay. *J. Dent. Res.* **67**:1267–1270.

Morotomi, M., T. Ohno, and M. Mutai. 1988. Rapid and correct identification of intestinal *Bacteroides* spp. with chromosomal DNA probes by whole-cell dot blot hybridization. *Appl. Environ. Microbiol.* **54**:1158–1162.

Morrissey, D.V. and M.L. Collins. 1989. Nucleic acid hybridization assays employing dA-tailed capture probes. Single capture methods. *Mol. Cell. Probes* **3**:189–207.

Moseley, S.L., I. Huq, A.R.M.A. Alim, M. So, M. Samapour-Motalebi, and S. Falkow. 1980. Detection of enterotoxigenic *Escherichia coli* by DNA colony hybridization. *J. Infect. Dis.* **142**:892–898.

Mullis, K.B. and F.A. Faloona. 1987. Specific synthesis of DNA *in vitro* via a polymerase catalysed chain reaction. *Methods Enzymol.* **155**:335–350.

Murray, P.A. and C.K. French. 1989. DNA probe detection of periodontal pathogens. *In: New Biotechnology in Oral Research*, H.M. Myers, ed., pp. 33–53. Basel: Karger.

Murray, R.G.E., D.J. Brenner, R.R. Colwell, P. De Vos, M. Goodfellow, P.A.D. Grimont, N. Pfennig, E. Stackebrandt, and G.A. Zavarzin. 1990. Report of the ad hoc committee on approaches to taxonomy within the proteobacteria. *Int. J. Syst. Bacteriol.* **40**:213–215.

Phillips, I. 1990. New methods for identification of obligate anaerobes. *Rev. Infect. Dis.* **12**:S127–S132.

Pollard-Knight, D., A.C. Simmonds, A.P. Schaap, H. Akhavan, and M.A.W. Brady. 1990b. Nonradioactive DNA detection on Southern blots by enzymatically triggered chemiluminescence. *Anal. Biochem.* **185**:353–358.

Pollard-Knight, D., C.A. Read, M.J. Downes, L.A. Howard, M.R. Leadbetter, S.A. Pheby, E. McNaughton, A. Syms, and M.A.W. Brady. 1990a. Nonradioactive nucleic acid detection by enhanced chemiluminescence using probes directly labeled with horseradish peroxidase. *Anal. Biochem.* **185**:84–89.

Poxton, I.R. 1988. Methods for the immunological analysis of anaerobes. *In: Anaerobes Today*, J.M. Hardie and S.P. Borriello, eds., pp. 151–158. Chichester: Wiley.

Quivey, R.G., Jr. 1987. The use of DNA probes in dental diagnosis and therapy. *Adv. Dent. Res.* **1**:99–108.

Ranki, M., A. Palva, M. Virtanen, M. Laaksonen, and H. Söderlund. 1983. Sandwich hybridization as a convenient method for the detection of nucleic acids in crude samples. *Gene* (Amst.) **21**:77–85.

Roberts, M.C., B. Moncla, and G.E. Kenny. 1987. Chromosomal DNA probes for identification of *Bacteroides* species. *J. Gen. Microbiol.* **133**:1423–1430.

Roberts, M.C., S.L. Hillier, F.D. Schoenknecht, and K.K. Holmes. 1984. Nitrocellulose filter blots for species identification of *Mobiluncus curtissii* and *Mohiluncus mulieris*. *J. Clin. Microbiol.* **20**:826–827.

Roberts, M.C., S.L. Hillier, F.D. Schoenknecht, and K.K. Holmes. 1985. Comparison of Gram stain, DNA probe, and culture for the identification of species of *Mobiluncus* in female genital specimens. *J. Infect. Dis.* **152**:74–77.

Romond, C. 1989. New diagnostic methods for anaerobic bacteria. *Scand. J. Infect. Dis.* (Suppl.) **62**:35–40.

Saiki, R.K., P.S. Walsh, C.H. Levenson, and H.A. Erlich. 1989. Genetic analysis of amplified DNA with immobilized sequence-specific oligonucleotide probes. *Proc. Natl. Acad. Sci. USA* **86**:6230–6234.

Saiki, R.K., D.H. Gelfand, S. Stoffel, S.J. Scharf, R. Higuchi, G.T. Horn, K.B. Mullis, and H.A. Erlich. 1988. Primer-directed enzymatic amplification of DNA with a thermostable DNA polymerase. *Science* **239**:487–494.

Salyers, A.A., S.P. Lynn, and J.F. Gardner. 1983. Use of randomly cloned DNA fragments for identification of *Bacteroides thetaiotaomicron*. *J. Bacteriol.* **154**:287–293.

Savitt, E.D., M.N. Strzempko, K.K. Vaccaro, W.J. Peros, and C.K. French. 1988. Comparison of cultural methods and DNA probe analyses for the detection of *Actinobacillus actinomycetemcomitans*, *Bacteroides gingivalis*, and *Bacteroides intermedius* in subgingival plaque samples. *J. Periodontol.* **59**:431–438.

Schaap, A.P., H. Akhavan, and L.J. Romano. 1989. Chemiluminescent substrates for alkaline phosphatase: application to ultrasensitive enzyme-linked immunoassays and DNA probes. *Clin. Chem.* **35**:1863–1864.

Shah, H.N. and M.D. Collins. 1988. Proposal for reclassification of *Bacteroides asaccharolyticus*, *Bacteroides gingivalis* and *Bacteroides endodontalis* in a new genus, *Porphyromonas*. *Int. J. Syst. Bacteriol.* **38**:128–131.

Shah, H.N. and D.M. Collins. 1990. *Prevotella*, a new genus to include *Bacteroides meleninogenicus* and related species formerly classified in the genus *Bacteroides*. *Int. J. Syst. Bacteriol.* **40**: 205–208.

Smith, G.L.F., S.S. Socransky, and C.M. Smith. 1989. Non-isotopic DNA probes for the identification of subgingival microorganisms. *Oral Microbiol. Immunol.* **4**:41–46.

Stahl, D.A., B. Flesher, H.R. Mansfield, and L. Montgomery. 1988. Use of phylogenetically based hybridization probes for studies of ruminal microbial ecology. *Appl. Environ. Microbiol.* **54**:1079–1084.

Steinman, C.R. 1975. Specific detection and semiquantitation of micro-organisms in tissue nucleic acid hybridization. I. Characterization of the method and application to model systems. *J. Lab. Clin. Med.* **86**:164–174.

Strzempko, M.N., S.L. Simon, C.K. French, J.A. Lippke, F.F. Raia, E.D. Savitt, and K.K. Vaccaro. 1987. A cross-reactivity study of whole genomic DNA probes for *Haemophilus actinomycetemcomitans*, *Bacteroides intermedius*, and *Bacteroides gingivalis*. *J. Dent. Res.* **66**:1543–1546.

Sutter, V.L., D.M. Citron, M.A.C. Edelstein, and S.M. Finegold. 1985. *Wadsworth Anaerobic Bacteriology Manual*, 4th Ed. Belmont, California: Star Publications.

Tanner, A., H.D. Bouldin, and M.F.J. Maiden. 1989. Newly delineated periodontal pathogens with special reference to *Selenomonas* species. *Infection* **17**:182–187.

Tenover, F.C. 1988. Diagnostic deoxyribonucleic acid probes for infectious diseases. *Clin. Microbiol. Rev.* **1**:82–101.

Thompson, J.D., S. Decker, D. Haines, R.S. Collins, M. Feild, and M. Gillespie. 1989. Enzymatic amplification of RNA purified from crude cell lysate by reversible target capture. *Clin. Chem.* **35**:1878–1881.

Urdea, M.S., J.A. Running, T. Horn, J. Clyne, L. Ku, and B.D. Warner. 1987. A novel method for the rapid detection of specific nucleotide sequences in crude biological samples without blotting or radioactivity; application to the analysis of hapatitis B virus in human serum. *Gene* (Amst.) **61**:253–264.

Urdea, M.S., B.D. Warner, J.A. Running, M. Stempien, J. Clyne, and T. Horn. 1988. A comparison of non-radioisotopic hybridization assay methods using fluorescent, chemiluminescent and enzyme labeled synthetic oligodesoxynucleotide probes. *Nucleic Acids Res.* **16**:4937–4956.

Willis, A.T. 1989. History. *In*: *Anaerobic Infections in Humans*, S.M. Finegold and W.L. George, eds., pp. 1–22. San Diego: Academic Press.

Wilson, K.H., R. Blitchington, P. Shah, G. McDonald, R.D. Gilmore, and L.P. Mallavia. 1989. Probe directed at a segment of *Rickettsia rickettsii* rRNA amplified with polymerase chain reaction. *J. Clin. Microbiol.* **27**:2692–2696.

Woese, C.R. 1987. Bacterial evolution. *Microbiol. Rev.* **51**:221–271.

Wu, D.Y. and R.B. Wallace. 1989. The ligation amplification reaction (LAR)-amplification of specific DNA sequences using sequential rounds of template-dependent ligation. *Genomics* **4**:560–569.

Yolken, R.H. 1988. Nucleic acids or immunoglobulins: which are the molecular probes of the future? *Mol. Cell. Probes* **2**:87–96.

Young, B.D. and M.L.M. Anderson. 1985. Quantitative analysis of solution hybridisation. *In*: *Nucleic Acid Hybridisation*, B.D. Hames and S.J. Higgins, eds., pp. 47–71. Oxford: IRL Press.

Zappa, U., M. Reinking-Zappa, H. Graf, R. Gmür, and E. Savitt. 1990. Comparison of serological and DNA probe analysis for detection of suspected periodontal pathogens in subgingival plaque samples. *Arch. Oral Biol.* **35** Suppl.: 161–164.

45

Molecular Biology of Bile Acid 7α-Dehydroxylation in an Intestinal *Eubacterium* Species

Darrell H. Mallonee and Phillip B. Hylemon

45.1 Introduction

The gastrointestinal tract of adult humans contains as many as 10^{14} bacteria, which may represent 400 to 500 different species (Moore and Holdeman, 1974; Holdeman et al., 1976). More than 99% of these bacteria are obligate anaerobes. The majority (two-thirds) belong to the following species: *Bacteroides* spp., *Eubacterium aerofaciens*, *E. eligens*, *E. biforme*, *E. rectale*, *Fusobacterium prausnitzii*, *Bifidobacterium adolescentis*, *B. longum*, *Peptostreptococcus productus*, and *Gemmiger formicilis* (Moore and Holdeman, 1974; Holdeman et al., 1976). These bacteria are capable of carrying out a wide variety of biotransformations of both endogenous and exogenous compounds (Goldman, 1978). In contrast to oxidative and conjugative biotransformations that are carried out in the liver, the intestinal microflora tend to hydrolyze and reduce compounds that enter the gastrointestinal tract.

Bile acids are synthesized from cholesterol in the liver and represent a major pathway for elimination of cholesterol from the body (Vlahcevic et al., 1990). The two primary bile acids that are synthesized from cholesterol in man are cholic acid and chenodeoxycholic acid (Figure 45.1). These bile acids are then conjugated to either glycine or taurine before secretion into the biliary bile (Vlahcevic et al., 1990). Bile salts are primarily absorbed by active transport systems in the terminal ileum and returned to the liver via the portal blood. However, approximately 20% to 25% of the biliary bile

acid pool is deoxycholic acid, which is a secondary bile acid generated from cholic acid by intestinal bacteria (Figure 45.1). Indeed, the final composition of biliary bile acids in man is determined by liver biosynthetic enzymes and intestinal bacterial biotransformations.

During the enterohepatic circulation of bile salts, several hundred milligrams per day escape and enter the large bowel where they are metabolized into a variety of products (Figure 45.2). Known microbial biotransformations include the hydrolysis of the glycine and taurine conjugates yielding free bile acids, the oxidation of hydroxy groups at C-3, C-7, and C-12, and the reduction of oxo bile acids yielding either α- or β-hydroxy groups (Figure 45.2). Quantitatively the most important bile acid biotransformation is the 7α-dehydroxylation of cholic acid and chenodeoxycholic acid yielding deoxycholic acid and lithocholic acid, respectively. Certain members of the intestinal microflora are also capable of the direct 7β-dehydroxylation of ursodeoxycholic acid (Figure 45.2) and capable of transforming 3-sulfated bile acids into a number of metabolites (Hylemon, 1985).

The literature on bile acid metabolism by the intestinal microflora has been reviewed (Hylemon and Glass, 1983; Hylemon, 1985; Macdonald et al., 1983). The major fecal bile acids are deoxycholic acid and lithocholic acid (Ali et al., 1966). Moreover, most fecal bile acids are in the unconjugated form (Eneroth et al., 1966). Most intestinal bacteria that are capable of the 7α-dehydroxylation of primary bile acids have

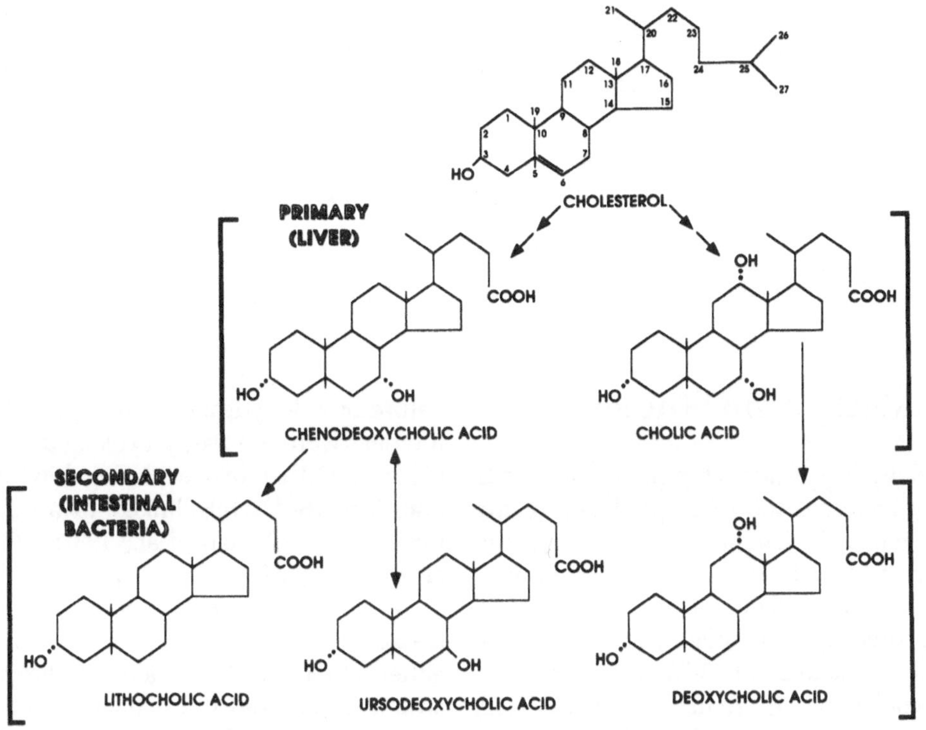

FIGURE 45.1 Structures of bile acids in man and   indicated sites of biosynthesis.

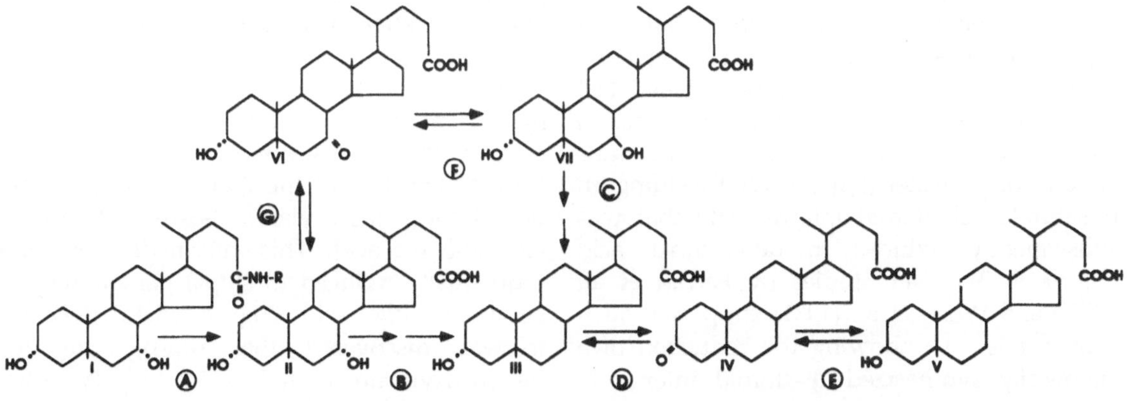

FIGURE 45.2 Bile salt biotransformations carried out by intestinal microflora. Abbreviations for enzymes (letters) and steroids (numbers): A, bile salt hydrolase; B, 7α-dehydroxylase; C, 7β-dehydroxylase; D, 3α-hydroxysteroid dehydrogenase; E, 3β-hydroxysteroid dehydrogenase; F, 7β-hydroxysteroid dehydrogenase; G, 7α-hydroxysteroid dehydrogenase; I, conjugated chenodeoxycholic acid; II, chenodeoxycholic acid; III, lithocholic acid; IV, 3-oxo-lithocholic acid; V, isolithocholic acid; VI, 7-oxo-chenodeoxycholic acid; VII, ursodeoxycholic acid.

been identified as members of the genera *Clostridium* or *Eubacterium* (Hylemon and Glass, 1983). However, the number of intestinal bacteria capable of 7α-dehydroxylation represents a relatively small fraction of the total intestinal microflora (10^3 to 10^5 bacteria per gram wet weight feces) (Stellwag and Hylemon, 1979; Ferrari et al., 1980).

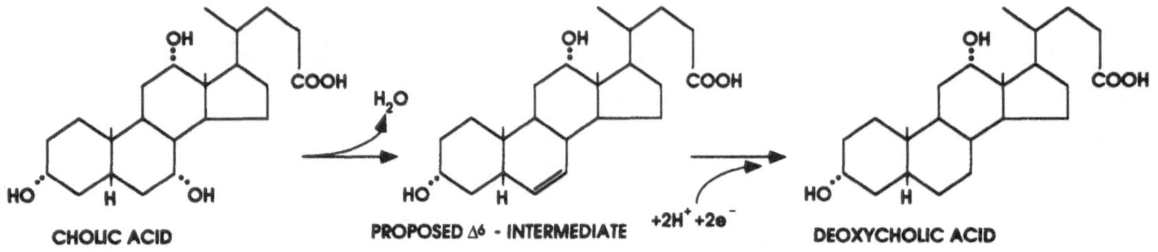

FIGURE 45.3 Proposed reaction mechanism for 7α-dehydroxylation of cholic acid (Samuelsson, 1960).

45.2 Mechanism Studies

The 7α-dehydroxylation of primary bile acids markedly increases the hydrophobicity of the bile acid molecule. There is a decrease in the solubility and an alteration of the critical micellar concentration (Vlahcevic et al., 1990). In addition, hydrophobic bile acids are more powerful regulators of 3-hydroxy-3-methylglutaryl-coenzyme A reductase (HMG-CoA reductase) and cholesterol 7α-hydroxylase, the rate-limiting enzymes in the cholesterol and bile acid biosynthetic pathways, respectively (Heuman et al., 1988).

The mechanism of 7α-dehydroxylation of cholic acid was first investigated in vivo by Samuelsson (1960) using doubly labeled (^{3}H and ^{14}C) cholic acid fed to rats. The reaction was proposed to proceed by a diaxial trans elimination of water (6βH, 7αOH) yielding an unsaturated Δ^6-steroid intermediate that was subsequently reduced to deoxycholic acid (Figure 45.3). Later studies by Ferrari et al. (1977) and White et al. (1981) supported such a mechanism by showing the reduction of a chemically synthesized Δ^6-steroid intermediate (3α, 12α-dihydroxy-5β-chol-6-enoic acid) by whole cells or cell extracts of *Clostridium bifermentans* and *Eubacterium* sp. VPI 12708, respectively. Studies by White et al. (1982) demonstrated that *Eubacterium* sp. VPI 12708 was also capable of the direct 7β-dehydroxylation of ursodeoxycholic acid (see Figure 45.2).

Additional studies showed that both 7α- and 7β-dehydroxylation activities were stimulated by NAD+ and inhibited by NADH in dialyzed cell extracts of *Eubacterium* sp. VPI 12708. These results were surprising in view of the reductive

nature of 7-dehydroxylation (Figure 45.3), and the data were not easily explained until it was discovered that bile acid ring biotransformations occurred while linked to an adenosine nucleotide (Coleman et al., 1987a). Gas liquid chromatography-mass spectrophotometry showed the accumulation of oxidized bile acids (3-oxo-12α-hydroxy-4-cholenoic acid) when either cholic acid or deoxycholic acid was added to dialyzed cell extracts prepared from cultures induced by primary bile acids. This finding and the data that suggest that 3-oxo-7α-hydroxy-Δ^4-steroids easily lose the 7α-hydroxy group led to a new proposed mechanism of 7α-dehydroxylation in *Eubacterium* sp. VPI 12708 (Figure 45.4).

The new mechanism is hypothesized to start by an initial oxidation of the 3α-hydroxy group and of the C4-C5 bond that may require NAD+. The loss of the 7α-hydroxy group by dehydration then yields 3-oxo-12α-hydroxy-4,6-cholenoic acid. This intermediate is then sequentially reduced (3 steps) generating the secondary bile acid (Figure 45.4). Recently it was discovered that during the 7α-dehydroxylation of cholic acid there is an accumulation of allo-deoxycholic acid (3α-12α-dihydroxy-5α-cholanoic acid) in cell extracts of *Eubacterium* sp. VPI 12708 (Hylemon et al., 1991). These results show that the reduction of 3-oxo-12α-hydroxy-4-cholenoic acid can yield either the 5β or 5α configuration. These data would suggest induction of specific 5α- and 5β-Δ^4-steroid oxidoreductases by cholic acid in this organism. Moreover, the formation of allo-deoxycholic acid may be a branch of the cholic acid 7α-dehydroxylation pathway.

Support for the new mechanism of cholic

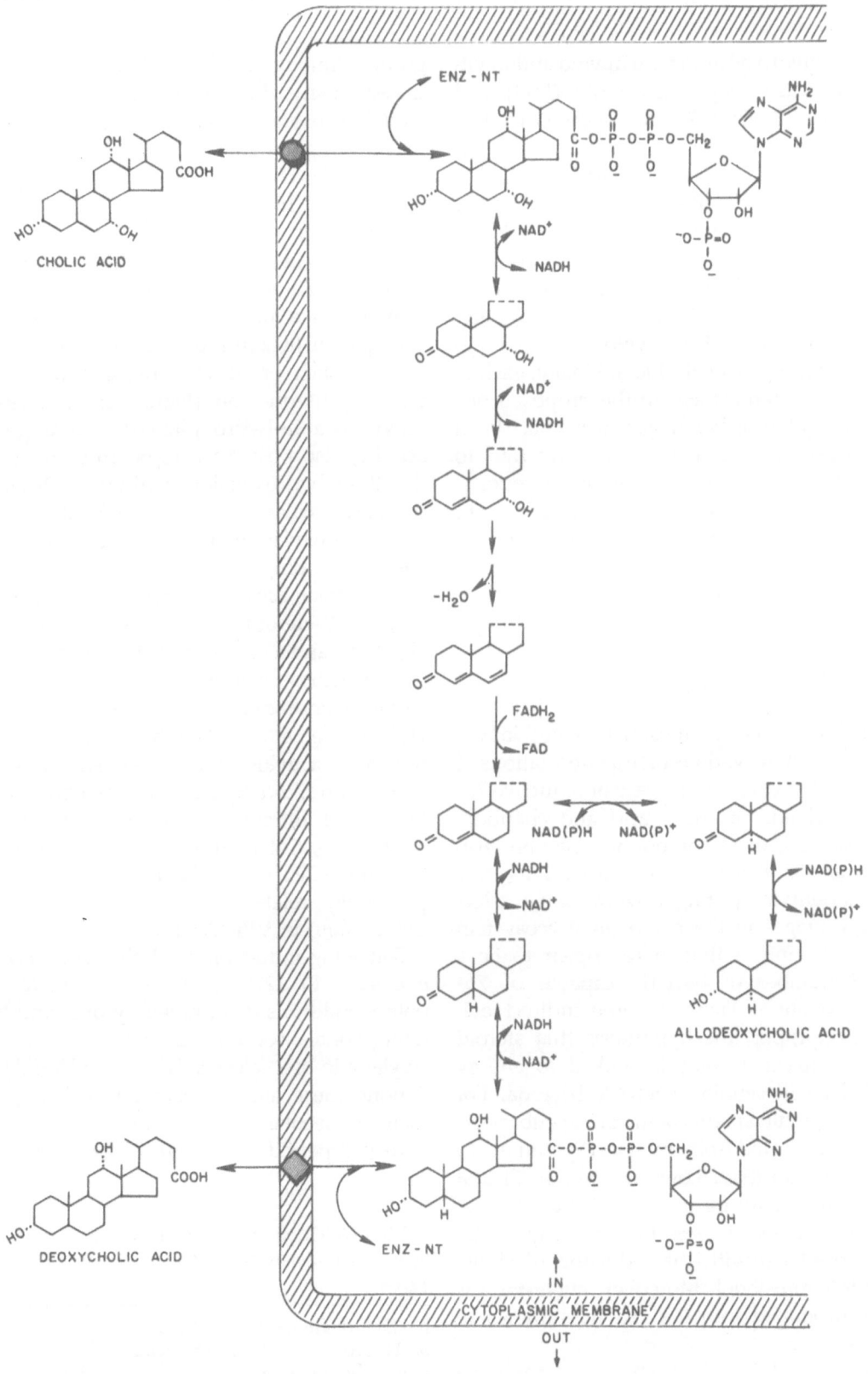

FIGURE 45.4 Proposed mechanism for bile acid 7α-dehydroxylation in *Eubacterium* sp. VPI 12708

(Coleman et al., 1988; Hylemon et al., 1991).

acid 7α-dehydroxylation both in vivo and in vitro was gained through the use of [5β-^{3}H]- and [^{14}C]-labeled cholic acid. It was shown that during the conversion of a mixture of [5β-^{3}H]-cholic acid and [24-^{14}C]-cholic acid to deoxycholic acid there was a differential loss (>80%) of the [5β-^{3}H] indicating the formation of a 3-oxo-Δ^4-steroid intermediate (Figure 45.4). The loss of the [5β-^{3}H] of cholic acid occurred in cell extracts or whole cells of all 7α-dehydroxylating bacteria tested as well as in vivo when given to humans (Björkhem et al., 1989). Finally, each of the various intermediates in the proposed bile acid 7α-dehydroxylation pathway has been synthesized and shown to be converted to deoxycholic acid in whole cells or cell extracts of Eubacterium sp. VPI 12708 (Hylemon et al., 1991). Therefore, cholic acid appears to induce several enzymes that are involved in the pathway of bile acid 7α-dehydroxylation.

45.3   Physiology

It is not yet clear why intestinal anaerobic bacteria carry out the various biotransformations of bile acids. However, in the case of reductive 7α-dehydroxylation of cholic acid and chenodeoxycholic acid, these compounds may be serving as ancillary electron acceptors. This may give 7α-dehydroxylating bacteria some selective advantages in the competitive ecosystem of the intestine in that there appear to be a limited number of bacteria capable of 7α-dehydroxylation. There is some indirect evidence to support the hypothesis that steroid reductive reactions may be linked to energy metabolism in certain anaerobic bacteria. For example, certain species of intestinal eubacteria have been isolated from animals (Eyssen et al., 1973) and man (Sadzikowski et al., 1977) that require cholesterol for growth. The amount of growth of many of these bacteria is quantitatively correlated with the reduction of cholesterol to coprostanol. Moreover, cholesterol is not incorporated into the cell membranes of these bacteria.

The bile acid 7α-dehydroxylation activity expressed by Eubacterium sp. VPI 12708 is induced by culturing in the presence of unconjugated C-24 bile acids that possess a 7α-hydroxyl group. Among the bile acids promoting induced activity, cholic acid was shown to induce 7α-dehydroxylation activity to a level three-to fivefold higher than chenodeoxycholic acid when using whole-cell assays with the Eubacterium strain (White et al., 1980).

Eubacterium sp. VPI 12708 displays several other bile acid-inducible activities that may be related to 7α-dehydroxylation. One of these activities is a bile acid-adenosine nucleotide conjugation (ligation) activity (Coleman et al., 1987a) which produces a reactive adenylated bile acid intermediate (Figure 45.4). A second activity is a 3α-hydroxysteroid dehydrogenase activity, also postulated to be an early step in the 7α-dehydroxylation pathway. Other inducible activities which may be involved in the 7α-dehydroxylation pathway include a primary bile acid binding activity, an NADH:flavin oxidoreductase activity (Lipsky and Hylemon, 1980), a Δ^6-steroid reductase activity (White et al., 1981), and a series of Δ^4-steroid oxidoreductase activities including the 5α- and 5β-Δ^4-steroid oxidoreductase activities (Figure 45.4; Hylemon et al., 1991). The ligation and Δ^6-reductase activities have been shown to co-elute with 7α-dehydroxylase activity during HPLC gel filtration chromatography (White et al., 1983; Coleman et al., 1987a). Table 45.1 presents a list of bile acid-inducible activities potentially related to 7α-dehydroxylation in Eubacterium sp. VPI 12708.

Following induction by cholic acid, Eubacterium sp. VPI 12708 synthesizes several new polypeptides, as determined by one- and two-dimensional sodium dodecyl sulfate-polyacrylamide gel electrophoresis (SDS-PAGE). Among these are polypeptides with approximate molecular weights of 77,000, 56,000 (two polypeptides), 45,000, 27,000 and 23,500

Table 45.1 Cholic acid-inducible enzyme activities associated with Eubacterium sp. VPI 12708

NADH: flavin oxidoreductase
3α-Hydroxysteroid oxidoreductase
Bile acid-CoA ligation (conjugation) activity
Primary bile acid binding activity
Δ^6-Steroid reductase
Δ^4-Steroid oxidoreductases

(Paone and Hylemon, 1984; White et al., 1981). Four of these cholic acid-inducible polypeptides have been purified to homogeneity. These include a 27,000-M_r polypeptide purified by Coleman et al. (1987b), a 45,000-M_r polypeptide purified by White et al. (1988a), a 23,500-M_r polypeptide purified by Mallonee et al. (1990), and a 72,000-M_r polypeptide purified by Franklund and Hylemon (unpublished data). Recent unpublished data has shown that the 27,000 M_r polypeptide has 3α-dehydrogenase activity and that the 72,000-M_r polypeptide is an NADH:flavin oxidoreductase. All four of these polypeptides react on Western blots to antibodies produced to HPLC fractions containing 7α-dehydroxylation activity (Paone and Hylemon, 1984; White et al., 1988a). The genes coding for these three polypeptides have also been cloned and sequenced; the results are discussed in the next section.

45.4   Genetics

Cloning of the *baiA* Gene Family

The lack of a genetic transfer system for *Eubacterium* sp. VPI 12708 has necessitated the use of a molecular cloning and sequencing approach to the study of the genetic organization of elements related to bile acid 7α-dehydroxylation. The genes coding for several cholic acid-inducible polypeptides from the *Eubacterium* strain have been cloned and sequenced. Included among these are the genes coding for three copies of the 27,000-M_r bile acid-inducible polypeptide, which have been cloned on three separate DNA fragments. The first copy (*baiA1*) was cloned on overlapping 2.2-kb *EcoRI* and 1.1-kb *TaqI* restriction fragments, utilizing an initial oligonucleotide probe made from sequence information derived from the purified 27,000-M_r polypeptide (Coleman et al., 1987b, 1988). Southern blot analysis of DNA restriction fragments from the *Eubacterium* strain using a 23-mer oligonucleotide probe made to a sequence internal to the *baiA1* gene revealed the presence of three hybridizing bands of similar intensity. A second copy of the *baiA* gene family (*baiA2*) was cloned on a 2.9-kb *EcoRI* restriction fragment (White et al., 1988b). The third copy (*baiA3*) was cloned on an 11.2-kb DNA fragment (Gopal-Srivastava et al., 1990). Restriction maps of the three *baiA* genes and surrounding areas are presented in Figure 45.5.

The *baiA1* and *baiA3* genes share 100% nucleotide sequence identity within the coding region of 747 bases (Gopal-Srivastava et al., 1990). This sequence identity remains for 930 bases in the upstream (5') direction from the initiation codons for the *baiA* genes and for at least 325

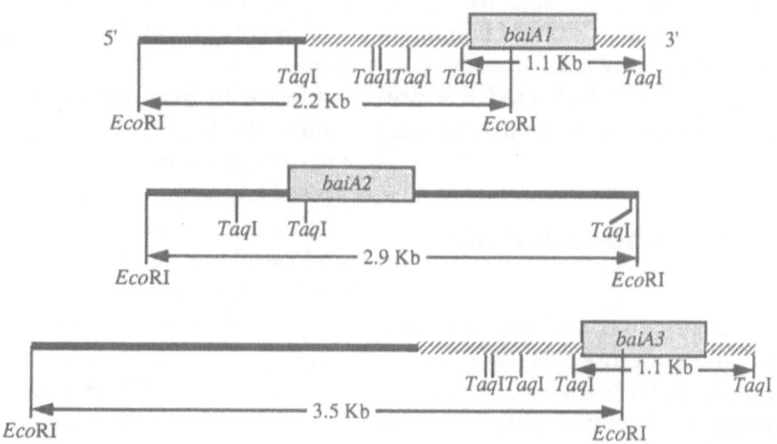

FIGURE 45.5   Partial restriction maps for the *baiA1*, *baiA2*, and *baiA3* genes and surrounding regions from *Eubacterium* sp. VPI 12708 (Coleman et al., 1988; White et al., 1988b; Gopal-Srivastava et al., 1990). The *baiA* genes are boxed; region of homology between the *baiA1* and *baiA3* clones is highlighted.

bases in the downstream (3') direction from the stop codons. Included in the identical 5' regions are the proposed promoter structures for the *baiA* genes and a second open-reading frame.

Although the point of sequence divergence was identified in the 5' direction from the *baiA1* and *baiA3* genes, the 3' point of divergence could not be determined despite several attempts to obtain clones in this region. The *baiA2* gene shares 81% nucleotide sequence identity with the *baiA1* and *baiA3* genes, while the polypeptide encoded by the *baiA2* gene shares 92% amino acid sequence identity with the other two 27,000-M_r polypeptides (White et al., 1988b; Gopal-Srivastava et al., 1990). The nucleotide sequences surrounding the *baiA2* gene share no homology with the sequences surrounding the other two *baiA* genes, except for a 15-bp pair 5' segment containing the putative ribosome-binding sites for the genes. Southern blots were inconclusive in determining the relative locations on the chromosome for the three copies of the *baiA* gene, and the mechanism of duplication for the two identical copies (*baiA1* and *baiA3*) could not be determined. However, duplication of the *baiA1* and *baiA3* genes was not affected by culturing the bacteria in the presence or absence of cholic acid (Gopal-Srivastava et al., 1990).

Northern blot analysis of transcripts containing the *baiA1* and *baiA3* coding regions revealed small (~950 bases) monocistronic mRNAs (Coleman et al., 1988). However, the *baiA2* gene is part of a large bile acid-inducible operon. Northern blot analysis of the transcript containing the *baiA2* gene has indicated the presence of an mRNA species more than 6-kb long (White et al., 1988a).

Cloning of a Bile Acid-Inducible Operon

The 2.9-kb *Eco*RI fragment containing the *baiA2* gene also contained an open-reading frame coding for a polypeptide whose N-terminal sequence matched the sequence obtained from the purified 45,000-M_r bile acid-inducible polypeptide (White et al., 1988a). This open-reading frame extended from a point just downstream from the *baiA2* gene to the 3' end

of the *Eco*RI fragment. However, the 3' end of the gene coding for the 45,000-M_r polypeptide was not present on the *Eco*RI fragment. An open-reading frame potentially coding for a polypeptide having a calculated M_r of 19,514 was found immediately upstream from the *baiA2* gene (Mallonee et al., 1990). The N-terminal sequence of this hypothetical polypeptide matched the N-terminal sequence obtained from the purified 23,500-M_r (as determined by SDS-PAGE) bile acid-inducible polypeptide. The purified polypeptide and the hypothetical polypeptide encoded by the open-reading frame were therefore judged to be identical. Immediately upstream from the open reading frame coding for the 19,500-M_r polypeptide was what appeared to be the 3' end of a fourth open-reading frame.

The remaining part of the bile acid-inducible operon in the 5' direction was cloned on three additional overlapping restriction fragments (Figure 45.6; Mallonee et al., 1990). Proceeding in the 5' direction from the 2.9-kb *Eco*RI fragment, the clones obtained included a 300-bp *Nru*I fragment, a 3.0-kb *Kpn*I fragment, and finally a 2.4-kb *Eco*RI fragment. Difficulties in obtaining large clones in this region prevented the cloning of a large DNA fragment that could potentially contain the entire operon. Sequence analysis of these four overlapping fragments revealed the presence of a fourth, fifth, and sixth open reading frame on the operon. These open-reading frames could potentially code for polypeptides with calculated M_rs of (5' to 3') 59,513, 58,272, and from 9,099 to 11,447 depending on the correct initiation codon.

Upstream from the open-reading frame encoding the hypothetical 58,000-M_r polypeptide was a putative promoter structure (Mallonee et al., 1990). This putative promoter was very similar to the putative promoters for the *baiA1* and *baiA3* genes, sharing more than 69% nucleotide sequence identity over a stretch of 130 bases. Preliminary data suggest a protein-binding operator region is associated with the promoter of this bile acid-inducible operon.

Because it was difficult to obtain clones for the 3' end of the operon, amplification of a 300-bp overlapping *Hae*III fragment by inverse polymerase chain reaction was performed (Mallonee et al., 1990). Amplified fragments

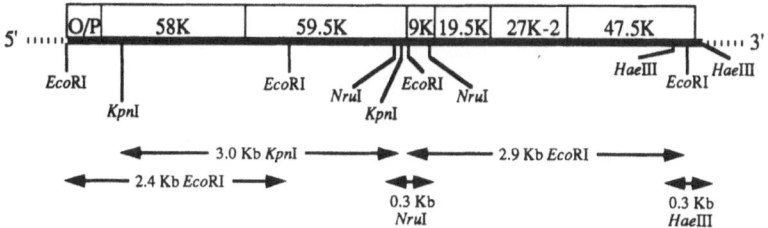

FIGURE 45.6   Partial restriction map and locations of open reading frames for the proposed bile acid-inducible operon from *Eubacterium* sp. VPI 12708 (Mallonee et al., 1990). Operator-promoter region, O/P; arrows indicate regions used for cloning or PCR sequencing.

were then used for direct sequencing, which revealed the 3' end of the gene encoding the 45,000-M_r polypeptide and resulted in an open-reading frame coding for a polypeptide with a calculated M_r of 47,448. Sequence information obtained was insufficient to determine if an additional open-reading frame existed downstream (3') from the gene encoding the 47,500-M_r polypeptide.

The open-reading frames on the operon have been labeled as bile acid-inducible (*bai*) genes as with the *BaiA* genes (Gopal-Srivastava et al., 1990; Mallonee et al., 1990). The genes coding for the hypothetical 58,000, 59,500, and 9,000-M_r polypeptides are designated *baiB*, *baiC*, and *baiD*, respectively. The genes coding for the 19,500 and 47,500-M_r bile acid-inducible polypeptides are designated *baiE* and *baiF*, respectively.

It is hypothesized that most or all of these bile acid-inducible polypeptides are involved in the multistep 7α-dehydroxylation pathway (see Figure 45.4). Antibodies prepared against the 27,000-M_r polypeptides have been shown to inhibit 7α-dehydroxylation activity (Paone and Hylemon, 1984). Antibodies prepared against HPLC fractions containing 7α-dehydroxylation activity (Paone and Hylemon, 1984) have also been shown to react against the 19,500-, 27,000-, and 47,500-M_r polypeptides, and possibly against the 58,000- and 59,500-M_r polypeptides (Coleman et al., 1987b; White et al., 1988a; Mallonee et al., 1990).

Data bank searches for amino acid sequences similar to the polypeptides encoded by the *bai* genes revealed no significantly similar sequences to the 59,500-, 47,500-, 19,500-, or 9,000- to 11,500-M_r polypeptides (Mallonee et al., 1990). The amino acid sequence for the 27,000-M_r polypeptides showed significant homology to several alcohol/polyol/sugar dehydrogenases, especially in the proposed pyridine nucleotide binding domains (Coleman et al., 1988; White et al., 1988b). The 27,000-M_r polypeptides catalyze the oxidation of the 3α-hydroxy group of bile acid substrates in the 7α-dehydroxylation pathway (see Figure 45.4). A comparison of the N-terminal amino acid sequences of the three 27,000-M_r polypeptides and several homologous polypeptides is shown in Figure 45.7.

Significant homology was also obtained between the amino acid sequence for the 58,000-M_r polypeptide and several polypeptides involved in the adenylation of cyclic carboxylated compounds (Mallonee et al., 1990). These homologous polypeptides include 4-coumarate:CoA ligase from *Petroselinum crispum* (parsley; Lozoya et al., 1988), tyrocidine synthetase 1 (*tycA* gene product) from *Bacillus brevis* (Weckermann et al., 1988), luciferase from *Photinus pyralis* (North American firefly; de Wet et al., 1987), and a polypeptide (*entE* gene product) involved in the activation of 2,3-dihydroxybenzoate during enterobactin (enterochelin) synthesis in *Escherichia coli* (Liu et al., 1989). The sequence homologies for these polypeptides range from 19% to 31% amino acid sequence identity over a span of 118 to 521 amino acids with optimum alignment, including gaps. It was therefore postulated that the 58,000-M_r polypeptide may be involved in the formation of the bile acid-adenosine nucleotide described by Coleman et al. (1987a). Recent data has shown that the 58,000-M_r polypeptide is a subunit of an α2 enzyme that catalyzes the formation of bile acid-coenzyme A conjugates (D.H. Mallonee et al., unpublished data). Table

βA

```
ADH1  1  Ser Phe Thr Leu Thr Asn Lys Asn Val Ile Phe Val Ala Gly Leu Gly Gly Ile Gly Leu Asp
ADH2  1      Met Ala Ile Ala Asn Lys Asn Ile Ile Phe Val Ala Gly Leu Gly Gly Ile Gly Phe Asp
GDHA  2  Tyr Thr Asp Leu Lys Asp Lys Val Val Val Ile Thr Gly Gly Ser Thr Gly Leu Gly Arg Ala
GDHB  2  Tyr Lys Asp Leu Glu Gly Lys Val Val Val Ile Thr Gly Ser Ser Thr Gly Leu Gly Lys Ser
RDH   9  Asn Thr Ser Leu Ser Gly Lys Val Ala Ala Ile Thr Gly Ala Ala Ser Gly Ile Gly Leu Gly
HSDH  1      Met Arg Leu Lys Asp Lys Val Ile Leu Val Thr Ala Ser Thr Arg Gly Ile Gly Leu Ala
27-1  1  Met Lys Leu Gln Asp Lys Ile Thr Ile Ile Thr Gly Gly Thr Arg Gly Ile Gly Phe Ala
27-2  1  Met Asn Leu Val Gln Asp Lys Val Thr Ile Ile Thr Gly Gly Thr Arg Gly Ile Gly Phe Ala
27-3  1  Met Lys Leu Val Gln Asp Lys Ile Thr Ile Ile Thr Gly Gly Thr Arg Gly Ile Gly Phe Ala
```

βB

```
ADH1  22  Thr Ser Lys Glu Leu Leu Lys Arg Asp Leu Lys Asn Leu Val Ile Leu Asp Arg Ile Glu Asn
ADH2  21  Thr Ser Arg Glu Ile Val Lys Ser Gly Pro Lys Asn Leu Val Ile Leu Asp Arg Ile Glu Asn
GDHA  23  Met Ala Val Arg Phe Gly Gln Glu Glu Ala Lys Val Val Ile Asn Tyr Tyr Asn Asn Glu Glu
GDHB  23  Met Ala Ile Arg Phe Ala Thr Glu Lys Ala Lys Val Val Val Asn Tyr Arg Ser Lys Glu Asp
RDH   30  Cys Ala Arg Thr Leu Leu Gly Ala Gly Ala Lys Val Val Leu Ile Asp Arg Glu Gly Glu Lys
HSDH  21  Ile Ala Gln Ala Cys Ala Lys Glu Gly Ala Lys Val Tyr Met Gly Ala Arg Asn Leu Glu Arg
27-1  22  Ala Ala Lys Leu Phe Ile Glu Asn Gly ALa Lys Val Ser Ile Phe Gly Glu Thr Gln Glu Glu
27-2  22  Ala Ala Lys Ile Phe Ile Asp Asn Gly ALa Lys Val Ser Ile Phe Gly Glu Thr Gln Glu Glu
27-3  22  Ala Ala Lys Leu Phe Ile Glu Asn Gly ALa Lys Val Ser Ile Phe Gly Glu Thr Gln Glu Glu
```

FIGURE 45.7 Comparison of amino acid sequences for polypeptides encoded by three *baiA* genes from *Eubacterium* sp. VPI 12708 (27-1, 27-2, and 27-3) with sequences for several dehydrogenases from short non-zinc alcohol/polyol/sugar class of dehydrogenases (Jornvall et al., 1984). These sequences contain proposed pyridine nucleotide binding sites for the enzymes (Jornvall et al., 1984; Wierenga et al., 1986). Boxed residues are composed of functionally similar amino acids as described by Wierenga et al. (1986). Numbers indicate residue number of first amino acid on line. Abbreviations: ADH1, *Drosophila melanogaster* alcohol dehydrogenase (Thatcher, 1980); ADH2, *Drosophila mojavensis* alcohol dehydrogenase (Atkinson et al., 1988); GDHA, *Bacillus megaterium* glucose dehydrogenase A (Heilmann et al., 1988); GDHB, *B. megaterium* glucose dehydrogenase B (Jany et al., 1984); RDH, *Klebsiella aerogenes* ribitol dehydrogenase (Morris et al., 1974); HSDH, 7α-hydroxysteroid dehydrogenase from *Eubacterium* sp. VPI 12708 (Baron et al., 1991; Franklund et al., 1990).

45.2 presents a list of proposed cholic acid-inducible polypeptides from *Eubacterium* sp. VPI 12708.

45.5 Summary and Conclusions

The inducible bile acid 7α-dehydroxylation activity in *Eubacterium* sp. VPI 12708 proceeds by a complex multistep pathway. Significant progress has been made in recent years in elucidating the various steroid intermediates in this pathway. Many of the bile acid-inducible genes coding for the polypeptides involved in the pathway have apparently been cloned and sequenced. However, the process of relating these polypeptides to specific catalytic func-

Table 45.2 Cholic acid-inducible polypeptides from *Eubacterium* sp. VPI 12708

Molecular weight	N-terminal amino acid sequence obtained	Gene cloned	Proposed enzymatic activity
9,000[a]	—	X	UK[b]
19,500	X	X	UK
27,000-1	X	X	3α-HSDH[c]
27,000-2	—	X	3α-HSDH
27,000-3	X	X	3α-HSDH
47,500	X	X	UK
58,000	—	X	Ligation[d]
59,500[a]	—	X	UK
72,000	X	—	FOR[e]

[a] Proposed from open reading frame on cholic acid-inducible operon.
[b] Unknown.
[c] 3α-Hydroxysteroid dehydrogenase.
[d] Bile acid-CoA ligase.
[e] NADH:flavin oxidoreductase.

tions has only begun. Expression of the various enzymes in *Escherichia coli* shows promise, and some of the catalytic activities of the bile acid-inducible polypeptides will undoubtedly be analyzed in this host species.

A full understanding of the pathway and the enzymes involved, however, would be greatly expedited by the development of a transformation system for *Eubacterium* sp. VPI 12708 and by the development of a selection procedure for mutations in the 7α-dehydroxylation pathway. Future studies will also be directed toward understanding the control mechanisms for expression of the bile acid-inducible operon. In addition, the characterization of proteins involved in bile acid transport and secretion should be a fruitful area of research.

References

Ali, S.S., A. Kuksis, and J.M.R. Beveridge. 1966. Excretion of bile acids by three men on a fat-free diet. *Can. J. Biochem.* **44**:957–969.

Atkinson, P.W., L.E. Mills, W.T. Stammer, and D.T. Sullivan. 1988. Structure and evolution of the *Adh* genes of *Drosophila mojavensis*. *Genetics* **120**:713–723.

Baron, S.F., C.V. Franklund, and P.B. Hylemon. 1991. Cloning, sequencing, and expression of the gene coding for bile acid 7α-hydroxysteroid dehydrogenase from *Eubacterium* sp. strain VPI 12708. *J. Bacteriol.* **173**:4558–4569.

Björkhem, I., K. Einarsson, P. Melone, and P. Hylemon. 1989. Mechanism of intestinal formation of deoxycholic acid from cholic acid in humans: evidence for a 3-oxo-Δ⁴-steroid intermediate. *J. Lipid Res.* **30**:1033–1040.

Coleman, J.P., W.B. White, B. Egestad, J. Sjövall, and P.B. Hylemon. 1987a. Biosynthesis of a novel bile acid nucleotide and mechanism of 7α-dehydroxylation by an intestinal *Eubacterium* species. *J. Biol. Chem.* **262**:4701–4707.

Coleman, J.P., W.B. White, and P.B. Hylemon. 1987b. Molecular cloning of bile acid 7-dehydroxylase from *Eubacterium* sp. strain VPI 12708. *J. Bacteriol.* **169**:1516–1521.

Coleman, J.P., W.B. White, M. Lijewski, and P.B. Hylemon. 1988. Nucleotide sequence and regulation of a gene involved in bile acid 7α-dehydroxylation in *Eubacterium* sp. strain VPI 12708. *J. Bacteriol.* **170**:2070–2077.

de Wet, J.R., K.V. Wood, M. Deluca, D.R. Helinski, and S. Subramani. 1987. Firefly luciferase

gene: structure and expression in mammalian cells. *Mol. Cell. Biol.* **7**:725–737.

Eneroth, P., B. Gordon, R. Ryhage and J. Sjövall. 1966. Identification of mono- and dihydroxy bile acids in human feces by gas-liquid chromatography and mass spectrometry. *J. Lipid Res.* **7**:511–523.

Eyssen, H.J., G.G. Parmentier, F.C. Compernolle, G. DePauw, and M. Piessens-Denef. 1973. Biohydrogenation of sterols by *Eubacterium* ATCC 21408-Nova species. *Eur. J. Biochem.* **36**:411–421.

Ferrari, A., C. Scolastinco, and I. Beretta. 1977. On the mechanism of cholic acid 7α-dehydroxylation by a *Clostridium bifermentans* cell free extract. *FEBS Lett.* **75**:166–168.

Ferrari, A., N. Pacini, E. Canzi, and F. Bruno. 1980. Prevalence of oxygen intolerant microorganisms in primary bile acid 7-dehydroxylating mouse intestinal microflora. *Curr. Microbiol.* **4**:257–260.

Franklund, C.V., P. de Prada, and P.B. Hylemon. 1990. Puification and characterization of a microbial, NADP-dependent bile acid 7α-hydroxysteroid dehydrogenase. *J. Biol. Chem.* **265**:9842–9849.

Goldman, P. 1978. Biochemical pharmacology of the intestinal flora. *Annu. Rev. Pharmacol. Toxicol.* **18**:523–539.

Gopal-Srivastava, R., D.H. Mallonee, W.B. White, and P.B. Hylemon. 1990. Multiple copies of a bile acid-inducible gene in *Eubacterium* sp. strain VPI 12708. *J. Bacteriol.* **172**:4420–4426.

Heilmann, H.J., H.J. Mägert, and Hans G. Gassen. 1988. Identification and isolation of glucose dehydrogenase genes of *Bacillus megaterium* M1286 and their expression in *Escherichia coli*. *Eur. J. Biochem.* **174**:485–490.

Heuman, D.M., Z.R. Vlahcevic, M.L. Bailey, and P.B. Hylemon. 1988. Regulation of bile acid synthesis II. Effect of bile acid feeding on enzymes regulating hepatic cholesterol and bile acid synthesis in the rat. *Hepatology* (Baltimore) **8**:892–897.

Holdeman, L.V., I.J. Good, and W.E.C. Moore. 1976. Human fecal flora: variation in bacterial composition within individuals and a possible effect of emotional stress. *Appl. Environ. Microbiol.* **31**:359–375.

Hylemon, P.B. 1985. Metabolism of bile acids in intestinal microflora. In: *New Comprehensive Biochemistry, Vol. 12, Sterols and Bile Acids*, H. Danielson and J. Sjovall, eds., pp. 331–343. New York: Elsevier.

Hylemon, P.B. and T.L. Glass. 1983. Biotransformation of bile acids and cholesterol by the intestinal microflora. In: *Human Intestinal Microflora in Health and Disease*, D.J. Henteg, ed., pp. 189–213. New York: Academic Press.

Hylemon, P.B., P.D. Melone, C.V. Franklund, E. Lund, and I. Björkhem. 1991. Mechanism of intestinal 7α-dehydroxylation of cholic acid: evidence that allo-deoxycholic acid is an inducible side-product. *J. Lipid Res.* **32**:89–96.

Jany, K.D., W. Ulmer, M. Fröschle, and G. Pfleiderer. 1984. Complete amino acid sequence of glucose dehydrogenase from *Bacillus megaterium. FEBS Lett.* **165**:6–10.

Jörnvall, H., H. von Bahr-Lindström, K.D. Jany, W. Ulmer, and M. Fröschle. 1984. Extended super family of short alcohol-polyol-sugar-dehydrogenases: structural similarities between glucose and ribitol dehydrogenases. *FEBS Lett.* **165**:190–196.

Lipsky, R.H. and P.B. Hylemon. 1980. Characterization of a NADH:flavin oxidoreductase induced by cholic acid in a 7α-dehydroxylating intestinal *Eubacterium* species. *Biochim. Biophys. Acta* **612**:328–336.

Liu, J., K. Duncan, and C.T. Walsh. 1989. Nucleotide sequence of a cluster of *Escherichia coli* enterobactin biosynthesis genes: identification and purification of its product 2,3-dihydro-2,3-dihydroxybenzoate dehydrogenase. *J. Bacteriol.* **171**:791–798.

Lozoya, E., H. Hoffmann, C. Douglas, W. Schulz, D. Scheel, and K. Hahlbrock. 1988. Primary structures and catalytic properties of isoenzymes encoded by the two 4-coumarate:CoA ligase genes in parsley. *Eur. J. Biochem.* **176**:661–667.

Macdonald, I.A., V.D. Bokkenheuser, J. Winter, A.M. McLernon, and E.H. Mosbach. 1983. Degradation of steroids in the human gut. *J. Lipid Res.* **24**:675–700.

Mallonee, D.H., W.B. White, and P.B. Hylemon. 1990. Cloning and sequencing of a bile acid-inducible operon from *Eubacterium* sp. strain VPI 12708. *J. Bacteriol.* **172**:7011–7019.

Moore, W.E.C. and L.V. Holdeman. 1974. Human fecal flora: the normal flora of 20 Japanese-Hawaiians. *Appl. Microbiol.* **27**:961–979.

Morris, H.R., D.H. Williams, G.G. Midwinter, and B.S. Hartley. 1974. A mass-spectrometric sequence study of the enzyme ribitol dehydrogenase from *Klebsiella aerogenes. Biochem. J.* **141**:701–713.

Paone, D.A.M. and P.B. Hylemon. 1984. HPLC purification and preparation of antibodies to cholic acid-inducible polypeptides from *Eubacterium* sp. VPI 12708. *J. Lipid Res.* **25**:1343–1349.

Sadzikowski, M.R., J.F. Sperry, and T.D. Wilkins. 1977. Cholesterol-reducing bacterium from human feces. *Appl. Environ. Microbiol.* **34**:355–362.

Samuelsson, B. 1960. Bile acids and steroids: on the mechanism of the biological formation of deoxycholic acid from cholic acid. *J. Biol. Chem.* **235**:361–366.

Stellwag, E.J. and P.B. Hylemon. 1979. 7α-dehydroxylation of cholic acid and chenodeoxycholic acid by *Clostridium leptum. J. Lipid Res.* **20**:325–333.

Thatcher, D.R. 1980. The complete amino acid sequence of three alcohol dehydrogenase alleloenzymes (*Adh*[N-11], *Adh*[S] and *Adh*[UF]) from the fruitfly *Drosophila melanogaster. Biochem. J.* **187**:875–886.

Vlahcevic, Z.R., D.M. Heuman, and P.B. Hylemon. 1990. Physiology and pathophysiology of enterohepatic circulation of bile acids. *In: Hepatology. A Textbook of Liver Disease*, D. Zabim and T.D. Boyer, eds., pp. 341–377. Philadelphia: Saunders.

Weckermann, R., R. Fürbass, and M.A. Marahiel. 1988. Complete nucleotide sequence of the *tycA* gene coding the tyrocidine synthetase 1 from *Bacillus brevis. Nucleic Acids Res.* **16**:11841.

White, B.A., A.F. Cacciapuoti, R.J. Fricke, T.R. Whitehead, E.H. Mosbach, and P.B. Hylemon. 1981. Cofactor requirements for 7α-dehydroxylation of cholic and chenodeoxycholic acid in cell extracts of the intestinal anaerobic bacterium, *Eubacterium* species V.P.I. 12708. *J. Lipid Res.* **22**:891–898.

White, B.A., R.J. Fricke, and P.B. Hylemon. 1982. 7β-Dehydroxylation of ursodeoxycholic acid by whole cells and cell extracts of the intestinal anaerobic bacterium, *Eubacterium* species VPI 12708. *J. Lipid Res.* **23**:145–153.

White, B.A., R.L. Lipsky, R.J. Fricke, and P.B. Hylemon. 1980. Bile acid induction specificity of 7α-dehydroxylase activity in an intestinal *Eubacterium* species. *Steroids* **35**:103–109.

White, B.A., D.A.M. Paone, A.F. Cacciapuoti, R.J. Fricke, E.H. Mosbach, and P.B. Hylemon. 1983. Regulation of bile acid 7-dehydroxylase activity by NAD⁺ and NADH in cell extracts of *Eubacterium* species V.P.I. 12708. *J. Lipid Res.* **24**:20–27.

White, W.B., J.P. Coleman, and P.B. Hylemon. 1988a. Molecular cloning of a gene encoding a 45,000-dalton polypeptide associated with bile acid 7-dehydroxylation in *Eubacterium* sp. strain VPI 12708. *J. Bacteriol.* **170**:611–616.

White, W.B., C.V. Franklund, J.P. Coleman, and P.B. Hylemon. 1988b. Evidence for a multigene family involved in bile acid 7-dehydroxylation in *Eubacterium* sp. strain VPI 12708. *J. Bacteriol.* **170**:4555–4561.

Wierenga, R.K., P. Terpstra, and W.G.L. Hol. 1986. Prediction of the occurrence of the ADP-binding βαβ-fold in proteins, using an amino acid sequence fingerprint. *J. Mol. Biol.* **187**:101–107.

46

Cloning and Expression in *Escherichia coli* of Three Amylase Genes of a Strictly Anaerobic Thermophile, *Dictyoglomus thermophilum*, and Their Nucleotide Sequences

Sueharu Horinouchi and Teruhiko Beppu

46.1 Introduction

In our search for microorganisms that produce heat-stable amylases, we have targeted extremely thermophilic microorganisms or caldoactive bacteria capable of growing at temperatures above 70°C, based on the concept that a thermophile produces thermostable enzymes. This screening led to a novel bacterlum that we have named *Dictyoglomus thermophilum* (Saiki et al., 1985), which produces an amylase complex with acidic pH optimum useful for industrial starch saccharification (Kobayashi et al., 1988). Difficulties encountered in purifying and characterizing each of the amylases in the complex prompted us to clone and express the amylase genes in *Escherichia coli* and then to examine their properties. This chapter briefly describes the taxonomic properties of *D. thermophilum* and the cloning of three amylase genes from this strain into *E. coli*.

46.2 Some Taxonomic Properties of *Dictyoglomus thermophilum*

During the past several years, extremely thermophilic and anaerobic chemoorganotrophs, including *Thermoanaerobium brockii* (Zeikus et al., 1979), *Thermoanaerobacter ethanolicus* (Wiegel and Ljungdahl, 1981), and *Thermobacteroides acethoethylicus* (Ben-Bassat and Zeikus, 1981), have been isolated·from hot springs and volcanic environments. Our screening of hot springs in Japan for thermophilic and anaerobic bacteria allowed us to isolate a species whose taxonomic properties did not fit those of any previously described.

This species, *Dictyoglomus thermophilum* (type strain, ATCC 35947), is a gram-negative rod 0.4 to 0.6 μm wide by 5 to 20 μm long with neither motility nor spore-forming ability. The temperature range for growth is between 50° and 80°C, with optimum growth at 78°C; the pH range for growth is between 5.9 and 8.3. The doubling time at 73°C and pH 7.2 is about 2.5 h. The guanine-plus-cytosine content of DNA is 29 mol% as determined by its melting temperature. This result clearly contradicts a widely accepted but poorly understood assumption about thermophiles, i.e., that their DNA should have a high G + C content. It is also interesting that this strain often forms bundles and large spherical bodies called "rotund bodies" (Figure 46.1). The bundles and large spherical bodies consist of two to several cells and several dozen cells, respectively, and are surrounded by a common outer cell mem-

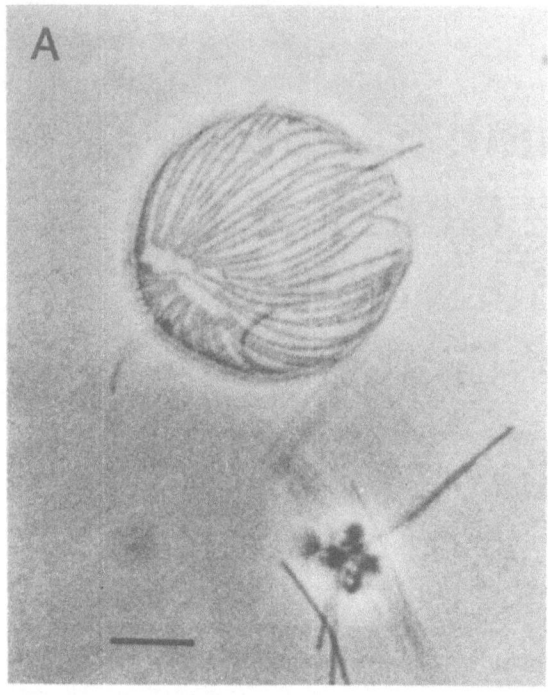

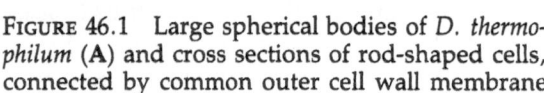

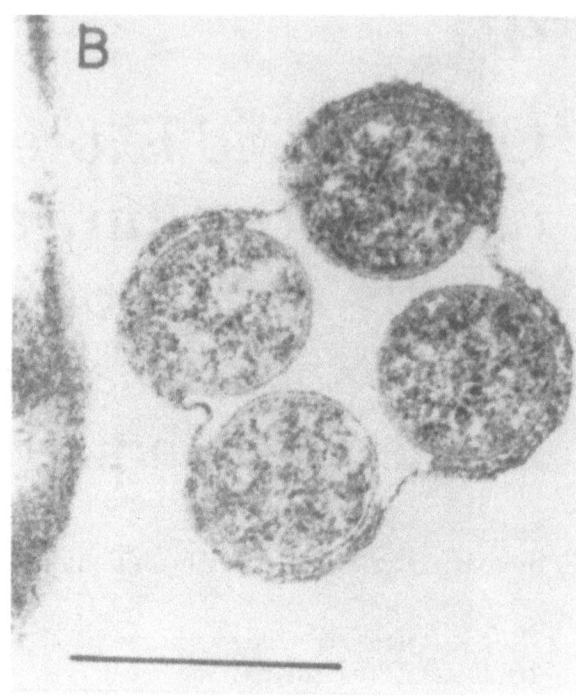

FIGURE 46.1 Large spherical bodies of *D. thermo-phiulum* (**A**) and cross sections of rod-shaped cells, connected by common outer cell wall membrane (**B**). In **B**, four cells are connected to form bundle. Bars = 0.5 μm. (Photographs were kindly provided for this chapter by Dr. T. Saiki.)

brane. The spherical bodies formed by the association of many filamentous cells frequently grow to 50 to 100 μm in diameter, depending on the number of cells. Similar large spherical bodies are also observed in the case of an archaebacterium, *Thermus aquaticus* (Brock and Edwards, 1970).

46.3 Amylase Production by *Dictyoglomus thermophilum*

Dictyoglomus thermophilum was found to produce highly heat-stable amylases when a crude extract of the cells was assayed by the amylose-iodine method (Kobayashi et al., 1988). Almost all the activity was detected in the culture medium. Glucose-containing medium (0.5%) supported significantly less enzyme production than starch-containing medium. The pH and temperature optima for the amylase activity were 5.0 and 90°C, respectively; activity was stabilized in the presence of $CaCl_2$.

Our attempts to purify the amylase using a DEAE-Toyopearl column showed that amylase activities were eluted in four different fractions. Preliminary studies on each fraction suggested that this strain produced at least three different amylases, yielding differing patterns of products from starch and pH optima. At this stage, instead of attempting to purify each amylase in the mixture, we concentrated on employing recombinant DNA techniques for cloning the individual amylase genes in *E. coli*. After cloning and expression of the genes, *E. coli* transformants should contain a single species of the amylase, facilitating their purification. Another application of recombinant DNA techniques should also be an improvement in their yields.

46.4   Cloning in *E. coli* and Nucleotide Sequences of Three Amylase Genes

Shotgun cloning and subcloning

We used well-developed *E. coli* systems (Maniatis et al., 1982) for cloning the amylase genes from *Dictyoglomus thermophilum*. The plasmid vector we used, pYEJ001, contained an ampicillin resistance gene as a selectable marker and an *E. coli* consensus promoter for expression of foreign genes. Chromosomal DNA prepared from *D. thermophilum* was digested with *Hind*III and the mixture was ligated into the *Hind*III site located just downstream from the *E. coli* promoter. Amylase-producing *E. coli* transformants were selected as colonies forming a halo on a I$_2$-stained starch plate. One transformant containing a 5.2-kb insert in pYEJ001 was obtained. Reintroduction of the plasmid into *E. coli* caused all the transformants to form a halo. We named this initially isolated amylase gene *amyA*.

Similar shotgun-cloning experiments using *Sau*3AI instead of *Hind*III for obtaining chromosomal fragments yielded about 100 colonies capable of forming a I$_2$-stained halo on starch plates among about 100,000 ampicillin-resistant transformants. Southern blot hybridization using the 5.2-kb fragment containing the *amyA* gene as the probe, together with restriction mapping of the plasmids in the transformants, revealed two additional amylase genes different from *amyA*. These genes were named *amyB* and *amyC*. Judging from the number of *E. coli* transformants that were obtained and assayed for amylase activity, we may assume that the genomic bank accounted for virtually all amylase genes in *D. thermophilum* that could be cloned by the selection method used.

We next subcloned the three amylase genes. The 5.2-kb fragment containing the *amyA* gene appeared not to contain a promoter that was functional in *E. coli*, because the DNA fragment was unable to confer amylase activity when inserted in pYEJ001 in the opposite orientation. Subcloning of the amylase genes was, therefore, carried out in such a way that the orientation in various plasmid constructions was the same as that in the originally isolated plasmid with respect to the *E. coli* promoter of pYEJ001. As a result, *amyA*, *amyB*, and *amyC* were localized to 2.6-kb, 2.5-kb, and 1.8-kb fragments, respectively. In all cases, decreasing the distance between the consensus promoter in pYEJ001 and the translation initiation codons of the amylases resulted in 8- to 10-fold increases in amylase production in *E. coli*. However, the yields of these amylases were still considerably lower than those from amylase genes of other bacterial sources also expressed in *E. coli*. The low yields may be explained in terms of an unusual codon usage pattern of the *D. thermophilum* genes, as described next.

Localization of the amylase produced in recombinant *E. coli* cells

In amylolytic microorganisms, most of the amylase activity is generally found in culture media. This is consistent with the function of amylase in degrading a high molecular mass polymer into small units for transport into the cell. Likewise, in *D. thermophilum*, almost all amylase activity was found in the culture medium (Kobayashi et al., 1988). To localize the amylase produced in recombinant *E. coli* cells, we cultured *E. coli* transformants containing each of the amylase genes on vector plasmids and separated them into intracellular, periplasmic, and extracellular fractions by the cold osmotic shock method (Cornelis et al., 1982). Amylase activity, together with β-lactamase activity (specified by the vector), was measured for each fraction. Under the experimental conditions in which more than 90% of β-lactamase activity was found in the periplasmic fraction, indicative of successful fractionation, all the amylase activities for AmyA and AmyB were found in the intracelllular fractions and hardly any was detected in the periplasmic or extracellular fraction. For AmyC, however, a very small portion of the total amylase activity was detected in the media. The intracellular localization of these amylases when produced in *E.*

coli is discussed next in relationship to the absence of a typical signal sequence for secretion.

Nucleotide sequences of the amylase genes

The nucleotide sequences of the subcloned fragments containing the amylase genes were determined by the M13-dideoxynucleotide method (Yanisch-Perron et al., 1985). As for the *amyA* gene, an open-reading frame starting with ATG (methionine) and terminating with TAA, which consisted of 686 amino acids (M_r, 81,200), was present. The deduced NH_2-terminal amino acid sequence coincided with that determined from purified amylase protein by an automated Edman degradation procedure. Likewise, the *amyB* and *amyC* genes encode 562-amino-acid (M_r, 67,000) and 498-amino acid (M_r, 59,000) proteins, respectively. Potential ribosome-binding sequences were located 5 bp upstream from the translation initiation codons for *amyA* and *amyC*, AAAG-GAGGT and GGAGGT, respectively. For *amyB*, GGAGGT is also present 8 bp upstream from the ATG start codon. The nucleotide sequences of the 3' end of 16S rRNA in procaryotes, which are important for translation initiation by forming nucleotide pairs with mRNAs, are well conserved. Therefore, these sequences having perfect complementarity to the 3' end of *E. coli* 16S rRNA must also be serving as ribosome-binding sites in *D. thermophilum*.

As mentioned, the cloned fragments contained no detectable promoter activities. Consistent with this, procaryotic promoter "consensus" sequences (TTGACA for −35 and TATAAT for −10; Rosenberg and Court, 1979) were not present upstream from any of the amylase translation initiation codons. It is probable that transcriptional control sequences of *D. thermophilum* with an extremely low G + C content in its DNA (29 mol%) are so different from the other procaryotic consensus sequences that the *E. coli* transcriptional machinery is unable to recognize them. The alternative situation is illustrated by *Streptomyces* genes, which have an extremely high G + C content of DNA (about 73 mol%); almost none of the *Streptomyces* genes are transcribed in *E. coli* because of a great deviation of the −35 and −10 sequences from the consensus sequences (Bibb and Cohen, 1982; Horinouchi et al., 1987).

An additional point we would like to note with respect to the amino acid sequences deduced from their nucleotide sequences is that the primary translation products of these three amylases lack a typical signal sequence for secretion. This is consistent with the observation that almost all the amylase activity was found in the intracellular fractions of recombinant *E. coli* cells, as described. All the bacterial amylases so far characterized have a signal sequence at the NH_2 termini of the primary translation product, which is cleaved during secretion into the media. A typical signal sequence has two or three basic amino acids just next to the translational start Met, a long stretch of hydrophobic amino acids, and a cleavage site recognized by signal peptidases (von Heijne, 1984; Pugsley and Schwartz, 1985).

The inability to secrete AmyA and AmyB into the periplasmic space of *E. coli* cells can be ascribed to the absence of any signal peptide-like sequences at their NH_2 termini. To explain the low level of AmyC secretion, we speculate that its NH_2-terminal amino acid sequence, two Lys residues followed by hydrophobic amino acids, is recognized to a small extent as a signal peptide by the *E. coli* secretory machinery. At any rate, these findings suggest the presence of some unknown mechanism of protein secretion in this gram-negative thermophile.

Codon usage pattern of *D. thermophilum* genes

The codon usage pattern of the amylase genes of *D. thermophilum*, which has one of the lowest G + C composition of any living organism, shows a characteristic feature; the G + C content of the first letter of the codons is significantly high, while that of the third letter is somewhat low (Figure 46.2). Of 20 amino acids, 18 allow all four bases, T, C, A and G, in the third codon position, which is thus called a "wobble" position. For the first letter of the codons, such a choice is possible only for Arg (C or A) and leu (T or C); no such choice is allowed for the second codon position. One can therefore expect that genes of procaryotes with highly biased G + C contents might possess characteristic codon usage patterns to minimize a possible bias of the amino acid composition.

We can safely conclude that *D. thermophilum* genes coding for proteins have such a codon usage, that a higher proportion of G + C in the first position of codons reduces a bias of the amino acid composition, and that a lower proportion of G + C in the third position accounts for the extremely low overall G + C content. The G + C composition of the second position reflects the average G + C content of DNA of this strain, which presents a striking contrast to genes with extremely high G + C contents.

The codon usage of genes of the genus *Streptomyces*, which has an average composition of 73 mol% G + C, is highly biased; the high G + C content is reflected exclusively in the third codon position (Thompson and Gray, 1983; Bibb et al., 1984). For example, the G + C contents of the third position of the α-amylase genes from *S. limosus* (Long et al., 1987) and *S. hygroscopicus* (Hoshiko et al., 1987) are 97.2 mol% and 95.6 mol%, respectively (see Figure 46.2). The G + C composition of the second position is dramatically low, and that of the first position reflects the average G + C content of DNA of this genus. All the *Streptomyces* genes so far characterized have the same tendency. The same holds for genes of the genus *Thermus* with a G + C content of 68% (Kagawa et al., 1984).

In connection with the nonrandom distribution of G and C at each codon position, Kagawa et al. (1984) formulated an explanation for the increased G + C in the third position of the *Thermus leuB* (3-isopropylmalate dehydrogenase) gene. Their view is that the high G + C in the third position may stabilize the dynamic structure formed during transcription and translation at extremely high temperatures. However, it is clear that this is not applicable to thermophilic microorganisms in general.

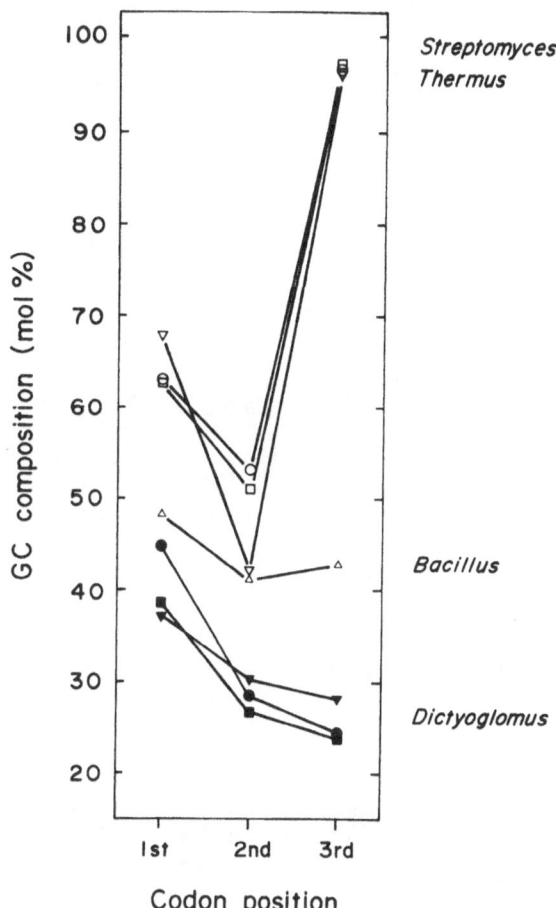

FIGURE 46.2 G + C base composition at three different codon positions for several genes having various overall G + C contents: *Dictyoglomus thermophilum* amylases *amyA* (●); *amyB* (■); *amyC* (▼); *Bacillus subtilis* α-amylase gene, *amyEm*[+] (△) (Yamazaki et al., 1983); *Streptomyces limosus* α-amylase, *aml* (○) (Long et al., 1987); *Streptomyces hygroscopicus* α-amylase gene, *amy* (□) (Hoshiko et al., 1987); *Thermus flavus* malate dehydrogenase gene *mdh* (▽) (Nishiyama et al., 1986). (From Horinouchi et al., 1988, with permission.)

46.5 Enzymatic Properties of the *Dictyoglomus thermophilum* Amylases

Comparison of the amino acid sequences with those of other amylases

α-Amylases are widely distributed among both procaryotes and eucaryotes. A comparison of

AmyB MIYDDKIFGDLCHKEFLVEREVKKLEEIYLEEVLPEDPKPEDEIEFTFNCPLKFHITSGKIVKDNREIYTFNIQERKTQWNDSIFNFSEIIKIKIPPLKE

AmyB NGLYQIHLYEMNEKIYEQYLSIDS-FEAPLW-SEESIIYHIFIDRFAKDEKEVE-YS--ENLKEKLGGNLKGILSRLDYIENLGINTIWISPIFK----
TAA   ATP---AD-WRSQ-S-IYFLLTDRFARTDGSTTATCNTADQKYC-GGTWQGIIDKLDYIQGMGFTAIWITPVTAQLPQD----
AmyC M--KIKFFIKR-TLIFILVTFLTYIHG-YNEPWYKNAIFY-EVFVRS-FADSDGDRVGDLNGLIDKLDYFKNLNITALWLMPIFP-----

AmyB ---STSYHGYDIEDYFEIDPIWGTKDELKKLVREA-FNRGIRIILDFVPNHMSYKNPIFQKALKDKNLRSWFI-FKGEDYET-F-FGVKSMPKINLKNKE
TAA   CAYGDAYTGYWQTDIYSLNENYGTADDLKALSS-ALHERGMYLMVDVVANHMGYDGAG--SSV-DY---SVFKDFSSQDYFHPFCFIQNYEDTQVEDCW
AmyC ---SVSYHGYDVTDYYDIHPGYGTNEDFENLIRKA-HEKNIKIILDLVVNHTSSRHPWFVSSA------SSYN-SPYIDYYIWSTEKPEKNSNLWYK-KP

AmyB AIDYIIN-------------------AAKYWIREFGISGYRMDHATGP--DI-N--FWSIFY-YNLKSEFPETFY-FGEIV-ET-PKETKKYVGKFDG
TAA   LGDNTV---SL-PDLDTTKDVVKNEWYDWVGSLVSNYSIDGLRIDTVKHVQKD----FWPG--YN-K--AAGVYCIGE-VLDGDPAYTCPYQNVMDG
AmyC TGYYALFWSEMPDLNFDNPKVREEVKKIAKFWI-EKGVDGFRLDAAKHIYDDSKNIQWWKEFYSY-LKSIKPD-VVLVGE-VWDNEYKIAEYYKGLPSN

                        AmyA LAKYDESNH

AmyB TLDFYLFKIIRDF-FIGKRWSTKEFVK--MIDLEEKFYGNKFKR-ISFLENHDSNRFLWVAK---DKKLLRLA-SIFQ-FSI---NAI-PIIYNGQEMGC
TAA   VLN-YPIYYPLLNAF-KSTSGSMDDLYNMINTV-KSDCPDSTLLGTFVENHDQVRVRTFFGGSIDKSI---LAGSI----YTLAGNTF---IYYGEEIGM
AmyC F-NF-PLSDKIMNSSS-KSKRLRNYRISRLKRLF-GENNTDFADAI-FLRNHDQVRVRTFFGGSIDKSI---LAGSI----YTLAGNTF---IYYGEEIGM

                        AmyA LFYDNHRR

AmyB SQYR-DILEGNRT--LHEHARLPIPWSDD-KQDKELIDFYIQLVKIRKSHP-ALYKGTF-IPIFSDM-ISFIKETQEES-ILVLIN-IEDKEE-IFNLNG
TAA   AGGN-DPA--NREAT-WLSGY-PTD-SELYKLIASAN--AIRNYAISKDTGFVTYKNP-YIKDDTT--IAMRKGTDG-SQI-VTIL-S-NKGASGDSYTL
AmyC EGSKPDEY---IREPFKWTDDMKSKYQT--YWIIPRYN--LPGNG-IALDTEEKD-PNSIYNHYKKLLEIRV-KCRAL-SNGK--IERIKTQDRSILAYKL

AmyB TYRDLFSGNIYTNSLKLGPMSAHLLLRI---DH 562 aa
TAA   SLSGA-S---YTAGQQLTEVIGCTTVTVGS-DGNVPVPMAGGLPRVLYPTEKLAGS-KICSDSS 478 aa
AmyC ELEDEKIMVVH-N-LNRIENTFNFNNEIKEKDILYIRN-AKTKENKIILGPYSTVIVKIP 498 aa

FIGURE 46.3 Alignment of amino acid sequences of *AmyB*, *AmyC*, and Taka-amylase A. Symbols: identical amino acids in Taka-amylase A (TAA) and either AmyB or AmyC (*); catalytic residues, Glu and Asp, of Taka-amylase A (▼); amino acids participating in substrate binding and locating in active center cleft (▽). Although no significant homology between AmyA and other amylases was found, two parts of the AmyA sequences are also aligned with well conserved regions of amylases. (From Horinouchi et al, 1988, with permission.)

the amino acid sequences of amylases from a wide range of species shows the presence of four conserved regions positioned at similar intervals along the length of each protein (Nakajima et al., 1986). On the basis of the studies on Taka-amylase A, at least three of these are suggested to be involved in substrate binding or contribute components of the active site (Toda et al., 1982; Matsuura et al., 1984). Using dot-matrix analysis to determine similarities in the amino acid sequence (Novotny, 1982) among the three amylases from *D. thermophilum*, we found considerable homology between AmyB and AmyC at the protein level. Figure 46.3 shows alignment of the two amino acid sequences. However, no significant homology between AmyA and either AmyB or AmyC was predicted. In addition, a computer-aided similarity search was unable to predict any protein having a similarity to AmyA. AmyB and AmyC, in resembling each other, show significant homology in terms of entire sequence to Taka-amylase A, the most extensively studied amylase. It was noted especially that these related amylases contain four regions with a considerable degree of amino acid sequence identity to the conserved regions, which probably participate in substrate binding (Figure 41.2). We can then reliably predict that the catalytic amino acid residues for both AmyB and AmyC are Glu and Asp, as indicated in Figure 46.3.

Purification and characterization of the amylases produced by *E. coli* transformants

Purification. AmyA and AmyB were purified to homogeneity, each giving a single band with a molecular mass of 75,000 and 67,000, respectively, on sodium dodecyl sulfate-polyacrylamide gel electrophoresis (SDS-PAGE). These apparent M_r are in good agreement with those predicted from the nucleotide sequences. Because these amylases are extremely heat stable, the crude extracts of the recombinant *E. coli* cells were heat-treated at 75°C for 30 min before several purification steps of chromatography. This heat treatment and subsequent centrifugation were very useful in removing a considerable portion of the *E. coli* proteins. We

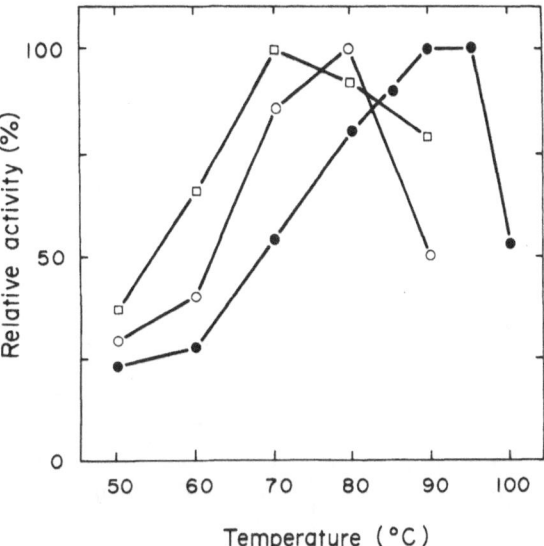

FIGURE 46.4   Effect of temperature on thermophilic amylase activities of AmyA, AmyB, and AmyC. Reaction mixture containing 0.4% soluble starch, 40 mM sodium acetate, pH 5.5, 5 mM CaCl$_2$ and 2 to 3 U/ml of each enzyme was incubated for 60 min at the indicated temperatures. Symbols: AmyA (●); AmyB (○); AmyC (□). (From Horinouchi et al., 1988, with permission.)

used a heat-treated crude extract from an *E. coli* transformant to characterize AmyC. Activity staining of the crude extract showed that the M_r of AmyC was 60,000, in good agreement with that predicted by the nucleotide sequence.

Optimum temperature. The optimum temperatures for AmyA, AmyB, and AmyC as determined by the decrease in the intensity of the blue color of amylose–iodine complexes were 90°C, 80°C, and 70°C, respectively (Figure 46.4).

Optimum pH. All the three amylases had pH optimum at 5.5. Figure 46.5 shows the effects of pH on amylase activities of AmyB and AmyC; AmyA showed a pH dependence similar to AmyC.

Mode of starch degradation. The products of the action of the amylases on soluble starch were determined by following the time course of the decrease in blue color along with the increase in the amount of reducing sugar. AmyA rapidly decreased the blue color, which suggested that this amylase was a highly liq-

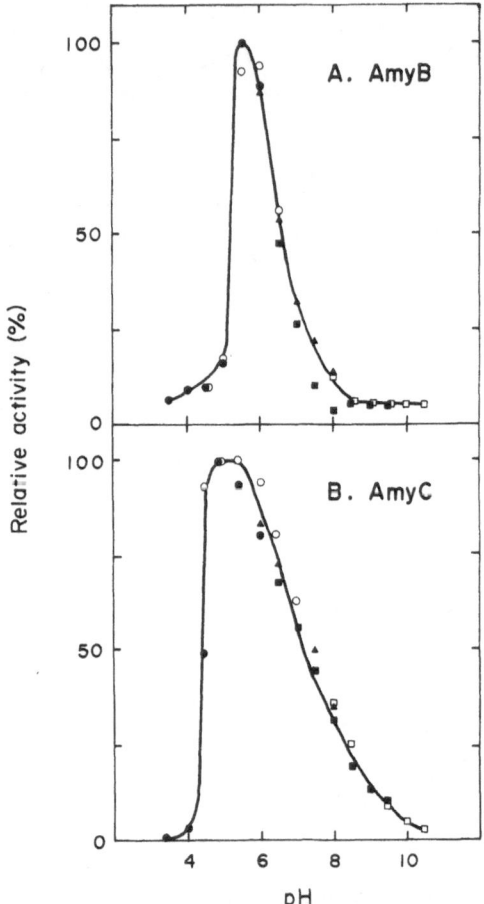

FIGURE 46.5 Effect of pH on thermophilic activities of AmyB and AmyC. Buffers: 40 mM sodium
acetate (●), 40 mM Mes (○), 40 mM Mops (▲), 40
mM Bicine (■), and 40 mM Caps (□). Activity was
measured after incubation for 60 min at 70°C.
(From Horinouchi et al., 1988, with permission.)

phy (Figure 46.7) indicated that AmyB hydrolyzes soluble starch almost completely into glucose and maltose. AmyC showed intermediate
characteristics between those of AmyA and
AmyB. Thin-layer chromatographic analysis
showed that AmyC produced significant
amounts of maltose and maltotriose with short
chain lengths. The ratio of oligosaccharides that
remained in the reaction mixture after prolonged incubation was also intermediate when
compared with ratios obtained with AmyA and
AmyB.

46.6   Conclusions

Dictyoglomus thermophilum, which was isolated
from a hot spring, is a gram-negative, obligately anaerobic, and extremely thermophilic
bacterium that forms a characteristic cell-
association structure called a rotund body and
possesses DNA of an extremely low G + C content. This strain produces at least three different amylases with an extreme heat stability
and an acidic pH optimum that will be potentially useful in industrial starch saccharification.
Recombinant DNA techniques enabled us to
clone three different amylase genes and characterize each of the amylases produced in recombinant *E. coli* cells. Nucleotide sequencing
revealed their primary structures, and two of
the amylases showed considerable similarities
to Taka-amylase A over the length of amino
acid sequences.

These findings should provide insight for
elucidating the structure–function relationships of amylases, by site-directed mutagenesis for example, although the yields of
these amylases in *E. coli* cells are still low.
The absence of a typical signal sequence for
secretion at the NH_2 termini of the primary
translation products strongly suggests the presence of some unknown mechanism of protein
secretion in this organism. The codon usage in
these amylase genes is highly biased, reflecting
the fact that the G + C content of DNA is 29
mol%. The distribution of G and C at each position of the codons was nonrandom; the G + C
content at the first codon position is significantly high, whereas that for the third position is

uefying type (Figure 46.6A). Consistent with
this was the analysis of the degradation products by thin-layer chromatography on silica
gel (Figure 46.6B); oligomers appeared first, followed by a gradual increase in the amounts of
maltotriose, maltose, and glucose on prolonged
incubation. On the other hand, AmyB produced larger amounts of reducing sugar,
although the blue color decreased more slowly
(Figure 46.7A). At the point where the blue col-
or decreased almost to zero, the amount of reducing sugar in the reaction mixture reached
about 50% of the total sugars. Analysis of the
hydrolysis products by thin-layer chromatogra-

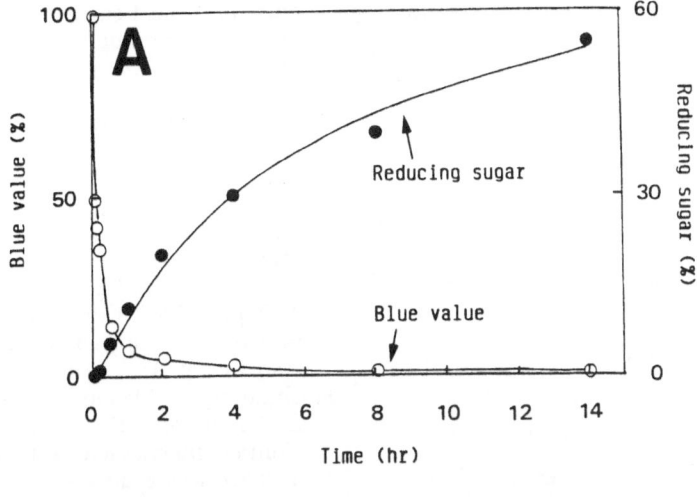

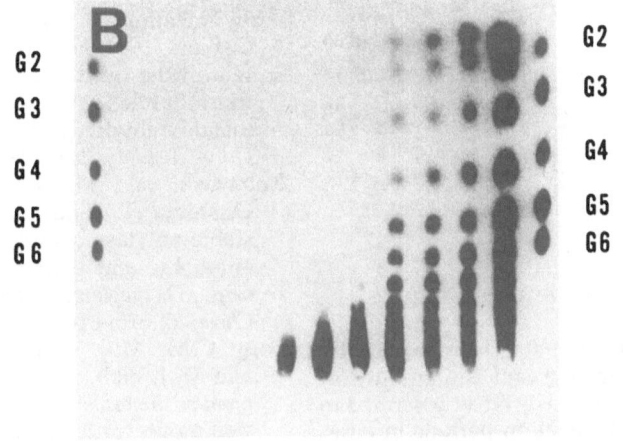

FIGURE 46.6 Time course of AmyA reaction on soluble starch by measuring blue color and production of reducing sugars (**A**) and by thin-layer chromatography (**B**). **A**. Assay mixtures, containing 10 m*M* sodium phosphate, pH 7.0, 1% soluble starch, and 10 U/ml of AmyA, were incubated at 80°C for indicated times. Blue color and production of reducing sugars were measured by the I₂/KI and 3,5-dinitrosalicylic acid method, respectively. **B**. Assay mixtures of **A** were incubated at 80°C for indicated times; products were analyzed by silica gel thin-layer chromatography; S, standard mixture for oligomers, e.g., G2 = (G1c)₂. (From Fukusumi et al., 1988, with permission.)

somewhat low. In addition, codons consisting only of A and T were preferentially used in this thermophile. Further studies on *D. thermophilum* genes will reveal interesting mechanisms of regulation of gene expression that may generally be functional in the thermophiles having extremely low G + C contents.

References

Ben-Bassat, A. and J.G. Zeikus. 1981. *Thermobacteroides acetoethylicus* gen. nov. and sp. nov., a new chemoorganotrophic, anaerobic, thermophilic bacterium. *Arch. Microbiol.* **128**:365–370.

Bibb. M.J. and S.N. Cohen. 1982. Gene expression in *Streptomyces*: construction and application of

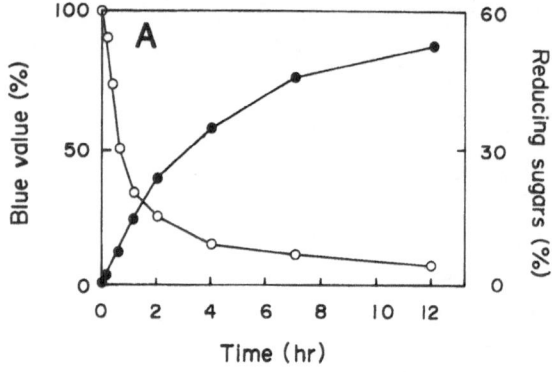

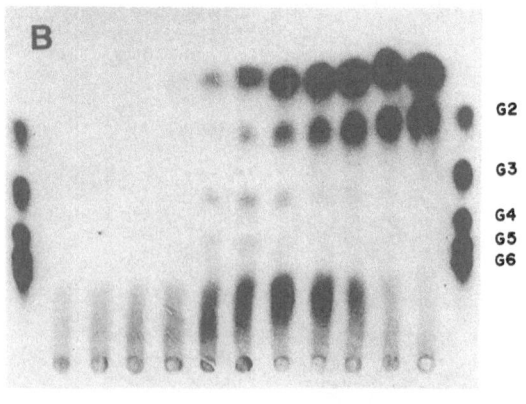

FIGURE 46.7 Time course of starch hydrolysis by AmyB. Blue color values (○) and the amount of reducing sugars (●) were measured as described in legend of Figure 46.6. Incubation periods in lanes 1 to 11 were 0, 0.5, 3, 10, 30, 60, 120, 240, 420, 720, and 1200 min, respectively. G2 to G6 indicates $(Glc)_2$ to $(Glc)_6$. (From Horinouchi et al., 1988, with permission.)

promoter-probe plasmid vectors in *Streptomyces lividans. Mol. Gen. Genet.* **187**:265–277.

Bibb, M.J., P.R. Findlay, and M.W. Johnson. 1984. The relationship between base composition and codon usage in bacterial genes and its use for the simple and reliable identification of protein-coding sequence. *Gene* (Amst.) **30**:157–166.

Brock, T.D. and M.R. Edwards. 1970. Fine structure of *Thermus aquaticus*, an extreme thermophile. *J. Bacteriol.* **104**:509–517.

Cornelis, P., C. Digneffe, and K. Willemot. 1982. Cloning and expression of a *Bacillus coagulans* amylase in *Escherichia coli. Mol. Gen. Genet.* **186**:507–511.

Fukusumi, S., A. Kamizono, S. Horinouchi, and T. Beppu. 1988. Cloning and nucleotide sequ-

ence of a heat-stable amylase gene from an anaerobic thermophile, *Dictyoglomus thermophilum. Eur. J. Biochem.* **174**:15–21.

Horinouchi, S., S. Fukusumi, T. Ohshima, and T. Beppu. 1988. Cloning and expression in *Escherichia coli* of two additional amylase genes of a strictly anaerobic thermophile, *Dictyoglomus thermophilum*, and their nucleotide sequences with extremely low guanine-plus-cytosine contents. *Eur. J. Biochem.* **176**:243–253.

Horinouchi, S., M. Nishiyama, A. Nakamura, and T. Beppu. 1987. Construction and characterization of multicopy expression-vectors in *Streptomyces* spp. *Mol. Gen. Genet.* **210**:468–475.

Hoshiko, S., O. Makabe, C. Nojiri, K. Katsumata, E. Satoh, and K. Nagaoka. 1987. Molecular cloning and characterization of the *Streptomyces hygroscopicus* α-amylase. *J. Bacteriol.* **169**:1029–1036.

Kagawa, Y., H. Nojima, N. Nukiwa, M. Ishizuka, T. Nakajima, T. Yasuhara, T. Tanaka, and T. Oshima. 1984. High guanine plus cytosine content in the third letter of codons of an extreme thermophile. DNA sequence of the isopropylmalate dehydrogenase gene of *Thermus thermophilus. J. Biol. Chem.* **259**:2956–2960.

Kobayashi, Y., M. Motoike, S. Fukusumi, T. Ohshima, T. Saiki, and T. Beppu. 1988. Heat-stable amylase complex produced by a strictly anaerobic and extremely thermophilic bacterium, *Dictyoglomus thermophilum. Agric. Biol. Chem.* **42**:615–616.

Long, C.M., M.-J. Virolle, S.-Y. Chang, S. Chang, and M. J. Bibb. 1987. α-Amylase gene of *Streptomyces limosus*: nucleotide sequence, expression motifs, and amino acid sequence homology to mammalian and invertebrate α-amylases. *J. Bacteriol.* **169**:5745–5754.

Maniatis, T., E.F. Fritsch, and J. Sambrook. 1982. *Molecular cloning—A Laboratory Manual.* Cold Spring Harbor, New York: Cold Spring Harbor Laboratory.

Matsuura, Y., M. Kusunoki, W. Harada, and M. Kakudo. 1984. Structure and possible catalytic residues of Taka-amylase A. *J. Biochem.* **95**:697–702.

Nakajima, R., T. Imanaka, and S. Aiba. 1986. Comparison of amino acid sequences of eleven different α-amylases. *Appl. Microbiol. Biotechnol.* **23**:355–360.

Nishiyama, M., N. Matsubara, K. Yamamoto, S. Iijima, T. Uozumi, and T. Beppu. 1986. Nucleotide sequence of the malate dehydrogenase gene of *Thermus flavus* and its mutation directing an increase in enzyme activity. *J. Biol. Chem.* **261**:14178–14183.

Novotny, J. 1982. Matrix program to analyze primary structure homology. *Nucleic Acids Res.* **10**:127–131.

Pugsley, A.P. and M. Schwartz. 1985. Export and secretion of proteins by bacteria. *FEMS Microbiol. Rev.* **32**:3–38.

Rosenberg, M. and D. Court. 1979. Regulatory sequences involved in promotion and termination of RNA transcription. *Annu. Rev. Genet.* **13**:319–353.

Saiki, T., Y. Kobayashi, K. Kawagoe, and T. Beppu. 1985. *Dictyoglomus thermophilum* gen. nov., sp. nov., a chemoorganotrophic, anaerobic, thermophilic bacterium. *Int. J. Syst. Bacteriol.* **35**:253–259.

Thompson, C.J. and G.S. Gray. 1983. Nucleotide sequence of a streptomycete aminoglycoside phosphotransferase gene and its relationship to phosphotransferases encoded by resistance plasmids. *Proc. Natl. Acad. Sci. USA* **80**:5190–5194.

Toda, H., K. Kondo, and K. Narita. 1982. The complete amino acid sequence of Taka-amylase A. *Proc. Jpn. Acad.* **58**:208–212.

von Heijne, G. 1984. How signal sequences maintain cleavage specificity. *J. Mol. Biol.* **173**:243–251.

Yamazaki, H., K. Ohmura, A. Nakayama, Y. Takeichi, K. Otozai, M. Yamasaki, G. Tamura, and K. Yamane. 1983. α-Amylase genes (*amyR2* and *amyE*[+]) from an α-amylase-hyperproducing *Bacillus subtilis* strain: molecular cloning and nucleotide sequences. *J. Bacteriol.* **156**:327–337.

Yanisch-Perron, C., J. Vieira, and J. Messing. 1985. Improved M13 phage cloning vectors and host strains: nucleotide sequences of M13mp18 and pUC19 vectors. *Gene* (Amst.) **33**:103–119.

Wiegel, J. and L.G. Ljungdahl. 1981. *Thermoanaerobacter ethanolicus* gen. nov., sp. nov., a new, extreme thermophilic, anaerobic bacterium. *Arch. Microbiol.* **128**:343–348.

Zeikus, J.G., P.W. Hegge, and M.A. Anderson. 1979. *Thermoanaerobium brockii* gen. nov. and sp. nov., a new chemoorganotrophic caldoactive, anaerobic bacterium. *Arch. Microbiol.* **122**:41–48.

47

A Novel Class of Industrially Important Debranching Enzymes: The Thermoanaerobic Amylopullulanases

Robert D. Coleman

47.1 Introduction

This chapter highlights the more significant work on industrially important carbohydrases, with special attention to the multifaceted debranching enzymes. A major focus of the chapter is on a new class of debranching enzymes (amylopullulanases) that have dual specificity: (1) hydrolysis of $\alpha - 1 \rightarrow 6$ linkages in pullulan and (2) cleavage primarily of $\alpha - 1 \rightarrow 4$ glucosidic bonds in starch. These unique enzymes have high molecular weights (100–450 kDa), are usually glycosylated, and are largely found in thermoanaerobes. Work involving the few known attempts to increase amylopullulanase production and stability via heterologous gene cloning is discussed, particularly the expression of the *Thermoanaerobium brockii* amylopullulanase ("debranching enzyme") gene in *Escherichia coli* and *Bacillus subtilis* and enzyme secretion. Continuing efforts, either in gene transfer to other host systems or in genetic manipulation of the native host, should be rewarding. Some of the more thermostable amylopullulanases may have excellent potential for increasing the production of maltose or specialty syrups made from starch and other starch-like raw materials.

47.2 The Demand for Highly Thermostable Carbohydrases

Since thermophilic bacteria were discovered more than 100 years ago, there has been a fascination with and a curiosity about the adaptations that allow these microorganisms to survive in their native habitats of springs, pools, and creeks warmed by geothermal activity. Industrial demand for alternate microorganisms and enzymes that can thrive at higher temperatures has been exceptional, especially for application in product areas such as industrial enzymes and sweeteners where profit margins can be thin. For many years, industrial and academic laboratories have invested considerable effort in screening for novel meso- and thermophilic microbial strains that are multifunctional or can tolerate extreme environments such as elevated temperatures, minimal nutrients, high salinity, high hydrostatic pressure, or high or low pH.

Greater thermostabilities and wider pH optima are often highly preferred traits for industrial enzymes. The enzymatic hydrolysis of starch has been used for centuries in the production of beer and alcohol. More recently, the conversions of starch into glucose, fructose, maltose, and other syrups have required enzymes with improved properties. When dextrose and high-fructose corn syrup are produced from starch, two processes are initially carried out. First, starch is liquefied with a highly temperature-stable α-amylase (EC 3.2.1.1). Then, during saccharification, a pullulanase (EC 3.2.1.41), a debranching enzyme, can be added with glucoamylase (EC 3.2.1.3) to yield more than 94%–95% dextrose. During saccharification, both enzymes must be compatible at pH 4.3 to 5.0 and must survive for days at 60°C. Glucoamylase acts as an exodegrading carbohydrase, while pullulanase debranches $\alpha - 1 \rightarrow 6$ linkages. The starch bioconversion processes are operated at higher temperatures to minimize microbial contamination and to improve starch solubility.

In 1986, the U.S. enzyme market for the food industry was about $155 million; about 60% of that was for carbohydrases, including α-amylase ($15 million), glucoamylase ($25 million), glucose isomerase ($45 million), and other enzymes (about $5 million) (Hester, 1986).

47.3 Carbohydrases and Their Substrates Broadly Defined

Although polysaccharide-modifying enzymes such as carbohydrases have traditionally been limited to producing starch-related products, the development of various microbial polysaccharides and the generally increasing demand for new or improved industrial polysaccharides have prompted interest in enzymes that are responsible for the biosynthesis and degradation of a much wider variety of materials and uses (Yalpani and Desrochers, 1987).

Starch-degrading enzymes have been categorized under a wide range of definitions, some more specific than others. For example, *polysaccharide-modifying enzymes* usually imply those enzymes that hydrolyze starch. However, the category should include catalysts of both synthetic and degradative activities, as characterized by Yalpani and Desrochers (1987). The *synthetic enzymes* relate to biosynthesis and all modifications. Some of the hydrolytic enzymes such as cellulases, chitinases, 1,4-β-D-glucanase, endo-1,3,1,4-β-D-glucanase, glucoamylase, pectinase, pullulanase, xanthanase, and xylanase could be defined as *degradative polysaccharases*. *Alpha-glucosidases* occur ubiquitously in nature (in bacteria, fungi, plants, and animals), catalyze the hydrolysis of alpha-glucosidic linkages, and are among the most significant carbohydrate-splitting enzymes. Their substrates are, depending on their specificity, oligo- and polysaccharides (Truscheit et al., 1981). Finally, *carbohydrases* such as α-amylases, β-amylases, glucoamylases, pullulanases, amylopullulanases, and isoamylases hydrolyze amylose or amylopectin and, in general, can be included under the definitions given earlier.

Amylose and amylopectin are the unbranched and branched polysaccharides, respectively, constituting starch, which is 15%–25% amylose and 75%–85% amylopectin. Amylose is a linear (1-4)-α-glucan consisting of helically arranged chains of polysaccharides with an average degree of polymerization of about 1,000 glucose residues. Amylopectin, on the other hand, is a branched polysaccharide that has a much higher molecular weight (to about 10^7). The molecules have a treelike structure consisting of two types of $\alpha - 1 \rightarrow 4$-glucan chains with average lengths of 20 and more than 50 D-glucose residues that are connected, respectively, by $\alpha - 1 \rightarrow 6$ linkages. A molecule of amylopectin can contain up to 100,000 D-glucose residues and 4,000 to 5,000 interchain $\alpha - 1 \rightarrow 6$-glucosidic linkages (and the same number of individual chains) (Truscheit et al., 1981).

Until recently, the enzymes that degrade starch were generally divided into two main groups, those that specifically hydrolyzed the $\alpha - 1 \rightarrow 6$ interchain linkages (debranching enzymes) and those that split $\alpha - 1 \rightarrow 4$ bonds. The $\alpha - 1 \rightarrow 4$ glucosidases were further subdivided into those that cause random, or internal, cleavage (endoenzymes) and those that act from chain ends (exoenzymes). This chapter describes a unique pullulanase activity, termed amylopullulanase (Saha et al., 1988), that cleaves $\alpha - 1 \rightarrow 4$ linkages in starch and $\alpha - 1 \rightarrow 6$ linkages in pullulan, a linear glucan of about 480 maltotriosyl units linked through $\alpha - 1 \rightarrow 6$ glucosidic linkages. One of the thermoanaerobes known to produce an amylopullulanase is *Thermoanaerobium brockii* (Coleman et al., 1987).

Saha and Zeikus (1989) categorized enzymes that hydrolyze pullulan into four types: (1) *neopullulanase*, hydrolyzing $\alpha - 1 \rightarrow 4$ bonds but producing panose instead of isopanose; (2) *glucoamylase*, degrading pullulan from nonreducing termini and producing glucose; (3) *isopullulanase*, cleaving $\alpha - 1 \rightarrow 4$ glucosidic linkages and producing isopanose; and (4) *pullulanase*, hydrolyzing $\alpha - 1 \rightarrow 6$ glucosidic bonds to form maltotriose.

47.4   Neopullulanase: A Unique Maltogenic α-Amylase

A neopullulanase from *Bacillus stearothermophilus* TRS40 hydrolyzed pullulan efficiently (45%) but starch very slightly (13%) (Kuriki et al., 1988; Imanaka and Kuriki, 1989). The principal degradation product from pullulan hydrolysis was panose, with small amounts of glucose and maltose simultaneously produced at a final molar ratio of 3:1:1, respectively (Kuriki et al., 1988).

These products indicate that both $\alpha - 1 \rightarrow 4$ and $\alpha - 1 \rightarrow 6$ linkages are cleaved in pullulan. Thus, this particular neopullulanase has dual specificity, at least when pullulan is the substrate. The neopullulanase gene has been successfully cloned and expressed in *Bacillus subtilis* (Kuriki et al., 1988). A discussion relating to amylopullulanases, which also have dual specificity, is provided later. Earlier work by Suzuki and Imai (1985) characterized a pullulan hydrolase from *B. stearothermophilus* KP1064 that bioconverts amylose and pullulan into maltose and panose, respectively. Although the authors suggested an enzyme assignment as a unique type of maltogenic α-amylase (1,4-α-D-glucan glucanhydrolase, EC 3.2.1.1), the enzyme could also be broadly identified as a maltogenic neopullulanase, but one with only $\alpha - 1 \rightarrow 4$ linkage specificity for both starch and pullulan.

47.5   The Quest for Greater Glucoamylase Thermostability: From *Aspergillus* to Thermoanaerobes

Glucoamylase (EC 3.2.1.3) is an exoacting carbohydrase that produces glucose by cleaving $\alpha - 1 \rightarrow 4$ bonds consecutively from the non-reducing termini of starch. The various applications and properties of glucoamylase has been reviewed by Fogarty and Kelly (1979, 1980). The primary industrial use of glucoamylase is to make glucose syrups that are subsequently used for fermentation, for production of crystalline glucose, or as a feedstock in the production of high-fructose syrups. Glucoamylase is also used in combination with β-amylases or maltogenic fungal β-amylase for improving the yield of maltose (Hyun and Zeikus, 1985b). The enzyme is made both by bacteria and by eucaryotes such as yeast and fungi; from the latter source, especially from *Aspergillus niger*, much of the glucoamylase used for commercial applications is obtained.

Structural homologies have been observed for glucoamylases. Two glycosylated forms of glucoamylases (E.C.3.2.1.3), designated GAI and GAaII, are secreted by *A. niger* and *A. awamori*. A single gene encodes both forms, which have identical amino acid sequences and hydrolyze $\alpha - 1 \rightarrow 4$ and $\alpha - 1 \rightarrow 6$ linkages but differ in their primary structures and in the lengths of their carboxyl-terminated amino acid sequences. It has been suggested that these differences arise because of differential RNA splicing (Yalpani and Desrochers, 1987).

Because saccharification temperatures greater than 60°C would provide better substrate solubility and minimize interference from contaminating microorganisms, alternate thermostable glucoamylases (and pullulanases) have been sought. Before 1985, the temperature optima of most known glucoamylases were 40° to 60°C, except for *Humicola lanuginosa* glucoamylase, which has optimal activity near 65°C (Fogarty and Kelly, 1979, 1980). In 1985, Hyun and Zeikus (1985b, 1985c) characterized unique glucoamylase and pullulanase activities in *Clostridium thermohydrosulfuricum* 39E, a strain known to ferment starch to ethanol with a 2-h doubling time. The pullulanase and glucoamylase activities were stable and optimally active at 85°C and 75°C, respectively. Because both activities had similar pH optima, pH 5.5–6.0 for the pullulanase and pH 4.0–6.0 for the glucoamylase, they could probably be used in a single step to efficiently transform liquefied starch into a glucose syrup. However, for secreted enzyme in the range of grams per liter, the genes might first have to be transferred and expressed in an alternate host such as *B. subtilis*. Both pullulanase and glucoamylase activities in *C. thermohydrosulfuricum* are membrane bound.

47.6 Increased Production of Thermostable α- and β-Amylases via Heterologous Cloning

Amylase synthesis appears to be regulated and is a rate-limiting step during the growth of thermoanaerobes on starch (Hyun and Zeikus, 1985c). Both glucoamylase and pullulanase synthesis are induced by maltose and are subject to catabolite repression in *Clostridium thermohydrosulfuricum*. Hyperproductive, catabolite repression-resistant amylase mutants generated via nitrosoguanidine treatment have increased starch metabolism and ethanol production. Ethanol-resistant mutants of *C. thermohydrosulfuricum* (Lovitt et al., 1984) have also been made.

Amylolytic enzymes that split only internal, $\alpha - 1 \rightarrow 4$ bonds are α-amylases or $\alpha - 1,4$-glucan-4-glucanohydrolases (EC 3.2.1.1). These enzymes can hydrolyze starch, glycogen, and related $\alpha - 1,4$-glucans. When amylopectin is degraded by an α-amylase, the α-dextrins formed contain the original interchain $\alpha - 1 \rightarrow 6$ glucosidic linkages not cleaved by α-amylases (Truscheit et al., 1981).

A wealth of molecular biology information relating to α-amylase heterologous gene cloning is known (Shinomiya et al., 1981; Palva, 1982; Aiba et al., 1983; Joyet et al., 1984; Tsukagoshi et al., 1984; Vehamaanpera and Korbola, 1986; Long et al., 1987; Yalpani and Desrochers, 1987; Haeckel and Bahl, 1989; Verhasselt et al., 1989). Of relevance to this chapter are the cloning and expression of thermophilic α-amylase genes from *Bacillus stearothermophilus* into *Escherichia coli* (Tsukagoshi et al., 1984) and into *B. subtilis* and *B. stearothermophilus* (Aiba et al., 1983). The last host, containing its recombinant plasmid, produced about fivefold more α-amylase than the wild-type *B. stearothermophilus* strain. Other recombinant systems

have shown rather large increases in gene expression and protein secretion over native strains (Dean and Kaelbling, 1981). Recent work using donor α-amylase genes from anaerobes cloned into *E. coli* revealed that 99% of the amylase activity in an *E. coli* host carrying the thermostable α-amylase gene from *C. thermosulfurogenes* DSM 3896 (Haekel and Bahl, 1989) was cell associated. However, Verhasselt et al. (1989) conjectured that the extracellular α-amylase activity of *E. coli* containing the *C. acetobutylicum* amylase gene might be secreted amylase or perhaps activity resulting from cell lysis.

β-Amylases (1,4-α-D-glucan maltohydrolase, EC 3.2.1.2) hydrolyze starch by splitting successive maltose residues from amylopectin and amylose at the nonreducing termini, producing β-maltose. These exoenzymes are found in plants; most of the industrial β-amylases are obtained from soybeans. No mammals and very few microorganisms are known to produce β-amylases.

Several species of the genus *Bacillus*, such as *B. polymyxa* (Kawazu et al., 1987) and *B. cereus* var. *mycoides* (Takasaki, 1976a, 1976b), produce β-amylases. Hyun and Zeikus (1985a) initially reported that the type strain of *C. thermosulfurogenes* produces an extracellular β-amylase. Hyun and Zeikus (1985d) reported that constitutive or derepressed mutants for β-amylase production were obtained. More recently, a patent issued to Udaka et al. (1989) described a thermophilic β-amylase secreted from *C. thermosulfurogenes* (ATCC 33733), a thermophilic obligate anaerobe. *Bacillus subtilis* was transformed with plasmid pNK1 containing the β-amylase gene, and the transformed *B. subtilis* cells produced β-amylase having the same thermostability as that obtained from *C. thermosulfurogenes*. The enzyme has an optimum temperature of 75°C and is stable to heat even at 85°C. Although the optimum pH range is 5.5 to 6.0, the enzyme is stable in the pH range of 3.5 to 6.5. One favorable property of this β-amylase is that its thermostability is particularly superior to that of the enzyme derived from soybeans and other sources. The enzyme is expected to be very useful for industrial production of maltose and other sugars from starch.

47.7　The Multifaceted Debranchers

Debranching enzymes (EC 3.2.1.) hydrolyze branch points or the $\alpha - 1 \rightarrow 6$ glucosidic interchain linkages in certain α-D-glucans such as amylopectin and glycogen molecules. The enzymes have been known for more than 20 years and have been quite useful for the detailed structural analysis of amylopectins and glycogens and of various dextrins derived from these polysaccharides (Evans et al., 1979). The $\alpha - 1 \rightarrow 6$ glucosidic linkages inhibit the action of glucoamylases and maltogenic β-amylases. Therefore, the debranching enzymes used in glucose, maltose, and high-fructose syrup production from starch hydrolysis improve the saccharification reaction by incorporating a specific amylopectin-debranching enzyme into the system. Criteria for a suitable debranching enzyme for addition during saccharification (Ostergaard et al., 1982) include the following:

1. Heat stability at 60°C
2. Acidophilic character (operative at pH 4.5)
3. Suitability for food production.

Novo Industri in Denmark, after an extensive screening program, succeeded in isolating a species of *Bacillus* that produces an enzyme with the desired properties (Norman, 1983). The enzyme, marketed under the name Promozyme, has been applied in the production of glucose and high-fructose corn syrup.

Debranching enzymes have largely been classified as either pullulanases (EC 3.2.1.41) or isoamylases (EC 3.2.1.68). Nearly 20 years ago Harada et al. (1972) examined the debranching action of isoamylase (or amylopectin 6-glucanohydrolase) from *Pseudomonas amyloderamosa* SB15 and compared it with the pullulanase from *Aerobacter aerogenes* (ATCC 9621), the latter microorganism now known as *Klebsiella aerogenes* or *K. pneumoniae* and the most studied source of pullulanase. The analysis of Harada et al. revealed that isoamylase hydrolyzed both inner and outer branching linkages of amylopectin, while pullulanase cleaved outer linkages well but scarcely affected inner linkages.

An abundance of information on numerous debranching enzymes, especially pullulanases, is now available.

Although isoamylases will catalyze the hydrolysis of $\alpha - 1 \rightarrow 6$ glucosidic linkages in amylopectin and glycogen, they have little or no debranching activity on pullulan. Urlaub and Wober (1975) defined isoamylases as being similar to pullulanases but having no activity on pullulan. More recently, Amemura et al. (1988) cloned the gene (*iam*) coding for *P. amyloderamosa* SB-15, determined its nucleotide sequence, and compared the amino acids of its catalytic site with those of other amylolytic enzymes. Later, Fujita et al. (1989) examined the size of the transcripts of *iam*, the transcription initiation sites, and the regulation of transcription by maltose and glucose. The synthesis of amylases is regulated differently in gram-negative and gram-positive bacteria, such as *E. coli* and *B. subtilis*, respectively. Whereas the maltose regulon of *E. coli*, including the genes coding for certain amylases, is induced by maltose and repressed by glucose, expression of the α-amylase gene in *B. subtilis* is repressed by glucose and not induced by maltose.

Isoamylase activity was found in *Pseudomonas amyloderamosa* strain SB-15 grown in medium containing maltose or starch but not in medium containing glucose (Harada et al., 1968). This finding prompted interest in comparing *iam* gene regulation in strain SB-15 to the regulation of other amylase genes. Besides strain SB-15, isoamylases have also been found in a *Flavobacterium* sp. (Evans et al., 1979), in yeast (Kobayashi, 1956), and in *Bacillus amyloliquefaciens* (Urlaub and Wober, 1975).

Distinct from either pullulanase- or isoamylase-type activity are other debranching enzymes that are not fully understood. An example is a debranching enzyme from *E. coli* K12 that was purified 312 fold and that hydrolyzed $\alpha - 1 \rightarrow 6$ glucosidic linkages in phosphorylase and β-amylase limit dextrins prepared from glycogen and amylopectin. Amylopectin was completely hydrolyzed, but the enzyme showed a very low activity with glycogen as the substrate (Jeanningros et al., 1976). The authors concluded that the enzyme cannot be classified as a pullulanase because it has practically no activity with pullulan. It also differs from the bacterial isoamylases because of its inability to hydrolyze glycogen.

Pullulanases (pullulan 6-glucanohydrolases) that have been found in various microorganisms cleave pullulan to maltotriose, hydrolyze $\alpha - 1 \rightarrow 6$ linkages in amylopectin and have limited effect on glycogen. Morgan et al. (1979) surveyed 297 strains representing 26 species of the genus *Bacillus* and found that 46% (138) of the strains could hydrolyze pullulan. Some of these activities may be one or more of the pullulan-hydrolyzing enzymes listed earlier. However, because pullulanase is involved in branched α-glucan metabolism in bacteria (Norman and Wober, 1975), many amylolytic species of *Bacillus* probably do secrete pullulanase. Related studies on pullulanases from *Bacillus* species have demonstrated an alkaline pullulanase from *Bacillus* no. 202-1 (Nakamura et al., 1975), thermophilic pullulanases from *Thermoactinomyces thalpophilus* (Odibo and Obi, 1988) and *B. acidopullulyticus* (Schulein and Hojer-Pedersen, 1984), a pullulanase from *B. cereus* var. *mycoides* (Takasaki, 1976a, 1976b), and Novo's Promozyme (Ostergaard et al., 1982; Norman, 1983). Pullulanases have also been reported in *Aspergillus niger* (Sakano et al., 1971) and *Streptomyces flavochromogenes* (Ohba et al., 1978).

Many microorganisms that produce debranching enzymes such as pullulanases also appear to produce other carbohydrases with unique properties. One might expect that use of glucose (or maltose) as a carbon source for some of these microorganisms could more readily be achieved with combined $\alpha - 1 \rightarrow 4$ and $\alpha - 1 \rightarrow 6$ linkage hydrolysis of starch or starchlike substrates catalyzed by extracellular carbohydrases. Simultaneous production of extracellular β-amylase and pullulanase by *B. cereus* var. *mycoides* has been observed (Takasaki, 1976a, 1976b). The thermoanaerobe *C. thermohydrosulfuricum* contains both pullulanase and glucoamylase activities, albeit cell-bound (Hyun and Zeikus, 1985b). Finally, production of α-amylase, pullulanase, and α-glucosidase by *C. thermosulfurogenes* (DSM 3896) (Hyun and Zeikus, 1985a) results in an obvious change in the cell envelope structure. When high levels of

extracellular proteins were found in the culture medium as noted earlier, degradation of the cell envelope, formation of blebs, and a high number of free vesicles were observed (Haeckel and Bahl, 1989). Transfer of the thermostable α-amylase gene from *C. thermosulfurogenes* (DSM 3896) into *E. coli* produced an enzyme indistinguishable from the native enzyme (Haeckel and Bahl, 1989).

The pullulanase from *Klebsiella pneumoniae*, a gram-negative organism, was one of the first debranching enzymes studied (Bender and Wallenfels, 1961). An excellent review of pullulanase from *K. pneumoniae* is presented by d'Enfert et al. (1988). The enzyme is excreted from the cells into the culture broth. Production is induced by a substrate of the enzyme, pullulan, or by a product of the enzyme, maltose (Takizawa and Murooka, 1985). In addition, maltotriose can induce the enzyme, but glucose represses it (Michaelis et al., 1985). The 145,000-dalton pullulanase is first excreted to the outer membrane of the cell (Wohner and Wober, 1978) and then released into the growth medium at the end of the exponential growth phase. Pullulanase in either the outer membrane or the growth medium could allow *K. pneumoniae* to use pullulan as a carbon source. Because protein secretion by gram-negative bacteria is still not well understood, pullulanase production by *K. pneumoniae* is a good model for study.

The pullulanase gene (*pul*) of *K. pneumoniae* has been cloned into *E. coli*, with the *pul*[+] recombinant producing about three- to sevenfold more pullulanase than did the wild-type strain *K. pneumoniae* W70 (Takizawa and Murooka, 1985). When the *E. coli pul*[+] plasmid was transferred back into the parent strain, W70, approximately a 20- to 40-fold increase in pullulanase production (with 100- to 150-fold higher intracellular pullulanase) was seen. Michaelis et al. (1985) noted that the *pul* gene is controlled by *mal T*, the positive regulatory gene of the maltose regulon. They also showed that when *E. coli* was transformed with the *pul* gene, the enzyme was distributed in both the inner and outer membranes and that no enzyme was released into the growth medium. Characterization of the region 5' to the *pul* (or *pul A*) gene and of the beginning of the gene revealed that (1) the K. pneumoniae pullulanase is likely to be a lipoprotein; (2) an additional mal T-controlled promoter (the mal X promoter) lies adjacent to the pul A promoter and is oriented in the opposite direction; (3) the pul A and mal X promoters, like the three mal T-controlled promoters, have the consensus sequence 5'-GGAGA 35 base pairs upstream from the transcription initiation site; and (4) both the pul A and mal X promoters appear to lack any detectable upstream binding sites for the cyclic AMP-binding protein (Chapon and Raibaud, 1985).

A cell-associated pullulanase (80% soluble and about 20% membrane associated) from Bacteroides thetaiotaomicron, recently characterized by Smith and Salyers (1989), appears to be one of two pullulanases produced by the gram-negative obligate anaerobe. After directed insertional mutagenesis was used to inactivate the Bacteroides thetaiotaomicron pullulanase gene, the pulluanase-specific activity of the mutant was found to be approximately 45% that of the wild-type strain. A conclusion was that the remaining pullulan-degrading activity was not caused by multiple chromosomal copies of the inactivated gene (Smith and Salyers, 1989). This conclusion was based on gene probe analysis of the parent and mutant strains of Bacteroides thetaiotaomicron. Cloning the gene into E. coli produced a cell-associated and soluble pullulanase with a molecular mass of 72 kDa, similar to the pullulanase from the native strain. Another Bacteroides pullulanase gene, for chondroitinase II, was transferred into E. coli, but expression levels were very low (Guthrie et al., 1985).

An example of greatly enhanced pullulanase activity is seen in the enzyme from the Clostridium sp. strain EM1 (Antranikian et al., 1987a; Madi et al., 1987). Large increases in both the thermostable α-amylase and pullulanase activities from strain EM1 were observed during continuous culture under substrate limitation versus batch culture; the activity per volume increased to 12 fold. Growth-limiting amounts of starch led to overproduction of enzyme, partial disintegration of the cell surface, and a proliferation of the cytoplasmic membrane that is associated with the formation of blebs

and extracellular vesicles (Antranikian et al., 1987a, 1987b), similar to the morphological changes seen for C. thermosulfurogenes (DSM 3896) during the production of extracellular carbohydrates (Haeckel and Bahl, 1989). Optimal pullulanase and α-amylase activities of 2800 and 1450 U/liter, respectively, were achieved from strain EM1 growing on 1% (wt/vol) starch in a chemostat. The amounts of pullulanase and α-amylase excreted were 70% and 55%, respectively. This process for the excretion of thermostable amylolytic enzymes from thermoanaerobes has been filed with the German Patent Office (No. 3639267.7, November 1986).

47.8 The Amylopullulanases

Although pullulanases are known to hydrolyze $\alpha - 1 \rightarrow 6$ linkages in pullulan, $\alpha - 1 \rightarrow 4$ bonds in pullulan and starch are unaffected. If $\alpha - 1 \rightarrow 4$ bonds in pullulan are cleaved to produce panose or isopanose, the enzymes are designated as neopullulanase or isopullulanase, respectively. However, carbohydrases with specificity towards $\alpha - 1 \rightarrow 6$ and $\alpha - 1 \rightarrow 4$ linkages in pullulan and starch, respectively, represent an entirely new class of pullulanases, tentatively called amylopullulanases by Saha and Zeikus (1989).

The tentative designation seems appropriate because amylopullulanases debranch pullulan and also act as amylolytic activities in the presence of starch. Earlier work (Sakano et al., 1982) revealed that Thermoactinomyces vulgaris α-amylase attacks some of the $\alpha - 1 \rightarrow 6$ linkages in partial hydrolysates of pullulan as well as $\alpha - 1 \rightarrow 4$ bonds in starch and pullulan. Subsequent studies by Sakano et al. (1983) suggested that the Thermoactinomyces vulgaris enzyme can hydrolyze $\alpha - 1 \rightarrow 4$ glucosidic linkages in starch and pullulan at the same catalytic site. However, the Thermoactinomyces vulgaris enzyme primarily attacks $\alpha - 1 \rightarrow 4$ bonds in pullulan, producing panose (Fukushima et al. 1982), whereas most of the more recently discovered pullulanases having dual specificity (Table 47.1) hydrolyze only $\alpha - 1 \rightarrow 6$ linkages in pullulan to produce maltotriose. Therefore, further discussion of amylopullulanases includes only enzymes that act as true pullula-

Table 47.1 Amylopullulanases and their properties

Microorganism	Reference	Amylopullulanase			Degradation products from hydrolysis of	
		Weight (kDa)	Glycosylation	Production	Pullulan	Starch
Thermoanaerobium Tok6-B1	Plant et al., 1987	120	?	Extracellular	Maltotriose	Malto-oligo-saccharides
T. brockii (ATCC 33075)	Coleman et al., 1987	105	Yes	Extracellular	Maltotriose	Various oligomers
Clostridium thermo-hydrosulfuricum Z21-109	Saha et al., 1988	136.5	Yes	Intracellular	Maltotriose	DP2-DP4
Thermoanaerobacter sp. B6A	Saha et al., 1990a	450	Yes	Extra and intracellular	Maltotriose	Maltose
Bacillus strain 3183	Saha and Zeikus, 1989	116–205	?	Extracellular	Maltotriose	DP2 and higher
Clostridium thermo-hydrosulfuricum E101-69	Melasniemi, 1987, 1988	330, 370	Yes	Extracellular	Maltotriose	Malto-oligo-saccharides
Bacillus subtilis Tu	Takasaki, 1987	450	?	Extracellular	Maltotriose	Maltotriose, maltose

nases when pullulan is the substrate (that cleave only $\alpha - 1 \rightarrow 6$ bonds) but can also hydrolyze $\alpha - 1 \rightarrow 4$ linkages in starch.

The data on products of starch hydrolysis (Table 47.1) suggest that some if not most of the amylopullulanases also have $\alpha - 1 \rightarrow 6$ specificity in starch. *Clostridium thermohydrosulfuricum* Z21–109 produces an intracellular enzyme that cleaves both $\alpha - 1 \rightarrow 4$ and $\alpha - 1 \rightarrow 6$ bonds in starch to yield DP2-DP4 products (Saha et al., 1988). In addition, starch is hydrolyzed to maltose by a single enzyme isolated from *Thermoanaerobacter* sp. B6A (Saha et al., 1990a). Using nuclear magnetic resonance analysis of degradation products generated by the digestion of starch with the *T. brockii* debranching enzyme, Coleman et al. (1987) found that the $\alpha - 1 \rightarrow 6$ bonds were reduced from 4% (control, no enzyme) to 3.02% (with enzyme). Therefore, the amylopullulanases do appear to have strict specificity with pullulan but not with starch as the substrate.

All the known amylopullulanases have been reported during the last several years (Table 47.1). Five of the seven activities are produced by thermoanaerobes, and only one enzyme is associated with a mesophile, *B. subtilis* TU grown at 30°C), which secretes an enzyme with temperature optima of 60°C and 50°C for the pullulanase and amylase functions, respectively (Takasaki, 1987). All the enzymes are greater than 100 kda; the amylopullulanases from *Thermoanaerobacter* sp. B6A, *C. thermohydrosulfuricum* E101–69, and *B. subtilis* TU are very large (> 300 kd) and are likely to have two or more subunits. For certain amylopullulanases (Table 47.1), glycosylation has been found to occur. Glycosylation may stabilize the enzyme, but this has yet to be demonstrated. Calcium, however, does appear to enhance or stabilize the amylopullulanases derived from *Thermoanaerobium* Tok6-B1, *C. thermohydrosulfuricum* E101–69, and *B. subtilis* TU.

Because most of the thermostable and multifunctional amylopullulanases are secreted, they are of obvious industrial interest. Many amylopullulanases with suitable pH and temperature ranges and optima have excellent promise for producing specialty corn syrups. Saha and Zeikus (1989) evaluated a thermostable amylo-

pullulanase from thermophilic *Bacillus* sp. 3183 for potential use with glucoamylase in the saccharification of liquefied starch. Although the amylopullulanase from *B. subtilis* TU (Takasaki, 1987) increased dextrose yields by 1% to 3% versus glucoamylase alone, the amylopullulanase from *Bacillus* sp. 3183 did not increase the glucose yield with glucoamylase or reduce the overall saccharification time.

Often only low amounts of enzyme are assayed in the cell-free broth of a thermoanaerobic fermentation. Therefore, the gene for the enzyme can become a prime candidate for transfer into a host such as *B. subtilis* that might produce higher levels of extracellular enzyme. Of the genes expressing amylopullulanases (see Table 47.1), only those from *T. brockii* (Coleman et al., 1987) and *C. thermohydrosulfuricum* (Saha et al., 1990b) have been expressed in other host systems. The author is unaware of attempts to clone other amylopullulanase genes. *Escherichia coli* DH 5a, with pUC 18, was the heterologous cloning system used to express the amylopullulanase gene from *C. thermohydrosulfuricum*. This work helped to confirm the enzyme's dual substrate specificity.

Coleman et al. (1987) transferred the amylopullulanase gene from its native host (*Thermoanaerobium brockii*) into *E. coli* and *B. subtilis*. After partial *Hind*III hydrolysis fragments of *T. brockii* genomic DNA were ligated with *Hind*III-cut pBR322, recombinants were transferred to *E. coli* RR1, and $Amp^r Tet^s$ colonies were selected and then grown on pullulan plates (PM) (Morgan et al., 1979) containing D-cycloserine to screen for colonies with intracellular amylopullulanase. The colony producing the largest zone was designated as RR1(pCPC901). The 15.5-kb plasmid (pCPC901, not shown), containing 11.2 kb of *T. brockii* DNA, was subsequently trimmed to a 9.9-kb plasmid designated pCPC902 ($Amp^r Tet^s DE^+$) by ligation of partial *Hinf*I fragments from pCPC901 and transfer of chimeric plasmids into *E. coli* RR1. For cloning foreign DNA into *B. subtilis*, it is generally advantageous to avoid vector instability by inserting smaller segments into host vectors. The reduction in size of pCPC901 and pCPC902 (see Figure 47.1) did not appear to affect the amount of

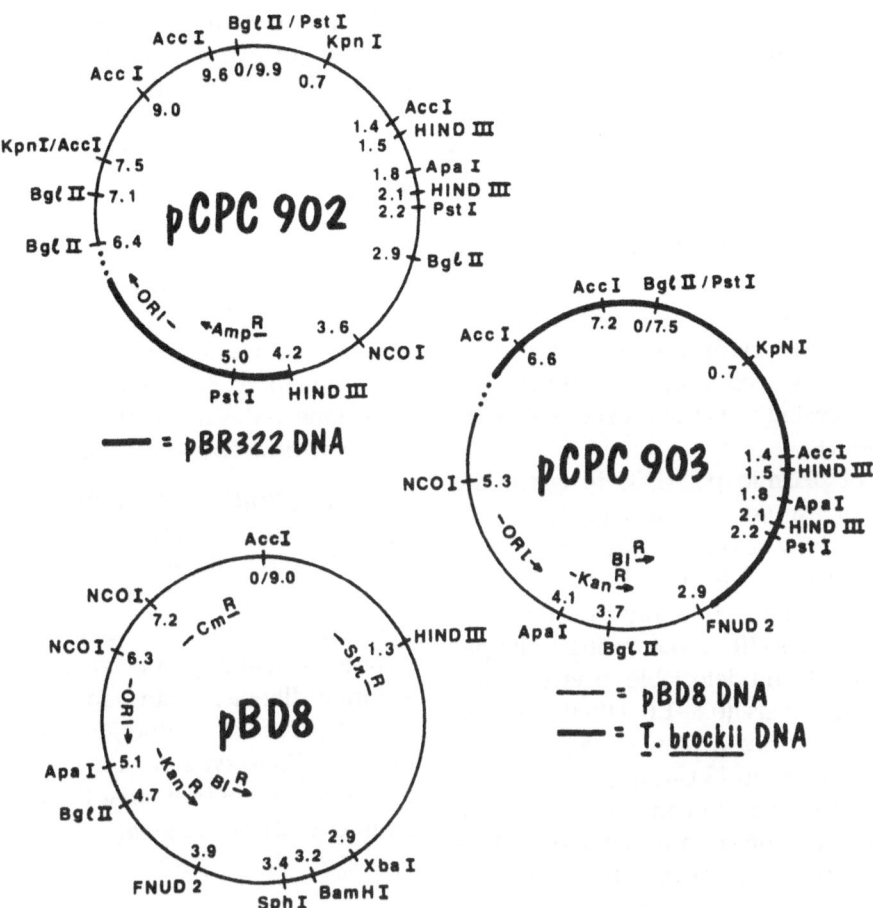

FIGURE 47.1 Cleavage maps with restriction endonucleases for pBD8, pCPC902, and pCPC903. Numerals inside circles indicate molecular size in kilobases. Reprinted from Coleman (1987) with permission of the publisher.

amylopullulanase produced in *E. coli*. Zones around RR1(pCPC901) and RR1(pCPC902) colonies grown on PM plates were the same.

Because of its high frequency of transfer into *B. subtilis* (Gryczan and Dubnau, 1978) and its availability of *Hin*fI sites in nonessential segments of the plasmid (McKenzie et al., 1986), pBD8, encoding kanamycin, chloramphenicol, and streptomycin resistances, was chosen as a vector for introducing the amylopullulanase gene into *B. subtilis*. Partial digests of pBD8 and pCPC902 were made with *Hin*fI, ligation of fragments, and transfer of the new constructs into competent cells (*B. subtilis* B1–20 or B1–109). The pullulan reactive (PRR) plate assay (Yang and Coleman, 1987) was used to distinguish debranching from α-amylase activities and to eliminate the ethanol washes required

in the PM plate assay. PRR plates contained 1.5% agar and 2% PRR in 100mM sodium acetate (pH 5.0). The procedure of Rinderknecht et al. (1967) was modified to prepare PRR; i.e., pullulan and reactive red were used to substitute for starch and Remazol brilliant blue. A Kan[r]CM[s] transformant of *B. subtilis* B1–20 had the largest zone of clearing on a PRR plate and was designated B1–20(pCPC903). pCPC903 (7.4 kb), as seen in Figure 47.1. Most of the gene coding region is in a 2.2-kb *Pst*I fragment that is common to the *E. coli* and *B. subtilis* chimeric vectors pCPC902 and cPCP903, respectively (Figure 47.1).

Although the *T. brockii* amylopullulanase secreted from *B. subtilis* was unglycosylated and had less thermostability, more enzyme was secreted from *B. subtilis* (0.80–1.0 U/ml) than from *T. brockii* (0.23 U/ml). *E. coli* did not export any measurable enzyme. From the fermentation broth of *B. subtilis* containing pCPC903, three active species of the amylopullulanase enzyme were separated; two species may be protease digestion products of the larger protein (105 kDa). The enzyme can cleave all the $\alpha - 1 \rightarrow 6$ glucosidic linkages (and none of the $\alpha - 1 \rightarrow 4$ bonds) in pullulan, but it hydrolyzed mostly $\alpha - 1 \rightarrow 4$ and very few of the $\alpha - 1 \rightarrow 6$ linkages in starch. Hydrolysis of pullulan by the enzyme produced only maltotriose, but starch was digested to various-sized oligomers (see Table 47.1).

Stability of plasmid pCPC903 in either the sporogenic or asporogenic host (B1-20 or B1-109, respectively) was consistently greater than 90% through 40 to 50 h in shake-flask experiments. The plasmid was highly transformable back into *B. subtilis* (10^5 transforming units per μg of DNA) with no detectable in vivo alterations. Enzyme levels (0.8–1.0 U/ml) secreted from either *B. subtilis* host were higher than those from *T. brockii* (0.23 U/ml).

It will be of interest to compare the regulatory region of a gene from an anaerobic thermophile with that of *B. subtilis*. To our knowledge, the work of Coleman et al. (1987) was the first example of such a gene (from a thermophilic anaerobe) being expressed in, and the gene product being exported from, *B. subtilis*. Efstathiou and Truffaut (1986) showed that the *Clostridium acetobutylicum* gene from β-isopropyl malate dehydrogenase could also be expressed and a genetic defect could be complemented in *B. subtilis*.

47.9 Some Final Thoughts

Much of the known or partially understood molecular biology and genetics for anaerobic bacteria is reviewed in this volume. The success of heterologous cloning of various genes from anaerobes into *E. coli*, *B. subtilis*, and other host systems bodes well for future industrial applications. Although there is still much to be accomplished, the developments to date are quite encouraging.

Because of the nature of their secreted enzymes, the thermoanaerobes have become par-

ticularly attractive. However, with the recent discovery of extreme thermophilic anaerobes, bacteria that tolerate and grow in temperatures above 80°C, an even more unique and challenging area of research and development is apparent. How are these microorganisms effectively transformed with foreign DNA? Will the recognition signals for genes in extreme thermophiles be functional in meso- or thermophilic microorganisms? Is glycosylation a prerequisite for enzyme thermostability? Many other questions remain to be answered. However, the work done in this fascinating area should prove to be beneficial and intriguing.

Acknowledgments. The author wishes to acknowledge the support of CPC International, Summit-Argo, Illinois, and his collaboration with S.S. Yang, and M.P. McAlister in the work that is cited in this chapter. The author's current work at Argonne National Laboratory, Argonne, Illinois, is supported in part by the U.S. Department of Energy, Assistant Secretary for Conservation and Renewable Resources, Office of Industrial Technology, under contract W-31-109-Eng-38.

References

Aiba, S., K. Kitai, and T. Imanaka. 1983. Cloning and expression of thermostable alpha-amylase gene from *Bacillus stearothermophilus* in *Bacillus stearothermophilus* and *Bacillus subtilis*. *Appl. Environ. Microbiol.* **46**:1059–1065.

Amemura, A., R. Chakraborty, M. Fujita, M. Noumi, and M. Futai. 1988. Cloning and nucleotide sequence of the isoamylase gene from *Pseudomonas amyloderamosa* SB–15. *J. Biol. Chem.* **263**:9271–9275.

Antranikian, G., C. Herzberg, and G. Gottschalk. 1987a. Production of thermostable alpha-amylase, pullulanase and alpha glucosidase in continuous culture by a new *Clostridium* isolate. *Appl. Environ. Microbiol.* **53**:1668–1673.

Antranikian, G., C. Herzberg, G. Gottschalk, and F. Mayer. 1987b. Changes in the cell envelope structure of *Clostridium* sp. strain EM1 during massive production of alpha-amylase and pullulanase. *FEMS Microbiol. Lett.* **41**:193–197.

Bender, H. and K. Wallenfels. 1961. Untersuchungen an Pullulan. II. Spezifisher Abbau durch ein Bacterielles. *Enzyme Biochem. Z.* **334**:79–95.

Chapon, C. and O. Raibaud. 1985. Structure of two divergent promoters located in front of the gene encoding pullulanase in *Klebsiella pneumoniae* and positively regulated by the malt product. *J. Bacteriol.* **164**:639–645.

Coleman, R., S. Yang, and M. McAlister. 1987. Cloning of the debranching-enzyme gene from *Thermoanaerobium brockii* into *Escherichia coli* and *Bacillus subtilis*. *J. Bacteriol.* **169**:4302–4307.

Dean, D. and M. Kaelbling. 1981. A genetic engineering manifesto for the genus *Bacillus*. *Ann. N.Y. Acad Sci.* **369**:23–32.

d'Enfert, C., I. Reyss, A. Ryter, and A.P. Pugsley. 1988. Pullulanase: a new specific secretion pathway in *Escherichia coli*. *In: Membrane Biogenesis*, NATO ASI Series, Vol. H16, J.A.F. Op den Kamp, ed., pp. 419–428. Berlin: Springer-Verlag.

Efstathiou, I. and N. Truffaut. 1986. Cloning of *Clostridium acetobutylicum* genes and their expression in *Escherichia coli* and *Bacillus subtilis*. *Mol. Gen. Genet.* **204**:317–321. (Erratum; *Mol. Gen. Genet.* **205**:384, 1986.)

Evans, R., D. Manners, and J. Stark. 1979. Partial purification and properties of a bacterial isoamylase. *Carbohydr. Res.* **76**:203–213.

Fogarty, W. and C. Kelly. 1979. Starch-degrading enzymes of microbial origins. *Prog. Indian Microbiol.* **15**:87–150.

Fogarty, W. and C. Kelly. 1980. Amylases, amyloglucosidases and related glucanases. *In Microbial Enzymes and Bioconversions*, A. Rose, ed., pp. 115–169. New York: Academic Press.

Fujita, M., A. Amemura, and M. Futai. 1989. Transcription of the isoamylase gene (*iam*) in *Pseudomonas amyloderamosa* SB-15. *J. Bacteriol.* **171**:4320–4325.

Fukushima, J., Y. Sakano, H. Iwai, Y. Itoh, M. Tanura, and T. Kobayashi. 1982. Hydrolysis of the $\alpha - 1 \rightarrow 6$ glucosidic linkages by an alpha-amylase from *Thermoactinomyces vulgaris*. *Agric. Biol. Chem.* **46**:1423–1424.

Gryczan, T. and D. Dubnau. 1978. Construction and properties of chimeric plasmids in *Bacillus subtilis*. *Proc. Nat. Acad. Sci. USA* **75**:1428–1432.

Guthrie, E., N. Shoemaker, and A. Salyers. 1985. Cloning and expression in *Ecsherichia coli* of a gene coding for a chondroitin lyase from *Bacteriodes thetaiotaomicron*. *J. Bacteriol.* **164**:510–515.

Haeckel, K. and H. Bahl. 1989. Cloning and expression of the thermostable alpha-amylase gene from *Clostridium thermosulfurogenes* (DSM 3896) in *Escherichia coli*. *FEMS Microbiol.* **60**:333–338.

Harada, T., K. Yokobayashi, and A. Misaki. 1968. Formation of isoamylase by *Pseudomonas*. *Appl. Microbiol.* **16**:1439–1444.

Harada, T., A. Misaki, H. Akai, K. Yokobayashi, and K. Sugimoto. 1972. Characterization of *Pseudomonas* isoamylase by its actions on amylopectin and glycogen: Comparison with *Aerobacter* pullulanase. *Biochim. Biophy. Acta* **268**:497–505.

Hester, A. 1986. Market for recombinant enzymes for the food industry. *Gen. Technol. News*, August, p. 6.

Hyun, H. and J. Zeikus. 1985a. General biochemical characterization of thermostable beta-amylase from *Clostridium thermosulfurogenes*. *Appl. Environ. Microbiol.* **49**:1168–1173.

Hyun, H. and G. Zeikus. 1985b. General biochemical characterization of thermostable pullulanase and glucoamylase from *Clostridium thermohydrosulfuricum*. *J. Bacteriol.* **164**:1162–1170.

Hyun, H. and G. Zeikus. 1985c. Regulation and genetic enhancement of glucoamylase and pullulanase production in *Clostridium thermohydrosulfuricum*. *J. Bacteriol.* **164**:1146–1152.

Hyun, H. and G. Zeikus. 1985d. Regulation and genetic enhancement of beta-amylase production in *Clostridium thermosulfurogenes*. *J. Bacteriol.* **164**:1162–1170.

Imanaka, T. and T. Kuriki. 1989. Pattern of action of *Bacillus stearothermophilus* neopullulanase on pullulan. *J. Bacteriol.* **171**:369–374.

Jeanningros, R., N. Creuzet-Sigal, C. Frixon, and J. Cattaneo. 1976. Purification and properties of a debranching enzyme from *Escherichia coli*. *Biochim. Biophys. Acta* **438**:186–199.

Joyet, P., M. Guerineau, and H. Heslot. 1984. Cloning of a thermostable alpha-amylase gene from *Bacillus licheniformis* and its expression in *Escherichia coli* and *Bacillus subtilis*. *FEMS Microbiol. Lett.* **21**:353–358.

Kawazu, T., Y. Nakanishi, N. Uozmi, T. Sasaki, H. Yamagata, N. Tsukagashi, and S. Udaka. 1987. Cloning and nucleotide sequence of the gene coding for enzymatically active fragments of the *Bacillus polymyxa* beta-amylase. *J. Bacteriol.* **169**:1564–1570.

Kobayashi, T. 1956. Studies on isoamylase. *Bull. Agric. Chem. Soc. Jpn.* **19**:163–166.

Kuriki, T., S. Okada and T. Imanaka. 1988. New type of pullulanase from *Bacillus stearothermophilus* and molecular cloning and expression in *Bacillus subtilis*. *J. Bacteriol.* **170**:1554–1559.

Long, C., M. Virolle, S.-Y. Chang, S. Chang, and M. Bibb. 1987. Alpha amylase gene of *Streptomyces limosus*: nucleotide sequence, expression motifs, and amino acid sequence homology to mammalian and invertebrate alpha amylase. *J. Bacteriol.* **169**:5745–5754.

Lovitt, R., R. Longin, and G. Zeikus. 1984. Ethanol production by thermophilic bacteria: physiological comparison of solvent effects on parent and alcohol-tolerant strains of *Clostridium thermohydrosulfuricum*. *Appl. Environ. Microbiol.* **48**:171–177.

McKenzie, T., T. Hoshino, T. Tanaka, and N. Sueoka. 1986. The nucleotide sequence of pUB110: some salient features in relation to replication and its regulation. *Plasmid* **15**:93–103.

Madi, E., G. Antranikian, K. Ohmiya, and G. Gottschalk. 1987. Thermostable amylolytic enzymes from a new *Clostridium* isolate. *Appl. Environ. Microbiol.* **53**:1661–1667.

Melasniemi, H. 1987. Characterization of alpha-amylase and pullulanase activities of *Clostridium thermohydrosulfuricum. Biochem. J.* **246**:193–197.

Melasniemi, H. 1988. Purification and some properties of the extracellular alpha-amylase-pullulanase produced by *Clostridium thermohydrosulfuricum. Biochem. J.* **250**:813–818.

Michaelis, S., C. Chapon, C. D'Enfert, A. Pugsley, and M. Schwartz. 1985. Characterization and expression of the structural gene for pullulanase, a maltose-inducible secreted protein of *Klebsiella pneumoniae. J. Bacteriol.* **164**:633–638.

Morgan, F., K. Adams, and F. Priest. 1979. A cultural method for the detection of pullulan-degrading enzymes in bacteria and its application to the genus *Bacillus. J. Appl. Bacteriol.* **46**:291–294.

Nakamura, N., K. Watanabe and K. Horikoshi. 1975. Purification and some properties of alkaline pullulanase from a strain of *Bacillus* no. 202-1, an alkalophilic microorganism. *Biochim. Biophys. Acta* **397**:188–193.

Norman, B. 1983. A novel *Bacillus* pullulanase—its properties and application in the glucose syrups industry. *J. Jpn. Soc. Starch Sci.* **30**:200–211.

Norman, J. and G. Wober. 1975. Comparative biochemistry of alpha-glucan utilization in *Pseudomonas amyloderamosa* and *Pseudomonas saccharophila. Arch. Microbiol.* **102**:253–260.

Odibo, F. and S. Obi. 1988. Purification and characterization of a thermostable pullulanase from *Thermoactinomyces thalpophilus. J. Ind. Microbiol.* **3**:343–350.

Ohba, R., H. Chaen, S. Hayashi, and S. Veda. 1978. Immobilization of *Streptomyces flavochromogenes* pullulanase on tannic acid and TEAE-cellulose. *Biotechnol. Bioeng.* **20**:665–676.

Ostergaard, J., E. Berhmans, and B. Norman. 1982. Novel debranching enzyme for glucose and high fructose corn syrup production. *In*: *American Association of Cereal Chemists Annual Meeting*, October 24–28, 1982, San Antonio, Texas.

Palva, I. 1982. Molecular cloning of alpha-amylase gene from *Bacillus amyloliquefaciens* and its expression in *Bacillus subtilis. Gene* (Amst.) **19**:81–87.

Plant, R., R. Clemens, H. Morgan, and R. Daniel. 1987. Active site and substrate specificity of *Thermoanaerobium* Tok6-B1 pullulanase. *Biochem. J.* **246**:537–541.

Rinderknecht, H.P., P. Wilding, and B.J. Haverback. 1967. A new method for the determination of alpha-amylase. *Experientia* (Basel) **23**:805.

Saha, B. and G. Zeikus. 1989. Novel highly thermostable pullulanase from thermophiles. *Trends Biotechnol.* **7**:234–239.

Saha, B., S. Mathupala, and G. Zeikus. 1988. Purification and characterization of a highly thermostable novel pullulanase from *Clostridium thermohydrosulfuricum. Biochem. J.* **252**:343–348.

Saha, B., R. Lamed, C.-Y. Lee, S. Mathupala, and G. Zeikus. 1990a. Characterization of an endoacting amylopullulanase from *Thermoanaerobacter* strain B6A. *Appl. Environ. Microbiol.* **56**:881–886.

Saha, B., S. Mathupala, and G. Zeikus. 1990b. Thermostable amylopullulanase from a thermoanaerobe—a new enzyme for starch bioprocessing. *In*: *Bridging the Gap: Case Studies of University Technology Transfer to Industry*, April 30-May 1, 1990, East Lansing, Michigan.

Sakano, Y., N. Masuda, and T. Kobayashi. 1971. Hydrolysis of pullulan by a novel enzyme from *Aspergillus niger. Agric. Biol. Chem.* **35**:971–973.

Sakano, Y., J. Fukushima, and T. Kobayashi. 1983. Hydrolysis of the $\alpha - 1 \rightarrow 4$ and $\alpha - 1 \rightarrow 6$ glucosidic linkages in trisaccharides by the *Thermoactinomyces vulgaris* α-amylase. *Agric. Biol. Chem.* **47**:2211–2216.

Sakano, Y., S. Hiraiwa, J. Fukushima, and T. Kobayashi. 1982. Enzymatic properties and action patterns of *Thermoactinomyces vulgaris. Agric. Biol. Chem.* **46**:1121–1129.

Schulein, M. and B. Hojer-Pedersen. 1984. Characterization of a new class of thermophilic pullulanases from *Bacillus acidopullulyticus*: comparison with *Aerobacter aerogenes* pullulanase; compatibility with *Aspergillus niger* glucoamylase. *Ann. N.Y. Acad. Sci.* **434**:271–274.

Shinomiya, S., K. Yamane, T. Mizukami, F. Kawamura, and H. Saito. 1981. Cloning of thermostable alpha-amylase gene using *Bacillus subtilis* phage p11 as a vector. *Agric. Biol. Chem.* **45**:1733–1735.

Smith, K. and A. Salyers. 1989. Cell-associated pullulanase from *Bacteriodes thetaiotaomicron*: cloning, characterization and insertional mutagenesis to determine role in pullulan utilization. *J. Bacteriol.* **171**:2116–2123.

Suzuki, Y. and T. Imai. 1985. *Bacillus stearothermophilus* KP 1064 pullulanhydrolase. *Appl. Microbiol. Biotechnol.* **21**:20–26.

Takasaki, Y. 1976a. Productions and utilizations of beta-amylase and pullulanase from *Bacillus cereus* var. *mycoides. Agric. Biol. Chem.* **40**:1515–1522.

Takasaki, Y. 1976b. Purifications and enzymatic properties of beta-amylase and pullulanase from *Bacillus cereus* var. *mycoides. Agric. Biol. Chem.* **40**:1515–1522.

Takasaki, Y. 1987. Pullulanase-amylase complex enzyme from *Bacillus subtilis. Agric. Biol. Chem.* **51**:9–16.

Takizawa, N. and Y. Murooka. 1985. Cloning of the pullulanase gene and overproduction of pullulanase in *Escherichia coli* and *Klebsiella* aerogenes. *Appl. Environ. Microbiol.* **49**:294–298.

Truscheit, E., W. Frommer, B. Junge, L. Muller, D. Schmidt, and W. Wingender. 1981. Chemistry and biochemistry of microbial alpha-glucosidase inhibitors. *Angew. Chem.* **20**:744–761.

Tsukagoshi, N., H. Ihara, H. Yamagata, and S. Vdaka. 1984. Cloning and expression of a thermostable alpha-amylase gene from *Bacillus stearothermophilus* in *Escherichia coli. Mol. Gen. Gen.* **193**:58–63.

Udaka, S., H. Yamagato, K. Noriyuki, T. Kato, and N. Tsukagoshi. 1989. Beta-amylase gene derived from *Clostridium thermosulfurogenes* and expressed in *Bacillus subtilis*. European Patent number EP 337090, October 18, 1989.

Urlaub, H. and G. Wober. 1975. Identification of isoamylase, a glycogen-debranching enzyme from *Bacillus amyloliquefaciens. FEBS Lett.* **57**:1–4.

Vehamaanpera, J. and M. Korbola. 1986. Stability of the recombinant plasmid carrying the *Bacillus*

amyloliquefaciens alpha-amylase gene in *Bacillus subtilis. Appl. Microbiol. Biotechnol.* **23**:456–461.

Verhasselt, P., F. Poncelet, K. Vits, K. Gool, and J. Vanderleyden. 1989. Cloning and expression of *Clostridium acetobutylicum* alpha-amylase gene in *Escherichia coli. FEMS Microbiol. Lett.* **59**:135–140.

Wohner, G. and G. Wober. 1978. Pullulanase, an enzyme of starch catabolism, is associated with the outer membrane of *Klebsiella. Arch. Microbiol.* **116**:303–310.

Yalpani, M. and M. Desrochers. 1987. Developments and prospects in enzymatic biopolymer modification. *In: Industrial Polysaccharides: Genetic Engineering, Structure/Property Relations and Applications*, M. Yalpani, ed. pp. 7–34. Amsterdam: Elsevier.

Yang, S.S. and R.D. Coleman. 1987. Detection of pullulanase in polyacrylamide gels using pullulan reactive red agar plates. *Anal. Biochem.* **160**:480–482.

Appendix:
A List of Strict Anaerobes

J. Gregory Zeikus, Monique Hermann, Michel Magot, and Madeleine Sebald

Material Covered

The list that is given in the Appendix covers all species of strict anaerobes, and the genera in which they are included. When in rare cases a genus includes both facultative and strict anaerobic species, only the latter are quoted (i.e., *Streptococcus*). In a few instances, the strictly anaerobic versus facultative character is debatable at the genus level or at the species level by reason of paucity of information on the physiology or strain variation (i.e., *Actinomyces*, *Propionibacterium*, *Treponema*). Microaerophilic anaerobes have been excluded from the list (e.g., *Campylobacter*). Nevertheless, we included the *Capnocytophaga* although these are capnophilic rather than strictly anaerobic.

The List

An alphabetic list of genera is given separately for *Archaebacteria* and *Eubacteria*, and the type species of each genus, when designated, is indicated by an asterisk.

The species designation is followed by naming the type strain, when available. The type strain numbers refer to the following collections: ATCC, American Type Culture Collection, Rockville, Maryland, U.S.A.; AUCM, All-Union Culture of Microorganisms, Moscow, U.S.S.R.; CECT, Coleccion Española de Culti-vos Tipo, Burjasot, Valencia, Spain; DSM, Deutsche Sammlung von Mikroorganismen und Zellkulturen GmbH, Braunschweig, F.R.G.; FERM, Fermentation Research Institute, Tsukuba, Japan; IAM, Institute of Applied Microbiology, University of Tokyo, Tokyo, Japan; IMUSSR, Institute of Microbiology of the U.S.S.R. Academy of Sciences, Moscow, U.S.S.R.; JCM, Japan Collection of Microorganisms, Riken, Japan; NCDO, National Collection of Dairy Organisms, Shinfield, Reading, U.K.; NCIB, National Collection of Industrial Bacteria, Torry Research Station, Aberdeen, Scotland, U.K.; NCTC, National Collection of Type Cultures, Central Public Health Laboratories, London, U.K.; NRCC, National Research Council Collection, Ottawa, Ontario, Canada; VPI, Virginia Polytechnic Institute, Blacksburg, Virginia, U.S.A.

When not deposited in a collection, the strain number of the original description is noted.

References

We provide references only for those genera or species that have been described or redescribed since the current edition of *Bergey's Manual of Systematic Bacteriology* (1984 to 1989). The reference of a new genus is given with the corresponding new type species.

Archaebacteria

Archaeglobus
    A. fulgidus.   TS: DSM 4304.
    A. profundus.   Burggraf et al., 1990b; TS: DSM 5631.

Caldococcus
    C. litoralis.   Svetlichnyi et al., 1988; TS: Z-1301.

Desulfurococcus
    D. amylolyticus.   Bonch-Osmolovskaya et al., 1988; TS: DSM 3822.
    D. mobilis.   TS: DSM 2161.
    D. mucosus.   TS: DSM 2162.

Halomethanococcus
    H. doii.   Yu and Kawamura, 1987; TS: ATCC 43619 (to be reclassified).

Hyperthermus
    H. butylicus.   Zillig et al., 1991; TS: DSM 5496.

Methanobacterium
    M. alcaliphilum.   TS: DSM 3387.
    M. bryantii.   TS: DSM 863.
    M. espanolae.   Patel et al., 1990; TS: NRC 5912.
    M. formicicum.   TS: DSM 1535.
    M. ivanovii.   Jain et al., 1987; TS: DSM 2611.
    M. palustre.   Zellner et al., 1989b; TS: DSM 3108.
    M. thermoaggregans.   TS: DSM 3266.
    M. thermoalcaliphilum.   TS: DSM 3267.
    M. thermoautotrophicum.
                 TS: DSM 1053.
    M. thermoformicicum.   TS: DSM 3720.
    M. thermophilum.   Laurinavichyus et al., 1988; strain M.
    M. uliginosum.   TS: DSM 2956.
    M. wolfei.   TS: DSM 2970.

Methanobrevibacter
    M. arboriphilicus.   TS: DSM 1125.
    M. ruminantium.   TS: DSM 1093.
    M. smithii.   TS: DSM 861.

Methanococcoides
    M. methylutens.   TS: DSM 2657.
    M. euhalobius.   Obraztsova et al., 1987; TS: strain 283.

Methanococcus
    M. aeolicus.   The type strain has not been designated.
    M. igneus.   Burggraf et al., 1990a; TS: DSM 5666.
    M. jannaschii.   TS: DSM 2661.
    M. maripaludis.   TS: DSM 2067.
    M. thermolithotrophicus.
                 TS: DSM 2095.
    M. vannielii.   TS: DSM 1224.
    M. voltae.   TS: DSM 1537.

Methanocorpusculum
    M. aggregans.   Xun et al., 1989; TS: DSM 3027.
    M. bavaricum.   Zellner et al., 1989c; TS: DSM 4179.
    M. labreanum.   Zhao et al., 1989; TS: DSM 4855.
    M. parvum.   Zellner et al., 1987; TS: DSM 3823.
    M. sinense.   Zellner et al., 1989c; TS: DSM 4274.

Methanoculleus.   Maestrojuan et al., 1990.
    M. bourgense.   Maestrojuan et al., 1990; TS: DSM 3045.
    M. marisnigri.   TS: DSM 1498.
    M. olentangyi.   TS: DSM 2772.
    M. thermophilicus.   TS: DSM 2373.

Methanogenium
    M. cariaci.   TS: DSM 1497.
    M. frittonii.   Harris et al., 1984; TS: DSM 2832.
    M. liminatans.   Zellner et al., 1990; TS: DSM 4140.
    M. organophilum.   Widdel et al., 1988; TS: DSM 3596.
    M. tationis.   TS: DSM 2702.

Methanohalobium
    M. evestigatus.   Zhilina and Zavarzin; 1985; TS: DSM 3721.

Methanohalophilus
    M. mahii.   Paterek and Smith, 1988; TS: ATCC 35705.
    M. oregonense.   Liu et al., 1990; TS: 115VP
    M. zhilinae.   Mathrani et al., 1988; TS: DSM 4017.

Methanolacinia
    M. paynteri.   Zellner et al., 1989a; TS: DSM 2545.

Methanolobus
    M. siciliae.   TS: DSM 3028.
    M. tindarius.   TS: DSM 2278.

M. vulcani.   TS: DSM 3029.
Methanomicrobium
   *M. mobile.   TS: DSM 1539.
Methanoplanus
   M. endosymbiosus.   TS: DSM 3599.
   *M. limicola.   TS: DSM 2279.
Methanopyrus.   Huber et al., 1989a.
   No species description.
Methanosaeta
   M. concilii.   Patel and Sprott, 1990; TS:
   DSM 3671.
   M. thermoacetophila.   Patel and Sprott,
   1990; TS: DSM 4774.
Methanosarcina
   M. acetivorans.   TS: DSM 2834.
   M. alcaliphilum.   Nakatsugawa   and
   Horikoshi, 1989.
   *M. barkeri.   TS: DSM 800.
   M. frisia.   Blotevogel   and   Fischer,
   1989; TS: DSM 3318.
   M. mazei.   TS: DSM 2053.
   M. thermophila.   TS: DSM 1825.
   M. vacuolata.   TS: DSM 1232.
Methanosphaera
   M. cuniculi.   Biavati et al., 1988; TS:
   DSM 4103.
   *M. stadtmanae.   TS: DSM 3091.
Methanospirillum
   *M. hungatei.   TS: DSM 864.
Methanothermus
   *M. fervidus.   TS: DSM 2088.
   M. sociabilis.   TS: DSM 3496.
Methanothrix**
   M. soehngenii.   TS: DSM 2139.
Pyrobaculum
   *P. islandicum.   Huber et al., 1987; TS:
   DSM 4184.
   P. organotrophum.   Huber et al., 1987;
   TS: DSM 4185.
Pyrococcus
   *P. furiosus.   TS: DSM 3638.
   P. woesei.   TS: DSM 3773.
Pyrodictium
   P. brockii.   TS: DSM 2708.
   *P. occultum.   TS: DSM 2709.
Staphylothermus
   *S. marinus.   TS: DSM 3639.

**see also: Methanosaeta.

Thermococcus
   *T. celer.   TS: DSM 2476.
   T. litoralis.   Neuner et al., 1990; TS:
   DSM 5473.
   T. stetteri.   Miroshnichenko et al.,
   1989; TS: DSM 5263.
Thermofilum
   *T. pendens.   TS: DSM 2475.
Thermoproteus
   T. neutrophilus.   TS: DSM 2338.
   *T. tenax.   TS: DSM 2078.
   T. uzoniensis.   Bonch-Osmolovskaya et
   al., 1990a; TS: DSM 5263 (=Z605).

Eubacteria

Acetivibrio
   *A. cellulolyticus.   TS: ATCC 33288.
   A. cellulosolvens.   Khan et al., 1984; TS:
   NRCC 2936.
   A. ethanolgignens.   TS: ATCC 33324.
   A. multivorans.   Tanaka et al., 1991; TS:
   DSM 6139.
Acetoanaerobium
   A. noterae.   Sleat et al., 1985; TS: ATCC
   35199.
Acetobacterium
   A. carbinolicum.   Eichler and Schink,
   1984; TS: DSM 2925.
   A. malicum.   Tanaka   and   Pfennig,
   1988; TS: DSM 4132.
   A. wieringae.   TS: DSM 1911.
   *A. woodii.   TS: DSM 1030.
Acetobacteroides
   A. glycinophilus.   Phelps and Zeikus (in
   manuscript); TS: 3078.
Acetofilamentum
   *A. rigidum.   Dietrich et al., 1988b; TS:
   DSM 20769.
Acetogenium
   *A. kivui.   Leigh et al., 1981; TS: DSM
   2030.
Acetomicrobium
   A. faecalis.   Winter et al., 1987; TS:
   DSM 20678.
   *A. flavidum.   Soutschek et al., 1984;
   TS: DSM 20664.
Acetothermus
   *A. paucivorans.   Dietrich et al., 1988a;
   TS: DSM 20768.

Acidaminobacter
  *A. hydrogenoformans. Stams and Hansen, 1984; TS: DSM 2784.
Acidaminococcus
  *A. fermentans. TS: ATCC 25085.
Actinomyces
  *A. bovis. TS: ATCC 13683.
  A. denticolens. TS: NCTC 11490.
  A. georgiae. Johnson et al., 1990; TS: ATCC 49285.
  A. gerencseriae. Johnson et al., 1990; TS: ATCC 23860.
  A. hordeovulneris. TS: ATCC 35275.
  A. howellii. TS: NCTC 11636.
  A. israelii. TS: ATCC 12102.
  A. meyeri. TS: ATCC 35568.
  A. naeslundii. TS: DSM 43013.
  A. odontolyticus. TS: ATCC 17929.
  A. pyogenes. TS: ATCC 19411.
  A. slackii. Dent and Williams, 1986; TS: NCTC 11923.
  A. viscosus. TS: ATCC 15987.
Anaerobacter
  *A. polyendosporus. Duda et al., 1987; TS: PS-1, IM USSR
Anaeroplasma
  *A. abactoclasticum. TS: ATCC 27879.
  A. bactoclasticum. TS: ATCC 27112.
  A. intermedium. Robinson and Freundt, 1987; TS: ATCC 43166.
  A. varium. Robinson and Freundt, 1987; TS: ATCC 43167.
Anaerorhabdus
  *A. furcosus. Shah and Collins, 1986; TS: ATCC 25662.
Anaerobiospirillum
  *A. succinoproducens. TS: ATCC 29305.
Anaerovibrio
  *A. lipolytica. TC: ATCC 33276.
Asteroleplasma
  *A. anaerobium. Robinson and Freundt, 1987; TS: ATCC 27880.
Bacteroides. Shah and Collins, 1989**.
  B. cacae. Johnson et al., 1986; TS: ATCC 43185.

**See also: *Anaerorhabdus; Dichelobacter; Fibrobacter; Megamonas;Mitsuokella; Prevotella; Porphyromonas; Rikenella; Ruminobacter; Sebaldella; Tissierella; Wolinella.*

B. cellulosolvens. Murray et al., 1984; TS: NRCC 2944.
B. coagulans. TS: ATCC 29798.
B. distasonis. TS: ATCC 8503.
B. eggerthii. TS: ATCC 27754.
B. forsythus. Tanner et al., 1986; TS: ATCC 43037.
*B. fragilis. TS: ATCC 25285.
B. galacturonicus. Jensen and Canale-Parola, 1986; TS: ATCC 43244.
B. gracilis. TS: ATCC 33236.
B. helcogenes. Benno et al., 1983; TS: DSM 20613.
B. heparinolyticus. Okuda et al., 1985; TS: ATCC 35895.
B. levii. TS: ATCC 29147.
B. macacae. TS: ATCC 33141.
B. merdae. Johnson et al., 1986; TS: ATCC 43184.
B. ovatus. TS: ATCC 8483.
B. pectinophilus. Jensen and Canale-Parola, 1986; TS: ATCC 43243.
B. pneumosintes. TS: ATCC 33048.
B. polypragmatus. TS: NRCC GP4.
B. putredenis. TS: ATCC 29800.
B. pyogenes. Benno et al., 1983; TS: DSM 20611.
B. salivosus. Love et al., 1987; TS: NCTC 11632.
B. splancnicus. TS: ATCC 29572.
B. stercoris. Johnson et al., 1986; TS: ATCC 43183.
B. suis. Benno et al., 1983; TS: DSM 20612.
B. tectum. Love et al., 1986; TS: NCTC 11853.
B. thetaiotaomicron. TS: ATCC 29148.
B. uniformis. TS: ATCC 8492.
B. ureolyticus. TS: NCTC 10941.
B. vulgatus. TS: ATCC 8482.
B. xylanolyticus. Scholten-Koerselman et al., 1986; TS: DSM 3808
Bifidobacterium
  B. adolescentis. TS: ATCC 15703.
  B. angulatum. TS: ATCC 27535.
  B. animalis. TS: ATCC 25527.
  B. asteroides. TS: ATCC 25910.
  *B. bifidum. TS: ATCC 29521.
  B. boum. TS: ATCC 27917.
  B. breve. TS: ATCC 15700.

B. catenulatum. TS: ATCC 27539.
B. choerinum. TS: ATCC 27686.
B. cuniculi. TS: ATCC 27916.
B. coryneforme. TS: ATCC 25911.
B. dentium. TS: ATCC 27534.
B. gallicum. Lauer, 1990; TS: DSM 20093.
B. gallinarum. Watabe et al., 1983; TS: ATCC 33777.
B. globosum. TS: ATCC 25865.
B. indicum. TS: ATCC 25912.
B. infantis. TS: ATCC 15697.
B. longum. TS: ATCC 15707.
B. magnum. TS: ATCC 27540.
B. merycicum. Biavati and Mattarelli, 1991; TS: ATCC 49391.
B. minimum. TS: ATCC 27538.
B. pseudocatenulatum. TS: ATCC 27919.
B. pseudolongum. TS: ATCC 25526.
B. pullorum. TS: ATCC 27685.
B. ruminantium. Biavati and Mattarelli, 1991; TS: ATCC 49390.
B. subtile. TS: ATCC 27537.
B. suis. TS: ATCC 27533.
B. thermophilum. TS: ATCC 25525.

Bilophila
B. wadsworthia. Baron et al., 1989; TS: ATCC 49260.

Butyribacterium
B. methylotrophicum. TS: DSM 3468.

Butyrivibrio
B. crossotus. TS: ATCC 29175.
B. fibrosolvens. TS: ATCC 19171.

Capnocytophaga
C. canimorsus. Brenner et al., 1989; TS: ATCC 35979.
C. cynodegmi. Brenner et al., 1989; TS: ATCC 49044.
C. gingivalis. TS: ATCC 33624.
C. ochracea. TS: ATCC 27872.
C. sputigena. TS: ATCC 33612.

Centipeda
C. periodontii. Lai et al., 1983; TS: DSM 2778.

*Clostridium***
C. absonum. TS: ATCC 2755.
C. aceticum. TS: ATCC 35044.

**See also: *Anaerobacter, Sporohalobacter, Syntrophospora.*

C. acetobutylicum. TS: ATCC 824.
C. acidiurici. TS: ATCC 7906.
C. aerotolerans. Van Gylswyk and van der Toorn, 1987; TS: ATCC 43524
C. aldrichii. Yang et al., 1990; TS: ATCC49358.
C. aminovalericum. TS: ATCC 13725.
C. arcticum. TS: "Jordan and McNicol no. III."
C. argentinense. Suen et al., 1988; TS: ATCC 27322.
C. aurantibutyricum. TS: ATCC 17777.
C. barati. TS: ATCC 27638.
C. barkeri. TS: ATCC 25849.
C. beijerinckii. TS: ATCC 25752.
C. bifermentans. TS: ATCC 638.
C. botulinum. type A: TS: ATCC 25763; proteolytic type B: TS: ATCC 7949; nonproteolytic type B: TS: ATCC 25765; type C: TS: ATCC 25766; type D: TS: ATCC 25767; type E: TS: ATCC 9564; type F: TS: ATCC 27321. (See also *C. argentinense.*)
C. butyricum. TS: ATCC 19398.
C. cadaveris. TS: ATCC 25783.
C. carnis. TS: ATCC 25777.
C. celatum. TS: ATCC 27791.
C. celerecrescens. Palop et al., 1989; TS: CECT 954.
C. cellobioparum. TS: ATCC 15832.
C. cellulovorans. TS: ATCC 35296.
C. chartatabidum. Kelly et al., 1987; TS: 163.
C. chauvoei. TS: ATCC 10092.
C. clostridiiforme. TS: ATCC 25537.
C. colinum. TS: ATCC 27770.
C. coccoides. TS: ATCC 29236.
C. cochlearium. TS: ATCC 17787.
C. cocleatum. TS: ATCC 29902.
C. collagenovorans. Jain and Zeikus, 1988; TS: DSM 3089.
C. cylindrosporum. TS: DSM 605.
C. difficile. TS: ATCC 9689.
C. disporicum. Horn, 1987; TS: NCIB 12424.
C. durum. TS: ATCC 27763.
C. fallax. TS: NCTC 8380.
C. felsineum. TS: ATCC 17788.
C. fervidus. Patel et al., 1987; TS: ATCC 43204.

C. formicaceticum. TS: DSM 92.

C. ghonii. TS: ATCC 25757.

C. glycolicum. TS: ATCC 14880.

C. haemolyticum. TS: ATCC 9650.

C. halophilum. Fendrich et al., 1990; TS: DSM 5387.

C. hastiforme. TS: ATCC 33268.

C. histolyticum. TS: NCTC 503.

C. homopropionicum. Dörner and Schink, 1990; TS: DSM 5847.

C. indolis. TS: ATCC 25771.

C. innocuum. TS: ATCC 14501.

C. intestinalis. Lee et al., 1989; TS: ATCC 49213.

C. irregulare. TS: ATCC 25756.

C. josui. Sukhumasavi et al., 1988; TS: FERM P-9684.

C. kluyveri. TS: NCIB 10680.

C. lentocellum. Murray et al., 1986; TS: NCIB 11756.

C. leptum. TS: ATCC 29065.

C. limosum. TS: ATCC 25620.

C. litorale. Fendrich et al., 1990; TS: DSM 5388.

C. lituseburense. TS: ATCC 25759.

C. longisporum. Varel, 1989; TS: DSM 605.

C. magnum. TS: DSM 2767.

C. malenominatum. TS: ATCC 25776.

C. mangenotii. TS: ATCC 25761.

C. methylpentosum. Himelbloom and Canale-Parola, 1989; TS: ATCC 43829.

C. nexile. TS: ATCC 27757.

C. novyi. type A: TS: ATCC 17861; type B: ATCC 25758; type C: ATCC 27323.

C. oceanicum. TS: ATCC 25647.

C. oroticum. TS: ATCC 13619.

C. oxalicum. Dehning and Schink, 1989a; TS: Alt 0 × 1.

C. papyrosolvens. TS: DSM 2782.

C. paraputrificum. TS: ATCC 25780.

C. pasteurianum. TS: ATCC 6013.

C. perfringens. TS: ATCC 13124.

C. pfennigii. Krumholz and Bryant, 1985; TS: DSM 3222.

C. polysaccharolyticum. TS: ATCC 33142.

C. propionicum. TS: ATCC 25522.

C. proteolyticum. Jain and Zeikus, 1988; TS: DSM 3090.

C. puniceum. TS: NCIB 11596.

C. purinolyticum. TS: DSM 1384.

C. putrefaciens. TS: ATCC 25786.

C. putrificum. TS: NCIB 10677.

C. quercicolum. TS: ATCC 25974.

C. ramosum. TS: ATCC 25582.

C. rectum. TS: ATCC 25751.

C. roseum. TS: ATCC 17797.

C. saccharolyticum. TS: DSM 2544.

C. sardiniensis. TS: ATCC 33455.

C. sartagoformum. TS: ATCC 25778.

C. scatologenes. TS: ATCC 25775.

C. scindens. Morris et al., 1985; TS: ATCC 35704.

C. septicum. TS: ATCC 12464.

C. sordellii. TS: ATCC 9714.

C. sphenoides. TS: DSM 632.

C. spiroforme. TS: ATCC 29900.

C. sporogenes. TS: ATCC 3584.

C. sporosphaeroides. TS: ATCC 25781.

C. stercorarium. TS: NCIB 11754.

C. sticklandii. TS: ATCC 12662.

C. subterminale. TS: ATCC 25774.

C. symbiosum. TS: ATCC 14940.

C. tertium. TS: ATCC 14753.

C. tetani. TS: ATCC 19406.

C. tetanomorphum. Wilde et al., 1989; TS: DSM 4474.

C. thermoaceticum. TS: DSM 521.

C. thermoauto-trophicum. TS: DSM 1974.

C. thermobutyricum. Wiegel et al., 1989; TS: DSM 4928.

C. thermocellum. TS: ATCC 27405.

C. thermocopriae. Jin et al., 1988; TS: IAM 13577.

C. thermohydrosulfur-icum. TS: DSM 567.

C. thermolacticum. Le Ruyet et al., 1985; TS: DSM 2910.

C. thermopalmarium. Lawson Anani Soh et al., 1991; TS: DSM 5974.

C. thermosaccharo-lyticum. TS: ATCC 7956.

C. thermosuccinogenes. Drent et al., 1991; TS: DSM 5807.

C. thermosulfurogenes. TS: ATCC 33743.

C. tyrobutyricum. TS: ATCC 25755.

C. villosum. TS: DSM 1645.

C. xylanolyticum. Rogers and Baecker, 1991; TS: ATCC 49623.

Coprococcus

    C. catus.   TS: ATCC 27761.

    C. comes.   TS: ATCC 27758.

    *C. eutactus.   TS: ATCC 27759.

Coriobacterium

    *C. glomerans. Haas and König, 1988; TS: DSM 20642.

Desulfobacter

    D. curvatus. Widdel, 1987; TS: DSM 3379.

    D. hydrogenophilus. Widdel, 1987; TS: DSM 3380.

    D. latus. Widdel, 1987; TS: DSM 3381.

    *D. postgatei.   TS: DSM 2034.

Desulfobacterium

    D. anilini. Schnell et al., 1989; TS: DSM 4660.

    *D. autotrophicum. Brysch et al., 1987; TS: DSM 3382.

    D. catecholicum. Szewzyk and Pfennig, 1987; TS: DSM 3882.

    D. indolicum. Bak and Widdel, 1986a; TS: DSM 3383.

    D. macestii. Gogotova and Vainshtein, 1989; TS: AUCCM-B1598.

    D. niacini (Imhoff-Stuckle and Pfenning, 1983). Devereux et al., 1989; TS: DSM 2650.

    D. phenolicum. Bak and Widdel, 1986b; TS: DSM 3384.

    D. vacuolatum. Schauder et al., 1986; TS: DSM 3385.

Desulfobulbus

    D. elongatus. Samain et al., 1984; TS: DSM 2908.

    *D. propionicus.   TS: DSM 2032.

Desulfococcus

    D. biacutus. Platten et al., 1990; TS: DSM 5651.

    *D. multivorans.   TS: DSM 2059.

Desulfohalobium

    D. retbaense. Ollivier et al., 1991; TS: DSM 5692.

Desulfomicrobium

    D. aspheronum. Rozanova et al., 1988; TS: AUCCM 1105.

    D. baculatus. Rozanova et al., 1988; TS: AUCCM 1378.

Desulfomonas

    *D. pigra. TS: ATCC 29098. To be reclassified as *Desulfovibrio.* Devereux et al., 1989.

Desulfomonile

    D. tiedjei. DeWeerd et al., 1990; TS: ATCC 49306.

Desulfonema

    *D. limicola.   TS: DSM 2076.

    D. magnum.   TS: DSM 2077.

Desulfosarcina

    *D. variabilis.   TS: DSM 2060.

Desulfotomaculum

    D. acetoxidans.   TS: DSM 771.

    D. antarcticum.   TS: IAM 64.

    D. geothermicum. Daumas et al., 1988; TS: DSM 3669.

    D. guttoideum. Gogotova and Vainshtein, 1983; TS: DSM 4024.

    D. kuznetsovii. Nazina et al., 1988; TS: AUCCM 17.

    *D. nigrificans.   TS: ATCC 19998.

    D. orientis.   TS: ATCC 19365.

    D. ruminis.   TS: ATCC 23193.

    D. sapomandens. Cord-Ruwish and Garcia, 1985; TS: DSM 3223

    D. thermoacetoxidans. Min and Zinder, 1990; TS: CAMZ.

Desulfovibrio

    D. africanus.   TS: NCIB 8401.

    D. alcoholovorans. Qatibi et al., 1991; TS: DSM 5433.

    D. baarsii. TS: DSM 2075. To be reclassified. Devereux et al., 1989.

    D. baculatus.   TS: DSM 1741.

    D. carbinolicus. Nanninga and Gottschal, 1987; TS: DSM 3852.

    D. desulfuricans subsp. TS: NCIB 9335. *aestuarii.*

    *D. desulfuricans subsp. TS: NCIB 8307. *desulfuricans.*

    D. fructosovorans. Ollivier et al., 1988; TS: DSM 3604.

    D. furfuralis. Folkerts et al., 1989; TS: DSM 2590.

    D. giganteus. Esnault et al., 1988; TS: DSM 4123.

    D. gigas.   TS: NCIB 9332.

    D. halophilus. Caumette et al., 1991; TS: DSM 5663.

D. *salexigens*.　　　　TS: ATCC 14822.

D. *sapovorans*.　TS: DSM 2055. To be re-classified. Devereux et al., 1989.

D. *simplex*. Zellner et al., 1989; TS: DSM 4141.

D. *sulfodismutans*. Bak and Pfennig, 1987; TS: DSM 3696.

D. *termitidis*. Trinkel et al., 1990; TS: DSM 5308.

D. *vulgaris* subsp.　　TS: NCIB 9442.
oxamicus.

D. *vulgaris* subsp.　　TS: DSM 644.
vulgaris.

Desulfurella

*D. acetivorans. Bonch-Osmolovskaya et al., 1990b; TS: DSM 5264.

Desulfuromonas

*D. *acetoxidans*.　　TS: DSM 684.

Dichelobacter　　Dewhirst et al., 1990; TS:
D. *nodosus*.　　　　ATCC 25549.

Dictyoglomus

*D. thermophilum. Saiki et al., 1985; TS: ATCC 35947.

Eubacterium

E. *acidaminophilum*. Zindel et al., 1988; TS: ATCC 3953.

E. *aerofaciens*.　　TS: ATCC 25986.

E. *alactolyticum*.　　TS: ATCC 23263.

E. *angustum*. Beuscher and An-dreesen, 1984; TS: DSM 1989.

F. *biforme*.　　　　TS: ATCC 27806.

E. *brachy*.　　　　TS: ATCC 33089.

E. *budayi*.　　　　TS: ATCC 25541.

E. *cellulosolvens*. Van Gylswyk and van der Toorn, 1986; TS: 6.

E. *callanderi*. Mountfort et al., 1988; TS: DSM 3662.

E. *combesii*.　　　　TS: ATCC 25545.

E. *contortum*.　　　TS: ATCC 25540.

E. *cylindroides*.　　TS: ATCC 27803.

E. *desmolans*. Morris et al., 1986; TS: ATCC 43058.

E. *dolichum*.　　　　TS: ATCC 29143.

E. *eligens*.　　　　TS: ATCC 27750.

E. *fissicatena*.　　　TS: DSM 3598.

E. *formicigenerans*.　TS: ATCC 27755.

E. *fossor*. Bailey and Love, 1986; TS: NCTC 11919.

E. *hadrum*.　　　　TS: ATCC 29173.

E. *hallii*.　　　　TS: DSM 3353.

E. *lentum*.　　　　TS: ATCC 25559

*E. *limosum*.　　　TS: ATCC 8486.

E. *moniliforme*.　　TS: ATCC 25546.

E. *multiforme*.　　　TS: ATCC 2552.

E. *nitritogenes*.　　TS: ATCC 25547.

E. *nodatum*.　　　TS: ATCC 33099.

E. *oxidoreducens*.　Krumholz and Bryant, 1986b; TS: DSM 3217.

E. *plautii*.　　　　TS: ATCC 29863.

E. *plexicaudatum*.　TS: VPI 7582.

E. *ramulus*.　　　　TS: ATCC 29099.

E. *rectale*.　　　　TS: ATCC 33656.

E. *ruminantium*.　　TS: ATCC 17233.

E. *saburreum*.　　　TS: ATCC 33271.

E. *siraeum*.　　　　TS: ATCC 29066.

E. *suis*.　　　　　TS: ATCC 33144.

E. *tarantellus*.　　TS: ATCC 29255.

E. *tenue*.　　　　TS: ATCC 25553.

E. *timidum*.　　　TS: ATCC 33092.

E. *tortuosum*.　　　TS: ATCC 25548.

E. *tucumanense*. Diaz et al., 1989; TS: ATCC 49281.

E. *uniforme*. Van Gylswyk and van der Toorn, 1985; TS: ATCC 35992.

E. *ventriosum*.　　TS: ATCC 27560.

E. *xylanophilum*. Van Gylswyk and van der Toorn, 1985; TS: ATCC 35991.

E. *yurii* subsp. *margaretiae*. Margaret and Krywolap, 1986; TS: ATCC 43715.

E. *yurii* subsp. *schtika*. subsp. nov-Margaret and Krywolap, 1988; TS: ATCC 43716.

E. *yurii* subsp. *yurii*. Margaret and Krywolap, 1986; TS: ATCC 43714.

Fervidobacterium

*F. nodosum. Patel et al., 1985; TS: ATCC 35602.

F. *islandicum*. Huber et al., 1990; TS: DSM 5733.

Fibrobacter

F. *intestinalis*. Montgomery et al., 1988; TS: ATCC 43854.

*F. *succinogenes* subsp. *succinogenes*. Montgomery et al., 1988; TS: ATCC 19169.

F. *succinogenes* subsp. *elongata*. Mont-gomery et al., 1988; TS: ATCC 43856.

Flexistipes
  F. sinusarabici. Fiala et al., 1990; TS: DSM 4947.
Fusobacterium
  F. alocis. Cato et al., 1985; TS: ATCC 35896.
  F. gonidiaformans. TS: ATCC 25563.
  F. mortiferum. TS: ATCC 25557.
  F. naviforme. TS: ATCC 25832.
  F. necrophorum. TS: ATCC 25286.
  F. nucleatum subsp. *animalis.* Gharbia and Shah, 1990; TS: LH 37.
  F. nucleatum subsp. *fusiforme.* Gharbia and Shah, 1990; TS: NCTC 11362.
  F. nucleatum subsp. *nucleatum.* Dzink et al., 1990; TS: ATCC 25586
  F. nucleatum subsp. *polymorphum.* Dzink et al., 1990; TS: ATCC 10953.
  F. nucleatum subsp. *vincentii.* Dzink et al., 1990; TS: ATCC 49256.

  F. necrogenes. TS: ATCC 25556.
  F. perfoetens. TS: ATCC 29250.
  F. periodonticum. Slots et al., 1983; TS: ATCC 33693.
  F. prausnitzii. TS: ATCC 27768.
  F. pseudonecrophorum. Shinjo et al., 1990; TS: Fn 521.
  F. russii. TS: ATCC 25533.
  F. simiae. TS: ATCC 33568.
  F. sulci. Cato et al., 1985; TS: ATCC 35585.
  F. ulcerans. Adriaans and Shah, 1988; TS: NCTC 12111.
  F. varium. TS: ATCC 8501.

Gemmiger
  G. formicilis. TS: 206.
Haloanaerobium
  H. praevalens. Zeikus et al., 1983b; TS: ATCC 33744.
Halobacteroides
  H. acetoethylicus. Rengpipat et al., 1988; TS: ATCC 43120.
  H. halobius. Oren et al., 1984; TS: ATCC 35273.
Ilyobacter
  I. delafieldii. Janssen and Harfoot, 1990; TS: DSM 5704.
  I. polytropus. Stieb and Schink, 1984; TS: DSM 2926.

  I. tartaricus. Schink, 1984b; TS: DSM 2382.
Lachnospira
  L. multipara. TS: ATCC 19207.
Leptotrichia
  L. buccalis. TS: DSM 1135.
Malonomonas
  M. rubra. Dehning and Schink, 1989b; TS: DSM 5091.
Megamonas
  M. hypermegas. TS: ATCC 25560.
Megasphaera
  M. cerevisiae. Engelmann and Weiss, 1985; TS: DSM 20462.
  M. elsdenii. TS: ATCC 25940.
Mitsuokella
  M. dentalis. Haapasalo et al., 1986; TS: DSM 3688.
  M. multiacidus. TS: ATCC 27723.
Mobiluncus
  M. curtisii subsp. *curtisii.* Spiegel and Roberts, 1984; TS: ATCC 35241.
  M. curtisii subsp. *holmesii.* Spiegel and Roberts, 1984; TS: ATCC 35242.
  M. mulieris. Spiegel and Roberts, 1984; TS: ATCC 35243.
Oxalobacter
  O. formigenes. Allison et al., 1985; TS: ATCC 35274.
  O. vibrioformis. Dehning and Schink, 1989a; TS: Wo O × 3.
Pectinatus
  P. cerevisiiphilus. TS: ATCC 29359.
  P. frisingensis. Schleifer et al., 1990; TS: ATCC 33332.
Pelobacter
  P. acetylenicus. Schink, 1985; TS: DSM 3246.
  P. acidigallici. Schink and Pfennig, 1982a; TS: DSM 2377.
  P. carbinolicus. Schink, 1984a; TS: DSM 2380.
  P. propionicus. Schink, 1984a; TS: DSM 2379.
  P. venetianus. Schink and Stieb, 1983; TS: DSM 2394.
Peptococcus
  P. niger. TS: ATCC 27731.

Peptostreptococcus
*P. anaerobius.   TS: ATCC 27337.
P. asacharolyticus.   TS: ATCC 14963.
P. barnesae. Schiefer-Ullrich and Andreesen, 1985; TS: DSM 3244.
P. heliotrinreducens.   TS: ATCC 29202.
P. hydrogenalis. Ezaki et al., 1990; TS: JCM 7635.
P. indolicus.   TS: ATCC 29427.
P. magnus.   TS: ATCC 15794.
P. micros.   TS: ATCC 33270.
P. prevotii.   TS: ATCC 9321.
P. productus.   TS: ATCC 27340.
P. tetradius.   TS: ATCC 35098.
Porphyromonas. Shah and Collins, 1988.
*P. asacharolytica.   TS: ATCC 25260.
P. gingivalis.   TS: ATCC 33277.
P. endodontalis. Van Steenbergen et al., 1984; TS: ATCC 35406.
Prevotella. Shah and Collins, 1990.
P. biviae.   TS: ATCC 29303.
P. buccae.   TS: ATCC 33574.
P. buccalis.   TS: NCDO 2354.
P. corporis.   TS: ACC 33547.
P. denticola.   TS: NCDO 2352.
P. disiens.   TS: ATCC 29426.
P. heparinolytica.   TS: ATCC 35895.
P. intermedia.   TS: ATCC 25611.
P. loescheii.   TS: ATCC 15930.
*P. melaninogenica.   TS: ATCC 25845.
P. oralis.   TS: ATCC 33269.
P. oris.   TS: ATCC 33573.
P. oulora. Shah et al., 1985; TS: NCTC 11871.
P. ruminicola subsp.   TS: ATCC 19188.
  brevis.
  subsp. ruminicola.   TS: ATCC 19189.
P. veroralis.   TS: ATCC 33779.
P. zoogleoformans.   TS: ATCC 33285.
Propionibacterium
P. acidipropionici.   TS: ATCC 25562.

P. acnes.   TS: ATCC 6919.
P. avidum.   TS: ATCC 25577.
P. granulosum.   TS: ATCC 25564.
*P. freudenreichii.   TS: ATCC 6207.
P. jensenii.   TS: ATCC 4868.
P. lymphophilum.   TS: ATCC 27520.
P. propionicum. Charfreitag et al., 1988; TS: DSM 43307.

P. thoenii.   TS: ATCC 4874.
Propionigenium
P. modestum. Schink and Pfennig, 1982b; TS: DSM 2376
Propionispira
P. arboris. Schink et al., 1982; TS: ATCC 33732.
Propionivibrio
P. dicarboxylicus. Tanaka et al., 1990; TS: JCM 7784.
Rikenella
*R. microfusus. Collins et al., 1985; TS: NCTC 11190.
Ruminobacter
*R. amylophilus. Stackebrandt and Hippe, 1986; TS: ATCC 29744.
Ruminococcus
R. albus.   TS: ATCC 27210.
R. bromii.   TS: ATCC 27255.
R. callidus.   TS: VPI 57-31.
*R. flavefaciens.   TS: ATCC 19208.
R. gnavus.   TS: ATCC 29149.
R. lactaris.   TS: ATCC 29176.
R. obeum.   TS: ATCC 29174.
R. pasteurii. Schink, 1984b; TS: DSM 2381.
R. torques.   TS: ATCC 27756.
Sarcina
S. maxima.   TS: DSM 316.
*S. ventriculi.   TS: ATCC 19633.
Sebaldella
*S. termitidis. Collins and Shah, 1986a; TS: NCTC 11300.
Selenomonas
S. artemidis. Moore et al., 1987; TS: ATCC 43528.
S. dianae. Moore et al., 1987; TS: ATCC 43527.
S. flueggei. Moore et al., 1987; TS: ATCC 43531.
S. infelix. Moore et al., 1987; TS: ATCC 43532.
S. lacticifex. Schleifer et al., 1990; TS: DSM 20757.
S. noxia. Moore et al., 1987; TS: ATCC 43542.
S. ruminantium.
  subsp lactilytica.   TS: ATCC 19205.
  subsp. ruminantium.   TS: ATCC 12561.
*S. sputigena. Johnson et al., 1985; Neo

TS: ATCC 35185.

Serpula. Stanton et al., 1991.
- *S. hyodysenteriae.*   TS: ATCC 2168.
- *S. innocens.*   TS: ATCC 29796.

Spirochaeta
- *S. aurantia* subsp.   TS: DSM 1902.
  aurantia.
- *S. isovalerica.*   TS: DSM 2461.
- *S. litoralis.*   TS: DSM 2029.
- *S. stenostrepta.*   TS: DSM 2028.
- *S. zuelzerae.*   TS: DSM 1903.

Sporohalobacter
- **S. lortetii.* Oren et al., 1987; TS: ATCC 35059.
- *S. marismortui.* Oren et al., 1987; TS: ATCC 35420.

Sporomusa
- *S. acidovorans.* Ollivier et al., 1985a; TS: DSM 3132.
- *S. malonica.* Dehning et al., 1989; TS: DSM 5090.
- *S. ovata.* Möller et al., 1984; TS: DSM 2662.
- *S. paucivorans.* Hermann et al., 1987; TS: DSM 3697.
- **S. sphaeroides.* Moller et al., 1984; TS: DSM 2875.
- *S. termitida.* Breznak et al., 1988; TS: DSM 4440.

Streptococcus
includes three anaerobic species:
- *S. hansenii.*   TS: ATCC 27752.
- *S. pleomorphus.*   TS: NCTC 11087.
- *S. parvulus.*   TS: ATCC 33793.

Succinomonas
- **S. amylolytica.*   TS: ATCC 19206.

Succinovibrio
- **S. dextrinosolvens.*   TS: ATCC 19716.

Syntrophobacter
- *S. wolinii.* Boone and Bryant, 1980; TS: DSM 2805.

Syntrophococcus
- **S. sucromutans.* Krumholz and Bryant, 1986a; TS: DSM 3224.

Syntrophomonas
- *S. sapovorans.* Roy et al., 1986; TS: DSM 3441.
- **S. wolfei.* McInerney et al., 1981; TS: DSM 2245.

Syntrophospora
- **S. bryantii.* Zhao et al., 1990; TS: DSM 3014.

Syntrophus
- **S. buswelli.* Mountfort et al., 1984; TS: DSM 2612.
- *S. massiliense.* Heitz et al., in manuscript; TS: DSM 4156.

Thermoanaerobacter
- **T. ethanolicus.*   TS: ATCC 31550.
- *T. finnii.* Schmid et al., 1986; TS: DSM 3389.

Thermoanaerobium
- **T. brockii.*   TS: ATCC 33075.
- *T. lactoethylicum.* Kondratieva et al., 1989; TS: ZE1.

Thermobacteroides
- **T. acetoethylicus.* Ben-Bassat and Zeikus, 1981; TS: ATCC 33265.
- *T. leptospartum.* Toda et al., 1988; TS: IAM 13499.
- *T. proteolyticus.* Ollivier et al., 1985b; TS: ATCC 35242.

Thermodesulfobacterium
- **T. commune.* Zeikus et al., 1983a; TS: ATCC 33708.
- **T. mobile.* Rozanova and Pivovarova, 1988; TS: DSM 1276.

Thermosipho
- **T. africanus.* Huber et al., 1989b; TS: DSM 5309.

Thermotoga
- **T. maritima.* Huber et al., 1986; TS: DSM 3109.
- *T. neapolitana.* Jannasch et al., 1988; TS: DSM 4359.
- *T. thermarum.* Windberger et al., 1989; TS: DSM 5069.

Tissierella
- **T. praeacuta.* Collins and Shah, 1986b; TS: ATCC 25539.

Treponema**
includes following strict anaerobes:
- *T. bryantii.*   TS: ATCC 33254.
- *T. saccharophilum.* Paster and Canale-Parola, 1985; TS: DSM 2985.

** See also: *Serpula.*

T. socranskii subsp. *socranskii*. Smibert et al., 1984; TS: ATCC 35536.

T. socranskii subsp. *buccale*. Smibert et al., 1984; TS: ATCC 35534.

T. socranskii subsp. *paredis*. Smibert et al., 1984; TS: ATCC 35535.

Veillonella

V. atypica	TS: ATCC 17744.
V. caviae.	TS: ATCC 33540.
V. criceti.	TS: ATCC 17747.
V. dispar.	TS: ATCC 17748.
*V. parvula.	TS: ATCC 10790.
V. ratti.	TS: ATCC 17746.
V. rodentium.	TS: ATCC 17743.

Wolinella

W. curva.	TS: ATCC 35224.
W. recta.	TS: ATCC 33238.
*W. succinogenes.	TS: ATCC 29543.

Zymomonas

*Z. mobilis.	TS: DSM 424.

Zymophilus

Z. paucivorans. Schleifer et al., 1990; TS: DSM 20756.

*Z. raffinosivorans. Schleifer et al., 1990; TS: DSM 20765.

References

Adriaans, B. and H. Shah. 1988. *Fusobacterium ulcerans* sp. nov. from tropical ulcers. *Int. J. Syst. Bacteriol.* **38**:447–448.

Allison, M.J., K.A. Dawson, W.R. Mayberry, and J.G. Foss. 1985. *Oxalobacter formigenes* gen. nov. sp. nov.: oxalate-degrading anaerobes that inhabit the gastrointestinal tract. *Arch. Microbiol.* **141**:1–7.

Bailey, G.D. and D.N. Love. 1986. *Eubacterium fossor* sp. nov., an agar-corroding organism from normal pharynx and oral and respiratory tract lesions of horses. *Int. J. Syst. Bacteriol.* **36**:383–387.

Bak, F. and N. Pfennig. 1987. Chemolitotrophic growth of *Desulfovibrio sulfodismutans* sp. nov. by disproportionation of inorganic sulfur compounds. *Arch. Microbiol.* **147**:184–189.

Bak, F. and F. Widdel. 1986a. Anaerobic degradation of indolic compounds by sulfate-reducing enrichment cultures, and description of *Desulfobacterium indolicum* gen. nov., sp. nov. *Arch. Microbiol.* **146**:170–176.

Bak, F. and F. Widdel. 1986b. Anaerobic degradation of phenol and phenol derivatives by *Desul-*fobacterium phenolicum sp. nov. *Arch. Microbiol.* **146**:177–180.

Baron, E.J., P. Summanen, J. Downes, M.C. Roberts, H. Wexler and S.M. Finegold. 1989. *Bilophila wadsworthia* gen. nov. and sp. nov., a unique gram-negative anaerobic rod recovered from appendicitis specimens and human faeces. *J. Gen. Microbiol.* **135**:3405–3411.

Ben-Bassat, A. and J.G. Zeikus. 1981. *Thermobacteroides acetoethylicus* gen. nov. and spec. nov., a new chemoorganotrophic, anaerobic, thermophilic bacterium. *Arch. Microbiol.* **128**:365–370.

Benno, Y., W. Watabe, and T. Mitsuoka. 1983. *Bacteroides pyogenes* sp. nov., *Bacteroides suis* sp. nov., and *Bacteroides helcogenes* sp. nov., new species from abscesses and feces of pigs. *Syst. Appl. Microbiol.* **4**:396–407.

Bergey's Manual of Systematic Bacteriology. 1984. Vol. 1: N.R. Krieg (editor); 1986. Vol. 2: P.H.A. Sneath (ed.); 1989. Vol. 3: J.T. Staley (ed.) and Vol. 4: S.T. Williams (ed.). Baltimore: Williams & Wilkins.

Beuscher, H.U. and J.R. Andreesen. 1984. *Eubacterium angustum* sp. nov., a gram-positive anaerobic, non-sporeforming, obligate purine fermenting organism. *Arch. Microbiol.* **140**:2–8.

Biavati, B. and P. Mattarelli. 1991. *Bifidobacterium ruminantium* sp. nov. and *Bifidobacterium merycicum* sp. nov. from the rumens of cattle. *Int. J. Syst. Bacteriol.* **41**:163–165.

Biavati, B., M. Vasta, and J.G. Ferry. 1988. Isolation and characterization of *Methanosphaera cuniculi* sp. nov. *Appl. Environ. Microbiol.* **54**: 768–771.

Blotevogel, K.-H. and U. Fischer. 1989. Transfer of *Methanococcus frisius* to the genus *Methanosarcina* as *Methanosarcina frisia* comb. nov. *Int. J. Syst. Bacteriol.* **39**:91–92.

Bonch-Osmolovskaya, E.A., M.L. Miroschichenko, N.A. Kostrikina, N.A. Chernych, and G.A. Zavarzin. 1990a. *Thermoproteus uzoniensis* sp. nov., a new extremely thermophilic archaebacterium from Kamchatka continental hot springs. *Arch. Microbiol.* **154**:556–559.

Bonch-Osmolovskaya, E.A., T.G. Sokolova, N.A. Kostrikina, and G.A. Zavarzin. 1990b, *Desulfurella acetivorans* gen. nov. and sp. nov.—a new thermophilic sulfur-reducing eubacterium. *Arch. Microbiol.* **153**:151–155.

Bonch-Osmolovskaya, E.A., A.I. Sleserev, M.L. Miroshnichnenko, T.P. Svetlichnaya, and V.A. Alekseev. 1988. Characteristics of *Desulfurococcus amylolyticus* n. sp.—a new extremely thermophilic archaebacterium isolated from thermal springs of Kamchatka and Kunashir Island. *Microbiology* (English translation of *Mikrobiologiya*) **57**:78–85.

Boone, D.R. and M.P. Bryant. 1980. Propionatedegrading bacterium *Syntrophobacter wolinii* sp.

nov., gen. nov. from methanogenic ecosystems. *Appl. Environ. Microbiol.* **40**:626–632.

Brenner, D.J., D.G. Hollis, G.R. Fanning, and R.E. Weaver. 1989. *Capnocytophaga canimorsus* sp. nov. (formerly CDC Group DF-2), a cause of septicemia following dog bite, and *C. cynodegmi* sp. nov., a cause of localized wound infection following dog bite. *J. Clin. Microbiol.* **27**:231–235.

Breznak, J.A., J.M. Switzer, and H.J. Seitz. 1988. *Sporomusa termitida* sp. nov., an H_2/CO_2-utilizing acetogen isolated from termites. *Arch. Microbiol.* **150**:282–288.

Brysch, K., C. Schneider, G. Fuchs, and F. Widdel. 1987. Lithoautotrophic growth of sulfate-reducing bacteria, and description of *Desulfobacterium autotrophicum* gen. nov., sp. nov. *Arch. Microbiol.* **148**:264–274.

Burggraf, S., H. Fricke, A. Neuner, J. Kristjansson, P. Rouvier, L. Mandelco, C.R. Woese, and K.O. Stetter. 1990a. *Methanococcus igneus* sp. nov., a novel hyperthermophilic methanogen from a shallow submarine hydrothermal system. *Syst. Appl. Microbiol.* **13**:263–269.

Burggraf, S., H.W. Jannasch, B. Nicolaus, and K.O. Stetter. 1990b. *Archaeglobus profundus* sp. nov. represents a new species with the sulfate-reducing archaebacteria. *Syst. Appl. Microbiol.* **13**:24–28.

Cato, E.P., L.V.H. Moore, and W.E.C. Moore. 1985. *Fusobacterium alocis* sp. nov. and *Fusobacterium sulci* sp. nov. from the human gingival sulcus. *Int. J. Syst. Bacteriol.* **35**:475–477.

Caumette, P., Y. Cohen, and R. Matheron. 1991. Isolation and characterization of *Desulfovibrio halophilus* sp. nov., a halophilic sulfate-reducing bacterium isolated from Solar Lake (Sinaï). *Syst. Appl. Microbiol.* **13**:33–38.

Charfreitag, O., M.D. Collins, and E. Stackebrandt. 1988. Reclassification of *Arachnia propionica* as *Propionibacterium propionicus* comb. nov. *Int. J. Syst. Bacteriol.* **38**:354–357.

Collins, M.D. and H.N. Shah. 1986a. Reclassification of *Bacteroides termitidis* Sebald (Holdeman and Moore) in a new genus *Sebaldella*, as *Sebaldella termitidis* comb. nov. *Int. J. Syst. Bacteriol.* **36**:349–350.

Collins, M.D. and H.N. Shah. 1986b. Reclassification of *Bacteroides praeacutus* Tissier (Holdeman and Moore) in a new genus, *Tissierella*, as *Tissierella praeacuta* comb nov. *Int. J. Syst. Bacteriol.* **36**:461–463.

Collins, M.D., H.N. Shah, and T. Mitsuoka. 1985. Reclassification of *Bacteroides microfusus* (Kaneuchi and Mitsuoka) in a new genus *Rikenella*, as *Rikenella microfusus* comb. nov. *Syst. Appl. Microbiol.* **6**:79–81.

Cord-Ruwisch, R. and J.L. Garcia. 1985. Isolation and characterization of an anaerobic benzoate-degrading spore-forming sulfate-reducing bac-

terium, *Desulfotomaculum sapomandens* sp. nov. *FEMS Microbiol. Lett.* **29**:325–330.

Daumas, S., R. Cord-Ruwisch, and J.L. Garcia. 1988. *Desulfotomaculum geothermicum* sp. nov., a thermophilic, fatty acid-degrading, sulfate-reducing bacterium isolated with H_2 from geothermal ground water. *Antonie Leeuwenhoek J. Microbiol.* **54**:165–178.

Dehning, I. and B. Schink. 1989a. Two new species of anaerobic oxalate-fermenting bacteria, *Oxalobacter vibrioformis* sp. nov. and *Clostridium oxalicum* sp. nov., from sediment samples. *Arch. Microbiol.* **153**:79–84.

Dehning, I. and B. Schink. 1989b. *Malonomonas rubra* gen. nov. sp. nov., a microaerotolerant anaerobic bacterium growing by decarboxylation of malonate. *Arch. Microbiol.* **151**:427–433.

Dehning, I., M. Stieb, and B. Schink. 1989. *Sporomusa malonica*, sp. nov., a homoacetogenic bacterium growing by decarboxylation of malonate or succinate. *Arch. Microbiol.* **151**:421–426.

Dent, V.E. and R.A.D. Williams. 1986. *Actinomyces slackii* sp. nov. from dental plaque of dairy cattle. *International Journal of Systematic Bacteriology* **36**:392–395.

Devereux, R., M. Delaney, F. Widdel, and D.A. Stahl. 1989. Natural relationships among sulfate-reducing eubacteria. *J. Bacteriol.* **171**:6689–6695.

DeWeerd, K.A., M. Mandelco, R.S. Tanner, C.R. Woese, and J.M. Sulfita. 1990. *Desulfomonile tiedjei* gen. nov. and sp. nov., a novel anaerobic, dehalogenating, sulfate-reducing bacterium. *Arch. Microbiol.* **154**:23–30.

Dewhirst, F.E., B.J. Paster, S. La Fontaine and J.I. Rood. 1009. Transfer of *Kingella indologenes* (Snell and Lapage 1976) to the genus *Suttonella* gen. nov. as *Suttonella indologenes* comb. nov.; transfer of *Bacteroides nodosus* (Beveridge 1941) to the genus *Dichelobacter* gen. nov. as *Dichelobacter nodosus* comb. nov.; and assignment of the genera *Cardiobacterium*, *Dichelobacter* and *Suttonella* to *Cardiobacteriaceas* fam. nov. in the gamma subdivision of *Proteobacteria* based upon 16S ribosomal ribonucleic acid sequence comparisons. *International Journal of Systematic Bacteriology* **40**:426–433.

Diaz, H.F., C.G. Nunez, and F. Sineriz. 1989. *Eubacterium tucumanense* sp. nov.: an anaerobic gram-positive non-sporeformer isolated from an anaerobic digester. *J. Gen. Microbiol.* **135**:2537–2541.

Dietrich, G., N. Weiss, and J. Winter. 1988a. *Acetothermus paucivorans*, gen. nov., sp. nov., a strictly anaerobic, thermophilic bacterium from sewage sludge, fermenting hexoses to acetate, CO_2 and H_2. *Syst. Appl. Microbiol.* **10**:174–179.

Dietrich, G., N. Weiss, F. Fiedler, and J. Winter. 1988b. *Acetofilamentum rigidum* gen. nov. sp. nov., a strictly anaerobic bacterium from sewage sludge. *Syst. Appl. Microbiol.* **10**:273–278.

Dörner, C. and B. Schink. 1990. *Clostridium homopropionicum* sp. nov., a new strict anaerobe growing with 2-, 3-, or 4-hydroxybutyrate. *Arch. Microbiol.* **154**:342–348.

Drent, W.J., G.A. Lahpor, W.M. Wiegant, and J.C. Gottschal. 1991. Fermentation of inulin by *Clostridium thermosuccinogenes* sp. nov., a thermophilic anaerobic bacterium isolated from various habitats. *Appl. Environ. Microbiol.* **57**:455–462.

Duda, V.l., A.V. Lebedinsky, M.S. Mushegjan, and L.L. Mitjushina. 1987. A new anaerobic bacterium, forming up to five endospores per cell—*Anaerobacter polyendosporus* gen. et spec. nov. *Arch. Microbiol.* **148**:121–127.

Dzink, J.L., M.T. Sheenan, and S.S. Socransky. 1990. Proposal of three subspecies of *Fusobacterium nucleatum* Knorr 1922: *Fusobacterium nucleatum* subsp. nov., comb. nov.; *Fusobacterium nucleatum* subsp. *polymorphum* subsp. nov., nom. rev., comb. nov.; and *Fusobacterium nucleatum* subsp. *vicentii* subsp. nov., nom. rev., comb. nov. *Int. J. Syst. Bacteriol.* **40**:74–78.

Eichler, B. and B. Schink. 1984. Oxidation of primary aliphatic alcohols by *Acetobacterium carbinolicum* sp. nov., a homoacetogenic anaerobe. *Arch. Microbiol.* **140**:147–152.

Engelmann, U. and N. Weiss. 1985. *Megasphaera cerevisiae* sp. nov.: a new gram-negative obligately anaerobic coccus isolated from spoiled beer. *Syst. Appl. Microbiol.* **6**:287–290.

Esnault, G., P. Caumette, and J.L. Garcia. 1988. Characterisation of *Desulfovibrio giganteus* sp. nov., a sulfate-reducing bacterium isolated from a brackish coastal lagoon. *Syst. Appl. Microbiol.* **10**:147–151.

Ezaki, T., S.-L. Liu, Y. Yashimoto, and E. Yabuuchi. 1990. *Peptostreptococcus hydrogenalis* sp. nov. from human fecal and vaginal flora. *Int. J. Syst. Bacteriol.* **40**:305–306.

Fendrich, L., H. Hippe, and G. Gottschalk. 1990. *Clostridium halophilum* sp. nov. and *C. littorale* sp. nov., an obligate halophilic and a marine species degrading betaine in the Stickland reaction. *Arch. Microbiol.* **154**:127–132.

Fiala, G., C.R. Woese, T.A. Langworthy, and K.O. Stetter. 1990. *Flexistipes sinusarabici*, a novel genus and species of eubacteria occuring in the Atlantis II deep brines of the Red Sea. *Arch. Microbiol.* **154**:120–126.

Folkerts, M., U. Ney, H. Kneifel, E. Stackebrandt, E.G. Witte, H. Förstel, S.M. Schoberth, and H. Sahm. 1989. *Desulfovibrio furfuralis* sp. nov., a furfural degrading strictly anaerobic bacterium. *Syst. Appl. Microbiol.* **11**:161–169.

Gharbia, S.E. and H.N. Shah. 1990. Heterogeneity within *Fusobacterium nucleatum*, proposal of four subspecies. *Lett. Appl. Microbiol.* **10**:105–108.

Gogotova, G.l. and M.B. Vainshtein. 1983. Spore forming, sulfate-reducing bacterium *Desulfotomaculum guttoideum* sp. nov. *Microbiology* (English translation of *Mikrobiologiya*) **52**:618–622.

Gogotova, G.l. and M.B. Vainshtein. 1989. Description of a sulfate-reducing bacterium, *Desulfobacterium macestii* sp. nov., which is capable of autotrophic growth. *Mikrobiologiya* **58**:76–80.

Haapasalo, M., H. Ranta, H. Shah, K. Ranta, K. Lounatmaa, and R.M. Kroppenstedt. 1986. *Mitsuokella dentalis* sp. nov. from dental root canals. *Int. J. Syst. bacteriol.* **36**:566–568.

Haas, F. and H. König. 1988. *Coriobacterium glomerans* gen. nov., sp. nov. from the intestinal tract of the red soldier bug. *Int. J. Syst. Bacteriol.* **38**:382–384.

Harris, J.E., P.A. Pinn, and R.P. Davis. 1984. Isolation and characterization of a novel thermophilic, freshwater methanogen. *Appl. Environ. Microbiol.* **48**:1123–1128.

Hermann, M., M.R. Popoff, and M. Sebald. 1987. *Sporomusa paucivorans* sp. nov., a methyltrophic bacterium that forms acetic acid from hydrogen and carbon dioxide. *Int. J. Syst. Bacteriol.* **37**:93–101.

Himelbloom, B.H. and E. Canale-Parola. 1989. *Clostridium methylpentosum* sp. nov.: a ring-shaped intestinal bacterium that ferments only methylpentoses and pentoses. *Arch. Microbiol.* **151**:287–293.

Horn, N. 1987. *Clostridium disporicum* sp. nov., a saccharolytic species able to form two spores per cell, isolated from a rat cecum. *Int. J. Syst. Bacteriol.* **37**:398–401.

Huber, R., J.K. Kristjansson, and K.O. Stetter. 1987. *Pyrobaculum* gen. nov., a new genus of neutrophilic, rod-shaped archaebacteria from continental solfataras growing optimally at 100°C. *Arch. Microbiol.* **149**:95–101.

Huber, R., M. Kurr, H.W. Jannasch, and K.O. Stetter. 1989a. A novel group of abyssal methanogenic archaebacteria (*Methanopyrus*) growing at 110°C. *Nature* (London) **342**:833–834.

Huber, R., C.R. Woese, T.A. Langworthy, H. Fricke, and K.O. Stetter. 1989b. *Thermosipho africanus* gen. nov. represents a new genus of thermophilic eubacteria within the "Thermotogales." *Syst. Appl. Microbiol.* **12**:32–37.

Huber, R., C.R. Woese, T.A. Langworthy, J.K. Kristjansson, and K.O. Stetter. 1990. *Fervidobacterium islandicum* sp. nov., a new extremely thermophilic eubacterium belonging to the "Thermotogales." *Arch. Microbiol.* **154**:105–111.

Huber, R., T.A. Langworthy, H. König, M. Thomm, C.R. Woese, U.B. Sleytr, and K.O. Stetter. 1986. *Thermotoga maritima* sp. nov. represents a new genus of unique extremely thermophilic eubacteria growing up to 90°C. *Arch. Microbiol.* **144**:324–333.

Imhoff-Stuckle, D. and N. Pfennig. 1983. Isolation and characterization of a nicotinic acid-degrading sulfate-reducing bacterium, *Desulfococcus niacini* sp. nov. *Arch. Microbiol.* **136**:194–198.

Jain, M.K. and J.G. Zeikus. 1988. Taxonomic distinction of two new protein specific, hydrolytic anaerobes: isolation and characterization of *Clostridium proteolyticum* sp. nov. and *Clostridium collagenovorans* sp. nov. *Syst. Appl. Microbiol.* **10**:134–141.

Jain, M.K., T.E. Thompson, E. Conway de Macario, and J.G. Zeikus. 1987. Speciation of *Methanobacterium* strain Ivanov as *Methanobacterium ivanovii* sp. nov., *Syst. Appl. Microbiol.* **9**:77–82.

Jannasch, H.W., R. Huber, S. Belkin, and K.O. Stetter. 1988. *Thermotoga neapolinata* sp. nov. of the extremely thermophilic, eubacterial genus *Thermotoga. Arch. Microbiol.* **150**:103–104.

Janssen, P.H. and C.G. Harfoot. 1990. *Ilyobacter delafieldii* sp. nov., a metabolically restricted anaerobic bacterium fermenting PHB. *Arch. Microbiol.* **154**:253–259.

Jensen, N.S. and E. Canale-Parola. 1986. *Bacteroides pectinophilus* sp. nov. and *Bacteroides galacturonicus* sp. nov.: two pectinolytic bacteria from the human intestinal tract. *Appl. Environ. Microbiol.* **52**:880–887.

Jin, F., K. Yamasato, and K. Toda. 1988. *Clostridium thermocopriae* nov. sp., a cellulolytic thermophile from animal feces compost soil and a hot spring in Japan. *Int. J. Syst. Bacteriol.* **38**:1279–1281.

Johnson, J.L., L.V. Holdeman, and W.E.C. Moore. 1985. Replacement of the type strain of *Selenomonas sputigena* under rule 18 g. Request for an opinion. *International Journal of Systematic Bacteriology* **35**:371–374.

Johnson, J.L., W.E.C. Moore, and L.V.H. Moore. 1986. *Bacteroides caccae* sp. nov., *Bacteroides merdae* sp. nov., and *Bacteroides stercoris* sp. nov. isolated from human feces. *Int. J. Syst. Bacteriol.* **36**:499–501.

Johnson, J.L., L.V.H. Moore, B. Kaneko, and W.E.C. Moore. 1990. *Actinomyces georgiae* sp. nov., *Actinomyces gerencseriae* sp. nov., designation of two genospecies of *Actinomyces naeslundii*, and inclusion of *A. naeslundii* serotype II and III and *Actinomyces viscosus* serotype II in *A. naeslundii* genospecies 2. *Int. J. Syst. Bacteriol.* **40**:273–286.

Kelly, W.J., R.V. Asmundson, and D.H. Hoperoft. 1987. Isolation and characterization of a strictly anaerobic, cellulolytic spore former: *Clostridium chartatabidum* sp. nov. *Arch. Microbiol.* **147**:169–173.

Khan, A.W., E. Meek, L.C. Sowden, and J.R. Colvin. 1984. Emendation of the genus *Acetivibrio* and description of *Acetivibrio cellulosolvens* sp. nov., a nonmotile cellulolytic mesophile. *Int. J. Syst. Bacteriol.* **34**:419–422.

Kilpper-Bälz, R. and K.R. Schleifer. 1988. Transfer of *Streetococcus morbillorum* to the genus *Gemella* as *Gemella morbillorum* comb. nov. *Int. J. Syst. Bacteriol.* **38**:442–443.

Kondratieva, E.N., E.V. Zacharova, V.I. Duda, and V.V. Krivenko. 1989. *Thermoanaerobium lactoethylicum* spec. nov., a new anaerobic bacterium from a hot spring of Kamchatka. *Arch. Microbiol.* **151**:117–122.

Krumholz, L.R. and M.P. Bryant. 1985. *Clostridium pfennigii* sp. nov. uses methoxyl groups of monobenzenoids and produces butyrate. *Int. J. Syst. Bacteriol.* **35**:454–456.

Krumholz, L.R. and M.P. Bryant. 1986a. *Syntrophococcus sucromutans* sp. nov. gen. nov. uses carbohydrates as electron donors and formate, methoxymonobenzenoids or *Methanobrevibacter* as electron acceptor systems. *Arch. Microbiol.* **143**:313–318.

Krumholz, L.R. and M.P. Bryant. 1986b. *Eubacterium oxidoreducens* sp. nov. requiring H_2 or formate to degrade gallate, pyrogallol, phloroglucinol and quercitin. *Arch. Microbiol.* **144**:8–14.

Lai, C.-H., B.M. Males, P.A. Dougherty, P. Berthold, and M.A. Listgarten. 1983. *Centipeda periodontii* gen. nov., sp. nov. from human periodontal lesions. *Int. J. Syst. Bacteriol.* **33**:628–635.

Lauer, E. 1990. *Bifidobacterium gallicum* sp. nov. isolated from human feces. *Int. J. Systematic Bacteriology* **40**:100–102.

Laurinavichyus, K.S., S.V. Kotel'nikova, and A.Y Obraztsova. 1988. New species of thermophilic methane-producing bacteria, *Methanobacterium thermophilum. Microbiology* (English translation of *Mikrobiologiya*) **57**:832–838.

Lawson Anani Soh A., H. Ralambotiana, B. Ollivier, G. Prensier, E. Tine, and J.-L. Garcia. 1991. *Clostridium thermopalmarium* sp. nov., a moderately thermophilic butyrate-producing bacterium isolated from palm wine in Senegal. *Syst. Appl. Microbiol.* **14**:135–139.

Lee, W.-K., T. Fujisawa, S. Kawamura, K. Itoh, and T. Mitsuoka. 1989. *Clostridium intestinalis* sp. nov., an aerotolerant species isolated from the feces of cattle and pigs. *Int. J. Syst. Bacteriol.* **39**:334–336.

Leigh, J.A., F. Mayer, and R.S. Wolfe. 1981. *Acetogenium kivui*, a new thermophilic hydrogen-oxidizing, acetogenic bacterium. *Arch. Microbiol.* **129**:275–280.

Le Ruyet, P., H.C. Dubourguier, G. Albagnac, and G. Prensier. 1985. Characterization of *Clostridium thermolacticum* sp. nov., a hydrolytic thermophilic anaerobe producing high amount of lactate. *Syst. Appl. Microbiol.* **6**:196–202.

Liu, Y., D.R. Boone, and C. Choy. 1990. *Methanohalophilus oregonense* sp. nov., a methyltrophic methanogen from an alkaline, saline aquifer. *Int. J. Syst. Bacteriol.* **40**:111–116.

Love, D.N., J.L. Johnson, R.F. Jones, and A. Calverley. 1987. *Bacteroides salivosus* sp. nov., an asacharolytic, black-pigmented species from cats. *Int. J. Syst. Bacteriol.* **37**:307–309.

Love, D.N., J.L. Johnson, R.F. Jones, M. Bailey, and A. Calverley. 1986. *Bacteroides tectum* sp. nov. and characteristics of other nonpigmented *Bacteroides* isolated from soft-tissue infections from cats and dogs. *Int. J. Syst. Bacteriol.* **36**: 123–128.

Maestrojuan, G.M., D.R. Boone, L. Xun, R.A. Mah, and L. Zhang. 1990. Transfer of *Methanogenium bourgense*, *Methanogenium marisnigri*, *Methanogenium olentangyi*, and *Methanogenium thermophilicum* to the genus *Methanoculleus* gen. nov., emendation of *Methanoculleus marisnigri* and *Methanogenium* and description of new strains of *Methanoculleus bourgense* and *Methanoculleus marisnigri*. *Int. J. Syst. Bacteriol.* **40**:117–122.

Margaret, B.S. and G.N. Krywolap. 1986. *Eubacterium yurii* subsp. *yurii* sp. nov. and *Eubacterium yurii* subsp. *margaretiae* subsp. nov.: test tube brush bacteria from subgingival dental plaque. *Int. J. Syst. Bacteriol.* **36**:145–149.

Margaret, B.S. and G.N. Krywolap. 1988. *Eubacterium yurii* subsp. *schtitka* subsp. nov.: test tube brush bacteria from subgingival dental plaque. *Int. J. Syst. Bacteriol.* **38**:207–208.

Mathrani, I.M., D.R. Boone, R.A. Mah, G.E. Fox, and P.P. Lau. 1988. *Methanohalophilus zhilinae* sp. nov., an alkaliphilic, halophilic, methylotrophic methanogen. *Int. J. Syst. Bacteriol.* **38**: 139–142.

McInerney, M.J., M.P. Bryant, R.B. Hespell, and J.W. Costerton. 1981. *Syntrophomonas wolfei* gen. nov., sp. nov., an anaerobic, syntrophic, fatty acid-oxidizing bacterium. *Appl. Environ Microbiol.* **41**:1029-1039.

Min, H. and S.H. Zinder. 1990. Isolation and characterization of a thermophilic sulfate-reducing bacterium *Desulfotomaculum thermoacetoxidans* sp. nov. *Arch. Microbiol.* **153**:399–404.

Miroshnichenko, M.L., E.A. Bonch-Osmolovskaya, A. Neuner, N.A. Kostrikina, N.A. Chernych, and V.A. Alekseev. 1989. *Thermococcus stetteri* sp. nov., a new extremely thermophilic marine sulfur-metabolizing archaebacterium. *Systematic and Applied Microbiology* **12**:257–262.

Möller, B., R. Ossmer, B.H. Howard, G. Gottschalk, and H. Hippe. 1984. *Sporomusa*, a new genus of Gram-negative anaerobic bacteria including *Sporomusa sphaeroides* spec. nov. and *Sporomusa ovata* spec. nov. *Arch. Microbiol.* **139**:388–396.

Montgomery, L., B. Flesher, and D. Stahl. 1988. Transfer of *Bacteroides succinogenes* (Hungate) to *Fibrobacter* gen. nov. as *Fibrobacter succinogenes* comb. nov. and description of *Fibrobacter intesti-*

nalis sp. nov. *Int. J. Syst. Bacteriol.* **38**:430–435.

Moore, L.V.H., J.L. Johnson, and W.E.C. Moore. 1987. *Selenomonas noxia* sp. nov., *Selenomonas flueggei* sp. nov., *Selenomonas infelix* sp. nov., *Selenomonas dianae* sp. nov., and *Selonomonas artemidis* sp. nov., from the human gingival crevice. *Int. J. Syst. Bacteriol.* **37**:271–280.

Morris, G.N., J. Winter, E.P. Cato, A.E. Ritchie, and V.D. Bokkenheuser. 1985. *Clostridium scindens* sp. nov., a human intestinal bacterium with desmolytic activity on corticoids. *Int. J. Syst. Bacteriol.* **35**:478–481.

Morris, G.N., J. Winter, E.P. Cato, A.E. Ritchie, and V.D. Bokkenheuser. 1986. *Eubacterium desmolans* sp. nov., a steroid desmolase-producing species from cat fecal flora. *Int. J. Syst. Bacteriol.* **36**:183–186.

Mountfort, D.O., W.J. Brulla, L.R. Krumholz, and M.P. Bryant. 1984. *Syntrophus buswellii* gen. nov. sp. nov.: a benzoate catabolizer from methanogenic ecosystems. *Int. J. Syst. Bacteriol.* **34**:216–217.

Mountfort, D.O., W.D. Grant, R. Clarke, and R.A. Asher. 1988. *Eubacterium callanderi* sp. nov. that demethoxylates O-methoxylated aromatic acids to volatile fatty acids. *Int. J. Syst. Bacteriol.* **38**:254–258.

Murray, W.D., L.C. Sowden, and J.R. Colvin. 1984. *Bacteroides cellulosolvens* sp. nov., a cellulolytic species from sewage sludge. *Int. J. Syst. Bacteriol.* **34**:185–187.

Murray, W.D., L. Hofmann, N.L. Campbell, and R.H. Madden. 1986. *Clostridium lentocellum* sp. nov., a cellulolytic species from river sediment containing paper-mill waste. *Syst. Appl. Microbiol.* **8**:181–184.

Nakatsugawa, N. and J.P. Horikoshi. 1989. Alkalophilic, methanogenic bacteria (*Methanosarcina alcaliphilum*) and fermentation method for the fast production of methane. Research Development Corporation of Japan, No. 33134, European Patent.

Nanninga, H.J. and J.C. Gottschal. 1987. Properties of *Desulfovibrio carbinolicus* sp. nov. and other sulfate-reducing bacteria isolated from an anaerobic purification plant. *Appl. Environ. Microbiol.* **53**:802–809.

Nazina, T.N., A.E. Ivanova, L.P. Kanchaveli, and E.P. Rozanova. 1988. A new sporoforming thermophilic methylotrophic sulfate-reducing bacterium, *Desulfotomaculum kuznetzovii*. *Mikrobiology* (English translation of *Microbiologiya*) **57**:659–663.

Neuner, A., H.W. Jannasch, S. Belkin, and K.O. Stetter. 1990. *Thermococcus litoralis* sp. nov.: a new species of extremely thermophilic marine archaebacteria. *Arch. Microbiol.* **152**:205–207.

Obraztsova, A.Y., O.V. Shipin, L.V. Bezrukova, and S.S. Belyaev. 1987. Properties of the coccoid methylotrophic methanogen, *Methanococ-*

coides euhalobius sp. nov. *Microbiology* (English translation of *Mikrobiologiya*) 56:523–527.

Okuda, K., T. Kato, J. Shiozu, 1. Takazoe, and T. Nakamura. 1985. *Bacteroides heparinolyticus* sp. nov. isolated from humans with periodontitis. *Int. J. Syst. Bacteriol.* 35:438–442.

Ollivier, B., R. Cord-Ruwisch, E.C. Hatchikian, and J.L. Garcia. 1988. Characterization of *Desulfovibrio fructosovorans* sp. nov. *Arch. Microbiol.* 149:447–450.

Ollivier, B., R. Cord-Ruwisch, A. Lombardo, and J.-L. Garcia. 1985a. Isolation and characterization of *Sporomusa acidovorans* sp. nov., a methyltrophic homoacetogenic bacterium. *Arch. Microbiol.* 142:307–310.

Ollivier, B.M., R.A. Mah, T.J. Ferguson, D.R. Boone, J.L. Garcia, and R. Robinson. 1985b. Emendation of the genus *Thermobacteroides*: *Thermobacteroides proteolyticus* sp. nov., a proteolytic acetogen from a methanogenic enrichment. *Int. J. Syst. Bacteriol.* 35:425–428.

Ollivier, B., C.E. Hatchikian, G. Prensier, J. Guezennec, and J.-L. Garcia. 1991. *Desulfohalobium retbaense* gen. nov., sp. nov., a halophilic sulfate-reducing bacterium from sediments of a hypersaline lake in Senegal. *Int. J. Syst. Bacteriol.* 41:74–81.

Oren, A., H. Pohla, and E. Stackebrandt. 1987. Transfer of *Clostridium lortetii* to a new genus *Sporohalobacter* gen. nov. as *Sporohalobacter lortetii* comb. nov., and description of *Sporohalobacter marismortui* sp. nov. *Syst. Appl. Microbiol.* 9:239–246.

Oren, A., W.G. Weisburg, M. Kessel, and C.R. Woese. 1984. *Halobacteroides halobius* gen. nov., sp. nov., a moderately halophilic anaerobic bacterium from the bottom sediments of the Dead Sea. *Syst. Appl. Microbiol.* 5:58–70.

Palop, M.L., S. Valles, F. Pinaga, and A. Flors. 1989. Isolation and characterization of an anaerobic cellulolytic bacterium *Clostridium celerecrescens* nov. sp. *Int. J. Syst. Bacteriol.* 39: 68–71.

Paster, B.J. and E. Canale-Parola. 1985. *Treponema saccharophilum* sp. nov., a large pectinolytic spirochete from the bovine rumen. *Appl. Environ. Microbiol.* 50:212–219.

Patel, G.B. and G.D. Sprott. 1990. *Methanosaeta concilii* gen. nov., sp. nov. ("*Methanothrix concilii*") and *Methanosaeta thermoacetophila* nom. rev., comb. nov. *Int. J. Syst. Bacteriol.* 40:79–82.

Patel, B.K.C., H.W. Morgan, and R.M. Daniel. 1985. *Fervidobacterium nodosum*, gen. nov., sp., nov., a new chemoorganotrophic caldoactive anaerobic bacterium. *Arch. Microbiol.* 141:63–69.

Patel, B.K.C., C. Monk, H. Littleworth, H.W. Morgan, and R.M. Daniel. 1987. *Clostridium fervidus* sp. nov. a new chemoorganotrophic ace-

togenic thermophile. *Int. J. Syst. Bacteriol.* 37:123–126.

Patel, G.B., G.D. Sprott, and J.E. Fein. 1990. Isolation and characterization of *Methanobacterium espanolae* sp. nov., a mesophilic, moderately acidiphilic methanogen. *Int. J. Syst. Bacteriol.* 40:12–18.

Paterek, J.R. and P.H. Smith. 1988. *Methanohalophilus mahii* gen. nov., sp. nov., a methylotrophic halophilic methanogen. *Int. J. Syst. Bacteriol.* 38:122–123.

Platen, H., A. Temmes, and B. Schink. 1990. Anaerobic degradation of acetone by *Desulfococcus biacutus* spec. nov. *Arch. Microbiol.* 154:355–361.

Qatibi, A.l., V. Nivière, and J.-L. Garcia. 1991. *Desulfovibrio alcoholovorans* sp. nov., a sulfate reducing bacterium able to grow on glycerol 1,2- and 1,3-propanediol. *Arch. Microbiol.* 155:143–148.

Rengpipat, S., T.A. Langworthy, and J.G. Zeikus. 1988. *Halobacteroides acetoethylicus* sp. nov., a new obligately anaerobic halophile isolated from deep subsurface hypersaline environments. *Syst. Appl. Microbiol.* 11:28–35.

Robinson, l.M. and E.A. Freundt. 1987. Proposal for an amended classification of anaerobic Mollicutes. *Int. J. Syst. Bacteriol.* 37:78–81.

Rogers, G.A. and A.A.W. Baecker. 1991. *Clostridium xylanolyticum* sp. nov., an anaerobic xylanolytic bacterium from decayed *Pinus patula* wood chips. *Int. J. Syst. Bacteriol.* 41:140–143.

Roy, F., E. Samain, H. C. Dubourguier, and G. Albagnac. 1986. *Syntrophomonas sapovorans* sp. nov., a new obligately proton reducing anaerobe oxidizing saturated and unsaturated long chain fatty acids. *Arch. Microbiol.* 145:142–147.

Rozanova, E.P. and T.A. Pivovarova. 1988. Reclassification of *Desulfovibrio thermoehilus* (Rozanova, Khudyakova, 1974). *Mikrobiology* (English translation of *Mikrobiologiya*) 57:85–89.

Rozanova, E.P., T.N. Nazina, and A.S. Galushko. 1988. Isolation of a new genus of sulfate-reducing bacteria and description of a new species of this genus, *Desulfomicrobium aspheronum* gen. nov., sp. nov. *Mikrobiology* (English translation of *Mikrobiologiya*) 57:514–520.

Saiki, T., Y. Kobayashi, K. Kawagoe, and T. Beppu. 1985. *Dictyoglomus thermophilum* gen. nov., sp. nov., a chemoorganotrophic, anaerobic, thermophilic bacterium. *Int. J. Syst. Bacteriol.* 35:253–259.

Samain, E., H.C. Dubourguier, and G. Albagnac. 1984. Isolation and characterization of *Desulfobulbus elongatus* new species from a mesophilic industrial digester. *Syst. Appl. Microbiol.* 5:391–401.

Schauder, R., B. Eikmanns, R.K. Thauer, F. Widdel, and G. Fuchs. 1986. Acetate oxidation to CO_2 in anaerobic bacteria via a novel pathway not involving reactions of the citric acid cycle. *Arch. Microbiol.* **145**:162–172.

Schiefer-Ullrich, H. and J.R. Andreesen. 1985. *Peptostreptococcus barnesae* sp. nov., a grampositive, anaerobic, obligately purine utilizing coccus from chicken feces. *Arch. Microbiol.* **143**:26–31.

Schink, B. 1984a. Fermentation of 2,3-butanediol by *Pelobacter carbinolicus* sp. nov. and *Pelobacter propionicus* sp. nov., and evidence for propionate formation from C_2 compounds. *Arch. Microbiol.* **137**:33–41.

Schink, B. 1984b. Fermentation of tartrate enantiomers by anaerobic bacteria, and description of two new species of strict anaerobes, *Ruminococcus pasteurii* and *Ilyobacter tartaricus*. *Arch. Microbiol.* **139**:409–414.

Schink, B. 1985. Fermentation of acetylene by an obligate anaerobe, *Pelobacter acetylenicus* sp. nov. *Arch. Microbiol.* **142**:295–301.

Schink, B. and N. Pfennig. 1982a. Fermentation of trihydroxybenzenes by *Pelobacter acidigallici* gen. nov. sp. nov., a new strictly anaerobic, non-sporeforming bacterium. *Arch. Microbiol.* **133**:195–201.

Schink, B. and N. Pfennig. 1982b. *Propionigenium modestum* gen. nov. sp. nov., a new strictly anaerobic nonsporing bacterium growing on succinate. *Arch. Microbiol.* **133**:209–216.

Schink, B. and M. Stieb. 1983. Fermentative degradation of polyethylene glycol by a new strictly anaerobic, Gram negative, nonsporeforming bacterium, *Pelobacter venetianus* sp. nov. *Appl. Environ. Microbiol.* **45**:1905-1913.

Schink, B., T.E. Thompson, and J.G. Zeikus. 1982. Characterization of *Propionispira arboris* gen. nov. sp. nov., a nitrogen-fixing anaerobe common to wetwoods of living trees. *J. Gen. Microbiol.* **128**:2771–2779.

Schleifer, K.H., M. Leuteritz, N. Weiss, W. Ludwig, G. Kirchhof, and H. Seidel-Rufer. 1990. Taxonomic study of anaerobic, gram-negative, rod-shaped bacteria from breweries: emended description of *Pectinatus cerevisiiphilus* and description of *Pectinatus frisingensis* sp. nov., *Selenomonas lacticifex* sp. nov., *Zymophilus raffinosivorans* gen. nov., sp. nov., and *Zymophilus paucivorans* sp. nov. *Int. J. Syst. Bacteriol.* **40**:19–27.

Schmid, U., H. Giesel, S.M. Schoberth, and H. Sahm. 1986. *Thermoanaerobacter finnii* sp. nov., a new ethanologenic sporogenous bacterium. *Syst. Appl. Microbiol.* **8**:80–85.

Schnell, S., F. Bak, and N. Pfennig. 1989. Anaerobic degradation of aniline and dihydroxybenzenes by newly isolated sulfate-reducing bacteria and description of *Desulfobacterium anilini*. *Arch. Microbiol.* **152**:556–563.

Scholten-Koerselman, 1., F. Houwaard, P. Janssen, and A.J.B. Zehnder. 1986. *Bacteroides xylanolyticus* sp. nov., a xylanolytic bacterium from methane producing cattle manure. *Antonie Leeuwenhoek J. Microbiol.* **52**:543–554.

Shah, H.N., and M.D. Collins. 1986. Reclassification of *Bacteroides furcosus* Veillon and Zuber (Hauduroy, Ehringer, Urbain, Guillot and Magrou) in a new genus *Anaerorhabdus*, as *Anaerorhabdus furcosus* comb. nov. *Syst. Appl. Microbiol.* **8**:86–88.

Shah, H.N. and M.D. Collins. 1988. Proposal for reclassification of *Bacteroides asaccharolyticus*, *Bacteroides gingivalis* and *Bacteroides endodontalis* in a new genus, *Porphyromonas*. *Int. J. Syst. Bacteriol.* **38**:128–131.

Shah, H.N. and M.D. Collins. 1989. Proposal to restrict the genus *Bacteroides* (Castellani and Chalmers) to *Bacteroides fragilis* and closely related species. *Int. J. Syst. Bacteriol.* **39**:85–87.

Shah, H.N. and M.D. Collins. 1990. *Prevotella*, a new genus to include *Bacteroides melaninogenicus* and related species formerly classified in the genus *Bacteroides*. *Int. J. Syst. Bacteriol.* **40**:205–208.

Shah, H.N., M.D. Collins, J. Watabe, and T. Mitsuoka. 1985. *Bacteroides oulorum* sp. nov., a nonpigmented saccharolytic species from the oral cavity. *Int. J. Syst. Bacteriol.* **35**:193–197.

Shinjo, T., K. Hizaiwa, and S. Miyazato. 1990. Recognition of Biovar C of *Fusobacterium necrophorum* (Flügge) Moore and Holdeman as *Fusobacterium pseudonecrophorum* sp. nov., nom. rev. (ex Prévot 1940). *Int. J. Syst. Bacteriol.* **40**:71–73.

Sleat, R., R.A. Mah, and R. Robinson. 1985. *Acetoanaerobium noterae* gen. nov., sp. nov.: an anaerobic bacterium that forms acetate from H_2 and CO_2. *Int. J. Syst. Bacteriol.* **35**:10–15.

Slots, J., T.V. Potts and P.A. Mashimo. 1983. *Fusobacterium periodonticum*, a new species from the human oral cavity. *Journal of Dental Research* **62**:960–963.

Smibert, R.M., J.L. Johnson, and R.R. Ranney. 1984. *Treponema socranskii* sp. nov., *Treponema socranskii* subsp. *socranskii* subsp. nov., *Treponema socranskii* subsp. *buccale* subsp. nov., and *Treponema socranskii* subsp. *paredis* subsp. nov. isolated from the human periodontia. *Int. J. Syst. Bacteriol.* **34**:457–462.

Soutschek, E., J. Winter, F. Schindler, and O. Kandler. 1984. *Acetomicrobium flavidum*, gen. nov., sp. nov., a thermophilic, anaerobic bacterium from sewage sludge, forming acetate, CO_2 and H_2 from glucose. *Syst. Appl. Microbiol.* **5**:377–390.

Spiegel, C.A. and M. Roberts. 1984. *Mobiluncus* gen. nov., *Mobiluncus curtisii* subsp. *curtisii* sp.

nov., *Mobiluncus curtisii* subps. *holmesii* subsp. nov., and *Mobiluncus mulieris* sp. nov., curved rods from the human vagina. *Int. J. Syst. Bacteriol.* **34**:177–184.

Stackebrandt, E. and H. Hippe. 1986. Transfer of *Bacteroides amylophilus* to a new genus *Ruminobacter* gen. nov., nom. rev. as *Ruminobacter amylophilus* comb. nov. *Syst. Appl. Microbiol.* **8**:204–207.

Stams, A.J.M. and T.A. Hansen. 1984. Fermentation of glutamate and other compounds by *Acidaminobacter hydrogenoformans* gen. nov. sp. nov., an obligate anaerobe isolated from black mud. Studies with pure culture and mixed cultures with sulfate-reducing and methanogenic bacteria. *Arch. Microbiol.* **137**:329–337.

Stanton, T.B., N.S. Jensen, T.A. Casey, L.A. Tordoff, F.E. Dewhirst, and B.J. Paster. 1991. Reclassification of *Treponema hyodysenteriae* and *Treponema innocens* in a new genus, *Serpula*, gen. nov., as *Serpula hyodysenteriae* comb. nov. and *Serpula innocens*, comb. nov. *Int. J. Syst. Bacteriol.* **41**:50–58.

Stieb, M. and B. Schink. 1984. A new 3-hydroxybutyrate fermenting anaerobe, *Ilyobacter polytropus*, gen. nov. sp. nov., possessing various fermentation pathways. *Arch Microbiol.* **140**:139–146.

Suen, J.C., C.L. Hatheway, A.G. Steigerwalt, and D.J. Brenner. 1988. *Clostridium argentinense* sp. nov.: a genetically homogeneous group composed of all strains of *Clostridium botulinum* toxin type G and some non toxigenic strains previously identified as *Clostridium subterminale* or *Clostridium hastiforme*. *Int. J. Syst. Bacteriol.* **38**:375–381.

Sukhumavasi, J., K. Ohmiya, S. Shimizu, and K. Ueno. 1988. *Clostridium josui* sp. nov., a cellulolytic moderate thermophilic species from Thai compost. *Int. J. Syst. Bacteriol.* **38**:179–182.

Svetlichnyi, V.A., A.I. Slesarev, T.P. Svetlichnaya, and G.A. Zavarzin. 1988. *Caldococcus litoralis* gen. nov. sp. nov. A new marine, extremely thermophilic, sulfur-reducing Archaebacterium. *Microbiology* **56**:658–664 (English translation of *Mikrobiologiya* 1987;56:831–838).

Szewzyk, R. and N. Pfennig. 1987. Complete oxidation of catechol by the strictly anaerobic sulfate-reducing *Desulfobacterium catecholicum* sp. nov. *Arch. Microbiol.* **147**:163–168.

Tanaka, K. and N. Pfennig. 1988. Fermentation of 2-methoxy-ethanol by *Acetobacterium malicum* sp. nov. and *Pelobacter venetianus*. *Arch. Microbiol.* **149**:181–187.

Tanaka, K., N. Nakamura, and E. Mikami. 1990. Fermentation of maleate by a gram negative strictly anaerobic non-spore-former, *Propionivibrio dicarboxylicus* gen. nov., sp. nov. *Arch. Microbiol.* **154**:323–328.

Tanaka, K., K. Nakamura, and E. Mikami. 1991. Fermentation of cinnamate by a mesophilic strict anaerobe, *Acetovibrio multivorans* sp. nov. *Arch. Microbiol.* **155**:120–124.

Tanner, A.C.R., M.A. Listgarten, J.L. Ebersole, and M.N. Strzempko. 1986. *Bacteroides forsythus* sp. nov., a slow-growing, fusiform *Bacteroides* sp. from the human oral cavity. *Int. J. Syst. Bacteriol.* **36**:213–221.

Toda, Y., T. Saiki, T. Uozumi, and T. Beppu. 1988. Isolation and characterization of a protease-producing, thermophilic, anaerobic bacterium, *Thermobacteroides leptospartum* sp. nov. *Agric. Biol. Chem.* **52**:1339–1344.

Trinkel, M., A. Breunig, R. Schauder, and H. Konig. 1990. *Desulfovibrio termitidis* sp. nov., a carbohydrate-degrading sulfate-reducing bacterium from the hindgut of a termite. *Syst. Appl. Microbiol.* **13**:372–377.

Van Gylswyk, N.O and J.J.T.K. van der Toorn. 1985. *Eubacterium uniforme* sp. nov. and *Eubacterium xylanophilum* sp. nov., fiber-digesting bacteria from the rumina of sheep fed corn stover. *Int. J. Syst. Bacteriol.* **35**:323–326.

Van Gylswyk, N.O and J.J.T.K. van der Toorn. 1986. Description and designation of a neotype strain of *Eubacterium cellulosolvens* (*Cillobacterium cellulosolvens* Bryant, Small, Bouma and Robinson) Holdeman and Moore. *Int. J. Syst. Bacteriol.* **36**:275–277.

Van Gylswyk, N.O and J.J.T.K. van der Toorn. 1987. *Clostridium aerotolerans* sp. nov., a xylanolytic bacterium from corn stover and from the rumina of sheep feed corn stover. *Int. J. Syst. Bacteriol.* **37**:102–105.

Van Steenbergen, T.J.M., A.J. Van Winkelhoff, D. Mayrand, D. Grenier, and J. de Graaff. 1984. *Bacteroides endodontalis* sp. nov., an asacharolytic black-pigmented *Bacteroides* species from infected dental root canals. *Int. J. Syst. Bacteriol.* **34**:118–120.

Varel, V.H. 1989. Reisolation and characterization of *Clostridium longisporum*, a ruminal sporoforming cellulolytic anaerobe. *Arch. Microbiol.* **152**:209–214.

Watabe, J., Y. Benno, and T. Mitsuoka. 1983. *Bifidobacterium gallinarum* sp. nov.: a new species isolated from the ceca of chickens. *Int. J. Syst. Bacteriol.* **33**:127–132.

Widdel, F. 1987. New types of acetate-oxidizing, sulfate-reducing *Desulfobacter* species, *D. hydrogenophilus* sp. nov., *D. latus* sp. nov., and *D. curvatus* sp. nov. *Arch. Microbiol.* **148**:286–291.

Widdel, F., P.E. Rouviere, and R.S. Wolfe. 1988. Classification of secondary alcohol-utilizing methanogens including a new thermophilic isolate. *Arch. Microbiol.* **150**:477–481.

Wiegel, J., S.-U. Kuk, and G.W. Kohring; 1989. *Clostridium thermobutyricum* sp. nov., a moderate thermophile isolated from a cellulolytic culture that produces butyrate as the major product. *Int. J. Syst. Bacteriol.* **39**:199–204.

Wilde, E., H. Hippe, N. Tosunoglu, G. Schallehn, K. Herwig, and G. Gottschalk. 1989. *Clostridium tetanomorphum* sp. nov., nom. rev. *Int. J. Syst. Bacteriol.* **39**:127–134.

Windberger, E., R. Huber, A. Trincone, H. Fricke, and K.O. Stetter. 1989. *Thermotoga thermarum* sp. nov. and *Thermotoga neapolitana* occurring in African continental solfataric springs. *Arch. Microbiol.* **151**:506–512.

Winter, J., E. Braun, and H.-P. Zabel. 1987. *Acetomicrobium faecalis* spec. nov., a strictly anaerobic bacterium from sewage sludge, producing ethanol from pentoses. *Syst. Appl. Microbiol.* **9**:71–76.

Xun, L., D.R. Boone, and R.A. Mah. 1989. Deoxyribonucleic acid hybridization study of *Methanogenium* and *Methanocorpusculum* species, emendation of the genus *Methanocorpusculum* and transfer of *Methanogenium aggregans* to the genus *Methanocorpusculum* as *Methanocorpusculum aggregans* comb. nov. *Int. J. Syst. Bacteriol.* **39**:109–111.

Yang, J.C., D.P. Chynoweth, D.S. Williams, and A. Li. 1990. *Clostridium aldrichii* sp. nov., a cellulolytic mesophile inhabiting a wood-fermenting anaerobic digester. *Int. J. Syst. Bacteriol.* **40**:268–272.

Yu, I.K. and F. Kawamura. 1987. *Halomethanococcus doii*, new genus new species, an obligately halophilic methanogenic bacterium from solar salt ponds. *J. Gen. Appl. Microbiol.* **33**:303–310.

Zeikus, J.G., M.A. Dawson, T.E. Thompson, K. Ingvorsen, and E.C. Hatchikian. 1983a. Microbial ecology of volcanic sulfidogenesis. Isolation and characterization of *Thermodesulfobacterium commune* new genus, new species. *J. Gen. Microbiol.* **129**:1159–1170.

Zeikus, J.G., P.W. Hegge, T.E. Thompson, T.J. Phelps, and T.A. Langworthy. 1983b. Isolation and description of *Haloanaerobium praevalens* gen. nov. and sp. nov., an obligately anaerobic halophile common to Great Salt Lake sediments. *Curr. Microbiol.* **9**:225–234.

Zellner, G., P. Messner, H. Kneifel.and J. Winter. 1989. *Desulfovibrio simplex* spec. nov., a new sulfate-reducing bacterium from a sour whey digester. *Arch. Microbiol.* **152**:329–334.

Zellner, G., C. Alten, E. Stackebrandt, E. Conway de Macario, and J. Winter. 1987. Isolation and characterization of *Methanocorpusculum parvum*, gen. nov., spec. nov., a new tungsten-requiring, coccoid methanogen. *Arch. Microbiol.* **147**:13–20.

Zellner, G., U.B. Sleytr, P. Messner, H. Kneifel, and J. Winter. 1990. *Methanogenium liminatans* spec. nov., a new coccoid, mesophilic methanogen able to oxidize secondary alcohols. *Arch. Microbiol.* **153**:287–293.

Zellner, G., P. Messner, H. Kneifel, B.J. Tindall, J. Winter, and E. Stackebrandt. 1989a. *Methanolacinia* gen. nov., incorporating *Methanomicrobium paynteri* as *Methanolacinia paynteri* comb. nov. *J. Gen. Appl. Microbiol.* **35**:185–202.

Zellner, G., K. Bleicher, E. Braun, H. Kneifel, B.J. Tindall, E. Conway de Macario, and J. Winter. 1989b. Characterization of a new mesophilic, secondary alcohol utilizing methanogen, *Methanobacterium palustre* spec. nov. from a peat bog. *Arch. Microbiol.* **151**:1–9.

Zellner, G., E. Stackebrandt, P. Messner, B.J. Tindall, E. Conway de Macario, H. Kneifel, and U.B. Sleytr. 1989c. *Methanocorpusculaceae* fam. nov., represented by *Methanocorpusculum parvum*, *Methanocorpusculum sinense* spec. nov. and *Methanocorpusculum bavaricum* spec. nov. *Arch. Microbiol.* **151**:381–390.

Zhao, H., D. Yang, C.R. Woese, and M.P. Bryant. 1990. Assignment of *Clostridium bryantii* to *Syntrophospora bryantii* gen. nov., comb. nov. on the basis of a 16S rRNA sequence analysis of its crotonate-grown pure culture. *Int. J. Syst. Bacteriol.* **40**:40–44.

Zhao, Y., D.R. Boone, R.A. Mah, J.E. Boone, and L. Xun. 1989. Isolation and characterization of *Methanocorpusculum labreanum* sp. nov. from the LaBrea Tar Pits. *Int. J. Syst. Bacteriol.* **39**:10–13.

Zhilina, T.N. and G.A. Zavarzin. 1985. New methanogenic bacteria. *Priroda* **7**:103–105.

Zillig, W., I. Holz, and S. Wunderl. 1991. *Hyperthermus butylicus* gen. nov., sp. nov., a hyperthermophilic, anaerobic, peptide-fermenting, facultatively H_2S-generating archaebacterium. *Int. J. Syst. Bacteriol.* **41**:169–170.

Zindel, U., W. Freudenberg, M. Rieth, J.R. Andreesen, J. Schnell, and F. Widdel. 1988. *Eubacterium acidaminophilum* sp. nov., a versatile amino acid-degrading anaerobe producing or utilizing H_2 or formate. Description and enzymatic studies. *Arch. Microbiol.* **150**:254–266.

Index

Brock/Springer Series in Contemporary Bioscience

(Continued from page ii)